Undergraduate Texts in Physics

Undergraduate Texts in Physics (UTP) publishes authoritative texts covering topics encountered in a physics undergraduate syllabus. Each title in the series is suitable as an adopted text for undergraduate courses, typically containing practice problems, worked examples, chapter summaries, and suggestions for further reading. UTP titles should provide an exceptionally clear and concise treatment of a subject at undergraduate level, usually based on a successful lecture course. Core and elective subjects are considered for inclusion in UTP.

UTP books will be ideal candidates for course adoption, providing lecturers with a firm basis for development of lecture series, and students with an essential reference for their studies and beyond.

Fabian Cadiz · Arnaud Couairon

Classical Electrodynamics

Fundamentals and Applications

 Springer

Fabian Cadiz
Laboratoire de Physique de la Matière
Condensée
CNRS, Ecole Polytechnique, Institut
Polytechnique de Paris
Palaiseau, France

Arnaud Couairon
Centre de Physique Théorique
CNRS, Ecole Polytechnique, Institut
Polytechnique de Paris
Palaiseau, France

ISSN 2510-411X   ISSN 2510-4128   (electronic)
Undergraduate Texts in Physics
ISBN 978-3-031-86784-2   ISBN 978-3-031-86785-9   (eBook)
https://doi.org/10.1007/978-3-031-86785-9

Preface

Classical electrodynamics is a cornerstone of physics and engineering education. Its principles underpin a vast array of technologies and serve as a foundation for advanced topics such as quantum mechanics, statistical physics, and relativity.

This book aims to address the unique challenges faced by students from diverse backgrounds, particularly those transitioning to advanced studies. We aim to provide a unified treatment of classical electrodynamics and its applications, incorporating mathematical rigor, clear explanations of symmetry and invariance principles, and concrete connections to modern technologies. By covering this vast domain, we hope this book will provide a comprehensive and engaging learning experience for students, empowering them to delve deeper into the fascinating world of electrodynamics.

This book is organized into twenty chapters, covering topics ranging from Coulomb's law and electrostatics to electromagnetic waves and special relativity. Each chapter is enriched with illustrations of contemporary applications, from the working principles of electric motors and magnetic levitation trains to medical imaging technologies like MRI. These real-world examples not only ground the theory in practical contexts but also inspire readers to see the relevance of electromagnetism in everyday life. Additionally, throughout the text, we provide brief notes on the most influential scientists who have shaped the laws and principles discussed, offering historical context to the physics.

To support learning, we have included fully solved exercises at the end of each chapter. These problems vary in difficulty and are designed to reinforce theoretical concepts and build problem-solving skills. We believe this approach will appeal to a wide audience, from undergraduate students taking their first steps in electromagnetism to graduate students seeking a deeper understanding of the subject. We hope this book will inspire students to approach the study of classical electromagnetism with curiosity, enthusiasm, and confidence.

We drew inspiration from a variety of sources, including the excellent course by Jean–Jacques Greffet at the Institut d'Optique (Palaiseau, France), Feynman's Lectures on Physics, and classic textbooks by John David Jackson, David J. Griffiths, and Andrew Zangwill. We also benefited from Jorge Krause's insightful lecture notes at the Pontificia Universidad Católica (Chile) and Jimmy Roussel's excellent notes on geometrical optics. We extend our immense gratitude to our colleagues Sébastien Corde, François Hache, Fabien Bretenaker, and Antoine Browaeys at École Polytechnique, with whom we have had the privilege of teaching electromagnetism courses for the past eight years and who have inspired us in writing this book. Their lectures and tutorials have significantly influenced this book.

We welcome feedback from readers and instructors and hope this book will become a valuable resource for classrooms worldwide.

Palaiseau, France   Fabian Cadiz
December 2024   Arnaud Couairon

Competing Interests The authors have no competing interests to declare that are relevant to the content of this manuscript.

Contents

Chapter 1
Coulomb's Law, Electric Field, and the Superposition Principle

Abstract This chapter introduces the fundamental principles governing electrostatic phenomena, laying the groundwork for the study of classical electrodynamics. It begins by establishing the concept of electric charge as an intrinsic property of matter, analogous to mass in gravitational interactions, and elucidates the nature of attractive and repulsive forces between charges. A central focus is *Coulomb's law*, which quantitatively describes the inverse-square force between two stationary point charges. Building upon this, the chapter rigorously develops the *superposition principle*, demonstrating how the total electrostatic force or electric field due to a system of multiple charges is the vector sum of individual contributions. This principle is extended to introduce the crucial concept of the *electric field* as a vector field that mediates electrostatic interactions, providing a framework to analyze forces without direct action-at-a-distance. The chapter also explores the visualization of electric fields through *electric field lines*. Furthermore, it addresses the practical transition from *discrete point charges to continuous charge* distributions (volume, surface, and line charge densities), providing integral formulations for calculating electric fields in such scenarios. The mathematical tool of the *Dirac delta distribution* is introduced to unify the treatment of discrete and continuous charge distributions within a single framework. Finally, the chapter discusses the *asymptotic behavior of the electric field* at large distances from various charge configurations, emphasizing how the field's decay rate is influenced by the overall charge neutrality of the system. Through these foundational concepts, the chapter provides a comprehensive introduction to the static electric interactions.

Keywords Coulomb's law · Electrostatics · Superposition principle

1.1 Introduction

The electromagnetic force is one of the four fundamental forces through which all particles that constitute nature interact. It relies on one of the elementary properties of matter: electric charge. Charge plays a role analogous to that of mass in the gravitational force. It is the electromagnetic force that confines electrons around

© The Author(s), under exclusive license to Springer Nature Switzerland AG 2025

F. Cadiz and A. Couairon, *Classical Electrodynamics*, Undergraduate Texts in Physics,

https://doi.org/10.1007/978-3-031-86785-9_1

nuclei to form atoms and, in turn, allows atoms to bond together to form molecules. At the macroscopic level, it governs most everyday phenomena, such as a person sitting on a chair or a car remaining at rest on a slope. Both examples represent macroscopic manifestations of electromagnetic forces. Electromagnetic interactions also allow us to observe the world around us, as our ability to see is due to the sensitivity of our eyes to certain oscillations of electromagnetic fields, commonly known as light. In this chapter, we establish the two principles that govern electrostatic phenomena: Coulomb's law and the superposition principle. We also introduce the concept of the electric field, a vector field that encapsulates all the information about the charge distribution at a given instant.

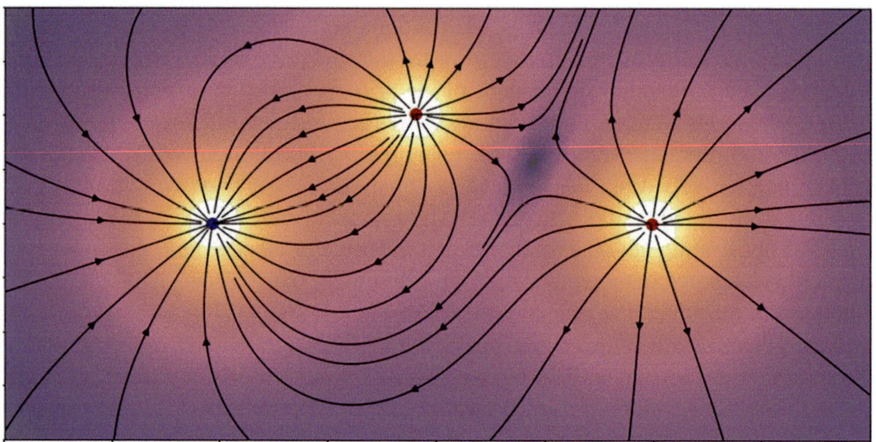

1.2   The Electric Charge

The particles that make up matter (electrons, protons, and neutrons) have a fundamental property called electric charge, analogous to the role that mass plays in gravitational interaction. As discovered by Charles Dufay in 1733, there are two types of charges: positive and negative. In fact, and in contrast with the gravitational force, the force between two charged objects can be repulsive or attractive depending on the relative sign of their charges. In the international system (S.I.) the unit of charge is the coulomb (C), and the fundamental charge is that of the electron, given by

$$q_e = -e = -1.602176634 \times 10^{-19}\,\mathrm{C}\,.$$

The charge of a proton has the same magnitude as that of the electron, but an opposite sign, that is, $q_p = +e = 1.6 \times 10^{-19}\mathrm{C}$. The neutron is a particle with zero net charge. Consequently, the charge Q of every macroscopic object, which is necessarily composed of an integer number of electrons and protons, is

$$Q = ne, \quad n \in \mathbb{Z}\,.$$

It is determined from experiments that, for an isolated system, the total charge is a conserved quantity.

1.3 Force Between Two Point Charges: Coulomb's Law (1785)

Consider a system of two point charges q and q', in a reference frame where both are at rest, separated by a distance r in vacuum, as shown in Fig. 1.1. The electrostatic force, strictly valid when the charges are not in relative motion, was studied experimentally by Dufay, Cavendish, and Charles Augustin de Coulomb (Fig. 1.2).

Coulomb's law (1785)
The force exerted by a charge q' located at $\mathbf{x}'$ on another charge q located at $\mathbf{x}$ is given by Coulomb's law:

$$\mathbf{F}_q = \frac{qq'}{4\pi\epsilon_0|\mathbf{x}-\mathbf{x}'|^3}(\mathbf{x}-\mathbf{x}') = \frac{qq'}{4\pi\epsilon_0 r^2}\mathbf{u}_r \; . \tag{1.1}$$

That is to say, as with Newton's inverse-square law for gravitation, the electrostatic force is inversely proportional to the square of the distance $r = |\mathbf{x}-\mathbf{x}'|$. This is a central force: its direction is along the line that joins both charges $\mathbf{u}_r = (\mathbf{x}-\mathbf{x}')/|\mathbf{x}-\mathbf{x}'|$. Note that the direction of the force on q depends on the sign of qq'. The proportionality constant in Coulomb's law is approximately equal to

$$\frac{1}{4\pi\epsilon_0} = 8.9875 \times 10^9 \, \mathrm{N\,m^{-2}C^{-2}} \; ,$$

where ϵ_0 is known as the *vacuum permittivity*:

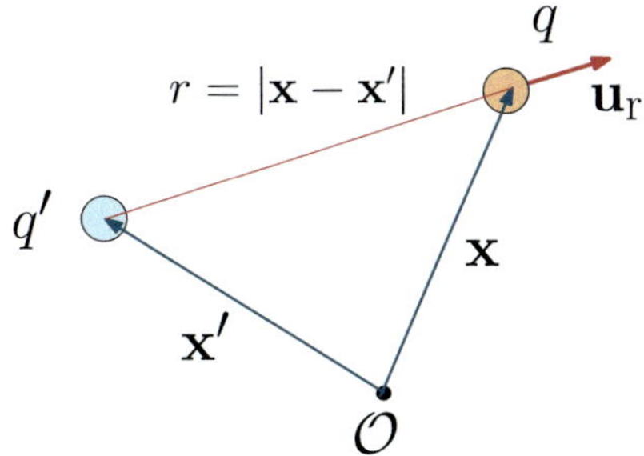

Fig. 1.1 Two point charges q and q' separated by a distance r

Fig. 1.2 Charles–Augustin de Coulomb (1736–1806), a French engineer and physicist who made significant contributions to mechanics and electrostatics. He formulated the laws of solid friction and invented the torsion balance

$$\epsilon_0 = 8.854 \times 10^{-12}\,\mathrm{N}^{-1}\mathrm{m}^{-2}\mathrm{C}^2 \ .$$

It is clear that the force on q' exerted by q is given by $\mathbf{F}_{q'} = -\mathbf{F}_q$, as it should be by Newton's third law. Note that this simple system of two charges cannot remain at rest unless the presence of other forces keeps them in place. Strictly speaking, if the charges are not held at rest by external forces, they will be set in relative motion due to their mutual interaction, and then one should add another force that takes into account this movement. Later, we will see that this will give rise to a new manifestation of the electromagnetic interaction, which we call magnetism.

It is an empirical fact that electrostatic forces can be attractive or repulsive (see Fig. 1.3), depending on the type of charges involved. The nature of the force is determined by the product of the charges: if both have the same sign, the force will be repulsive, while if they are of opposite sign, the force will be attractive. Consequently, the sign we assign to the two types of existing charges is absolutely arbitrary. By historical convention, the charge of the electron is negative ($q_e = -e$), and that of the proton is positive ($q_e = +e$). The sign of the electron and proton charges could have been perfectly inverted, with the force between two charged particles (and consequently all experimental observations) being absolutely insensitive to this change.

Fig. 1.3 Unlike the mass (an always positive quantity), the charges of two particles do not always have the same sign. Charges of different sign are attracted, whereas charges of the same sign repel each other

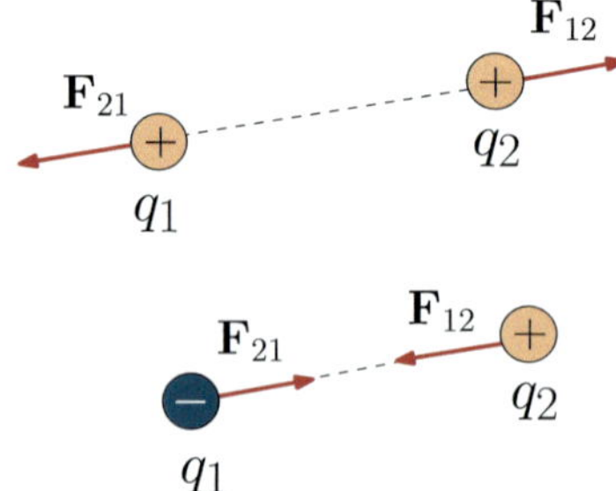

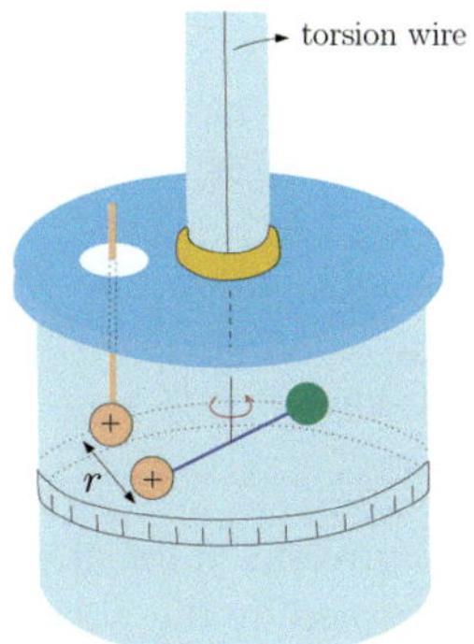

Fig. 1.4 Torsion balance used by Coulomb to study the electrostatic force between charged objects. Picture taken at the Museum of Ecole Polytechnique (Mus'X)

The device used by Coulomb consists of a torsion balance, depicted in Fig. 1.4, which is able to measure small forces with precision. Two metallic spheres are joined by an insulating rod, which is suspended by a fiber. A sphere of positive charge approaches a distance r from one of the spheres of the balance. If the latter is also positively charged, the two spheres repel, and consequently there is a rotation of the rod and a twist of the suspension fiber at an angle proportional to the force experienced by the charge. Using this device, Coulomb discovered that the electrostatic force scales as $1/r^2$ and that its repulsive or attractive nature depends on the relative sign of both charges.

Example 1.1—The electric force dominates at the atomic level
The electrostatic force at the atomic scale is much more intense than the gravitational force. As an example, consider the hydrogen atom, formed by the interaction between an electron (charge $-e$) and a proton (charge $+e$), separated by a distance r (see Fig. 1.5).

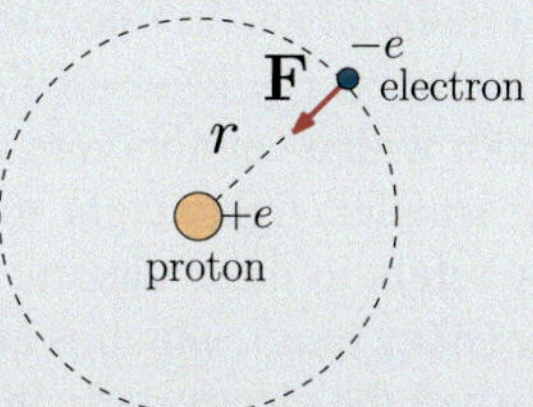

Fig. 1.5 An electron describing a circular orbit around a proton

The ratio between the magnitudes of the electric and gravitational forces is

$$\frac{F_e}{F_g} = \frac{e^2}{4\pi\epsilon_0 r^2}\frac{r^2}{Gm_e m_p},$$

where $G = 6.7 \times 10^{-11}\,\text{m}^3\,\text{kg}^{-1}\,\text{s}^{-2}$ is the universal gravitational constant. The masses of the electron and the proton are, respectively, $m_e = 9.1 \times 10^{-31}\,\text{kg}$ and $m_p \simeq 2000\,m_e$. Hence,

$$\frac{F_e}{F_g} = \frac{e^2}{4\pi\epsilon_0 G m_e m_p} \simeq 2 \times 10^{39} \ .$$

The magnitude of the electrostatic interaction is in this case 39 orders of magnitude larger than the gravitational interaction; the electric interaction predominates in an atomic system.

Even if the electric interaction governs at the atomic scale, it is well known that at large scales the universe is governed by the gravitational force. It is gravity and not the electrostatic force that determines the movement of the Earth around the Sun. The reason is that at large scales matter tends to be electrically neutral. For example, the hydrogen atom seen from distances much larger than the separation distance between the electron and the proton is practically electrically neutral. Since charges of opposite sign attract each other, they end up remaining together to form matter, generating a screening effect which damps large-scale electrostatic interactions.

1.3.1   Why are There Charged Objects?

Atoms in their stable state are electrically neutral; the nucleus has the same number of protons as electrons that are bound to it. In this way, in principle, all macroscopic objects are neutral. However, there may be charge transfer mechanisms that can cause an imbalance of charge in certain objects. For example, static electricity is produced when certain materials are rubbed against each other, such as wool against plastic or shoe soles against a carpet, where the rubbing process causes the electrons to be removed from the surface of a material and to relocate onto the surface of the other material that offers more favorable energy levels. This generates an electric imbalance in each object, leaving both with a net non-zero charge. These charged bodies are then capable of attracting smaller, lightweight materials such as small pieces of paper. The word *electricity* refers to these electric bodies, and derives from the ancient Greek word for amber (ἤλεχτρον), which is an electric body when rubbed.

Naturally, Coulomb established the electrostatic interaction law by studying the resultant force between charged macroscopic objects, and not between two fundamental particles in vacuum. Coulomb's law (1.1) between fundamental particles is a postulate which has never been contradicted by experiments. It was derived for macroscopic objects, but it can be inferred from the atomic spectra that Coulomb's law is true for length scales as small as the hydrogen atom, 1×10^{-10} m. It has also been verified from experiments involving nuclear physics, extending its validity to scales at least as small as the proton radius, 1×10^{-15} m.

1.3.2 Insulators and Conductors

Matter can be roughly classified into two categories. The first one is that of insulators (or dielectrics): those materials in which charges cannot move easily, since electrons are strongly bound to the host atoms; the transfer of charge to or from an insulating object is not favorable (although possible). Classic examples are glasses and plastics. The second category is that of conductors (typically metals): electrons in these materials can move freely through the medium; it is easy to add or remove charges from this type of object.

When a charged insulator is brought near to a conductor, the charges on the surface of the latter will be redistributed due to the presence of the charges in the insulator. As an example, if the insulator is positively charged, then due to the Coulomb attraction, negative charges will accumulate in the region of the conductor closest to the insulator, leaving a deficit of negative charges (and therefore a positive charge) in its other end, as shown in Fig. 1.6. This phenomenon is called polarization by induction. Although the total charge of the conductor remains zero, it is not evenly distributed on its surface, leaving areas with positive and negative charge.

1.3.3 How to Detect Charged Objects?

The electroscope is an instrument capable of determining if an object is electrically charged. Figure 1.7 shows a conductive sphere that is connected by a rod (equally conductive) to two very thin metal sheets. When an electrified object is brought close to the sphere, the latter is polarized by induction, transmitting to the plates a charge of the same sign as the object to be studied. This charge is equally distributed over both sheets, and as charges of the same sign are repelled, the electrostatic force will

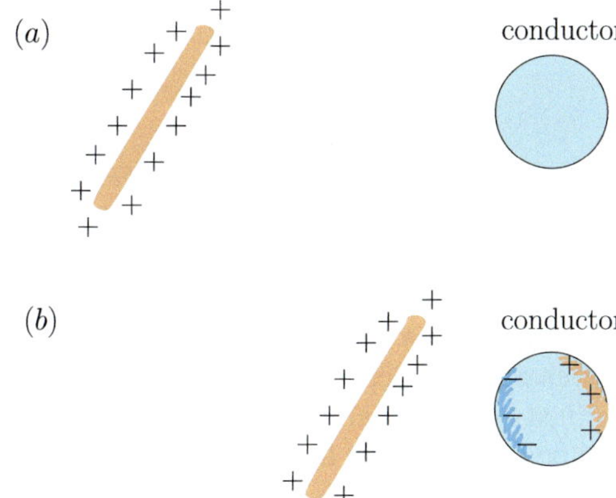

Fig. 1.6 a A positively charged rod slowly approaches an electrically neutral conductive sphere. **b** When both objects are close enough, the electrons in the metal will be attracted to the rod, leaving an electron deficit (positive charge) in the area opposite to it

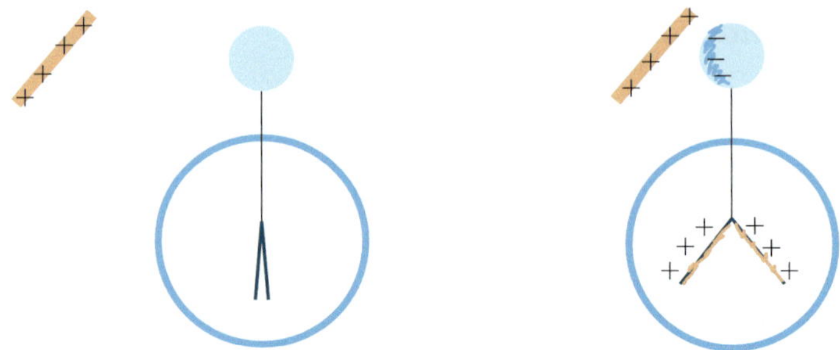

Fig. 1.7 A conductive sphere is connected to two thin sheets (equally conductive). When a charged rod is brought closer to the sphere, a charge of opposite sign is concentrated by induction on the upper part of the sphere. This implies that the charge of the same sign as that of the rod accumulates in both sheets, which separate due to Coulombian repulsion

cause a separation of both sheets, with this separation being a measure of the amount of charge they have received.

1.4 Force Between More Than Two Charges: Superposition Principle

Coulomb's law describes the interaction between two point charges, but it does not tell us what occurs when more than two charges are present. Experimentally, it has been observed that nature obeys the superposition principle stated below.

The superposition principle.
Suppose there are N charges q_j with $j \in \{1, 2, \ldots N\}$, and that the position of the jth charge is $\mathbf{x}_j$. When placing a charge q at the position $\mathbf{x}$, the force on q is given by the superposition principle:

$$\mathbf{F}_q = \sum_{j=1}^{N} \frac{q q_j}{4\pi \epsilon_0 |\mathbf{x} - \mathbf{x}_j|^3} (\mathbf{x} - \mathbf{x}_j) \,. \tag{1.2}$$

That is, the total force on q is the vector sum of Coulomb forces exerted on q by the other charges q_j. The superposition principle then states that the force between two charges is independent of the presence of other charges. The fact that this law is verified experimentally allowed the determination of Coulomb's law (valid between two point charges) in the first place, since it was obtained from the study of forces between two charged macroscopic objects on a torsion balance (where an extremely large number of elementary charges is involved). Indeed, the force between two elementary particles or between two macroscopic objects follows the same law of $1/r^2$ (provided the separation distance is much greater than the characteristic size of both objects).

1.5 The Electric Field

The electrostatic force, like the gravitational force, acts at a distance. Suppose we have two point charges, q and q_0, where the position of q is fixed. We can measure the force that q exerts on the test charge q_0 as the latter is placed at different positions in space. The charge q_0 will experience a force $\mathbf{F}_{q_0}(\mathbf{x})$ proportional to q_0 at every position $\mathbf{x}$ where it is located. The force per unit charge exerted on q_0 therefore solely depends on $\mathbf{x}$; it is independent of q_0. In this sense, we can say that the charge q generates a field at every point $\mathbf{x}$ in space that is independent of whether or not we place the charge q_0. It is said that an electric charge generates an *electric field* (mathematically described by a vector field), which is capable of acting on other charges. From this point of view, it is the electric field that is the mediator of the electrostatic interaction and transmits the electrostatic force at a distance.

Electric field
The electric field $\mathbf{E}$ at a point $\mathbf{x}$ is defined as the force experienced by a point charge q_0 at $\mathbf{x}$ divided by q_0:

$$\mathbf{E}(\mathbf{x}) = \frac{\mathbf{F}_{q_0}(\mathbf{x})}{q_0}. \tag{1.3}$$

In this way, the electric field is independent of the charge q_0, it is intrinsic to the system defined by all the other charges that are present. It is a force per unit charge, and so its unit is NC^{-1}.

As an example, Fig. 1.8 illustrates the electric field $\mathbf{E}$ generated by a negative charge at a distance r from it. A charge q_0 placed at this position will feel a force $\mathbf{F} = q_0\mathbf{E}$. The direction of the force will be given by both $\mathbf{E}$ and the sign of q_0.

Under this new perspective, we then say that charges are sources of electric fields. When a charge q_0 is in a region where there is an electric field $\mathbf{E}(\mathbf{x})$, the force on q_0 is $\mathbf{F}_{q_0} = q_0\mathbf{E}(\mathbf{x})$. Using this definition of the electric field and Coulomb's law, we conclude that the electric field at a distance r from a point charge q is given by

$$\boxed{\mathbf{E}(\mathbf{r}) = \frac{q}{4\pi\epsilon_0 r^2}\mathbf{u}_r \, ,} \tag{1.4}$$

Fig. 1.8 The figure shows the field at distance r from a negative charge $-q$. The force on a charge q_0 is simply $q_0\mathbf{E}$ (in this example, $q_0 > 0$)

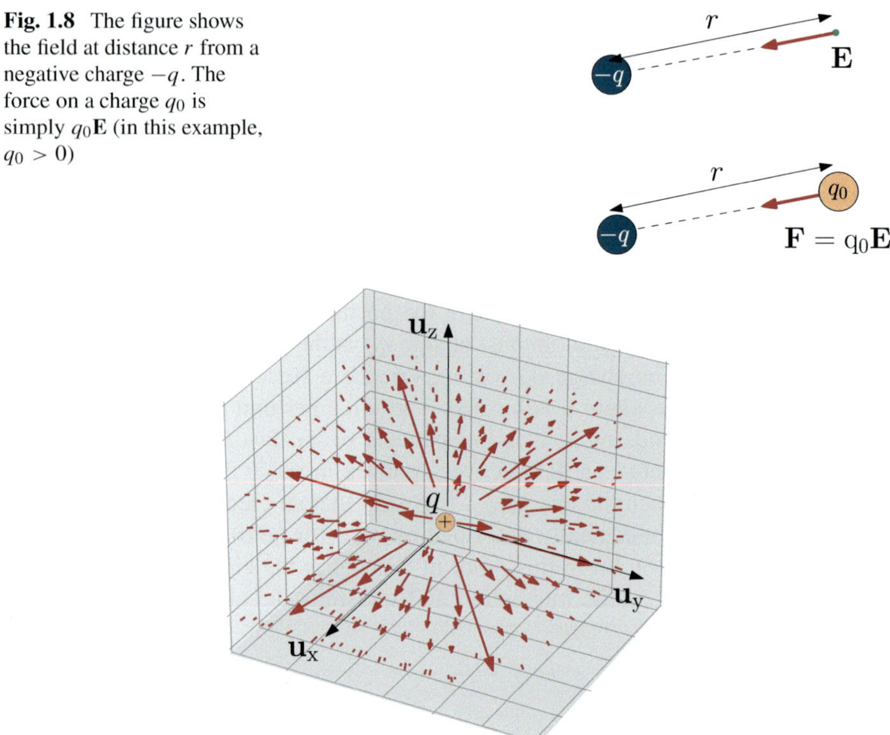

Fig. 1.9 Graphic representation of the electric field generated by a positive charge at the origin

where $\mathbf{u}_r$ is the unit vector in the radial direction in spherical coordinates with the origin at the position of the charge q. To graphically represent the electric field, one could draw the vector $\mathbf{E}(\mathbf{r})$ for every node $\mathbf{r}$ of a grid. Figure 1.9 shows the electric field representation for a positive point charge at the origin.

Instead of representing the electric field of a point charge in three dimensions, we can benefit from the fact that this is a radial field. The point charge at the origin generates a field such that there is no preferred direction in space. If we consider any plane that passes through the charge, then at every point on the plane the electric field is within the plane. We can therefore restrict ourselves to a simpler two-dimensional representation, as shown in Fig. 1.10, which shows separately the electric field generated by a positive and a negative point charge in a plane that intersects the charge.

Note that this representation has important limitations related to the fact that the electric field magnitude rapidly decays as the distance increases. If we plot the electric field too close to the charges, the magnitude of the arrows would be so large that it would overlap with all the other arrows. In addition, we rapidly lose track of the field behavior when the size of the field gets smaller than the arrow tip. In Sect. 1.5.1, we will see a more convenient way to represent electric fields.

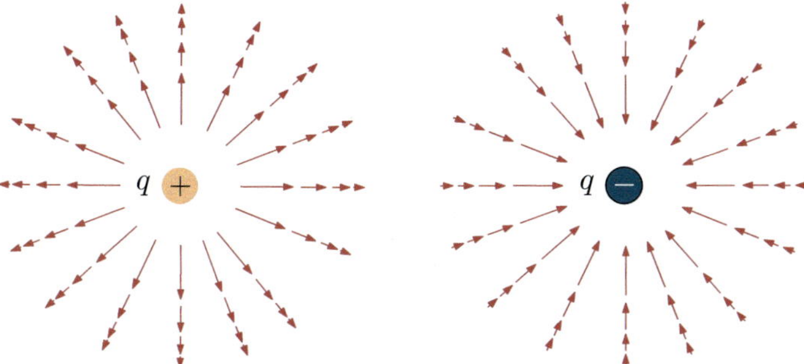

Fig. 1.10 Graphic representation of the electric field **E** generated by a positive (left) and negative (right) charge

Fig. 1.11 The superposition principle states that the electric field at a point P will be the vector sum of the fields generated separately by all charges except the charge located at P

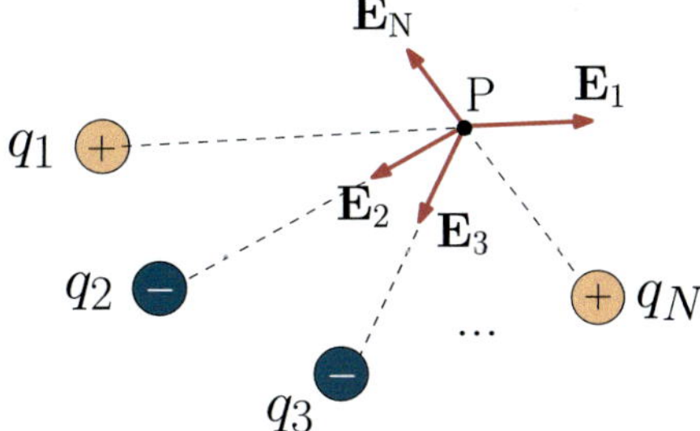

By construction, the superposition principle is also valid for the electric field, so that the field at $\mathbf{x}$ generated by a discrete distribution of N point charges located at $\mathbf{x}_j$, with $j \in \{1, 2, \ldots N\}$, can be written as

$$\mathbf{E}(\mathbf{x}) = \sum_{j=1}^{N} \frac{1}{4\pi\epsilon_0} \frac{q_j(\mathbf{x} - \mathbf{x}_j)}{|\mathbf{x} - \mathbf{x}_j|^3} \,.$$

Coulomb's law and the superposition principle are the *fundamental* empirical laws of electrostatics. Figure 1.11 shows the electric fields produced at point P by each point charge q_i present in the system. The total electric field at this point will be the vector sum of all these fields.

Example 1.2—Field of a dipole
An electric dipole is a system of two charges of equal magnitude but opposite signs, close to each other. It is very important in electrostatics since it is the

simplest model one can think of for a globally neutral molecule or atom. Consider two charges $\pm Q$ located at $x = \pm l$, respectively, as shown in Fig. 1.12. Let us calculate the total electric field on the x-axis.

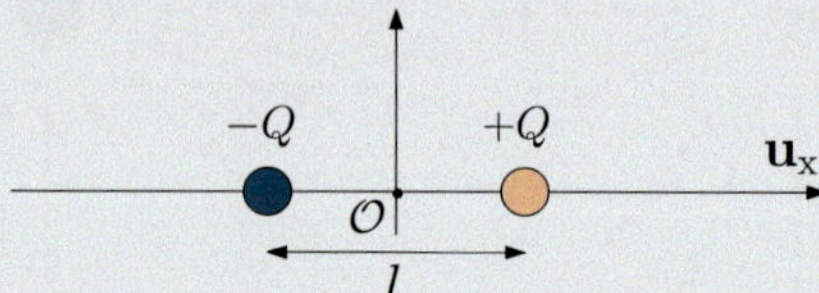

Fig. 1.12 Electric dipole

By the superposition principle, we obtain

$$\mathbf{E(x)} = \frac{1}{4\pi\epsilon_0}\left(\frac{Q}{(x+l)^2} - \frac{Q}{(x-l)^2}\right)\mathbf{u}_x , \quad x < -l ,$$

$$\mathbf{E(x)} = \frac{1}{4\pi\epsilon_0}\left(-\frac{Q}{(x+l)^2} - \frac{Q}{(x-l)^2}\right)\mathbf{u}_x , \quad -l < x < l ,$$

$$\mathbf{E(x)} = \frac{1}{4\pi\epsilon_0}\left(-\frac{Q}{(x+l)^2} + \frac{Q}{(x-l)^2}\right)\mathbf{u}_x , \quad x > l .$$

Figure 1.13 shows a graph of the field amplitude $E(x)$ along the x-axis.

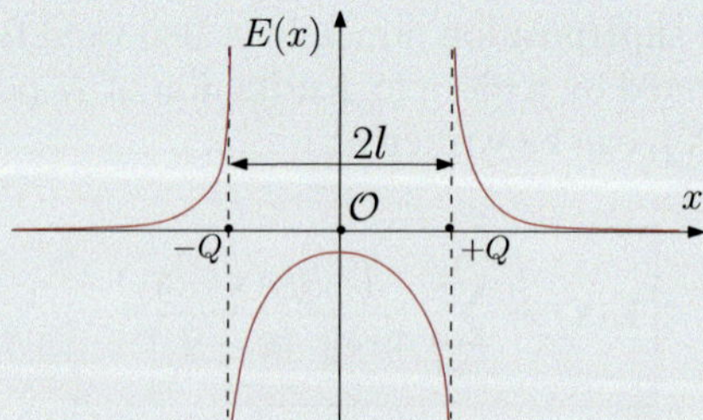

Fig. 1.13 Electric field magnitude along the axis of an electric dipole

It diverges at the position of the two point charges, and exhibits even parity ($E(x)=E(-x)$). We will see in the following chapter that this reflects the way the charges are distributed with respect to the $x = 0$ plane. In the limit when $x \gg l$, we can write up to first order in l/x

$$(x \pm l)^{-2} \simeq x^{-2}\left(1 \mp \frac{2l}{x}\right)$$

so that for $x \gg l$

$$\mathbf{E}(\mathbf{x}) \simeq \frac{Q}{4\pi\epsilon_0 x^2}\left(-1 + \frac{2l}{x} + 1 + \frac{2l}{x}\right)\mathbf{u}_x = \frac{lQ}{\pi\epsilon_0 x^3}\,\mathbf{u}_x\ .$$

We see that at large distances the field generated by a dipole decays as $1/r^3$, much faster than the field of a single charge due to the mutual screening of both charges.

1.5.1 Electric Field Lines

A field line is a curve Γ that is parallel to the electric field at every point $\mathbf{x} \in \Gamma$, i.e., $\mathbf{E}(\mathbf{x}) \parallel d\mathbf{x}$, where $d\mathbf{x}$ is an infinitesimal displacement vector, tangent to the curve at $\mathbf{x}$, see Fig. 1.14. To find a mathematical expression for the field lines, one can look for a curve parametrization such that, at every point $\mathbf{x}$ on the curve, $d\mathbf{x} \times \mathbf{E}(\mathbf{x}) = \mathbf{0}$. Since the electric field is uniquely defined at every point, only one field line passes through any given point in space; this means that the field lines do not intersect. In addition, in electrostatics the electric field lines never close on themselves; this property will be shown later in Chap. 3. The field lines start from positive charges (or infinity) and end at negative charges (or infinity).

Let us find as an example the field lines for the electric field of a point charge at the origin. As stated in Sect. 1.5, we can restrict the study to a plane containing the charge. In polar coordinates, any given point $\mathbf{x}$ is defined by its distance r to the origin and the angle θ that the vector $\mathbf{x}$ makes with respect to the x-axis. An infinitesimal change in coordinates $r + dr$ and $\theta + d\theta$ leads to a new position $\mathbf{x} + d\mathbf{x}$. As shown in Fig. 1.15, one can write such displacement as $d\mathbf{x} = dr\,\mathbf{u}_r + r\,d\theta\,\mathbf{u}_\theta$. At point $\mathbf{x}$, the electric field is of the form $\mathbf{E}(\mathbf{x}) = E(r)\mathbf{u}_r$, so that the field line passing through $\mathbf{x}$

Fig. 1.14 Electric field lines are curves such that the electric field is tangent to the curve at every point

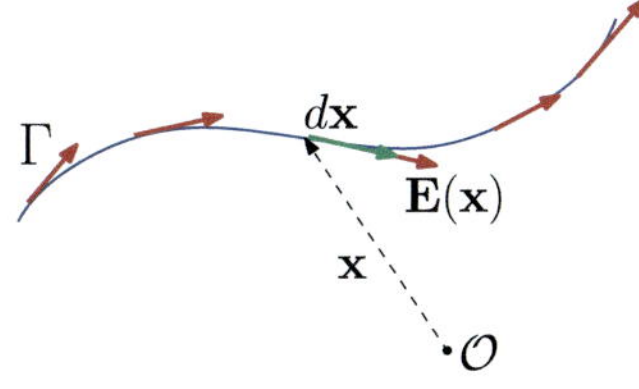

Fig. 1.15 An infinitesimal displacement $d\mathbf{x}$ in polar coordinates

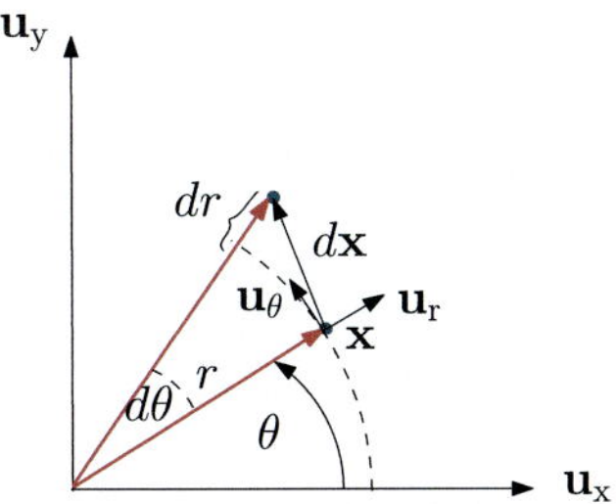

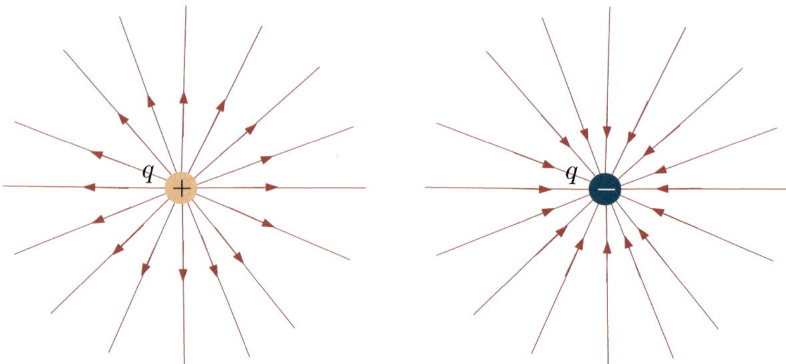

Fig. 1.16 Electric field lines of a positive (left) and negative (right) point charge

must be a solution of the equation

$$dx \times \mathbf{E}(\mathbf{x}) = -rE(r)d\theta \mathbf{u}_z = \mathbf{0} \,,$$

that is, $d\theta = 0$, whose solution is simply $\theta = \text{Cst}$. For each value of Cst, the solution is a radial line passing through the origin. The set of field lines for a point charge therefore consists of all the radial lines passing through the charge. Note that the sign of the charge will determine whether the field is radially pointing toward or away from the charge. It is customary to plot arrows along the field lines to explicitly show the direction of the electric field. Figure 1.16 shows the electric field lines for a positive and a negative point charge.

In example 1.2, the field in the axis of a dipole was calculated. A more detailed calculation can be performed to obtain the field at every point in space, as will be shown in Sect. 3.4. Figure 1.17 shows the field lines generated by such an electric dipole in a plane intersecting both charges. Note that these field lines can be inferred from a superposition of the field lines of the positive and negative charges that constitute the dipole, at least qualitatively. The advantage of such representation is that it is easy to see the direction of the electric field, even at points where its magnitude is small. The magnitude of the electric field is represented by the field line density, that is, the number of field lines per unit perpendicular area.

1.6 From Discrete to Continuous Charge Distributions

Any piece of matter is made of elementary particles, and so is described by a discrete set of point charges separated by vacuum. In a solid, for example, neighboring atoms are separated by a distance of the order of $a \sim 1 \times 10^{-10}$ m, so that the electric field exhibits rapid variations over distances of the order of a. However, when dealing with macroscopic objects, we are not capable of resolving such rapid variations at

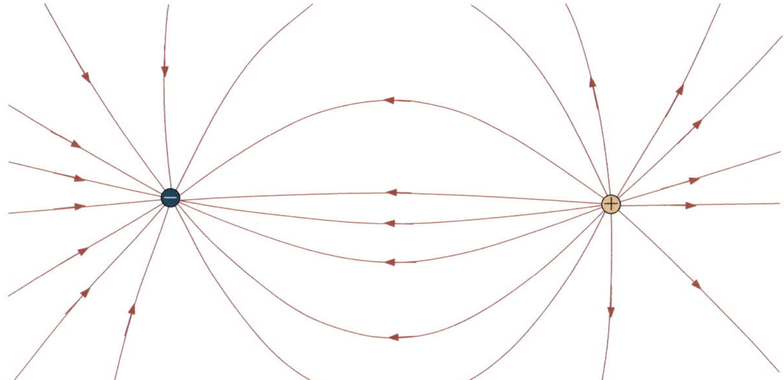

Fig. 1.17 Graphic representation of the electric field lines of an electric dipole

the atomic scale. As a consequence, we are only interested in averaged values of both the charge spatial distribution and the resulting electric field. The idea is therefore to replace a discrete set of charges by a continuous medium in which a smoothly varying charge density can be defined at every point.

Let Ω be the volume occupied by a macroscopic object. At any point $\mathbf{x}' \in \Omega$, consider a sub-volume d^3x' around $\mathbf{x}'$ such that $a^3 \ll d^3x' \ll \Omega$, containing a very large number N of point charges (see Fig. 1.18). Within d^3x', there is a total charge $dq(\mathbf{x}') = \sum_{i=1}^{N} q_i$.

The idea is to consider d^3x' as being a homogeneously charged volume, that is, we associate with $\mathbf{x}'$ a certain average charge density per unit volume $\varrho(\mathbf{x}')$ (Cm^{-3}) such that

$$dq(\mathbf{x}') = \varrho(\mathbf{x}')d^3x' \ .$$

At this stage, the real charge density has been replaced by its spatial average over a length scale much larger than a. The field generated at a point $\mathbf{x}$ by $dq(\mathbf{x}')$ will be approximately that of a point charge located at $\mathbf{x}'$:

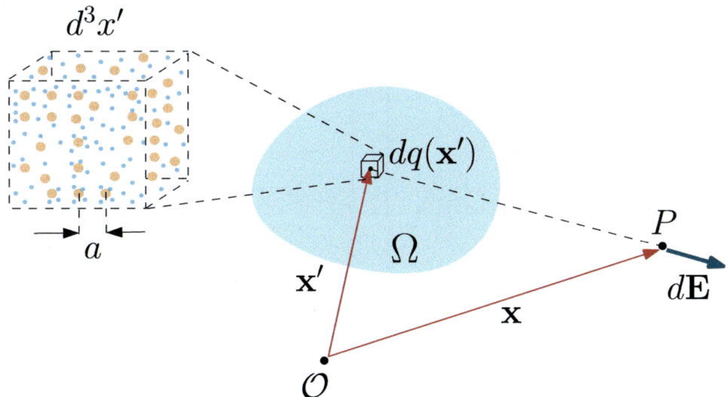

Fig. 1.18 A macroscopic object is here represented by a volume Ω

$$d\mathbf{E}(\mathbf{x}) = \frac{\varrho(\mathbf{x}')d^3x'}{4\pi\epsilon_0}\frac{\mathbf{x}-\mathbf{x}'}{|\mathbf{x}-\mathbf{x}'|^3} \, .$$

To obtain the field generated by the whole charge distribution, it is enough, according to the superposition principle, to add the individual contributions of each elementary volume within Ω. We therefore sum over the possible values of $\mathbf{x}' \in \Omega$ and the following integral formula is obtained in the limit $d^3x' \to 0$:

$$\mathbf{E}(\mathbf{x}) = \frac{1}{4\pi\epsilon_0}\iiint_\Omega \frac{\varrho(\mathbf{x}')}{|\mathbf{x}-\mathbf{x}'|^3}(\mathbf{x}-\mathbf{x}')d^3x' \, . \tag{1.5}$$

When one of the dimensions of Ω is much smaller than the other ones (for example, when the charge is concentrated over a very small thickness), one may go further and consider a two-dimensional charge distribution over a surface S, so that at each $\mathbf{x}' \in S$ we associate a surface charge density, denoted σ (Cm^{-2}) to distinguish it from a volume charge density, and

$$\mathbf{E}(\mathbf{x}) = \frac{1}{4\pi\epsilon_0}\iint_S \frac{\sigma(\mathbf{x}')}{|\mathbf{x}-\mathbf{x}'|^3}(\mathbf{x}-\mathbf{x}')dS' \, .$$

Finally, for a one-dimensional charge distribution over a curve Γ, we consider a line charge density, typically denoted λ (Cm^{-1}), so that the field at $\mathbf{x}$ is given by

$$\mathbf{E}(\mathbf{x}) = \frac{1}{4\pi\epsilon_0}\int_\Gamma \frac{\lambda(\mathbf{x}')}{|\mathbf{x}-\mathbf{x}'|^3}(\mathbf{x}-\mathbf{x}')dl' \, .$$

A more rigorous justification of this method of spatial averaging is given in the Appendix A.3, where it is shown that the field given by (1.5) is a smoothed version of the field generated by point charges. Figure 1.19 (left) illustrates an example of the field lines produced by a set of point charges. One can associate with this discrete

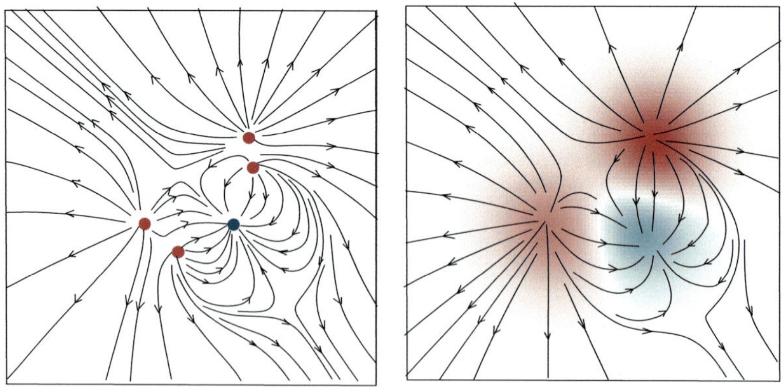

Fig. 1.19 Left: electric field lines of a discrete set of point charges. Right: electric field lines of a smoothed version of the charge density, now a continuous function of space

charge density a continuous charge density, ϱ, which represents a smoothed version of the true density. The resulting field generated by this continuous density, shown in Fig. 1.19 (right), is likewise smoothed relative to the true field.

Example 1.3—Uniformly charged ring
Consider a ring of radius R with uniform line charge density λ. Let us calculate the electric field generated by this charge distribution at any point of its axis of symmetry, at distance z from the center of the ring.

A differential length element on the ring is given, in cylindrical coordinates, by $dl = Rd\theta$. The associated charge is $dq = \lambda dl = \lambda Rd\theta$, so that the electric field at a point $\mathbf{x}$ (see Fig. 1.20) due to this infinitesimal charge is given by

$$d\mathbf{E}(\mathbf{x}) = \frac{dq}{4\pi\epsilon_0}\frac{\mathbf{x}-\mathbf{x}'}{|\mathbf{x}-\mathbf{x}'|^3} = \frac{d\theta\, R\lambda}{4\pi\epsilon_0}\frac{\mathbf{x}-\mathbf{x}'}{|\mathbf{x}-\mathbf{x}'|^3}\,,$$

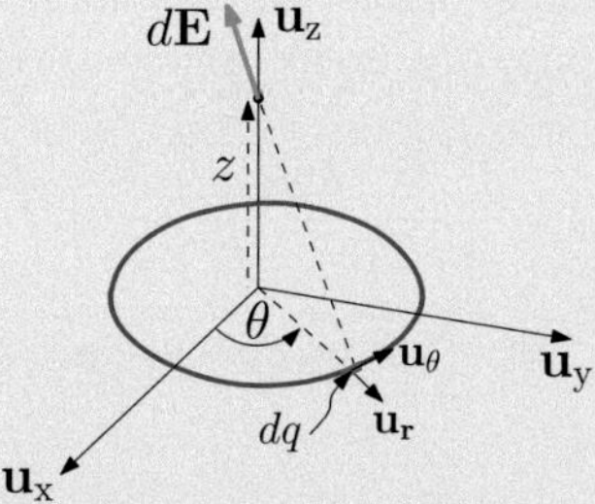

Fig. 1.20 A uniformly charged ring

where $\mathbf{x}' = R\mathbf{u}_r$ is the position of this charge element, so that the total field at a point $\mathbf{x} = z\mathbf{u}_z$ on the axis of symmetry of the ring is obtained by integrating over the entire ring:

$$\mathbf{E}(\mathbf{x} = z\mathbf{u}_z) = \frac{R\lambda}{4\pi\epsilon_0}\int_0^{2\pi} d\theta\,\frac{z\mathbf{u}_z - R\mathbf{u}_r}{(z^2 + R^2)^{3/2}}\,,$$

where we used $|\mathbf{x}-\mathbf{x}'| = (z^2 + R^2)^{1/2}$. Of these two integrals (along $\mathbf{u}_z$, $\mathbf{u}_r$), only the integral along $\mathbf{u}_z$ is non-zero. This is easy to anticipate by appealing to the symmetry of the charge distribution: every infinitesimal charge element dq in the ring generates an electric field $d\mathbf{E}$ and has a diametrically opposite charge element $dq' = dq$ generating an electric field $d\mathbf{E}'$ whose radial component is opposite to that of $d\mathbf{E}'$, so that the sum of both the electric fields is a vector along $\mathbf{u}_z$ (see Fig. 1.21).

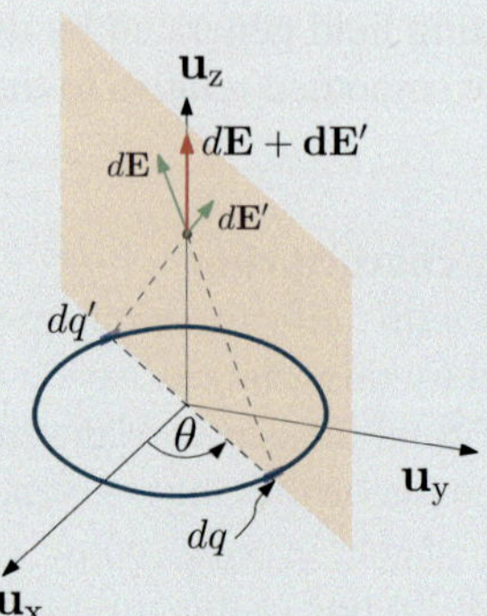

Fig. 1.21 Two diametrically opposite charge elements produce a vertical electrical field on the axis of the ring

We find a strong relationship between the symmetry of the charge distribution and the spatial orientation of the electric field. In the next chapter, we will see that this is a consequence of Curie's principle. We have

$$\mathbf{E}(z\mathbf{u}_z) = \frac{Rz\lambda}{4\pi\epsilon_0(z^2 + R^2)^{3/2}}\mathbf{u}_z \int_0^{2\pi} d\theta$$

$$= \frac{Q}{4\pi\epsilon_0} \frac{z}{(z^2 + R^2)^{3/2}}\mathbf{u}_z \, ,$$

where the result was expressed as a function of the total charge of the ring, $Q = 2\pi R\lambda$. Using $\cos\alpha = z/(z^2 + R^2)^{1/2}$, $\sin\alpha = R/(z^2 + R^2)^{1/2}$, this result can be rewritten as

$$\mathbf{E}(z\mathbf{u}_z) = \frac{Q}{4\pi\epsilon_0 R^2} \cos\alpha \sin^2\alpha\, \mathbf{u}_z = \frac{Q}{4\pi\epsilon_0 R^2}(c - c^3)\mathbf{u}_z,$$

with $c = \cos\alpha$. The magnitude of the electric field is maximum when $c - c^3$ is maximum, i.e., when $c = \cos\alpha = 1/\sqrt{3}$, that is, for $z = R/\sqrt{2}$.

Remarks

- At $z = 0$, $\mathbf{E} = 0$, because the fields generated by diametrically opposite charges cancel each other.
- If z is sufficiently large compared to the dimensions of the ring ($z \gg R$), then at lowest order in (R/z)

$$\mathbf{E}(z\mathbf{u}_z) \simeq \frac{Q}{4\pi\epsilon_0 z^2}\mathbf{u}_z,$$

which corresponds to the Coulomb field of a point charge Q at distance z. Far from the ring, the field generated by the ring of total charge Q is the same as the Coulomb field generated by a point charge Q located at the origin.

1.7   The Dirac Distribution, the Mathematical Problem of Dealing with Point Charges

As seen in Sect. 1.6, performing a spatial average of discrete charge distributions has allowed us to model a charged object Ω by a function $\varrho : \Omega \to \mathbb{R}$ which is continuous inside Ω and that represents the averaged charge density per unit volume at all points of Ω. This density allows the electric field to be calculated in all space by means of the integral (1.5). Let us now reverse the problem: can we express a point charge via a charge density ϱ? If yes, this would mean that we can consider (1.5) as a completely general expression of the electric field, whether it is generated by a discrete set of charges or a continuous charge distribution.

Indeed, in classical physics, it is frequent to resort to the notion of a point object to deal, for example, with fundamental particles such as the electron or with objects whose dimensions are negligible with respect to the relevant length scales of the problem. Here, our objective is to assign to a charge q located at point $\mathbf{x}_0$ a charge density ϱ. Assuming that this charge density exists, it must vanish everywhere except at x_0, that is $\varrho(\mathbf{x}, \mathbf{x_0}) = 0$ if $\mathbf{x} \neq \mathbf{x_0}$, and its integral over the whole space must represent the total charge equal to q, $\iiint_{\mathbb{R}^3} \varrho(\mathbf{x}, \mathbf{x_0})d^3x = q$. In addition, by replacing ϱ in the integral (1.5), we should obtain the electric field of a point charge:

$$\frac{1}{4\pi\epsilon_0} \iiint_{\mathbb{R}^3} \varrho(\mathbf{x'}, \mathbf{x_0}) \frac{(\mathbf{x} - \mathbf{x'})}{|\mathbf{x} - \mathbf{x'}|^3} d^3x' = \frac{q}{4\pi\epsilon_0} \frac{(\mathbf{x} - \mathbf{x_0})}{|\mathbf{x} - \mathbf{x_0}|^3} . \tag{1.6}$$

We see that the role of $\varrho(\mathbf{x'}, \mathbf{x_0})$ in this integral is tantamount to extracting the value of the integrand at $\mathbf{x'} = \mathbf{x_0}$. The property (1.6) cannot be fulfilled by a classical function ϱ that is zero everywhere except at a single point, as it would make the integral in (1.6) equal to zero.

However, Eq. (1.6) is satisfied by writing $\varrho(\mathbf{x}, \mathbf{x_0}) = q\delta(\mathbf{x} - \mathbf{x_0})$, where $\delta(\mathbf{x} - \mathbf{x_0})$ is not a function but a generalized function called the Dirac *distribution* centered at $\mathbf{x_0}$, named after Paul Dirac (Fig. 1.22). The theory of distributions was properly established by the French mathematician Laurent Schwartz (1915–2002), providing the mathematical framework to generalize the classical notion of functions. While Schwartz's theory will not be treated in this textbook, we will use the Dirac distribution as a practical tool to deal with discrete distributions of charges. Intuitively, the Dirac distribution acts as weight used only under an integral. It can be seen as the limit of a sequence of functions $(\delta_n)_{n\in\mathbb{N}}$, such that $\lim_{n\to\infty} \delta_n(\mathbf{x}) = 0$ for $\mathbf{x} \neq \mathbf{0}$ and such that $\lim_{n\to\infty} \iiint_{\mathbb{R}^3} \delta_n(\mathbf{x})\varphi(\mathbf{x})d^3x = \varphi(\mathbf{0})$ for every regular function φ. For

Fig. 1.22 Paul Dirac (1902–1984), a British physicist and one of the founders of quantum mechanics. His famous Dirac equation describes the quantum behavior of relativistic electrons and allowed him to predict the existence of antimatter

example, a possible candidate is $\delta_n(\mathbf{x}) = n$ for $|\mathbf{x}| < 1/n^{1/3}$ and $\delta_n(\mathbf{x}) = 0$ otherwise. By means of an abuse of notation, we can then define the Dirac *distribution* as $\delta(\mathbf{x}) \equiv \lim_{n\to\infty} \delta_n(\mathbf{x})$. It satisfies the required properties, namely, $\delta(\mathbf{x} - \mathbf{x}_0) = 0$ for $\mathbf{x} \neq \mathbf{x}_0$ and

$$\iiint_{\mathbb{R}^3} \delta(\mathbf{x} - \mathbf{x}_0)\varphi(\mathbf{x})d^3x = \lim_{n\to\infty} \iiint_{\mathbb{R}^3} \delta_n(\mathbf{x} - \mathbf{x}_0)\varphi(\mathbf{x})d^3x = \varphi(\mathbf{x}_0) \; . \tag{1.7}$$

The integral on the left should always be interpreted as the limit shown on the right of Eq. (1.7), although for simplicity we will never write this limit again.

The Dirac distribution and charge density of a point charge
The Dirac distribution centered at $\mathbf{x}_0$ is noted $\delta(\mathbf{x} - \mathbf{x}_0)$ and is defined by

$$\iiint_{\mathbb{R}^3} \delta(\mathbf{x} - \mathbf{x}_0)\varphi(\mathbf{x})d^3x = \varphi(\mathbf{x}_0) \; . \tag{1.8}$$

for every regular function φ. The charge density associated to a point charge q located at $\mathbf{x}_0$ writes

$$\varrho(\mathbf{x}, \mathbf{x}_0) = q\delta(\mathbf{x} - \mathbf{x}_0) \; . \tag{1.9}$$

Sometimes it is tempting to manipulate δ as if it were a function. As an example, a fundamental identity of electrostatics involving the Dirac distribution found in most textbooks is the following:

$$\boxed{\nabla^2\left(\frac{1}{|\mathbf{x}|}\right) = -4\pi\delta(\mathbf{x}) \; ,} \tag{1.10}$$

where ∇^2 is the Laplace operator. However, one must be careful with the interpretation of (1.10), since the left term is a function whereas the right term is not. A more appropriate way to write (1.10) would be:

$$\iiint_{\mathbb{R}^3} \varphi(\mathbf{x}) \nabla^2 \left(\frac{1}{|\mathbf{x}|} \right) d^3x = -4\pi \varphi(0) , \tag{1.11}$$

where φ is a sufficiently regular function. Indeed, from (1.8) we see that $\nabla^2 \left(\frac{1}{|\mathbf{x}|} \right)$ behaves (behind an integral) as the distribution $-4\pi \delta(\mathbf{x})$. Writing (1.10) instead of (1.11) is clearly more convenient despite its lack of mathematical rigor, and this is the way most physicists would write the identity. A brief discussion of the Delta distribution and its main properties are summarized in the Appendix A.2.

1.8 Asymptotic Behavior of the Electric Field

From the integral (1.5), we see that the magnitude of the electric field satisfies

$$|\mathbf{E}(\mathbf{x})| = \frac{1}{4\pi\epsilon_0} \left| \iiint_{\mathbb{R}^3} \frac{\varrho(\mathbf{x}')(\mathbf{x}-\mathbf{x}')}{|\mathbf{x}-\mathbf{x}'|^3} d^3x' \right| \leq \frac{1}{4\pi\epsilon_0} \iiint_{\mathbb{R}^3} \frac{|\varrho(\mathbf{x}')|}{|\mathbf{x}-\mathbf{x}'|^2} d^3x' ,$$

so that

$$|\mathbf{x}|^2 |\mathbf{E}(\mathbf{x})| \leq \frac{1}{4\pi\epsilon_0} \iiint_{\mathbb{R}^3} |\varrho(\mathbf{x}')| \frac{|\mathbf{x}|^2}{|\mathbf{x}-\mathbf{x}'|^2} d^3x' .$$

Then,

$$\lim_{|\mathbf{x}| \to \infty} |\mathbf{x}|^2 |\mathbf{E}(\mathbf{x})| \leq \frac{1}{4\pi\epsilon_0} \iiint_{\mathbb{R}^3} |\varrho(\mathbf{x}')| d^3x' .$$

We see that if ϱ is an integrable function ($\iiint_{\mathbb{R}^3} |\varrho(\mathbf{x}')| d^3x' < \infty$) the electric field $\mathbf{E}$ defined by the integral (1.5) decays as or faster than $1/|\mathbf{x}|^2$ at infinity:

$$\iiint_{\mathbb{R}^3} |\varrho(\mathbf{x}')| d^3x' < \infty \Rightarrow \lim_{|\mathbf{x}| \to \infty} |\mathbf{x}|^2 |\mathbf{E}(\mathbf{x})| < \infty .$$

There are some common examples of charge distributions for which ϱ is not integrable (an infinite line of charge, or an infinite charged plane). In those cases, the electric field decays slower than $1/r^2$ at infinity (the field of the charge line decays as $1/r$, whereas the field of an infinite plane is uniform in space). These examples are limit cases that should only be considered as approximations for points sufficiently close to a charge line of finite length or to a finite plane, respectively. Since every real charge distribution is necessarily bounded, the electric field always decays at least as $1/r^2$ far from its sources.

1.9   Summary and Essential Formulas

- Electric charges generate an electric field $\mathbf{E}(\mathbf{x})$, which is a vector field defined over all space. Physically, $\mathbf{E}(\mathbf{x})$ represents the force per unit of charge that a charge placed at position $\mathbf{x}$ would experience.
- The force experienced by a charge q in the presence of an electric field $\mathbf{E}$ is

$$\boxed{\mathbf{F}_q = q\mathbf{E}\,.}$$

- The electric field generated by a charge density $\varrho(\mathbf{x})$, which is a scalar field defined over all space, is given by the Coulomb integral (1.5)

$$\boxed{\mathbf{E}(\mathbf{x}) = \frac{1}{4\pi\epsilon_0} \iiint_{\mathbb{R}^3} \frac{\varrho(\mathbf{x}')(\mathbf{x}-\mathbf{x}')}{|\mathbf{x}-\mathbf{x}'|^3} d^3x'\,.}$$

This equation is not only valid when ϱ represents the charge density of a continuous medium, but also in the case of a discrete distribution of charges. Thanks to the Dirac distribution, the charge density associated with a point charge q located at $\mathbf{x}_0$ can be written as $\varrho(\mathbf{x}) = q\delta(\mathbf{x}-\mathbf{x}_0)$. In this way, we obtain the electric field of a point charge (Coulomb's law):

$$\mathbf{E}(\mathbf{x}) = \frac{1}{4\pi\epsilon_0} \iiint_{\mathbb{R}^3} q\delta(\mathbf{x}'-\mathbf{x}_0) \frac{(\mathbf{x}-\mathbf{x}')}{|\mathbf{x}-\mathbf{x}'|^3} d^3x'\,,\ \text{ i.e. } \boxed{\mathbf{E}(\mathbf{x}) = \frac{q}{4\pi\epsilon_0} \frac{(\mathbf{x}-\mathbf{x}_0)}{|\mathbf{x}-\mathbf{x}_0|^3}\,.}$$

- Far from any charged volume Ω of finite size, the electric field decays at least as rapidly as $1/r^2$. If the system is globally neutral (total charge in Ω equals zero), the electric field will decay at least as rapidly as $1/r^3$ (case of an electric dipole).

Problems

1.1 Operation of an electroscope
Consider two massless rods of length l. Each rod carries a charge q of mass m at its end. The rods form an angle θ with the vertical axis. Express the charge q as a function of m, l, θ and the gravitational constant g. Is it possible to determine the sign of this charge? Assuming that $l = 1\,\text{m}$, $m = 0.5\,\text{kg}$, and that the precision in the measurement of θ is 1 deg, what is the minimum charge that can be measured with this electroscope?

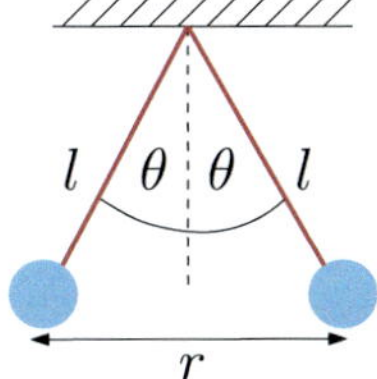

1.2 Superposition of forces

Four equal positive charges Q are placed at the corners of a square of side length L in a horizontal plane. As shown below, a positive charge q of mass m is placed at a height h above the center of the square. Find h such that the charge q will be in equilibrium.

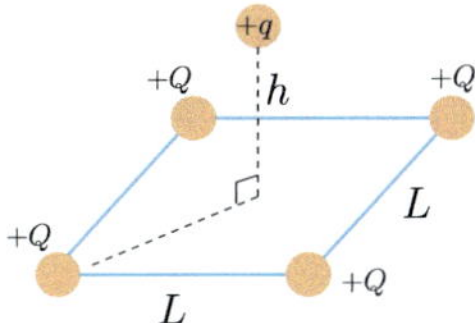

1.3 Field of a quadrupole

Consider the distribution of point charges. It is globally neutral. Calculate the electric field at a point P located at distance r from the origin on the axis that passes through the central charge $+2Q$. How does the magnitude of the electric field vary with the distance r when $r \gg d$?

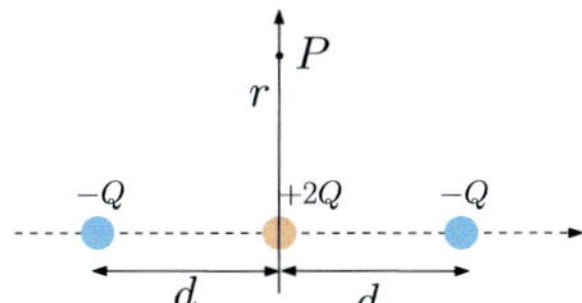

1.4 Uniformly charged disk

Consider a disk of radius R, uniformly charged with surface charge density σ. Calculate the electric field on the axis of the disk at distance z from its center. Calculate the field in the limit $R \to \infty$ (which would correspond to the case of a charged infinite plane).

1.5 Superposition of two known distributions

Consider an infinite plane of charge with surface density $\sigma > 0$. A circular hole of radius R is made, as shown below.

(a) Calculate the electric field at any point belonging to the axis of the hole (the z-axis), perpendicular to the plane.

(b) Along the axis of the hole is placed a charge line of length a (line segment $0 \le z \le a$), charge density $\lambda > 0$ per unit length and whose nearest point is located at a distance μ from the center of the hole. Calculate the strength of the repulsion experienced by the charge line.

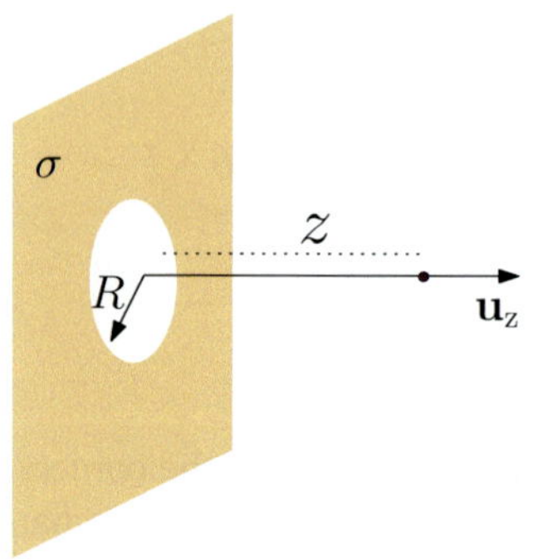

1.6 Field on the axis of a charged line

Consider a homogeneous line of charge with linear charge density λ, and finite length a. Calculate the electric field at point P located at distance x from the end.

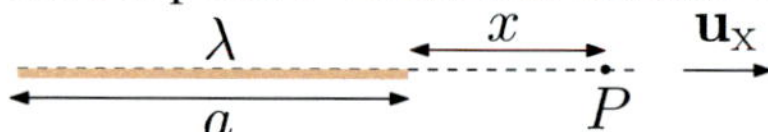

1.7 Off-axis field of a charged line

Consider a rectilinear charge distribution with uniform density λ. A point P is at distance r from the distribution, and its projection on the axis of the charged line lies at distances l_1 and l_2 from its ends.

(a) Calculate the electric field at P.
(b) What happens when r is much larger than l_1 and l_2?
(c) Make an approximation when r is much smaller than l_1 and l_2. What happens when l_1 and l_2 tend to infinity?

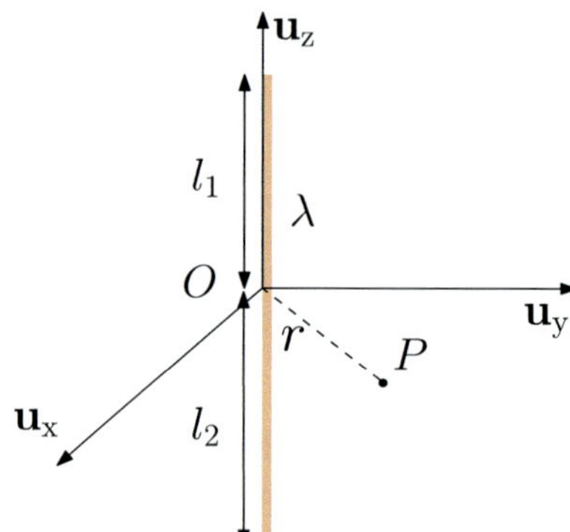

1.8 Force between a charged ribbon and a charged line

Calculate the force per unit length that a very long surface distribution of width b and surface charge density σ exerts on an equally long wire carrying a charge density λ per unit length. The wire is placed in the same plane as the ribbon, a distance r away from the bottom edge of the ribbon, as shown below.

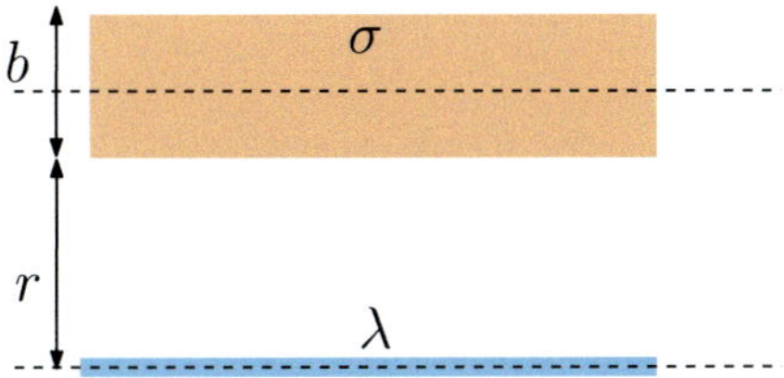

1.9 Asymmetric rod

A non-conducting rod is bent into a semicircle of radius R. A charge $+q$ is uniformly distributed along the rod in the upper half and a charge $-q$ is evenly distributed along the rod in the lower half, as shown below. Determine the magnitude and direction of the electric field at the center of the semicircle.

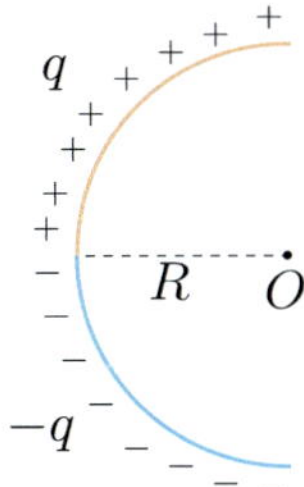

1.10 Circle arc

A thin bar with uniform charge density per unit length λ is bent into the shape of an arc of a circle of radius R. The angle subtended by the arc is equal to $2\theta_0$, symmetric with respect to the x-axis, as shown below. What is the electric field $\mathbf{E}$ at the origin O? What happens when $\theta_0 \to \pi$ and when $\sin\theta_0 \ll 1$?

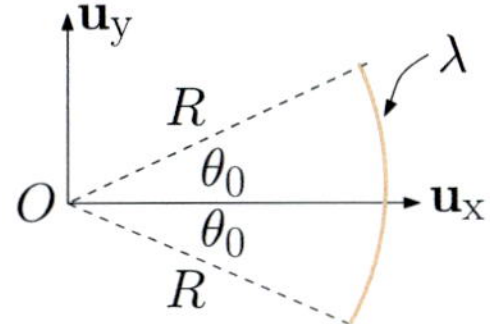

1.11 Charged hemisphere

A hemisphere of radius a carries a total charge Q evenly distributed on its surface. Find the electric field at the center of curvature.

1.12 Discrete distribution as a volume density

Show that a discrete distribution of charges q_i, located at points $\mathbf{x}_i$, with $i \in \{1, 2, \ldots N\}$, is consistent with a volume charge distribution given by

$$\rho(\mathbf{x}) = \sum_{i=1}^{N} q_i \delta(\mathbf{x} - \mathbf{x}_i)$$

Chapter 2
Gauss's Law and Symmetry Properties

Abstract This chapter presents *Gauss's law*, a fundamental principle in electrostatics that provides an alternative and often simpler method for calculating electric fields, particularly for charge distributions exhibiting spatial symmetries. Building upon the foundational concepts of Coulomb's law and the superposition principle introduced previously, the chapter first defines the *flux of a vector field through a surface*, both for planar and arbitrary closed surfaces. The core of the chapter is the formal introduction of *Gauss's law*, presenting both its integral and differential forms. It highlights the profound relationship between the electric flux through a closed surface and the enclosed charge, demonstrating how this law is a direct consequence of Coulomb's law. A significant emphasis is placed on leveraging *symmetry arguments in electrostatics* through *Curie's principle*, which dictates that the symmetries of a charge distribution are reflected in its resulting electric field. This principle is then applied to analyze invariances under spatial translation, rotation, and mirror symmetries (and antisymmetries), providing powerful tools to constrain the direction and spatial dependence of the electric field. Practical applications of Gauss's law are illustrated through detailed examples, such as determining the electric field generated by a uniformly charged sphere and an infinite plane of charge. The chapter also includes a proof of Gauss's law and a summary of essential formulas, reinforcing its utility as a powerful analytical tool in electrostatics, especially when dealing with highly symmetric charge configurations.

Keywords Gauss's law · Electrostatics · Curie's principle · Symmetry · Invariances

2.1 Introduction

In Chap. 1, we introduced the fundamental laws of electrostatics. Coulomb's law for a point charge and the superposition principle led to the integral Eq. (1.5) which allows us to calculate the electric field at any point in space, provided the charge distribution $\varrho : \mathbb{R}^3 \mapsto \mathbb{R}$ is known:

© The Author(s), under exclusive license to Springer Nature Switzerland AG 2025

F. Cadiz and A. Couairon, *Classical Electrodynamics*, Undergraduate Texts in Physics,
https://doi.org/10.1007/978-3-031-86785-9_2

$$E(x) = \frac{1}{4\pi \epsilon_0} \iiint_{\mathbb{R}^3} \frac{\varrho(x')(x - x')}{|x - x'|^3} d^3 x'. \tag{2.1}$$

Equation (2.1) fully resolves the problem of electrostatics, provided the spatial distribution ϱ is known. However, in many situations, the spatial charge distribution is not known a priori, rendering (2.1) insufficient in practice. Generally, the charge density ϱ depends on the electric field, while the electric field, in turn, depends on the charge distribution. This creates a self-consistent problem that requires additional mathematical tools and laws for resolution.

In this chapter, we introduce Gauss's law, which establishes a profound relationship between the spatial variation of the electric field and the charge density. This law is a direct consequence of the fundamental Eq. (2.1). To formulate Gauss's law mathematically, we first introduce the concept of the flux of a vector field through a surface.

We will demonstrate how Gauss's law simplifies the calculation of the electric field when the charge distribution exhibits invariances and symmetries. For this, we also introduce Curie's principle, which asserts that the symmetries of the causes are reflected in their effects. This principle proves highly useful in electrostatics, as it allows us to constrain the orientation of the electric field and the dependencies of its components solely based on symmetry arguments.

2.2   Flux of a Vector Field Through a Surface

2.2.1   Flux Through a Plane Surface

Consider the plane surface of Fig. 2.1 with area S and normal n (unit vector perpendicular to the surface at all points). If a uniform vector field E is present, the flux of E through S is defined as

$$\Phi_{S,E} = E \cdot n S = |E| S \cos \theta,$$

where θ denotes the angle between E and n. The flux of a vector field through a surface is thus a scalar quantity proportional to the field component normal to the surface. The sign of the flux depends on the relative orientation of E and the normal n. By defining the vector surface $S = Sn$, the flux of a uniform vector field through a plane surface is simply written as a scalar product

$$\Phi_{S,E} = E \cdot S.$$

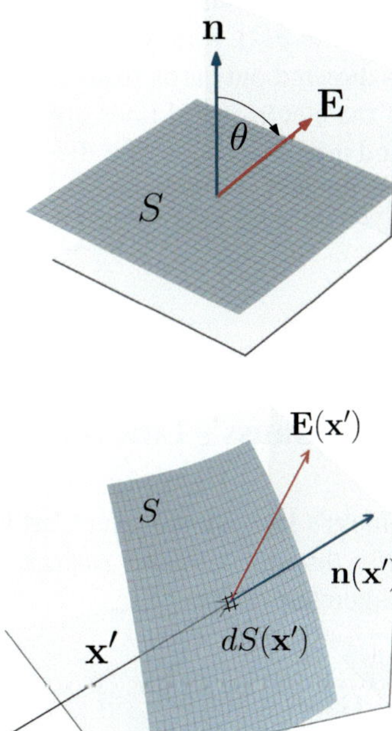

Fig. 2.1 A plane surface is defined by its area S and its normal **n**. The flux of **E** through the surface is $\mathbf{E} \cdot \mathbf{n} S$

Fig. 2.2 An arbitrary surface S can be partitioned into infinitesimal surface elements so that one can define a local flux at each point $\mathbf{x}' \in S$

2.2.2   Flux Through an Arbitrary Surface

In the general case, the surface S will be curved, and the electric field **E** will not be uniform but will vary over the surface. The flux of the field through an arbitrary surface S can be obtained by partitioning the latter into a sum of infinitesimal surface elements. Thus, at position $\mathbf{x}' \in S$, we will have a locally flat surface element $dS(\mathbf{x}')$ with a normal $\mathbf{n}(\mathbf{x}')$ (see Fig. 2.2). The elementary flux through dS will then be

$$\Delta \Phi_{dS,\mathbf{E}} = \mathbf{E}(\mathbf{x}') \cdot \mathbf{n}(\mathbf{x}') dS(\mathbf{x}') = \mathbf{E}(\mathbf{x}') \cdot d\mathbf{S}(\mathbf{x}') \,.$$

The total flux through the surface S is obtained by adding all the surface elements and then taking the limit $dS(\mathbf{x}') \to 0$. We then obtain the flux for a general surface.

Flux of a vector field through a surface
The flux of a vector field **E** through a surface S is defined as the integral:

$$\Phi_{S,\mathbf{E}} = \iint_S \mathbf{E}(\mathbf{x}') \cdot d\mathbf{S}(\mathbf{x}') \,,$$

where $d\mathbf{S}(\mathbf{x}') = dS(\mathbf{x}')\mathbf{n}(\mathbf{x}')$.

When the surface S is closed (i.e., when it encloses a bounded volume $\Omega \in \mathbb{R}^3$, so that $S = \partial\Omega$), we say that the surface is positively oriented if the normal at every point is directed outwards from Ω. In this way, the flux of a field $\mathbf{E}$ through $\partial\Omega$ is positive if the electric field lines are leaving the volume Ω through $\partial\Omega$ ($\mathbf{E}(\mathbf{x}') \cdot \mathbf{n}(\mathbf{x}') > 0$), and it will be negative if the lines enter through $\partial\Omega$ ($\mathbf{E}(\mathbf{x}') \cdot \mathbf{n}(\mathbf{x}') < 0$). The notation for the flux through a closed surface is the following:

$$\Phi_{\partial\Omega,\mathbf{E}} = \oiint_{\partial\Omega} \mathbf{E}(\mathbf{x}') \cdot d\mathbf{S}(\mathbf{x}') .$$

2.3 Gauss's Law for the Electric Flux (1835)

Gauss's law, named after Carl Friedrich Gauss (see Fig. 2.3), states that the electric flux through a closed surface is proportional to the charge enclosed by it. More concretely:

Gauss's law

For every positively oriented, closed surface $\partial\Omega$, the flux of the electric field through $\partial\Omega$ is proportional to $Q(\Omega)$, the total charge enclosed by $\partial\Omega$:

$$\Phi_{\partial\Omega,\mathbf{E}} = \oiint_{\partial\Omega} \mathbf{E}(\mathbf{x}') \cdot d\mathbf{S}(\mathbf{x}') = \frac{Q(\Omega)}{\epsilon_0} . \qquad (2.2)$$

Fig. 2.3 Carl Friedrich Gauss (1777–1855), a German mathematician and physicist who made profound contributions to numerous fields including mathematics, astronomy, and physics

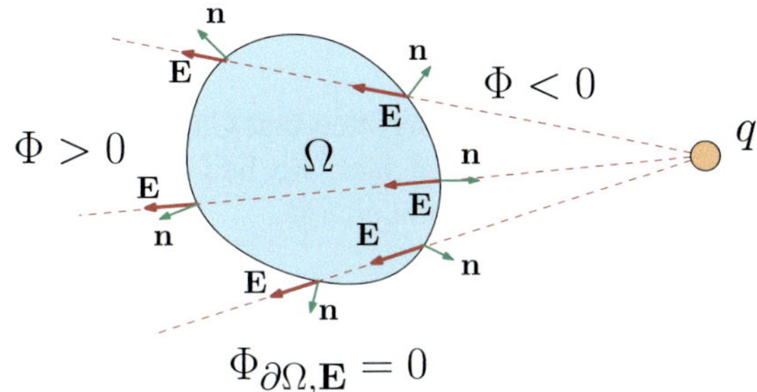

Fig. 2.4 The total electric flux through a closed surface $\partial\Omega$ due to a charge outside Ω is always zero

Writing explicitly the charge enclosed by $\partial\Omega$ as an integral of the charge density ϱ over Ω, Gauss's law is rewritten as

$$\oiint_{\partial\Omega} \mathbf{E}(\mathbf{x}') \cdot d\mathbf{S}(\mathbf{x}') = \frac{1}{\epsilon_0} \iiint_{\Omega} \varrho(\mathbf{x}')d^3x'.$$

Gauss's law can be very useful for calculating the electric field of charge distributions having spatial symmetries. Note also that, according to Gauss's law, electric fields created by charges outside of Ω do not contribute to the flux $\Phi_{\partial\Omega,\mathbf{E}}$. The reason is that for each point charge outside Ω, the incoming flux is always perfectly compensated with the outgoing flux, as can be seen schematically in Fig. 2.4. This property will be demonstrated in Sect. 2.5.

The flux through a closed surface only depends on the charge enclosed by it, and it is independent of the surface shape as long as the total charge inside remains the same.

Remarks

- Gauss's law is a consequence of Coulomb's law. It is natural to wonder if, given Gauss's law, one can demonstrate Coulomb's law. If we wish to be mathematically rigorous, the answer is no. Indeed, as we shall see later, part of the information contained in Coulomb's law is not found in Gauss's law, with the missing part being the fact that the electric field lines cannot form closed loops, a result that will be demonstrated in Chap. 3. However, using Gauss's law and assuming that the field of a point charge has spherical symmetry, then Coulomb's law can be demonstrated as follows:

  - Given a point charge q, we consider a spherical surface S of radius r centered on the charge.
  - At every point on the surface S, the electric field has the same magnitude, which we call $E(r)$. In addition, its direction coincides with the normal to S, $\mathbf{E}(r) = E(r)\mathbf{u}_r = E(r)\mathbf{n}$. The flux through S is then the area of the sphere multiplied by the magnitude of the electric field $\oiint_S \mathbf{E}(\mathbf{x}') \cdot \mathbf{n}(\mathbf{x}')dS(\mathbf{x}') = 4\pi r^2 E(r)$.
  - Finally, Gauss's law equates the calculated flux with q/ϵ_0. Coulomb's law is then retrieved:

$$\mathbf{E}(\mathbf{r}) = \frac{q}{4\pi\epsilon_0 r^2}\mathbf{u}_r.$$

- It should not be forgotten that Gauss's law alone is not enough to demonstrate Coulomb's law. In fact, we had to assume the radial symmetry of the field generated by a point charge.

• Later we will see that when the charges are in motion, Coulomb's law ceases to be valid, but Gauss's law, on the other hand, will continue to be fulfilled. This is why Gauss's law is associated with one of the four fundamental Maxwell equations of electromagnetism.

Example 2.1—Flux through different surfaces
The flux of the electric field generated by a charge Q through any of the three closed surfaces represented in two dimensions in Fig. 2.5 is the same, $\Phi_E = Q/\epsilon_0$. Note that Q can be a point charge, a set of discrete point charges of total charge Q, or the total charge Q of a continuous charge distribution contained in S_1. The flux of $\mathbf{E}$ through a surface only depends on the total charge enclosed by the surface, but not on the way the charges are distributed inside it.

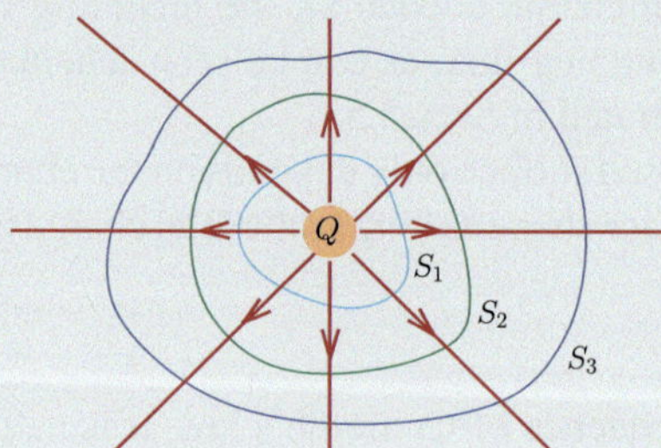

Fig. 2.5 The electric flux through the three depicted surfaces is the same, since they all enclose the same amount of charge

2.3.1 *Differential Form of Gauss's Law*

It is possible, from Gauss's law, to establish a relationship between the spatial derivatives of the electric field and the charge density at each point of space. Indeed, combining Gauss's law and Green–Ostrogradsky theorem (see Sect. A.1.6), we have

$$\oiint_{\partial\Omega} \mathbf{E}(\mathbf{x}) \cdot d\mathbf{S}(\mathbf{x}) = \iiint_{\Omega} \nabla \cdot \mathbf{E}(\mathbf{x})d^3x = \frac{1}{\epsilon_0}\iiint_{\Omega} \varrho(\mathbf{x})d^3x.$$

We conclude that, for every $\Omega \subseteq \mathbb{R}^3$

$$\iiint_{\Omega} \left(\mathbf{\nabla} \cdot \mathbf{E}(\mathbf{x}) - \frac{\varrho(\mathbf{x})}{\epsilon_0} \right) d^3x = 0 \,.$$

At any point $\mathbf{x} \in \mathbb{R}^3$, we can always choose an infinitesimally small volume dV such that the integrand does not vary significantly over dV. The integral then simply equals the product of the integrand by dV. This means that the integrand should be zero, so that at any point $\mathbf{x}$ we have:

$$\boxed{\mathbf{\nabla} \cdot \mathbf{E}(\mathbf{x}) = \frac{\varrho(\mathbf{x})}{\epsilon_0} \,.} \tag{2.3}$$

Equation (2.3) is the differential (also called local) form of Gauss's law. Note that if we multiply Eq. (2.3) by an infinitesimal volume dV around $\mathbf{x}$, we get $\mathbf{\nabla} \cdot \mathbf{E}(\mathbf{x})dV = \varrho dV/\epsilon_0$. The right-hand term is the charge inside dV divided by ϵ_0 and so $\mathbf{\nabla} \cdot \mathbf{E}(\mathbf{x})dV$ is the flux through the surface enclosing dV. The divergence of the electric field represents therefore an infinitesimal flux per unit volume. The divergence at $\mathbf{x}$ is positive where the electric field lines are diverging outwards from $\mathbf{x}$, and it is negative where the field lines are converging toward $\mathbf{x}$. We already know that this must be the case when a charge is at position $\mathbf{x}$ (see Sect. 1.5.1), confirming Eq. (2.3).

Example 2.2—Electric flux through an infinitesimal cube
Consider an infinitesimal volume defined in Cartesian coordinates by $[x, x + dx] \times [y, y + dy] \times [z, z + dz]$, as shown in Fig. 2.6. Let us assume that the electric field $\mathbf{E}(x, y, z)$ is known and calculate the electric flux through its boundary S.

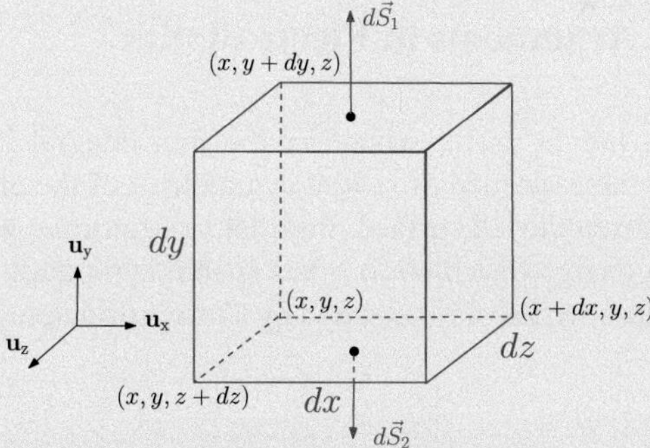

Fig. 2.6 Infinitesimal cube used to calculate the electric flux

The flux can be calculated as the sum of the flux over each of the six planar surfaces that define the cube,

$$\Phi_{S,\mathbf{E}} = \sum_{i=1}^{6} \mathbf{E}_i \cdot d\mathbf{S}_i.$$

Consider, for example, the top and bottom surfaces shown in Fig. 2.6. We have $d\mathbf{S}_1 = dxdz\mathbf{u}_y$ and $d\mathbf{S}_2 = -d\mathbf{S}_1$, and so

$$\mathbf{E}_1 \cdot d\mathbf{S}_1 + \mathbf{E}_2 \cdot d\mathbf{S}_2 \approx \left(E_y(x, y + dy, z) - E_y(x, y, z)\right) dxdz.$$

Since for an infinitesimal dy we can write $E_y(x, y + dy, z) - E_y(x, y, z) \approx \frac{\partial E(x,y,z)}{\partial y} dy$:

$$\mathbf{E}_1 \cdot d\mathbf{S}_1 + \mathbf{E}_2 \cdot d\mathbf{S}_2 = \frac{\partial E(x, y, z)}{\partial y} dxdydz.$$

An analogous calculation for the other four surfaces yields

$$\Phi_{S,\mathbf{E}} = \left(\frac{\partial E(x, y, z)}{\partial x} + \frac{\partial E(x, y, z)}{\partial y} + \frac{\partial E(x, y, z)}{\partial z}\right) dxdydz$$
$$= \nabla \cdot \mathbf{E}(x, y, z)dxdydz,$$

where we recognized the expression for the divergence in Cartesian coordinates. We conclude that divergence of the electric field at x, y, z is indeed the flux per unit volume around that point.

2.4 Symmetry Arguments in Electrostatics

In practice, Gauss's law is useful whenever the flux integral in Eq. (2.2) can be simplified thanks to the existence of spatial symmetries of the electric field and the use of a properly chosen closed surface. In order to determine whether the electric field generated by a charge distribution ϱ has spatial symmetries, it is sufficient to study the symmetry properties of ϱ, as stated by Curie's principle, named after Pierre Curie (see Fig. 2.7).

Curie's principle (1894)
When some causes (here, a charge density ϱ) produce some effects (here, an electric field $\mathbf{E}$), the symmetries of the causes are found in the effects.

Note that the reciprocal is not true, because the effects may have more symmetries than their causes. Of course, this principle can be demonstrated in the particular case

Fig. 2.7 Pierre Curie (1859–1906), a French physicist, made groundbreaking contributions to crystallography, magnetism, and radioactivity. He discovered piezoelectricity and formulated key laws governing the properties of ferromagnetic and paramagnetic materials

of electrostatics from the Coulomb integral (2.1). A demonstration can be found in the Appendix A.4.

As will be shown below, the study of the symmetries of the charge distribution will allow us to

1. Reduce the number of variables required to describe the spatial variations of $\mathbf{E}$.
2. Constrain the orientation of the field $\mathbf{E}$.

2.4.1 Invariance by Spatial Translation Along an Axis

Consider the case of a charge distribution that is invariant under translation along an axis (the z-axis, for example). This means, in Cartesian coordinates, that for every $a \in \mathbb{R}$,

$$\varrho(x, y, z + a) = \varrho(x, y, z) ,$$

so that the charge density is independent of the z-coordinate, $\varrho(x, y, z) = \varrho(x, y)$. This translates directly into an invariance of the electric field upon this coordinate:

$$\mathbf{E}(x, y, z) = \mathbf{E}(x, y) .$$

2.4.2 Invariance by Spatial Rotation Around an Axis

Consider the case of a charge distribution that is invariant under rotation around an axis (the z-axis, for example). This means, in cylindrical coordinates, that for every $\varphi \in [0, 2\pi]$,

$$\varrho(r, \theta + \varphi, z) = \varrho(r, \theta, z) ,$$

so that the charge density is independent of θ, $\varrho(r, \theta, z) = \varrho(r, z)$, and this translates into an invariance of the electric field upon this coordinate:

$$\mathbf{E}(r, \theta, z) = \mathbf{E}(r, z) \, .$$

Example 2.3—Cylindrical/spherical symmetries

Two particular cases of major importance in electrostatics are those of a charge distribution having either cylindrical or spherical symmetry.

In the first case, the charge density ϱ is invariant by spatial translation along—and also by spatial rotation around—the z-axis. It is therefore independent of z and θ in cylindrical coordinates and one has

$$\varrho(r, \theta, z) = \varrho(r) \quad \Rightarrow \quad \mathbf{E}(r, \theta, z) = \mathbf{E}(r) \, .$$

In the case of spherical symmetry, the charge density ϱ is invariant under rotation around any axis passing through the origin. It is then independent of θ and ϕ in spherical coordinates and this translates into

$$\varrho(r, \theta, \phi) = \varrho(r) \quad \Rightarrow \quad \mathbf{E}(r, \theta, \phi) = \mathbf{E}(r) \, .$$

Remark

In conclusion, invariances of the charge distribution allow us to reduce the number of variables required to describe the spatial variation of $\mathbf{E}$. However, it does not tell us anything about the direction of the electric field. To constrain its direction, one can use the mirror symmetry or antisymmetry of the charge distribution.

2.4.3   Mirror Symmetries

Let $\mathbf{\Pi}$ be a plane in space. For any point $P \in \mathbb{R}^3$ let us consider the reflection P' of P with respect to the plane $\mathbf{\Pi}$ (see Fig. 2.8). In other words, P' is the symmetric point of P with respect to the plane $\mathbf{\Pi}$:

$$P' = \mathrm{sym}_{\mathbf{\Pi}} P \, .$$

To find P', first we find the point $\Gamma \in \mathbf{\Pi}$ which is closest to P by means of an orthogonal projection. This determines a line passing through Γ and P, and P' is the point on this line which is at the same distance to Γ than P.

Fig. 2.8 The point P' is the symmetric (reflection) of P with respect to the plane Π

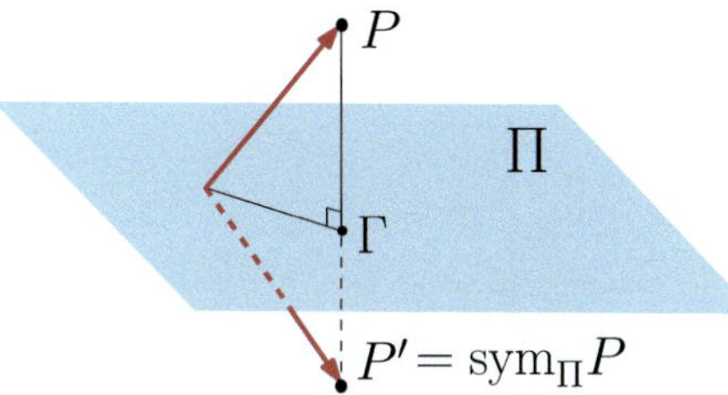

Mirror symmetry of a charge distribution
We say that Π is a plane of symmetry for the charge distribution ϱ if ϱ satisfies the mirror symmetry with respect to Π:

$$\varrho(P') = \varrho(P) \quad \text{with} \quad P' = \mathrm{sym}_\Pi P$$

for every point P in space. By Curie's principle, we conclude that the electric field generated by ϱ exhibits this mirror symmetry as well (see Fig. 2.9)

$$\mathbf{E}(P') = \mathrm{sym}_\Pi \mathbf{E}(P).$$

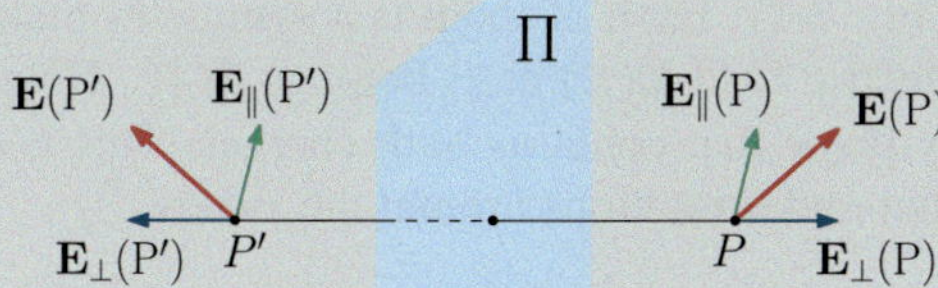

Fig. 2.9 The electric field is symmetric with respect to a plane of symmetry of the charge distribution

The mirror symmetry for a vector quantity implies $\mathbf{E}_\perp(P) = -\mathbf{E}_\perp(P')$ and $\mathbf{E}_\parallel(P) = \mathbf{E}_\parallel(P')$. In particular, for a point $P \in \Pi$, one has $P' = P$ and therefore $\mathbf{E}_\perp(P) = -\mathbf{E}_\perp(P) = 0$. In other words, for any point P belonging to a symmetry plane Π, the electric field at P is within the plane Π.

Consider, for example, two identical charges $q > 0$ located on the y-axis, for which the electric field lines are shown in Fig. 2.10 (left). Despite the simplicity of this charge distribution, the electric field is complex and depends on all three Cartesian coordinates. However, we can still predict the field orientation at specific points in

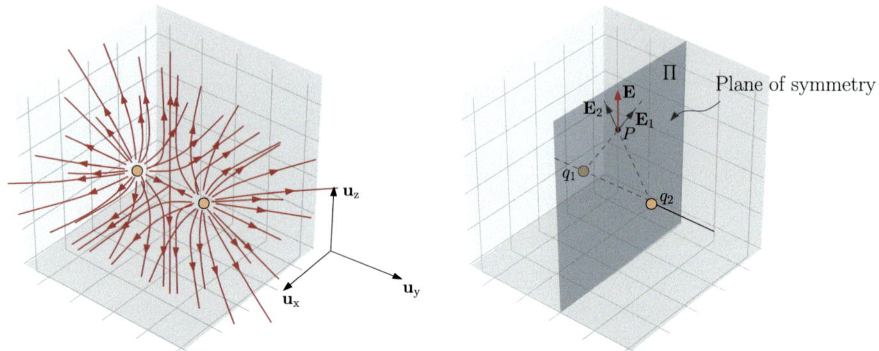

Fig. 2.10 Left: Electric field lines generated by a pair of equal and positive charges. Right: A plane Π equidistant to the two charges is a symmetry plane.The electric field at any point on this plane belongs to it

space. Indeed, this charge distribution has mirror symmetry with respect to the plane Π which is equidistant to both charges. At any point $P \in \Pi$, the superposition of the two electric fields generated separately by each charge gives indeed a total field within the plane, the normal components cancel each other out. This is illustrated in Fig. 2.10 (right). In consequence, any field line passing through a point in Π must be included in the plane.

Moreover, we can consider now a plane Π' that contains both charges. Two such planes are shown in Fig. 2.11. Each of them is a symmetry plane, and so for every point P in Π', the electric field $\mathbf{E}(P)$ at P belongs to Π'. Note that the field lines look the same in any plane that contains both charges. This is because the charge distribution is invariant under rotations around the y-axis.

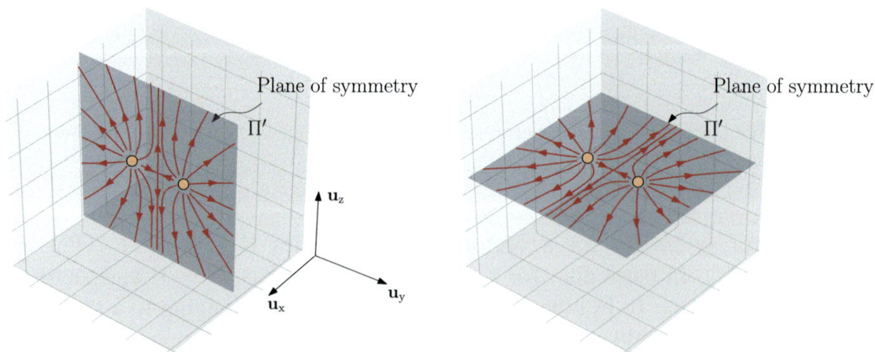

Fig. 2.11 The planes that contain the two charges are also symmetry planes for the charge distribution. At any point on these planes, the electric field belongs to the plane

Mirror antisymmetry

We say that Π^* is a plane of antisymmetry for the charge distribution if ϱ is antisymmetric with respect to Π^*:

$$\varrho(P') = -\varrho(P) \quad \text{with} \quad P' = \text{sym}_{\Pi^*} P$$

for every P. By Curie's principle, the electric field generated by ϱ exhibits this mirror antisymmetry as well,

$$\mathbf{E}(P') = -\text{sym}_{\Pi^*}\mathbf{E}(P)\,.$$

The mirror antisymmetry implies $\mathbf{E}_\perp(P) = \mathbf{E}_\perp(P')$ and $\mathbf{E}_\parallel(P) = -\mathbf{E}_\parallel(P')$ (see Fig. 2.12). In particular, $P' = P$ for every point $P \in \Pi^*$ and therefore, $\mathbf{E}_\parallel(P) = -\mathbf{E}_\parallel(P) = 0$. In other words, for a point P belonging to an antisymmetry plane Π^*, the electric field at P is perpendicular to the plane Π^*.

Consider, for example, the plane Π^* equidistant to two charges of equal strength but of opposite sign (electric dipole) shown in Fig. 2.13. This charge distribution has mirror antisymmetry with respect to Π^*. At any point in this plane, the superposition of the two electric fields generated separately by each charge gives a total field that is perpendicular to this plane; the parallel components cancel out.

As in the case of two identical charges, any plane Π that contains the charges is a symmetry plane, so that in any point $P \in \Pi$, $\mathbf{E}(P) \in \Pi$. In Fig. 2.13, one of those planes is shown, together with the electric field lines. Note that once again the electric field lines look the same in every plane that contains the charges, due to the invariance of the charge distribution under rotations around the y-axis.

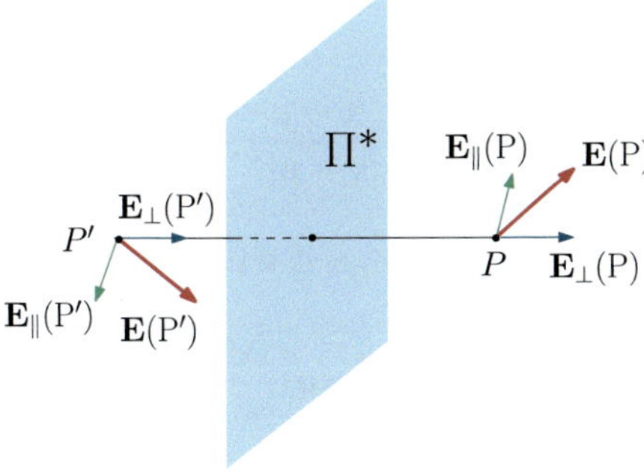

Fig. 2.12 The electric field must exhibit mirror antisymmetry with respect to an antisymmetry plane

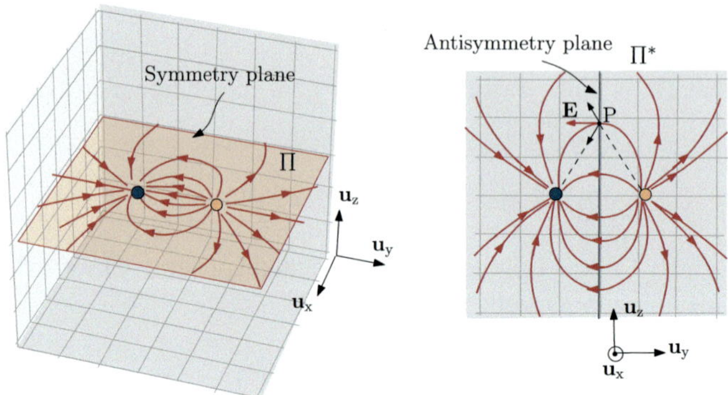

Fig. 2.13 Any plane Π containing the charges is a symmetry plane. The plane Π^* equidistant to two charges of opposite sign is an antisymmetry plane

In conclusion, to constrain the orientation of the electric field at any point P, one can look for a symmetry (antisymmetry) plane of the charge distribution containing P. If such a plane exists, the electric field at P will be within (perpendicular) to this plane.

2.4.4   *Justification of the Mirror Symmetries of the Electric Field*

The mirror symmetry properties stated above are not additional laws, since they can be demonstrated directly from the Coulomb integral (1.5). Any plane Π partitions $\mathbb{R}^3$ into two disjoint regions D_1 and D_2. As such, any point $\mathbf{u}' \in D_2$ can be written as the reflection of a point $\mathbf{u} \in D_1$ with respect to the plane Π, as seen in Fig. 2.14. Then, the electric field at an arbitrary position $\mathbf{x}$ can be written as an integral over D_1 only:

$$\mathbf{E}(\mathbf{x}) = \frac{1}{4\pi\epsilon_0} \iiint_{D_1} \left(\varrho(\mathbf{u}) \frac{\mathbf{x} - \mathbf{u}}{|\mathbf{x} - \mathbf{u}|^3} + \varrho(\mathbf{u}') \frac{\mathbf{x} - \mathbf{u}'}{|\mathbf{x} - \mathbf{u}'|^3} \right) d^3 u \,.$$

Similarly, the electric field at $\mathbf{x}' = \mathrm{sym}_\Pi \mathbf{x}$ is written as

$$\mathbf{E}(\mathrm{sym}_\Pi \mathbf{x}) = \mathbf{E}(\mathbf{x}') = \frac{1}{4\pi\epsilon_0} \iiint_{D_1} \left(\varrho(\mathbf{u}) \frac{\mathbf{x}' - \mathbf{u}}{|\mathbf{x}' - \mathbf{u}|^3} + \varrho(\mathbf{u}') \frac{\mathbf{x}' - \mathbf{u}'}{|\mathbf{x}' - \mathbf{u}'|^3} \right) d^3 u \,.$$

Noting that $|\mathbf{x} - \mathbf{u}| = |\mathbf{x}' - \mathbf{u}'|$, $|\mathbf{x} - \mathbf{u}'| = |\mathbf{x}' - \mathbf{u}|$ and more generally that $\mathbf{a} - \mathbf{b} = \mathrm{sym}_\Pi(\mathbf{a}' - \mathbf{b}')$ for any pair of vectors $\mathbf{a}$ and $\mathbf{b}$:

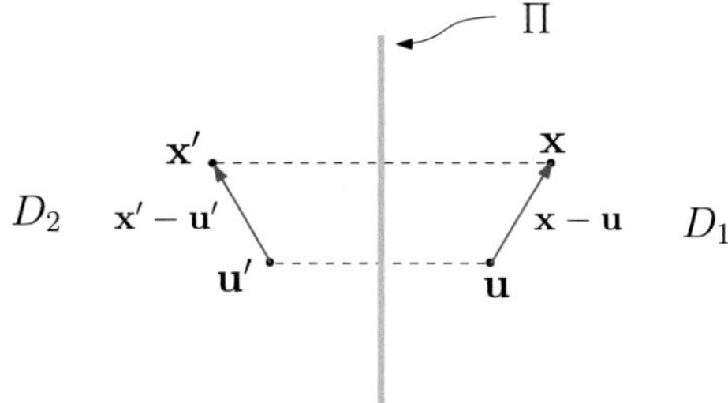

Fig. 2.14 A plane Π permits to partition the whole space into two disjoint regions D_1 and D_2. To any point $\mathbf{u} \in D_1$, one can consider $\mathbf{u}' \in D_2$, its symmetric with respect to Π

$$\mathbf{E}(\mathrm{sym}_\Pi \mathbf{x}) = \mathbf{E}(\mathbf{x}') = \mathrm{sym}_\Pi \frac{1}{4\pi\epsilon_0} \iiint_{D_1} \left(\varrho(\mathbf{u}) \frac{(\mathbf{x} - \mathbf{u}')}{|\mathbf{x} - \mathbf{u}'|^3} + \varrho(\mathbf{u}') \frac{(\mathbf{x} - \mathbf{u})}{|\mathbf{x} - \mathbf{u}|^3} \right) d^3u .$$

From here, it is clear that

1. If ϱ has mirror symmetry with respect to Π, $\varrho(\mathbf{u}) = \varrho(\mathbf{u}')$ then $\mathbf{E}$ as well:

$$\mathbf{E}(\mathbf{x}') = \mathrm{sym}_\Pi \frac{1}{4\pi\epsilon_0} \iiint_{D_1} \varrho(\mathbf{u}) \left(\frac{(\mathbf{x} - \mathbf{u}')}{|\mathbf{x} - \mathbf{u}'|^3} + \frac{(\mathbf{x} - \mathbf{u})}{|\mathbf{x} - \mathbf{u}|^3} \right) d^3u = \mathrm{sym}_\Pi \mathbf{E}(\mathbf{x}) .$$

2. If ϱ has mirror antisymmetry with respect to $\Pi^* = \Pi$, $\varrho(\mathbf{u}) = -\varrho(\mathbf{u}')$ and:

$$\mathbf{E}(\mathbf{x}') = \mathrm{sym}_{\Pi^*} \frac{1}{4\pi\epsilon_0} \iiint_{D_1} \varrho(\mathbf{u}) \left(\frac{(\mathbf{x} - \mathbf{u}')}{|\mathbf{x} - \mathbf{u}'|^3} - \frac{(\mathbf{x} - \mathbf{u})}{|\mathbf{x} - \mathbf{u}|^3} \right) d^3u = -\mathrm{sym}_{\Pi^*} \mathbf{E}(\mathbf{x}) .$$

Example 2.4—Field of a uniformly charged sphere
Consider a homogeneously distributed charge inside a sphere of radius R. Let us find an expression for the electric field in all space generated by this charge distribution:

$$\varrho(\mathbf{x}') = \begin{cases} \varrho_0 & \text{if } |\mathbf{x}'| \le R \\ 0 & \text{if } |\mathbf{x}'| > R \end{cases}$$

A way to solve this problem is to evaluate the Coulomb integral (1.5). A much more convenient method is to use Gauss's law. For this, one needs to wisely choose the closed surface S to calculate the electric flux. The first step is then to establish the possible symmetries of the electric field. In spherical coordinates the charge density only depends on r, thus the electric field as well by Curie's principle, $\mathbf{E}(r, \theta, \phi) = \mathbf{E}(r)$. In addition, for any point P, there is an infinite number of symmetry planes containing P and the center O of the sphere. Figure 2.15 shows two such planes.

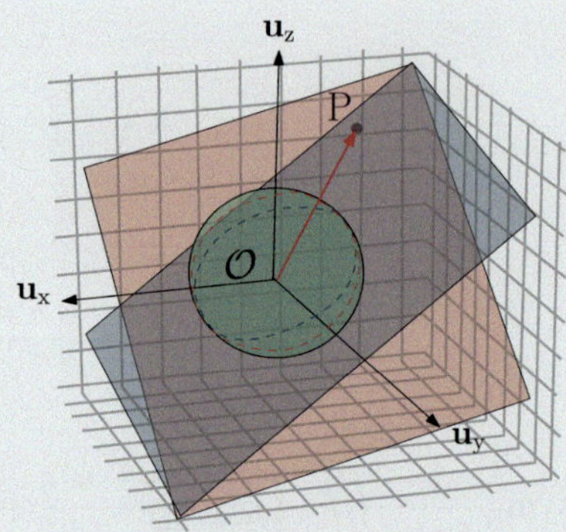

Fig. 2.15 For any point P one can find symmetry planes that contain P and the center of the sphere. They all have in common the direction $\mathbf{u}_r$

Their intersection is along OP, which coincides with the radial direction $\mathbf{u}_r$. Since $\mathbf{E}(P)$ belongs to all these planes, one concludes that the orientation of $\mathbf{F}$ is along $\mathbf{u}_r$ and therefore $\mathbf{E}(r, \theta, \phi) = E(r)\mathbf{u}_r$. In this way, to obtain the field in the region outside the charge distribution $(r > R)$, we choose a spherical surface of radius $r > R$ to apply Gauss's law (see Fig. 2.16).

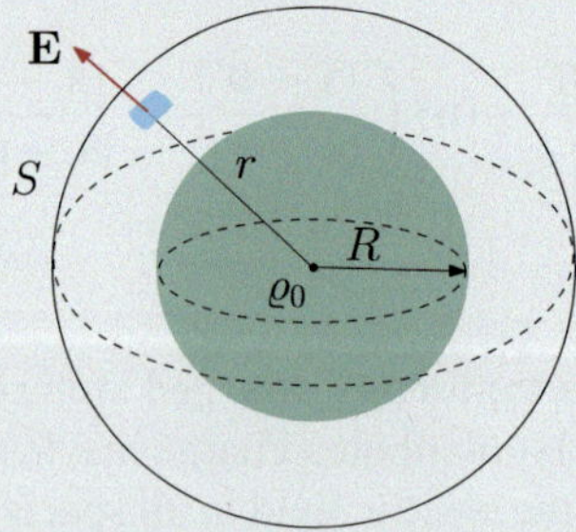

Fig. 2.16 A uniformly charged sphere surrounded by a closed, concentrical spherical surface

At any point on this surface, the normal is $\mathbf{n} = \mathbf{u}_r$ and the electric field has a constant magnitude $E(r)$ and direction $\mathbf{u}_r$, thus parallel to the normal. The electric field flux through this surface is

$$
\Phi_{S,\mathbf{E}} = \oiint_S \mathbf{E}(\mathbf{x}') \cdot \mathbf{n}(\mathbf{x}')\, dS(\mathbf{x}') = \int_0^\pi d\theta \int_0^{2\pi} d\phi\, r^2 \sin\theta\, E(r) \underbrace{\mathbf{u}_r \cdot \mathbf{u}_r}_{=1}
$$

$$
= r^2 E(r) \int_0^\pi d\theta \sin\theta \int_0^{2\pi} d\phi = E(r) 4\pi r^2
$$

and using Gauss's law:

$$\Phi_{S,\mathbf{E}} = E(r)4\pi r^2 = \frac{Q}{\epsilon_0} ,$$

where $Q = \frac{4}{3}\pi R^3 \varrho_0$ is the total charge of the sphere. Finally:

$$\mathbf{E}(r) = E(r)\mathbf{u}_r = \frac{Q}{4\pi\epsilon_0 r^2}\mathbf{u}_r ,$$

that is, the field at $r > R$ is the same as the field of a point charge Q located at the origin. In terms of the charge density ϱ_0:

$$\mathbf{E}(r) = \frac{\varrho_0 R^3}{3\epsilon_0 r^2}\mathbf{u}_r .$$

Now, for the field inside the sphere, we again choose a spherical surface S of radius $r < R$ (see Fig. 2.17). Again the flux is

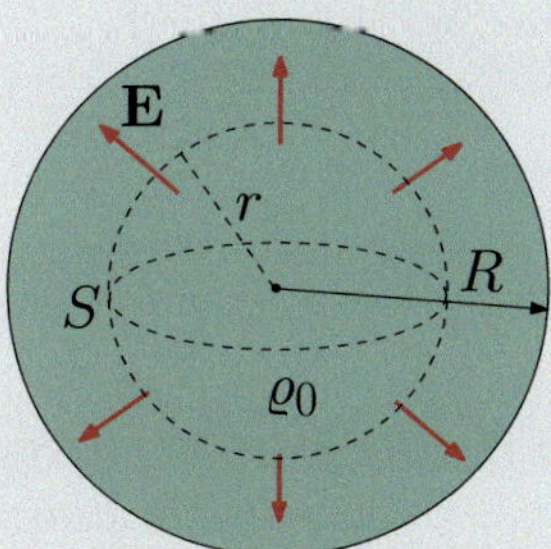

Fig. 2.17 A spherical closed surface inside the uniformly charged sphere

$$\Phi_{S,\mathbf{E}} = \oiint_S d\mathbf{S}(\mathbf{x}') \cdot \mathbf{E}(\mathbf{x}') = E(r)4\pi r^2 = \frac{Q(r)}{\epsilon_0} ,$$

where $Q(r)$ is the charge enclosed by the surface S, which in this case is equal to

$$Q(r) = \frac{4}{3}\pi r^3 \varrho_0 ,$$

so that

$$E(r)4\pi r^2 = \frac{4\pi r^3 \varrho_0}{3\epsilon_0} , \quad \text{i.e.}$$

$$E(r) = \frac{r\varrho_0}{3\epsilon_0} .$$

The magnitude of the electric field within the charged sphere increases linearly with r. Finally:

$$\mathbf{E}(\mathbf{x}) = \begin{cases} \dfrac{\varrho_0 r}{3\epsilon_0}\mathbf{u}_r = \dfrac{Qr}{4\pi\epsilon_0 R^3}\mathbf{u}_r & \text{if } r \leq R, \\[2mm] \dfrac{\varrho_0 R^3}{3\epsilon_0 r^2}\mathbf{u}_r = \dfrac{Q}{4\pi\epsilon_0 r^2}\mathbf{u}_r & \text{if } r > R. \end{cases}$$

Note that the field is continuous across the sphere at $|\mathbf{x}| = R$ despite the discontinuity of ϱ. Figure 2.18 shows the dependence of $E(r)$ as a function of the distance r to the origin.

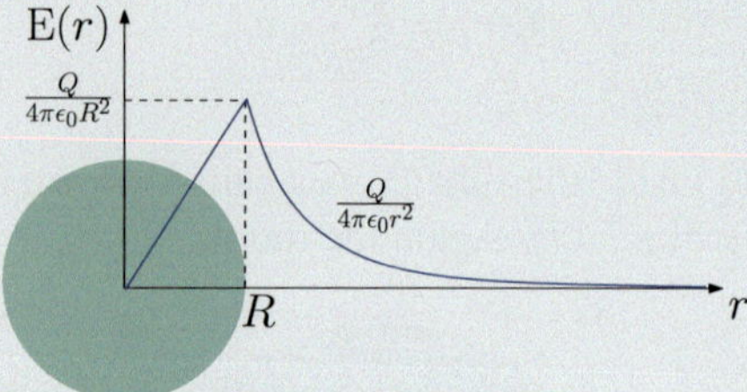

Fig. 2.18 Graph of the electric field magnitude as a function of the distance with respect to the center of the sphere

Example 2.5—Field of an infinite plane of charge

Consider the infinite plane $z = 0$ charged with a uniform surface density $\sigma > 0$ (Fig. 2.19). We will determine, using Gauss's law, the electric field that it generates in all space.

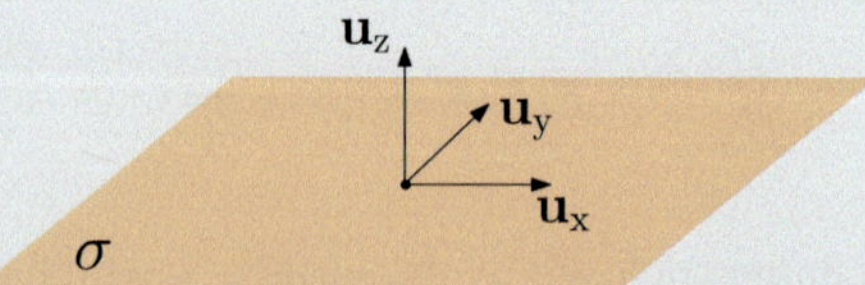

Fig. 2.19 An infinite plane of charge

In Cartesian coordinates, the charge distribution of the plane is invariant under translations along the x- and y-axes. By Curie's principle, the electric field depends only on the z coordinate, $\mathbf{E}(x, y, z) = \mathbf{E}(z)$. In addition, for any point P, one can find an infinite number of symmetry planes containing P and (parallel to) the z-axis. In Fig. 2.20, two of them are shown.

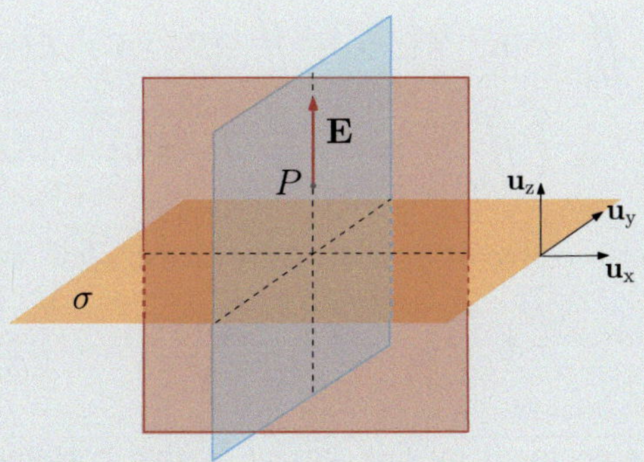

Fig. 2.20 Two planes of symmetry containing the point P

It follows that the orientation of the electric field at P is along the z-axis and one has $\mathbf{E}(x, y, z) = E(z)\mathbf{u}_z$. Finally, note that the plane Oxy containing the charges is a plane of symmetry, so that $\mathbf{E}$ has a mirror symmetry with respect to this plane, meaning that $\mathbf{E}(-z) = -E(z)\mathbf{u}_z$.

To calculate $E(z)$, we use Gauss's law, where the closed surface S that we consider is a cylinder of height $2z$ that intersects the charged plane symmetrically, as illustrated in Fig. 2.21.

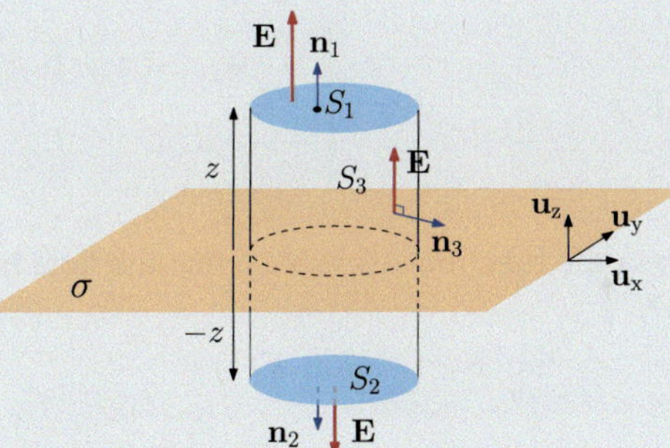

Fig. 2.21 Gauss surface chosen to calculate the electric field

The cylinder can be divided into 3 surfaces: two basis surfaces S_1 and S_2 (both of area A), and the lateral surface S_3. The flux of $\mathbf{E}$ through the cylinder is then:

$$\Phi_{S,\mathbf{E}} = \oiint_S dS(\mathbf{x}')\mathbf{n}(\mathbf{x}') \cdot \mathbf{E}(\mathbf{x}') = \iint_{S_1} dS_1\mathbf{n}_1 \cdot \mathbf{E} + \iint_{S_2} dS_2\mathbf{n}_2 \cdot \mathbf{E} + \iint_{S_3} dS_3\mathbf{n}_3 \cdot \mathbf{E}.$$

The flux through S_3 is zero, since at every point of this surface the normal is perpendicular to $\mathbf{E}$, $\mathbf{E}(\mathbf{x}) \cdot \mathbf{n}_3(\mathbf{x}) = 0$, $\forall \mathbf{x} \in S_3$. In addition, the magnitude of $\mathbf{E}$ on both basis surfaces is constant and equal to $E(z)$ (note that on S_2 the field points along $-\mathbf{u}_z$, and the normal $\mathbf{n}_2$ as well). With this:

$$\Phi_{S,\mathbf{E}} = \iint_{S_1} dS_1 \mathbf{u}_z \cdot E(z)\mathbf{u}_z + \iint_{S_2} dS_2(-\mathbf{u}_z) \cdot E(z)(-\mathbf{u}_z)$$

$$= E(z)\left(\iint_{S_1} dS_1(\mathbf{x}') + \iint_{S_2} dS_2(\mathbf{x}')\right) = 2AE(z),$$

where A is the surface area of S_1 and S_2. By Gauss's law, the electric flux is given by

$$\Phi_{S,\mathbf{E}} = 2AE(z) = \frac{Q(S)}{\epsilon_0},$$

where the charge enclosed by this surface is that contained on a disk of area A,

$$Q(S) = \sigma A,$$

so that the magnitude of the field is uniform in space:

$$E(z) = \frac{\sigma}{2\epsilon_0} = E.$$

Finally, taking into account the mirror symmetry of the electric field with respect to $z = 0$:

$$\mathbf{E}(\mathbf{x}) = \frac{\sigma z}{2\epsilon_0 |z|}\mathbf{u}_z = \begin{cases} \dfrac{\sigma}{2\epsilon_0}\mathbf{u}_z & \text{if } z > 0\,, \\[2mm] -\dfrac{\sigma}{2\epsilon_0}\mathbf{u}_z & \text{if } z < 0\,. \end{cases}$$

Figure 2.22 shows the function $E(z)$. Note that the field has a discontinuity at $z = 0$, given by

$$\lim_{z \to 0^+} \mathbf{E}(z) - \lim_{z \to 0^-} \mathbf{E}(z) = \frac{\sigma}{\epsilon_0}\,.$$

due to the presence of a charged surface at $z = 0$. This discontinuity is removed if one considers a more realistic model in which the charge distribution has a non-zero thickness. Finally, since the total charge density is not integrable, there is no guarantee of a field decay as $1/r^2$ (or faster) far from the plane. In fact, the field does not decay at all as a function of the distance z.

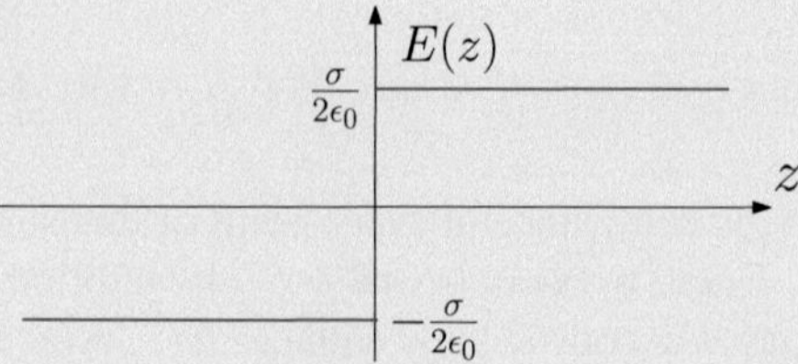

Fig. 2.22 Electric field as a function of the distance to the infinite plane

Of course, an infinite plane of charge does not exist in reality. For a finite plane, the result obtained here will be a good approximation for all points away from the edges and located at a distance z much smaller than the characteristic size of the plane. Moreover, since any smooth surface can be approximated locally by a flat surface, the result obtained in this problem also holds for the electric field generated by a surface at a point infinitesimally close to it. That is, in the vicinity of every point $\mathbf{x}$ of a charged surface, the field due to this surface reads

$$\mathbf{E} = \pm \frac{\sigma(\mathbf{x})}{2\epsilon_0} \mathbf{n}(\mathbf{x})$$

where $\sigma(\mathbf{x})$ and $\mathbf{n}(\mathbf{x})$ are the charge density and the normal to the surface at point $\mathbf{x}$, respectively. The sign depends on whether the field is calculated above (in the direction of $\mathbf{n}(\mathbf{x})$) or below the surface.

2.5 Proof of Gauss's Law

Gauss's law can be proved by different means. An elegant and short proof involving the Dirac distribution can be found in the Appendix A.2.2.4. Here, purely geometrical arguments will be used instead. For this, it is convenient to introduce the notion of solid angle, which corresponds to a two-dimensional angle representing the apparent size of a surface seen from a point O. For example, consider two concentric spheres of radii r_1 and r_2 respectively, as shown in Fig. 2.23.

The two spherical portions S_1 and S_2, seen from O, have the same apparent size. This is because both subtend the same solid angle. Mathematically, a surface element in spherical coordinates is written $dS = r^2 \sin\theta\, d\theta\, d\phi$. The elementary solid angle is

Fig. 2.23 The two spherical portions S_1 and S_2 have the same apparent size when seen from O. They subtend the same solid angle

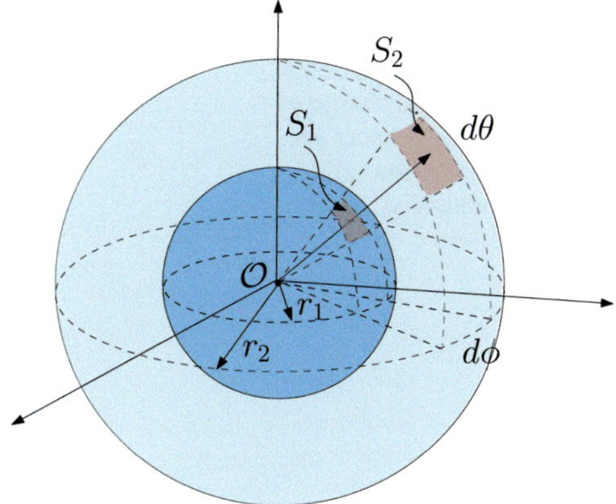

defined as $d\Omega = \sin\theta\, d\theta\, d\phi$. It is clear that the two surfaces of the figure are described by the same intervals of θ and of ϕ, and then both subtend the same total solid angle $\Omega = \int d\Omega$. The surfaces S_1 and S_2 then read:

$$S_1 = r_1^2\Omega, \qquad S_2 = r_2^2\Omega.$$

The solid angle Ω is measured in steradians, and can vary between 0 and 4π. For the more general case of an arbitrary surface element $d\mathbf{S} = dS\,\mathbf{n}$ (not necessarily a portion of a sphere) at distance r from O, the solid angle $d\Omega$ that it subtends will be the same as that of its projection on a unit sphere, as shown in Fig. 2.24.

If $d\mathbf{S}$ is at distance r from O, $r^2 d\Omega$ corresponds to the projected surface of $d\mathbf{S}$ on a sphere of radius r, i.e., $r^2 d\Omega = |d\mathbf{S} \cdot \mathbf{u}_r|$. Now, to prove Gauss's law, consider a point charge q and a closed surface S containing q. The flux of the electric field on a surface element $d\mathbf{S}$ located at distance r from the charge (see Fig. 2.25) will be

$$d\Phi_{dS,\mathbf{E}} = \mathbf{E} \cdot d\mathbf{S} = \frac{q}{4\pi\epsilon_0 r^2}\underbrace{\mathbf{u}_r \cdot d\mathbf{S}}_{=r^2 d\Omega} = \frac{q}{4\pi\epsilon_0}d\Omega,$$

with $d\Omega$ the solid angle that subtends $d\mathbf{S}$ when viewed from the charge. This is a consequence of the radial character and the $1/r^2$ dependence of the electric field generated by an elementary charge. Then, the electric flux through the entire surface

Fig. 2.24 The solid angle $d\Omega$ subtended by an infinitesimal surface $d\mathbf{S}$ seen from O is the same as the one of the projection of $d\mathbf{S}$ on a unit sphere centered at O

Fig. 2.25 The electric field flux generated by a charge q over the surface element $d\mathbf{S}$ is proportional to the solid angle $d\Omega$ subtended by $d\mathbf{S}$ when seen from q

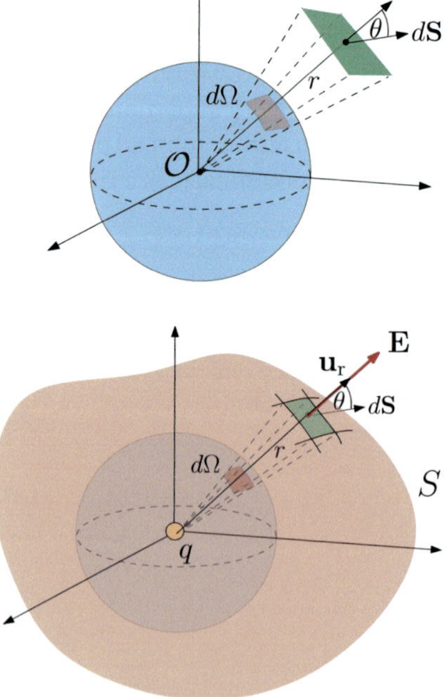

will be

$$\Phi_{S,\mathbf{E}} = \frac{q}{4\pi\epsilon_0} \underbrace{\oiint_S d\Omega}_{=4\pi} = \frac{q}{\epsilon_0},$$

since the total solid angle subtended by any closed surface containing the charge is 4π.

Consider now the case in which the charge q lies outside the closed surface S. As shown in Fig. 2.26, for every surface element $d\mathbf{S}_1$ on the region of S farther from the charge, there is another surface element $d\mathbf{S}_2$ that subtends the same solid angle $d\Omega_1$ but such that $\mathbf{u}_r \cdot d\mathbf{S}_2 < 0$. The total electric flux over dS_1 and dS_2 reads:

$$d\Phi_{dS_1,\mathbf{E}} + d\Phi_{dS_2,\mathbf{E}} = \frac{q}{4\pi\epsilon_0} \left(\frac{\mathbf{u}_r \cdot d\mathbf{S}_1}{r_1^2} + \frac{\mathbf{u}_r \cdot d\mathbf{S}_2}{r_2^2} \right) = \frac{q}{4\pi\epsilon_0} (d\Omega_1 - d\Omega_1) = 0.$$

If we decompose the surface S into such elementary pairs, we conclude that the total electric flux through S is zero. Finally, for a set of charges q_i, the total electric field will be the superposition of the fields $\mathbf{E}_i$ generated individually by each charge, according to the superposition principle. Gauss's law can be applied individually to each charge and we then obtain:

$$\Phi_{S,\mathbf{E}} = \oiint_S \mathbf{E} \cdot d\mathbf{S} = \frac{1}{\epsilon_0} \sum_{\text{charges inside } S} q_i,$$

where only the charges contained in S are included in the sum. For a continuous distribution of charges, the sum is replaced by an integral over the volume delimited by S.

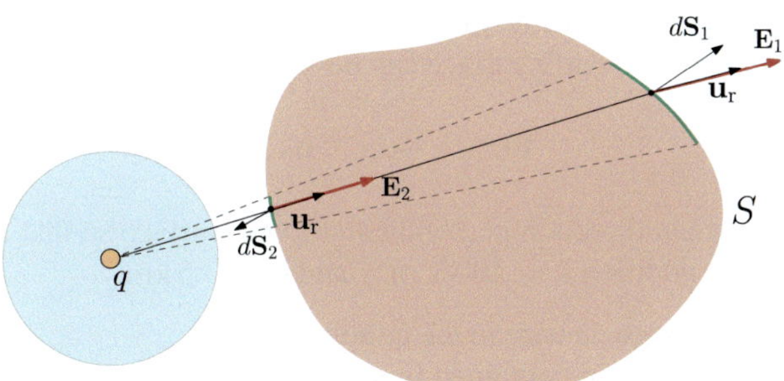

Fig. 2.26 The electric flux over S generated by a charge outside S is always zero

2.6   Summary and Essential Formulas

- The electric field generated by a charge density $\varrho : \mathbb{R}^3 \to \mathbb{R}$ is, according to Coulomb's law and the superposition principle, given by the integral (2.1):

$$\mathbf{E}(\mathbf{x}) = \frac{1}{4\pi \epsilon_0} \iiint_{\mathbb{R}^3} \frac{\varrho(\mathbf{x}')(\mathbf{x} - \mathbf{x}')}{|\mathbf{x} - \mathbf{x}'|^3} d^3x'.$$

- Gauss's law is a consequence of Coulomb's law (2.1) and states that the flux of $\mathbf{E}$ over any closed surface $\partial\Omega$ is proportional to the charge contained in the volume Ω enclosed by $\partial\Omega$ (see Eq. (2.2)):

$$\oiint_{\partial\Omega} \mathbf{E}(\mathbf{x}') \cdot \mathbf{n}(\mathbf{x}') dS(\mathbf{x}') = \frac{1}{\epsilon_0} \iiint_{\Omega} \varrho(\mathbf{x}') d^3x' = \frac{Q(\Omega)}{\epsilon_0} \Leftrightarrow \nabla \cdot \mathbf{E} = \frac{\varrho}{\epsilon_0}.$$

- Knowing ϱ, and from Gauss's law (2.2), it is not possible to deduce Coulomb's law, unless one assumes the radial symmetry of the field generated by a point charge. In this sense, Gauss's law alone contains less information than Coulomb's integral (2.1). This lack of information can be completed by symmetry arguments that constrain the spatial dependence of the electric field when the charge distribution allows for it. When the charge distribution does not have any particular symmetry, Gauss's law alone is not sufficient to determine $\mathbf{E}$.

Problems

2.1   Charge density in the atmosphere

The electric field near the Earth surface can be written as

$$\mathbf{E}(z) = -ae^{-\alpha z}\mathbf{u}_z,$$

where $\mathbf{u}_z$ is the vertical direction perpendicular to the surface so that z is the height from the ground. In addition $a = 100 \, \text{V m}^{-1}$ and $\alpha = 3.5 \, \text{km}$.

(a) Determine the charge density in the atmosphere.
(b) What is the total charge contained in a vertical column of section S extending from $z = 0$ to infinity? What is the total charge in the atmosphere? (the Earth radius is $R_T = 6400 \, \text{km}$).

2.2   Electric flux through a square surface

Calculate the electric flux generated by a point charge q through a square surface Σ of side L. The charge is located at distance $d = L/2$ from the center of the surface on an axis perpendicular to it, as shown below.

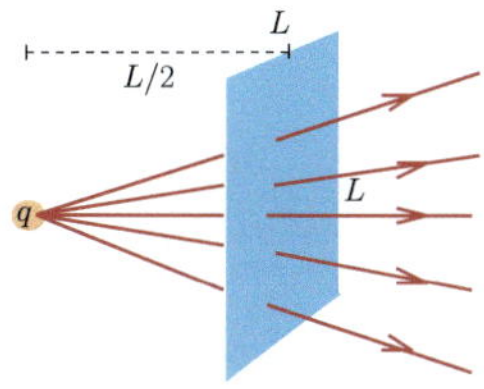

2.3 Field of an infinite charged line

Calculate the electric field of an infinite line of charge using Gauss's law. Check that the result is the same as that previously calculated in Problem 1.7 using the Coulomb integral.

2.4 A coaxial cable

Consider a very long coaxial cable, consisting of a solid inner cylinder of radius a with a uniform volume charge density ϱ and a hollow outer cylinder of radius b ($b > a$) that carries a surface charge density σ such that the cable in its entirety is electrically neutral. Find the electric field produced by the cable in all space.

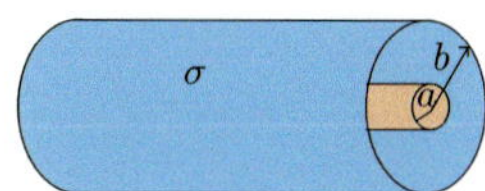

2.5 Superposition of spheres

Consider two nonconcentric spheres of radius R. The first sphere carries a volume charge density ϱ and the second $-\varrho$. The centers of the spheres are at a distance smaller than $2R$. Let **d** be the vector that goes from the center of the positively charged sphere to the center of the negatively charged sphere, as shown below. Show that the electric field within the intersection of the spheres is constant, and find its value.

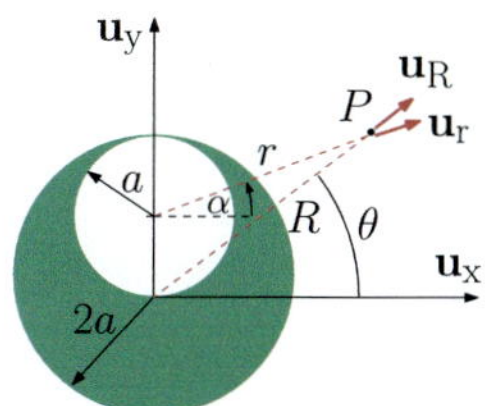

2.6 Sphere with cavity

A sphere centered at the origin, of radius $2a$, carries a homogeneous charge density ϱ. Material is extracted from this sphere so as to form a spherical cavity of radius a. The cavity center is on the y-axis and at distance a from the origin, as shown below. The rest of the sphere carries the homogeneous charge density ϱ.

(a) Show that the electric field at point P outside the sphere is equal to

$$\mathbf{E} = \frac{\varrho}{3\epsilon_0} \left(\frac{8a^3}{r^2 + a^2 + 2ar \sin \alpha} \mathbf{u}_R - \frac{a^3}{r^2} \mathbf{u}_r \right),$$

where $\mathbf{u}_R = \mathbf{R}/R$, $\mathbf{u}_r = \mathbf{r}/r$.

(b) Now consider the case where the point P is within the cavity, at a distance $r < a$ from its center. Show that the x-component of the electric field at this point is zero. Find E_y.

Chapter 3
Electric Potential, Electrostatic Energy, and the Circulation Law

Abstract This chapter introduces the fundamental concept of the *electric potential* (V) as a scalar field from which the electrostatic field ($\mathbf{E}$) can be derived, emphasizing the relationship $\mathbf{E} = -\nabla V$. This formulation, combined with Gauss's law, provides a complete and often more convenient framework for solving electrostatic problems. The chapter details the definition of electric potential, including considerations for absolute potential (relative to infinity) and potential relative to ground. A significant portion is dedicated to the relationship between *work, electric potential, and energy*. It establishes that the work done by the electric field on a charge is path-independent, leading to the definition of *electrostatic potential energy* for both discrete and continuous charge distributions. The concept of *electrostatic energy density* is introduced, demonstrating that the total energy can be interpreted as localized in regions where the electric field is non-zero. The chapter further explores the *multipole expansion*, a powerful technique for approximating the electric potential of an arbitrary charge distribution at large distances. This expansion introduces the concepts of monopole, dipole, and quadrupole moments, highlighting how the far-field behavior of the potential and electric field is dominated by the lowest non-zero moment. Special attention is given to the *electric dipole*, including its definition, the potential and field it generates, and its behavior (potential energy, force, and torque) in an external electric field. Finally, the chapter formalizes the two fundamental laws of electrostatics: *Gauss's law* (in its differential form, relating divergence of $\mathbf{E}$ to charge density) and the *circulation law* (stating that the curl of $\mathbf{E}$ is zero, implying the conservative nature of the electrostatic field). It concludes by demonstrating, through *Helmholtz's decomposition theorem*, that these two differential laws are sufficient to uniquely determine the electric field, thereby retrieving Coulomb's law.

Keywords Electric potential · Electrostatic energy · Circulation law · Electrostatics

F. Cadiz and A. Couairon, *Classical Electrodynamics*, Undergraduate Texts in Physics,
https://doi.org/10.1007/978-3-031-86785-9_3

3.1 Introduction

In this chapter we will show that the electrostatic field is conservative, meaning it derives from a scalar potential V, called the electric potential, such that $\mathbf{E} = -\nabla V$. This relationship, together with Gauss's law, forms a complete system of equations in electrostatics. Indeed, given a charge distribution ϱ, there is a unique field $\mathbf{E}$ that satisfies Gauss's law, decays to zero at infinity, and derives from a potential V.

In many cases, it is more convenient to calculate the electric potential first and then determine the electric field by taking its gradient. This approach is particularly useful when analyzing the field far from an electric dipole, a system consisting of two charges of opposite sign. Such a system can approximate any charge distribution that is globally neutral. We demonstrate that the electric potential represents potential energy per unit charge and provides a basis for defining the electrostatic energy of a system of charges.

3.2 An Important Identity in Electrostatics

Let us begin by demonstrating the following identity, of crucial importance in electrostatics

$$\nabla \frac{1}{|\mathbf{x} - \mathbf{x}'|} = -\frac{\mathbf{x} - \mathbf{x}'}{|\mathbf{x} - \mathbf{x}'|^3} \, . \tag{3.1}$$

Proof A very convenient way to demonstrate this identity is to use spherical coordinates with origin in $\mathbf{x}'$, in which case $\mathbf{x} - \mathbf{x}' = \mathbf{r} = r\mathbf{u}_r$ and $|\mathbf{x} - \mathbf{x}'| = r$, so that

$$\nabla \frac{1}{|\mathbf{x} - \mathbf{x}'|} = \nabla \frac{1}{r} = -\frac{1}{r^2} \nabla r = -\frac{1}{r^2} \frac{\partial}{\partial r}(r\mathbf{u}_r) = -\frac{1}{r^2}\mathbf{u}_r \, .$$

Finally,

$$\nabla \frac{1}{|\mathbf{x} - \mathbf{x}'|} = -\frac{1}{r^3}\mathbf{r} = -\frac{\mathbf{x} - \mathbf{x}'}{|\mathbf{x} - \mathbf{x}'|^3} \, ,$$

which demonstrates (3.1).

This identity has a direct consequence for the electrostatic field, since the latter can always be written as the integral (1.5)

$$\mathbf{E}(\mathbf{x}) = \frac{1}{4\pi\epsilon_0} \iiint_{\mathbb{R}^3} \varrho(\mathbf{x}') \frac{(\mathbf{x} - \mathbf{x}')}{|\mathbf{x} - \mathbf{x}'|^3} d^3 x' \, .$$

Using identity (3.1) and the linearity of the gradient operator,

$$\mathbf{E}(\mathbf{x}) = -\frac{1}{4\pi\epsilon_0}\iiint_{\mathbb{R}^3}\varrho(\mathbf{x}')\nabla\frac{1}{|\mathbf{x}-\mathbf{x}'|}d^3x' = -\nabla\left\{\frac{1}{4\pi\epsilon_0}\iiint_{\mathbb{R}^3}\varrho(\mathbf{x}')\frac{1}{|\mathbf{x}-\mathbf{x}'|}d^3x'\right\}.$$

We obtain that the electric field is the gradient of a scalar quantity that will be called the electric potential.

3.3 The Electric Potential

Electric potential

The electric potential V of a charge distribution ϱ is defined by the integral

$$V(\mathbf{x}) = \frac{1}{4\pi\epsilon_0}\iiint_{\mathbb{R}^3}\frac{\varrho(\mathbf{x}')}{|\mathbf{x}-\mathbf{x}'|}d^3x' + C, \tag{3.2}$$

where C is an arbitrary constant. The electric field generated by ϱ then writes

$$\mathbf{E}(\mathbf{x}) = -\nabla V(\mathbf{x}). \tag{3.3}$$

Remarks

1. The potential of a point charge q located at $\mathbf{x}_0$ is given by

$$V(\mathbf{x}) = \frac{1}{4\pi\epsilon_0}\frac{q}{|\mathbf{x}-\mathbf{x}_0|} + C.$$

2. If the charge density is integrable $\left(\iiint_{\mathbb{R}^3}|\varrho(\mathbf{x}')|d^3x' < \infty\right)$, then:

$$\lim_{|\mathbf{x}|\to\infty}|\mathbf{x}|V(\mathbf{x}) < \infty$$

so that the potential decays at least as fast as $1/|\mathbf{x}|$ at infinity.

3.3.1 Absolute Potential and Earth's Potential

The electric field is insensitive to the choice of the constant C used in (3.2) to define the potential. In practice, two natural choices are used for this constant:

- **The absolute potential (with respect to infinity).** We choose the constant so that the potential is zero at any point located infinitely far from any charge. The potential defined this way, also called the absolute potential, is given by the choice $C = 0$:

$$V(\mathbf{x}) = \frac{1}{4\pi\epsilon_0} \iiint_{\mathbb{R}^3} \frac{\varrho(\mathbf{x}')}{|\mathbf{x} - \mathbf{x}'|} d^3 x'$$

and verifies $\lim_{|\mathbf{x}|\to\infty} V(\mathbf{x}) = 0$ if the charge density is integrable.

- **Potential with respect to the ground (Earth).** The absolute potential of the surface of the Earth, V_G, has two remarkable properties:

1. V_G is the same at every point on the Earth's surface, since the latter is a conductor (this property will be demonstrated in Chap. 4). One can show that the absolute potential of the Earth is

$$V_G = \frac{1}{4\pi\epsilon_0} \frac{Q_T}{R_T} ,$$

   where Q_T is the total charge on the surface of the Earth and R_T is its radius.

2. In many situations, Earth can be considered as an isolated system, V_G is time-independent and therefore it is natural to assume that its surface charge Q_T is a constant. The space- and time-invariant ground potential can then be used as a reference for the potential by attributing to it a value of 0. This means we choose $C = -V_G$ in (3.2) and the potential with respect to the Earth at point $\mathbf{x}$ is

$$V(\mathbf{x}) = \frac{1}{4\pi\epsilon_0} \iiint_{\mathbb{R}^3} \frac{\varrho(\mathbf{x}')}{|\mathbf{x} - \mathbf{x}'|} d^3 x' - V_G .$$

Unless stated otherwise, we will use mainly the absolute potential (i.e., $V = 0$ at infinity). Sometimes, we will make no distinction between ground potential and absolute potential. This means that if the electrostatic system under investigation is small compared to the Earth, *infinity* can be considered as a point on the ground rather than true infinity, which would be infinitely far away from the Earth.

Example 3.1—Potential at the center of a conductive sphere
Consider a conductive sphere S of radius R having a uniform charge density σ on its surface. By definition, the absolute potential at any point $\mathbf{x}$ is given by:

$$V(\mathbf{x}) = \frac{\sigma}{4\pi\epsilon_0} \oiint_S dS(\mathbf{x}') \frac{1}{|\mathbf{x} - \mathbf{x}'|} .$$

At the center of the sphere, $\mathbf{x} = \mathbf{0}$, and so

$$V(0) = \frac{\sigma}{4\pi\epsilon_0} \oiint_S dS(\mathbf{x}') \frac{1}{|\mathbf{x}'|} = \frac{\sigma}{4\pi\epsilon_0} \int_0^{2\pi} d\phi \int_0^{\pi} \sin\theta d\theta R^2 \frac{1}{R} = \frac{\sigma}{\epsilon_0} R \ .$$

The total charge on the sphere is $Q = 4\pi R^2 \sigma$, and then:

$$V(0) = \frac{Q}{4\pi\epsilon_0 R} \ .$$

We will see in Chap. 4 that the potential of a conductive sphere, in fact, has the same value $V = V(0)$ at any point inside the sphere.

3.3.2 *Equipotential Surfaces*

Equipotential surface
An equipotential surface is a set of points in space for which the potential has the same value $V = V_0$. In other words, if $\mathbf{x}$ belongs to an equipotential surface, and if $d\mathbf{x}$ is an infinitesimal displacement along the tangent at point $\mathbf{x}$ to this surface, one has

$$dV = \nabla V(\mathbf{x}) \cdot d\mathbf{x} = 0 \ ,$$

that is to say,

$$\mathbf{E}(\mathbf{x}) \cdot d\mathbf{x} = 0 \ .$$

The electric field lines are, therefore, perpendicular at all points to the equipotential surfaces.

Figure 3.1 shows three equipotentials generated by a positive (left) and a negative (right) point charge. They correspond to surfaces for which $1/r =$ constant, therefore concentric spheres centered on the charge. Also shown are the field lines which are perpendicular to the equipotentials.

The potential decreases in the direction of the field lines, such that, in Fig. 3.1, $V_1 > V_2 > V_3$ for a positive charge, whereas $V_1 < V_2 < V_3$ for a negative charge. Equation (3.2) says that the potential of an arbitrary charge distribution is the superposition of the potentials created by each individual charge in the system. Figure 3.2 shows the equipotentials together with the electric field lines of a system of two equal positive charges, in a plane containing these charges. Each equipotential corresponds to a different value V_0 of the potential, which is represented here

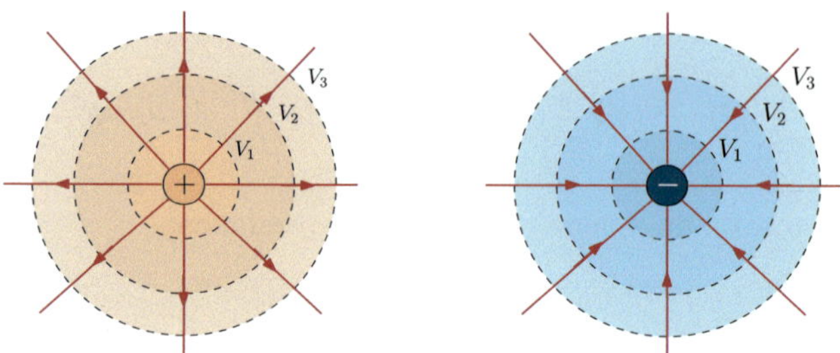

Fig. 3.1 The equipotentials surfaces for a point charge are spheres centered on the charge

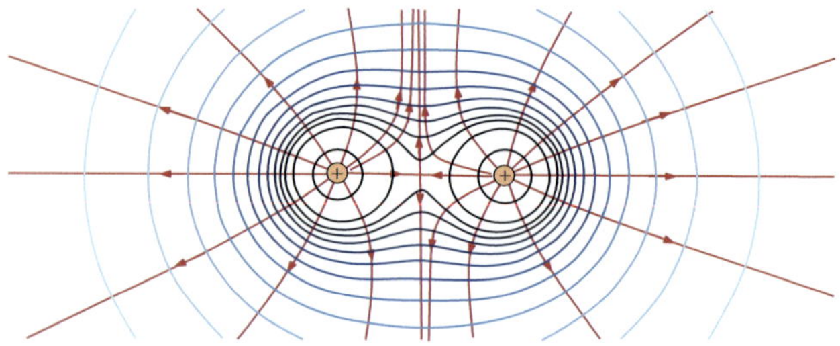

Fig. 3.2 Electric field lines (red) and equipotentials for two positive charges

by a color. It can be seen that the electric field lines are always perpendicular to the equipotentials.

3.3.3   *Potential and Mirror Symmetries*

The absolute electric potential obeys the same symmetry properties as those obeyed by the charge distribution. A demonstration of Curie's principle, as applied to the potential, can be found in the Appendix A.4. Following the arguments of Sect. 2.4.4 for the mirror symmetries of the electric field, consider a plane Π which partitions the whole space into two disjoint domains D_1 and $D_2 = \{\mathbf{u}' = \mathrm{sym}_\Pi \mathbf{u}, \mathbf{u} \in D_1\}$. The absolute potential ($C = 0$ in (3.2)) can be written as:

$$V(\mathbf{x}) = \frac{1}{4\pi \epsilon_0} \iiint_{\mathbf{u} \in D_1} \left(\frac{\varrho(\mathbf{u})}{|\mathbf{x} - \mathbf{u}|} + \frac{\varrho(\mathbf{u}')}{|\mathbf{x} - \mathbf{u}'|} \right) d^3 u \ .$$

If ϱ has either mirror symmetry or antisymmetry with respect to Π, $\varrho(\mathbf{u}') = \pm\varrho(\mathbf{u})$ and

$$V(\mathbf{x}) = \frac{1}{4\pi \epsilon_0} \iiint_{\mathbf{u} \in D_1} \varrho(\mathbf{u}) \left(\frac{1}{|\mathbf{x} - \mathbf{u}|} \pm \frac{1}{|\mathbf{x} - \mathbf{u}'|} \right) d^3 u \ . \tag{3.4}$$

Now, let $\mathbf{x}' = \mathrm{sym}_\Pi \mathbf{x}$, then

$$V(\mathbf{x}') = \frac{1}{4\pi\epsilon_0} \iiint_{\mathbf{u}\in D_1} \varrho(\mathbf{u}) \left(\frac{1}{|\mathbf{x}' - \mathbf{u}|} \pm \frac{1}{|\mathbf{x}' - \mathbf{u}'|} \right) d^3u \, ,$$

but $|\mathbf{x}' - \mathbf{u}| = |\mathbf{x} - \mathbf{u}'|$ and $|\mathbf{x}' - \mathbf{u}'| = |\mathbf{x} - \mathbf{u}|$ so that

$$V(\mathbf{x}') = \frac{1}{4\pi\epsilon_0} \iiint_{\mathbf{u}\in D_1} \varrho(\mathbf{u}) \left(\frac{1}{|\mathbf{x} - \mathbf{u}'|} \pm \frac{1}{|\mathbf{x} - \mathbf{u}|} \right) d^3u \, . \tag{3.5}$$

Comparing (3.4) and (3.5), we see that

- If ϱ has mirror symmetry with respect to Π ($\varrho(\mathbf{u}) = \varrho(\mathbf{u}')$), the same holds for the absolute potential

$$V(\mathbf{x}' = \text{sym}_\Pi \mathbf{x}) = V(\mathbf{x}) \, .$$

- If ϱ has mirror antisymmetry with respect to Π ($\varrho(\mathbf{u}) = -\varrho(\mathbf{u}')$), the same holds for the absolute potential

$$V(\mathbf{x}' = \text{sym}_\Pi \mathbf{x}) = -V(\mathbf{x}) \, .$$

Note that adding an arbitrary constant C to the potential will break the mirror anti-symmetry of the potential, while it will preserve the mirror symmetry. In Fig. 3.2, one sees that the plane equidistant to the two point charges is a plane of symmetry for the charge distribution, as well as any plane that contains the charges. In consequence, the potential has mirror symmetry with respect to these planes. This is confirmed by the shape of the equipotentials in Fig. 3.2.

3.4 Potential and Field Generated by an Electric Dipole

Dipole moment

An electric dipole is the system formed by two charges of equal magnitude q but of opposite sign, separated by a distance d. If $q > 0$ is at position $\mathbf{x}_1$, and $-q$ at position $\mathbf{x}_2$, the dipole moment $\mathbf{p}$ of this system is defined as

$$\mathbf{p} = q\mathbf{x}_1 - q\mathbf{x}_2 = q(\mathbf{x}_1 - \mathbf{x}_2) \, .$$

Writing $d = |\mathbf{x}_2 - \mathbf{x}_1|$ and defining $\mathbf{u} = (\mathbf{x}_1 - \mathbf{x}_2)/d$ as the unit vector in the direction that joins the negative to the positive charge (see Fig. 3.3), the dipole moment can be written as

$$\boxed{\mathbf{p} = qd\mathbf{u}} \, . \tag{3.6}$$

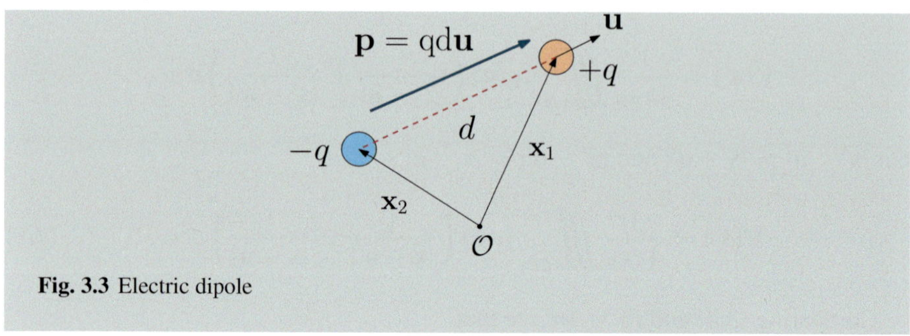

Fig. 3.3 Electric dipole

Figure 3.4 shows two examples of electric dipoles found in nature. The first one is the hydrogen atom, in which an electron orbits around the proton at a distance equal to the Bohr radius, $d \approx 5.29 \times 10^{-11}$ m. The dipole moment of this atom points toward the nucleus and has a magnitude of $|\mathbf{p}| = de = 8.46 \times 10^{-30}$ Cm.

Another example is the water molecule. It is a globally neutral system, however, the electrons are more likely to surround the oxygen atom, leaving an excess of positive charge near the two hydrogen atoms. This molecule can therefore be modeled as a dipole moment of magnitude $|\mathbf{p}| = 6.2 \times 10^{-30}$ Cm. Molecules having a net dipole moment are called polar and are, therefore, subjected to electric forces between them, as in the case of water.

The concept of electric dipole is therefore as important as that of a point charge. Any arbitrary charge distribution that is globally neutral indeed constitutes an electric dipole, provided its dipole moment is non-zero, as will be shown in Sect. 3.5. It is the case for atoms, molecules, or even neutral conductors in the presence of an electric field (see Sect. 4.2.8).

Now, let us consider an electric dipole oriented along the $\mathbf{u}_z$ direction, as shown in Fig. 3.5. The charge distribution is invariant under rotation around the z-axis. In spherical coordinates (we choose the origin at the midpoint between the two charges), the potential is therefore independent of ϕ, that is $V(r, \theta, \phi) = V(r, \theta)$. This means that, for the calculation of the potential, we can restrict ourselves to any plane containing the dipole.

Fig. 3.4 Two examples of electric dipoles: hydrogen atom (top), water molecule (bottom)

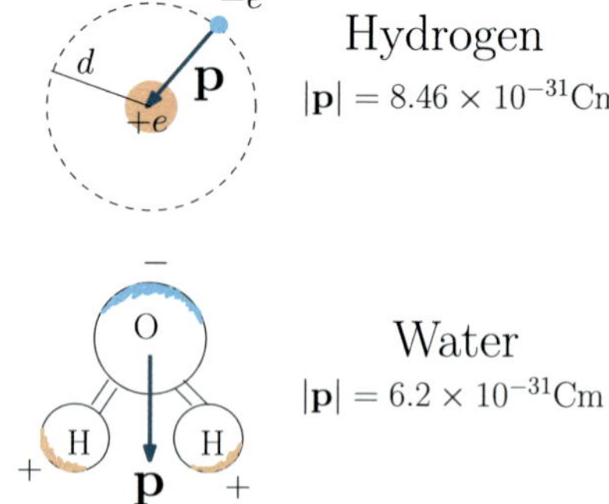

Fig. 3.5 Electric dipole oriented along the $\mathbf{u}_z$ axis

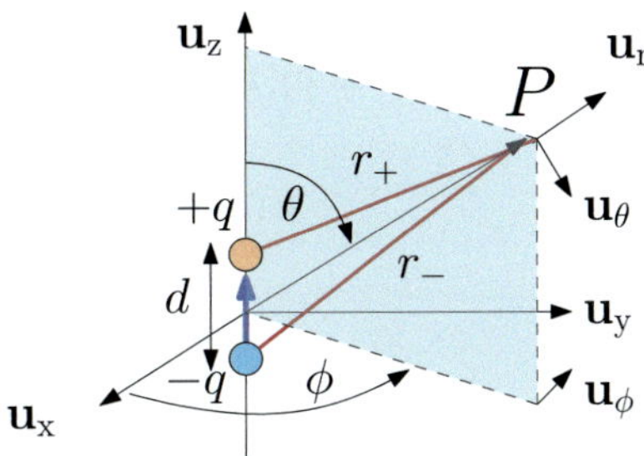

By the superposition principle, the potential at any point P is given by

$$V(P) = \frac{1}{4\pi\epsilon_0}\left(\frac{q}{r_+} - \frac{q}{r_-}\right) , \tag{3.7}$$

where the distances $r_\pm$ can be expressed as

$$r_+^2 = r^2 - dr\cos\theta + (d/2)^2 ,$$
$$r_-^2 = r^2 + dr\cos\theta + (d/2)^2 ,$$

and so,

$$r_\pm = r\sqrt{1 \mp (d/r)\cos\theta + (d/2r)^2} .$$

Far from the dipole, $d \ll r$, and we can use the Taylor series expansion to first order in d/r:

$$\frac{1}{r_\pm} = \frac{1}{r}\left(1 \mp \frac{d}{r}\cos\theta + \cdots\right) .$$

With this expansion and (3.7), the potential far from the dipole takes the form

$$V(P) = V(r,\theta) \approx \frac{q}{4\pi\epsilon_0 r}\frac{d\cos\theta}{r} = \frac{p\cos\theta}{4\pi\epsilon_0 r^2} ,$$

where $p = qd$ denotes the magnitude of the dipole moment. Thus, the potential far from a dipole decays as $1/r^2$ and the equipotential surfaces can be found by the parametric equation $r^2 = C\cos\theta$, where C is a constant for each surface. In particular, in the xy-plane $\Pi = (O, \mathbf{u}_x, \mathbf{u}_y)$ (i.e., $\theta = \pi/2$) the (absolute) potential is zero. This can be expected from the mirror antisymmetry of the charge distribution (and therefore, of the potential) with respect to this plane. For any point P in an antisymmetry plane, $V(P) = -V(P) = 0$.

Finally, in terms of the dipole moment $\mathbf{p} = p\mathbf{u}_z = p\cos\theta\,\mathbf{u}_r - p\sin\theta\,\mathbf{u}_\theta$, the potential can be written as

$$\boxed{V(\mathbf{x}) = \frac{\mathbf{p}\cdot\mathbf{u}_r}{4\pi\epsilon_0 r^2} = \frac{\mathbf{p}\cdot\mathbf{x}}{4\pi\epsilon_0|\mathbf{x}|^3} .} \tag{3.8}$$

The electric field can be obtained as $\mathbf{E} = -\nabla V$. Recalling the invariance of the charge distribution by rotation around the z-axis:

$$\mathbf{E}(r, \theta) = -\frac{\partial V(r, \theta)}{\partial r}\mathbf{u}_r - \frac{1}{r}\frac{\partial V(r, \theta)}{\partial \theta}\mathbf{u}_\theta ,$$

so that

$$E_r = -\frac{\partial V(r, \theta)}{\partial r} = \frac{p \cos \theta}{2\pi \epsilon_0 r^3} ,$$

$$E_\theta = -\frac{1}{r}\frac{\partial V(r, \theta)}{\partial \theta} = \frac{p \sin \theta}{4\pi \epsilon_0 r^3} .$$

We retrieve in the radial component the result of Example 1.2 for the field in the axis of the dipole ($\theta = 0$). Finally,

$$\mathbf{E}(r, \theta) = \frac{p}{4\pi \epsilon_0 r^3}(2 \cos \theta\, \mathbf{u}_r + \sin \theta\, \mathbf{u}_\theta) . \tag{3.9}$$

Equation (3.9) can be rewritten in the form of a compact expression that depends only on the dipole moment and on vector $\mathbf{x}$:

$$\mathbf{E}(\mathbf{x}) = \frac{1}{4\pi \epsilon_0 r^3}(3(\mathbf{p} \cdot \mathbf{u}_r)\mathbf{u}_r - \mathbf{p}) = \frac{1}{4\pi \epsilon_0 |\mathbf{x}|^5}(3(\mathbf{p} \cdot \mathbf{x})\mathbf{x} - |\mathbf{x}|^2\mathbf{p}) . \tag{3.10}$$

The electric field far from the dipole decays as $1/r^3$. The field lines can be obtained by solving $\mathbf{E} \times d\mathbf{x} = \mathbf{0}$, which in this case gives $r = C \sin^2 \theta$, where C is a constant for each field line. Figure 3.6 shows the electric field lines and the equipotentials in any plane containing both charges.

3.5 The Multipole Expansion

Let us suppose that the total charge distribution of a system lies inside a finite, bounded region of space Ω near the origin (see Fig. 3.7). Here the goal is to obtain an approximate expression for the electric potential at any point P far from Ω, in order to generalize what we have done for the electric dipole in Sect. 3.4.

In electrostatics, the expression for the electric potential reads

$$V(\mathbf{x}) = \frac{1}{4\pi \epsilon_0} \iiint_\Omega \frac{\varrho(\mathbf{x}')}{|\mathbf{x} - \mathbf{x}'|} d^3 x' ,$$

where $|\mathbf{x} - \mathbf{x}'| = \left(|\mathbf{x}|^2 - 2\mathbf{x} \cdot \mathbf{x}' + |\mathbf{x}'|^2\right)^{1/2}$, so that

Fig. 3.6 Equipotential
surfaces and electric field
lines for a dipole

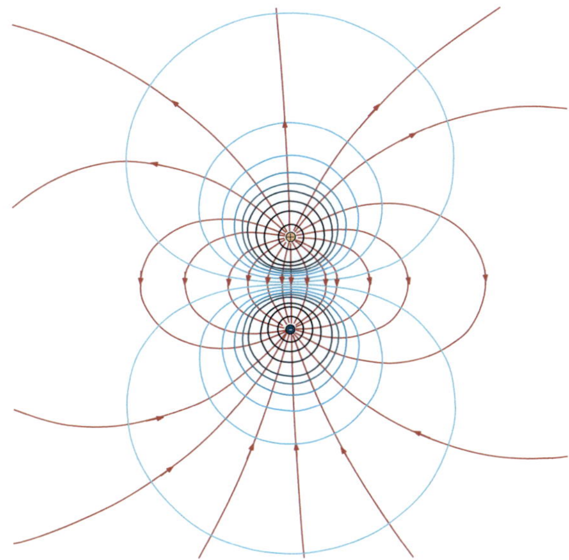

Fig. 3.7 An arbitrary charge
distribution inside a bounded
volume Ω

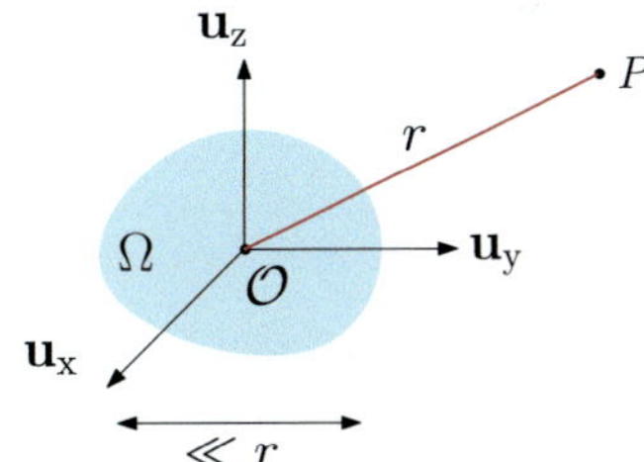

$$\frac{1}{|\mathbf{x}-\mathbf{x}'|} = \frac{1}{|\mathbf{x}|}\left(1 - 2\,\frac{\mathbf{x}\cdot\mathbf{x}'}{|\mathbf{x}|^2} + \frac{|\mathbf{x}'|^2}{|\mathbf{x}|^2}\right)^{-1/2} .$$

At distances $r = |\mathbf{x}|$ much larger than the typical size of Ω, we have $|\mathbf{x}'|/|\mathbf{x}| \ll 1$.
Using the Taylor expansion

$$(1+z)^{-1/2} = 1 - \frac{1}{2}z + \frac{3}{8}z^2 + O(z^3)$$

valid for $z \ll 1$, we find

$$\frac{1}{|\mathbf{x}-\mathbf{x}'|} \approx \frac{1}{|\mathbf{x}|}\left\{1 + \frac{\mathbf{x}\cdot\mathbf{x}'}{|\mathbf{x}|^2} + \frac{3(\mathbf{x}\cdot\mathbf{x}')^2 - |\mathbf{x}|^2|\mathbf{x}'|^2}{2|\mathbf{x}|^4} + O\left(\frac{|\mathbf{x}'|}{|\mathbf{x}|}\right)^3\right\} .$$

Therefore, the potential at second order in $|\mathbf{x}'|/|\mathbf{x}|$ is given by

$$V(\mathbf{x}) = \frac{1}{4\pi\epsilon_0|\mathbf{x}|} \iiint_\Omega \varrho(\mathbf{x}')d^3x' + \frac{\mathbf{x}}{4\pi\epsilon_0|\mathbf{x}|^3} \cdot \iiint_\Omega \rho(\mathbf{x}')\mathbf{x}'d^3x'$$
$$+ \frac{1}{8\pi\epsilon_0|\mathbf{x}|^5} \iiint_\Omega \left(3(\mathbf{x}\cdot\mathbf{x}')^2 - |\mathbf{x}'|^2|\mathbf{x}|^2\right)\rho(\mathbf{x}')d^3x' . \tag{3.11}$$

Writing the scalar products as $\mathbf{x}\cdot\mathbf{x}' = \mathbf{x}^T\mathbf{x}'$ and $|\mathbf{x}|^2 = \mathbf{x}^T\mathbf{x}$, the integrand in the third term of (3.11) can be rewritten

$$\left(3(\mathbf{x}\cdot\mathbf{x}')^2 - |\mathbf{x}'|^2|\mathbf{x}|^2\right) = \mathbf{x}^T\left(3\mathbf{x}'\mathbf{x}'^T - |\mathbf{x}'|^2\right)\mathbf{x} .$$

In the first term of the expansion (3.11), we recognize the total charge of the distribution

$$Q = \iiint_\Omega \varrho(\mathbf{x}')d^3x' ,$$

whereas the dipole moment of the charge distribution appears in the second term.

Dipole moment of an arbitrary charge distribution
The dipole moment $\mathbf{p}$ of an arbitrary charge distribution ϱ writes

$$\mathbf{p} = \iiint_{\mathbb{R}^3} \mathbf{x}\varrho(\mathbf{x})d^3x . \tag{3.12}$$

Note that for an electric dipole composed of two point charges $\pm q$ separated by a distance d, (3.12) reduces to $\mathbf{p} = q(\mathbf{x}_1 - \mathbf{x}_2) = qd\mathbf{u}$, where $\mathbf{u}$ is the unit vector oriented from the negative to the positive charge.

Finally, by defining the quadrupole moment as the matrix

$$D = \iiint_\Omega \varrho(\mathbf{x})\left\{3\mathbf{x}'\cdot\mathbf{x}'^T - |\mathbf{x}'|^2\right\}d^3x' , \tag{3.13}$$

the multipole expansion writes

$$V(\mathbf{x}) = \frac{Q}{4\pi\epsilon_0|\mathbf{x}|} + \frac{\mathbf{p}\cdot\mathbf{x}}{4\pi\epsilon_0|\mathbf{x}|^3} + \frac{1}{8\pi\epsilon_0}\frac{\mathbf{x}^T D\mathbf{x}}{|\mathbf{x}|^5} + \cdots , \tag{3.14}$$

which is a good approximation far from the charge distribution, that is, for $|\mathbf{x}|$ much larger than the typical length scale of Ω.

Remarks

1. The first term, that decays as $1/r$, corresponds to the potential of a point charge Q at the origin. Considering only this term in the expansion is the simplest approximation that one can make, provided that $Q \neq 0$.
2. The second term in the expansion, decaying as $1/r^2$, corresponds to the potential of an electric dipole $\mathbf{p}$ located at the origin. Far from any neutral charge distribution $(Q = 0)$, the potential is mainly that of a dipole moment (if $\mathbf{p} \neq \mathbf{0}$).
3. The third term decays as $1/r^3$, and corresponds to the potential of a quadrupole. One can construct a simple quadrupole with a set of four charges such that the total charge and total dipole moment are zero.

3.6 Work, Electric Potential, and Energy

The electrostatic potential plays a a role in electromagnetism analogous to that of the gravitational potential in classical mechanics. Suppose that an external operator carries a charge q from point $\mathbf{x}_1$ to $\mathbf{x}_2$ along the path Γ, in the presence of an electric field $\mathbf{E}$ (see Fig. 3.8). To move this charge q, the operator has to exert a force $-\mathbf{F}_e$ that opposes the electric force $\mathbf{F}_e = q\mathbf{E}$ acting on the charge.

The work provided by the operator is then given by the path integral

$$W = -q \int_{\mathbf{x}_1}^{\mathbf{x}_2} \mathbf{E}(\mathbf{x}) \cdot d\mathbf{x} \ . \tag{3.15}$$

In general, the value of this integral depends on the particular path Γ taken by the charge q. However, since an electrostatic field derives from a potential, introduction of the relation $\mathbf{E}(\mathbf{x}) = -\nabla V(\mathbf{x})$ into (3.15) writes

$$W = q \int_{\mathbf{x}_1}^{\mathbf{x}_2} \nabla V(\mathbf{x}) \cdot d\mathbf{x}$$

and we obtain the integral of an exact differential $\nabla V(\mathbf{x}) \cdot d\mathbf{x} = dV$:

$$W = q \int_{\mathbf{x}_1}^{\mathbf{x}_2} dV = q\left(V(\mathbf{x}_2) - V(\mathbf{x}_1)\right) \ .$$

Fig. 3.8 A charge q moved by an operator against the effect of an electric field

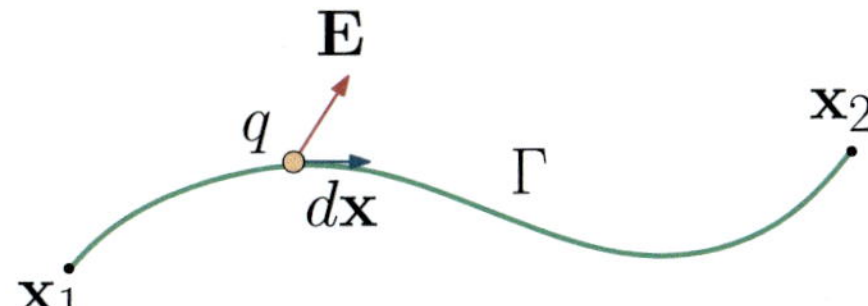

The work is independent of the path between $\mathbf{x}_1$ and $\mathbf{x}_2$.

$$W = -q \int_{\Gamma} \mathbf{E}(\mathbf{x}) \cdot d\mathbf{x} = q\left(V(\mathbf{x}_2) - V(\mathbf{x}_1)\right) \ . \tag{3.16}$$

From (3.16), we see that qV has units of energy, and corresponds to the potential energy of a charge q in the presence of an electric potential V. The unit of electric potential is the volt, with $1\,\mathrm{V} = 1\,\mathrm{J\,C^{-1}}$.

Electrostatic energy of a point charge

The electrostatic potential energy of a charge q located at $\mathbf{x}$ is

$$U(\mathbf{x}) = qV(\mathbf{x}) \ ,$$

where $V(\mathbf{x})$ is an external electrostatic potential at $\mathbf{x}$. Indeed, the force associated with this potential reads

$$\mathbf{F}(\mathbf{x}) = -\nabla U(\mathbf{x}) = q\mathbf{E}(\mathbf{x}) \ .$$

Using the work-energy theorem, we conclude that for a particle in an electric potential V, the total mechanical energy

$$E = K + U = \frac{1}{2}m\left(\frac{d\mathbf{x}}{dt}\right)^2 + qV(\mathbf{x})$$

is a constant.

Example 3.2—Energy of an electron in a hydrogen atom

In Example 1.1 we saw that the electron and the proton are held together by the electrostatic interaction, since gravity is negligible at these atomic scales. Let us suppose a circular orbit of the electron around the nucleus, with the orbit radius $r = a_0 \approx 5.29 \times 10^{-11}\,\mathrm{m}$ (Bohr's radius).

The electric potential generated by the proton at the position of the electron is

$$V(a_0) = \frac{e}{4\pi\epsilon_0 a_0} = 27.25\,\mathrm{V} \ ,$$

where $e = 1.602 \times 10^{-19}\,\mathrm{C}$ is the proton charge. The electrostatic potential energy of the electron is therefore:

$$U = q_e V(a_0) = -eV(a_0) = -4.36 \times 10^{-18}\,\mathrm{J} \ .$$

In atomic physics it is more convenient to use a different unit for the energy, the electronvolt (eV), which is defined as the magnitude of the energy of an electron placed in a potential of 1 Volt ($1\,\text{eV} = 1.6 \times 10^{-19}\,\text{J}$). With this definition, the electrostatic energy of the electron in a hydrogen atom is simply:

$$U = -27.25\,\text{eV} .$$

The kinetic energy of the electron is given by $K = \frac{1}{2}m_e v^2$, with $m_e = 9.1 \times 10^{-31}\,\text{kg}$ the electron mass, and v the modulus of its velocity. The equilibrium of forces in the radial direction gives

$$\frac{m_e v^2}{a_0} = \frac{q_e^2}{4\pi\epsilon_0 a_0^2} ,$$

then

$$K = \frac{1}{2}m_e v^2 = \frac{1}{2}\frac{q_e^2}{4\pi\epsilon_0 a_0} = -\frac{1}{2}U .$$

Finally, the total energy of the electron in a hydrogen atom is

$$E = K + U = \frac{1}{2}U = -13.6\,\text{eV} ,$$

which corresponds to the well-known ionization energy of hydrogen.

Example 3.3—Potential of a uniformly charged sphere
Let us calculate the absolute potential generated by a sphere of radius R and charge Q uniformly distributed in its volume. We start from the electric field generated by this charge distribution (see Example 2.4), and then use the relation $\mathbf{E} = -\nabla V$ or, equivalently

$$V(r) = \int_r^\infty \mathbf{E}(\mathbf{x}) \cdot d\mathbf{x} .$$

By choosing a radial trajectory, $d\mathbf{x} = dr\,\mathbf{u}_r$, we get for $r > R$:

$$V(r) = \int_r^\infty E(r)dr = \int_r^\infty \frac{Q\,dr}{4\pi\epsilon_0 r^2} = \frac{Q}{4\pi\epsilon_0}\int_r^\infty \frac{dr}{r^2} = \frac{Q}{4\pi\epsilon_0 r} ,$$

whereas for $r < R$, we find:

$$V(r) = \int_R^\infty E(r)dr + \int_r^R E(r)dr$$

$$= \int_R^\infty \frac{Q}{4\pi\epsilon_0 r^2}dr + \int_r^R \frac{Qr}{4\pi\epsilon_0 R^3}dr = \frac{Q(3R^2 - r^2)}{8\pi\epsilon_0 R^3} \ .$$

Finally,

$$V(r) = \begin{cases} \frac{Q(3R^2 - r^2)}{8\pi\epsilon_0 R^3} & \text{if } r \le R, \\ \frac{Q}{4\pi\epsilon_0 r} & \text{if } r > R \ . \end{cases}$$

Figure 3.9 shows the variation of the potential with respect to the distance r. Note that the potential is continuous and both the electric field and the potential outside the sphere are identical to the ones of a point charge located at the origin.

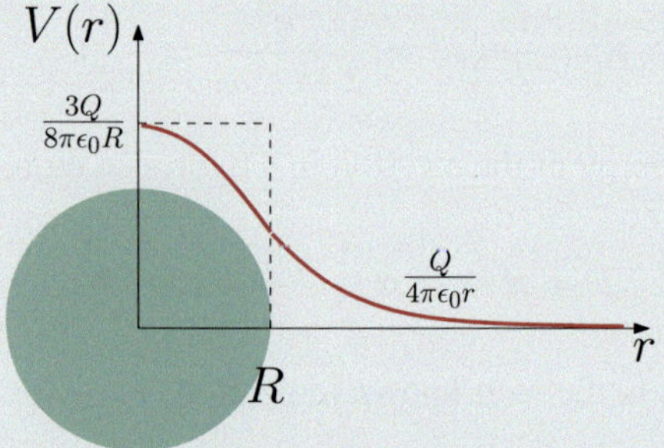

Fig. 3.9 Potential generated by a uniformly charged sphere

Example 3.4—Electrical breakdown of air

Under normal conditions, air is a good electric insulator because air molecules (mostly dinitrogen) are neutral. However, whenever a sufficiently high electric field builds up, a free electron present in air can be accelerated to an energy such that it can ionize a dinitrogen molecule by impact. This impact results in two free electrons, which will be in turn accelerated by the electric field and produce four more electrons by impact with two other dinitrogen molecules. This chain reaction ultimately ionizes the air, causing a spark to appear.

Let us estimate the order of magnitude of the breakdown electric field, considering that an electron in air travels a distance $\lambda = 5\,\mu\text{m}$ before colliding with a N_2 molecule. The kinetic energy that the electron acquires between the impacts is $\Delta K = q\Delta V$, where ΔV is the potential difference between two points separated by a distance λ. If the electric field is constant, then $\Delta V = E\lambda$ and

$$\Delta K = q\,\Delta V = q E \lambda\ .$$

The ionization energy of dinitrogen is $\sim 15\,\mathrm{eV}$. For an electron to be able to ionize a N_2 molecule during a collision, one must have

$$\Delta K = q E \lambda \geq 15\,\mathrm{eV}\ .$$

Therefore, the order of magnitude of the electric breakdown of air is

$$E = \frac{15}{\lambda} = 3 \times 10^6\,\mathrm{V\,m^{-1}}\ .$$

3.6.1 *Potential Energy of a Charge Distribution*

Let us now suppose that there is a discrete charge distribution of point charges q_j located at position $\mathbf{x}_j$, with $j = 1, 2, \ldots, N$. The potential energy of this system is equal to interaction energy of the charges. It corresponds to the work that would be required to bring all the charges from infinity, where they do not interact, to their current positions where each charge interacts with the field generated by all other charges. To calculate this work, we proceed in steps as follows:

1. First, we bring the charge q_1 from infinity to the position $\mathbf{x}_1$. The work W_1 required for this is zero since there is no other electric field apart from that created by q_1 itself. Thus, $W_1 = 0$ and at this stage the system (single charge isolated from the rest of the charges) has no potential energy:

$$U_1 = W_1 = 0\ .$$

2. Now, we bring the charge q_2 from infinity to position $\mathbf{x}_2$, as illustrated in Fig. 3.10. For this, one must work against the electric field created by q_1, requiring a work W_2 given by:

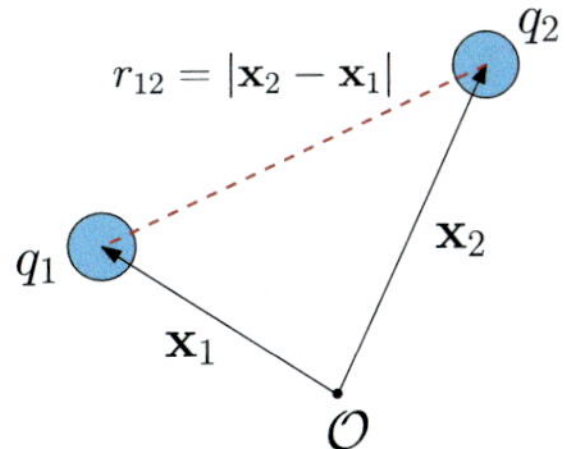

Fig. 3.10 A charge q_2 is brought from infinity at a distance r_{12} from charge q_1

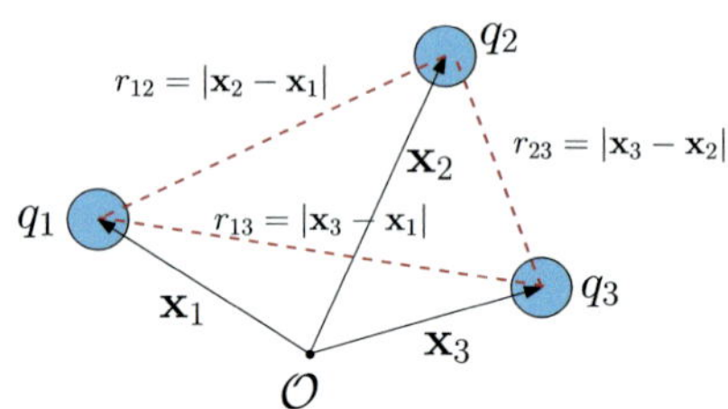

Fig. 3.11 A charge q_3 is brought from infinity to its position in the system of charges

$$W_2 = V(\mathbf{x}_2)q_2 = \left(\frac{1}{4\pi \epsilon_0} \frac{q_1}{r_{12}} \right) q_2 \, .$$

In this configuration, the system possesses a total potential energy of

$$U_2 = W_1 + W_2 = \frac{1}{4\pi \epsilon_0} \frac{q_1 q_2}{r_{12}} \, .$$

3. In the same way, we bring the charge q_3 from infinity (see Fig. 3.11). For this, one must work against the combined electric field created by q_2 and q_1. The work W_3 needed for this is

$$W_3 = q_3 V(\mathbf{x}_3) = \frac{1}{4\pi \epsilon_0} \left(\frac{q_1}{r_{13}} + \frac{q_2}{r_{23}} \right) q_3 \, .$$

The potential energy of the system at this stage is

$$U_3 = W_1 + W_2 + W_3 = \frac{1}{4\pi \epsilon_0} \left(\frac{q_1}{r_{12}} q_2 + \frac{q_1}{r_{13}} q_3 + \frac{q_2}{r_{23}} q_3 \right) \, .$$

Since $r_{ij} = r_{ji}$, the energy U_3 can be rewritten as

$$U_3 = \frac{1}{2} \frac{1}{4\pi \epsilon_0} \left\{ \left(\frac{q_2}{r_{21}} + \frac{q_3}{r_{31}} \right) q_1 + \left(\frac{q_1}{r_{12}} + \frac{q_3}{r_{32}} \right) q_2 + \left(\frac{q_1}{r_{13}} + \frac{q_2}{r_{23}} \right) q_3 \right\} \, .$$

Finally,

$$U_3 = \frac{1}{2} \left\{ V(\mathbf{x}_1)q_1 + V(\mathbf{x}_2)q_2 + V(\mathbf{x}_3)q_3 \right\} \, ,$$

where $V(\mathbf{x}_i)$ is the potential at the position of charge q_i due to all other charges. This result can be now easily extended to the case of an arbitrary number of charges.

Potential energy of an arbitrary charge distribution

For a system of N point charges q_j located at $\mathbf{x}_j$, with $j = 1, 2, \ldots, N$, the potential energy is the interaction energy

$$U = \frac{1}{2} \sum_{i=1}^{N} q_i V(\mathbf{x}_i) \,. \tag{3.17}$$

For the case of a continuous charge distribution ϱ, the expression for the potential energy is generalized to the integral

$$U = \frac{1}{2} \iiint_{\mathbb{R}^3} dq(\mathbf{x}) V(\mathbf{x}) = \frac{1}{2} \iiint_{\mathbb{R}^3} \varrho(\mathbf{x}) V(\mathbf{x}) d^3 x \,. \tag{3.18}$$

Example 3.5—Energy of a uniformly charged sphere and the classical electron radius

Let us calculate the total electrostatic energy of a uniformly charged sphere of radius R and charge Q. In Example 3.3, the potential at a distance r from the center of the sphere was calculated:

$$V(r) = \begin{cases} \frac{Q(3R^2 - r^2)}{8\pi\epsilon_0 R^3} & \text{if } r \leq R, \\ \frac{Q}{4\pi\epsilon_0 r} & \text{if } r > R. \end{cases}$$

Then, the total energy of the charge distribution is given by the integral

$$U = \frac{1}{2} 4\pi \int_0^R r^2 V(r) \varrho(r) dr = \frac{1}{2} 4\pi \varrho \int_0^R r^2 V(r) dr \,,$$

where the charge density ϱ is such that $Q = \frac{4}{3}\pi R^3 \varrho$, and so,

$$U = \frac{3Q^2}{16\pi\epsilon_0 R^6} \int_0^R r^2 (3R^2 - r^2) dr = \frac{3Q^2}{16\pi\epsilon_0 R^6} \frac{4R^5}{5} \,.$$

Finally,

$$U = \frac{3}{5} \frac{Q^2}{4\pi\epsilon_0 R} \,. \tag{3.19}$$

Now, assuming that the electron is a uniformly charged sphere of radius r_e and charge $-e = -1.6 \times 10^{-19}$ C, what would this radius be? The rest energy E of the electron and its mass $m_e = 9 \times 10^{-31}$ kg are related by Einstein's equation $E = m_e c^2$, where $c = 3 \times 10^8$ m s^{-1} is the speed of light. The energy U can be interpreted as the electron's self-energy due to its electric charge, representing the energy required to hold the electron's charge together. For consistency, we must have

$$U = \frac{3}{5}\frac{e^2}{4\pi \epsilon_0 r_e} \leq E = m_e c^2 \ .$$

Solving for the electron radius gives

$$r_e \geq \frac{3}{5}\frac{e^2}{4\pi \epsilon_0 m_e c^2} = 1.69 \times 10^{-15} \text{m} \ ,$$

indicating that its radius must be at least larger than half the radius of a proton. Today, the electron is believed to have no physical size. However, Quantum Field Theory tell us that, at very short length scales, the electron is constantly surrounded by a cloud of electron-positron pairs, so at the end the concept of a single pointlike charge breaks down. The electron can be seen instead as a charge distribution with a finite size a, given by the reduced Compton wavelength:

$$a = \frac{\hbar}{m_e c} \sim 4 \times 10^{-13} \text{ m} \ .$$

The classical radius r_e calculated here is, in fact, smaller than the length scales at which quantum field theory must be applied to describe electromagnetic interactions.

3.6.2 Electrostatic Energy Density

From Eq. (3.18), we see that the electrostatic energy of a charge distribution ϱ can be obtained as an integral of the form

$$U = \iiint_{\mathbb{R}^3} \frac{\varrho(\mathbf{x}) V(\mathbf{x})}{2} d^3 x \ .$$

The differential form of Gauss's law (2.3) allows us to rewrite the charge density as $\varrho = \epsilon_0 \mathbf{\nabla} \cdot \mathbf{E}$, and so,

$$U = \iiint_{\mathbb{R}^3} \frac{\epsilon_0}{2} V(\mathbf{x})(\mathbf{\nabla} \cdot \mathbf{E}(\mathbf{x})) d^3 x \ .$$

In addition, $V(\nabla \cdot \mathbf{E}) = \nabla \cdot (V\mathbf{E}) - \mathbf{E} \cdot \underbrace{\nabla V}_{-\mathbf{E}} = \nabla \cdot (V\mathbf{E}) + |\mathbf{E}|^2$. Finally, the electrostatic energy can be written

$$U = \frac{\epsilon_0}{2} \iiint_{\mathbb{R}^3} \left(\nabla \cdot (V\mathbf{E}) + |\mathbf{E}|^2 \right) d^3x .$$

The first term corresponds to a surface integral (using Green–Ostrogradsky formula, as in (A.15) at infinity). It is zero since V and $\mathbf{E}$ both decay to zero at infinity if ϱ is integrable. Thus, the total electrostatic energy becomes

$$U = \frac{\epsilon_0}{2} \iiint_{\mathbb{R}^3} |\mathbf{E}(\mathbf{x})|^2 d^3x . \tag{3.20}$$

Electrostatic energy density

The electrostatic energy density (in Jm^{-3}) writes

$$\boxed{u_E(\mathbf{x}) = \frac{\epsilon_0}{2}|\mathbf{E}(\mathbf{x})|^2 .} \tag{3.21}$$

The electrostatic energy can therefore be interpreted as being localized in the regions of space where the electric field is non-zero. The total electrostatic energy U of a system is the integral

$$U = \iiint_{\mathbb{R}^3} u_E(\mathbf{x}) d^3x .$$

Example 3.6—Energy density of an homogeneously charged sphere

In Example 2.4, the electric field of a uniformly charged sphere of radius R and charge Q was obtained. It is given by

$$\mathbf{E}(r) = \begin{cases} \dfrac{Qr}{4\pi\epsilon_0 R^3}\mathbf{u}_r & \text{if } r \leq R , \\[2ex] \dfrac{Q}{4\pi\epsilon_0 r^2}\mathbf{u}_r & \text{if } r > R . \end{cases}$$

The electrostatic energy density is, according to (3.21),

$$u_E(r) = \frac{\epsilon_0}{2}|\mathbf{E}(\mathbf{x})|^2 = \begin{cases} \dfrac{\epsilon_0 Q^2 r^2}{2(4\pi\epsilon_0)^2 R^6} & \text{if } r \leq R, \\[2ex] \dfrac{\epsilon_0 Q^2}{2(4\pi\epsilon_0)^2 r^4} & \text{if } r > R , \end{cases}$$

and the total energy of the sphere is

$$
U = 4\pi \int_0^\infty u_E(r) r^2 dr = 4\pi \left(\int_0^R u_E(r) r^2 dr + \int_R^\infty u_E(r) r^2 dr \right)
$$

$$
= \frac{Q^2}{2(4\pi\epsilon_0)} \left(\int_0^R \frac{r^4}{R^6} dr + \int_R^\infty \frac{1}{r^2} dr \right)
$$

$$
= \frac{Q^2}{2(4\pi\epsilon_0)} \left(\frac{1}{5R} + \frac{1}{R} \right) = \frac{3Q^2}{5(4\pi\epsilon_0) R} \ .
$$

This is the same result obtained in Example 3.5.

3.6.3   *Potential Energy of a Dipole in an External Electric Field*

Suppose that an electric dipole $\mathbf{p} = q\mathbf{d}$ is in a region where there is an external electric field $\mathbf{E}$. Since the electric dipole is a neutral system, one may expect the net electric force acting on it to be zero if the electric field is homogeneous. However, this is, in general, not true in the presence of a spatial gradient of the electric field along the direction of the dipole. To show this, let us calculate the potential energy U_p associated with the dipole in this field. Here, we will not consider the internal potential energy of the dipole itself, which is simply $-\dfrac{q^2}{4\pi\epsilon_0 d}$.

According to (3.17) and Fig. 3.12, the potential energy of the dipole in the external electric field reads

$$
U_p = -q V(\mathbf{x}) + q V(\mathbf{x} + \mathbf{d}) \ ,
$$

with $\mathbf{E} = -\nabla V$. By using a Taylor expansion for the potential up to first order in $|\mathbf{d}|/|\mathbf{x}|$ ($|\mathbf{d}| \ll |\mathbf{x}|$), we get

Fig. 3.12 Electric dipole in the presence of an external electric field

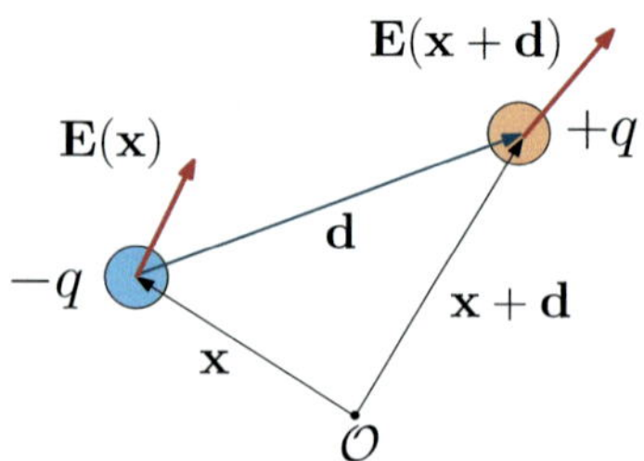

$$V(\mathbf{x} + \mathbf{d}) = V(\mathbf{x}) + \nabla V(\mathbf{x}) \cdot \mathbf{d} \, ,$$

so that the potential energy reads

$$U_p(\mathbf{x}) = q\mathbf{d} \cdot \nabla V(\mathbf{x}) = \mathbf{p} \cdot \nabla V(\mathbf{x}) \, ,$$

or, equivalently,

$$\boxed{U_p(\mathbf{x}) = -\mathbf{p} \cdot \mathbf{E}(\mathbf{x}) \, .} \tag{3.22}$$

This result is valid for an arbitrary distribution of charge with density $\varrho(\mathbf{x})$ that constitutes a dipole of moment $\mathbf{p} = \iiint_\Omega \varrho(\mathbf{x}')(\mathbf{x}' - \mathbf{x})d^3x'$ with zero total charge $\iiint_\Omega \varrho(\mathbf{x})d^3x = 0$, where Ω is a volume bounding the charge distribution. Indeed, for a charge distribution in an external electric potential $V(\mathbf{x})$, the potential energy reads

$$U_p = \iiint_\Omega \varrho(\mathbf{x}')V(\mathbf{x}')d^3x' \, .$$

Now define the position $\mathbf{x}$ of the dipole as the position of an arbitrarily chosen reference point within the small volume Ω. A Taylor expansion of the potential reads

$$V(\mathbf{x}') = V(\mathbf{x}) + \nabla V(\mathbf{x}) \cdot (\mathbf{x}' - \mathbf{x}) = V(\mathbf{x}) - \mathbf{E}(\mathbf{x}) \cdot (\mathbf{x}' - \mathbf{x})$$

and can be introduced in the expression for the potential energy,

$$U_p = \iiint_\Omega \varrho(\mathbf{x}')V(\mathbf{x})d^3x' - \iiint_\Omega \varrho(\mathbf{x}')\mathbf{E}(\mathbf{x}) \cdot (\mathbf{x}' - \mathbf{x})d^3x' \, .$$

The first term simply reads $V(\mathbf{x}) \iiint_\Omega \varrho(\mathbf{x}')d^3x'$ and is zero because the total charge of the dipole is zero. The result is therefore

$$U_p(\mathbf{x}) = -\left(\iiint_\Omega \varrho(\mathbf{x}')(\mathbf{x}' - \mathbf{x})d^3x' \right) \cdot \mathbf{E}(\mathbf{x}) = -\mathbf{p} \cdot \mathbf{E}(\mathbf{x}) \, .$$

3.6.4 Force and Torque Exerted by an External Electric Field on a Dipole

The potential energy (3.22) is a function of the position and orientation of the dipole (six degrees of freedom). It is minimum when the dipole moment $\mathbf{p}$ aligns with the electric field $\mathbf{E}$.

In order to derive the force $\mathbf{F}$ and torque $\mathbf{N}$ acting on the dipole due to the electric field, consider an elementary displacement $d\mathbf{x}$ and an elementary rotation $d\boldsymbol{\Omega}$ from a state of mechanical equilibrium of the dipole. The related change in potential energy reads

$$dU_p = -\mathbf{p} \cdot d\mathbf{E} - d\mathbf{p} \cdot \mathbf{E}(\mathbf{x})$$

$$= -\mathbf{p} \cdot \underbrace{\left(\frac{\partial \mathbf{E}}{\partial x} dx + \frac{\partial \mathbf{E}}{\partial y} dy + \frac{\partial \mathbf{E}}{\partial z} dz \right)}_{d\mathbf{E}} - \underbrace{(d\boldsymbol{\Omega} \times \mathbf{p})}_{d\mathbf{p}} \cdot \mathbf{E} .$$

This expression must be identified with the elementary work

$$dW = -\mathbf{F} \cdot d\mathbf{x} - \mathbf{N} \cdot d\boldsymbol{\Omega}$$

performed by an external operator acting on the dipole against the electric field, i.e., with a force $-\mathbf{F}$ and a torque $-\mathbf{N}$ opposite to those due to the electric field so as to maintain mechanical equilibrium. Finally comparing dW and dU_p, the components of the force acting on the dipole are

$$F_x = \mathbf{p} \cdot \frac{\partial \mathbf{E}}{\partial x}, \quad F_y = \mathbf{p} \cdot \frac{\partial \mathbf{E}}{\partial y}, \quad F_z = \mathbf{p} \cdot \frac{\partial \mathbf{E}}{\partial z} , \tag{3.23}$$

and the torque is obtained by applying a circular shift to the scalar triple product $(d\boldsymbol{\Omega} \times \mathbf{p}) \cdot \mathbf{E} = (\mathbf{p} \times \mathbf{E}) \cdot d\boldsymbol{\Omega}$:

$$\mathbf{N} = \mathbf{p} \times \mathbf{E} .$$

The force acting on the dipole could have been obtained directly from a Taylor expansion of the field. Consider for instance the x-component of the force

$$\begin{aligned} F_x &= -q E_x(\mathbf{x}) + q E_x(\mathbf{x} + \mathbf{d}) \\ &= -q E_x(\mathbf{x}) + q \left[E_x(\mathbf{x}) + (\mathbf{d} \cdot \boldsymbol{\nabla}) E_x(\mathbf{x}) \right] \\ &= (q\mathbf{d} \cdot \boldsymbol{\nabla}) E_x(\mathbf{x}) = (\mathbf{p} \cdot \boldsymbol{\nabla}) E_x(\mathbf{x}) . \end{aligned}$$

A similar calculation can be done for the other components, which yields $F_y = (\mathbf{p} \cdot \boldsymbol{\nabla}) E_y(\mathbf{x})$, $F_z = (\mathbf{p} \cdot \boldsymbol{\nabla}) E_z(\mathbf{x})$ and finally, a compact expression for the electric force applied to the dipole reads

$$\boxed{\mathbf{F}(\mathbf{x}) = (\mathbf{p} \cdot \boldsymbol{\nabla}) \mathbf{E}(\mathbf{x}) .} \tag{3.24}$$

As for the potential energy, expression (3.24) is valid for a dipole of moment $\mathbf{p}$ defined by a continuous distribution of charge $\varrho(\mathbf{x})$. Choosing a reference point $\mathbf{x}$ that represents the position of the dipole and performing a Taylor expansion of the electric field around $\mathbf{x}$, the x-component of the force indeed writes

$$\begin{aligned} F_x &= \iiint_\Omega \varrho(\mathbf{x}') E_x(\mathbf{x}') d^3 x' = \iiint_\Omega \varrho(\mathbf{x}') \left[E_x(\mathbf{x}) + ((\mathbf{x}' - \mathbf{x}) \cdot \boldsymbol{\nabla}) E_x(\mathbf{x}) \right] d^3 x' \\ &= \underbrace{\iiint_\Omega \varrho(\mathbf{x}') d^3 x'}_{0} E_x(\mathbf{x}) + \iiint_\Omega \varrho(\mathbf{x}')((\mathbf{x}' - \mathbf{x}) \cdot \boldsymbol{\nabla}) E_x(\mathbf{x}) d^3 x' . \end{aligned}$$

The first term is zero because the total charge is zero for the dipole and taking out of the integral the field gradient that is evaluated at $\mathbf{x}$, the expression for the dipole moment can be recognized in the second term, which reads

$$F_x = (\mathbf{p} \cdot \nabla) E_x(\mathbf{x}) \ .$$

A similar calculation for F_y and F_z gives (3.24) again.

At this stage, careful readers have noticed that we obtained two different expressions, (3.23) and (3.24), for the force acting on the dipole. It is not obvious to see if they give the same values. Consider for instance the x-component of the force in (3.23) and (3.24): the two expressions are equal only if

$$(\mathbf{p} \cdot \nabla) E_x(\mathbf{x}) = \mathbf{p} \cdot \frac{\partial \mathbf{E}(\mathbf{x})}{\partial x} \ .$$

Expanding the above equation in Cartesian coordinates, we find

$$p_x \frac{\partial E_x}{\partial x} + p_y \frac{\partial E_x}{\partial y} + p_z \frac{\partial E_x}{\partial z} = p_x \frac{\partial E_x}{\partial x} + p_y \frac{\partial E_y}{\partial x} + p_z \frac{\partial E_z}{\partial x} \ ,$$

or simply

$$- p_y \underbrace{\left(\frac{\partial E_y}{\partial x} - \frac{\partial E_x}{\partial y} \right)}_{(\nabla \times \mathbf{E})_z} + p_z \underbrace{\left(\frac{\partial E_x}{\partial z} - \frac{\partial E_z}{\partial x} \right)}_{(\nabla \times \mathbf{E})_y} = 0 \ ,$$

which is true when $\nabla \times \mathbf{E} = \mathbf{0}$, i.e., for electrostatics (to be shown in Sect. 3.7), but not true in general. The origin of this difference lies in the different assumptions made in the derivation of (3.23) and (3.24): the force expression (3.23) was derived from the potential energy of the dipole whose expression is valid only in the context of electrostatics, while the direct calculation of the force expression (3.24) remains valid beyond the context of electrostatics since it relies on the expression for the electric (Lorentz) force and a Taylor expansion, performed in both cases, for the electric field or its potential.

The most general expression for the force acting on the dipole is finally (3.24). This implies that a dipole will be pushed toward regions with larger electric fields. In the case where $\mathbf{p}$ does not depend on the electric field, we can also write

$$\mathbf{F} = -\nabla U_p = \nabla (\mathbf{p} \cdot \mathbf{E}) \ ,$$

that is to say, the force is conservative.

Finally, note that the torque acts on a dipole even in the absence of a net force there. Indeed, neglecting the spatial gradient of $\mathbf{E}$, one has a torque with respect to the center of gravity of the dipole given by a direct calculation:

$$\mathbf{N} = \frac{1}{2} \left(\mathbf{d} \times q\mathbf{E} - \mathbf{d} \times (-q\mathbf{E}) \right) ,$$

which corresponds, in that case, to the expression found by using the electrostatic interaction energy:

$$\boxed{\mathbf{N} = \mathbf{p} \times \mathbf{E} .} \tag{3.25}$$

3.7   The Two Fundamental Laws of Electrostatics

So far, we have seen that from the fundamental principles of electrostatics, that is, Coulomb's law and the superposition principle, one can demonstrate Gauss's law (2.2), whose differential form (2.3) relates the divergence of the electric field to the charge distribution:

$$\nabla \cdot \mathbf{E} = \frac{\varrho}{\epsilon_0} .$$

3.7.1   The Circulation Law and the Curl of E

The existence of the electrostatic potential V implies that the electric field is conservative. This means that its circulation over a closed path Γ is zero. Indeed, from (3.16), we see that the path integral over a curve which starts at a and ends at a (see Fig. 3.13) is

$$\boxed{\oint_\Gamma \mathbf{E}(\mathbf{x}) \cdot d\mathbf{x} = -(V(a) - V(a)) = 0 .} \tag{3.26}$$

Since the circulation law (3.26) holds for any closed path, it also applies to an infinitesimal loop. Consider, for example, the square contour in the xy plane shown in Fig. 3.14.

The line integral along the horizontal segment from (x, y) to $(x + dx, y)$ is approximately:

$$\int_{x,y}^{x+dx,y} \mathbf{E} \cdot d\mathbf{l} \approx E_x(x, y)dx$$

Fig. 3.13 The circulation of the electric field around a closed path is zero

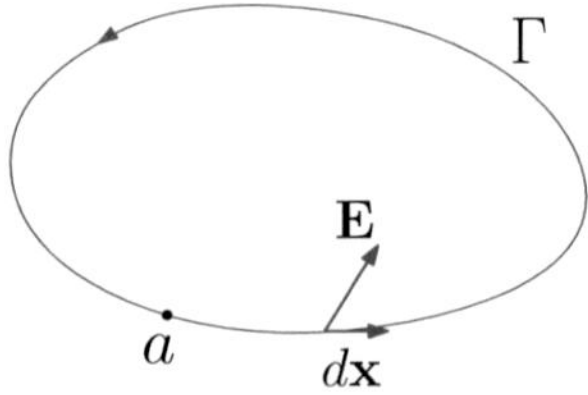

Fig. 3.14 An infinitesimal square contour

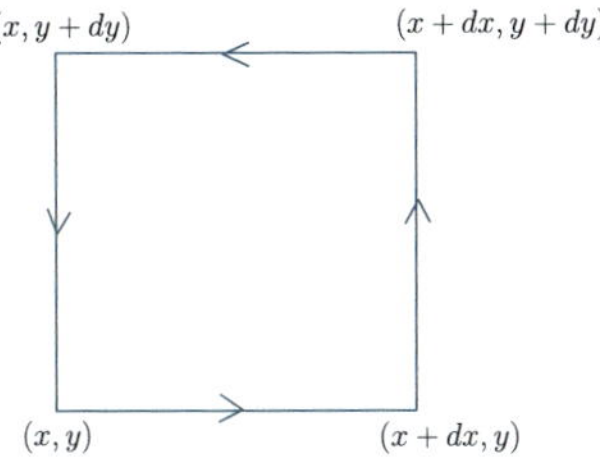

since $d\mathbf{l} = \mathbf{u}_x$. On the opposite horizontal segment, where $d\mathbf{l} = -\mathbf{u}_x$, the contribution becomes

$$\int_{x+dx,y+dy}^{x,y+dy} \mathbf{E} \cdot d\mathbf{l} \approx -E_x(x, y + dy)dx \ .$$

Similarly, for the vertical segments along the y-axis, we can compute their contributions. Adding the four segments, the total circulation becomes

$$\oint \mathbf{E} \cdot d\mathbf{l} \approx (E_x(x, y) - E_x(x, y + dy))dx + (E_y(x + dx, y) - E_y(x, y))dy$$

Using the first-order approximations $E_x(x, y) - E_x(x, y + dy) \approx -\frac{\partial E_x}{\partial y}(x, y)dy$ and $E_y(x + dx, y) - E_y(x, y) \approx \frac{\partial E_y}{\partial x}(x, y)dx$, the circulation simplifies to

$$\oint \mathbf{E} \cdot d\mathbf{l} = \left(\frac{\partial E_y(x, y)}{\partial x} - \frac{\partial E_x(x, y)}{\partial y} \right) dxdy = (\nabla \times \mathbf{E})_z (x, y)dxdy \ .$$

Thus, the ith component of the curl of $\mathbf{E}$ at a point represents the circulation per unit area around that point in the plane perpendicular to $\mathbf{u}_i$. In general, the curl quantifies how much a field circulates around a point, as illustrated in Fig. 3.15.

Fig. 3.15 The curl (in red) of a vector field (in blue) at point $\mathbf{x}$ is non-zero when the field circulates around $\mathbf{x}$

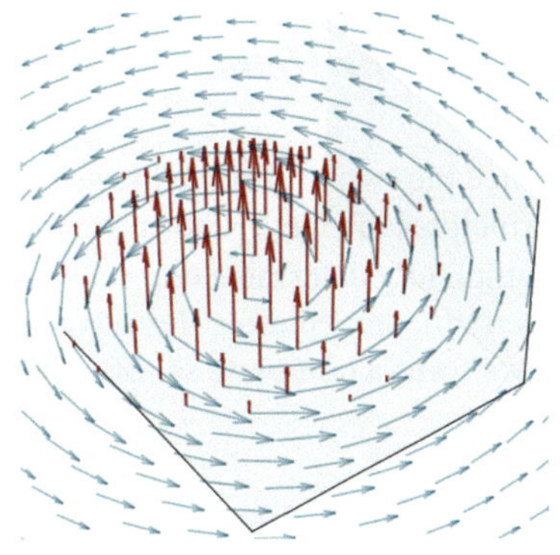

By choosing infinitesimal contours in the yz and xz planes, we conclude that the curl of the electrostatic field must be zero. In other words, the electric field is irrotational:

$$\boxed{\nabla \times \mathbf{E}(\mathbf{x}) = 0 \,.}$$

(3.27)

This result can also be derived immediately from the circulation law (3.26) and Stokes's theorem (A.17):

$$\oint_{\Gamma} \mathbf{E}(\mathbf{x}) \cdot d\mathbf{x} = \iint_{S(\Gamma)} (\nabla \times \mathbf{E}(\mathbf{x})) \cdot d\mathbf{S}(\mathbf{x}) = 0 \,,$$

where Γ is any closed curve and $S(\Gamma)$ is any surface bounded by Γ. Since the surface integral is zero for any S bounded by a contour, we deduce (3.27), the differential form of the circulation law (3.26).

3.7.1.1 Analogy with Fluid Mechanics

The term irrotational originates from fluid mechanics. Suppose the vector field represents the velocity of a fluid. A non-zero curl (or rotational field) corresponds to the presence of vorticity. For instance, if the fluid rotates like a rigid body with angular velocity $\mathbf{\Omega}$ around the origin, the velocity field is given by $\mathbf{v}(\mathbf{x}) = \mathbf{\Omega} \times \mathbf{x}$. In this case:

$$\nabla \times \mathbf{v}(\mathbf{x}) = 2\mathbf{\Omega} \,.$$

Here, the curl is uniform and directly proportional to the angular velocity of the fluid. Therefore, the curl of a field quantifies the rotational behavior of the system. In the case of the electrostatic field, there can be no vorticity; the field is irrotational. Consequently, the electric field lines cannot close on themselves, as this would imply a non-zero circulation and hence a non-zero curl.

3.7.2 The Helmholtz's Decomposition

Since the two differential laws (2.3) and (3.27) were obtained from the fundamental principles of electrostatics, we may ask ourselves whether they alone are sufficient to completely determine the electric field. This is indeed the case: thanks to Helmholtz's theorem, named after Hermann von Helmholtz (see Fig. 3.16), these two differential laws form a complete set of equations for the electric field.

Fig. 3.16 Hermann von Helmholtz (1821–1894), a German physicist with significant contributions in physics, physiology, and mathematics. He is renowned for his work on the conservation of energy, the Helmholtz free energy in thermodynamics, and the Helmholtz equation in wave theory

Helmholtz's–Hodge decomposition

Let $\mathbf{F}$ be a vector field that is continuously differentiable and that decays to zero at infinity as $1/|\mathbf{x}|^2$ or faster. Then, $\mathbf{F}$ can be written as the sum of a divergence-free term ($\nabla \times \mathbf{A}$) and a curl-free term (∇V), i.e., there is a vector field $\mathbf{A}$ and a scalar potential V such that $\mathbf{F}$ can be written

$$\mathbf{F} = \nabla \times \mathbf{A} - \nabla V \, ,$$

with

$$V(\mathbf{x}) = \frac{1}{4\pi} \iiint_{\mathbb{R}^3} \frac{\nabla' \cdot \mathbf{F}(\mathbf{x}')}{|\mathbf{x} - \mathbf{x}'|} d^3 x',$$

$$\mathbf{A}(\mathbf{x}) = \frac{1}{4\pi} \iiint_{\mathbb{R}^3} \frac{\nabla' \times \mathbf{F}(\mathbf{x}')}{|\mathbf{x} - \mathbf{x}'|} d^3 x' \, .$$

Note that this decomposition is not unique for V and $\mathbf{A}$ ($V' = V + C$ and $\mathbf{A}' = \mathbf{A} + \nabla \phi$, where C is a constant and ϕ an arbitrary scalar field define the same vector field $\mathbf{F}$ as V and $\mathbf{A}$).

Corollaries

- If $\mathbf{F}$ is an irrotational field ($\nabla \times \mathbf{F} = \mathbf{0}$), also called a conservative or a curl-free field, then it derives from a scalar potential. That is, there is a scalar field V such that

$$\mathbf{F} = -\nabla V \, .$$

Note that this confirms a well-known result from Poincaré (A.1.7):

$$\nabla \times \mathbf{E} = \mathbf{0} \quad \Leftrightarrow \quad \mathbf{E} = -\nabla V \, . \tag{3.28}$$

- If $\mathbf{F}$ is solenoidal ($\nabla \cdot \mathbf{F} = 0$), also called a divergence-free field, then it derives from a vector potential $\mathbf{A}$ such that

$$\mathbf{F} = \nabla \times \mathbf{A} \ .$$

The two differential laws (2.3) and (3.27) thus form a complete theory. They contain all the information of electrostatics since there is a unique vector field $\mathbf{E}$ that decays as $1/|\mathbf{x}|^2$ (or faster) such that:

$$\nabla \cdot \mathbf{E} = \varrho/\epsilon_0 \quad \text{and} \quad \nabla \times \mathbf{E} = \mathbf{0} \ .$$

Indeed, the Helmholtz theorem states that $\mathbf{E}$ is completely determined by the differential-integral representation:

$$\mathbf{E}(\mathbf{x}) = -\nabla \frac{1}{4\pi} \iiint_{\mathbb{R}^3} \frac{\nabla' \cdot \mathbf{E}(\mathbf{x}')}{|\mathbf{x} - \mathbf{x}'|} d^3 x' = -\nabla \underbrace{\frac{1}{4\pi\epsilon_0} \iiint_{\mathbb{R}^3} \frac{\varrho(\mathbf{x}')}{|\mathbf{x} - \mathbf{x}'|} d^3 x'}_{V(\mathbf{x})} \ ,$$

where we recognize the absolute electric potential (3.2). Using identity (3.1),

$$\mathbf{E}(\mathbf{x}) = \frac{1}{4\pi\epsilon_0} \iiint_{\mathbb{R}^3} \frac{\varrho(\mathbf{x}')(\mathbf{x} - \mathbf{x}')}{|\mathbf{x} - \mathbf{x}'|^3} d^3 x' \ . \tag{3.29}$$

Thus, we retrieve Coulomb's law for a continuous distribution of charge as obtained from the superposition principle.

3.8   Summary and Essential Formulas

- The electrostatic field derives from a scalar potential V such that $\mathbf{E} = -\nabla V$. In terms of the charge distribution, this potential writes

$$\boxed{V(\mathbf{x}) = \frac{1}{4\pi\epsilon_0} \iiint_{\mathbb{R}^3} \frac{\varrho(\mathbf{x}')}{|\mathbf{x} - \mathbf{x}'|} d^3 x' + C \ ,}$$

where C is a constant. If the charge density is integrable $\left(\iiint_{\mathbb{R}^3} |\varrho(\mathbf{x})| d^3 x < \infty\right)$, one has $C = \lim_{|\mathbf{x}| \to \infty} V(\mathbf{x})$.
- The work W done by the electric field on a charge q that moves from A to B along a path Γ depends solely on the extrema of the path, and writes

$$W = q \int_{\mathbf{x}_1}^{\mathbf{x}_2} \mathbf{E} \cdot d\mathbf{x} = q(V(\mathbf{x}_1) - V(\mathbf{x}_2)) \ .$$

In particular: (i) The quantity qV represents the potential energy of a charge q in an external potential V. (ii) The work done over a closed path is zero. The electric field is therefore conservative, or equivalently, irrotational. This circulation law can be written in either its integral or differential form:

$$\oint \mathbf{E} \cdot d\mathbf{x} = 0 \;\Leftrightarrow\; \nabla \times \mathbf{E} = \mathbf{0} \,.$$

- The electrostatic energy of a charge distribution ϱ is given by

$$U = \frac{1}{2} \iiint_{\mathbb{R}^3} \rho(\mathbf{x}')V(\mathbf{x}')d^3x' = \iiint_{\mathbb{R}^3} u_E(\mathbf{x}')d^3x' \,,$$

where the energy density is

$$u_E = \frac{1}{2}\epsilon_0 |\mathbf{E}|^2 \,.$$

- An atom, a molecule, or in general any globally neutral charge distribution generates an electric field that decays as $1/r^3$ at infinity. This field is characterized by the dipole moment $\mathbf{p}$ of the charge distribution, such that the electric potential far from the charge distribution is that of an electric dipole

$$V(\mathbf{x}) = \frac{\mathbf{p} \cdot \mathbf{x}}{4\pi\epsilon_0|\mathbf{x}|^3} \,.$$

The moments of the charge distribution are defined as

$$\text{Monopole moment:} \quad Q = \iiint_{\Omega} \varrho(\mathbf{x}')d^3x'.$$

$$\text{Dipole moment:} \quad \mathbf{p} = \iiint_{\Omega} \varrho(\mathbf{x}')\mathbf{x}'d^3x'.$$

$$\text{Quadrupole moment:} \quad D = \iiint_{\Omega} \left\{3\mathbf{x}' \cdot \mathbf{x}'^T - |\mathbf{x}'|^2\right\} \varrho(\mathbf{x}')d^3x' \,.$$

- Far from a charge distribution, that is for $|\mathbf{x}| \gg a$ where a denotes the largest dimension of the distribution, we can perform a multipole expansion of the potential, which is an expansion in powers of $a/|\mathbf{x}|$.

$$V(\mathbf{x}) = \underbrace{\frac{Q}{4\pi\epsilon_0|\mathbf{x}|}}_{\text{monopole}} + \underbrace{\frac{\mathbf{p} \cdot \mathbf{x}}{4\pi\epsilon_0|\mathbf{x}|^3}}_{\text{dipole}} + \underbrace{\frac{\mathbf{x}^T D\mathbf{x}}{8\pi\epsilon_0|\mathbf{x}|^5}}_{\text{quadrupole}}$$

The expansion is usually truncated at the lowest non-zero term, thereby characterizing the distribution as a monopole $Q \neq 0$, dipole ($Q = 0$ and $\mathbf{p} \neq \mathbf{0}$), quadrupole ($Q = 0$, $\mathbf{p} = \mathbf{0}$ and $D \neq 0$), etc.

- An electric dipole $\mathbf{p}$ in an electric field $\mathbf{E}$ has a potential energy $U_p = -\mathbf{p} \cdot \mathbf{E}$. In consequence, it is subjected to a force

$$\mathbf{F} = (\mathbf{p} \cdot \nabla)\mathbf{E}$$

and a torque

$$\mathbf{N} = \mathbf{p} \times \mathbf{E} \ .$$

The force is non-zero only for non-uniform electric fields and tends to pull the dipoles toward regions of high-fields. The torque tends to align the dipoles parallel to the fields.

- The circulation law, together with Gauss's law, contains all the information of electrostatics. Indeed, by using Helmholtz's theorem, it is possible to show that the only field $\mathbf{E}$ that decays to zero at infinity and that satisfies both Coulomb's law and the circulation law is given by the Coulomb integral (1.5).

Problems

3.1 Potential difference between two charged planes

A small charge q of mass m is suspended by a wire between two infinite vertical planes with surface charge densities σ and $-\sigma$, separated by a distance d. If the wire forms an angle θ with respect to the vertical direction:

(a) Determine the value of σ.
(b) What is the potential difference between the planes?

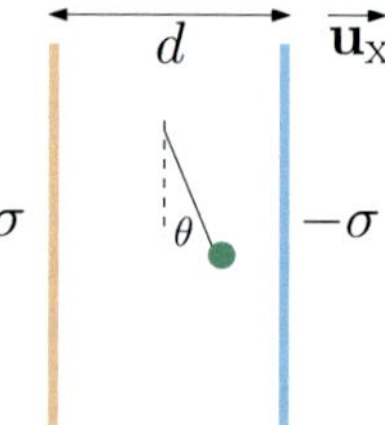

3.2 Superposition of two spheres

Two spheres with centers O_1 and O_2, separated by a distance d, have the same radius R and are uniformly charged in their volume with densities $-\rho$ and ρ, respectively. For the region outside the spheres, which electrostatic system is the set of spheres equivalent to? Calculate the potential and the modulus of the electric field at any point outside the spheres in the limit $|\mathbf{d}| \ll r$.

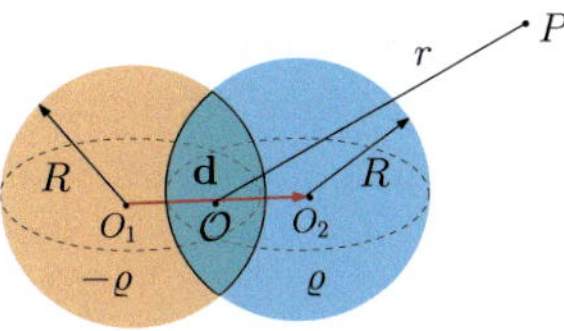

3.3 Linear particle accelerator

In a linear accelerator, electrons are accelerated in a vacuum tube thanks to a series of gates g_n through which the electrons can pass. Even-numbered gates are grounded, whereas odd-numbered gates are brought to a potential $V(t)$ that switches from V to $-V$ periodically in time with a switching period of $2T$. In this way, an electron always sees a negative field on its trajectory, and hence gets accelerated all the way. The length l_k is such that the position of the electron is $x = g_k$ at $t = kT$.

(a) The gates are wide enough for the electric field between them to be considered uniform. At $t = 0$, the electrons enter the gate g_0 with a velocity $v_0 = 0$ and $V(0) = V$. Determine the electric field $\mathbf{E}_k$ that an electron sees while it moves between g_k and g_{k+1}; express the result in terms of l_k.

(b) What is the work W_k done by the field on the electron between those two gates? From this, determine the electron velocity v_k as it passes g_k. What is the kinetic energy K_k?

(c) For k large enough, give an approximation of l_k considering that the velocity does not vary much between g_k and g_{k+1}.

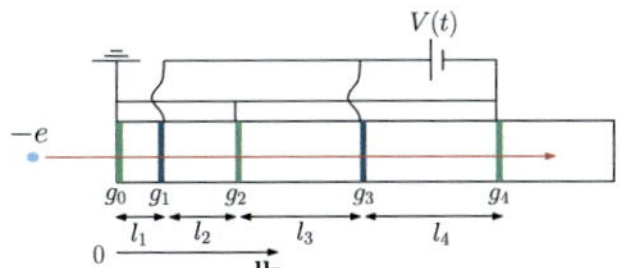

3.4 Potential on the axis of a charged ring

Consider a ring of radius R that is charged uniformly with a linear charge density λ, as shown below. What is the electric potential on its axis at a distance z from the center of the ring? Use this result to obtain the potential on the axis of a hollow disk of internal radius a, external radius b, and uniform surface charge density σ. Finally, what is the potential of a uniformly charged infinite plane?

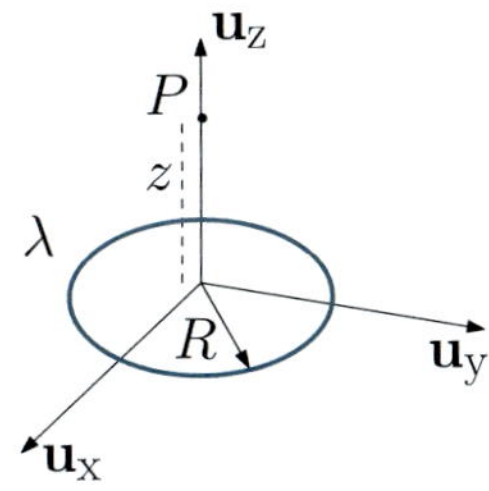

3.5 Potential at a point outside the axis of a charged ring

Express the potential at any point of space generated by a charged ring of radius R and charge Q in terms of the complete elliptic integral of the first kind, defined by

$$K(\lambda) = \int_0^{\pi/2} \frac{d\theta}{\sqrt{1 - \lambda^2 \sin^2 \theta}}, \qquad \sqrt{2\lambda}\, K(\lambda) = \int_0^{\pi} \frac{d\theta}{\sqrt{b - \cos \theta}}, \qquad \lambda = \frac{2}{1 + b}.$$

3.6 Ion between two cylindrical electrodes

Ions of charge $q > 0$ are accelerated from rest in a region between two electrodes with potential difference V_0 between them. Then, they enter the region between two infinitely long, concentric semi-cylindrical electrodes of radii a and b, respectively. The ions travel in a semi-circular path at constant speed in between the half-cylinders before leaving. Find the potential difference $V_b - V_a$ between the cylinders.

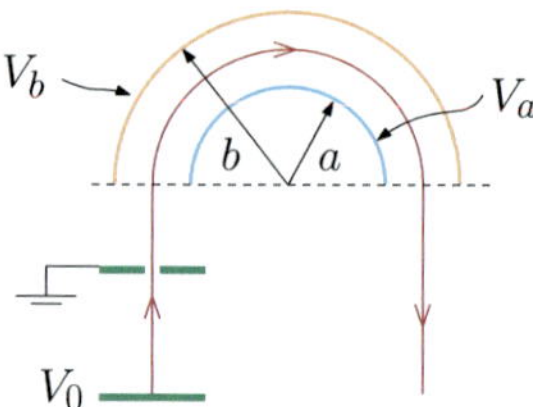

3.7 Infinite plane and point charge

A negative point charge $-q$ is at distance a from an infinite plane with uniform charge density $\sigma > 0$, as illustrated below.

(a) Find the electric potential at any point $0 < x < a$ of the x-axis, which is perpendicular to the plane and passing through the charge $-q$. Use the potential in the plane ($x = 0$) as a reference.

(b) A particle of mass m and charge $-e$ is free to move in between the plane and the charge $-q$. If it starts from rest at $x = a/2$ on the x-axis, with what velocity will it impact on the charge plane?

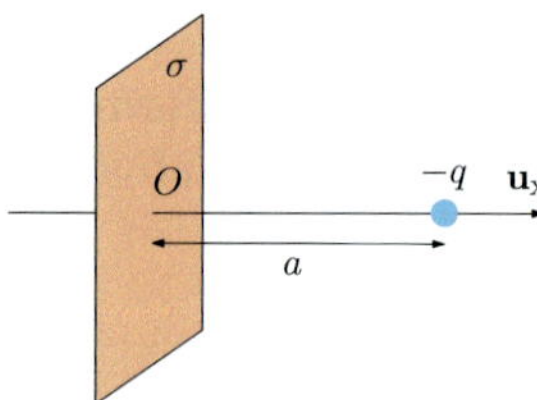

3.8 Superposition of infinite wires

Two infinite and parallel wires have a linear charge density of λ and $-\lambda$, respectively. They are at a distance a from the origin and parallel to the z-axis, as shown below.

(a) Find the potential at any point (x, y, z).
(b) Find the equipotential surfaces.

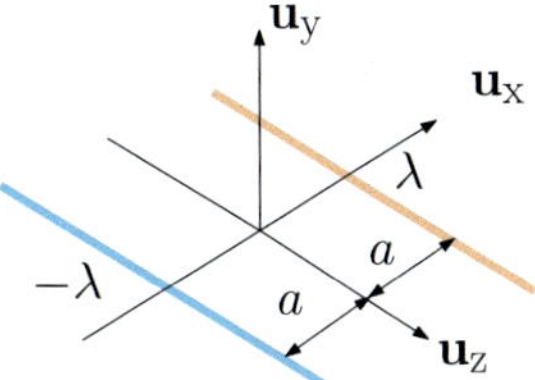

3.9 Force between a dipole and a charged line

An electric dipole $\mathbf{p}$ is at a distance r from a very thin and infinitely long line of charge with homogeneous charge density $\lambda > 0$. Assume that the dipole is oriented along the electric field generated by the line, as shown below.

(a) If the distance d between the two charges of the dipole is such that $d \ll r$, determine the force acting on the dipole, up to first order in d/r.

(b) Show that in this limit, the force can be expressed as

$$\mathbf{F}(\mathbf{x}) = (\mathbf{p} \cdot \nabla)\,\mathbf{E}(\mathbf{x}) \ .$$

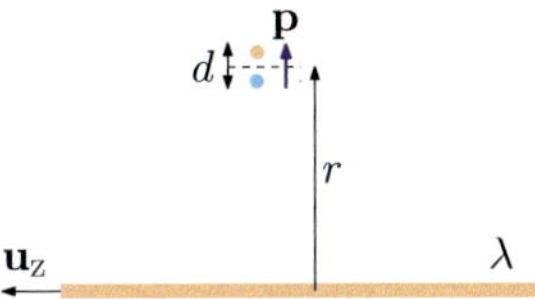

3.10 Dipole–dipole interaction

An electric dipole of moment $\mathbf{p}$ is placed at the origin of a system of spherical coordinates.

(a) Express the electric field generated by the dipole at $M(\mathbf{r})$ as a function of $\mathbf{p}$ and $\mathbf{r}$.

(b) Another dipole of moment $\mathbf{p}'$ is placed at point $M(\mathbf{r})$ and is arbitrarily oriented with respect to $\mathbf{p}$. Express the potential energy of the system of dipoles as a function of $\mathbf{p}$, $\mathbf{p}'$ and $\mathbf{r}$.

(c) Show that the force $\mathbf{F}_{\mathbf{p}\to\mathbf{p}'}$ acting on $\mathbf{p}'$ due to $\mathbf{p}$ is expressed as

$$\mathbf{F}_{\mathbf{p}\to\mathbf{p}'} = \frac{1}{4\pi\epsilon_0 r^5}\left(3(\mathbf{p}\cdot\mathbf{r})\mathbf{p}' + 3(\mathbf{p}'\cdot\mathbf{r})\mathbf{p} + 3(\mathbf{p}\cdot\mathbf{p}')\mathbf{r} - 15(\mathbf{p}\cdot\mathbf{r})(\mathbf{p}'\cdot\mathbf{r})\frac{\mathbf{r}}{r^2}\right) \ .$$

(d) Is there a force acting on $\mathbf{p}$?

Chapter 4
Conductors at Equilibrium and Poisson's Equation

Abstract This chapter focuses on the behavior of *conductors in electrostatic equilibrium* and introduces *Poisson's equation* ($\nabla^2 V = -\varrho/\epsilon_0$) as a powerful mathematical tool for determining the electric potential. It begins by establishing key properties of conductors in static electric fields: the electric field vanishes inside a conductor, all excess charge resides on its surface, and the electric field at the surface is always normal to it. These properties lead to *Coulomb's theorem*, which quantifies the discontinuity of the electric field across a conductor's surface, and the constant nature of the electric potential throughout the conductor and on its surface. The concept of *electrostatic energy of a conductor* and the *electrostatic pressure* exerted on its surface are also discussed. The chapter then introduces Poisson's equation as the governing partial differential equation for the electric potential, which simplifies to Laplace's equation ($\nabla^2 V = 0$) in charge-free regions. It emphasizes that solving this equation, subject to appropriate *boundary conditions*, is the primary method for determining the electric potential and, consequently, the electric field in complex scenarios where charge distributions are not known a priori. Two main types of boundary conditions are detailed: *Dirichlet's boundary condition* (specifying potential values on the boundary) and *Neumann's boundary condition* (specifying the normal component of the electric field on the boundary). Advanced techniques for solving Poisson's and Laplace's equations are presented, including the *method of images* for problems involving conductors with simple geometries (like infinite grounded planes) and *expansion in Legendre polynomials* for problems with azimuthal symmetry. Finally, the chapter introduces the concept of *Green's functions* as a formal solution to Poisson's equation with given boundary conditions, providing a general framework for addressing complex electrostatic problems.

Keywords Conductors · Poisson's equation · Laplace's equation · Electrostatics

4.1 Introduction

Matter can be broadly classified into two categories: conductors and insulators. This chapter focuses on the behavior of conductive materials in the presence of static electric fields. We demonstrate that the electric field must vanish at every point inside a conductor for the charges to reach equilibrium. Additionally, we show that the electric potential inside and on the surface of a conductor is constant. This straightforward mathematical condition for the potential provides a strong motivation to introduce Poisson's equation, a partial differential equation that the electric potential satisfies. Various methods for solving Poisson's equation are presented. The study of insulators is deferred to Chap. 6.

4.2 Conductors in Equilibrium

Inside a conductor, there is a large density of free electrons. The word *free* means that these electrons are not bound to any particular atom of the host crystal. In consequence, if an electric field is present inside a conductor, electrons will be accelerated, thus generating a flow of charges, that is to say, an electric current. Two possibilities arise:

1. The conductor is isolated: In this case, the current will eventually stop once the displacement of electrons in the medium generates an electric field that cancels the external field, and the system reaches a static equilibrium.
2. The electric field is imposed by an external source: This is the case, for example, when a battery imposes a constant potential difference between two points of a conductor. A stationary electric current develops, according to what is known as Ohm's law. The study of these transport phenomena is the subject of Chap. 7.

In the following we will establish the main properties of conductors in electrostatic equilibrium (case 1 above), which constrains how the electric field and potential vary in space.

4.2.1 The Electric Field Vanishes Inside a Conductor at Equilibrium

Since an electron will be accelerated as soon as a non-zero electric field is present, the only acceptable solution for static equilibrium is to have a zero electric field at every point inside a conductor. If Ω represents the volume of a conductive material, then

$$\mathbf{E}(\mathbf{x}) = \mathbf{0} \quad \forall \ \mathbf{x} \in \Omega \,,$$

which can be written in terms of the potential as

$$-\nabla V(\mathbf{x}) = 0 \quad \forall \mathbf{x} \in \Omega \,.$$

4.2.2 The Charge of a Conductor at Equilibrium is Distributed on Its Surface

A consequence of the previous property is that every charge in a conductor Ω must reside on its surface $\partial\Omega$. To show this, imagine an arbitrary closed surface $S \in \Omega$, as shown in Fig. 4.1. Since the electric field is zero in Ω, the electric flux over S is zero, and according to Gauss's law this implies that the charge Q_S contained in the volume inside S must be zero:

$$\Phi_{S,\mathbf{E}} = \oiint_S \mathbf{E} \cdot d\mathbf{S} = \frac{Q(S)}{\epsilon_0} = 0 \ .$$

Since S is an arbitrary surface contained in Ω, it can be chosen as an infinitesimal surface bounding a volume element localized around each point in Ω. It follows that the charge density ϱ is zero everywhere inside a conductor,

$$\boxed{\varrho(\mathbf{x}) = 0 \quad \forall \ \mathbf{x} \in \Omega \ .}$$

Note that a much more direct way to show this is to use the differential form of Gauss's law (2.3), since from $\mathbf{E} = 0$ inside a conductor we conclude that $\nabla \cdot \mathbf{E} = \frac{\varrho}{\epsilon_0} = 0$ and so $\varrho = 0$. In summary, all the charge of a conductor Ω at equilibrium must be distributed on its surface. A conductor at equilibrium is therefore completely characterized by its surface charge density σ (in C m^2) on $\partial\Omega$. In the absence of any particular symmetry, the surface charge density is non-uniform.

4.2.3 The Electric Field is Normal to the Surface of a Conductor at Equilibrium

For the electrons on the surface of a conductor to be at rest, the electric field must be normal to the surface at every point. Otherwise, a non-zero tangential component of the field will set the electrons on the surface in motion. In fact, since the electric field is zero in Ω, it is necessarily perpendicular to the surface $\partial\Omega$. To show this, let us consider a point $\mathbf{x}$ on the surface ($\mathbf{x} \in \partial\Omega$) and decompose the electric field into two perpendicular components, normal and parallel to the surface, $\mathbf{E}(\mathbf{x}) = \mathbf{E}_{\perp}(\mathbf{x}) + \mathbf{E}_{\parallel}(\mathbf{x})$.

Fig. 4.1 Gauss's law applied to a closed surface S inside a conductor

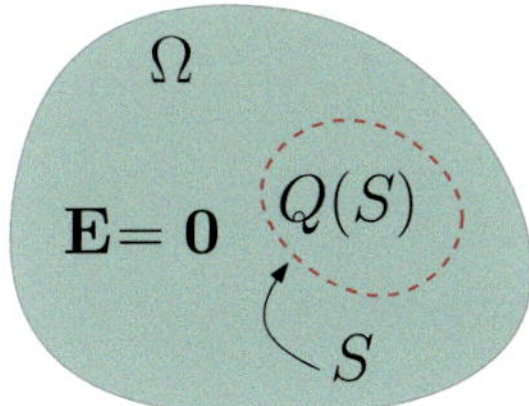

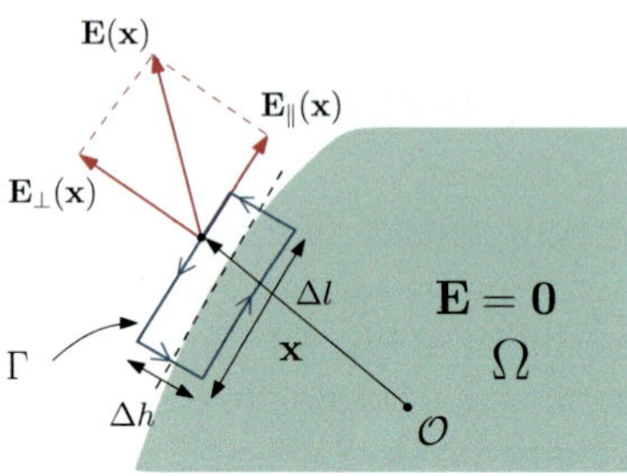

Fig. 4.2 Circulation law applied to a small loop at the interface between vacuum and a conductor

We apply the circulation law for the electric field,

$$\oint_\Gamma \mathbf{E} \cdot d\mathbf{l} = 0 \; ,$$

where Γ is the infinitesimal planar contour around $\mathbf{x}$, with two sides parallel and two sides perpendicular to the surface, as shown in Fig. 4.2.

The dimensions of the contour Γ being infinitesimal, we neglect the spatial variation of the electric field over the contour sides that are parallel to the surface. The contribution of the parallel segment above the surface is then $-|\mathbf{E}_\parallel(\mathbf{x})|\Delta l$. In addition, we consider the limiting case of the circulation law in which the length Δh of the contour sides normal to the surface tends to zero. In this limit, the contribution of these segments to the circulation is zero and we are left with the contribution of the segment above the surface,

$$\oint_\Gamma \mathbf{E} \cdot d\mathbf{l} \underset{\Delta h \to 0}{\approx} -|\mathbf{E}_\parallel(\mathbf{x})|\Delta l = 0 \; ,$$

which means that, for every $\mathbf{x} \in \partial\Omega$, $\mathbf{E}_\parallel(\mathbf{x}) = \mathbf{0}$ and the electric field is normal to the surface:

$$\boxed{\mathbf{E}(\mathbf{x}) = E(\mathbf{x})\mathbf{n}(\mathbf{x}) \quad \forall \; \mathbf{x} \in \partial\Omega \; .}$$

This also means that the surface of a conductor is an equipotential, since for any pair of points A and B on the surface, we have

$$V(A) - V(B) = \int_A^B \mathbf{E} \cdot d\mathbf{l}$$

and choosing a path from A to B contained in $\partial\Omega$

$$V(A) - V(B) = \int_A^B \underbrace{\mathbf{E} \cdot d\mathbf{l}}_{\mathbf{E}_\parallel \cdot d\mathbf{l} = 0} = 0 \; .$$

4.2.4 Coulomb's Theorem: Electric Field Discontinuity Across the Surface of a Conductor at Equilibrium

There is a relationship between the electric field $\mathbf{E}$ and the charge density σ on the surface of a conductor at equilibrium. We will establish it by applying Gauss's law to a closed cylindrical surface S around a point $\mathbf{x}$ on the conductor surface $\partial\Omega$, as illustrated in Fig. 4.3. The base area A of the cylinder is supposed to be small enough for the field on the top surface to be uniform. For the same reason, the charge density at $\mathbf{x}$ can be considered uniform on the section of $\partial\Omega$ within the cylinder.

Since the field is zero inside the conductor, the electric flux is non-zero only through the top surface and possibly through the lateral surface of the cylinder lying above the conductor. The charge contained in S is the charge σA present on the surface of the conductor inside S. We consider the limit case where the height of the cylinder tends to zero. The electric flux through the lateral surface then tends to zero. The electric field on the top base surface tends to the field $E(\mathbf{x}^+)$ just above the surface of the conductor (the electric field $E(\mathbf{x}^-) = 0$ just below the surface). Gauss's law then reads

$$\oiint_S \mathbf{E} \cdot d\mathbf{S} = E(\mathbf{x}^+)A = \frac{\sigma(\mathbf{x})A}{\epsilon_0}$$

and so, we obtain the result known as Coulomb's theorem for the electric field just above the surface of a conductor,

$$\boxed{\mathbf{E}(\mathbf{x}^+) = \frac{\sigma(\mathbf{x})}{\epsilon_0}\mathbf{n} \qquad \mathbf{x} \in \partial\Omega \,.} \tag{4.1}$$

4.2.5 Electrostatic Energy of a Conductor

Since all the charge in a conductor Ω is distributed on its surface $\partial\Omega$ and since the potential V is constant on $\partial\Omega$, the electrostatic energy of the conductor, according to (3.18), is written as

Fig. 4.3 Gauss's law applied to an infinitesimal cylinder whose axis is perpendicular to the surface of a conductor

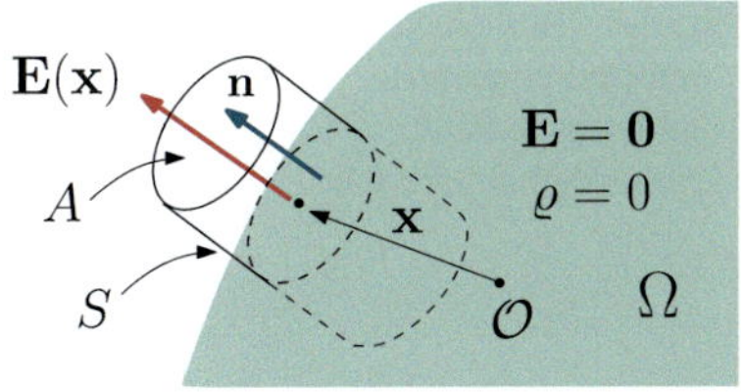

$$U = \frac{1}{2} \oiint_{\partial\Omega} V(\mathbf{x}')dq(\mathbf{x}') = \frac{1}{2}V_0 \underbrace{\oiint_{\partial\Omega} \sigma(\mathbf{x}')dS(\mathbf{x}')}_{Q} = \frac{1}{2}V_0 Q \; ,$$

where V_0 is the potential in the conductor and Q the total charge on its surface. Finally,

$$\boxed{U = \frac{1}{2}V_0 Q \; .} \tag{4.2}$$

4.2.6   Cavendish's Experiment and Testing the $1/r^2$ Law

In 1773, before Coulomb published the $1/r^2$ law for the force between charges, Henry Cavendish conducted an experiment involving a spherical shell consisting of two concentric charged spheres. His goal was to study the dependence of the electric force on the distance between charges. As illustrated in Fig. 4.4, this was done by surrounding a conductive sphere of radius R_1 and charge Q_1 by a conductive concentrical shell of radius R_2 and charge Q_2.

Due to the spherical symmetry of the system, the charges must be homogeneously distributed on the two surfaces. Since, according to Coulomb's law, the electric force between point charges varies as $1/r^2$, it is easy to show that the outer sphere produces no electric field inside. No force acts on the charges of the inner sphere. Indeed, consider an arbitrary point P inside the outer sphere, and define the two infinitesimal portions of the sphere dS_1 and dS_2 that subtend the same solid angle $d\Omega$ when viewed from P, as shown in Fig. 4.5. That is, $dS_1 = d\Omega r_1^2$ and $dS_2 = d\Omega r_2^2$. Since the charge density of the sphere is uniform, and because of the inverse square law, $dS_1/r_1^2 = dS_2/r^2 = d\Omega$. The forces that dS_1 and dS_2 exert on a charge q at P exactly cancel each other. Note that this is exactly the same kind of argument used to demonstrate Gauss's law in Sect. 2.5. Gauss's law is a consequence of the $1/r^2$ law.

Now, suppose the shell and the inner sphere are connected by a very thin conductive wire. Connecting both spheres will result in a charge transfer between them in order for the conductors to have the same electrical potential. The new charges will be Q_1' and Q_2', and the field between the two is only given by the charge of the inner shell $\mathbf{E}(r) = Q_1'/(4\pi\epsilon_0 r^2)\mathbf{u}_r$ and so,

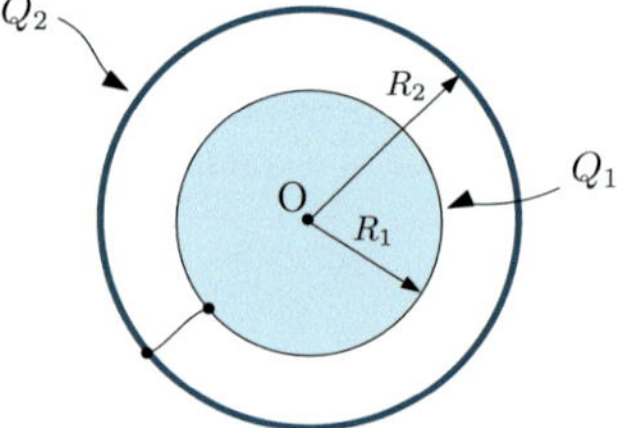

Fig. 4.4 Two electrically connected concentric conductors used in Cavendish's experiment

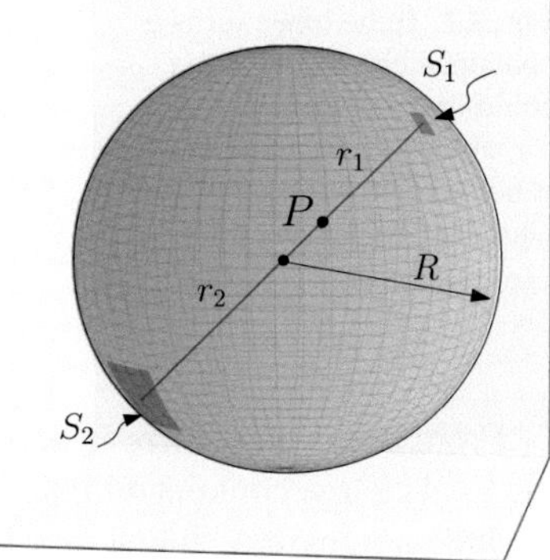

Fig. 4.5 The surfaces S_1 and S_2 subtend the same solid angle when viewed from P

$$V(R_1) - V(R_2) = \int_{R_1}^{R_2} E(r)\,dr = \frac{Q_1'}{4\pi\epsilon_0}\left(\frac{1}{R_1} - \frac{1}{R_2}\right) = 0\,.$$

One concludes that all the charge initially contained in the inner sphere must be transferred to the outer shell, $Q_1' = 0$. After disconnecting both conductors, Cavendish removed the outer shell and measured the charge in the sphere with an electroscope. It indeed showed zero charge, and he concluded that if the electric force deviated from a r^{-2} law, for example in the form of $1/r^{2+\delta}$, the deviation δ could not be larger than 0.02.

Modern experiments based on this idea have determined that $\delta \leq 10^{-16}$ and from this we know that the photon mass cannot be larger than $1 \times 10^{-50} kg$! The effect of a non-zero photon mass on the electric potential is the subject of Exercise 4.12.

4.2.7 Hollow Conductors and Faraday's Cages

Experiments such as the one mentioned in example Sect. 4.2.6 can be used to test Coulomb's inverse squared law with great accuracy. One of the reasons for that is that the conductive shell does not need to be a sphere, it can have any shape. Consider any hollow conductor, that is, a conductor with a cavity inside. As shown in Fig. 4.6, let S be a closed surface, completely inside the conductor, and enclosing the cavity. If the conductor is at equilibrium, the electric field is zero everywhere inside and by Gauss's law we conclude that the total charge enclosed by S must be zero.

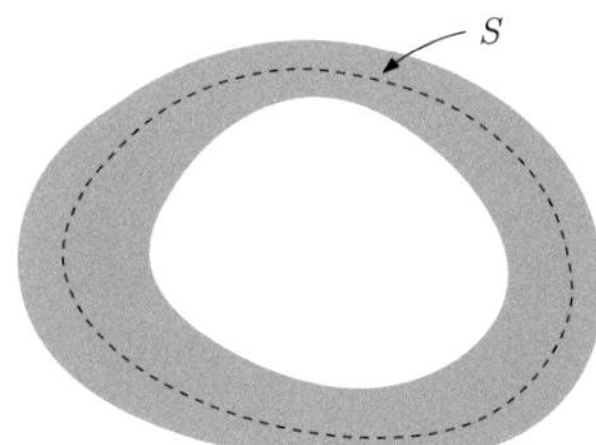

Fig. 4.6 Gauss's law applied to a surface enclosing the cavity of a hollow conductor

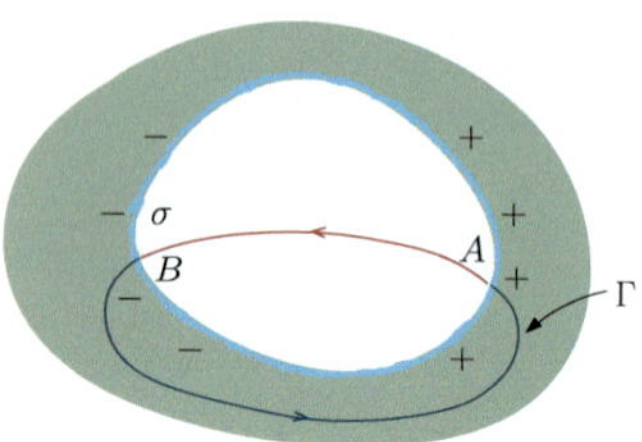

Fig. 4.7 If the inner surface contained charge, then one could find a closed curve Γ for which the circulation of **E** is not zero, in contradiction with the circulation law

From this, and if there is no charge inside the cavity, we can only conclude that the total charge contained in the inner surface of the conductor is zero. This does not exclude a non-zero charge density σ such that its integral over the inner surface is zero. But, if there were a non-zero charge density, then there would be an electric field inside the cavity, going from the positive to the negative charges. As shown in Fig. 4.7, one could find a closed contour Γ that joins two points A and B along an electric field line, and then closes on itself by taking a path inside the conductor, where $\mathbf{E} = \mathbf{0}$. This would lead to a non-zero circulation of the electric field over a closed loop, in contradiction with the circulation law (3.26).

The fact that no charge is present in the inner surface can be experimentally verified by doing the experiment illustrated in Fig. 4.8. Imagine a spherical conductor that has been initially electrified with positive charge. The conductor has a small aperture allowing a small metallic sphere to be placed inside. If the small ball touches the outer surface, it becomes positively charged, and produces a repulsion of the electroscope's sheets. If the experiment is repeated by touching the inner surface of the sphere, no charge is transferred to the small sphere.

This is the principle of the Faraday's cage, used to shield a region of space from any external electric field. Enclosing that region with a solid conducting cage will completely screen any possible influence of fields outside the cage. It turns out that even if the cage is not solid, but made instead of a mesh of conductive wires, it provides protection from electrostatic fields. This is illustrated in Fig. 4.9.

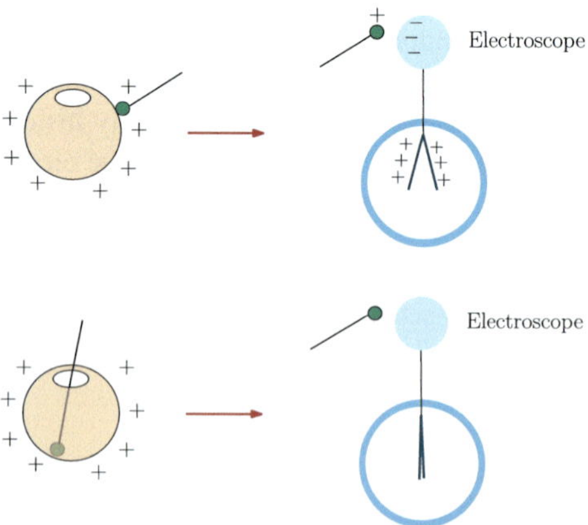

Fig. 4.8 An initially discharged sphere can only gain charge from the outer surface of a charged conductor

Fig. 4.9 A Faraday's cage
cancels the influence of
external charges inside it

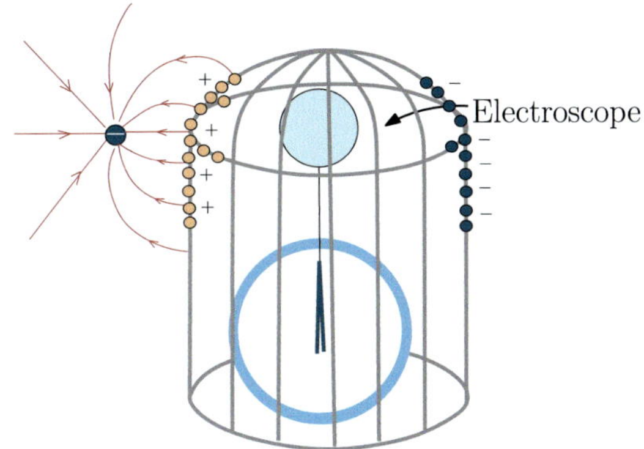

Example 4.1—Point charge surrounded by a conductive spherical shell
Suppose a point charge q is placed inside a neutral and conductive spherical
shell, as shown in Fig. 4.10.

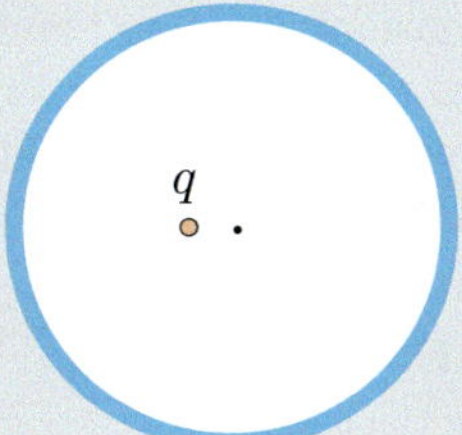

Fig. 4.10 Point charge q surrounded by a conductive shell

Due to the phenomenon of electrostatic induction, the charge q will produce
a charge separation in the metallic sphere. Let Q_{int} and Q_{ext} be the charges
induced on the inner and outer surfaces of the shell, respectively. Since the
sphere is neutral, its total charge must be zero:

$$Q_{\text{int}} + Q_{\text{ext}} = 0 \, .$$

To determine Q_{int}, we can apply Gauss's law to a closed surface S
completely contained in the region inside the shell, as shown in Fig. 4.11.

The total charge contained in the volume inside S is $Q_{\text{int}} + q$, and since the
electric field is zero at every point of S, we obtain

$$\oiint \mathbf{E}(\mathbf{x}) \cdot \mathbf{n}(\mathbf{x}) dS(\mathbf{x}) = \frac{(Q_{\text{int}} + q)}{\epsilon_0} = 0 \, ,$$

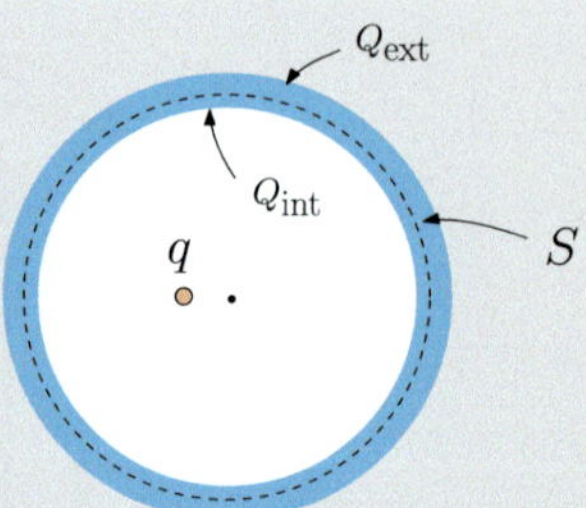

Fig. 4.11 The surface S is fully contained within the spherical shell

which means $Q_{\text{int}} = -q$, so that $Q_{\text{ext}} = q$. Unless the charge q is located at the center of the shell, the charge distribution on the inner surface will be inhomogeneous. A larger charge of opposite sign to that of q must accumulate near q to cancel its effect inside the shell. As for the charge on the outer surface of the shell, this will be, in contrast, always homogeneously distributed. This can be understood by noting that the electric potential is constant at every point inside the shell. Since the superposition of the point charge q and the charge on the inner surface of the shell generates a null electric field inside the shell, the corresponding potential is constant in that region. Then, the potential on the outer surface must also be constant, equal to that inside the shell, and this can only be achieved if the charge Q_{ext} is homogeneously distributed. The electric field outside the shell is therefore equal to the field generated by a point charge at the origin, and in spherical coordinates reads

$$\mathbf{E}(r) = \frac{1}{4\pi\epsilon_0} \frac{q}{r^2} \mathbf{u}_r \ .$$

Figure 4.12 shows the electric field lines inside and outside the shell.

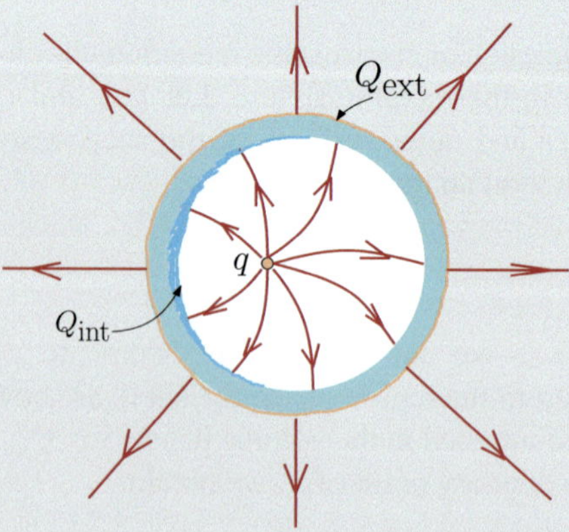

Fig. 4.12 Electric field lines generated by the charge q and the rearrangement of charge on the surface of the shell

If a second point charge q' is brought near to the shell from the outside, the charge on the outer surface will redistribute in order to cancel the electric field generated by q' inside the shell and keep a constant potential inside it. The charge distribution on the inner shell will therefore not change, since the charge q' is already completely screened inside the shell. This is known as a Faraday shield; any electric field outside the shell is completely blocked inside it by the rearrangement of the charge on the outer surface of the shell. The region enclosed by the shield is therefore completely decoupled from any external electric field. Figure 4.13 shows a sketch of the field lines for the case $q > 0$ and $q' < 0$.

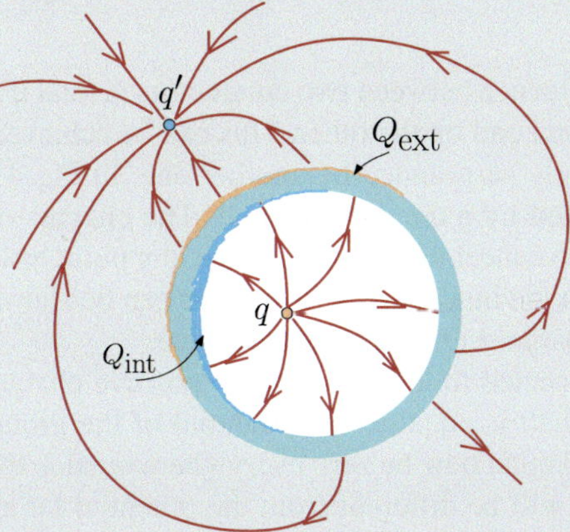

Fig. 4.13 The electric field outside the shell is the superposition of the field generated by q' and the Coulomb field generated by a point charge q, exactly as if it were located at the origin

4.2.8 Interaction Between Conductors

When two or more conductors are present, and if at least one of them is charged, the charges on their surfaces must be distributed in a way that the electric field is zero inside every conductor and, according to Coulomb's theorem (4.1), equal to $\mathbf{E}(\mathbf{x}) = \sigma(\mathbf{x})/\epsilon_0 \mathbf{n}(\mathbf{x})$ above its surface, where $\mathbf{n}(\mathbf{x})$ is the normal at point $\mathbf{x}$. This is the main reason why, when dealing with conductors, additional mathematical tools beyond the Coulomb integral (1.1) are necessary to determine the electric field: the charge densities are not known a priori.

This redistribution of charges in a conductor can be easily interpreted as an attraction of negative charges toward the positive charges on the other conductors, leaving an equal amount of positive charge on the opposite side. This is illustrated in Fig. 4.14.

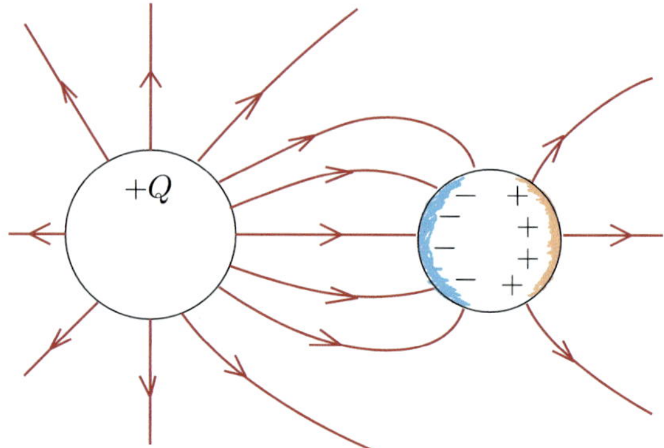

Fig. 4.14 Electrostatic induction on a conductor due to the presence of a charged conductor nearby

We say that the influence between two conductors is total if all the field lines start from one conductor and end on the other. This can be achieved if, for example, one conductor is completely surrounded by another one. In Fig. 4.15, a conductor with charge Q is surrounded by a conductive shell. The charges on the surfaces of the shell rearrange so as to cancel the electric field at any point inside it. This means that the induced charge on the inner surface is $-Q$, as can be shown easily with Gauss's law, thus leaving a charge $+Q$ on the outer surface.

If the latter is connected to the ground, this positive charge is evacuated so that the potential of the shell is equal to the potential of the ground. This is necessary since the electric field must now be zero everywhere outside the shell, otherwise the potential of the shell will be different from the potential far away (equal to that of the ground). Gauss's law then implies that the total charge contained in the system must be zero, the shell is now negatively charged with a charge $-Q$, and all the field lines connect the two conductors, which are therefore under total influence.

4.2.9 Force on a Charged Conductor: The Electrostatic Pressure

Consider a conductor Ω in electrostatic equilibrium in the presence of an arbitrary distribution of charges. At any point $\mathbf{x}'$ outside Ω yet infinitely close to its surface, there is an electric field normal to $\partial\Omega$.

Therefore, a force is exerted on a differential surface element $dS(\mathbf{x}')$ around $\mathbf{x}'$ since it contains a charge $dq(\mathbf{x}') = \sigma(\mathbf{x}')dS(\mathbf{x}')$ (see Fig. 4.16). However, the total electric field at $\mathbf{x}'$ is not only due to the charge on the surface element $dS(\mathbf{x}')$ itself, but also to the external charge in the rest of the system. We can then write

$$\mathbf{E}(\mathbf{x}') = \mathbf{E}_{dS}(\mathbf{x}') + \mathbf{E}_{\text{ext}}(\mathbf{x}') \,,$$

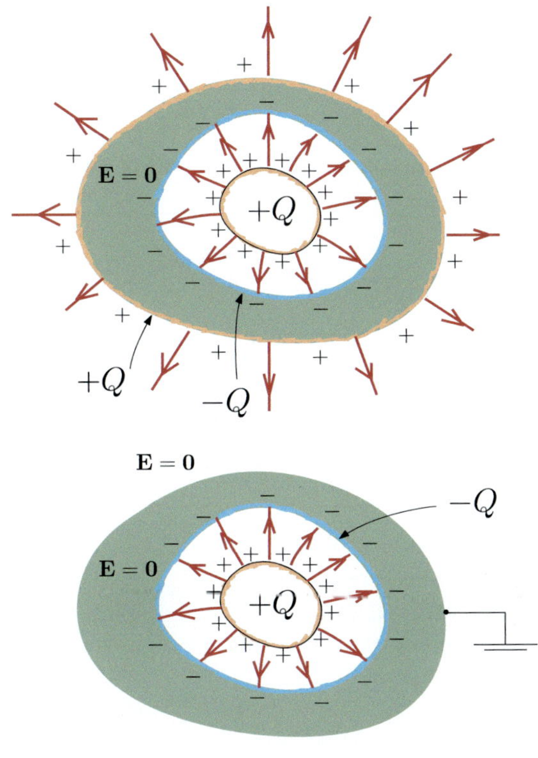

Fig. 4.15 Two conductors brought to total influence. First, the conductor inside the shell is charged (top), and the shell is then connected to the ground (bottom)

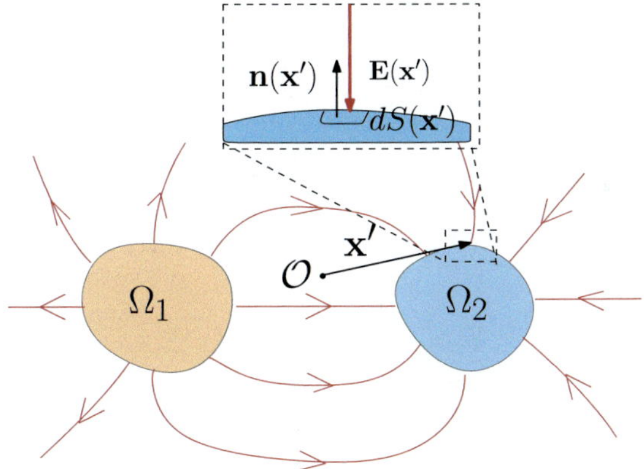

Fig. 4.16 The electric field will produce a force on an infinitesimal surface element dS on the conductor

where $\mathbf{E}_{dS}(\mathbf{x}') = \dfrac{\sigma(\mathbf{x}')}{2\epsilon_0}\mathbf{n}'(\mathbf{x}')$ is the electric field due to the charge on $dS(\mathbf{x}')$ (identical to the field generated by a charged infinite plane), while $\mathbf{E}_{\text{ext}}(\mathbf{x}')$ is the field generated by all the other charges present in the system. By Coulomb's theorem (4.1), we know the field at $\mathbf{x}'$,

$$\mathbf{E}(\mathbf{x}') = \frac{\sigma(\mathbf{x}')}{\epsilon_0}\mathbf{n}(\mathbf{x}') = \frac{\sigma(\mathbf{x}')}{2\epsilon_0}\mathbf{n}(\mathbf{x}') + \mathbf{E}_{\text{ext}}(\mathbf{x}') ,$$

so that the external field acting on $dS(\mathbf{x}')$ can be written as

$$\boxed{\mathbf{E}_{\text{ext}}(\mathbf{x}') = \frac{\sigma(\mathbf{x}')}{2\epsilon_0}\mathbf{n}(\mathbf{x}') .} \tag{4.3}$$

The electrostatic force acting on the charge $dq(\mathbf{x}')$ contained in $dS(\mathbf{x}')$ is therefore

$$d\mathbf{F}_e = \underbrace{dq(\mathbf{x}')}_{\sigma(\mathbf{x}')dS(\mathbf{x}')} \mathbf{E}_{\text{ext}}(\mathbf{x}') = \frac{\sigma^2(\mathbf{x}')}{2\epsilon_0}dS(\mathbf{x}')\mathbf{n}(\mathbf{x}') .$$

The electrostatic pressure

An electrostatic force dF_e acts on any differential surface element dS of a charged conductor Ω. For every point on the surface of the conductor, this force is given by

$$d\mathbf{F}_e(\mathbf{x}') = P_e(\mathbf{x}')dS(\mathbf{x}')\mathbf{n}(\mathbf{x}') ,$$

where P_e is the electrostatic pressure

$$\boxed{P_e(\mathbf{x}') = \frac{\sigma^2(\mathbf{x}')}{2\epsilon_0} .} \tag{4.4}$$

The pressure P_e is always positive regardless of the sign of the charge on the surface element, meaning that the force dF_e is always pointing out of the conductor. The total electrostatic force on the conductor is given by

$$\mathbf{F}_e = \oiint_{\partial\Omega} P_e(\mathbf{x}')\mathbf{n}(\mathbf{x}')dS(\mathbf{x}') = \oiint_{\partial\Omega} \frac{\sigma(\mathbf{x}')^2}{2\epsilon_0}\mathbf{n}(\mathbf{x}')dS(\mathbf{x}') .$$

An external agent must then provide a force that cancels the electrostatic force in order to keep the charged conductors at fixed positions.

4.3 Poisson's Equation

We know that the electric charge on a conductor must reside on its surface. If symmetries do not guarantee a uniform charge distribution, we face a fundamental difficulty, arising from the fact that the electric field and the surface charge densities are interdependent, both are related by Coulomb's theorem (4.1). There is, luckily, a proper way of mathematically stating the problem. Recall that

- The electric field is completely determined by the electric potential V, since $\mathbf{E} = -\nabla V$.
- The potential is constant on the surface of every conductor.
- The electric field, in other words, the potential gradient is normal to the surface of every conductor.

It is then convenient to find an equation that V must satisfy at every point in space. The presence of conductors then translates into simple boundary conditions for the potential. Once the potential is determined, so is the electric field. The charge density at the surface of every conductor can then be found by Coulomb's theorem.

We saw that the differential form of the circulation law (3.27), that is $\nabla \times \mathbf{E}(\mathbf{x}) = 0$, guarantees the existence of a scalar potential V such that

$$\mathbf{E}(\mathbf{x}) = -\nabla V(\mathbf{x}) \ .$$

Combining this with the differential form of Gauss's law (2.3), we obtain

$$\nabla \cdot \underbrace{\mathbf{E}(\mathbf{x})}_{-\nabla V} = -\nabla^2 V(\mathbf{x}) = \frac{\varrho(\mathbf{x})}{\epsilon_0} \ ,$$

where $\nabla^2 = \nabla \cdot \nabla$ is the Laplace operator. In Cartesian's coordinates,

$$\nabla^2 V = \frac{\partial^2 V}{\partial x^2} + \frac{\partial^2 V}{\partial y^2} + \frac{\partial^2 V}{\partial z^2} = \frac{-\varrho(x, y, z)}{\epsilon_0} \ .$$

Poisson's equation (1813)
The electric potential V satisfies Poisson's equation

$$\nabla^2 V(\mathbf{x}) = -\frac{\varrho(\mathbf{x})}{\epsilon_0} \tag{4.5}$$

which, in regions free of charges, reduces to Laplace's equation

$$\boxed{\nabla^2 V(\mathbf{x}) = 0\,.}\qquad\qquad (4.6)$$

These equations are named after Siméon Denis Poisson and Pierre-Simon Laplace (see Fig. 4.17)

Fig. 4.17 Left: Siméon Denis Poisson (1781–1840), a French mathematician and physicist, he made major contributions to the modern study of electromagnetism, elasticity, and fluid mechanics. Right: Pierre-Simon Laplace (1749–1827), a French mathematician and astronomer, was a central figure in celestial mechanics and mathematical physics. He formulated the Laplace transform, widely used in engineering and physics

Remark

Note that for a point charge q at $\mathbf{x}'$, we can write the charge density in terms of the Dirac distribution as $\varrho(\mathbf{x}) = q\delta(\mathbf{x} - \mathbf{x}')$. We also know that the potential generated by this charge is simply $V(\mathbf{x}) = \frac{q}{4\pi\epsilon_0|\mathbf{x}-\mathbf{x}'|}$. Poisson's equation then becomes

$$\nabla^2 V = \frac{q}{4\pi\epsilon_0}\nabla^2\frac{1}{|\mathbf{x}-\mathbf{x}'|} = -\frac{q\delta(\mathbf{x}-\mathbf{x}')}{\epsilon_0}$$

and we retrieve the fundamental Eq. (1.10), valid in the framework of distributions

$$\nabla^2\frac{1}{|\mathbf{x}-\mathbf{x}'|} = -4\pi\delta(\mathbf{x}-\mathbf{x}')\,.\qquad\qquad (4.7)$$

Example 4.2 - Potential generated by a uniformly charged sphere

The potential of a uniformly charged sphere was obtained in Example 3.3. It can also be retrieved by solving Poisson's equation. Due to the spherical symmetry of the charge distribution, the potential only depends on r in spherical coordinates. Poisson's equation takes the form:

$$\nabla^2 V(r) = \frac{1}{r^2}\frac{\partial}{\partial r}\left(r^2\frac{\partial V(r)}{\partial r}\right) = -\frac{\varrho(r)}{\epsilon_0}\,,$$

where $\varrho(r) = \varrho$ for $r \leq R$ and $\varrho(r) = 0$ otherwise.

Outside the sphere, the potential satisfies Laplace's equation:

$$\frac{\partial}{\partial r}\left(r^2 \frac{\partial V(r)}{\partial r}\right) = 0$$

and so, $r^2 \frac{\partial V(r)}{\partial r} = A$, where A is a constant. Assuming that V tends to zero at infinity (absolute potential), one concludes that $V(r) = -\frac{A}{r}$. At infinity, the potential should tend to that of a point charge $Q = \frac{4}{3}\pi R^3 \varrho$, so that $A = -Q/(4\pi\epsilon_0)$.

Inside the sphere, the potential satisfies

$$\frac{\partial}{\partial r}\left(r^2 \frac{\partial V(r)}{\partial r}\right) = -\frac{r^2 \varrho}{\epsilon_0} = -\frac{3r^2 Q}{4\pi\epsilon_0 R^3}\,.$$

Integrating with respect to r, we find

$$r^2 \frac{\partial V(r)}{\partial r} = -\frac{r^3 Q}{4\pi\epsilon_0 R^3} + B$$

and a final integration gives

$$V(r) = -\frac{r^2 Q}{8\pi\epsilon_0 R^3} - \frac{B}{r} + C\,.$$

A finite solution for $r \to 0$ is obtained only if $B = 0$. In fact, according to (1.10), $\nabla^2(1/r) = -4\pi\delta(r)$ so that keeping $B \neq 0$ would imply the presence of a point charge at the origin. Finally, the continuity of the potential at $r = R$ gives

$$-\frac{Q}{8\pi\epsilon_0 R} + C = \frac{Q}{4\pi\epsilon_0 R}\,.$$

This yields $C = \frac{3Q}{8\pi\epsilon_0 R}$ and the potential inside the sphere:

$$V(r) = \frac{Q}{8\pi\epsilon_0 R}\left(3 - \frac{r^2}{R^2}\right)\,.$$

We obtain the same result as in example 3.3.

4.3.1 Poisson's Equation in a Finite Volume

Often, we are only interested in determining the solution to Poisson's equation inside a finite volume $\Omega \subset \mathbb{R}^3$. Poisson's integral equation, whose demonstration can be found in the Appendix A.5, will be helpful in understanding the boundary conditions that one must impose.

Poisson's integral equation If V satisfies Poisson's equation (4.5) in Ω, then, for any point $\mathbf{x}$ in Ω,

$$V(\mathbf{x}) = \frac{1}{4\pi\epsilon_0} \iiint_\Omega \frac{\varrho(\mathbf{x}')d^3x'}{|\mathbf{x} - \mathbf{x}'|} - \frac{1}{4\pi} \oiint_{\partial\Omega} \frac{d\mathbf{S}(\mathbf{x}') \cdot \mathbf{E}(\mathbf{x}')}{|\mathbf{x} - \mathbf{x}'|}$$
$$+ \frac{1}{4\pi} \oiint_{\partial\Omega} d\mathbf{S}(\mathbf{x}') \cdot \frac{\mathbf{x}' - \mathbf{x}}{|\mathbf{x}' - \mathbf{x}|^3} V(\mathbf{x}') . \quad (4.8)$$

It is interesting to interpret this formula. The first term in (4.8) corresponds to the superposition of the potentials generated by the charges contained in Ω. The two extra terms tell us that all the information about the charges outside Ω is contained in V and $\mathbf{E}$ on the boundary $\partial\Omega$. Note that (4.8) is not a practical formula for solving Poisson's equation, since we still need to know the charge distribution ϱ and the potential itself at the boundary to evaluate the integrals. Yet, (4.8) provides us with a strong mathematical result: a proper boundary condition is enough to encode all the information about the fields outside Ω.

In particular, one can choose $\Omega = \mathbb{R}^3$ in (4.8). For an integrable charge density, the electric field decays at least as $1/r^2$, the second integral tends to zero and we obtain the well known result:

$$V(\mathbf{x}) = \frac{1}{4\pi\epsilon_0} \iiint_{\mathbb{R}^3} \frac{\varrho(\mathbf{x}')d^3x'}{|\mathbf{x} - \mathbf{x}'|} + C$$

with C the value of the potential at infinity. Indeed, one can take for $\partial\Omega$ a sphere of radius r centered at $\mathbf{x}$, and then take the limit $r \to \infty$, so that the third term in (4.8) is

$$\begin{aligned}
C &= \lim_{r\to\infty} \frac{1}{4\pi} \oiint_{\partial\Omega} d\mathbf{S}(\mathbf{x}') \cdot \frac{\mathbf{x}' - \mathbf{x}}{|\mathbf{x}' - \mathbf{x}|^3} V(\mathbf{x}') \\
&= \lim_{r\to\infty} \frac{1}{4\pi} \oiint_{\partial\Omega} r^2 \sin\theta \, d\theta \, d\phi \, \mathbf{u}_r \cdot \frac{\mathbf{u}_r}{r^2} V(r, \theta, \phi) \\
&= \lim_{r\to\infty} \frac{1}{4\pi r^2} \oiint_{\partial\Omega} V(r, \theta, \phi) r^2 \sin\theta \, d\theta \, d\phi = \lim_{r\to\infty} \langle V \rangle ,
\end{aligned}$$

with $\langle V \rangle$ the average of the potential over the sphere of radius r.

Example 4.3 - Application to the case of a conductor
Let us apply Poisson's integral equation (4.8) to the case of a conductor of volume Ω in the presence of an arbitrary charge distribution ϱ outside Ω, see Fig. 4.18. First, we choose an arbitrary point $\mathbf{x} \in \Omega$ inside the conductor, and use Poisson's integral equation.

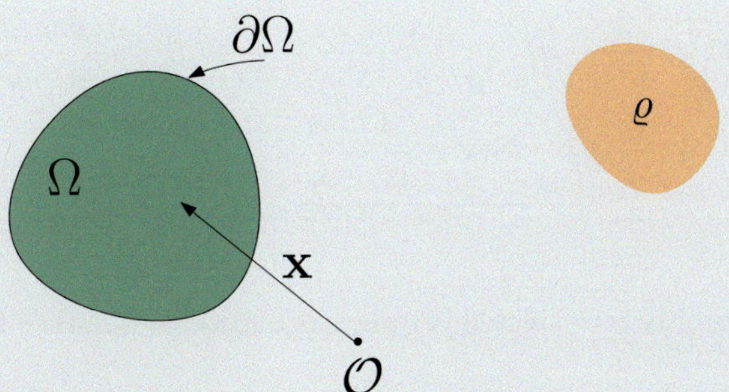

Fig. 4.18 A conductor Ω in the presence of a charge distribution ϱ

Since $\varrho(\mathbf{x}') = 0$, $\mathbf{E}(\mathbf{x}') = \mathbf{0}$ at any point $\mathbf{x}'$ inside Ω and $V(\mathbf{x}') = V_0$ at any point on $\partial\Omega$,

$$V(\mathbf{x}) = V_0 \frac{1}{4\pi} \oiint_{\partial\Omega} d\mathbf{S}(\mathbf{x}') \cdot \frac{\mathbf{x}' - \mathbf{x}}{|\mathbf{x}' - \mathbf{x}|^3} \ .$$

The surface integral on the right-hand side, multiplying V_0, is equivalent to the electric flux through $\partial\Omega$ produced by a charge $q \equiv \epsilon_0$ located at $\mathbf{x}$. By Gauss's law, this flux equals $q/\epsilon_0 \equiv 1$, so that Poisson's integral equation tells us that the potential at any point inside Ω is equal to the potential at its boundary

$$V(\mathbf{x}) = V_0, \quad \mathbf{x} \in \Omega \ .$$

Now, consider a point $\mathbf{x}$ outside Ω and apply Poisson's integral equation to the volume $\mathbb{R}^3/\Omega$ (entire space excluding Ω). Its boundary is again $\partial\Omega$ and infinity, except that this time the normal at any point $\mathbf{x}'$ in $\partial\Omega$ points inward, as shown in Fig. 4.19.

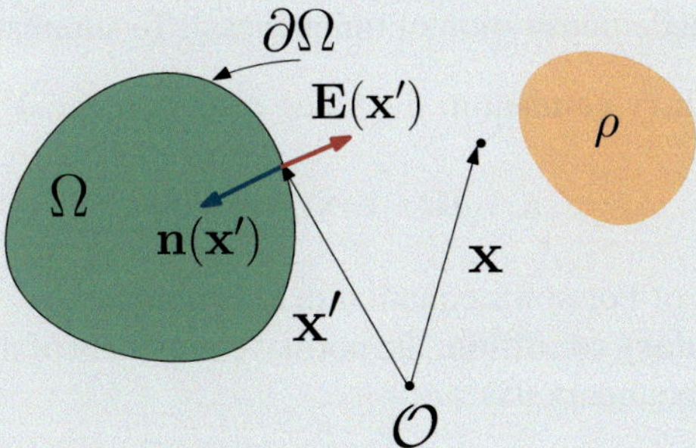

Fig. 4.19 Now the point $\mathbf{x}$ is outside Ω

Coulomb's theorem gives the electric field at any point $\mathbf{x}'$ in $\mathbb{R}^3/\Omega$ infinitely close to $\partial\Omega$, $\mathbf{E}(\mathbf{x}') = -\dfrac{\sigma(\mathbf{x}')}{\epsilon_0}\mathbf{n}(\mathbf{x}')$, and we obtain

$$V(\mathbf{x}) = \frac{1}{4\pi\epsilon_0} \iiint_{\mathbb{R}^3/\Omega} \frac{\varrho(\mathbf{x}')d^3x'}{|\mathbf{x} - \mathbf{x}'|} + \frac{1}{4\pi\epsilon_0} \oint_{\partial\Omega} \frac{\sigma(\mathbf{x}')dS(\mathbf{x}')}{|\mathbf{x} - \mathbf{x}'|}$$

$$+ \frac{V_0}{4\pi} \underbrace{\oiint_{\partial\Omega} dS(\mathbf{x}') \cdot \frac{\mathbf{x}' - \mathbf{x}}{|\mathbf{x}' - \mathbf{x}|^3}}_{0} .$$

The last integral is zero since this time $\mathbf{x}$ is outside Ω (Gauss's law). Finally the potential

$$V(\mathbf{x}) = \frac{1}{4\pi\epsilon_0} \iiint_{\mathbb{R}^3/\Omega} \frac{\varrho(\mathbf{x}')d^3x'}{|\mathbf{x} - \mathbf{x}'|} + \frac{1}{4\pi\epsilon_0} \oiint_{\partial\Omega} \frac{\sigma(\mathbf{x}')dS(\mathbf{x}')}{|\mathbf{x} - \mathbf{x}'|}$$

is simply the superposition of the potential generated by the charge distribution ϱ and that generated by the surface charge density σ on the conductor.

4.4 Boundary Conditions for Poisson's Equation

From (4.8), we see that it is possible to determine the potential at any point $\mathbf{x} \in \Omega$ if one knows ϱ in Ω as well as the normal component of the electric field and the potential at the boundary $\partial\Omega$. However, in general there is no solution for an arbitrary choice of these two quantities. Indeed, they are not independent since $\mathbf{E} = -\nabla V$. In practice, it is enough to specify either the value of the potential at $\partial\Omega$, or the normal component of the electric field at $\partial\Omega$. These are two common boundary conditions, for which the solution of Poisson's equation is unique (the reader can refer to Appendices A.5.2 and A.5.3 for a demonstration of uniqueness). To summarize:

- **Dirichlet's boundary condition**: the value of the potential V is imposed at the boundary $\partial\Omega$:
$$\mathbf{x} \in \partial\Omega : V(\mathbf{x}) = V_D(\mathbf{x}) .$$

  Then, the solution of Poisson's equation in Ω is unique.
- **Neumann's boundary condition**: the normal component of the electric field $\mathbf{E}(\mathbf{x})$ is specified at the boundary $\partial\Omega$
$$\mathbf{x} \in \partial\Omega : \mathbf{E}(\mathbf{x}) \cdot \mathbf{n}(\mathbf{x}) = E_N(\mathbf{x}) .$$

Then, the solution of Poisson's equation in Ω is unique, up to a constant.

Example 4.4 - Potential generated by a conductive sphere

Let $\Omega = \{r \leq R\}$ be the interior of a conductive sphere of radius R, and V_0 the absolute potential at its surface $\partial\Omega$. Then, a constant $V(r) = V_0$ is a solution of Poisson (Laplace)'s equation in Ω and satisfies Dirichlet's boundary condition $V = V_0$ at the surface $\partial\Omega$. It is therefore the only possible solution of this Poisson–Dirichlet's problem according to Sect. 4.4. For the region $\widetilde{\Omega} = \{r > R\}$ outside the sphere, $V(r) = \frac{V_0 R}{r}$ is a solution of Poisson (Laplace)'s equation and satisfies the Dirichlet boundary conditions $V = V_0$ at $r = R$ and $V = 0$ at infinity. It is the only solution of this Poisson–Dirichlet's problem in $\widetilde{\Omega} = \{r > R\}$.

On the other hand, let us suppose we don't know the potential V_0 of the sphere but instead know its total charge Q. By symmetry, the charge density must be uniform on its surface $\sigma = Q/(4\pi R^2)$. Coulomb's theorem (4.1) states that the electric field in $\widetilde{\Omega}$ normal to the sphere is simply

$$E_N = \frac{\sigma}{\epsilon_0} = \frac{Q}{4\pi R^2 \epsilon_0} ,$$

whereas the electric field tends to zero at infinity. Note that

$$V(r) = \frac{Q}{4\pi \epsilon_0 r}$$

is a solution of Poisson (Laplace)'s equation in $\widetilde{\Omega}$, and it gives an electric field just outside the sphere whose normal component to the sphere reads

$$-\nabla V \cdot \mathbf{u}_r | \, (r = R) = \frac{Q}{4\pi \epsilon_0 R^2} = E_N .$$

We conclude that $V(r) = \frac{Q}{4\pi \epsilon_0 r}$ is the only solution (up to a constant) of Poisson–Neumann's problem outside the sphere. Comparing the Poisson–Dirichlet's and Poisson–Neumann's solutions in $\widetilde{\Omega}$ yields the relationship between the charge Q of the sphere and its absolute potential V_0,

$$V_0 = \frac{Q}{4\pi \epsilon_0 R} .$$

4.4.1 The Method of Images

The method of images helps us to easily find the solution of Poisson's equation when the potential satisfies Dirichlet boundary conditions. Let us consider a point charge q located at $\mathbf{x}' \in \Omega$, in the presence of a conductor represented by the volume $\widetilde{\Omega} = \mathbb{R}^3/\Omega$, as shown in Fig. 4.20.

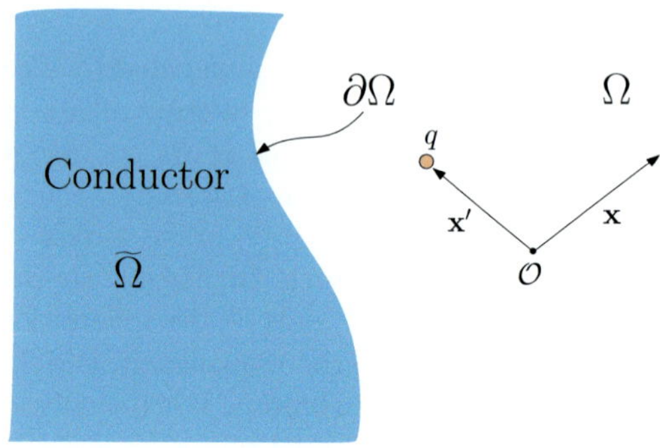

Fig. 4.20 A charge q in the presence of a conductor of boundary $\partial\Omega$

Suppose that the potential at the surface of the conductor is known to be V_D. We want to solve Poisson–Dirichlet's problem

$$\begin{cases} \mathbf{x} \in \Omega, \quad \nabla^2 V(\mathbf{x}) = -\dfrac{q}{\epsilon_0}\delta(\mathbf{x} - \mathbf{x}') , \\ \mathbf{x} \in \partial\Omega, \ V(\mathbf{x}) = V_D. \end{cases}$$

The method of images consists in replacing the conductor $\widetilde{\Omega}$ by a set of point charges such that the potential at any point $\mathbf{x}$ in Ω is the same as in the original problem. More specifically, the method consists in finding, if the geometry of the problem allows it, a set of point charges $q'_i = q'_i(q, \mathbf{x}')$ (also called image charges) located at the positions $\mathbf{x}'_i = \mathbf{x}'_i(q, \mathbf{x}')$ in $\widetilde{\Omega}$, that would give the same boundary conditions at $\partial\Omega$ as in the case where $\widetilde{\Omega}$ is a conductor. Note that all image charges are outside Ω, implying that they do not contribute to the charge density ϱ in Ω and so, the potential at any point $\mathbf{x}$ in Ω still satisfies the original Poisson's equation $\nabla^2 V = -(q/\epsilon_0)\delta(\mathbf{x} - \mathbf{x}')$.

Example 4.5 - Point charge in the presence of an infinite grounded plane
Consider a point charge q at distance d from an infinite conductor plane at potential $V = 0$, as shown in Fig. 4.21.

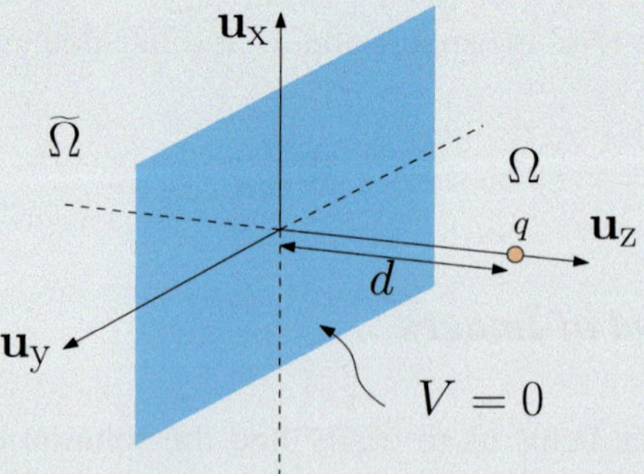

Fig. 4.21 A point charge q at distance d from a grounded, infinite plane at $z = 0$

We want to solve the Poisson–Dirichlet's problem for the potential in the region $\Omega = \{\mathbf{x} \mid z > 0\}$:

$$
\begin{cases}
\mathbf{x} \in \Omega : & \nabla^2 V(\mathbf{x}) = -\dfrac{q}{\epsilon_0}\delta(\mathbf{x} - \mathbf{x}_0), \\[2mm]
\mathbf{x} \in \partial\Omega : & V(\mathbf{x}) = 0,
\end{cases}
$$

where $\mathbf{x}_0 = d\mathbf{u}_z$ is the position of the point charge q. This problem can be solved with the method of images if we replace the conductive plane by a point charge q' located at $\mathbf{x}' = -d'\mathbf{u}_z \in \widetilde{\Omega}$, as illustrated in Fig. 4.22.

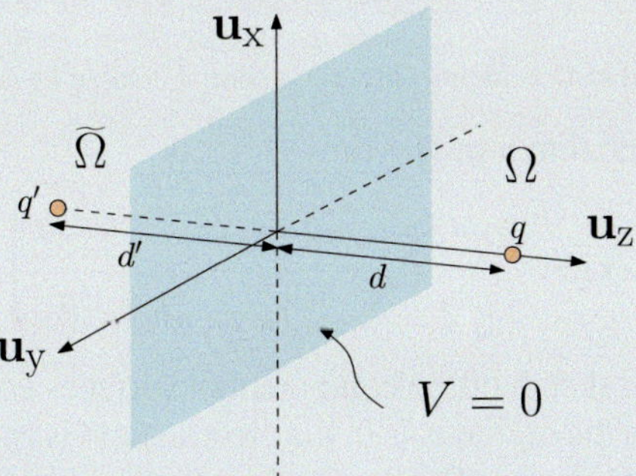

Fig. 4.22 The grounded plane is replaced by a point charge q' located in $\widetilde{\Omega}$

The potential in Ω due to both charges is simply

$$
V(\mathbf{x}) = \frac{q}{4\pi\epsilon_0 |\mathbf{x} - d\mathbf{u}_z|} + \frac{q'}{4\pi\epsilon_0 |\mathbf{x} + d'\mathbf{u}_z|} \, ,
$$

which still satisfies

$$
\nabla^2 V(\mathbf{x}) = -\frac{q}{\epsilon_0}\delta(\mathbf{x} - \mathbf{x}') \qquad \mathbf{x} \in \Omega
$$

since $\delta(\mathbf{x} - \mathbf{x}') = 0$ for every $\mathbf{x}$ in Ω and $\mathbf{x}'$ in $\widetilde{\Omega}$. We choose q' and d' such that the potential is zero at the surface $z = 0$, that is

$$
V(z = 0) = \frac{1}{4\pi\epsilon_0}\left\{ \frac{q}{\sqrt{r^2 + d^2}} + \frac{q'}{\sqrt{r^2 + d'^2}} \right\} = 0 \, ,
$$

with $r^2 = x^2 + y^2$. This can be obtained by taking $q' = -q$ and $d' = d$. We see that the potential in Ω is then identical to that generated by an electric dipole

$\mathbf{p} = 2qd\mathbf{u}_z$ centered at the origin. The plane $z = 0$ is indeed an equipotential for which $V = 0$ at every point, see Fig. 4.23.

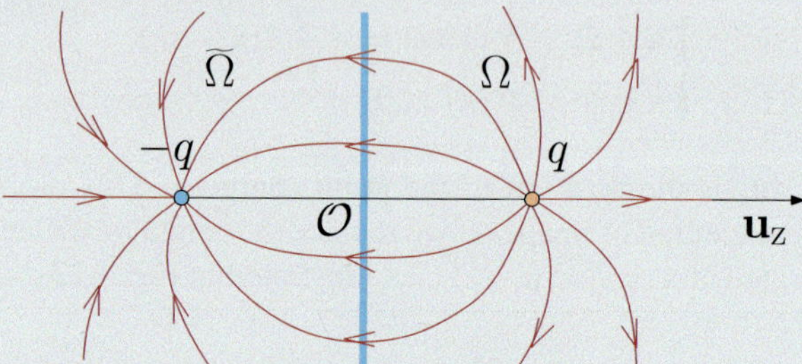

Fig. 4.23 The potential in Ω is identical to the potential created by an electric dipole

Finally the potential at $\mathbf{x}$ in Ω writes

$$V(\mathbf{x}) = \frac{q}{4\pi\epsilon_0} \left\{ \frac{1}{|\mathbf{x} - d\mathbf{u}_z|} - \frac{1}{|\mathbf{x} + d\mathbf{u}_z|} \right\} .$$

Figure 4.24 shows a sketch of the equipotential surfaces and field lines for the original problem. In the region $z < 0$, the electric field is zero and the potential is constant and equal to zero. For $z > 0$, the potential and the electric field are identical to those of an electric dipole.

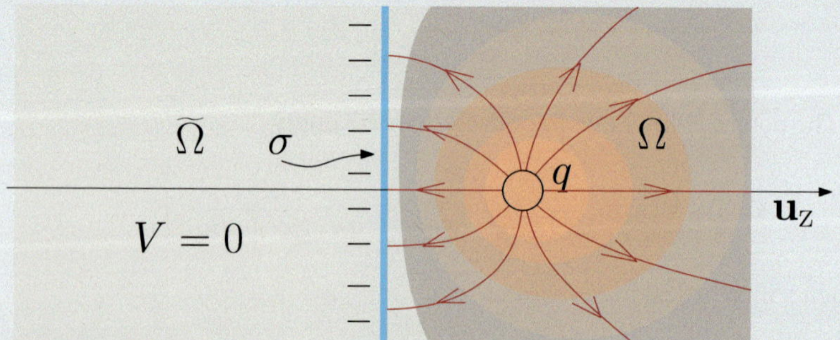

Fig. 4.24 Electric field lines and equipotential surfaces generated by the charge q and the grounded plane

Once we have obtained the potential, we can calculate the charge density and the electric field. The charge density σ at the surface of the conductor plane is given by Coulomb's theorem (4.1):

$$\sigma(\mathbf{x}) = \epsilon_0 \mathbf{E}(\mathbf{x}) \cdot \mathbf{n} \qquad \mathbf{x} \in \partial\Omega ,$$

which writes in this particular case

$$\sigma(\mathbf{x}) = -\epsilon_0 \left. \frac{\partial V(\mathbf{x})}{\partial z} \right|_{z=0} .$$

Since

$$\frac{\partial}{\partial z} \frac{1}{|\mathbf{x} - \mathbf{x}'|} = \mathbf{u}_z \cdot \boldsymbol{\nabla}\left(\frac{1}{|\mathbf{x} - \mathbf{x}'|} \right) = -\mathbf{u}_z \cdot \frac{\mathbf{x} - \mathbf{x}'}{|\mathbf{x} - \mathbf{x}'|^3} ,$$

then

$$\frac{\partial}{\partial z} \frac{1}{|\mathbf{x} - d\mathbf{u}_z|} = -\frac{z - d}{|\mathbf{x} - d\mathbf{u}_z|^3}$$

and

$$\frac{\partial}{\partial z} \frac{1}{|\mathbf{x} + d\mathbf{u}_z|} = -\frac{z + d}{|\mathbf{x} + d\mathbf{u}_z|^3} ,$$

so that

$$\left. \frac{\partial V(\mathbf{x})}{\partial z} \right|_{z=0} = -\frac{\sigma(x, y)}{\epsilon_0} = \frac{2qd}{4\pi\epsilon_0(r^2 + d^2)^{3/2}}$$

and the charge density induced on the conductor plane reads

$$\sigma(\mathbf{x}) = -\frac{1}{2\pi} \frac{qd}{\left(r^2 + d^2\right)^{3/2}} .$$

The total induced charge can be calculated by integrating σ in cylindrical coordinates

$$Q = \int_0^{2\pi} d\phi \int_0^\infty r\sigma(r)dr = -\int_0^\infty \frac{rqd}{\left(r^2 + d^2\right)^{3/2}} dr = -q .$$

Not surprisingly, the induced charge corresponds to the image charge $q' = -q$.

4.4.2 *Expansion in Legendre's Polynomials*

Let us suppose a region Ω in space free of charges ($\varrho = 0$), so that the potential satisfies Laplace's equation

$$\nabla^2 V(\mathbf{x}) = 0 \quad \text{for } \mathbf{x} \text{ in } \Omega .$$

If the charge distribution is invariant under rotations with respect to an axis (azimuthal symmetry), the general solution to Laplace's equation may be written as a series of well-known polynomial functions, called Legendre's polynomials. To show this, let us use spherical coordinates, and suppose that the potential is invariant under rotations around the z-axis. This means that $V(r, \theta, \phi) = V(r, \theta)$. Now look for a solution to Laplace's equation of the form

$$V(r, \theta) = \frac{R(r)}{r} \Theta(\theta) \ .$$

Injecting into Laplace's equation yields

$$\frac{\Theta}{r} \frac{d^2 R}{dr^2} + \frac{R}{r^3 \sin \theta} \frac{d}{d\theta} \left(\sin \theta \frac{d\Theta}{d\theta} \right) = 0$$

and multiplying the latter equation by $r^3/(R\Theta)$, we obtain

$$\frac{r^2}{R} \frac{d^2 R}{dr^2} = -\frac{1}{\Theta \sin \theta} \frac{d}{d\theta} \left(\sin \theta \frac{d\Theta}{d\theta} \right) \ .$$

This is an equation with separate variables: for all values of r and θ, the left-hand side depends on r only while the right-hand side depends on θ only, which requires both sides to be equal to a constant, that we call λ. We then obtain an equation for Θ:

$$\frac{1}{\sin \theta} \frac{d}{d\theta} \left(\sin \theta \frac{d\Theta}{d\theta} \right) + \lambda \Theta = 0$$

whose general solution is written in terms of Legendre's polynomials P_l.

The regularity of Θ at $\theta = 0$ and $\theta = \pi$ imposes the condition $\lambda = l(l + 1)$ where l is a positive integer. Then,

$$\Theta(\theta) = P_l(\cos \theta)$$

and the equation for the radial part of the potential writes

$$\frac{d^2 R}{dr^2} - \frac{l(l + 1)}{r^2} R = 0 \ ,$$

whose solution is written

$$R_l(r) = A_l r^{l+1} + B_l r^{-l} \ ,$$

where A_l and B_l are constants.

Fig. 4.25 Adrien–Marie
Legendre (1752–1833) was a
French mathematician who
made foundational
contributions to number
theory, geometry, statistics,
and mathematical analysis.
The figure represents the first
five Legendre's polynomials

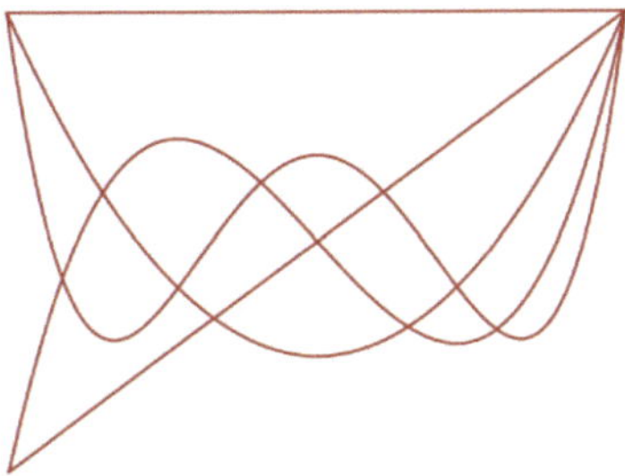

General solution of Laplace's equation with azimuthal symmetry

Let $\Omega \subseteq \mathbb{R}$ be a region in space in which the potential satisfies Laplace's equation. If the potential is invariant under rotations around the z-axis, the potential can be written, in spherical coordinates, as a series involving the Legendre polynomials P_l:

$$V(r, \theta) = \sum_{l=0}^{\infty} \left(A_l r^l + B_l r^{-(l+1)} \right) P_l(\cos \theta) . \tag{4.9}$$

The Legendre polynomials are illustrated in Fig. 4.25 and their main properties can be found in Appendix A.6.

Example 4.6 - Potential generated by a uniformly charged ring

Let us consider a uniformly charged ring of radius R and charge Q. Let us find the absolute potential at any point P in space, as shown in Fig. 4.26.

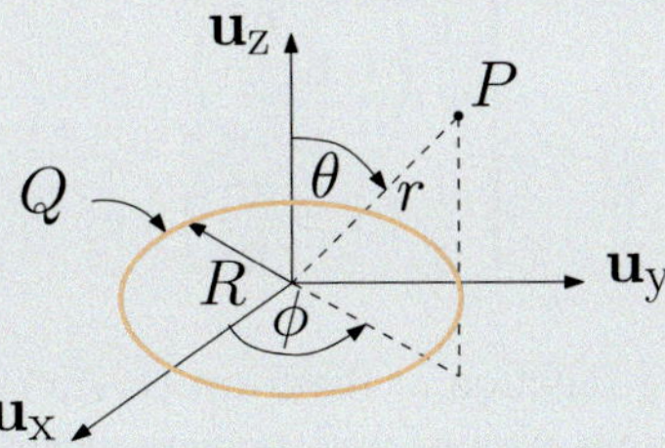

Fig. 4.26 Uniformly charged ring

The charge density being invariant under rotations around the z-axis, it follows that the electric potential is independent of ϕ in spherical coordinates. Everywhere outside the ring, the potential is a solution to Laplace's equation, which writes in its most general form as (4.9):

$$V(r,\theta) = \sum_{l=0}^{\infty} \left(A_l r^l + B_l r^{-(l+1)}\right) P_l(\cos\theta), \quad r < R \ ,$$

$$V(r,\theta) = \sum_{l=0}^{\infty} \left(C_l r^l + D_l r^{-(l+1)}\right) P_l(\cos\theta), \quad r \geq R \ .$$

Since the charge density is integrable, we have $\lim_{r\to\infty} V(r,\theta) = 0$ so that $C_l = 0$ for $l = 0, 1, 2, \ldots$, a non-negative integer. In addition, for the potential to remain finite in $r = 0$ we must have $B_l = 0$ for $l = 0, 1, 2, \ldots$. In Problem 3.5, it is shown that the potential on the axis of the ring is given by

$$V(r,0) = \frac{1}{4\pi\epsilon_0} \frac{Q}{\sqrt{r^2 + R^2}}$$

so that we must have

$$V(r,0) = \sum_{l=0}^{\infty} A_l r^l = \frac{1}{4\pi\epsilon_0} \frac{Q}{\sqrt{r^2 + R^2}} \qquad r < R \ ,$$

$$V(r,0) = \sum_{l=0}^{\infty} D_l r^{-(l+1)} = \frac{1}{4\pi\epsilon_0} \frac{Q}{\sqrt{r^2 + R^2}} \qquad r \geq R \ ,$$

since $P_l(1) = 1$ for every l. Now, we can write

$$\frac{1}{\sqrt{r^2 + R^2}} = \begin{cases} \dfrac{1}{r}\left(1 + \dfrac{R^2}{r^2}\right)^{-1/2} & r < R \ , \\[2ex] \dfrac{1}{R}\left(1 + \dfrac{r^2}{R^2}\right)^{-1/2} & r \geq R \ . \end{cases}$$

In both cases we have a function of the form $(1+x)^{-1/2}$ with $x \leq 1$, which therefore has the following expansion in powers of x:

$$(1+x)^{-1/2} = \sum_{l=0}^{\infty} \frac{x^l}{l!} \left(\frac{d^l}{dx^l}(1+x)^{-1/2}\right)\Bigg|_{x=0} \ .$$

By iteration, one can show that

$$\left. \left(\frac{d^l}{dx^l}(1+x)^{-\frac{1}{2}} \right) \right|_{x=0} = \frac{(-1)^l(2l-1)!!}{2^l} \quad \text{for } l \geq 1 \,,$$

with $(2l-1)!! = 1 \times 3 \times 5 \times \ldots \times (2l-1)$. In this way,

$$(1+x)^{-1/2} = 1 + \sum_{l=1}^{\infty} \frac{x^l(-1)^l}{l!2^l}(2l-1)!!$$

and we find

$$V(r,0) = \sum_{l=0}^{\infty} A_l r^l = \frac{Q}{4\pi\epsilon_0}\left(\frac{1}{R} + \sum_{l=1}^{\infty} \frac{(-1)^l}{2^l}\frac{(2l-1)!!}{l!}\frac{r^{2l}}{R^{2l+1}} \right) \qquad r < R \,,$$

$$V(r,0) = \sum_{l=0}^{\infty} D_l r^{-(l+1)} = \frac{Q}{4\pi\epsilon_0}\left(\frac{1}{r} + \sum_{l=0}^{\infty} \frac{(-1)^l}{2^l}\frac{(2l-1)!!}{l!}\frac{R^{2l}}{r^{2l+1}} \right) \qquad r > R \,.$$

We conclude that $A_l = D_l = 0$ for every odd l. In addition, $A_0 = \dfrac{Q}{4\pi\epsilon_0 R}$, $D_0 = \dfrac{Q}{4\pi\epsilon_0 r}$ and

$$A_{2l} = \frac{(-1)^l}{2^l}\frac{(2l-1)!!}{l!R^{2l+1}}, \qquad l \geq 1 \,,$$

$$D_{2l} = \frac{(-1)^l}{2^l}\frac{(2l-1)!!R^{2l}}{l!}, \qquad l \geq 1 \,.$$

Finally, the potential is given by

$$V(r,\theta) = \frac{Q}{4\pi\epsilon_0} \times \left(\frac{1}{R} + \sum_{l=1}^{\infty} \frac{(-1)^l(2l-1)!!}{2^l l!\, R^{2l+1}} r^{2l} P_{2l}(\cos\theta) \right) \quad r < R \,,$$

$$= \frac{Q}{4\pi\epsilon_0} \times \left(\frac{1}{r} + \sum_{l=1}^{\infty} \frac{(-1)^l(2l-1)!!R^{2l}}{2^l l!\, r^{2l+1}} P_{2l}(\cos\theta) \right) \quad r \geq R \,.$$

Figure 4.27 shows a sketch of the equipotential surfaces together with the field lines in a plane of symmetry passing through the center of the ring and containing the z-axis.

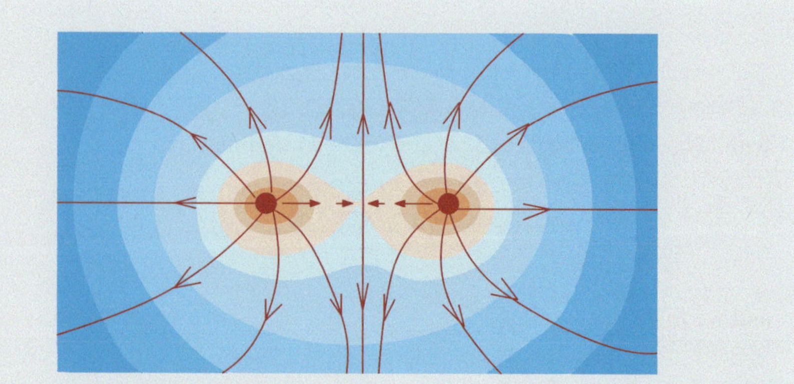

Fig. 4.27 Electric field lines and equipotential surfaces generated by a uniformly charged ring

4.5 Green's Function and Formal Solution to Poisson's Equation with Boundary Conditions

In order to solve Poisson's equation for an arbitrary charge distribution ϱ, it is sufficient to determine the solution when the charge distribution is a point charge at an arbitrary position in space. This results from the fact that, within the framework of distributions, any charge density ϱ can be written as a linear superposition of point charges such that

$$\varrho(\mathbf{x}) = \iiint_{\mathbb{R}^3} \varrho(\mathbf{x}')\delta(\mathbf{x} - \mathbf{x}')d^3x' \ .$$

The linearity of Poisson's equation ensures that solving the problem with a Dirac charge distribution and the proper boundary conditions are sufficient to find the solution for an arbitrary density ϱ by simply using the superposition principle.

Green's function

A Green's function G in $\Omega \subseteq \mathbb{R}^3$ is a solution to the Poisson's equation in Ω for a point charge density located at $\mathbf{x}'$, that is to say,

$$\nabla^2 G(\mathbf{x}, \mathbf{x}') = -4\pi\,\delta(\mathbf{x} - \mathbf{x}') \quad \text{for } \mathbf{x} \in \Omega \ . \tag{4.10}$$

More generally, a Green function, named after George Green (see Fig. 4.28), represents the impulse response of a physical system.

$$\oint_C P dx + Q dy = \iint_D \left(\frac{\partial Q}{\partial x} - \frac{\partial P}{\partial y} \right) dx dy$$

Fig. 4.28 George Green (1793–1841), an English mathematician and physicist, made pioneering contributions to mathematical physics. He is responsible for Green's theorem shown here and for introducing the concept of a potential function, essential for modern electromagnetism, fluid dynamics, and elasticity

Remark

Let $L(\mathbf{x})$ be any solution of Laplace's equation in Ω, that is,

$$\nabla^2 L(\mathbf{x}) = 0, \quad \text{for } \mathbf{x} \in \Omega .$$

Then, since $\nabla^2 \dfrac{1}{|\mathbf{x} - \mathbf{x}'|} = -4\pi \delta(\mathbf{x} - \mathbf{x}')$, we infer that the function

$$G(\mathbf{x}, \mathbf{x}') = \frac{1}{|\mathbf{x} - \mathbf{x}'|} + L(\mathbf{x})$$

is a Green function in Ω.

Using Green's second identity (A.42) with V and G, the potential writes in terms of the Green function as:

$$V(\mathbf{x}) = \frac{1}{4\pi \epsilon_0} \iiint_\Omega \varrho(\mathbf{x}') G(\mathbf{x}', \mathbf{x}) d^3 x'$$

$$+ \frac{1}{4\pi} \oiint_{\partial\Omega} d\mathbf{S}(\mathbf{x}') \cdot \left(G(\mathbf{x}', \mathbf{x}) \nabla' V(\mathbf{x}') - V(\mathbf{x}') \nabla' G(\mathbf{x}', \mathbf{x}) \right) . \quad (4.11)$$

This result allows us to identify a formal solution to the Dirichlet–Poisson's and Neumann–Poisson's problems, which consists in finding the appropriate Green's function for each case.

4.5.1   *Formal Solution to Poisson–Dirichlet Problem*

The problem consists in finding a solution of Poisson's equation in Ω given the potential V_D at its boundary $\partial\Omega$,

$$\begin{cases} \mathbf{x} \in \Omega : \quad \nabla^2 V(\mathbf{x}) = -\dfrac{\varrho(\mathbf{x})}{\epsilon_0} , \\[2mm] \mathbf{x} \in \partial\Omega : V(\mathbf{x}) = V_D(\mathbf{x}) . \end{cases}$$

For this, we solve the associated Green–Dirichlet problem

$$\begin{cases} \mathbf{x} \in \Omega : \ \nabla^2 G_D(\mathbf{x}, \mathbf{x}') = -4\pi \delta(\mathbf{x} - \mathbf{x}') \, , \\ \mathbf{x} \in \partial\Omega : G_D(\mathbf{x}, \mathbf{x}') = 0 \, . \end{cases}$$

Writing the Green function as

$$G_D(\mathbf{x}, \mathbf{x}') = \frac{1}{|\mathbf{x} - \mathbf{x}'|} + L_D(\mathbf{x}) \, ,$$

the problem is reduced to finding L_D such that

$$\begin{cases} \mathbf{x} \in \Omega : \ \nabla^2 L_D(\mathbf{x}, \mathbf{x}') = 0 \, , \\ \mathbf{x} \in \partial\Omega : L_D(\mathbf{x}, \mathbf{x}') = -\dfrac{1}{|\mathbf{x} - \mathbf{x}'|} \, . \end{cases}$$

Replacing in (4.11), the formal solution for the potential is therefore

$$V(\mathbf{x}) = \frac{1}{4\pi\epsilon_0} \iiint_\Omega \varrho(\mathbf{x}') G_D(\mathbf{x}, \mathbf{x}') d^3 x' - \frac{1}{4\pi} \oiint_{\partial\Omega} V_D(\mathbf{x}') d\mathbf{S}(\mathbf{x}') \cdot \nabla' G_D(\mathbf{x}', \mathbf{x}) \qquad (4.12)$$

4.5.2 Formal Solution to Poisson–Neumann Problem

The problem consists in solving Poisson's equation in Ω given the normal component of the electric field at its boundary $\partial\Omega$,

$$\begin{cases} \mathbf{x} \in \Omega : \ \nabla^2 V(\mathbf{x}) = -\dfrac{\varrho(\mathbf{x})}{\epsilon_0} \, , \\ \mathbf{x} \in \partial\Omega : \nabla V(\mathbf{x}) \cdot \mathbf{n}(\mathbf{x}) = -E_N(\mathbf{x}) \, . \end{cases}$$

The associated Green–Neumann problem is

$$\begin{cases} \mathbf{x} \in \Omega : \ \nabla^2 G_N(\mathbf{x}, \mathbf{x}') = -4\pi \delta(\mathbf{x} - \mathbf{x}') \, , \\ \mathbf{x} \in \partial\Omega : \mathbf{n}(\mathbf{x}) \cdot \nabla G_N(\mathbf{x}, \mathbf{x}') = -\dfrac{4\pi}{S} \, , \end{cases}$$

where S is the total surface area of $\partial\Omega$. Again, we look for G_N in the form

$$G_N(\mathbf{x}, \mathbf{x}') = \frac{1}{|\mathbf{x} - \mathbf{x}'|} + L_N(\mathbf{x})$$

and so, L_N must satisfy

$$\begin{cases} \mathbf{x} \in \Omega : \ \nabla^2 L_N(\mathbf{x}, \mathbf{x}') = 0 \, , \\ \mathbf{x} \in \partial\Omega : \mathbf{n}(\mathbf{x}) \cdot \nabla L_N(\mathbf{x}, \mathbf{x}') = -\dfrac{4\pi}{S} - \mathbf{n}(\mathbf{x}) \cdot \nabla \dfrac{1}{|\mathbf{x} - \mathbf{x}'|} \, , \end{cases}$$

from which we see that the formal solution of the Poisson–Neumann's problem writes

$$V(\mathbf{x}) = \frac{1}{4\pi\epsilon_0} \iiint_\Omega \varrho(\mathbf{x}')G_N(\mathbf{x}',\mathbf{x})d^3x' - \frac{1}{4\pi} \oiint_{\partial\Omega} dS(\mathbf{x}')E_N(\mathbf{x}')G_N(\mathbf{x}',\mathbf{x}) + \langle V\rangle_S \ , \qquad (4.13)$$

where $\langle V\rangle_S$ is the mean value of the potential over the surface $\partial\Omega$,

$$\langle V\rangle_S = \frac{1}{S} \oiint_{\partial\Omega} dS(\mathbf{x})V(\mathbf{x}) \ .$$

Example 4.7 - The mean value theorem

In a region without charges, the electric potential at any point $\mathbf{x}$ is equal to its mean value over any sphere centered at $\mathbf{x}$. To show this, let us consider the Green–Dirichlet function that satisfies

$$\nabla^2 G_D(\mathbf{x},\mathbf{x}') = -4\pi\delta(\mathbf{x}-\mathbf{x}') \quad \mathbf{x}\in\Omega \ ,$$

$$\mathbf{G}_D(\mathbf{x},\mathbf{x}') = 0, \quad \mathbf{x}\in\partial\Omega \ .$$

If Ω is a region free of charges, then (4.12) reduces to

$$V(\mathbf{x}) = -\frac{1}{4\pi} \oiint_{\partial\Omega} V(\mathbf{x}')d\mathbf{S}(\mathbf{x})\cdot\nabla' G_D(\mathbf{x},\mathbf{x}') \ .$$

We can now choose Ω as the volume inside a sphere of radius a centered at $\mathbf{x}$, such that the Green–Dirichlet function becomes

$$G_D(\mathbf{x},\mathbf{x}') = \frac{1}{|\mathbf{x}-\mathbf{x}'|} - \frac{1}{a}$$

so that

$$\nabla' G_D(\mathbf{x},\mathbf{x}') = \frac{\mathbf{x}-\mathbf{x}'}{|\mathbf{x}-\mathbf{x}'|^3} \ .$$

Then, for $\mathbf{x}'$ in $\partial\Omega$

$$\mathbf{n}(\mathbf{x}')\cdot\nabla' G_D(\mathbf{x},\mathbf{x}') = \frac{\mathbf{x}'-\mathbf{x}}{|\mathbf{x}'-\mathbf{x}|}\cdot\frac{\mathbf{x}-\mathbf{x}'}{|\mathbf{x}-\mathbf{x}'|^3} = -\frac{1}{|\mathbf{x}-\mathbf{x}'|^2} = -\frac{1}{a^2}$$

and we conclude that at any point $\mathbf{x}$ inside the sphere, the potential is equal to its mean value over its surface,

$$V(\mathbf{x}) = \frac{1}{4\pi a^2} \oiint_{\partial\Omega} dS(\mathbf{x}')V(\mathbf{x}') = \langle V\rangle_S \ .$$

In particular, this can be applied to a spherical conductor to show that the potential anywhere inside is equal to the potential at the surface of the conductor.

4.6 Summary and Essential Formulas

- For a conductor Ω in electrostatic equilibrium, the electric field and the charge density vanish at every point inside Ω: $\varrho = 0$, $\mathbf{E} = 0$. In the region outside the conductor, at a point infinitely close to the surface $\partial\Omega$, the electric field is given by Coulomb's theorem (4.1)

$$
\mathbf{E}(\mathbf{x}) = \frac{\sigma(\mathbf{x})}{\epsilon_0}\,\mathbf{n}(\mathbf{x}) \quad \text{for } \mathbf{x} \text{ in } \partial\Omega \,,
$$

where σ is the charge density on the surface of the conductor and $\mathbf{n}(\mathbf{x})$ the normal direction pointing outward.

- The electric potential is constant at any point inside a conductor in equilibrium, and equal to its value at its surface V_0. The electrostatic energy of a conductor with total charge Q writes

$$
U = \frac{1}{2} V_0 Q \,.
$$

- The electrostatic force $\mathbf{F}_e$ acting on a conductor Ω at equilibrium is the integral over its surface $\partial\Omega$ of the electrostatic pressure P_e, defined as

$$
P_e = \frac{\sigma^2}{2\epsilon_0}
$$

so that

$$
\mathbf{F}_e = \oiint_{\partial\Omega} P_e(\mathbf{x}')\mathbf{n}(\mathbf{x}')dS(\mathbf{x}') \,.
$$

- The electric potential V satisfies Poisson's equation

$$
\nabla^2 V(\mathbf{x}) = -\frac{\varrho(\mathbf{x})}{\epsilon_0} \,,
$$

which, in regions free of charges, reduces to Laplace's equation $\nabla^2 V = 0$.
- A unique solution to Poisson's equation exists in a region Ω with

  - Dirichlet's boundary condition, which specifies the value of the potential $V = V_D$ at any point of the boundary $\partial\Omega$.

- Neumman's boundary condition, which specifies the component of the electric field normal to the surface at any point in $\partial\Omega$, that is $\mathbf{E} \cdot \mathbf{n} = E_N$. In that case, the solution is unique up to an additive constant but its gradient, that is the electric field, does not depend on this constant.

- For a charge distribution that is invariant under rotations around the z-axis, the solution to Laplace's equation writes in spherical coordinates as a series involving the Legendre polynomials,

$$V(r, \theta) = \sum_{l=0}^{\infty} \left(A_l r^l + B_l r^{-(l+1)} \right) P_l(\cos\theta) .$$

- Mathematically, a formal solution to Poisson's equation is obtained by solving an associated Green–Poisson problem:

$$\nabla^2 G(\mathbf{x}, \mathbf{x}') = -4\pi \delta(\mathbf{x} - \mathbf{x}') \quad \mathbf{x} \in \Omega ,$$

where the choice of the boundary condition for G in $\partial\Omega$ depends on the boundary conditions of the original problem. For Dirichlet's boundary condition, one imposes

$$\mathbf{x} \in \partial\Omega : G(\mathbf{x}, \mathbf{x}') = 0$$

and the potential writes

$$V(\mathbf{x}) = \frac{1}{4\pi \epsilon_0} \iiint_\Omega \varrho(\mathbf{x}') G(\mathbf{x}, \mathbf{x}') d^3 x' - \frac{1}{4\pi} \oiint_{\partial\Omega} V_D(\mathbf{x}') d\mathbf{S}(\mathbf{x}') \cdot \nabla' G(\mathbf{x}', \mathbf{x}) ,$$

whereas for Neumann's boundary condition, one imposes

$$\mathbf{x} \in \partial\Omega : \mathbf{n}(\mathbf{x}) \cdot \nabla G(\mathbf{x}, \mathbf{x}') = -\frac{4\pi}{S} ,$$

where S is the surface area of $\partial\Omega$. The potential then writes

$$V(\mathbf{x}) = \frac{1}{4\pi \epsilon_0} \iiint_\Omega \varrho(\mathbf{x}') G(\mathbf{x}', \mathbf{x}) d^3 x' - \frac{1}{4\pi} \oiint_{\partial\Omega} dS(\mathbf{x}') E_N(\mathbf{x}') G(\mathbf{x}', \mathbf{x}) + \langle V \rangle_S ,$$

where $\langle V \rangle_S$ is the mean value of the potential over the surface $\partial\Omega$.

Problems

4.1 Conductive sphere and sparks

Consider a conductive sphere of radius $R = 30\,\mathrm{cm}$. What is the maximum charge than can be accumulated in the sphere before a spark forms between the sphere and a second, small sphere connected to the ground at a distance $d = 1\,\mathrm{m}$ from it? What is then the maximal potential difference (with respect to the ground) that can be applied to the sphere? Recall that the breakdown field of air is $E = 3 \times 10^6 \, \text{Vm}^2$.

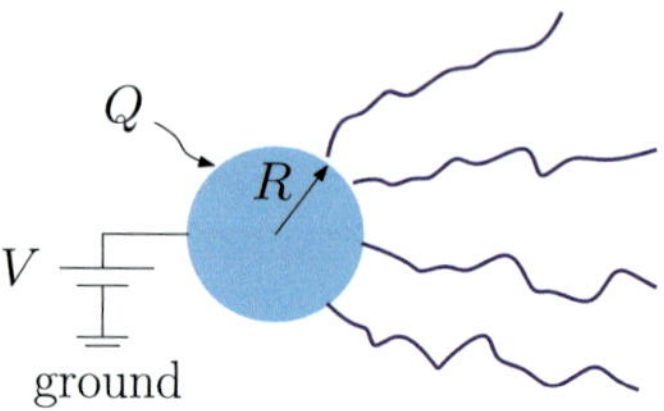

4.2 Screening length in semiconductors

Semiconductors are materials in which the electrical conductivity can be tuned by several orders of magnitude by, for example, introducing small amounts of foreign atoms (impurities), also called dopants. Two types of charge carriers coexist in such materials: free electrons, each of charge $-e$ at a density $n(\mathbf{x})$, and free *holes*, each of charge $+e$ at a density $p(\mathbf{x})$. Far from any impurity, charge neutrality imposes $n = p$. Suppose that now a positively ionized impurity is introduced at $r = 0$, creating a charge imbalance in the crystal. This local charge imbalance will be partially compensated by the fact that electrons (holes) will be attracted to (repelled from) this ion, screening the electric field of the ion at larger scales. The goal of this exercise is to determine the characteristic screening length beyond which charge neutrality is recovered. We will admit that the densities n and p obey, at every point, a Boltzmann distribution law:

$$n(\mathbf{x}) = n_0 e^{eV(\mathbf{x})/k_B T}, \quad p(\mathbf{x}) = n_0 e^{-eV(\mathbf{x})/k_B T} \,,$$

where V is the electric potential, T the temperature, and k_B is Boltzmann's constant.

(a) Assuming that the system possesses spherical symmetry around the positive ion at the origin, what is the differential equation satisfied by the potential?
(b) Assuming that $k_B T \gg eV$, reduce the previous differential equation to a linear equation.
(c) Integrate the linear equation found previously by using the change of variable $U(r) = rV(r)$. Introduce a typical length scale before integrating.
(d) Interpret the solution.

4.3 Charge density of the hydrogen atom

Consider an hydrogen atom centered at the origin. In a simple model, the time-averaged electric potential is given in spherical coordinates by

$$V(r) = \frac{q}{4\pi\epsilon_0} \frac{e^{-\alpha r}}{r} \left(1 + \frac{\alpha r}{2}\right)$$

with q the magnitude of the electron charge, and $\alpha = a_0/2$, where a_0 is Bohr's radius. Find the charge distribution associated with this potential.

4.4 Charge between cylinders

The region between two concentric metallic cylinders of radius a and b, respectively, is filled with a homogeneous charge density ϱ, as shown below. Neglecting edge effects, determine the potential and the electric field between the cylinders, as well as the charge density on their surfaces, when both are connected to the ground.

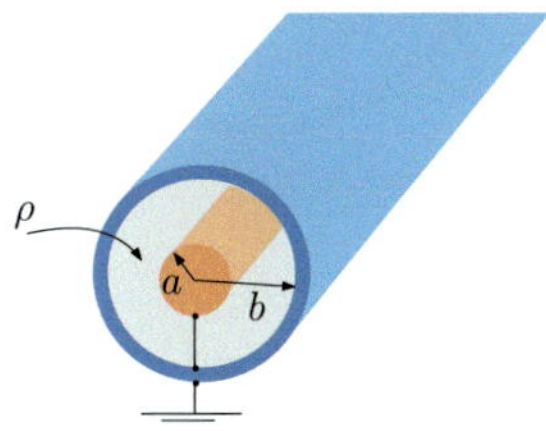

4.5 System of concentric conductors

A metallic sphere of radius R and charge Q is surrounded by a neutral, concentric spherical shell, also metallic, of inner radius $a > R$ and outer radius b, as shown below.

(a) Find the charge density on each surface.
(b) Find the absolute potential at the common center of both conductors.
(c) If the outer shell is connected to the ground, how does the previous answers change?

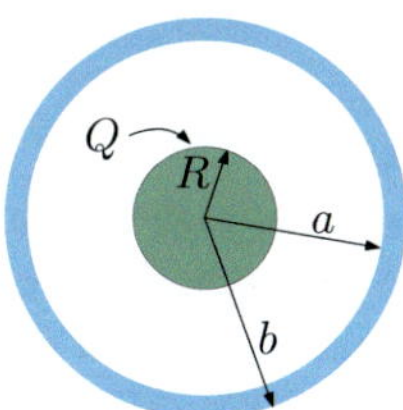

4.6 Three conductive spheres

Consider a system of three concentric conductive spherical shells as shown below. The inner shell, of radius a is initially discharged, the middle one, of radius b, has a charge $-Q$ and the outer one, of radius c, a charge $+Q$. The thickness of each shell is negligible.

(a) Find the electrostatic potential on each of the conductive shells.
(b) If the inner and outer shell are connected by a very thin wire, isolated from the central shell, what is the new value of the electric potential and the charge on each sphere?

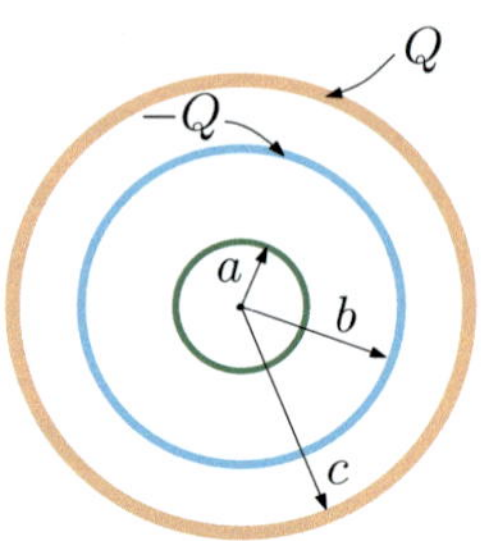

4.7 Cavities in a conductor

Two point charges q_a and q_b are each placed at the center of a spherical cavity of radius a and b, respectively. The cavities lie inside a conductive and neutral sphere of radius $R > a, b$ as shown below.

(a) Find the charge density in the conductive sphere.
(b) What is the electric field at any point outside the sphere?
(c) What is the electric field inside each cavity?

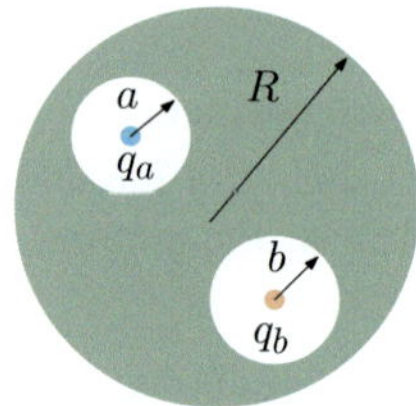

4.8 Charge on the Earth surface

The Earth surface (radius $R_E = 6400\,\text{km}$) and the ionosphere (at a height $h \sim 100\,\text{km}$) are assumed to be two spherical conductive shells forming a neutral system, illustrated below. The typical potential difference between the ground and the ionosphere is on the order of $350\,\text{kV}$. Since the air is partially conductive, the electric field decreases exponentially as a function of the height z with respect to the ground ($E = E_0 e^{-z/H}$ with $H = 3.5\,\text{km}$), as seen in problem 2.1.

(a) What is the magnitude E_0 of the electric field at the Earth surface?
(b) What is the total charge accumulated on the Earth surface?

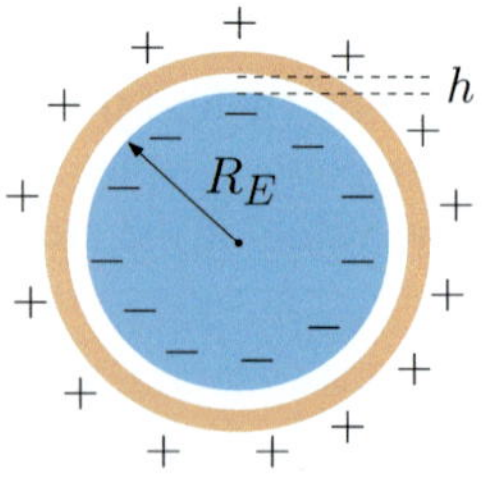

4.9 Point charge in the presence of a grounded conductive sphere

Consider a point charge q located at a distance d from the center of a conductive sphere of radius $a < d$ that is kept at zero potential, as shown below. Determine the electrostatic potential at any point $\mathbf{x}$ outside the sphere, as well as the induced charge in the sphere and the force between the point charge and the sphere. The notation $\alpha = a/d$ can be introduced to facilitate calculations.

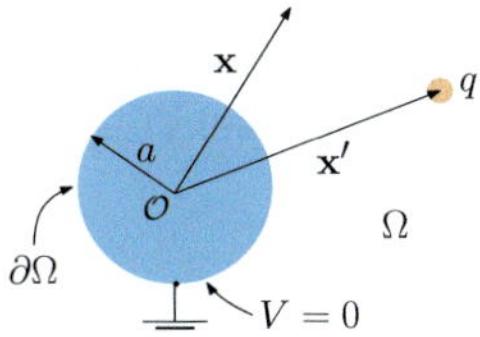

4.10 Dipole inside a spherical shell

A dipole $\mathbf{p}$ is located at the origin and oriented along the z-axis. The dipole is surrounded by a spherical shell of radius R connected to the ground. Find the potential everywhere in space, as well as the charge density induced on the sphere.

4.11 A corollary of the mean value theorem

Show that the electric field in vacuum satisfies the relations

$$\oiint_{\partial\Omega} \frac{d\mathbf{S}(\mathbf{x}') \cdot \mathbf{E}(\mathbf{x}')}{|\mathbf{x} - \mathbf{x}'|} = 0 \quad \text{and} \quad \iiint_{\Omega} \mathbf{E}(\mathbf{x}') \cdot \frac{\mathbf{x} - \mathbf{x}'}{|\mathbf{x} - \mathbf{x}'|^3} d^3 x' = 0 \, ,$$

where $\partial\Omega$ is a sphere of arbitrary radius centered at $\mathbf{x}$.

4.12 Electrostatics with massive photons

Suppose a point charge at the origin, of density $\varrho(\mathbf{r}) = q\delta(\mathbf{r})$. In exercise 13.4 it is shown that if the photon had a non-zero mass, Poisson's equation should be replaced by

$$\nabla^2 V - \mu^2 V = -\frac{\varrho}{\epsilon_0}$$

where μ is a constant. We recall that in spherical coordinates, for a system that does not depend on the spherical angles ϕ and θ, the expression for the Laplacian is

$$\Delta V = \frac{1}{r} \frac{d^2(rV)}{dr^2} \, .$$

(a) Give the expression for the potential $V(r)$ generated by the point charge. In order to find integration constants, consider the limit case when $\mu = 0$.

(b) Find the expression for the electric field at any point in space.

(c) Consider now a charge distribution of density $\varrho(\mathbf{r})$ occupying a finite volume Ω. What are the expressions for the potential and the electric field at any point of space?

(d) Apply the results to calculate the potential at any point inside a sphere of radius a, that carries a uniformly distributed charge Q on its surface. Hint: consider spherical coordinates r, θ, ϕ. Use symmetry arguments first. Consider the integration variable s such that $s^2 = a^2 + r^2 - 2ar\cos\theta$.

(e) In 1755, Benjamin Franklin attempted to measure the interaction force between two charged objects and observed that an object placed inside a (hollow) charged sphere was not attracted by the shell. From the expression of the potential obtained in the previous question, comment on Franklin's observation. Can we measure the photon mass from Franklin's experiment?

Chapter 5
Capacitors

Abstract This chapter introduces the fundamental concept of *capacitance*, a geometric property quantifying the ability of a conductor to store charge for a given electric potential. It begins by defining the capacitance of an isolated conductor and its relationship to electrostatic energy, then generalizes this to a system of multiple conductors through *induction coefficients*. The core of the chapter focuses on the *capacitor*, a ubiquitous electronic component consisting of two conductors carrying equal and opposite charges. It defines the capacitance of such a system in terms of the potential difference between the conductors and derives the *electrostatic energy stored* within a capacitor. Practical methods for *determining capacitance* are illustrated through common geometries, such as parallel plate and cylindrical capacitors. The chapter also details the rules for calculating *equivalent capacitance* when multiple capacitors are connected in series or parallel. A significant portion is dedicated to the *forces acting on capacitors*, distinguishing between isolated systems (fixed charge) and systems connected to a voltage source (fixed potential difference). This leads to the introduction of *thermodynamic potentials*. Specifically, the *Helmholtz free energy* is introduced for isolated systems and the *Gibbs free energy* for systems at constant potential. The chapter demonstrates how these free energies serve as powerful tools to systematically derive electrostatic forces and torques on conductors, ensuring consistency in their calculation under various operating conditions.

Keywords Capacitors · Capacitance · Electrostatics

5.1 Introduction

In this chapter, we apply the results from Chap. 4 on conductors at equilibrium to define a geometric property known as capacitance. This quantity quantifies how much charge a conductor can hold on its surface for a given absolute potential. We also introduce the capacitor, a critical component in electronic applications, which, broadly speaking, is any system composed of two conductors under mutual influence.

We extend the analysis to the case of N conductors in equilibrium, deriving expressions for the force and torque on a conductor due to interactions with the rest

© The Author(s), under exclusive license to Springer Nature Switzerland AG 2025   129
F. Cadiz and A. Couairon, *Classical Electrodynamics*, Undergraduate Texts in Physics,
https://doi.org/10.1007/978-3-031-86785-9_5

of the system. The applications of capacitors in electronics are vast, ranging from energy storage to signal processing and even smartphone touchscreens, to name just a few.

5.2 Capacitance

5.2.1 Capacitance of an Isolated Conductor

The capacitance of a conductor in electrostatic equilibrium is defined from the relation existing between its electric potential V and the charge Q on its surface. Let Ω be the volume occupied by a conductor, then

$$Q = \oiint_{\partial\Omega} \sigma(\mathbf{x}')dS(\mathbf{x}'),$$

$$V = \frac{1}{4\pi\epsilon_0} \oiint_{\partial\Omega} \frac{\sigma(\mathbf{x}')dS(\mathbf{x}')}{|\mathbf{x}-\mathbf{x}'|} \quad \text{for } \mathbf{x} \in \Omega .$$

We see that if the charge density is scaled as $\sigma' = \alpha\sigma$, then necessarily:

$$V' = \alpha V \quad \text{and} \quad Q' = \alpha Q .$$

In consequence, $Q/V = Q'/V'$ is a constant that does not depend on V or Q.

It is always possible to normalize the potential to its value V at the boundary $\partial\Omega$, thus, writing $V(\mathbf{x}) = V\phi(\mathbf{x})$, the normalized potential satisfies $\nabla^2\phi(\mathbf{x}) = 0$ everywhere except on the boundary, where $\phi(\mathbf{x} \in \partial\Omega) = 1$. At infinity, $\phi(\mathbf{x} \to \infty) = 0$. For a given conductor, the shape of the boundary $\partial\Omega$ is fixed and the normalized potential $\phi(\mathbf{x})$ is therefore unique. Coulomb's theorem then yields

$$\sigma(\mathbf{x}' \in \partial\Omega) = -\epsilon_0 V \nabla\phi(\mathbf{x}') \cdot \mathbf{n}(\mathbf{x}') ,$$

where the gradient is calculated on the outer side of the conductor. The total charge on the conductor then reads

$$Q = V \oiint_{\partial\Omega} \left(-\epsilon_0 \nabla\phi(\mathbf{x}') \cdot \mathbf{n}(\mathbf{x}')\right) dS(\mathbf{x}') ,$$

which shows that Q/V, like the normalized potential, only depends on the geometry of the conductor Ω.

Capacitance of a conductor

The capacitance of a conductor in equilibrium is a constant that only depends on its geometry, and is defined as

$$C = Q/V \, ,$$

(5.1)

where V is the absolute potential of the conductor when the latter has a total charge Q on its surface. In SI units, the unit for the capacitance is the farad (F), with $1\,\mathrm{F} = 1\,\mathrm{C}\,\mathrm{V}^{-1}$. In electronics, capacitances are rarely larger than one microfarad ($1\,\mu\mathrm{F} = 1 \times 10^{-6}\,\mathrm{F}$).

5.2.2 Capacitance and Electrostatic Energy

In Sect. 4.2, the electrostatic energy of a charged conductor was shown to be

$$U = \frac{1}{2} Q V \, .$$

In terms of the capacitance ($Q = CV$), this expression can be rewritten as

$$U = \frac{1}{2} C V^2 = \frac{Q^2}{2C} \, ,$$

(5.2)

showing that the electrostatic energy of a conductor is a quadratic function of its potential, or of its charge.

Example 5.1 Capacitance of a charged conductive sphere

Consider a conductive sphere of radius R carrying a total charge Q. As shown in Example 4.4, its absolute potential is given by

$$V = \frac{Q}{4\pi \epsilon_0 R}$$

and therefore, its capacitance reads

$$C = Q/V = 4\pi \epsilon_0 R \, .$$

5.2.3   Generalization: Induction Coefficient Between Conductors

Suppose a system of N conductors is in equilibrium. Conductor i carries a charge Q_i and is at potential V_i, with $i = 1, 2, \ldots, N$.

Charges and potentials satisfy

$$Q_i = \oiint_{\partial \Omega_i} \sigma_i(\mathbf{x}') dS(\mathbf{x}'),$$

$$V_i = \sum_{j=1}^{N} \frac{1}{4\pi \epsilon_0} \oiint_{\partial \Omega_j} \frac{\sigma_j(\mathbf{x}') dS(\mathbf{x}')}{|\mathbf{x}_i - \mathbf{x}'|} \quad \text{for } \mathbf{x}_i \in \Omega_i .$$

This shows clearly the linear relationship between charges and potentials. As a consequence, we can write in the most general form

$$Q_i = \sum_{j=1}^{N} C_{ij} V_j .$$

Induction coefficient

The induction coefficient between the conductors i and j is defined as

$$C_{ij} = \frac{\partial Q_i}{\partial V_j} . \qquad (5.3)$$

Generalizing the case of an isolated conductor, the self-capacitance coefficient of the conductor i in the presence of the $N - 1$ other conductors is defined as

$$C_{ii} = \frac{\partial Q_i}{\partial V_i} , \qquad (5.4)$$

whereas the C_{ij} coefficients with $i \neq j$ are called the mutual capacitance coefficients. It follows that $C_{ij} = C_{ji}$, which is demonstrated in Sect. 5.4.

5.3   The Capacitor

A capacitor is a system of two conductors under total influence, that is, carrying charges of equal magnitude but of opposite sign, as shown in Fig. 5.1. This system is globally neutral, so according to the multipole expansion (3.11), far away it generates an electric field equivalent to that of an electric dipole. A significant fraction of

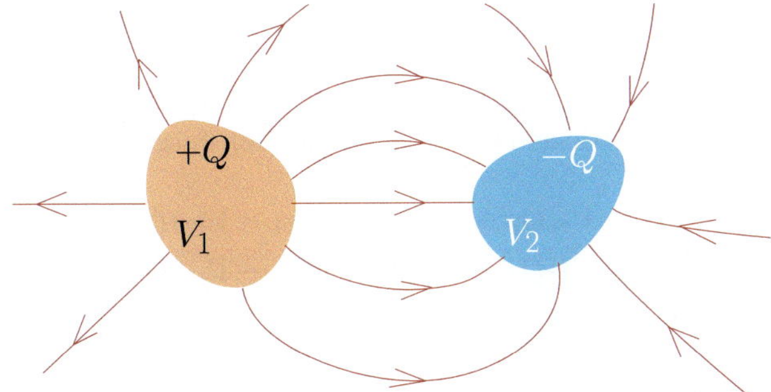

Fig. 5.1 Two conductors of opposite charge constitutes a capacitor

the field lines is confined in the region between the two conductors, and so is the electrostatic energy density of the system.

5.3.1 Capacitance of a Capacitor

Under ideal and typical operating conditions, the plates of a capacitor always have exactly equal and opposite charges. This is a fundamental principle of how capacitors function in a standard electrical circuit, rooted in the principle of charge conservation. Consider the two conductors of a capacitor carrying charges $Q_1 = Q$ and $Q_2 = -Q$, at potentials V_1 and V_2, respectively. In terms of the induction coefficients, we have

$$Q_1 = Q = C_{11} V_1 + C_{12} V_2 \, ,$$
$$Q_2 = -Q = C_{21} V_1 + C_{22} V_2 \, ,$$

and since $C_{12} = C_{21}$, this can be rewritten as

$$Q = C_{11} V_1 + C_{12} V_2 \, ,$$
$$-Q = C_{12} V_1 + C_{22} V_2 \, .$$

We look for a linear relationship between the charge Q and the potential difference $\Delta V = V_1 - V_2$ between the conductors

$$Q = C(V_1 - V_2) = C \Delta V \, , \tag{5.5}$$

where C is, by definition, the capacitance of the capacitor:

$$\boxed{C = \frac{Q}{\Delta V} = \frac{Q}{V_1 - V_2}} \tag{5.6}$$

and represents its capacity to produce a charge separation for a given potential difference.

Now, if the capacitance matrix (C) of coefficients C_{ij} is invertible, the inverse matrix (Γ) is also symmetric and the system is written equivalently as

$$V_1 = \Gamma_{11} Q_1 + \Gamma_{12} Q_2 = (\Gamma_{11} - \Gamma_{12})Q \ ,$$
$$V_2 = \Gamma_{12} Q_1 + \Gamma_{22} Q_2 = (\Gamma_{12} - \Gamma_{22})Q \ ,$$

where

$$\Gamma_{11} = \frac{C_{22}}{C_{11}C_{22} - C_{12}^2}, \quad \Gamma_{12} = \Gamma_{21} = \frac{-C_{12}}{C_{11}C_{22} - C_{12}^2}, \quad \Gamma_{22} = \frac{C_{11}}{C_{11}C_{22} - C_{12}^2} \ .$$

Now calculating the potential difference,

$$V_1 - V_2 = (\Gamma_{11} + \Gamma_{22} - 2\Gamma_{12})Q \ ,$$

we can insert the Γ_{ij} coefficients and express the capacitance in terms of the influence coefficients:

$$C = \frac{Q}{\Delta V} = \frac{1}{\Gamma_{11} + \Gamma_{22} - 2\Gamma_{12}} = \frac{C_{11}C_{22} - C_{12}^2}{C_{11} + C_{22} + 2C_{12}} \ .$$

In the case of a parallel plate capacitor, the capacitance matrix coefficients is not invertible since $C_{11} = C_{22} = -C_{12}$. Nevertheless, the capacitance is still defined by relation (5.6) and its value is simply given by the unique coefficient $C = C_{11} = C_{22}$.

5.3.2   Electrostatic Energy of a Capacitor

A general system of conductors has, according to (4.2), a total electrostatic energy of

$$U = \frac{1}{2} \sum_i Q_i V_i \ .$$

In the case of a capacitor under total influence, we then find

$$U = \frac{1}{2}(QV_1 - QV_2) = \frac{1}{2} Q \underbrace{(V_1 - V_2)}_{\Delta V}$$

and the electrostatic energy stored in the capacitor is

$$U = \frac{1}{2}Q\Delta V = \frac{1}{2}\frac{Q^2}{C} = \frac{1}{2}C\Delta V^2 \qquad (5.7)$$

showing that the potential energy of a capacitor is a quadratic function of its charge or the potential difference.

5.3.3 Determination of a Capacitance

In order to determine the capacitance of a capacitor under total influence, the following general procedure can be applied:

1. Assume a charge Q on one of the conductors, and therefore a charge $-Q$ on the other one.
2. The electric field $\mathbf{E}$ between the conductors can be then determined. For instance, the superposition principle can be applied.
3. The potential difference is given by

$$\Delta V = V_1 - V_2 = \int_1^2 \mathbf{E} \cdot d\mathbf{l} .$$

The most appropriate path to calculate this line integral will depend on the symmetries of the electric field.
4. Finally, the capacitance is given by $C = Q/\Delta V$, which is independent of charge Q or potential difference $V_1 - V_2$.

Example 5.2 The parallel plate capacitor
Consider a capacitor made of two parallel plates of area A separated by a distance $d \ll A$, as shown in Fig. 5.2. In order to get rid of edge effects, we will consider that both plates are of infinite extension (this is a good approximation at any point between the plates away from the edges since $d \ll A$).

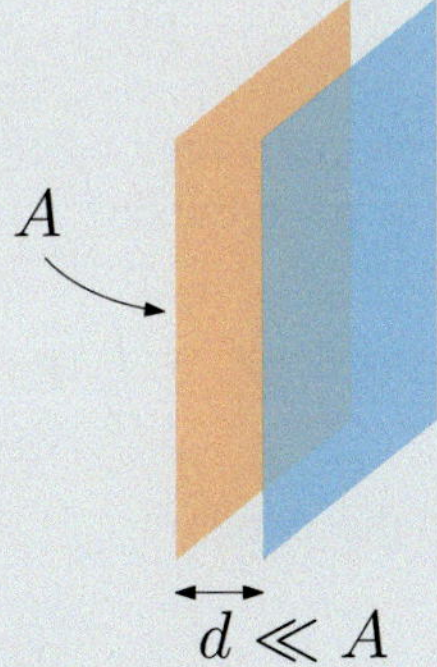

Fig. 5.2 Two conductive parallel plates separated by a distance d

Suppose that the left and right plates are homogeneously charged with surface densities σ and $-\sigma$, respectively. The x-axis is chosen perpendicular to the plates with the origin O at the center of the left plate, as shown in Fig. 5.3.

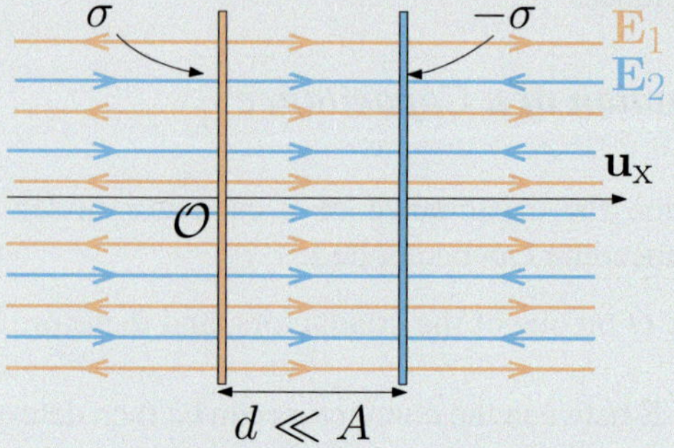

Fig. 5.3 Electric field lines generated by each of the two infinite plates

The electric field will be the superposition of the fields $\mathbf{E}_1$ (left plate) and $\mathbf{E}_2$ (right plate) generated by each plate separately. Recalling the electric field generated by an infinite plane carrying a uniform surface charge density (Example 2.5), we have

$$\mathbf{E}_1(\mathbf{x}) = \frac{\sigma}{2\epsilon_0} \frac{x}{|x|} \mathbf{u}_x ,$$

$$\mathbf{E}_2(\mathbf{x}) = \frac{-\sigma}{2\epsilon_0} \frac{x-d}{|x-d|} \mathbf{u}_x .$$

The superposition of both fields vanishes everywhere except for the region between the plates, so that the plates are under total influence and the field between them is given by

$$\mathbf{E}(\mathbf{x}) = \mathbf{E}_1(\mathbf{x}) + \mathbf{E}_2(\mathbf{x}) = \begin{cases} \frac{\sigma}{\epsilon_0} \mathbf{u}_x & \text{if } 0 < x < d, \\ \mathbf{0} & \text{otherwise} . \end{cases}$$

The field lines are thus confined in the region between the plates and so is the electrostatic energy density. The potential difference between the plates writes

$$\Delta V = V_1 - V_2 = \int_0^d \mathbf{E} \cdot d\mathbf{x} .$$

By choosing a path parallel to the x-axis,

$$\Delta V = \int_0^d \frac{\sigma}{\epsilon_0} dx = \frac{\sigma d}{\epsilon_0} .$$

If the area of each plate is A, then the total charge on the left plate is given by $Q = A\sigma$ and the capacitance reads

$$C = \frac{Q}{\Delta V} = \frac{\epsilon_0 A}{d} .$$

For a capacitor made of square plates of lateral size of $10\,\text{cm}$, we have $A = 1 \times 10^{-2}\text{m}^2$, and supposing a separation of $d = 1\,\text{cm}$, the capacitance is $C = 9 \times 10^{-12}\text{F}$. If a potential difference of $\Delta V = 1.5\,\text{V}$ is applied, the charge on the plates will be $Q = C\Delta V = 1.5 \times 9 \times 10^{-12}\text{C} = 1.35 \times 10^{-11}\text{C}$ and the electrostatic energy stored is $U = \frac{1}{2}Q\Delta V = 1.01 \times 10^{-12}\text{J}$.

Example 5.3 Cylindrical capacitor
Consider a capacitor made of two cylindrical and concentric conductors, of radius a and $b > a$, and length $L \gg (b - a)$. The capacitance may be obtained by supposing that the inner conductor carries a charge Q and the outer one $-Q$. Considering the two cylinders as virtually infinite, we can assume that the charge distribution has cylindrical symmetry and the electric field writes, in cylindrical coordinates, $\mathbf{E}(\mathbf{x}) = E(r)\mathbf{u}_r$. This is true far from the edges of the capacitor. Choosing as a closed Gauss surface S a cylinder of radius r, with $a < r < b$, concentric to the conductors and of length h (see Fig. 5.4), Gauss's law writes

$$\oiint_S \mathbf{E}(\mathbf{x}') \cdot \mathbf{n}(\mathbf{x}')dS(\mathbf{x}') = E(r)2\pi r h = \frac{Q(S)}{\epsilon_0} .$$

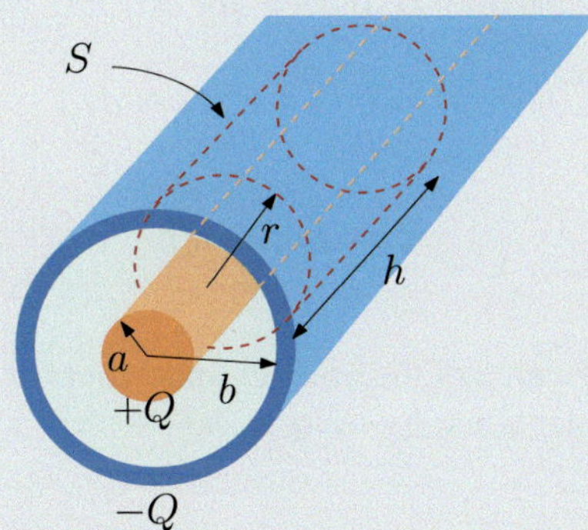

Fig. 5.4 Two cylindrical concentric conductors of radius a and b

The charge $Q(S)$ inside the surface S is $h\lambda = hQ/L$, and so

$$E(r)2\pi r h = \frac{Qh}{\epsilon_0 L} \ .$$

Hence, the electric field writes

$$\mathbf{E}(\mathbf{x}) = \frac{Q}{2\pi \epsilon_0 r L}\mathbf{u}_r \ .$$

Note that for $r < a$ and $r > b$ the charge $Q(S)$ inside S is zero; the electric field is confined between the conductors. The potential difference is

$$V(a) - V(b) = \int_a^b \mathbf{E} \cdot d\mathbf{l} = \int_a^b E(r)dr \ ,$$

where a radial path has been used ($d\mathbf{l} = dr\,\mathbf{u}_r$). Then,

$$\Delta V = \frac{Q}{2\pi \epsilon_0 L} \int_a^b \frac{dr}{r} = \frac{Q}{2\pi \epsilon_0 L} \ln\left(\frac{b}{a}\right) \ .$$

Finally, the capacitance is

$$C = \frac{Q}{\Delta V} = \frac{2\pi \epsilon_0 L}{\ln\left(\dfrac{b}{a}\right)} \ .$$

Note that in the limit $d = b - a \ll a, b$, we should retrieve the result obtained for the parallel plate capacitor (2). Indeed, for $d \ll a$,

$$\ln(b/a) = \ln(1 + d/a) \approx d/a$$

and so,

$$C \approx \frac{2\pi \epsilon_0 L a}{d} = \frac{\epsilon_0 A}{d} \ ,$$

where $A = 2\pi L a$ is the area of the inner cylinder, approximately equal to the area of the outer cylinder when $b - a \ll a, b$.

Example 5.4 Touchscreens

Capacitive touchscreens are often found in tablets and smartphones. Many different technologies exist, but the principle relies on the fact that the human body is a relatively good electrical conductor. One way to build a touchscreen is to put, above the display panel, a layer of thin (and transparent) electrodes covered by a layer of glass. When a human finger touches the screen, it has an impact on the capacitance measured between two thin electrodes. As shown in Fig. 5.5, the finger draws some electric field lines from the left electrode and this decreases the capacitance between this electrode and another *sensor* electrode.

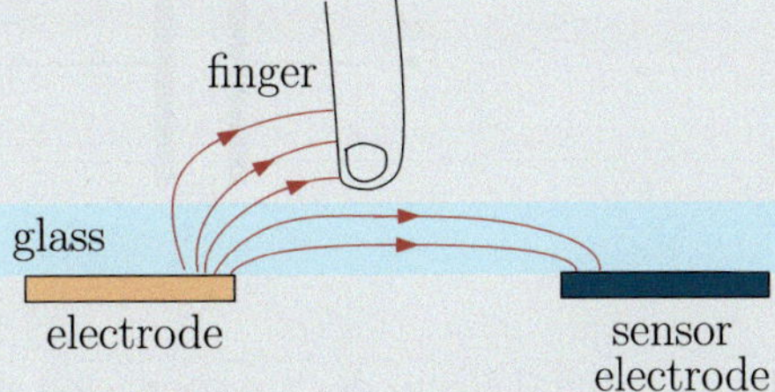

Fig. 5.5 A finger changes the capacitance between two electrodes underneath the glass film

The position of the finger when it touches the screen can be retrieved, for example, by using sensor electrodes at the four corners of the screen. More accurate positioning can be achieved by the use of a grid of sensor wires. Of course, the same principles can be used to construct simple proximity sensors. Due to this working principle, some of these screens will not work at all if the user is wearing insulating gloves, for instance during a cold day.

5.3.4 *Connection of Two or More Capacitors*

In electronics, a capacitor is commonly represented by the symbol shown in Fig. 5.6 on the left, which depicts two parallel plates connected to two wires. The plates are used to represent the two conductors of the capacitor, also called electrodes, regardless of their actual shape and geometry. Some capacitors have specific positive and negative polarities, as will be discussed in Chap. 6, and special care must be taken since applying the reverse polarity may result in permanent damage of the capacitor. These polarized capacitors are represented by the symbol shown in Fig. 5.6 on the right.

Two capacitors may be connected together either in series, or in parallel. A series connection is shown in Fig. 5.7, where two capacitors of capacitance C_1 and C_2 are connected to one another at point c. This ensemble can be thought of as a two-terminal device, where an external source could be connected to a and b. Suppose that the system is neutral and that initially no electrode is charged. If a potential difference $\Delta V = V(a) - V(b)$ is now imposed, a charge transfer must occur between the electrode connected to a and that connected to b, leaving a charge $+Q$ on the first and

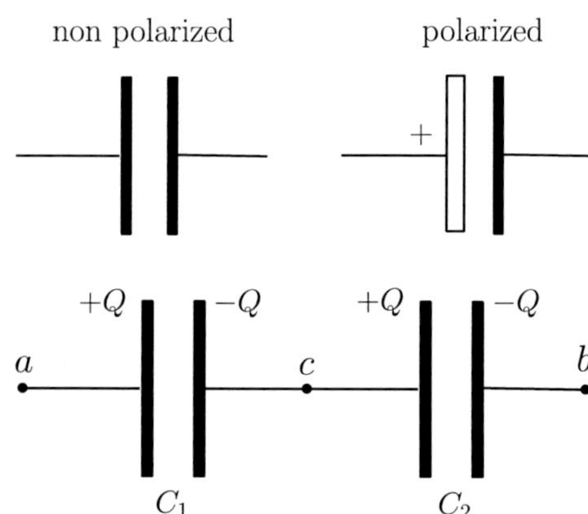

Fig. 5.6 Symbol used to represent a non polarized capacitor (left) and a polarized capacitor (right), for which the polarity is important

Fig. 5.7 Two capacitors connected in series

$-Q$ on the second. The two central electrodes are connected at c, and so they must be at the same potential $V(c)$ once equilibrium is established. For this condition to be realized, the electric field must be zero between the two central electrodes, which is achieved if their charge produce a field that neutralizes the field generated by the external electrodes. We conclude that each capacitor remains electrically neutral as shown in Fig. 5.7.

Now we can write

$$Q = C_1(V(a) - V(c)) \quad \text{and} \quad Q = C_2(V(c) - V(b)) \,,$$

that is to say,

$$\frac{Q}{C_1} + \frac{Q}{C_2} = (V(a) - V(c)) + (V(c) - V(b)) = \Delta V$$

and so,

$$Q\left(\frac{1}{C_1} + \frac{1}{C_2}\right) = \Delta V = \frac{Q}{C_{\mathrm{eq}}} \,.$$

From this we see that the two capacitors in series act as a single capacitor of equivalent capacitance C_{eq} given by

$$\frac{1}{C_{\mathrm{eq}}} = \frac{1}{C_1} + \frac{1}{C_2} \,.$$

More generally, a system of N capacitors in series is equivalent to a single capacitor of capacitance C_{eq} given by

$$\boxed{C_{\mathrm{eq}}^{-1} = \sum_{i=1}^{N} \frac{1}{C_i} \quad \text{for capacitors in series .}} \tag{5.8}$$

Consider now two capacitors C_1 and C_2 connected in parallel, as shown in Fig. 5.8. When connected to an external source that imposes a potential difference $\Delta V = V(a) - V(b)$, a charge redistribution must occur between the electrodes at potential $V(a)$ and those at $V(b)$. The two capacitors remain electrically neutral, but do not carry necessarily the same charge, which we call Q_1 and Q_2.
Here, we can write

$$Q_1 = C_1(V(a) - V(b)) \quad \text{and} \quad Q_2 = C_2(V(a) - V(b)) .$$

The total charge stored on the electrodes connected at a is $Q = Q_1 + Q_2$, so that

$$Q = C_1(V(a) - V(b)) + C_2(V(a) - V(b)) = (C_1 + C_2)\Delta V$$

and the capacitances simply add up. For a system of N capacitors in parallel, the equivalent capacitance is the sum of the individual capacitances,

$$\boxed{C_{\mathrm{eq}} = \sum_{i=1}^{N} C_i \quad \text{for capacitors in parallel .}} \tag{5.9}$$

Fig. 5.8 Two capacitors connected in parallel

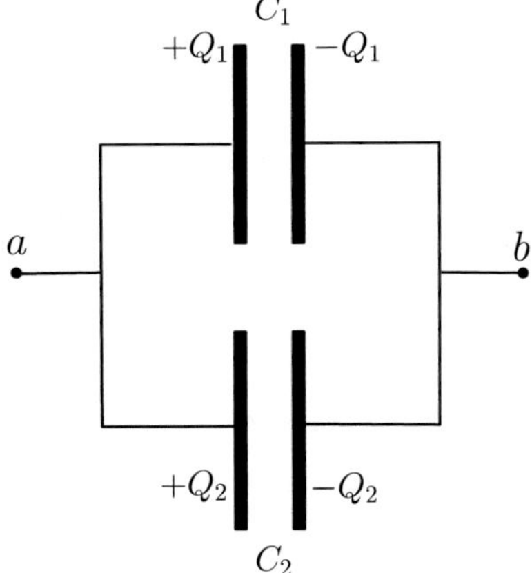

5.3.5 *Force on a Capacitor*

The two conductors of a capacitor carry opposite charges and therefore, attract each other. This electrostatic attraction is characterized by a macroscopic force $\mathbf{F}_e$ exerted by one of the conductors on the other. In order to determine this force, consider the case of a capacitor made of a fixed conductor Ω_1 and a mobile conductor Ω_2, whose center of mass is located at position $\mathbf{x}$, as shown in Fig. 5.9. Let us assume that an external operator moves in a quasi-static way the mobile conductor by a quantity $d\mathbf{x}$, that is, the displacement is infinitely slow so that all the work done by the operator to move the system corresponds to a gain (or loss) of potential energy for the conductor. This is achieved if the operator exerts a force $\mathbf{F} = -\mathbf{F}_e$ that cancels the electrostatic attraction between the conductors, thus producing a work $\delta W = -\mathbf{F}_e \cdot d\mathbf{x}$.

According to Eq. (5.7), the electrostatic energy of the system writes

$$U = \frac{1}{2} Q \Delta V = \frac{Q^2}{2C} \, ,$$

where C is the capacitance. This energy will depend on the position $\mathbf{x}$ of the mobile conductor.

Isolated system, fixed charge

If the system is isolated, there is no possible charge transfer between the conductors, so that Q is a constant independent of $\mathbf{x}$ and the electrostatic energy reads

$$U(\mathbf{x}) = \frac{1}{2} Q \Delta V(\mathbf{x}) = \frac{Q^2}{2C(\mathbf{x})} \, .$$

Writing that the variation of the system electrostatic energy equals the work of the operator, $dU = \delta W$, we find

$$dU = -\frac{Q^2}{2C(\mathbf{x})^2} \nabla C(\mathbf{x}) \cdot d\mathbf{x} = -\mathbf{F}_e \cdot d\mathbf{x}$$

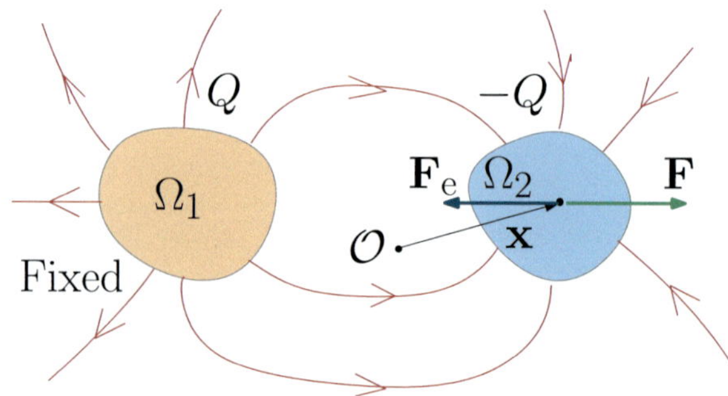

Fig. 5.9 The conductor Ω_1 is kept fixed while the conductor Ω_2 is mobile

and the electrostatic force on the mobile conductor writes

$$\mathbf{F}_e = -\,\nabla U|_Q = -\frac{Q}{2}\nabla(\Delta V(\mathbf{x})) = \frac{Q^2}{2C(\mathbf{x})^2}\nabla C(\mathbf{x})\,, \qquad (5.10)$$

where the vertical bar and subscript $|_Q$ indicates differentiation assuming constant Q.

Example 5.5 Force on a parallel plane capacitor
Consider a parallel-plate capacitor in which the plates, of area S, are charged with a charge density $\pm\sigma$. Let us calculate the electrostatic force acting on the top plate when both plates are separated by a distance x, as shown in Fig. 5.10.

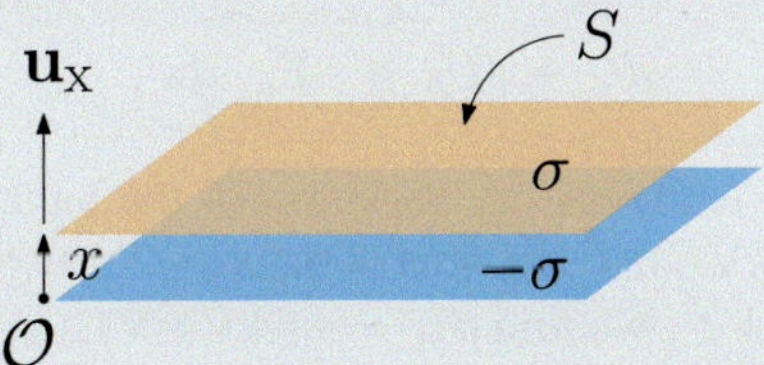

Fig. 5.10 Two conductive parallel plates separated by a variable distance x

According to Eq. (4.4), an electrostatic pressure $P_e = \dfrac{\sigma^2}{2\epsilon_0}$ acts on the top plate so that the total electrostatic force acting on it is

$$\mathbf{F}_e = -P_e S\mathbf{u}_x = -\frac{\sigma^2 S}{2\epsilon_0}\mathbf{u}_x\,.$$

On the other hand, the result can also be obtained from Eq. (5.10), the expression for the electrostatic force established for an electrically isolated system. If the top plate is at distance x from the bottom plate, this force is given by

$$\mathbf{F}_e = -\,\nabla U|_Q = \frac{Q^2}{2C(x)^2}\nabla C(\mathbf{x})$$

where $Q = \sigma S$ is the charge on the top capacitor and $C(x) = \dfrac{\epsilon_0 S}{x}$ the capacitance. We find

$$\mathbf{F}_e = \frac{\sigma^2 S^2 x^2}{2(\epsilon_0 S)^2}\nabla\left(\frac{\epsilon_0 S}{x}\right) = \frac{\sigma^2 S x^2}{2\epsilon_0}\left(-\frac{1}{x^2}\mathbf{u}_x\right) = -\frac{\sigma^2 S}{2\epsilon_0}\mathbf{u}_x\,,$$

which coincides with the result obtained from the electrostatic pressure.

Fixed potential difference

Now consider a more subtle case in which the potential difference between the conductors is kept constant. This can be achieved if the capacitor is connected to a battery or a voltage source imposing a difference ΔV between the two conductors, as shown in Fig. 5.11 (a discussion on the working principles of batteries is found in Chap. 7).

If the mobile conductor is displaced by an external operator by a quantity $d\mathbf{x}$, a charge transfer between the conductors, provided by the battery, must occur in order to keep ΔV fixed.

Hence, the two conductors alone do not constitute a closed system. When considering the energy variation of the two conductors, we must include the energy provided by the battery. The work of the external operator, $\delta W = -\mathbf{F}_e \cdot d\mathbf{x}$, is thus equal to the energy variation of the new closed system, which consists in the set of conductors and the battery. Therefore,

$$dU + dU_b = -\mathbf{F}_e \cdot d\mathbf{x} \,,$$

where $dU_b = -\Delta V dQ$ is the energy variation of the battery for a charge transfer $dQ = \Delta V dC(\mathbf{x}) = \Delta V \nabla C(\mathbf{x}) \cdot d\mathbf{x}$, that is to say, $dU_b = -(\Delta V)^2 \nabla C(\mathbf{x}) \cdot d\mathbf{x}$.

Since $U = \frac{1}{2} C(\mathbf{x})(\Delta V)^2$, we find $dU = \frac{1}{2}(\Delta V)^2 \nabla C(\mathbf{x}) \cdot d\mathbf{x}$ and finally,

$$\mathbf{F}_e = \frac{1}{2}(\Delta V)^2 \nabla C(\mathbf{x}) \,.$$

We note that using $Q = C\Delta V$ shows that the expression for the force is the same as that calculated for an isolated system with fixed charge. However, it also corresponds to the plus rather than the minus gradient of the potential energy:

$$\boxed{\mathbf{F}_e = \nabla U|_{\Delta V} = \frac{1}{2}\Delta V \nabla Q(\mathbf{x}) = \frac{(\Delta V)^2}{2}\nabla C(\mathbf{x}) \,.} \tag{5.11}$$

The reason for this asymmetry in the formulas (5.10) and (5.11) lies in the fact that here the capacitor is not a closed system. Considering instead the potential

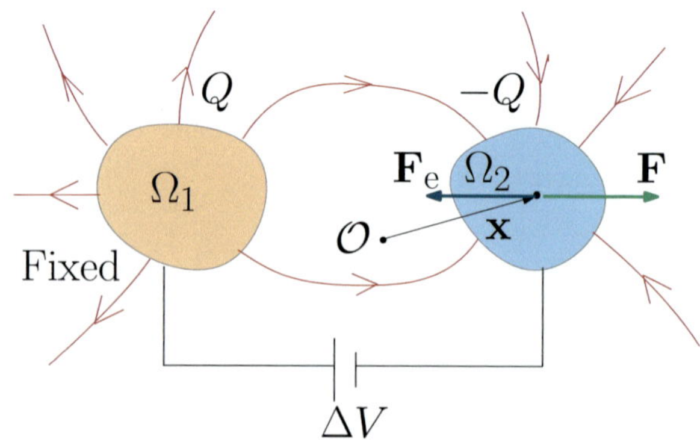

Fig. 5.11 Two conductors kept at a fixed potential difference

$\tilde{U} = U + U_b$ corresponding to the complete system {conductors and battery}, we retrieve the conventional form

$$\mathbf{F}_e = - \nabla\tilde{U}\Big|_{\Delta V} \; .$$

Example 5.6 Force between parallel planes at fixed potential
Suppose that two square, conductive and parallel plates of lateral size a are kept at a potential difference V_0, and that the left plate is free to move in the $x-z$ plane, as shown in Fig. 5.12.

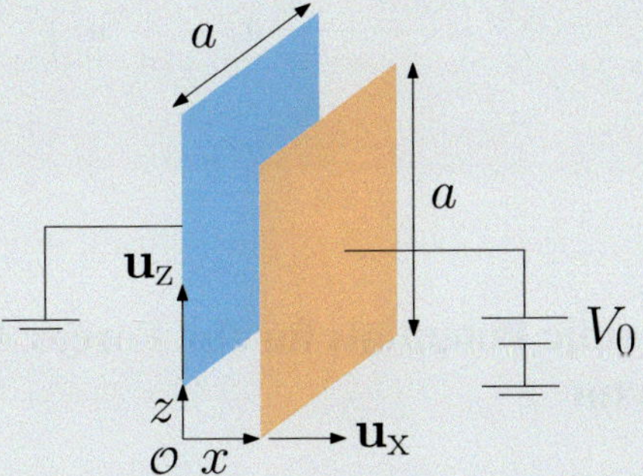

Fig. 5.12 A plane capacitor is maintained at a fixed potential difference

When the plates are at distance x and the bottom left corner of the left plate is at position $z\mathbf{u}_z$, the electrostatic energy of the system writes

$$U(x, z) = \frac{1}{2}C(x, z)V_0^2 \; .$$

Neglecting edge effects, we may assume that the electric field is homogeneous and that the field lines, confined to the region of area $a(a-z)$ between the plates, are perpendicular to both plates. This is illustrated in Fig. 5.13.

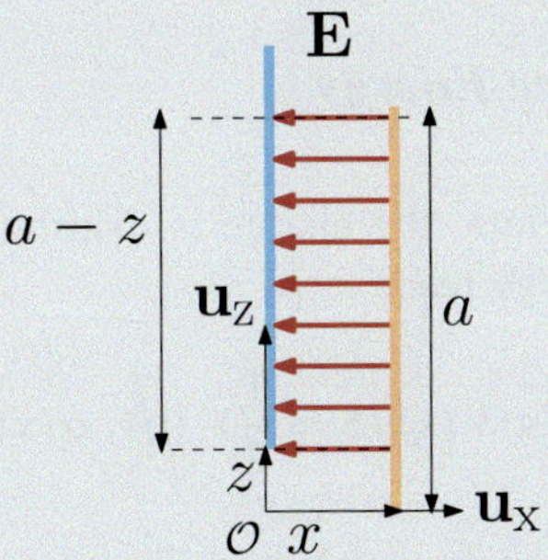

Fig. 5.13 The electric field is assumed to be uniform in the region between the plates

The capacity is then

$$C(x, z) \approx \frac{\epsilon_0 a (a - z)}{x} \, .$$

The force, opposite to the attraction force between the plates, that an external operator must exert to maintain the left plate at this position is then given by

$$\mathbf{F} = - \nabla U|_{\Delta V} = - \frac{\partial U}{\partial x} \mathbf{u}_x - \frac{\partial U}{\partial z} \mathbf{u}_z$$
$$= \frac{V_0^2}{2} \left(\frac{\epsilon_0 (a - z) a}{x^2} \mathbf{u}_x + \frac{\epsilon_0 a}{x} \mathbf{u}_z \right) \, .$$

5.4 Thermodynamic Potentials for the Forces Acting on a Conductor

In this section, we extend our analysis to the case of a set of N conductors at equilibrium, which describes common situations in circuit electricity and electronics. For instance, numerous metallic contacts are necessary in integrated circuits. We generalize the expressions for the electrostatic energy and for the force acting on a conductor in a set of conductors. As for capacitors, the force is shown to derive from an energy function in both cases of an isolated system and a system of conductors connected to batteries so that their potential is constant. The reason is that the energy function considered for each case is analogous to a *thermodynamic potential* that can be identified to a free energy: the *Helmholtz free energy* when the system is isolated, or the *Gibbs free energy* when the potentials are constant.

5.4.1 Helmholtz's Free Energy

For a conductor at equilibrium, the charge is distributed on the surface and the potential is uniform. Its energy reads

$$F = \frac{1}{2} \oiint_{\partial \Omega} \sigma(\mathbf{x}) V(\mathbf{x}) dS = \frac{1}{2} V \oiint_{\partial \Omega} \sigma(\mathbf{x}) dS = \frac{1}{2} Q V \, ,$$

where Q is the total charge on the conductor. The energy of a system of N isolated conductors at equilibrium,

$$F = \frac{1}{2} \sum_{i=1}^{N} Q_i V_i \tag{5.12}$$

simply corresponds to the constitution energy for the set of charges Q_i, $i = 1, 2, \ldots, N$, where conductor i is at potential V_i and carries the charge Q_i. This constitution energy was obtained by calculating the amount of work performed by a virtual external operator to bring every charge Q_i from a reference state where it lies at zero potential to its final state on conductor i at potential V_i. The basic assumption is that the operator exerts a force on each charge element dQ_i that opposes at any time the electrostatic interaction force due to the charges already set in their final state. In other words, the operator is assumed to work slowly enough so that the system can be constituted in a quasi-static and reversible way. In thermodynamics, this corresponds to the situation where there is no heat dissipation and the system stores all the work of the operator. For the present case, the operator's work is stored in the form of electrostatic energy that can be restored or transformed into work again. In thermodynamics, energy that is available (free) for doing some work is called *free energy*. More precisely, the change of *Helmholtz's free energy*, denoted as F (or A) quantifies the amount of reversible work that can be done on or extracted from a system at constant temperature. The work that can be obtained from the system is always smaller than or equal to the variation of the system free energy. In the case of N conductors at equilibrium, the reversible work done by the virtual operator to assemble the system in its final state from the reference state corresponds precisely to the Helmholtz free energy.

The Helmholtz free energy is a *state function*: it does only depend on the state of the system, not on its history, in keeping with the fact that Eq. (5.12) does only depend on the charges and potentials of the N conductors in their final equilibrium state, not in the way the system was assembled by the virtual operator. In this expression, the charge Q_i and the potential V_i constitute a pair of conjugate variables, similarly to temperature and entropy, or to pressure and volume in thermodynamics. Potentials and charges are not independent of each other but linked by a linear relation playing the same role as an equation of state in thermodynamics, which reads

$$Q_i = \sum_{j=1}^{N} C_{ij} V_j . \tag{5.13}$$

Quantities C_{ij} are the coefficients of the capacitance matrix, which can be inverted[1] as a matrix $(\Gamma) = (C)^{-1}$, called the elastance matrix, with coefficients Γ_{ij}:

$$V_i = \sum_{j=1}^{N} \Gamma_{ij} Q_j . \tag{5.14}$$

---

[1] If there is no charge at ∞.

The Helmholtz free energy can then be written as a function of charges Q_i, $i = 1, 2, \ldots, N$, only:

$$F = \frac{1}{2} \sum_{i=1}^{N} \sum_{j=1}^{N} Q_i \Gamma_{ij} Q_j \,, \tag{5.15}$$

or as a function of potentials V_i, $i = 1, 2, \ldots, N$, only:

$$F = \frac{1}{2} \sum_{i=1}^{N} \sum_{j=1}^{N} V_i C_{ij} V_j \,.$$

The fact that the free energy is a state function means that F admits a total differential dF that represents the external work δW_{ext} to bring the charges dQ_i, $i = 1, 2, \ldots, N$, from their reference state at zero potential to their final state at potential V_i. Since the work of the operator reads

$$\delta W = \sum_i V_i dQ_i \,,$$

we readily obtain an expression for dF:

$$dF = \sum_i V_i dQ_i, \quad \text{where } V_i = \left. \frac{\partial F}{\partial Q_i} \right|_{Q_{j,j \neq i}} \tag{5.16}$$

and the vertical bar with subscript $|_{Q_{j,j \neq i}}$ indicates differentiation with respect to Q_i assuming constant Q_j for all $j \neq i$.

A total differential satisfies the Schwarz integrability condition

$$\frac{\partial}{\partial Q_j} \left(\frac{\partial F}{\partial Q_i} \right) = \frac{\partial}{\partial Q_i} \left(\frac{\partial F}{\partial Q_j} \right) \,.$$

Using the far right part of Eq. (5.16) and identifying the coefficients of the elastance matrix, $\frac{\partial V_j}{\partial Q_i} = \Gamma_{ji}$ from Eq. (5.14), the Schwarz integrability condition shows the symmetry of the elastance matrix:

$$\Gamma_{ij} = \frac{\partial V_i}{\partial Q_j} = \frac{\partial V_j}{\partial Q_i} = \Gamma_{ji} \quad \text{for } i \neq j \,.$$

Therefore, its inverse, the capacitance matrix, is symmetric as well:

$$C_{ij} = C_{ji} \quad \text{for } i \neq j \,.$$

Now, it is clear that an expression for the free energy state function $F(Q_1, \ldots, Q_N)$ can be obtained by rewriting Eq. (5.16) using the symmetry of the elastance matrix

$$dF = \sum_{i=1}^{N}\sum_{j=1}^{N}\Gamma_{ij}Q_j dQ_i = \sum_{i=1}^{N}\Gamma_{ii}Q_i dQ_i + \sum_{i=1}^{N}\sum_{j=i+1}^{N}\Gamma_{ij}(Q_j dQ_i + Q_i dQ_j)$$

and integration yields an expression for the Helmholtz free energy

$$F(Q_1,\ldots,Q_N) = \frac{1}{2}\sum_{i=1}^{N}\left(\Gamma_{ii}Q_i^2 + 2\sum_{j>i}^{N}\Gamma_{ij}Q_i Q_j\right), \qquad (5.17)$$

which is an explicit quadratic function of charges. It gives the same numerical value as Eq. (5.12). However, Eq. (5.12) involves potentials and charges, which are conjugate variables linked by the equation of state (5.13). In thermodynamics, the Helmholtz free energy is a relevant potential to study systems at constant temperature because the temperature and volume are two independent variables for the Helmholtz free energy. Similarly, Expression (5.17), or alternatively (5.15), involves only independent charge variables and is therefore the relevant energy function to study isolated systems with invariant charges.

5.4.2 Forces on a Conductor by Means of the Free Energy

In this section, we consider an isolated system of N conductors. Our goal is to determine forces and torques that apply on each conductor. Conductor i carries a charge Q_i, is at potential V_i, has a center of mass at position $\mathbf{x}_i$ and an orientation given by Euler's angles $\boldsymbol{\alpha}_i$. All conductors can virtually move in all directions around their center of mass and change orientation. Without loss of generality, the force and torque on a conductor can be obtained by assuming a system of 2 conductors where only conductor 2 is mobile while conductor 1 represents all other fixed conductors, as shown in Fig. 5.14. Conductor 2 has 6 degrees of freedom, corresponding to three coordinates for its position $\mathbf{x} = (x, y, z)$ and three angles $\boldsymbol{\alpha} = (\alpha, \beta, \gamma)$ for its orientation.

The charges on the conductors exert the force $\mathbf{f}$ and torque $\boldsymbol{\tau}$ on the mobile conductor. If a virtual operator moves the mobile conductor in a reversible way by a quantity $d\mathbf{x}$ and changes its orientation by a quantity $d\boldsymbol{\alpha}$, thus exerting a force

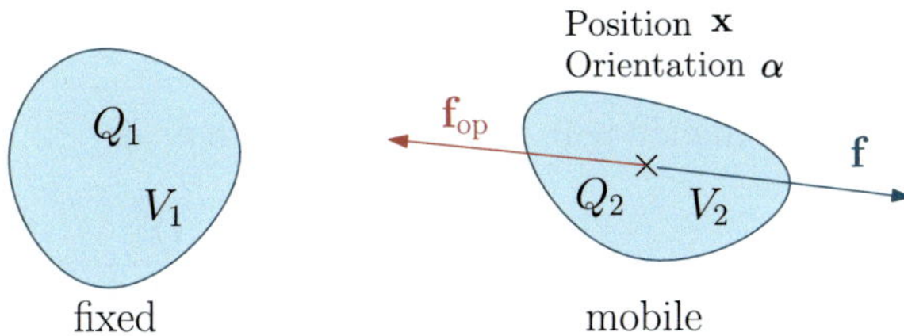

Fig. 5.14 A conductor of charge Q_2 and potential V_2 is free to move in the presence of a fixed conductor of charge Q_1 and potential V_1

$\mathbf{f}_{op} = -\mathbf{f}$ and a torque $\boldsymbol{\tau}_{op} = -\boldsymbol{\tau}$ that are opposite to the force and torque exerted by the charges, the energy received by the conductor corresponds to the work of the operator $\delta W_{op} = -\mathbf{f} \cdot d\mathbf{x} - \boldsymbol{\tau} d\boldsymbol{\alpha}$.

The change of the Helmholtz free energy of the system $F(Q_1, \ldots, Q_N, \mathbf{x}, \boldsymbol{\alpha})$, now a function of the charges and the position and orientation variables of the mobile conductor under investigation, corresponds to the work of the virtual operator and reads

$$dF = \sum_{i=1}^{N} V_i dQ_i - \mathbf{f} \cdot d\mathbf{x} - \boldsymbol{\tau} \cdot d\boldsymbol{\alpha} , \tag{5.18}$$

showing that $(\mathbf{f}, \mathbf{x})$ and $(\boldsymbol{\tau}, \boldsymbol{\alpha})$ are pairs of conjugate variables. The fact that the Helmholtz free energy is a state function then translate into the mathematical property:

$$\boxed{\mathbf{f} = -\, \nabla_{\mathbf{x}} F|_{Q_1 \ldots Q_N, \boldsymbol{\alpha}} , \qquad \boldsymbol{\tau} = -\, \nabla_{\boldsymbol{\alpha}} F|_{Q_1 \ldots Q_N, \mathbf{x}} , }$$

which correspond to expressions for the force and torque acting on the mobile conductor. In summary, to find the electrostatic force or torque acting on a mobile isolated conductor, the Helmholtz free energy of the system of conductors must be first expressed as a function of the charges on each conductor. A dependency upon the position and orientation coordinates of the mobile conductor is obtained via the dependency of the elastance (or capacitance) matrix coefficients on these variables. The force component along a given direction x is the derivative of the free energy with respect to the corresponding coordinate, holding constant all other variables:

$$f_x = -\, \frac{\partial F}{\partial x}\bigg|_{Q_1 \ldots Q_N, y, z, \boldsymbol{\alpha}} .$$

The torque around a given axis is the derivative of the free energy with respect to the corresponding angular coordinate, holding constant all other variables:

$$f_\alpha = -\, \frac{\partial F}{\partial \alpha}\bigg|_{Q_1 \ldots Q_N, \mathbf{x}, \beta, \gamma} .$$

For instance, the electrostatic force acting on the mobile plate (along the x-axis) of a plane capacitor can be obtained from the Helmholtz free energy

$$F(Q, x) = \frac{1}{2} \frac{Q^2}{C(x)} ,$$

by differentiation with respect to x, which reads

$$f_x = -\, \frac{\partial F}{\partial x}\bigg|_Q = \frac{1}{2} \frac{dC}{dx} \frac{Q^2}{C^2(x)} = \frac{1}{2} \frac{dC}{dx}(\Delta V)^2 .$$

If the two plates have a surface area S and are separated by a distance x, the capacitance is $C(x) = \epsilon_0 S/x$, leading to an attractive force $f_x = -\dfrac{1}{2}\dfrac{Q^2}{\epsilon_0 S} < 0$, as expected. A careless derivation using the expression for the Helmholtz free energy as a function of potential rather than charge, $F = C(x)(\Delta V)^2/2$, may lead to a wrong repulsive force $f_x = -\frac{1}{2}\frac{dC}{dx}(\Delta V)^2$ if one forgets that the potential difference ΔV is a function of x as well via the equation of state $\Delta V = Q/C(x)$. This emphasizes the importance of properly expressing the Helmholtz free energy as a function of charges for an isolated system (invariant charges) before performing any differentiation.

5.4.3 The Gibbs Free Energy

When the system of conductors is no longer isolated but connected to generators holding the potentials constant, as shown in Fig. 5.15, the relevant state function to calculate forces or torques acting on a mobile conductor is the *Gibbs free energy*, also called the *free enthalpy* in thermodynamics and chemistry.

The *Gibbs free energy* G is defined from the Helmholtz free energy by a common operation in classical mechanics, called a *Legendre transformation* which amounts to subtracting the product of a pair of conjugated variable. We have seen that the differential for the Helmholtz free energy reads

$$dF = \sum_{i=1}^{N} V_i dQ_i - \mathbf{f} \cdot d\mathbf{x} - \boldsymbol{\tau} \cdot d\boldsymbol{\alpha} ,$$

allowing us to identify Q_i and V_i as conjugate variables and F as a function the independent variables that represent the charge Q_i, the position $\mathbf{x}$ and the orientation $\boldsymbol{\alpha}$ for each conductor. The Legendre transformation that defines the Gibbs free energy reads

$$G(V_1, \ldots, V_N, \mathbf{x}, \boldsymbol{\alpha}) = F(Q_1, \ldots, Q_N, \mathbf{x}, \boldsymbol{\alpha}) - \sum_{i=1}^{N} Q_i V_i . \tag{5.19}$$

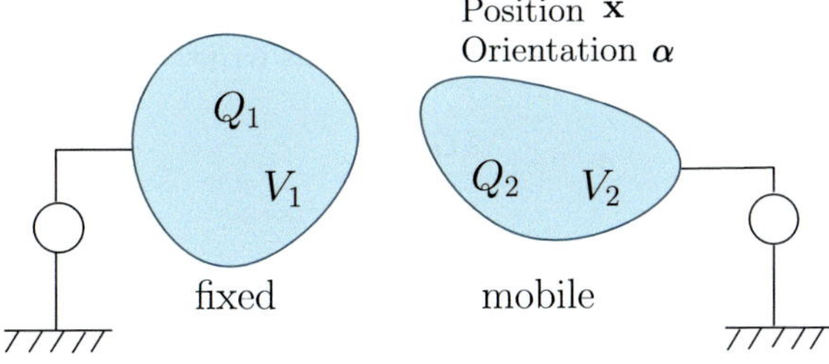

Fig. 5.15 A mobile conductor in the presence of a fixed conductor. The potential difference is kept constant

Introducing Eq. (5.12) shows that the numerical value for the Gibbs free energy of the system of N conductors at fixed potentials is just the opposite of the Helmholtz free energy of the system of isolated conductors:

$$G(V_1, \ldots, V_N, \mathbf{x}, \boldsymbol{\alpha}) = -\frac{1}{2} \sum_{i=1}^{N} Q_i V_i \ ,$$

which can also be expressed as a function of potentials using Eq. (5.13) and the capacitance matrix:

$$G(V_1, \ldots, V_N, \mathbf{x}, \boldsymbol{\alpha}) = -\frac{1}{2} \sum_{i=1}^{N} \sum_{i=1}^{N} V_i C_{ij}(\mathbf{x}, \boldsymbol{\alpha}) V_j \ . \tag{5.20}$$

An equivalent expression using the symmetry of the capacitance matrix reads

$$G(V_1, \ldots, V_N, \mathbf{x}, \boldsymbol{\alpha}) = -\frac{1}{2} \sum_{i=1}^{N} \left(C_{ii}(\mathbf{x}, \boldsymbol{\alpha}) V_i^2 + 2 \sum_{j>i}^{N} C_{ij}(\mathbf{x}, \boldsymbol{\alpha}) V_i V_j \right) \ . \tag{5.21}$$

It is a matter of calculus to obtain the differential of the Gibbs free energy by differentiating Eq. (5.19):

$$dG = dF - \sum_{i=1}^{N} (Q_i dV_i + V_i dQ_i)$$

$$= - \sum_{i=1}^{N} Q_i dV_i - \mathbf{f} \cdot d\mathbf{x} - \boldsymbol{\tau} \cdot d\boldsymbol{\alpha} \ .$$

The above expression makes it clear that independent variables for the Gibbs free energy are the potentials, the positions, and orientations of each conductor. In addition, we can derive the force and torque applied to a conductor at position $\mathbf{x}$ and with orientation $\boldsymbol{\alpha}$ from an expression for the Gibbs free energy as a function of its independent variables as in Eq. (5.20).

$$\boxed{\mathbf{f} = -\nabla_{\mathbf{x}} G|_{V_1 \ldots V_N, \boldsymbol{\alpha}}, \qquad \boldsymbol{\tau} = -\nabla_{\boldsymbol{\alpha}} G|_{V_1 \ldots V_N, \mathbf{x}} \ .}$$

Component-wise, these expressions write

$$f_x = -\frac{\partial G}{\partial x}\bigg|_{V_1, \ldots, V_N, y, z, \alpha} \ , \qquad \tau_\alpha = -\frac{\partial G}{\partial \alpha}\bigg|_{V_1, \ldots, V_N, \mathbf{x}, \beta, \gamma} \ .$$

For example, the Gibbs free energy for a plane plate capacitor connected to a battery with distance x between its plates writes

$$G(\Delta V, x) = -\frac{1}{2}C(x)(\Delta V)^2$$

with $C(x) = \epsilon_0 S/x$. From this expression, we can derive the component along the x-axis of the force acting on the mobile plate:

$$f_x = -\left.\frac{\partial G}{\partial x}\right|_V = \frac{1}{2}\frac{\partial C}{\partial x}(\Delta V)^2 = -\frac{1}{2}\frac{\epsilon_0 S}{x^2}(\Delta V)^2 \,.$$

The final result is the same as that obtained using the Helmholtz free energy; the force is attractive as it must be. The reason for the difference between F and G is due to the fact that the system of conductors connected to batteries is no longer a closed system. Every conductor receives charges from the battery to which it is connected. Now the total energy in the system of conductors and batteries is preserved. The Gibbs free energy accounts for the free energy that is available in the system {conductors and batteries}.

5.5 Summary and Essential Formulas

- In a system of N conductors in electrostatic equilibrium, the charge Q_i at the surface of the conductor i depends linearly on the potentials of the N conductors. One can therefore write

$$Q_i = \sum_{j=1}^{N} C_{ij} V_j \,,$$

  where $C_{ij} = \frac{\partial Q_i}{\partial V_j}$ are the coefficients of the capacitance matrix, which is symmetric: $C_{ij} = C_{ji}$. The self-capacitance of conductor i corresponds to

$$C_{ii} = \frac{\partial Q_i}{\partial V_i}$$

  and its unit is the farad ($1\,\text{F} = 1\,\text{C}\,\text{V}^{-1}$).
- A neutral system of two conductors ($N = 2$ and $Q_1 = -Q_2 = Q$) is called capacitor, and the capacitance is defined as

$$\boxed{C = \frac{Q}{\Delta V}}$$

  where ΔV is the potential difference between the conductors. The electrostatic energy associated with such a system is given by

$$U = \frac{1}{2}Q\Delta V = \frac{1}{2}\frac{Q^2}{C} = \frac{1}{2}C(\Delta V)^2 \,.$$

- For capacitors connected in series, the equivalent capacitance is given by

$$C_{\text{eq}}^{-1} = \sum_{i=1}^{N} \frac{1}{C_i} ,$$

  whereas for capacitors connected in parallel, the equivalent capacitance is

$$C_{\text{eq}} = \sum_{i=1}^{N} C_i .$$

- For an isolated capacitor (of constant charge), the electrostatic force acting on the conductor at position $\mathbf{x}$ reads

$$\mathbf{F}_e = \frac{Q^2}{2C(\mathbf{x})^2} \nabla C(\mathbf{x}) .$$

- For a capacitor connected to a battery that imposes a fixed potential difference ΔV, the electrostatic force acting on the conductor at position $\mathbf{x}$ is

$$\mathbf{F}_e = \frac{1}{2} \Delta V \nabla Q(\mathbf{x})$$

  and since $Q(\mathbf{x}) = C(\mathbf{x}) \Delta V$,

$$\mathbf{F}_e = \frac{1}{2} \Delta V^2 \nabla C(\mathbf{x}) = \frac{Q^2}{2 C^2} \nabla C(\mathbf{x}) .$$

- For a system of N conductors where conductor i is at potential V_i and carries a charge Q_i, the electric force and torque applied on one of the conductors located at position $\mathbf{x}$ and having an orientation given by angular coordinates $\boldsymbol{\alpha}$ can be calculated from

  – the Helmholtz free energy

$$F(Q_1, \ldots, Q_N) = \frac{1}{2} \sum_{i=1}^{N} \sum_{j=1}^{N} Q_i \Gamma_{ij}(\mathbf{x}, \boldsymbol{\alpha}) Q_j$$

  if the conductors are isolated. Quantities Γ_{ij} are the coefficients of the inverse of the capacitance matrix. The force $\mathbf{F}_e$ and torque $\boldsymbol{\tau}_e$ read

$$\mathbf{F}_e = -\nabla_{\mathbf{x}} F|_{Q_i, \boldsymbol{\alpha}} , \qquad \boldsymbol{\tau}_e = -\nabla_{\boldsymbol{\alpha}} F|_{Q_i, \mathbf{x}} ,$$

  where $\nabla_{\mathbf{x}}$ and $\nabla_{\boldsymbol{\alpha}}$ denote the gradient with respect to coordinates of $\mathbf{x}$ and $\boldsymbol{\alpha}$, respectively.

– the Gibbs free energy

$$G(V_1, \ldots, V_N) = -\frac{1}{2} \sum_{i=1}^{N} \sum_{j=1}^{N} V_i C_{ij}(\mathbf{x}, \boldsymbol{\alpha}) V_j$$

if the conductors are held at constant potential. The force $\mathbf{F}_e$ and torque $\boldsymbol{\tau}_e$ read

$$\mathbf{F}_e = -\nabla_{\mathbf{x}} G|_{V_i,\boldsymbol{\alpha}}, \qquad \boldsymbol{\tau}_e = -\nabla_{\boldsymbol{\alpha}} G|_{V_i,\mathbf{x}} .$$

Problems

5.1 Spherical capacitor

Consider a spherical capacitor made of two concentric conductive spheres of radius a and b, shown below. Determine its capacitance.

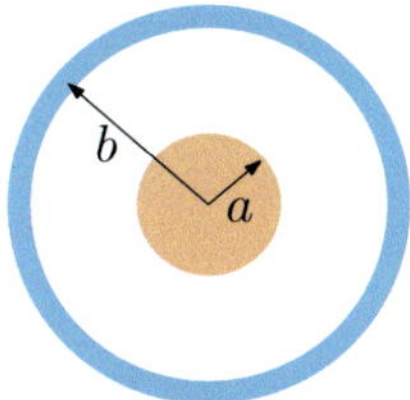

5.2 Capacitance between the Earth surface and the ionosphere

Based on the results of Exercise 4.8. What is the capacitance of the Earth-ionosphere system? Compare it to the one of a spherical capacitor with the same amount of charge.

5.3 Capacitance of three concentric spheres

Consider the capacitor formed by two concentric and conductive spheres of radius a and b. In the region between the spheres, a spherical, conductive and neutral shell is introduced, with inner radius c and outer radius d. What is the capacitance of this system?

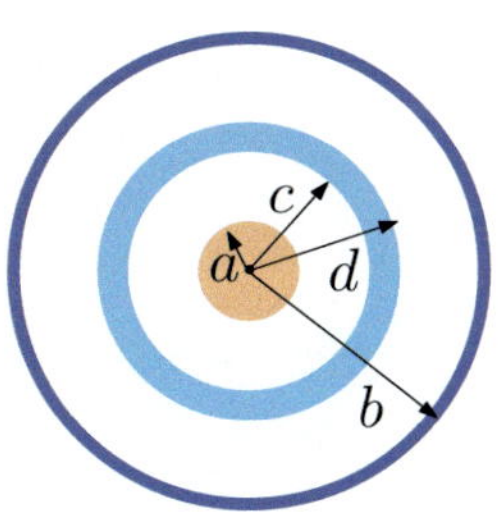

5.4 Plates at an angle

Consider two metallic plates forming an angle θ_0 as shown below. The bottom plate is grounded, whereas the top one is held at a potential V_0. Consider that the plates are long enough in the z-direction so that edge effects can be neglected.

(a) Determine the potential and the electric field in the region between the plates.
(b) Determine the capacity if the total length of the capacitor in the z-direction is L.

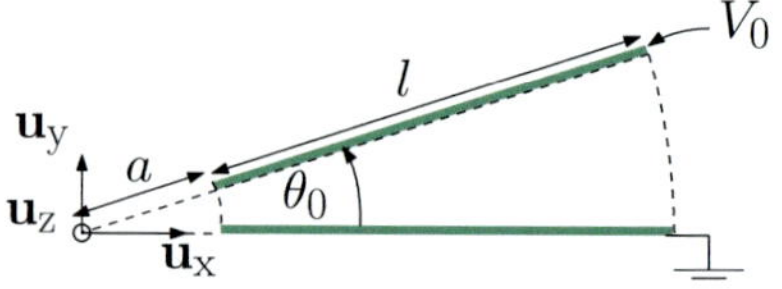

5.5 Conical capacitor

Two conductive and concentric cones are described in spherical coordinates by $\theta = \theta_1$ and $\theta = \theta_2 > \theta_1$, respectively. They are of infinite extension and separated by an infinitesimal distance at the origin O. If the inner conductor is grounded and the outer one at a potential V_0, determine the potential and the electric field between the cones.

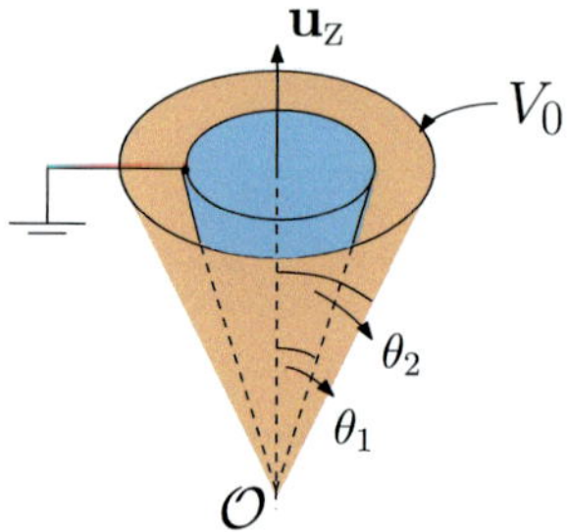

5.6 Two conductive hemispheres

Consider a sphere of radius R composed of two distinct hemispheres, shown below such that the potential at its surface is piecewise continuous and has the form

$$V(R, \theta) = \begin{cases} V_0 & \text{if } 0 \leq \theta \leq \pi/2 \\ -V_0 & \text{if } \pi/2 \leq \theta \leq \pi \end{cases}$$

Find the potential everywhere in space, as well as the surface charge density on each hemisphere. What is the capacitance of the system?

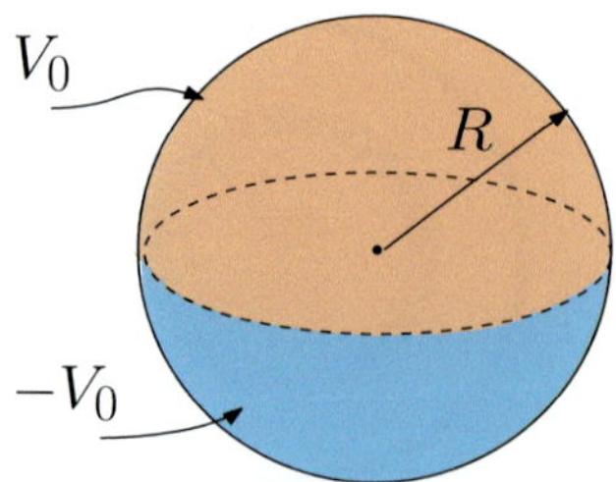

5.7 Electrostatic force from Helmholtz's and Gibb's free energies

Show that the expression for the electrostatic force acting on a conductor in a set of N conductors

$$\mathbf{F}_e = -\nabla_{\mathbf{x}} F|_{Q_1,\dots,Q_N} ,$$

where F denotes the Helmholtz free energy of the system yields the same result as its counterpart

$$\mathbf{F}_e = -\nabla_{\mathbf{x}} G|_{V_1,\dots,V_N} ,$$

where G denotes the Gibbs free energy of the system.

5.8 Separation of capacitor plates

A charge Q is placed on a capacitor of capacitance C_1. One of the terminals is connected to ground and the other is insulated at potential V_1. The separation between the plates is now increased and the capacitance becomes C_2 ($C_2 < C_1$). What happens to the potential on the isolated plate when the plates are separated? Express the potential V_2 as a function of the potential V_1.

5.9 Energy in a capacitor

(a) Consider two capacitors in series, the rigid center section of length a being mobile along the vertical axis. The area of each plate is S. Show that the capacitance of the series combination is independent of the position of the center section and is given by $C = S\epsilon_0/(b-a)$.
(b) If the voltage difference between the outside plates is held constant at V_0, what is the change in the energy stored in the capacitors if the center section is removed?

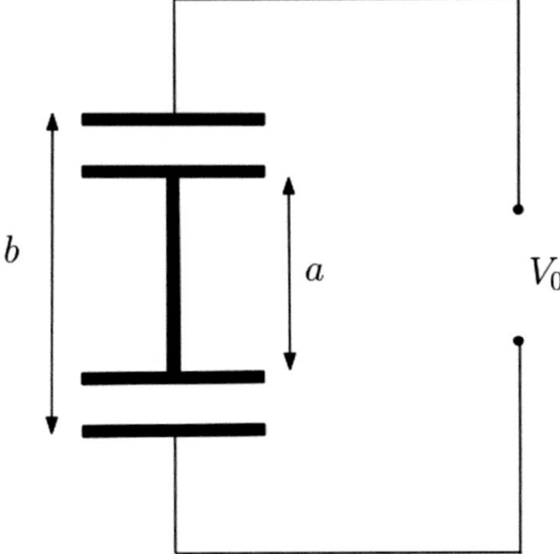

5.10 Work to separate plates of a charged capacitor

A parallel-plate capacitor is charged to a potential V and then disconnected from the charging circuit. The plates are circular with radius r. The separation between the plates is $d \ll r$. How much work is done by slowly changing the separation of the plates from d to $d' \neq d$?

5.11 System of three conductors

Three small, identical conductive spheres of radius r, are placed in a vacuum at the three vertices of an equilateral triangle. They are isolated and carry charges Q_1, Q_2 Q_3, respectively. Let a be the side of the equilateral triangle, with $a \gg r$.

(a) If we write $Q_i = \sum_{j=1}^{3} C_{ij} V_j$, with $i = 1, 2, 3$, where V_j is the potential of the sphere carrying the charge Q_j, calculate the coefficients C_{ij} as a function of r and $x \equiv r/a$. Limit Taylor expansions in x to second order.

(b) We do the following operation: the sphere A is grounded for a sufficient time for the electrostatic equilibrium to be reestablished, while the other spheres remain isolated. What is the potential of sphere A? Then, the connection to ground is cut. Express the charge Q_1' after this operation as a function of x, Q_1, Q_2, and Q_3. We do the same operation again with B, then with C once the connection to ground for sphere B is cut. What are the charges Q_1', Q_2' Q_3' of the three spheres after the third sphere has been disconnected from the ground? Give their expressions as function of x, Q_1, Q_2, and Q_3.

(c) Express the initial energy U_1 of the system, and the final energy U_2 after the three operations of question (b), in the case where $Q_1 = Q_2 = Q_3 = Q$.

(d) Calculate the values of C_{ij}, Q_1', Q_2', Q_3', U_2/U_1 in the case where $Q = 10^{-4}$ C, $a = 1$ m, $r = 5$ cm. Data: $\kappa \equiv 1/(4\pi\epsilon_0) = 9 \times 10^9$ m F^{-1}.

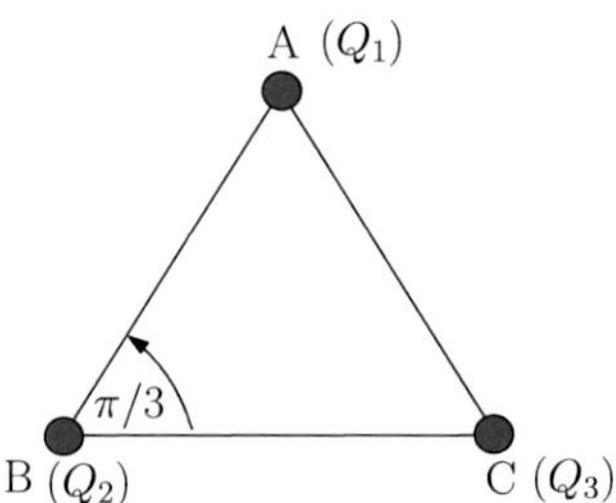

Chapter 6
The Field in Dielectric Media

Abstract This chapter explores the behavior of *dielectric materials* (insulators) in electric fields, contrasting them with conductors. It introduces the *polarization vector* (**P**) as a macroscopic quantity representing the average electric dipole moment density within a medium, which arises from the distortion or alignment of molecular dipoles. The chapter demonstrates how a polarized dielectric can be effectively modeled as a distribution of *bound charges* (volume and surface densities) that contribute to the total electric field. A key concept introduced is the *electric displacement field* (**D**), defined as $\mathbf{D} = \epsilon_0 \mathbf{E} + \mathbf{P}$. The chapter shows that sources for **D** are solely *free charges*, simplifying Gauss's law in dielectric media. It examines *linear, homogeneous, and isotropic (LHI) media*, where **P** is linearly related to **E**, leading to the definition of *electric susceptibility* (χ) and the *dielectric constant* (ϵ_r). The physical interpretation of ϵ_r is emphasized: it quantifies the reduction of the electric field within the dielectric due to the screening effect of bound charges. The chapter also discusses various *polarization mechanisms*, including electronic, ionic, and orientational polarization, and derives the *Clausius–Mossotti relation*, which connects microscopic polarizability to the macroscopic dielectric constant. Crucially, it establishes the *boundary conditions* for electric fields and displacement fields at the interface between different dielectric media. Finally, the chapter examines the impact of dielectrics on *capacitance*, demonstrating how inserting a dielectric material increases the ability of a capacitor to store charge. It also introduce the *electrostatic free energy* for polarized media and the *forces and torques* exerted on dielectrics in electric fields, providing a comprehensive understanding of their macroscopic behavior.

Keywords Dielectrics · Insulators · Dielectric constant · Polarization vector · Electrostatics

F. Cadiz and A. Couairon, *Classical Electrodynamics*, Undergraduate Texts in Physics, https://doi.org/10.1007/978-3-031-86785-9_6

6.1   Introduction

In this chapter, we examine the behavior of insulating materials, also known as dielectrics. In contrast to conductors, where electrons can move freely, electrons within insulators are strongly bound to their atoms or molecules. Since these are typically electrically neutral, one might initially assume that insulators do not significantly interact with external electrostatic fields. However, as demonstrated by Michael Faraday when he filled the space between two conductors with a dielectric, this assumption is incorrect. Molecules and atoms in dielectric materials do contribute to the total electric field experienced by a free charge within the dielectric.

We show that a dielectric material is fully characterized by its dipole moment density P, which encapsulates the material's response to an external electric field. Because dipoles generate their own electric field, they must be accounted for in determining the total electric field. In linear and isotropic media, this effect is quantified by the dielectric constant—a dimensionless quantity that renormalizes the electric field relative to the field that would exist in the absence of the dielectric.

We discuss the various mechanisms that lead to polarization in a medium and derive the boundary conditions that the electric field must satisfy at the interface between two dielectric materials. Finally, we explore how dielectrics influence capacitance, a phenomenon first observed by Faraday.

6.2   The Polarization Vector

In a polarized medium, where molecules or atoms have their dipole moments aligned in the same direction, every infinitesimal volume element possesses a corresponding infinitesimal net electric dipole moment $d\mathbf{p}$.

Polarization

The polarization vector $\mathbf{P}(\mathbf{x})$ at a point $\mathbf{x}$ in a medium represents the average electrical dipole moment density at that location. Within an infinitesimal volume d^3x surrounding $\mathbf{x}$, the total electric dipole moment $d\mathbf{p}(\mathbf{x})$ is given by

$$\boxed{\mathbf{P}(\mathbf{x})d^3x = d\mathbf{p}(\mathbf{x})\,.}$$

The polarization vector has units of cm^2.

By treating a polarized material as a continuous medium, we assume that the polarization vector $\mathbf{P}$ varies smoothly with position $\mathbf{x}$. To justify this approximation, we consider infinitesimal volume elements d^3x around each point $\mathbf{x}$, which are macroscopically small but contain a significant number of molecules. A rigorous derivation of this averaging process will be presented in Sect. 6.4.

6.2.1 *Potential Generated by a Dielectric Medium*

Consider a volume Ω of dielectric material where the polarization vector $\mathbf{P}(\mathbf{x}')$ is defined at each point $\mathbf{x}'$ within the volume, as depicted in Fig. 6.1. Each infinitesimal volume element d^3x' effectively acts as a macroscopic electric dipole with a dipole moment $d\mathbf{p}(\mathbf{x}') = \mathbf{P}(\mathbf{x}')d^3x'$. The potential at point $\mathbf{x}$, due to this volume element, is equivalent to the potential generated by a point electric dipole,

$$dV(\mathbf{x}) = \frac{d\mathbf{p}(\mathbf{x}') \cdot (\mathbf{x} - \mathbf{x}')}{4\pi\epsilon_0|\mathbf{x} - \mathbf{x}'|^3} = \frac{\mathbf{P}(\mathbf{x}') \cdot (\mathbf{x} - \mathbf{x}')d^3x'}{4\pi\epsilon_0|\mathbf{x} - \mathbf{x}'|^3}\ .$$

The total potential at point $\mathbf{x}$ is calculated by integrating the contributions from each infinitesimal volume element d^3x' throughout the entire volume Ω

$$V(\mathbf{x}) = \frac{1}{4\pi\epsilon_0} \iiint_\Omega \frac{\mathbf{P}(\mathbf{x}') \cdot (\mathbf{x} - \mathbf{x}')d^3x'}{|\mathbf{x} - \mathbf{x}'|^3}\ . \tag{6.1}$$

6.2.2 *Bound Charge Densities*

The potential given by (6.1) can be written as the potential generated by a charged volume Ω. Using the identity

$$\nabla \frac{1}{|\mathbf{x}|} = -\frac{\mathbf{x}}{|\mathbf{x}|^3}\ ,$$

we can write

$$\frac{(\mathbf{x} - \mathbf{x}')}{|\mathbf{x} - \mathbf{x}'|^3} = \nabla' \frac{1}{|\mathbf{x} - \mathbf{x}'|}\ ,$$

Fig. 6.1 An infinitesimal volume element d^3x' within a dielectric medium possesses a total dipole moment $\mathbf{P}(x')d^3x'$

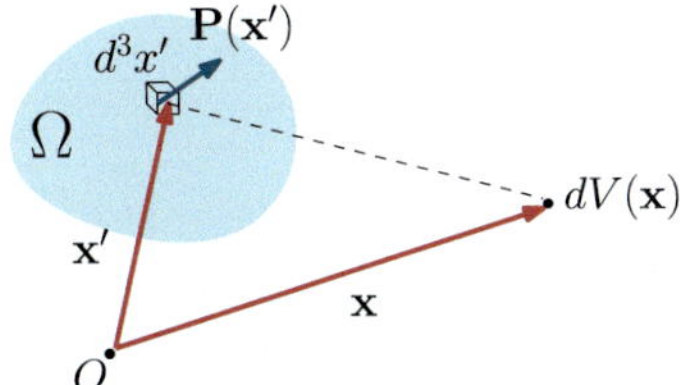

where $\mathbf{\nabla}'$ denotes the gradient operator with respect to the coordinates of $\mathbf{x}'$. The potential can then be rewritten as

$$V(\mathbf{x}) = \frac{1}{4\pi\epsilon_0} \iiint_\Omega \mathbf{P}(\mathbf{x}') \cdot \mathbf{\nabla}' \frac{1}{|\mathbf{x} - \mathbf{x}'|} d^3 x' \,.$$

Now consider the identity: $\mathbf{\nabla} \cdot (f\mathbf{F}) = f\mathbf{\nabla} \cdot \mathbf{F} + \mathbf{F} \cdot \mathbf{\nabla} f$, where f is a scalar field and $\mathbf{F}$ a vector field. Taking $f = 1/|\mathbf{x} - \mathbf{x}'|$ and $\mathbf{F} = \mathbf{P}(\mathbf{x}')$, we obtain

$$\mathbf{P}(\mathbf{x}') \cdot \mathbf{\nabla}' \frac{1}{|\mathbf{x} - \mathbf{x}'|} = \mathbf{\nabla}' \cdot \frac{\mathbf{P}(\mathbf{x}')}{|\mathbf{x} - \mathbf{x}'|} - \frac{1}{|\mathbf{x} - \mathbf{x}'|} \mathbf{\nabla}' \cdot \mathbf{P}(\mathbf{x}') \,.$$

By introducing this relation in the expression for the potential, we obtain

$$V(\mathbf{x}) = \frac{1}{4\pi\epsilon_0} \iiint_\Omega \mathbf{\nabla}' \cdot \frac{\mathbf{P}(\mathbf{x}')}{|\mathbf{x} - \mathbf{x}'|} d^3 x' - \frac{1}{4\pi\epsilon_0} \iiint_\Omega \frac{1}{|\mathbf{x} - \mathbf{x}'|} \mathbf{\nabla}' \cdot \mathbf{P}(\mathbf{x}') d^3 x' \,.$$

By applying the Green–Ostrogradsky theorem (A.15), the first volume integral in the expression for the potential can be transformed into a surface integral over the boundary of the region, $\partial\Omega$, leading to

$$\boxed{V(\mathbf{x}) = \oiint_{\partial\Omega} \frac{\mathbf{P}(\mathbf{x}') \cdot \mathbf{n}(\mathbf{x}') dS(\mathbf{x}')}{4\pi\epsilon_0 |\mathbf{x} - \mathbf{x}'|} + \iiint_\Omega \frac{-\mathbf{\nabla}' \cdot \mathbf{P}(\mathbf{x}') d^3 x'}{4\pi\epsilon_0 |\mathbf{x} - \mathbf{x}'|} \,.} \qquad (6.2)$$

The potential generated by a dielectric material, as given by (6.2), is equivalent to the potential generated by a distribution of bound charges. These bound charges have a volume density $\varrho_P(\mathbf{x}) = -\mathbf{\nabla} \cdot \mathbf{P}(\mathbf{x})$ (Cm^{-3}) within the dielectric and a surface density $\sigma_P(\mathbf{x}) = \mathbf{P}(\mathbf{x}) \cdot \mathbf{n}(\mathbf{x})$ (Cm^{-2}) on its surface. These charges arise from the polarization of the dielectric material, which can be thought of as a collection of neutral dipole pairs. In a crystal, for example, electrons are bound to the positive ions in the lattice, forming such dipoles.

Bound charge density

A polarized dielectric Ω can be seen as a medium containing a bound charge (to distinguish it from the free charge density) with volume density ϱ_P in Ω and surface density σ_P in $\partial\Omega$. The volume density of bound charges ϱ_P is a measure of the non-uniformity of $\mathbf{P}$ within the material

$$\boxed{\varrho_P(\mathbf{x}) = -\mathbf{\nabla} \cdot \mathbf{P}(\mathbf{x}) \qquad \mathbf{x} \in \Omega \,,} \qquad (6.3)$$

while the surface density of bound charges is given by the component of $\mathbf{P}$ that is normal to the surface,

$$\boxed{\sigma_P(\mathbf{x}) = \mathbf{P}(\mathbf{x}) \cdot \mathbf{n}(\mathbf{x}) \qquad \mathbf{x} \in \partial\Omega \,.} \qquad (6.4)$$

Remark

The total bound charge in a dielectric is zero, as demonstrated by the divergence theorem (A.15). This is consistent with the fact that dielectrics are globally neutral.

$$Q_P = \iiint_\Omega -\nabla' \cdot \mathbf{P}(\mathbf{x}')d^3x' + \oiint_{\partial\Omega} \mathbf{P}(\mathbf{x}') \cdot \mathbf{n}(\mathbf{x}')dS(\mathbf{x}') = 0 \,.$$

The appearance of a dipole moment at the molecular level arises from a localized nonzero charge distribution, which is reflected in the bound charge densities ϱ_P and σ_P.

In summary, the potential due to the dielectric material is

$$\boxed{V(\mathbf{x}) = \oiint_{\partial\Omega} \frac{\sigma_P(\mathbf{x}')dS(\mathbf{x}')}{4\pi\epsilon_0|\mathbf{x} - \mathbf{x}'|} + \iiint_\Omega \frac{\varrho_P(\mathbf{x}')d^3x'}{4\pi\epsilon_0|\mathbf{x} - \mathbf{x}'|} \,.} \qquad (6.5)$$

The electric field is obtained by using the identity $\mathbf{E} = -\nabla V$. Naturally, we obtain a Coulomb integral with the appropriate charge distributions:

$$\mathbf{E}_{\mathrm{d}}(\mathbf{x}) = \frac{1}{4\pi\epsilon_0} \oiint_{\partial\Omega} \frac{\sigma_P(\mathbf{x}')(\mathbf{x} - \mathbf{x}')dS(\mathbf{x}')}{|\mathbf{x} - \mathbf{x}'|^3} + \frac{1}{4\pi\epsilon_0} \iiint_\Omega \frac{\varrho_P(\mathbf{x}')(\mathbf{x} - \mathbf{x}')d^3x'}{|\mathbf{x} - \mathbf{x}'|^3} \,.$$

In a dielectric material without free charges, the electric field created by the polarization charges is known as the *depolarizing field*. If the polarization of the dielectric is induced by an external electric field, $\mathbf{E}_{\mathrm{ext}}$, the total electric field within the dielectric is the combination of the external field and the depolarizing field:

$$\mathbf{E} = \mathbf{E}_{\mathrm{ext}} + \mathbf{E}_{\mathrm{d}} \,.$$

The depolarizing field acts in the opposite direction to the external field. This counteracts the external field, resulting in a total electric field that is generally weaker than it would be in a material without polarization.

6.3 Gauss's Law in a Dielectric

Suppose that we now have a certain distribution of *free* charges ϱ_{free} submerged in a dielectric medium, that is, charges that are not already counted in as polarization charges.[1] These charges generate an electric field that will polarize the molecules in the medium surrounding them. In addition to these free charges, a polarized dielectric medium will also contribute to the total electric field via its bound charge density ϱ_P. Gauss's law remains valid in a dielectric, but we must include all the charge density in the right-hand term (free and bound) to write

$$\nabla \cdot \mathbf{E}(\mathbf{x}) = \frac{\varrho_{\text{free}}(\mathbf{x}) + \varrho_P(\mathbf{x})}{\epsilon_0} \, .$$

The bound charge density can be written in terms of the polarization vector as $\varrho_P = -\nabla \cdot \mathbf{P}$ according to (6.3), so that

$$\nabla \cdot \mathbf{E}(\mathbf{x}) = \frac{\varrho_{\text{free}}(\mathbf{x}) - \nabla \cdot \mathbf{P}(\mathbf{x})}{\epsilon_0}$$

and rearranging the terms, we obtain

$$\nabla \cdot (\epsilon_0 \mathbf{E}(\mathbf{x}) + \mathbf{P}(\mathbf{x})) = \varrho_{\text{free}} \, .$$

We see that the electric displacement field, defined as $\mathbf{D} = \epsilon_0 \mathbf{E} + \mathbf{P}$, has its sources solely due to free charges. In other words, bound charges do not directly contribute to the divergence of the electric displacement field.

Electric displacement and generalized Gauss's law in a dielectric
The electric displacement is the vector field defined as

$$\boxed{\mathbf{D}(\mathbf{x}) = \epsilon_0 \mathbf{E}(\mathbf{x}) + \mathbf{P}(\mathbf{x}) \, ,} \tag{6.6}$$

which satisfies the differential Gauss law:

$$\boxed{\nabla \cdot \mathbf{D}(\mathbf{x}) = \varrho_{\text{free}}(\mathbf{x})} \tag{6.7}$$

whose integral form is

$$\boxed{\oiint_S \mathbf{D}(\mathbf{x}') \cdot d\mathbf{S}(\mathbf{x}') = Q_{\text{free}}(S) \, .} \tag{6.8}$$

The flux of the electric displacement field $\mathbf{D}$ through any closed surface S is equal to the total free charge enclosed by that surface. The physical unit of

---

[1] The term *free* can be misleading since ϱ_{free} may not only contain charges that are free to move but also immobile charges such as ions or charged impurities that are not free to move but are not bound to any other charge of opposite sign.

the displacement vector is the same as that of the polarization vector, which is cm^{-2}.

Remarks

1. Equation (6.7) is a more convenient way of writing (2.3), because the displacement field $\mathbf{D}$ has the advantage of only conserving free charges as sources, making it easier to deal with (6.7) instead of (2.3).
2. All the information about the polarization of the medium is contained in $\mathbf{D}$. In vacuum, we have $\mathbf{P} = \mathbf{0}$, $\mathbf{D} = \epsilon_0 \mathbf{E}$ and (6.8) is equivalent to Gauss's law in vacuum.
3. There is no fundamental distinction between free and bound charges at the atomic scale. Gauss's law in a dielectric (6.7) and the distinction between ϱ_{free} and ϱ_P arises after averaging both the charge density and the resulting electric field so as to smooth their rapid variations at the atomic scale. This will be detailed in Sect. 6.4.

6.4 Gauss's Law: From Vacuum to a Dielectric

Here, we will show that the generalized Gauss law for a dielectric (6.8) can be obtained by performing a spatial average of both charge density and electric field. As discussed in the Appendix A.3, the average of a quantity f at position $\mathbf{x}$ is given by

$$\langle f(\mathbf{x}) \rangle = \iiint_{\mathbb{R}^3} f(\mathbf{x}')\varphi(\mathbf{x} - \mathbf{x}')d^3x' , \tag{6.9}$$

where φ is a function centered at the origin with a spatial extension d and unit integral. In A.3, we show that by averaging the charge density of an atom of size a over a scale $d \gg a$, we can justify the use of the Coulomb integral for a charged volume $\Omega \gg d^3$. However, this approximation is somehow *crude* in the sense that φ was assumed to be constant within the atom. We will show now that by considering the first order variations of φ at the atomic scale, we can retrieve the behavior of dielectrics.

Consider an atom centered at $\mathbf{x}_k$, consisting of N charges q_i located at positions $\mathbf{x}_i$ relative to $\mathbf{x}_k$, as illustrated in Fig. 6.2. The average charge density of this atom is given by

$$\langle \varrho_{\text{at}}(\mathbf{x}) \rangle = \sum_{i=1}^{N} q_i \varphi(\mathbf{x} - \mathbf{x}_k - \mathbf{x}_i) .$$

Instead of assuming that the weighting function φ is constant within an atom (as in A.3), we will consider a first-order approximation:

$$\varphi(\mathbf{x} - \mathbf{x}_k - \mathbf{x}_i) \approx \varphi(\mathbf{x} - \mathbf{x}_k) - \nabla\varphi(\mathbf{x} - \mathbf{x}_k) \cdot \mathbf{x}_i .$$

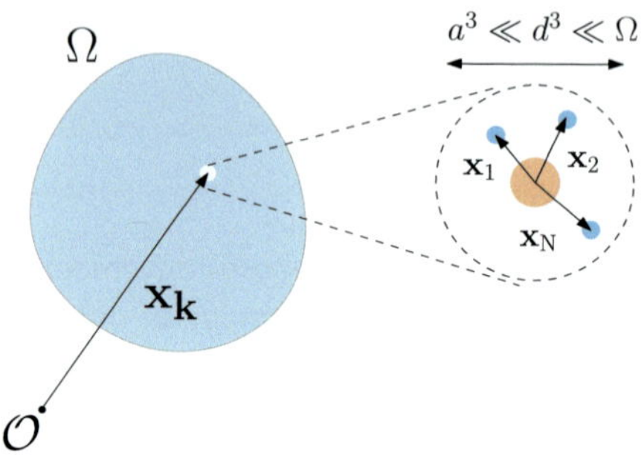

Fig. 6.2 We consider an atom at position $\mathbf{x}'$ and its surrounding electrons

This approximation leads to the average charge density of an atom:

$$\langle \varrho_{\mathrm{at}}\rangle(\mathbf{x}) = \underbrace{\left(\sum_{i=1}^{N} q_i\right)}_{Q_{\mathrm{at}}} \varphi(\mathbf{x} - \mathbf{x}_k) - \underbrace{\left(\sum_{i=1}^{N} q_i \mathbf{x}_i\right)}_{\mathbf{p}_{\mathrm{at}}} \cdot \nabla \varphi(\mathbf{x} - \mathbf{x}_k)$$

$$= Q_{\mathrm{at}}\varphi(\mathbf{x} - \mathbf{x}_k) - \mathbf{p}_{\mathrm{at}} \cdot \nabla \varphi(\mathbf{x} - \mathbf{x}_k) \,,$$

where Q_{at} and $\mathbf{p}_{\mathrm{at}}$ are the total charge and dipole moment of the atom, respectively. While neutral atoms have $Q_{\mathrm{at}} = 0$, ions or charged materials can have a nonzero total charge due to an excess or lack of electrons per atom. By summing over all atoms in the dielectric, we obtain the total average charge distribution:

$$\langle \varrho\rangle(\mathbf{x}) = \sum_{k} Q_k \varphi(\mathbf{x} - \mathbf{x}_k) - \sum_{k} \mathbf{p}_k \cdot \nabla \varphi(\mathbf{x} - \mathbf{x}_k) \,.$$

Since $\mathbf{p}_k$ is a constant vector (not a function of $\mathbf{x}$), we can apply the product rule for the divergence:

$$\mathbf{p}_k \cdot \nabla \varphi(\mathbf{x} - \mathbf{x}_k) = \nabla \cdot (\mathbf{p}_k \varphi(\mathbf{x} - \mathbf{x}_k)) \,.$$

Now let us define the polarization vector $\mathbf{P}$ as the averaged dipole moment density

$$\mathbf{P}(\mathbf{x}) = \sum_{k} \mathbf{p}_k \varphi(\mathbf{x} - \mathbf{x}_k) \,.$$

Identifying $\sum_k Q_k \varphi(\mathbf{x} - \mathbf{x}_k)$ as the free charge density (which can include immobile charges such as fixed ions or charged impurities), we obtain

$$\langle \varrho\rangle(\mathbf{x}) = \langle \varrho_{\mathrm{free}}\rangle(\mathbf{x}) - \nabla \cdot \mathbf{P}(\mathbf{x}) \,.$$

We recognize the bound charge density $\varrho_P = -\nabla \cdot \mathbf{P}$. By spatially averaging Gauss's law in vacuum, we obtain

$$\langle \mathbf{\nabla} \cdot \mathbf{E} \rangle (\mathbf{x}) = \frac{\langle \varrho(\mathbf{x}) \rangle}{\epsilon_0} = \frac{\langle \varrho_{\text{free}} \rangle (\mathbf{x}) - \mathbf{\nabla} \cdot \mathbf{P}(\mathbf{x})}{\epsilon_0} \, .$$

Since φ vanishes at infinity, we can use integration by parts (Divergence theorem with a surface at infinity) to show that

$$\langle \mathbf{\nabla} \cdot \mathbf{E}(\mathbf{x}) \rangle = \iiint_{\mathbb{R}} [\mathbf{\nabla}' \cdot \mathbf{E}(\mathbf{x}')] \varphi(\mathbf{x} - \mathbf{x}') d^3 x' = - \iiint_{\mathbb{R}} \mathbf{E}(\mathbf{x}') \cdot \mathbf{\nabla}' \varphi(\mathbf{x} - \mathbf{x}') d^3 x'$$

$$= \iiint_{\mathbb{R}} \mathbf{E}(\mathbf{x}') \cdot \mathbf{\nabla} \varphi(\mathbf{x} - \mathbf{x}') d^3 x' = \mathbf{\nabla} \cdot \left(\iiint_{\mathbb{R}} \mathbf{E}(\mathbf{x}') \varphi(\mathbf{x} - \mathbf{x}') d^3 x' \right) = \mathbf{\nabla} \cdot \langle \mathbf{E} \rangle (\mathbf{x}) \, .$$

Therefore, the spatially averaged electric field satisfies

$$\mathbf{\nabla} \cdot \langle \mathbf{E} \rangle = \frac{\langle \varrho_{\text{free}} \rangle - \mathbf{\nabla} \cdot \mathbf{P}}{\epsilon_0} \, . \tag{6.10}$$

By defining the displacement vector as $\mathbf{D} = \epsilon_0 \langle \mathbf{E} \rangle + \mathbf{P}$,
we recover Gauss's law for dielectrics:

$$\mathbf{\nabla} \cdot \mathbf{D} = \varrho_{\text{free}} \, . \tag{6.11}$$

6.4.1 Electric Susceptibility and Dielectric Constant

Consider a dielectric medium with a known distribution of free charges $\varrho_{\text{free}}(\mathbf{x})$. To determine the electric field $\mathbf{E}$ and displacement field $\mathbf{D}$ everywhere, we must solve the following equations:

$$\mathbf{\nabla} \times \mathbf{E}(\mathbf{x}) = \mathbf{0}, \tag{6.12}$$

$$\mathbf{\nabla} \cdot \mathbf{D}(\mathbf{x}) = \varrho_{\text{free}}(\mathbf{x}) \, . \tag{6.13}$$

The first equation implies that the electric field is conservative, meaning $\mathbf{E} = -\mathbf{\nabla} V$. In a vacuum, where there are no polarization charges, $\mathbf{D} \equiv \epsilon_0 \mathbf{E}$ and $\varrho(\mathbf{x}) = \varrho_{\text{free}}(\mathbf{x})$.

Therefore, the second equation simplifies to

$$\mathbf{\nabla} \cdot \epsilon_0 \mathbf{E}(\mathbf{x}) = \varrho_{\text{free}}(\mathbf{x}) \, .$$

This leads to Poisson equation

$$\nabla^2 V = -\frac{\varrho}{\epsilon_0} \, .$$

To find the scalar potential and electric field in a dielectric, we must solve Poisson equation and use the relation $\mathbf{E} = -\mathbf{\nabla} V$. However, unlike in a vacuum, the system of Eqs. (6.12) and (6.13) is incomplete. A constitutive relation is needed to establish a connection between the electric field $\mathbf{E}$ and the displacement field $\mathbf{D}$. This relation, which is specific to the dielectric material, describes how the medium responds to an applied electric field, determining the polarization of the medium. We will now examine various types of constitutive relations, expressed as $\mathbf{D}(\mathbf{E})$ or equivalently $\mathbf{P}(\mathbf{E})$. These relations characterize the material properties of different dielectrics.

6.4.2   Linear Homogeneous and Isotropic (LHI) Media

For many materials, the properties are *homogeneous*, meaning they are uniform throughout the dielectric volume and do not depend on position $\mathbf{x}$.

In many cases, the polarization density $\mathbf{P}$ is *linearly* related to the electric field. This means the displacement vector $\mathbf{D} = \epsilon_0 \mathbf{E} + \mathbf{P}$ is also linear in $\mathbf{E}$. In a Cartesian basis $\mathbf{u}_x, \mathbf{u}_y, \mathbf{u}_z$, the relationship between the components of $\mathbf{D}$ and $\mathbf{E}$ can be expressed as

$$
\begin{pmatrix} D_x \\ D_y \\ D_z \end{pmatrix} = \underbrace{\begin{bmatrix} \epsilon_{xx} & \epsilon_{xy} & \epsilon_{xz} \\ \epsilon_{yx} & \epsilon_{yy} & \epsilon_{yz} \\ \epsilon_{zx} & \epsilon_{zy} & \epsilon_{zz} \end{bmatrix}}_{\text{permittivity matrix}} \begin{pmatrix} E_x \\ E_y \\ E_z \end{pmatrix} .
$$

The matrix $[\epsilon]$ is called the *permittivity matrix*. Its nine coefficients are assumed to be independent of $\mathbf{x}$ for homogeneous media, while the electric and displacement fields can vary with position $\mathbf{x}$. We will show later that the permittivity matrix is *symmetric*, meaning $\epsilon_{xy} = \epsilon_{yx}$, $\epsilon_{xz} = \epsilon_{zx}$ and $\epsilon_{yz} = \epsilon_{zy}$. This reduces the number of independent coefficients to six. Furthermore, any symmetric matrix can be diagonalized. This means there exists a specific basis $(\mathbf{u}_1, \mathbf{u}_2, \mathbf{u}_3)$ for which the permittivity matrix is diagonal:

$$
\begin{pmatrix} D_x \\ D_y \\ D_z \end{pmatrix} = \begin{bmatrix} \epsilon_1 & 0 & 0 \\ 0 & \epsilon_2 & 0 \\ 0 & 0 & \epsilon_3 \end{bmatrix} \begin{pmatrix} E_1 \\ E_2 \\ E_3 \end{pmatrix} .
$$

The axes corresponding to this diagonal basis are called the *principal axes* of the dielectric medium. If $\epsilon_1 \neq \epsilon_2 \neq \epsilon_3$, the material is anisotropic. Birefringent crystals like calcite are examples of anisotropic materials refracting optical rays in two different directions, as shown in Fig. 6.3. This is known as double refraction.

In an isotropic medium, where all directions are equivalent, the diagonal coefficients of the permittivity matrix are equal: $\epsilon_1 = \epsilon_2 = \epsilon_3 \equiv \epsilon$. This simplifies the linear relationship between $\mathbf{D}$ and $\mathbf{E}$ to:

$$
\mathbf{D}(\mathbf{x}) = \epsilon \mathbf{E}(\mathbf{x}) ,
$$

where the permittivity ϵ is a scalar quantity that is independent of position $\mathbf{x}$ for a homogeneous dielectric.

Fig. 6.3 Double refraction
by a birefringent crystal.
Picture taken at the Museum
of Ecole Polytechnique
(Mus'X)

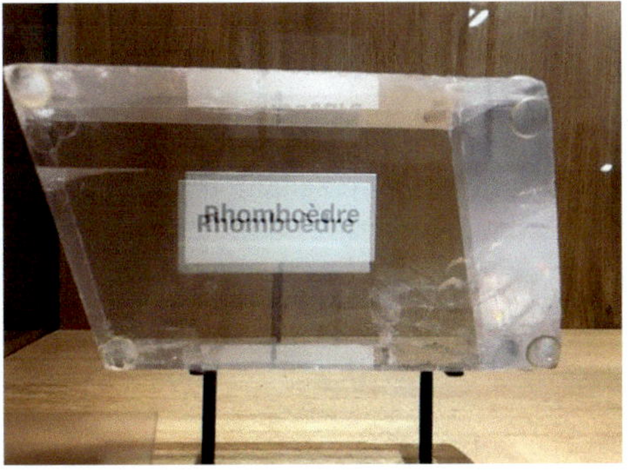

6.4.3 Electric Susceptibility and Dielectric Constant

For a linear and isotropic medium (even if it is not homogeneous), the polarization
vector **P** can be expressed as

$$\mathbf{P}(\mathbf{x}) = \epsilon_0 \chi(\mathbf{x})\mathbf{E}(\mathbf{x}) \, ,$$

where $\chi(\mathbf{x})$ is the *electric susceptibility*, a dimensionless quantity that characterizes
the response of the material to an electric field. Using the relation $\mathbf{D} = \epsilon_0 \mathbf{E} + \mathbf{P}$ from
(6.6), we can write the displacement vector as

$$\mathbf{D}(\mathbf{x}) = \epsilon_0 (1 + \chi(\mathbf{x}))\mathbf{E}(\mathbf{x}) = \epsilon(\mathbf{x})\mathbf{E}(\mathbf{x}) \, . \tag{6.14}$$

Dielectric constant
We define the *permittivity* of the medium as

$$\epsilon = \epsilon_0 (1 + \chi) \, ,$$

where ϵ_0 is the permittivity of free space and χ is the electric susceptibility. For
an isotropic and homogeneous material (where χ is constant), the permittivity
is also a constant.

The *dielectric constant* (also known as the *relative permittivity*) is defined
as the dimensionless quantity

$$\epsilon_r = \frac{\epsilon}{\epsilon_0} \, . \tag{6.15}$$

This allows us to express the displacement vector **D** as

$$\mathbf{D} = \epsilon \mathbf{E} = \epsilon_0 \epsilon_r \mathbf{E} \ .$$

Remarks

1. The dielectric constant ϵ_r is a dimensionless constant that characterizes the electrical properties of a material. It has the following properties:

   a. For *all materials*: $\epsilon_r \geq 1$.
   b. For *gases*: ϵ_r is approximately equal to 1.
   c. For *metals and conductors*: ϵ_r is much greater than 1.

2. The polarization vector can be expressed in terms of the permittivity as

$$\mathbf{P}(\mathbf{x}) = (\epsilon - \epsilon_0)\mathbf{E}(\mathbf{x}) \ .$$

In a dielectric with no free charges ($\varrho_{\text{free}} = 0$), the volume density of bound charge ϱ_P is given by

$$\varrho_P = -\nabla \cdot \mathbf{P} = -\nabla \cdot [(\epsilon - \epsilon_0)\mathbf{E}] = -\frac{\epsilon_0}{\epsilon}(\nabla \epsilon) \cdot \mathbf{E} \ .$$

This equation shows that the bound charge density is zero unless there is a gradient of the permittivity in the direction of the electric field.[2] In a homogeneous dielectric, where the permittivity is constant, the bound charge density is zero. This implies that the bound charges are confined to the surface of the dielectric.

3. The dielectric constant ϵ_r is always greater than or equal to 1. In a linear, homogeneous, and isotropic dielectric medium, the differential form of generalized Gauss's law can be written as

$$\boxed{\nabla \cdot \mathbf{E}(\mathbf{x}) = \frac{\varrho_{\text{free}}(\mathbf{x})}{\epsilon} = \frac{\varrho_{\text{free}}(\mathbf{x})}{\epsilon_r \epsilon_0} \ .} \tag{6.16}$$

This means that the laws of electrostatics in such dielectric medium are equivalent to those in a vacuum, provided we replace the vacuum permittivity ϵ_0 with $\epsilon = \epsilon_r \epsilon_0$ or, equivalently, replace the free charge density with $\varrho_{\text{free}}/\epsilon_r$.

6.4.4 *Physical Interpretation of the Dielectric Constant*

When a medium is polarized by an external electric field, it generates an internal electric field that opposes the external field. This effect reduces the total electric field by a factor of ϵ_r compared to its value in a vacuum. This occurs because the bound

---

[2] We have used Gauss's law in the absence of free charge: $\nabla \cdot \mathbf{E} = \varrho_P/\epsilon_0$.

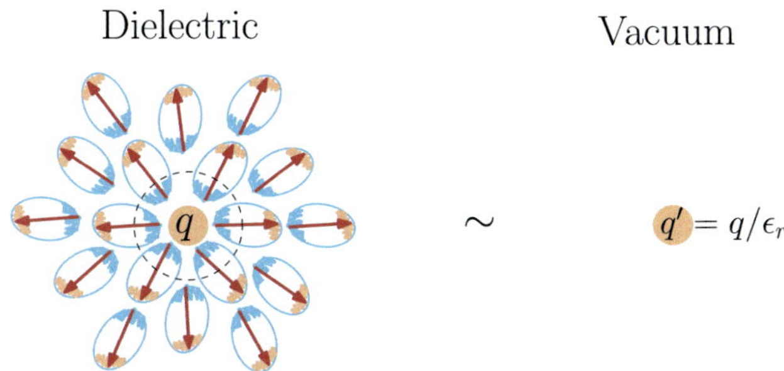

Fig. 6.4 A charge q placed inside a dielectric material is effectively screened by the polarization of the material. This screening effect can be represented by considering the charge q as an effective charge $q' = q/\epsilon_r$ in a vacuum

charges in the dielectric tend to screen the free charges. As illustrated in Fig. 6.4, a positive charge q in a dielectric is surrounded by a cloud of negative bound charges, effectively reducing its net charge. The electric field generated within the dielectric is equivalent to the field of a charge q/ϵ_r in a vacuum. A dielectric with a very high relativeenlargethispage24pt permittivity $\epsilon_r \gg 1$ behaves similarly to a conductor. This is because the electric field generated by the bound charges tends to cancel the external electric field. The electric field generated by the polarization **P** is often referred to as the *depolarizing electric field*.

Table 6.1 provides examples of dielectric constants.

Table 6.1 Examples of dielectric constants

Material	Dielectric constant	Material	Dielectric constant
Vacuum	1.0000	Quartz, fused	3.8
Air	1.0006	Wood (Maple)	4.4
PTFE, FEP (Teflon)	2.0	Glass	4.9–7.5
Polypropylene	2.20–2.28	Castor oil	5.0
ABS resin	2.4–3.2	Wood (Birch)	5.2
Polystyrene	2.45–4.0	Mica, muscovite	5.0–8.7
Waxed paper	2.5	Glass-bonded mica	6.3–9.3
Transformer oil	2.5–4	Porcelain, Steatite	6.5
Hard rubber	2.5–4.80	Alumina	8.0–10.0
Wood (Oak)	3.3	Distilled water	80.0
Silicones	3.4–4.3	Barium-strontium-titanite	7500
Bakelite	3.5–6.0		

Example 6.1 - Charged sphere in a dielectric

Consider a sphere of radius R with a uniformly distributed free charge q, immersed in a homogeneous dielectric medium with a constant relative permittivity ϵ_r. Due to the spherical symmetry of the free charge distribution and the isotropic nature of the dielectric, the bound charge distribution and the electric and displacement fields will also exhibit spherical symmetry. In spherical coordinates (with the origin at the center of the sphere), the displacement field $\mathbf{D}$ can be expressed as

$$\mathbf{D} = D(r)\mathbf{u}_r \ .$$

Applying Gauss's law for dielectrics (6.8) to a spherical surface S of radius r $(r > R)$ concentric with the sphere of charge q, we write:

$$\oiint_S \mathbf{D}(\mathbf{x}') \cdot d\mathbf{S}(\mathbf{x}') = 4\pi r^2 D(r) = q \ .$$

From this, we can determine the displacement field as

$$\mathbf{D}(r) = \frac{q}{4\pi r^2}\mathbf{u}_r \ .$$

The electric field and polarization can be easily calculated from $\mathbf{D}$:

$$\mathbf{E}(r) = \frac{\mathbf{D}(r)}{\epsilon_r \epsilon_0} = \frac{q}{4\pi \epsilon_r \epsilon_0 r^2}\mathbf{u}_r \ .$$

The dielectric medium attenuates the magnitude of the electric field by a factor of ϵ_r compared to its value in a vacuum. Using the relation $\mathbf{P} = \epsilon_0(\epsilon_r - 1)\mathbf{E}$, we can find the polarization vector:

$$\mathbf{P}(r) = \frac{q(\epsilon_r - 1)}{4\pi \epsilon_r r^2}\mathbf{u}_r \ .$$

In this case, the bound charge density $\rho_P = -\nabla \cdot \mathbf{P}$ is zero everywhere within the dielectric except at the interface with the spherical charge. This is because the polarization vector is divergence-free within the homogeneous dielectric: $\nabla \cdot \mathbf{P}(\mathbf{x}) = (\epsilon - \epsilon_0)\nabla \cdot \mathbf{E}(\mathbf{x}) = 0$. The divergence of the electric field $\mathbf{E}$ is zero due to its inverse-square dependence on distance (characteristic of Coulomb's law).

At the surface of the dielectric $(r = R)$, the surface charge density of the bound charge is

$$\sigma_P = \mathbf{P} \cdot \mathbf{n}|_{r=R} = \mathbf{P}(R) \cdot (-\mathbf{u}_r) = -\frac{(\epsilon_r - 1)q}{4\pi \epsilon_r R^2} \ .$$

The total bound charge on the surface is

$$Q_S = -4\pi R^2 \frac{(\epsilon_r - 1)q}{4\pi \epsilon_r R^2} = -\frac{(\epsilon_r - 1)}{\epsilon_r}q \; .$$

This bound charge partially shields the free charge q. Note that the dielectric remains globally neutral, so an equal and opposite bound charge $-Q_S$ appears on the outer boundary of the dielectric but does not affect the electric field outside the sphere of radius R.

To interpret the effect in terms of the depolarizing field, note that the point charge q placed in a vacuum would produce an electric field following the inverse square law:

$$\mathbf{E}_{\text{ext}} = \frac{q}{4\pi \epsilon_0 r^2}\mathbf{u}_r \; .$$

When a dielectric sphere is placed around the charge, this external field induces a polarization. The polarization charges on the surface of the sphere create a depolarizing field $\mathbf{E}_d$. Using Gauss's law, we can find that $\mathbf{E}_d$ is equal to $-\mathbf{P}/\epsilon_0$:

$$\mathbf{E}_d = \frac{Q_S}{4\pi \epsilon_0 r^2}\mathbf{u}_r \; .$$

This depolarizing field acts in the opposite direction to the field created by the point charge, effectively shielding it.

Combining the external field and the depolarizing field, and using $Q_S + q = q/\epsilon_r$, we retrieve the total electric field,

$$\mathbf{E} = \mathbf{E}_{\text{ext}} + \mathbf{E}_d = \frac{q}{4\pi \epsilon_0 \epsilon_r r^2}\mathbf{u}_r \; .$$

This equation shows that the presence of the dielectric reduces the overall electric field compared to the case without the dielectric.

6.5 The Local Field

Consider a dielectric medium with no free charges. Imagine a single particle (e.g., an electron) within an atom or molecule of this dielectric. We will study the electrostatic influence of all other charges (bound and free) on this particle. The electric field generated at the particle location $\mathbf{x}$ by all charges except the particle itself is called the local field $\mathbf{E}_l$. If the polarization is uniform throughout the dielectric region Ω, there is no volume polarization charge ($\varrho_P = 0$). In this case, the macroscopic

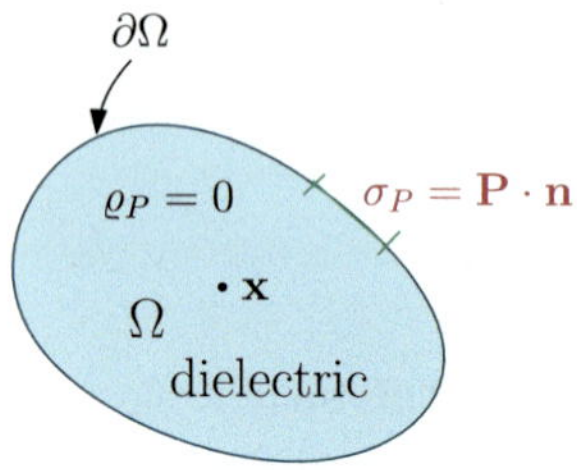

Fig. 6.5 A dielectric medium Ω with no volume polarization charge

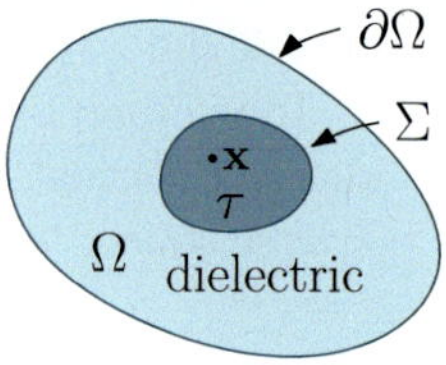

Fig. 6.6 A small volume τ centered around point **x**

depolarizing field $\mathbf{E}_d$ induced by $\mathbf{P}$ is solely due to the surface charges σ_p on the boundary $\partial\Omega$ (as illustrated in Fig. 6.5).

To define the local field acting on the particle at **x**, consider a small volume τ centered around **x**. This volume should be significantly larger than the average distance between particles, atoms, or molecules (typically around 1×10^{-10} m), as illustrated in Fig. 6.6.

The local electric field at **x** consists of two components:

- *The external field* $\mathbf{E}_{\text{ext}}(\mathbf{x})$ generated by charges outside the dielectric domain Ω.
- *The internal field* generated by the bound charges within the domain Ω, excluding the particle at **x**.

These two components together contribute to the total local electric field at the particle location. To evaluate the contribution of the internal field to the local electric field, we apply the superposition principle to the domain Ω. We can divide Ω into two regions:

1. A small volume τ centered around point **x**.
2. The remaining domain Ω' bounded by the inner surface Σ of τ and the outer surface $\partial\Omega$ of Ω.

We can calculate the electric field at **x** due to the charges in these two regions separately.

The domain Ω' is a part of Ω and is therefore also uniformly polarized. It does not contribute to the volume polarization charge ($\varrho_P = 0$). The surface charges on the boundaries $\partial\Omega$ and Σ do contribute to the electric field at **x**. These surface charges are due to the polarization of the dielectric.

The local field at **x** is the sum of the following contributions:

1. *External field*: $\mathbf{E}_{\text{ext}}(\mathbf{x})$ generated by charges outside the dielectric domain Ω.

2. Field from the outer surface: $\mathbf{E}_{\partial\Omega}(\mathbf{x})$ generated by the surface charges on the outer boundary $\partial\Omega$.
3. Field from the inner surface: $\mathbf{E}_{\Sigma}(\mathbf{x})$ generated by the surface charges on the inner boundary Σ.
4. Field from the volume: $\mathbf{E}_{\tau}(\mathbf{x})$ generated by the charges within the small volume τ around $\mathbf{x}$.

In summary, the local field at $\mathbf{x}$ is given by

$$\mathbf{E}_l(\mathbf{x}) = \mathbf{E}_{\text{ext}}(\mathbf{x}) + \mathbf{E}_{\partial\Omega}(\mathbf{x}) + \mathbf{E}_{\Sigma}(\mathbf{x}) + \mathbf{E}_{\tau}(\mathbf{x}) \ .$$

6.5.1 Calculation of $\mathbf{E}_{\partial\Omega}(\mathbf{x})$

The electric field at $\mathbf{x}$ generated by the polarization charges on $\partial\Omega$ can be calculated using the following integral:

$$\mathbf{E}_{\partial\Omega}(\mathbf{x}) = \oiint_{\partial\Omega} \frac{\sigma_P(\mathbf{x}')(\mathbf{x} - \mathbf{x}')}{4\pi\epsilon_0 |\mathbf{x} - \mathbf{x}'|^3} dS = \oiint_{\partial\Omega} \frac{\mathbf{P}\cdot\mathbf{n}\,(\mathbf{x}')(\mathbf{x} - \mathbf{x}')}{4\pi\epsilon_0 |\mathbf{x} - \mathbf{x}'|^3} dS \ .$$

However, an actual integration is not necessary: since the polarization is uniform in Ω, we have $\nabla \cdot \mathbf{P} = 0$. This implies that there is no volume polarization charge, and therefore, the depolarizing field $\mathbf{E}_d$ created by polarization charges in Ω is solely due to the surface polarization charges on $\partial\Omega$. In other words, the field $\mathbf{E}_{\partial\Omega}(\mathbf{x})$ is identical to the depolarizing field $\mathbf{E}_d()$.

$$\mathbf{E}_{\partial\Omega}(\mathbf{x}) = \mathbf{E}_d(\mathbf{x}) \ .$$

6.5.2 Evaluation of $\mathbf{E}_{\tau}(\mathbf{x})$

To evaluate the electric field $\mathbf{E}_{\tau}(\mathbf{x})$ generated by the charges within the small volume τ, we need to exclude the particle at $\mathbf{x}$ itself. Note that the surface polarization charges on the boundary Σ of τ contribute to the field $\mathbf{E}_{\Sigma}(\mathbf{x})$, not to the field $\mathbf{E}_{\tau}(\mathbf{x})$.

We will limit our calculations to the following two cases:

1. *Polar Liquid*: In a polar liquid, the medium is isotropic. We choose the volume τ to be a sphere centered at $\mathbf{x}$, as shown in Fig. 6.7 (left). Due to the spherical symmetry, the field generated by the charges within τ at the center $\mathbf{x}$ is zero.
2. *Isotropic Cubic Crystal*: In an isotropic cubic crystal, we choose the surface Σ to be a cube centered at $\mathbf{x}$, with faces parallel to the crystal lattice, as shown in Fig. 6.7 (right). The cubic symmetry of the crystal lattice ensures that the field at $\mathbf{x}$ is also zero in this case.

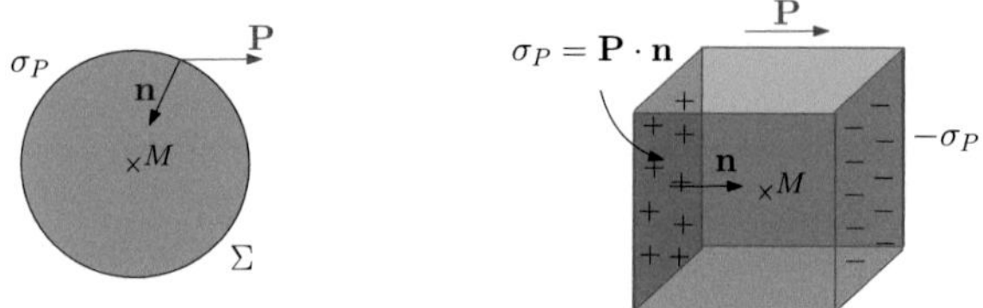

Fig. 6.7 Within an isotropic medium, Σ can be chosen to be a sphere centered at $\mathbf{x}$ (left) or a cube centered at $\mathbf{x}$, aligned with the crystal lattice of a cubic material (right)

Therefore, in both cases (polar liquid and isotropic cubic crystal), the charges within the small volume τ do not contribute to the local field at $\mathbf{x}$:

$$\mathbf{E}_\tau(\mathbf{x}) = \mathbf{0} .$$

6.5.3   Calculation of $\mathbf{E}_\Sigma(\mathbf{x})$

Consider the two cases where Σ is either a sphere or a cube centered at $\mathbf{x}$. In both cases, we will evaluate the electric field $\mathbf{E}_\Sigma(\mathbf{x})$ generated at $\mathbf{x}$ by the polarization charge lying on the surface Σ.

For the sphere, since the polarization is uniform, the surface polarization charge on the spherical surface is $\sigma_P = \mathbf{P} \cdot \mathbf{n}$, where $\mathbf{n}$ is the unit normal vector to the sphere pointing outward with respect to the dielectric or equivalently, inward with respect to the sphere. In spherical coordinates with the z-axis aligned with $\mathbf{P}$, this becomes $\sigma_P = -P \cos\theta$. Due to the symmetry of the charge distribution, any plane containing the z-axis is a plane of symmetry. This means that the electric field at the center of the sphere, $\mathbf{x}$, must lie along the z-axis. To calculate the electric field, we can divide the sphere into a series of thin rings centered on the z-axis. Each ring has a polar angle range from θ to $\theta + d\theta$. The surface area of a ring is $d\Sigma = 2\pi r^2 \sin\theta d\theta$, and the charge on this element is $dq = \sigma_P dS$. The electric field contribution from each ring element at the center is directed along $-\mathbf{u}_r$. Projecting this onto the z-axis, we find that the z-component of the field element is

$$dE_z = \frac{dq}{4\pi\epsilon_0 r^2}(-\mathbf{u}_r \cdot \mathbf{u}_z) = \frac{P\cos^2\theta \sin\theta d\theta}{2\epsilon_0} .$$

Integrating this expression over all polar angles from 0 to π gives us the total z-component of the electric field at the center:

$$E_z = \frac{P}{2\epsilon_0}\int_0^\pi \cos^2\theta \sin\theta d\theta = -\frac{P}{2\epsilon_0}\left[\frac{\cos^3\theta}{3}\right]_0^\pi = \frac{P}{3\epsilon_0} .$$

Therefore, the electric field at the center of the sphere due to the surface charge distribution is

$$\mathbf{E}_\Sigma(\mathbf{x}) = \frac{\mathbf{P}}{3\epsilon_0} \ .$$

If the surface Σ is a cube with a uniform polarization vector $\mathbf{P}$, the four faces parallel to $\mathbf{P}$ will have no surface polarization charge. The remaining two faces will have surface charges of $\sigma_P = \pm P$, depending on their orientation. To calculate the electric field at the center of the cube, we can focus on the two charged faces. The field at the center will be along the z-axis, which connects the centers of these faces.

Consider a single face, say the left face with charge density $\sigma_P = P$. A small surface element $d\Sigma$ on this face contributes to the electric field at the center. The z-component of this field is

$$dE_z = d\mathbf{E} \cdot \mathbf{u}_z = \frac{\sigma_0 d\Sigma}{4\pi\epsilon_0 r^2}\mathbf{u} \ .$$

Here, r is the distance from the surface element to the center, and $\mathbf{u}$ is the unit vector pointing from the element to the center. We can simplify this expression by noting that $\dfrac{d\Sigma}{r^2}\mathbf{u} \cdot \mathbf{u}_z$ is equal to the solid angle $d\Omega$ subtended by $d\Sigma$ at the center. Therefore:

$$dE_z = \frac{\sigma_P}{4\pi\epsilon_0}d\Omega \ .$$

The solid angle subtended by the entire left face at the center is $\Omega = \dfrac{4\pi}{6}$. Integrating over the face, we find the total z-component of the electric field due to this face:

$$E_z = \frac{P}{4\pi\epsilon_0}\Omega \ .$$

As for the negatively charged plate of a capacitor, the right face contributes to the z-component of the electric field in the same direction, even though it has the opposite charge. So, the total electric field at the center due to both faces is

$$\mathbf{E} = 2 \times \frac{P}{4\pi\epsilon_0}\Omega\,\mathbf{u}_z = \frac{P}{3\epsilon_0}\mathbf{u}_z \ .$$

The electric field at the center of the cube due to the surface charges is the same as that for a sphere with the same polarization:

$$\mathbf{E}_\Sigma(\mathbf{x}) = \frac{\mathbf{P}}{3\epsilon_0} \ .$$

6.5.4 Final Result for the Local Field

Gathering all contributions to the local field, we have

$$\underbrace{\mathbf{E}_l(\mathbf{x})}_{\text{local}} = \underbrace{\mathbf{E}_{\text{ext}}(\mathbf{x})}_{\text{external}} + \underbrace{\mathbf{E}_d(\mathbf{x})}_{\text{depolarizing}} + \frac{\mathbf{P}}{3\epsilon_0} + \mathbf{0} \,.$$

Since the sum of the first two terms on the right-hand side is, by definition, the macroscopic electric field $\mathbf{E}$ appearing in Maxwell's equations, we obtain

$$\boxed{\; \mathbf{E}_l(\mathbf{x}) = \underbrace{\mathbf{E}(\mathbf{x})}_{\text{macroscopic field}} + \underbrace{\frac{\mathbf{P}}{3\epsilon_0}}_{\text{corrective field}} \;}\,.$$

Note that the local field $\mathbf{E}_l(\mathbf{x})$ is different from the macroscopic field $\mathbf{E}(\mathbf{x})$. This means that the motion of a particle at $\mathbf{x}$ should be described using the local field in Newton's law: $\mathbf{f} = q\mathbf{E}_l(\mathbf{x})$. The corrective field term $\frac{\mathbf{P}}{3\epsilon_0}$ is significant for polar liquids and crystals. It is less relevant for dilute media.

For a LHI medium, the polarization is linked to the electric field: $\mathbf{P} = \epsilon_0 \chi \mathbf{E}$. This allows us to express the local field as a function of the macroscopic field:

$$\mathbf{P} = \epsilon_0 \chi \mathbf{E} \quad \Rightarrow \quad \mathbf{E}_l = \left(1 + \frac{\chi}{3}\right)\mathbf{E}, \quad \chi = \epsilon_r - 1 \,.$$

Clearly, the local field differs from $\mathbf{E}$ if the susceptibility χ cannot be neglected, that is if the relative permittivity significantly differs from 1 ($\epsilon_r \neq 1$), which is precisely the case for dense media: liquids and crystals. For dilute media (gases), $\chi \ll 1 \Rightarrow \mathbf{E}_l(\mathbf{x}) \sim \mathbf{E}(\mathbf{x})$.

6.6 Polarization Mechanisms

6.6.1 Electronic Polarization

Under the action of the local field $\mathbf{E}_l$, the electron cloud in an atom undergoes a distortion that is accompanied by the generation of a dipole moment $\mathbf{p}$. This is illustrated in Fig. 6.8.

Consider a simplified model of an atom where the charge distribution is spherically symmetrical. Inside a sphere of radius R, the electron cloud, carrying a total charge of $-Ze$, has a uniform average charge density given by

$$\varrho^- = -3Ze/4\pi R^3 \,.$$

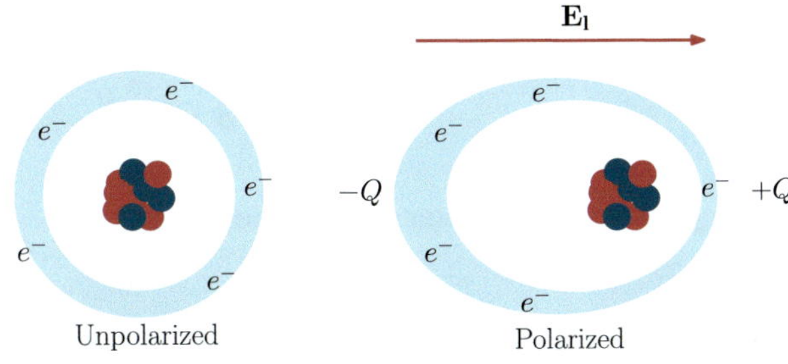

Fig. 6.8 The local field polarizes an atom, generating a non-zero dipole moment **p**

When subjected to a local electric field, the electron cloud shifts slightly to the left by a distance d, which is much smaller than the atomic radius R. This displacement generates an electric dipole moment:

$$\mathbf{p} = Ze\mathbf{d} \, ,$$

as illustrated in Fig. 6.9.

The positive charge of the nucleus experiences two fields:

1. The local electric field, $\mathbf{E}_l$.
2. The electric field, $\mathbf{E}_-$, generated by the displacement of the negative charge distribution.

As demonstrated in Example 4, applying Gauss's law to a uniformly charged sphere yields:

$$\mathbf{E}_- = \frac{\varrho^-\mathbf{d}}{3\epsilon_0} \, ,$$

where ϱ^- is the average charge density of the electron cloud and $\mathbf{d}$ is the displacement of the electron cloud.

For the positive charge to be in equilibrium, the sum of these two electric fields must be zero:

$$\mathbf{E}_l + \mathbf{E}_- = 0 \, .$$

Fig. 6.9 An electron cloud displaced by **d** with respect to the nucleus

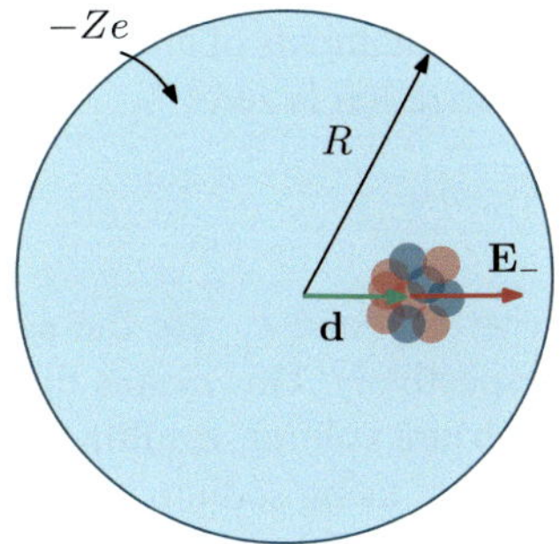

Therefore, we can express the local electric field as

$$\mathbf{E}_l = -\frac{\varrho^-}{3\epsilon_0}\mathbf{d} = \frac{\mathbf{p}}{4\pi\epsilon_0 R^3} \ ,$$

where $\mathbf{p}$ is the dipole moment of the atom.

Polarizability

We define the *polarizability*, α, as the proportionality constant that relates the dipole moment of an atom to the local electric field acting upon it:

$$\mathbf{p} = \epsilon_0\alpha\mathbf{E}_l \ .$$

The polarizability has the units of volume and can be thought of as a rough measure of the size of the atom. The permittivity of free space, ϵ_0, ensures dimensional consistency.

From the equation relating the dipole moment, $\mathbf{p}$, to the local electric field, $\mathbf{E}_l$, we can deduce the polarizability, α:

$$\mathbf{p} = \epsilon_0\alpha\mathbf{E}_l = 4\pi\epsilon_0 R^3\mathbf{E}_l \ .$$

Comparing the coefficients of $\mathbf{E}_l$, we find that the polarizability of an atom in this simplified model is given by

$$\boxed{\alpha = 4\pi R^3} \ .$$

6.6.2 Ionic Polarization

In a crystal without a permanent dipole moment, the ions have charges $q_i = \pm e$ and are located at their equilibrium positions $\mathbf{r}_i$ when there is no external electric field. The polarization, $\mathbf{P}$, within a unit volume, $\mathcal{U}$, is defined as the sum of the dipole moments of all the ions in that volume. In the absence of an external field, the polarization is zero:

$$\mathbf{P} = \sum_{i\in\mathcal{U}} e(\mathbf{r}_i^+ - \mathbf{r}_i^-) = \mathbf{0} \ ,$$

where $\mathbf{r}_i^+$ and $\mathbf{r}_i^-$ are the equilibrium positions of the positive and negative ions, respectively. This means that the positive and negative charges are balanced within each unit volume, resulting in a net dipole moment of zero. Figure 6.10 illustrates this concept using sodium chloride (NaCl) as an example. In the NaCl crystal structure, each pair of neighboring Na^+ and Cl^- ions forms a dipole. However, the dipoles

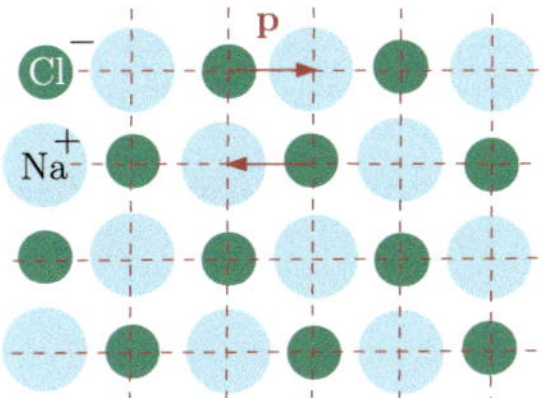

Fig. 6.10 Non-permanently polarized crystal

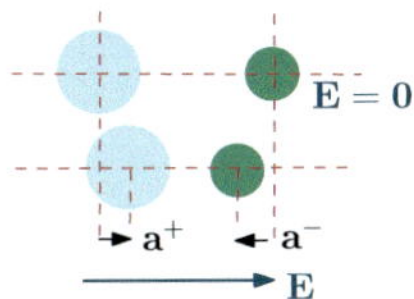

Fig. 6.11 The local electric field causes a distortion of the crystal lattice. This distortion results from the displacement of the ions within the crystal structure

of neighboring pairs are oriented in opposite directions, effectively canceling each other out. This results in a net zero dipole moment for the entire unit volume.

When a local electric field, $\mathbf{E}_l$, is applied to a crystal, the ions within the crystal are displaced slightly from their original equilibrium positions. This displacement, denoted by $\mathbf{a}_i$ for each ion, results in a distortion of the crystal lattice, as depicted in Fig. 6.11.

This distortion induces a polarization, $\mathbf{P}$, expressed as

$$\mathbf{P} = \sum_{i \in \mathcal{U}} e(\mathbf{r}_i^+ + \mathbf{a}_i^+ - \mathbf{r}_i^- - \mathbf{a}_i^-) \, ,$$

where $\mathbf{a}_i^+$ and $\mathbf{a}_i^-$ are the displacements of the positive and negative ions, due to the field. Considering the example of sodium chloride and assuming that all positive (Na^+ and negative (Cl^- ions experience the same displacement, $\mathbf{a}^+$ and $\mathbf{a}^-$, respectively, we can simplify the equation to:

$$\mathbf{P} = \sum_{i \in \mathcal{U}} e(\mathbf{a}_i^+ - \mathbf{a}_i^-) = ne(\mathbf{a}^+ - \mathbf{a}^-) \, ,$$

where n is the number of atoms per unit volume. This expression for the polarization is based on the assumption that the crystal is non-polarized in the absence of an electric field.

To determine the polarizability, we will show that difference in displacements, $\mathbf{a}^+ - \mathbf{a}^-$, is directly proportional to the local electric field, $\mathbf{E}_l$ by considering the equilibrium of a cation. According to Newton's second law, the electric force due to the local field must balance the restoring force that arises from the displacement of the cation relative to the anion. This restoring force can be expressed as the product of the stress and the surface area, $|\mathbf{r_i}^+ - \mathbf{r_i}^-|^2$, between the cation and anion. We can use Hooke's law for the stress, $\sigma = Y\epsilon$, where Y is the Young modulus of the crystal and $\epsilon = \dfrac{|\mathbf{a}^+ - \mathbf{a}^-|}{|\mathbf{r}_i^+ - \mathbf{r}_i^-|}$ is the strain or proportional deformation. Along the x-axis, we have the following equation:

$$eE_l - Y|\mathbf{r}_i^+ - \mathbf{r}_i^-||\mathbf{a}^+ - \mathbf{a}^-| = 0 \ .$$

Solving for the displacement difference, $|\mathbf{a}^+ - \mathbf{a}^-| = (eE_l)/(Y|\mathbf{r}_i^+ - \mathbf{r}_i^-|)$, and substituting this into the expression for polarization, we find:

$$\mathbf{P} = ne(\mathbf{a}^+ - \mathbf{a}^-) = n\frac{e^2}{Y|\mathbf{r}_i^+ - \mathbf{r}_i^-|}\mathbf{E}_l \ .$$

Finally, we can identify the polarizability α by comparing this expression to the definition of polarization in terms of the electric field:

$$\mathbf{P} = n\epsilon_0\alpha\mathbf{E}_l \ .$$

This leads to the expression for the polarizability:

$$\boxed{\alpha = \frac{e^2}{\epsilon_0 Y|\mathbf{r}_i^+ - \mathbf{r}_i^-|} \ .}$$

6.6.3   Polarization by Dipolar Orientation

In this section, we will examine dielectric media composed of polar molecules, such as water (as illustrated in Fig. 6.12). These molecules possess a permanent dipole moment with a magnitude of p_0. Upon application of an external electric field, the dipole moments of these molecules tend to align with the field. This alignment results in the material acquiring a polarization. We will focus on this type of polarization, which is caused by the orientation of the molecules.

In the absence of an external electric field, the dipole moments of the molecules within the medium are randomly oriented. This means that the dipole moments point in various directions, leading to a net dipole moment of zero within a unit volume. Even though each individual molecule has a permanent dipole moment, the random orientation cancels out their collective effect. This phenomenon is illustrated in Fig. 6.13.

When a local electric field, $\mathbf{E}_l$, is applied, the dipole moments of the molecules tend to align parallel to the field. However, thermal agitation counteracts this alignment, causing the molecules to randomly rotate. As a result, the average alignment of the dipole moments with the local field is only partial, as illustrated in Fig. 6.14.

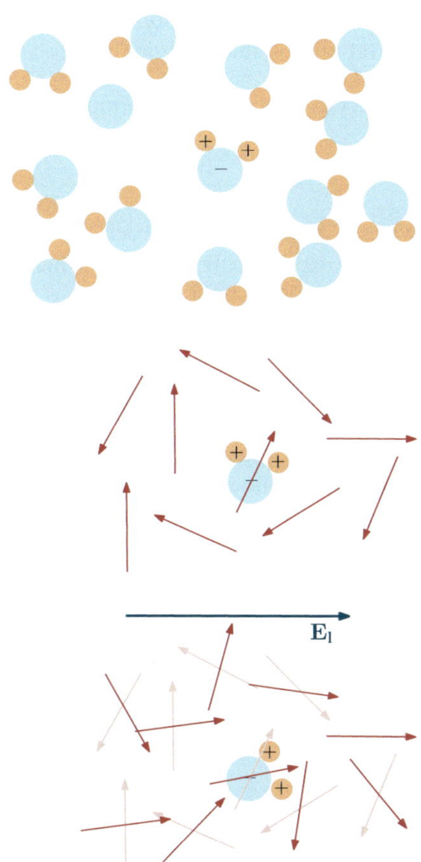

Fig. 6.12 A dielectric medium that contains polar molecules (here H_2O)

Fig. 6.13 In the absence of an external electric field, the molecular dipoles are randomly oriented

Fig. 6.14 The molecular dipoles tend to align in the presence of an external electric field, but are also subjected to thermal agitation

When summing the dipole moments of all molecules within a unit volume, we obtain the polarization, **P**. Due to the random orientation of the dipoles in the absence of an external field, the magnitude of the polarization, $|\mathbf{P}|$, falls within the range of 0 to np_0, where n is the number density of molecules and p_0 is the magnitude of the individual dipole moments.

To calculate the polarization, we must consider the competition between the alignment of dipoles with the field and the thermal agitation that tends to disorient them. Thermal agitation is described by Boltzmann's law, which gives the probability $d\mathcal{P}$ of a particle having an energy within a specific range $[W, W + dW]$ at a temperature T:

$$d\mathcal{P} = \mathcal{P}_0 e^{-\frac{W}{k_B T}} dW \,,$$

where:

- $k_B = 1.38 \times 10^{-23} \mathrm{Jk}^{-1}$ is the Boltzmann constant.
- W is the energy of the particle.

- The constant $\mathcal{P}_0$ is a normalization factor that ensures that the sum of probabilities over all possible energies equals 1.

In this case, the energy of a molecular dipole in the local field is given by

$$W = -\mathbf{p} \cdot \mathbf{E}_l = -p_0 E_l \cos\theta, \quad 0 < \theta < \pi ,$$

where θ is the angle between the dipole moment and the electric field. For convenience, we define $W_0 = p_0 E_l$, which represents the maximum energy a dipole can have in the local field.

The contribution of a single molecule to the polarization is given by $p_0 \cos\theta\, \mathbf{u}\, d\mathcal{P}$. In this expression, $\mathbf{u}$ is a unit vector pointing in the direction of the local electric field and $d\mathcal{P}$ is the probability of the molecule having an energy W. To calculate the total polarization, $\mathbf{P}$, we must integrate the contribution of each molecule over all possible energy states or, equivalently, over all possible molecular orientations. This leads to the following expression:

$$\mathbf{P} = \mathbf{u}\, n\, p_0 \frac{\displaystyle\int_{-W_0}^{W_0} \cos\theta\, \mathcal{P}_0 e^{-\frac{W}{kT}}\, dW}{\displaystyle\int_{-W_0}^{W_0} \mathcal{P}_0 e^{-\frac{W}{kT}}\, dW} .$$

The calculation can be performed analytically by introducing the following change of variables:

$$\cos\theta = -\frac{W}{W_0}, \qquad X = -\frac{W}{kT} .$$

Using these substitutions, the polarization can be expressed as

$$\mathbf{P} = n p_0 \mathbf{u} \frac{kT}{W_0} \frac{\displaystyle\int_{-X_0}^{X_0} X e^X dX}{\displaystyle\int_{-X_0}^{X_0} e^X dX} ,$$

where $X_0 = +\dfrac{W_0}{kT}$. By evaluating the integrals, we arrive at

$$\mathbf{P} = n p_0 \mathbf{u} \left(\frac{1}{\tanh X_0} - \frac{1}{X_0} \right) = n p_0 \mathcal{L}(X_0) \mathbf{u} .$$

The function $\mathcal{L}(X) = 1/\tanh(X) - 1/X$ is known as the *Langevin function* and is plotted on Fig. 6.15.

- *Low Temperatures*: When X_0 is large (which occurs at low temperatures T), the Langevin function approaches saturation ($\mathcal{L}(X_0) \to 1$). In this limit, all molecules

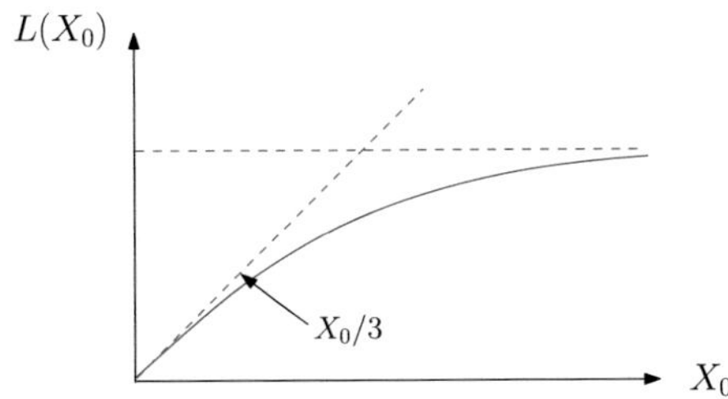

Fig. 6.15 Langevin's function $\mathcal{L}(X)$

tend to align perfectly with the field. This leads to a maximum polarization:

$$\mathbf{P} = np_0\mathbf{u} .$$

- *High Temperatures and Weak Fields*: When X_0 is small (which occurs at high temperatures or weak fields), the Langevin function can be approximated as $\mathcal{L}(X_0) \simeq X_0/3$. In this case, the polarization as a function of the local field becomes

$$\mathbf{P} = np_0\mathcal{L}(X_0)\frac{\mathbf{E}_l}{E_l} \quad \text{i.e.} \quad \mathbf{P} = np_0\frac{p_0}{3kT}\mathbf{E}_l .$$

This expression allows us to identify the polarizability for molecular orientation:

$$\boxed{\alpha = \frac{p_0^2}{3\epsilon_0 kT}} .$$

- *Overall Polarization*: For molecular media such as water, both electronic polarizability and molecular orientation contribute to the total polarization. Therefore, the total polarization can be expressed as

$$\mathbf{P} = \epsilon_0 n\left(\alpha_e + \frac{p_0^2}{3\epsilon_0 kT}\right)\mathbf{E}_l ,$$

where α_e is the electronic polarizability.

6.6.4 Clausius–Mossotti Relation

In previous sections, we discussed several physical mechanisms that contribute to polarization. These mechanisms led us to expressions for the polarizability, α, which relates the macroscopic polarization density, $\mathbf{P}$, to the local field, $\mathbf{E}_l$:

$$\mathbf{P} = \epsilon_0 n\alpha\mathbf{E}_l .$$

In Sect. 6.4, we defined the dielectric constant or relative permittivity, ϵ_r, which connects the macroscopic polarization to the macroscopic electric field, $\mathbf{E}$:

$$\mathbf{P} = \epsilon_0 \chi \mathbf{E} = \epsilon_0(\epsilon_r - 1)\mathbf{E} ,$$

where χ is the *electric susceptibility*.

The Clausius–Mossotti relation, named after Ottaviano–Fabrizio Mossotti and Rudolf Clausius (see Fig. 6.16), also known as the Lorentz-Lorenz relation, connects the polarizability of a dielectric medium to its dielectric constant, ϵ_r, or susceptibility, $\chi = \epsilon_r - 1$. In a linear, homogeneous, and isotropic medium, the local field can be expressed as a function of the macroscopic field and the polarization density:

$$\mathbf{E}_l = \mathbf{E} + \frac{\mathbf{P}}{3\epsilon_0} .$$

By substituting the local field into the expression for the polarization density and rearranging terms, we arrive at

$$\left(1 - \frac{n\alpha}{3}\right) \mathbf{P} = n\epsilon_0 \alpha \mathbf{E} .$$

Using the relation $\mathbf{P} = \epsilon_0(\epsilon_r - 1)\mathbf{E}$, we obtain

$$n\epsilon_0 \alpha = \epsilon_0(\epsilon_r - 1) \left(1 - \frac{n\alpha}{3}\right) .$$

Finally, rearranging this equation yields the Clausius–Mossotti relation:

$$\boxed{\frac{n\alpha}{3} = \frac{\epsilon_r - 1}{\epsilon_r + 2}} \quad \text{Clausius–Mossotti relation} \qquad (6.17)$$

Fig. 6.16 Left: Ottaviano-Fabrizio Mossotti (1791–1863), an Italian physicist and astronomer. Right: Rudolf Clausius (1822–1888), a German physicist and one of the founders of thermodynamics

Alternatively, we can express the susceptibility, χ, in terms of the polarizability, α:

$$\chi = \frac{n\alpha}{1 - \dfrac{n\alpha}{3}} \, . \tag{6.18}$$

The Clausius–Mossotti relation (6.17) indicates that the dielectric constant, ϵ_r, is a function of both temperature and pressure. This dependence arises through the temperature dependence of the polarizability, α, and the pressure dependence of the number density of molecules, n. This provides a very good model to predict the dielectric constant of liquids. Note that (6.18) gives a negative dielectric constant for $n\alpha > 3$, and diverges for $n\alpha = 3$. The latter corresponds to an unstable situation where the polarization increases strongly, creating a very large local field, which in turn polarizes the atoms or molecules even more. Of course, at some point, the proportionality between the polarization and the external field breaks down. Very close to this situation, a material can build up an spontaneous, permanent polarization sustained by its own field. This polarization will subsist even if the external electric field is switched off. This is at the origin of ferroelectricity in ionic crystals.

6.7 Boundary Conditions at the Interface Between Dielectric Media

The conditions that the electric field must satisfy at the interface between two dielectric media are derived from the following differential laws:

$$\nabla \cdot \mathbf{D}(\mathbf{x}) = \rho(\mathbf{x}) \, ,$$
$$\nabla \times \mathbf{E}(\mathbf{x}) = 0 \, ,$$

where $\mathbf{D}$ is the electric displacement vector, ρ is the free charge density and $\mathbf{E}$ is the electric field.

6.7.1 Normal Component of $\mathbf{D}$ at a Charged Interface

Consider two regions of space separated by an interface. Choose a point $\mathbf{x}$ on the interface. A closed cylinder can be constructed as shown in Fig. 6.17, with one cap in region 2 and the other cap in region 1. The normal vector $\mathbf{n}(\mathbf{x})$ points from region 2 to region 1.

Fig. 6.17 Gauss law applied
to a small cylinder at the
interface between two media

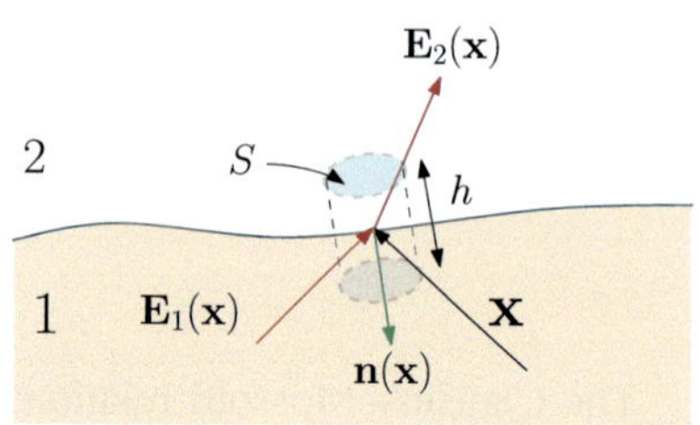

Gauss's law can be written as

$$\oiint_S \mathbf{D}(\mathbf{x}) \cdot d\mathbf{S}(\mathbf{x}) = \iiint_{\Omega(S)} \varrho(\mathbf{x}) d^3 x \ ,$$

where $\Omega(S)$ is the volume of the cylinder enclosed by the surface S, with height
h and radius R. The lateral surface area of the cylinder is $2\pi h R$. As the height
h approaches zero, the flux of $\mathbf{D}$ through the lateral surface becomes negligible.
Therefore, we only need to consider the flux through the top and bottom caps, each
with an area $A = \pi R^2$:

$$\lim_{h \to 0} \oiint_S \mathbf{D}(\mathbf{x}') \cdot d\mathbf{S}(\mathbf{x}') = \mathbf{n}(\mathbf{x}) \cdot (\mathbf{D}_1(\mathbf{x}) - \mathbf{D}_2(\mathbf{x})) \, A \ ,$$

where: $\mathbf{D}_1$ and $\mathbf{D}_2$ are the electric displacement vectors in regions 1 and 2, respectively.

If we assume that the interface carries a surface charge density $\sigma(\mathbf{x})$, then as
the height h approaches zero, the enclosed free charge within the Gaussian cylinder
becomes $A\sigma(\mathbf{x})$.

Applying Gauss's law, we get

$$\mathbf{n}(\mathbf{x}) \cdot (\mathbf{D}_1(\mathbf{x}) - \mathbf{D}_2(\mathbf{x})) \, A = \sigma(\mathbf{x}) A \ .$$

Dividing both sides by A, we find that the normal component of the electric
displacement vector is discontinuous at the interface:

$$\boxed{\mathbf{n} \cdot (\mathbf{D}_1 - \mathbf{D}_2)|_{\mathbf{x}} = \sigma(\mathbf{x})} \tag{6.19}$$

where $\mathbf{n}$ is the normal vector pointing from medium 2 to medium 1.

For linear, homogeneous, and isotropic dielectric media, this equation can be
rewritten in terms of the electric field as

$$\mathbf{n} \cdot (\epsilon_1 \mathbf{E}_1 - \epsilon_2 \mathbf{E}_2)|_{\mathbf{x}} = \sigma(\mathbf{x}) \tag{6.20}$$

where ϵ_1 and ϵ_2 are the permittivities of the two media.

Fig. 6.18 Circulation law applied to a small planar loop at the interface between two media

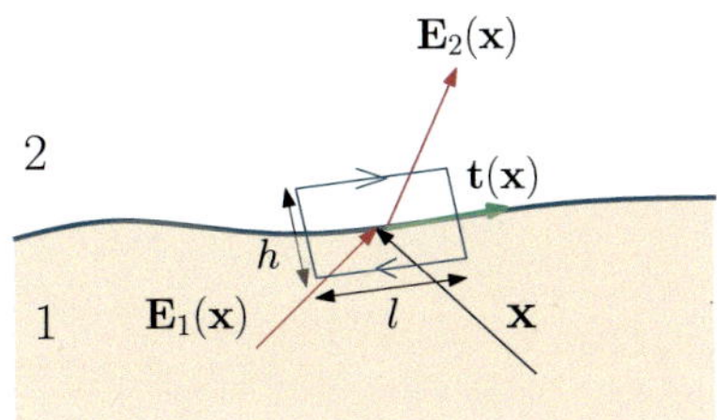

6.7.2 Continuity of the Tangential Component of $\mathbf{E}$

To demonstrate that the tangential component of the electric field, $\mathbf{E}$, is continuous at the interface between two dielectric media, we will integrate the following equation over a closed rectangular path, Γ:

$$\nabla \times \mathbf{E}(\mathbf{x}) = 0 \ .$$

This equation is a consequence of the fact that the electric field is conservative. Integrating over the closed path Γ results in the law of circulation for the electrostatic field:

$$\oint_\Gamma \mathbf{E}(\mathbf{x}) \cdot d\mathbf{x} = 0 \ .$$

This equation states that the line integral of the electric field around a closed loop is zero (Fig. 6.18). In the limit as the height h approaches zero, the circulation integral reduces to the sum of the line integrals along the two segments parallel to the tangential direction $\mathbf{t}(\mathbf{x})$ at the interface. These segments are assumed to be sufficiently small that the electric field, $\mathbf{E}$, can be considered approximately constant along each segment.

$$l(\mathbf{E}_1(\mathbf{x}) - \mathbf{E}_2(\mathbf{x})) \cdot \mathbf{t}(\mathbf{x}) = 0$$

where l is the length of each segment. Since this equation must hold for any value of l, we conclude that the tangential component of the electric field is continuous at the interface:

$$\boxed{\mathbf{t} \cdot (\mathbf{E}_1 - \mathbf{E}_2)|_{\mathbf{x}} = 0 \ .} \tag{6.21}$$

Example 6.2 - Hemisphere filled with a dielectric
Consider two concentric conductive spheres with radii a and b. The inner sphere carries a charge of $+Q$, while the outer sphere has a charge of $-Q$. Suppose that half of the space between them is filled with a dielectric material having a dielectric constant $\epsilon_r = \epsilon/\epsilon_0$, as illustrated in Fig. 6.19.

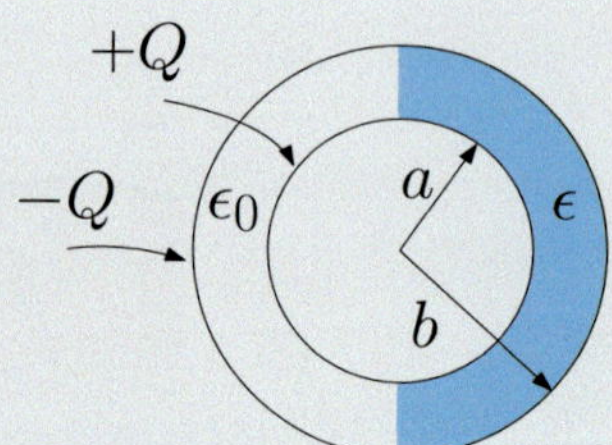

Fig. 6.19 A hemispherical region (half of the space) between the two concentric conductive spheres is filled with a dielectric material

Given that both conductive spheres are equipotential surfaces, the total charge density (comprising both free and bound charges) must be uniformly distributed on the surfaces of the spheres. As a result, the electric field exhibits spherical symmetry, meaning it can be expressed as

$$\mathbf{E}(\mathbf{x}) = E(r)\mathbf{u}_r \ ,$$

where $E(r)$ is the magnitude of the electric field and $\mathbf{u}_r$ is the radial unit vector in spherical coordinates with the origin at the center of the spheres. Now, consider a spherical surface S with radius r, concentric with both conductive spheres. The radius r should satisfy $a < r < b$ to lie between the two spheres. Applying Gauss's law for dielectrics to this surface, we obtain

$$\oiint_S \mathbf{D}(\mathbf{x}) \cdot d\mathbf{S}(\mathbf{x}) = Q$$

where $\mathbf{D}$ is the electric displacement vector and Q is the enclosed charge.

We can divide the spherical surface S into two hemispheres:

- S_1: Defined by $0 \le \phi \le \pi$ (the right hemisphere in the dielectric region).
- S_2: Defined by $\pi \le \phi \le 2\pi$ (the left hemisphere in vacuum).

Using this division, the surface integral in Gauss's law becomes

$$\oiint_S \mathbf{D}(\mathbf{x}) \cdot d\mathbf{S}(\mathbf{x}) = \iint_{S_1} \mathbf{D}(\mathbf{x}) \cdot d\mathbf{S}(\mathbf{x}) + \iint_{S_2} \mathbf{D}(\mathbf{x}) \cdot d\mathbf{S}(\mathbf{x}) = Q \ .$$

Substituting the electric displacement vector $\mathbf{D} = \epsilon_0 \epsilon_r \mathbf{E}$ in the dielectric region (S_1) and $\mathbf{D} = \epsilon_0 \mathbf{E}$ in the vacuum region (S_2), we get

$$\oiint_S \mathbf{D}(\mathbf{x}) \cdot d\mathbf{S}(\mathbf{x}) = \iint_{S_1} \epsilon_0 \epsilon_r \mathbf{E}(\mathbf{x}) \cdot d\mathbf{S}(\mathbf{x}) + \iint_{S_2} \epsilon_0 \mathbf{E}(\mathbf{x}) \cdot d\mathbf{S}(\mathbf{x}) \ .$$

Since the electric field is spherically symmetric, its magnitude E(r) is constant over each hemisphere. Evaluating the surface integrals, we obtain

$$\oiint_S \mathbf{D}(\mathbf{x}) \cdot d\mathbf{S}(\mathbf{x}) = 2\pi r^2 E(r)\epsilon_0 \left(1 + \epsilon_r\right) ,$$

where ϵ_r is the dielectric constant of the dielectric region S_1. Finally, solving for the electric field $\mathbf{E}(\mathbf{x})$, we find the electric field between the two concentric spheres:

$$\mathbf{E}(\mathbf{x}) = \frac{Q}{2\pi r^2(\epsilon_0 + \epsilon)}\mathbf{u}_r \quad \text{for } a \le r \le b .$$

To determine the free charge density on the sphere of radius $r = a$, we can use the boundary condition (6.20) for the normal component of the electric field, with with a zero electric field inside the sphere:

$$\mathbf{E}(\mathbf{x}) \cdot \mathbf{n}(\mathbf{x}) = \frac{\sigma(\mathbf{x})}{\epsilon(\mathbf{x})}, \quad \mathbf{x} \in S_a : \{r = a\} .$$

Therefore, the free charge density at $r = a$ is different for the two hemispheres:

$$\sigma_2 = \epsilon_0 E(a) = \frac{Q\epsilon_0}{2\pi a^2 (\epsilon + \epsilon_0)} ,$$

$$\sigma_1 = \epsilon E(a) = \frac{Q\epsilon}{2\pi a^2 (\epsilon + \epsilon_0)} .$$

To determine the bound charge density induced on the dielectric surface, we use the following relation:

$$\mathbf{P}(\mathbf{x}) = (\epsilon - \epsilon_0)\,\mathbf{E}(\mathbf{x}) .$$

The bound surface charge density, σ_P, can then be calculated as

$$\sigma_P = \mathbf{P} \cdot \mathbf{n}\big|_{\mathbf{x}\in S_a}$$

where $\mathbf{n}$ is the unit normal vector pointing outward from the hemispherical region (i.e., $\mathbf{n} = -\mathbf{u}_r$ for $r = a$). For the hemisphere filled with the dielectric:

$$\sigma_{P1} = -(\epsilon - \epsilon_0)\,E(r)\big|_{r=a} = \left(\frac{\epsilon_0 - \epsilon}{\epsilon_0 + \epsilon}\right)\frac{Q}{2\pi a^2} .$$

For the hemisphere filled with vacuum, $\sigma_{P2} = 0$.

The total charge density on the inner sphere ($r = a$) is uniform and equal for both hemispheres:

$$\sigma_T = \sigma + \sigma_P = \frac{Q\epsilon_0}{2\pi a^2(\epsilon + \epsilon_0)} \, .$$

6.8   Capacitors with Dielectrics

When a dielectric material is inserted between the two plates of a capacitor, the capacitance increases by a factor of ϵ_r, a phenomenon discovered by Michael Faraday in the 19th century. To illustrate this, consider a parallel-plate capacitor with area A, plate separation d, and a charge Q. Now, introduce a dielectric material with a thickness t (where $t \leq d$) between the plates, as depicted in Fig. 6.20.

To calculate the capacitance, we will first determine the potential difference between the plates. In the absence of a dielectric, the magnitude of the electric field between the plates (assuming that the plates are large compared to their separation d) is given by

$$E_0 = \frac{\sigma}{\epsilon_0} = \frac{Q}{\epsilon_0 A} \, ,$$

where σ is the surface charge density on the plates, ϵ_0 is the permittivity of free space, and A is the area of the plates. When a dielectric is inserted, Gauss's law for dielectrics, (6.8), tells us that the free charges are the sources of the electric displacement field, $\mathbf{D}$. Therefore, solving for the electric field in the dielectric is equivalent to solving for the electric field in vacuum with modified charges (multiplied by ϵ_0). The electric displacement field between the plates is homogeneous and equal to

$$\mathbf{D} = \sigma \mathbf{u}_x \, ,$$

where $\mathbf{u}_x$ is the unit vector in the x-direction. Outside the dielectric region, where $\epsilon_r = 1$, we have

$$\mathbf{D} = \epsilon_0 \mathbf{E}$$

which means the electric field remains the same as in the case without a dielectric:

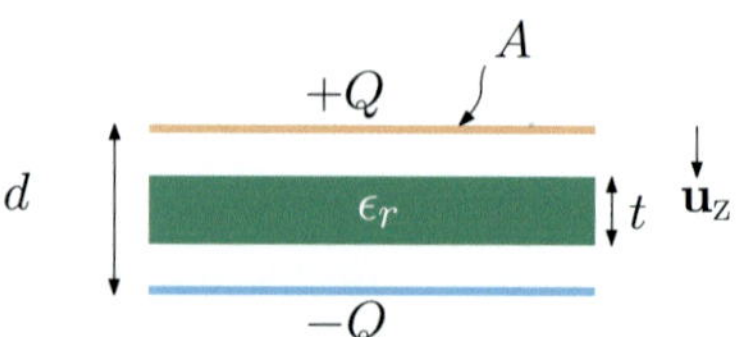

Fig. 6.20 A parallel-plate capacitor with a dielectric slab inserted between its plates

$$\mathbf{E}_0 = (\sigma/\epsilon_0)\mathbf{u}_x \ .$$

Inside the dielectric, we have

$$\mathbf{D} = \epsilon_r\epsilon_0\mathbf{E} \ ,$$

which means the electric field is weakened by a factor of ϵ_r compared to its value in vacuum: $\mathbf{E}_D = \mathbf{E}_0/\epsilon_r$. Since the electric field $\mathbf{E}_D$ inside the dielectric is a uniform field, the polarization charges only appear on the surfaces of the dielectric. This is because the orientation of the electric dipoles within the dielectric aligns with the electric field generated by the capacitor plates, creating a shielding effect that reduces the field inside the dielectric, as illustrated in Fig. 6.21. To find the potential difference between the plates, we can integrate the electric field along a vertical path (parallel to $\mathbf{u}_x$) that connects the upper plate to the lower plate:

$$\Delta V = \int_0^d E(x)dx = E_0(d-t) + E_Dt \ .$$

Substituting the expressions for E_0 and E_D, we get

$$\Delta V = \frac{Q}{A\epsilon_0}(d-t) + \frac{Q}{A\epsilon_0\epsilon_r}t = \frac{Q}{A\epsilon_0}\left(d-t\left(1-\frac{1}{\epsilon_r}\right)\right) \ .$$

Finally, the capacitance is given by

$$C = \frac{Q}{\Delta V} = \frac{\epsilon_0 A}{d - t\left(1 - \dfrac{1}{\epsilon_r}\right)} \ .$$

When the thickness of the dielectric, t, is equal to the distance between the plates, d, the entire space between the plates is filled with the dielectric. In this case, the capacitance becomes

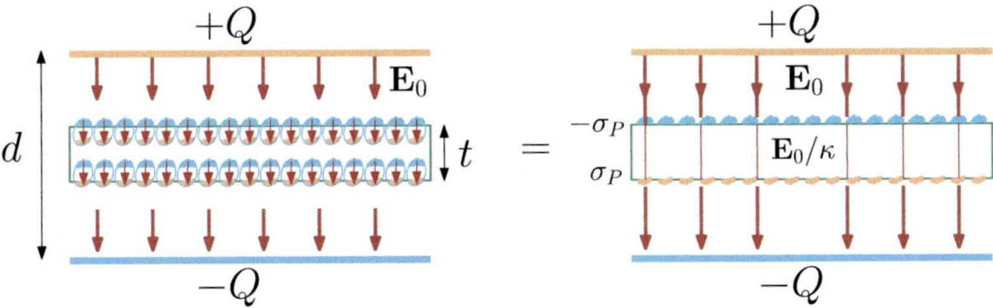

Fig. 6.21 A dielectric material placed in an external electric field becomes polarized. This polarization results in the formation of a surface charge density on the surface of the dielectric

$$C = \frac{\epsilon_r \epsilon_0 A}{d} = \epsilon_r C_0 \, ,$$

where C_0 is the capacitance without the dielectric. Therefore, inserting a dielectric material with a high dielectric constant between the plates of a capacitor is an effective way to increase its capacitance without modifying its physical dimensions.

6.8.1 Capacitance Coefficients in a Dielectric Medium

Consider a system of N conductors immersed in a linear, homogeneous, and isotropic medium. As an example, Fig. 6.22 illustrates a system with $N = 2$ conductors. We have previously established the relationship between the charges Q_i carried by the conductors and their corresponding potentials V_i when they are in a vacuum. This relationship led to the definition of the capacitance matrix. The question is: Does the capacitance matrix change when the conductors are surrounded by a dielectric medium instead of a vacuum?

We have previously seen that for a linear, homogeneous, and isotropic dielectric inserted into a capacitor, the capacitance increases by a factor of the relative permittivity (ϵ_r). We can expect a similar change in the capacitance coefficients for a system of N conductors. Figure 6.22 compares a system of $N = 2$ conductors in vacuum with the same system immersed in a dielectric medium. Both systems have the same conductor positions and potentials (V_1 and V_2).

- In the *vacuum* case, the potential at any point $\mathbf{x}$ between the conductors must satisfy Laplace's equation:

$$\nabla^2 V = 0 \tag{6.22}$$

with the boundary conditions

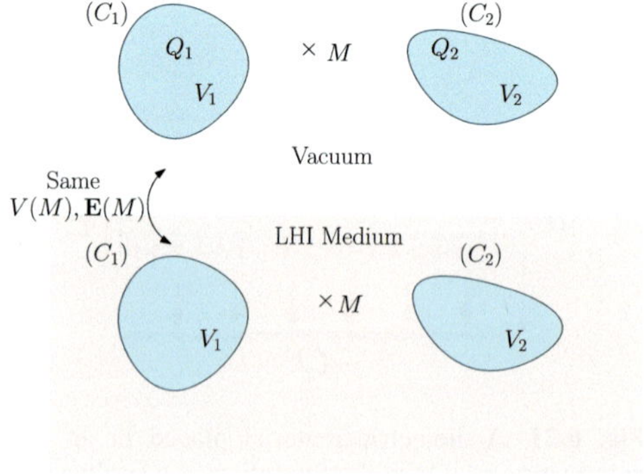

Fig. 6.22 System of conductors in vacuum (top) or in a dielectric medium (bottom)

$$V(\mathbf{x}) = V_1 \text{ for } \mathbf{x} \in (S_1) , \quad V(\mathbf{x}) = V_2 \text{ for } \mathbf{x} \in (S_2) ,$$

where S_1 and S_2 are the surfaces of the conductors. The electric field can then be obtained from the potential using:

$$\mathbf{E} = -\nabla V .$$

- If a *linear, homogeneous, and isotropic dielectric medium* is present between the conductors, the electric displacement field satisfies the Maxwell–Gauss equation with no free charge:

$$\nabla \cdot \mathbf{D} = 0 .$$

Since $\mathbf{D} = \epsilon \mathbf{E}$, where ϵ is the permittivity of the dielectric, the Maxwell–Gauss equation becomes

$$\epsilon \nabla \cdot \mathbf{E} = 0 .$$

Given that the electric field is conservative, $\mathbf{E} = -\nabla V$, we find that the potential, V, satisfies Poisson's equation (6.22). The boundary conditions for the potential remain the same as in the vacuum case. Therefore, the potential at any point between the conductors is identical to the potential in a vacuum. Consequently, the electric field between the conductors is also the same as in the vacuum case.

The primary difference between the two cases (vacuum versus dielectric) lies in the free charge carried by the conductors.

In a vacuum, Coulomb's law states that at the surface of a conductor:

$$\epsilon_0 \mathbf{E} = \sigma_0 \mathbf{n} ,$$

where $\mathbf{E}$ is the electric field on the surface of a conductor, $\mathbf{n}$ is the unit normal vector to the conductor surface, and σ_0 is the free surface charge density.

When a dielectric is present, the boundary condition for the normal component of the electric displacement field becomes

$$\mathbf{D} - \mathbf{0} = \epsilon \mathbf{E} = \sigma \mathbf{n} .$$

Since the electric field and the normal vector are the same as in the vacuum case, we find

$$\sigma = \epsilon \frac{\sigma_0}{\epsilon_0} = \epsilon_r \sigma_0 ,$$

This shows that the free charge density on the conductors is multiplied by the relative permittivity, ϵ_r. Consequently, the total charges on the conductors (Q_1 and Q_2) are also multiplied by ϵ_r. The relation between charges and potentials involving the capacitance matrix is given by

$$\begin{pmatrix} Q_1 \\ Q_2 \end{pmatrix} = \begin{bmatrix} C_{11} & C_{12} \\ C_{21} & C_{22} \end{bmatrix} \begin{pmatrix} V_1 \\ V_2 \end{pmatrix} \, .$$

When a dielectric medium is present, potentials remain the same as in vacuum, charges increase by a factor of ϵ_r. Therefore, every coefficient of the capacitance matrix is multiplied by ϵ_r.

Example 6.3 - Spherical capacitor with dielectric
Consider a spherical capacitor of inner radius a and outer radius b. The space between the inner and outer spheres is filled with a dielectric material having a dielectric constant ϵ_r, as illustrated in Fig. 6.23.

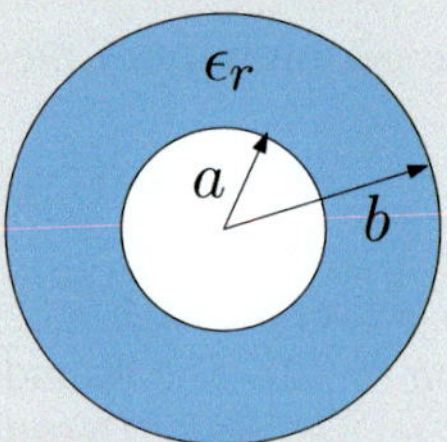

Fig. 6.23 Spherical capacitor filled with dielectric

Assume that the surface $r = a$ carries a charge Q, while the surface $r = b$ carries a charge $-Q$. To calculate the capacitance, we can determine the potential difference between these surfaces. Since both surfaces are equipotential, the charge density must be uniformly distributed on each sphere. Applying Gauss's law for dielectrics (6.8) to a spherical surface with radius r (where $a < r < b$), we get

$$\oiint_S d\mathbf{S}(\mathbf{x}) \cdot \mathbf{D}(\mathbf{x}) = Q \, .$$

The electric displacement vector, $\mathbf{D}$, inherits the spherical symmetry of the charge distribution. Therefore, it can be expressed as $\mathbf{D}(\mathbf{x}) = D(r)\mathbf{u}_r$, where $D(r)$ is a function of the radial distance r and $\mathbf{u}_r$ is the radial unit vector. Substituting this expression into Gauss's law and evaluating the surface integral, we get

$$\oiint_S d\mathbf{S}(\mathbf{x}) \cdot \mathbf{D}(\mathbf{x}) = 4\pi D(r)r^2 = Q \, .$$

Solving for $D(r)$, we obtain

$$\mathbf{D}(\mathbf{x}) = \frac{Q}{4\pi r^2}\mathbf{u}_r, \quad a \leq r \leq b \, .$$

Since $\mathbf{D}(\mathbf{x}) = \epsilon_0 \epsilon_r \mathbf{E}(\mathbf{x})$, we can find the electric field $\mathbf{E}(\mathbf{x})$ between the concentric spheres:

$$\mathbf{E}(\mathbf{x}) = \frac{Q}{4\pi \epsilon_0 \epsilon_r r^2} \mathbf{u}_r, \quad a \leq r \leq b .$$

To calculate the potential difference between the spherical surfaces, we can integrate the electric field along a radial path:

$$\Delta V = V(a) - V(b) = \int_a^b \mathbf{E}(\mathbf{x}) \cdot d\mathbf{x} ,$$

where $\mathbf{x} = r\mathbf{u}_r$ and $d\mathbf{x} = dr\mathbf{u}_r$. Substituting the expression for $\mathbf{E}(\mathbf{x})$, we get

$$\Delta V = \int_a^b \frac{dr\, Q}{4\pi \epsilon_0 \epsilon_r r^2} = \frac{Q}{4\pi \epsilon_0 \epsilon_r} \left(\frac{1}{a} - \frac{1}{b} \right) .$$

Finally, the capacitance is given by

$$C = \frac{Q}{\Delta V} = \frac{4\pi \epsilon_0 \epsilon_r ab}{(b - a)} .$$

Comparing this result to the capacitance in a vacuum, we see that the capacitance of a spherical capacitor filled with a dielectric is increased by a factor of ϵ_r.

6.8.2 Types of Capacitors

Ceramic capacitors are widely used in electronics due to their small size and high capacitance values. One common type is the ceramic disc capacitor, which consists of a ceramic disc coated with a metal layer on both sides (typically silver). This configuration is shown in Fig. 6.24 (left).

Another type is the multilayer ceramic chip capacitor (MCCC). These capacitors are composed of hundreds of thin ceramic layers interleaved with alternating metal layers. This structure effectively creates a large number of capacitors connected in parallel, resulting in a high capacitance value despite the small volume. Figure 6.24(right) illustrates an MCCC. Ceramic capacitors typically have capacitances ranging from 1 picofarad (1pF) to 100 microfarads (100μF). An important characteristic of ceramic capacitors is that they do not have a polarity. This means they can be connected in either direction without affecting their performance.

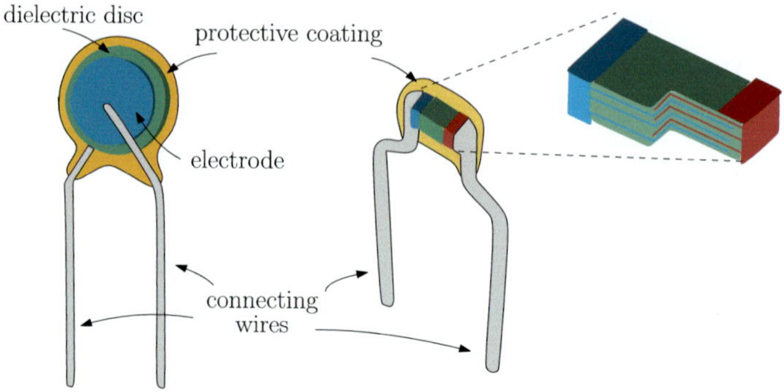

Fig. 6.24 Ceramic disc capacitor (left) and multilayer ceramic chip capacitor (right)

Electrolytic capacitors are another type of capacitor where the anode (positive plate) is made of a metal like aluminum or tantalum with a very thin layer of oxide. This oxide layer serves as the dielectric. The cathode is treated with an electrolyte paste or solution, which is a medium that allows ions to move freely. When a positive voltage is applied between the anode and cathode, an electrochemical reaction occurs, causing the oxide layer to become even more insulating. This increased insulation leads to a higher capacitance, typically ranging from 1 microfarad (1μF) to a few tens of millifarads (mF).

Figure 6.25 illustrates the typical internal construction of an electrolytic capacitor. The anode and cathode are foils separated by a paper spacer saturated with an electrolyte. These foils are then rolled into a cylindrical shape to increase the capacitance by a factor of two. This rolling configuration allows both sides of each foil to act as capacitor plates, effectively creating two parallel-plate capacitors in parallel. Additionally, the foils are roughened to increase their surface area, further enhancing the capacitance.

Fig. 6.25 An electrolytic capacitor

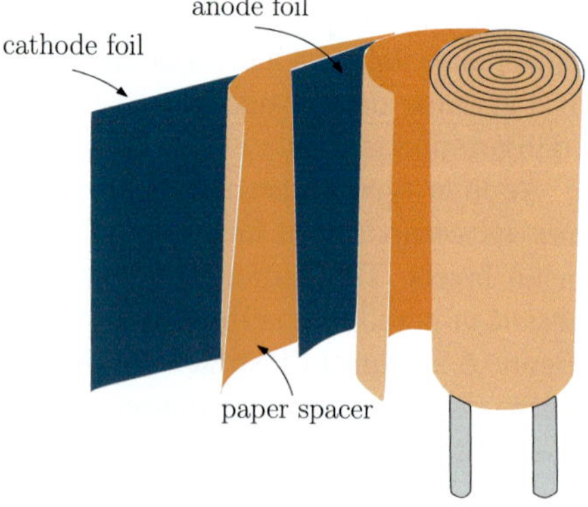

Due to their operating principle, electrolytic capacitors are polarized. This means that they must be connected in a specific polarity (positive to anode, negative to cathode). If the reverse polarity is applied, the oxide layer can be damaged. This damage can lead to a short circuit, which will cause the capacitor to heat up dramatically and potentially cause the electrolyte to evaporate.

In addition to traditional dielectric capacitors, there is a category of capacitors known as supercapacitors. These devices utilize electrochemical phenomena to store electric charge, achieving exceptionally high capacitances that can reach thousands of farads. One such phenomenon is the Helmholtz double layer. This occurs when a conductor is in contact with an electrolyte. At the interface between the conductor and electrolyte, two layers of opposite polarity form: one on the surface of the electrode and the other within the electrolyte. These layers are separated by an atomically thin layer of solvent molecules.

6.9   The Electrostatic Free Energy for a Polarized Medium

In Chap. 3, we defined and calculated the electrostatic potential energy of a charge distribution as the work required by an external agent to bring the charges from infinity (where the potential is zero) to their current positions in a quasistatic manner. In Chap. 5, we identified this work as the Helmholtz free energy, F, of the system. The free energy is a state function that depends solely on the final state of the system. The question we now consider is how the free energy is modified when polarization charges are present.

6.9.1   Electrostatic Free Energy Density for a Dielectric Inserted in a Plane Plate Capacitor

Consider a dielectric material that is exposed to an external electric field or is in proximity to external (free) charges. For example, we can imagine a dielectric placed between the plates of a capacitor, where one plate, at potential V, carries a charge Q and the other is grounded, as illustrated in Fig. 6.26.

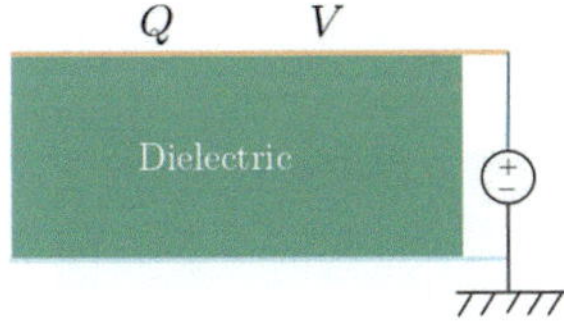

Fig. 6.26   A dielectric between two conducting plates

The free energy of the system corresponds to the work performed by an external agent, such as a generator, that supplies charges to polarize the dielectric medium. This work is assumed to be done quasistatically, without dissipating energy as heat. We know that the Helmholtz free energy, F, is a state function that depends only on the final state of the system. In Chap. 5, we derived that to supply a charge dQ to the positive plate of a capacitor, starting from an equilibrium state with potential V and charge Q, the work done by the generator (and thus the infinitesimal change in free energy) is

$$dF = VdQ \ .$$

In this equation, the potential V and the charge Q are conjugate variables. Therefore, as demonstrated in Chap. 5, the variation of the Helmholtz free energy must be expressed as a function of Q only if we want to integrate the free energy $F(Q)$. However, in this context, we are interested in the dependence of the free energy on the polarization charges. To achieve this, we will express dF as a function of the polarization $\mathbf{P}$ using the electric field $\mathbf{E}$ and the electric displacement field $\mathbf{D}$.

The electric displacement field can be determined using the jump condition for its normal component at the interface between the dielectric and the metal plate. Since the electric field is zero inside the metal, the normal component of the electric displacement field at the interface is equal to the surface free charge density, σ, on the metal plate:

$$\mathbf{D} \cdot \mathbf{n} = \sigma \ ,$$

where $\mathbf{n}$ is the unit normal vector pointing from the metal plate into the dielectric, as illustrated in Fig. 6.27.

Differentiating this equation, we obtain

$$dD = d\sigma = \frac{dQ}{S} \ ,$$

where S is the surface area of the metal plate. Since $V = Ee$, where e is the thickness of the dielectric, we can express the variation in the Helmholtz free energy as

$$dF = VdQ = (Ee)(Sd\sigma) = (eS)Ed\sigma = (eS)(EdD) = (eS)(\mathbf{E} \cdot d\mathbf{D}) \ ,$$

where eS is the volume of the dielectric. Therefore, the differential of the Helmholtz free energy density is

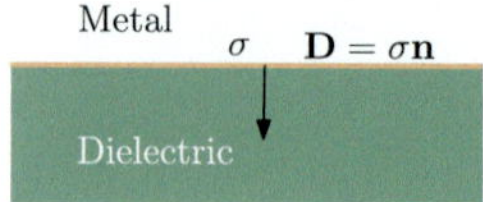

Fig. 6.27 Displacement vector in the dielectric infinitely close to the interface

$$\boxed{df = \mathbf{E} \cdot d\mathbf{D}} \; .$$

The free energy density is an exact differential, implying that $\mathbf{E}$ and $\mathbf{D}$ are conjugate variables. To find an expression for the state function $f(\mathbf{D})$, the differential df must be expressed as a function of $\mathbf{D}$ only before integration.

6.9.2 Free Energy Density for a Dielectric Medium: General Case

The result obtained for the capacitor, where the electric and displacement fields are uniform between the plates, can be generalized to any polarized dielectric medium in the presence of an electric field $\mathbf{E}(\mathbf{x})$. By considering $\mathbf{E}(\mathbf{x})$ and $\mathbf{D}(\mathbf{x})$ as locally uniform over an infinitesimal volume d^3x, we can write

$$df = \mathbf{E}(\mathbf{x}) \cdot d\mathbf{D}(\mathbf{x}) \; .$$

Integrating this expression over the volume Ω of the dielectric yields the variation in free energy:

$$\boxed{dF = \iiint_\Omega \mathbf{E}(\mathbf{x}) \cdot d\mathbf{D}(\mathbf{x}) d^3x} \; .$$

This expression is an exact differential, since F is a state function.

The presence of two infinitesimal elements, $d\mathbf{D}$ and d^3x, in the integral might seem unusual. However, it is correct because the volume integral will eliminate d^3x, resulting in a single infinitesimal element on each side of the equation.

Using the relation between the electric displacement vector and the electric field:

$$\mathbf{D} = \epsilon_0 \mathbf{E} + \mathbf{P} \; ,$$

we can differentiate both sides to get

$$d\mathbf{D} = \epsilon_0 d\mathbf{E} + d\mathbf{P} \; .$$

Substituting this into the expression for df, we obtain

$$df = \epsilon_0 \mathbf{E} \cdot d\mathbf{E} + \mathbf{E} \cdot d\mathbf{P} \; .$$

The first term, $\epsilon_0 \mathbf{E} \cdot d\mathbf{E}$, represents the energy required to establish the electric field, i.e., to change the electric field from $\mathbf{E}$ to $\mathbf{E} + d\mathbf{E}$. This term would be present even in the absence of a dielectric medium.

The second term, $\mathbf{E} \cdot d\mathbf{P}$, represents the energy required to establish the polarization in the medium, i.e., to change the polarization from $\mathbf{P}$ to $\mathbf{P} + d\mathbf{P}$ while keeping the electric field $\mathbf{E}$ constant.

6.9.3   Helmholtz's Free Energy Density for a Linear Homogeneous and Isotropic Medium

To find the Helmholtz free energy density of a polarized dielectric, we must integrate the differential $df = \mathbf{E} \cdot d\mathbf{D}$. However, $\mathbf{E}$ and $\mathbf{D}$ are related through the equation of state of the medium. For a linear, homogeneous, and isotropic medium, we have

$$\mathbf{D} = \epsilon \mathbf{E} \, ,$$

where ϵ is the permittivity of the medium. Substituting this into the expression for df, we get

$$df = \epsilon \mathbf{E} \cdot d\mathbf{E} \, .$$

Integrating this expression, we find that the Helmholtz free energy density is given by

$$f = \epsilon \frac{\mathbf{E}^2}{2} \, .$$

Using the relation $\mathbf{D} = \epsilon \mathbf{E}$, we can obtain equivalent expressions for the Helmholtz free energy density:

$$f = \frac{\mathbf{E} \cdot \mathbf{D}}{2} = \frac{\mathbf{D}^2}{2\epsilon} \, .$$

These expressions show that the Helmholtz free energy density for a linear, homogeneous, and isotropic medium can be expressed as a function of either the electric field, $\mathbf{E}$ or the electric displacement field, $\mathbf{D}$, or a combination of both. Using the relation $\epsilon = \epsilon_0(1 + \chi)$, where χ is the electric susceptibility, we can express the Helmholtz free energy density as a sum of the energy stored in the electric field and the energy stored in the polarization of the dielectric:

$$f = \epsilon_0 \frac{\mathbf{E}^2}{2} + \epsilon_0 \chi \frac{\mathbf{E}^2}{2} \, .$$

To find the total Helmholtz free energy of the dielectric, we integrate the free energy density over the volume Ω:

$$F = \iiint_\Omega f(\mathbf{x}) d^3x = \iiint_\Omega \epsilon \frac{\mathbf{E}^2(\mathbf{x})}{2} d^3x \, .$$

This expression represents the total electrostatic energy stored in the dielectric.

An alternative approach to deriving the free energy density involves starting from a reference state with zero electric and displacement fields ($\mathbf{E} = \mathbf{0}$ and $\mathbf{D} = \mathbf{0}$) and then integrating the free energy up to the final state characterized by $\mathbf{E}$ and $\mathbf{D}$. For a linear, homogeneous, and isotropic medium, we have $\mathbf{D} = \epsilon \mathbf{E}$. Along the integration path, we consider intermediate states characterized by electric fields and displacements of the form $\alpha \mathbf{E}$ and $\alpha \mathbf{D}$, respectively, where $0 \leq \alpha \leq 1$.

The variation of the free energy to reach a neighboring state $((\alpha + d\alpha)\mathbf{E}, (\alpha + d\alpha)\mathbf{D})$ from the current state $(\alpha \mathbf{E}, \alpha \mathbf{D})$ is given by

$$df = \alpha \mathbf{E} \cdot (d\alpha)\mathbf{D} = \mathbf{E} \cdot \mathbf{D}\alpha d\alpha \ .$$

Integrating this expression from $\alpha = 0$ to $\alpha = 1$, we obtain

$$f = \int_0^1 \mathbf{E} \cdot \mathbf{D}\alpha d\alpha = \mathbf{E} \cdot \mathbf{D} \int_0^1 \alpha d\alpha = \frac{1}{2}\mathbf{E} \cdot \mathbf{D} \ .$$

Therefore, the free energy density for a linear, homogeneous, and isotropic medium is

$$\boxed{f = \frac{1}{2}\mathbf{E} \cdot \mathbf{D} = \frac{1}{2\epsilon}\mathbf{D}^2 = \epsilon\frac{\mathbf{E}^2}{2} = \epsilon_0\frac{\mathbf{E}^2}{2} + \frac{\mathbf{E} \cdot \mathbf{P}}{2} \ .}$$

6.9.4 Gibb's Free Energy

Another thermodynamic state function that is often useful is the *Gibbs free energy*, also known as the *free enthalpy*. It is obtained from the Helmholtz free energy through a Legendre transformation. Working with the free energy densities, this Legendre transformation is

$$g = f - \mathbf{E} \cdot \mathbf{D} \ . \tag{6.23}$$

Differentiating this definition, we get the differential for the Gibbs free energy density:

$$\boxed{dg = -\mathbf{D} \cdot d\mathbf{E} \ .}$$

This equation suggests that it is natural to express the electric displacement $\mathbf{D}$ as a function of the electric field $\mathbf{E}$ to find an expression for the Gibbs free energy density as a function of $\mathbf{E}$. For example, for a linear, homogeneous, and isotropic dielectric, we substitute the relation $\mathbf{D} = \epsilon \mathbf{E}$ into the expression (6.23) for g, together with the expression for the Helmholtz free energy density $f = \epsilon \mathbf{E}^2/2$; we obtain

$$g(\mathbf{E}) = -\frac{1}{2}\epsilon \mathbf{E}^2 \ .$$

This expression is the same as that obtained through direct integration of the differential dg and shows that the Gibbs free energy density can be expressed as a function of the electric field $\mathbf{E}$.

To illustrate the use of the Gibbs free energy, consider a linear, homogeneous, and possibly anisotropic dielectric medium with a permittivity matrix $[\epsilon]$. The equation of state relating the electric displacement vector $\mathbf{D}$ to the electric field $\mathbf{E}$ can be expressed in a Cartesian basis as

$$\begin{pmatrix} D_x \\ D_y \\ D_z \end{pmatrix} = \begin{bmatrix} \epsilon_{xx} & \epsilon_{xy} & \epsilon_{xz} \\ \epsilon_{yx} & \epsilon_{yy} & \epsilon_{yz} \\ \epsilon_{zx} & \epsilon_{zy} & \epsilon_{zz} \end{bmatrix} \begin{pmatrix} E_x \\ E_y \\ E_z \end{pmatrix} .$$

The differential of the Gibbs free energy is

$$dg = -\mathbf{D} \cdot d\mathbf{E} = -D_x dE_x - D_y dE_y - D_z dE_z .$$

Since f is a state function, df is an exact differential and must satisfy the Schwarz property:

$$\left. \frac{\partial D_x}{\partial E_y} \right|_{E_x} = \left. \frac{\partial D_y}{\partial E_x} \right|_{E_y} .$$

Using the equation of state, we can show that this implies

$$\epsilon_{xy} = \epsilon_{yx} .$$

Similarly, we can show by considering the two other pairs of variables (E_x, E_z) and (E_y, E_z), that $\epsilon_{xz} = \epsilon_{zx}$ and $\epsilon_{yz} = \epsilon_{zy}$. Therefore, the symmetry of the permittivity matrix is a direct consequence of the fact that the Gibbs free energy is a state function. This elegant approach demonstrates how thermodynamic principles can be used to infer properties of materials.

6.10   Forces Applied to a Dielectric Medium

6.10.1   Force and Torque Density

From Eq. (3.23), we know that the force $\mathbf{F}$ exerted on a dipole moment $\mathbf{p}$ in an electric field $\mathbf{E}$ can be expressed component-wise in a Cartesian basis as

$$F_x = \mathbf{p} \cdot \nabla E_x ,$$
$$F_y = \mathbf{p} \cdot \nabla E_y ,$$
$$F_z = \mathbf{p} \cdot \nabla E_z .$$

The torque on the dipole moment is given by

$$\mathbf{T} = \mathbf{p} \times \mathbf{E} .$$

For a volume element d^3x in a dielectric medium, the dipole moment is $d\mathbf{p}(\mathbf{x}) = \mathbf{P}(\mathbf{x})d^3x$. The infinitesimal force exerted on this dipole moment can be expressed component-wise as

$$d F_x = \mathbf{P}d^3x \cdot \nabla E_x ,$$
$$d F_y = \mathbf{P}d^3x \cdot \nabla E_y ,$$
$$d F_z = \mathbf{P}d^3x \cdot \nabla E_z .$$

This leads to the force density $\mathbf{f} = \dfrac{d\mathbf{F}}{d^3x}$, whose components are $f_x = (\mathbf{P} \cdot \nabla)E_x$, $f_y = (\mathbf{P} \cdot \nabla)E_y$, $f_z = (\mathbf{P} \cdot \nabla)E_z$. A compact expression for the force density writes

$$\boxed{\mathbf{f} = (\mathbf{P} \cdot \nabla)\mathbf{E} .}$$

The torque density exerted on a dielectric medium can be similarly obtained:

$$\boxed{\tau = \mathbf{P} \times \mathbf{E} .}$$

These results lead to the following observations:

- *Uniform Electric Field*: The force components f_x, f_y, and f_z are zero if the electric field $\mathbf{E}$ is uniform. Forces arise in regions where the electric field is non-uniform.
- *Linear homogeneous and isotropic medium*:

    - In such a medium, the torque density $\tau = \mathbf{0}$ because $\mathbf{P}$ is parallel to $\mathbf{E}$.
    - Using $\mathbf{P} = \epsilon_0 \chi \mathbf{E}$, the force density becomes

$$\mathbf{f} = \frac{\epsilon_0}{2}\nabla(\chi \mathbf{E}^2) = \frac{1}{2}\nabla(\mathbf{E} \cdot \mathbf{P}) .$$

    It is important to note that this expression differs from the force density on a permanent dipole moment which does not depend on $\mathbf{E}$: $\mathbf{f} = \nabla(\mathbf{P} \cdot \mathbf{E})$.
    These distinctions highlight the unique behavior of induced dipole moments in a dielectric medium compared to permanent dipole moments.

6.10.2   *Electrostatic Pressure for a Linear Homogeneous Isotropic Medium*

We will now extend the concept of electrostatic pressure to the case of a conductor surrounded by a linear, homogeneous, and isotropic (LHI) dielectric medium with a permittivity ϵ, as illustrated in Fig. 6.28.

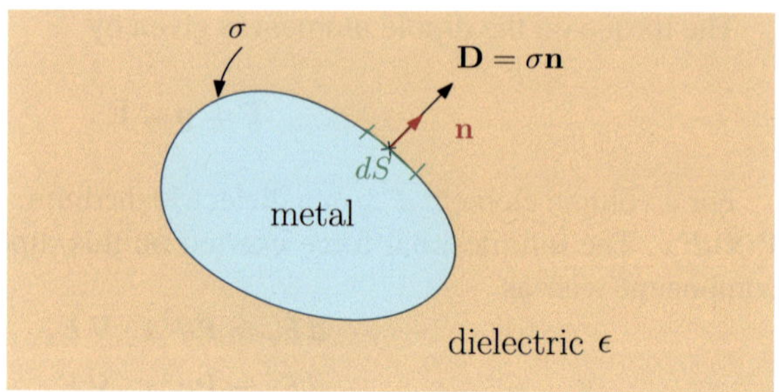

Fig. 6.28 A conductor surrounded by a LHI dielectric medium

Consider a conductor with a surface charge density σ. To derive an expression for the electrostatic pressure, we focus on an infinitesimal surface element dS on the conductor and the rest of the system, denoted as R. We aim to determine the force $d\mathbf{f}$ exerted on the charge $dq = \sigma\, dS$ on dS by the electrostatic forces due to charges in R.

$$d\mathbf{f} = dq\, \mathbf{E}_R(\mathbf{x}) = \sigma\, dS\, \mathbf{E}_R(\mathbf{x}) ,$$

where $\mathbf{x}$ is a point on the surface of the conductor and $\mathbf{E}_R(\mathbf{x})$ is the electric field at $\mathbf{x}$ generated by charges in R (excluding the charges on dS itself).

Coulomb's theorem allows us to relate the electric displacement field $\mathbf{D}$ to the surface charge density σ:

$$\mathbf{D} = \sigma\mathbf{n} ,$$

where $\mathbf{n}$ is the unit normal vector pointing outward from the conductor. The corresponding electric field writes

$$\mathbf{E} = \frac{\sigma}{\epsilon}\mathbf{n} .$$

The total electric field at point $\mathbf{x}$ can be expressed as the sum of the fields produced by the charges on dS and the charges in R:

$$\mathbf{E}(\mathbf{x}) = \mathbf{E}_{dS}(\mathbf{x}) + \mathbf{E}_R(\mathbf{x}) .$$

From the perspective of point $\mathbf{x}$, the infinitesimal surface element dS behaves like an infinite plane carrying a surface charge density σ. Therefore, the field $\mathbf{E}_{dS}(\mathbf{x})$ is the same as the field produced by an infinite plane:

$$\mathbf{E}_{dS}(\mathbf{x}) = \frac{\sigma}{2\epsilon}\mathbf{n} .$$

Combining these equations, we can find the field $\mathbf{E}_R(\mathbf{x})$:

$$\mathbf{E}_R(\mathbf{x}) = \mathbf{E}(\mathbf{x}) - \mathbf{E}_{dS}(\mathbf{x}) = \frac{\sigma}{2\epsilon}\mathbf{n} .$$

The force exerted on dS can be expressed as the product of the surface area and the electrostatic pressure:

$$d\mathbf{f}(\mathbf{x}) = \sigma\, dS \frac{\sigma}{2\epsilon}\mathbf{n} = p_e(\mathbf{x})\, dS\mathbf{n} .$$

This leads to the expression for the electrostatic pressure:

$$\boxed{\; p_e(\mathbf{x}) = \frac{\sigma^2(\mathbf{x})}{2\epsilon} \;} .$$

We see that the electrostatic pressure has the same form as in a vacuum, but with the permittivity replaced by the permittivity of the medium. This indicates that the electrostatic pressure is reduced by a factor of ϵ_r. This is consistent with the increase in capacitance by a factor of ϵ_r when a dielectric medium is inserted into a capacitor.

6.11   Summary and Essential Formulas

- A material medium, composed of molecules or neutral atoms, is characterized at the macroscopic scale by its polarization vector $\mathbf{P}$, which represents the average dipole moment per unit volume.
- A polarized medium Ω for which $\mathbf{P} \neq \mathbf{0}$ can be seen as a charge distribution in Ω, consisting of a volume density ϱ_P and a surface density σ_P on $\partial\Omega$, and given by

$$\boxed{\; \varrho_P = -\nabla \cdot \mathbf{P} \quad \text{and} \quad \sigma_P = \mathbf{P} \cdot \mathbf{n} , \;}$$

where $\mathbf{n}$ is the normal to the surface $\partial\Omega$, pointing outward.
- The potential at any point $\mathbf{x}$ generated by a polarized dielectric carrying both a distribution of free charges of density ϱ_{free} and polarization charges of density ϱ_P in the volume Ω and σ_P on its surface $\partial\Omega$ writes

$$V(\mathbf{x}) = \frac{1}{4\pi\epsilon_0}\left[\iiint_{(\Omega)} \frac{\varrho_{\text{free}}(\mathbf{x}') + \varrho_P(\mathbf{x}')}{|\mathbf{x} - \mathbf{x}'|}d^3x' + \iint_{\partial\Omega} \frac{\sigma_P(\mathbf{x}')}{|\mathbf{x} - \mathbf{x}'|}dS'\right] .$$

- Gauss's law in a material medium is written

$$\boxed{\; \oiint_{\partial\Omega} \mathbf{D} \cdot d\mathbf{S} = Q_{\text{free}} \quad \Leftrightarrow \quad \nabla \cdot \mathbf{D} = \varrho_{\text{free}} , \;}$$

where $\mathbf{D}$ is the displacement vector, defined by

$$\boxed{\; \mathbf{D} = \epsilon_0\mathbf{E} + \mathbf{P} . \;}$$

- The displacement field exhibits a discontinuity of its normal component across an interface

$$(\mathbf{D}_1 - \mathbf{D}_2) \cdot \mathbf{n} = \sigma \, ,$$

  where σ denotes the surface density of free charge at the interface and $\mathbf{n}$ the normal to the interface pointing from medium 2 to medium 1.
- The tangential component of the electric field is continuous across an interface between two dielectric media

$$(\mathbf{E}_1 - \mathbf{E}_2) \cdot \mathbf{t} = 0 \, ,$$

  where $\mathbf{t}$ is the unit tangent vector to the interface.
- The differential for the volume density of electrostatic free energy writes

$$df = \mathbf{E} \cdot d\mathbf{D} \, .$$

  The differential for the electrostatic free energy writes

$$dF = \iiint_\Omega \mathbf{E}(\mathbf{x}) \cdot d\mathbf{D}(\mathbf{x}) d^3 x \, .$$

  Integration to obtain the free energy F requires an equation of state $\mathbf{D}(\mathbf{E})$.
- The force exerted on a polarized medium in an electric field is expressed via the force density:

$$\mathbf{f} = (\mathbf{P} \cdot \nabla)\mathbf{E} \, .$$

  The density of torque exerted on the dielectric medium is

$$\boldsymbol{\tau} = \mathbf{P} \times \mathbf{E}$$

  and is non-zero only if $\mathbf{P}$ and $\mathbf{E}$ are not colinear.
- In a linear, homogeneous, and isotropic medium, the dipole moments are aligned according to the orientation of the electric field $\mathbf{E}$, and the polarization density $\mathbf{P}$ is proportional to $\mathbf{E}$, $\mathbf{P} = \epsilon_0 \chi \mathbf{E}$, with $\chi > 0$. The displacement vector can be written

$$\mathbf{D} = \epsilon \mathbf{E} \, ,$$

  where $\epsilon = \epsilon_0(1 + \chi) > \epsilon_0$ is the permittivity of the medium. If the medium is homogeneous, everything occurs as in vacuum, except that the permittivity of vacuum is multiplied by the relative permittivity, or dielectric constant $\epsilon_r = 1 + \chi \geq 1$. In particular, the fundamental equations of electrostatics in matter are then written

$$\nabla \cdot \mathbf{E} = \frac{\varrho_{\text{free}}}{\epsilon} \quad \text{and} \quad \nabla \times \mathbf{E} = \mathbf{0} \, ,$$

where ϱ_{free} represents the density of free charges. The contribution of the bound (polarization) charges is thus included in ϵ. The dielectric medium tends to decrease the magnitude of the electric field generated by free charges, because the induced bound charges tend to shield the latter.

- In a linear, homogeneous, and isotropic medium, the volume density of free energy is

$$f(\mathbf{D}) = \frac{\mathbf{D}^2}{2\epsilon}.$$

- The local field felt by an atom or molecule in a medium is given by $\mathbf{E}_l = \mathbf{E} + \dfrac{\mathbf{P}}{3\epsilon_0}$.
- The polarizability α of an atom/molecule is a dimensionless constant defined from the relation between the dipole moment and the electric field, $\mathbf{p} = \epsilon_0 \alpha \mathbf{E}_l$.
- Clausius–Mossotti relation applies to diluted media:

$$\frac{n\alpha}{3} = \frac{\epsilon_r - 1}{\epsilon_r + 2},$$

where ϵ_r is the dielectric constant, α the polarizability, and n the density of dipoles in the medium.

- The capacity of the plane plate capacitor of thickness d with a dielectric of permittivity ϵ between the plates reads

$$C = \frac{\epsilon S}{d}.$$

More generally, in a LHI medium the capacitance matrix coefficients are multiplied by ϵ_r.

- The expression for the electrostatic pressure on a surface carrying a charge of density σ is $p = \dfrac{\sigma^2}{2\epsilon}$.

Problems

6.1 Dielectric shell

Consider a dielectric spherical shell of inner radius a, outer radius b, and permittivity ϵ. A point charge q is located at the center of the shell, as shown below. Determine the electric field at all points in space, both inside and outside the shell. Additionally, calculate the bound charge densities on the inner and outer surfaces of the shell.

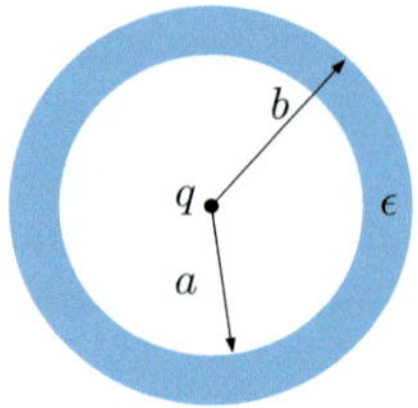

6.2 Conductive sphere surrounded by a dielectric shell

A conductive sphere of radius a carries a charge Q. The sphere is surrounded by a nonconductive concentric shell of internal radius $b > a$, external radius c, and relative dielectric constant ϵ_r. The entire system, as illustrated below, is isolated from its surroundings.

(a) Calculate the electric field at all points in space.
(b) Find the potential difference between the surface of the conducting sphere and infinity.
(c) What is the energy needed to move a charge q from infinity to the outer surface of the dielectric shell?

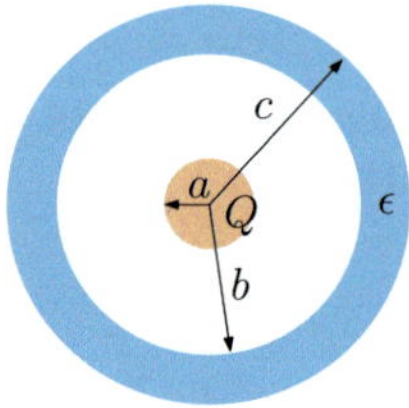

6.3 Concentric spheres

Consider three concentric conductive spheres of radii a, b, and c. The space between the first two spheres is filled with a dielectric of constant ϵ_r. Initially the sphere of radius a is not charged and the spheres of radii b and c carry a charge q_1 and q_2, respectively. The inner sphere of radius a is connected to the sphere of radius c by a thin insulated wire, as shown below.

(a) Calculate the free charge on each of the three spheres.
(b) Calculate the bound charge on the external surface ($r = b$) and on the internal surface ($r = a$) of the dielectric.

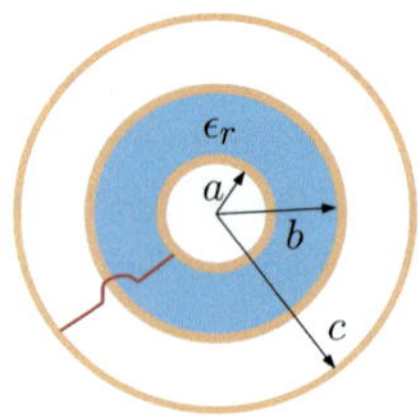

6.4 Cylindrical capacitor filled with an inhomogeneous dielectric

Consider two coaxial conducting cylinders. The inner cylinder of radius r_1 carries a uniformly distributed charge, q. The outer cylinder, with a radius $r_2 > r_1$, carries a uniformly distributed charge, $-q$. The space between the cylinders is filled with a dielectric material whose relative permittivity, $\epsilon_r(r)$, is a function of the radial distance, r, from the central axis. Both cylinders have a length l, which is significantly greater than r_1 and r_2, allowing us to assume cylindrical symmetry for the electric fields, as if the conductors were infinitely long.

(a) Find an expression for $\epsilon_r(r)$ that generates an electrostatic field that is radial and independent of the distance to the axis.

(b) Calculate the capacity of the corresponding cylindrical capacitor.

(c) Calculate the surface density of bound charges on the surface of the dielectric material at $r = r_1$ and $r = r_2$, for the function $\epsilon_r(r)$ calculated in question a.

(d) In cylindrical coordinates, for a vector $\mathbf{A}$ that only depends on r, the expression for the divergence is $\nabla \cdot \mathbf{A} = \dfrac{1}{r}\dfrac{\partial}{\partial r}(r A_r)$. Calculate the volume density of polarization charges in the dielectric.

(e) Calculate the total amount of bound charges, including charges on the surface and in the volume of the dielectric.

(f) Calculate the electrostatic potential energy stored in the capacitor.

(g) Assuming that the dielectric material between the two cylinders can be easily slid out along the central axis without encountering friction, calculate the work required to remove this dielectric from the region between the plates of the cylindrical capacitor.

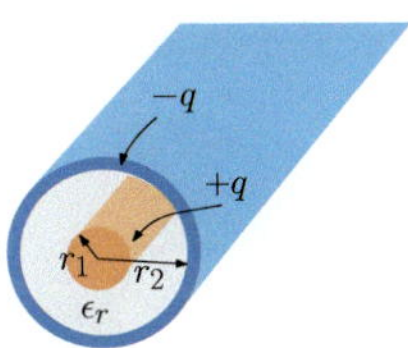

6.5 Dielectric sphere in a uniform field

A dielectric sphere of radius a and permittivity $\epsilon = \epsilon_r\epsilon_0$ is placed in an initially uniform electric field $\mathbf{E} = E_0\mathbf{u}_z$, as shown below. Find the potential and the electric field in all space. Show that the polarization vector is given by

$$\mathbf{P} = \frac{3\epsilon_0(\epsilon_r - 1)}{\epsilon_r + 2}\mathbf{E}_0$$

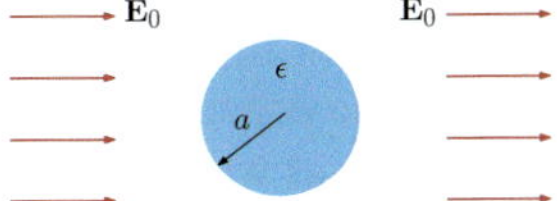

6.6 Classical model for the Van der Waals force

A dipole $\mathbf{p}_0$ is located at position $\mathbf{d}$ from the center of a dielectric sphere of radius R and dielectric constant ϵ_r. We assume $d = |\mathbf{d}| \gg R$.

(a) Express the electric field $\mathbf{E}_0$ at the center of the sphere due to the dipole $\mathbf{p}_0$.
(b) Considering $\mathbf{E}_0$ as approximately constant in the vicinity of the sphere, express the dipole moment $\mathbf{p}$ of the sphere as a function of ϵ_r, $\mathbf{E}_0$, and R. *Hint: Use the result of Exercise* 6.5.
(c) Using the result of Exercise 3.10, show that the force between the two dipoles reads

$$\mathbf{F}_{\mathbf{p}\to\mathbf{p}_0} = -\frac{3}{4\pi\epsilon_0}\left(\frac{\epsilon_r - 1}{\epsilon_r + 2}\right)\frac{R^3}{d^7}[4(\mathbf{p}_0 \cdot \hat{\mathbf{r}})^2\hat{\mathbf{r}} + p_0^2\hat{\mathbf{r}} - (\mathbf{p}_0 \cdot \hat{\mathbf{r}})\mathbf{p}_0]$$

where $\hat{\mathbf{r}} = \mathbf{d}/d$ is the unit radial vector.
(d) Express the force acting on the dielectric sphere, averaged over all possible directions of $\mathbf{p}_0$ and comment the result.

6.7 Point charge and dielectric sphere

A point charge q is in vacuum at a distance d from the center of a dielectric sphere of radius $a < d$ and dielectric constant $\epsilon_r = \epsilon/\epsilon_0$.

(a) Find the potential at all points in space.
(b) Calculate the Cartesian components of the electric field near the center of the sphere.
(c) Verify that in the limit $\epsilon \to \infty$, the solution approaches for a conducting sphere.

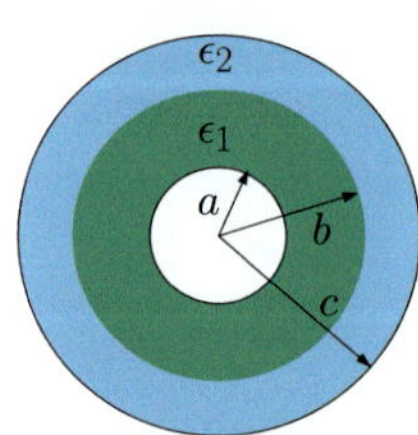

6.8 Spherical capacitor with two dielectrics

Consider a conductive sphere of inner radius a and outer radius c. The space between the two surfaces is filled with two different dielectrics, so that the dielectric constant is ϵ_{r_1} between a and b, and ϵ_{r_2} between b and c. Determine the capacitance of the system.

6.9 Force on a dielectric bar

Consider a parallel-plate capacitor with square plates of side length a. The plates are separated by a distance $d \ll a$. The capacitor carries a total charge Q. Calculate the force exerted on a dielectric material when it is placed at a distance x from the origin O, as illustrated below.

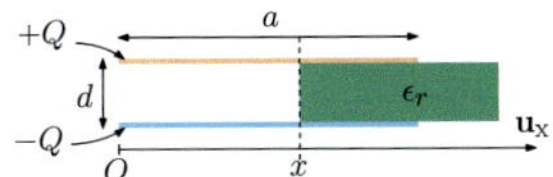

6.10 Capacitor with different dielectrics

Calculate the capacity of the capacitor displayed below.

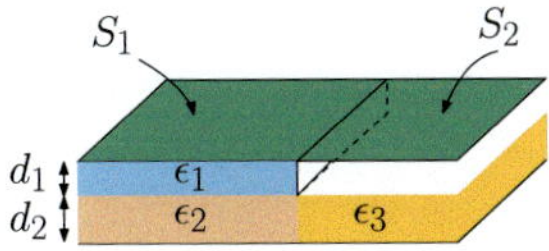

Chapter 7
Electric Currents, Ohm's Law and Electrical Networks

Abstract This chapter transitions from electrostatics to *electrodynamics*, focusing on the flow of electric charge as *electric currents*. It begins with a *microscopic description of conductivity* using *Drude's model*, explaining how free electrons in a conductor acquire a drift velocity under an applied electric field, leading to a steady current. The concepts of carrier lifetime and mobility are introduced to characterize this microscopic behavior. The chapter then derives *Ohm's law* in both its local and macroscopic forms, establishing the fundamental relationship between current density, electric field, and the material *conductivity (or resistivity)*. It explores the implications of time-varying electric fields on conductivity and discusses the phenomenon of charge neutrality in conductors, introducing the concept of *dielectric relaxation time*. A brief overview of *conductors, semiconductors, and superconductors* highlights their distinct electrical properties, particularly their temperature dependence. The latter part of the chapter examines in detail *DC circuit elements* and *electrical networks*. It defines bipoles (two-terminal devices) and introduces the *passive sign convention* for consistent circuit analysis. The concept of *electrical power* in circuit components, including the *Joule effect* (power dissipated as heat in resistors), is thoroughly discussed. Various types of bipoles, such as resistors, diodes (including PN junction, LED, Zener, and tunnel diodes), and ideal sources (voltage and current), are characterized by their *current-voltage (I-V) characteristics*. Finally, the chapter lays out the foundational rules for analyzing complex electrical networks: *Kirchhoff's Current Law (KCL)*, based on charge conservation at nodes, and *Kirchhoff's Voltage Law (KVL)*, derived from the conservative nature of the electric field in steady states. It introduces powerful circuit simplification theorems, including the *Superposition theorem* for linear circuits with multiple sources, and *Thévenin's and Norton's theorems* for reducing complex two-terminal networks to simpler equivalent circuits. The chapter concludes with a discussion of *voltmeters and ammeters* and their operating principles.

Keywords Electricity · Conductivity · Drude's model · Resistance · Resistivity · Ohm's law · Electrical networks · Kirchhoff's laws

   215
F. Cadiz and A. Couairon, *Classical Electrodynamics*, Undergraduate Texts in Physics, https://doi.org/10.1007/978-3-031-86785-9_7

7.1 Introduction

So far, we have only considered the case of static equilibrium in conductors, where charges are at rest on average and do not move. Locally, however, each electron experiences forces from electric fields generated by nuclei, other electrons, and charged impurities, causing them to constantly move and interact. While electrons in a metal are never truly at rest, their movements average out to produce no net flow of charge. This situation changes when a conductor is connected to a battery or, more generally, to a voltage source, which imposes an electric field that drives electrons through the circuit. This controlled flow of electrons, known as electric current, is ubiquitous and powers our lights, modern electronic devices, and even our muscles.

In this chapter, we introduce Drude's model of metallic conductivity and derive the relationship between current density and the electric field inside a conductor, which, in the simplest case, leads to Ohm's law. We define and discuss the concepts of conductivity and resistivity and provide a brief overview of the electrochemical principles underlying the operation of batteries.

We then establish the notation and conventions used for analyzing stationary circuits, which consist of interconnected electrical components such as resistors, diodes, capacitors, and voltage or current sources. This analysis is grounded in Kirchhoff's laws. Finally, we introduce useful mathematical theorems for simplifying linear circuits.

7.2 Microscopic Description of Conductivity. Drude's Model (1900)

A rigorous microscopic description of the electrical conductivity of a conductor requires tools from statistical physics and quantum mechanics. However, it is possible to understand the origin of conductivity from *phenomenological* arguments based solely on classical physics, as carried out in the Drude model, proposed by Paul Drude (see Fig. 7.1). This section reviews his derivation.

In a crystal such as copper, positive ions will only slightly shift from their average position, forming a periodic lattice that undergoes weak perturbations from perfect periodicity. This small amplitude motion is due to their large mass. Each atom of the metal liberates at least one electron from the atom's outer shell. These electrons are mobile throughout the entire volume of the metal. They are called *charge carriers* because they are mobile. This is illustrated in Fig. 7.2.

At equilibrium, electrons are constantly colliding with each other and with the positive ions. After each collision, the electron momentum acquires a new, random direction so that in the absence of an external electric field the electron's motion describes a random walk, similar to what happens to a molecule in a perfect gas. In equilibrium, the average velocity of an electron in a conductor is zero, $\langle \mathbf{v} \rangle = \mathbf{0}$ as well as the average electric field. Despite its very large velocity between collisions,

Fig. 7.1 Paul Drude (1863–1906), a German physicist known for his pioneering works in solid-state physics and optics. He contributed significantly to the study of light-matter interactions and electromagnetic theory

Fig. 7.2 In a metal, free electrons move in the presence of a fixed periodic lattice of positive ions

the electron does not advance in any particular direction. To get an idea of how fast electrons move in metals, recall that for a classical gas the average kinetic energy is $1/2 m_e \langle \mathbf{v}^2 \rangle = 3/2 k_B T$, where k_B is Boltzmann's constant and T is the temperature. At room temperature ($T = 292$ K), we have $\sqrt{\langle \mathbf{v}^2 \rangle} = \overline{v} \sim 10^5$ m/s. The quantity $\overline{v}$ is called the root mean square (rms) velocity. This value is however an underestimation, since electrons in metals behave according to quantum laws. In copper, for example, the velocity between collisions can reach 10^6 m/s due to the Pauli exclusion principle.

Let us now consider an out-of equilibrium situation, in which an external electric field $\mathbf{E}$ is present inside a conductor. Due to this field, in between two collisions the electron will be accelerated in the direction opposite to the electric field. After multiple collisions, the electron has drifted in this direction so that the electron acquires a non-zero average velocity $\langle \mathbf{v}_d \rangle$, called drift velocity. This is illustrated in Fig. 7.3. Typically, it is orders of magnitude smaller than the thermal velocity between collisions, but it is this velocity, always in the same direction, that is responsible for the appearance of an electrical current.

Fig. 7.3 In the presence of
an electric field, an electron
will drift due to its
acceleration

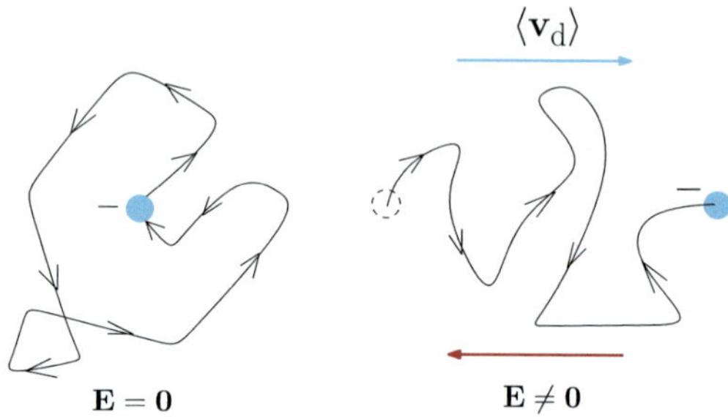

7.2.1 *Carrier Lifetime*

The collision of an electron is defined as the start time of the electron's motion.
While the electron does not undergo another collision, the time t after the collision is
defined as the *age* of the electron. Let us suppose that for an electron born at $t = 0$,
the probability for a collision to occur during the time interval dt between times t
and $t + dt$ is dt/τ, where τ is independent of the carrier's age t and of the velocity
that the electron had before the collision. Let us calculate the probability $p(t)$ that
an electron reaches age t, that is, no collision occurs between 0 and t. To do this, we
will link $p(t + dt)$ to $p(t)$. If no collisions occurred between 0 and $t + dt$, it means
that no collisions occurred between 0 and t, and no collisions occurred between the
remaining time interval dt, so $p(t + dt)$ is equal to the product of probabilities that
no collision occurs

$$\underbrace{p(t + dt)}_{\text{No coll. btw. 0 and } t + dt} = \underbrace{p(t)}_{\text{0 and } t} \; \underbrace{(1 - dt/\tau)}_{t \text{ and } t + dt} \;\; \Rightarrow \;\; \frac{dp}{p} = -\frac{dt}{\tau} \; .$$

After simple algebra, we obtain a well known ordinary differential equation for $p(t)$
whose solution is

$$p(t) = p_0 e^{-t/\tau} \; ,$$

where p_0 is determined from the normalization constraint

$$\int_0^{+\infty} p(t)\,dt = 1 \;\Rightarrow\; p_0 = 1/\tau \; .$$

The age of the electron will lie between t and $t + dt$ with probability $p(t)$. Hence,
we can calculate the average carrier's lifetime

$$\langle t \rangle = \int_0^{\infty} t \, p(t)\,dt = \int_0^{\infty} \frac{t}{\tau} e^{-t/\tau}\,dt = \tau \; .$$

We find an average lifetime equal to the constant τ introduced earlier.

$$\boxed{\langle t \rangle = \tau = \Lambda/\overline{v}} \quad \text{(mean free path/rms velocity) .}$$

We can thus interpret this time constant as the time between two collisions, which is the ratio of the mean free path by the rms velocity.

7.2.2 Carrier Mobility

In the absence of an electric field, the velocity $\mathbf{v}_i$ of the ith electron is randomly distributed and when averaged spatially, we obtain $\langle \mathbf{v}_i \rangle = \mathbf{0}$. Under the influence of an electric field $\mathbf{E}$ applied at $t = 0$, the motion of carriers is governed by Newton's second law:

$$m_e \frac{d\mathbf{v}_i}{dt} = -e\mathbf{E} . \tag{7.1}$$

The velocity $\mathbf{v}_i$ of an electron of mass m_e and charge $-e$, that will be accelerated in the direction of the electric field satisfies:

$$\mathbf{v}_i(t) = \mathbf{v}_i(0) - \frac{et}{m_e}\mathbf{E} .$$

Electrons cannot accelerate indefinitely. Of course, the above equation, applied to a single electron, is only valid between two collisions, since an electron will give away part of its kinetic energy during each collision. The Newton equation must then be modified to take into account the fact that sooner or later, a collision will occur. In order to take the effect of collisions into account, let us average the electron velocities, and divide the electrons into two families:

- The fraction dt/τ of electrons that suffer a collision between t and $t + dt$. The contribution of these electrons to the average velocity at $t + dt$ is zero.
- The fraction $(1 - dt/\tau)$ of electrons that have not participated in any collision between t and $t + dt$. Their velocity has been increased by the electric field $\mathbf{v}_i(t + dt) = \mathbf{v}_i(t) - e\mathbf{E}dt/m_e$. Their contribution to the average velocity is therefore

$$\langle \mathbf{v}_i(t + dt) \rangle = \left(1 - \frac{dt}{\tau}\right)\left(\langle \mathbf{v}_i(t) \rangle - \frac{e\mathbf{E}dt}{m_e}\right) .$$

Rewriting this at first order in dt, we have:

$$\langle \mathbf{v}_i(t + dt) \rangle - \langle \mathbf{v}_i(t) \rangle = \left(-\frac{e}{m_e}\mathbf{E} - \frac{1}{\tau}\langle \mathbf{v}_i(t) \rangle\right)dt . \tag{7.2}$$

In the limit $dt \to 0$ we recognize Newton's equation for the velocity of the gas of electrons $\mathbf{v}(t) = \langle \mathbf{v}_i(t) \rangle$:

$$m_e \frac{d\mathbf{v}}{dt} = -e\mathbf{E} - m_e \frac{\mathbf{v}}{\tau} , \tag{7.3}$$

where a friction term proportional to the average velocity $\mathbf{v}$ of the electron gas, with a characteristic time τ, describes the effect of collisions. This term, describing the interaction between each electron and the rest of the crystal, represents energy dissipation as heat (the Joule effect). Equation (7.3) admits a steady-state solution for the velocity

$$\mathbf{v} = -\frac{e\tau}{m_e}\mathbf{E} = -\mu_e\mathbf{E} ,$$
(7.4)

where $\mu_e = e\tau_e/m_e$ is the electron mobility. This quantity determines the proportionality constant between the external electric field and the steady state average velocity of the electrons. It is this average velocity that supports the steady current flowing in a conductor whenever an electric field is applied.

Before describing macroscopic properties of this current, note the connection between the average time between collisions, introduced as a microscopic feature of the metal, and the trend of the macroscopic electron gas to return to equilibrium: If we suppose that, starting from a steady state, the electric field is turned off at $t = 0$, the averaged velocity will then decay exponentially to zero according to $\mathbf{v}(t) = \mathbf{v}_0\exp(-t/\tau)$, with a typical characteristic time τ. The latter quantity then represents the time it takes for the system to come back to the equilibrium, and it is therefore called relaxation time. The relaxation time is equal to the average time between two collisions, in keeping with the intuitive idea that the only way for the electron gas to reach equilibrium is by means of collisions.

7.3   Electric Currents

Current density
Let n be the density of mobile charges, each of charge q, in a conductor and $\mathbf{v}$ the average velocity of these carriers. The current density is defined as

$$\mathbf{j} = qn\mathbf{v}$$
(7.5)

whose unit in the S.I. system is $\mathrm{C\,s^{-1}\,m^{-2}}$.

Remark Note that a current density may exist inside a neutral medium. It is the case of a conductor, in which the total charge density is generally zero (equal amounts of free electrons and positive ions per unit volume). However, only electrons are mobile in metals and therefore the density of mobile charges equals the density of free electrons. A current may develop in a metal, even if locally, at every point, it remains a neutral material.

The flux of the current density $\mathbf{j}$ through a surface represents the amount of charge per unit time going through that surface. Indeed, consider an arbitrary infinitesimal

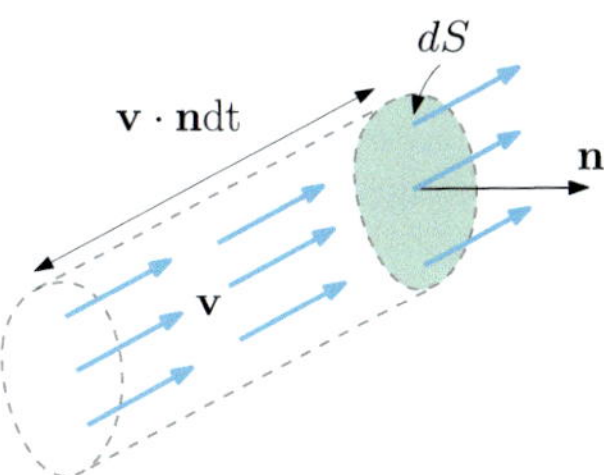

Fig. 7.4 The charge flux through dS during a time dt is $q\mathbf{v}\cdot\mathbf{n}\,dS\,dt$

surface dS with normal $\mathbf{n}$ as shown in Fig. 7.4. During a time interval dt, all the charges that are contained inside a volume $dt\mathbf{v}\cdot\mathbf{n}\,dS$ have gone through dS.

Therefore, the charge dq that goes through $d\mathbf{S}$ during the time interval dt is

$$dq = qn\,dS\,\mathbf{v}\cdot\mathbf{n}\,dt = \underbrace{qn\mathbf{v}}_{\mathbf{j}}\cdot d\mathbf{S}\,dt\,,$$

and the charge going through $d\mathbf{S}$ per unit time is

$$dq/dt = \mathbf{j}\cdot\mathbf{n}\,dS = \Phi_{dS,\mathbf{j}}\,.$$

Note that for a current density of electrons, $qn < 0$ so that the average drift velocity $\mathbf{v}$ and the associated current density $\mathbf{j}$ are of opposite sign. It is purely a matter of convention then that a positive flux of electrons through a surface corresponds to a negative current.

Electric current
The current I through a surface S of a conductor is defined as the total amount of charge going through S per unit time. According to the definition of the current density $\mathbf{j}$, this can be written as the flux of $\mathbf{j}$ through the surface S:

$$\boxed{I = \frac{dq}{dt} = \iint_S dS(\mathbf{x}')\mathbf{n}(\mathbf{x}')\cdot\mathbf{j}(\mathbf{x}') = \Phi_{S,\mathbf{j}}} \tag{7.6}$$

The unit for the electric current is the Ampere, with $1\,A = 1\,C/s$.

7.3.1 Charge Conservation Law—Integral Form

Since all the fundamental interactions between elementary particles seem to conserve the electric charge, it results that the total charge in the universe must be a conserved

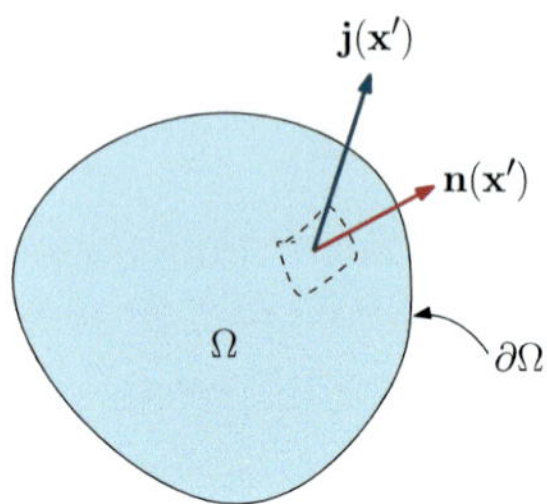

Fig. 7.5 A current flux through the boundary of Ω must be accompanied by a change in the total charge inside

quantity. We will see that the charge density ϱ and the current density $\mathbf{j}$ are related by an equation that assures the local conservation of charge at every point. Let us consider a volume Ω and its positively oriented surface $\partial\Omega$ as shown in Fig. 7.5. At time t, the current going through $\partial\Omega$ is given by

$$I(t) = \Phi_{S,\mathbf{j}(t)} = \oiint_{\partial\Omega} d\mathbf{S}(\mathbf{x}') \cdot \mathbf{j}(\mathbf{x}', t) .$$

Since charge is conserved, a positive (resp. negative) $I(t)$ implies that there must be a decrease (resp. increase) of the total charge $Q(t)$ contained in Ω. That is $dQ(t)/dt = -I(t)$ or, equivalently:

$$\frac{dQ}{dt} = \frac{d}{dt} \iiint_{\Omega} \varrho(\mathbf{x}', t)d^3x' = - \oiint_{\partial\Omega} d\mathbf{S}(\mathbf{x}') \cdot \mathbf{j}(\mathbf{x}', t) ,$$

which gives the integral form of the charge conservation law

$$\iiint_{\Omega} \frac{\partial\varrho(\mathbf{x}', t)}{\partial t}d^3x' + \oiint_{\partial\Omega} d\mathbf{S}(\mathbf{x}') \cdot \mathbf{j}(\mathbf{x}', t) = 0 . \qquad (7.7)$$

7.3.2 Charge Conservation Law—Differential Form

Using Green–Ostrogradsky theorem in (7.7):

$$\iiint_{\Omega} \left(\boldsymbol{\nabla}' \cdot \mathbf{j}(\mathbf{x}', t) + \frac{\partial\varrho(\mathbf{x}', t)}{\partial t} \right) d^3x' = 0 \quad \forall\, \Omega$$

from which we obtain the differential form of the charge conservation law

$$\boldsymbol{\nabla} \cdot \mathbf{j}(\mathbf{x}, t) + \frac{\partial\varrho(\mathbf{x}, t)}{\partial t} = 0 . \qquad (7.8)$$

This equation ensures the local conservation of charge at every point in space and time. From this, we see that a current density compatible with a steady state regime ($\partial \varrho / \partial t = 0$) must satisfy

$$\nabla \cdot \mathbf{j}(\mathbf{x}) = 0 \tag{7.9}$$

which means that the total current across any closed surface must be zero in the steady state. This simple law is very useful when dealing with electric circuits, as will be shown in Sect. 7.10.

7.3.3 Ohm's Law—Microscopic Form

The current density $\mathbf{j}$ in a conductor having a density n of free electrons may be written as $\mathbf{j} = -ne\langle \mathbf{v}_i \rangle$. According to Drude's model, using (7.4) we see that at steady state the current density is proportional to the electric field $\mathbf{E}$:

$$\boxed{\mathbf{j} = ne\mu_e \mathbf{E} = \frac{ne^2 \tau_e}{m_e} \mathbf{E} = \sigma \mathbf{E} \, .} \tag{7.10}$$

Equation (7.10) is the microscopic form of Ohm's law. The quantity $\sigma = ne^2 \tau_e / m$ is the electrical conductivity of the material, whose units in the S.I. system is the Siemens per meter (S/m), where a Siemens corresponds to $1\,\mathrm{S} = 1\,\mathrm{C\,m^{-1}\,s^{-1}\,V^{-1}}$. The resistivity ρ (not to be confused with a charge density) is defined as the inverse of the conductivity, $\rho = 1/\sigma$. A good insulator (glass, plastics, dry air) have a conductivity of the order of $\sigma = 10^{-12} - 10^{-16}\,\mathrm{S\,m^{-1}}$, whereas good conductors (gold, copper, silver) have conductivities around $\sigma = 1 \times 10^8\,\mathrm{S\,m^{-1}}$. The electrical conductivity is one of the physical properties with the largest variation in nature - spanning 24 orders of magnitude between a good conductor and an insulator.

Remark Note that the steady state current density must satisfy $\nabla \cdot \mathbf{j} = 0$ and, according to Ohm's law, $\sigma \nabla \cdot \mathbf{E} = 0$ (if the conductivity is homogeneous). From Gauss's law, this implies $\varrho = 0$ and so the charge density inside a conductor must vanish at steady state. This property extends the known result for conductors at equilibrium and implies that the electric potential in a metal satisfies Laplace's equation

$$\nabla^2 V = 0 \, .$$

Methods of potential theory to find the potential, the field $\mathbf{E} = -\nabla V$ and the current $\mathbf{j} = \sigma \mathbf{E}$ apply in a homogeneous ohmic conductor.

Example 7.1 - Time between collisions in copper

Consider copper, for which the conductivity is $\sigma = 6.54 \times 10^7$ S m^{-1}. It has a face-centered cubic structure with lattice parameter $a = 3.61 \times 10^{-10}$ m. Assuming that each atom of copper contributes with one free electron, and that each elementary cell contains 4 atoms, we deduce that the density of free electrons is

$$n = 4a^{-3} = 8.5 \times 10^{28} \text{m}^{-3}$$

and using the electron charge and mass, the microscopic Ohm law (7.10) gives us the time between collisions

$$\tau = 3 \times 10^{-14}\text{s} = 30\,\text{fs} \ .$$

7.3.4   Ohm's Law in the Case of a Time-Dependent Electric Field

The microscopic form of Ohm law $\mathbf{j} = \sigma \mathbf{E}$ is valid at steady state. In the more general case in which the electric field varies in time, we will see that the conductivity σ of the medium depends on how fast the field oscillates. In other words, the conductivity becomes frequency-dependent. In order to show this, let us suppose that the electric field has a sinusoidal time-dependence of the form

$$\mathbf{E}(t) = \mathbf{E}_0 \cos(\omega t)$$

and let us find a solution for the equation of motion of an electron accelerated by this electric field

$$\frac{d\mathbf{v}}{dt} = -\frac{e}{m_e}\mathbf{E}_0 \cos(\omega t) - \frac{\mathbf{v}}{\tau} \ .$$

To solve it, it is convenient to write the electric field and the electron velocity as

$$\mathbf{E} = \mathrm{Re}\{\underline{\mathbf{E}}\} \qquad \mathbf{v} = \mathrm{Re}\{\underline{\mathbf{v}}\} \ ,$$

where the complex electric field and velocity are given by $\underline{\mathbf{E}} = \mathbf{E}_0 e^{i\omega t}$ and $\underline{\mathbf{v}} = \mathbf{v}_0 e^{i\omega t}$, respectively. Since the equation of motion is linear, we can write

$$\frac{d\underline{\mathbf{v}}}{dt} = -\frac{e}{m_e}\underline{\mathbf{E}} - \frac{\underline{\mathbf{v}}}{\tau}$$

so that the real part of the latter equation (on both sides) gives the original equation of motion. We can therefore solve the equation for the complex velocity, which is particularly simple since $\dfrac{d\mathbf{v}}{dt} = i\omega\mathbf{v}_0 e^{i\omega t}$ so that

$$i\omega\mathbf{v}_0 e^{i\omega t} = -\frac{e}{m_e}\mathbf{E}_0 e^{i\omega t} - \frac{\mathbf{v}_0 e^{i\omega t}}{\tau} .$$

The complex exponentials cancel out on both sides and we obtain for the complex amplitude of the velocity

$$\mathbf{v}_0 = -\frac{e\tau/m_e}{(1 + i\omega\tau)}\mathbf{E}_0 .$$

Since the complex current density is given by $\underline{\mathbf{j}} = -en\underline{\mathbf{v}}$, we get the generalized form of Ohm law for a sinusoidal regime:

$$\underline{\mathbf{j}} = \underline{\sigma}(\omega)\underline{\mathbf{E}} \tag{7.11}$$

where the complex conductivity is given by

$$\underline{\sigma}(\omega) = \frac{\sigma_0}{1 + i\omega\tau} = \frac{ne^2\tau/m_e}{1 + i\omega\tau} \tag{7.12}$$

where $\sigma_0 = ne^2\tau/m_e$ is the static conductivity. From this, the real current density writes

$$\mathbf{j} = \mathrm{Re}\{\underline{\mathbf{j}}\} = \mathrm{Re}\left\{\frac{ne^2\tau/m_e}{(1 + i\omega\tau)}\mathbf{E}_0 e^{i\omega t}\right\} = \frac{\sigma_0}{\sqrt{1 + (\omega\tau)^2}}\mathbf{E}_0 \cos\left(\omega t - \tan^{-1}(\omega\tau)\right) .$$

This current does not respond instantaneously to the electric field due to the phase term $\tan^{-1}(\omega\tau)$. Only if $\omega \ll \frac{1}{\tau}$, in other words if the period of the sinusoidal field is much longer than the time between collisions, one may write a static form of Ohm's law at each time, $\mathbf{j} \approx \sigma_0 \mathbf{E}_0 \cos(\omega t) = \sigma_0\mathbf{E}(t)$.

7.3.5 *A Few Comments on the Classical Drude Model*

The Drude model is a simple way to describe conductivity and Ohm's law in terms of collisions between classical particles, but it has its limitations. A more complete, quantitative description of conductivity requires taking into account the quantum behavior of electrons in solids. In reality, not all free electrons can participate in conductivity. Electrons with small kinetic energies are excluded from any collision mechanism due to Pauli exclusion principle, and for the same reason these electrons may not accelerate under the presence of an electric field. Moreover, one may think that an electron's velocity will be modified each time it encounters an ion from the crystal lattice. However, the experimental values for the time between collisions show that the mean free path is much larger than the distance between atoms in the crystal, and this cannot be explained with a classical picture. In reality, electrons behave like waves that are not perturbed at all by the periodic arrangement of atoms of the lattice. Instead, these waves are only perturbed by deviations from a perfect crystal (atomic vibrations, impurities, etc.).

Finally, there is a whole class of materials relevant for modern technologies called semiconductors, for which the conductivity is intermediate between that of an insulator and that of a conductor, and which that can be tuned by several orders of magnitude by small perturbations (light excitation, electric fields, incorporation of minute proportions of chemical impurities). This versatility is at the heart of the major technological revolution of the 20th century (transistors, diodes, LEDs, solar cells, and semiconductor lasers are a few examples of devices based on semiconducting materials). In solids in general, an electron may not have an arbitrary energy, but instead only energies that fall within bounded intervals (called bands) separated by energy gaps where no possible states are allowed for the electrons. In a semiconductor, a gap exists between non-conductive and conductive states that is of the order of 1eV. This explains why, for example, increasing the temperature (i.e., the available thermal energy) increases the conductivity of a semiconductor, which is opposite to what is observed in metals. The existence of this energy gap can only be explained by quantum mechanics.

7.4   Ohm's Law—Macroscopic Form (1827)

Let us now consider a wire made of a conductive material of length l, section S, and conductivity σ. A difference of electric potential $\Delta V = V_A - V_B$ is imposed between the two ends of the wire, as shown in Fig. 7.6. This potential difference generates an electric field $\mathbf{E}$, and in consequence, a current density $\mathbf{j} = \sigma \mathbf{E}$. At steady state, $\nabla \cdot \mathbf{j} = 0$ so that the current entering the wire through A equals the output current at B. If the section S is constant, the current density $\mathbf{j}$ as well as the electric field $\mathbf{E}$ are uniform inside the conductor. Then

$$\Delta V = V_A - V_A = \int_A^A \mathbf{E(x)} \cdot d\mathbf{x} = El \ .$$

Fig. 7.6 An imposed electric field is accompanied by a steady current density inside a conductor

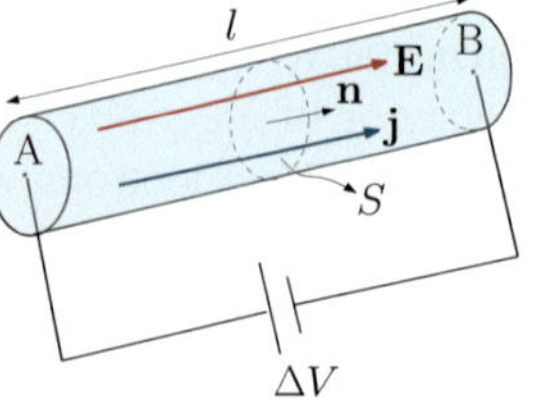

The current density therefore writes

$$\mathbf{j} = \sigma \mathbf{E} = \sigma \frac{\Delta V}{l} \mathbf{n} \,,$$

where $\mathbf{n}$ is the direction normal to the transverse section S and along the electric field. The current through the wire is then

$$I = \mathbf{j} \cdot \mathbf{n} S = \Delta V \frac{\sigma S}{l} \,,$$

which leads to the macroscopic form of Ohm's law, named after Georg Simon Ohm (see Fig. 7.7).

> **Ohm's law (1827)**
> The potential difference ΔV between two extremities A and B of a conductor of section S, length l, and conductivity σ is proportional to the current I going through it:
>
> $$\boxed{\Delta V = \frac{l}{\sigma S} I = R I}$$ (7.13)
>
> where $R = \frac{l}{\sigma S} = \frac{l \rho}{S}$ is the *resistance* of the conductor between A and B. In the S.I. system, the unit for the resistance is the Ohm (Ω), with $1\,\Omega = 1\,\mathrm{V\,A^{-1}}$.

Fig. 7.7 Georg Simon Ohm (1789–1854), a German physicist and mathematician. His works became crucial to the development of electrical science and engineering

7.4.1   Relation Between Resistance and Capacity

Consider a capacitor defined by two equipotential surfaces, S_1 at potential V_1 and S_2 at potential V_2, with a conducting material in between. The surfaces S_1 and S_2 are sections of a current tube, that is, the lateral surface of the tube is tangent to the current vector at each point of the surface (Fig. 7.8).

Due to Ohm's law, the current is proportional to the electric field, hence, a current tube is also a field tube. The lateral surface of the tube is tangent to the electric field at each point of the surface. The resistance of this capacitor is defined as the ratio $R = (V_1 - V_2)/I$ where

$$\left.\begin{aligned} V_1 - V_2 &= \int_{A_1}^{A_2} \mathbf{E} \cdot d\mathbf{l} \\ I &= \iint_{S_1} \mathbf{j} \cdot \mathbf{n} dS, \quad \mathbf{j} = \sigma \mathbf{E} \end{aligned}\right\} \quad R = \frac{\displaystyle\int_{A_1}^{A_2} \mathbf{E} \cdot d\mathbf{l}}{\displaystyle\iint_{S_1} \sigma \mathbf{E}_1 \cdot \mathbf{n}_1 dS_1} \ .$$

Note that charge conservation, i.e., $\nabla \cdot \mathbf{j} = 0$ implies that the current I can be calculated as the flux of the current vector through any section of the field tube (here through S_1; it would yield the same result if calculated through S_2). The capacity between the two surfaces S_1 at potential V_1 and S_2 at potential V_2 reads $C = Q/(V_1 - V_2)$ where Q denotes the free charge on surface S_1, that is

$$Q = \iint_{S_1} \sigma_1 dS_1 \ ,$$

where the surface density of free charge σ_1 is obtained from Coulomb's theorem:

$$\sigma_1 = \epsilon_0 \mathbf{E}_1 \cdot \mathbf{n}_1 \ .$$

Thus,

$$C^{-1} = \frac{V_1 - V_2}{Q} = \frac{\displaystyle\int_{A_1}^{A_2} \mathbf{E} \cdot d\mathbf{l}}{\displaystyle\iint_{S_1} \epsilon_0 \mathbf{E}_1 \cdot \mathbf{n}_1 dS_1} \ .$$

From the resistance and the capacity of this capacitor, we obtain

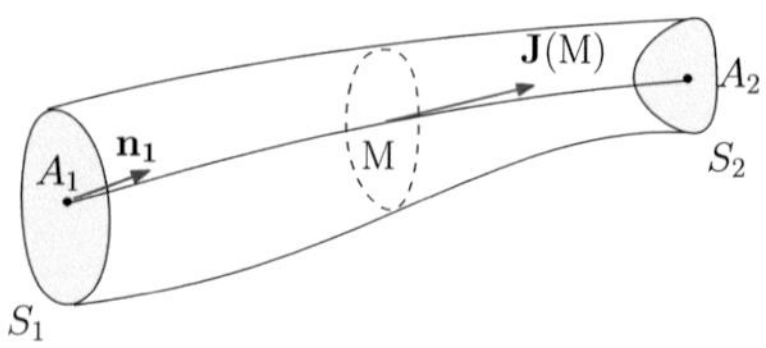

Fig. 7.8 A capacitor filled with a material of resistance R

$$\sigma R = \frac{\epsilon_0}{C} \quad \text{i.e.,} \quad \boxed{RC = \rho\epsilon_0 = \frac{\epsilon_0}{\sigma}} \, . \tag{7.14}$$

The product of the resistance by the capacity of a conductor is equal to the product of the metal resistivity by vacuum's permittivity.

7.4.2 Charge Neutrality in a Conductor

We have seen that inside a conductor in electrostatic equilibrium, the electric field must vanish, so that $\mathbf{E} = \mathbf{0}$ and the charge density ϱ are zero at every point. We will show now that more generally, when a time-varying electric field is present, the charge neutrality in a conductor may or may not be respected depending on the frequency of the electric field with respect to the characteristic timescales at which charge densities may oscillate in the metal. Let us first suppose a sinusoidal electric field of the form

$$\mathbf{E}(t) = \mathbf{E}_0 \cos(\omega t) = \text{Re}\{\underline{\mathbf{E}}\}$$

for which the current density is given by Ohm's law (7.11) in the sinusoidal regime

$$\mathbf{j}(t) = \text{Re}\left\{\sigma(\omega)\underline{\mathbf{E}}\right\} = \text{Re}\left\{\frac{\sigma_0}{1 + \omega\tau}\underline{\mathbf{E}}\right\} \, .$$

Now, recall that the charge density ϱ must satisfy the charge-conservation law (7.8)

$$\frac{\partial \varrho(\mathbf{x}, t)}{\partial t} + \nabla \cdot \mathbf{j}(\mathbf{x}, t) = 0$$

and writing $\varrho = \text{Re}\{\underline{\varrho}\} = \text{Re}\{\varrho_0 e^{i\omega t}\}$, we may instead look for the equation that ϱ_0 must satisfy

$$\frac{\partial \underline{\varrho}(\mathbf{x}, t)}{\partial t} + \nabla \cdot \underbrace{\underline{\mathbf{j}}(\mathbf{x}, t)}_{\underline{\sigma}(\omega)\underline{\mathbf{E}}} = 0 \, ,$$

which becomes:

$$i\omega\underline{\varrho} + \underline{\sigma}(\omega)\nabla \cdot \underline{\mathbf{E}} = 0$$

and according to Gauss's law, $\nabla \cdot \underline{\mathbf{E}} = \underline{\varrho}/\epsilon_0$

$$i\omega\varrho_0 + \frac{\underline{\sigma}(\omega)}{\epsilon_0}\varrho_0 = i\omega\varrho_0 + \frac{\sigma_0}{(1 + i\omega\tau)\epsilon_0}\varrho_0 = 0$$

or, equivalently

$$-\omega^2 \varrho_0 + i\frac{\omega}{\tau}\varrho_0 + \underbrace{\frac{\sigma_0}{\epsilon_0 \tau}}_{\omega_p^2}\varrho_0 = 0 \, ,$$

where we define the plasma frequency as

$$\omega_p = \sqrt{\frac{\sigma_0}{\epsilon_0 \tau}} = \sqrt{\frac{ne^2}{m_e \epsilon_0}}$$

and the latter equation rewrites

$$\left(\omega_p^2 - \omega^2 + i\frac{\omega}{\tau}\right)\varrho_0 = 0 \, .$$

Typically, in metals, we have $1/\tau \ll \omega_p$ so that $\omega = \omega_p$ is a solution of the above equation for which $\varrho \neq 0$. Thus, an electric field oscillating at the plasma frequency can break the charge neutrality of the conductor, maintaining charge waves that naturally oscillate in phase with the electric field.

In the general case $\omega \neq \omega_p$, we see that the charge density satisfies the equation of motion of a damped harmonic oscillator of eigenfrequency ω_p

$$\boxed{\frac{\partial^2 \varrho}{\partial t^2} + \frac{1}{\tau}\frac{\partial \varrho}{\partial t} + \omega_p^2 \varrho = 0 \, .} \tag{7.15}$$

At low frequencies ($\omega \ll \frac{1}{\tau} \ll \omega_p$), the first term may be neglected with respect to the other two and the equation writes

$$\frac{\partial \varrho}{\partial t} + \underbrace{\tau \omega_p^2}_{=\sigma_0 \epsilon_0} \varrho = 0 \, .$$

In this regime, if a perturbation breaks the charge neutrality in a conductor at $t = 0$, the charge density will evolve as

$$\varrho(\mathbf{x}, t) = \varrho_0(\mathbf{x})e^{-t/\tau_d}$$

so that the medium recovers its neutrality exponentially fast, as e^{-t/τ_d}. The characteristic decay time $\tau_d = \sigma_0/\epsilon$ is called the dielectric relaxation time. Any charge imbalance excited by a low-frequency perturbation is therefore neutralized within a time τ_d. Note that from (7.14), the dielectric relaxation time can also be written as $\tau_d = RC$. In metals, this time can be extremely short, of the order of $\tau_d = 1.3 \times 10^{-18}$s, whereas in an insulator such as quartz, we have $\tau \sim 10^5$s, which is of the order of one day.

7.5 Conductors, Semiconductors, Superconductors

The resistivity of a metal, as described by the Drude model, is given by $\rho = 1/\sigma = \frac{m}{ne^2\tau_e}$. At high temperatures, resistivity increases linearly with temperature, $\rho(T) \propto T$, because thermal energy enhances lattice vibrations, reducing the lifetime $\tau_e \propto 1/T$ between electron collisions. This behavior is well-verified at high temperatures and is relatively insensitive to the purity of the metal, making resistivity a reliable measure of temperature. Many thermometers utilize this principle.

In contrast, a distinct class of materials, known as semiconductors, exhibits the opposite behavior: resistivity decreases as temperature increases. This phenomenon, first observed by Michael Faraday, is strikingly different from the behavior of metals. Not only can temperature increase the conductivity of a semiconductor, but also absorption of light, leading to a phenomenon called photoconductivity. An electric field can also modify the material's conductivity, which is the principle behind field effect transistors. All these properties are due to the fact that electrons in semiconductors are not free to move, but can become mobile if a minimum amount of energy, called the bandgap, is supplied. This can be understood in the framework of quantum mechanics.

Now let us discuss what happens at low temperatures. In many metals, such as copper, silver, platinum, and gold, resistivity approaches a constant value ρ_0 at cryogenic temperatures, as shown in Fig. 7.9. At these low temperatures, of a few degrees Kelvin, lattice vibrations no longer dominate electron collisions; instead, resistivity is limited by impurities or defects in the crystal lattice. The higher the concentration of impurities or defects, the larger the value of ρ_0. Even in exceptionally pure metals, the gain in conductivity compared to room temperature is limited to about six orders of magnitude.

The physicist Heike Kammerlingh Onnes (Fig. 7.10), a pioneer in low-temperature physics and a specialist in the liquefaction of gases, achieved the liquefaction of helium in 1911. Helium becomes liquid at 4.2 K, enabling Onnes to study the conductivity of metals at extremely low temperatures. His groundbreaking experiments revealed that the resistivity of mercury drops abruptly by several orders of magnitude at a critical temperature $T_c = 4.18 K$. In superconductors, resistivity is at least 18 orders of magnitude lower than that of copper at room temperature. This dramatic

Fig. 7.9 Resistivity versus temperature for a non-superconducting metal

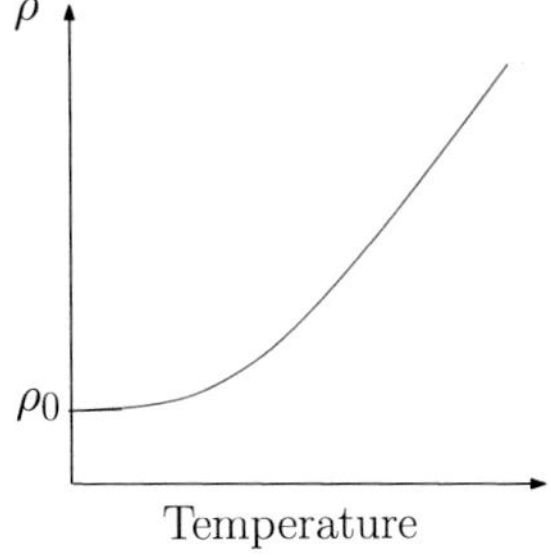

Fig. 7.10 Heike Kamerlingh Onnes (1853–1926), a Dutch physicist and pioneer of low-temperature physics. He achieved the first liquefaction of helium, significantly advancing the field of cryogenics

behavior results from a phase transition, wherein electrons cease to behave as in normal metals. In this superconducting state, electric currents flow without any heat dissipation.

This extraordinary phenomenom, where currents flow without resistance, is reminiscent of the mystery physicists faced before the advent of quantum mechanics regarding the stability of the hydrogen atom. According to classical physics, the hydrogen atom should be unstable, with the electron radiating energy in the form of electromagnetic waves. Quantum mechanics resolved this puzzle by demonstrating that the electron in a hydrogen atom occupies a stationary state, where its energy is well-defined, and no radiation occurs. Similarly, in a superconductor, the macroscopic ensemble of electrons occupies a global stationary state, allowing currents to flow without energy loss. The way superconductors respond to a magnetic field is discussed in Sect. 11.4.3.

7.6 Batteries

An electric battery is a device capable of providing electric energy through the energy released by chemical reactions. It was Alessandro Volta who invented the first electric battery in 1799, the famous voltaic pile depicted in Fig. 7.11. One way to construct such a pile is through a stacking of alternating copper and zinc discs separated by a cardboard soaked in brine. This brine solution constitues an electrolyte: a substance through which ions (not electrons) can move. Each individual element Cu/electrolyte/Zn generates an **electromotive force** (or emf) which can put electrons in motion. Despite its name, the emf is a work per unit charge, and so it is measured in volts and for one cell, it is around 0.76V. The stacking produces a larger total

Fig. 7.11 Left: Alessandro Volta (1745–1827), Italian physicist and chemist. He demonstrated that electricity could be generated chemically, laying the foundation for modern electrochemistry and electrical engineering. He also discovered methane and conducted early research on electrical capacitance. Right: The voltaic pile invented by Alessandro Volta (1799). Picture taken at the Museum of Ecole Polytechnique (Mus'X)

emf. Before the invention of the electric generator by the end of the 19th-century, the whole electrical industry was powered with batteries.

7.6.1 Daniell's Cell (1836)

The first voltaic piles triggered a significant amount of research and allowed for very important discoveries, such as the electrolysis of water and the isolation of several chemical elements. In practice, a major drawback was that they could not provide a large current for a sustained period of time. John Frederic Daniell invented in 1836 an electrochemical cell that was a significant improvement with respect to volta's pile. It became the first practical source of electricity, and was widely adopted by the industry. We will briefly describe the working principles of such battery.

This battery can be viewed as the junction of two electrochemical cells, as shown in Fig. 7.12. The first one consists on a piece of solid Copper (Cu) immersed in a solution of Copper sulfate ($CuSO_4$). This salt dissolves in water into its ions Cu^{2+} and SO_4^{2-}, forming an electrolyte.

In this cell, an equilibrium is established between the reduction (gain of electrons) of Copper ions ($Cu^{2+}_{(aq)} + 2e^- \rightarrow Cu_{(s)}$) and the oxidation (loss of electrons) of Copper ($Cu_{(s)} \rightarrow Cu^{2+}_{(aq)} + 2e^-$), which we write

$$Cu \rightleftharpoons Cu^{2+}_{aq} + 2e^- \, .$$

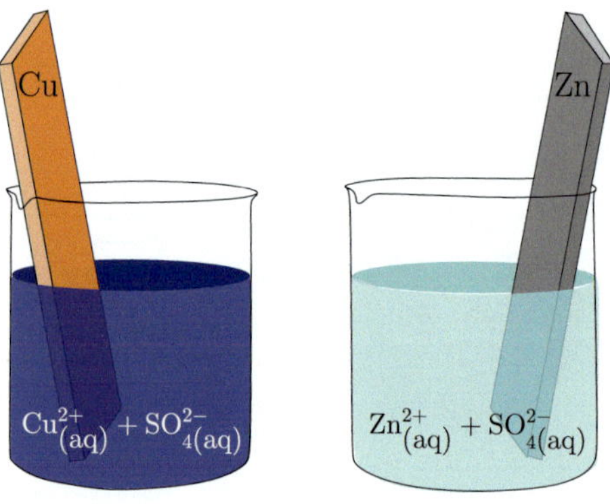

Fig. 7.12 In each cell, a metal M is in contact with an electrolyte containing M^{2+} ions

Similarly, the second cell contains solid Zinc (Zn) in a solution of Zinc sulfate ($ZnSO_4$). The equilibrium between the reduction of Zinc ions and the oxidation of Zinc is:

$$Zn \rightleftharpoons Zn^{2+}_{aq} + 2e^- \ .$$

If an electrical connection is made between the two electrolytes, a potential difference between the Copper and the Zinc electrodes of around 1.1 V appears under normal conditions. This potential difference can be explained in the framework of thermodynamics applied to the different oxidation-reduction processes involved. Such connection between the cells can be made with a salt bridge, that is, a tube filled with an inert electrolyte solution, such as potassium nitrate or potassium chloride. A porous diaphragm helps to contain the electrolyte within the tube while allowing the passage of ions to the cells. This is illustrated in Fig. 7.13.

When the circuit is closed through a wire connecting both metals, a spontaneous oxidation-reduction reaction occurs. The oxidation of Zinc liberates electrons from the Zinc electrode ($Zn_{(S)} \rightarrow Zn^{2+}_{(aq)} + 2e^-$) and Zinc ions into the electrolyte. The Zinc electrode is thus the anode, the liberated electrons are pushed away by it since it is negatively charged. These electrons will then flow from the Zinc cell into the Copper cell, where Copper ions precipitate and are reduced at the Copper electrode by capturing these electrons($Cu^{2+}_{(aq)} + 2e^- \rightarrow Cu_{(S)}$). The Copper electrode is therefore the cathode, electrons are consumed in it and it is positively charged. During the functioning of the battery, solid copper is accumulated at the cathode and Zinc corrodes into the solution. Eventually, this will lead to the depletion of reactants and the battery will stop working (unless the battery is recharged by inversing the electrochemical reaction).

Note that ions in the salt bridge move to each cell in order to maintain the charge neutrality of the electrolytes. This is illustrated in Fig. 7.14. In summary, a continuous flow of charge is established and electrical energy is generated. This energy comes from the chemical energy released through the oxidation and reduction reactions that take place. Note that the total reaction can be written as:

Fig. 7.13 The two cells are electrically connected thanks to the use of a salt bridge

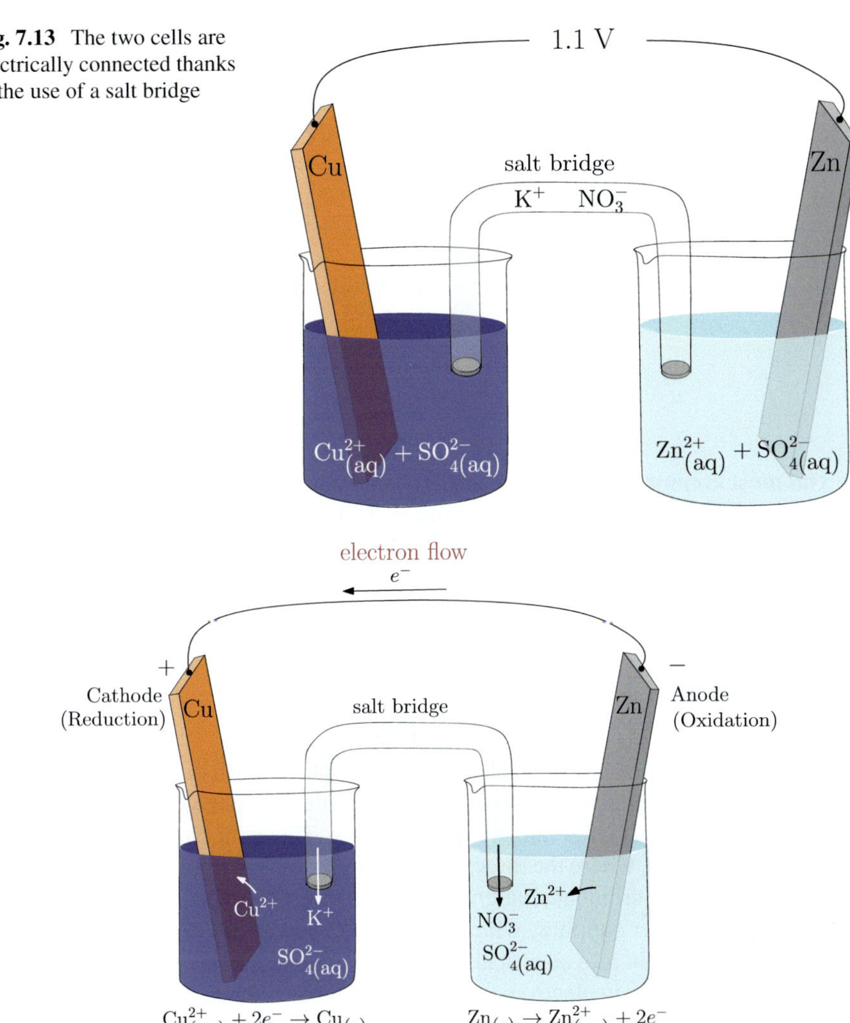

Fig. 7.14 Flow of charges on a Daniell cell

$$Cu^{2+}_{(aq)} + Zn_{(s)} \rightarrow Cu_{(s)} + Zn^{2+}_{(aq)} \ .$$

Nowadays, standard commercial batteries are largely dominated by Zinc-carbon or alkaline batteries. The latter are represented in Fig. 7.15. The electrodes can be connected to a circuit through two pieces of metal called current collectors. This kind of battery provides typically 1.5V and rely on a non-reversible electrochemical reaction, therefore they are not rechargeable.

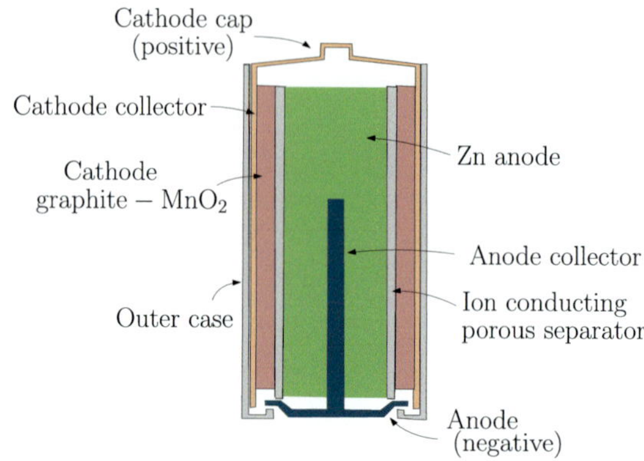

Fig. 7.15 Schematics of an alkaline battery

The most commonly employed rechargeable batteries are lithium-ion, in which both electrodes are compounds containing Lithium (Li), and Lithium ions are exchanged between the electrodes through the electrolyte which is a solution of Lithium salt.

7.7 DC Circuit Elements

A direct current (DC) is a type of electrical current that flows in one direction only in a steady state. DC circuits provide a steady and reliable flow of electrical energy, making them suitable for a wide range of applications. This is in contrast to alternating current (AC) circuits, where the direction of current flow reverses periodically. Common sources of DC current are batteries, which convert chemical energy into electrical energy, and power supplies, which convert AC power to DC power. Solar cells, which convert sunlight into electrical energy, are another example. DC currents have numerous applications in electronic devices (e.g., smartphones, laptops), automotive systems, industrial equipment and battery-powered devices.

7.7.1 Bipoles

A bipole refers to a two-terminal electrical component or device. It is a fundamental building block in circuit analysis and design. Common examples of bipoles include:

- Resistors, which resist the flow of electric current according to Ohm's law.
- Capacitors, which store electrical energy in an electric field.
- Inductors, which store electrical energy in a magnetic field, as discussed in Chap. 11.

- Diodes, which are components that allow current to flow in only one direction.
- Transistors, which can amplify or switch electronic signals.

The latter two examples rely on the physics of semiconductors and are at the heart of modern microelectronics.

A bipole $\mathcal{B}$ has two terminals. These terminals are the points where the bipole connects to other components in a circuit. The representation that will be used here is shown in Fig. 7.16. They may be labeled by anchors or with symbols like $+$ and $-$ to indicate the polarity of voltage or current flow. For DC currents, the amplitude of the current entering through A is equal to that exiting at B. This is due to the charge conservation law. The voltage across the bipole is the potential difference between the terminals,

$$V = V_{AB} = V_A - V_B \ .$$

By combining bipoles in various configurations, we can create a wide range of electronic circuits. For the analysis of such circuits, it is crucial to establish a consistent convention for assigning voltage and current directions. The convention that we will use, known as the *passive sign convention*, helps to ensure accurate analysis and calculations. It consists of assigning a direction for the current and voltage arbitrarily, but in opposite directions, as illustrated in Fig. 7.17.

- Current direction: using an arrow, we arbitrarily assign a direction for the current flow through the component. If the real current entering terminal A of the component flows in this direction (corresponding to a motion of electrons in the opposite direction), its value is positive. Otherwise, its value is negative.
- Voltage orientation: using an arrow, we arbitrarily assign a reference direction to the voltage: $V = V_{AB}$ represents the voltage difference between head and tail. Here, $V = V_A - V_B$. The value of V is positive if A is the terminal with the highest potential, and B, the terminal with the lowest potential. This defines A as the *positive* voltage terminal of the bipole. In the passive sign convention, the positive current orientation is defined as flowing into the bipole through the positive voltage terminal A (arrows point in opposite directions).

The current I and voltage V are therefore algebraic quantities: their value can be positive or negative. The chosen directions may therefore be different from the directions of the actual current flow and voltage.

Remark Consistency is a key point: the chosen convention should be consistent throughout the circuit analysis. The initial assignments of voltage polarities and

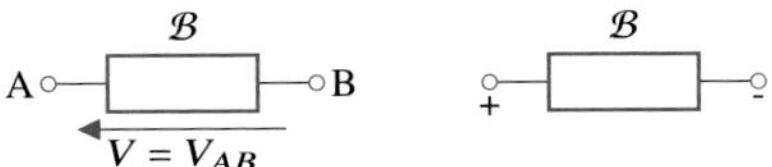

Fig. 7.16 Convention used to represent a two-terminal device, or bipole

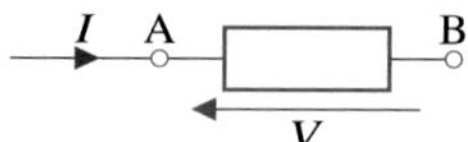

Fig. 7.17 Passive sign convention for a bipole: positive current flows through the bipole into the positive voltage terminal

current directions can be arbitrary. The final results, given by the physical laws, will be correct regardless of the initial choices. Unless otherwise stated, we will use the passive sign convention throughout the book.

7.7.2 Electrical Power in a Component

Consider the bipole AB in Fig. 7.17 with a potential difference $V = V_A - V_B$ between the terminals A and B. The potential energy of a free charge carrier q at terminal A is $q V_A$, where V_A is the potential at A. The power entering at terminal A is expressed as the amount of energy entering at A per unit time, which is the product of the potential energy of a single carrier at A and the number of carriers entering at A per unit time: $\frac{dN}{dt} q V_A = I V_A$. The same reasoning applies to terminal B: the product of the energy $q V_B$ of a charge carrier at terminal B and the rate of charge carrier exiting at B per unit time yields the power exiting terminal B: $\frac{dN}{dt} q V_B = I V_B$. The electrical power received by the component AB (passive component) is then

$$P = I V_A - I V_B \implies \boxed{P = V I \,.}$$

Remarks

- The passive sign convention (PSC) ensures that the power absorbed by a component, $P = V I$, is positive if the current enters the positive terminal (with the highest potential). If the current enters the negative terminal (with the lowest potential), $P < 0$ and the component is supplying power. This is the case for a battery.
- An alternative convention exists in circuit electricity: the active sign convention considers the direction of positive current as an arrow pointing to the negative (lowest potential) terminal of a component. In other words, the current and the voltage-difference arrow point in the same direction, as shown in Fig. 7.18.

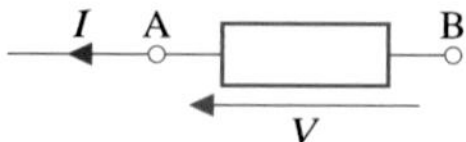

Fig. 7.18 Active sign convention for a bipole: positive current flows through the bipole into the negative voltage terminal

However, a positive current entering the negative terminal is equivalent to a negative current entering the positive terminal. This means that the constitutive relation between current flowing through and voltage across a component is obtained in the active sign convention by changing I to $-I$ in the PSC formulation. For instance, in the active sign convention, the power received by a component is written as $P = -VI$. If both the current I and the voltage V are positive, the power received by the component is negative, or equivalently, the positive power VI is generated by the component. The PSC ensures that a bipole receives a power $P = VI$, which is positive if the electric power effectively flows into the bipole, or negative if the power flows out of the bipole and back into the circuit.

7.7.3 Joule Effect (1860)—Power Dissipated in a Resistance

At steady state, the power absorbed by a resistance equals the amount of power dissipated in the form of heat. Microscopically, this is due to the relaxation of the electron kinetic energy inside the crystal during collisions. Part of this energy is transferred to the vibrational modes of the atoms in the crystal, increasing its temperature. The electric energy thus generates heat inside the conductor. The power dissipated must equal the power $P = IV$ supplied, so that according to Ohm's law (7.13)

$$\boxed{P = IV = I^2R = V^2/R \,.}$$
(7.16)

We see that the energy dissipated is proportional to the square of the current. One may apply this principle to generate heat in an oven, to generate hot air or to heat an incandescent filament in a light bulb. For some applications, this Joule effect, named after James Prescott Joule (see Fig. 7.19) must be minimized in order to maintain good energy efficiency. This is the case for a transmission line, where any loss of energy between a power plant and a city must be minimized. This is why transmission lines usually operate at very high voltages, thus minimizing the current for a given electrical power.

7.7.4 Current-Voltage Characteristics

The relationship between the voltage across the terminals (V) and the current flowing through the terminals (I) is often referred to as the voltage-current characteristic or the IV characteristic of the bipole, illustrated as a graph of current I as a function of voltage V. In the example of Fig. 7.20 (left), I_0 is the cut-off current for which the voltage across the terminals of the bipole is zero, while V_0 is the voltage in an open circuit ($I = 0$). A linear bipole is a bipole whose characteristic is a straight line as shown in Fig. 7.20 (right).

Fig. 7.19 James Prescott
Joule (1818–1889) was an
English physicist and
mathematician. His works on
the relationship between heat
and mechanical energy lead
to the formulation of the first
law of thermodynamics

Fig. 7.20 IV characteristics
of a bipole (left) and a linear
bipole (right)

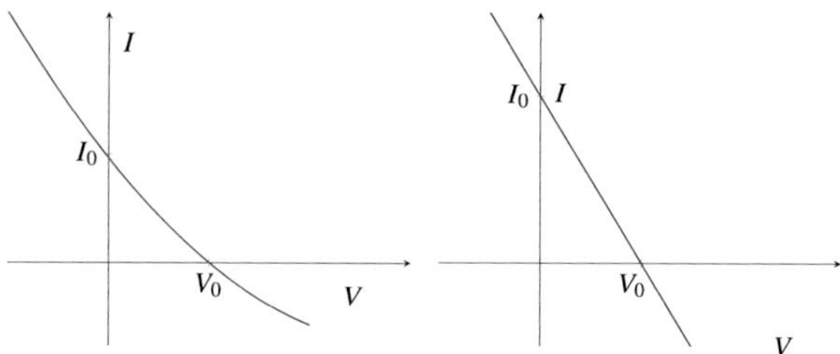

7.7.5 Resistors

- *Perfect resistors*: This is a resistor that satisfies Ohm's law

$$V = RI \ ,$$

  where R is the resistance of the component. In other words, the voltage across
  a resistor is directly proportional to the current flowing through it. For a perfect
  resistor, the IV characteristic is a straight line passing through the origin, with a
  slope equal to the inverse of the resistance.
- *Voltage dependent resistor (VDR)*: this is a resistor whose resistance depends on
  the voltage across its terminals. It is also known as a varistor.
- *Photoresistor*: This is a resistor whose voltage depends on the flux ϕ of photons
  it receives. The higher the photon flux, the lower the resistance. This effect can
  be achieved with semiconducting materials which exhibit photoconductivity, a
  phenomenon in which absorption of light generates free carriers in their conduction
  band.

The IV characteristics for these kinds of resistor bipoles are shown in Fig. 7.21.

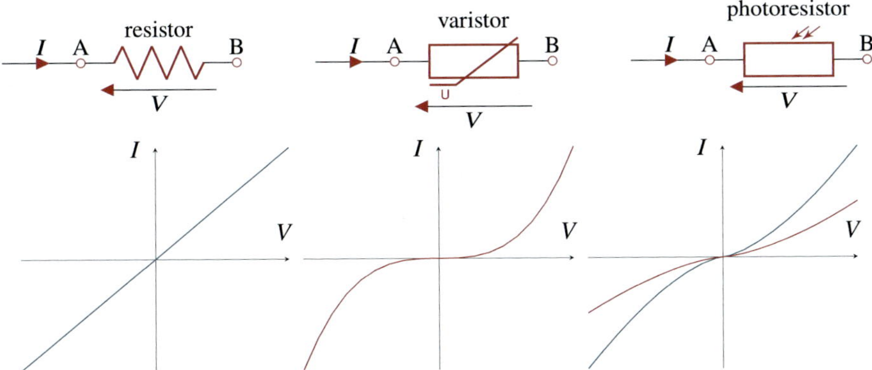

Fig. 7.21 IV characteristics for a resistor (left), a voltage dependent resistor (center), and a photoresistor (right), for which the IV curve depends on the incident photon flux

7.7.6 Diodes

A diode is a two-terminal electronic component that allows current to flow in only one direction. It is typically created by joining two types of doped semiconductors. Doping involves intentionally introducing impurities into the crystal lattice of the semiconductor material.

In an n-type semiconductor, an impurity with an extra valence electron is introduced. This extra electron becomes free to move through the lattice, significantly increasing the conductivity of the material. Conversely, in a p-type semiconductor, an impurity with one fewer valence electron is added. This deficiency creates a *hole* in the lattice, representing the absence of a free electron. Interestingly, these holes behave as positive charge carriers because electrons from neighboring atoms move to fill the vacancies, effectively causing the holes to move within the crystal. In p-type semiconductors, the primary charge carriers are holes, which carry a positive charge.

A p-n junction is illustrated in Fig. 7.22, where the terminal *A* is connected to the p-type semiconductor, and the terminal *B* to the n-type semiconductor.

This kind of bipole does not behave the same way under positive (forward) and negative (reverse) polarity of the applied voltage. At the interface between the two semiconductors, a highly resistive region known as the depletion zone forms. The

Fig. 7.22 A p-n junction

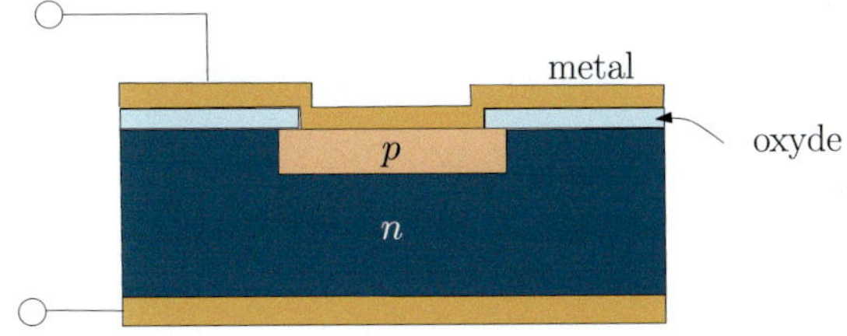

width of this zone can be adjusted by the application of a potential difference. Under forward bias, the width of the depletion zone decreases, leading to an exponential increase in the forward current. In contrast, under reverse bias, the depletion zone widens, making the diode even more resistive so that practically no current flows. As a result, current con only flow in one direction. This enables for the conversion of an alternating current to a direct current in a circuit called rectifier. This property was utilized shortly after the invention of the p-n junction in the early 1940s to enhance the reliability of wireless communications.

In 1947, the first semiconductor transistor was invented at Bell Labs. It was based on two p-n junctions made of germanium. This lead to the replacement of old vacuum tubes and allowed for the miniaturization of electrical circuits. The first integrated circuit was invented a decade later. Nowadays, microprocessors contain billion of transistors based on silicon. The p-n junction can also be used to emit light, the principle relies in the radiative recombination of electron and holes in the p-n junction. These are called Light Emitting Diodes (LEDs) and are the most energetically efficient emitters of light. Other applications of p-n junctions include semiconductor laser diodes, phototransistors and photovoltaic cells. Figure 7.23 shows the IV characteristics of different types of diodes.

For diodes under forward bias, the anode is at a higher potential than the cathode. The anode and the cathode correspond to the A and B terminals shown in Fig. 7.24.

- PN junction diode (Fig. 7.23a):

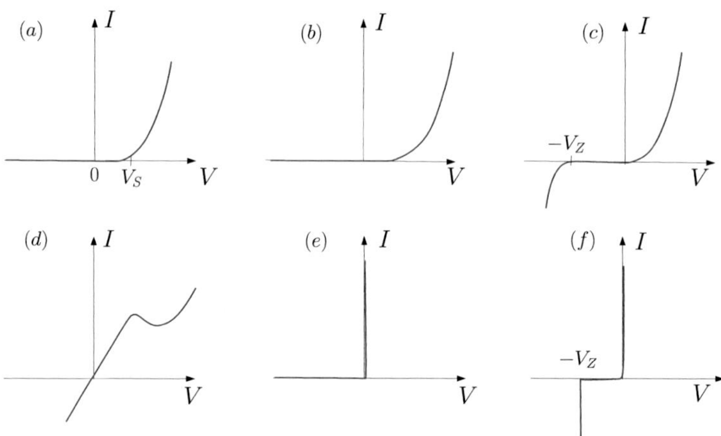

Fig. 7.23 Voltage-current characteristics of different types of diodes: **a** PN junction diode; **b** Light-Emitting Diode; **c** Zener diode; **d** Tunnel diode; **e** ideal PN junction diode; **f** Ideal Zener diode

Fig. 7.24 PN junction diode

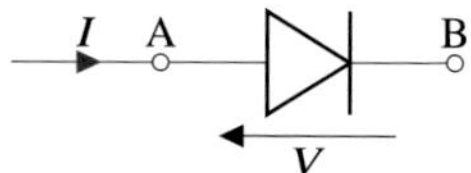

- Forward Bias: $V > 0$, the diode conducts. The IV characteristic shows a rapid increase in current with a small increase in voltage, after a certain threshold voltage V_S (typically 0.7V for silicon diodes).
- Reverse Bias: $V < 0$, the diode blocks current flow. The IV characteristic shows a very small leakage current until a certain reverse breakdown voltage is reached, at which point the diode conducts heavily before being damaged.

- Light-Emitting Diode (LED) (Figs. 7.23b and 7.25):

  - Forward Bias: when forward-biased, a LED emits light. The IV characteristic is similar to a PN junction diode, but with a higher forward voltage drop.
  - Reverse Bias: similar to a PN junction diode, it blocks current flow in reverse bias and there is no light emission.

- Zener Diode (Figs. 7.23c and 7.26):

  - Forward Bias: similar to a PN junction diode, it conducts in the forward bias region.
  - Reverse Bias: unlike a PN junction diode, it allows a controlled amount of current to flow in the reverse bias region when the voltage across it exceeds a specific breakdown voltage (the Zener voltage $-V_Z$). This characteristic is used for voltage regulation.

- Tunnel diode (Figs. 7.23d and 7.27):
  A tunnel diode, also known as an Esaki diode, is a special type of semiconductor diode that exhibits a unique voltage-current characteristic with a region of negative differential resistance. This characteristic arises from a quantum mechanical phenomenon known as quantum tunneling: In a heavily doped PN junction, the energy barrier between the conduction band of the n-type region and the valence band of the p-type region is very narrow. This allows electrons to tunnel directly through the barrier, even when the applied voltage is insufficient to overcome the barrier classically. The IV characteristic of a tunnel diode is distinctive:

  - Forward Bias: as the forward bias voltage increases, the current initially increases rapidly due to tunneling. At a certain peak voltage, the current reaches a maximum. Beyond the peak voltage, the current decreases with increasing

Fig. 7.25 Light-Emitting diode

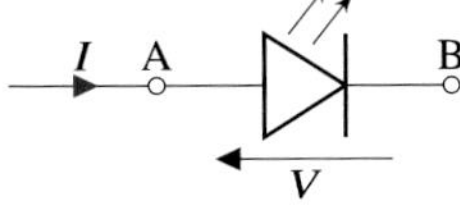

Fig. 7.26 Zeener diode

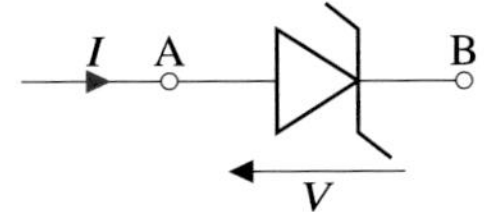

Fig. 7.27 Tunnel diode

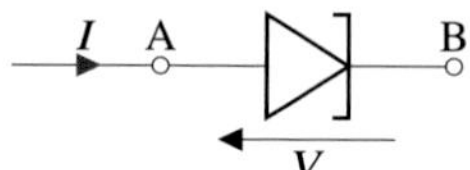

voltage, resulting in a region of negative differential resistance. Finally, the current starts to increase again due to conventional forward biasing.
– Reverse Bias: the diode behaves like a conventional PN junction diode, with a small reverse leakage current.

- Ideal diodes (Fig. 7.23e):
  An ideal diode is a theoretical device that perfectly conducts current in one direction (forward bias) and completely blocks current in the other direction (reverse bias). It is a simplified model idealizing the limitations of real-world diodes. It is used to analyze and design circuits. The key characteristics of an ideal diode are:

  – Zero forward voltage drop: When forward-biased, an ideal diode offers zero resistance, allowing current to flow without any voltage drop.
  – Infinite reverse resistance: When reverse-biased, an ideal diode offers infinite resistance, completely blocking the flow of current.
  – Zero switching time: An ideal diode can switch instantaneously between the conducting and non-conducting states.

- Ideal Zener diode (Fig. 7.23f):
  An ideal Zener diode is a model for analyzing circuits by neglecting the limitations of real-world Zener diodes. Its IV characteristic is similar to that of a PN junction diode when forward biased. When reverse biased, an ideal Zener diode remains blocked until the voltage across it exceeds the Zener voltage. The diode then enters in the breakdown region, in which the voltage across the diode remains nearly constant at the Zener voltage, even as the current through the diode increases.

7.7.7   *Utility Range of a Bipole*

A bipole can usually support a maximum power P_{max} before being damaged or before a significant performance degradation occurs. The validity range of the bipole characteristic law is limited by this power, as illustrated in Fig. 7.28.

This typically involves a range of voltages and currents that the device can handle. Various factors affect the utility range: the maximum power that a bipole can safely dissipate without overheating, or the maximum voltage and current that a bipole can withstand without breaking down. The operating temperature range may also affect the performance and reliability of the component.

Fig. 7.28 The utility range of a bipole is the region of the IV characteristic limited by $I = P_{\max}/V$ (dashed curves)

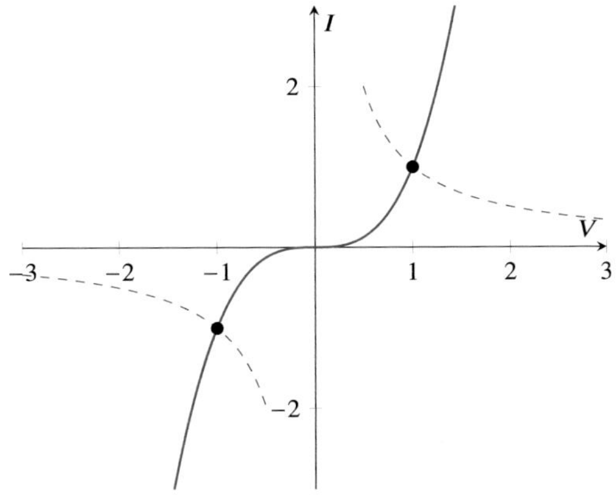

7.7.8 *Classification of Bipoles*

Bipoles can be classified based on several key features:

1. Linearity: A bipole is linear if its voltage-current relationship is directly proportional. This means that doubling the voltage doubles the current. This the case of ideal resistors. In contrast, if its voltage-current relationship is not directly proportional, the bipole is nonlinear. This is the case of diodes, transistors, and many other electronic components.
2. Receptor/Generator: A passive bipole (receptor) cannot generate energy. It can only dissipate or store energy. Its voltage-current characteristic passes through the origin $(I, V) = (0, 0)$. Examples: resistors, capacitors, and inductors (see Chap. 11). In contrast, an active bipole (generator) can generate energy. Its voltage-current characteristic does not pass through the origin. It is polarized, with two distinguishable terminals. Examples: batteries, power supplies, and transistors.
3. Symmetry: A bipole is symmetric if its voltage-current characteristic remains the same when the terminals are reversed. Examples: resistors, capacitors, and inductors. In contrast, a bipole is asymmetric if its voltage-current characteristic changes when the terminals are reversed. Examples: diodes and transistors.

These classifications facilitates the analysis and design of complex electrical circuits. For instance, linear bipoles can be analyzed using linear circuit analysis techniques and linear algebra, while nonlinear bipoles require more advanced techniques.

7.8 Connection of Bipoles

When two bipoles $\mathcal{B}_1$ and $\mathcal{B}_2$ are in series or in parallel, they form a bipole $\mathcal{B}$ with a IV characteristic that can be graphically obtained as a function of those of the constitutive bipoles.

7.8.1 *Connecting Bipoles in Series*

Series connection is one of the fundamental ways to connect electrical components. When components are connected in series, they share a single node, meaning the same current flows through each component as shown in Fig. 7.29.

Figure 7.30a illustrates the IV characteristics C_k of bipoles $\mathcal{B}_k$ connected in series. The current I flowing through each component is the same. A key characteristics of the series connection is voltage division: the total voltage V across the series combination is equal to the sum $V_1 + V_2$ of the individual voltage drops across each component. This yields the IV characteristic C: $(I, V = V_1 + V_2)$ for the bipole equivalent to the series connection.

7.8.1.1   Connecting Bipoles in Parallel

Parallel connection is another fundamental way to connect electrical components. In a parallel connection, the terminals of two or more components are connected together at both ends, as shown in Fig. 7.31.

Figure 7.30b illustrates the IV characteristics C_k of bipoles $\mathcal{B}_k$ connected in parallel. The voltage V across each component is the same. A key characteristic of parallel connections is current division: the total current I flowing into the parallel combination is divided among the components. This yields the IV characteristic C of the bipole equivalent to the parallel connection: $(I = I_1 + I_2, V)$.

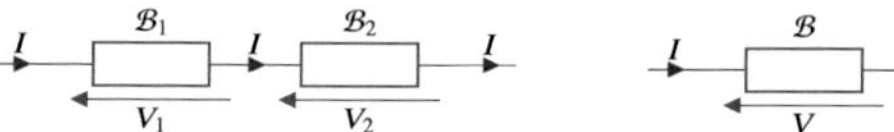

Fig. 7.29 Two bipoles connected in series (left) and its equivalent bipole (right)

Fig. 7.30 Voltage-current characteristics of two bipoles connected **a** in series; **b** in parallel

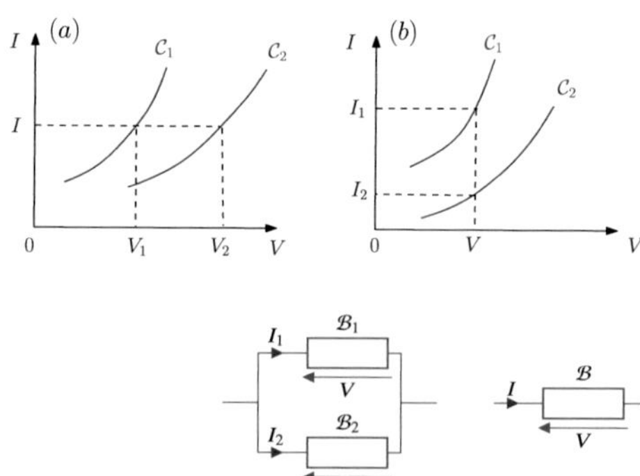

Fig. 7.31 Two bipoles connected in parallel (left) and its equivalent bipole (right)

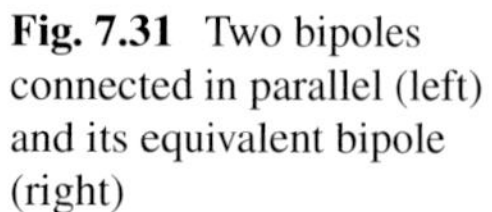

7.8.2 *Connecting Resistors in Series and in Parallel*

For resistors R_1 and R_2 connected in series, the equivalent resistance is the sum of the individual resistances. The same amount of current I flows through each resistor. Applying twice Ohm's law, we find the voltage across each resistor: $V_1 = R_1 I$ and $V_2 = R_2 I$. The voltage difference at the terminals of the equivalent bipole is $V = V_1 + V_2 = (R_1 + R_2)I$. Since it must also satisfy Ohm's law, $V = RI$, we conclude that resistance R is the sum of the resistances wired in series:

$$R = R_1 + R_2 \ .$$

For resistors connected in parallel, the reciprocal of the equivalent resistance is equal to the sum of the reciprocals of the individual resistances. Indeed, Ohm's law applied to each resistor writes $V = R_1 I_1$ and $V = R_2 I_2$ where the voltage V is the same for the two resistors as the terminals of R_1 are connected to those of R_2. The current I is the sum of the currents flowing in each resistor $I = V/R_1 + V/R_2$. Ohm's law must be valid for the resistor equivalent to the parallel connection, $I = V/R$. We conclude that

$$\frac{1}{R} = \frac{1}{R_1} + \frac{1}{R_2} \ .$$

The inverse of a resistance is called a conductance:

$$G = \frac{1}{R}$$

and its unit is the Siemens, with symbol S $= \Omega^{-1}$. The conductance of a bipole is the sum of the conductances wired in parallel.

Example 7.2 - A diode and resistor in series
When an ideal diode and a resistor are connected in series, their combined IV characteristic is a combination of their individual characteristics. The voltage across the diode and the resistor must add up to the total applied voltage. For $V > 0$ (forward bias) the IV-characteristic is a straight line passing through the origin, with a slope equal to the reciprocal resistance of the resistor. For $V < 0$ (reverse bias), the IV-characteristic is a horizontal line along the voltage axis, indicating zero current flow for any applied reverse voltage. The resistance behavior dominates in forward bias, whereas the diode behavior dominates in reverse bias. This is illustrated by the red curve in Fig. 7.32a.

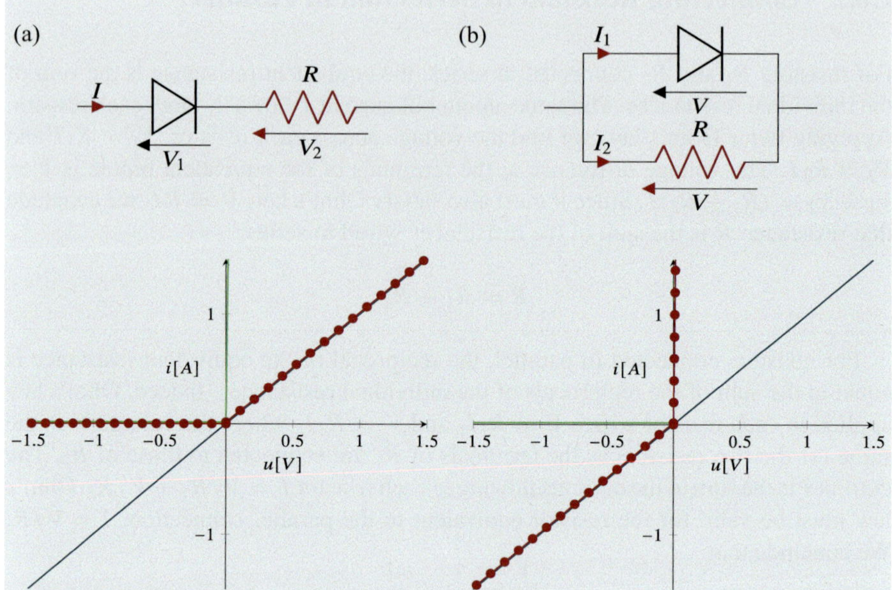

Fig. 7.32 IV-characteristics (red dotted curve) for an ideal diode (green curve) and a resistor (blue curve) connected in series (**a**); parallel (**b**)

Now if the diode and the resistor are connected in parallel, the current across each element must add up to the total current. For $V > 0$ (Forward bias) the IV characteristic is a vertical line along the current axis, indicating that the diode behavior dominates the circuit response. For $V < 0$ (Reverse bias) the IV characteristic is a straight line passing through the origin, with a slope equal to the inverse of the resistance of the resistor, indicating that the resistor's behavior dominates in reverse bias. This is illustrated by the red curve in Fig. 7.32b.

7.9   Linear Bipoles

Linear bipoles are two-terminal electrical components whose voltage-current relationship is linear. This means that this relationship is a straight line: the current flowing through the component is expressed in the form of a generalized Ohm law,

$$V = RI - e \, ,$$

where R is the resistance and e is the algebraic electromotive force (emf) of the bipole. Equivalently, the IV characteristic can be written in terms of conductance (G), the reciprocal of resistance:

Fig. 7.33 IV characteristic
of a linear bipole

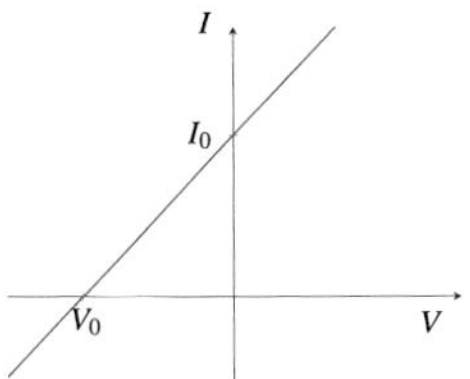

$$I = GV + I_0 \,,$$

where $G = 1/R$ is the conductance and I_0 is the short circuit current, also known as electromotive current, which is the current that flows through the component when its terminals are short-circuited. This is illustrated in Fig. 7.33.

7.9.1 Examples of Linear Bipoles

- Ideal resistors: their IV characteristics is given by Ohm's law: $V = RI$.
- Ideal generator (emf source): it is a bipole whose voltage is constant for any current crossing it. Batteries, generators, solar cells are emf sources.
  Symbols representing emf sources are given together with a voltage, E, indicating the value of the potential difference across the terminals: $E = V_+ - V_- = V_A - V_B$.

  - Battery: For a single cell (left symbol in Fig. 7.34) or a multiple cell battery (right symbol in Fig. 7.34), the large bar represents the positive terminal, also known as cathode for an active component.
  - Voltage source: the American symbol (left in Fig. 7.35) indicates polarity while the European symbol (right in Fig. 7.35) does not. For this reason, we will use the American symbols for voltage (and current) sources.

  The IV characteristic for the ideal voltage source is a straight line of equation $V = E$, where the current I is assumed to be oriented to enter by the positive terminal of the voltage source, as illustrated in Fig. 7.36.
  Real-world sources of emf such as batteries, generators, solar cells have an internal resistance, which reduces the terminal voltage when current flows. The relationship

Fig. 7.34 Symbols to
represent batteries with (left)
single cell or (right) multiple
cells

Fig. 7.35 Symbols used to
represent a voltage source
(left) American symbol.
(right) European symbol

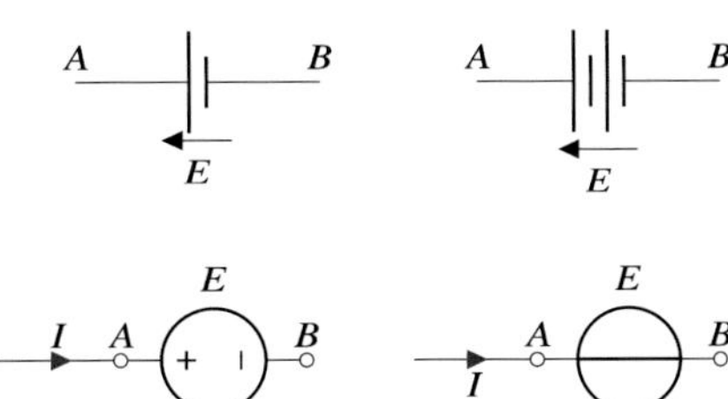

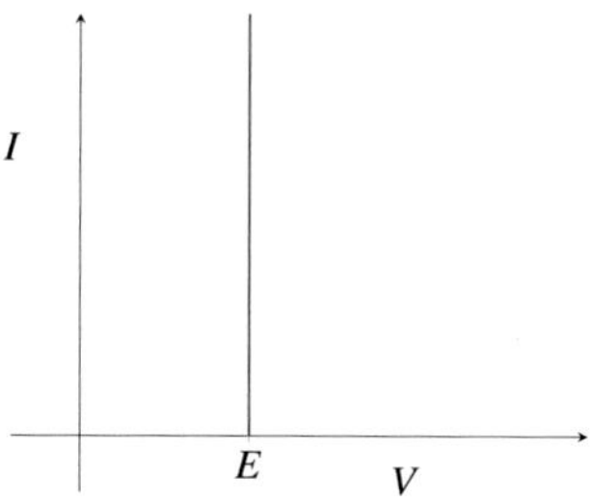

Fig. 7.36 IV characteristic of an ideal voltage source

Fig. 7.37 This circuit is equivalent to a real emf bipole with internal resistance r

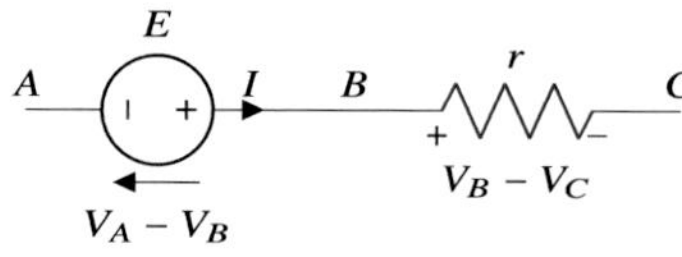

Fig. 7.38 Symbols used to represent a current source: (left) American symbol. (right) European symbol

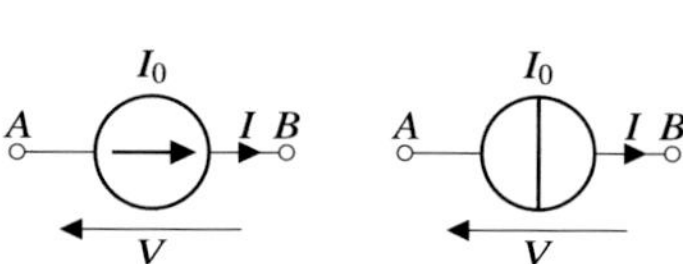

between emf (E), terminal voltage (V), internal resistance (r), and current (I) is given by $V = rI - E$. In a circuit, it can be represented as a series connection of an ideal voltage source and an ideal resistor, as illustrated in Fig. 7.37. The series connection allows us to easily check the relation

$$V = V_A - V_C = (V_A - V_B) + (V_B - V_C) = -E + rI .$$

The concept of an ideal voltage source is approached by a battery with a weak internal resistance.

- Ideal current source: an ideal current source is a model of bipole crossed by a constant current for any voltage difference across its terminals. The American convention (left symbol in Fig. 7.38) indicates the positive real direction of current I_0, while the European convention (right symbol in Fig. 7.38) does not. With the chosen passive sign convention marked in Fig. 7.38, the equation for the IV characteristics is $I = I_0$. The voltage across an ideal current source is completely determined by the circuit it is connected to.

7.9.2   Connection of Linear Bipoles

- In series: assume a series connection of N bipoles $\mathcal{B}_k$, $k = 1, \ldots N$. The IV relation for each bipole writes

$$V_k = R_k I - E_k ,$$

where V_k is the voltage across the terminals, R_k is the internal resistance and E_k is the emf for the bipole $\mathcal{B}_k$. The same current I crosses each bipole. The potential difference at the terminals of the equivalent bipole is $V = \sum_{k=1}^{N} V_k$. Introducing the IV relations, this voltage becomes

$$V = \left(\sum_{k=1}^{N} R_k \right) I - \sum_{k=1}^{N} E_k = RI - E \ , \quad \text{with } R = \sum_{k=1}^{N} R_k \quad \text{and } E = \sum_{k=1}^{N} E_k \ .$$

- In parallel: assume a parallel connection of the N bipoles.
Now each bipole has the same voltage V across its terminals. The IV relation for each bipole writes
$$I_k = G_k V + I_{0k} \ ,$$

where I_k is the current through bipole $\mathcal{B}_k$, G_k is its internal reciprocal resistance and I_{0k} is its short circuit current. The total current flowing in the equivalent bipole is $I = \sum_{k=1}^{N} I_k$. Introducing the IV relations, this current becomes

$$I = \left(\sum_{k=1}^{N} G_k \right) V + \sum_{k=1}^{N} I_{0k} = GV + I_0 \ , \quad \text{with } G = \sum_{k=1}^{N} G_k \quad \text{and } I_0 = \sum_{k=1}^{N} I_{0k} \ .$$

7.10 Electric Networks, Kirchhoff's Laws

Electric circuits are interconnected networks of electrical components, such as resistors, diodes, capacitors, inductors, and voltage and current sources. These components are connected by conductive wires, which are assumed to have negligible resistance. Currents meet or drain away at the nodes of the network. An example is shown in Fig. 7.39. In electric networks:

- *Nodes* are points where two or more circuit elements connect.
- *Branches* are paths between two nodes, consisting of one or more circuit elements.
- *Loops* are closed paths within a network, formed by a sequence of branches passing only once through a given node.

Fig. 7.39 Example of electric network

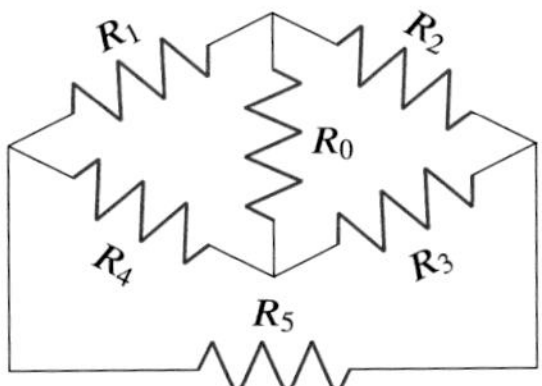

Kirchhoff's laws, named after Gustav Kirchhoff (see Fig. 7.40), are the rules linking currents at junctions of network branches or voltages around network loops.

Kirchhoff's current law (KCL)
The algebraic sum of currents entering a node (junction) in a circuit is equal to the algebraic sum of currents leaving the node.

For example, consider a node with five branches shown in Fig. 7.41. If currents I_1, I_2, and I_3 are flowing into the node, and current I_4 and I_5 are flowing out, then according to KCL:

$$I_1 + I_2 + I_3 - I_4 - I_5 = 0 .$$

Consider a node with N branches connected to it. Let us assign a current I_k to each branch k, with the direction of the current arbitrarily chosen. Kirchhoff's Current Law (KCL) can be expressed mathematically as

$$\boxed{\sum_{k}^{N} \epsilon_k I_k = 0 ,}$$

Fig. 7.40 Gustav Kirchhoff (1824–1887) was a German physicist who made significant contributions to circuit theory, spectroscopy, and thermodynamics

Fig. 7.41 Node with five branches, illustrating Kirchhoff's current law

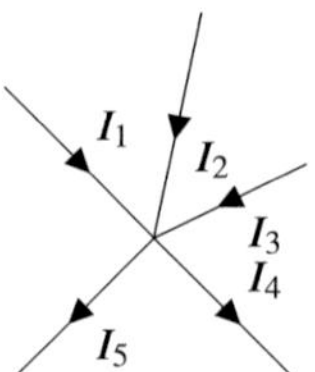

where $\epsilon_k = +1$ if the current is pointing toward the node, $\epsilon_k = -1$ if the current is pointing away from the node. Of course, this is based on the principle of conservation of charge. The rate of change of the charge at a node per unit time is the algebraic sum of the currents. $dQ/dt = \sum_k^N \epsilon_k I_k$. In the regime of steady currents there is no charge build up in the node, justifying KCL

Kirchhoff's Voltage Law (KVL)
The sum of the voltages across all branches along a loop is zero.

This law is valid in the absence of electromagnetic induction (see Chap. 11), in other words in the time-independent regime. In that case the electric field is a conservative field, which is expressed by the circulation law

$$\mathbf{\nabla} \times \mathbf{E} = \mathbf{0} .$$

The integral form of this equation can be written for any closed contour C as

$$\oint_C \mathbf{E} \cdot d\mathbf{l} = 0 .$$

For any loop in an electrical circuit, such as the one shown in Fig. 7.42, the integral can be written as a sum over the branches forming the contour:

$$\oint_C \mathbf{E} \cdot d\mathbf{l} = \int_A^B \mathbf{E} \cdot d\mathbf{l} + \int_B^C \mathbf{E} \cdot d\mathbf{l} + \cdots + \int_E^A \mathbf{E} \cdot d\mathbf{l} = 0 .$$

Since $\mathbf{E}$ is conservative, $\mathbf{E} \cdot d\mathbf{l} = -\mathbf{\nabla} V \cdot d\mathbf{l} = -dV$. Therefore, between two nodes, we have $\int_A^B \mathbf{E} \cdot d\mathbf{l} = V_A - V_B$. Now identifying the potential difference $V_A - V_B = V_1$ and applying similar arguments to the other branches, we find

$$V_1 + V_2 + V_3 + V_4 + V_5 = 0 .$$

Fig. 7.42 A loop interconnecting five linear bipoles

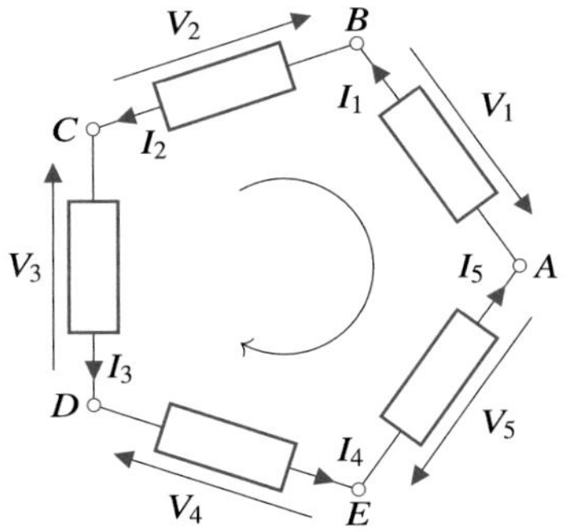

The sum of the voltages across all branches along the loop is zero.

Consider now a loop with N branches. Let us assign a voltage V_k to each branch k, with an arbitrarily chosen direction. Currents in each branch are assigned from the passive sign convention. It is also useful to mark a loop arrow, say in the clockwise direction. Kirchhoff's Voltage Law (KVL) can be expressed mathematically as

$$\sum_{k}^{N} \epsilon_k V_k = 0 \,,$$

where $\epsilon_k = +1$ if the voltage V_k is pointing in the direction of the loop arrow and $\epsilon_k = -1$ if it is pointing in the opposite direction.

For a linear network, the voltage in each branch can be expressed using the IV relation for linear bipoles,

$$V_k = \epsilon_k R_k I_k - \epsilon_k' E_k \,,$$

where $\epsilon_k = +1$ if the chosen current orientation points in the opposite direction to the voltage arrow, $\epsilon_k = -1$ if they point in the same direction, $\epsilon_k' = +1$ if the current arrow exits the voltage source by the positive terminal and $\epsilon_k' = -1$ otherwise.

As an example, let us express the voltage $V = V_A - V_B$ across terminals A and B of the portion of circuit illustrated in Fig. 7.43, comprising four branches with a predefined orientation of positive current in each branch.

The voltage across each element between A and B is added to the total with a plus sign if the current is oriented from A to B in a resistor or enters the positive terminal of a voltage source, and a minus sign otherwise. In this example:

$$V = (R_1 I_1) + (R_2 I_2 - E) + (-R_3 I_3 + E') + (-R_4 I_4) \,.$$

7.11 Superposition Theorem for Steady Currents

The superposition theorem is a powerful tool used in circuit analysis to determine the voltage across or the current through any element in a linear circuit with multiple independent sources.

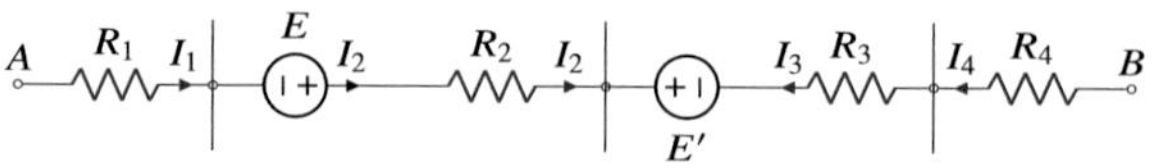

Fig. 7.43 Portion of circuit with four branches, each branch representing a linear bipole

Superposition theorem
In a linear circuit containing multiple independent sources, the response (voltage or current) in any branch can be determined by considering the effect of each independent source acting alone, with all other independent sources replaced by their internal resistances.

This is possible because Kirchhoff's laws are linear and linear circuits obey the superposition principle, which means that the response to multiple inputs is the sum of the responses to each input applied individually. The superposition theorem is applicable only to linear circuits. Applying the superposition theorem consists in the following steps:

1. Determine all the independent voltage and current sources in the circuit.
2. For each independent source, deactivate all other sources. That is

   - Replace voltage sources with short circuits.
   - Replace current sources with open circuits.

3. Analyze the simplified circuit with only one active source to determine the voltage or current across the desired element.
4. Repeat steps 2 and 3 for each independent source.
5. Algebraically add the contributions from each source to find the total voltage or current.

A key point is the independence of sources: The output of one source must not be affected by the output of another source.

As an illustration, let us apply the superposition theorem to find the currents I_1, I_2, I_3 in the circuit illustrated in Fig. 7.44.

We will find a first state by deactivating the voltage source E_2, i.e., replacing it by a short circuit (see Fig. 7.45 left). The current I_1' in the resulting circuit flows in a parallel association of resistors R_2 and R_3, which is a resistor with resistance $R' = R_2 R_3/(R_2 + R_3)$ (Fig. 7.45 right).

Applying KVL to the equivalent circuit on the right, where I_1' flows in the series connection of resistors R and R_1, we have

$$E_1 = (R' + R_1)I_1' .$$

Fig. 7.44 A circuit composed of two voltage sources and three resistors

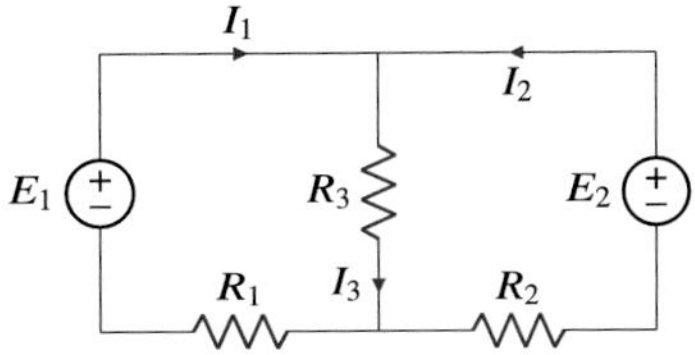

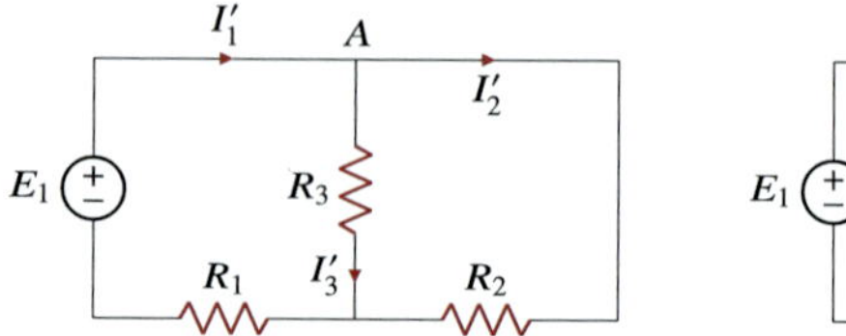

Fig. 7.45 Left: the circuit when the source E_2 has been turned off. Right: equivalent circuit

Applying KCL to node A on the one hand, and KVL to the loop with resistors R_2 and R_3 on the other hand, we have

$$I_1' = I_3' + I_2' ,$$
$$R_3 I_3' = R_2 I_2' .$$

Solving in I_1', I_2', I_3', we find

$$I_1' = \frac{E_1}{(R' + R_1)}, \quad I_2' = \frac{R_3}{(R_2 + R_3)} \frac{E_1}{(R' + R_1)}, \quad I_3' = \frac{R_2}{(R_2 + R_3)} \frac{E_1}{(R' + R_1)} .$$

Replacing R', we obtain

$$I_1' = \frac{E_1(R_2 + R_3)}{R^2}, \quad I_2' = \frac{E_1 R_3}{R^2}, \quad I_3' = \frac{E_1 R_2}{R^2} ,$$

where $R^2 \equiv R_2 R_3 + R_1 R_2 + R_1 R_3$. Now we consider the circuit of Fig. 7.46 where E_1 is deactivated and E_2 is the only source. We could do the calculation again but we clearly have a circuit that is the same as in the previous case if we replace E_1 by E_2, R_1 by R_2, I_1' by I_2'', and I_2' by I_1''. The quantity R^2 is not changed in this transformation.

Therefore, the solution for the currents I_k'' is

$$I_1'' = \frac{E_2 R_3}{R^2}, \quad I_2'' = \frac{E_2(R_1 + R_3)}{R^2}, \quad I_3'' = \frac{E_2 R_1}{R^2} .$$

The superposition theorem now gives the solution for the complete circuit:

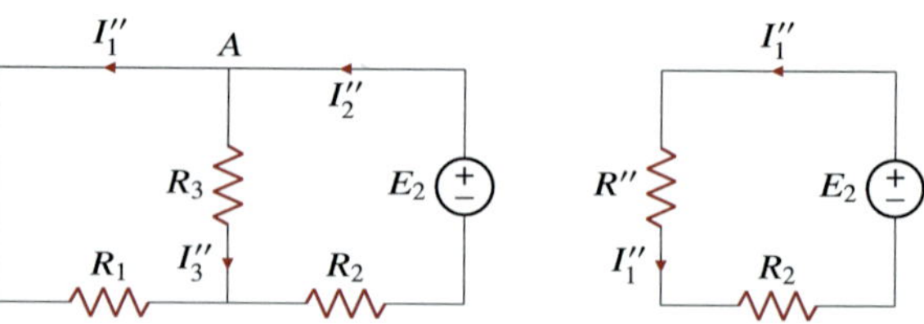

Fig. 7.46 Left: the circuit when the source E_1 is turned off. Right: equivalent circuit

$$I_1 = I_1' + I_1'' = \frac{E_1(R_2 + R_3) + E_2 R_3}{R^2},$$

$$I_2 = -I_2' - I_2'' = -\frac{E_1 R_3 + E_2(R_1 + R_3)}{R^2},$$

$$I_3 = I_3' + I_3'' = \frac{E_1 R_2 + E_2 R_1}{R^2}.$$

7.12 Bipolar Networks, Thévenin's and Norton's Theorems

A *bipolar network* is a type of electrical network that has two terminals. This means it can be connected to other components in a circuit to form larger networks. Bipolar networks can be complex combinations of resistors, voltage and current sources, diodes, and transistors.

A *linear bipolar network*, illustrated in Fig. 7.47, lumps linear bipoles (E_k, R_k) and (I_{0j}, G_j), characterized by their electromotive force E_k and resistance R_k, and their short-circuit current I_{0j} and reciprocal resistance G_j, respectively. A linear bipole network has two points of connection to other components and is equivalent to a single linear bipole which can be described as a series connection of an ideal voltage source and an internal resistance (Thévenin's theorem), or as a parallel connection of an ideal current source and an internal resistance (Norton's theorem). This equivalence between complex circuits to much simpler ones helps in analyzing circuit behavior and calculating voltages and currents. It aids in designing circuits by allowing for easier calculations and simulations.

7.12.1 Thévenin's Theorem

Thévenin's theorem is a powerful tool in circuit analysis that allows us to simplify complex linear circuits into a simpler equivalent circuit. This equivalent circuit consists of a single voltage source (V_{Th}) in series with a single resistor (R_{Th}), as shown in Fig. 7.48.

Finding the Thévenin equivalent bipole across two load terminals A and B consists in the following steps:

1. Calculate the Thévenin voltage (V_{Th}):

Fig. 7.47 A linear bipolar network

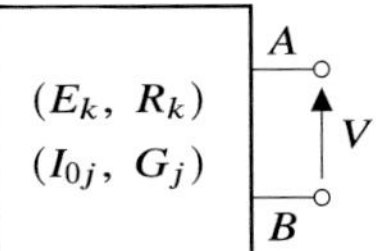

Fig. 7.48 Thévenin
equivalent circuit

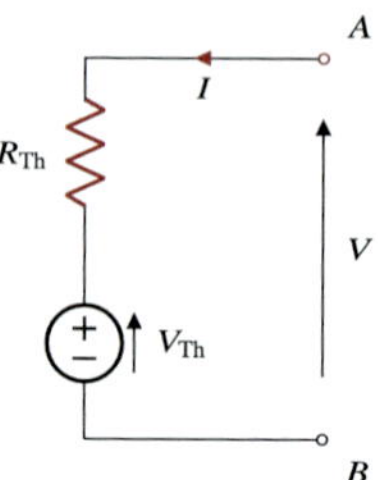

Remove any load from the circuit and calculate the open-circuit voltage V_1
across the terminals (with $I_1 = 0$). Thévenin's emf is $V_{\text{Th}} = V_1$. This is shown in
Fig. 7.49.

2. Calculate the Thévenin resistance (R_{Th}):
   Deactivate all independent sources (replace voltage sources with short circuits,
   i.e., $E_k = 0$ and current sources with open circuits, i.e., $I_{0j} = 0$). Then, calculate
   the equivalent resistance seen from the load terminals, using $V_2 = R_{\text{Th}} I_2$ where
   V_2 is any arbitrary potential. This is shown in Fig. 7.50.

Thévenin's theorem results directly from the superposition theorem applied to
these two states:

$$I = I_1 + I_2 \equiv I_2, \qquad V = V_1 + V_2 \equiv V_{\text{Th}} + R_{\text{Th}} I .$$

Now, connecting the load resistor to the circuit between A and B result in a simple
circuit shown in Fig. 7.51. The current flowing through the load resistor is obtained
by applying KVL to this circuit: $V_{\text{Th}} + R_{\text{Th}} I + R_{\text{load}} I = 0$. Solving for I, we find

$$I = -\frac{V_{\text{Th}}}{R_{\text{Th}} + R_{\text{load}}} .$$

Fig. 7.49 The Thévenin
voltage is the open-circuit
voltage when nothing is
connected to the bipolar
network

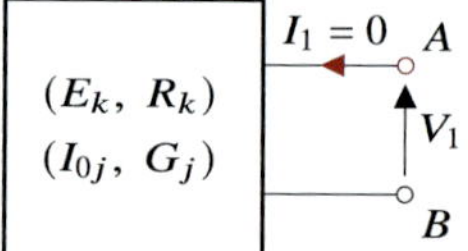

Fig. 7.50 The Thévenin
resistance is the equivalent
resistance of the bipolar
network when all
independent sources are
turned off

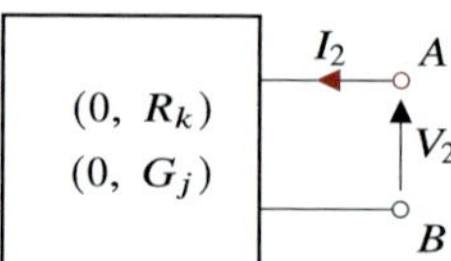

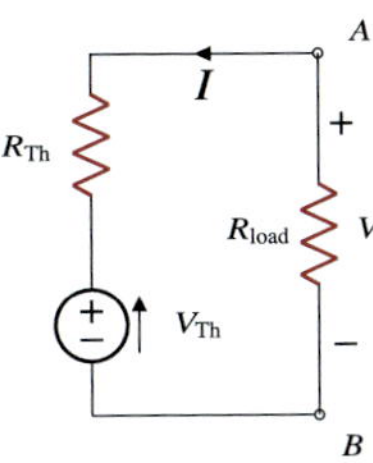

Fig. 7.51 A load connected to a bipolar network is equivalent to connect the load to the Thévenin equivalent circuit of the network

7.12.2 Norton's Theorem

Norton's Theorem is another powerful tool in circuit analysis, similar to Thevenin's theorem. It allows us to simplify a complex linear circuit into a simpler equivalent circuit consisting of a current source (I_{No}) in parallel with a resistance (R_{No}), shown in Fig. 7.52.

Finding the Norton equivalent circuit across two load terminals A and B consists in the following steps:

1. Calculate the Norton Current (I_{No}): Short-circuit the load terminals and calculate the short-circuit current I_1 flowing through the short circuit. This current is $-I_{No}$, as shown in Fig. 7.53.
2. Calculate the Norton resistance (R_{No}): Deactivate all independent sources (replace voltage sources with short circuits, i.e., $E_k = 0$ and current sources with open circuits, i.e., $I_{0j} = 0$). Then, calculate the equivalent resistance seen from the load terminals by writing Ohm's law, $I_2 = G_{No}V_2$ where V_2 is any arbitrary voltage, as shown in Fig. 7.54. The equivalent resistance is $R_{No} = G_{No}^{-1}$.

Norton's theorem directly follows from the superposition of these two states:

$$V = V_1 + V_2 \equiv V_2, \quad I = I_1 + I_2 \equiv G_{No}V - I_{No} .$$

Thévenin and Norton equivalent circuits are closely related. The Thévenin resistance is equal to the Norton resistance: $R_{Th} = R_{No}$. The Thévenin voltage is related to the Norton current by the equation: $V_{Th} = R_{No}I_{No}$. Now, connecting a load resistor between terminals A and B of a linear bipolar network is equivalent to connecting it to either the Thévenin equivalent circuit (Fig. 7.51) or the Norton equivalent circuit (Fig. 7.55). Applying KCL to node A, we have $I_{No} = G_{No}V + G_{load}V$, where

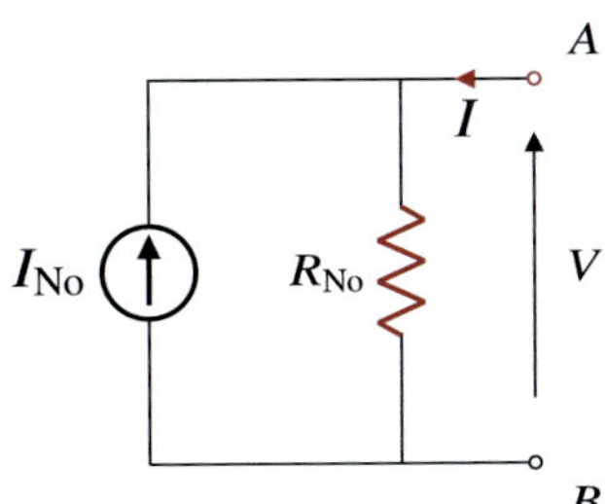

Fig. 7.52 Norton equivalent circuit for a bipolar network

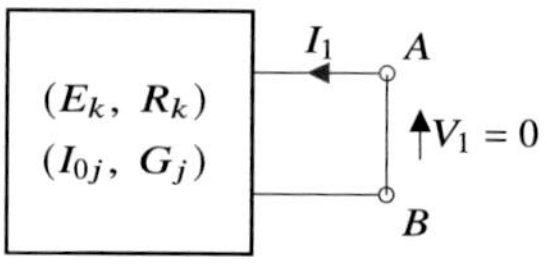

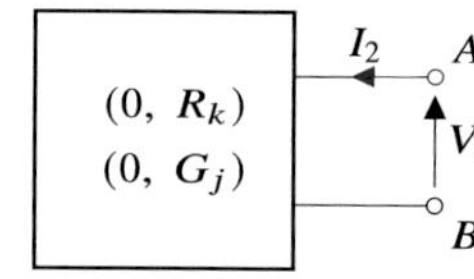

Fig. 7.53 The Northon current is obtained by short-circuiting the terminals A and B

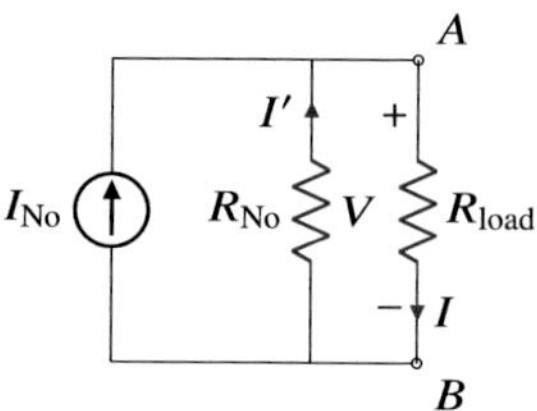

Fig. 7.54 The Northon resistance is obtained by turning off all the independent sources of the network

Fig. 7.55 A load connected to a bipolar network is equivalent to its connection to the equivalent Norton circuit of the system

$G_{\text{load}} = R_{\text{load}}^{-1}$. Solving for V, we find

$$V = \frac{I_{\text{No}}}{G_{\text{No}} + G_{\text{load}}} \, .$$

The current is $I = G_{\text{load}} V$.

7.13   Voltmeters and Ammeters

Analog voltmeters do not measure voltage directly. Instead, they connect the two electrodes to a resistor and measure the current through it. Then the voltage is calculated according to Ohm's law. The principle relies on the galvanometer: a resistor that has the shape of a moving coil is put in a strong, fixed magnetic field. This field produces a torque on the coil that is proportional to the current, as will be shown in Chap. 9. This torque moves a needle that is used as a pointer to read the current, as shown in Fig. 7.56. The galvanometer constitutes an ammeter, since it measures directly the current through the coil.

Note that an ammeter should be connected in series with the circuit being measured and should be able to support the current that is to be measured. Its internal resistance should therefore be as small as possible in order to avoid perturbing the current

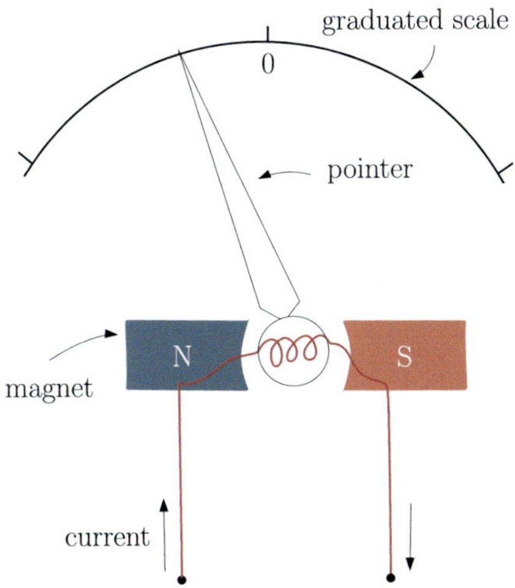

Fig. 7.56 The galvanometer can be used to measure a current

through the circuit. Typically, small currents are enough to provide large deflections of the pointer. In order to increase the range of the device (i.e., to measure larger currents), one can put a shunt resistor in parallel with the coil of the galvanometer, which is a resistance much smaller than that of the galvanometer itself. Most of the current will therefore flow through the shunt resistor, and only a small fraction of the current will go through the galvanometer. The ratio of both resistances allows one to calibrate the measurement.

Figure 7.57 shows a different type of galvanometer, invented by Leopoldo Nobili in 1825. The fixed frame is traversed by the current to be detected, which produces a magnetic field that influences two magnetized needles that are oriented in opposite directions. This reduces the impact of the Earth's magnetic field on the measurement. Under the influence of the current's magnetic field, the lower magnetized needle rotates and drives the upper needle. One can read the angle of rotation on a graduated disc. The deflection increases with the intensity of the current circulating in the frame. It is thanks to the extreme sensitivity of Nobili's galvanometer that Carlo Matteucci was able to detect, in the early 1840s, the electrical current generated by muscles.

Voltmeters should be connected in parallel with the circuit being measured. It can be made by adding an internal resistance R in series with a galvanometer so as to draw as little current as possible from the circuit that is being measured. This is illustrated in Fig. 7.58

Modern digital voltmeters rely on more complex electronics to measure voltage more directly, with the measurement being displayed directly as a digital signal. There are several kinds of digital voltmeters. One example is that of a ramp type voltmeter, which compares the input analog voltage of the voltmeter V_{input} to a ramp voltage with a negative slope. This voltmeter includes a clocked circuit capable of counting the time interval between the moment the ramp voltage coincides with V_{input} and the moment it crosses 0 V. The time interval corresponds to an integer number

Fig. 7.57 Nobili's galvanometer. Picture taken at the Museum of Ecole Polytechnique (Mus'X)

Fig. 7.58 An analog voltmeter can be made by using a high resistance R in series with a galvanometer G. It should be connected in parallel to the circuit to minimize the current I_g being drawn

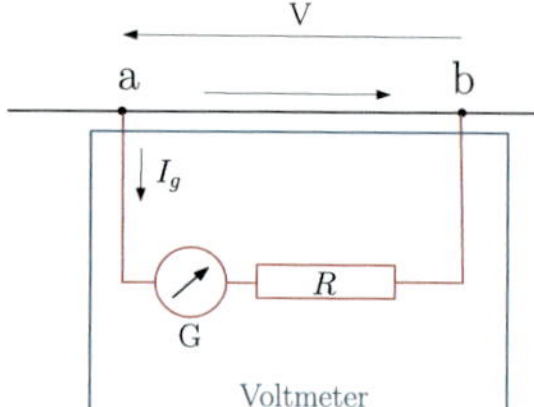

of clock pulses, providing a digital count representing the input analog voltage. The measurement can be repeated periodically by sending a cycle of ramps, as illustrated in Fig. 7.59.

In a digital ammeter, the current is actually converted into a calibrated voltage across a shunt resistor, which is then converted to a digital signal.

7.14   Summary and Essential Formulas

- Electrons in a conductor will acquire a non-zero average velocity **v** in the presence of an external electric field. According to the Drude model, the velocity is given by

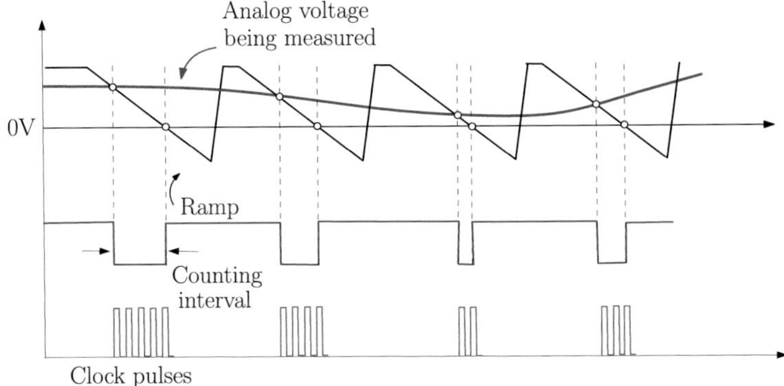

Fig. 7.59 A ramp type digital voltmeter uses comparison between a ramp waveform with the input voltage

$$\mathbf{v} = \frac{-e\tau}{m_e}\mathbf{E} = -\mu\mathbf{E},$$

where $-e$ is the electron charge, τ the mean time between collisions, and m_e the electron mass. This allows to define the mobility $\mu = \frac{e\tau}{m_e} > 0$. As a result, if the density of free electrons is n, a current density $\mathbf{j}$ will appear, which is given by Ohm's law

$$\mathbf{j} = -ne\mathbf{v} = \sigma\mathbf{E} ,$$

where:

$$\sigma = \frac{ne^2\tau}{m_e}$$

is the conductivity, whose unit is Siemens per meter ($[\sigma] = Sm^{-1}$). The current I through a surface S is the flux of $\mathbf{j}$ over S, and represents the total amount of charge per unit time crossing S

$$I = \Phi_{dS,\mathbf{j}} = \iint_S \mathbf{j}(\mathbf{x}') \cdot \mathbf{n}(\mathbf{x}')dS(\mathbf{x}').$$

- Since the electric charge is a conserved quantity, the charge density ϱ and the current density $\mathbf{j}$ are related, at every point, by the charge conservation law

$$\frac{\partial\varrho}{\partial t} + \nabla\cdot\mathbf{j} = 0 .$$

- Ohm's law in its macroscopic form states that the potential difference between points A and B of a conductor reads

$$V_A - V_B = R_{AB}I \, ,$$

where I is the current flowing from A to B, and R_{AB} is the total resistance of the segment AB.

- For a rectilinear wire of cross-section S and length l, the resistance is given by

$$R = \frac{\rho l}{S} \, ,$$

where the resistivity ρ is given by the inverse of the conductivity, $\rho = 1/\sigma$. The resistance and the capacitance are linked by the relation

$$\boxed{RC = \frac{\epsilon_0}{\sigma} \, .}$$

- Electric circuits are interconnected networks of electrical components, such as resistors, diodes, capacitors, inductors, and voltage and current sources. These components are connected by conductive wires, which are assumed to have negligible resistance compared to the components.
  Kirchhoff's current law (KCL) states that the algebraic sum of currents entering a node (junction) in a circuit is equal to the algebraic sum of currents leaving the node. It derives from the charge conservation law in the steady state regime.
  Kirchhoff's voltage law (KVL) states that the sum of the voltages across all branches along a loop is zero. It follows from the fact that the electric field is conservative in electrostatics.
- The superposition theorem states that, in a linear circuit containing multiple independent sources, the response (voltage or current) in any branch can be determined by considering the effect of each independent source acting alone, with all other independent sources replaced by their internal resistances.
- A linear bipole network is any network made of linear bipoles that has two points of connection to other components. It is equivalent to a simple linear bipole, which can be either a series connection of an ideal voltage source with an internal resistance (Thévenin's theorem), or a parallel connection of an ideal current source with an internal resistance (Norton's theorem).

Problems

7.1 Measurement of the resistance of a generator with a Wheatstone bridge

Prove that when the Wheatstone bridge is balanced (i.e., $R_a R = R_b r$), no current flows through the ammeter A regardless of the state of the switch K (open or closed). In other words, demonstrate that the current through the ammeter is independent of the resistance R' and the current flowing through it.

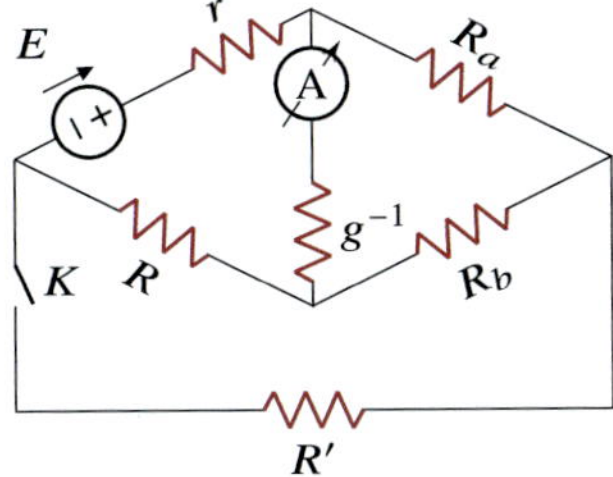

7.2 Electrical network

Given: $E_0 = 30V$, $E_1 = 6V$, $E_3 = 8V$, $E = 2V$ $R_0 = 2\Omega$, $R_1 = 20\Omega$, $R_2 = 3\Omega$, $R_3 = 20\Omega$, $R_4 = 4\Omega$, $R = 4\Omega$.

Calculate the currents in the different branches of the network. Give the values of the voltage differences $V_A - V_D$, $V_B - V_C$, $V_A - V_C$, $V_B - V_D$

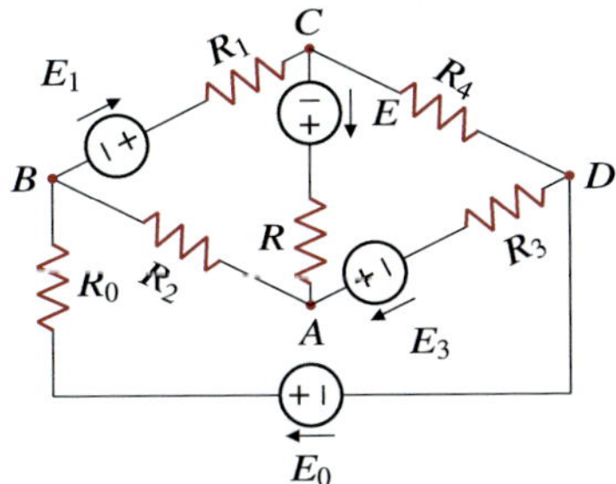

7.3 Equivalent resistance of a rectangular grid

This rectangular network is composed of six squares. If r is the resistance of each side of a square, what is the resistance of this network between points A and A'?

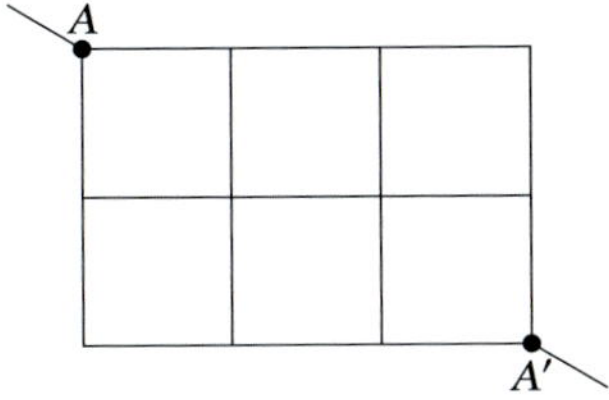

7.4 Equivalent resistance of a triangular grid

This network consists of four equilateral triangles. If the resistance of each side is r, calculate the resistance of the network between points A and A'.

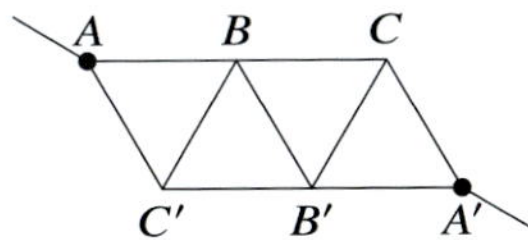

7.5 Equivalent resistance

Calculate the equivalent resistance of this network between points A and B.

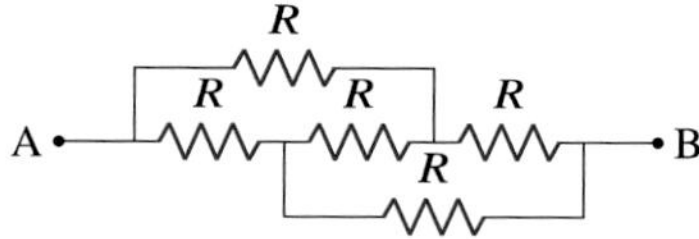

7.6 Application of Thévenin's theorem

Calculate the current in the branch AB using the Thévenin equivalent generator of the two-terminal network to which the branch is connected. Data: $r_1 = 2\Omega$, $R_1 = 3\Omega$, $r_2 = 1\Omega$, $R_2 = 9\Omega$, $R_3 = 6\Omega$, $E_1 = 4\text{V}$, $E_2 = 3\text{V}$.

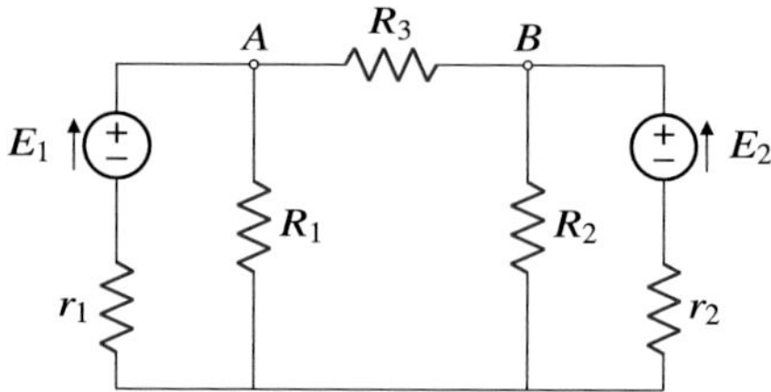

7.7 Thévenin's theorem vs Norton's theorem

Calculate the current flowing through R_3 using (i) Thévenin's theorem; (ii) Norton's theorem.

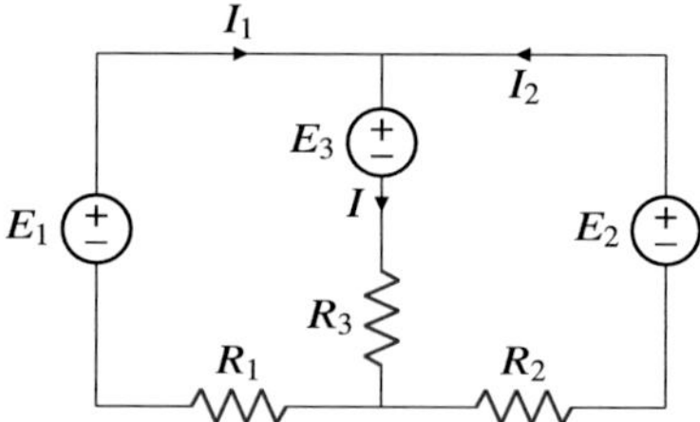

7.8 Variable resistor and equivalent generator

A variable resistor is connected across a voltage source with negligible internal resistance.

(a) Find the equivalent voltage (emf) and internal resistance of the portion of the circuit between points A and C, as a function of the total resistance R (between A and B) and the variable portion x (between A and C).

(b) A standard battery (with known emf and internal resistance) is connected to the variable resistor, with a galvanometer in between. The two voltage sources are connected in opposition. Find the value of x (denoted as x_0) that makes the current through the galvanometer zero. Calculate the current through the galvanometer for any value of x that is not equal to x_0.

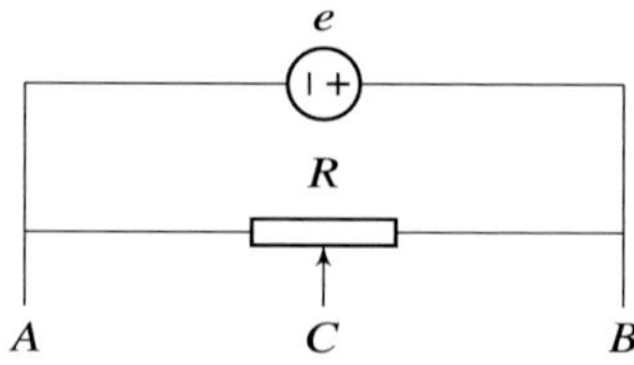

Chapter 8
Magnetostatics, Biot–Savart Law

Abstract This chapter introduces the fundamental principles of *magnetostatics*, the study of magnetic fields produced by stationary electric currents. It begins by establishing the empirical observation that moving electric charges exert a *magnetic interaction* on each other, distinct from the electrostatic force. This leads to the definition of the *magnetic field* (**B**) as a vector field that mediates these interactions. A central concept is the *Lorentz force*, which describes the total force experienced by a charged particle in the presence of both electric and magnetic fields ($\mathbf{F} = q(\mathbf{E} + \mathbf{v} \times \mathbf{B})$). The chapter presents the dynamics of charged particles under the influence of uniform magnetic fields, including *circular motion (cyclotron motion)* and *helical trajectories*, and discusses applications in *particle accelerators* (cyclotrons and synchrotrons) and *magnetic confinement*. The *Hall effect* is presented as a direct consequence of the Lorentz force, explaining the generation of a transverse electric field in current-carrying conductors subjected to a magnetic field, and its various technological applications. The chapter also introduces the *Penning trap*, a device that uses combined electric and magnetic fields to confine charged particles in three dimensions. The core of magnetostatics is then formalized with *Biot–Savart law*, which provides a method for calculating the magnetic field generated by continuous current distributions (volume currents) and, more specifically, by *current-carrying wires*. The chapter illustrates how to visualize magnetic fields through *magnetic field lines*, emphasizing their characteristic closed-loop nature. Finally, the *Laplace force* is derived, describing the force exerted by an external magnetic field on a current-carrying wire, and its implications for closed current loops in uniform magnetic fields.

Keywords Magnetostatics · Biot–Savart law · Lorentz force · Laplace's force · Cyclotron

8.1 Introduction

The word magnetism derives from 'magnesian stone' (μαγνήτηςλίσθος in ancient greek). In both China and Greece, it was observed that a naturally occurring iron oxide, today called magnetite (Fe_3O_4), can possess 'magnetic' properties. When

F. Cadiz and A. Couairon, *Classical Electrodynamics*, Undergraduate Texts in Physics, https://doi.org/10.1007/978-3-031-86785-9_8

magnetite is naturally magnetized, it is referred to as lodestone, such as those found by the Greeks in the region of Magnesia.

This mineral could attract other iron-containing substances, which could then become magnetized themselves. One of the earliest applications of magnetism was in navigation. The compass, invented centuries ago, uses a magnetized iron needle that aligns with the magnetic field of the Earth, providing a means of determining the direction to the geographical poles. Over time, numerous magnetic materials composed of various mineral compounds have been discovered. These materials have found countless applications across various fields. In modern electronics, magnets are used to store data in hard drives and to convert electrical signals into sound in speakers. They are also essential components of magnetic resonance imaging (MRI) machines, electric motors, and electric power generation plants.

In 1820, the Danish physicist Hans Christian Ørsted (Fig. 8.1) made a groundbreaking discovery: electric currents generate magnetic fields. This was demonstrated by observing that a compass needle deflected in the presence of an electric current, as illustrated in Fig. 8.2. This finding highlighted the strong connection between electricity and magnetism. Following Ørsted's discovery, French physicists Biot, Savart, and Ampère developed the mathematical laws governing static magnetic phenomena, known as magnetostatics. These laws describe the behavior of magnetic fields that do not change over time.

In this chapter, we introduce the interaction law between charges in relative motion. Drawing an analogy with electrostatics, we define the magnetic field as a vector field that encapsulates the contributions of all moving charges in the system. We present the Lorentz force, which governs the dynamics of a charged particle in the presence of both electric and magnetic fields, and solve the equations of motion for specific cases. Phenomena such as the Hall effect and the electromagnetic trapping of charged particles are also discussed. We then introduce Biot–Savart law, which provides an expression for the magnetic field generated by an arbitrary current distribution. Finally, we derive the Laplace force, which describes the force acting on a current-carrying wire in the presence of an external magnetic field.

Fig. 8.1 Hans Christian Ørsted (1777–1851), a Danish physicist and chemist

Fig. 8.2 An electric current changes the orientation of a compass needle in its surroundings

8.2 The Magnetic Interaction Law

It is an experimental fact that when two electric charges move relative to each other, they experience, in addition to the electric force, a so-called magnetic interaction. Consider two charges q and q', located at positions $\mathbf{x}$ and $\mathbf{x}'$, respectively, with velocities $\mathbf{v}$ and $\mathbf{v}'$, as illustrated in Fig. 8.3.

Magnetic force

The magnetic force exerted on a charge q by another charge q' is approximately given by

$$\mathbf{F}_q = \frac{\mu_0}{4\pi} \frac{qq'}{|\mathbf{x} - \mathbf{x}'|^3} \mathbf{v} \times \left(\mathbf{v}' \times (\mathbf{x} - \mathbf{x}')\right) , \tag{8.1}$$

where μ_0 is the vacuum magnetic permeability, with a numerical value in SI units of $\mu_0 = 1.256 \times 10^{-7}\,\mathrm{m\ kg\ C^{-2}} \approx 4\pi \times 10^{-7}\,\mathrm{m\ kg\ C^{-2}}$.

Remarks

- The relative magnitudes of the magnetic and electric forces between two point charges can be expressed as

Fig. 8.3 The magnetic interaction between two point charges depends on their relative positions and velocities

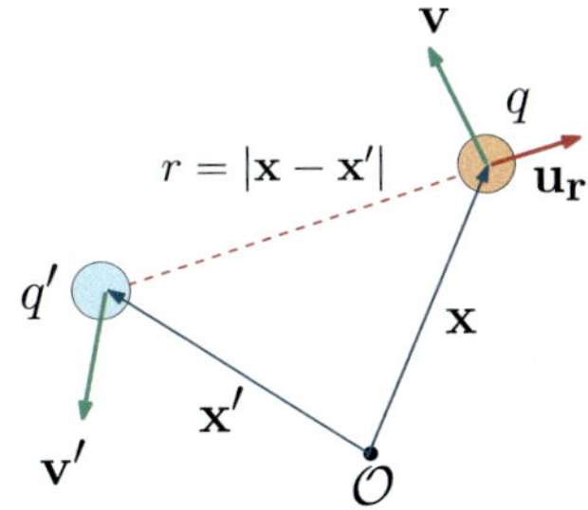

$$\frac{|\mathbf{F}_{\text{magnetic}}|}{|\mathbf{F}_{\text{electric}}|} = \frac{\dfrac{\mu_0}{4\pi}\dfrac{qq'}{|\mathbf{x}-\mathbf{x}'|^2}|\mathbf{v}||\mathbf{v}'|}{\dfrac{1}{4\pi\epsilon_0}\dfrac{qq'}{|\mathbf{x}-\mathbf{x}'|^2}} = \epsilon_0\mu_0 vv' = \frac{vv'}{c^2}\,,$$

where c is a constant representing a velocity, given by

$$c = \frac{1}{\sqrt{\epsilon_0\mu_0}} = 3 \times 10^8\,\text{m s}^{-1}\,. \tag{8.2}$$

As will be shown in Chap. 13, c is the speed of light. The magnetic force is significantly smaller than the electric force when the velocities of the charges are much smaller than the speed of light ($vv'/c^2 \ll 1$).

- Since a particle's velocity depends on the observer's frame of reference, the magnetic force (8.1) is not the same in all Galilean reference frames. For instance, in the frame $\mathcal{R}'$ where q' is at rest, the force experienced by q is purely electrostatic, governed by Coulomb's law. However, in any other frame moving relative to $\mathcal{R}'$, the magnetic interaction must be included to reconcile observations with those in $\mathcal{R}'$. This demonstrates that magnetism naturally arises from the principle of relativity and represents a different aspect of the unified electromagnetic force. The distinction between electric and magnetic fields depends on the chosen reference frame.

  Equation (8.1) is an approximation valid when the velocities are small compared to c. The exact law can be derived using the laws of special relativity (see Exercise 20.5). This derivation shows that the magnetic force emerges as a consequence of transforming the electromagnetic force between reference frames.

- Unlike the electrostatic force, the magnetic force is not a central force. The magnetic force $\mathbf{F}_q$ acting on q is not directed along the line connecting the charges but is perpendicular to both the relative position vector $\mathbf{x} - \mathbf{x}'$ and the velocity $\mathbf{v}$. Consequently, the magnetic force does no work on the charges.

Example 8.1—Correction to Coulomb's law due to particle motion
When changing from one reference frame to another, not only the magnetic force changes, but also the electric force. Consider two charges, q, separated by a distance d, at rest in an inertial reference frame $\mathcal{R}$. In this frame, the only force between the charges is the electrostatic Coulomb force $\mathbf{F}_e$. Now consider another frame $\mathcal{R}'$ moving at a constant velocity $-\mathbf{v}$ along the x-axis relative to $\mathcal{R}$. In $\mathcal{R}'$, the charges appear to move with velocity $\mathbf{v}$ in the x-direction, as shown in Fig. 8.4.

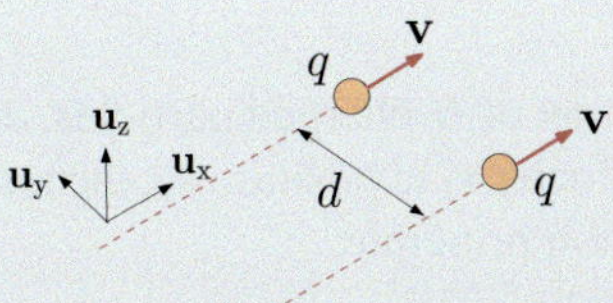

Fig. 8.4 Two identical charges at rest in one frame appear to move with velocity **v** in another

An observer in $\mathcal{R}'$ concludes that a magnetic interaction exists between the charges. The magnetic force acting on the charge on the right in $\mathcal{R}'$ is given, according to (8.1), by

$$\mathbf{F}'_m = \frac{q^2 \mu_0}{4\pi d^2} v\mathbf{u}_x \left(v\mathbf{u}_x \times -\mathbf{u}_y\right) = \frac{q^2 \mu_0}{4\pi d^2} v^2 \mathbf{u}_y \ .$$

To ensure that the total force experienced by a charge is independent of the reference frame, we must admit that Coulomb's law takes different forms in $\mathcal{R}$ and $\mathcal{R}'$. Let $\mathbf{F}'_e$ be the electric force expressed in frame $\mathcal{R}'$, in which the charges are moving with velocity **v**. For the charge on the right, the total force $\mathbf{F}'_e + \mathbf{F}'_m$ must equal the Coulomb force $\mathbf{F}_e$ observed in $\mathcal{R}$:

$$\frac{q^2}{4\pi d^2} \mu_0 v^2 \mathbf{u}_y + \mathbf{F}'_e = \mathbf{F}_e = -\frac{q^2}{4\pi \epsilon_0 d^2} \mathbf{u}_y \ .$$

Substituting $1/c^2 = \epsilon_0 \mu_0$, we find

$$\mathbf{F}'_e = -\frac{q^2}{4\pi \epsilon_0 d^2}(1 + v^2/c^2)\mathbf{u}_y = \mathbf{F}_e(1 + v^2/c^2) \ . \tag{8.3}$$

Thus, in any frame where the charges are in motion, Coulomb's law requires a correction proportional to $(v/c)^2$. For small velocities compared to c, this correction is negligible, and the standard Coulomb law suffices. However, Eq.(8.3) is not exact because we have already made the approximation of small velocities in (8.1) and assumed that $\mathbf{F}' = \mathbf{F}$, which is not true in general for relativistic velocities. In fact, it is the equation of the dynamics (given by $\mathbf{F} = d\mathbf{p}/dt$), that should lead to the same conclusions in both frames. The exact transformation for this particular case writes:

$$\mathbf{F}'_e = \frac{1}{\sqrt{1 - (v/c)^2}}\mathbf{F}_e \approx \mathbf{F}_e \left(1 + \frac{v^2}{2c^2}\right) \quad |v| \ll c \ . \tag{8.4}$$

This result is a particular case of the general field transformation laws (20.10) in special relativity (see Chap. 20).

8.2.1 Superposition Principle for Magnetic Forces

If a charge q interacts with a set of N charged particles, denoted by $\{q_1, q_2, \ldots, q_N\}$, where the ith charge has a velocity $\mathbf{v}_i$ at position $\mathbf{x}_i$, the total magnetic force on charge q is given by the superposition principle:

$$\mathbf{F}_q = \sum_{i=1}^{N} \mathbf{F}_q(i) = \frac{\mu_0}{4\pi} \frac{q q_i}{|\mathbf{x} - \mathbf{x}_i|^3} \mathbf{v} \times (\mathbf{v}_i \times (\mathbf{x} - \mathbf{x}_i)) \ . \tag{8.5}$$

This equation indicates that the total magnetic force on charge q is the vector sum of the individual magnetic forces exerted by each of the other charges.

8.3   The Magnetic Field

Magnetic phenomena can be described using a magnetic field $\mathbf{B}$. Similar to the approach in electrostatics, the goal is to express the magnetic force acting on a charge q due to the influence of other charges as

$$\boxed{\mathbf{F}_q = q\mathbf{v} \times \mathbf{B}(\mathbf{x}) \ .} \tag{8.6}$$

Remarks

- The magnitude of the magnetic force is directly proportional to the magnitude of the velocity, $|\mathbf{v}|$, and the sine of the angle θ between $\mathbf{v}$ and $\mathbf{B}$. Mathematically, $|\mathbf{F}| = qvB \sin\theta$.
- The magnetic force is always perpendicular to the velocity vector $\mathbf{v}$ (see Fig. 8.5). This means that the magnetic force does no work on the charged particle. As a result, a particle moving in a magnetic field will experience a change in direction but will maintain a constant speed.

By comparing equations (8.6) and (8.1), we can identify the magnetic field at point $\mathbf{x}$ that is generated by a single point charge q'.

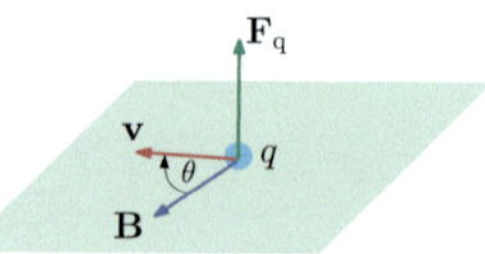

Fig. 8.5 The magnetic force acting on a charge q is perpendicular to both the magnetic field, $\mathbf{B}$, and the charge velocity, $\mathbf{v}$

Magnetic field of a point charge

The magnetic field at point $\mathbf{x}$ generated by a point charge q' moving with velocity $\mathbf{v}'$ at position $\mathbf{x}'$ is given by

$$\boxed{\mathbf{B}(\mathbf{x}) = \frac{\mu_0}{4\pi} q' \frac{\mathbf{v}' \times (\mathbf{x} - \mathbf{x}')}{|\mathbf{x} - \mathbf{x}'|^3}.} \tag{8.7}$$

If $\mathbf{E}'(\mathbf{x})$ is the electrostatic field generated by q' at point $\mathbf{x}$ (that is, the field generated in the frame where q' is at rest), then we can express the magnetic field generated by the charge's motion as

$$\mathbf{B}(\mathbf{x}) = \mu_0 \epsilon_0 \mathbf{v}' \times \mathbf{E}'(\mathbf{x}) = \frac{\mathbf{v}' \times \mathbf{E}'}{c^2}. \tag{8.8}$$

This is true provided $|\mathbf{v}| \ll c$. The exact result writes

$$\mathbf{B} = \frac{1}{\sqrt{1 - (v'/c)^2}} \frac{\mathbf{v}' \times \mathbf{E}'}{c^2}$$

and is demonstrated in Chap. 20.

From this definition, it is evident that the magnetic field at a point $\mathbf{x}$ generated by a moving charge is not static, as the vector $\mathbf{x} - \mathbf{x}'(t)$ varies with time. However, a stationary electric current, composed of a large ensemble of charges in motion such that a steady flow of charge is established, can create a static magnetic field.

The SI unit for $\mathbf{B}$ is the *tesla*, (T), where $1\,\text{T} = 1\,\text{N}\,\text{A}^{-1}\,\text{m}^{-1}$. Another commonly used unit is the *gauss* (G), with $1\,\text{G} = 1 \times 10^{-4}\,\text{T}$. The magnetic field of Earth typically has a magnitude of $0.2\,\text{G}$ or $20\,\mu\text{T}$. In comparison, electromagnets can generate fields ranging from $0.1\,\text{T}$ to $1\,\text{T}$ while superconducting electromagnets can produce fields up to $50\,\text{T}$.

8.3.1 *Where Does the Magnetic Field of a Magnet Originate?*

We have observed a connection between moving charges (currents) and magnetic fields. However, certain materials, such as magnetite, exhibit magnetic properties even in the absence of any macroscopic electric current flow at equilibrium. In a classical model of matter, each atom can be considered as a positive nucleus surrounded by orbiting electrons, which effectively create microscopic currents. If these microscopic currents were aligned in the same direction, they could collectively produce a macroscopic magnetic field. In reality, the magnetism of a magnet does not originate from the orbital motion of electrons. Instead, it arises from an intrinsic property of electrons known as spin. Spin is a quantum mechanical property that has no classical equivalent. The magnetism of magnetic materials is therefore a quantum phenomenon, which will be explored in more detail in Chap. 10.

8.3.2 The Lorentz Force

When a particle of charge q is located in a region where both electric and magnetic fields, $\mathbf{E}$ and $\mathbf{B}$, are present, the net force acting on the particle is given by

$$\boxed{\mathbf{F} = q\,(\mathbf{E} + \mathbf{v} \times \mathbf{B})\,.} \tag{8.9}$$

This force is known as the Lorentz force, named after Hendrik Lorentz, who derived it in 1895.

8.4 The Discovery of the Electron (1897)

In the late 19th century, it was discovered that gases, typically excellent insulators, could become conductive when subjected to extremely high voltages. This phenomenon was studied using discharge tubes, which consisted of two electrodes inside a sealed glass tube filled with a low-pressure gas, as illustrated in Fig. 8.6. By applying a sufficiently high voltage, the gas could be ionized, allowing an electric current to flow through it. This process also resulted in the glass glowing at the end opposite the cathode, forming a clear shadow of the anode. These observations led scientists to believe that rays of an unknown nature were emitted from the cathode toward the anode. These rays were subsequently named cathode rays. It is now understood that cathode rays are actually beams of electrons accelerated from the

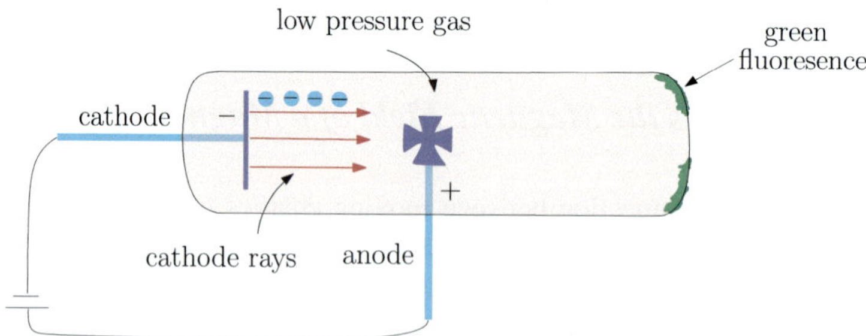

Fig. 8.6 Cathode rays are now understood to be streams of electrons that travel from the cathode to the anode within a discharge tube

cathode towards the anode. When these electrons strike the glass behind the anode, they cause it to fluoresce, emitting a green glow.

Sir William Crookes observed that cathode rays could be deflected by a magnet, similar to the behavior of charged particles. In France, Jean Perrin discovered that a metal plate placed in the path of these rays acquired a negative electric charge. These findings suggested that cathode rays were composed of negatively charged particles moving through the tube. Sir Joseph John Thomson (Fig. 8.7) conducted experiments to measure the mass and charge of these particles, leading to the discovery of corpuscles, now known as electrons, within atoms. One method involved using a device like the one depicted in Fig. 8.8 to first measure the charge-to-mass ratio of the electron. In this device, electrons are accelerated from the cathode (C) to the anode (A) and then enter a region with both an electric field $\mathbf{E} = E\mathbf{u}_y$ and a magnetic field $\mathbf{B} = B\mathbf{u}_z$.

Let $\Delta V = V_A - V_C$ be the potential difference between the electrodes. By applying the principle of energy conservation, the kinetic energy gained by electrons as they accelerate from C to A is:

Fig. 8.7 J.J. Thomson (1856–1940), a British physicist, pioneered mass spectroscopy and was the first to separate isotopes of a stable element, revolutionizing atomic and molecular science

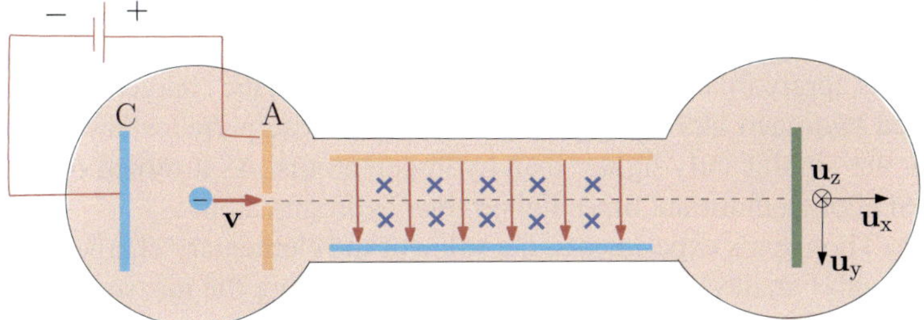

Fig. 8.8 By subjecting electrons to both electric and magnetic fields, it is possible to determine their charge-to-mass ratio, e/m

$$\frac{1}{2}m_e v^2 = -e\Delta V \, ,$$

where m_e is the mass of an electron and e its charge. The speed of the electrons as they enter the region with the magnetic field is

$$v = \sqrt{\frac{-2e\Delta V}{m_e}} \, .$$

In the presence of uniform, perpendicular electric and magnetic fields, $\mathbf{E} = E\mathbf{u}_y$ and $\mathbf{B} = B\mathbf{u}_z$, there exists a specific velocity $\mathbf{v} = v\mathbf{u}_x$ for which a charged particle will experience no net force. This velocity is given by

$$\mathbf{v} = \frac{1}{B^2}\mathbf{E} \times \mathbf{B} = \frac{E}{B}\mathbf{u}_x \, .$$

If a particle moves with this velocity, its trajectory will remain a straight line, as the Lorentz force acting on it will be zero:

$$\mathbf{F} = q\,(\mathbf{E} + \mathbf{v} \times \mathbf{B}) = q\left(\mathbf{E} + \frac{1}{B^2}(\mathbf{E} \times \mathbf{B}) \times \mathbf{B}\right) = \mathbf{0} \, ,$$

obtained by using the vector triple product identity (BAC − CAB rule) to expand $(\mathbf{E} \times \mathbf{B}) \times \mathbf{B} = -B^2\mathbf{E}$.

By carefully adjusting the values of ΔV, E and B, it is possible to ensure that the electrons pass through a hole centered on the axis of the tube and follow a straight-line trajectory. Under these conditions, the charge-to-mass ratio of the electron can be measured:

$$\frac{-e}{m_e} = \frac{E^2}{2\Delta V B^2} \, .$$

This experiment yielded a value of:

$$\frac{e}{m_e} = -1.7588 \times 10^{11} \, \mathrm{C\,kg^{-1}} \, .$$

This value is approximately a thousand times larger than the charge-to-mass ratio of the ionized hydrogen atom, suggesting that either these particles have a very high charge or are significantly lighter than hydrogen atoms. As it turned out, electrons are indeed about a thousand times lighter than hydrogen atoms.

Prior to Thomson's experiments, the value of the elementary charge e was determined through Faraday's electrolysis experiments. Using the measured e/m_e ratio, Thomson was able to calculate the mass of the electron:

$$m_e = 9.11 \times 10^{-31} \, \mathrm{kg} \, .$$

Cathode rays were found to be composed of negatively charged particles, which are approximately two thousand times lighter than the lightest atom (hydrogen). These particles, known as electrons, are fundamental constituents of all atoms and play a crucial role in chemical bonding and electronics.

8.5 Dynamics of a Charged Particle in Uniform, Perpendicular Electric and Magnetic Fields

Consider a point charge q with mass m subjected to the Lorentz force in a simplified scenario where both the electric and magnetic fields are uniform and mutually perpendicular, given by $\mathbf{E} = E\mathbf{u}_x, \mathbf{B} = B\mathbf{u}_z$, as depicted in Fig. 8.9. The equation of motion for the particle is:

$$m\frac{d\mathbf{v}}{dt} = q\,(\mathbf{E} + \mathbf{v} \times \mathbf{B}) = \frac{q}{m}\,(E\mathbf{u}_x + \mathbf{v} \times B\mathbf{u}_z)\ ,$$

The equation of motion can be rewritten as

$$\frac{d\mathbf{v}}{dt} = \frac{q}{m}\left(E\mathbf{u}_x - \frac{dx}{dt}B\mathbf{u}_y + \frac{dy}{dt}B\mathbf{u}_x\right)\ .$$

Assuming an initial velocity $\mathbf{v}_0 = (v_{0x}, v_{0y}, v_{0z})$, we obtain the following set of differential equations describing the motion of the particle:

$$\frac{d^2z}{dt^2} = 0\ ,$$

$$\frac{d^2y}{dt^2} = -\frac{qB}{m}\frac{dx}{dt}\ ,$$

$$\frac{d^2x}{dt^2} = \frac{q}{m}\left(E + \frac{dy}{dt}B\right)\ .$$

The solution to the first differential equation is

$$z(t) = z_0 + v_{0z}t\ .$$

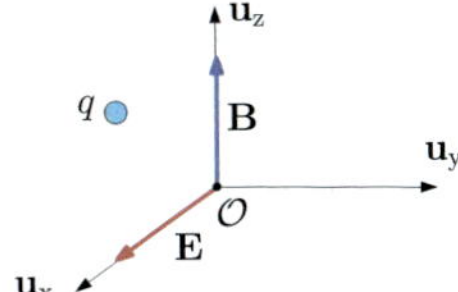

Fig. 8.9 A charged particle q subjected to uniform and mutually perpendicular electric and magnetic fields

This represents a uniform motion along the z-axis, parallel to the magnetic field. Defining the frequencies $\omega_c = \dfrac{qB}{m}$ and $\omega_1 = \dfrac{qE}{m}$, we can rewrite the equations of motion in the xy-plane as

$$\frac{d^2 y}{dt^2} = -\omega_c \frac{dx}{dt} \, ,$$

$$\frac{d^2 x}{dt^2} = \omega_1 + \omega_c \frac{dy}{dt} \, .$$

The solution to the equations of motion is

$$x(t) = -\frac{v_{0y}}{\omega_c} \cos \omega_c t + \frac{v_{0x}}{\omega_c} \sin \omega_c t + \frac{\omega_1}{\omega_c^2}(1 - \cos \omega_c t) + \frac{v_{0y}}{\omega_c} + x_0 \, ,$$

$$y(t) = \frac{v_{0y}}{\omega_c} \sin \omega_c t + \frac{v_{0x}}{\omega_c} \cos \omega_c t - \frac{\omega_1}{\omega_c}\left(t - \frac{1}{\omega_c} \sin \omega_c t\right) - \frac{v_{0x}}{\omega_c} + y_0 \, .$$

For a charge initially at the origin ($x_0 = y_0 = 0$) with a non-zero initial velocity component along the z-axis, the trajectory will be a spiral. This spiral results from the combination of a circular motion in the xy-plane and a drift along the z- and y-axes, as illustrated in the Fig. 8.10.

It is important to note that the charge does not drift in the direction of the electric field, as it would if the magnetic field were zero. The combined presence of the electric and magnetic fields causes the projection of the particle's position on the x-axis to oscillate back and forth. This phenomenon results in a current flowing in the y-direction, perpendicular to the electric field. This current is known as the Hall current.

8.5.1 Circular Motion in a Uniform Magnetic Field

When $\mathbf{E} = \mathbf{0}$ and $\mathbf{B} = B\mathbf{u}_z$, the trajectory of the charge is given by

$$x(t) = -\frac{v_{0y}}{\omega_c} \cos \omega_c t + \frac{v_{0x}}{\omega_c} \sin \omega_c t + \frac{v_{0y}}{\omega_c} + x_0 \, ,$$

$$y(t) = \frac{v_{0y}}{\omega_c} \sin \omega_c t + \frac{v_{0x}}{\omega_c} \cos \omega_c t - \frac{v_{0x}}{\omega_c} + y_0 \, ,$$

Fig. 8.10 Trajectory of a charged particle with an initial velocity component along the z-axis

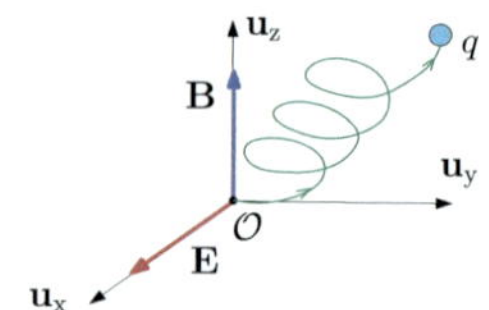

$$z(t) = z_0 + v_{0z}t .$$

In this case, the particle trajectory is a helix. The projection of this helix onto the xy-plane describes a uniform circular motion with an angular frequency

$$\omega_c = \frac{qB}{m} .\tag{8.10}$$

This frequency is known as the cyclotron frequency. The trajectory is illustrated in Fig. 8.11.

If the initial velocity component along the z-axis is zero, the particle's trajectory will be a circle confined to the xy-plane, as illustrated in Fig. 8.12. Let v and R be the speed and radius of this circular motion, respectively. Then,

$$v = \omega_c R .$$

Rearranging this equation, we find the radius of the circle,

$$R = \frac{mv}{qB} = \frac{v}{\omega_c} .\tag{8.11}$$

It is noteworthy that the cyclotron frequency, ω_c, is independent of both the particle's velocity and the radius of its orbit. Instead, it depends solely on the charge-to-mass ratio, q/m. Therefore, measuring the cyclotron frequency offers an alternative method for determining the q/m value of charged particles.

Figure 8.13 shows the trajectory of an electron subjected to a magnetic field. Electrons are emitted from a hot filament inside a tube filled with a low pressure Neon gas. The electrons are then accelerated by an electric field, and filtered through a circular aperture, which selects those with a well-defined velocity. In the presence of a magnetic field, electrons then follow a circular trajectory if their initial velocity is perpendicular to the field, as shown in Fig. 8.13a. The trajectory is visualized using the fluorescence of Neon atoms inside the tube, which are excited by the electron beam. When the tube is slightly rotated, the component of the electron's initial

Fig. 8.11 Trajectory of a charged particle with an initial velocity component along the z-axis, subjected to a uniform magnetic field

Fig. 8.12 Cyclotron motion of a charge in the presence of a uniform magnetic field

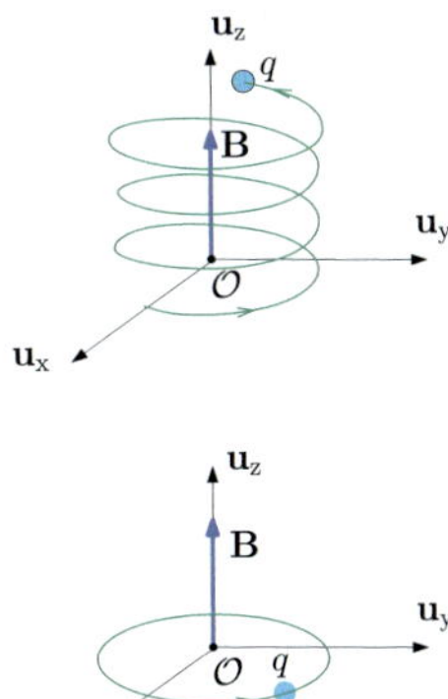

Fig. 8.13 Cyclotron motion of electrons in the presence of a uniform magnetic field

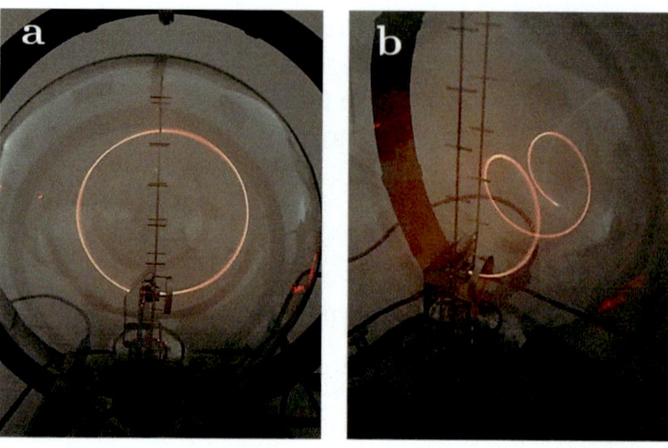

velocity perpendicular to the magnetic field remains unchanged, and the trajectory transforms into a helix, as depicted in Fig. 8.13b.

Magnetic fields can be used to confine the trajectories of charged particles in space. This principle, known as magnetic confinement, is crucial in nuclear fusion reactors, where plasma must be maintained at extremely high temperatures–on the order of hundreds of millions of degrees Celsius. Since no solid material can withstand such extreme conditions, magnetic fields are employed to contain the plasma particles along paths confined to a donut-shaped surface. This confinement results from the combined effects of the magnetic field generated by the current flowing through the plasma itself and an externally applied toroidal magnetic field.

8.6　Particle Accelerators

Particle accelerators have a wide variety of applications. They allow for the study of high energy physics, for example at the Large Hadron Collider. This is essential to test the theories of particle physics. Some accelerators are conceived to maximize the amount of X-rays that the charged particles emit, which can then be used to peform experiments for material science. For medical applications, accelerators can be use to sterilize equipment, for treating tumors and cancer.

Linear accelerators use electric fields to accelerate charges along a line. This is treated in Exercise 3.3. For many applications these accelerators excel, but for achieving very high energies, they become extremely long, requiring significant space and infrastructure.

A cyclotron is a particle accelerator that has the advantage of using magnetic fields to confine the charge's trajectory to a limited volume. They consist of two hollow, semicircular electrodes placed within a uniform magnetic field $\mathbf{B} = B\mathbf{u}_z$, which is perpendicular to the plane of the electrodes. An oscillating potential difference, $\pm\Delta V$, is applied between the electrodes, creating an oscillating electric field in the gap between them. Ions exiting one electrode are accelerated towards the other by the electric field. Within the second electrode, the ions follow a semicircular trajectory under the influence of the magnetic field.

Remarkably, in the non-relativistic case, the time required for the ions to complete a semicircular path inside an electrode is independent of their velocity. This is because the time depends solely on the cyclotron frequency, $\omega_c = qB/m$. The electric field is synchronized to reverse its polarity with a period $T = \pi/\omega_c$, ensuring that the ions are always accelerated as they cross the gap between the electrodes. The trajectory of the ions, schematically shown in Fig. 8.14, is a series of arcs with increasing radii. This process continues until the ions reach the edge of the apparatus at high speeds.

The final kinetic energy of the ions is limited by the maximum radius R of the orbit before they escape the accelerator and by the strength of the magnetic field. Since $v = qBR/m$, the final energy is given by

$$E = \frac{1}{2}mv^2 = \frac{q^2 B^2 R^2}{2m} .$$

Notably, the final energy is independent of the strength of the electric field. The synchrotron accelerator is an evolution of the cyclotron. As shown in Fig. 8.15, it combines smaller magnets with linear sections where particles are accelerated by electric fields. The key advantage of the synchrotron design is that the particle trajectory remains fixed, defined by a closed loop. This allows the vacuum chamber to be a large, thin toroidal structure. These accelerators are named "synchrotrons" because they synchronize the accelerating electric field with the increasing velocity of the particles and dynamically adjust the magnetic field strength to keep the particles

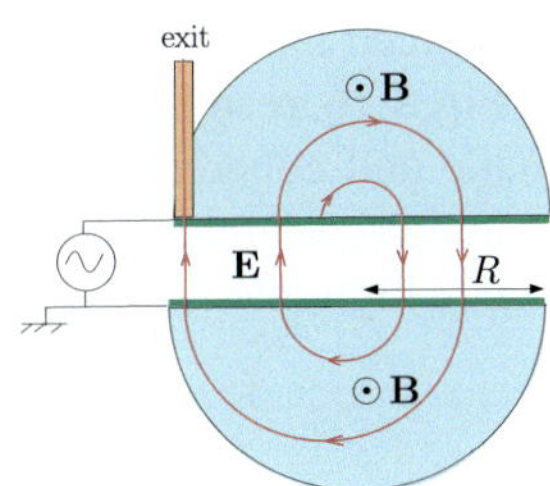

Fig. 8.14 The cyclotron particle accelerator

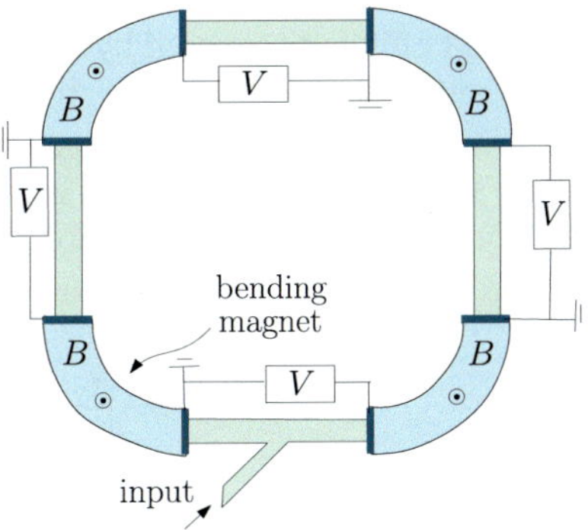

Fig. 8.15 The synchrotron particle accelerator

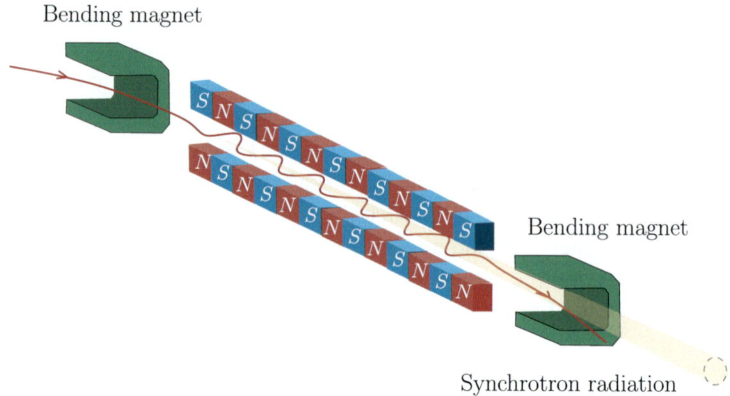

Fig. 8.16 Relativistic particles in the presence of oscillating magnetic fields produce intense, highly collimated radiation

confined to a circular path as they gain energy. Synchrotrons can accelerate particles to speeds close to the speed of light.

Some synchrotron facilities, such as SOLEIL near Paris, are specifically designed as light sources. Accelerated electrons emit electromagnetic radiation—a phenomenon explained in Chap. 19. These facilities incorporate undulators and wigglers, which consist of a series of alternating magnetic poles arranged in a periodic structure. As relativistic electrons pass through these oscillating magnetic fields, they are forced into narrow sinusoidal oscillations or large-amplitude wiggles, as shown in Fig. 8.16, resulting in highly collimated radiation. Undulators produce monochromatic and coherent X-rays used for imaging and probing the atomic structure of materials. Wigglers provide intense, broadband radiation suitable for applications where high intensity over a wide wavelength range is needed.

8.6.1 The Hall Effect (1879)

The Hall effect, discovered by Edwin Hall in 1879, is a consequence of the Lorentz force acting on charge carriers. It occurs when a current flows through a conductor in the presence of an external magnetic field **B**. A transverse electric field, perpendicular to the velocity of the electrons, is generated as a result.

If **v** is the average velocity of an electron in the conductor, we can modify the Drude model equation of motion (see Chap. 7) to include the magnetic force:

$$m\frac{d\mathbf{v}}{dt} + \frac{m}{\tau}\mathbf{v} = -e\mathbf{E} - e\mathbf{v} \times \mathbf{B}\,.$$

In a steady-state condition, this equation becomes

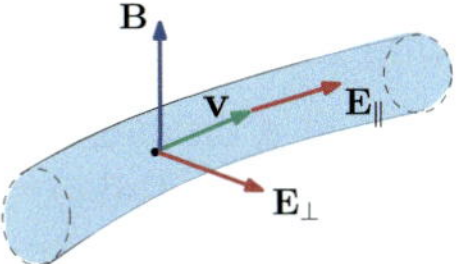

Fig. 8.17 The electric field is decomposed into its components parallel and perpendicular to the velocity of the electrons

$$\mathbf{v} = -\frac{e\tau}{m}\mathbf{E} - \frac{e\tau}{m}\mathbf{v} \times \mathbf{B} .$$

The electric field $\mathbf{E}$ can be decomposed into two components: $\mathbf{E}_\parallel$, the component parallel to the velocity of the electrons and $\mathbf{E}_\perp$, the component perpendicular to the velocity of the electrons. This decomposition is illustrated in Fig. 8.17.

Projecting the equation of motion onto the direction parallel to the velocity yields:

$$\mathbf{v} = -\frac{e\tau}{m}\mathbf{E}_\parallel .$$

This leads to the familiar Ohm law,

$$\mathbf{j} = \frac{ne^2\tau}{m}\mathbf{E}_\parallel = \sigma\mathbf{E}_\parallel ,$$

where n is the density of free electrons and σ is the conductivity.

Projecting the equation of motion onto the direction perpendicular to the velocity imposes

$$\mathbf{0} = -\frac{e\tau}{m}\mathbf{E}_\perp - \frac{e\tau}{m}\mathbf{v} \times \mathbf{B} .$$

The transverse field $\mathbf{E}_\perp$, also known as the Hall field, is given by

$$\boxed{\mathbf{E}_\perp = \mathbf{E}_\mathrm{H} = -\mathbf{v} \times \mathbf{B} .} \tag{8.12}$$

The establishment of this Hall field is necessary to ensure that the electron's trajectory is not deflected by the external magnetic field in the steady state. This field arises from an accumulation of electrons at the surface of the conductor. Combining the parallel and perpendicular components of the electric field, we obtain

$$\mathbf{E} = \mathbf{E}_\parallel + \mathbf{E}_\perp = \frac{\mathbf{j}}{\sigma} - \frac{\mathbf{j}}{-ne} \times \mathbf{B} .$$

This equation can be written as

$$\mathbf{E} = \varrho\mathbf{j} - R_H\mathbf{j} \times \mathbf{B} ,$$

where $\varrho = \dfrac{1}{\sigma}$ is the resistivity of the material and $R_H = -\dfrac{1}{ne}$ is the Hall coefficient.

Historically, the Hall effect played a crucial role in understanding the conduction properties of semiconductor materials. Today, it finds widespread applications in various everyday devices and modern technologies. Hall effect probes enable the direct measurement of magnetic fields by utilizing the relationship between the current flowing through a conductor and the perpendicular magnetic field applied to it. This principle allows for the construction of compasses that measure magnetic field strength based on voltage differences. The Hall effect can also be utilized to create contactless switches for powering various objects. For instance, in the automotive industry, Hall effect switches are used in engine ignition systems. Hall effect probes have diverse applications, including:

- Rotation speed detection: Used in bicycle wheels, automotive speedometers, electronic ignition systems, and gear tooth rotation control.
- Movement detection: Found in smartphones, paintball guns, and certain GPS systems.
- Position and motion sensors: Primarily used in DC motors, especially brushless types.
- Automotive industry: Applications in fuel injection, ignition systems, and wheel rotation sensors for anti-lock braking systems.
- Spacecraft propulsion: Hall-effect thrusters, such as the one shown in Fig. 8.18, are employed to propel spacecraft after they reach orbit.

These are just a few examples of the many applications of Hall effect probes.

Fig. 8.18 Hall Effect thruster. Reproduced from: B.A. Jorns et al., J. Appl. Phys. **136**, 053302 (2024); doi: 10.1063/5.0205985; licensed under a Creative Commons Attribution (CC BY) license

8.7 The Penning Trap

We have seen that a uniform magnetic field is able to confine a charge's trajectory to a circle in the plane perpendicular to the field, but it does not act on the charge's motion in the axis parallel to the field. A Penning trap is a device invented by Hans Georg Dehmlet (1922–2017) that uses both electric and magnetic fields to confine charged particles to a bounded trajectory in all three dimensions within the trap. They are used, for example, at CERN to store and study anti-matter, and are central to the realization of quantum computation based on trapped ions. To create a Penning trap, one needs a static magnetic field $\mathbf{B} = B\mathbf{u}_z$ and a quadrupolar electric field which, in cylindrical coordinates, is given by:

$$\mathbf{E} = \alpha \left(\frac{r}{2}\mathbf{u}_r - z\mathbf{u}_z \right) . \tag{8.13}$$

The electrostatic potential is therefore of the form $V(r, z) = \alpha z^2 - \frac{\alpha r^2}{2}$. The equipotentials, defined by $z^2 - \frac{r^2}{2} = C$ where C is a constant, are hyperbolas. Figure 8.19 shows a pair of equipotentials $V = \pm V_0$.

Therefore, three hyperbolic conductors with the shape of these equipotentials can be connected as shown in Fig. 8.19. Both the top and bottom conductors are at a pottential difference of V_0 with respect to the central conductor. This will produce exactly the field (8.13). Indeed, in the region between the conductors the potential is a solution of Laplace's equation with the boundary conditions $V = \pm V_0$. This problem has a unique solution given by $V(r, z) = \alpha z^2 - \frac{\alpha r^2}{2}$.

A simpler geometry that generates a similar potential consists of the superposition of a uniformly charged ring of density λ and two spheres of charge Q, located at $z = \pm a$ along the axis of the ring, as shown in Fig. 8.20.

If $Q = -\frac{3\pi a\lambda}{8}$, the field close to the center of the ring is approximately given by (8.13) with $\alpha = \frac{7Q}{3\pi \epsilon_0 a^3}$. A demonstration can be found in A.7. Figure 8.21 shows the calculated electric field on a symmetry plane of the charge distribution. It shows that (8.13) is a good description of the field close to the ring center. If $\alpha > 0$, a positive charge will feel a restoring force along the z-axis, but will be pushed radially towards the ring. No trapping is possible with a purely electrostatic field. Adding a magnetic

Fig. 8.19 Equipotentials
of the quadrupole field

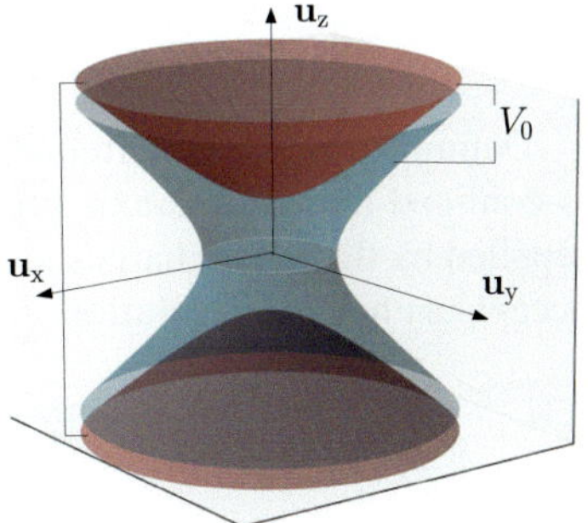

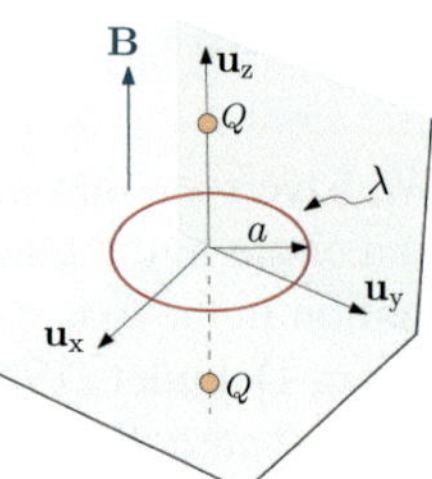

Fig. 8.20 A simple version of the Penning trap

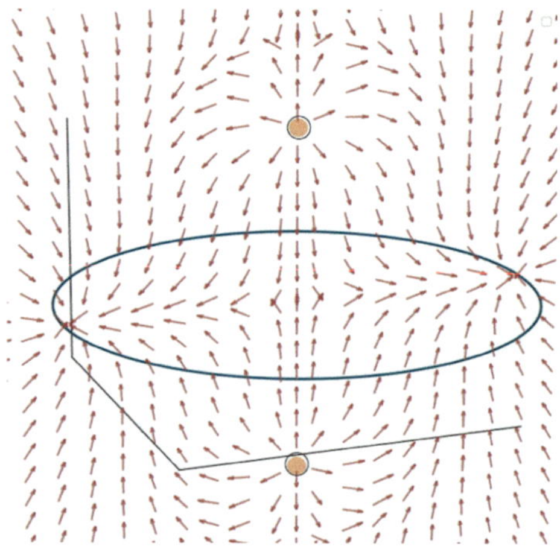

Fig. 8.21 Electric field produced by a negatively charged ring of radius a and two identical positive charges at $z = \pm a$ on the axis

field along z is what permits the trajectory of such a charge to be confined in the plane as well, by forcing a circular motion in the xy plane.

Indeed, the charge dynamics is given by $m\frac{d\mathbf{v}}{dt} = q\left(\alpha\frac{r}{2}\mathbf{u}_r - \alpha z\mathbf{u}_z + \mathbf{v}\times B\mathbf{u}_z\right)$. Projecting along the three cartesian coordinates

$$m\frac{dv_x}{dt} = q\alpha\frac{x}{2} + qBv_y \, ,$$

$$m\frac{dv_y}{dt} = q\alpha\frac{y}{2} - qBv_x \, ,$$

$$m\frac{dv_z}{dt} = -q\alpha z \, .$$

Defining the cyclotron frequency $\omega_c = qB/m$ and $\omega_z^2 = \frac{q\alpha}{m}$, we see that the charge is confined along the z-axis provided $\alpha q > 0$, meaning that the charge q must be repelled by the point charges at $z = \pm a$. In that case the equation of motion along z accepts an harmonic solution

$$\frac{d^2 z}{dt^2} + \omega_z^2 z = 0 \quad \rightarrow \quad z(t) = z_0 \cos(\omega_z t + \theta_z) \, .$$

For the in-plane movement, we define the variable $Z = x + iy$ so that

$$\begin{aligned}
\frac{d^2 Z}{dt^2} &= \frac{d^2 x}{dt^2} + i\frac{d^2 y}{dt^2} \\
&= \frac{\omega_z^2}{2}x + \omega_c\frac{dy}{dt} + i\left(\frac{\omega_z^2}{2}y - \omega_c\frac{dx}{dt}\right) \\
&= \frac{\omega_z^2}{2}Z - i\omega_c\frac{dZ}{dt}\,.
\end{aligned}$$

The general solution of this equation writes

$$Z(t) = Ae^{-i\omega_+ t} + Be^{-i\omega_- t}$$

with

$$\omega_\pm = \frac{\omega_c \pm \sqrt{\omega_c^2 - 2\omega_z^2}}{2}\,.$$

Finally, going back to the original cartesian coordinates, we have

$$x(t) = \mathrm{Re}(Z(t)) = |A|\cos(\omega_+ t + \varphi) + |B|\cos(\omega_- t + \phi)\,,$$

$$y(t) = \mathrm{Im}(Z(t)) = -|A|\sin(\omega_+ t + \varphi) - |B|\sin(\omega_- t + \phi)\,.$$

We see that the motion in the plane is the superposition of two circular motions of angular frequencies ω_+ and ω_-, respectively, provided that $\omega_c^2 - 2\omega_z^2 > 0$. In other words, the magnitude of the magnetic field should satisfy $|B| > \sqrt{\frac{2m\alpha}{q}}$. Many experiments are performed in the regime $|\omega_c| \gg \omega_z$ so that, for a positive charge, $\omega_+ \approx \omega_c$ and $\omega_- \approx \frac{\omega_z^2}{2\omega_c} \ll \omega_+$. For a negative charge like an electron, $\omega_c < 0$ and in this case $\omega_- \approx \omega_c < 0$ and $\omega_+ \approx \frac{\omega_z^2}{2\omega_c} < 0$.

Figure 8.22 shows a positive charge's trajectory for $\omega_c \gg \omega_z$. The trajectory is the superposition of a sinusoidal motion at frequency ω_z in the z direction, a fast circular motion close to the cyclotron frequency ω_c, called reduced cyclotron motion, and a slow circular motion at the so-called magnetron frequency $\omega_m = \frac{\omega_z^2}{2\omega_c}$. For large enough velocities, the radius associated with the magnetron motion is $R \sim v/\omega_m$, whereas that of the reduced cyclotron motion is $r \sim v/\omega_c \ll R$.

Trapping electrons in such traps has been crucial for conducting sensitive experiments that test our understanding of nature. One of the greatest successes of quantum electrodynamics is the remarkable agreement between the theoretical predictions and the experimentally measured value of the so-called anomalous magnetic moment of the electron, accurate to more than 10 significant digits.

Fig. 8.22 A negative charge's trajectory close to the center of the Penning trap

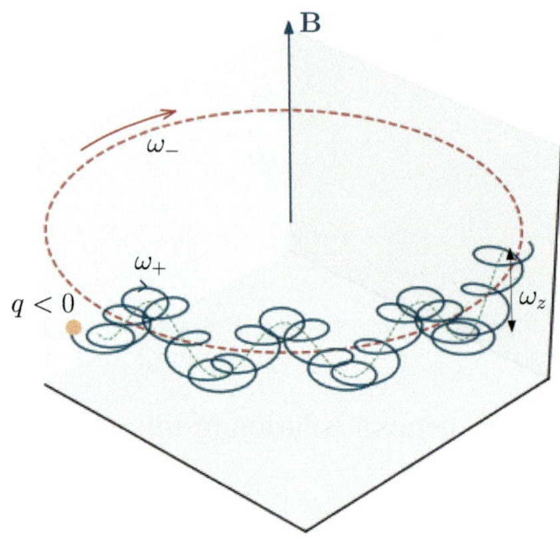

8.8 The Biot–Savart Law (1820): Magnetic Fields from Stationary Currents

In 1820, French physicists Jean-Baptiste Biot and Félix Savart experimentally established the law that now bears their names. By varying the distance between a stationary current in a wire and a magnetized object, they determined the mathematical formula describing the magnetic field generated by the current. The Biot–Savart law is considered the fundamental law of magnetostatics, analogous to Coulomb's law in electrostatics. In the following section, we will derive the Biot–Savart law from the definition of the magnetic field generated by a moving point particle (see (8.7)).

8.8.1 Magnetic Field Generated by a Continuous Current Distribution

Consider a fixed region of space, Ω, containing a distribution of mobile charges with a density ρ. These charges are moving within Ω, creating a current density that we assume is constant over time (stationary current). A small volume element d^3x' around a point $\mathbf{x}'$ within Ω contains an infinitesimal charge $dq(\mathbf{x}') = \varrho(\mathbf{x}')d^3x'$. This mobile element also has a current density $\mathbf{j}(\mathbf{x}') = \varrho(\mathbf{x}')\mathbf{v}(\mathbf{x}')$, where $\mathbf{v}$ is the velocity field. This scenario is illustrated in Fig. 8.23.

Based on (8.7) and the superposition principle, the magnetic field at $\mathbf{x}$ generated by the motion of charges in the volume element d^3x' is

$$d\mathbf{B}(\mathbf{x}) = \frac{\mu_0}{4\pi}dq(\mathbf{x}')\mathbf{v}(\mathbf{x}') \times \frac{\mathbf{x} - \mathbf{x}'}{|\mathbf{x} - \mathbf{x}'|^3} = \frac{\mu_0}{4\pi}d^3x'\mathbf{j}(\mathbf{x}') \times \frac{\mathbf{x} - \mathbf{x}'}{|\mathbf{x} - \mathbf{x}'|^3} \; . \qquad (8.14)$$

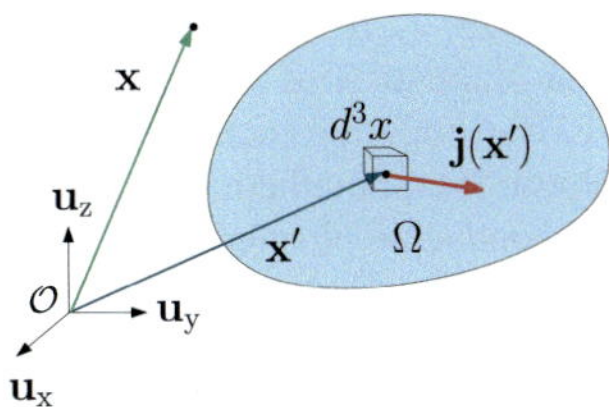

Fig. 8.23 A volume Ω containing an arbitrary current density $\mathbf{J}$

To find the total magnetic field at $\mathbf{x}$, we integrate this expression over the entire volume Ω:

$$\mathbf{B}(\mathbf{x}) = \frac{\mu_0}{4\pi} \iiint_\Omega \frac{\mathbf{j}(\mathbf{x}') \times (\mathbf{x} - \mathbf{x}')}{|\mathbf{x} - \mathbf{x}'|^3} d^3 x' \ . \tag{8.15}$$

8.8.2 Magnetic Field from a Current-Carrying Wire

Now consider a current I circulating through a wire with a small, constant cross-section dS. In this case, we can approximate the current flow as occurring along a one-dimensional path Γ, referred to as a line of current, that follows the shape of the circuit. This is illustrated in Fig. 8.24.

Since the cross-section dS is constant along the wire, the magnitude of the current density $|\mathbf{j}| = I/dS = j$ is also constant. Consider a point $\mathbf{x}'$ inside the conductor and take the volume element $d^3 x' = dl'(\mathbf{x}')dS$, where dl' is the differential line element at that point. We have

$$\mathbf{j}(\mathbf{x}')d^3 x' = \mathbf{j}(\mathbf{x}')dS\,dl'(\mathbf{x}') = j\,dS\,dl(\mathbf{x}')\,\mathbf{t}(\mathbf{x}') \ ,$$

where $\mathbf{t}(\mathbf{x}')$ denotes the direction of the current density, tangent to the curve Γ. Since $j\,dS = I$ is the total current through the wire, we can write the vector line element as $d\mathbf{l} = dl\,\mathbf{t}$. Using (8.14), we obtain the magnetic field generated by the current element $I d\mathbf{l}$:

$$d\mathbf{B}(\mathbf{x}) = \frac{\mu_0}{4\pi} \frac{I d\mathbf{l}(\mathbf{x}') \times (\mathbf{x} - \mathbf{x}')}{|\mathbf{x} - \mathbf{x}'|^3} \ .$$

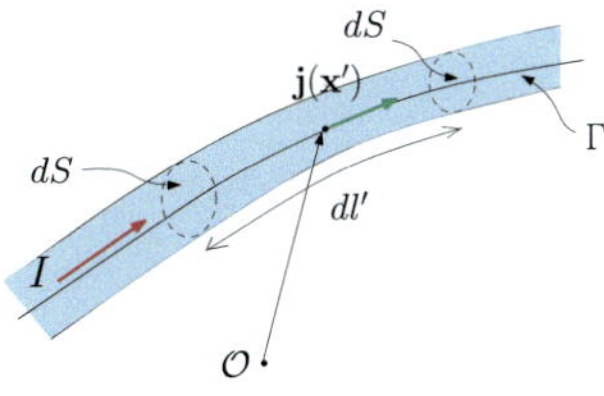

Fig. 8.24 A wire with a small cross-section can be approximated as a one-dimensional current flowing along the path Γ

Fig. 8.25 Left: Jean-Baptiste Biot (1774–1862), a french physicist and mathematician. He conducted pioneering research in circular birefringence. Right: Felix Savart (1791–1841), a French physicist who made significant contributions to acoustics, including the invention of the Savart wheel for measuring sound frequencies

Integrating this expression over the curve Γ yields the Biot–Savart law for a line of current, named after Jean-Baptiste Biot and Félix Savart (see Fig. 8.25).

The Biot–Savart law (1820)

A line of current is described by a curve Γ through which a current I circulates. The magnetic field at $\mathbf{x}$ generated by this circuit is given by the Biot–Savart law:

$$\mathbf{B}(\mathbf{x}) = \frac{\mu_0 I}{4\pi} \int_{\Gamma} d\mathbf{l}(\mathbf{x}') \times \frac{(\mathbf{x} - \mathbf{x}')}{|\mathbf{x} - \mathbf{x}'|^3} . \tag{8.16}$$

Example 8.2—Magnetic field at the axis of a circular current loop

Consider a circular loop in the xy-plane carrying a current I in the counterclockwise direction. A differential current element located at point $\mathbf{x}' = R\left(\cos\theta\,\mathbf{u}_x + \sin\theta\,\mathbf{u}_y\right) = R\mathbf{u}_r(\theta)$ can be written as $I d\mathbf{x}'$, where $d\mathbf{x}' = Rd\theta\mathbf{u}_\theta$ and $\mathbf{u}_\theta = (-\sin\theta\,\mathbf{u}_x + \cos\theta\,\mathbf{u}_y)$. This is illustrated in Fig. 8.26.

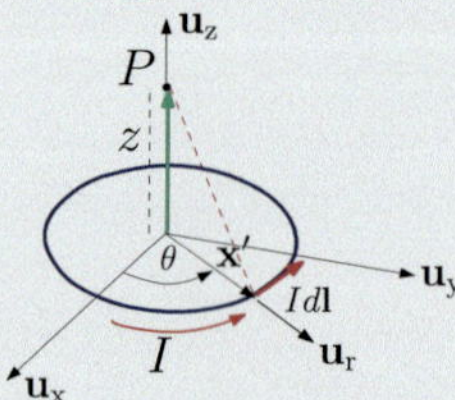

Fig. 8.26 A differential current element $I dl$ along a circular loop

Using the Biot–Savart law, we have:

$$d\mathbf{B}(\mathbf{x}) = \frac{\mu_0 I}{4\pi} \frac{d\mathbf{x}' \times (\mathbf{x} - \mathbf{x}')}{|\mathbf{x} - \mathbf{x}'|^3}$$

with $|\mathbf{x} - \mathbf{x}'| = (R^2 + z^2)^{1/2}$. Therefore,

$$d\mathbf{B}(\mathbf{x}) = \frac{\mu_0 I}{4\pi} \frac{R d\theta\, \mathbf{u}_\theta \times (z\mathbf{u}_z - R\mathbf{u}_r)}{(R^2 + z^2)^{3/2}}$$

$$= d\theta \frac{\mu_0 I}{4\pi} \frac{(R^2\mathbf{u}_z + Rz\mathbf{u}_r)}{(R^2 + z^2)^{3/2}} .$$

Integrating this expression over the loop, we find

$$\mathbf{B}(\mathbf{x}) = \frac{\mu_0 I}{4\pi} \int_0^{2\pi} d\theta \frac{[R^2\mathbf{u}_z + Rz(\sin\theta\, \mathbf{u}_y + \cos\theta\, \mathbf{u}_x)]}{(R^2 + z^2)^{3/2}} .$$

The integrals along $\mathbf{u}_x$ and $\mathbf{u}_y$ vanish, consistent with symmetry arguments (see Sect. 9.3). Finally, the magnetic field at a distance z from the loop is

$$\mathbf{B}(\mathbf{x}) = \frac{\mu_0 I}{4\pi} \int_0^{2\pi} d\theta \frac{R^2}{(R^2 + z^2)^{3/2}}\mathbf{u}_z = \frac{\mu_0 I}{2} \frac{R^2}{(R^2 + z^2)^{3/2}}\mathbf{u}_z .$$

8.8.3 Magnetic Field Lines

The magnetic field lines are defined in the same way as the electric field lines. They represent the set of curves Γ such that at any point $\mathbf{x}$ on Γ, the tangent to the curve aligns with the direction of $\mathbf{B}(\mathbf{x})$ at that point. These lines can be found by solving the equation

$$\mathbf{B} \times d\mathbf{x} = \mathbf{0} ,$$

where $d\mathbf{x}$ is an infinitesimal displacement along the curve.

Reminder: If we have the expressions for the components of the magnetic field as functions of spatial coordinates, we can determine the network of field lines by solving the ordinary differential equation derived from the condition $\mathbf{B} \times d\mathbf{x} = \mathbf{0}$.

In Cartesian coordinates, this condition is equivalent to

$$\frac{dx}{B_x(x, y, z)} = \frac{dy}{B_y(x, y, z)} = \frac{dz}{B_z(x, y, z)} .$$

By using separation of variables (if possible) or numerical methods, we can integrate this equation to obtain the equation for the field lines. The integration constants are determined by specifying the initial conditions, such as requiring a field line to pass through a particular point.

A similar procedure applies to other coordinate systems. For example, in cylindrical coordinates, the condition $\mathbf{B} \times d\mathbf{x} = \mathbf{0}$ becomes

$$\frac{dr}{B_r(r, \theta, z)} = \frac{r\,d\theta}{B_\theta(r, \theta, z)} = \frac{dz}{B_z(r, \theta, z)} .$$

In magnetostatics, magnetic field lines encircle the electric currents that produce them, ultimately forming closed loops. Figure 8.27 illustrates the magnetic field lines for an infinitely long wire. Due to the cylindrical symmetry of the system, the magnetic field lines are concentric circles perpendicular to the wire axis. These results will be demonstrated in Chap. 9 using Ampère's law.

A practical way to visualize the magnetic field lines in a plane is to place a rectangular array of magnetic needles that can freely rotate around the axis perpendicular to the plane. In the presence of a magnetic field, they experience a torque and tend to align along the field. This will be demonstrated in Chap. 9. Figure 8.28 shows the result for a magnet bar. These field lines will be calculated in Sect. 9.6.

Fig. 8.27 Magnetic field lines around a current-carrying wire

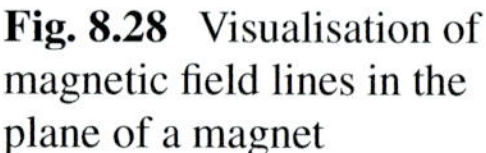

Fig. 8.28 Visualisation of magnetic field lines in the plane of a magnet

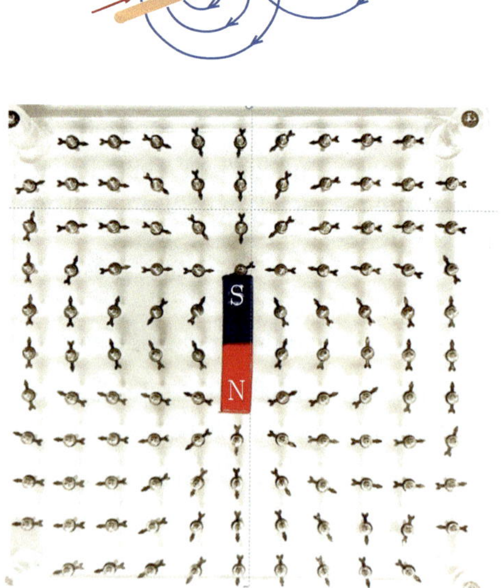

8.9 The Laplace Force

A conductor carrying a current experiences a force $\mathbf{F}_L$, known as the Laplace force, when placed in an external magnetic field $\mathbf{B}$. This force arises due to the Hall effect, which is the generation of a transverse electric field $\mathbf{E}_H$ perpendicular to the velocity $\mathbf{v}$ of the electrons. The Hall field is given by

$$\mathbf{E}_H = -\mathbf{v} \times \mathbf{B} \ .$$

While the Hall field cancels out the magnetic force acting on the electrons, it exerts a force on the positive ions of the host crystal, which are not involved in the current flow. The electric force acting on a positive ion with charge $+e$ is

$$\mathbf{f} = e\mathbf{E}_H = -e\mathbf{v} \times \mathbf{B} \ .$$

If n is the density of ions (and electrons), the force acting on a differential volume d^3x' around $\mathbf{x}'$ is

$$d\mathbf{f}(\mathbf{x}') = \mathbf{f}d^3x' = -en\mathbf{v} \times \mathbf{B}d^3x' = \mathbf{j} \times \mathbf{B}d^3x' \ .$$

For a wire conductor with a cross-section $\Delta S(\mathbf{x}')$, we can write $d^3x' = \Delta S(\mathbf{x}')dl(\mathbf{x}')$, where dl is a differential length element. The total force on d^3x' can be rewritten as

$$d\mathbf{f}(\mathbf{x}') = \Delta S(\mathbf{x}')dl(\mathbf{x}')\mathbf{j}(\mathbf{x}') \times \mathbf{B}(\mathbf{x}') \ .$$

Defining $d\mathbf{l}(\mathbf{x}')$ as the vector with magnitude $dl(\mathbf{x}')$ and direction coinciding with the current density, we obtain the Laplace force acting on a current element $Id\mathbf{l}$,

$$\boxed{d\mathbf{F}_L(\mathbf{x}') = Id\mathbf{l}(\mathbf{x}') \times \mathbf{B}(\mathbf{x}') \ .} \tag{8.17}$$

The Laplace force
The Laplace force on a linear conductor described by the curve Γ is given by

$$\boxed{\mathbf{F}_L = \int_\Gamma Id\mathbf{l} \times \mathbf{B}(\mathbf{x}) \ .} \tag{8.18}$$

where I is the current carried by the conductor and $\mathbf{B}$ is the magnetic field.

Example 8.3—Force on a current carrying loop in a uniform magnetic field

Consider a conductor carrying a current I in the presence of a uniform magnetic field $\mathbf{B}$, as illustrated in Fig. 8.29.

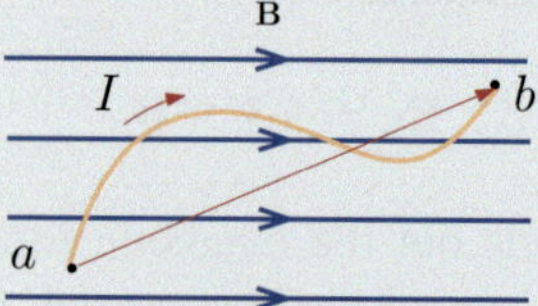

Fig. 8.29 A wire carrying a current I between points a and b in the presence of a uniform magnetic field $\mathbf{B}$

The force on this conductor is given by (8.18)

$$\mathbf{F}_L = I \left(\int_a^b d\mathbf{x} \right) \times \mathbf{B}(\mathbf{x}) = I \left(\mathbf{x}_b - \mathbf{x}_a \right) \times \mathbf{B}(\mathbf{x}) \ .$$

Let $\mathbf{l} = \mathbf{x}_b - \mathbf{x}_a$ be the vector directed from a to b. Using this notation

$$\mathbf{F}_L = I\,\mathbf{l} \times \mathbf{B} \ .$$

For a closed circuit of arbitrary shape, the force becomes

$$\mathbf{F}_L = I \left(\oint_\Gamma d\mathbf{x} \right) \times \mathbf{B}$$

However, $\oint_\Gamma d\mathbf{x} = \mathbf{0}$ for a closed loop. Therefore, the net Laplace force on a closed circuit in a uniform magnetic field is zero, $\mathbf{F}_L = \mathbf{0}$.

8.10 Summary and Essential Formulas

- Moving electric charges generate a magnetic field $\mathbf{B}$, which is a vector field. There exists a profound relationship between electric and magnetic fields. In a reference frame where a charge is stationary, an observer will perceive only an electrostatic field. However, in any other reference frame, the observer will see the charge in motion, resulting in both an electric field and a magnetic field. This demonstrates that electricity and magnetism are interconnected aspects of a single phenomenon.

- The magnetic field generated by a current density $\mathbf{j}$ is given by the integral equation (8.15)

$$\mathbf{B}(\mathbf{x}) = \frac{\mu_0}{4\pi} \iiint_{\mathbb{R}^3} \frac{\mathbf{j}(\mathbf{x}') \times (\mathbf{x} - \mathbf{x}')}{|\mathbf{x} - \mathbf{x}'|^3} d^3 x' \ .$$

- The field generated by a line current defined by the curve Γ is given by the Biot–Savart law, which is a special case of (8.15):

$$\mathbf{B}(\mathbf{x}) = \frac{\mu_0 I}{4\pi} \int_\Gamma d\mathbf{l}(\mathbf{x}') \times \frac{(\mathbf{x} - \mathbf{x}')}{|\mathbf{x} - \mathbf{x}'|^3} \ .$$

- The force experienced by a charge q in the presence of a magnetic field $\mathbf{B}$ is

$$\mathbf{F}_q = q\mathbf{v} \times \mathbf{B} \ .$$

This force is responsible for a wide variety of phenomena. A free charge in a uniform magnetic field that is perpendicular to the charge velocity describes a circular path of radius $R = \dfrac{mv}{qB}$ and of frequency $\omega = v/R = \dfrac{qB}{m}$. In a conductive medium through which electrons of velocity $\mathbf{v}$ flow, it leads to the Hall effect: In the presence of a magnetic field $\mathbf{B}$, the electric field in a conductor carrying a steady-state current density $\mathbf{j}$ reads

$$\mathbf{E} = \varrho\mathbf{j} + R_H \mathbf{B} \times \mathbf{j}$$

where the Hall coefficient is

$$R_H = -1/ne \ .$$

- The Laplace force is the electromagnetic force exerted by a magnetic field $\mathbf{B}(\mathbf{x})$ on a current-carrying conductor. It is the macroscopic manifestation of the Lorentz force, which acts on individual charged particles. The Laplace force is perpendicular to both the direction of the current and the magnetic field, and is given by the following equation:

$$\mathbf{F}_L = \int_\Gamma I\, d\mathbf{x} \times \mathbf{B}(\mathbf{x}) \ ,$$

where I is the current flowing through the conductor, Γ is the path followed by the conductor within the magnetic field and $d\mathbf{x}$ is is an infinitesimal element of the path Γ. For this element, the Laplace force element $d\mathbf{F}_L = I\,d\mathbf{x} \times \mathbf{B}(\mathbf{x})$ is perpendicular to both the direction of the current and the magnetic field.

Problems

8.1 Mass spectrometer

A mass spectrometer is a device used to measure the mass of charged particles. Its key components are illustrated below. An ion with mass m and charge q is initially at rest and is then accelerated by a potential difference ΔV. The ion then enters a region with a uniform magnetic field $\mathbf{B} = -B\mathbf{u}_z$ and a uniform electric field $\mathbf{E} = E\mathbf{u}_x$, which are perpendicular to each other. By adjusting the fields, the velocity $\mathbf{v}$ of the ions can be set so that they pass through this region in a straight line. These particles then enter a second region with a different magnetic field $\mathbf{B}_0 = -B_0\mathbf{u}_z$. In this region, the ions follow a circular path and eventually strike a detector at a distance $2r$ from the entry point.

(a) Determine the velocity $\mathbf{v}$ of the particles as they enter the selector.
(b) Calculate the magnetic field $\mathbf{B}$ that should be applied so that the particles pass through the selector without being deflected in the presence of $\mathbf{E}$ and $\mathbf{B}$.
(c) Once the value of r is measured, derive an expression for the mass m of the particle.

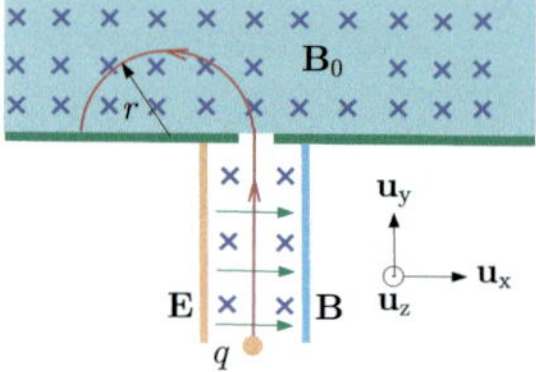

8.2 Proton in a uniform magnetic field

A proton with charge q and speed v enters a region of uniform magnetic field $\mathbf{B} = -B\mathbf{u}_z$ The angle of incidence is θ.

(a) Determine the angle of departure, θ'?
(b) Calculate the distance d.

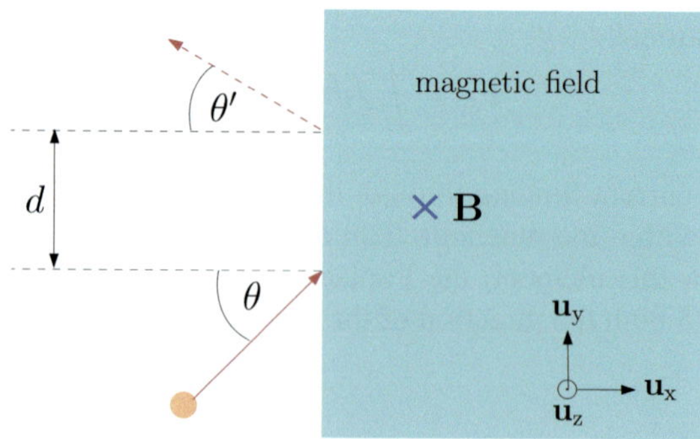

8.3 Electron in a uniform magnetic field

An electron (charge $-e$) is launched with initial velocity $\mathbf{v} = v\mathbf{u}_x$ midway between two parallel plates. A constant magnetic field $\mathbf{B} = -B\mathbf{u}_z$ exists between the plates. Neglecting gravitational effects:

(a) Determine the direction in which the electron is deflected.
(b) Calculate the magnitude of the velocity if the particle strikes the end of one of the plates.

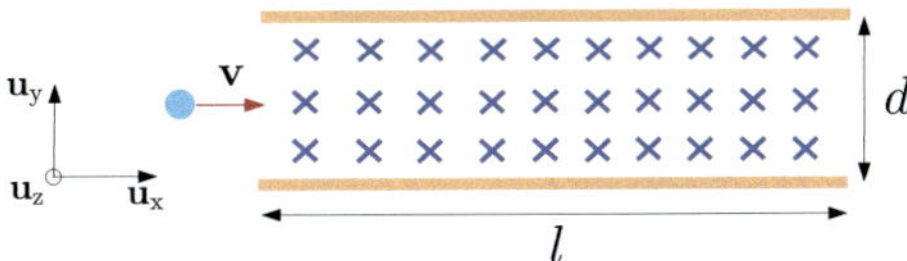

8.4 Charge in a uniform magnetic field

A charged particle with mass m and charge q enters a uniform magnetic field $\mathbf{B}$ with a velocity v, as shown in the figure below. Determine the time elapsed between the particle entering and exiting the magnetic field region.

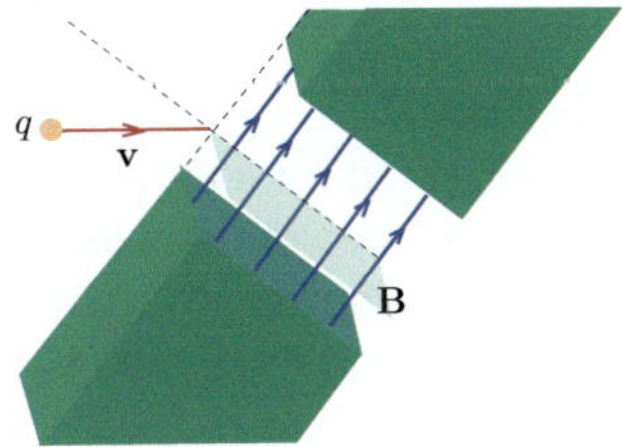

8.5 Proton in a cyclotron accelerator

A proton is accelerated in a cyclotron with a radius $R = 2\,\text{m}$, an oscillating potential difference $\Delta V = 50\sin(2\pi f t + \varphi)\text{kV}$, and a magnetic field $B = 1\,\text{T}$. Determine the oscillation frequency f of the electric field. Ignoring relativistic effects, calculate the maximum kinetic energy that the proton can acquire. How many turns are required to achieve this maximum kinetic energy?

8.6 The Hall effect

Consider a conductive plate of length l and rectangular cross-section with dimensions a and b. The plate carries a uniform current density $\mathbf{j}$ parallel to its length (along the y-axis). The conductor has n free charges (q) per unit volume.

(a) What is the effect of the magnetic field $\mathbf{B}$ on the free charges? What forces do they experience, and what occurs at the side walls $x = 0$ and $x = a$? On which wall do charges q accumulate?
(b) The accumulation or deficit of charges on the walls creates an electric field $\mathbf{E}_H$ (The Hall field). How does this field affect the free charges?
(c) For what value of the $\mathbf{E}_H$ field is equilibrium achieved, canceling the effect of the magnetic field $\mathbf{B}$? What is the trajectory of the free charges under this condition?

(d) What is the Hall voltage V_H that appears between the walls in terms of the Hall constant $R_H = -1/nq$? How can the Hall effect be used to measure an unknown magnetic field (Hall effect probe)?

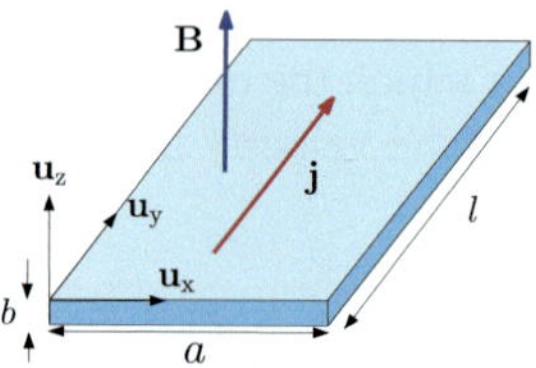

8.7 Application of the Hall effect I

Consider a copper band of length $a = 2\,\text{cm}$ and thickness $b = 1\,\text{mm}$, carrying a current of 50 A. The band is subjected to a uniform magnetic field of 2 T. The free charges are electrons ($q = -e = -1.6 \times 10^{-19}$ C), and the number of free electrons per cubic meter is $n = 8 \times 10^{28}\text{m}^{-3}$.

(a) Calculate the displacement speed of the electrons.
(b) Calculate the value of the Hall field and the Hall voltage V_H.

8.8 Application of the Hall effect II

A Hall field of $2.5\,\text{mV m}^{-1}$ is measured for a current density of $1 \times 10^7\,\text{A m}^{-2}$ in a magnetic field of $B = 1$ T. Calculate the volume density of free electrons in sodium and compare it to the number of sodium atoms. The atomic mass of sodium 23g mol^{-1}, its density is 0.97g cm^{-3}, and the Avogadro number is 6.02×10^{23}.

8.9 Magnetic field from a straight current-carrying wire

Consider a thin, rigid wire carrying a current I along the x-axis. Calculate the magnetic field at point P and analyze the specific case where the wire is symmetric with respect to the y-axis. What occurs when the wire is infinitely long? Determine the current required for a compass located 10 cm from the wire to experience a magnetic field 10 times stronger than the Earth's magnetic field ($B_T = 0.2 \times 10^{-4}$ T).

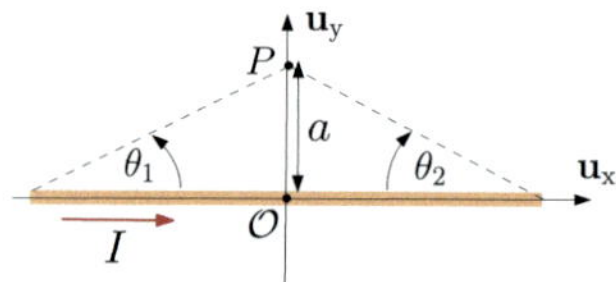

8.10 Magnetic field of a rotating charged disc

Consider a very thin disc with a uniformly distributed surface charge density σ. The disc begins rotating at an angular speed ω around its axis. Determine the magnetic field on the disc axis.

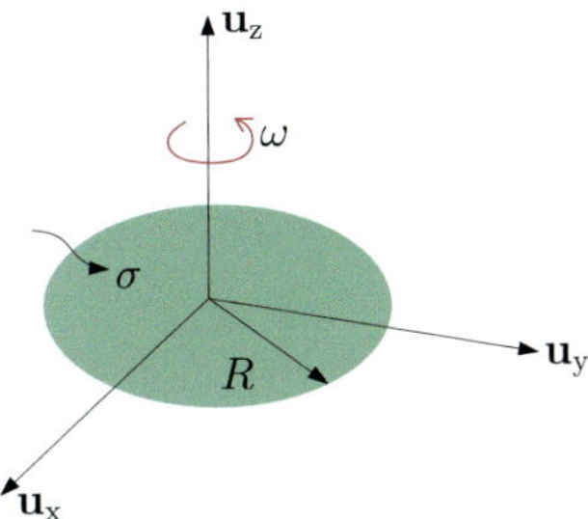

8.11 Magnetic field from an annular quadrant

Consider the current loop formed by radial segments and circular arc segments, as illustrated in the figure below. Assuming the two horizontal segments are infinitesimally close, determine the magnetic field $\mathbf{B}$ at point P.

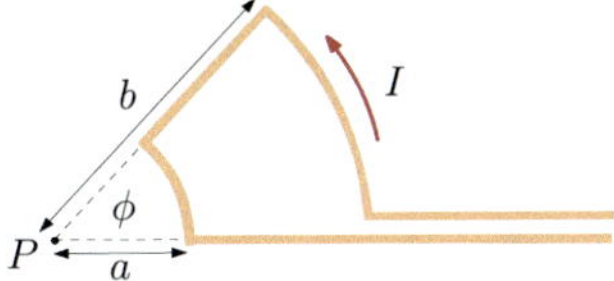

8.12 Field near a corner

A rectilinear conductor with current I goes down the y-axis to the origin and then continues along the horizontal direction. Show that the magnetic field at point (x, y) with $x > 0$, $y > 0$ is given by

$$\mathbf{B}(x, y) = \frac{\mu_0 I}{4\pi} \left(\frac{1}{x} + \frac{1}{y} + \frac{r}{xy} \right) \mathbf{u}_z$$

with $r = \sqrt{x^2 + y^2}$.

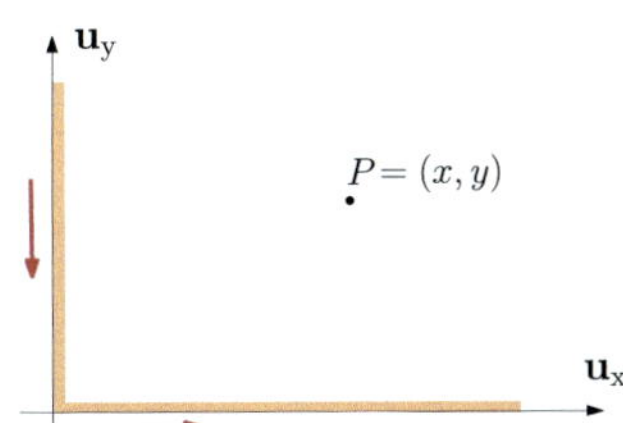

8.13 Helmholtz coils

Consider two coils, each with N turns and radius R, arranged perpendicular to the z-axis. The centers of the coils are located at $z = l$ and $z = -l$. Both coils carry a constant current I in the same direction, as shown in the figure. Determine the magnetic field on the axis at a distance z from the origin. Verify that the first derivative of the field at the midpoint is zero.

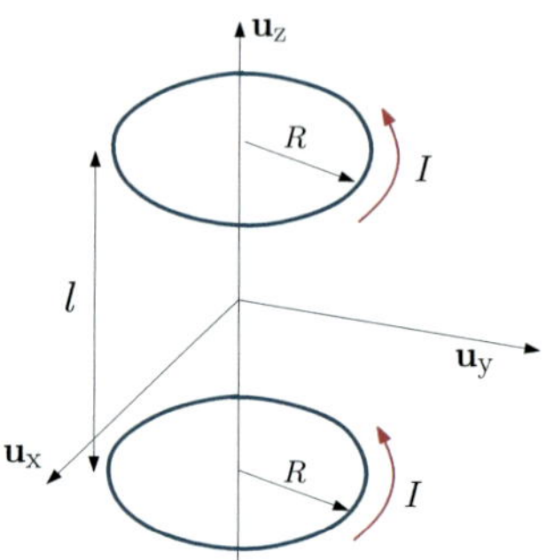

8.14 Force between two parallel currents

Consider two parallel wires of length l separated by a distance $a \ll l$. The wires carry currents I_1 and I_2 in the y-direction. Calculate the force between the conductors.

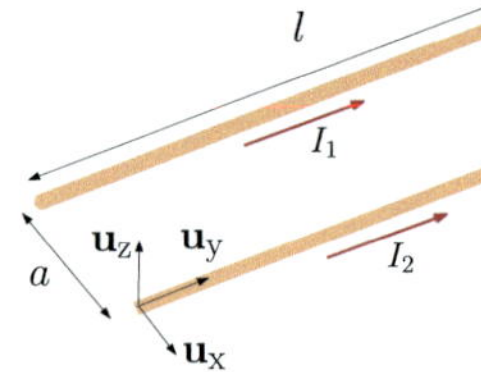

Chapter 9
Ampère's Law and Magnetic Dipole

Abstract This chapter presents *Ampère's law*, a cornerstone of magnetostatics that provides a powerful method for calculating magnetic fields, especially for current distributions with high symmetry. It presents both the *differential and integral forms of Ampère's law*, demonstrating how the circulation of the magnetic field around a closed loop is directly proportional to the enclosed current. The chapter emphasizes the application of *symmetry arguments* (invariance under translation, rotation, and mirror symmetry/antisymmetry) to simplify magnetic field calculations. A significant portion is dedicated to the *magnetic vector potential* ($\mathbf{A}$), defined such that $\mathbf{B} = \nabla \times \mathbf{A}$. The chapter shows how $\mathbf{A}$ can be calculated from current distributions and discusses its *non-uniqueness*, leading to the concept of *gauge transformations* and the choice of the *Coulomb gauge* ($\nabla \cdot \mathbf{A} = 0$), under which $\mathbf{A}$ satisfies a Poisson-like equation. The profound physical significance of the vector potential, even in regions where the magnetic field is zero, is highlighted through the *Aharonov–Bohm effect*, an experimental confirmation of quantum mechanical principles. The chapter then formalizes the *two fundamental laws of magnetostatics: Gauss's law for magnetism* ($\nabla \cdot \mathbf{B} = 0$), which signifies the absence of magnetic monopoles, and *Ampère's law*. These two laws, together with the Helmholtz decomposition theorem, are shown to uniquely determine the magnetic field. Finally, the chapter introduces the *magnetic dipole*, defining its *magnetic moment* ($\mathbf{m}$) for both current loops and arbitrary current distributions. It demonstrates that the magnetic field far from a localized current distribution resembles that of a magnetic dipole and derives the expressions for the *forces and torques* experienced by a magnetic dipole in an external magnetic field.

Keywords Ampere's law · Magnetic dipole · Magnetostatics · Invariances · Symmetry

9.1 Introduction

The discovery that electrical currents affect the needle of a compass led physicists to investigate the mathematical relationships between currents and magnetic fields. André-Marie Ampère (see Fig. 9.1) was the first to establish a formula for the force

Fig. 9.1 André-Marie Ampère (1775–1836), a French physicist and mathematician, and one of the founders of classical electromagnetism. He was the first to postulate the existence of the electron. The unit of electric current is named in his honor

of interaction between two current-carrying wires. In this chapter we will formulate a law, now known as Ampère's law, that plays a role analogous to Gauss's law in electrostatics, as it simplifies the calculation of magnetic fields when the current distribution exhibits invariances and symmetries. In its differential form, Ampère's law states that a current density $\mathbf{j}$ is the source of a rotational field $\mathbf{B}$.

We show how Curie's principle can be applied to the study of magnetic fields and use Ampère's law to calculate the magnetic field for several classic examples. We also introduce the concept of the vector potential $\mathbf{A}$, defined such that $\mathbf{B} = \nabla \times \mathbf{A}$, which arises from the absence of magnetic monopoles. Together with Ampère's law, this forms a complete set of equations for magnetostatics. Furthermore, we define the magnetic dipole as an elementary loop of current and show that, at points far from a confined current density, the magnetic field can be approximated by that of a magnetic dipole. Finally, we derive the expressions for the force and torque experienced by a magnetic dipole in the presence of an external magnetic field.

9.2 Ampère's Law

9.2.1 Ampère's Law—Differential Form

From (8.15), let us calculate the curl of an arbitrary magnetic field generated by a current density inside a volume $\Omega \subseteq \mathbb{R}^3$:

$$\nabla \times \mathbf{B}(\mathbf{x}) = \nabla \times \frac{\mu_0}{4\pi} \iiint_\Omega \frac{\mathbf{j}(\mathbf{x}') \times (\mathbf{x} - \mathbf{x}')}{|\mathbf{x} - \mathbf{x}'|^3} \, d^3x'$$

$$= \frac{\mu_0}{4\pi} \iiint_\Omega \nabla \times \left(\mathbf{j}(\mathbf{x}') \times \frac{\mathbf{x} - \mathbf{x}'}{|\mathbf{x} - \mathbf{x}'|^3} \right) d^3x' \, .$$

Now, we use the vector identity (A.9) in Sect. A.1.5 giving the curl of the cross product of the two vector fields $\mathbf{j}$ and $\mathbf{E}$:

$$\nabla \times (\mathbf{j} \times \mathbf{E}) = \mathbf{j}(\nabla \cdot \mathbf{E}) - \mathbf{E}(\nabla \cdot \mathbf{j}) + (\mathbf{E} \cdot \nabla)\mathbf{j} - (\mathbf{j} \cdot \nabla)\mathbf{E} .$$

Since $\mathbf{j}(\mathbf{x}')$ is independent of $\mathbf{x}$, we have $\mathbf{E}(\nabla \cdot \mathbf{j}) = 0$ and $(\mathbf{E} \cdot \nabla)\mathbf{j} = 0$. Setting

$$\mathbf{E}_{\mathbf{x}'}(\mathbf{x}) = \frac{\mathbf{x} - \mathbf{x}'}{|\mathbf{x} - \mathbf{x}'|^3} ,$$

which represents the electrostatic field generated at $\mathbf{x}$ by a point charge q located at $\mathbf{x}'$, with $\dfrac{q}{4\pi\epsilon_0} = 1$ V m, we obtain

$$\nabla \times \mathbf{B}(\mathbf{x}) = \frac{\mu_0}{4\pi} \left\{ \iiint_\Omega \mathbf{j}(\mathbf{x}') \, (\nabla \cdot \mathbf{E}_{\mathbf{x}'}(\mathbf{x})) \, d^3x' - \iiint_\Omega (\mathbf{j}(\mathbf{x}') \cdot \nabla)\mathbf{E}_{\mathbf{x}'}(\mathbf{x}) \, d^3x' \right\} .$$

In the first integral on the right-hand side, we recognize the divergence of the electrostatic field $\mathbf{E}_{\mathbf{x}'}(\mathbf{x})$ satisfying the differential form of Gauss's law (2.3),

$$\nabla \cdot \mathbf{E}_{\mathbf{x}'}(\mathbf{x}) = 4\pi\delta(\mathbf{x} - \mathbf{x}') .$$

Therefore, the integral of the first term yields

$$\frac{\mu_0}{4\pi} \iiint_\Omega \mathbf{j}(\mathbf{x}') \, (\nabla \cdot \mathbf{E}_{\mathbf{x}'}(\mathbf{x})) \, d^3x' = \mu_0 \iiint_\Omega \mathbf{j}(\mathbf{x}')\delta(\mathbf{x} - \mathbf{x}') \, d^3x' = \mu_0\mathbf{j}(\mathbf{x}) .$$

We will now demonstrate that the second integral is zero:

$$\mathcal{I} = \iiint_\Omega (\mathbf{j}(\mathbf{x}') \cdot \nabla)\mathbf{E}_{\mathbf{x}'}(\mathbf{x}) \, d^3x' = 0 .$$

To integrate $(\mathbf{j}(\mathbf{x}') \cdot \nabla)\mathbf{E}_{\mathbf{x}'}(\mathbf{x})$, we apply the vector identity (A.8),

$$\nabla(\mathbf{j} \cdot \mathbf{E}_{\mathbf{x}'}) = (\mathbf{j} \cdot \nabla)\mathbf{E}_{\mathbf{x}'} + (\mathbf{E}_{\mathbf{x}'} \cdot \nabla)\mathbf{j} + \mathbf{j} \times (\nabla \times \mathbf{E}_{\mathbf{x}'}) + \mathbf{E}_{\mathbf{x}'} \times (\nabla \times \mathbf{j}) ,$$

where the second and fourth terms on the right-hand side vanish because $\mathbf{j}(\mathbf{x}')$ does not depend on $\mathbf{x}$. The third term also vanish since the electrostatic field $\mathbf{E}_{\mathbf{x}'}$ has zero curl. Therefore, we obtain

$$\mathcal{I} = \iiint_\Omega \nabla\,(\mathbf{j}(\mathbf{x}') \cdot \mathbf{E}_{\mathbf{x}'}(\mathbf{x})) \, d^3x' = \nabla \left(\iiint_\Omega \mathbf{j}(\mathbf{x}') \cdot \mathbf{E}_{\mathbf{x}'}(\mathbf{x}) \, d^3x' \right) ,$$

where the last step is allowed because the gradient acts on coordinates $\mathbf{x}$. Now we can write the electrostatic field as a function of $\mathbf{x}'$ and use the property that it is conservative, $\mathbf{E}_{\mathbf{x}'}(\mathbf{x}) = -\mathbf{E}_{\mathbf{x}}(\mathbf{x}') = \nabla' V_{\mathbf{x}}(\mathbf{x}')$. Here $V_{\mathbf{x}}(\mathbf{x}') = 1/|\mathbf{x} - \mathbf{x}'|$ and ∇' denotes the gradient with respect to coordinates $\mathbf{x}'$. The current density is stationary, $\nabla' \cdot \mathbf{j}(\mathbf{x}') = 0$, allowing us to write

$$\mathbf{j}(\mathbf{x}') \cdot \mathbf{E}_{\mathbf{x}'}(\mathbf{x}) = \mathbf{j}(\mathbf{x}') \cdot \nabla' V_{\mathbf{x}}(\mathbf{x}') = \nabla' \cdot \big(V_{\mathbf{x}}(\mathbf{x}')\mathbf{j}(\mathbf{x}')\big) \ .$$

Finally, using the divergence theorem (A.15), we obtain

$$\mathcal{I} = \nabla \left(\iiint_{\Omega} \nabla' \cdot \big(V_{\mathbf{x}}(\mathbf{x}')\mathbf{j}(\mathbf{x}')\big) \ d^3 x' \right) = \nabla \left(\oiint_{\partial\Omega} V_{\mathbf{x}}(\mathbf{x}')\mathbf{j}(\mathbf{x}') \cdot \mathbf{n}(\mathbf{x}')dS \right) \ .$$

The current density is assumed to be localized inside the volume Ω, and cannot have a perpendicular component to the surface, therefore, $\mathbf{j}(\mathbf{x}') \cdot \mathbf{n}(\mathbf{x}') = 0$. This completes our demonstration that $\mathcal{I} = 0$.

We have obtained the differential form of Ampère's law.

Ampère's law (1826)—differential form

Given a stationary current density $\mathbf{j}$ ($\nabla \cdot \mathbf{j} = 0$) localized within a volume $\Omega \subseteq \mathbb{R}^3$, Ampère's law states

$$\boxed{\nabla \times \mathbf{B}(\mathbf{x}) = \mu_0 \mathbf{j}(\mathbf{x}) \ .} \tag{9.1}$$

This law states that the current density $\mathbf{j}$ generates, at every point $\mathbf{x} \in \mathbb{R}^3$, a rotational magnetic field $\mathbf{B}$ around $\mathbf{j}$.

9.2.2 Ampère's Law—Integral Form

Let Γ be a closed curve in $\mathbb{R}^3$ and $S(\Gamma)$ be a surface bounded by the curve Γ. Stokes's theorem (A.20) states

$$\oint_{\Gamma} \mathbf{B}(\mathbf{x}) \cdot d\mathbf{x} = \iint_{S(\Gamma)} (\nabla \times \mathbf{B}(\mathbf{x})) \cdot \mathbf{n}(\mathbf{x})dS(\mathbf{x}) \ .$$

The right-hand side can be rewritten by using the differential form of Ampère's law,

$$\oint_{\Gamma} \mathbf{B}(\mathbf{x}) \cdot d\mathbf{x} = \mu_0 \iint_{S(\Gamma)} \mathbf{j}(\mathbf{x}) \cdot \mathbf{n}(\mathbf{x})dS(\mathbf{x}) = \mu_0 \Phi_{S(\Gamma),\mathbf{j}} \ .$$

This shows that the right-hand side is proportional to the flux of $\mathbf{j}$ over $S(\Gamma)$, or equivalently, the current passing through $S(\Gamma)$. For simplicity, we will denote this current as I_{Γ}.

Fig. 9.2 According to Stokes's theorem, the flux of **j** is the same through any oriented surface bounded by Γ

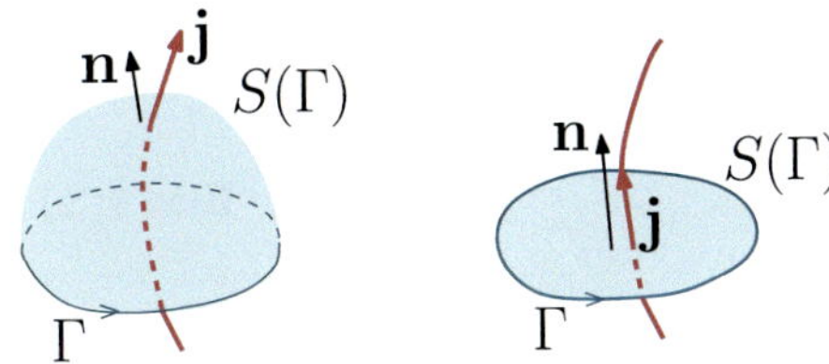

Ampère's law (1826)—Integral form
The circulation of the magnetic field around a closed curve $\Gamma \in \mathbb{R}^3$ is proportional to the current $I(\Gamma)$ flowing through any surface $S(\Gamma)$ bounded by the curve Γ:

$$\oint_{\Gamma} \mathbf{B}(\mathbf{x}) \cdot d\mathbf{x} = \mu_0 I_{\Gamma} \tag{9.2}$$

Remarks
1. For Stokes's theorem to hold, Γ and $S(\Gamma)$ must have consistent orientations. If you are looking at $S(\Gamma)$ such that the normal vector points toward you, then Γ should be oriented counterclockwise.
2. Given a curve Γ, an infinity of surfaces $S(\Gamma)$ can be chosen that are bounded by Γ. However, the flux of **j** over $S(\Gamma)$ is independent of this choice. Figure 9.2 illustrates two different possibilities, where the current passing through $S(\Gamma)$ remains the same in both cases. As seen in Chap. 7, this result follows from the conservation of charge, $\nabla \cdot \mathbf{j} = 0$

9.3 Symmetry Arguments in Magnetostatics

Ampère's law is particularly useful when the sources of the magnetic field exhibit spatial symmetries. According to Curie's principle, to determine the symmetries of the magnetic field (the effect), it is sufficient to study the symmetries of the current density **j** (the cause).

9.3.1 Invariance Under Translation Along an Axis

Consider a current density **j** that remains invariant under translation along the z-axis. In Cartesian coordinates, this means

$$\mathbf{j}(x, y, z + a) = \mathbf{j}(x, y, z) \quad \text{for all } a .$$

Therefore, the current density is independent of the z-coordinate, $j = j(x, y)$. This directly implies that the magnetic field $\mathbf{B}$ is also invariant under translations along the z-axis:

$$\mathbf{B}(x, y, z) = \mathbf{B}(x, y) \ .$$

9.3.2 Invariance Under Rotation Around an Axis

Consider a current density $\mathbf{j}$ that remains invariant under rotations around the z-axis. In cylindrical coordinates, this means:

$$\mathbf{j}(r, \theta + a, z) = \mathbf{j}(r, \theta, z) \quad \text{for all } a \ .$$

Therefore, the current density is independent of θ, $\mathbf{j}(r, \theta, z) = \mathbf{j}(r, z)$. Consequently, the magnetic field also does not depend on this coordinate:

$$\mathbf{B}(r, \theta, z) = \mathbf{B}(r, z) \ .$$

9.3.3 Mirror Symmetry

To constrain the orientation of the magnetic field, we need to determine if any mirror symmetry is present. To demonstrate mirror symmetries in magnetostatics, we can follow a similar approach to that used in electrostatics. Any plane Π divides $\mathbb{R}^3$ into two disjoint regions, D_1 and D_2. Any point $\mathbf{u}' \in D_2$ can be written as the reflection of a point $\mathbf{u} \in D_1$ in the plane Π. Then, the magnetic field at position $\mathbf{x}$ can be expressed as an integral over D_1 using Eq. (8.15):

$$\mathbf{B}(\mathbf{x}) = \frac{\mu_0}{4\pi} \iiint_{D_1} \left(\mathbf{j}(\mathbf{u}) \times \frac{\mathbf{x} - \mathbf{u}}{|\mathbf{x} - \mathbf{u}|^3} + \mathbf{j}(\mathbf{u}') \times \frac{\mathbf{x} - \mathbf{u}'}{|\mathbf{x} - \mathbf{u}'|^3} \right) d^3 u \ ,$$

where a prime denotes the symmetric of the corresponding quantity with respect to Π. Similarly, the magnetic field at $\mathbf{x}' = \mathrm{sym}_\Pi \mathbf{x}$ is

$$\mathbf{B}(\mathrm{sym}_\Pi \mathbf{x}) = \mathbf{B}(\mathbf{x}') = \frac{\mu_0}{4\pi} \iiint_{D_1} \left(\mathbf{j}(\mathbf{u}) \times \frac{\mathbf{x}' - \mathbf{u}}{|\mathbf{x}' - \mathbf{u}|^3} + \mathbf{j}(\mathbf{u}') \times \frac{\mathbf{x}' - \mathbf{u}'}{|\mathbf{x}' - \mathbf{u}'|^3} \right) d^3 u \ .$$

Since $|\mathbf{x} - \mathbf{u}| = |\mathbf{x}' - \mathbf{u}'|$, $|\mathbf{x} - \mathbf{u}'| = |\mathbf{x}' - \mathbf{u}|$, and $\mathbf{a} - \mathbf{b} = \mathbf{a}' - \mathbf{b}'$ for any vectors $\mathbf{a}$ and $\mathbf{b}$, we have

$$\mathbf{B}(\mathrm{sym}_\Pi \mathbf{x}) = \mathbf{B}(\mathbf{x}') = \frac{\mu_0}{4\pi} \iiint_{D_1} \left(\mathbf{j}(\mathbf{u}) \times \frac{(\mathbf{x} - \mathbf{u}')'}{|\mathbf{x} - \mathbf{u}'|^3} + \mathbf{j}(\mathbf{u}') \times \frac{(\mathbf{x} - \mathbf{u})'}{|\mathbf{x} - \mathbf{u}|^3} \right) d^3 u \ .$$

From the previous analysis, it is clear that

1. If $\mathbf{j}$ has mirror symmetry with respect to $\mathbf{\Pi}$, then for any vector $\mathbf{b}$, we have $\mathbf{j}(\mathbf{u}') \times \mathbf{b}' = -\mathrm{sym}_\mathbf{\Pi}\mathbf{j}(\mathbf{u}) \times \mathbf{b}$. This implies that $\mathbf{B}$ has mirror antisymmetry with respect to $\mathbf{\Pi}$:

$$\mathbf{B}(\mathbf{x}') = \frac{\mu_0}{4\pi} \iiint_{D_1} - \left(\mathbf{j}(\mathbf{u}') \times \frac{\mathbf{x} - \mathbf{u}'}{|\mathbf{x} - \mathbf{u}'|^3} + \mathbf{j}(\mathbf{u}) \times \frac{\mathbf{x} - \mathbf{u}}{|\mathbf{x} - \mathbf{u}|^3} \right)' d^3u = -\,\mathrm{sym}_\mathbf{\Pi}\mathbf{B}(\mathbf{x}) \ .$$

Therefore mirror symmetry for $\mathbf{j}$ leads to mirror antisymmetry for $\mathbf{B}$.

$$\boxed{\text{Mirror symmetry for } \mathbf{j} \quad \rightarrow \quad \mathbf{B}(\mathbf{x}') = -\mathrm{sym}_\mathbf{\Pi}\mathbf{B}(\mathbf{x}) \ .}$$

The mirror symmetry of $\mathbf{j}$ implies $\mathbf{B}_\perp(P) = \mathbf{B}_\perp(P')$ and $\mathbf{B}_\parallel(P) = -\mathbf{B}_\parallel(P')$ (see Fig. 9.3). For a point $P \in \mathbf{\Pi}$, since $P' = P$, we have $\mathbf{B}_\parallel(P) = -\mathbf{B}_\parallel(P) = 0$. Therefore, for a point P on a symmetry plane $\mathbf{\Pi}$, the magnetic field is perpendicular to $\mathbf{\Pi}$.

2. If $\mathbf{j}$ has mirror antisymmetry with respect to $\mathbf{\Pi}$, we denote this plane as $\mathbf{\Pi}^*$. For any vector $\mathbf{b}$, we have $\mathbf{j}(\mathbf{u}') \times \mathbf{b}' = \mathrm{sym}_{\mathbf{\Pi}^*}\mathbf{j}(\mathbf{u}) \times \mathbf{b}$. This implies that $\mathbf{B}$ has mirror symmetry with respect to $\mathbf{\Pi}^*$ (Fig. 9.4):

$$\mathbf{B}(\mathbf{x}') = \frac{\mu_0}{4\pi} \iiint_{D_1} \left(\mathbf{j}(\mathbf{u}') \times \frac{\mathbf{x} - \mathbf{u}'}{|\mathbf{x} - \mathbf{u}'|^3} + \mathbf{j}(\mathbf{u}) \times \frac{\mathbf{x} - \mathbf{u}}{|\mathbf{x} - \mathbf{u}|^3} \right)' d^3u = \mathrm{sym}_{\mathbf{\Pi}^*}\mathbf{B}(\mathbf{x}) \ .$$

Therefore, mirror antisymmetry for $\mathbf{j}$ leads to mirror symmetry for $\mathbf{B}$.

$$\boxed{\text{Mirror antisymmetry for } \mathbf{j} \quad \rightarrow \quad \mathbf{B}(\mathbf{x}') = \mathrm{sym}_{\mathbf{\Pi}^*}\mathbf{B}(\mathbf{x}) \ .}$$

This mirror antisymmetry of $\mathbf{j}$ implies $\mathbf{B}_\perp(P) = -\mathbf{B}_\perp(P')$ and $\mathbf{B}_\parallel(P) = \mathbf{B}_\parallel(P')$ (see Fig. 9.4). For a point $P \in \mathbf{\Pi}^*$, since $P' = P$, we have $\mathbf{B}_\perp(P) = -\mathbf{B}_\perp(P) =$

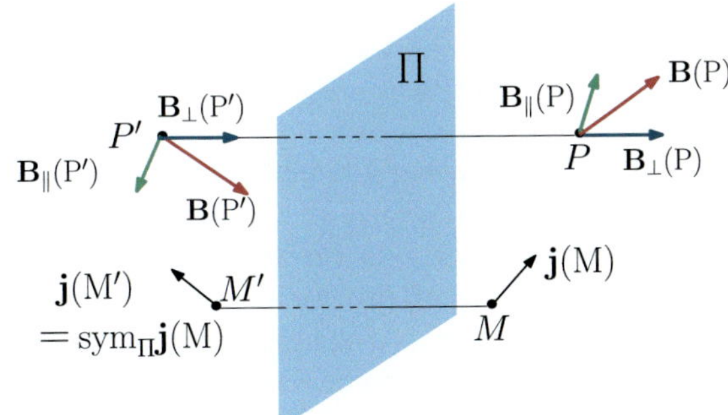

Fig. 9.3 If $\mathbf{\Pi}$ is a symmetry plane for the current distribution $\mathbf{j}$, then $\mathbf{B}$ has mirror antisymmetry with respect to $\mathbf{\Pi}$

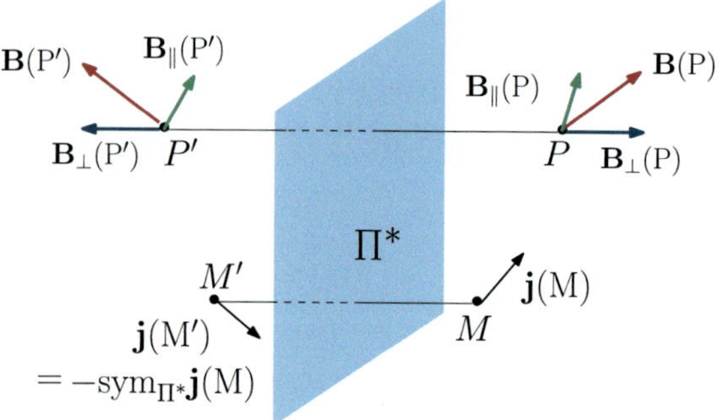

Fig. 9.4 If $\mathbf{\Pi}^*$ is an antisymmetry plane for the current distribution $\mathbf{j}$, then $\mathbf{B}$ has mirror symmetry with respect to $\mathbf{\Pi}$

$\mathbf{0}$. Therefore, for any point P on an antisymmetry plane $\mathbf{\Pi}^*$, the magnetic field lies within $\mathbf{\Pi}^*$.

Remark The magnetic field is said to be a pseudovector because it acquires a minus sign upon reflection by a symmetry plane due to its general expression (8.15), which involves a cross product.

Example 9.1—Magnetic field of an infinitely long straight wire
Calculating the magnetic field of current distributions with cylindrical symmetry is straightforward using Ampère's law. To illustrate this, let us calculate the magnetic field of an infinitely long straight wire, as shown in Fig. 9.5.

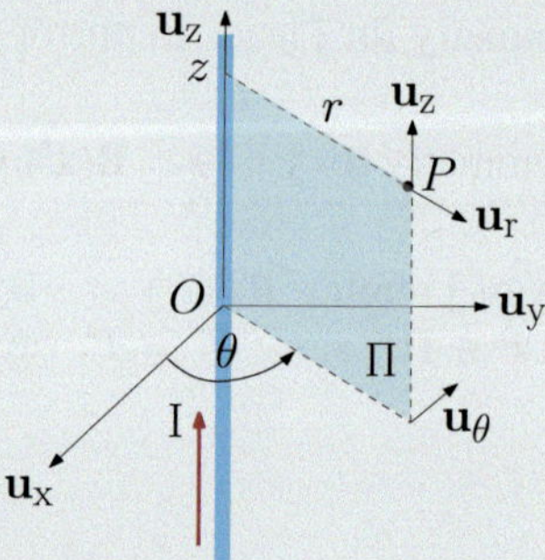

Fig. 9.5 An infinitely long straight wire

First, by choosing a cylindrical coordinate system with the z-axis as the wire axis, the invariance of the current density under translation along and rotation around the z-axis results in the magnetic field being independent of θ and z, $\mathbf{B}(r, \theta, z) = \mathbf{B}(r)$. For any point P, the plane $\mathbf{\Pi}$ passing through P and containing the wire is a symmetry plane for $\mathbf{j}$. This implies that at point P, the magnetic field is perpendicular to $\mathbf{\Pi}$ and therefore along $\mathbf{u}_\theta$. These symmetries constrain the magnetic field to vary as $\mathbf{B}(r, \theta, z) = B(r)\mathbf{u}_\theta$. To apply Ampère's law, we can choose Γ to be a circle of radius r concentric with the wire axis, such that $d\mathbf{x} = r\,d\theta\,\mathbf{u}_\theta$. This curve is shown in Fig. 9.6.

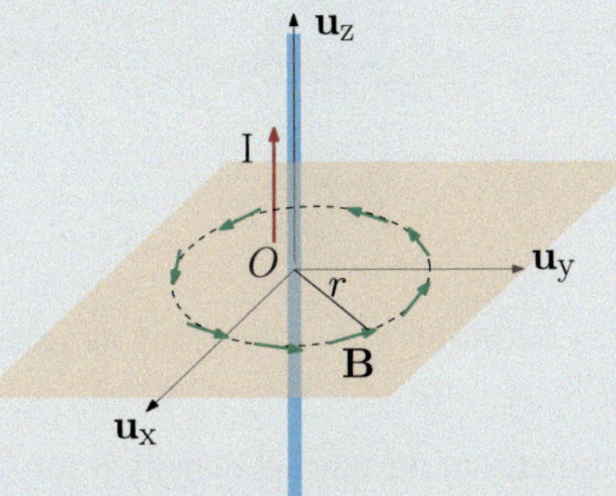

Fig. 9.6 The magnetic field has constant magnitude and is always parallel to the tangent of a circle perpendicular to the axis of the current

The current through any surface bounded by the contour Γ is simply I. Therefore, Ampère's law gives

$$\oint_\Gamma \mathbf{B}(\mathbf{x}) \cdot d\mathbf{x} = \int_0^{2\pi} B(r)r\,d\theta\,\mathbf{u}_\theta \cdot \mathbf{u}_\theta = 2\pi r B(r) = \mu_0 I \ .$$

Finally,

$$\mathbf{B}(r) = \frac{\mu_0 I}{2\pi r}\mathbf{u}_\theta \ .$$

The magnetic field thus circulates around the current.

Example 9.2—Magnetic field of a solenoid

A solenoid consists of a wire wound in the shape of a circular helix, as shown in Fig. 9.7. The radius R is typically much smaller than the total length of the solenoid. Let us calculate the magnetic field produced by an infinitely long, ideal solenoid with and n turns per unit length, carrying a current I.

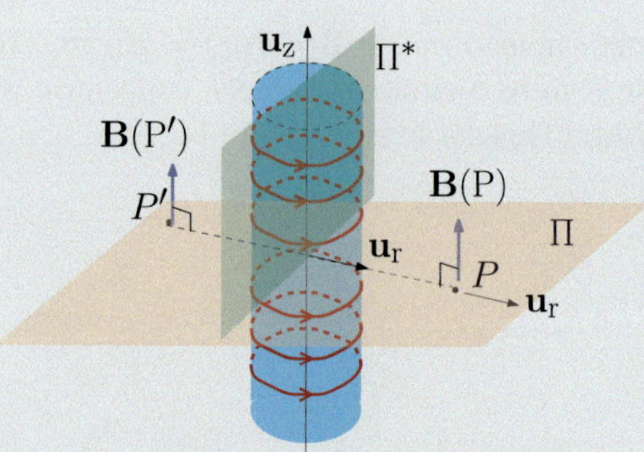

Fig. 9.7 An infinitely long solenoid is shown along with its symmetry and antisymmetry planes of the current density

The cylindrical symmetry of the current density results in the magnetic field being independent of θ and z in cylindrical coordinates, $\mathbf{B}(r, \theta, z) = \mathbf{B}(r)$. For any point P, the plane Π containing P and perpendicular to the z-axis is a symmetry plane for $\mathbf{j}$. This implies that the magnetic field is perpendicular to Π, oriented along $\mathbf{u}_z$, and $\mathbf{B}(r, \theta, z) = B(r)\mathbf{u}_z$. To determine the magnetic field outside the solenoid, we apply Ampère's law using the plane curve Γ shown in Fig. 9.8. For the two segments perpendicular to the $\mathbf{u}_z$ axis, $\mathbf{B}(\mathbf{x}) \cdot d\mathbf{x} = 0$. Therefore, the circulation of $\mathbf{B}$ along Γ is due solely to the two segments parallel to the $\mathbf{u}_z$ axis:

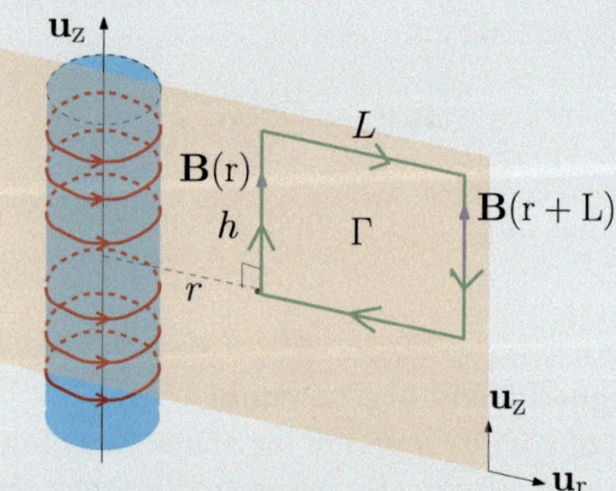

Fig. 9.8 Planar curve Γ for applying Ampère's law to determine the magnetic field outside the solenoid

$$\oint_{\Gamma} \mathbf{B}(\mathbf{x}) \cdot d\mathbf{x} = h(B(r) - B(r + L)) \,.$$

Taking the limit $L \to \infty$, we have $B(r + L) \to 0$. Since no current is enclosed by Γ, Ampère's law becomes

$$hB(r) = 0 \,.$$

This implies that the magnetic field at any point outside the solenoid is zero. To determine the magnetic field inside the solenoid, consider the plane curve Γ shown in Fig. 9.9.

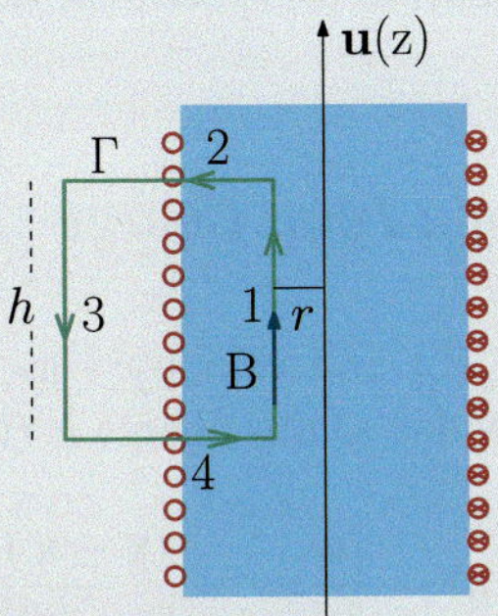

Fig. 9.9 Planar curve Γ for determining the magnetic field at a point inside the solenoid

In segments 2 and 4, $\mathbf{B}(\mathbf{x})$ is perpendicular to $d\mathbf{x}$, so these segments do not contribute to the circulation. In segment 3, which is outside the solenoid, the magnetic field is zero. Therefore, only the circulation along segment 1 contributes, and we obtain

$$\oint_{\Gamma} \mathbf{B}(\mathbf{x}) \cdot d\mathbf{x} = B(r)h \; .$$

The current enclosed by Γ is I multiplied by the number of loops within Γ, which equals nh. Ampère's law then gives

$$B(r)h = \mu_0 I n h \; .$$

This shows that the magnetic field inside the solenoid is uniform and given by $B = \mu_0 I n$. In conclusion,

$$\mathbf{B}(\mathbf{x}) = \begin{cases} \mu_0 I n \, \mathbf{u}_z & \text{if } |\mathbf{x}| \leq R \; , \\ \mathbf{0} & \text{if } R \leq |\mathbf{x}| \; . \end{cases}$$

9.4   The Vector Potential

In electrostatics, the electric potential V is defined by the relation $\mathbf{E} = -\nabla V$. Similarly, in magnetostatics, we can define a vector potential $\mathbf{A}$ such that the magnetic field $\mathbf{B}$ can be obtained as the curl of this potential, $\mathbf{B} = \nabla \times \mathbf{A}$.

9.4.1   Vector Potential of a Point Charge

Consider a particle with charge q and velocity $\mathbf{v}$ located at position $\mathbf{x}'$ at a given instant. The magnetic field at $\mathbf{x}$ due to this charge is (from (8.7)):

$$\mathbf{B}(\mathbf{x}) = \frac{\mu_0}{4\pi} q \mathbf{v} \times \frac{(\mathbf{x} - \mathbf{x}')}{|\mathbf{x} - \mathbf{x}'|^3} \, .$$

Using the identity $\dfrac{(\mathbf{x} - \mathbf{x}')}{|\mathbf{x} - \mathbf{x}'|^3} = -\nabla \left(\dfrac{1}{|\mathbf{x} - \mathbf{x}'|} \right)$, the magnetic field can be written as

$$\mathbf{B}(\mathbf{x}) = \frac{\mu_0}{4\pi} \left(\nabla \left(\frac{1}{|\mathbf{x} - \mathbf{x}'|} \right) \times q\mathbf{v} \right) \, .$$

Since $q\mathbf{v}$ is independent of $\mathbf{x}$, applying the vector calculus identity (A.6),

$$\nabla \times (a\mathbf{A}) = a \left(\nabla \times \mathbf{A} \right) + \nabla a \times \mathbf{A} \, ,$$

where a is a scalar and $\mathbf{A}$ is a vector, we get

$$\nabla \left(\frac{1}{|\mathbf{x} - \mathbf{x}'|} \right) \times q\mathbf{v} = \nabla \times \left(\frac{q\mathbf{v}}{|\mathbf{x} - \mathbf{x}'|} \right) \, .$$

Therefore,

$$\mathbf{B}(\mathbf{x}) = \nabla \times \left(\frac{\mu_0}{4\pi} \frac{q\mathbf{v}}{|\mathbf{x} - \mathbf{x}'|} \right) \, .$$

Vector potential of a point charge
The vector potential generated by a point charge q at $\mathbf{x}'$, moving with velocity $\mathbf{v}$, is given by

$$\mathbf{A}(\mathbf{x}) = \frac{\mu_0}{4\pi} \frac{q\mathbf{v}}{|\mathbf{x} - \mathbf{x}'|} \, . \tag{9.3}$$

The magnetic field generated by this charge verifies

$$\mathbf{B}(\mathbf{x}) = \nabla \times \mathbf{A}(\mathbf{x}) .$$

9.4.2 *Vector Potential of an Arbitrary Current Distribution*

The vector potential can be generalized for an arbitrary current distribution. The Biot–Savart law can be rewritten as

$$\mathbf{B}(\mathbf{x}) = \frac{\mu_0}{4\pi} \iiint_\Omega \mathbf{j}(\mathbf{x}') \times \nabla \left(-\frac{1}{|\mathbf{x} - \mathbf{x}'|} \right) d^3 x' = \nabla \times \left\{ \frac{\mu_0}{4\pi} \iiint_\Omega \frac{\mathbf{j}(\mathbf{x}')}{|\mathbf{x} - \mathbf{x}'|} d^3 x' \right\} .$$

Vector potential of an arbitrary current distribution
For an arbitrary current density localized inside a volume $\Omega \subseteq \mathbb{R}^3$, the vector potential is given by the integral

$$\mathbf{A}(\mathbf{x}) = \frac{\mu_0}{4\pi} \iiint_{\mathbb{R}^3} \frac{\mathbf{j}(\mathbf{x}')}{|\mathbf{x} - \mathbf{x}'|} d^3 x' . \tag{9.4}$$

For the case of a line of current, this reduces to

$$\mathbf{A}(\mathbf{x}) = \frac{\mu_0 I}{4\pi} \int_\Gamma \frac{d\mathbf{l}(\mathbf{x}')}{|\mathbf{x} - \mathbf{x}'|} . \tag{9.5}$$

9.4.3 *The Vector Potential Is Not Unique*

Even though (9.4) defines a unique vector potential, there are actually infinitely many possible vector potentials that satisfy

$$\nabla \times \mathbf{A} = \mathbf{B} . \tag{9.6}$$

If $\mathbf{A}$ is defined by (9.4), we can always add the gradient of an arbitrary scalar field to $\mathbf{A}$:

$$\mathbf{A}' = \mathbf{A} + \nabla\psi \ .$$

Since $\nabla \times \nabla\psi = 0$, both $\mathbf{A}$ and $\mathbf{A}'$ define the same magnetic field,

$$\mathbf{B} = \nabla \times \mathbf{A} = \nabla \times \mathbf{A}' \ .$$

From the Helmholtz theorem seen in Sect. 3.7.2, to uniquely determine a vector field that decays to zero at infinity, we must specify its curl and its divergence. The vector potential defined by Eq. (9.4) satisfies

$$\nabla \times \mathbf{A} = \mathbf{B} \quad \text{and} \quad \nabla \cdot \mathbf{A} = 0 \ .$$

This corresponds to a particular choice among all possible vector fields that satisfy $\nabla \times \mathbf{A} = \mathbf{B}$ and decay to zero at infinity. This choice is known as the Coulomb gauge.

9.4.4 Poisson's Equation for the Vector Potential

Similar to how the electrostatic potential satisfies Poisson's equation, the vector potential in the Coulomb gauge satisfies

$$\boxed{\nabla^2 \mathbf{A}(\mathbf{x}) = -\mu_0 \mathbf{j}(\mathbf{x}) \ .} \tag{9.7}$$

In other words, each component A_i of $\mathbf{A}$ satisfies Poisson's equation with a source term $-\mu_0 j_i$.

The demonstration is the following: Using the definition of the vector potential, $\nabla \times \mathbf{A} = \mathbf{B}$, and applying the vector identity A.4 to $\mathbf{A}$, we have $\nabla \times (\nabla \times \mathbf{A}) = \nabla(\nabla \cdot \mathbf{A}) - \nabla^2 \mathbf{A}$, leading to

$$\nabla \times \mathbf{B} = \nabla(\nabla \cdot \mathbf{A}) - \nabla^2 \mathbf{A} \ .$$

Using Ampère's law (9.1) and the Coulomb gauge condition ($\nabla \cdot \mathbf{A} = 0$), we obtain Eq. (9.7), $\nabla^2 \mathbf{A} = -\mu_0 \mathbf{j}$.

Example 9.3—Vector potential of a uniform magnetic field and of a solenoid

Suppose a uniform magnetic field $\mathbf{B} = B_0 \mathbf{u}_z$. In cylindrical coordinates, the curl of $\mathbf{A}$ will have a non-zero component along z if, for example, we choose $\mathbf{A} = A_\theta(r)\mathbf{u}_\theta$.

By choosing Γ as an anti-clockwise (when looked from above) circular loop of radius r in the xy plane, we get

$$\oint_{\Gamma} \mathbf{A} \cdot d\mathbf{l} = 2\pi r A_{\theta}(r) = \iint_{S(\Gamma)} (\nabla \times \mathbf{A}) \cdot d\mathbf{S}$$

where $S(\Gamma)$ is the planar circular surface enclosed by the loop. Since $\nabla \times \mathbf{A} = \mathbf{B} = B_0 \mathbf{u}_z$, we must have

$$\oint_{\Gamma} \mathbf{A} \cdot d\mathbf{l} = 2\pi r A_{\theta}(r) = \pi r^2 B_0 \ .$$

Finally, we found that the vector potential $\mathbf{A} = \frac{r}{2} B_0 \mathbf{u}_{\theta}$ generates a uniform magnetic field along z. It can be written as

$$\mathbf{A}(\mathbf{x}) = \mathbf{B} \times \frac{r\mathbf{u}_r}{2} \ .$$

The same procedure can be applied to find the vector potential associated with an infinitely long solenoid of radius R. The field is zero outside and uniform inside the solenoid, along its axis, $\mathbf{B} = B_0 \mathbf{u}_z$. Again, we can choose the vector potential to be along $\mathbf{u}_{\theta}$ and calculate its circulation over a closed circular loop of radius r, as shown in Fig. 9.10.

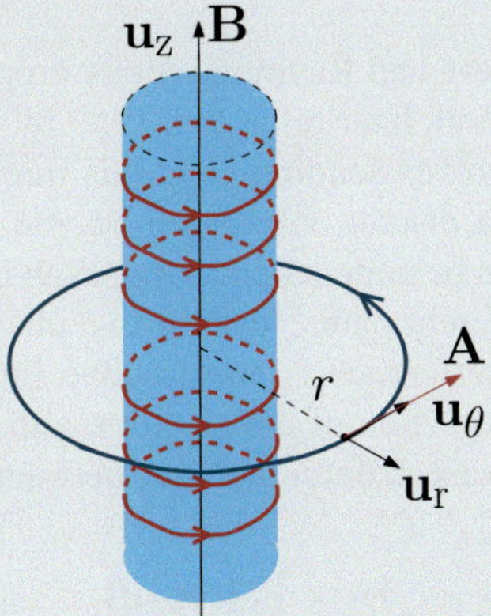

Fig. 9.10 We can choose $\mathbf{A}$ so that its circulation around a circular loop is simply $2\pi r A(r)$

We have

$$\oint_{\Gamma} \mathbf{A} \cdot d\mathbf{l} = 2\pi r A_{\theta}(r) = \iint_{S(\Gamma)} \mathbf{B} \cdot d\mathbf{S} = \Phi_{S,\mathbf{B}}$$

the magnetic flux is constant for $r > R$ and given by $\pi R^2 B_0$. For $r \leq R$, the flux is $\pi r^2 B_0$. The vector potential is therefore

$$\mathbf{A}(\mathbf{x}) = \begin{cases} \dfrac{B_0 r^2}{2} \mathbf{u}_\theta & \text{if } r \leq R \text{ ,} \\[2mm] \dfrac{B_0 R^2}{2r^2} \mathbf{u}_\theta & \text{if } r > R \text{ .} \end{cases}$$

9.4.5 The Aharonov–Bohm Effect

From the Lorentz force $\mathbf{F} = q\,(\mathbf{E} + \mathbf{v} \times \mathbf{B})$, it appears that the vector potential $\mathbf{A}$ has no direct effect on a particle's dynamics. For instance, if a particle moves in a region where $\mathbf{B} = \mathbf{0}$, no magnetic force acts on the particle even if $\mathbf{A} \neq \mathbf{0}$. In classical physics, therefore, the vector potential is treated merely as a mathematical tool, without any assigned physical reality. Quantum mechanics, however, fundamentally changes this perspective. From the inception of quantum theory, it became clear that the wave equation describing an electron's evolution must explicitly involve the vector potential $\mathbf{A}$. This dependence arises from the need for the physics to remain invariant under gauge transformations. In quantum mechanics, the vector potential cannot be ignored.

In 1949, Werner Ehrenberg and Raymond Siday proposed a thought experiment to test this quantum prediction, later refined in 1959 by Yakir Aharonov and David Bohm. The experiment involves sending electrons through a double slit, creating an interference pattern on a distant observation screen, similar to the behavior of light waves. This pattern can be understood as the result of interference between two waves that travel along different paths, leading to a phase difference ϕ. This phase determines whether the interference is constructive (bright fringe) or destructive (dark fringe). When the interaction of the electron with an electromagnetic field is considered, an additional phase difference arises between the two paths. This phase difference is given by

$$\delta\phi = \frac{q}{\hbar} \oint_\Gamma \mathbf{A} \cdot d\mathbf{l} \tag{9.8}$$

where Γ is the closed contour formed by path C_1 from A to the observation point P, and C_2 returning from P to A, as illustrated in Fig. 9.11.

By placing a long solenoid between the two paths, a non-zero vector potential $\mathbf{A}$ exists outside the solenoid, even though the magnetic field is $\mathbf{B} = \mathbf{0}$ along the paths. Remarkably, the phase difference $\delta\phi$ is non-zero, leading to a measurable shift in the interference pattern on the screen. This prediction has been experimentally verified.

Fig. 9.11 The Aharonov–Bohm effect: a solenoid placed between the two paths creates a phase difference in the electron's wavefunction, despite $\mathbf{B} = \mathbf{0}$ along the paths

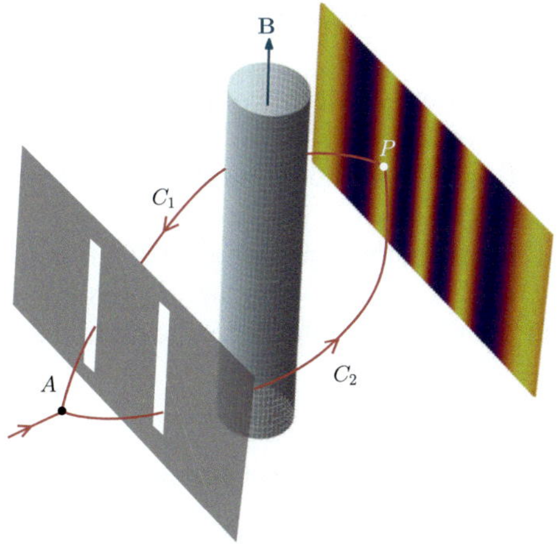

In classical physics, the vector potential is not uniquely defined. Similarly, in quantum mechanics, the phase shift remains invariant under a gauge transformation of $\mathbf{A}$. For instance, if $\mathbf{A}' = \mathbf{A} + \boldsymbol{\nabla}\varphi$, the resulting phase difference is

$$\delta\phi' = \frac{q}{\hbar}\oint_\Gamma \mathbf{A}' \cdot d\mathbf{l} = \frac{q}{\hbar}\oint_\Gamma \mathbf{A} \cdot d\mathbf{l} + \frac{q}{\hbar}\underbrace{\oint_\Gamma \boldsymbol{\nabla}\varphi \cdot d\mathbf{l}}_{=0} = \delta\phi \ .$$

Thus, the phase difference depends only on the physical electromagnetic field configuration, not on the specific choice of gauge. The phase difference can also be expressed in terms of the magnetic flux $\Phi_{S,\mathbf{B}}$ through any surface $S(\Gamma)$ bounded by Γ:

$$\delta\phi = \frac{q}{\hbar}\oint_\Gamma \mathbf{A} \cdot d\mathbf{l} = \frac{q}{\hbar}\iint_{S(\Gamma)} \underbrace{\boldsymbol{\nabla} \times \mathbf{A}}_{\mathbf{B}} \cdot d\mathbf{S} = \frac{q}{\hbar}\Phi_{S,\mathbf{B}} \ . \tag{9.9}$$

This is remarkable because, even though the phase shift $\delta\phi$ can be written in terms of $\mathbf{B}$, it occurs in a region where $\mathbf{B} = \mathbf{0}$ everywhere the particle is likely to be found. Here, the action at a distance is mediated by the vector potential $\mathbf{A}$, rather than the magnetic field $\mathbf{B}$, which is a purely quantum mechanical effect. Classically, such a phenomenon is impossible. It has also been predicted, though not yet experimentally verified, that a similar effect could arise from the scalar potential. In this case, the phase shift would result from a difference in scalar potential in a region where the electric field is zero.

The Aharonov–Bohm effect is regarded as experimental proof of the gauge principle that governs modern physics. It illustrates that only quantities invariant under gauge transformations, such as the phase shift $\delta\phi$, correspond to physical phenomena. This principle underpins much of quantum field theory and modern particle physics.

9.5 The Two Fundamental Laws of Magnetostatics

We will now state the two fundamental laws of magnetostatics. We have already seen that a vector potential $\mathbf{A}$ can be defined such that the magnetic field $\mathbf{B}$ is the curl of $\mathbf{A}$: $\mathbf{B} = \nabla \times \mathbf{A}$. Consequently, the divergence of the magnetic field is always zero: $\nabla \cdot \mathbf{B} = 0$.

Together, this equation and the differential form of Ampère's law form a complete set of equations that uniquely determine the magnetic field at every point in space (as per the Helmholtz decomposition theorem in Sect. 3.7.2).

9.5.1 Gauss's Law for Magnetism

The divergence of a curl is always zero:

$$\nabla \cdot (\nabla \times \mathbf{A}) = 0 \,.$$

Therefore, the differential equation for the divergence of $\mathbf{B}$ is simply

$$\boxed{\nabla \cdot \mathbf{B} = 0 \,.} \tag{9.10}$$

The magnetic field is thus solenoidal. The integral form of (9.10), obtained from the divergence theorem (A.15), is

$$\boxed{\oiint_S \mathbf{B}(\mathbf{x}') \cdot \mathbf{n}(\mathbf{x}')dS(\mathbf{x}') = \iiint_{\Omega(S)} \nabla \cdot \mathbf{B}(\mathbf{x}')d^3x' = 0 \,.} \tag{9.11}$$

This means that the magnetic flux through any closed surface is zero. As we will see later, this remains true even for time-varying fields. This fundamental equation implies the non-existence of magnetic monopoles.

In summary, the two fundamental laws of magnetostatics are

1. Gauss's law for magnetism, also known as Maxwell–Thomson equation:

$$\oiint_{\partial\Omega} \mathbf{B}(\mathbf{x}) \cdot \mathbf{n}(\mathbf{x})dS(\mathbf{x}) = 0 \quad \Leftrightarrow \quad \nabla \cdot \mathbf{B} = 0 \,.$$

   The magnetic flux over any closed surface is zero or, equivalently, the divergence of the magnetic field is zero.
2. Ampère's law:

$$\oint_\Gamma \mathbf{B}(\mathbf{x}) \cdot d\mathbf{x} = \mu_0 I_\Gamma \quad \Leftrightarrow \quad \nabla \times \mathbf{B} = \mu_0 \mathbf{j} \,.$$

The circulation of the magnetic field over any closed curve Γ is proportional to the current I_Γ passing through any surface $S(\Gamma)$ bounded by the curve Γ.

These differential equations form a complete set of equations for magnetostatics, as per the Helmholtz theorem. In a region free of currents, $\nabla \times \mathbf{B} = \mathbf{0}$. This allows us to define a scalar potential ϕ such that $\mathbf{B} = -\nabla\phi$. Since $\nabla \cdot \mathbf{B} = 0$, this scalar potential must satisfy Laplace's equation, $\nabla^2\phi = 0$. We can then apply the methods discussed in Chap. 4.

9.5.2 *Completeness of Magnetostatics*

For a current density $\mathbf{j}$ decaying as $1/|\mathbf{x}|^2$ at infinity, there exists a unique magnetic field $\mathbf{B}$ that also decays faster than $1/|\mathbf{x}|^2$ at infinity and satisfies:

$$\nabla \cdot \mathbf{B} = 0 \quad \text{and} \quad \nabla \times \mathbf{B} = \mu_0\mathbf{j} \, .$$

The Helmholtz theorem guarantees that $\mathbf{B}$ is uniquely determined by the following integral-differential representation:

$$\mathbf{B}(\mathbf{x}) = \nabla \times \frac{1}{4\pi} \iiint_{\mathbb{R}^3} \frac{\nabla' \times \mathbf{B}(\mathbf{x}')}{|\mathbf{x} - \mathbf{x}'|} d^3x' = \nabla \times \frac{\mu_0}{4\pi} \iiint_{\mathbb{R}^3} \frac{\mathbf{j}(\mathbf{x}')}{|\mathbf{x} - \mathbf{x}'|} d^3x' \, .$$

Using the vector identity (A.6), we have $\nabla \times (f\mathbf{j}(\mathbf{x}')) = f\nabla \times \mathbf{j}(\mathbf{x}') + \nabla f \times \mathbf{j}(\mathbf{x}')$ with $f = 1/|\mathbf{x} - \mathbf{x}'|$, and noting that $\nabla \times \mathbf{j}(\mathbf{x}') = \mathbf{0}$, we obtain the Biot–Savart law,

$$\mathbf{B}(\mathbf{x}) = \frac{\mu_0}{4\pi} \iiint_{\mathbb{R}^3} \frac{\mathbf{j}(\mathbf{x}') \times (\mathbf{x} - \mathbf{x}')}{|\mathbf{x} - \mathbf{x}'|^3} d^3x' \, . \tag{9.12}$$

Therefore, the two fundamental differential equations, (9.10) and (9.1), are sufficient to determine the magnetic field.

9.6 The Magnetic Dipole

In Sect. 8.8.2, we saw that the magnetic field on the axis of a circular current loop decays as $1/r^3$. This is reminiscent of the $1/r^3$ dependence of the electric field generated by an electric dipole. In magnetostatics, a small current loop defines a magnetic dipole. In this section, we will show that the field lines of a magnetic dipole are equivalent to those of an electric dipole.

Consider a square loop of side length l carrying a current I in the xy-plane, oriented counterclockwise, as shown in Fig. 9.12. Let us calculate the magnetic field at a point P, located at a distance r from the origin, which is much larger than the

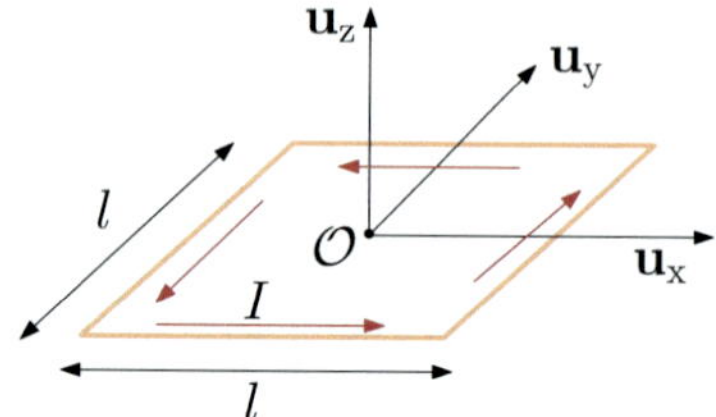

Fig. 9.12 A square current loop

loop size. We first compute the vector potential at point P due to the four current segments of the loop. The two segments parallel to the y-axis contribute:

$$\mathbf{A}_1(\mathbf{x}) \approx \frac{\mu_0 I}{4\pi} \frac{(-l)}{d_1}\mathbf{u}_y \ , \qquad \mathbf{A}_2(\mathbf{x}) \approx \frac{\mu_0 I}{4\pi} \frac{l}{d_2}\mathbf{u}_y \ ,$$

where (Fig. 9.13)

$$d_1 = \sqrt{(x + l/2)^2 + y^2 + z^2} = \sqrt{r^2 + lx + l^2/4} \ ,$$
$$d_2 = \sqrt{(x - l/2)^2 + y^2 + z^2} = \sqrt{r^2 - lx + l^2/4} \ .$$

The y-component of the vector potential is given by the superposition

$$A_y(\mathbf{x}) = \frac{\mu_0 I l}{4\pi}\left(\frac{1}{d_2} - \frac{1}{d_1}\right) \approx \frac{\mu_0 I l^2 x}{4\pi r^3} \ .$$

Similarly, the x-component is

$$A_x(\mathbf{x}) \approx -\frac{\mu_0 I l^2 y}{4\pi r^3} \ .$$

Defining the magnetic moment of the loop as $\mathbf{m} = I S \mathbf{n}$, where $S = l^2$ is the surface area of the loop and n is the unit normal to the loop, the vector potential becomes

$$\mathbf{A}(\mathbf{x}) = \frac{\mu_0 I S}{4\pi r^3}(x\mathbf{u}_y - y\mathbf{u}_x) = \frac{\mu_0}{4\pi}\frac{\mathbf{m} \times \mathbf{x}}{|\mathbf{x}|^3} \ .$$

Fig. 9.13 We consider a point P, far away from the loop, whose distance to the origin is much larger than the loop size

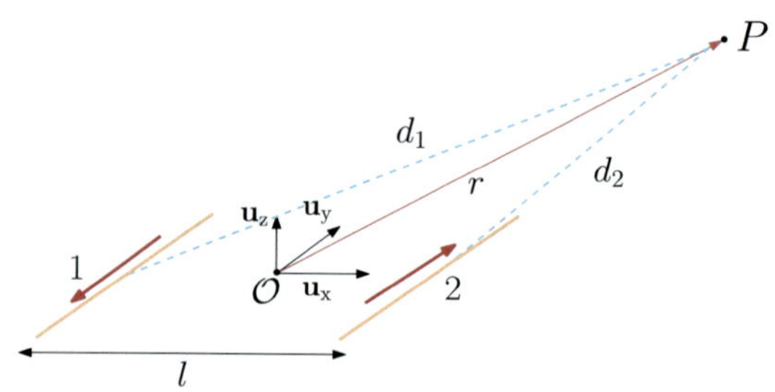

The magnetic field far from the magnetic moment is

$$\mathbf{B}(\mathbf{r}) = \nabla \times \mathbf{A}(\mathbf{r}) = \nabla \times \left(\frac{\mu_0}{4\pi} \frac{\mathbf{m} \times \mathbf{x}}{|\mathbf{x}|^3} \right) .$$

Using the vector identity (A.9) for the curl of the cross product of two vectors, we have $\nabla \times (\mathbf{m} \times \mathbf{x}/|\mathbf{x}|^3) = \mathbf{m}(\nabla \cdot \mathbf{x}/|\mathbf{x}|^3) - (\mathbf{m} \cdot \nabla)\mathbf{x}/|\mathbf{x}|^3$. Noting that $\nabla \cdot \mathbf{x}/|\mathbf{x}|^3 = 0$ for $\mathbf{x} \neq \mathbf{0}$ and $\mathbf{m}$ is constant, we get

$$\nabla \times \left(\frac{\mathbf{m} \times \mathbf{x}}{|\mathbf{x}|^3} \right) = -\nabla \left(\mathbf{m} \cdot \frac{\mathbf{x}}{|\mathbf{x}|^3} \right) .$$

This leads to the remarkable result:

$$\mathbf{B}(\mathbf{x}) = -\nabla \left(\frac{\mu_0}{4\pi} \frac{\mathbf{m} \cdot \mathbf{x}}{|\mathbf{x}|^3} \right) = -\nabla \psi . \tag{9.13}$$

The scalar field ψ is identical to the electric potential of an electric dipole with moment $\mathbf{p} = \mathbf{m}$, after replacing μ_0 with $1/\epsilon_0$ (see (3.8)). Therefore, far from the source, the magnetic field of a magnetic dipole has the same form as the electric field of an electric dipole,

$$\mathbf{E}(\mathbf{x}) = -\nabla \left(\frac{1}{4\pi\epsilon_0} \frac{\mathbf{p} \cdot \mathbf{x}}{|\mathbf{x}|^3} \right) .$$

The field lines of both systems are identical, as shown in Fig. 9.14.

Fig. 9.14 The magnetic field lines far from a magnetic dipole are topologically equivalent to the electric field lines far from an electric dipole

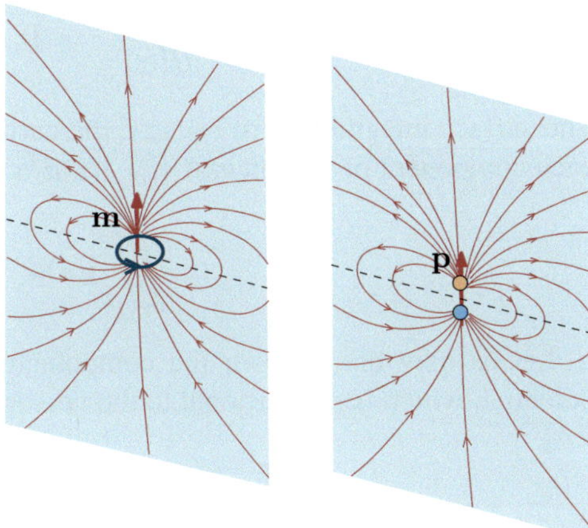

9.7   Multipole Expansion for the Magnetic Field

Let us consider a current distribution confined to a bounded volume Ω that contains the origin. The vector potential is given by

$$\mathbf{A}(\mathbf{x}) = \frac{\mu_0}{4\pi} \iiint_\Omega \frac{\mathbf{j}(\mathbf{x}')}{|\mathbf{x} - \mathbf{x}'|} d^3x' .$$

For a point $\mathbf{x}$ far away from Ω, $|\mathbf{x}'| \ll |\mathbf{x}|$, we can approximate[1]

$$\frac{1}{|\mathbf{x} - \mathbf{x}'|} \approx \frac{1}{|\mathbf{x}|} + \frac{\mathbf{x} \cdot \mathbf{x}'}{|\mathbf{x}|^3} .$$

Substituting this approximation into the vector potential, we get

$$\mathbf{A}(\mathbf{x}) \approx \frac{\mu_0}{4\pi |\mathbf{x}|} \iiint_\Omega \mathbf{j}(\mathbf{x}') \, d^3x' + \frac{\mu_0}{4\pi |\mathbf{x}|^3} \iiint_\Omega (\mathbf{x} \cdot \mathbf{x}') \, \mathbf{j}(\mathbf{x}') \, d^3x' .$$

In this expansion, the first term, which would correspond to a magnetic monopole, is zero. To show this, we need to show that each component of the integral $\iiint_\Omega j_i(\mathbf{x}')d^3x'$ is zero, where $\mathbf{x}'$ is here a dummy variable. Consider the product rule (A.7) for the divergence: $\nabla \cdot (x_i \mathbf{j}) = \mathbf{j} \cdot \nabla x_i + x_i \nabla \cdot \mathbf{j}$. For a stationary current, $\nabla \cdot \mathbf{j} = 0$ and we have $\nabla x_i = \mathbf{u}_i$, the unit vector along the x_i-axis, leading to $j_i(\mathbf{x}) = \mathbf{j} \cdot \nabla x_i = \nabla \cdot (x_i \mathbf{j})$. Integrating over Ω and applying the divergence theorem (A.15), we get

$$\iiint_\Omega j_i(\mathbf{x}) \, d^3x = \iiint_\Omega \nabla \cdot (x_i \mathbf{j}) \, d^3x = \oiint_{\partial\Omega} x_i \mathbf{j} \cdot \mathbf{n} dS = 0 .$$

The surface integral is zero because the current is localized within Ω, showing that each component of the integral $\iiint_\Omega \mathbf{j}(\mathbf{x})d^3x$ is zero. Therefore,

$$\iiint_\Omega \mathbf{j}(\mathbf{x})d^3x = \mathbf{0} .$$

Let us now break down the components of the dominant term of the vector potential, which is proportional to the vector

$$\iiint_\Omega (\mathbf{x} \cdot \mathbf{x}') \, j_i(\mathbf{x}') \, d^3x' \quad \text{for} \quad i = x, y, z .$$

---

[1] We can write

$$|\mathbf{x} - \mathbf{x}'|^{-1} = \left(|\mathbf{x}|^2 + |\mathbf{x}'|^2 - 2\mathbf{x} \cdot \mathbf{x}' \right)^{-1/2} = \frac{1}{|\mathbf{x}|} \left(1 + \frac{|\mathbf{x}'|^2}{|\mathbf{x}|^2} - \frac{2\mathbf{x} \cdot \mathbf{x}'}{|\mathbf{x}|^2} \right)^{-1/2} .$$

A Taylor expansion at first order in $|\mathbf{x}'|/|\mathbf{x}|$ gives the required approximation.

We can rewrite $j_i(\mathbf{x}') = \mathbf{j}(\mathbf{x}') \cdot \nabla' x_i'$, where ∇' applies to coordinates $\mathbf{x}'$. Using the Einstein's summation convention (repeated indices are summed over), we have $\mathbf{x} \cdot \mathbf{x}' = \sum_k x_k x_k' = x_k x_k'$. Therefore,

$$(\mathbf{x} \cdot \mathbf{x}') \, j_i(\mathbf{x}') = (x_k x_k')(\mathbf{j}(\mathbf{x}') \cdot \nabla' x_i') = x_k \{\mathbf{j}(\mathbf{x}') \cdot (x_k' \nabla' x_i')\} \, .$$

Applying the product rule (A.5) for the gradient, we find

$$(\mathbf{x} \cdot \mathbf{x}') j_i(\mathbf{x}') = x_k \left\{ \mathbf{j}(\mathbf{x}') \cdot \left[\nabla'(x_k' x_i') - x_i' \nabla' x_k' \right] \right\} \, .$$

Applying now the product rule (A.7) for the divergence and using $\nabla' \cdot \mathbf{j}(\mathbf{x}') = 0$, we get

$$(\mathbf{x} \cdot \mathbf{x}') \, j_i(\mathbf{x}') = x_k \left\{ \nabla' \cdot (x_i' x_k' \mathbf{j}(\mathbf{x}')) - x_i' \mathbf{j}(\mathbf{x}') \cdot \nabla' x_k' \right\} \, .$$

The second term simplifies to $x_k x_i' \, \mathbf{j}(\mathbf{x}') \cdot \nabla' x_k' = x_i' \, x_k \, j_k(\mathbf{x}') = x_i'(\mathbf{j}(\mathbf{x}') \cdot \mathbf{x})$, leading to

$$(\mathbf{x} \cdot \mathbf{x}') \, j_i(\mathbf{x}') = x_k \, \nabla' \cdot (x_i' x_k' \mathbf{j}(\mathbf{x}')) - x_i'(\mathbf{j}(\mathbf{x}') \cdot \mathbf{x}) \, .$$

Integrating over Ω and using the divergence theorem, the first term vanishes due to the localization of the current ($\mathbf{j}(\mathbf{x}') = \mathbf{0}$ on $\partial\Omega$). This leaves us with

$$\iiint_\Omega (\mathbf{x} \cdot \mathbf{x}') j_i(\mathbf{x}') \, d^3 x' = - \iiint_\Omega x_i'(\mathbf{j}(\mathbf{x}') \cdot \mathbf{x}) \, d^3 x' \, .$$

Combining all components, we obtain

$$\iiint_\Omega (\mathbf{x} \cdot \mathbf{x}') \mathbf{j}(\mathbf{x}') \, d^3 x' = - \iiint_\Omega (\mathbf{j}(\mathbf{x}') \cdot \mathbf{x}) \, \mathbf{x}' \, d^3 x' \, .$$

Using the vector triple product identity $(\mathbf{x}' \times \mathbf{j}(\mathbf{x}')) \times \mathbf{x} = (\mathbf{x}' \cdot \mathbf{x}) \mathbf{j}(\mathbf{x}') - (\mathbf{x} \cdot \mathbf{j}(\mathbf{x}'))\mathbf{x}'$, we can rewrite the integral as

$$\iiint_\Omega \mathbf{j}(\mathbf{x}')(\mathbf{x}' \cdot \mathbf{x}) d^3 x' = \frac{1}{2} \iiint_\Omega (\mathbf{x}' \times \mathbf{j}(\mathbf{x}')) \times \mathbf{x} \, d^3 x' \, .$$

Finally, the vector potential becomes

$$\mathbf{A}(\mathbf{x}) \approx \frac{\mu_0}{4\pi} \frac{1}{2} \left(\iiint_\Omega \mathbf{x}' \times \mathbf{j}(\mathbf{x}') \, d^3 x' \right) \times \frac{\mathbf{x}}{|\mathbf{x}|^3} \, .$$

The quantity $\left(\iiint_\Omega \mathbf{x}' \times \mathbf{j}(\mathbf{x}') \, d^3 x' \right)$ is the magnetic dipole moment $\mathbf{m}$ of the current distribution. Thus, the vector potential far from the distribution is

$$\mathbf{A}(\mathbf{x}) \approx \frac{\mu_0}{4\pi} \mathbf{m} \times \frac{\mathbf{x}}{|\mathbf{x}|^3} \, .$$

Magnetic moment of an arbitrary current distribution
The magnetic moment $\mathbf{m}$ of an arbitrary current density $\mathbf{j}$ contained in a volume Ω is defined as

$$\mathbf{m} = \frac{1}{2} \iiint_\Omega \mathbf{x}' \times \mathbf{j}(\mathbf{x}')\, d^3x' \; .$$

For a line current I flowing along a closed curve Γ, $\mathbf{j}(\mathbf{x}')d^3x' = I\,d\mathbf{l}$, where $d\mathbf{l}$ is a tangent vector to the curve Γ. Thus, the magnetic moment $\mathbf{m}$ of a current I flowing in a circuit defined by the closed curve Γ is

$$\boxed{\mathbf{m} = \frac{I}{2} \oint_\Gamma \mathbf{x}' \times d\mathbf{l}(\mathbf{x}') \; .} \tag{9.14}$$

For a planar current loop, such as the one illustrated in Fig. 9.15, it is shown below that the magnetic moment (9.14) simplifies to $\mathbf{m} = I S\mathbf{n}$, where S is the area enclosed by the loop and $\mathbf{n}$ is the unit normal vector to the plane of the loop (pointing in the direction determined by the right-hand rule). In the SI system, the unit of magnetic moment is Am^2.

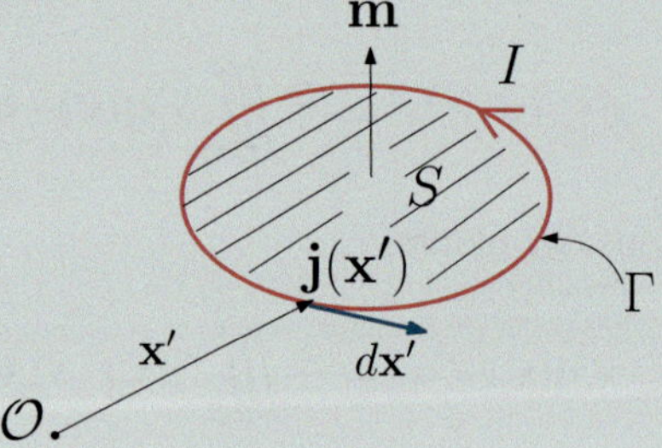

Fig. 9.15 A planar current loop defines a magnetic moment $\mathbf{m} = I S\mathbf{n}$

Remarks

- Unlike the electric field, where the first-order approximation for an arbitrary charge distribution includes a monopole term, the first-order approximation for the magnetic field of an arbitrary current distribution is a magnetic dipole. This distinction highlights the absence of magnetic monopoles in magnetostatics, a conclusion supported by extensive experimental evidence. To date, the only known sources of magnetism are electric currents or intrinsic magnetic moments. However, as discussed in Sect. 9.7.2, our current theoretical framework does not rule out the existence of magnetic monopoles. Their discovery would profoundly reshape our understanding of electromagnetism and the fundamental symmetries of nature.
- For a close contour Γ, any surface bounded by Γ is characterized by a surface vector defined as

$$\mathbf{S} = \iint_S \mathbf{n}(\mathbf{x})\,dS(\mathbf{x}) = \frac{1}{2}\oint_\Gamma \mathbf{x}' \times d\mathbf{l}(\mathbf{x}') \ .$$

Here, $\mathbf{n}(\mathbf{x})$ is the unit normal vector to the surface at point $\mathbf{x}$. To show this, we may take the scalar product of the integral (9.14) with an arbitrary unit vector $\mathbf{u}$:

$$\mathbf{m} \cdot \mathbf{u} = \frac{I}{2}\oint_\Gamma (\mathbf{x} \times d\mathbf{l}) \cdot \mathbf{u} = \frac{I}{2}\oint_\Gamma (\mathbf{u} \times \mathbf{x}) \cdot d\mathbf{l}$$

where we have performed a circular shift of the scalar triple product. Using Stoke's theorem (A.20)

$$\mathbf{m} \cdot \mathbf{u} = \frac{I}{2}\iint_{S(\Gamma)} \nabla \times (\mathbf{u} \times \mathbf{x}) \cdot \mathbf{n}\,dS$$

where $S(\Gamma)$ is any surface bounded by Γ. Using the identity (A.9) and the fact that $\mathbf{u}$ is a constant vector, $\nabla \times (\mathbf{u} \times \mathbf{x}) = (\nabla \cdot \mathbf{x})\mathbf{u} - (\mathbf{u} \cdot \nabla)\mathbf{x} = 3\mathbf{u} - \mathbf{u} = 2\mathbf{u}$. Finally

$$\mathbf{m} \cdot \mathbf{u} = I \iint_{S(\Gamma)} \mathbf{n}\,dS \cdot \mathbf{u} \ .$$

Since $\mathbf{u}$ is arbitrary, we conclude that a general expression for the dipole moment of a planar or non-planar current-carrying loop is

$$\mathbf{m} = I \iint_{S(\Gamma)} \mathbf{n}\,dS = I\mathbf{S}$$

where I is the current flowing through the loop.

9.7.1 Earth's Magnetic Field

In Earth's outer core, convection currents of molten iron and nickel generate a magnetic field that extends thousands of kilometers into space. The Earth's rotation provides the kinetic energy necessary to drive these currents, while the convective motions are sustained by heat transfer from the internal energy of the core. Together with Faraday's law of induction (see Chap. 11), these processes form the basis for the dynamo mechanism, which explains how Earth's magnetic field is maintained over geological timescales.

At first approximation, this field can be modeled as a magnetic dipole, akin to a bar magnet, as illustrated schematically in Fig. 9.16. The best-fitting dipole defines the geomagnetic poles, which are the antipodal points where the axis of this equivalent magnet intersects Earth's surface. Notably, the north geomagnetic pole corresponds

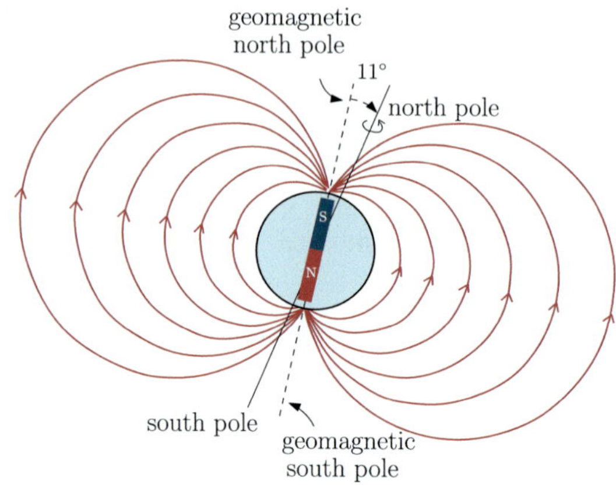

Fig. 9.16 The Earth's magnetic field at its surface resembles that of a magnetic dipole

to the south pole of Earth's equivalent magnet. The axis of this equivalent dipole is tilted at an angle of approximately 11° relative to Earth's rotational axis.

Near Earth's surface, the magnetic field strength was first measured by Gauss in 1832. It ranges between 25 and 65 μT, which is strong enough to align a small compass needle, one of the earliest practical applications of magnetism in navigation. In fact, the very definition of the *north pole* of a magnet is the end that points toward Earth's North Magnetic Pole when the magnet is suspended freely, such as in a compass.

However, Earth's magnetic field is not static. It evolves over geological timescales, with significant changes occurring approximately every several hundred thousand years. This field acts as a shield, protecting the planet from harmful solar wind and cosmic radiation. Without it, Earth's upper atmosphere would be stripped away. Mars, for instance, is believed to have lost its atmosphere billions of years ago after its magnetic field shut down, leaving it vulnerable to solar wind.

9.7.2 *Magnetic Monopoles*

One of the four fundamental equations of electromagnetism, Eq. (9.10), states that there are no magnetic charges, or monopoles. Despite a long search, magnetic monopoles have never been observed. This absence raises an intriguing question: why is there such asymmetry between the electric and magnetic laws? Could magnetic monopoles exist after all? In 1931, Paul Dirac proposed a quantum theory of the magnetic charge. Remarkably, his work demonstrated that magnetic monopoles could theoretically exist, but only with specific, quantized values of the magnetic charge. To understand this, let us consider a point-like magnetic monopole at the origin. Analogous to Gauss's law for electric charges, such a monopole would create a radial magnetic field of the form:

$$\mathbf{B}_m = \frac{q_m}{4\pi r^2}\mathbf{u}_r$$

where q_m is its magnetic charge. Defined this way, q_m has the same units as a magnetic flux. The problem, however, is that the introduction of a magnetic charge violates the condition $\nabla \cdot \mathbf{B} = 0$, making it impossible to define a vector potential $\mathbf{A}$ such that $\nabla \times \mathbf{A} = \mathbf{B}$ everywhere. Nevertheless, by excluding the origin, it is possible to define two potentials: $\mathbf{A}^N$, valid for the northern hemisphere ($\theta \in [0, \pi/2]$ in spherical coordinates), and $\mathbf{A}^S$, valid for the southern hemisphere ($\theta \in (\pi/2, \pi]$). These potentials are related through a gauge transformation and satisfy $\nabla \times \mathbf{A} = \mathbf{B}$ in their respective regions. Thus, despite the singularity at the origin, the magnetic field can still be expressed as the curl of a vector field everywhere else.

As discussed in Sect. 9.4.5, the phase acquired by an electric charge q moving along a closed loop Γ in the presence of a magnetic field is given by

$$\delta\phi = \frac{q}{\hbar}\oint_\Gamma \mathbf{A}\cdot d\mathbf{l} = \frac{q}{\hbar}\iint_{S(\Gamma)} \mathbf{B}\cdot d\mathbf{S}$$

where the surface $S(\Gamma)$ bounded by Γ can be chosen arbitrarily. This phase shift, measurable experimentally, leads to the Aharonov–Bohm effect. For consistency, $\delta\phi$ must be independent of the chosen surface, implying:

$$\frac{q}{\hbar}\iint_{S_1(\Gamma)} \mathbf{B}\cdot d\mathbf{S} - \frac{q}{\hbar}\iint_{S_2(\Gamma)} \mathbf{B}\cdot d\mathbf{S} = 2n\pi \quad n \in \mathbb{Z}.$$

If we consider two surfaces S_1 and S_2 sharing the same boundary Γ as shown in Fig. 9.17, the difference of these two integrals equals $q/\hbar$ times the magnetic flux through the closed surface formed by S_1 and $-S_2$. According to Gauss's law, this flux equals the magnetic charge q_m enclosed. Thus, we arrive at the Dirac quantization condition:

Fig. 9.17 A magnetic charge q_m would create a non-zero magnetic flux over a closed surface that contains the charge

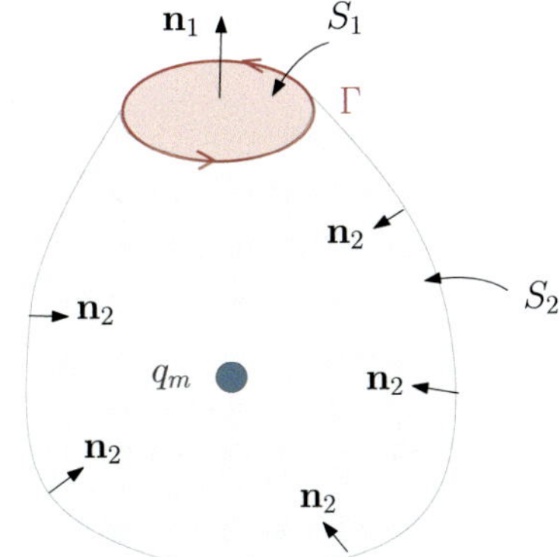

$$qq_m = 2\hbar n\pi \quad n \in \mathbb{Z} \,. \tag{9.15}$$

This condition implies that if magnetic monopoles exist, their magnetic charge cannot have any value. Furthermore, since electric charge is known to be quantized[2]— an integer multiple of the elementary charge e—then the magnetic charge is also quantized, with the smallest possible magnetic charge given by

$$\frac{2\pi\hbar}{e} = \Phi_0$$

where Φ_0 is called the quantum of flux. The existence of magnetic monopoles remains one of the most tantalizing open questions in modern physics. Many theories that try to extend the standard model predict that they should exist. Their discovery would revolutionize our understanding of electromagnetism, quantum mechanics, and even the fundamental structure of the universe. A number of experiments around the world are trying to detect them.

9.8 Forces and Torques on a Magnetic Dipole in an External Magnetic Field

Consider a magnetic dipole located at position $\mathbf{x}$ in an external magnetic field $\mathbf{B}$, as shown in Fig. 9.18. Similar to how an electric field exerts a force and torque on an electric dipole, the external magnetic field will exert a force $\mathbf{F}$ and a torque $\boldsymbol{\tau}$ on the magnetic dipole.

Fig. 9.18 A magnetic dipole in an external magnetic field

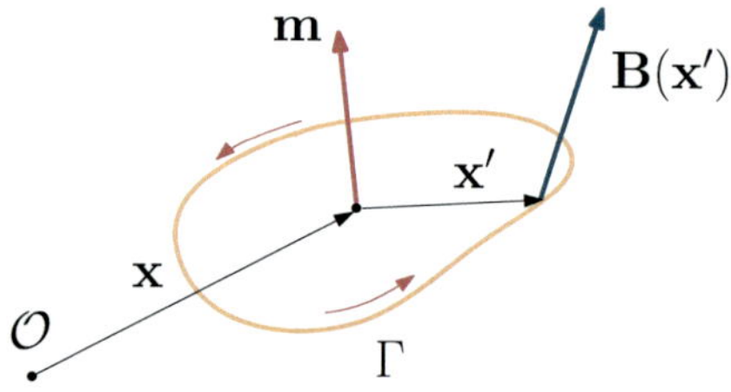

9.8.1   *Force on a Current Loop in an Inhomogeneous Magnetic Field*

For a current-carrying wire loop Γ in a magnetic field $\mathbf{B}$, the force on a small segment of the wire is given by the Laplace force,

$$\mathbf{F} = \oint_\Gamma I \, dl(\mathbf{x}') \times \mathbf{B}(\mathbf{x}') \ .$$

If the magnetic field is uniform, $\mathbf{B}$ can be pulled out of the integral and since $\oint_\Gamma dl(\mathbf{x}') = \mathbf{0}$, the total force on the loop is zero.

For a non-uniform magnetic field, consider the x-component of the force,

$$F_x = \mathbf{F} \cdot \mathbf{u}_x = \oint_\Gamma I \, dl(\mathbf{x}') \times \mathbf{B}(\mathbf{x}') \cdot \mathbf{u}_x = \oint_\Gamma I \, \mathbf{B}(\mathbf{x}') \times \mathbf{u}_x \cdot dl(\mathbf{x}') \ ,$$

where a circular permutation was used in the scalar triple product. Using the Stokes theorem (A.20), we can convert the line integral into a surface integral:

$$F_x = \iint_S I \, \nabla' \times (\mathbf{B}(\mathbf{x}') \times \mathbf{u}_x) \cdot \mathbf{n}(\mathbf{x}') dS \ .$$

Using the vector identity[3]

$$\nabla \times (\mathbf{B} \times \mathbf{u}_x) = \frac{\partial \mathbf{B}}{\partial x} \ ,$$

we get

$$F_x = \iint_S I \, \frac{\partial \mathbf{B}(\mathbf{x}')}{\partial x'} \cdot \mathbf{n}(\mathbf{x}') dS \ .$$

Approximating $\dfrac{\partial \mathbf{B}(\mathbf{x}')}{\partial x'} = \dfrac{\partial \mathbf{B}(\mathbf{x})}{\partial x}$ for points $\mathbf{x}'$ close to $\mathbf{x}$[4] and pulling $\dfrac{\partial \mathbf{B}(\mathbf{x})}{\partial x}$ out the integral, we get

$$F_x = \iint_S I \, \frac{\partial \mathbf{B}(\mathbf{x})}{\partial x} \cdot \mathbf{n}(\mathbf{x}') dS = I \iint_S \mathbf{n}(\mathbf{x}') dS \cdot \frac{\partial \mathbf{B}(\mathbf{x})}{\partial x} \ .$$

Recognizing the magnetic dipole moment $\mathbf{m} = I \iint_S \mathbf{n}(\mathbf{x}') dS$, we obtain

---

[3] This identity can be demonstrated using (A.9), Gauss's law for magnetism ($\nabla \cdot \mathbf{B} = 0$) and the fact that $\mathbf{u}_x$ is a constant vector.

[4] This is justified by a Taylor expansion at leading order of $\dfrac{\partial \mathbf{B}(\mathbf{x}')}{\partial x'}$, considering $\mathbf{x}' = \mathbf{x} + (\mathbf{x}' - \mathbf{x})$:

$$\frac{\partial \mathbf{B}(\mathbf{x}')}{\partial x'} \sim \frac{\partial \mathbf{B}(\mathbf{x})}{\partial x} \ .$$

$$\boxed{F_x = \mathbf{m} \cdot \frac{\partial \mathbf{B}(\mathbf{x})}{\partial x}} \; . \qquad (9.16)$$

Similarly, we can derive expressions for F_y and F_z. Combining these, we get the general expression for the force on a magnetic dipole in an inhomogeneous magnetic field

$$F_i = \mathbf{F} \cdot \mathbf{u}_i = \mathbf{m} \cdot \frac{\partial \mathbf{B}}{\partial x_i} \; .$$

The Stern–Gerlach experiment, which demonstrated the quantization of angular momentum, utilized the force on magnetic dipoles in an inhomogeneous magnetic field.

If $\mathbf{m}$ is a constant vector (independent of $\mathbf{B}$), then the force can be written as

$$\mathbf{F} = -\nabla(-\mathbf{m} \cdot \mathbf{B}) \; .$$

This allows us to define a potential energy function U_m for a permanent magnetic dipole $\mathbf{m}$ in a magnetic field $\mathbf{B}$,

$$U_m = -\mathbf{m} \cdot \mathbf{B} \; .$$

There is an obvious analogy between the potential energy $U_m = -\mathbf{m} \cdot \mathbf{B}$ for a magnetic dipole and the potential energy $U_p = -\mathbf{p} \cdot \mathbf{E}$ for an electric dipole in an electric field. However, there are subtle differences in the forces acting on these dipoles. The expression $F_x = \mathbf{m} \cdot \dfrac{\partial \mathbf{B}(\mathbf{x})}{\partial x}$ was derived from the Laplace force, which is valid even for time-dependent magnetic fields $\mathbf{B}(\mathbf{x}, t)$. Therefore, the force exerted by a magnetic field on a magnetic dipole remains the same, regardless of whether the field is static or time-varying.

In the static regime, the external magnetic field satisfies Ampère's law,

$$\nabla \times \mathbf{B} = \mu_0 \mathbf{j} \; .$$

Since the source current $\mathbf{j}$ is located far away from the dipole, $\mathbf{j}(\mathbf{x}) = \mathbf{0}$ and $\nabla \times \mathbf{B} = \mathbf{0}$. The x-component of the force on the dipole can be written as

$$F_x = m_x \frac{\partial B_x}{\partial x} + m_y \frac{\partial B_y}{\partial x} + m_z \frac{\partial B_z}{\partial x} = m_x \frac{\partial B_x}{\partial x} + m_y \frac{\partial B_x}{\partial y} + m_z \frac{\partial B_x}{\partial z} \; .$$

In vector notation, this becomes

$$F_x = (\mathbf{m} \cdot \nabla) B_x \; , \qquad \mathbf{F} = (\mathbf{m} \cdot \nabla) \mathbf{B} \; .$$

This expression for the force on a magnetic dipole in a static magnetic field is different from the general expression (9.16) derived earlier. While both expressions coincide in the static regime, the general expression (9.16) is valid for both static and time-varying magnetic fields. It is interesting to note the analogy between the force on a magnetic dipole in a static magnetic field and the force on an electric dipole in a static electric field: $\mathbf{F} = (\mathbf{p} \cdot \nabla)\mathbf{E}$. However, this analogy does not hold for time-varying fields.

9.8.2 *Torque on a Magnetic Dipole*

The torque on a magnetic dipole in a magnetic field can be derived by considering the torque on each infinitesimal current element of the dipole. For a current loop Γ, the torque is given by

$$\boldsymbol{\tau}(\mathbf{x}) = \oint_{\Gamma} \mathbf{x}' \times (I d\mathbf{x}' \times \mathbf{B}(\mathbf{x}')) .$$

Using the $BAC - CAB$ rule and assuming a uniform magnetic field[5] $\mathbf{B}(\mathbf{x}') \simeq \mathbf{B}(\mathbf{x})$, we can simplify the expression:

$$\boldsymbol{\tau}(\mathbf{x}) = \oint_{\Gamma} \mathbf{x}' \times [I d\mathbf{x}' \times \mathbf{B}(\mathbf{x})] = \oint_{\Gamma} I[\mathbf{x}' \cdot \mathbf{B}(\mathbf{x})]d\mathbf{x}' - \oint_{\Gamma} I(\mathbf{x}' \cdot d\mathbf{x}')\,\mathbf{B}(\mathbf{x}) .$$

The second term vanishes since $\mathbf{B}(\mathbf{x})$ can be pulled out of the integral and we have $\int_{\Gamma} \mathbf{x}' \cdot d\mathbf{x}' = (\mathbf{x}_0'^2 - \mathbf{x}_0'^2)/2 = 0$, where $\mathbf{x}_0'$ is an origin on Γ. Consider the x-component of the torque,

$$\tau_x = \boldsymbol{\tau} \cdot \mathbf{u}_x = \int_{\Gamma} I[\mathbf{x}' \cdot \mathbf{B}(\mathbf{x})]\mathbf{u}_x \cdot d\mathbf{x}' .$$

Using the Stokes theorem, we can convert the line integral into a surface integral:

$$\tau_x = \iint_{S} I\left\{\nabla' \times [(\mathbf{x}' \cdot \mathbf{B}(\mathbf{x}))\mathbf{u}_x]\right\} \cdot \mathbf{n}(\mathbf{x}')dS .$$

---

[5] This is justified by a Taylor expansion of the magnetic field

$$\mathbf{B}(\mathbf{x}') = \mathbf{B}(\mathbf{x}) + [(\mathbf{x}' - \mathbf{x}) \cdot \nabla]\mathbf{B}(\mathbf{x}) + \cdots ,$$

limited to its leading order.

Simplifying the curl,[6] then applying a circular permutation to the resulting triple product and finally pulling $\mathbf{B}(\mathbf{x})$ and $\mathbf{u}_x$ out of the integral, we obtain

$$\tau_x = \iint_S I\,[\mathbf{B}(\mathbf{x}) \times \mathbf{u}_x] \cdot \mathbf{n}(\mathbf{x}')dS = \left(\iint_S I\mathbf{n}(\mathbf{x}')dS\right) \times \mathbf{B}(\mathbf{x}) \cdot \mathbf{u}_x \;.$$

Recognizing the magnetic dipole moment $\mathbf{m} = I \iint_S \mathbf{n}(\mathbf{x}')dS$, the x-component of the torque can be written as $\tau_x = (\mathbf{m} \times \mathbf{B}(\mathbf{x})) \cdot \mathbf{u}_x$. In vector notation, this becomes

$$\boxed{\boldsymbol{\tau}(\mathbf{x}) = \mathbf{m} \times \mathbf{B}(\mathbf{x}) \;.} \tag{9.17}$$

This expression shows that a magnetic dipole experiences a torque that tends to align its magnetic moment with the external magnetic field. This behavior is analogous to the torque experienced by an electric dipole $\mathbf{p}$ in an electric field: $\boldsymbol{\tau} = \mathbf{p} \times \mathbf{E}$.

When two magnets are brought close to each other, they experience both a torque and a force due to their magnetic moments interacting. If their axes coincide, the magnets tend to align such that their moments are parallel and in the same direction. This alignment minimizes their magnetic energy. Specifically:

- Unlike poles (north and south) attract each other. The magnetic field between the poles becomes strong, and the force pulls the magnets together. If there is no friction or another opposing force, the magnets will stick to each other.
- Like poles (north-north or south-south) repel each other, creating a weaker field between them and causing the magnets to push apart.

This behavior is illustrated in Fig. 9.19, where the magnetic field pattern between the poles of two magnets is visualized thanks to magnetic particles suspended in a fluid that align with the field and attract each other. Between the north and south poles, the field is strong and the force is attractive. Between two like poles (south-south in the figure), the field is weak and the magnets repel each other.

---

[6] Using (A.6), we can show that

$$\nabla' \times [(\mathbf{x}' \cdot \mathbf{B}(\mathbf{x}))\mathbf{u}_x] = (\mathbf{x}' \cdot \mathbf{B}(\mathbf{x}))\nabla' \times \mathbf{u}_x + \left[\nabla'(\mathbf{x}' \cdot \mathbf{B}(\mathbf{x}))\right] \times \mathbf{u}_x \;.$$

The first term is zero since $\mathbf{u}_x$ is a constant vector. In the second term, $\mathbf{B}(\mathbf{x})$ is a constant vector for the gradient ∇'. Using Cartesian coordinates, it is easy to show that $\nabla'(\mathbf{x}' \cdot \mathbf{B}(\mathbf{x})) = \mathbf{B}(\mathbf{x})$. Therefore,

$$\nabla' \times [(\mathbf{x}' \cdot \mathbf{B}(\mathbf{x}))\mathbf{u}_x] = \mathbf{B}(\mathbf{x}) \times \mathbf{u}_x \;.$$

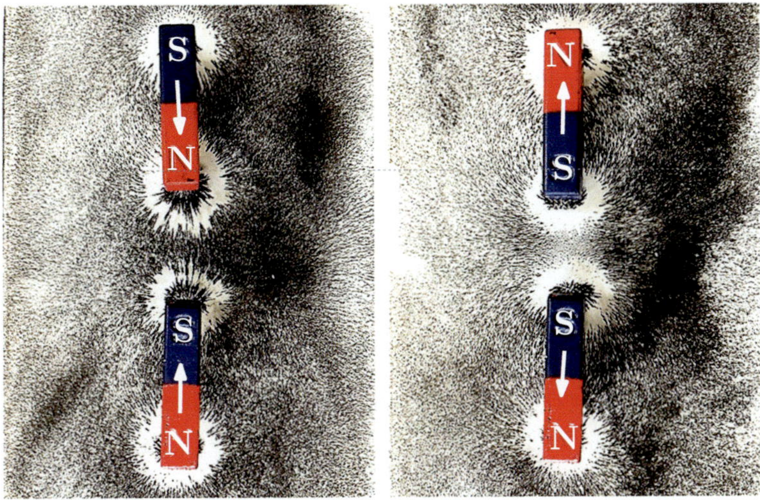

Fig. 9.19 Magnetic powder particles suspended in a fluid align with the magnetic field of the nearby magnets

9.9 Summary and Essential Formulas

- Ampère's law states that the circulation of the magnetic field **B** around any closed curve Γ is proportional to the total current I_Γ flowing through any surface $S(\Gamma)$ bounded by Γ:

$$\oint_\Gamma d\mathbf{x} \cdot \mathbf{B}(\mathbf{x}) = \mu_0 I_\Gamma \ .$$

- In differential form, Ampère's law can be written as

$$\boxed{\nabla \times \mathbf{B} = \mu_0 \mathbf{j} \ .}$$

- The magnetic field **B** can be derived from a vector potential **A** as

$$\mathbf{B} = \nabla \times \mathbf{A}$$

this implies that the divergence of the magnetic field is zero,

$$\boxed{\nabla \cdot \mathbf{B} = 0 \ .}$$

Equivalently, the magnetic flux through any closed surface is zero:

$$\oiint_S d\mathbf{S} \cdot \mathbf{B} = 0 \ .$$

- The vector potential $\mathbf{A}$ is not unique. Given a scalar field ψ, $\mathbf{A}' = \mathbf{A} + \nabla\psi$ defines the same magnetic field $\mathbf{B}$. This gauge invariance allows us to choose a specific gauge, such as the Coulomb gauge ($\nabla \cdot \mathbf{A} = 0$), which simplifies calculations. In this gauge, the vector potential satisfies Poisson's equation,

$$\nabla^2 \mathbf{A}(\mathbf{x}) = -\mu_0 \mathbf{j}(\mathbf{x}) \ .$$

- A magnetic moment $\mathbf{m}$ can be associated with a closed current loop carrying a current I. For a planar loop enclosing a surface S with unit normal vector $\mathbf{n}$, the magnetic moment is

$$\boxed{\mathbf{m} = I S \mathbf{n} \ .}$$

For a non-planar loop Γ bounding a surface S, the magnetic moment is

$$\boxed{\mathbf{m} = I\mathbf{S} \ , \quad \text{with} \quad \mathbf{S} = \frac{1}{2}\oint_\Gamma \mathbf{x}' \times d\mathbf{x}' = I \iint_{S(\Gamma)} \mathbf{n}(\mathbf{x}')dS \ .}$$

- A magnetic dipole $\mathbf{m}$ in an external magnetic field $\mathbf{B}$ experiences a torque τ and a force $\mathbf{F}$ given by

$$\tau = \mathbf{m} \times \mathbf{B} \ ,$$

$$\mathbf{F} = \nabla \left(\mathbf{m} \cdot \mathbf{B} \right) \ .$$

Here, we assume that the magnetic moment $\mathbf{m}$ is constant (independent of $\mathbf{B}$). If $\mathbf{m}$ is not constant, the torque remains the same, but the force components are given by

$$F_i = \mathbf{m} \cdot \frac{\partial \mathbf{B}}{\partial x_i} \ .$$

The torque tends to align the dipole with the magnetic field and the force tends to move it toward regions of stronger field.

- For a point far away from an arbitrary current distribution, the vector potential can be approximated as that of a magnetic dipole,

$$\mathbf{A}(\mathbf{x}) \approx \frac{\mu_0}{4\pi} \frac{\mathbf{m} \times \mathbf{x}}{|\mathbf{x}|^3} \ .$$

The magnetic field far from the dipole is given by

$$\mathbf{B}(\mathbf{x}) = \nabla \times \left(\frac{\mu_0}{4\pi} \frac{\mathbf{m} \times \mathbf{x}}{|\mathbf{x}|^3} \right) \ .$$

$\mathbf{B}(\mathbf{x})$ is similar to the electric field of an electric dipole,

$$\mathbf{B}(\mathbf{x}) = -\nabla \left(\frac{\mu_0}{4\pi} \frac{\mathbf{m} \cdot \mathbf{x}}{|\mathbf{x}|^3} \right).$$

Problems

9.1 Force between a rectangular loop and a straight wire

Consider a rectangular loop carrying a clockwise current I_2 placed a distance a away from an infinitely long, straight wire carrying a current I_1, as shown below. Both the loop and the wire are in the same plane. Calculate the net force on the current loop I_2.

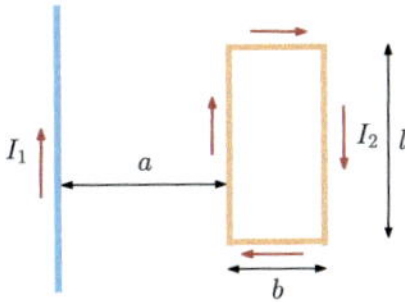

9.2 Magnetic field of a coaxial cable

Consider a coaxial cable consisting of two infinitely long, concentric cylindrical conductors. A current I_0 flows through the inner conductor, while a current $-I_0$ flows through the outer conductor. Determine the magnetic field at all points in space.

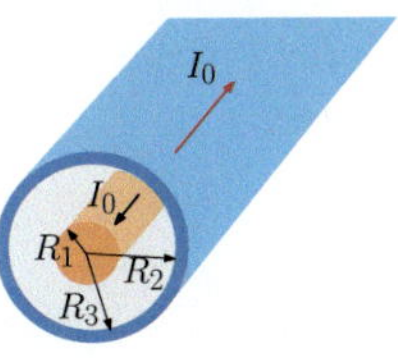

9.3 Magnetic field of a non-uniform cylindrical current

Consider an infinitely long cylindrical conductor of radius R carrying a current I. The current density within the conductor varies with the radial distance r from the axis of the conductor and is given by $J(r) = \alpha r$, where α is a constant. Determine the magnetic field at all points in space.

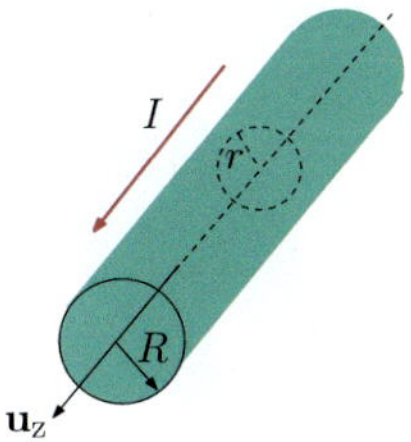

9.4 Magnetic field of a plane ribbon conductor

Consider a flat, infinitely long ribbon conductor of width w, lying in the xy-plane. The current I flows along the x-axis through the ribbon. Determine the magnetic field at a point P in the xy-plane, located at a distance h from the conductor.

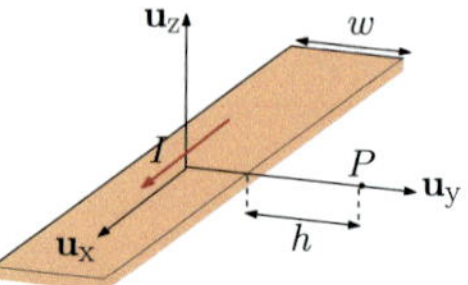

9.5 Magnetic field of a current sheet

Consider two infinite, parallel planes separated by a distance b. A uniform current density $\mathbf{j} = j_0\mathbf{u}_x$ flows between the planes. Determine the magnetic field everywhere in space.

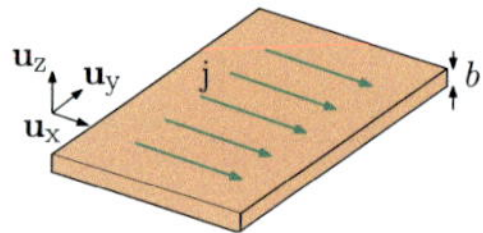

9.6 Superposition of magnetic fields

Two parallel currents flow along the x-axis. The first current is due to a uniform current density $j_1\mathbf{u}_x$ flowing between two infinite parallel planes separated by a distance b $(-b/2 \leq z \leq b/2)$. The second current $j(r)\mathbf{u}_x$ is due to an infinitely long cylindrical conductor of radius R with a non-uniform current density $j(r) = j_0\left(1 - \frac{r}{R}\right)$ (for $r \leq R$) flowing along its axis. Determine the value of j_0 in terms of j_1 such that the resulting magnetic field in the z-direction is zero at a point P located at a distance $R/2$ from the cylinder axis.

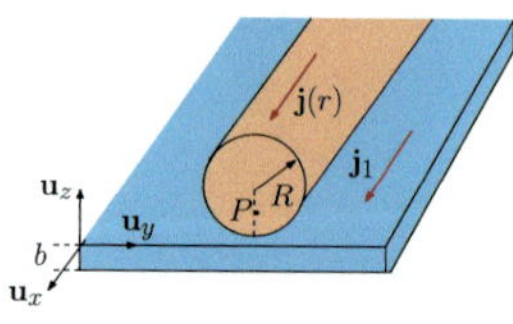

9.7 Wire with a cavity

A cylindrical wire of radius R carries a uniform current density $\mathbf{j} = j_0\mathbf{u}_z$. There is a cylindrical current-free cavity within the wire, whose axis is parallel to the wire axis and at a distance b from it. Determine the magnetic field inside the cavity.

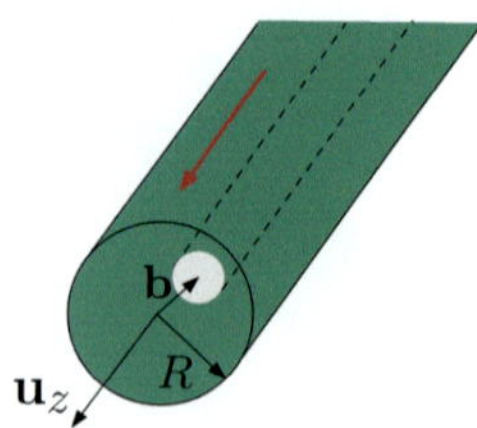

9.8 Magnetic force between a cylinder and a ribbon conductor

An infinitely long cylindrical conductor of radius R carries a current I. It is parallel to an infinitely long, flat ribbon conductor of width a, carrying a current I'. The two conductors are coplanar, and the axis of the cylinder is a distance b from the edge of the ribbon.

(a) Determine the magnetic field on the x-axis for $x > a$.

(b) Calculate the force per unit length between the two conductors.

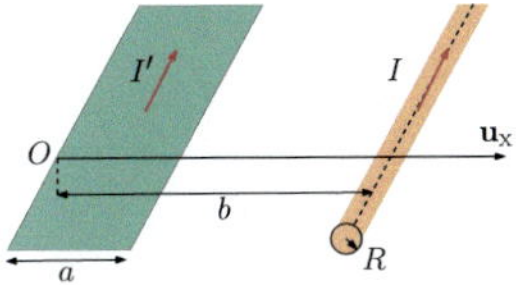

Chapter 10
The Origin of Magnetism in Matter

Abstract This chapter explores the *microscopic origins of magnetism in matter*, classifying materials into *diamagnetic, paramagnetic, and ferromagnetic* categories based on their response to external magnetic fields. It emphasizes that magnetism is fundamentally a quantum mechanical phenomenon, primarily arising from the *orbital and spin magnetic moments of electrons*. The concept of the *Bohr magneton* as the fundamental unit of magnetic moment is introduced. The chapter then presents the *dynamics of a magnetic moment in a magnetic field*, explaining *Larmor precession* and the subsequent relaxation processes that lead to the alignment of magnetic moments with an external field. This discussion highlights the principles behind *Electron Spin Resonance (ESR)* and *Nuclear Magnetic Resonance (NMR)*, including their applications in fields like medical imaging (MRI). The macroscopic magnetic properties of materials are quantified by the *magnetization vector* (**M**), representing the average magnetic dipole moment per unit volume. A key development is the representation of a magnetized material by equivalent *magnetization current densities* (volume and surface currents), which contribute to the total magnetic field. This leads to the introduction of the *magnetic excitation field* (**H**), defined as $\mathbf{H} = \mathbf{B}/\mu_0 - \mathbf{M}$, which simplifies Ampére's law in magnetic media by relating it directly to free currents. The chapter establishes the *constitutive relations* between **B**, **H**, and **M**, defining *magnetic susceptibility* (χ_m) and *magnetic permeability* (μ) for linear materials, and discussing the complex, nonlinear behavior of ferromagnets, including *hysteresis* and the distinction between soft and hard magnetic materials. Finally, the chapter outlines the *boundary conditions* for magnetic fields (**B** and **H**) at the interface between different magnetic media, demonstrating the continuity of the normal component of **B** and the discontinuity of the tangential component of **H** in the presence of surface currents. These principles provide a comprehensive framework for understanding and analyzing magnetic phenomena in various materials.

Keywords Magnetism · Ferromagnetism · Paramagnetism · Diamagnetism · Spin · Hysteresis · Permeability

© The Author(s), under exclusive license to Springer Nature Switzerland AG 2025
F. Cadiz and A. Couairon, *Classical Electrodynamics*, Undergraduate Texts in Physics,
https://doi.org/10.1007/978-3-031-86785-9_10

10.1 Introduction

This chapter delves into the microscopic origins of magnetism and explores its three primary types: ferromagnetism, paramagnetism, and diamagnetism. We discuss how these phenomena are intrinsically quantum mechanical, strongly influenced by the electron spin—an intrinsic form of angular momentum associated with an inherent magnetic moment.

We examine the dynamics of a magnetic moment in a uniform magnetic field, leading to the phenomenon of Larmor's precession, which is fundamental to technologies such as magnetic resonance imaging (MRI). Additionally, we present the fundamental laws of magnetostatics in magnetic media, characterized by a magnetic dipole density $\mathbf{M}$. The dependence of $\mathbf{M}$ on an external magnetic field varies significantly across different types of magnetism.

Finally, we establish the boundary conditions that the magnetic field must satisfy at the interface between two media, providing a comprehensive framework for understanding magnetism in various contexts.

10.2 Magnetic Media—Paramagnetism, Diamagnetism, Ferromagnetism

10.2.1 The Different Types of Magnetism

In electrostatics, materials are broadly classified into two categories: insulators (dielectrics) and conductors. In contrast, magnetic materials are categorized into three primary groups based on their response to external magnetic fields:

- **Ferromagnetism**: Ferromagnetic materials are strongly attracted to external magnetic fields. They exhibit spontaneous magnetization, meaning they retain their magnetic properties even after the external field is removed. This property makes them crucial for various applications in industry, such as motors, hard drives, and generators.
  Examples: Iron (Fe), Cobalt (Co), Nickel (Ni), Iron Oxide (Fe_2O_3).
- **Paramagnetism**: Paramagnetic materials are weakly attracted to external magnetic fields. However, they lose their magnetic properties when the external field is removed.
  Examples: Aluminum (Al), Calcium (Ca), Sodium (Na).
- **Diamagnetism**: Diamagnetic materials are weakly repelled by external magnetic fields. While all materials exhibit diamagnetic properties at the atomic level, these effects are often masked by stronger paramagnetic or ferromagnetic behaviors.
  Examples: Bismuth (Bi), Silver (Ag), Copper (Cu), Mercury (Hg), Lead (Pb), water (H_2O).

10.2.2 *Microscopic Origin of Magnetism*

The magnetic properties of matter primarily arise from the magnetic moments **m** associated with electrons in atoms. These magnetic moments have two origins:

- **Orbital magnetic moment**: Consider the classical model of the hydrogen atom, where an electron orbits a proton at a distance r. This orbital motion of the electron constitutes a tiny current loop, generating an orbital magnetic moment, as shown in Fig. 10.1.

  The angular momentum of the electron is given by $\mathbf{L} = \mathbf{r} \times \mathbf{p} = r\mathbf{u}_r \times m_e\mathbf{v}$, where m_e is the electron mass and $r\mathbf{u}_r$ is its position, using a spherical coordinate system centered at the proton. Assuming a circular orbit in the xy-plane, this simplifies to $\mathbf{L} = rm_ev\mathbf{u}_z$. The electron orbital motion can be considered as a tiny current loop, generating a magnetic moment $\mathbf{m}_L = IS\mathbf{u}_z$. Here, $S = \pi r^2$ is the area of the loop, and $I = \frac{-ev}{2\pi r}$ is the current due to the electron motion. Combining these expressions, we obtain

$$\mathbf{m}_L = \frac{-evr}{2}\mathbf{u}_z = -\frac{e}{2m_e}\mathbf{L} \; . \tag{10.1}$$

  The magnetic moment m_L associated with the electron orbital motion is proportional to its angular momentum **L**. Quantum mechanics dictates that the electron angular momentum is quantized. This quantization restricts the possible values of $|\mathbf{L}|$ to integer multiples of the reduced Planck constant $\hbar = 1.054 \times 10^{-34}$Js: $L = n\hbar$, where n is an integer. As a result, the orbital magnetic moment (10.1) is also quantized:

$$\boxed{\mathbf{m}_L = -\frac{\mu_B}{\hbar}\mathbf{L} \;\Rightarrow\; |\mathbf{m}_L| = n\mu_B \; .} \tag{10.2}$$

  Here, $\mu_B = \frac{1}{2}e\hbar/m_e \approx 9 \times 10^{-24}$Am^2 is the Bohr magneton, the fundamental unit of magnetic moment. The orbital magnetic moment is therefore an integer multiple of μ_B, the *quantum* of magnetic moment.

- **Spin magnetic moment**: Even when not orbiting, an electron possesses an intrinsic angular momentum **S**, known as spin. This quantum mechanical property has no classical analog. The projection of the spin along any axis can only take two values,

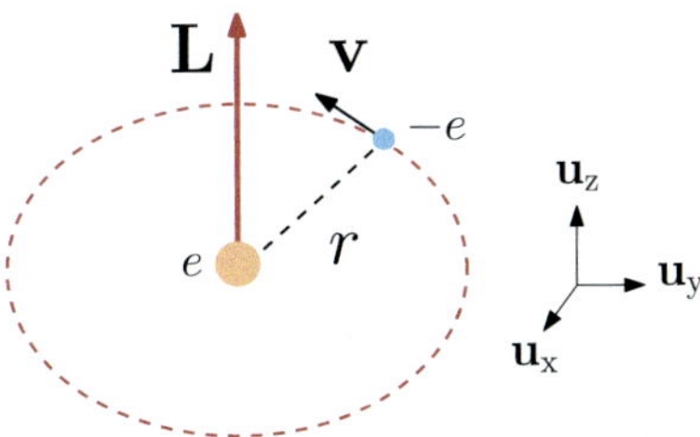

Fig. 10.1 The orbital motion of an electron bound in a hydrogen atom generates a magnetic moment

$S = \pm \hbar/2$. The spin contributes to magnetism through the spin magnetic moment:

$$\boxed{\mathbf{m}_S = -g\frac{\mu_B}{\hbar}\mathbf{S} \;\Rightarrow\; m_S \approx \pm\mu_B \;.}$$ (10.3)

The dimensionless g-factor, approximately equal to the value of 2 predicted by Dirac's equation, accounts for the effects of relativity and quantum mechanics. This quantity can be experimentally measured, for instance, by trapping electrons in a Penning trap (see Sect. 8.7). The most precise experimental value known to date has a relative uncertainty of one ten-trillionth:

$$g = 2.00231930436092 \pm (3.6 \times 10^{-13}) \;.$$

This represents one of the most precise measurements in science. Our current theoretical understanding relies on quantum field theory, which predicts the value of g in terms of the fine-structure constant $\alpha = \frac{e^2}{4\pi\epsilon_0\hbar c}$. The fine-structure constant can be measured independently with a relative uncertainty of 10^{-10}. The theoretical value of g, derived using the measured value of α, agrees with the experimental result to 10 significant figures. This is one of the greatest triumphs in the history of theoretical physics.

We will recall that the magnitude of the electron spin magnetic moment is, therefore, approximately equal to the Bohr magneton, $|\mathbf{m}_S| \approx \mu_B$.

- **Atom magnetic moment**: An atom magnetic moment arises from the combined contributions of the orbital and spin magnetic moments of its electrons. While nucleons (protons, neutrons) also possess intrinsic magnetic moments, they are significantly smaller (about 2000 times smaller) than those of electrons, $m_n \sim 10^{-3} \; \mu_B$. Therefore, the magnetic properties of solids are primarily determined by the behavior of electrons. However, the interaction between nuclear and electron spins can lead to observable effects, as seen in techniques like nuclear magnetic resonance (NMR) allowing the detection, at the atomic scale, of different environments of the nuclei in matter.

10.2.3   Dynamics of a Magnetic Moment in a Magnetic Field—The Larmor Precession

A particle with angular momentum $\mathbf{J}$ (resulting in an atom from both orbital and spin contributions) possesses a magnetic moment

$$\mathbf{m} = \gamma\mathbf{J} \;,$$

where $\gamma = -g\mu_B/\hbar$ is the *gyromagnetic ratio*. For an electron, $\gamma \approx -e/m_e$. In a uniform magnetic field $\mathbf{B}$, the magnetic moment experiences a torque $\tau = \mathbf{m} \times \mathbf{B}$, according to (9.17). This torque causes the angular momentum to precess about the direction of the magnetic field. The dynamics of the angular momentum $\mathbf{J}$ is governed by the equation of equation of motion

$$\frac{d\mathbf{J}}{dt} = \tau = \mathbf{m} \times \mathbf{B} .$$

The equation of motion for the magnetic moment $\mathbf{m}$ is

$$\boxed{\frac{d\mathbf{m}}{dt} = \gamma \mathbf{m} \times \mathbf{B} .} \tag{10.4}$$

This equation describes the Larmor precession of the magnetic moment around the magnetic field.

Key properties of Larmor's precession:

- Conservation of the parallel component: The component of the magnetic moment parallel to the field, $\mathbf{m} \cdot \mathbf{B}$, remains constant, as shown by taking the scalar product with $\mathbf{B}$ on both sides of (10.4).
- Conservation of the magnitude: The magnitude of the magnetic moment, $|\mathbf{m}|$, remains constant, as shown by taking the scalar product with $\mathbf{m}$ on both sides of (10.4).
- Circular motion: In a uniform magnetic field $\mathbf{B} = B_0 \mathbf{u}_z$, the projection of $\mathbf{m}$ onto the xy-plane, (m_x, m_y) satisfies

$$\frac{dm_x(t)}{dt} = -\omega_L m_y ,$$
$$\frac{dm_y(t)}{dt} = \omega_L m_x .$$

The solution to this system[1] shows that (m_x, m_y) rotates with the Larmor frequency $\omega_L = -\gamma B_0$, as illustrated in Fig. 10.2.

This precession phenomenon is fundamental to many physical phenomena, including nuclear magnetic resonance (NMR) and electron spin resonance (ESR).

Remark: The interaction energy between a magnetic moment and a magnetic field is given by $W = -\mathbf{m} \cdot \mathbf{B}$. This energy is minimized when the magnetic moment aligns with the magnetic field in an energetically stable configuration. While the Larmor precession equation (10.4) suggests perpetual precession preventing $\mathbf{m}$ to truly aligns with the magnetic field, this is for an isolated system where

---

[1] The complex variable $\underline{m}_\perp = m_x + i m_y$ satisfies

$$\frac{d\underline{m}_\perp}{dt} = i\omega_L \underline{m}_\perp .$$

The solution is $\underline{m}_\perp = \underline{m}_0 \exp(i\omega_L t)$.

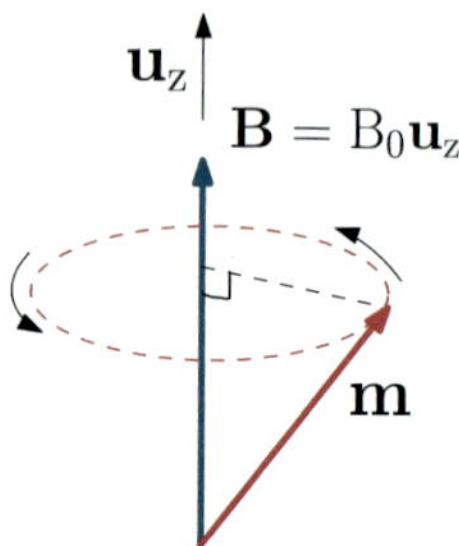

Fig. 10.2 Larmor's precession of a magnetic dipole around the magnetic field **B**

$\mathbf{m} \cdot \mathbf{B} = m_z B_0$ remains constant. To reach the thermodynamical equilibrium state, where the magnetic moment aligns with the field, energy exchange with the environment is necessary.

To account for this relaxation process (return to equilibrium of the magnetic moment), we can introduce a damping term in the equation of motion (10.4):

$$\frac{d\mathbf{m}}{dt} = \gamma \mathbf{m} \times \mathbf{B} - \frac{\mathbf{m} - \mathbf{m}_{\mathrm{eq}}(\mathbf{B})}{T_1} \ . \tag{10.5}$$

Here, T_1 is the relaxation time, which characterizes the rate at which the magnetic moment approaches its equilibrium orientation $\mathbf{m}_{\mathrm{eq}}(\mathbf{B})$.

For example, consider a uniform magnetic field $\mathbf{B} = B_0 \mathbf{u}_z$. In this case, the equilibrium orientation of the magnetic moment is $\mathbf{m}_{\mathrm{eq}}(\mathbf{B}) = m_s \mathbf{u}_z$. Projecting the equation of motion (10.5) onto the z-axis, we get

$$\frac{dm_z(t)}{dt} = -\frac{m_z(t) - m_s}{T_1} \ .$$

Given an initial condition $m_z(0) = m_s \cos\theta$, the solution for the z-component of the magnetic moment is

$$m_z(t) = m_s \left(1 - e^{-t/T_1}\right) + m_s \cos\theta \, e^{-t/T_1} \ .$$

In addition to the Larmor precession around the magnetic field, the magnetic moment spirals toward its equilibrium position, as illustrated in Fig. 10.3. In a non-magnetic material, the magnetic moments of atoms are randomly oriented due to thermal fluctuations. Applying an external magnetic field causes these moments to partially align with the field, leading to magnetization. This alignment process is characterized by a relaxation time T_1, during which the magnetic moments reach their equilibrium orientation.

Fig. 10.3 A magnetic moment will relax to its energetically favorable equilibrium position through interactions with its environment

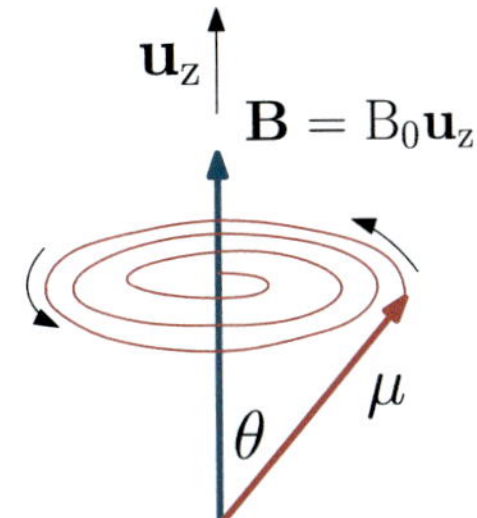

Example 10.1 - Electron Spin Resonance (ESR) and Nuclear Magnetic Resonance (NMR)

Electron Spin Resonance (ESR)

When a strong, static magnetic field is applied to a material containing unpaired electrons, the electron spins precess around the field direction. Electron Spin Resonance (ESR) is a technique that exploits the selective absorption of a weak microwave radiation by these precessing spins. This absorption occurs only when the frequency of the microwave radiation matches the natural Larmor precession frequency of the spins.

By varying the strength of the static magnetic field and monitoring the absorption of microwave radiation, a resonance spectrum can be measured. ESR spectroscopy can be used to identify paramagnetic substances, study the nature of chemical bonds within molecules by identifying unpaired electrons, and investigate molecular structure through their interaction with the immediate surroundings. This technique has applications in various fields, including biology, chemistry, physics, and medicine.

Nuclear Magnetic Resonance (NMR)

Nuclear Magnetic Resonance is a similar technique applied to nuclei with non-zero spin. In this case, the atomic electron cloud tends to screen the effective magnetic field experienced by the nuclei. As a result, the nuclear spin resonance frequency depends on the electron density distribution in the corresponding molecular orbitals.

NMR spectroscopy is widely used to determine the structure of organic molecules in solution, study molecular physics, and analyze crystalline and non-crystalline materials. Additionally, it forms the basis of magnetic resonance imaging (MRI), a powerful medical imaging technique.

10.2.4   The Magnetization Vector

Every atom with a non-zero magnetic moment behaves like a tiny magnet, interacting with external magnetic fields. The macroscopic magnetic properties of a material depend on the collective behavior of these atomic magnetic moments. If these
moments are exhibit a net alignment, as shown in Fig. 10.4, the material possesses a
net magnetic moment. Conversely, if the moments are randomly oriented, the material net magnetic moment is negligible, and it is considered demagnetized, or simply
non-magnetic.

> **Magnetization**
> Consider a volume Ω containing a very large number of atoms. A *mesoscopic*
> volume d^3x within Ω contains a large number N of atoms, each with a magnetic
> moment $\mathbf{m}_i$. The magnetization $\mathbf{M}$ at point $\mathbf{x}$ is defined as the average magnetic
> moment per unit volume,
>
> $$\mathbf{M}(\mathbf{x})d^3x = \sum_{i=1}^{N} \mathbf{m}_i , \qquad (10.6)$$
>
> where the sum is over all atoms within d^3x.

10.2.5   Microscopic Origin of Diamagnetism

Most insulating materials exhibit weak diamagnetic behavior. This arises from the
fact that atoms in insulators have filled electron shells, leading to a cancellation of
individual electron magnetic moments (orbital and spin magnetic moments cancel
each other). A purely quantum mechanical effect,[2] known as diamagnetism, occurs

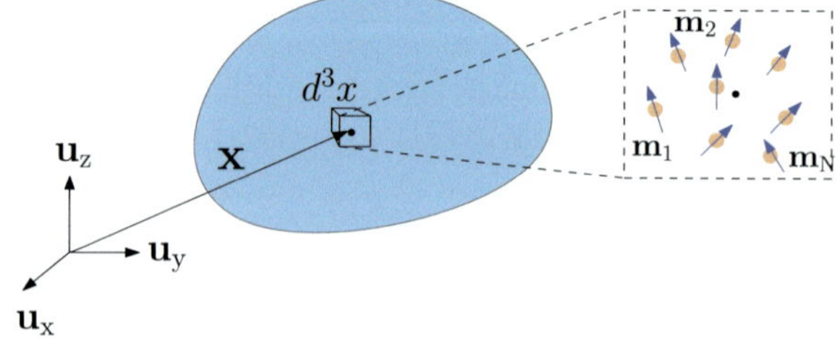

Fig. 10.4 A magnetic material can be described by a non-zero net magnetic moment within each infinitesimal volume

---

[2] Although a semi-classical, qualitative model due to Langevin allows us to understand diamagnetism in terms of Faraday's law of induction, which is the main subject of Chap. 11.

when an external magnetic field modifies the atomic orbitals. Essentially, a diamagnetic material responds to an external magnetic field by developing a magnetization **M** that opposes the applied field. This results in a weak repulsive force, which can lead to phenomena like magnetic levitation. Superconductors, which exhibit perfect diamagnetism (Meissner effect), completely expel magnetic fields from their interior.

10.2.6 Microscopic Origin of Paramagnetism and Ferromagnetism

In metals, free electrons contribute to paramagnetism, known as *Pauli's paramagnetism*. Each electron possesses a spin, and in the presence of an external magnetic field, one spin configuration becomes energetically favorable. This leads to a small net magnetization $\mathbf{M} \neq 0$ aligned with the external field at thermal equilibrium.

In solids with ions containing unpaired electrons, a stronger form of paramagnetism, known as *Curie's paramagnetism*, occurs. The magnetic moments of these ions align with the external field, further enhancing the material magnetic response.

In both cases, illustrated in Fig. 10.5, magnetization is temporary and disappears due to thermal fluctuations when the external field is turned off.

As temperature decreases, Curie's paramagnetism becomes more pronounced. Below a critical temperature T_C, known as the Curie temperature (typically of several hundred kelvins), the interaction between neighboring magnetic moments becomes significant. This interaction leads to spontaneous magnetization even without magnetic field. The alignment of magnetic moments forms magnetic domains known as Weiss domains, generally separated by crystalline defects. The material transitions from a paramagnetic to a ferromagnetic state, as illustrated in Fig. 10.6.

These magnetic domains remain aligned due to their mutual interactions, resulting in a *remanent magnetization*. The magnetic behavior of a material, whether paramagnetic or ferromagnetic, is influenced by energy scales and temperature.

While dipole-dipole interactions might seem to explain ferromagnetic order, they predict critical temperatures much lower (of the order of $T_C = 0.1$ K) than observed

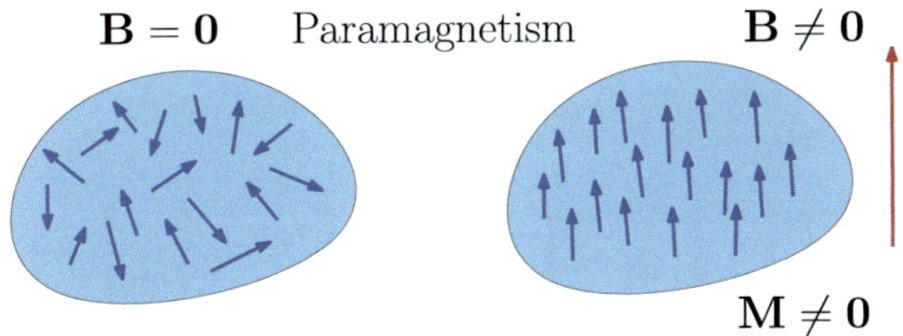

Fig. 10.5 In a paramagnetic material at thermal equilibrium, the magnetic moments of atoms are randomly oriented. However, when an external magnetic field is applied, these moments tend to align with the field

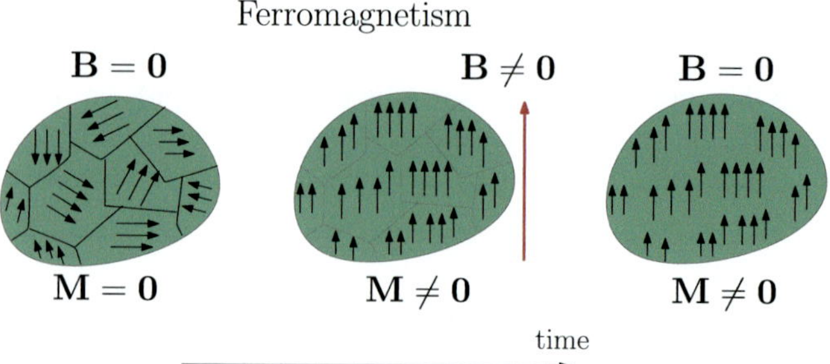

Fig. 10.6 Applying an external magnetic field aligns the magnetic moments. Below a critical temperature, this alignment persists even after the field is turned off. This is known as ferromagnetism

values (for example, in the case of iron, $T_C = 1043$ K). The true mechanism behind ferromagnetism is a quantum mechanical effect known as exchange interaction. This interaction arises from the combination of Pauli's exclusion principle and Coulomb's repulsion between electrons, and it strongly depends on the relative orientation of electron spins.

Other forms of collective magnetism exist: In *antiferromagnetic materials*, neighboring magnetic moments align antiparallel to each other, as shown in Figure Fig. 10.7. This often occurs in crystal lattices with two equivalent sublattices. If both sublattices have equal magnetic moments, the net magnetization is zero. However, if the sublattices have different magnetic moments, a net magnetization can arise under an antiferromagnetic arrangement, leading to ferrimagnetism. Ferrites (Fe_2O_3, Fe_3O_4) are examples of *ferrimagnetic materials*.

Example 10.2 - From electronics to spintronics
The electron spin, a fundamental property responsible for ferromagnetism, is now utilized in hard drives for storing information. Tiny ferromagnetic domains within the hard drive can be magnetized in specific directions to represent binary data (0s and 1s). Reading this data involves a phenomenon known as giant magnetoresistance, where the electrical resistance between ferromagnetic

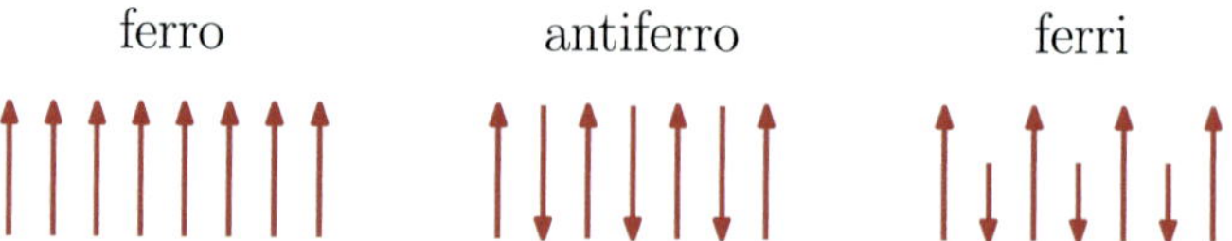

Fig. 10.7 Besides ferromagnetism, other forms of magnetic order, such as antiferromagnetism and ferrimagnetism, can also occur

layers varies depending on the relative alignment of their magnetizations. *Spintronics*, an emerging field of nanotechnology, aims to leverage the electron spin to develop advanced optoelectronic devices with enhanced performance.

10.3 Magnetic Field Generated by a Magnetized Material

Consider a volume Ω with a magnetic dipole moment density $\mathbf{M}$. This means that a small volume element d^3x' around a point $\mathbf{x}'$ has a total dipole moment $d\mathbf{m}(\mathbf{x}') = \mathbf{M}(\mathbf{x}')d^3x'$, as shown in Fig. 10.8.

The infinitesimal vector potential at a point $\mathbf{x}$ due to an infinitesimal magnetic dipole moment is given by

$$dA(\mathbf{x}) = \frac{\mu_0}{4\pi} \frac{d\mathbf{m}(\mathbf{x}') \times (\mathbf{x} - \mathbf{x}')}{|\mathbf{x} - \mathbf{x}'|^3} \; .$$

Substituting $d\mathbf{m}(\mathbf{x}') = \mathbf{M}(\mathbf{x}')d^3x'$, we get

$$dA(\mathbf{x}) = \frac{\mu_0}{4\pi} \frac{\mathbf{M}(\mathbf{x}') \times (\mathbf{x} - \mathbf{x}')}{|\mathbf{x} - \mathbf{x}'|^3} d^3x' \; .$$

To find the total vector potential at $\mathbf{x}$, we integrate over the entire volume Ω:

$$\mathbf{A}(\mathbf{x}) = \frac{\mu_0}{4\pi} \iiint_\Omega \frac{\mathbf{M}(\mathbf{x}') \times (\mathbf{x} - \mathbf{x}')}{|\mathbf{x} - \mathbf{x}'|^3} d^3x' \; .$$

Using the vector identity (A.6), we get

$$\mathbf{M}(\mathbf{x}') \times \frac{(\mathbf{x} - \mathbf{x}')}{|\mathbf{x} - \mathbf{x}'|^3} = \mathbf{M}(\mathbf{x}') \times \nabla' \frac{1}{|\mathbf{x} - \mathbf{x}'|} = \frac{\nabla' \times \mathbf{M}(\mathbf{x}')}{|\mathbf{x} - \mathbf{x}'|} - \nabla' \times \left(\frac{\mathbf{M}(\mathbf{x}')}{|\mathbf{x} - \mathbf{x}'|} \right) \; .$$

We can rewrite the integral as

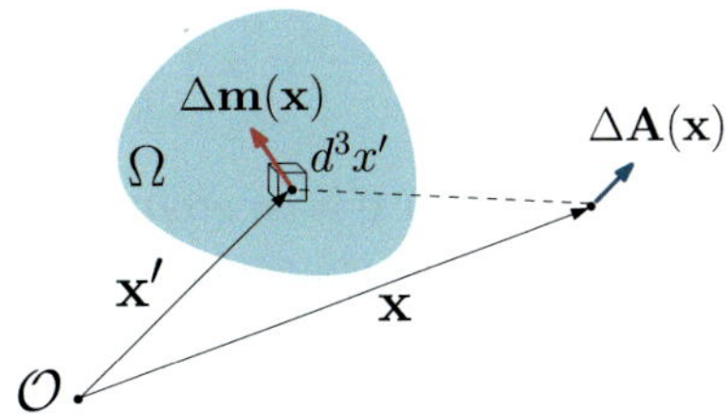

Fig. 10.8 The vector potential at point $\mathbf{x}$ can be calculated by summing the contributions from each infinitesimal dipole within the volume Ω

$$\mathbf{A}(\mathbf{x}) = \frac{\mu_0}{4\pi} \iiint_\Omega \frac{\nabla' \times \mathbf{M}(\mathbf{x}')}{|\mathbf{x} - \mathbf{x}'|} d^3x' - \frac{\mu_0}{4\pi} \iiint_\Omega \nabla' \times \left(\frac{\mathbf{M}(\mathbf{x}')}{|\mathbf{x} - \mathbf{x}'|}\right) d^3x' .$$

Applying the curl formula (A.19) to the second integral (see details below), we obtain

$$\mathbf{A}(\mathbf{x}) = \frac{\mu_0}{4\pi} \iiint_\Omega \frac{\nabla' \times \mathbf{M}(\mathbf{x}')}{|\mathbf{x} - \mathbf{x}'|} d^3x' + \frac{\mu_0}{4\pi} \oiint_{\partial\Omega} \frac{\mathbf{M}(\mathbf{x}') \times \mathbf{n}(\mathbf{x}')}{|\mathbf{x} - \mathbf{x}'|} dS(\mathbf{x}') . \tag{10.7}$$

Equation (10.7) resembles the general solution to Poisson's equation for the vector potential in magnetostatics, which is generated by a volume current density $\mathbf{j}(\mathbf{x}')$ and a surface current density $\mathbf{j}_s(\mathbf{x}')$:

$$\mathbf{A}(\mathbf{x}) = \frac{\mu_0}{4\pi} \iiint_\Omega \frac{\mathbf{j}(\mathbf{x}')}{|\mathbf{x} - \mathbf{x}'|} d^3x' + \frac{\mu_0}{4\pi} \oiint_{\partial\Omega} \frac{\mathbf{j}_s(\mathbf{x}')dS(\mathbf{x}')}{|\mathbf{x} - \mathbf{x}'|} .$$

This similarity allows us to introduce the concept of *magnetization currents*, as discussed below.

To complete the derivation of Eq. (10.7), we will focus on the second term on the right-hand side, $\mathbf{A}_2$. Consider a constant unit vector $\mathbf{u}$. We will show that the transformation into a surface integral can be applied to the projection of this term onto $\mathbf{u}$. Pushing the scalar product with $\mathbf{u}$ under the integral, we have

$$\mathbf{A}_2 \cdot \mathbf{u} = - \iiint_\Omega \frac{\mu_0}{4\pi} \left[\nabla' \times \left(\frac{\mathbf{M}(\mathbf{x}')}{|\mathbf{x} - \mathbf{x}'|}\right)\right] \cdot \mathbf{u}\, d^3x' .$$

Using the vector identity (A.10), we can rewrite the integrand as

$$\left(\nabla \times \frac{\mathbf{M}(\mathbf{x}')}{|\mathbf{x} - \mathbf{x}'|}\right) \cdot \mathbf{u} = \nabla \cdot \left(\frac{\mathbf{M}(\mathbf{x}')}{|\mathbf{x} - \mathbf{x}'|} \times \mathbf{u}\right) .$$

Applying the divergence theorem (A.15) to convert the volume integral into a surface integral, we get

$$\mathbf{A}_2 \cdot \mathbf{u} = - \iiint_\Omega \frac{\mu_0}{4\pi} \nabla' \cdot \left(\frac{\mathbf{M}(\mathbf{x}')}{|\mathbf{x} - \mathbf{x}'|} \times \mathbf{u}\right) d^3x' = - \oiint_{\partial\Omega} \frac{\mu_0}{4\pi} \left(\frac{\mathbf{M}(\mathbf{x}')}{|\mathbf{x} - \mathbf{x}'|} \times \mathbf{u}\right) \cdot \mathbf{n}(\mathbf{x}')dS(\mathbf{x}') .$$

Rearranging the triple product (permutation of $\mathbf{u}$ and $\mathbf{n}$ with a change of sign), we find

$$\mathbf{A}_2 \cdot \mathbf{u} = \left(\oiint_{\partial\Omega} \frac{\mu_0}{4\pi} \frac{\mathbf{M}(\mathbf{x}')}{|\mathbf{x} - \mathbf{x}'|} \times \mathbf{n}(\mathbf{x}')dS(\mathbf{x}')\right) \cdot \mathbf{u} .$$

Considering that $\mathbf{u}$ is an arbitrary unit vector, this result is valid for all components of $\mathbf{A}_2$ and we can conclude that

$$\mathbf{A_2} = - \iiint_\Omega \frac{\mu_0}{4\pi} \left[\nabla' \times \left(\frac{\mathbf{M}(\mathbf{x}')}{|\mathbf{x} - \mathbf{x}'|} \right) \right] = \oiint_{\partial\Omega} \frac{\mu_0}{4\pi} \frac{\mathbf{M}(\mathbf{x}')}{|\mathbf{x} - \mathbf{x}'|} \times \mathbf{n}(\mathbf{x}') dS(\mathbf{x}') .$$

This completes the derivation of Eq. (10.7).

10.3.1 Magnetization Current Densities

By comparing Eq. (10.7) with the general expression for the vector potential of a current distribution (9.4),

$$\mathbf{A}(\mathbf{x}) = \iiint_\Omega \frac{\mathbf{j}(\mathbf{x}')}{|\mathbf{x} - \mathbf{x}'|} d^3 x' ,$$

we can identify equivalent current densities within the volume Ω and on its surface $\partial\Omega$:
Volume current density:

$$\mathbf{j}_V(\mathbf{x}) = \nabla \times \mathbf{M}(\mathbf{x}) .$$

Surface current density:

$$\mathbf{j}_S(\mathbf{x}) = \mathbf{M}(\mathbf{x}) \times \mathbf{n}(\mathbf{x}) .$$

With these equivalent current densities, the vector potential can be expressed as

$$\mathbf{A}(\mathbf{x}) = \frac{\mu_0}{4\pi} \iiint_\Omega \frac{\mathbf{j}_V(\mathbf{x}')}{|\mathbf{x} - \mathbf{x}'|} d^3 x' + \frac{\mu_0}{4\pi} \oiint_{\partial\Omega} \frac{\mathbf{j}_S(\mathbf{x}')}{|\mathbf{x} - \mathbf{x}'|} dS(\mathbf{x}') .$$

Magnetization currents
A magnetized material can be effectively modeled as a system of microscopic current loops. This equivalent current distribution can be described by a volume current density $\mathbf{j}_V$ and a surface current density $\mathbf{j}_S$ expressed in terms of the magnetization $\mathbf{M}$:
Volume current density:

$$\boxed{\mathbf{j}_V(\mathbf{x}) = \nabla \times \mathbf{M}(\mathbf{x}) .} \tag{10.8}$$

Surface current density:

$$\boxed{\mathbf{j}_S(\mathbf{x}) = \mathbf{M}(\mathbf{x}) \times \mathbf{n}(\mathbf{x}) ,} \tag{10.9}$$

where $\mathbf{n}$ is the unit normal to the surface of the magnetized material.

For a uniformly magnetized material, the internal microscopic current loops cancel each other, leaving only a surface current $\mathbf{j}_S = \mathbf{M} \times \mathbf{n}$, as shown in Fig. 10.9. However, for a non-uniformly magnetized material, both volume and surface currents contribute to the overall magnetic field, as shown in Fig. 10.10.

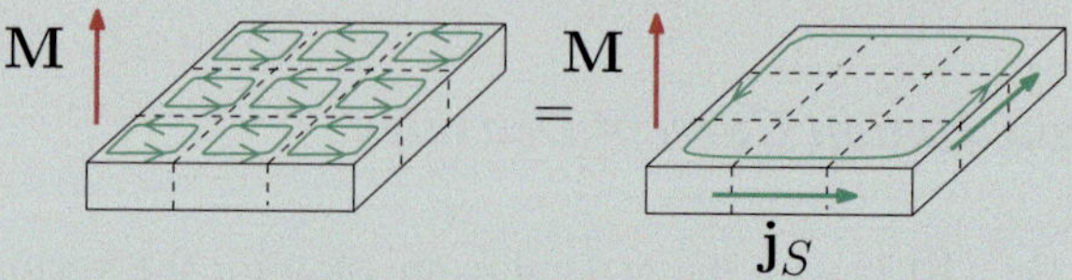

Fig. 10.9 A uniformly magnetized material can be effectively modeled as a surface current flowing on its boundary. The surface current density is $\mathbf{j}_S = \mathbf{M} \times \mathbf{n}$

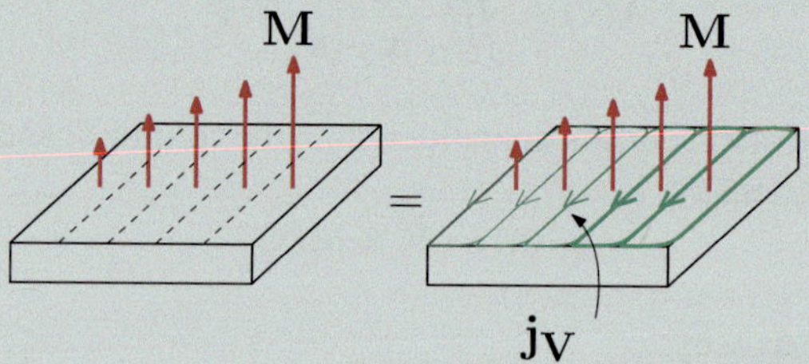

Fig. 10.10 In a non-uniformly magnetized medium, both surface and volume magnetization currents contribute to the overall magnetic field

These equivalent currents, often referred to as Amperian currents, provide a useful way to understand and calculate the magnetic fields generated by magnetized materials.

Note that the volume current density $\mathbf{j}_V$ is divergence-free ($\nabla \cdot \mathbf{j}_V(\mathbf{x}) = 0$), as expected for a steady-state current distribution. Similarly, the surface current density $\mathbf{j}_S$ is tangential to the surface ($\mathbf{n}(\mathbf{x}) \cdot \mathbf{j}_S(\mathbf{x}) = 0$).

In essence, the effect of an external magnetic field on a material can be understood as inducing equivalent current densities within the material.

It is important to note that, unlike polarization charges in dielectrics, which arise from the physical separation of positive and negative charges, magnetization currents are not associated with the actual flow of charges. For instance, the spin magnetic moment originates from a quantum mechanical property of the electron and does not involve classical charge motion.

10.4   Magnetostatics in Magnetic Media

In a magnetized medium, the total current density contributing to the magnetic field consists of two parts:

- *Free current density* ($\mathbf{j}_{\text{free}}(\mathbf{x})$): This is the current due to the motion of free charges, such as electrons in a conductor.
- *Magnetization current density* ($\mathbf{j}_M(\mathbf{x})$): This arises from the magnetization of the material and is given by $\mathbf{j}_M(\mathbf{x}) = \nabla \times \mathbf{M}(\mathbf{x})$.

Ampère's law for the magnetic field $\mathbf{B}$ in a magnetized medium can be written as

$$\nabla \times \mathbf{B}(\mathbf{x}) = \mu_0 \left(\mathbf{j}_{\text{free}}(\mathbf{x}) + \nabla \times \mathbf{M}(\mathbf{x}) \right) ,$$

where all quantities are spatially averaged fields. This means that the magnetization $\mathbf{M}$ characterizing the magnetic medium varies smoothly with $\mathbf{x}$. Equivalently, we can rewrite Ampère's law as

$$\nabla \times \left(\frac{\mathbf{B}(\mathbf{x})}{\mu_0} - \mathbf{M}(\mathbf{x}) \right) = \mathbf{j}_{\text{free}}(\mathbf{x}) .$$

This form highlights the role of the free current density as a source of magnetic field.

Magnetic excitation and Ampère's law
We define the *magnetic excitation field* $\mathbf{H}$ as

$$\mathbf{H}(\mathbf{x}) = \frac{\mathbf{B}(\mathbf{x})}{\mu_0} - \mathbf{M}(\mathbf{x}) . \tag{10.10}$$

This field satisfies Ampère's law in matter,

$$\nabla \times \mathbf{H}(\mathbf{x}) = \mathbf{j}_{\text{free}}(\mathbf{x}) . \tag{10.11}$$

While only the free current density appears explicitly as a source term, the magnetic excitation field $\mathbf{H}$ incorporates the effects of both free currents $\mathbf{j}_{\text{free}}$ and magnetization $\mathbf{M}$. It has the same SI unit as magnetization: A m^{-1}.

The fundamental equations of magnetostatics in a magnetized medium are

$$\nabla \cdot \mathbf{B}(\mathbf{x}) = 0 , \tag{10.12}$$

$$\nabla \times \mathbf{H}(\mathbf{x}) = \mathbf{j}_{\text{free}}(\mathbf{x}) . \tag{10.13}$$

Taking the divergence of (10.10) and using (10.12), the fundamental equations of magnetostatics in a magnetic medium can be written for the magnetic excitation field[3] $\mathbf{H}$ only as

$$\nabla \cdot \mathbf{H}(\mathbf{x}) = \rho_M(\mathbf{x}) = -\nabla \cdot \mathbf{M}(\mathbf{x}) \,,$$
$$\nabla \times \mathbf{H}(\mathbf{x}) = \mathbf{j}_{\text{free}}(\mathbf{x}) \,.$$

Here $\rho_M(\mathbf{x}) = -\nabla \cdot \mathbf{M}(\mathbf{x})$ represents a fictitious magnetic charge density. According to the Helmholtz decomposition theorem, the field $\mathbf{H}$ is uniquely specified by its divergence and curl. Therefore, both the free current density and the fictitious magnetic charge (or magnetization current) contribute to $\mathbf{H}$.

Equations (10.12) and (10.13) are analogous to the fundamental equations of electrostatics in a dielectric medium:

$$\nabla \cdot \mathbf{D}(\mathbf{x}) = \varrho_{\text{free}}(\mathbf{x}) \,,$$
$$\nabla \times \mathbf{E}(\mathbf{x}) = \mathbf{0} \,.$$

Taking the curl of the relation $\mathbf{D} = \epsilon_0 \mathbf{E} + \mathbf{P}$, this system becomes a set of equations for $\mathbf{D}$ only:

$$\nabla \cdot \mathbf{D}(\mathbf{x}) = \varrho_{\text{free}}(\mathbf{x}) \,,$$
$$\nabla \times \mathbf{D}(\mathbf{x}) = \mathbf{j}_P(\mathbf{x}) = \nabla \times \mathbf{P}(\mathbf{x}) \,.$$

The Helmholtz decomposition theorem shows that both the free charge density and the fictitious polarization current, defined by $\mathbf{j}_P(\mathbf{x}) = \nabla \times \mathbf{P}(\mathbf{x})$, contribute to $\mathbf{D}$.

In a linear, homogeneous, and isotropic dielectric, the electric displacement $\mathbf{D}$ is related to the electric field $\mathbf{E}$ by the constitutive relation $\mathbf{D} = \epsilon \mathbf{E}$. This relation allows us to express Maxwell's fundamental equations of electrostatics solely in terms of the electric field $\mathbf{E}$:

$$\nabla \cdot \mathbf{E}(\mathbf{x}) = \frac{\rho_{\text{free}}(\mathbf{x})}{\epsilon} \,,$$
$$\nabla \times \mathbf{E}(\mathbf{x}) = 0 \,.$$

The permittivity ϵ of the dielectric medium is larger than ϵ_0. It modifies the electric field generated by free charges, effectively reducing its magnitude.

Similarly, for magnetic materials, we need a constitutive relation between the magnetic field $\mathbf{B}$ and the magnetic excitation $\mathbf{H}$ to complete Maxwell's equations for magnetostatics in magnetic media. This relation will typically be more complex

---

[3] The simplified form of Ampère's law (10.11) in magnetic media explicitly shows that the magnetic excitation field $\mathbf{H}$ is directly generated by free currents applied and controlled by an external agent. According to the relation $\mathbf{B} = \mu_0 \mathbf{H} + \mathbf{M}$, the magnetization $\mathbf{M}$ of the material contributes to the total magnetic field $\mathbf{B}$ observed in the magnetic medium in response to the excitation field. For this reason, the term *magnetic field* was historically reserved for $\mathbf{H}$, while $\mathbf{B}$ was called the *magnetic induction*. A consensus in modern textbooks is to call $\mathbf{B}$ the magnetic field while the *auxiliary field* $\mathbf{H}$ is often called by its symbol.

than the linear relation between $\mathbf{D}$ and $\mathbf{E}$ in simple dielectrics, and it may exhibit nonlinear and hysteretic behavior.

10.4.1 Constitutive Relations of a Magnetic Material

In linear, homogeneous, and isotropic magnetic materials, the magnetic field generated by a free current can either increase (*paramagnetic* materials) or decrease (*diamagnetic* materials) compared to the magnitude it would have in vacuum. This behavior can be described by introducing an effective permeability μ, which is different from the vacuum permeability μ_0. However, for *ferromagnetic* materials, the relationship between $\mathbf{B}$ and $\mathbf{H}$ is more complex and depends on the material magnetization history (hysteresis). In this case, a simple linear relationship between $\mathbf{B}$ and $\mathbf{H}$ is not sufficient to describe the material magnetic behavior.

Magnetic susceptibility and permeability
In a *homogeneous* and *isotropic* medium, experience shows that the magnetization $\mathbf{M}$ depends on the magnetic excitation $\mathbf{H}$,

$$\mathbf{M}(\mathbf{x}) = \chi_m(\mathbf{H})\mathbf{H}(\mathbf{x}) .$$

where the dimensionless quantity $\chi_m(\mathbf{H})$ is the *magnetic susceptibility*. This leads to the following relation between $\mathbf{B}$ and $\mathbf{H}$:

$$\boxed{\mathbf{B} = \mu_0(1 + \chi_m(\mathbf{H}))\mathbf{H} = \mu(\mathbf{H})\mathbf{H} .} \tag{10.14}$$

Here $\mu(\mathbf{H})$ is the *magnetic permeability* of the material. For *linear* materials, the permeability μ is independent of $\mathbf{H}$, resulting in a linear relationship between $\mathbf{B}$ and $\mathbf{H}$.

For a linear, homogeneous, and isotropic magnetic medium, the fundamental equations of magnetostatics are

$$\nabla \cdot \mathbf{B}(\mathbf{x}) = 0 ,$$
$$\nabla \times \mathbf{B}(\mathbf{x}) = \mu \mathbf{j}_{\text{free}}(\mathbf{x}) .$$

These equations have the same form as those for vacuum, but with the permeability μ replacing the vacuum permeability μ_0.

Magnetic Susceptibility and Permeability
The magnetic susceptibility χ_m characterizes the linear relationship between the magnetization $\mathbf{M}$ and the magnetic field $\mathbf{H}$,

$$\mathbf{M} = \chi_m \mathbf{H} \,.$$

The permeability μ is related to the susceptibility by

$$\mu = \mu_0(1 + \chi_m) \,.$$

- *Diamagnetic Materials*: Diamagnetic materials have negative susceptibilities ($\chi_m < 0$), meaning they tend to weaken the applied magnetic field ($\mu < \mu_0$). This effect is typically very weak ($\chi_m \sim -10^{-5}$, compared to the susceptibility of superconductors $\chi_m \sim 1$) and independent of temperature, since diamagnetism is a property of the atomic orbitals occupied by electrons, specifically those in the lowest energy configuration.
- *Paramagnetic Materials*: Paramagnetic materials have positive susceptibilities ($\chi_m > 0$), meaning they enhance the applied magnetic field ($\mu > \mu_0$). The susceptibility can vary with temperature ($10^{-5} < \chi_m < 10^{-3}$).
- *Ferromagnetic Materials*: Ferromagnetic materials exhibit strong, nonlinear magnetic behavior. Their susceptibility depends on the applied field and the material history, leading to phenomena like hysteresis. Its value $\chi_m(\mathbf{H})$ can vary between 50 and 10^6. The response of a ferromagnet is therefore nonlinear, achieving very high values for the permeability ($\mu(\mathbf{H}) \gg \mu_0$).

Magnetic shielding and field concentration

A material with high permeability can be used to shield a region from external magnetic fields. For example, a spherical shell of high-permeability material can significantly reduce the magnetic field within the cavity. This is a magnetic shield, analogous to a Faraday cage for the electric field. Conversely, high-permeability materials can be used to concentrate magnetic field lines within a specific region, as illustrated in Fig. 10.11.

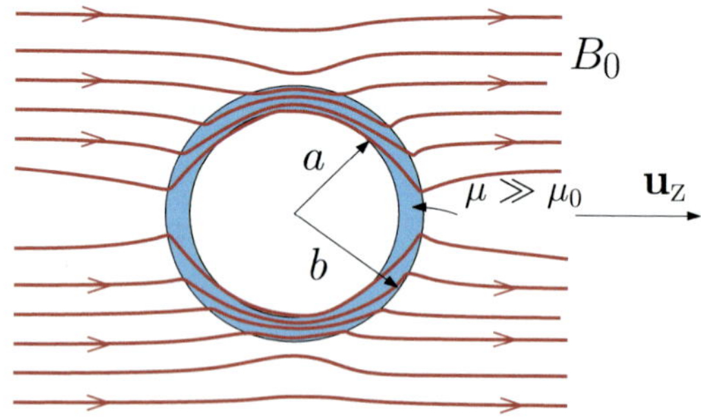

Fig. 10.11 Magnetic field lines around a high-permeability spherical shell in a uniform magnetic field

Example 10.3 - The magnetic excitation vector as a control parameter
The significance of the magnetic excitation vector $\mathbf{H}$ lies in its direct controllability through external currents. Consider a coil with N turns carrying a current I_{ext}, as shown in Fig. 10.12.

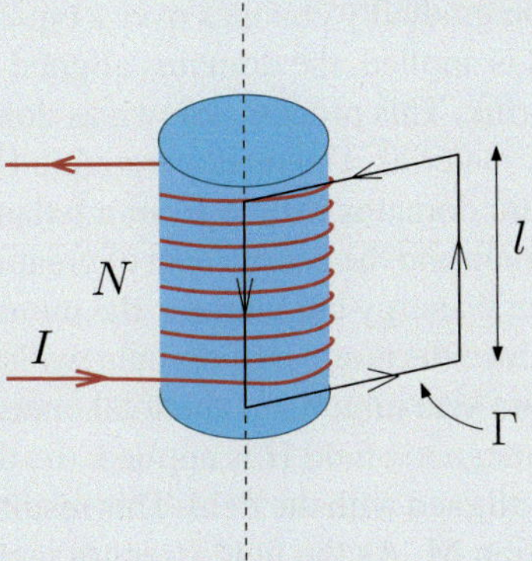

Fig. 10.12 A solenoid carrying a current I surrounding a magnetic material

Assuming an infinitely long coil, the symmetry of the current distribution $\mathbf{j}_{\text{ext}}$ implies a uniform $\mathbf{H}$ field inside the solenoid, parallel to the coil axis. Applying Ampère's law (10.11) to a square loop Γ of side length l within the solenoid, as illustrated in Fig. 10.12, we obtain

$$\oint_{\Gamma} \mathbf{H} \cdot d\mathbf{l} = Hl = \frac{Nl}{L} I_{\text{ext}},$$

where N is the total number of turns and L is the length of the solenoid. Introducing the turn density $n = N/L$, we get

$$H = n I_{\text{ext}}.$$

This equation shows that the magnetic excitation $\mathbf{H}$ inside the solenoid, is directly proportional to the external current I_{ext} and the turn density n. Therefore, the external current directly controls the magnetic excitation field $\mathbf{H}$, not $\mathbf{B}$.

10.4.2   *Hysteresis, Soft and Hard Ferromagnets*

Ferromagnetic materials exhibit a strong, *nonlinear* relationship between magnetization $\mathbf{M}$ and applied magnetic field $\mathbf{H}$. The magnitude of the magnetic field inside a ferromagnetic material can be considerable, up to 10^6 times larger than that of the applied external field. The relationship between **M** and **H** is characterized by *hysteresis*, meaning that the magnetization depends not only on the current field but also on the material previous magnetic history.

Initially, a ferromagnetic material is composed of small magnetic domains (Weiss domains with sizes between 10 and 100 μm), each with a uniform magnetization. These domains are separated by thin domain walls, known as Bloch walls, where the magnetization direction gradually changes over a typical width of 0.1 μm. When an external magnetic field is applied, the domains aligned with the field grow, while those aligned against it shrink. This process, known as domain wall motion, leads to a net magnetization of the material. This is illustrated in Fig. 10.13.

The size of the magnetic domains results from a balance between the magnetic energy stored within a domain and the energy cost of creating a domain wall. Smaller domains reduce the magnetic energy but increase the number of domain walls. Conversely, larger domains reduce the number of domain walls but increase the magnetic energy. The optimal domain size minimizes the total energy of the system.

When a weak external magnetic field **H** is applied, the domain walls move, favoring the growth of domains aligned with the field. This results in a linear and reversible increase in the magnetization **M**. As the field strength increases, domain walls may undergo sudden, irreversible displacements or disappear altogether. This process, known as domain wall motion, leads to a rapid increase in magnetization. Eventually, the magnetization reaches saturation M_s when all domains are aligned with the **H** field.

When the external field is reduced, the magnetization does not retrace its original path. Instead, it follows a different curve, resulting in a residual magnetization M_r even when the external field is zero. This is due to the pinning of domain walls to defects within the material. A ferromagnetic material with a residual magnetization becomes a permanent magnet.

To fully demagnetize the ferromagnetic material, a reverse magnetic field must be applied. When the field reaches a specific value, known as the coercive field H_c, the magnetization becomes zero. A hysteresis loop is obtained by subjecting the material to a cyclic variation of the magnetic field: $0 \rightarrow H \rightarrow -H \rightarrow H$. This loop illustrates the nonlinear and hysteretic behavior of ferromagnetic materials, which shows the relationship between magnetization and applied field.

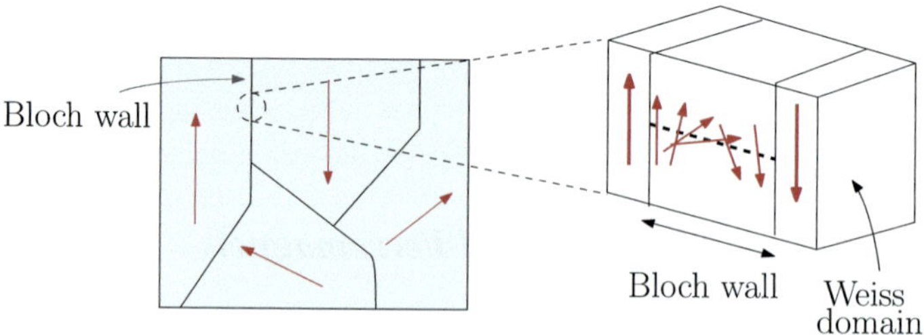

Fig. 10.13 Weiss magnetic domains and Bloch domain walls in ferromagnetic materials

Fig. 10.14 Hysteresis cycle

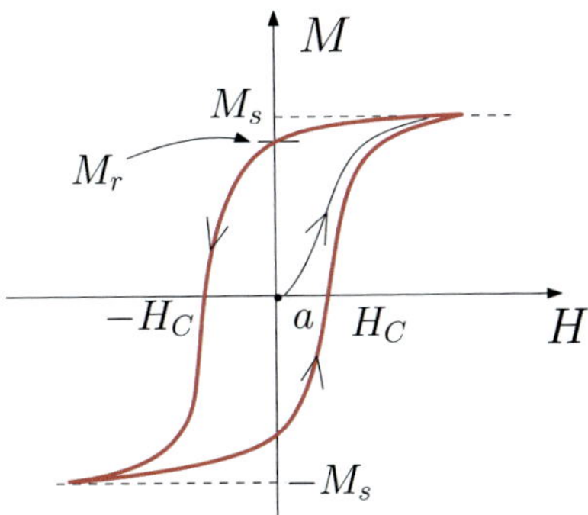

Figure 10.14 shows the hysteresis loop, which illustrates the variation of magnetization M with respect to the applied magnetic field **H**. Starting from an initially demagnetized state (point a), the magnetization increases as the field is applied. The key parameters characterizing the hysteresis loop are the remanence magnetization M_r and the coercive field H_c. These parameters are crucial for selecting ferromagnetic materials for specific applications.

There are two main categories of ferromagnetic materials:

- **Soft ferromagnetics**: These materials have a low coercive field and a high magnetic permeability. They are easy to magnetize and demagnetize, making them suitable for applications like electromagnets and transformers. Mu-metal, a Ni-Fe alloy with a permeability of around $10^5 \, \mu_0$, is a common example. It is used for shielding electronic equipment from static or slowly varying magnetic fields. In the absence of an external magnetic field, soft ferromagnets do not exhibit a significant magnetic field.
- **Hard ferromagnetics**: Also known as permanent magnets, these materials have a high coercive field and retain their magnetization even after the removal of the external field. They are used in applications like electric motors and generators. Rare-earth magnets, such as NdFeB, are among the strongest commercially available permanent magnets, with coercive fields on the order of $1 \times 10^6 \, \mathrm{A\,m^{-1}}$. They can generate significant magnetic fields at their surface, typically around 1 Tesla.

10.5 Boundary Conditions for the Magnetic Field

The fundamental laws of magnetostatics,

$$\nabla \cdot \mathbf{B}(\mathbf{x}) = 0 \,,$$

$$\nabla \times \mathbf{H}(\mathbf{x}) = \mathbf{j}(\mathbf{x}) \,,$$

provide information about the behavior of magnetic fields at the interface between two magnetic media with different permeabilities.

10.5.1   The Normal Component of **B** is Continuous Across an Interface Between Two Magnetic Media

Consider the situation shown in Fig. 10.15, where $\mathbf{x}$ is a point on the interface between two magnetic media (1) and (2). Let $\mathbf{B}_1(\mathbf{x})$ and $\mathbf{B}_2(\mathbf{x})$ be the magnetic fields at points infinitesimally close to $\mathbf{x}$ on either side of the interface. $\mathbf{n}$ is the unit normal vector to the interface, pointing from medium (2) to medium (1).

Consider a cylindrical Gaussian surface with its axis parallel to the normal vector $\mathbf{n}$. Applying the divergence theorem to this surface, we have

$$\nabla \cdot \mathbf{B} = 0 \rightarrow \oiint_S \mathbf{B}(\mathbf{x}) \cdot \mathbf{n}(\mathbf{x}) dS(\mathbf{x}) = 0 \,.$$

As the height h of the cylinder approaches zero, the flux through the curved surface vanishes. The remaining flux through the two end caps, each with area A, gives

$$A\,(\mathbf{B}_1 - \mathbf{B}_2) \cdot \mathbf{n} = 0 \,.$$

Here, we assume that the magnetic field is approximately constant over the small area of each cap. Therefore, at any point $\mathbf{x}$ on the interface,

$$\boxed{(\mathbf{B}_1 - \mathbf{B}_2) \cdot \mathbf{n}|_{\mathbf{x}} = 0 \,.}$$
(10.15)

This implies that the *normal* component of the magnetic field B is continuous across the interface.

10.5.2   The Tangential Component of **H** is Discontinuous Across a Surface Carrying a Surface Current Density

Now consider a closed rectangular loop Γ of length l and width h, oriented such that its normal vector $\mathbf{n}'$ is perpendicular to the interface and parallel to the horizontal

Fig. 10.15 A cylindrical Gaussian surface used to demonstrate the continuity of the normal component of the magnetic field across an interface

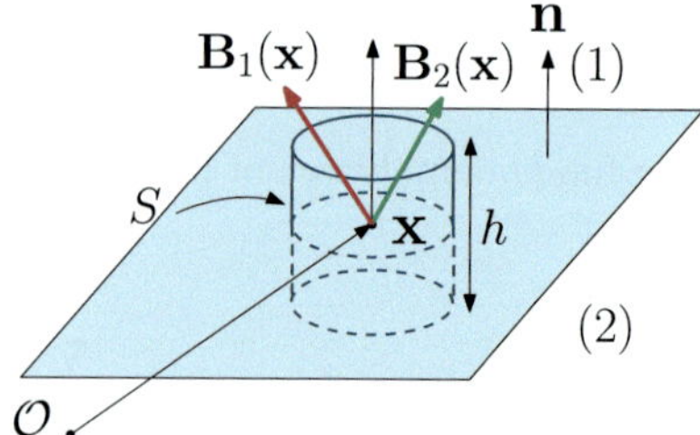

segment of the loop within region (1). This is illustrated in Fig. 10.16. Applying Ampère's law to this loop, we obtain

$$\oint_{\Gamma} \mathbf{H} \cdot d\mathbf{l} = \iint_{S(\Gamma)} \mathbf{j} \cdot \mathbf{n}' dS .$$

As h approaches zero, only the contributions from the horizontal segments of the loop remain significant. Assuming the length l is small enough for the magnetic field to remain approximately constant along these segments, we get

$$l\left(\mathbf{H}_1(\mathbf{x}) - \mathbf{H}_2(\mathbf{x})\right) \cdot \mathbf{t} = \lim_{h \to 0} \iint_{S(\Gamma)} \mathbf{j} \cdot \mathbf{n}' dS .$$

The right-hand side of the equation represents the total current flowing through the loop. In the limit as $h \to 0$, this current can be attributed to a surface current density $\mathbf{j}_S$ at the interface:

$$\lim_{h \to 0} \iint_{S(\Gamma)} \mathbf{j} \cdot \mathbf{n}' dS = l\, \mathbf{j}_S(\mathbf{x}) \cdot \mathbf{n}' .$$

Combining these results, we have

$$(\mathbf{H}_1(\mathbf{x}) - \mathbf{H}_2(\mathbf{x})) \cdot \mathbf{t} = \mathbf{j}_S(\mathbf{x}) \cdot \mathbf{n}' .$$

Since $\mathbf{t} = \mathbf{n}' \times \mathbf{n}$, where $\mathbf{n}$ is the unit vector normal to the interface pointing from medium (2) to (1), we can rewrite this equation as[4]

$$\{\mathbf{n} \times (\mathbf{H}_1(\mathbf{x}) - \mathbf{H}_2(\mathbf{x}))\} \cdot \mathbf{n}' = \mathbf{J}_j(\mathbf{x}) \cdot \mathbf{n}' .$$

Since this equation holds for any unit vector $\mathbf{n}'$ tangent to the interface, by rotation of the contour Γ around the vertical axis $\mathbf{n}$, we conclude that

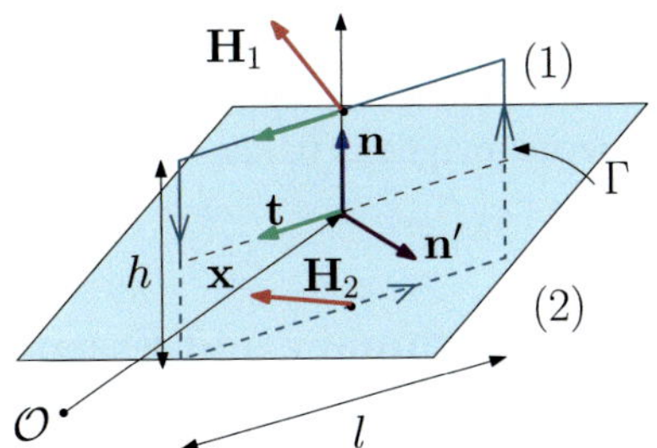

Fig. 10.16 A rectangular Amperian loop spanning the interface between two media

---

[4] Use the circular permutation in the scalar triple product $(\mathbf{H}_1 - \mathbf{H}_2) \cdot (\mathbf{n}' \times \mathbf{n}) = (\mathbf{n} \times (\mathbf{H}_1 - \mathbf{H}_2)) \cdot \mathbf{n}'$.

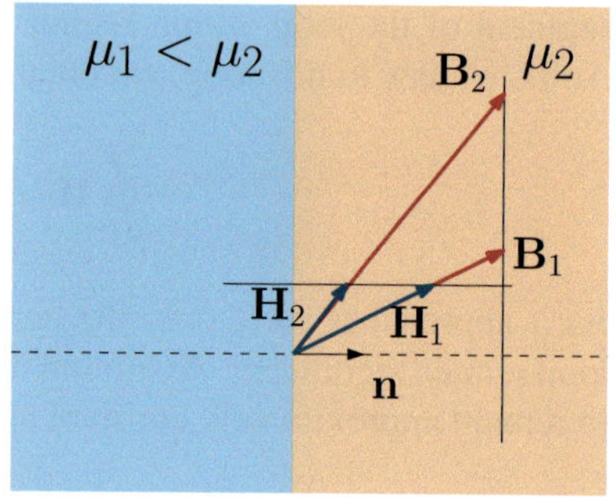

Fig. 10.17 Behavior of magnetic fields at an interface in the absence of surface currents

$$\boxed{\mathbf{n} \times (\mathbf{H}_1 - \mathbf{H}_2) = \mathbf{j}_S \,.} \tag{10.16}$$

This equation shows that the tangential component of the magnetic excitation field **H** is discontinuous across an interface carrying a surface current density $\mathbf{j}_S$. In the absence of a surface current, the tangential component of **H** is continuous. This case is illustrated in Fig. 10.17 for the case $\mu_2 > \mu_1$. As shown in the figure, the normal component of **B** is continuous across the interface, while the tangential component of **H** is also continuous. However, the tangential component of **B** is reduced in the region with higher permeability.

A consequence of the boundary conditions is that a high-permeability material can be used to confine magnetic fields. Consider the interface between air $(\mu_{\text{air}} = \mu_0)$ and a ferromagnetic material with a much higher permeability μ_{Fe}. In the absence of surface currents, the tangential component of **H** is continuous across the interface:

$$\mathbf{H}_\parallel^{\text{Fe}} = \mathbf{H}_\parallel^{\text{air}} \,.$$

Since $\mathbf{B} = \mu \mathbf{H}$, this implies:

$$|\mathbf{B}_\parallel^{\text{Fe}}| = \frac{\mu_{\text{Fe}}}{\mu_{\text{air}}} |\mathbf{B}_\parallel^{\text{air}}| \gg |\mathbf{B}_\parallel^{\text{air}}| \,.$$

Therefore, the tangential component of the magnetic field is significantly reduced as it enters the ferromagnetic material. This effect can be used to shield regions from external magnetic fields, similar to how a Faraday cage shields from electric fields.

10.6   Summary and Essential Formulas

- At the atomic scale, the orbital motion of electrons generates an orbital magnetic moment $\mathbf{m}_L = -(\mathbf{L}/\hbar)\mu_B$, where $L = n\hbar$ is the electron orbital angular momentum, n is an integer, $\mu_B \approx 9 \times 10^{-24}\,\text{Am}^2$ is Bohr's magneton and $\hbar$ is the reduced

Planck constant. Additionally, electrons possess an intrinsic spin angular momentum $\mathbf{S}$ leading to a spin magnetic moment $\mathbf{m}_S \approx -2\,(\mathbf{S}/\hbar)\,\mu_B$ such that $m_S = \pm\mu_B$. The total magnetic moment of an atom arises from the combined effects of these orbital and spin magnetic moments.

- In an external magnetic field $\mathbf{B}$, a magnetic moment undergoes Larmor's precession about the field direction. Simultaneously, the magnetic moment tends to align with the field through a relaxation process with a characteristic time T_1, corresponding to the time required for the moment to reach thermal equilibrium in the presence of $\mathbf{B}$. This alignment process leads to a net magnetization of the material.
- Magnetic materials can be classified into three main categories based on their response to an external magnetic field:

  - *Diamagnetics*: These materials have fully filled electron shells. The atoms do not possess a net magnetic moment, and their magnetic response arises from a modification of the atomic orbitals in the presence of a magnetic field. When exposed to an external magnetic field, diamagnetics develop a magnetization that opposes the field, leading to a slight repulsion.
  - *Paramagnetics*: In paramagnetic materials, the magnetic moments of atoms or free electrons with unsaturated electronic shells align with an external magnetic field to minimize their energy. This alignment results in a net magnetization that disappears when the field is removed due to thermal fluctuations. Paramagnetic materials are attracted to an external magnetic field.
  - *Ferromagnetics*: Ferromagnetic materials exhibit spontaneous magnetization, meaning they retain a magnetic moment even after the external field is removed. This behavior is due to a strong, purely quantum mechanical interaction between neighboring electron spins, known as exchange interaction: because of the Pauli principle, the Coulomb repulsion between two electrons depends on the relative orientation of their spins.

- A magnetic material is characterized by the magnetization vector $\mathbf{M}$, which represents the average magnetic dipole moment per unit volume.
- A magnetic medium Ω with magnetization $\mathbf{M}$ generates a magnetic field equivalent to that of a volume current density $\mathbf{j}_V$ within the volume Ω and a surface current density $\mathbf{j}_S$ on its boundary $\partial\Omega$:

$$\mathbf{j}_V(\mathbf{x}) = \nabla \times \mathbf{M}(\mathbf{x}) \quad \text{for } \mathbf{x} \in \Omega \,, \tag{10.17}$$

$$\mathbf{j}_S(\mathbf{x}) = \mathbf{M}(\mathbf{x}) \times \mathbf{n}(\mathbf{x}) \quad \text{for } \mathbf{x} \in \partial\Omega \,, \tag{10.18}$$

where $\mathbf{n}$ is the outward normal to the surface $\partial\Omega$.

- Ampère's law in a material medium is

$$\boxed{\oint_\Gamma \mathbf{H} \cdot d\mathbf{l} = I_{\text{free}} \quad \Leftrightarrow \quad \nabla \times \mathbf{H} = \mathbf{j}_{\text{free}} \,,}$$

where $\mathbf{H}$ is the magnetic excitation field, defined as

$$\mathbf{H} = \frac{\mathbf{B}}{\mu_0} - \mathbf{M} \, .$$

- In homogeneous and isotropic paramagnetic or diamagnetic media, the magnetization $\mathbf{M}$ is linearly proportional to the magnetic excitation $\mathbf{H}$:

$$\mathbf{M} = \chi_m \mathbf{H} \, ,$$

where χ_m is the magnetic susceptibility. For diamagnetic materials, χ_m is negative and typically on the order of -10^{-5}, while for paramagnetic materials, χ_m is positive and on the order of 10^{-3}. In ferromagnetic materials, the relationship between $\mathbf{M}$ and $\mathbf{H}$ is multivalued, nonlinear and depends on the material history, leading to the phenomenon of hysteresis.
- In a linear, homogeneous, and isotropic medium, the magnetic field $\mathbf{B}$ and the magnetic excitation $\mathbf{H}$ are related by

$$\mathbf{H} = \mu \mathbf{B} \, ,$$

where $\mu = \mu_0(1 + \chi_m)$ is the permeability of the medium. This relationship is analogous to the relation between the electric displacement $\mathbf{D}$ and the electric field $\mathbf{E}$ in a dielectric medium.

The fundamental equations of magnetostatics in a linear, homogeneous, and isotropic medium are

$$\nabla \times \mathbf{B} = \mu \mathbf{j}_{\text{free}} \, ,$$
$$\nabla \cdot \mathbf{B} = \mathbf{0} \, ,$$

where $\mathbf{j}_{\text{free}}$ represents the current density of free carriers (conduction current). These equations have the same form as those in vacuum, but with the permeability μ replacing the vacuum permeability μ_0.
- Boundary conditions at the interface between two magnetic media:

  - Continuity of the normal component of $\mathbf{B}$:

$$(\mathbf{B}_1 - \mathbf{B}_2) \cdot \mathbf{n} = 0 \, .$$

  The normal component of the magnetic field $\mathbf{B}$ is continuous across the interface. Here $\mathbf{n}$ is the unit normal to the interface.
  - Discontinuity of the tangential component of $\mathbf{H}$:

$$\mathbf{n} \times (\mathbf{H}_1 - \mathbf{H}_2) = \mathbf{j}_S \, .$$

  The tangential component of the magnetic excitation $\mathbf{H}$ is discontinuous across the interface if there is a surface current density $\mathbf{j}_S$ present. Here $\mathbf{n}$ is the unit normal to the interface pointing from medium (2) to medium (1).

Problems

10.1 Magnetic moment of a rotating charged sphere

Consider a uniformly charged sphere of radius a, total charge $-e$, and mass m_e rotating about the z-axis with angular velocity ω, as shown below. The goal of this exercise is to demonstrate that a classical model of the electron as a spinning sphere is not only inconsistent with experimental observations but also requires rotational speeds exceeding the speed of light.

(a) Express the charge density ϱ and mass density ρ_m in terms of e, m_e and a.

(b) Calculate the total angular momentum $\mathbf{L}$ and magnetic moment $\mathbf{m}$ of the rotating sphere. Use the following integral $\int_0^\pi \sin^3 \theta d\theta = 4/3$. Recall that the magnetic moment associated with a current density $\mathbf{j}$ in a volume Ω is given by

$$\mathbf{m} = \frac{1}{2} \iiint_\Omega \mathbf{x} \times \mathbf{j}(\mathbf{x}) d^3x \ .$$

(c) Write the magnetic moment as $\mathbf{m} = g\left(-\dfrac{e}{2m_e}\right)\mathbf{L}$ and determine the g-factor for this classical model. Compare this value to the experimentally determined g-factor for the electron spin, which is approximately 2.

(d) Assuming the electron is a uniformly charged sphere with radius $a = r_e = 1.69 \times 10^{-15}$ m, as determined in Example 3.5, calculate the rotational velocity at the equator required to produce the observed spin magnetic moment $|\mathbf{m}_S| \approx \mu_B$ where $\mu_B = 9 \times 10^{-24}$ A m^2 is Bohr's magneton. Compare this velocity to the speed of light $c = 3 \times 10^8$ m s^{-1}.

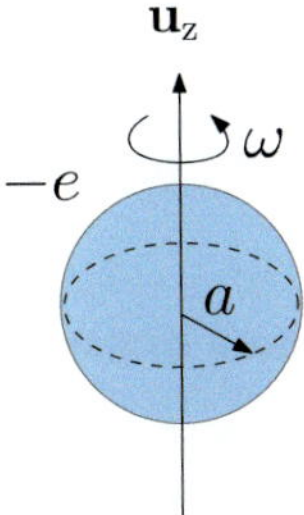

10.2 Magnetic spin resonance

Consider a spin magnetic moment $\mathbf{m}$ initially aligned along the z-axis due to a constant magnetic field $\mathbf{B}_0 = B_0 \mathbf{u}_z$. At time $t = 0$, a rotating magnetic field $\mathbf{B}_1$ is applied perpendicular to $\mathbf{B}_0$:

$$\mathbf{B}_1(t) = B_1(\cos \omega t \, \mathbf{u}_x + \sin \omega t \, \mathbf{u}_y) \ .$$

(a) Write the equation of motion for $\mathbf{m}$, considering the spin has reached thermal equilibrium in the field $\mathbf{B}_0$, corresponding to the initial condition $\mathbf{m}(0) = \mu_B \mathbf{u}_z$ when the rotating field $\mathbf{B}_1$ is applied. The spin relaxation time T_1.

(b) To simplify the analysis, we transform to a reference frame rotating with angular velocity ω around the z-axis. The unit vectors of the rotating frame are given in terms of those in the laboratory frame by

$$\mathbf{u}'_x(t) = \cos \omega t\, \mathbf{u}_x + \sin \omega t\, \mathbf{u}_y\,,$$
$$\mathbf{u}'_y(t) = -\sin \omega t\, \mathbf{u}_x + \cos \omega t\, \mathbf{u}_y\,,$$
$$\mathbf{u}'_z(t) = \mathbf{u}_z\,.$$

In this frame, the magnetic field $\mathbf{B}_1$ appears static. Show that the time derivative of the magnetic moment $\mathbf{m}'$ in the rotating frame is related to the time derivative of the magnetic moment $\mathbf{m}$ in the laboratory frame by

$$\frac{d\mathbf{m}}{dt} = \frac{d\mathbf{m}'}{dt} + \omega \mathbf{u}_z \times \mathbf{m}'\,.$$

(c) Using the previous result, derive the equation of motion for $\mathbf{m}'$ in the rotating frame in terms of $\omega_0 = -\gamma B_0$ and $\omega_1 = -\gamma B_1$, where γ is the gyromagnetic ratio. Show that in this frame, the spin precesses about a *static* effective magnetic field

$$\mathbf{B}_{\mathrm{eff}} = \frac{\omega - \omega_0}{\gamma}\mathbf{u}_z + B_1 \mathbf{u}'_x\,.$$

(d) Discuss the motion of the spin in the two limiting cases: $\omega_1 \ll |\omega_0 - \omega|$ and $\omega_1 \gg |\omega_0 - \omega|$, which case corresponds to resonance between the spin and the rotating magnetic field?

(e) Starting from thermal equilibrium in the presence of $\mathbf{B}_0$ ($\mathbf{m}(0) = \mu_B \mathbf{u}_z$), a resonant rotating pulse ($\mathbf{B}_1$ with $\omega \approx \omega_0$) of duration $T_\omega \ll T_1$ is applied. Determine the final state of the spin at $t = T_\omega$. What is the state of the spin for a $\pi/2$ pulse, corresponding to $T_\omega = \dfrac{\pi}{2\gamma B_1}$?

(f) Describe the subsequent evolution of the spin after a $\pi/2$ pulse.

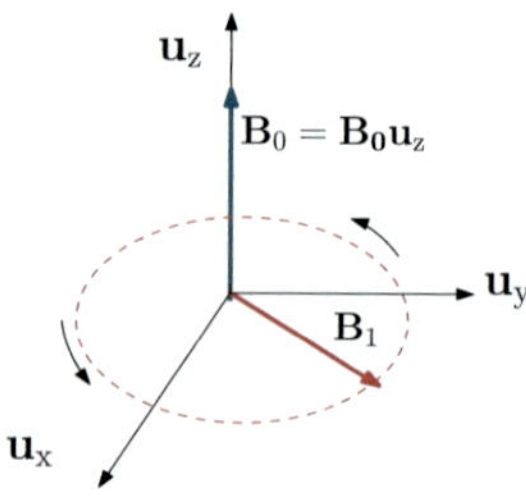

10.3 Magnetic field of a magnetized sphere

Consider a ferromagnetic sphere of radius a with a uniform magnetization $\mathbf{M} = M_0\mathbf{u}_z$. Calculate the magnetic field $\mathbf{B}$ and the magnetic excitation $\mathbf{H}$ everywhere in space.

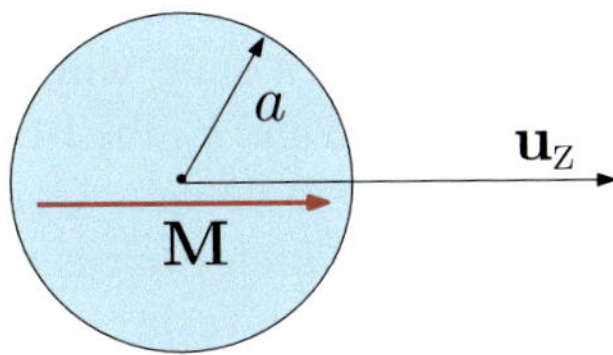

10.4 Magnetic shielding

Consider a spherical shell with inner radius a and outer radius b, made of a material with magnetic permeability μ. This shell is placed in a uniform magnetic field $\mathbf{B}(\mathbf{x}) = B_0\mathbf{u}_z$. Determine the scalar magnetic potential ϕ such that $\mathbf{B} = -\nabla\phi$ throughout space. What happens in the limit as μ approaches infinity?

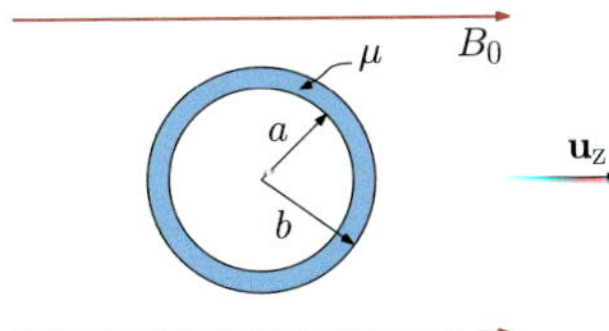

10.5 Field in the presence of a superconducting sphere

Consider a superconducting sphere of radius a placed in a uniform magnetic field $\mathbf{B}_0$. Due to the Meissner effect, the magnetic field is completely expelled from the interior of the sphere.

(a) Demonstrate that the magnetic field at the surface of the superconducting sphere is tangential to the sphere.
(b) Calculate the scalar magnetic potential ϕ such that $\mathbf{B} = -\nabla\phi$ for $r > a$.
(c) Show that the magnetic field outside the sphere can be expressed as

$$\mathbf{B} = \mathbf{B}_0 + \frac{\mu_0}{4\pi}\left\{\frac{3\mathbf{m}\cdot\mathbf{r}}{r^5}\mathbf{r} - \frac{\mathbf{m}}{r^3}\right\},$$

where the magnetic dipole moment $\mathbf{m}$ is given by

$$\mathbf{m} = \frac{-2\pi a^3}{\mu_0}\mathbf{B}_0.$$

10.6 Curie's paramagnetism of localized spins

Consider an ensemble of non-interacting localized spins. In the absence of an external magnetic field, the spins are oriented either parallel ($\mathbf{m} = +\mu_B\mathbf{u}_z$) or antiparallel ($\mathbf{m} = -\mu_B\mathbf{u}_z$) to the z-axis with equal probability, leading to zero net magnetization ($\mathbf{M} = 0$).

(a) Determine the energies of the two possible spin configurations ($\mathbf{m} = \pm\mu_B\mathbf{u}_z$) in the presence of an external magnetic field $\mathbf{B} = B_0\mathbf{u}_z$. Which configuration is energetically favorable?

(b) The spins tend to align with the magnetic field to minimize their energy, but this tendency competes with thermal fluctuations. Assuming thermal equilibrium, the probability of a spin occupying a state with energy E is proportional to e^{-E/k_BT}, where T is the temperature and k_B the Boltzmann constant.. Calculate the average magnetic moment of a single spin. Consider the two limiting cases $\mu_B B_0 \gg k_B T$ and $\mu_B B_0 \ll k_B T$. Which limit is relevant at room temperature for a magnetic field of approximately 1 Tesla?

(c) For a system with n spins per unit volume and $\mu_B B_0 \ll k_B T$, derive Curie's law for the magnetic susceptibility,

$$\chi_m \propto \frac{1}{T} \; .$$

Determine the proportionality constant. Estimate the order of magnitude of the magnetic susceptibility for $n = 1 \times 10^{21} \mathrm{cm}^{-3}$ at room temperature.

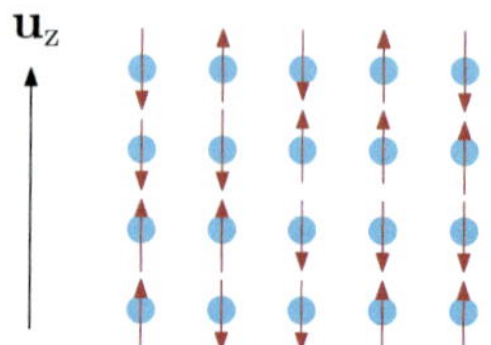

10.7   Transition between paramagnetism and ferromagnetism

Curie's law, $\chi_m \propto 1/T$, is valid at high temperatures where thermal fluctuations dominate. At lower temperatures, interactions between magnetic moments can lead to collective behavior, such as ferromagnetism.

The Weiss molecular field theory introduces a local magnetic field $\mathbf{B}_{\mathrm{loc}}$ felt by each spin:

$$\mathbf{B}_{\mathrm{loc}} = \mathbf{B} + \lambda\mathbf{M} \; ,$$

where λ is a constant.

(a) Using the results from Exercise 10.6, calculate the magnetization of an ensemble of spins under the influence of the local field $\mathbf{B}_{\mathrm{loc}}$ in an external field $\mathbf{B} = B_0\mathbf{u}_z$.

(b) For high temperatures, derive the Curie–Weiss law for the magnetic susceptibility:

$$\chi_m = \frac{C}{T - T_C} \; .$$

Determine the values of the constants C and T_C.

(c) Under what conditions can the magnetization **M** be non-zero in the absence of an external field **B**?

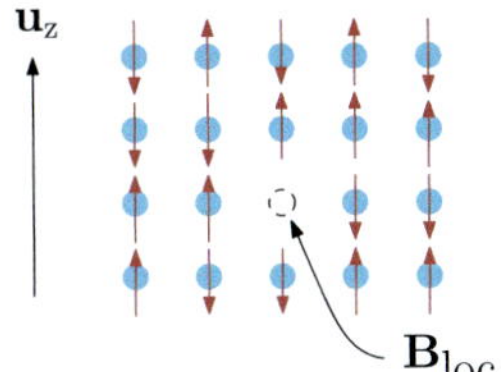

Chapter 11
Electromagnetic Induction, Faraday's Law and Magnetic Energy

Abstract This chapter introduces the pivotal concept of *electromagnetic induction*, building upon the established relationship between electricity and magnetism. It begins by detailing experimental observations of *induced currents* in the presence of changing magnetic fields or relative motion between circuits and magnetic fields. The core of the chapter is *Faraday's law of induction*, which quantitatively describes the *electromotive force (emf)* induced in a circuit as the negative time rate of change of the *magnetic flux* through the circuit. The chapter meticulously derives Faraday's law for both moving circuits in static magnetic fields (motional emf) and stationary circuits in time-varying magnetic fields (transformer emf), unifying these phenomena into a general formulation. The *differential form of Faraday's law* ($\nabla \times \mathbf{E} = -\partial \mathbf{B}/\partial t$) is presented as a fundamental Maxwell's equation, highlighting the non-conservative nature of the electric field in time-varying scenarios. Applications like *Foucault currents (eddy currents)* and the *Kelvin effect (skin depth)* are discussed, along with the unique inductive properties of *superconductors* (Meissner effect). The chapter then introduces *inductance*, a crucial property of circuits that quantifies their ability to oppose changes in current. Both *self-inductance* (due to the varying current within the circuit itself) and *mutual inductance* (magnetic coupling between circuits) are defined, with Neumann's formula for mutual inductance. The *energy stored in an inductor* is derived, leading to the concept of *magnetic energy density* in space. Finally, the chapter presents *magnetic potential energy* for current-carrying loops in external magnetic fields, providing a framework for calculating *forces and torques* on such circuits. It introduces the concepts of *Helmholtz free energy* and *Gibbs free energy* for systems of conductors and magnetic media, demonstrating how these thermodynamic potentials can be used to analyze equilibrium states and derive forces and torques while maintaining constant fluxes or currents, respectively.

Keywords Electromagnetic induction · Faraday's law · Inductance

F. Cadiz and A. Couairon, *Classical Electrodynamics*, Undergraduate Texts in Physics,
https://doi.org/10.1007/978-3-031-86785-9_11

"

11.1 Introduction

The interrelationship between electricity and magnetism was first established by physicists Ørsted, Ampère, Biot and Savart, who demonstrated that electric currents produce magnetic fields. A decade later, Michael Faraday discovered the inverse process: the generation of electricity with magnetism, a phenomenon known as electromagnetic induction. This principle forms the foundation of modern power generation.

Faraday's law of induction quantifies the relationship between the rate of change of magnetic flux and the induced electromotive force (emf) and is central to the operation of numerous devices. The first power plants, dating back to 1882, utilized steam turbines to rotate coils within magnetic fields, converting the mechanical energy of steam into electricity. This fundamental principle remains at the core of modern power generation, where energy sources such as water, wind, or nuclear power drive the turbines.

Beyond power generation, Faraday's law also underpins the operation of electric motors. By applying a current to a coil within a magnetic field, a torque is produced, causing mechanical rotation. This electromechanical energy conversion is pervasive in modern technology, powering devices ranging from small household appliances to large industrial machinery.

In this chapter, we delve into Faraday's law of induction and its applications in various electrical and electromechanical systems. We generalize Ohm's law and introduce the concept of inductance. We derive the magnetic potential energy for an arbitrary current distribution, enabling the definition of the free energy density stored in a magnetic field. Furthermore, we define the mutual inductance between filamentary circuits and derive expressions for the forces and torques exerted on a filamentary circuit within a system of circuits.

11.2 Induced Currents: Experimental Observations

Michael Faraday discovered the phenomenon of induction. He conducted experiments using two coils wound around a common iron core. When the current in one coil was suddenly changed, he observed an induced transient current in the second coil, even though the coils were electrically isolated. This phenomenon is illustrated in Fig. 11.1.

Fig. 11.1 A sudden change in current in one coil induces a transient current in the other

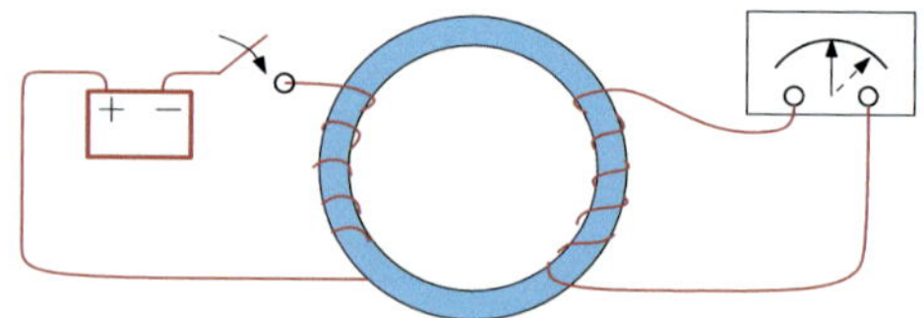

In a second series of experiments, Faraday observed that a current was induced in a conductor loop whenever there was relative motion between the loop and a magnetic field. This could be achieved by either moving a magnet near a stationary loop or moving the loop near a stationary magnet. These experiments exemplified the phenomenon of *electromagnetic induction* (Fig. 11.2).

Finally, electromagnetic induction can also occur due to changes in the geometry of a circuit within a magnetic field. Consider a closed circuit, part of which is a movable conductor, placed in a uniform, static magnetic field **B**, as shown in Fig. 11.3. As the conductor is moved, an induced current I is generated in the circuit. The direction of the induced current depends on the direction of motion of the conductor relative to the magnetic field. Importantly, no current is induced in the absence of a magnetic field.

All these phenomena are manifestations of the same fundamental principle: Faraday's law of electromagnetic induction.

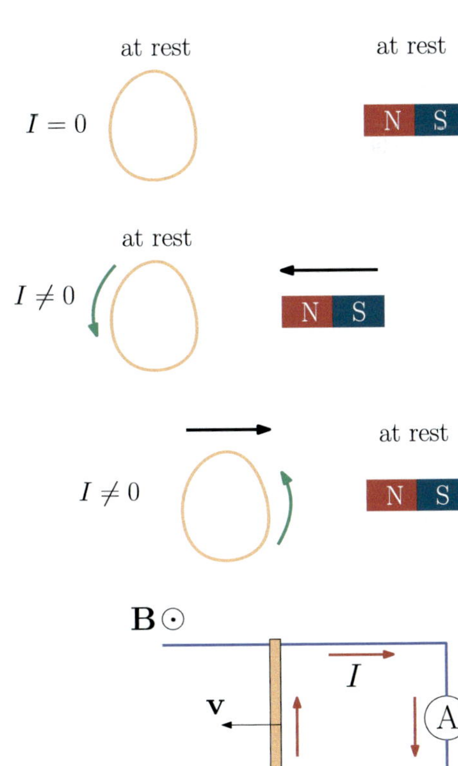

Fig. 11.2 The relative motion between the loop and the magnet induces a transient current in the loop

Fig. 11.3 Moving the conductive bar in a magnetic field induces a current in the circuit

11.3   Electromotive Force—Faraday's Law

In this section, we will establish Faraday's law of induction, a fundamental principle that unifies the various phenomena discovered by Faraday. In the case of a moving conductor in a static magnetic field, the induced current can be explained by the action of the Lorentz force on the charge carriers in the conductor. However, when the conductor is not moving and the magnetic field changes with time, the Lorentz force alone cannot account for the induced current. In this scenario, we must introduce the concept of an induced electric field that accompanies a time-varying magnetic field. This induced electric field is responsible for driving the current in the circuit.

11.3.1   Magnetic Flux Through a Moving Circuit

Faraday's law is concerned with the change in magnetic flux through a circuit. To explore this, consider a closed loop Γ centered around a point $\mathbf{x}(t)$ moving with velocity $\mathbf{v}$ in a reference frame $\mathcal{R}$. This loop is immersed in a time-varying magnetic field $\mathbf{B} = \mathbf{B}(\mathbf{x}, t)$, as depicted in Fig. 11.4.

Consider a surface $S(\Gamma)$ bounded by the curve Γ. The magnetic flux through this surface, $\Phi_{S(\Gamma),\mathbf{B}}$, will generally depend on both the position of the loop, $\mathbf{x}$, and time, t, if the magnetic field $\mathbf{B}(\mathbf{x}, t)$ is inhomogeneous or time-varying.

Using the chain rule, the total time derivative of the flux is given by

$$\frac{d\Phi_{S(\Gamma),\mathbf{B}}}{dt} = \frac{\partial}{\partial t}\Phi_{S(\Gamma),\mathbf{B}} + \underbrace{\frac{d\mathbf{x}}{dt}}_{\mathbf{v}} \cdot \nabla\Phi_{S(\Gamma),\mathbf{B}} \ .$$

The derivative of the loop position with respect to time is simply its velocity, $\mathbf{v}$. The term $(\mathbf{v} \cdot \nabla)\Phi_{S(\Gamma),\mathbf{B}}$ is known as the *motional* term and represents the change in magnetic flux due to the motion of the loop. Therefore, the total time derivative of the magnetic flux can be expressed as

$$\boxed{\frac{d\Phi_{S(\Gamma),\mathbf{B}}}{dt} = \left(\frac{\partial}{\partial t} + \mathbf{v} \cdot \nabla\right)\Phi_{S(\Gamma),\mathbf{B}} \ .}$$

(11.1)

Fig. 11.4 A moving circuit in a magnetic field

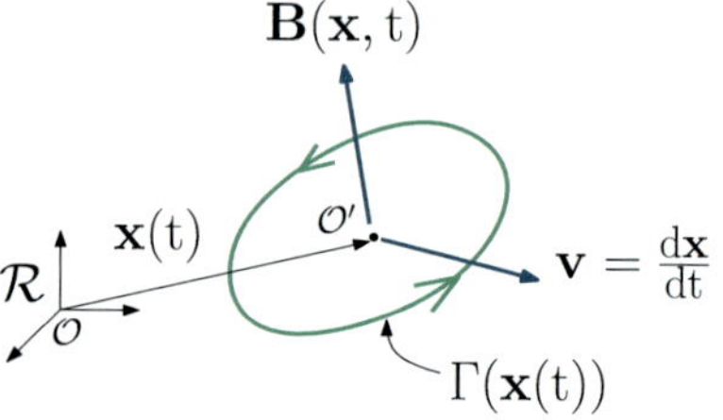

11.3.2 *Moving Circuit in a Static Magnetic Field* B

Consider a closed, rigid circuit Γ centered at point O' moving with a velocity $\mathbf{v} = \dfrac{d\mathbf{x}}{dt}$ through a static magnetic field $\mathbf{B}$, as shown in Fig. 11.5. In this scenario, the Lorentz force acting on the moving charges in the circuit induces an electromotive force (emf), resulting in an induced current.

At a given instant, a charge element δq located at $\mathbf{x}'$ relative to O' moves with a velocity $\mathbf{w} = \mathbf{v} + \mathbf{u}$, where $\mathbf{u}$ is the velocity of the charge relative to the circuit and is parallel to the infinitesimal line element $d\mathbf{l}(\mathbf{x}')$.

The magnetic force acting on the charge element is given by the Lorentz force law,

$$d\mathbf{F}(\mathbf{x}') = \delta q\,(\mathbf{w} \times \mathbf{B}) \ .$$

However, this force is perpendicular to the velocity of the charge, and thus, it does no work. Mathematically, this can be expressed as

$$0 = \delta q\,(\mathbf{w} \times \mathbf{B}) \cdot \mathbf{w}\,dt \ .$$

Expanding the cross product $\mathbf{w} \times \mathbf{B} = (\mathbf{v} + \mathbf{u}) \times \mathbf{B}$ and expanding the cross product with $\mathbf{w} = \mathbf{v} + \mathbf{u}$, we find that the resulting expression is written as the sum of only two non-zero terms:

$$0 = \delta q\,(\mathbf{v} \times \mathbf{B}) \cdot \mathbf{u}\,dt + \delta q\,(\mathbf{u} \times \mathbf{B}) \cdot \mathbf{v}\,dt.$$

The second term involves the Hall electric field, $\mathbf{E}_H = -\mathbf{u} \times \mathbf{B}$, which is responsible for the Laplace force. Given that $d\mathbf{l}(\mathbf{x}') = \mathbf{u}dt$, we can integrate over the contour Γ at a given instant to obtain

$$0 = \underbrace{\delta q \oint_{\Gamma(\mathbf{x})} (\mathbf{v} \times \mathbf{B}) \cdot d\mathbf{l}(\mathbf{x}')}_{\varepsilon} + \frac{\delta q}{dt} \oint_{\Gamma(\mathbf{x})} (d\mathbf{l}(\mathbf{x}') \times \mathbf{B}) \cdot \mathbf{v}dt \ . \qquad (11.2)$$

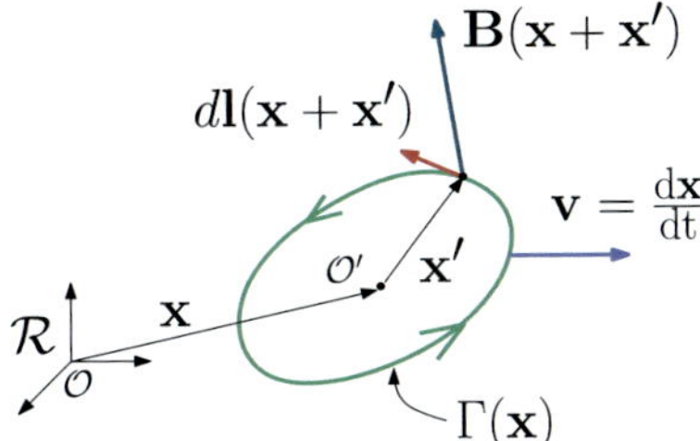

Fig. 11.5 Moving circuit in a static magnetic field

The first term can be identified as the product of the charge δq and the electromotive force (emf) ε induced in the loop:

$$\varepsilon = \oint_{\Gamma(\mathbf{x})} \mathbf{E}_{\text{emf}} \cdot d\mathbf{l}(\mathbf{x}') \ ,$$

where $\mathbf{E}_{\text{emf}} = \mathbf{v} \times \mathbf{B}$ is the electromotive field. It is important to note that $\mathbf{v}$ here refers to the velocity of the conductor as a whole, not the velocity of individual charge carriers within the conductor.

For an infinitesimal displacement $\delta\mathbf{x} = \mathbf{v}dt$ of the loop, the second term in Eq. (11.2) represents the work $\mathbf{F}_L \cdot \mathbf{v}dt$ done by the net Laplace force $\mathbf{F}_L$ acting on the loop:

$$\mathbf{F}_L = i \oint_{\Gamma(\mathbf{x})} (d\mathbf{l}(\mathbf{x}') \times \mathbf{B}) \ ,$$

where $i = \delta q/dt$ is the current flowing in the loop. Equation (11.2) can be interpreted as follows: to move the loop at velocity $\mathbf{v}$, an external agent must exert a force equal and opposite to the net Laplace force. This external agent does work (total work $\delta q\,\varepsilon$) on the charges in the loop, providing them with energy per unit charge equal to ε, the induced electromotive force (emf).

It is important to note that while ε is called an electromotive force, it is not a force in the traditional sense. Rather, it is the energy per unit charge supplied to the circuit, i.e., it has dimensions of a voltage.

The electromotive force induced in the circuit is generated by the electromotive field $\mathbf{E}_{\text{emf}} = \mathbf{v} \times \mathbf{B}$. This field, arising from the Lorentz magnetic force, drives the motion of charges within the conductor, resulting in the induced current. It is important to note that while $\mathbf{E}_{\text{emf}}$ acts like an electric field, it is fundamentally a manifestation of the magnetic force. By applying the Stokes's theorem, which relates the line integral of a vector field (circulation) to the surface integral (flux) of its curl, the electromotive force can be expressed as

$$\varepsilon = \iint_{S(\Gamma(\mathbf{x}))} \nabla' \times (\mathbf{v} \times \mathbf{B}(\mathbf{x} + \mathbf{x}')) \cdot \mathbf{n}(\mathbf{x}')dS(\mathbf{x}') \ .$$

Using the vector identity (A.9) to expand $\nabla' \times (\mathbf{v} \times \mathbf{B}(\mathbf{x} + \mathbf{x}'))$, and noting that the divergence of a magnetic field is always zero, the velocity $\mathbf{v}$ is a constant with respect to the spatial derivative ∇', we have

$$\nabla' \times (\mathbf{v} \times \mathbf{B}) = \mathbf{v}\underbrace{(\nabla' \cdot \mathbf{B})}_{0} - \mathbf{B} \cdot \underbrace{(\nabla' \cdot \mathbf{v})}_{0} + \underbrace{(\mathbf{B} \cdot \nabla')\mathbf{v}}_{0} - (\mathbf{v} \cdot \nabla')\mathbf{B} \ .$$

We can simplify the expression of the electromotive force to

$$\varepsilon = -\iint_{S(\Gamma(\mathbf{x}))} (\mathbf{v} \cdot \nabla')\mathbf{B}(\mathbf{x} + \mathbf{x}') \cdot \mathbf{n}(\mathbf{x}')dS(\mathbf{x}') \ .$$

Since $(\mathbf{v} \cdot \nabla')\mathbf{B}(\mathbf{x} + \mathbf{x}') = (\mathbf{v} \cdot \nabla)\mathbf{B}(\mathbf{x} + \mathbf{x}')$, we can rewrite the electromotive force as

$$\varepsilon = -(\mathbf{v} \cdot \nabla) \underbrace{\iint_{S(\Gamma(\mathbf{x}))} \mathbf{B}(\mathbf{x} + \mathbf{x}') \cdot \mathbf{n}(\mathbf{x}') dS(\mathbf{x}')}_{\Phi_{S(\Gamma(\mathbf{x})),\mathbf{B}}} .$$

Recognizing the integral as the magnetic flux $\Phi_{S(\Gamma(\mathbf{x})),\mathbf{B}}$, we have

$$\varepsilon = \oint_{\Gamma(\mathbf{x})} \mathbf{E}_{\mathrm{emf}} \cdot d\mathbf{l} = -(\mathbf{v} \cdot \nabla)\Phi_{S(\Gamma(\mathbf{x})),\mathbf{B}} . \tag{11.3}$$

We recognize in the right-hand side of Eq. (11.3) the motional term of the time derivative of the magnetic flux (11.1). Since the magnetic field is static, it corresponds as well to the total time derivative of the magnetic flux. Finally

$$\boxed{\varepsilon = \oint_{\Gamma} \mathbf{E}_{\mathrm{emf}} \cdot d\mathbf{l} = -\frac{d}{dt}\Phi_{S(\Gamma),\mathbf{B}} .} \tag{11.4}$$

This equation, known as Faraday's law of induction, states that the electromotive force induced in a circuit is equal to the negative time rate of change of the magnetic flux through the circuit.

The S.I. unit for the magnetic flux is the weber (see Fig. 11.6), with ($1\,\mathrm{Wb} = 1\,\mathrm{Tm}^2 = 1\,\mathrm{kg\,m^2\,s^{-2}\,A^{-1}}$), the induced electromotive force ε has units of $\mathrm{Wb\,s^{-1}}$, which is equivalent to volts (V). The negative sign in Faraday's law, Eq. (11.4), is a manifestation of Lenz's law, formulated by Heinrich Friedrich Emil Lenz in 1834, which states that the direction of the induced current is such that it opposes the change in magnetic flux. Note that this derivation of Faraday's law has been obtained using only the laws of magnetostatics. This underscores the power of these fundamental principles in explaining complex electromagnetic phenomena.

Remarks

- In a reference frame $\mathcal{R}'$ moving with velocity $\mathbf{v}$ relative to the laboratory frame $\mathcal{R}$, the circuit appears stationary. If the relative velocity $v = |\mathbf{v}|$ is much smaller than the speed of light, the Lorentz force acting on the charges in the circuit should be approximately the same in both reference frames. However, since the circuit is at rest in $\mathcal{R}'$, there must be an electric field $\mathbf{E}'$ present to account for the Lorentz force. This electric field is given by

$$\mathbf{E}' = \mathbf{v} \times \mathbf{B} .$$

  More generally, the electric and magnetic fields in the two reference frames are related by the following Galilean transformations (valid for velocities much smaller than the speed of light):

Fig. 11.6 Wilhelm Eduard
Weber (1804–1891), a
German physicist known for
his contributions to
electromagnetism and the
development of precise
measurement techniques. He
built the first electromagnetic
telegraph

$$\mathbf{E}' \approx \mathbf{E} + \mathbf{v} \times \mathbf{B} \, ,$$

$$\mathbf{B}' \approx \mathbf{B} \, ,$$

which are generalized to (20.10) in Chap. 20. These transformations ensure that
the Lorentz force remains invariant in both reference frames, highlighting the
relativistic nature of electromagnetism.

Therefore, the electromotive field $\mathbf{E}_{\text{emf}} = \mathbf{v} \times \mathbf{B}$ can be interpreted as either a
magnetic field (in the laboratory frame $\mathcal{R}$) or an electric field (in the moving frame
$\mathcal{R}'$). This underscores the interconnectedness of electric and magnetic fields and
their relative nature.

11.3.3 Circuit at Rest in a Time-Varying Magnetic Field

In this case, the laws of magnetostatics are insufficient to explain the observed phenomena. However, we can apply the principle of relativity to understand the situation
from a different reference frame. If we consider a moving reference frame in which
an observer sees the magnetic field as static, the problem reduces for the observer
to the case of a moving circuit in a static magnetic field, which we have already
analyzed.

In both cases, an electromotive force is induced in the circuit, given by Faraday's
law (see Fig. 11.7) ,

$$\varepsilon = -\frac{d\Phi_{S(\Gamma),\mathbf{B}}}{dt} \, ,$$

Here, the time-varying magnetic flux $\Phi_{S(\Gamma),\mathbf{B}}$ arises from the time-varying magnetic
field, rather than from the motion of the circuit. This induced emf can be attributed to

Fig. 11.7 Michael Faraday (1791–1867), an English scientist that made profound contributions to electromagnetism and electrochemistry, shaping modern physics. He introduced the concept of electric and magnetic field lines

an electric field $\mathbf{E}_{emf}$ that accompanies the time-varying magnetic field. This electric field drives the current in the circuit,

$$\varepsilon = \oint_{\Gamma} \mathbf{E}_{emf} \cdot d\mathbf{l} = \oint_{\Gamma} \mathbf{E} \cdot d\mathbf{l} = -\frac{d}{dt} \iint_{S(\Gamma)} \mathbf{B} \cdot d\mathbf{S} \ .$$

In essence, a time-varying magnetic field is always associated to an electric field, which in turn drives the current in the circuit.

Using the fact that the magnetic field $\mathbf{B}$ can be derived from a vector potential $\mathbf{A}$ such that $\mathbf{B} = \nabla \times \mathbf{A}$, we can write the electromotive force as

$$\oint_{\Gamma} \mathbf{E}_{emf} \cdot d\mathbf{l} = -\frac{d}{dt} \iint_{S(\Gamma)} (\nabla \times \mathbf{A}) \cdot d\mathbf{S} \ .$$

Applying the Stokes's theorem, we transform the surface integral into a line integral (circulation):

$$\oint_{\Gamma} \mathbf{E}_{emf} \cdot d\mathbf{l} = -\oint_{\Gamma} \frac{\partial \mathbf{A}}{\partial t} \cdot d\mathbf{l} \ .$$

Therefore, the electromotive field $\mathbf{E}_{emf}$ can be expressed in terms of the time derivative of the vector potential,

$$\mathbf{E}_{emf} = -\frac{\partial \mathbf{A}}{\partial t} \ .$$

Faraday's law of induction (1831)
In the most general case of a circuit Γ moving with velocity $\mathbf{v}$ in a time-varying magnetic field $\mathbf{B}$, an electromotive force (emf) is induced in the circuit. This emf is associated with an electromotive field $\mathbf{E}_{emf}$ given by

$$\boxed{\mathbf{E}_{\mathrm{emf}} = \mathbf{v} \times \mathbf{B} - \frac{\partial \mathbf{A}}{\partial t} \, .} \qquad (11.5)$$

Equation (11.5) combines the contributions from both the motion of the circuit in the magnetic field and the time-varying nature of the magnetic field itself.

The electromotive force around a closed loop Γ is given by the circulation of the electromotive field $\mathbf{E}_{\mathrm{emf}}$ around the loop. Faraday's law states that this emf is equal to the negative time rate of change of the magnetic flux through the surface $S(\Gamma)$ bounded by the loop:

$$\varepsilon = \oint_{\Gamma} \mathbf{E}_{\mathrm{emf}} \cdot d\mathbf{l} = -\left(\mathbf{v} \cdot \nabla + \frac{\partial}{\partial t}\right) \iint_{S(\Gamma)} \mathbf{B} \cdot d\mathbf{S} = -\frac{d}{dt} \Phi_{S(\Gamma),\mathbf{B}} \, ,$$

where $\Phi_{S(\Gamma),\mathbf{B}}$ is the magnetic flux through the surface $S(\Gamma)$. This equation,

$$\boxed{\varepsilon = -\frac{d}{dt} \iint_{S(\Gamma)} \mathbf{B} \cdot d\mathbf{S} = -\frac{d}{dt} \Phi_{S(\Gamma),\mathbf{B}} \, ,} \qquad (11.6)$$

known as Faraday's law of induction, is a fundamental law of electromagnetism.

11.4 Faraday's Law, Differential Form

As we have seen in electrostatics, a distribution of charges constitutes a source for a curl free electric field, $\mathbf{E}_{\mathrm{es}}$, satisfying $\nabla \times \mathbf{E}_{\mathrm{es}} = \mathbf{0}$. Faraday's law of induction states that a time-varying magnetic flux is always accompanied by an additional electric field $\mathbf{E}_{\mathrm{ind}}$. The total electric field is the sum of these two contributions,

$$\mathbf{E} = \mathbf{E}_{\mathrm{es}} + \mathbf{E}_{\mathrm{ind}} \, .$$

Consider a static, closed curve Γ. In this case $\mathbf{E}_{\mathrm{emf}} = \mathbf{E}_{\mathrm{ind}}$. Since the electrostatic field is conservative, its circulation around a closed loop is zero:

$$\oint_{\Gamma} \mathbf{E}_{\mathrm{es}}(\mathbf{x}, t) \cdot d\mathbf{l}(\mathbf{x}) = \iint_{S(\Gamma)} \underbrace{(\nabla \times \mathbf{E}_{\mathrm{es}}(\mathbf{x}, t))}_{0} \cdot d\mathbf{S}(\mathbf{x}) = \mathbf{0} \, .$$

This means that we can write $\varepsilon = \oint_{\Gamma} \mathbf{E}_{\mathrm{emf}}(\mathbf{x}, t) \cdot d\mathbf{l}(\mathbf{x}) = \oint_{\Gamma} \mathbf{E}(\mathbf{x}, t) \cdot d\mathbf{l}(\mathbf{x})$. By applying Stokes's theorem and Faraday's law of induction, we can relate the circulation of the total electric field $\mathbf{E}$ around this curve to the time rate of change of the magnetic flux through any surface $S(\Gamma)$ bounded by the curve:

$$\varepsilon = \oint_{\Gamma} \mathbf{E}(\mathbf{x}, t) \cdot d\mathbf{l}(\mathbf{x}) = \iint_{S(\Gamma)} (\nabla \times \mathbf{E}(\mathbf{x}, t)) \cdot d\mathbf{S}(\mathbf{x}) = - \iint_{S(\Gamma)} \frac{\partial}{\partial t} \mathbf{B}(\mathbf{x}, t) \cdot d\mathbf{S}(\mathbf{x}) .$$

This can be written as:

$$\iint_{S(\Gamma)} \left(\nabla \times \mathbf{E}(\mathbf{x}, t) + \frac{\partial}{\partial t} \mathbf{B}(\mathbf{x}, t) \right) \cdot d\mathbf{S}(\mathbf{x}) = 0 .$$

Since this equation holds for any arbitrary surface $S(\Gamma)$, the integrand must be zero everywhere. This leads to the differential form of Faraday's law of induction,

$$\boxed{\nabla \times \mathbf{E}(\mathbf{x}, t) = - \frac{\partial}{\partial t} \mathbf{B}(\mathbf{x}, t) .} \tag{11.7}$$

This equation is one of the four fundamental Maxwell's equations for time-varying electromagnetic fields. It generalizes the circulation law of electrostatics (3.27), which states that the curl of the electric field is zero for static electric fields.

A significant consequence of Faraday's law is that in the presence of time-varying fields, the electric field is no longer conservative. In other words, the line integral of the electric field around a closed loop is non-zero.

Using the vector potential $\mathbf{A}$, where $\mathbf{B} = \nabla \times \mathbf{A}$, we can rewrite Eq. (11.7) as

$$\nabla \times \mathbf{E} = - \frac{\partial \mathbf{B}}{\partial t} = - \frac{\partial}{\partial t} (\nabla \times \mathbf{A}) .$$

By rearranging terms, we obtain

$$\nabla \times \left(\mathbf{E} + \frac{\partial \mathbf{A}}{\partial t} \right) = 0 .$$

This equation suggests the introduction of a new vector field, often called the potential electric field, defined as

$$\mathbf{E}_{\text{pot}} = \mathbf{E} + \frac{\partial \mathbf{A}}{\partial t} .$$

This potential electric field is conservative, and its curl is zero. This allows us to introduce a scalar potential V such that $\mathbf{E}_{\text{pot}} = -\nabla V$. Therefore, the total electric field can be expressed as

$$\boxed{\mathbf{E} = -\nabla V - \frac{\partial \mathbf{A}}{\partial t} .} \tag{11.8}$$

The scalar potential V is the standard electrostatic potential, when $\mathbf{A}$ satisfies the Coulomb gauge condition $\nabla \cdot \mathbf{A} = 0$. Indeed, in that case the divergence of the electric field is given by

$$\mathbf{V} \cdot \mathbf{E} = -\mathbf{V}^2 V = \frac{\varrho}{\epsilon_0} \ .$$

The first term in the right-hand side of Eq. (11.8) represents therefore the conservative electric field $\mathbf{E}_{es}$ associated with charge separation ($\varrho \neq 0$), while the second term represents the non-conservative electric field $\mathbf{E}_{ind}$ that accompanies the time-varying magnetic field.

11.4.1 Foucault Currents

The experimental demonstration of induced currents was conducted by François Arago in 1824, before Faraday's discoveries, and later by Léon Foucault in 1851. The existence of such currents results in heat dissipation due to the Joule effect. Foucault demonstrated that the heat dissipated by a copper disk rotating between the poles of an electromagnet was equal to the work required to rotate the disk. This phenomenom is illustrated in Fig. 11.8, together with a picture of the Foucault induction machine. As the disk rotates, circulating currents - known as Foucault currents or eddy currents- are generated.

Applications of such currents include induction heating of cooking vessels and braking systems, where Foucault currents complement conventional brakes by dissipating the mechanical energy of a rotating wheel in the presence of an electromagnet. This provides the advantage of a contactless brake without friction. However, it only functions when the system is in motion and cannot bring the system to a complete stop on its own. While Foucault currents are beneficial in these applications, they can also lead to significant energy losses. In many situations minimizing these currents is crucial to improve system efficiency.

Fig. 11.8 Left: Foucault currents are generated in a conductive disk rotating in the presence of a magnetic field. Right: Foucault's induction machine. Picture taken at the Museum of Ecole Polytechnique (Mus'X)

11.4.2 Generalized Form of Ohm's Law

Consider a conducting wire moving with velocity $\mathbf{v}$ through an external magnetic field $\mathbf{B}$. The electrons in the wire experience a Lorentz force given by

$$\mathbf{F} = -e\mathbf{E} - e(\mathbf{v} + \mathbf{u}) \times \mathbf{B} \,,$$

where $\mathbf{u}$ is the drift velocity of the electrons relative to the wire. The equation of motion for an electron of mass m_e and charge $-e$ can be written as

$$m_e \frac{d}{dt}(\mathbf{v} + \mathbf{u}) + \frac{m_e}{\tau}\mathbf{u} = -e\mathbf{E} - e(\mathbf{v} + \mathbf{u}) \times \mathbf{B} \,.$$

The second term on the left-hand side represents a damping force due to collisions between electrons and ions in the conductor, with τ being the relaxation time. In the steady-state approximation, we neglect the acceleration term in the equation of motion, assuming that the electrons quickly reach a steady-state drift velocity. This approximation is valid for frequencies much lower than 10 GHz. Under this assumption, we have

$$\mathbf{u} = -\frac{e\tau}{m_e} \left(\mathbf{E} + (\mathbf{v} + \mathbf{u}) \times \mathbf{B}\right) \,.$$

The current density $\mathbf{j}$ in the wire is related to the electron drift velocity by

$$\mathbf{j} = -en\mathbf{u} \,.$$

Combining these equations, we obtain

$$\mathbf{j} = \sigma \left(\mathbf{E} + (\mathbf{v} + \mathbf{u}) \times \mathbf{B}\right) \,.$$

where $\sigma = \frac{ne^2\tau}{m_e}$ is the electrical conductivity of the material.

Projecting the equation onto directions parallel and perpendicular to $\mathbf{j}$, we obtain:

- Perpendicular component:

$$\mathbf{E}_\perp + (\mathbf{v} \times \mathbf{B})_\perp = -\mathbf{u} \times \mathbf{B} = \mathbf{E}_{\text{Hall}} \,.$$

  The left-hand side represents the total perpendicular electric field acting on the electrons in the reference frame of the conductor, which is equal to the Hall electric field.

- Parallel component:

$$\mathbf{j} = \sigma \left(\mathbf{E} + \mathbf{v} \times \mathbf{B}\right)_\| \,.$$

  We recognize the local form of a generalized Ohm's law.

Considering the electric field expression $\mathbf{E} = -\nabla V - \dfrac{\partial \mathbf{A}}{\partial t}$, we can rewrite the parallel component as

$$\mathbf{j} = \sigma(-\nabla V \underbrace{- \frac{\partial \mathbf{A}}{\partial t} + \mathbf{v} \times \mathbf{B}}_{\mathbf{E}_{\mathrm{emf}}})_{\|} \ .$$

Here, the term $-\frac{\partial \mathbf{A}}{\partial t} + \mathbf{v} \times \mathbf{B}$ represents the electromotive field $\mathbf{E}_{\mathrm{emf}}$. This shows that the induced current in the moving conductor is driven by the combined effect of the applied conservative electric field and the electromotive force.

Consider a wire of cross-sectional area S moving with velocity $\mathbf{v}$ through a magnetic field $\mathbf{B}$. To analyze the current flow in the wire, we can integrate the generalized form of Ohm's law along a path AB parallel to the current density $\mathbf{j}$ ($d\mathbf{l} \parallel \mathbf{j}$), as shown in Fig. 11.9:

$$\int_A^B \mathbf{j} \cdot d\mathbf{l} = -\sigma \int_A^B \nabla V \cdot d\mathbf{l} + \sigma \int_A^B \left(-\frac{\partial \mathbf{A}}{\partial t} + \mathbf{v} \times \mathbf{B} \right) \cdot d\mathbf{l} \ .$$

The current density $\mathbf{j}$ can be expressed as $\mathbf{j} = \dfrac{I}{S}\dfrac{d\mathbf{l}}{dl}$, where I is the current flowing from point A to point B. Integrating the generalized form of Ohm's law along the path AB, we obtain

$$I \underbrace{\int_A^B \frac{dl}{\sigma S}}_{R_{AB}} = V_A - V_B + \underbrace{\int_A^B \left(-\frac{\partial \mathbf{A}}{\partial t} + \mathbf{v} \times \mathbf{B} \right) \cdot d\mathbf{l}}_{\varepsilon_{AB}}.$$

The first term on the left-hand side, $\int_A^B \frac{dl}{\sigma S}$, represents the resistance R_{AB} between points A and B. The second term is the electromotive force ε_{AB} induced between points A and B. Thus, we arrive at the generalized Ohm's law,

$$\boxed{V_A - V_B = R_{AB} I - \varepsilon_{AB} \ .} \tag{11.9}$$

Remarks

- Since the electromotive field is not conservative, the potential difference $V_A - V_B$ between points A and B will generally depend on the path taken between the two points.
- For an open circuit ($I = 0$) or a conductor with negligible resistance ($R = 0$):

Fig. 11.9 A wire of section S moving at velocity $\mathbf{v}$ in a magnetic field

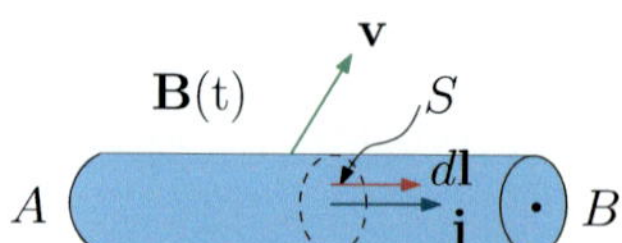

$$V_B - V_A = \varepsilon_{AB} \ .$$

The induced electromotive force creates a potential difference between points A and B, causing a charge separation in the circuit.

- For a wire forming a closed circuit Γ, $(A = B)$:

$$\varepsilon = \oint_\Gamma (\mathbf{E}_{\mathrm{emf}} \cdot d\mathbf{l}) = -\frac{d\Phi_{S(\Gamma),\mathbf{B}}}{dt} = RI \ ,$$

where R is the total resistance of the circuit.

- The power delivered to an infinitesimal charge element dq by the electromotive force is

$$dP_\varepsilon = dq\mathbf{E}_{\mathrm{emf}} \cdot \mathbf{u} \ ,$$

where $\mathbf{E}_{\mathrm{emf}}$ is the electromotive field and $\mathbf{u}$ is the velocity of the charge with respect to the circuit. Since $dq\,\mathbf{u} = (dq/dt)\,(dt\mathbf{u}) = I d\mathbf{l}$, the power delivered to the circuit by the electromotive force is given by:

$$P_\varepsilon = \oint_\Gamma \mathbf{E}_{\mathrm{emf}} \cdot I d\mathbf{l} = \varepsilon I \ .$$

Example 11.1 - A power plant generator
Consider a generator consisting of a permanent magnet producing a uniform magnetic field $\mathbf{B} = B\mathbf{u}_x$, and a rectangular coil of area A rotating with angular velocity ω about the y-axis, perpendicular to the magnetic field $\mathbf{B}$, as shown in Fig. 11.10.

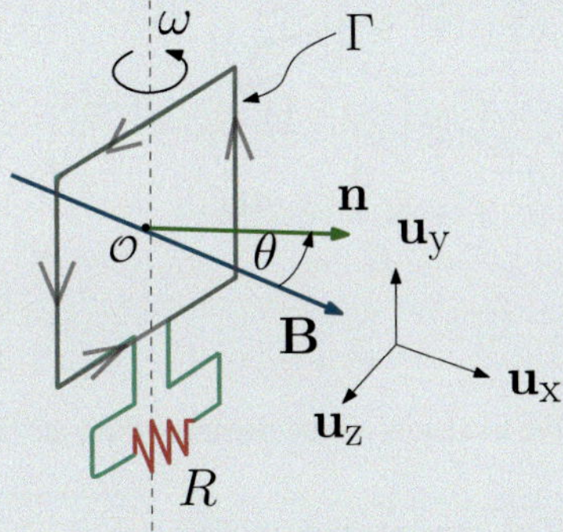

Fig. 11.10 Rotating coil in a static magnetic field

The closed loop Γ representing the rectangular coil is oriented as shown in Fig. 11.10. At a given instant, the normal vector $\mathbf{n}$ to the plane of the coil makes an angle θ with the x-axis.

Since the magnetic field is uniform, the magnetic flux through the surface $S(\Gamma)$ enclosed by the loop is given by

$$\Phi_{S(\Gamma),\mathbf{B}} = BA\cos\theta = BA\cos(\omega t + \theta_0) \,,$$

where and $\theta = \omega t + \theta_0$ is the angle between the normal vector to the loop and the magnetic field at time t.

The rate of change of magnetic flux with respect to time is

$$\frac{d\Phi_{S,\mathbf{B}}}{dt} = -BA\omega\sin(\omega t + \theta_0) \,.$$

Applying Faraday's law of induction, the induced electromotive force in the loop is

$$\varepsilon = -\frac{d\Phi_{S,\mathbf{B}}}{dt} = BA\omega\sin(\omega t + \theta_0) \,.$$

We see that ε varies sinusoidally with time. Since $\mathbf{B}$ is static, this electromotive force arises solely from the electromotive field $\mathbf{v} \times \mathbf{B}$ acting on the charge carriers in the two vertical segments ab and cd of the rotating coil shown in Fig. 11.11. Each point on these vertical segments moves in a circular path of radius r, where $2r = A^{1/2}$ is the length of the coil side.

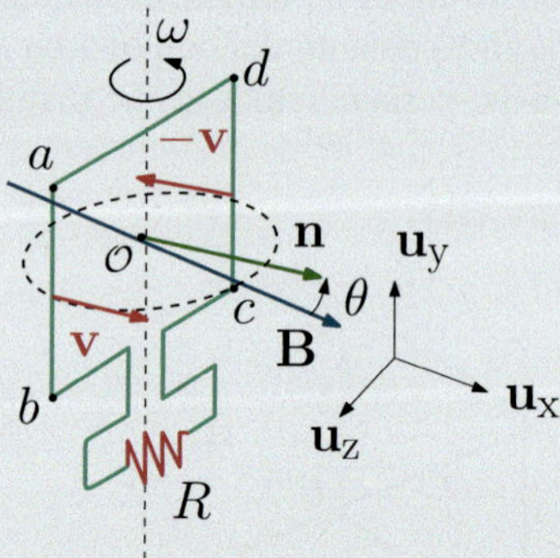

Fig. 11.11 The electromotive field acts on the carriers along the segments ab and cd

For the segment ab, the electromotive field is

$$\mathbf{E}_{\text{emf}} = \mathbf{v} \times \mathbf{B} = \omega r\mathbf{n} \times B\mathbf{u}_x = -\omega r B\sin\theta\mathbf{u}_y \,.$$

The induced emf in segment ab can be calculated by integrating the electromotive field along the length of the segment:

$$\varepsilon_{ab} = \int_a^b \mathbf{v} \times \mathbf{B} \cdot \underbrace{d\mathbf{l}}_{-dy\mathbf{u}_y} = \frac{AB\omega}{2}\sin(\omega t + \theta_0) \ .$$

For the segment cd, both the velocity $\mathbf{v}$ and the line element $d\mathbf{l}$ are reversed compare to segment ab. Therefore,

$$\varepsilon_{cd} = \varepsilon_{ab} \ .$$

The horizontal segments da and bc do not contribute since the electromotive field $\mathbf{v} \times \mathbf{B}$ is perpendicular to $d\mathbf{l}$. Therefore,

$$\varepsilon = \varepsilon_{ab} + \varepsilon_{cd} = 2\varepsilon_{ab} = AB\omega \sin(\omega t + \theta_0) \ .$$

This induced emf causes an alternating current to flow through the loop, which can be harnessed to power various devices.

If the ends of the loop are connected to a resistor R, the induced current through the circuit, according to Ohm's law, is

$$I = \frac{\varepsilon}{R} = \frac{AB\omega}{R}\sin(\omega t + \theta_0) \ .$$

This induced current generates a magnetic field that opposes the change in the original magnetic flux through the loop. This is in accordance with Lenz's law.

While there is no net force on the loop due to the uniform magnetic field, there is a torque exerted on the loop. This torque opposes the external torque that is causing the loop to rotate.

The power dissipated in the resistor is equal to the power supplied by the electromotive force,

$$P = \varepsilon I = \frac{(BA\omega)^2}{R}\sin^2(\omega t + \theta_0) \ .$$

On the other hand, the torque τ required to maintain the constant angular velocity ω must counteract the torque exerted by the magnetic field on the loop. This torque is given by

$$\tau = -\mathbf{m} \times \mathbf{B} = -IAB\sin(\omega t + \theta_0)\mathbf{u}_z \ ,$$

where $\mathbf{m} = m\mathbf{n} = IA\mathbf{n}$ is the magnetic dipole moment of the loop. Therefore, The mechanical power required to maintain the rotation of the loop is:

$$P_m = |\tau|\omega = \frac{A^2 B^2 \omega^2}{R}\sin^2(\omega t + \theta_0) \ .$$

This mechanical power is equal to the electrical power dissipated in the resistor, $P = \varepsilon I$, in accordance with the conservation of energy.

Example 11.2 - Langevin's model for diamagnetism

A classical model due to Paul Langevin explains diamagnetism as a consequence of Faraday's law of induction. It applies particularly well to dielectric materials in which electrons are strongly bound to atoms.

Consider a hydrogen atom with an electron orbiting a proton at a distance r (Fig. 11.12). This system possesses a magnetic dipole moment

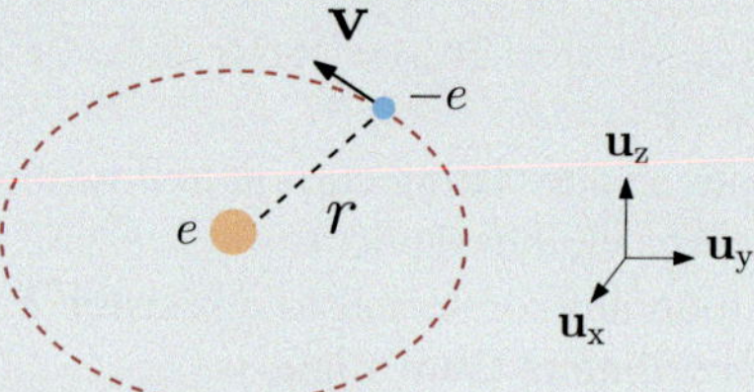

Fig. 11.12 An electron in an hydrogen atom

$$\mathbf{m} = I A \mathbf{u}_z = I \pi r^2 \mathbf{u}_z$$

where the current I associated with the electron orbital motion is

$$I = \frac{-ev}{2\pi r} \ .$$

Combining these equations, we get

$$\mathbf{m} = \frac{-evr}{2} \mathbf{u}_z \ .$$

Now, suppose an external time-varying magnetic field $\mathbf{B} = B(t)\mathbf{u}_z$ is applied. This time-varying field induces an electromotive force in the electron orbit. The work done per unit charge during one revolution is

$$\varepsilon = E_{\text{emf}} 2\pi r = -\frac{d}{dt}(B(t)\pi r^2) \ .$$

Assuming that the orbit radius remains constant, the induced electric field E_{emf} can be calculated as

$$E_{\text{emf}} = -\frac{r}{2}\frac{dB(t)}{dt} \ .$$

The induced electric field accelerates the electron, modifying its speed. The equation of motion for the electron is

$$- e E_{\text{emf}} = m_e \frac{dv}{dt} = \frac{er}{2} \frac{dB(t)}{dt} \, .$$

If the external magnetic field changes by ΔB, the electron speed changes by $\Delta v = er \Delta B/(2m_e)$. This change in speed leads to a change in the magnetic moment:

$$\Delta \mathbf{m} = \frac{-e \Delta v r}{2} \mathbf{u}_z = \frac{-e^2 r^2}{4m_e} \Delta B \, \mathbf{u}_z \, .$$

This change in the magnetic moment, the electron response, opposes the change in the external magnetic flux, consistent with Lenz's law.

Finally, if there are n atoms per unit volume, the magnetization $\mathbf{M}$ of the material is given by

$$\Delta \mathbf{M} = n \Delta \mathbf{m} = \frac{-ne^2 r^2}{4m_e} \Delta B = \chi_m \frac{\Delta B}{\mu_0} \, .$$

Assuming the material is sufficiently dilute, we can approximate the magnetic field experienced by each atom as the external magnetic field $\mathbf{H} \approx \dfrac{\mathbf{B}}{\mu_0}$, meaning that each atom solely responds to the external magnetic field and does not interact with other atoms. From this, we obtain a negative magnetic susceptibility,

$$\chi_m = -\mu_0 \frac{ne^2 r^2}{4m_e} \, .$$

This negative susceptibility is characteristic of diamagnetic materials.

For an atom with Z electrons, the diamagnetic susceptibility according to Langevin's model is

$$\chi_m = -\mu_0 \frac{Zne^2 r^2}{4m_e} \, .$$

For example, for graphite, which is composed of carbon atoms ($Z = 6$), with an average atomic radius $r \approx 0.07\,\text{nm}$ and a density $n = 9 \times 10^{22}\,\text{cm}^{-3}$, the estimated diamagnetic susceptibility is

$$\chi_m \sim -2 \times 10^{-5} \, .$$

This theoretical estimate compares reasonably well with the experimentally measured value of

$$\chi_m^{\text{exp}} = -1.4 \times 10^{-5} \, ,$$

even though graphite is a conductor.

Example 11.3 - Kelvin effect: penetration of a magnetic field inside a conductor

The Kelvin effect describes the penetration depth of a time-varying magnetic field into a conductor. Due to Faraday's law of induction, an induced current is generated within the conductor, which opposes the change in the external magnetic field.

Consider a medium with conductivity σ subjected to a time-varying magnetic field. Faraday's law of induction relates the induced electric field $\mathbf{E}$ to the time rate of change of the magnetic field $\mathbf{B}$:

$$\nabla \times \mathbf{E}(\mathbf{x}, t) = -\frac{\partial \mathbf{B}(\mathbf{x}, t)}{\partial t} \, .$$

This induced electric field, in turn, generates a current density $\mathbf{j}$ according to Ohm's law, $\mathbf{j}(\mathbf{x}, t) = \sigma \mathbf{E}(\mathbf{x}, t)$. This current density produces a magnetic field, as described by Ampère's law,

$$\nabla \times \mathbf{B}(\mathbf{x}, t) = \mu_0 \mathbf{j}(\mathbf{x}, t) = \mu_0 \sigma \mathbf{E}(\mathbf{x}, t) \, .$$

Combining this with Faraday's law, we obtain

$$\nabla \times (\nabla \times \mathbf{B}(\mathbf{x}, t)) = -\nabla^2 \mathbf{B}(\mathbf{x}, t) = \mu_0 \sigma \nabla \times \mathbf{E}(\mathbf{x}, t) = -\mu_0 \sigma \frac{\partial \mathbf{B}(\mathbf{x}, t)}{\partial t} \, .$$

Here, we have used the vector identity $\nabla \times (\nabla \times \mathbf{F}) = \nabla(\nabla \cdot \mathbf{F}) - \nabla^2 \mathbf{F}$ and the fact that the divergence of the magnetic field is zero, $\nabla \cdot \mathbf{B} = 0$. Finally, the magnetic field inside the conductor satisfies the following diffusion equation:

$$\nabla^2 \mathbf{B}(\mathbf{x}, t) - \mu_0 \sigma \frac{\partial \mathbf{B}(\mathbf{x}, t)}{\partial t} = 0 \, . \tag{11.10}$$

Assuming a time-harmonic variation of the fields at angular frequency ω, we can write the magnetic field as:

$$\mathbf{B}(\mathbf{x}, t) = \text{Re} \left\{ \mathbf{B}(\mathbf{x}) e^{i\omega t} \right\} \, ,$$

where the spatial component $\mathbf{B}(\mathbf{x})$ satisfies

$$\nabla^2 \mathbf{B}(\mathbf{x}) - i\omega\mu_0\sigma\mathbf{B}(\mathbf{x}) = 0 \ .$$

Considering a one-dimensional geometry where the magnetic field varies only with the x-coordinate and the conductor occupies the region $x > 0$, the diffusion equation simplifies to

$$\frac{d^2 B(x)}{dx^2} - i\omega\mu_0\sigma B(x) = 0 \ .$$

The solution to this equation is:

$$B(x, t) = \mathrm{Re}\left\{ B_0 e^{i\omega t} e^{-(1+i)x/\lambda_K} \right\} = B_0 \cos(\omega t - x/\lambda_K) e^{-x/\lambda_K} \ ,$$

where λ_K is the skin depth defined as

$$\lambda_K = \sqrt{\frac{2}{\mu_0\sigma\omega}} \ .$$

In conclusion, the magnetic field penetrates the conductor exponentially, with a characteristic penetration depth λ_K. This skin depth represents the distance from the surface at which the magnetic field amplitude is reduced by a factor of e.

For a good conductor like copper, with a conductivity $\sigma \sim 6 \times 10^7 \, \mathrm{S\,m^{-1}}$, at a frequency of $\omega = 3.8 \times 10^{15} \, \mathrm{s^{-1}}$ (corresponding to visible light), the skin depth is approximately $\lambda_K \sim 2.6\,\mathrm{nm}$. This indicates that the induced current is highly localized near the surface of the conductor, a phenomenon known as the skin effect.

11.4.3 Induction Effects in Superconductors

In Example 11.3, it was shown that a magnetic field can penetrate a conductor only within a distance from the surface given by the skin depth λ_K (11.4.2). This formula predicts that λ_K approaches zero for a perfect conductor, where the resistance becomes zero (infinite conductivity). However, the Drude model breaks down in that case because the electrons no longer acquire a drift velocity proportional to the electric field. Instead, the electrons are accelerated, as described by Newton's law of motion without friction. For an electron, this is expressed as:

$$m_0 \frac{d\mathbf{v}}{dt} = -e\mathbf{E}$$

As a result, the current density $\mathbf{j} = -ne\mathbf{v}$ in a perfect conductor satisfies:

$$\frac{d\mathbf{j}}{dt} = -ne\frac{d\mathbf{v}}{dt} = \frac{ne^2}{m_0}\mathbf{E} \qquad (11.11)$$

which is known as the first London equation. Combining the time derivative of Ampère's law with Faraday's law, along with the identity (A.4) and Eq. (11.11), leads to the following:

$$-\nabla^2 \frac{\partial \mathbf{B}}{\partial t} = \frac{\mu_0 ne^2}{m_0} \nabla \times \mathbf{E} = -\frac{\mu_0 ne^2}{m_0} \frac{\partial \mathbf{B}}{\partial t}$$

Defining the London penetration depth for a perfect conductor as $\lambda_L = \sqrt{\frac{m_0}{\mu_0 ne^2}}$, we find:

$$\left(\frac{\partial \mathbf{B}}{\partial t} - \lambda_L^2 \nabla^2 \frac{\partial \mathbf{B}}{\partial t} \right) = 0$$

This is a Helmholtz equation, whose solution, for a one-dimensional problem, is:

$$\frac{\partial B(x)}{\partial t} = \frac{\partial B(0)}{\partial t} e^{-x/\lambda_L}$$

where $x = 0$ corresponds to the surface of a conductor filling the region $x > 0$. This result shows that the magnetic field inside a perfect conductor is not necessarily zero. A perfect conductor opposes to a time-variation of the magnetic field. Any induced current will exist only near the surface of the conductor.

In 1933, Walther Meissner and Robert Ochsenfeld demonstrated that a superconductor behaves differently from a perfect conductor. While the resistivity of a superconductor drops to zero, experiments show that the magnetic field is completely expelled from its interior. This phenomenom, known as the Meissner effect, can be illustrated by considering a metallic sphere in a uniform magnetic field. As the sphere is cooled below the critical temperature T_c for superconductivity, the magnetic field is entirely expelled from its interior (see Fig. 11.13). This results in a repulsive force between the superconductor and nearby magnets, enabling effects such as magnetic levitation.

The expulsion of the magnetic field inside a superconductor is due to the emergence of superconducting currents near its surface, confined within the London penetration depth λ_L. These persistent currents flow indefinitely without dissipation, ensuring that the magnetic field remains zero inside. This demonstrates that a superconductor is not merely a perfect conductor; rather, it represents a macroscopic manifestation of a unique quantum phenomenom. Superconductors are widely used to generate the strong magnetic fields required for technologies like magnetic resonance

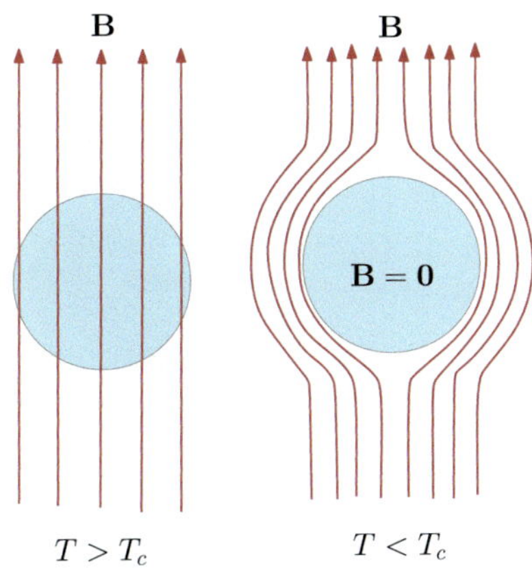

Fig. 11.13 The Meissner-Ochsenfeld effect

imaging (MRI). In MRI systems, the magnetic field is produced by superconducting electromagnets, commonly made of NbTi, cooled down with liquid helium to temperatures below T_c.

11.4.4 Magnetic Flux Through a Coil

A coil is formed by a wire wound into N turns around an axis. It is common to assume that the flux through the surface enclosed by the coil is simply N times the flux through the planar surface associated with a single turn, which is the projection of the coil onto the plane perpendicular to its axis. Typically, this surface is a square or a circle. To justify this, consider a helical coil of N turns and radius R, whose height increases linearly by d after every turn. The thickness of the coil is therefore Nd. Figure 11.14 illustrates the case of a coil with $N = 3$ turns and such that the initial and final points are joined by a line along the axis, supposed to be aligned with the z-axis.

Connecting both ends of the coil define a closed loop Γ, the electromotive force is determined by the flux of the magnetic field through any surface bounded by this loop. One possible choice of surface is the helical surface S_{helix}, shown in Fig. 11.14. It is parametrized by points $\mathbf{r}(r, \theta)$, given in cylindrical coordinates by

$$\mathbf{r}(r, \theta) = r\mathbf{u}_r + \frac{\theta}{2\pi}d\mathbf{u}_z \quad r \in [0, R], \theta \in [0, 2N\pi] .$$

The surface element is then computed as (see Sect. A.1.6.3)

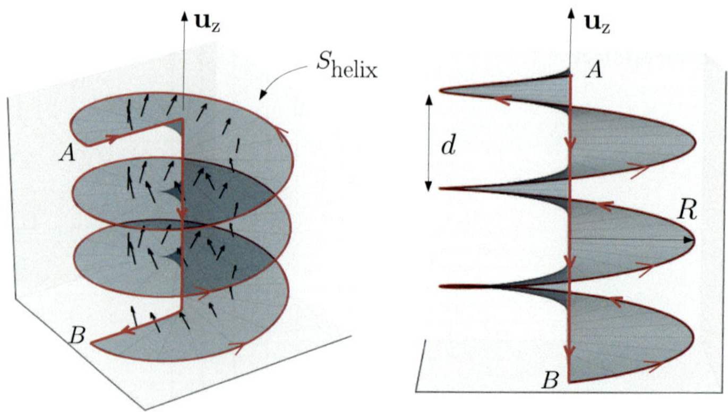

Fig. 11.14 The helical surface is bounded by the closed contour defined by the coil

$$dS = \frac{\partial \mathbf{r}}{\partial r} \times \frac{\partial \mathbf{r}}{\partial \theta} dr d\theta = \mathbf{u}_r \times \left(r \underbrace{\frac{\partial \mathbf{u}_r}{\partial \theta}}_{\mathbf{u}_\theta} + \frac{d}{2\pi}\mathbf{u}_z \right) dr d\theta$$

$$dS = \left(r\mathbf{u}_z - \frac{d}{2\pi}\mathbf{u}_\theta \right) dr d\theta$$

Suppose now a magnetic field invariant under translations along the coil's axis, i.e., $\mathbf{B} = \mathbf{B}(r, \theta)$. The flux of $\mathbf{B}$ through S_{helix} is given by

$$\Phi_{S_{\text{helix}},\mathbf{B}} = \int_0^{2N\pi} \int_0^R B_z(r, \theta) r dr d\theta - \frac{d}{2\pi} \int_0^{2N\pi} \int_0^R B_\theta(r, \theta) dr d\theta . \tag{11.12}$$

The first term in the right-hand side of (11.12) equals N times the flux of $\mathbf{B}$ over a circular surface of radius R in the xy plane. If $d \ll R$, the contribution of the second term becomes negligible (and exactly zero if the field is along z). This yields:

$$\Phi_{S_{\text{helix}},\mathbf{B}} = N\Phi_{S_{\text{circle}},\mathbf{B}} . \tag{11.13}$$

11.5   Application of Induction: The Transformer

A transformer is a device that alters the voltage magnitude of an alternating current (AC) power supply. It increases or decreases the voltage level, making it essential for efficient power transmission and distribution. For instance, electricity generated in power plants, typically at a few kilovolts, is stepped up to hundreds of kilovolts

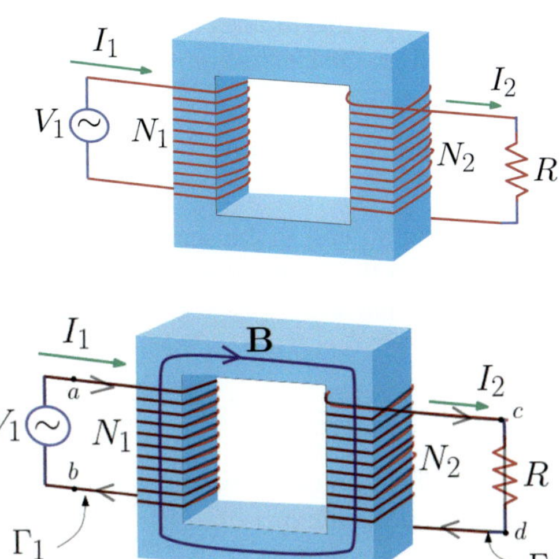

Fig. 11.15 A transformer consists of two coils wound around a common soft ferromagnetic core

Fig. 11.16 The first coil is connected to an alternating voltage source, while the second coil is connected to a resistor

for long-distance transmission. Before reaching homes, this high voltage is stepped down to safer levels, such as a few hundred volts for domestic use or tens of volts for charging devices.

A transformer consists of two coils wound around a common magnetic core made of a soft ferromagnetic material. The core garantees a good magnetic coupling between the coils. The primary coil, with N_1 turns, is connected to an AC voltage source $V_1 = V_0 \cos \omega t$. This voltage source generates an alternating current I_1 in the primary coil, which in turn produces a time-varying magnetic flux in the core. This changing magnetic flux comes with an electromotive force in the secondary coil, with N_2 turns, resulting in an induced current I_2.

The coils are typically wound around a ferromagnetic core to maximize the magnetic coupling between them. A simplified schematic of a transformer is shown in Fig. 11.15.

Consider the closed loops Γ_1 and Γ_2 corresponding to the primary and secondary coils, respectively, as shown in Fig. 11.16. For the primary coil, neglecting its internal resistance and assuming no additional emf sources, the generalized form of Ohm's law (11.9) gives

$$V(a) - V(b) = V_1(t) = -\varepsilon_{ab} = -\varepsilon_1 \ .$$

where ε_1 is the emf induced in the primary coil. The magnetic flux through one turn of the primary coil, of surface S, is given by $\Phi_{S,\mathbf{B}} = BS$, where $\mathbf{B}$ is the magnetic field. Applying Faraday's law of induction to the primary coil, we have

$$\varepsilon_1 = -N_1 \frac{d\Phi_{S,\mathbf{B}}}{dt} = -V_1(t) = -V_0 \cos \omega t \ .$$

Due to the high permeability of the iron core ($\mu \gg \mu_0$), the magnetic flux is effectively confined within the core, and the magnetic flux through each turn of the secondary coil has the same magnitude but opposite sign as the flux through a turn of the primary coil. Therefore, the electromotive force induced in the secondary coil is:

$$\varepsilon_2 = N_2 \frac{d\Phi_{S,\mathbf{B}}}{dt} = \frac{N_2}{N_1} V_0 \cos \omega t = \frac{N_2}{N_1} V_1 \, .$$

Assuming no emf in the segment cd of the secondary coil, by applying general Ohm's law to the segment dc (of negligible resistance), we find $V(d) - V(c) = -\varepsilon_{dc} = -\varepsilon_2$ and so

$$V_2(t) = V(c) - V(d) = \varepsilon_2 = \frac{N_2}{N_1} V_1$$

Hence, the ratio of the voltages across the primary and secondary coils is equal to the ratio of the number of turns:

$$\frac{V_2(t)}{V_1(t)} = \frac{N_2}{N_1} \, .$$

If this second coil is connected to a load resistor R, the induced current I_2 is given by

$$I_2(t) = \frac{\varepsilon_2(t)}{R} = \frac{N_2}{N_1} \frac{V_0}{R} \cos \omega t.$$

Assuming ideal conditions (negligible resistance in the coils), the power input to the primary coil must equal the power dissipated in the load resistance:

$$I_1(t) V_1(t) = I_2(t) V_2(t) \, .$$

This leads to the following relationship between the currents and voltages in the primary and secondary coils:

$$\frac{I_2(t)}{I_1(t)} = \frac{N_1}{N_2} = \frac{V_1(t)}{V_2(t)} \, .$$

To transmit electric power efficiently from a power plant to a city, transformers are used to step up the voltage and reduce the current. This significantly reduces power losses due to Joule heating, which is proportional to the square of the current ($P = RI^2$ for a current I in the transmission line, and R its resistance). For instance, increasing the voltage from 6 to 300 kV reduces the current exiting the transformer by a factor of 50, thereby reducing the power loss by a factor of 2500. To reduce power losses in the transformer itself, it is important not only to minimize the resistance of the coils, but also to suppress the eddy currents induced in the magnetic core. This is achieved by constructing the core from thin, electrically insulated sheets, which restrict eddy currents to small, isolated loops—thereby reducing the induced electromotive forces and associated energy losses. Alternatively, ferrites are often

used as core materials in high-frequency applications due to their high electrical
resistivity, which naturally limits eddy current formation.

11.6 Inductance

Faraday's law of induction describes the phenomenon of electromagnetic induction,
where a change in magnetic flux through a circuit induces an electromotive force.
This induced emf opposes the change in magnetic flux, a principle known as Lenz's
law. When a current flowing through a conductor changes, it creates a time-varying
magnetic field. This change in the magnetic field can induce an emf in the same
conductor, opposing the initial change in current.

11.6.1 Self-inductance

A closed circuit carrying a current I generates a magnetic field that permeates the
surrounding space. This magnetic field, in turn, produces a magnetic flux through
the circuit itself, as illustrated in Fig. 11.17.

Consider a closed loop Γ carrying a current I, and let $S(\Gamma)$ be a surface bounded
by Γ. Since the magnetic field $\mathbf{B}$ generated by the loop is proportional to the current
I, the magnetic flux $\Phi_{S(\Gamma),\mathbf{B}}$ through $S(\Gamma)$ is also proportional to I:

$$\Phi_{S(\Gamma),\mathbf{B}} = \iint_{S(\Gamma)} \mathbf{B}(\mathbf{x}) \cdot \mathbf{n}(\mathbf{x}) dS(\mathbf{x}) = LI \ .$$

The proportionality constant L is known as the self-inductance of the circuit, or
simply, inductance. It depends solely on the geometry of the circuit and is measured
in Henry (H), where $1\,\mathrm{H} = 1\,\mathrm{Wb\,A^{-1}}$ (Fig. 11.18).

According to Faraday's law of induction, a change in the current through a circuit
induces an electromotive force ε that opposes the change in current. This self-induced
emf is given by

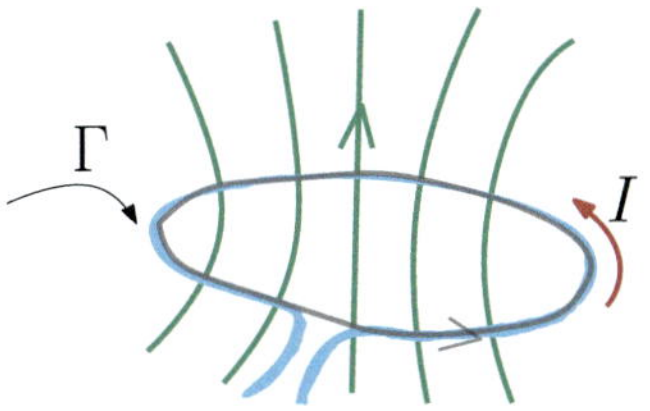

Fig. 11.17 A current
flowing through a loop
generates a magnetic field
that passes through the
surface enclosed by the loop,
creating a magnetic flux

Fig. 11.18 Joseph Henry (1797–1878), scientist from the United States who made significant contributions to the understanding of electromagnetic induction. The unit of inductance, the henry, is named in his honor

$$\varepsilon = -\frac{d\Phi_{S(\Gamma),\mathbf{B}}}{dt} = -\frac{d\Phi_{S(\Gamma),\mathbf{B}}}{dI}\frac{dI}{dt} = -L\frac{dI}{dt} \ .$$

The inductance L of the circuit quantifies its ability to oppose changes in current. A higher inductance value results in a larger self-induced emf for a given rate of change of current.

11.6.2   Inductance of a Solenoid

A solenoid, a coil wound into a tightly packed helix, generates a significant magnetic field within its core when a current flows through it. Consider a solenoid of length l, with N turns, and a radius R, where $R \ll l$. This configuration, illustrated in Fig. 11.19, allows us to neglect edge effects.

The magnetic field inside a solenoid is approximately uniform and parallel to the axis of the solenoid,

$$\mathbf{B} = B\mathbf{u}_z = \frac{\mu_0 N I}{l}\mathbf{u}_z \ ,$$

where $\mathbf{u}_z$ is a unit vector along the axis of the solenoid. The magnetic flux through a single turn of the solenoid is

Fig. 11.19 A solenoid with N turns, having a radius R and length l, where $R \ll l$

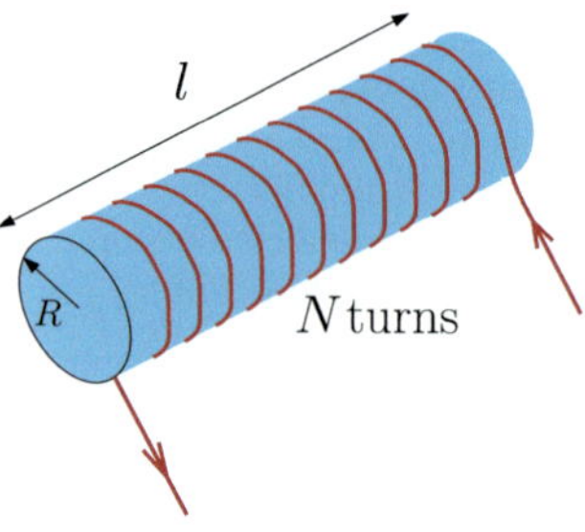

$$\Phi_B = B\pi R^2 = \frac{\mu_0 N I}{l}\pi R^2 \, .$$

The total magnetic flux through all N turns of the solenoid is

$$\Phi_{\text{total}} = N\Phi_B = \frac{\mu_0 \pi R^2 N^2 I}{l} \, .$$

The self-inductance of the solenoid, L, is defined as the ratio of the total magnetic flux to the current:

$$L = \frac{\Phi_{\text{total}}}{I} = \frac{\mu_0 \pi R^2 N^2}{l} \, . \tag{11.14}$$

11.6.3 Magnetic Coupling Between Current Loops

If a time-varying current flows through a circuit, it generates a time-varying magnetic field. This changing magnetic field can induce an electromotive force in a nearby circuit, as illustrated in Fig. 11.20 for two current loops.

Initially, the current in loop Γ_1 induces an emf in loop Γ_2. Conversely, the current induced in loop Γ_2 generates a magnetic field that can influence the current in loop Γ_1. Similar to the concept of capacitance between two conductors, we can define the mutual inductance between two circuits as a geometric coefficient that quantifies the efficiency of their magnetic coupling.

The magnetic flux through any surface $S(\Gamma_2)$ enclosed by the circuit Γ_2, due to the magnetic field $\mathbf{B}_1$ generated by Γ_1, is given by

$$\Phi_{S(\Gamma_2),\mathbf{B}_1} = \iint_{S(\Gamma_2)} \mathbf{B}_1 \cdot d\mathbf{S}_2 = \iint_{S(\Gamma_2)} (\nabla \times \mathbf{A}_1) \cdot d\mathbf{S}_2 \, .$$

Using the Stokes theorem (A.20), this can be rewritten as a line integral around the loop $S(\Gamma_2)$:

$$\Phi_{S(\Gamma_2),\mathbf{B}_1} = \oint_{\Gamma_2} \mathbf{A}_1(\mathbf{x}_2) \cdot d\mathbf{l}_2(\mathbf{x}_2) \, .$$

The vector potential $\mathbf{A}_1(\mathbf{x}_2)$ generated by the current I_1 in loop Γ_1 at a point $\mathbf{x}_2$ is given, in Coulomb's gauge, by

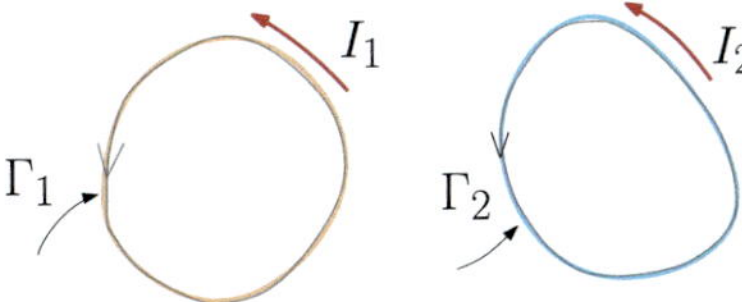

Fig. 11.20 Two loops will be magnetically coupled through the phenomenon of mutual induction

$$\mathbf{A}_1(\mathbf{x}_2) = \frac{\mu_0 I_1}{4\pi} \oint_{\Gamma_1} \frac{d\mathbf{l}_1(\mathbf{x}_1)}{|\mathbf{x}_1 - \mathbf{x}_2|} \,.$$

The magnetic flux through loop Γ_2 due to the magnetic field generated by loop Γ_1 is then

$$\Phi_{S(\Gamma_2),\mathbf{B}_1} = \iint_{S(\Gamma_2)} \mathbf{B}_1 \cdot d\mathbf{S}_2 = \oint_{\Gamma_2} \mathbf{A}_1(\mathbf{x}_2) \cdot d\mathbf{l}_2(\mathbf{x}_2) \,.$$

Substituting the expression for $\mathbf{A}_1(\mathbf{x}_2)$, we obtain

$$\Phi_{S(\Gamma_2),\mathbf{B}_1} = I_1 \underbrace{\frac{\mu_0}{4\pi} \oint_{\Gamma_1} \oint_{\Gamma_2} \frac{d\mathbf{l}_1(\mathbf{x}_1) \cdot d\mathbf{l}_2(\mathbf{x}_2)}{|\mathbf{x}_1 - \mathbf{x}_2|}}_{M} = M I_1,$$

The coefficient M is called the mutual inductance between the two circuits. It depends solely on the geometry of the two loops.

Mutual inductance The coefficient of mutual inductance, M, between two linear circuits Γ_1 and Γ_2 is defined by Neumann's formula:

$$\boxed{M = \frac{\mu_0}{4\pi} \oint_{\Gamma_2} \oint_{\Gamma_1} \frac{d\mathbf{l}_1(\mathbf{x}_1) \cdot d\mathbf{l}_2(\mathbf{x}_2)}{|\mathbf{x}_2 - \mathbf{x}_1|}} \,. \qquad (11.15)$$

The unit of mutual inductance is the henry (H), where $\mathrm{H} = \mathrm{Wb\,A^{-1}}$. The magnetic flux through circuit Γ_2 due to the current I_1 in circuit Γ_1 is

$$\Phi_{S(\Gamma_2),\mathbf{B}_1} = M I_1 \,.$$

Similarly, the magnetic flux through circuit Γ_1 due to the current I_2 in circuit Γ_2 is

$$\Phi_{S(\Gamma_1),\mathbf{B}_2} = M I_2 \,.$$

This symmetry in the mutual inductance is known as the reciprocity principle in electromagnetism.

Example 11.4 - Mutual inductance between non-coplanar loops
Consider two loops, as shown in Fig. 11.21. The smaller loop, of radius r, is centered on the axis of the larger loop, of radius R with $r \ll R$, at a distance z from its center. The planes of the two loops are inclined at an angle α to each other.

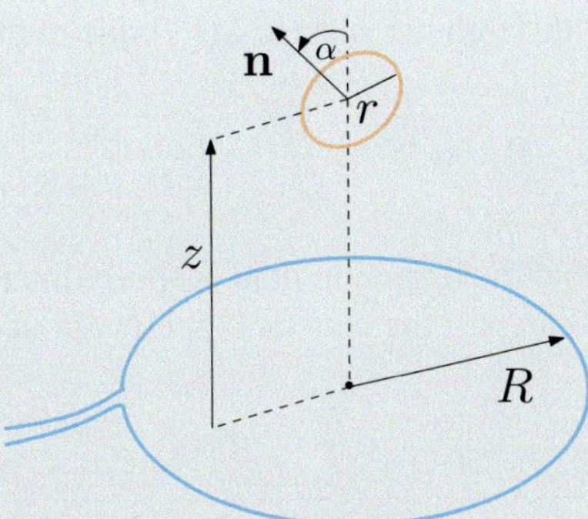

Fig. 11.21 Two loops whose normal vectors form an angle α with each other

One approach to calculate the mutual inductance between the two loops is to directly calculate the magnetic flux through loop Γ_1 due to a current in loop Γ_2. However, this calculation can be complex due to the inhomogeneous magnetic field generated by the smaller loop.

   A simpler approach is to calculate the magnetic flux through loop Γ_2 due to a current I_1 in loop Γ_1, as illustrated in Fig. 11.22. Since the radius r of the smaller loop is much smaller than the radius R of the larger loop, the magnetic field generated by the larger loop is approximately uniform over the area of the smaller loop. This simplifies the calculation significantly.

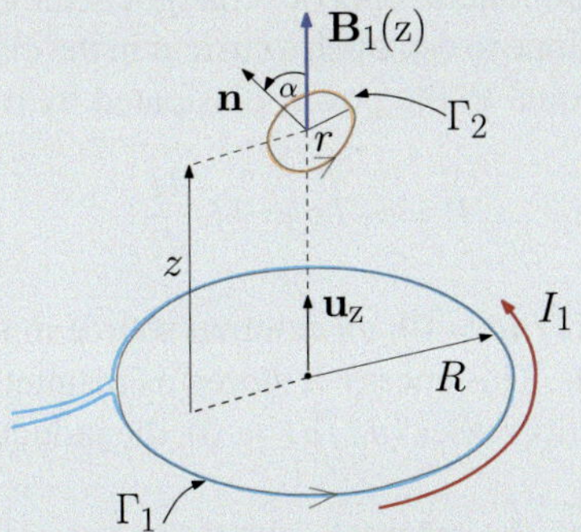

Fig. 11.22 The magnetic field generated by I_1 is approximately uniform over the area of the smaller loop Γ_2

The magnetic field on the axis of a circular loop of current, derived in Chap. 8, is given by

$$\mathbf{B}_1(z) = \frac{\mu_0 I_1 R^2}{2(z^2 + R^2)^{3/2}} \mathbf{u}_z \, .$$

The magnetic flux through the surface $S(\Gamma_2)$ due to this magnetic field is

$$\Phi_{S(\Gamma_2),\mathbf{B}_1} = \iint_{S(\Gamma_2)} \mathbf{B}_1 \cdot \mathbf{n}\,dS \approx \iint_{S(\Gamma_2)} \left(\frac{\mu_0 I R^2}{2(z^2 + R^2)^{3/2}} \right) \mathbf{u}_z \cdot \mathbf{n}\,dS .$$

Since $\mathbf{u}_z \cdot \mathbf{n} = \cos\alpha$ and the magnetic field is approximately constant over the small loop $S(\Gamma_2)$, we can pull the constant factors out of the integral:

$$\Phi_{S(\Gamma_2),\mathbf{B}_1} = \frac{\mu_0 I_1 R^2 \cos\alpha}{2(z^2 + R^2)^{3/2}} \iint_{S(\Gamma_2)} dS(\mathbf{x}) = \frac{\mu_0 \pi r^2 I_1 R^2 \cos\alpha}{2(z^2 + R^2)^{3/2}} .$$

Finally, the mutual inductance between the two loops is

$$M = \frac{\Phi_{S(\Gamma_2),\mathbf{B}_1}}{I_1} \approx \frac{\mu_0 \pi r^2 R^2 \cos\alpha}{2(z^2 + R^2)^{3/2}} .$$

11.6.4   Energy Stored in an Inductor

Due to its inductance L, a conductor opposes changes in the current flowing through it. Therefore, work must be done to establish a current in the circuit. The power supplied to the inductor must be equal to the power dissipated by the induced electromotive force:

$$P = -I\varepsilon = IL\frac{dI}{dt} .$$

If the current increases ($dI/dt > 0$), an external source must supply energy to overcome the self-induced emf. This energy is stored as magnetic energy in the inductor. Conversely, if the current decreases ($dI/dt < 0$), the inductor releases stored energy back to the circuit.

The total work done by the external source to increase the current from 0 to I over a time interval T is

$$W_{\text{ext}} = \int_0^T P(t)\,dt = \int_0^I LI'\,dI' = \frac{1}{2}LI^2 .$$

This work is stored as magnetic potential energy in the inductor:

$$\boxed{U_B = \frac{1}{2}LI^2 .} \tag{11.16}$$

This equation is analogous to the expression for the electric potential energy stored in a capacitor:

$$U_E = \frac{1}{2}\frac{Q^2}{C} \ .$$

Example 11.5 - Magnetic energy density in a solenoid

The inductance L of a solenoid with N turns, radius R, and length l is, according to (11.14),

$$L = \mu_0 \pi R^2 \frac{N^2}{l} \ .$$

The magnetic energy stored in this solenoid when a current I flows through it is

$$U_B = \frac{1}{2}LI^2 = \frac{\mu_0 \pi R^2 N^2 I^2}{2l} \ .$$

The magnetic field inside an infinitely long solenoid is uniform and has a magnitude of $B = \mu_0 N I / l$. Using this expression for the magnetic field, we can rewrite the magnetic energy stored in the solenoid as

$$U_B = \frac{\mu_0 \pi R^2 N^2}{2l}\frac{B^2 l^2}{\mu_0^2 N^2} = l\pi R^2 \frac{B^2}{2\mu_0} \ .$$

Since the volume of the solenoid is $V = \pi R^2 l$, the magnetic energy density u_B (energy per unit volume) is given by

$$u_B = \frac{U_B}{V} = \frac{1}{2\mu_0}B^2 \ .$$

This result is a general expression for the magnetic energy density in a region of space with a magnetic field $\mathbf{B}$.

11.7 Magnetic Potential Energy

11.7.1 *Energy Function for a Filamentary Circuit in an External Magnetic Field*

In electrostatics, we defined the potential energy of a charge q at position $\mathbf{x}$ in an electric potential $V(\mathbf{x})$ as the work done by an external agent to bring the charge from infinity to that position,

$$W = qV(\mathbf{x}) \ .$$

This work is done quasistatically to avoid energy dissipation. For a continuous charge distribution in a volume Ω with volume charge density $\varrho(\mathbf{x})$, the total electrostatic energy is

$$W = \frac{1}{2} \iiint_{\Omega} \varrho(\mathbf{x}) V(\mathbf{x}) d^3 x = \frac{\epsilon_0}{2} \iiint_{\mathbb{R}^3} \mathbf{E}^2(\mathbf{x}) d^3 x \ .$$

Analogously, we can define an energy function for a filamentary circuit Γ carrying a constant current I in an external magnetic field $\mathbf{B}(\mathbf{x})$. This energy represents the work done by an external agent to bring the circuit from infinity to its final position $\mathbf{x}$ and orient it in the external magnetic field $\mathbf{B}(\mathbf{x})$ (see Fig. 11.23). This process is assumed to be quasistatic, meaning it is slow enough to avoid energy dissipation. At infinity, where the magnetic field is negligible, no work is required to position the circuit. When the circuit is at position $\mathbf{x}$, the external magnetic field $\mathbf{B}(\mathbf{x})$ produces a magnetic flux $\Phi(\mathbf{x})$ through the surface bounded by the circuit.

The potential energy of a filamentary circuit is the work required to establish the current against the back emf induced by the changing magnetic field. We will show that this potential energy can be understood as the external work done by the external agent to move the circuit from infinity to its position $\mathbf{x}$ in the magnetic field,

$$\boxed{W = -I\Phi(\mathbf{x}) \ ,}$$

where $\Phi(\mathbf{x}) = \Phi_{S(\Gamma),\mathbf{B}} = \iint_{S(\Gamma)} \mathbf{B} \cdot \mathbf{n} dS$ is the magnetic flux through the circuit Γ at position $\mathbf{x}$.

In Chap. 8, we derived the expression for the force $\mathbf{f}$ exerted on a closed circuit (or a magnetic dipole) due to an external magnetic field $\mathbf{B}$. The x-component of this force is given by

$$f_x = \int_{\Gamma} I d\mathbf{l} \times \mathbf{B} \cdot \mathbf{u}_x \ .$$

Using a circular permutation on the triple product and the Stokes theorem (A.20), we can rewrite this as

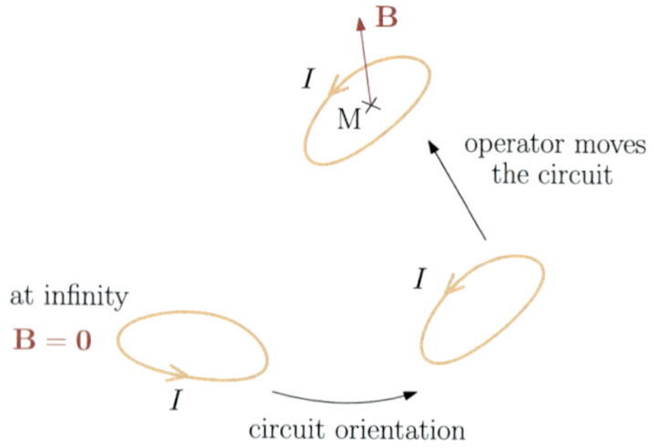

Fig. 11.23 The potential energy function of a filamentary circuit is the work done by an external agent to move the circuit quasistatically from infinity to its position in the magnetic field

$$f_x = \int_\Gamma I\mathbf{B} \times \mathbf{u}_x \cdot d\mathbf{l} = \iint_S I\nabla \times (\mathbf{B} \times \mathbf{u}_x) \cdot \mathbf{n}\,dS \ .$$

Since $\nabla \times (\mathbf{B} \times \mathbf{u}_x) = \dfrac{\partial \mathbf{B}}{\partial x}$, the force exerted on a circuit in the magnetic field $\mathbf{B}$ is

$$f_x = \iint_{S(\Gamma)} I\frac{\partial \mathbf{B}}{\partial x} \cdot \mathbf{n}\,dS \ .$$

To bring the circuit from infinity to its final position quasistatically, an external agent must exert a force $-\mathbf{f} = -f_x\mathbf{u}_x - f_y\mathbf{u}_y - f_z\mathbf{u}_z$, opposite to the magnetic force. For an infinitesimal displacement $d\mathbf{x} = dx\,\mathbf{u}_x + dy\,\mathbf{u}_y + dz\,\mathbf{u}_z$, the work done by the external agent is

$$\delta W = -\mathbf{f} \cdot d\mathbf{x} = -\iint_{S(\Gamma)} I\left(\frac{\partial \mathbf{B}}{\partial x}dx + \frac{\partial \mathbf{B}}{\partial y}dy + \frac{\partial \mathbf{B}}{\partial z}dz\right) \cdot \mathbf{n}\,dS \ .$$

This can be simplified as

$$\delta W = -\iint_{S(\Gamma)} I\,d\mathbf{B} \cdot \mathbf{n}\,dS = -I\,\Phi_{S(\Gamma),d\mathbf{B}} \ .$$

The vector $d\mathbf{B}(\mathbf{x})$ represents the change in the magnetic field at position $\mathbf{x}$ due to the infinitesimal displacement $d\mathbf{x}$ of the circuit, as illustrated in Fig. 11.24. The corresponding change in the magnetic flux through the circuit, $d\Phi_{S(\Gamma),\mathbf{B}}$ is linearly related to $d\mathbf{B}$: it is the magnetic flux $\Phi_{S(\Gamma),d\mathbf{B}}$. Therefore, the work done by the external agent during this infinitesimal displacement is:

$$\delta W = -I\,\Phi_{S(\Gamma),d\mathbf{B}} = -I\,d\Phi_{S(\Gamma),\mathbf{B}} \ .$$

This expression implies that the work done is an exact differential, meaning that the total work required to bring the circuit from infinity to its final position $\mathbf{x}$ is independent of the path taken. Hence, we can write the total work as

$$\boxed{W(\mathbf{x}) = -I\,\Phi(\mathbf{x}) \ .}$$

Fig. 11.24 The field experienced by the loop when moving from $\mathbf{x}$ to $\mathbf{x}'$ changes by the quantity $d\mathbf{B}$

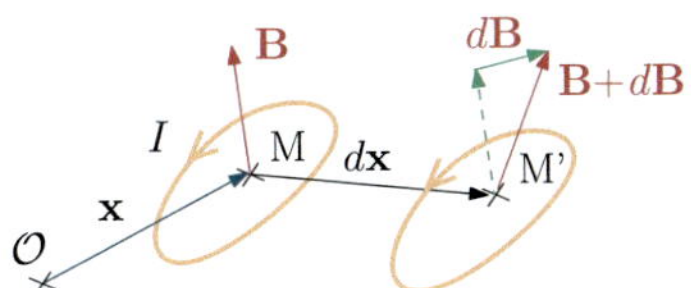

11.7.2 Forces and Torques on Current-Carrying Loops

The Laplace force acting on a line element $d\mathbf{l}$ of the circuit carrying a current I in a magnetic field $\mathbf{B}$ is given by

$$d\mathbf{f} = I d\mathbf{l} \times \mathbf{B} .$$

Integration over the circuit leads to a total force $\mathbf{f}$. Similarly, the magnetic field exerts a total torque $\boldsymbol{\tau} = \int_\Gamma (\mathbf{x} \times d\mathbf{f}(\mathbf{x}))$ on the circuit. To maintain its position and orientation, an external agent must exert an opposite force, $-\mathbf{f}$, and torque $-\boldsymbol{\tau}$. The work done by the external agent for an infinitesimal displacement $d\mathbf{x}$ and rotation $d\boldsymbol{\omega}$ is

$$dW = -\mathbf{f} \cdot d\mathbf{x} - \boldsymbol{\tau} \cdot d\boldsymbol{\omega} .$$

This work can be expanded as

$$dW = -f_x dx + -f_y dy + -f_z dz - \tau_\alpha d\alpha - \tau_\beta d\beta - \tau_\gamma d\gamma ,$$

where $d\alpha$, $d\beta$, $d\gamma$ are the components of the infinitesimal rotation $d\boldsymbol{\omega}$. The work function, as we have shown, is $W = -I\Phi$. The magnetic flux $\Phi(\mathbf{x}, \boldsymbol{\omega})$ depends on the position $\mathbf{x} = (x, y, z)$ of the circuit center of mass and its orientation $\boldsymbol{\omega} = (\alpha, \beta, \gamma)$.

Therefore, the components of the force $\mathbf{f}$ and torque $\boldsymbol{\tau}$ on the circuit can be obtained from the energy function using the following relations:

$$\boxed{f_x = -\frac{\partial W}{\partial x} = I\frac{\partial \Phi}{\partial x}} , \qquad \boxed{\tau_\alpha = -\frac{\partial W}{\partial \alpha} = I\frac{\partial \Phi}{\partial \alpha}} .$$

Similar expressions can be derived for the other components of the force and torque.

11.7.3 Magnetic Free Energy

In this section, we will derive expressions for the magnetic energy stored in a system of current-carrying conductors. These conductors interact with each other through the magnetic fields they generate. We will interpret this magnetic energy as the *free energy* of the system, treating it as a thermodynamic system that can exchange electrical work and heat with its environment. This free energy concept allows us to analyze the equilibrium properties and forces acting on the system. A brief review of thermodynamics, including the concept of free energy, is provided in Sect. A.9.

11.7.3.1 Continuous Distribution of Currents

Consider a general current distribution in a system of conductors where the initial current density $\mathbf{j}(\mathbf{x})$ and vector potential $\mathbf{A}(\mathbf{x})$ are zero. We aim to determine the energy required to establish a steady-state regime with a final current density $\mathbf{j}(\mathbf{x})$ and vector potential $\mathbf{j}(\mathbf{A})$. From Chap. **??**, we know that these quantities must satisfy Poisson's equation in the Coulomb gauge,

$$\nabla^2 \mathbf{A}(\mathbf{x}) = -\mu_0 \mathbf{j}(\mathbf{x}) \ .$$

To achieve this final state, we introduce a set of external generators connected to the conductors. These generators gradually increase the current in the conductors, ensuring that a steady-state equilibrium is maintained throughout the process.

The linearity of Poisson's equation implies that any intermediate state between the initial and final states can be described by a current density $\lambda \mathbf{j}(\mathbf{x})$ and a vector potential $\lambda \mathbf{A}(\mathbf{x})$, where $0 \leq \lambda \leq 1$. As the current density increases with time, the vector potential also increases, and an electromotive force is induced,

$$\mathbf{E}_{\text{emf}} = -\frac{d\lambda}{dt} \mathbf{A}(\mathbf{x}) \ .$$

To maintain the steady-state current distribution, the external generators must supply an electric field equal and opposite to this induced electric field.

The power per unit volume supplied by the generators to the conductors is

$$p = -\mathbf{E}_{\text{emf}} \cdot (\lambda \mathbf{j}) = \lambda \frac{d\lambda}{dt} \mathbf{A}(\mathbf{x}) \cdot \mathbf{j}(\mathbf{x}) \ .$$

The work done per unit volume by the generators to increase λ by $d\lambda$ is

$$dw = pdt = \mathbf{A}(\mathbf{x}) \cdot \mathbf{j}(\mathbf{x}) \lambda d\lambda \ .$$

Integrating this expression from $\lambda = 0$ to $\lambda = 1$, we obtain the total work density supplied to the conductors:

$$w(\mathbf{x}) = \frac{1}{2} \mathbf{A}(\mathbf{x}) \cdot \mathbf{j}(\mathbf{x}) \ .$$

In thermodynamics, the work done by an external agent during a quasistatic, reversible process is equal to the change in the free energy of the system: $\delta W = dF$ for an infinitesimal transformation, and by integration, $W = \Delta F = F_{\text{final}} - F_{\text{initial}}$. Therefore, the work density $w(\mathbf{x})$ we calculated earlier corresponds to the free energy density of the system.

The total free energy of the system, contained within a volume Ω, is given by

$$\boxed{F = \frac{1}{2} \iiint_{\Omega} \mathbf{A}(\mathbf{x}) \cdot \mathbf{j}(\mathbf{x}) d^3x \ .} \tag{11.17}$$

### 11.7.3.2	Case of Filamentary Circuits

Consider a system of n filamentary circuits Γ_l (with $l = 1, \ldots, n$), each carrying a current I_l, as shown in Fig. 11.25. For this system, the volume integral in (11.17) can be reduced to a line integral over the circuits:

$$F = \frac{1}{2} \sum_{l=1}^{n} \oint_{\Gamma_l} \mathbf{A}(\mathbf{x}') \cdot I_l d\mathbf{l}_l(\mathbf{x}') \, ,$$

where $d\mathbf{l}_l$ is an infinitesimal line element along the l-th circuit.

Applying the Green–Ostrogradsky theorem (A.15) to the line integrals, we can express the free energy as

$$F = \frac{1}{2} \sum_{l=1}^{n} I_l \iint_{S_l(\Gamma_l)} \mathbf{B}(\mathbf{x}') \cdot \mathbf{n}_l(\mathbf{x}') dS_l(\mathbf{x}') = \frac{1}{2} \sum_{l=1}^{n} I_l \Phi_l \, ,$$

where Φ_l is the magnetic flux through the l-th circuit. This magnetic flux consists of two components:

1. *Self-flux*: The flux produced by the current I_l in the l-th circuit itself, given by $L_l I_l$, where L_l is the self-inductance of the l-th circuit.
2. *Induced flux*: The flux produced by the currents in the other circuits, given by $\sum_{k \neq l} M_{lk} I_k$, where M_{lk} is the mutual inductance between circuits l and k.

In matrix form, the relationship between the fluxes and currents can be expressed as

$$\begin{bmatrix} \Phi_1 \\ \vdots \\ \Phi_n \end{bmatrix} = \underbrace{\begin{bmatrix} L_1 & \cdots & M_{1n} \\ \vdots & \ddots & \vdots \\ M_{n1} & \cdots & L_n \end{bmatrix}}_{\text{inductance matrix } [\mathcal{L}]} \begin{bmatrix} I_1 \\ \vdots \\ I_n \end{bmatrix} .$$

The matrix $[\mathcal{L}]$ is called the inductance matrix.

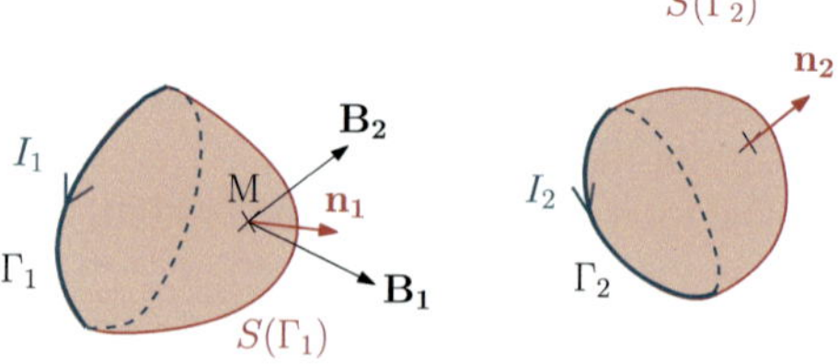

Fig. 11.25 Two current-carrying filamentary circuits

11.7.3.3 Properties of the Inductance Matrix

- The inductance matrix $[\mathcal{L}]$ is a symmetric matrix, meaning that $M_{lk} = M_{kl}$. This property is a direct consequence of Neumann's formula (11.15) and reflects the reciprocity principle of electromagnetic induction. For a system of two circuits, the inductance matrix takes the form

$$\begin{bmatrix} \Phi_1 \\ \Phi_2 \end{bmatrix} = \begin{bmatrix} L_1 & M \\ M & L_2 \end{bmatrix} \begin{bmatrix} I_1 \\ I_2 \end{bmatrix},$$

where $M_{12} = M_{21} = M$ is the mutual inductance between the two circuits.

- The diagonal elements (self-inductances) L_l are always positive. This is because a positive current I_l in the l-th circuit always produces a magnetic flux Φ_l that reinforces the current flow. This property ensures that the stored magnetic energy is always positive.

  This can also be understood from the orientation of the circuit. The direction of the current I_l defines a *north pole* and a *south pole* for the circuit Γ_l. The magnetic field $\mathbf{B}_l$ generated by the circuit points from the south pole to the north pole, which aligns with the orientation of the normal vector $\mathbf{n}_l$ to the surface S_l bounded by the circuit (see Fig. 11.26). As a result, the self-induced flux $\Phi_{S_l(\Gamma_l),\mathbf{B}_l}$ has the same sign as the current I_l. Since $\Phi_{S_l(\Gamma_l),\mathbf{B}_l} = L_l I_l$, it follows that $L_l > 0$ for all l.

- The inductance matrix $[\mathcal{L}]$ is a positive definite matrix. This means that its determinant is positive: $\det[\mathcal{L}] > 0$. For a system of two conductors, this condition implies an upper bound on the mutual inductance coefficient: $M^2 < L_1 L_2$.

- The sign of the mutual inductance M_{lk} depends on the relative orientation of the two circuits Γ_l and Γ_k. Figure 11.27 illustrates this point for a system of two circuits. In the first configuration, the magnetic flux $\Phi_{S_2,\mathbf{B}_1}$ through circuit Γ_2 due to the magnetic field $\mathbf{B}_1$ generated by Γ_1 is positive. This results in a positive mutual inductance:

$$\Phi_{S_2,\mathbf{B}_1} > 0, \quad \Phi_{S_2,\mathbf{B}_1} = M_{12} I_1 \quad \Rightarrow \quad M_{12} > 0.$$

In the second configuration, reversing the orientation of circuit Γ_2 leads to a negative magnetic flux $\Phi_{S_2,\mathbf{B}_1}$ and, consequently, a negative mutual inductance:

$$\Phi_{S_2,\mathbf{B}_1} < 0, \quad \Phi_{S_2,\mathbf{B}_1} = M_{12} I_1 \quad \Rightarrow \quad M_{12} < 0.$$

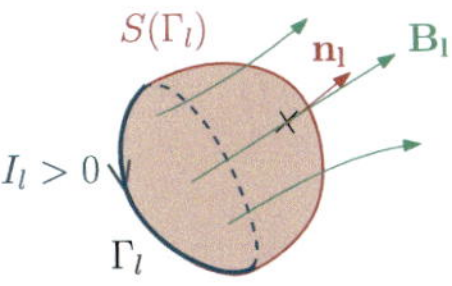

Fig. 11.26 The self-induced flux has the same sign as the current

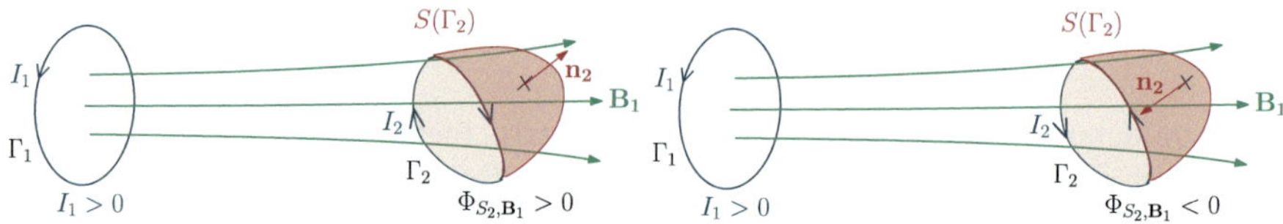

Fig. 11.27 Two current-carrying filamentary circuits

Remark: The concept of self-inductance is not strictly applicable to ideal, infinitely thin, filamentary circuits. This is because the Biot–Savart law, when applied to calculate the magnetic field at a point on the wire itself, leads to a divergent integral. However, in practical situations, wires have a finite radius. By considering a wire with a small but finite radius, we can avoid this singularity and calculate the self-inductance of a filamentary circuit. Note that the concept of mutual inductance between two filamentary circuits is well-defined and can be calculated without encountering such singularities.

11.7.4 Magnetic Energy Density in Terms of the Magnetic Field

From the expression (11.17) for the free energy, we can extend the integration volume to all of space $\mathbb{R}^3$ since the current density is zero outside the region Ω. Then, using Ampère's law $\mathbf{j} = \nabla \times \dfrac{\mathbf{B}}{\mu_0}$, we can rewrite the free energy as

$$F = \iiint_{\mathbb{R}^3} \frac{1}{2\mu_0} \mathbf{A}(\mathbf{x}) \cdot \nabla \times \mathbf{B}(\mathbf{x})\, d^3x \ .$$

Using the vector identity (A.10), $\nabla \cdot (\mathbf{A} \times \mathbf{B}) = -\mathbf{A} \cdot \nabla \times \mathbf{B} + \mathbf{B} \cdot \nabla \times \mathbf{A}$, and the fact that $\mathbf{B} = \nabla \times \mathbf{A}$, we obtain

$$F = -\frac{1}{2\mu_0} \iiint_{\mathbb{R}^3} \nabla \cdot (\mathbf{A} \times \mathbf{B})(\mathbf{x}) d^3x + \frac{1}{2\mu_0} \iiint_{\mathbb{R}^3} \mathbf{B}^2(\mathbf{x}) d^3x \ ,$$

Applying the divergence theorem (A.15) to the first integral and noting that the surface integral at infinity, $\oiint_{(S)\to\infty} (\mathbf{A} \times \mathbf{B})(\mathbf{x}) \cdot \mathbf{n} dS$, vanishes due to the decay of the fields, we arrive at

$$\boxed{F = \frac{1}{2\mu_0} \iiint_{\mathbb{R}^3} \mathbf{B}^2(\mathbf{x})\, d^3x \ .}$$

From this expression, we can identify the magnetic energy density,

$$u_B = \frac{\mathbf{B}^2}{2\mu_0} \, .$$ (11.18)

11.7.5 Helmholtz Free Energy and Gibbs Free Energy

Returning to the derivation of Eq. (11.17), we note that the transition from the steady state $(\mathbf{j}, \mathbf{A})$ to a neighboring state $(\mathbf{j} + d\mathbf{j}, \mathbf{A} + d\mathbf{A})$ involves the generation of an electric field $d\mathbf{A}/dt$ by external generators. This electric field opposes the electromotive force induced in the conductors. The resulting change in the free energy density is

$$df = \mathbf{j}(\mathbf{x}) \cdot d\mathbf{A}(\mathbf{x}),$$ (11.19)

This equation indicates that $\mathbf{j}$ and $\mathbf{A}$ are conjugate variables. A natural independent variable for the free energy density $f(\mathbf{A})$ is the vector potential $\mathbf{A}$.[1] Poisson's equation for the vector potential implies a linear relationship between the current density $\mathbf{j}$ and the vector potential $\mathbf{A}$. We can express this relationship as $\mathbf{A} = \lambda\mathbf{j}$, where λ is a constant. This linear relationship serves as an equation of state for the system.

To derive the free energy density, we can integrate the differential $df = \mathbf{j} \cdot d\mathbf{A}$, using the equation of state. Since $d\mathbf{A} = \lambda d\mathbf{j}$, we have

$$f = \int \lambda\mathbf{j} \cdot d\mathbf{j} = \lambda\frac{\mathbf{j}^2}{2} = \frac{1}{2}\mathbf{j} \cdot \mathbf{A} \, .$$

We could have integrated the differential $df = \mathbf{j} \cdot d\mathbf{A}$ with respect to $\mathbf{A}$ instead of $\mathbf{j}$. The key point is that the relationship between $\mathbf{j}$ and $\mathbf{A}$ is linear and can be expressed as an equation of state.[2] This equation of state allows us to express the differential df in terms of a single variable and integrate accordingly.

For a system of filamentary circuits, the free energy can be expressed as

$$F = \frac{1}{2}\sum_l I_l\Phi_l \, .$$ (11.20)

Here, the currents I_l and the magnetic fluxes Φ_l are conjugate variables, similar to the relationship between $\mathbf{j}$ and $\mathbf{A}$ for a distribution of volume current density. This means that the fluxes Φ_l are the natural independent variables for the Helmholtz free energy, as indicated by the differential

---

[1] In thermodynamics, for a gas undergoing a quasistatic process, the natural independent variables for the internal energy $U(V, S)$ are volume V and entropy S. This is expressed by the differential: $dU = -pdV + TdS$. The Helmholtz free energy $F(V, T)$ is a function of two independent variables: volume and temperature, with its differential given by: $dF = -pdV - SdT$.

[2] Similarly, in thermodynamics, pressure and volume are conjugate variables, linked by the equation of state for an ideal gas, $pV = nRT$. When calculating work done during a thermodynamic process, we express one variable in terms of the other using the equation of state.

$$dF = \sum_l I_l \, d\Phi_l \ .$$

The relationship between the fluxes and currents is given by the inductance matrix:

$$\Phi_l = \sum_{k=1}^{n} [\mathcal{L}]_{lk} I_k.$$

where $[\Phi_l]$ and $[I_k]$ are vectors of fluxes and currents, respectively. Inserting this relation in (11.20), we can express the Helmholtz free energy as a function of the currents,

$$F = \frac{1}{2} \sum_{l=1}^{n} \sum_{k=1}^{n} I_l [\mathcal{L}]_{lk} I_k \ .$$

Alternatively, by inverting the inductance matrix,[3] we can express the Helmholtz free energy as a function of its natural independent variables, the fluxes:

$$F(\Phi_1, \ldots, \Phi_n) = \frac{1}{2} \sum_{l=1}^{n} \sum_{k=1}^{n} \Phi_l [\mathcal{L}]_{lk}^{-1} \Phi_k \ .$$

This expression for the Helmholtz free energy can be used to derive the forces and torques acting on the elements of the circuit system.

Example 11.6 - Helmholtz free energy for two filamentary circuits
Consider two circuits connected to generators that maintain a steady-current flow. Any change in the current in one circuit will induce a change in the magnetic flux through the other circuit. This, in turn, induces an electromotive force $\varepsilon = -d\Phi/dt$ in the second circuit. To maintain the steady-state current, the generators must supply an emf ε_g equal and opposite to the induced emf: $\varepsilon_g = +d\Phi/dt$.

During a time interval dt, the generators provide an amount of energy given by

$$dF = \varepsilon_{g1} dt \, I_1 + \varepsilon_{g2} dt \, I_2 = I_1 d\Phi_1 + I_2 d\Phi_2.$$

Using the relationship between the fluxes and currents, $[\Phi_1, \Phi_2]^\dagger = [\mathcal{L}][i_1, i_2]^\dagger$, where $\dagger$ denotes transposition, we can express the differential of the free energy as

$$dF = I_1(L_1 dI_1 + M dI_2) + I_2(M dI_1 + L_2 dI_2)$$
$$= L_1 I_1 dI_1 + 2M d(I_1 I_2) + L_2 I_2 dI_2 \ .$$

---

[3] It is invertible as $[\mathcal{L}]$ is a positive definite matrix.

Integrating the differential dF, we obtain the final expression for the Helmholtz free energy

$$F = \frac{1}{2}L_1 I_1^2 + M I_1 I_2 + \frac{1}{2}L_2 I_2^2 \, .$$

The natural independent variables for the Helmholtz free energy F of a system of filamentary circuits are indeed the magnetic fluxes. While the Helmholtz free energy can be expressed as a function of currents, deriving forces and torques directly from this expression can be challenging, as it involves keeping the fluxes constant while taking derivatives with respect to the spatial coordinates or angular orientations, and varying the currents.

To address this, we can use a Legendre transform to obtain a new potential, where the currents are the independent variables. This new potential, known as the the Gibbs free energy, can be used to directly calculate forces and torques while keeping the currents I_l constant. The Legendre transform that defines the Gibbs free energy consists in subtracting products of pairs of conjugate variables:

$$G = F - \sum_l I_l \Phi_l \, .$$

The differential of the Gibbs free energy is

$$dG = \underbrace{dF}_{\sum_l i_l d\Phi_l} - \sum_l (I_l d\Phi_l + \Phi_l dI_l) \quad \Rightarrow \quad \boxed{dG = - \sum_l \Phi_l dI_l \, .}$$

This shows that the currents I_l are the independent variables of the Gibbs free energy.

For a distribution of volume current density $\mathbf{J}$, a Legendre transform of the Helmoltz free energy density, results in the Gibbs free energy density g:

$$g = f - \mathbf{A} \cdot \mathbf{j} \, .$$

Differentiating this relation, we find

$$dg = \underbrace{df}_{\mathbf{j} \cdot d\mathbf{A}} - \mathbf{A} \cdot d\mathbf{j} - \mathbf{j} \cdot d\mathbf{A} \quad \Rightarrow \quad \boxed{dg = -\mathbf{A} \cdot d\mathbf{j} \, ,}$$

which shows that the natural independent variable for the Gibbs free energy is indeed the current density $\mathbf{J}$.

Example 11.7 - Gibbs free energy for two filamentary circuits

For a set of two circuits, the Gibbs free energy G can be obtained from the Helmholtz free energy F through the Legendre transform

$$G(I_1, I_2) = F - I_1 \Phi_1 - I_2 \Phi_2 .$$

Differentiating and inserting the differential of the Helmholtz free energy, $dF = I_1 d\Phi_1 + I_2 d\Phi_2$, the differential of the Gibbs free energy is expressed as

$$dG = -\Phi_1 dI_1 - \Phi_2 dI_2 .$$

The exactness of the differential dG implies the following equality known as the Schwarz property:

$$\frac{\partial \Phi_1}{\partial I_2} = \frac{\partial \Phi_2}{\partial I_1} .$$

This condition is equivalent to the symmetry of the mutual inductance coefficients: $M_{12} = M_{21} = M$.

Inserting the equation of state $[\Phi_1, \Phi_2]^\dagger = [\mathcal{L}][I_1, I_2]^\dagger$ in the differential dG and integrating with respect to the currents, we obtain the expression for the Gibbs free energy:

$$G(I_1, I_2) = -\frac{1}{2} L_1 I_1^2 - M I_1 I_2 - \frac{1}{2} L_2 I_2^2 .$$

The components of the torque applied to circuit 1 can be obtained by taking the gradient of the Gibbs free energy with respect to orientation coordinates, α_i, while keeping the currents constant:

$$\tau_{\alpha_1} = -\left.\frac{\partial G}{\partial \alpha_1}\right|_{I_1, I_2} = \frac{1}{2}\frac{\partial L_1}{\partial \alpha_1} I_1^2 + \frac{\partial M}{\partial \alpha_1} I_1 I_2 + \frac{1}{2}\frac{\partial L_2}{\partial \alpha_1} I_2^2 .$$

To directly calculate the torque from the free energy expressed in terms of fluxes, we would need to express the fluxes as functions of the currents and their derivatives with respect to the orientation coordinates, α_i. This can be a complex task, especially for systems with multiple circuits.

Therefore, using the Gibbs free energy, where the currents are the independent variables, provides a more straightforward approach to calculating forces and torques. By taking the derivative of the Gibbs free energy with respect to the appropriate variable while keeping the currents constant, we can directly obtain the desired force or torque. This highlights the importance of choosing the appropriate potential function for a given problem, based on the variables that are held constant. In this case, the Gibbs free energy is the more convenient choice for calculating forces and torques while keeping the currents fixed.

11.8 The Magnetic Free Energy in Magnetized Media

We have previously derived the expression for the energy density of a magnetic field in vacuum, which represents the work required to establish a magnetic field $\mathbf{B}$ (or equivalently, a vector potential and current distribution). We will now extend this concept to magnetic media, where we aim to determine the energy required to establish a magnetic induction $\mathbf{H}$ and a magnetic field $\mathbf{B}$ within an initially unmagnetized medium.

11.8.1 Differential Expression for the Volume Density of Free Energy

The general expression for the vector potential in a magnetic medium is given by

$$\mathbf{A}(\mathbf{x}) = \iiint_{\Omega} \frac{\mu_0}{4\pi} \frac{\mathbf{j}(\mathbf{x}')}{|\mathbf{x} - \mathbf{x}'|} d^3 x' + \iiint_{\Omega} \frac{\mu_0}{4\pi} \frac{\mathbf{j}_M(\mathbf{x}')}{|\mathbf{x} - \mathbf{x}'|} d^3 x' + \iint_{\partial\Omega} \frac{\mu_0}{4\pi} \frac{\mathbf{j}_{s,M}(\mathbf{x}')}{|\mathbf{x} - \mathbf{x}'|} dS \ .$$

This equation reveals a nonlinear relationship between the vector potential $\mathbf{A}$ and the conduction current density $\mathbf{j}$, due to the presence of magnetization current densities $\mathbf{j}_M$ in Ω and $\mathbf{j}_{s,M}$ in $\partial\Omega$, which are themselves functions of the magnetic field.

If we have a steady-state regime with a current density $\mathbf{j}(\mathbf{x})$ and a corresponding vector potential $\mathbf{A}(\mathbf{x})$, a small perturbation in the current density, $\delta\mathbf{j}(\mathbf{x})$, will result in a small change in the vector potential, $\delta\mathbf{A}(\mathbf{x})$. This change in the vector potential is accompanied by an electromotive field given by

$$\mathbf{E}_{\text{emf}} = -\frac{\partial\mathbf{A}}{\partial t} \ .$$

The external operator must apply an electric field equal and opposite to the induced to maintain the steady-state current. This electric field does work on the free charges in the conductor. The work done on a charge element $\varrho d^3 x$ moving at velocity $\mathbf{v}$ during the time interval dt is

$$\delta W = (\varrho d^3 x)(-\mathbf{E}_{\text{emf}}) \cdot \mathbf{v}\, dt = \varrho \mathbf{v} \cdot \frac{\partial \mathbf{A}}{\partial t} dt\, (d^3 x) \,,$$

where ϱ is the charge density. Recognizing the current density $\mathbf{j} = \varrho \mathbf{v}$, the work done per per unit volume is given by

$$\frac{\delta W}{d^3 x} = \mathbf{j}(\mathbf{x}) \cdot \delta \mathbf{A}(\mathbf{x}) \,.$$

Integrating this expression over the entire volume Ω, we obtain the total work done, corresponding to the change in Helmholtz free energy:

$$dF = \iiint_\Omega \mathbf{j}(\mathbf{x}') \cdot \delta \mathbf{A}(\mathbf{x}') d^3 x' \,.$$

Since the current density is zero outside of Ω, we can extend the integration to all of space:

$$dF = \iiint_{\mathbb{R}^3} \mathbf{j}(\mathbf{x}') \cdot \delta \mathbf{A}(\mathbf{x}') d^3 x' \,.$$

Using the vector calculus identity:

$$\nabla \cdot (\mathbf{H} \times \delta \mathbf{A}) = -\mathbf{H} \cdot \nabla \times \delta \mathbf{A} + \delta \mathbf{A} \cdot \nabla \times \mathbf{H} \,,$$

and recognizing that $\nabla \times \delta \mathbf{A} = \delta \mathbf{B}$ and $\nabla \times \mathbf{H} = \mathbf{j}$, we can rewrite the expression for the work done as

$$dF = \underbrace{\iiint_{\mathbb{R}^3} \nabla \cdot (\mathbf{H} \times \delta \mathbf{A}) d^3 x'}_{\oiint_{S \to \infty} (\mathbf{H} \times \delta \mathbf{A}) \cdot \mathbf{n}\, dS = 0} + \iiint_{\mathbb{R}^3} \mathbf{H} \cdot \delta \mathbf{B} \,.$$

The first integral on the right-hand side can be converted to a surface integral using the divergence theorem. However, this surface integral vanishes as we integrate over all of space and the fields decay at infinity. Therefore, we are left with

$$dF = \iiint_{\mathbb{R}^3} \mathbf{H} \cdot \delta \mathbf{B} \,.$$

This leads to the following expression for the differential of the free energy density $f(\mathbf{B})$:

$$\boxed{df = \mathbf{H} \cdot d\mathbf{B} \,.}$$

This equation indicates that $\mathbf{H}$ and $\mathbf{B}$ are conjugate variables. Since the free energy is a state function, it can be expressed as a function of either $\mathbf{H}$ or $\mathbf{B}$.

To proceed, we need to introduce an equation of state that relates $\mathbf{H}$ and $\mathbf{B}$ for the specific magnetic medium. This equation of state will allow us to express the free energy as a function of a single variable.

11.8.2   Properties of Magnetic Free Energy

From the relation between the magnetic induction $\mathbf{B}$ and the magnetic field $\mathbf{H}$,

$$\mathbf{B} = \mu_0(\mathbf{H} + \mathbf{M}),$$

we obtain by differentiation:

$$d\mathbf{B} = \mu_0(d\mathbf{H} + d\mathbf{M}) .$$

Substituting this into the differential of the free energy density, we get

$$df = \mu_0 \mathbf{H} \cdot d\mathbf{H} + \mu_0 \mathbf{H} \cdot d\mathbf{M} .$$

The first term, $\mu_0 \mathbf{H} \cdot d\mathbf{H}$, represents the energy required to establish the magnetic excitation field $\mathbf{H}$, also called the magnetizing field, while the second term, $\mu_0 \mathbf{H} \cdot d\mathbf{M}$, represents the energy required to establish the magnetization in the medium.

In the case where the magnetization $\mathbf{M}$ is negligible compared to the magnetic field $\mathbf{H}$, we have $\mathbf{B} \simeq \mu_0 \mathbf{H}$ and the differential of the free energy density simplifies to

$$df = \mathbf{B} \cdot d\mathbf{H} .$$

In this case, only the energy required to establish the magnetizing field $\mathbf{H}$ contributes to the total free energy.

11.8.3   Helmholtz and Gibbs Free Energies

Having identified $\mathbf{B}$ and $\mathbf{H}$ as conjugate variables in the expression for the Helmholtz free energy density $df = \mathbf{H} \cdot d\mathbf{B}$, we can introduce the Gibbs free energy density g through a Legendre transform:

$$g = f - \mathbf{H} \cdot \mathbf{B} .$$

The differential of the Gibbs free energy density is

$$dg = -\mathbf{B} \cdot d\mathbf{H} .$$

This shows that the natural independent variable for the Gibbs free energy is $\mathbf{H}$.

To determine the specific form of the Gibbs free energy density, we need an equation of state that relates $\mathbf{H}$ and $\mathbf{B}$ for the given magnetic medium. This equation of state will allow us to express the Gibbs free energy density $g(\mathbf{H})$ as a function of $\mathbf{H}$ only. Similarly, we can express the Helmholtz free energy density $f(\mathbf{B})$ as a function of $\mathbf{B}$.

11.8.4 Free Energy for Linear, Homogeneous, Isotropic Media

For paramagnetic and diamagnetic materials, the relationship between the magnetic induction $\mathbf{B}$ and the magnetic field $\mathbf{H}$ is linear:

$$\mathbf{B} = \mu\mathbf{H} = \mu_0\mu_r\mathbf{H} \, ,$$

where μ is the permeability of the medium, μ_0 is the permeability of free space, μ_r is the relative permeability of the material.

From the relation between $\mathbf{B}$, $\mathbf{H}$ and the magnetization $\mathbf{M}$,

$$\frac{\mathbf{B}}{\mu_0} = \mathbf{H} + \mathbf{M} \, ,$$

we can derive the following relationship between the magnetization and the magnetic field:

$$\left(\frac{\mu}{\mu_0} - 1\right)\mathbf{H} = \mathbf{M} \, .$$

Introducing the magnetic susceptibility χ:

$$\boxed{\mu_r = 1 + \chi \, ,}$$

we obtain the equation of state $\mathbf{M}(\mathbf{H})$:

$$\boxed{\mathbf{M} = \chi\mathbf{H} \, .}$$

For paramagnetic and diamagnetic materials, the magnetic susceptibility χ is very small ($|\chi| \ll 1$, $\chi \sim 10^{-3}$ for paramagnetic materials; $\chi \sim -10^{-5}$ for diamagnetic materials). This implies that the magnetization $\mathbf{M}$ is much smaller than the magnetic field $\mathbf{H}$. As a result, we can approximate the relationship between $\mathbf{M}$, $\mathbf{B}$ and $\mathbf{H}$ as

$$\boxed{\mathbf{M} \approx \chi\mathbf{H} = \chi\frac{\mathbf{B}}{\mu_0} \, .}$$

This shows that the magnetization $\mathbf{M}$ is parallel or antiparallel to $\mathbf{B}$ and $\mathbf{H}$. The sign of the susceptibility determines the direction of the magnetization:

- Paramagnetic materials: $\chi > 0$, $\mathbf{M}$ is parallel to $\mathbf{H}$ and $\mathbf{B}$
- Diamagnetic materials: $\chi < 0$, $\mathbf{M}$ is antiparallel to $\mathbf{H}$ and $\mathbf{B}$

For a linear magnetic medium, where $\mathbf{B} = \mu\mathbf{H}$, the differential of the Helmholtz free energy density can be integrated to yield

$$\boxed{f = \frac{1}{2}\mathbf{H} \cdot \mathbf{B} \quad \text{or} \quad f(\mathbf{B}) = \frac{\mathbf{B}^2}{2\mu} \, .}$$

The total Helmholtz free energy of a magnetized domain is then

$$\boxed{F = \iiint_{\mathbb{R}^3} \frac{\mathbf{B}^2(\mathbf{x}')}{2\mu} d^3 x' \, .}$$

Applying the Legendre transform $g = f - \mathbf{H} \cdot \mathbf{B}$, we obtain the Gibbs free energy density and total Gibbs free energy density for a linear magnetic material:

$$g = -\frac{1}{2}\mathbf{H} \cdot \mathbf{B} = -\frac{\mu\mathbf{H}^2}{2} \, , \quad \text{and} \quad G = - \iiint_{\mathbb{R}^3} \frac{\mu\mathbf{H}^2}{2} d^3 x' \, .$$

11.9 Summary and Essential Formulas

- When a circuit (or part of it) moves with velocity $\mathbf{v}$ in a time-varying magnetic field $\mathbf{B}$, an electromotive force (emf) is induced in the circuit, generating an induced current. This emf is associated with an electromotive field $\mathbf{E}_{\text{emf}}$, given by:

$$\boxed{\mathbf{E}_{\text{emf}} = \mathbf{v} \times \mathbf{B} - \frac{\partial \mathbf{A}}{\partial t} \, .}$$

The first term, $\mathbf{v} \times \mathbf{B}$, arises from the motion of the conductor in the magnetic field. The second term, $-\frac{\partial \mathbf{A}}{\partial t}$, is due to the time-varying nature of the fields. This second term highlights a fundamental connection between time-varying electric and magnetic fields.

- The circulation of the electromotive field $\mathbf{E}_{\text{emf}}$ around a closed path Γ is known as the electromotive force ε. Faraday's law of induction relates this emf to the time rate of change of the magnetic flux $\Phi_{S(\Gamma),\mathbf{B}}$ through any surface $S(\Gamma)$ bounded by the path Γ:

$$\varepsilon = \oint_\Gamma \mathbf{E}_{\text{emf}} \cdot d\mathbf{l} = -\frac{d}{dt}\Phi_{S(\Gamma),\mathbf{B}} \, ,$$

where the magnetic flux is defined as

$$\Phi_{S(\Gamma),\mathbf{B}} = \iint_{S(\Gamma)} \mathbf{B} \cdot \mathbf{n}\, dS \, .$$

- The total time derivative of the magnetic flux $\Phi_{S(\Gamma),\mathbf{B}}$ is

$$-\frac{d}{dt}\Phi_{S(\Gamma),\mathbf{B}} = -\left(\mathbf{v} \cdot \nabla + \frac{\partial}{\partial t}\right)\Phi_{S(\Gamma),\mathbf{B}} \, .$$

Applying Stokes's theorem to Faraday's law, we can identify the contributions to the electromotive force:

1. *Motional emf*:

$$\oint_\Gamma (\mathbf{v} \times \mathbf{B}) \cdot d\mathbf{l} = -(\mathbf{v} \cdot \nabla)\Phi_{S(\Gamma),\mathbf{B}} \, .$$

   This term arises from the motion of the circuit in the magnetic field.
2. *Transformer emf*:

$$\oint_\Gamma \left(-\frac{\partial \mathbf{A}}{\partial t}\right) \cdot d\mathbf{l} = -\frac{\partial}{\partial t}\Phi_{S(\Gamma),\mathbf{B}} \, .$$

   This term arises from the time variation of the fields.

- The circulation law for the electric field is modified in the time-varying regime. Faraday's law in differential form is given by

$$\nabla \times \mathbf{E} = -\frac{\partial \mathbf{B}}{\partial t} \, .$$

This equation implies that the electric field is no longer conservative in the time-varying case. Using the relation $\mathbf{B} = \nabla \times \mathbf{A}$ and the Coulomb gauge condition $\nabla \cdot \mathbf{A} = 0$, we can write the electric field as

$$\mathbf{E} = -\nabla V - \frac{\partial \mathbf{A}}{\partial t} \, .$$

The first term, $-\nabla V$, is the conservative part of the electric field associated with the electrostatic potential V. The second term, $-\frac{\partial \mathbf{A}}{\partial t}$, is the non-conservative part of the electric field due to its time dependence.

- The generalized form of Ohm's law for a conductor in a non-conservative electric field is

$$\boxed{V_A - V_B = R_{AB} I - \varepsilon_{AB} \, ,}$$

where $V_A - V_B$ is the potential difference between points A and B, R_{AB} is the resistance of the conductor segment between A and B, I is the current flowing from A to B, ε_{AB} is the electromotive force induced between points A and B.

- The forces and torques acting on a circuit Γ carrying a steady current I in a magnetic field $\mathbf{B}$ can be derived from the energy function,

$$W = -I\Phi, \quad \text{where} \quad \Phi = \iint_{S(\Gamma)} \mathbf{B} \cdot \mathbf{n} \, dS \, .$$

The components of the force $\mathbf{f}$ and torque $\boldsymbol{\tau}$ on the circuit can be obtained from the energy function using the following relations:

$$f_x = -\frac{\partial W}{\partial x} = I\frac{\partial \Phi}{\partial x}, \quad \tau_\alpha = -\frac{\partial W}{\partial \alpha} = I\frac{\partial \Phi}{\partial \alpha} \, .$$

Similar expressions can be derived for the other components of the force and torque.

- For a current distribution with density $\mathbf{j}$ in a volume Ω, the Helmholtz free energy density and total Helmholtz free energy are given by

$$\boxed{f = \frac{1}{2}\mathbf{j} \cdot \mathbf{A} \quad \text{and} \quad F = \frac{1}{2} \iiint_\Omega \mathbf{j}(\mathbf{x}) \cdot \mathbf{A}(\mathbf{x}) d^3x \, .}$$

Alternatively, the free energy density and total free energy can be expressed in terms of the magnetic field $\mathbf{B}$

$$\boxed{f = \frac{\mathbf{B}^2}{2\mu_0} \quad \text{and} \quad F = \frac{1}{2} \iiint_{\mathbb{R}^3} \frac{\mathbf{B}^2(\mathbf{x})}{\mu_0} d^3x \, .}$$

For a linear magnetic material, the permeability of free space, μ_0, must be replaced in these expressions with the permeability of the material, μ.

- For a system of filamentary circuits carrying currents I_l, the relationship between the magnetic fluxes Φ_l and the currents I_l can be expressed in matrix form:

$$\begin{bmatrix} \Phi_1 \\ \vdots \\ \Phi_n \end{bmatrix} = \begin{bmatrix} L_1 & \cdots & M_{1n} \\ \vdots & \ddots & \vdots \\ M_{n1} & \cdots & L_n \end{bmatrix} \begin{bmatrix} I_1 \\ \vdots \\ I_n \end{bmatrix}$$

Here, the diagonal elements L_l represent the self-inductances of the individual circuits, while the off-diagonal elements M_{lk} represent the mutual inductances between pairs of circuits. The inductance matrix is symmetric and positive definite.

- Mutual inductance between two circuits is quantified by Neumann's formula:

$$M_{12} = \oint_{\Gamma_1} \oint_{\Gamma_2} \frac{\mu_0}{4\pi} \frac{d\mathbf{l}_2(\mathbf{x}_2) \cdot d\mathbf{l}_1(\mathbf{x}_1)}{|\mathbf{x}_2 - \mathbf{x}_1|} .$$

- In the context of interacting current-carrying circuits, the Helmholtz free energy is a state function that represents the work done by external forces to establish the set of currents. It is a positive quantity. For a system of two circuits, the differential of the Helmholtz free energy is given by

$$dF = I_1 d\Phi_1 + I_2 d\Phi_2 .$$

The corresponding expression for the Helmholtz free energy is

$$F = \frac{1}{2} L_1 I_1^2 + M I_1 I_2 + \frac{1}{2} L_2 I_2^2 .$$

For a single circuit carrying a current I, the stored magnetic energy is

$$F = \frac{1}{2} L I^2 .$$

- The forces and torques acting on a circuit Γ_k within a system of circuits carrying currents I_l (where $l = 1 \ldots n$) can be derived from the Helmholtz free energy $F(\Phi_l, x_l, \alpha_l, \ldots)$ using the following relations:

$$f_{x,\Gamma_k} = - \left. \frac{\partial F}{\partial x_k} \right|_{\Phi_l, \ldots} , \qquad \tau_\alpha = - \left. \frac{\partial F}{\partial \alpha_k} \right|_{\Phi_l \ldots} ,$$

where $\Phi_{l\ldots}$ means that all fluxes $\Phi_l, l = 1 \ldots n$ must be held constant. Alternatively, we can use the Gibbs free energy $G(I_l, x_l, \alpha_l, \ldots)$, defined by the Legendre transform

$$G = F - \sum_i I_i \Phi_i .$$

The forces and torques can then be obtained from the Gibbs free energy as follows:

$$f_{x,\Gamma_k} = - \left. \frac{\partial G}{\partial x_k} \right|_{I_l, \ldots} , \qquad \tau_\alpha = - \left. \frac{\partial G}{\partial \alpha_k} \right|_{I_l \ldots} .$$

Problems

11.1 Path dependence of the potential difference

Consider the circuit shown below. The voltmeters V_L and V_R each have an internal resistance of $R = 1\,\text{M}\Omega$ and measure the potential difference between points a and b. A time-varying magnetic field is present in the region enclosed by the circuit. At a certain instant, the voltmeter on the right measures $V_R = -0.2\,\text{V}$.

(a) What is the electromotive force?
(b) What does the voltmeter on the left read?

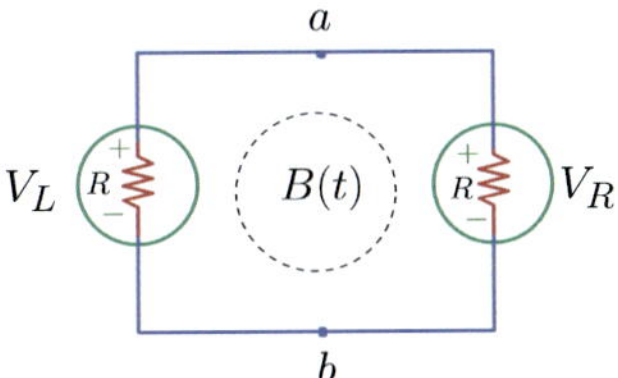

11.2 Mobile rail in a uniform magnetic field

Consider the circuit formed by a resistor R and a mobile conductive rail as shown below. The rail moves with a velocity $\mathbf{v}$ in a uniform magnetic field $\mathbf{B} = -B\mathbf{u}_z$.

(a) What is the relationship between the velocity $\mathbf{v}$ and the induced current I in the circuit?
(b) What is the force that must be applied to the rail to maintain its movement?

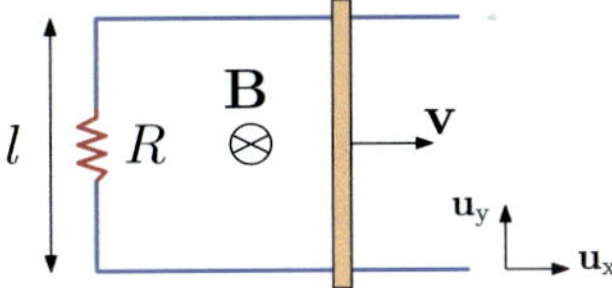

11.3 Rail on an inclined plane surface

A conductive bar of mass m slides without friction between two conductive and parallel rails separated by a distance l. The plane of the rails forms an angle θ with respect to the horizontal plane, the total resistance of the circuit is R and a uniform and vertical magnetic field is applied. The bar, initially at rest, slides downward due to the effect of gravity.

(a) Determine the induced current in the bar (magnitude and direction).
(b) What is the terminal velocity of the bar? (assume that the inclined plane is long enough for the bar to reach this velocity).
  Once the bar has reached its terminal velocity:
(c) What is the induced current?

(d) At what rate is energy dissipated in the circuit?
(e) What is the rate of work provided by gravity?

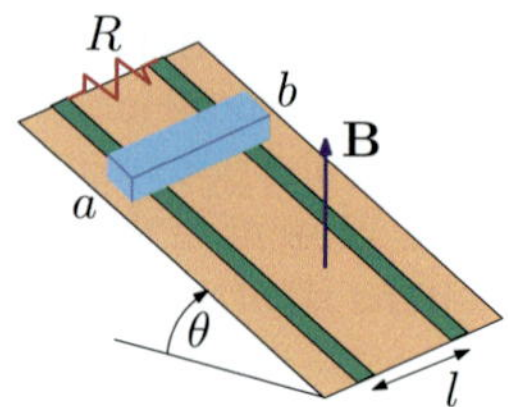

11.4 Induction by a time-varying magnetic field

A rectangular loop of length l and width w is coplanar to a very long and thin wire that carries a current I, at distance h to the closest side of the loop, as shown below.

(a) Determine the magnetic flux through the rectangular loop.
(b) Suppose that the current on the wire varies linearly with time as $I(t) = a + bt$. What is the electromotive force and the current induced in the loop if the latter has a total resistance R?

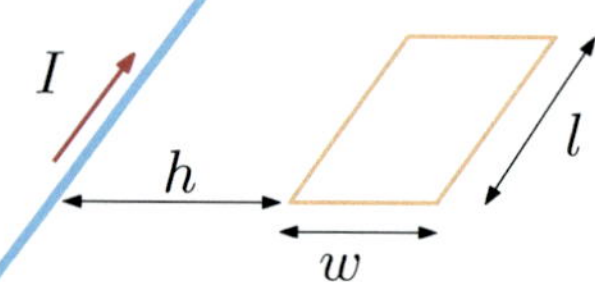

11.5 Relative motion between currents

A rectangular loop, of total resistance R, length l and width w moves at constant velocity $\mathbf{v}$ away from an infinitely long wire that carries a current I. Both the loop and the wire lie in the same plane. What is the induced current in the loop when the wire is a distance r away from its closest side?

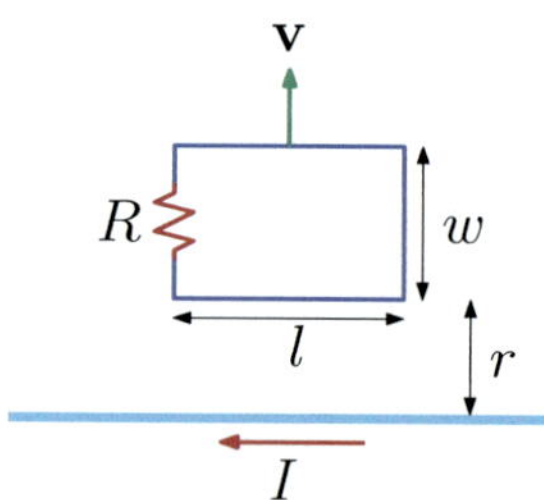

11.6 Current induced by gravity

A conductive bar of mass m and total resistance R slides on top of two conductive rails separated by a distance l. The bar is pulled in the horizontal direction by a cable of negligible mass that is connected, thanks to a pulley, to a block of mass M that moves vertically. A uniform magnetic field is applied in the vertical direction, and the bar starts from rest.

(a) What is the induced current trough the bar when it moves at speed v?
(b) Solve the equation of motion to find the velocity of the bar as a function of time.

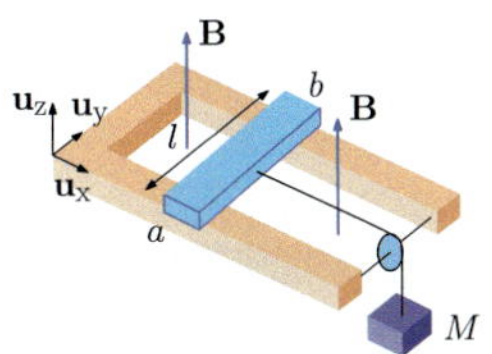

11.7 Rotation generated by induction

A disk has a charge q uniformly distributed on its outer border, a circle of radius $r = b$. It is suspended from a vertical wire attached to its center of gravity O. In the region $r < a$, there is a uniform magnetic field $\mathbf{B} = B\mathbf{u}_z$, perpendicular to the plane of the disk. If the magnetic field is turned off, what is the angular velocity acquired by the disk? The moment of inertia of the disk with respect to the z-axis is $\mathcal{J}$.

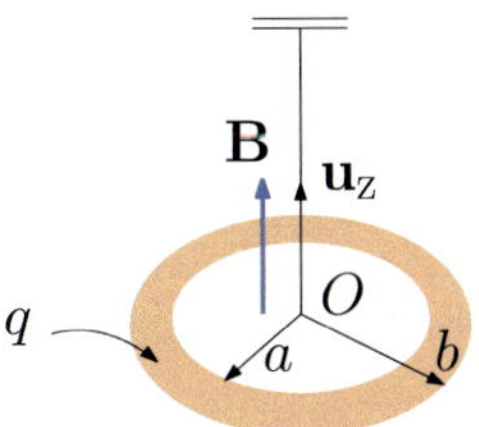

11.8 Mobile bar between two resistors

A conductive bar of length l slides without friction between two parallel and conductive rails. Two resistors R_1 and R_2 are connected at the extremities of the rails as shown below. An external agent moves the bar at constant velocity $\mathbf{v} = v\mathbf{u}_x$ in a uniform magnetic field $\mathbf{B} = -B\mathbf{u}_z$.

(a) Determine the electromotive force ε between the extremities a and b of the moving bar.
(b) Determine the current through both resistors.
(c) What is the total power dissipated in the resistors?
(d) What is the force required to maintain the movement of the bar at constant velocity?

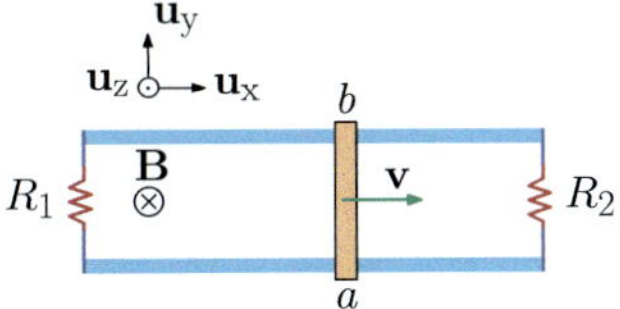

11.9 Induction in a toroid

Consider a toroid constituted of N square loops, each with a lateral size a and at a distance b from a very long and thin wire carrying a sinusoidal current $I = I_0 \sin \omega t$. Calculate the electromotive force induced in the terminals of the toroid

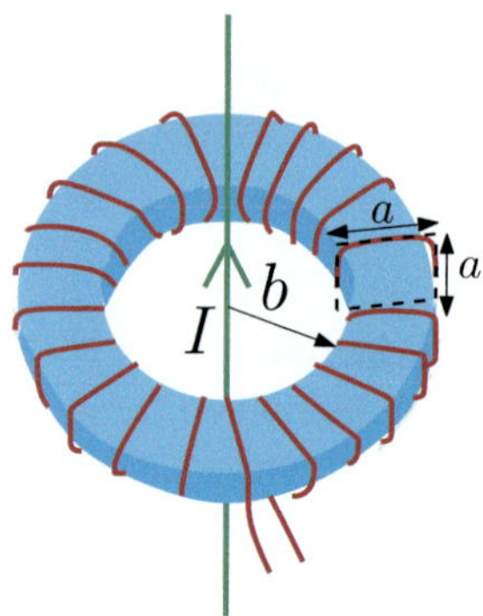

11.10 Solenoid moving in a magnetic field

A solenoid consisting of N turns of diameter D is in a uniform magnetic field $\mathbf{B} = B\mathbf{u}_z$ parallel to the axis of the solenoid. The two ends are connected to a device capable of measuring the total charge that goes trough the circuit, which initially is zero. The solenoid is then rotated upside down, and the total charge that went through the device is Q. If the total resistance of the circuit is R, what is the magnitude of the magnetic field?

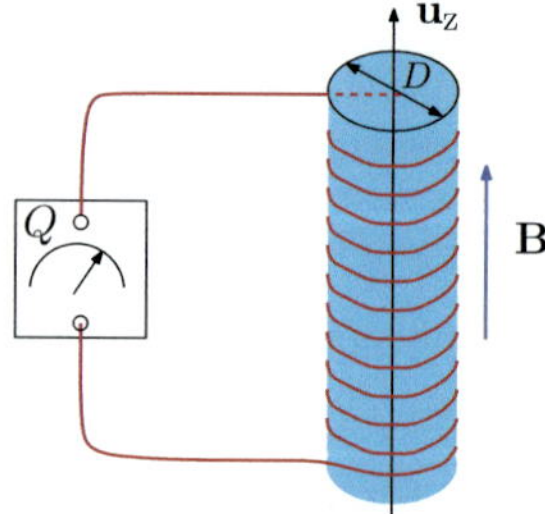

11.11 Magnetic brake

A train of mass m, height a and length l moves initially at a constant velocity $\mathbf{v} = v_0\mathbf{u}_x$ before entering a region of uniform magnetic field $\mathbf{B} = B\mathbf{u}_z$, as shown below. The train has a metallic frame made of a conductive loop of total resistance R, coplanar with the plane in which the train moves. Find the velocity of the train as a function of time. Assume that the train fully enters the magnetic field region.

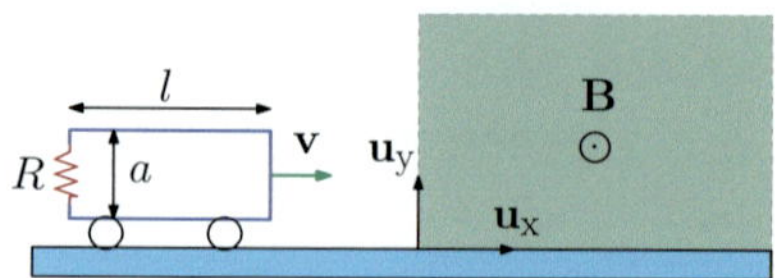

11.12 Magnetic parachute

Consider a square loop of size a and mass m, falling due to gravity in the Oyz plane in a magnetic field $\mathbf{B} = (B_0 - \alpha z)\mathbf{u}_x$, with $\alpha > 0$. Find the electromotive force and the induced current on the loop when the latter falls with a velocity $\mathbf{v} = -v\mathbf{u}_z$. What is the terminal velocity of the loop?

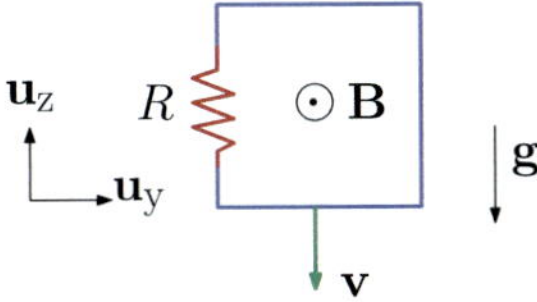

Chapter 12
Circuits in Transient Regimes

Abstract This chapter provides a comprehensive analysis of *electrical circuits in transient regimes* and their response to *alternating currents (AC)*. It begins by introducing the *quasi-static approximation*, a fundamental assumption in circuit theory that simplifies analysis by neglecting wave propagation effects, allowing for the application of Kirchhoff's laws to time-varying circuits. The chapter then characterizes the behavior of *passive circuit elements*—resistors, capacitors, and inductors—in the quasi-static regime. It derives their fundamental *current-voltage (I-V) relationships* and the expressions for *energy stored* in capacitors (electric field) and inductors (magnetic field). The analysis of *first-order circuits*, specifically *RC and RL circuits in series*, demonstrates their transient responses, including exponential charging/discharging and the concept of *time constants* ($\tau_{RC} = RC$ and $\tau_{RL} = L/R$). The energy transfer and dissipation in these circuits are also thoroughly examined. The chapter then extends to *second-order circuits*, focusing on the *series LC circuit* as an ideal oscillator with a characteristic *resonant frequency* ($\omega_0 = 1/\sqrt{LC}$)), and the more general *RLC circuit*, which exhibits overdamped, critically damped, and underdamped oscillatory responses depending on the damping factor and resonant frequency. The second part of the chapter transitions to the *forced sinusoidal regime* (AC currents). It introduces the powerful *complex representation of AC circuit components*, defining *impedance* (Z) as a complex quantity that generalizes resistance to include inductive and capacitive reactances. The rules for *series and parallel connection of impedances* are derived. The concept of a *transfer function* ($H(j\omega) = V_{\text{out}}/V_{\text{in}}$) is central to analyzing the *sinusoidal response of linear networks*, quantifying the gain ($G(\omega)$) and phase shift ($\phi(\omega)$) as functions of frequency. *Bode plots* are introduced as standard graphical tools for visualizing these frequency responses. Finally, the chapter applies these concepts to the design and analysis of *electronic filters* (low-pass, high-pass, band-pass, and band-stop filters), demonstrating how different RLC circuit configurations can achieve specific filtering characteristics. The *quality factor (Q-factor)* is introduced as a dimensionless parameter characterizing the sharpness of resonance and the selectivity of filters. The chapter concludes with a discussion of *power in AC circuits*, distinguishing between *active power* (useful work) and *reactive power* (energy oscillating between source and load), and introducing the concept of *power factor*.

© The Author(s), under exclusive license to Springer Nature Switzerland AG 2025

F. Cadiz and A. Couairon, *Classical Electrodynamics*, Undergraduate Texts in Physics, https://doi.org/10.1007/978-3-031-86785-9_12

Keywords Circuits · Transient regime · Sinusoidal regime · Impedance · Filters · Bode plot · Transfer function

12.1 Introduction

In circuit analysis, we often encounter two primary types of currents: steady-state currents and transient currents. A steady-state current is a current that remains constant over time. This occurs when the circuit has reached a stationary condition, typically after a sufficient amount of time has passed since the application of a constant voltage or current source. For the steady-state analysis of such circuits, one often assumes linearity of the elements, that their properties do not change in time and that the voltage and current sources supplying the circuit are constant. Chapter 7 presented methods for such steady-state analysis of electrical circuits, including Kirchhoff's laws, voltage-current relationships for components, superposition, and Thévenin's and Norton's theorems.

The transient current regime occurs when a circuit is subjected to a sudden change, such as the switching of a circuit element or the application of a time-varying input. During this transition period, the circuit variables (currents and voltages) change with time until a new steady-state condition is reached. For the analysis of the transient response of a circuit we will also assume linearity, invariance of the element's properties, and that the initial values of voltages across capacitors and currents through inductors at the moment of the change are known. The first part of this chapter focuses on methods for analyzing circuits in the transient regime. We derive the current-voltage relationships for capacitors and inductors and, using Kirchhoff's laws, analyze the behavior of various circuits composed of resistors, capacitors, and inductors. In the second part, we examine the forced sinusoidal regime, where the sources impose a sinusoidal variation in the voltages and currents within the circuit. We introduce the concepts of impedance and transfer functions, analyze the behavior of RLC circuits, and demonstrate how different types of filters can be constructed using these circuits.

12.2 Quasi-static Approximation in Circuit Theory

The quasi-static approximation is a fundamental assumption in circuit theory that simplifies the analysis of circuits with time-varying signals. It allows us to treat electromagnetic phenomena as if they were quasi-static, meaning that the fields change slowly enough that we can neglect the effects of wave propagation, that will be discussed in Chap. 13. This assumption is valid when the dimensions of the circuit are much smaller than the wavelength of the electromagnetic waves associated with the highest frequency component of the signal. It allows us to treat the propagation delay of signals within the circuit as negligible. The circuit components (resistors, capacitors, inductors) are also considered as lumped elements, meaning their physical dimensions are much smaller than the wavelength of the signal. This allows us to represent the circuit as a network of idealized components connected by ideal wires and ignore all phenomena like radiation, skin effect, and proximity effects, which become significant at higher frequencies.

The implications of the quasi-static approximation are a simplification of circuit analysis, making it a powerful tool for analyzing many circuits from simple resistive circuits to complex electronic systems: we can apply Kirchhoff's voltage and current laws to analyze the circuit, as if it were a DC circuit. The main limitations of the quasi-static approximation must be kept in mind for high-frequency circuits and for large circuits. As the frequency of the signal increases, the wavelength decreases, and the quasi-static approximation may break down. At high frequencies, distributed effects, such as transmission line effects, become significant. For very large circuits, the propagation delay of signals between different parts of the circuit can become non-negligible.

12.3 Passive Circuit Elements in the Quasi-static Regime

Components that do not require any external power source to operate are called passive elements. They can store or dissipate energy, but they cannot generate nor amplify it.

12.3.1 Resistors

A resistor is a first example of passive element. We recall its representation as a lump element, shown in Fig. 12.1.

Ohm's law is valid in the quasi-steady regime:

$$V(t) = RI(t) .$$

The power received by the resistor is

$$p(t) = V(t)I(t) = RI^2(t) .$$

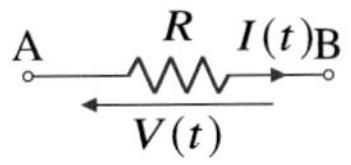

Fig. 12.1 A resistor

12.3.2 Capacitors

A second example is the capacitor consisting of one or more pairs of conductors separated by an insulator. A common example is the parallel-plate capacitor, composed of two parallel plates carrying equal and opposite charges. Capacitors in circuits are often represented by parallel plates, as shown in Fig. 12.2.

If the distance d between the plates is much smaller than the plate dimensions ($d \ll \sqrt{S}$, where S is the surface of the plates), we can approximate the capacitor as an ideal capacitor with infinite plates. In this case, the electric field between the plates is uniform and given by

$$\mathbf{E} = \frac{\sigma}{\epsilon_0}\mathbf{u}_x$$

where σ is the surface charge density on the plate carrying charge $+q$ and ϵ_0 is the permittivity of free space. Assuming negligible resistance in the connecting wires, the voltage V across the capacitor is

$$V = \int_A^B \mathbf{E} \cdot d\mathbf{l} = \frac{\sigma d}{\epsilon_0} = \frac{d}{\epsilon_0 S}q \, ,$$

where $q = \sigma S$ is the charge on the positively charged plate. This equation establishes a linear relationship between the voltage and the charge on the capacitor:

$$\boxed{q = CV \, ,} \tag{12.1}$$

where $C = \epsilon_0 S/d$ is the capacitance of the parallel-plate capacitor, determined solely by its geometric properties. If a current $I(t)$ flows through the wire, the charge $q(t)$ on the capacitor changes with time according to

$$\frac{dq}{dt} = I \, .$$

Differentiating (12.1) with respect to time, we obtain the current-voltage relationship for the capacitor:

$$\boxed{I(t) = C\frac{dV}{dt} \, .}$$

The energy W_E stored in a charged capacitor can be interpreted as the energy stored in the electric field between the plates. This energy can be calculated by integrating the energy density of the electric field over the volume of the capacitor:

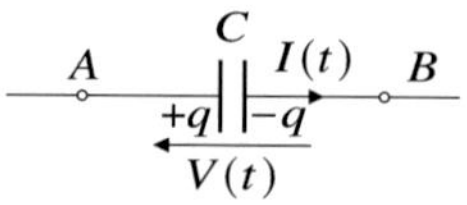

Fig. 12.2 A capacitor

$$W_E = \iiint \frac{1}{2}\epsilon_0 |\mathbf{E}|^2 d^3x = \frac{1}{2}\epsilon_0 \left(\frac{V}{d}\right)^2 Sd = \frac{1}{2}CV^2 ,$$

where $C = \epsilon_0 S/d$ is the capacitance of the capacitor. Therefore, the energy stored in a capacitor is

$$\boxed{W_E(t) = \frac{1}{2}CV^2(t) .}$$

The energy stored in a capacitor can also be calculated by considering the work done in charging the capacitor. At any instant t, the charge on the capacitor is $q(t)$ and the potential difference across it is $V(t)$. The work done in adding an infinitesimal charge dq to the capacitor is

$$dW = V(t)dq = \frac{q(t)}{C}dq .$$

Integrating this expression over the entire charging process, we get

$$W_E(t) = \int_0^{q(t)} \frac{q'}{C}dq' = \frac{q^2(t)}{2C} = \frac{1}{2}CV^2(t) .$$

This result is consistent with the power delivered to the capacitor, which is found by differentiating in time:

$$p(t) = \frac{dW_E}{dt} = CV(t)\frac{dV}{dt} = V(t)I(t) .$$

12.3.3 *Inductors*

A third example of a passive component is the inductor. It consists of a coil of wire wound into a specific shape, often a helix or solenoid. Figure 12.3 gives the symbols for an ideal inductor (top left) and a real-world inductor (right), which can be modeled as an ideal inductor in series with a resistor.

We have previously shown in Chap. 11 that the magnetic flux $\Phi_\mathbf{B}$ through the surface enclosed by a coil is proportional to the current I flowing through it,

Fig. 12.3 Left: Inductor with inductance L (top); equivalent source (bottom); Right: inductor with internal resistance r

$$\Phi_{\mathbf{B}} = \iint \mathbf{B} \cdot d\mathbf{S} = LI \ .$$

The proportionality constant L is called the inductance of the coil and depends only on its geometry. Differentiating the magnetic flux with respect to time, we obtain the induced electromotive force in a closed circuit that includes the inductor:

$$e(t) = -\frac{d\Phi_{\mathbf{B}}}{dt} = -L\frac{dI}{dt} \ .$$

Assuming negligible resistance in the wire and assuming that the magnetic field is negligible outside the coil, the emf is completely given by the line integral of the electromotive field from A to B. From generalized Ohm's law (11.9), the voltage across the coil terminals is equal to the induced emf, $V = -e = L\,dI/dt$, where passive sign convention was used as indicated in Fig. 12.3. This equation represents the current-voltage relationship for an inductor:

$$\boxed{V(t) = L\frac{dI}{dt} \ .} \tag{12.2}$$

Note that we can still apply Kirchhoff's voltage law (KVL) in circuits containing inductors, even though the electric field is not conservative in the presence of time-varying magnetic fields. In this case, the potential difference $V_A - V_B$ between the terminals of the inductor corresponds to the line integral from A to B of the conservative part of the electric field. Summing all the potential differences around a closed circuit still yields zero. However, this sum no longer corresponds to the line integral of the total electric field, which includes the induced field $\mathbf{E} = -\partial\mathbf{A}/\partial t$ in addition to the conservative field $-\nabla V$. It is this non-conservative part of the electric field that ensures the equality $V(t) = -e$ for an ideal inductor. Without it, the potential difference across the terminals of an inductor with no resistance would be zero. This, the presence of the induced field is very simply taken into account by writing (12.2).

The energy stored in an inductor, W_B, is associated with the magnetic field created by the current flowing through it. For an inductor modeled as a long solenoid with length d and coil surface S satisfying $d \gg \sqrt{S}$ (infinitely long solenoid), the magnetic field is uniform along the axis of the solenoid, with magnitude $B = \mu_0 n I$, where $n = N/d$ is the density of turns and N is the number of turns. The magnetic flux through the solenoid is $\Phi_B = NBS = N(\mu_0 n I)S$. This equation yields the inductance of the solenoid, $L = \mu_0 N^2 S/d$. The magnetic energy stored in the volume of the solenoid is

$$W_B = \iiint \frac{1}{2\mu_0}|\mathbf{B}|^2 d^3x = \frac{1}{2\mu_0}\left(\frac{\mu_0 N I}{d}\right)^2 Sd = \frac{1}{2}LI^2 \ .$$

This result is consistent with the rate of energy storage, or power, delivered to the inductor, which is given by

$$p(t) = \frac{dW_B}{dt} = V(t)I(t) = LI(t)\frac{dI}{dt} = \frac{d}{dt}\left(\frac{1}{2}LI^2(t)\right) .$$

We conclude after integrating in time that

$$W_B = \frac{1}{2}LI^2(t) .$$

12.3.4   Resistor-Capacitor Circuit (RC Circuit) in Series

The resistor-capacitor (RC) circuit shown in Fig. 12.4 consists of a voltage source, or battery, connected in series with a resistor of resistance R and a capacitor of capacitance C.

While the switch is open, there is no current in the circuit, and the capacitor is uncharged: $q = 0$ and $V = 0$. At time $t = 0$, the switch is closed. Applying Kirchhoff's voltage law, we obtain

$$E = RI + V .$$

Substituting the current-voltage relationship for the capacitor, $I = CdV/dt$, we arrive at the following first-order linear differential equation for the voltage across the capacitor:

$$\frac{dV}{dt} + \frac{V}{RC} = \frac{E}{RC} ,$$

with the initial condition $V(0) = 0$. Solving this differential equation yields

$$V(t) = E\left[1 - \exp\left(-\frac{t}{\tau_{RC}}\right)\right]$$

where $\tau_{RC} = RC$ is the time constant of the RC circuit. The voltage across the capacitor as a function of time is shown in Fig. 12.5a

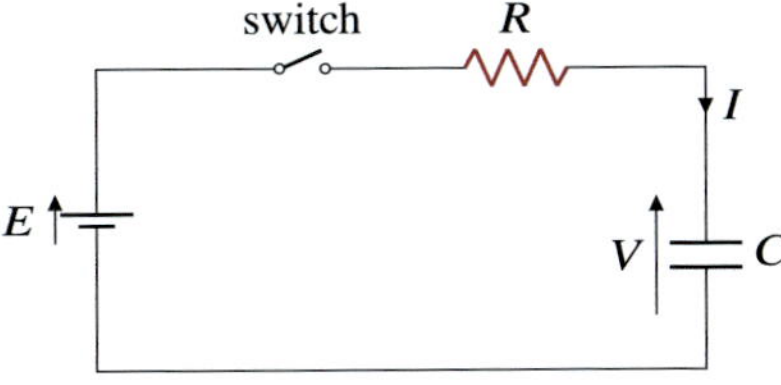

Fig. 12.4   RC circuit in series

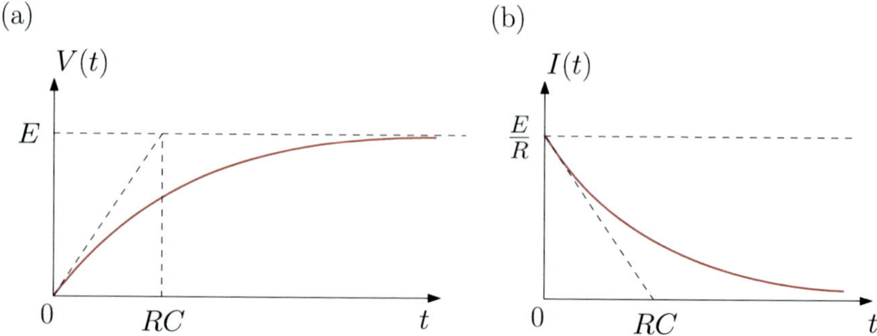

Fig. 12.5 RC circuit in series: **a** Voltage across the capacitor as a function of time. **b** Current through the resistor as a function of time

If the capacitor were to charge at a constant rate of E/τ_{RC}, it would indeed reach the voltage E at time $t = \tau_{RC}$. However, this is not the case. The charging process is exponential, and the rate of charging slows down as the capacitor approaches its final voltage. The time constant $\tau_{RC} = RC$ characterizes the rate of change of the voltage across the capacitor. After one time constant, the voltage reaches approximately 63% of its final value, not 100%. It takes approximately 5 time constants for the capacitor to be fully charged (within 1% of E). The current through the resistor, given by

$$I(t) = C\frac{dV}{dt} = \frac{E}{R}\exp\left(-\frac{t}{\tau_{RC}}\right),$$

decays exponentially with the same time constant τ_{RC}. Initially, the current is E/R, and it decreases to zero as the capacitor charges (See Fig. 12.5b).

12.3.5 Discharging of a Capacitor

Assume now that at time $t = T$, we open the switch. The capacitor is now charged to a voltage of E. We then remove the generator, short-circuit the capacitor (as shown in Fig. 12.6), and close the switch. Will the capacitor retain its charge and voltage?

No, the system will achieve equilibrium by becoming neutral, the capacitor will not hold its charge indefinitely. Once the switch is closed, the capacitor will discharge through the resistor. The charge on the capacitor will gradually decrease, and the voltage across it will decay exponentially with a time constant of $\tau = RC$.

Applying Kirchhoff's voltage law to the circuit, we obtain

$$0 = V + RI .$$

Fig. 12.6 A simple RC circuit consisting of a resistor and a capacitor connected in series

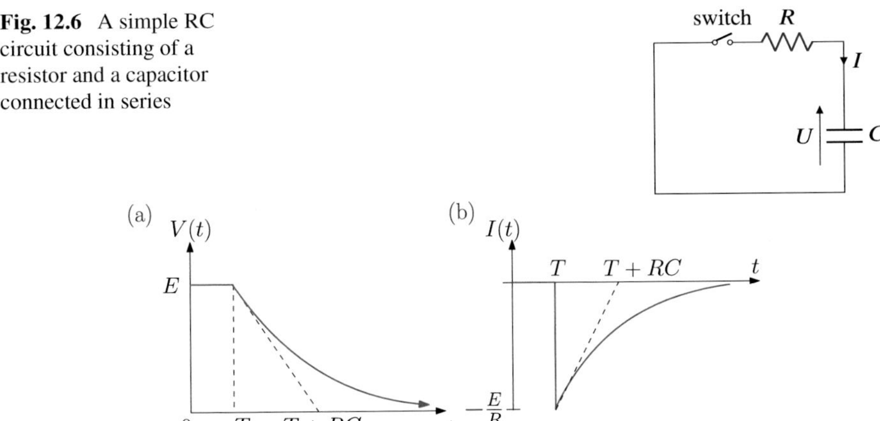

Fig. 12.7 Discharge of a Capacitor: **a** Voltage across the capacitor as a function of time. **b** Current through the resistor as a function of time

Substituting the current-voltage relationship for the capacitor, $I = C dV/dt$, we get

$$\frac{dV}{dt} + \frac{V}{RC} = 0 \, .$$

Solving this first-order differential equation with the initial condition $V(T) = E$, and defining the characteristic time $\tau_{RC} = RC$, we find the voltage across the capacitor as a function of time:

$$V(t) = E \exp\left(-\frac{(t - T)}{\tau_{RC}}\right) \, .$$

Differentiating this expression with respect to time, we obtain the current through the resistor,

$$I(t) = -\frac{E}{R} \exp\left(-\frac{(t - T)}{\tau_{RC}}\right) \, .$$

The capacitor discharges exponentially with a time constant $\tau_{RC} = RC$. The current through the resistor also decays exponentially with the same time constant (See Fig. 12.7a, b).

As demonstrated in Sect. 7.4.2, any charge imbalance in the volume of a conductor dissipates exponentially with a time constant $\tau_d = RC$, where R is the resistance and C is the capacitance of the conductor. This same time constant reappears in the behavior of an RC circuit, which is unsurprising since the RC circuit can be viewed as a lumped-element representation of a conductor that inherently exhibits both resistance and capacitance.

12.3.6 *Energy in the RC Circuit*

During the charging process of the capacitor, we established the relationship between the current through the resistor and the voltage $V = q/C$ across the capacitor:

$$RI + \frac{q}{C} = E \ .$$

Multiplying both sides by $dq = I dt$, we get

$$RI^2 dt + \frac{q dq}{C} = E dq \ .$$

Integrating both sides over time, or equivalently over the charge from 0 to $q = CE$, we obtain

$$\int_0^{+\infty} RI^2 dt + \int_0^{CE} \frac{q}{C} dq = E \int_0^{CE} dq \ .$$

Solving these integrals, we get

$$\int_0^{+\infty} RI^2 dt + \frac{1}{2} CE^2 = CE^2 \ .$$

The first term on the left-hand side represents the energy dissipated as heat in the resistor (Joule's heating). The second term represents the energy stored in the capacitor. The right-hand side represents the total energy supplied by the voltage source.

To analyze the discharge process, we start with the equation

$$RI + \frac{q}{C} = 0 \ .$$

Multiplying by $dq = I dt$ and integrating, we get

$$\int_0^{+\infty} RI^2 dt + \int_{CE}^{0} \frac{q}{C} dq = 0 \ .$$

Solving the second integral, we obtain

$$\int_0^{+\infty} RI^2 dt - \frac{1}{2} CE^2 = 0 \ .$$

This equation indicates that the energy dissipated as heat in the resistor (the first term) is equal to the initial energy, $\frac{1}{2} CE^2$, stored in the capacitor (the second term).

12.4 RL Circuit in Series

The resistor-inductor (RL) circuit shown in Fig. 12.8 consists of a voltage source or battery connected in series with a resistor of resistance R and an inductor of inductance L.

While the switch is open, there is no current in the circuit. At $t = 0$, the switch is closed, and initially, the current $I(0) = 0$. Applying Kirchhoff's voltage law to the circuit, we get

$$E = V(t) + RI(t) . \tag{12.3}$$

Using the voltage-current relationship for the inductor, $V(t) = L(dI/dt)$, we obtain the following first-order linear differential equation:

$$\frac{dI}{dt} + \frac{RI}{L} = \frac{E}{L} .$$

Solving this differential equation with the initial condition $I(0) = 0$, we get

$$\boxed{I(t) = \frac{E}{R}\left[1 - \exp\left(-\frac{t}{\tau_{RL}}\right)\right] \text{ with } \tau_{RL} = \frac{L}{R} .}$$

The current in the coil rises exponentially with a time constant $\tau_{RL} = L/R$ (See Fig. 12.9a).

The voltage across the coil, $V(t)$, can be obtained by differentiating the current with respect to time:

Fig. 12.8 RL circuit in series

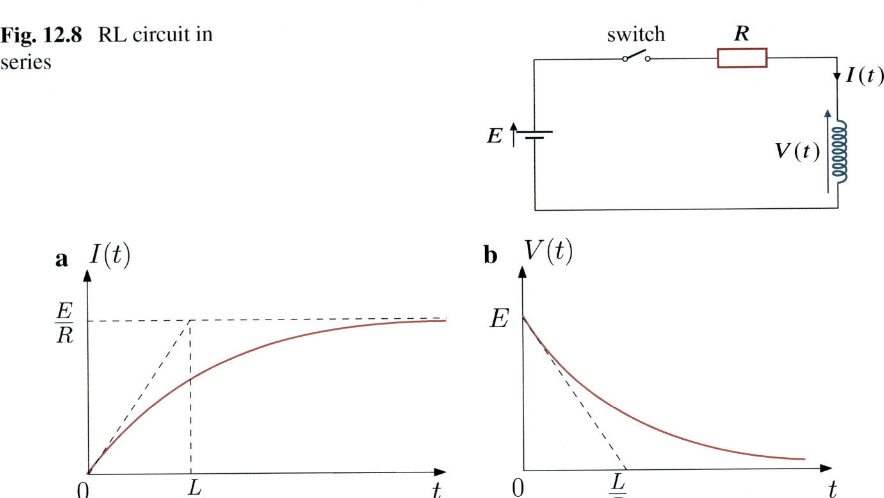

Fig. 12.9 **a** Current in the coil and **b** voltage across the coil for the *RL* circuit in series

$$V(t) = L\frac{dI}{dt} = E \exp\left(-\frac{t}{\tau_{RL}}\right) .$$

The voltage decays exponentially from an initial value of E with a time constant $\tau_{RL} = L/R$. (See Fig. 12.9b).

Now, assume that at time $t = T$, approximately $5\tau_{RL}$ after the switch is initially closed, the current through the inductor has reached a value very close to its steady-state value, E/R. At this point, we open the switch, remove the generator, and short-circuit the inductor (as shown in Fig. 12.10). When the switch is closed, the inductor, acting as a source of energy, will drive a current through the resistor. This current will gradually decrease as the inductor releases its stored energy. The voltage across the inductor will also decay exponentially.

Applying Kirchhoff's voltage law to the circuit, we obtain

$$L\frac{dI}{dt} + RI(t) = 0 . \tag{12.4}$$

Solving this first-order differential equation with the initial condition $I(T) = E/R$, and defining the characteristic time $\tau_{RC} = L/R$, we find the current through the inductor as a function of time:

$$\boxed{I(t) = \frac{E}{R} \exp\left(-\frac{(t - T)}{\tau_{RL}}\right) .}$$

The voltage across the inductor can be obtained by differentiating the current with respect to time:

$$V(t) = L\frac{dI}{dt} = -E \exp\left(-\frac{(t - T)}{\tau_{RL}}\right) .$$

Both the current and voltage decay exponentially with a time constant $\tau_{RL} = L/R$ (See Fig. 12.11a, b).

Fig. 12.10 A simple RL circuit consisting of a resistor and an inductor connected in series

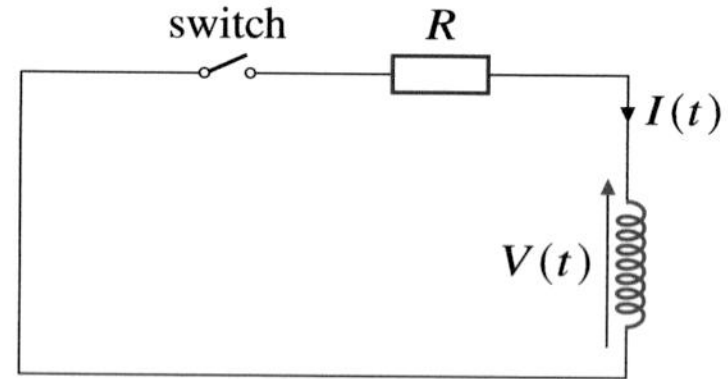

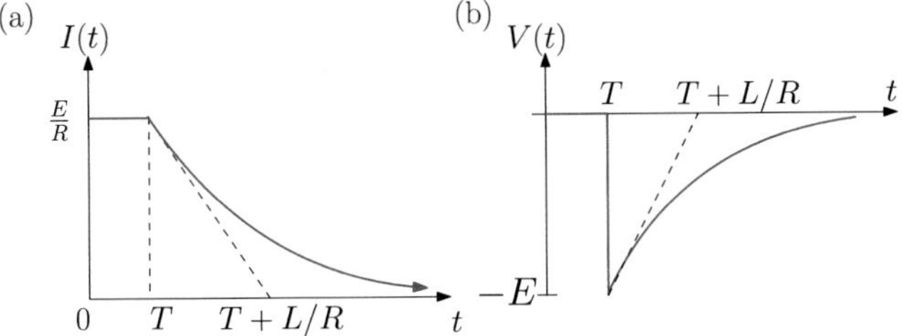

Fig. 12.11 Discharge of an inductor: **a** Current through the inductor as a function of time. **b** Voltage across the inductor as a function of time

12.4.1 Energy in the RL Circuit

During the energizing stage of the inductor, the governing equation for energy in the RL circuit can be obtained by multiplying (12.3) by $I(t)dt$:

$$LI dI + RI^2(t)dt = EI(t)dt .$$

Integrating both sides over time, from 0 to infinity, or equivalently over the current between 0 and its steady state value E/R, we get

$$\int_0^{E/R} d\left(\frac{1}{2}LI^2\right) + \int_0^{+\infty} RI^2(t)dt = \int_0^{+\infty} EI(t)dt .$$

The first term represents the energy stored in the inductor, the second term represents the energy dissipated as heat in the resistor, and the right-hand side represents the total energy supplied by the voltage source.

For the discharging process, the governing equation is obtained by multiplying (12.4) with $I dt$:

$$LI dI + RI^2(t)dt = 0 .$$

Integrating over time between 0 and $+\infty$ or equivalently, over the current between E/R and 0, we get

$$\int_{E/R}^0 d\left(\frac{1}{2}LI^2\right) + \int_0^{+\infty} RI^2(t)dt = 0 .$$

This equation shows that the energy initially stored in the inductor (the first term) is completely dissipated as heat in the resistor (the second term).

12.4.2 LC Circuit

An LC circuit is a simple electrical circuit consisting of an inductor and a capacitor connected in series shown in Fig. 12.12. This configuration allows for the oscillation of electrical energy between the inductor and capacitor.

This oscillatory behavior arises from the continuous exchange of energy between the inductor and capacitor. Initially, the capacitor is charged to a certain voltage $V_C = E$. This stores electrical energy in the capacitor electric field. Once the switch is closed, at $t = 0$, the capacitor discharges through the inductor. As the capacitor discharges, its stored electrical energy is transferred to the inductor, building up a magnetic field. When the capacitor is fully discharged, the current through the inductor starts to decrease, inducing a voltage across the inductor. This voltage causes the capacitor to charge again, but with opposite polarity. This process of energy transfer between the capacitor and the inductor repeats, resulting in sustained oscillations of the current and voltage in the circuit.

To quantify these oscillations, we can use Kirchhoff's voltage law, $V_L + V_C = 0$, and the constitutive equations for the inductor, $V_L = L\dfrac{dI}{dt}$:

$$L(dI/dt) + V_C = 0 .$$

Differentiating the current with respect to time and substituting the constitutive equation for the capacitor, $I = dq/dt = C\,dV_C/dt$, we get

$$\frac{d^2 V_C}{dt^2} + \frac{V_C}{LC} = 0 .$$

This is a second-order linear homogeneous differential equation with constant coefficients, which has a solution of the form:

$$V_C(t) = A\cos(\omega t) + B\sin(\omega t) ,$$

where $\omega = 1/\sqrt{LC}$ is the angular frequency of oscillation. The constants A and B are determined by the initial conditions $V_C(0) = E$ and $I(0) = 0$: We obtain $A = E$ and $B = 0$. Therefore, the voltage across the capacitor and the current through the inductor can be expressed as

Fig. 12.12 LC circuit in series

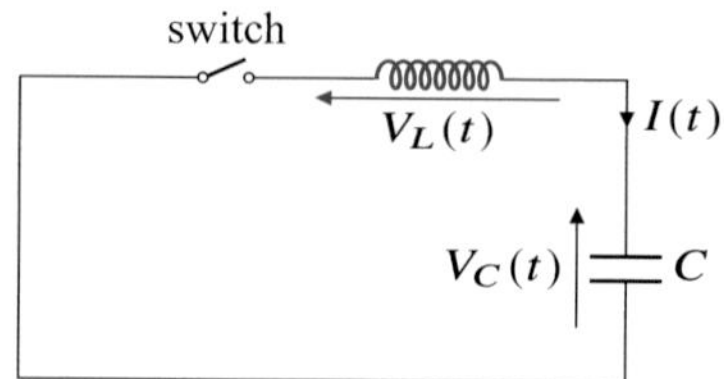

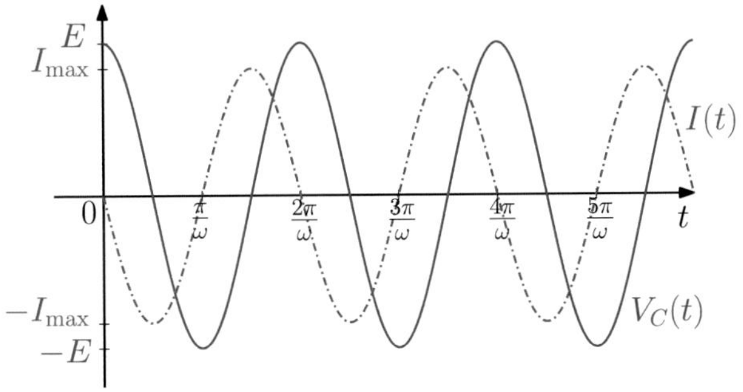

Fig. 12.13 Voltage across the terminals of the coil/capacitor (in blue) and current in the LC circuit in series (in red)

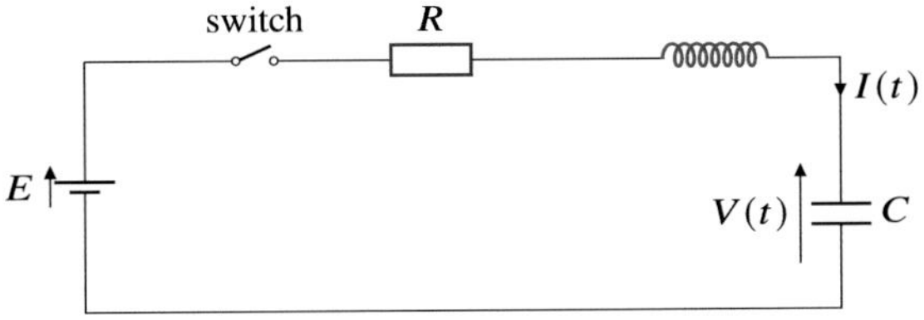

Fig. 12.14 A RLC circuit consisting of a resistor, an inductor, and a capacitor connected in series

$$V(t) = E\cos(\omega t) \, ,$$

$$I(t) = C\frac{dU}{dt} = -I_{\max}\sin(\omega t) \, ,$$

where $I_{\max} = \sqrt{\dfrac{C}{L}}\,E$. The voltage and current are illustrated in Fig. 12.13.

The LC circuit oscillates at a specific frequency, $\omega = 1/\sqrt{LC}$, known as the resonant frequency, which is solely determined by the values of the inductance and capacitance. LC circuits have various applications in electronics, including radio frequency tuners, where LC circuits are used to select specific frequencies in radios and televisions, and filters where the circuit is used to filter out specific frequencies from a signal. LC circuits are also used as oscillators to generate specific frequencies for various electronic devices.

12.5 RLC Circuit

The RLC circuit consists of a resistor, an inductor, and a capacitor connected in series, as illustrated on Fig. 12.14.

Initially, the switch is open, the capacitor is uncharged ($q(0) = 0$), and there is no current in the circuit ($I(0) = 0$). At $t = 0$ the switch is closed. Applying Kirchhoff's voltage law ($V_L + V_R + V = E$) and using the voltage-current relationships for the resistor ($V_R = RI$) and inductor ($V_C = LdI/dt$), as well as the relationships between the voltage and charge on the capacitor ($V = q/C$) and between the current and charge ($I = dq/dt$), we obtain the following second-order linear differential equation for the voltage across the capacitor:

$$\frac{d^2 V}{dt^2} + \frac{R}{L}\frac{dV}{dt} + \frac{V}{LC} = \frac{E}{LC} .$$

For convenience, we define the resonant frequency ω_0 and the damping time rate, ν, as

$$\omega_0^2 = \frac{1}{LC} ,$$

$$\nu = \frac{2R}{L} .$$

This allows us to rewrite the differential equation as

$$\frac{d^2 V}{dt^2} + 2\nu\frac{dV}{dt} + \omega_0^2 V = \omega_0^2 E . \tag{12.5}$$

The solution to (12.5) is the sum of the general solution to the homogeneous equation, $V_h(t)$, and a particular solution to the non-homogeneous equation,

$$V(t) = E + V_h(t) .$$

The initial conditions, $I(0) = 0$ and $V(0) = 0$, are used to determine the constants in the general solution. These conditions express the fact that both the current in the coil and the voltage across the capacitor are continuous when the switch is closed. The current in the circuit is determined by the IV relation for the capacitor, $I = C(dV/dt)$. Therefore the initial condition for the current is expressed as an initial condition on the time derivative of the voltage V. The nature of the solution to the homogeneous equation depends on the roots of the characteristic equation, $r^2 + 2\nu r + \omega_0^2 = 0$:

- Overdamped response: it occurs when occurs for $\nu > \omega_0$, for instance with a large resistance. The characteristic polynomial has two distinct real and negative roots:

$$r_\pm = -\nu \pm \sqrt{\nu^2 - \omega_0^2} .$$

The solution for the voltage, $V(t)$, and the current, $I(t)$, can be expressed as

$$V(t) = E + (Ae^{\nu_0 t} + Be^{-\nu_0 t})e^{-\nu t} ,$$

$$I(t) = C(A(\nu_0 - \nu)e^{\nu_0 t} - B(\nu_0 + \nu)e^{-\nu_0 t})e^{-\nu t} ,$$

where $\nu_0 = \sqrt{\nu^2 - \omega_0^2}$ and the constants A and B are determined by the initial conditions:

$$V(0) = 0 : \quad A + B = -E ,$$

$$I(0) = 0 : \quad A(\nu_0 - \nu) - B(\nu_0 + \nu) = 0$$

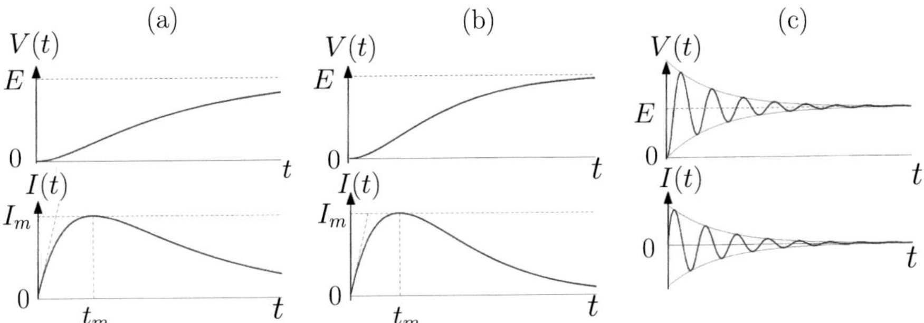

Fig. 12.15 Voltage across the terminals of the capacitor (First row) and current in the RLC circuit (Second row) for **a** the overdamped response; **b** the critically damped response and **c** the underdamped oscillatory response

Solving for A and B, we get

$$A = -\frac{E}{2}\left(1 + \frac{\nu}{\nu_0}\right) , \quad B = -\frac{E}{2}\left(1 - \frac{\nu}{\nu_0}\right) .$$

Substituting these values into the general solution, we obtain the voltage across the capacitor,

$$V(t) = E\left[1 - e^{-\nu t}\left(\cosh(\nu_0 t) + \frac{\nu}{\nu_0}\sinh(\nu_0 t)\right)\right] .$$

The current in the circuit can be obtained by differentiating the voltage with respect to time:

$$I(t) = CE\frac{\omega_0^2}{\nu_0}e^{-\nu t}\sinh(\nu_0 t) .$$

The voltage increases to E and the current reaches a maximum value of $I_m = CE\omega_0\left(\frac{\nu_0-\nu}{\omega_0}\right)^{-\nu/\nu_0}$ at time $t_m = \nu_0^{-1}\tanh^{-1}(\nu_0/\nu)$ before returning to zero, as shown in Fig. 12.15a.

- Critically damped response: it occurs for $\omega_0 = \nu$ (the resistance, inductance and capacitance satisfy $R = 2(L/C)^{1/2}$). The characteristic equation has a repeated real root:

$$r = -\nu .$$

The solution to (12.5) is expressed as

$$V(t) = E + (A + Bt)e^{-\nu t} .$$

The current in the circuit is found by the VI relation for the capacitor

$$I(t) = C\frac{dV}{dt} = C\left(-\nu(A + Bt)e^{-\nu t} + Be^{-\nu t}\right) .$$

The constants A and B are determined by the initial conditions and read

$$A = -E , \quad B = -\nu E .$$

The solutions for voltage and current are then read

$$V(t) = E\left(1 - (1 + \nu t)e^{-\nu t}\right) ,$$
$$I(t) = CE\nu^2 t e^{-\nu t} .$$

This situation corresponds to the fastest possible evolution toward the steady state regime with a charged capacitor. The current reaches its maximum value of $I_m = CE\nu/e$ at time $t_m = \nu^{-1}$ before decreasing back to zero, as shown in Fig. 12.15b.

- Underdamped oscillatory response: this situation occurs for $\nu < \omega_0$ (small resistance). The characteristic equation has complex conjugate roots

$$r_\pm = -\nu \pm i\omega , \quad \text{where } \omega = \sqrt{\omega_0^2 - \nu^2} .$$

The solution to (12.5) and the current in the circuit are expressed as

$$V(t) = E + e^{-\nu t} A \cos(\omega t - \phi) ,$$
$$I(t) = -CAe^{-\nu t} \left(\nu \cos(\omega t - \phi) + \omega \sin(\omega t - \phi)\right) ,$$

where the constants A and ϕ determined by initial conditions and read

$$\tan \phi = \frac{\nu}{\omega} , \quad A = -\frac{E}{\cos \phi} .$$

Substituting these values into the general solution, we obtain the voltage across the capacitor and current in the circuit:

$$V(t) = E\left[1 - e^{-\nu t}\left(\cos(\omega t) + \frac{\nu}{\omega}\sin(\omega t)\right)\right] ,$$
$$I(t) = CE\frac{\omega_0^2}{\omega}e^{-\nu t}\sin(\omega t) .$$

The voltage increases and exhibits damped oscillations around its steady state value E. The current also increases and returns to zero with damped oscillations as shown in Fig. 12.15c. The oscillation period is given by

$$T = \frac{2\pi}{\omega} = \frac{T_0}{\left(1 - \dfrac{v^2}{\omega_0^2}\right)^{1/2}} \; , \quad \text{where } T_0 = \frac{2\pi}{\omega_0} \; .$$

This shows that the period is larger than the oscillation period T_0 corresponding to the LC circuit.

12.5.1 *Energy in the RLC Circuit*

The governing equation for the RLC circuit,

$$L\frac{dI}{dt} + RI + \frac{q}{C} = E \; ,$$

allows us to analyze the energy transfer from the voltage source to the RLC circuit. Multiplying the governing equation by $dq = I\,dt$, we get

$$d\left(\frac{1}{2}LI^2\right) + RI^2 dt + d\left(\frac{1}{2}\frac{q^2}{C}\right) = EI\,dt \; .$$

Integrating in time, or between initial and final state, we find

$$\left[\frac{1}{2}LI^2\right]_0^{I_\infty} + \int_0^\infty RI^2 dt + \left[\frac{1}{2}\frac{q^2}{C}\right]_0^{CE} = \int_0^\infty EI\,dt \; .$$

The first term represents the energy stored in or released by the inductor. As $I_\infty = 0$ all the energy stored by the inductor during the charge is released to the circuit when the current decreases to zero. The second term represents the energy provided to the resistance and dissipated as heat. Using the expressions for $I(t)$, we can show that it is equal to $(1/2)CE^2$. The third term represents the energy stored in the capacitor, $(1/2)CE^2$. The term on the right-hand side is the energy provided by the voltage source: the integral can be calculated from $I(t)$, which yields CE^2. Whatever the values of the resistance and inductance, half of the energy of the source is flowing to the resistor and dissipated while the second half is stored in the capacitor.

12.6 AC Currents

Alternating Current (AC) is a type of electric current that periodically reverses direction and changes its magnitude sinusoidally with time. This is in contrast to Direct Current (DC) which flows in only one direction. Most electrical generators produce sinusoidal signals. AC sinusoidal currents are widely used in power transmission and distribution systems due to their ability to be easily transformed to different voltage levels. They also form the basis of many electronic devices and systems. AC sinusoidal signals are therefore of considerable practical importance. AC signals of sinusoidal waveform can be expressed as

$$f(t) = A \cos \omega t + B \sin \omega t = f_{\max} \cos(\omega t + \phi) \, . \tag{12.6}$$

Here, $f(t)$ represents a voltage, current, an emf, or any quantity that exhibit sinusoidal variation in time. This signal is characterized by its amplitude (peak value $f_{\max}$), frequency (cycles per second $v = \omega/(2\pi)$), and phase angle ϕ.

Complex numbers are often used to represent these quantities, making calculations easier. The signal $f(t)$ (Eq. (12.6)) can be expressed in complex notations as

$$F(t) = F_0 e^{j\omega t} \, ,$$

where

$$F_0 = f_{\max} e^{j\phi} \, .$$

Therefore, the relation between the real signal and the complex signal is

$$f(t) = \mathrm{Re}(F(t)) \, .$$

Using complex numbers to represent physical signals is a convenient way to facilitate calculations and easily identify amplitudes and phases. However, a voltage, a current, or any physical quantity corresponding to a sinusoidal signal is a real quantity. This means that after analyzing an AC circuit with complex quantities representing real signals, the physical interpretation of the output signal should be made by coming back to real quantities.

AC sinusoidal signals are also important because any periodic signal with period T can be written as a discrete sum of sinusoidal signals through the Fourier series:

$$f(t) = \frac{a_0}{2} + \sum_{n=1}^{+\infty} (a_n \cos \omega_n t + b_n \sin \omega_n t) \, ,$$

$$= \sum_{n=-\infty}^{+\infty} c_n e^{j\omega_n t} \, ,$$

where $\omega_n = n2\pi/T$ and a_n, b_n are the real amplitude coefficients of the signal component of angular frequency ω_n, whereas c_n is the complex valued coefficient of the component at ω_n. Moreover, any signal can be written as a continuous sum of AC signals through the Fourier transform:

$$f(t) = \frac{1}{2\pi} \int_{-\infty}^{+\infty} c(\omega) e^{j\omega t} \, d\omega \, ,$$

where $c(\omega)$ is the complex valued amplitude of the signal component with frequency ω. Therefore, analyzing an AC circuit for a sinusoidal excitation signal with angular frequency ω, and characterizing the circuit response in terms of frequency, allows us to reconstruct the circuit response to any excitation signal, by linear superposition of the responses to each frequency component in the excitation signal.

12.7 Complex Representation of AC Circuit Components

12.7.1 Generator, Voltages, Currents

Analyzing AC circuits involves considering the impedance of circuit components, which is a complex quantity combining resistance, inductive reactance, and capacitive reactance. These concepts quantify the opposition to the flow of AC current offered by resistors, inductors, and capacitors, respectively.

If a generator in the circuit is suddenly turned on and produces a sinusoidal voltage source, a permanent forced sinusoidal regime will be reached after a transient period. In this forced sinusoidal regime, all voltages and currents in the circuit are sinusoidal with the same frequency as the generator. For example, in the circuit represented in Fig. 12.16a, the sinusoidal voltage source is

$$e(t) = e_0 \cos \omega t$$

where e_0 is a real amplitude. The current in the inductor and voltage across the inductor are

$$i(t) = i_0 \cos(\omega t + \phi_i) \quad \text{and} \quad v(t) = v_0 \cos(\omega t + \phi_v) ,$$

where the peak current i_0 and peak voltage v_0 are real quantities. The phases ϕ_i and ϕ_v are the phase differences with respect to the voltage signal delivered by the generator, which is considered as the phase reference.

Figure 12.16b represents the same circuit using complex notations for the signal, which constitutes the standard in AC-circuit analysis. The voltage source E, current I, and voltage V across the inductor must be understood as the time dependent

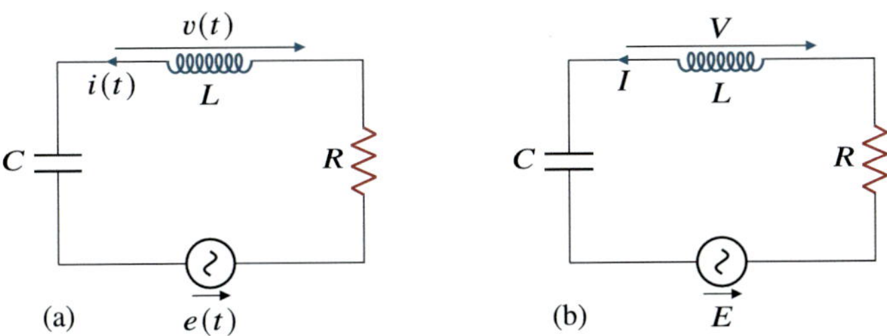

Fig. 12.16 An AC circuit with signals marked **a** in time dependent real notation; **b** with complex amplitudes

complex signals

$$E(t) = E_0 e^{j\omega t}, \quad I(t) = I_0 e^{j\omega t}, \quad V(t) = V_0 e^{j\omega t}$$

where E_0, I_0, and V_0 are the complex amplitudes of the corresponding signals. Their relation to the amplitude and phase of the real signals is

$$E_0 = e_0, \quad I_0 = i_0 e^{j\phi_i}, \quad V_0 = v_0 e^{j\phi_v} .$$

12.7.2 *Impedance of Passive Components*

- Consider any passive element, represented in Fig. 12.17. We define the impedance as the complex coefficient of proportionality between complex current and complex voltage amplitudes:

$$U = ZI .$$

This relation, similar to Ohm's law, shows that impedance is a generalization of the concept of resistance to AC circuits. The impedance is a complex number that represents the opposition to the flow of alternating current in an electrical circuit. We can derive the impedance of a bipole by means of the IV relationship for the bipole. To this aim, we will express this relationship for each bipole in terms of real signals, $v(t)$, the voltage across the bipole at time t, and $i(t)$, the current through the bipole at time t. Expressing this relationship in terms of the complex voltage V and current I allows us to obtain the impedance of the bipole.
- Impedance of a resistor.
  Ohm's law in the quasi-steady regime is expressed as $v(t) = Ri(t)$, where R is the resistance. Ohm's law in complex notations is expressed as

$$V = RI .$$

Therefore, the ratio V/I gives the impedance of a resistor,

$$\boxed{Z_R = R .}$$

The impedance of a resistor is purely real and equal to its resistance. This is illustrated in Fig. 12.18.

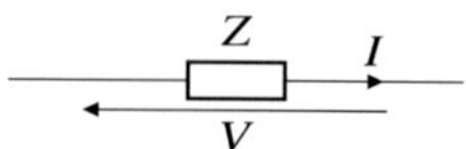

Fig. 12.17 A passive component in the sinusoidal regime is characterized by its impedance Z

Fig. 12.18 The impedance of a resistor is simply its resistance R

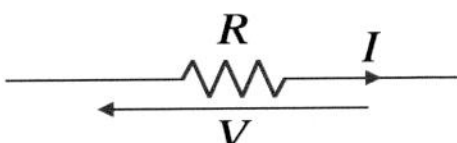

Fig. 12.19 The impedance of a capacitor is $Z = 1/j\omega C$

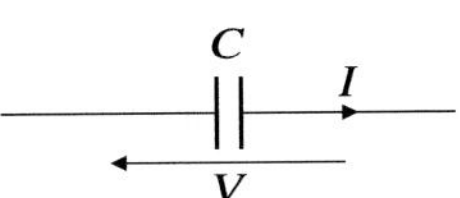

- Impedance of a capacitor

  For a capacitor in Fig. 12.19, the IV relationship is $i(t) = C\,dv(t)/dt$, where C is the capacitance. The complex representation of this linear relationship between $i(t)$ and $v(t)$ is

$$I = C\frac{dV}{dt} \, .$$

For a sinusoidal signal, substituting $V = V_0 e^{j\omega t}$ in this relation, we obtain

$$I = Cj\omega V_0 e^{j\omega t} = jC\omega V \, .$$

The ratio V/I gives the impedance for a capacitor,

$$\boxed{Z = \frac{1}{j\omega C}} \, .$$

The impedance of a capacitor is purely imaginary and inversely proportional to the frequency.

- Impedance of an inductor

  For an inductor, shown in Fig. 12.20, the IV relationship is expressed as $v(t) = L\,di(t)/dt$, where L is the inductance. Again, this relation is a linear relationship between $i(t)$ and $v(t)$ whose complex representation is

$$V(t) = L\frac{dI(t)}{dt} \, .$$

Substituting the complex current $I(t) = I_0 e^{j\omega t}$, we find

$$V(t) = LI_0 j\omega e^{j\omega t} = j\omega L I(t) \, .$$

The ratio V/I yields the impedance of an inductor:

$$\boxed{Z = j\omega L} \, .$$

Fig. 12.20 The impedance of an inductor is $Z = j\omega L$

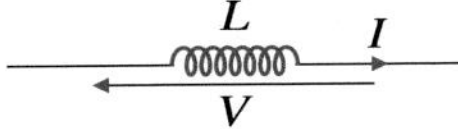

The impedance of an inductor is purely imaginary and directly proportional to the frequency.

12.7.3   Impedance Terminology

- **Resistance and reactance**
  For a general circuit element or bipole, the impedance is a complex number that measures the opposition of the bipole to the flow of alternating current due to a combination of capacitance, inductance, and resistance:

$$Z = R + jX \ .$$

  where R is the *resistance* (real part) and X is the *reactance* (imaginary part). Both the resistance and reactance are measured in ohm (Ω).
  A reactance can be of inductive or capacitive type depending on its relation to the signal angular frequency ω:

  - Inductive reactance: $X_L = \omega L$, where L is the inductance.
  - Capacitive reactance: $X_C = -1/(\omega C)$, where C is the capacitance.

- **Admittance**
  Admittance is the reciprocal of impedance, representing the ease with which alternating current flows through a circuit element.

$$Y = \frac{1}{Z} = G + jB \ .$$

  The real part, G, is known as the *conductance* while the imaginary part, B, is known as the *susceptance*. Both G and B are measured in siemens ($S \equiv \Omega^{-1}$).

12.7.4   Series and Parallel Connection of Impedances

Kirchhoff's laws are linear. Therefore they apply to complex current and voltages

- **Impedances add in series**: Consider two impedances connected in series, with complex current I flowing through the impedances, and complex voltages V_1 and V_2 across each impedance. Applying KVL and Ohm's law, we get

$$V = V_1 + V_2 = Z_1 I + Z_2 I \ .$$

  Equating the result to ZI allows us to identify the impedance of the series connection

$$Z = Z_1 + Z_2 \ .$$

Connecting impedances in series allows us to obtain a voltage divider: across each impedance, we have

$$V_1 = Z_1 I = Z_1 \frac{V}{Z} = \frac{Z_1}{Z_1 + Z_2} V \, ,$$

$$V_2 = Z_2 I = Z_2 \frac{V}{Z} = \frac{Z_2}{Z_1 + Z_2} V \, .$$

- **Admittances add in parallel**: Consider two impedances connected in parallel, with complex voltage V across the bipole and complex currents I_1 and I_2 flowing through each impedance. Applying KCL and Ohm's law, we get

$$I = I_1 + I_2 = \frac{V}{Z_1} + \frac{V}{Z_2} \, .$$

Equating the result to V/Z allows us to identify the admittance of the parallel connection:

$$\frac{1}{Z} = \frac{1}{Z_1} + \frac{1}{Z_2} \, ,$$

or equivalently:

$$Y = Y_1 + Y_2 \, .$$

Connecting impedances in parallel allows us to obtain a current divider: Through each impedance, we have

$$I_1 = Y_1 V = Y_1 \frac{I}{Y} = \frac{Y_1}{Y_1 + Y_2} I \, ,$$

$$I_2 = Y_2 V = Y_2 \frac{I}{Y} = \frac{Y_2}{Y_1 + Y_2} I \, .$$

12.8 Sinusoidal Response of a Linear Network

Figure 12.21 represents a linear network in a schematic format facilitating the analysis of the network response to AC currents. The three boxes represent the following functions:

- The source is a bipole (generator) providing the input voltage v_{in} and the input current i_{in}. We assume it is a linear bipole. It is useful to represent it as its equivalent Thévenin's voltage source in the case of a low internal resistance, or as its equivalent Norton's current source in the case of a large internal resistance.
- The transmission system has two input poles and two output poles. It is therefore a quadrupole, realizing a specific function. This means it transforms the input current and voltage into an output current i_{out} and voltage v_{out}.
- The load is a passive bipole. For instance, the load can be a simple resistor, or more generally a circuit characterized by its impedance.

Fig. 12.21 Schematic for a linear network modeled as a quadrupole. A source is connected to the two input poles. A load is connected to the two output poles. The quadrupole constitutes a transmission system

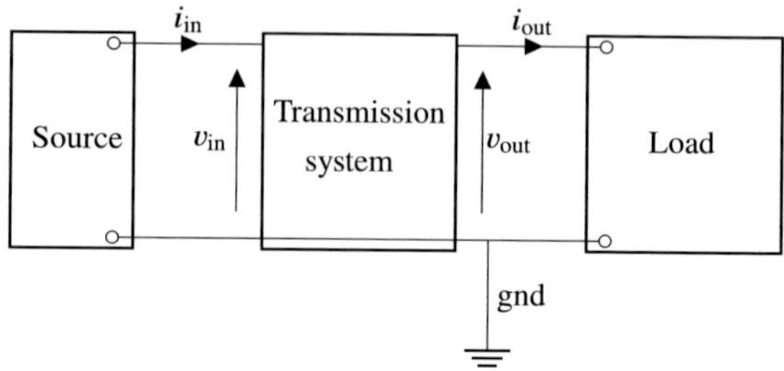

- There is usually a short circuit link between the bottom input and output nodes, represented by the ground connection. This means that the reference zero potential is taken at these nodes.

Analyzing the response of a network to AC currents consists in giving the relation between i_{out} and v_{in}, or between v_{out} and v_{in}.

12.8.1   Transfer Function

A transfer function is a mathematical representation of how a system responds to an input signal or, equivalently, how a system transforms an input signal into an output signal. In the context of AC currents, we look for the steady state response of a circuit to a sinusoidal excitation, which is the forced oscillation regime sustained by the sources after the transient response disappeared.

Consider a system with an input signal $v_{in}(t)$ and an output signal $v_{out}(t)$, The transfer function relating the input and output signals in the frequency domain is expressed as

$$H(j\omega) = \frac{V_{out}}{V_{in}}$$

where V_{out} and V_{in} denote the complex amplitudes of the output and input signals, respectively. The transfer function is a complex valued function of frequency:

$$H(j\omega) = G(\omega)e^{j\phi(\omega)} ,$$

where the modulus $G(\omega) = |H(j\omega)|$ represents the gain and the phase $\phi(\omega)$ represents the delay (or advance) of the output signal with respect to the input signal. The gain and the phase only depend on the circuit elements and on the angular frequency ω.

12.8.2   Bode Plots

A Bode plot is a graphical representation of the frequency response of a system, often used to analyze the behavior of transmission systems acting as filters. A Bode plot typically consists of two plots:

- The *Magnitude plot* shows the gain of the system, $G(\omega)$, as a function of frequency (ω). The gain is usually quantified by using a logarithmic scale, and is then measured in decibel (dB), a physical unit defined by the relation

$$G_{\mathrm{dB}} = 20 \log_{10} G \,,$$

where $\log_{10}$ denotes the decimal logarithm. From this relation and the definition of the gain, $G = |V_{\mathrm{out}}|/|V_{\mathrm{in}}|$, it is clear that

$$G_{\mathrm{dB}} = V_{\mathrm{out, dB}} - V_{\mathrm{in, dB}} \,.$$

The gain in dB corresponds to the voltage drop (difference) between output and input signals measured in decibel, $V_{\mathrm{out, dB}} = 20 \log_{10} |V_{\mathrm{out}}|$, and $V_{\mathrm{in, dB}} = 20 \log_{10} |V_{\mathrm{in}}|$. The coefficient of 20 means 20 dB. Therefore, a gain of $G_{\mathrm{dB}} = 20$ dB corresponds to a gain of $G = 10$, i.e., an amplification of the input signal by a factor of 10. Attenuation corresponds to a gain $G < 1$, or, equivalently, to a negative gain in decibel $G_{\mathrm{dB}} < 0$. A frequently used value is the gain $G_{\mathrm{dB}} = -3$ dB corresponding approximately to a drop in the gain by a factor of $\sqrt{2}$, i.e., $20 \log_{10}(1/\sqrt{2}) \approx -3$ dB. A gain of -3 dB means an attenuation of the signal by 3 dB or, equivalently, an attenuation of the signal power (square of the signal amplitude) by a factor of 2.
- The *phase plot* shows the phase shift, $\phi(\omega)$, (in degrees) between the input and output signals as a function of frequency.

As illustrated in Fig. 12.22, the key features of Bode plots must be indicated. In particular:

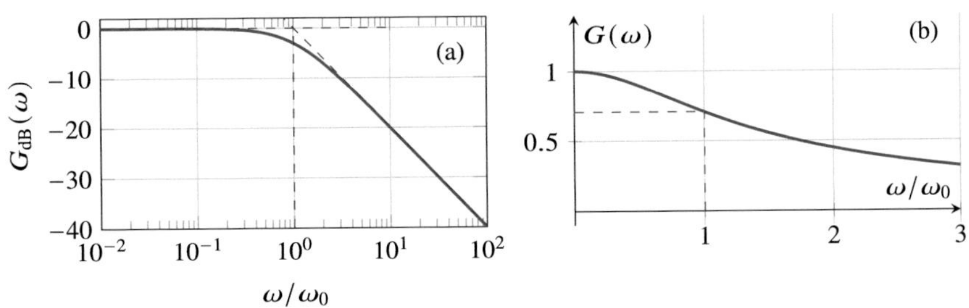

Fig. 12.22   Example of Bode plots: **a** Gain in decibel G_{dB} as a function of frequency (normalized to a reference frequency ω_0). The red dashed lines indicate the asymptotes for $\omega \ll \omega_0$ and $\omega \gg \omega_0$, intersecting at the reference frequency ω_0. **b** Gain G as a function of ω/ω_0. The red dashed lines indicate the frequency at which a gain of -3 dB is obtained ($G = 1/\sqrt{2}$)

- Asymptotic approximation: The Bode plot for the gain G_{dB} can be approximated by straight-line segments, which simplifies analysis. These asymptotes are obtained by studying the behavior of the gain at low frequency ($\omega \ll \omega_0$), large frequency ($\omega \gg \omega_0$), or in any other region of interest.
- The *corner frequency* is the frequency at which the gain starts to change rapidly.
- The *roll-off rate* is the rate at which the gain changes with frequency, often expressed in decibels per decade.

12.8.3   Filters

Electronic filters perform signal processing by removing unwanted frequencies in the signal and/or modifying its phase. Electronic filters can be passive and linear if using only resistances, inductances and capacitors. They are essential components in many electronic systems, from audio amplifiers to communication systems. They can be represented as a quadrupole electrical element. Figure 12.23 illustrates different types of filters:

- **Low-pass filters** pass low-frequency signals and attenuate high-frequency signals (Fig. 12.23a).
- **High-pass filters** pass high-frequency signals and attenuate low-frequency signals (Fig. 12.23b).
- **Band-pass filters** pass signals within a specific frequency range and attenuate signals outside that range (Fig. 12.23c).

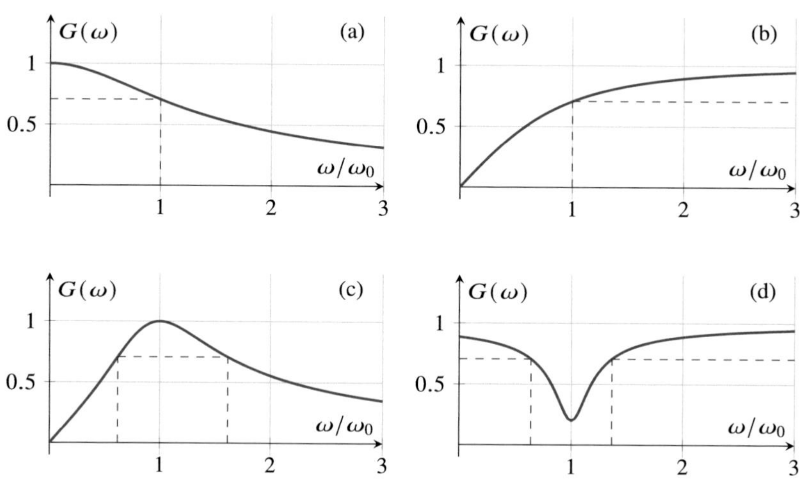

Fig. 12.23 Example of Bode plots representing filters: **a** A low-pass filter. **b** A high-pass filter. **c** A band-pass filter. **d** A band-stop filter. The horizontal dashed line indicates the range of frequencies such that $G \leq G_{max}/\sqrt{2}$.

- **Band-stop filters** (or notch filters) attenuate signals within a specific frequency range and pass signals outside that range.

By analyzing the Bode plot of the transfer function corresponding to a filter, we can determine the frequency response of the filter, its bandwidth, and phase shift characteristics. The band limits are usually determined by looking for cutoff frequencies. The cutoff frequency at $-3\,\mathrm{dB}$, ω_c, is defined as the frequency value for which the gain in decibel is reduced by $3\,\mathrm{dB}$, corresponding to $G(\omega_c) = G_{\mathrm{max}}/\sqrt{2}$:

$$G_{\mathrm{dB}}(\omega_c) = G_{\mathrm{max,dB}} - 3\,\mathrm{dB}\ .$$

12.9 Examples of Electronic Filters

12.9.1 RC Circuit as Low-Pass Filter

For the RC circuit, we have:

$$V = \frac{I}{jC\omega}\ , \quad E = \left(R + \frac{1}{jC\omega}\right) I\ .$$

$$H(j\omega) = \frac{V}{E} = \frac{1}{1 + jRC\omega}\ .$$

Defining $\omega_0 = 1/\sqrt{RC}$ and normalizing the frequency ω as $\tilde{\omega} = \omega/\omega_0$, we can express the gain and phase shift as

$$G(\tilde{\omega}) = \frac{1}{\sqrt{1 + \tilde{\omega}^2}}\ , \quad \phi(\tilde{\omega}) = -\tan^{-1}\tilde{\omega}\ .$$

The magnitude and phase of the transfer function can be plotted on a Bode plot to visualize the circuit frequency response. As shown in Fig. 12.24, the low-frequency signals are passed with minimal attenuation, while the high-frequency signals are attenuated. The circuit is a low-pass filter.

The gain in decibel is

$$G_{\mathrm{dB}}(\tilde{\omega}) = -20\log_{10}(\sqrt{1 + \tilde{\omega}^2})\ .$$

The cutoff frequency at $-3\,\mathrm{dB}$ corresponds to $\tilde{\omega}_c = 1$, or $\omega_c = \omega_0$. The asymptotic analysis yields the regimes summarized in the table below.

Validity range		$G(\tilde{\omega})$	$G_{\mathrm{dB}}(\tilde{\omega})$
$\omega \ll \omega_c$	$\tilde{\omega} \ll 1$	$G \approx 1$	$G_{\mathrm{dB}} \approx 0$
$\omega \ll \omega_c$	$\tilde{\omega} \gg 1$	$G \approx \tilde{\omega}^{-1}$	$G_{\mathrm{dB}} \approx -20\log_{10}\tilde{\omega}$

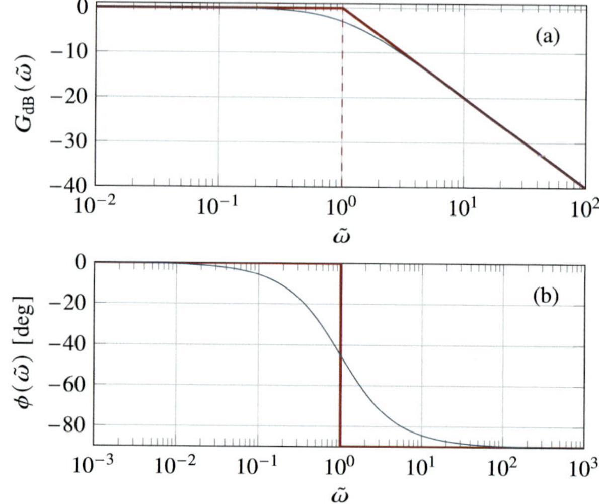

Fig. 12.24 Bode plots for a RC circuit (Low-pass filter): **a** Gain in decibel $G_{\mathrm{dB}}(\tilde{\omega})$ and **b** phase shift $\phi(\tilde{\omega})$ as function of the normalized frequency $\tilde{\omega}$. The red lines indicate the asymptotic regimes for $\tilde{\omega} \ll 1$ and $\tilde{\omega} \gg 1$

12.9.2   RLC Circuit

The RLC circuit is a fundamental circuit in electrical engineering and is used in various applications, including filters, oscillators, and power electronics. In a series RLC circuit, the components are connected end-to-end, so the same current flows through each component, as illustrated in Fig. 12.25. For analyzing a RLC circuit, it is convenient to represent the circuit as a quadrupole, powered by a source delivering input voltage $v_{\mathrm{in}}(t)$. The circuit realizes a transformation of these input signals into an output signal, usually the voltage across one of the components. In the example of Fig. 12.25, the output voltage, $v_{\mathrm{out}}(t)$, is that across the capacitor. It is related to the current $i(t)$ in the circuit.

Applying KVL to the circuit, we obtain the governing equation for the current $i(t)$

$$v_{\mathrm{in}}(t) = Ri(t) + L\frac{di(t)}{dt} + \frac{q(t)}{C} \, , \tag{12.7}$$

where $q(t)$ denotes the charge of the capacitor related to the current by

$$i = \frac{dq}{dt} \, .$$

Fig. 12.25 An RLC circuit represented as a quadrupole with output voltage across the capacitor

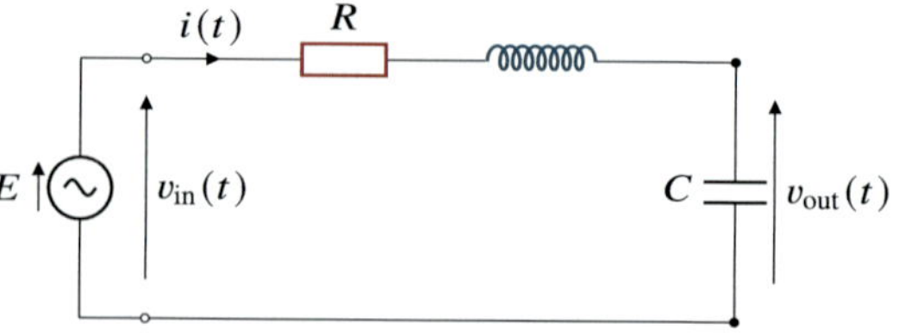

For the sinusoidal source, the input signal is given by

$$v_{\text{in}}(t) = v_{\text{max,in}} \cos \omega t$$

where ω is the angular frequency of the AC source and $v_{\text{max,in}}$ is the peak amplitude of the signal. As seen in previous sections, the current $i(t)$ is solution to a non-homogeneous linear differential equation. It is the superposition of the general solution to the homogeneous problem and a particular solution to the non-homogeneous problem. We have seem previously that the solution to the homogeneous problem is a transient current that decay to zero within a few characteristic times. After this transient stage, the solution is dominated by the particular solution to the non-homogeneous problem (12.7). We are looking for this solution, the steady state response of the circuit, which will exhibit forced sinusoidal oscillations. We look for $i(t)$ in the form

$$i(t) = i_{\text{max}} \cos(\omega t + \phi) .$$

To facilitate calculation, we systematically use complex notations for the input voltage (V_{in}) and current (I) to rewrite (12.7) as

$$V_{\text{in}} = RI + jL\omega I + \frac{1}{jC\omega} I .$$

Comparing this equation with Ohm's law, $V = Z_{RLC} I$, we can identify the complex impedance:

$$Z_{RLC} = \frac{V_{\text{in}}}{I} = R + j\left(L\omega - \frac{1}{C\omega}\right) .$$

This complex impedance can be written in terms of its amplitude and phase: $Z_{RLC} = |Z_{RLC}| e^{j\phi}$, where

$$|Z_{RLC}| = \sqrt{R^2 + \left(L\omega - \frac{1}{C\omega}\right)^2} \quad \text{and} \quad \tan\phi = \frac{L\omega - \dfrac{1}{C\omega}}{R} .$$

These expressions show that impedance of this circuit is purely resistive for a frequency $\omega_0 = 1/\sqrt{LC}$. At this frequency, the circuit will be resonant and the current will be maximum.

12.9.3 RLC Circuit, Output Voltage Across the Resistor

Consider the RLC circuit where the output signal is the voltage across the resistor, as illustrated in Fig. 12.26. The transfer function can be expressed using the voltage divider relation,

Fig. 12.26 A RLC circuit with output voltage across the resistor

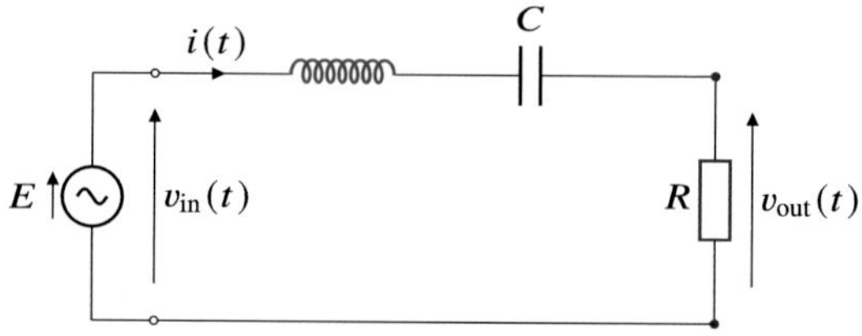

$$H(j\omega) = \frac{V_{out}}{V_{in}} = \frac{RI}{Z_{RLC}} .$$

Substituting the expression for Z_{RLC}, we obtain

$$H(j\omega) = \frac{R}{R + j\left(L\omega - \dfrac{1}{C\omega}\right)} .$$

From this relation, the gain and phase shift can be expressed as

$$G(\tilde{\omega}) = \frac{1}{\sqrt{1 + Q^2\left(\tilde{\omega} - \dfrac{1}{\tilde{\omega}}\right)^2}} ,$$

$$\tan(\phi(\tilde{\omega})) = -Q\left(\tilde{\omega} - \frac{1}{\tilde{\omega}}\right) ,$$

where the frequency is normalized with the resonant frequency ω_0 as

$$\tilde{\omega} = \frac{\omega}{\omega_0} , \qquad \text{where } \omega_0 = \frac{1}{\sqrt{LC}} ,$$

and the quantity Q-factor (quality factor) is given by

$$Q = \frac{1}{RC\omega_0} = \frac{L\omega_0}{R} .$$

Figure 12.27 shows the Bode plots for this circuit, which acts as a band-pass filter. The gain is maximum at frequency ω_0 with maximum gain $G_{max} = 1$. It is larger than $1/\sqrt{2}$ for frequencies within a band $\omega_{c,-} \leq \omega \leq \omega_{c,+}$, where the cutoff frequencies are given by

$$\omega_{c,\pm} = \omega_0\left(\sqrt{1 + \frac{1}{4Q^2}} \pm \frac{1}{2Q}\right) .$$

These frequencies define the bandwidth, the range of frequencies over which the circuit response is significant (larger gain than $-3\,\text{dB}$):

Fig. 12.27 Bode plots for a RLC circuit (Band-pass filter): **a** Gain in decibel $G_{dB}(\tilde{\omega})$ and **b** phase shift $\phi(\tilde{\omega})$ as function of the normalized frequency $\tilde{\omega}$. The dashed lines indicate the bandwidth. Quality factors are $Q = 1$ (green curves) and $Q = 3$) (blue curves)

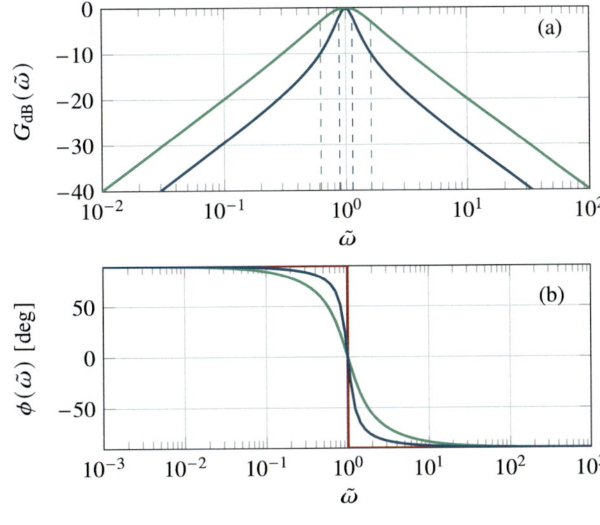

$$\Delta\omega = \omega_+ - \omega_- = \frac{\omega_0}{Q} \, .$$

This expression shows that a higher Q factor results in a narrower bandwidth. This is why the Q factor is called the *quality factor*: Q is a measure of the circuit selectivity. A higher quality factor means a sharper resonance peak.

As illustrated in Fig. 12.27b, at the resonant frequency, the phase angle between the voltage across the resistor and the current is zero. This indicates that the circuit behaves purely resistively at this frequency. Consequently, the impedance of the circuit is minimized, leading to maximum current flow. Since the resistor is in series with the current, it experiences the maximum voltage drop. In conclusion, when the output is taken across the resistor in a series RLC circuit, it functions as a band-pass filter. This configuration is highly selective, allowing only a narrow range of frequencies to pass through. This principle is exploited in various applications, including tuning circuits in radios and televisions, audio filters to enhance specific frequency ranges, and communication systems to isolate desired frequency bands.

12.9.4 Series RLC Circuit with Output Voltage Across the Capacitor

Consider now an RLC circuit where the output signal is the voltage across the capacitor, as illustrated in Fig. 12.28. The transfer function can again be expressed using the voltage divider relation,

$$H(j\omega) = \frac{V_{\text{out}}}{V_{\text{in}}} = \frac{(jC\omega)^{-1}I}{Z_{RLC}I} \, .$$

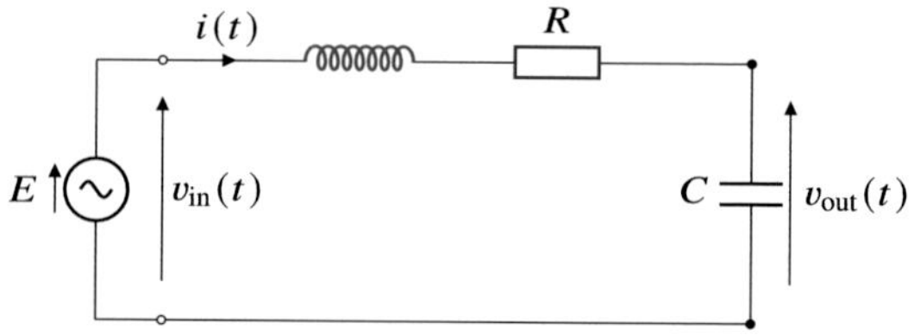

Fig. 12.28 An RLC circuit with output voltage across the capacitor

Substituting the expression for Z_{RLC}, we obtain

$$H(j\omega) = \frac{1}{jRC\omega - \left(LC\omega^2 - 1\right)} \, .$$

The gain and phase shift can be again expressed in terms of the normalized frequency $\tilde{\omega} = \omega/\omega_0$, where $\omega_0 = 1/\sqrt{LC}$ and quality factor $Q = 1/(RC\omega_0)$:

$$G(\tilde{\omega}) = \frac{1}{\sqrt{(1 - \tilde{\omega}^2)^2 + \dfrac{\tilde{\omega}^2}{Q^2}}} \, ,$$

$$\tan(\phi(\tilde{\omega})) = \frac{1}{Q\left(\tilde{\omega} - \dfrac{1}{\tilde{\omega}}\right)} \, .$$

Figure 12.29 illustrates the Bode plots for this RLC circuit with two different Q-factor values. The circuit behavior can transition from band-pass to low-pass filtering, depending on the specific Q-factor.

- Band-pass amplifier: $Q \geq 1/\sqrt{2}$.
  The gain curve exhibits a maximum (see blue curve in Fig. 12.27a) at a certain frequency $\omega_{r(C)}$. In this case, the maximum gain $G_{\max}$ and the frequency $\omega_{r(C)}$ maximizing the gain are

$$G_{\max} = \frac{Q}{\sqrt{1 - \dfrac{1}{2Q^2}}} \, , \quad \omega_{r(C)} = \omega_0 \sqrt{1 - \frac{1}{2Q^2}} \, .$$

  For large Q-factor values, the gain at resonance, $G(\omega_r) = Q/\tilde{\omega}_r(Q)$, is approximately equal to Q. As it is larger than 1, the circuit acts as an amplifier close to the resonant frequency.
  The cutoff frequencies at which the gain drops to $1/\sqrt{2}$ of its maximum value are

$$\omega_{c,\pm} = \omega_0 \left(1 - \frac{1}{2Q^2} \pm \frac{1}{Q}\sqrt{1 - \frac{1}{4Q^2}}\right)^{1/2} = \omega_{r(C)}\left(1 \pm \sqrt{\tilde{\omega}_{r(C)}^{-4} - 1}\right)^{1/2} \, .$$

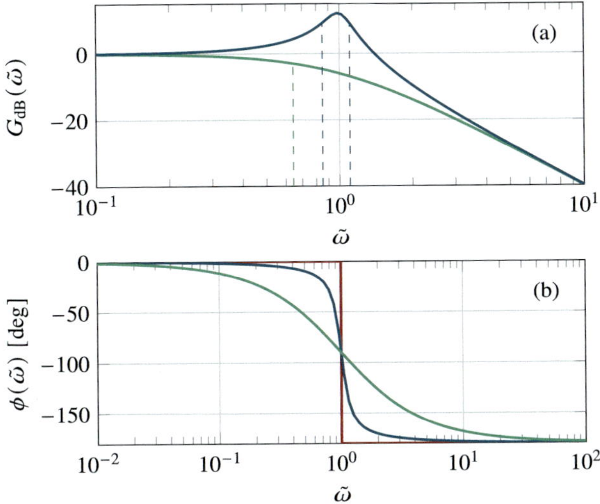

Fig. 12.29 Bode plots for a series RLC circuit with output voltage across the capacitor: **a** Gain in decibel $G_{\mathrm{dB}}(\tilde{\omega})$ and **b** phase shift $\phi(\tilde{\omega})$ as functions of the normalized frequency $\tilde{\omega}$. The dashed lines indicate the cutoff frequencies at $-3\,\mathrm{dB}$. The red lines indicate the asymptotes. Quality factors are $Q = 4$ (blue curves) and $Q = 0.5$) (green curves)

For large Q values, the bandwidth is approximately (by a Taylor expansion of $\omega_{c,\pm}$):

$$\Delta\omega = \omega_{c,+} - \omega_{c,-} \approx \frac{\omega_0}{Q} \quad \text{for } Q \gg 1 .$$

This expression is the same as in the case of the RLC circuit with output voltage across the resistor, indicating that a narrower bandwidth is obtained for a higher Q factor.

- Low-pass filter: $Q \leq 1/\sqrt{2}$.
  In this case, the maximum gain, $G_{\max} = 1$, is achieved at low frequencies ($\omega = 0$). The circuit acts as a low-pass filter (see green curve in Fig. 12.29a). The $-3\,\mathrm{dB}$ cutoff frequency is given by

$$\omega_c = \omega_0 \left(1 - \frac{1}{2Q^2} + \sqrt{\left(\frac{1}{2Q^2} - 1 \right)^2 + 1} \right)^{1/2} .$$

As illustrated in Fig. 12.29b, at the resonant frequency, $\omega = \omega_r$, the voltage across the capacitor lags the current by $\pi/2$ radians. This phase shift occurs because the inductive and capacitive reactances cancel each other out, leaving only the resistive component. As a result, the circuit exhibits maximum current flow. Since the capacitor and inductor are in series, they share the same current. However, the voltage across the capacitor is maximized due to the phase relationship between the current and voltage across a capacitor.

In conclusion, for large Q-factors, this series RLC circuit with capacitive output functions as a band-pass filter. It amplifies signals within a narrow frequency band centered around the resonant frequency. For small Q-factors, this circuit behaves as a low-pass filter. It attenuates high-frequency signals while allowing low-frequency signals to pass through. This versatile configuration finds applications in various electronic circuits, including filters, oscillators, and tuning circuits.

12.9.5   RLC Circuit with Output Voltage Across the Inductor

Consider finally the RLC circuit where the output signal is the voltage across the inductor, as illustrated in Fig. 12.30. The transfer function can again be expressed using the voltage divider relation,

$$H(j\omega) = \frac{V_{\text{out}}}{V_{\text{in}}} = \frac{(jL\omega)I}{Z_{RLC}I} \, .$$

Substituting the expression for Z_{RLC}, we obtain

$$H(j\omega) = \frac{jL\omega}{R + j\left(L\omega - \dfrac{1}{C\omega}\right)} \, .$$

Keeping the same definitions as previously for the resonant frequency, $\omega_0 = 1/\sqrt{LC}$, and quality factor, $Q = 1/(RC\omega_0)$, the gain and phase shift can be again expressed as

$$G(\tilde{\omega}) = \frac{1}{\sqrt{\dfrac{1}{Q^2\tilde{\omega}^2} + \left(1 - \dfrac{1}{\tilde{\omega}^2}\right)^2}} \, ,$$

$$\tan(\phi(\tilde{\omega})) = -\frac{1}{Q\left(\tilde{\omega} - \dfrac{1}{\tilde{\omega}}\right)} \, .$$

The expressions for the gain and phase shift are the same as those obtained for the RLC circuit with output voltage across the capacitor if we replace $\tilde{\omega}$ by $\tilde{\omega}^{-1}$.

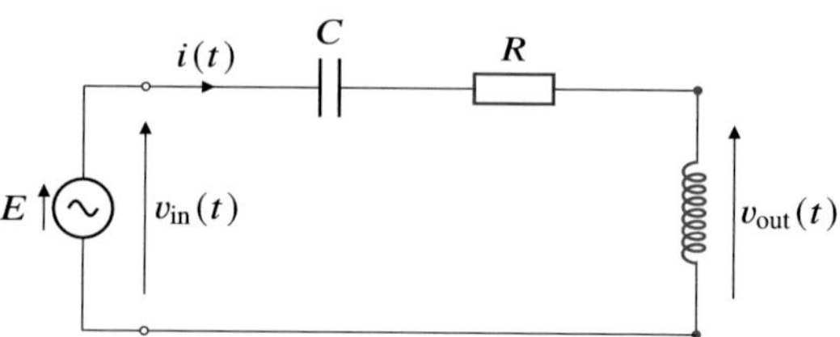

Fig. 12.30 A RLC circuit with output voltage across the inductor

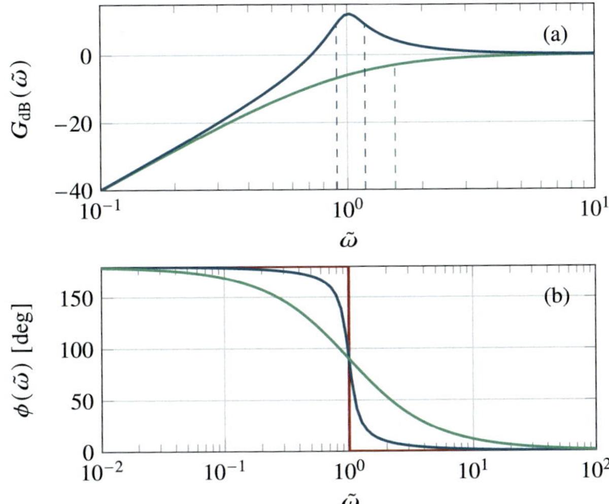

Fig. 12.31 Bode plots for a RLC circuit with output voltage at the inductor: **a** Gain in decibel $G_{dB}(\tilde{\omega})$ and **b** phase shift $\phi(\tilde{\omega})$ as function of the normalized frequency $\tilde{\omega}$. The dashed lines indicate the cutoff frequencies at $-3\,$dB. The red lines indicate the asymptotes. Quality factors are $Q = 4$ (blue curves) and $Q = 0.5)$ (green curves)

Figure 12.31 shows the Bode plots for this RLC circuit for two different values of the Q factor. Again, the circuit function depends on the Q-factor value. For small quality factor values, the circuit behaves as a high-pass filter. For large quality factor values, it behaves as a band-pass amplifier.

- Band-pass amplifyer: $Q \geq 1/\sqrt{2}$.
  The gain curve exhibits a maximum (see blue curve in Fig. 12.31a) at a certain frequency $\omega_{r(L)}$. The maximum gain $G_{\max}$ and the resonant frequency $\omega_{r(L)}$ maximizing the gain are given by

$$G_{\max} = \frac{Q}{\sqrt{1 - \dfrac{1}{4Q^2}}} \,, \quad \omega_{r(L)} = \frac{\omega_0}{\sqrt{1 - \dfrac{1}{2Q^2}}} \,.$$

The $-3\,$dB cutoff frequencies can be inferred from the expressions of cutoff frequencies obtained for the resonance of the RLC circuit with output voltage across the capacitor (using the transformation $\tilde{\omega} \to \tilde{\omega}^{-1}$):

$$\omega_{c,\pm} = \omega_0 \left(1 - \frac{1}{2Q^2} \mp \frac{1}{Q}\sqrt{1 - \frac{1}{4Q^2}} \right)^{-1/2} = \omega_{r(L)} \left(1 \mp \sqrt{\tilde{\omega}_{r(L)}^4 - 1} \right)^{-1/2} \,.$$

A Taylor expansion of these expressions for large Q-values yields again a simple result for the bandwidth

$$\Delta\omega = \omega_{c,+} - \omega_{c,-} \approx \frac{\omega_0}{Q} \quad \text{for } Q \gg 1 \,,$$

indicating as previously that a narrower bandwidth is obtained for a higher Q factor.

- High-pass filter: In the case $Q \leq 1\sqrt{2}$, the gain maximum is $G_{\max} = 1$, reached at high frequency, $\omega \gg \omega = r$. The circuit acts as a high-pass filter (see green curve in Fig. 12.31a). There is a single cutoff frequency at -3 dB given by

$$\omega_c = \omega_0 \left(1 - \frac{1}{2Q^2} + \sqrt{\left(\frac{1}{2Q^2} - 1 \right)^2 + 1} \right)^{-1/2} .$$

The phase angle is illustrated in Fig. 12.31b, indicating that he voltage across the inductor leads the current by $\pi/2$ degrees at resonance.

In conclusion, when the output voltage is taken across the inductor in a series RLC circuit, the circuit is in series resonance with inductive output. At the resonant frequency, the inductive reactance ($jL\omega$) and capacitive reactance ($1/(jC\omega)$) cancel each other out, leaving only the resistive component. This results in a maximum voltage across the inductor, as the entire source voltage is dropped across it. The circuit exhibits a high impedance at frequencies far from resonance. This configuration is often used in filter circuits, particularly band-pass filters, where the desired frequency band is selected based on the resonant frequency of the circuit.

12.10   Power in AC Circuits

12.10.1   AC Power in Real Notations

Consider a circuit with a sinusoidal voltage $v(t)$ across its terminals and a current $i(t)$ flowing through it. These signals can be expressed as

$$v(t) = V \cos(\omega t) , \quad i(t) = I \cos(\omega t - \phi) .$$

Here, ϕ is the phase difference between the current and voltage. The instantaneous power delivered to the circuit is given by

$$p(t) = v(t)i(t) = V I \cos \omega t \cos(\omega t - \phi) .$$

Using the trigonometric identity $\cos A \cos B = (1/2)[\cos(A + B) + \cos(A - B)]$, we can rewrite $p(t)$ as

$$p(t) = V I [\cos \phi + \cos(2\omega t - \phi)] .$$

This expression shows that the instantaneous power is a sinusoidal signal with frequency 2ω superimposed on a constant term.

The average power, P_a, is the time average of the instantaneous power over one period $T = 2\pi/\omega$:

$$P_a = \langle p(t) \rangle = \frac{1}{T} \int_0^T p(t)dt = \frac{1}{2} V I \cos \phi .$$

Comparing this result to the DC power expression, we observe an additional factor of 1/2. This arises from the time-averaging of sinusoidal signals. To establish a more direct connection to DC power, we introduce the concept of effective current and voltage:

$$I_{\text{eff}} = \sqrt{\langle i^2(t) \rangle} = \frac{I}{\sqrt{2}}, \qquad V_{\text{eff}} = \sqrt{\langle v^2(t) \rangle} = \frac{V}{\sqrt{2}} \; .$$

Using these effective values, the average power can be expressed as

$$P_a = V_{\text{eff}} I_{\text{eff}} \cos \phi \; .$$

The factor $\cos \phi$ is known as the power factor. For a given voltage, a higher power factor implies a lower current requirement to deliver the same power. This leads to reduced resistive losses in power transmission and distribution. Consequently, power companies often impose regulations to maintain a high power factor in domestic networks.

12.10.2 AC Power in Complex Notation

Since the instantaneous power is quadratic in the sinusoidal signals, care must be taken when using complex notation. To account for this, we introduce the concept of complex power, $\underline{P}$:

$$\underline{P} = \frac{1}{2} \underline{V} \underline{I}^* \; ,$$

where $\underline{I}^*$ denotes the complex conjugate of the current $\underline{I}$. Substituting the complex quantities $\underline{V} = V$, $\underline{I} = I \exp(-j\phi)$, and $\underline{I}^* = I \exp(j\phi)$, we get

$$\underline{P} = \frac{1}{2} V I (\cos \phi + j \sin \phi) = P_a + j P_r \; .$$

The real part of the complex power, $P_a = \frac{1}{2} V I \cos \phi$, represents the average power. The imaginary part, $P_r = \frac{1}{2} V I \sin \phi$, is known as the reactive power.

12.10.3 Active Versus Reactive Power

In real notations, the instantaneous power can be written as

$$p(t) = V I \cos \omega t \cos(\omega t - \phi) \; .$$

Using trigonometric identities, we can rewrite this as

$$p(t) = \frac{1}{2} V I \cos \phi \, [1 + \cos 2(\omega t - \phi)] - \frac{1}{2} V I \sin \phi \sin 2(\omega t - \phi) \; ,$$

or, in terms of powers P_a and P_r:

$$p(t) = P_a \left[1 + \cos 2(\omega t - \phi) \right] - P_r \sin 2(\omega t - \phi) \, .$$

The first term represents the instantaneous power received by the resistive parts of the circuit. It is modulated around the average P_u. This power is called *active power*. It represents the real power that performs useful work. It flows unidirectionally from the source to the load.

The second term represents the instantaneous power received by the reactive parts of the circuit (capacitors, inductors). Its average is zero. This power is called *reactive power*. It is essential for the operation of inductive and capacitive components, but it does not contribute to real work. It flows back and forth between the source and the load.

12.11   Summary and Essential Formulas

- The power received by a resistor with resistance R and crossed by a current I

$$P = RI^2 \, .$$

  This power is dissipated in the form of heat, a phenomenon known as *Joule heating*.
- For a capacitor with capacitance C with potential difference V across its terminals

  - The IV relationship is
    $$I = C \frac{dV}{dt} \, .$$

  - The energy stored in the electric field within the capacitor is

    $$W_E = \frac{1}{2} C V^2 \, .$$

- For an inductor with inductance L and crossed by a current I:

  - The IV relationship is
    $$V = L \frac{dI}{dt} \, .$$

  - The energy stored in the magnetic field in the inductor is

    $$W_B = \frac{1}{2} L I^2 \, .$$

- Studying the transient regime of an electric circuit amounts to applying Kirchhoff's laws together with the IV relationships of the circuit bipoles to find an ordinary differential equation (ode) governing the voltage across a particular bipole or current in a particular branch.

- Initial conditions must be satisfied: they are obtained by expressing continuity of the voltage across a capacitor and the current in a coil when a switch is closed to power the circuit.
- The solution to the governing ode satisfying the initial conditions gives rise to a transient regime involving characteristic times for the return to a steady state.
- For a RC circuit in series, the characteristic time is $\tau_{RC} = RC$.
- For a RL circuit in series, the characteristic time is $\tau_{RL} = L/R$.
- A LC circuit in series behaves as an oscillator with resonant frequency $\omega_0 = (LC)^{1/2}$.
- A RLC circuit exhibits different regimes with or without damped oscillations.

- The gain is a measure of the amplification or attenuation of a signal as it passes through a circuit. In the context of an RLC circuit, it is often expressed as the ratio of the output voltage to the input voltage. The gain of an RLC circuit depends on its configuration (series or parallel), component values, and the frequency of the input signal. At the resonant frequency, the gain is maximum for a series RLC circuit.
- The quality factor, or Q-factor, is a dimensionless parameter that characterizes the sharpness of the resonance peak in a resonant circuit. It is a measure of how underdamped an oscillatory system is. A higher Q-factor implies a narrower bandwidth and a sharper peak in the frequency response.
For a series RLC circuit, the Q factor is given by

$$Q = \omega_0 \frac{L}{R} = \frac{1}{\omega_0 RC}$$

where $\omega_0 = 1/\sqrt{LC}$ is the resonant frequency, L is the inductance, C is the capacitance and R is the resistance
A higher Q factor indicates a more selective circuit, meaning it will only respond strongly to frequencies near the resonant frequency. A higher Q-factor leads to a higher peak gain at the resonant frequency. A lower Q-factor results in a broader bandwidth and lower peak gain.

Problems

12.1 Kirchhoff's voltage law for a 2-loops RC circuit

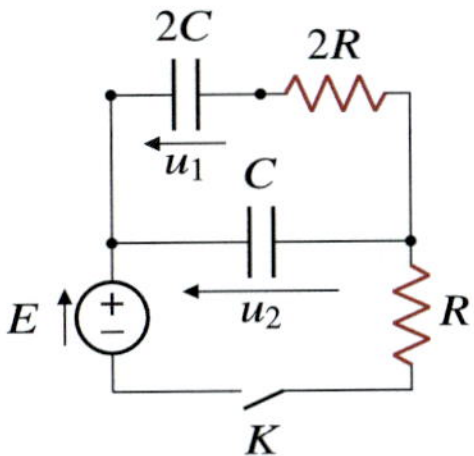

The quantities u_1 and u_2 denote the voltages across the capacitance $2C$ and C, respectively.

1. Write Kirchhoff's Voltage Laws (KVL) for the different loops in the circuit. Without integrating any equation, infer initial values for the currents if the capacitors are not charged at $t = 0$ when the switch K is closed. What are the steady state values for u_1 and u_2?
2. Find a differential equation satisfied by u_1. Solve it to express $u_1(t)$; assume the capacitors carry no charge at $t = 0$. Sketch the curve $u_1(t)$.

12.2 Neon lamp

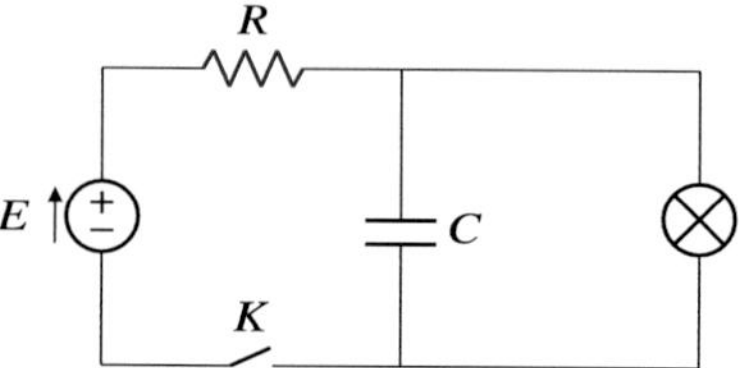

The neon lamp (symbol: cross in circle) has infinite resistance when it is off and resistance r when it is on. Let V_i denote the ignition voltage and V_e the extinction voltage for the lamp ($V_i > V_e$). At the initial time $t = 0$, the switch K is closed, the capacity carrying no charge. How does the system evolve?

12.3 Kirchhoff's laws in the transient regime

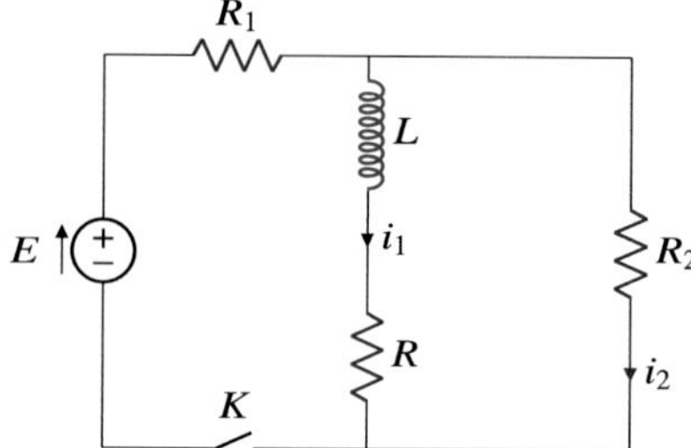

At $t = 0$, the switch K is closed. Express the currents in the coil and in the resistance R_2.

Answer: In the coil, $i_1 = \dfrac{E R_2}{R R_2 + R R_1 + R_1 R_2}(1 - e^{-t/\tau})$. In the resistance R_2,

$$i_2 = \frac{E R_2}{R R_2 + R R_1 + R_1 R_2}\left[\frac{R}{R_2} + \frac{R_1}{R_1 + R_2}e^{-t/\tau}\right].$$

12.4 RL and RC circuits in series

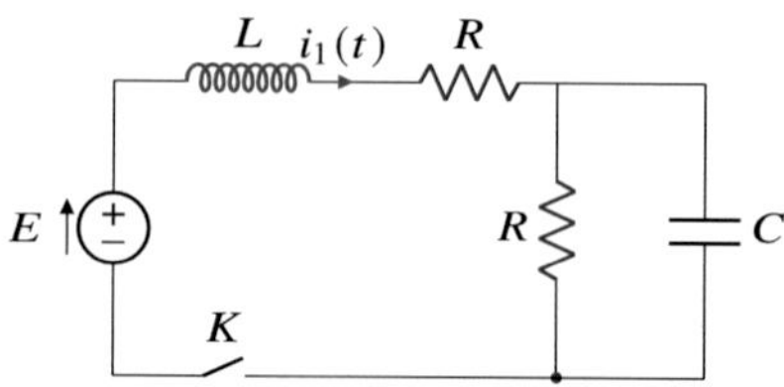

The (R, L) and (R, C) circuits have the same time constant τ. At $t = 0$, given that C is not charged, the switch K is closed. Express the current $i_1(t)$ in the coil.

$$\text{Answer: } i_1 = \frac{E}{2R}\left[1 + e^{-t/\tau}\left(\sin\frac{t}{\tau} - \cos\frac{t}{\tau}\right)\right].$$

12.5 AC Analysis of an LC circuit

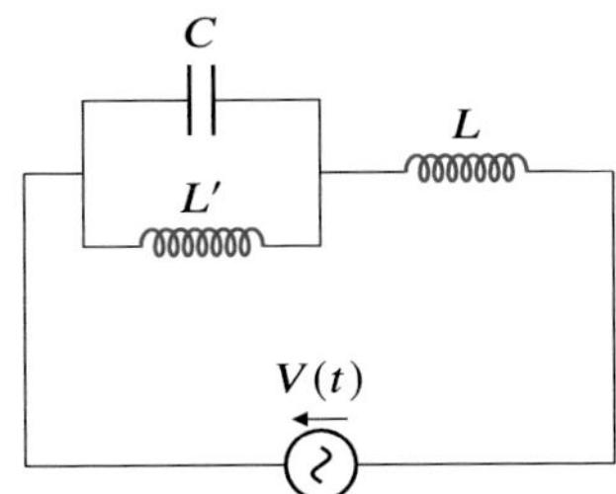

A sinusoidal voltage source, given by $V(t) = V_0 \exp(j\omega t)$ in complex notation, is applied to the circuit. Calculate the currents I and I' flowing through the inductors L and L', respectively, as functions of C, L, L', and the source voltage V_0. What is the value of I' when $L'C\omega^2 = 1$?

12.6 Current in a RLC circuit with two AC generators

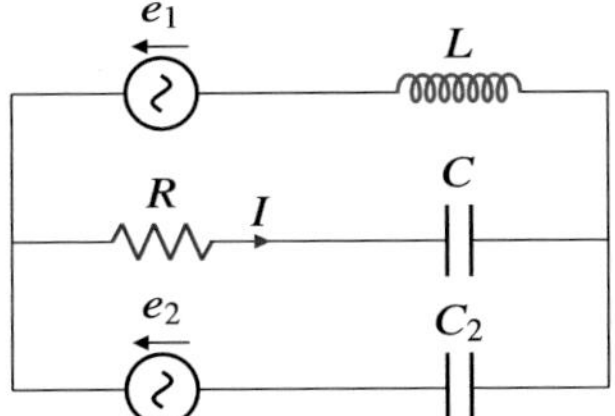

In this circuit, e_1 and e_2 are sinusoidal electromotive forces with an angular frequency of $\omega = 1 \times 5$ rad s^{-}1. The emf e_1 is used as a phase reference and has an amplitude of 2 V. The emf e_2 has an amplitude of $\sqrt{2}$ V and leads e_1 by a phase angle of $\pi/4$.
$C_2 = 10\,\text{nF}$, $C = 5\,\text{nF}$, $R = 1 \times 10^3\,\Omega$, $L = 20\,\text{mH}$.
Determine the current I flowing through the resistor.

12.7 Application of Thévenin's or Norton's theorem

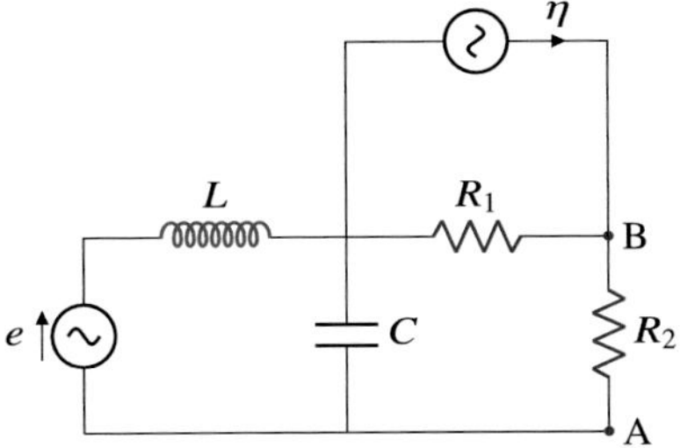

Given $R_1 = 50\,\Omega$, $R_2 = 100\,\Omega$, $L\omega = 150\,\Omega$, $1/C\omega = 75\,\Omega$, $e_{max} = 10\,\text{V}$ and $\eta_{max} = 0.1\,\text{A}$. e is a voltage source. η is a current source in phase with e. Assuming a steady-state sinusoidal regime with angular frequency ω, determine the voltage difference $V_B - V_A$ in both amplitude and phase form.

Answer: $V_B - V_A = 10\cos(\omega t - 3\pi/4)$.

12.8 High-pass RC filter

In the following RC circuit, a sinusoidal input signal $v_i(t)$ with a variable frequency ω is applied.

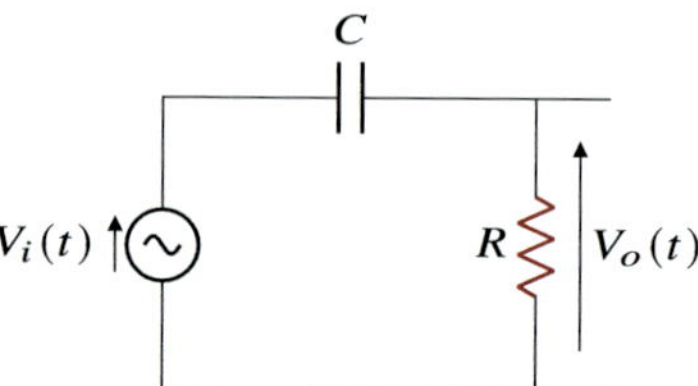

(a) Determine the transfer function $H(j\omega) = V_o/V_i$, where V_o is the complex output voltage across the resistor R. For simplicity, let $\omega_0 = 1/RC$.
(b) Plot the amplitude gain $G(\omega)$ and phase shift $\phi(\omega)$ of the transfer function.
(c) Plot the Bode diagram for the gain in decibels, $G_{dB} = 20\log_{10}(G(\omega/\omega_0))$.
(d) Given $R = 5\,\text{K}\Omega$ and $C = 40\,\text{nF}$, calculate the cutoff frequency of the filter.
(e) Keeping the capacitance constant at $C = 40\,\text{nF}$, determine the value of R required to attenuate the output signal by 10 dB for an input signal frequency of $f = 500\,\text{Hz}$?
(f) If a resistance R' is connected in parallel with the voltage source, show that the transfer function remains unchanged.

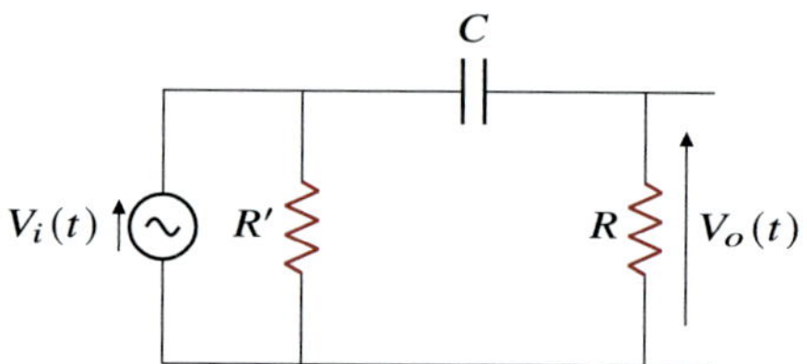

(g) Determine the input impedance Z_i of the filter. Express the magnitude of $|Z_i|$ as a function of angular frequency ω.

12.9 Second order filter

We add an inductor to the upper branch of the quadrupole circuit previously analyzed.

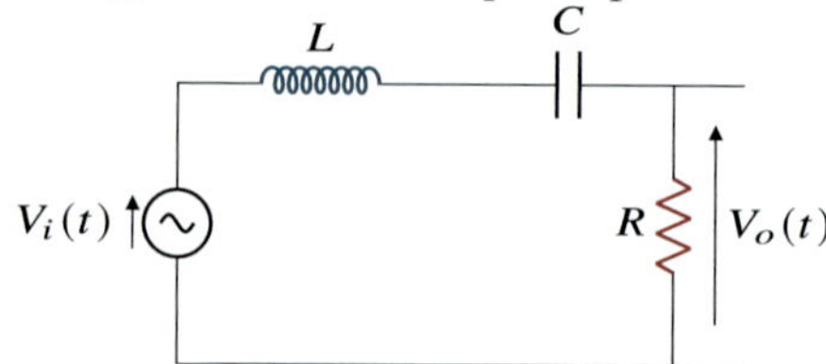

(a) Determine the transfer function $H(j\omega) = V_o/V_i$.
(b) Plot the amplitude gain $G(\omega)$ and phase shift $\phi(\omega)$ of the transfer function.
(c) Plot the Bode diagram for the gain in decibels.
(d) Identify the type of filter represented by the circuit.

12.10 T-Π transformation and filter

We aim to determine the conditions under which the two quadrupoles, T and Π, are equivalent.

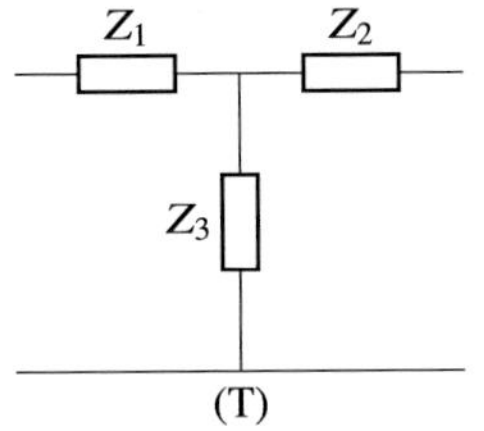

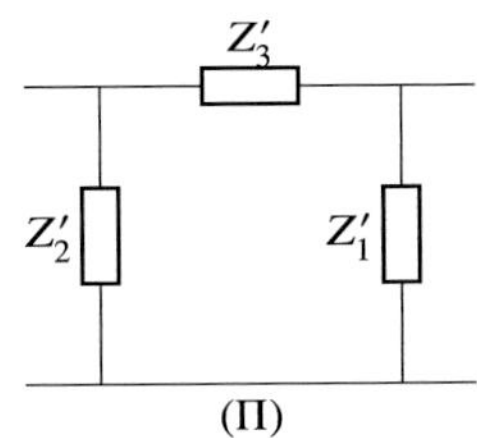

To this aim, we consider circuits (A) and (B).

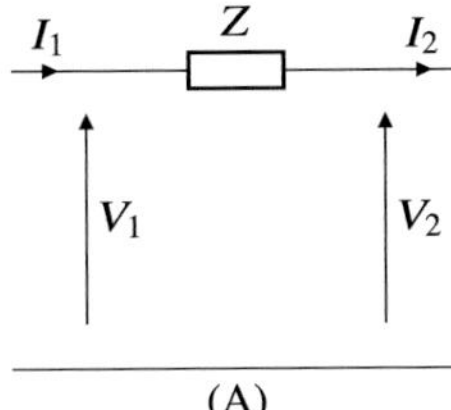

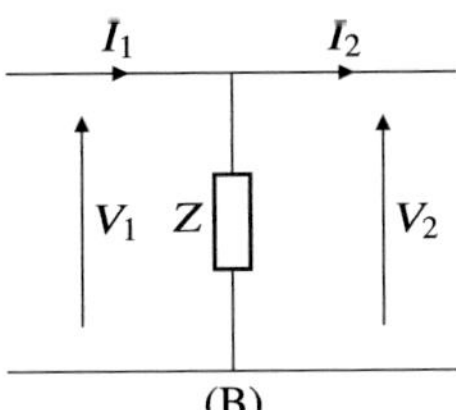

(a) For each quadrupole X (where $X \equiv A, B$), we seek a 2×2 matrix $\mathcal{M}_{(X)}$ such that

$$\mathcal{U}_2 = \mathcal{M}_{(X)}\mathcal{U}_1 \, ,$$

where $\mathcal{U}_i = \begin{pmatrix} V_i \\ I_i \end{pmatrix}$.

(b) Determine the matrices $\mathcal{M}_{(T)}$ and $\mathcal{M}_{(\Pi)}$ for quadrupoles T and Π, respectively.
(c) Find the conditions to be satisfied for the equivalence of T and Π.

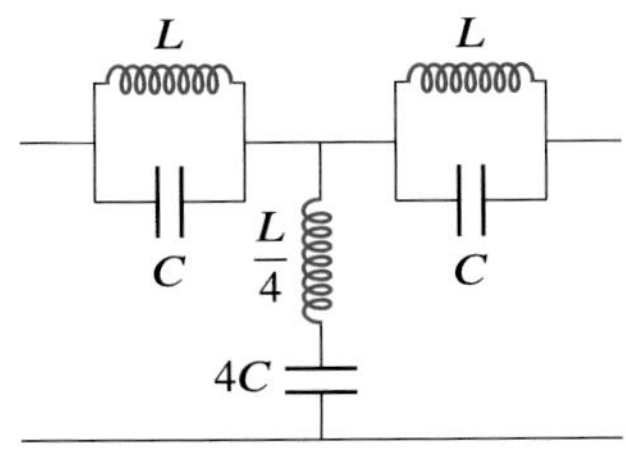

(d) Given the quadrupole shown above, we aim to determine the conditions for signal attenuation: A signal with complex amplitudes (V_1, I_1) is attenuated if $(V_2, I_2) = \lambda(V_2, I_2)$, where $|\lambda| < 1$. Find the conditions on the circuit parameters for this attenuation to occur.

(e) Analyze the frequency response of the quadrupole to determine its suitability for various filter applications.

Chapter 13
Maxwell's Equations, Electromagnetic Waves and Light

Abstract This chapter completes the study of classical electrodynamics by introducing *Maxwell's equations*, a unified set of four fundamental equations that describe all classical electromagnetic phenomena. It begins by addressing the inconsistency of Ampère's law in time-dependent regimes, leading to Maxwell's intuition of introducing the *displacement current* ($\mathbf{j}_D = \epsilon_0 \partial \mathbf{E}/\partial t$). This crucial addition completes Ampère's law, making it consistent with the *charge conservation law* and paving the way for the prediction of electromagnetic waves. The chapter then derives the *d'Alembert wave equation* for both electric and magnetic fields in vacuum, demonstrating that these fields propagate as waves at the speed of light ($c = 1/\sqrt{\epsilon_0 \mu_0}$). This prediction, experimentally confirmed by Hertz, established light as an electromagnetic phenomenon. The *structure of electromagnetic plane waves in vacuum* is analyzed in detail, highlighting their *transversality* (**E** and **B** fields are perpendicular to the propagation direction) and their mutual perpendicularity, forming a direct trihedron with the propagation vector. Various forms of *light polarization* are discussed, including linear, circular, and elliptical polarization, and their historical context and applications are briefly reviewed. Finally, the chapter introduces the concept of *electromagnetic energy conservation* through *Poynting's theorem*. This theorem identifies the *Poynting vector*, which describes the direction and rate of electromagnetic energy flow, and the electromagnetic energy density. The chapter demonstrates how energy is transported and quantifies the intensity of electromagnetic waves, providing a complete picture of energy dynamics in electromagnetic fields.

Keywords Maxwell's equations · d'Alembert's equation · Waves · Speed of light · Polarization · Poynting vector

13.1 Introduction

We begin this chapter by generalizing Ampère's law in order to make it compatible with the charge conservation law (7.8) in the time-dependent regime. This generalized law, known as the Ampère–Maxwell law, together with Gauss's law,

F. Cadiz and A. Couairon, *Classical Electrodynamics*, Undergraduate Texts in Physics,
https://doi.org/10.1007/978-3-031-86785-9_13

Faraday's law and Thomson's law (Gauss's law for magnetism) constitute a complete set of coupled equations between the electric and magnetic fields, known as Maxwell's equations in honor to James Clerk Maxwell who put them together[1] in 1864. Maxwell's equations not only synthesize all the known electrostatic and magnetostatic phenomena discussed so far, but establishes a profound relationship between them two. Indeed, they predict the existence of electric and magnetic disturbances which propagate together, called electromagnetic waves, propagating at the speed $c = 1/\sqrt{\epsilon_0 \mu_0} = 3 \times 10^8$ m/s, very close to the experimental values for the speed of light measured by Fizeau and Foucault, so that Maxwell proposed that light was an electromagnetic phenomenon.

The prediction of such waves was confirmed experimentally between 1886 and 1889 by Heinrich Hertz who demonstrated the existence of what is now called radio waves that propagate at the speed of light. This was a profound leap in our understanding of nature. Radio waves and light were both different forms of electromagnetic radiation obeying Maxwell's equations.

From Maxwell's equations, we will derive the propagation laws for electromagnetic waves in a vacuum and analyze the mathematical structure of their solutions, including their transversality and polarization. Additionally, we will derive the conservation law for electromagnetic energy within an arbitrary volume. As part of this discussion, we introduce the Poynting's vector, which describes the direction and rate of energy flow per unit surface area.

13.2   Ampère's Law and Displacement Current

So far, we have established the following laws that summarize the electrostatic, magnetostatic, and Faraday's induction phenomena:

$$
\boxed{
\begin{aligned}
\nabla \cdot \mathbf{E} &= \frac{\varrho}{\epsilon_0} & &\text{Gauss's law} \\[4pt]
\nabla \cdot \mathbf{B} &= 0 & &\text{Thomson's law} \\[4pt]
\nabla \times \mathbf{E} &= -\frac{\partial \mathbf{B}}{\partial t} & &\text{Faraday's law} \\[4pt]
\nabla \times \mathbf{B} &= \mu_0 \mathbf{j} & &\text{Ampère's law}
\end{aligned}
}
\tag{13.1}
$$

This set of equations is, however, incomplete. The problem comes from the fact that Ampère's law is inconsistent with the charge conservation law in the presence of time-dependent current densities. In order to illustrate this, let us consider the following example.

---

[1] In reality Maxwell wrote down 20 equations that were later reduced to only four by Oliver Heaviside.

Example 13.1—Problem with Ampère's law

The following is a problematic example when applying Ampère's law in a situation where the current is not stationary. Consider a parallel plate capacitor that is being charged with a current $I(t) = dQ(t)/dt$. The goal is to determine the magnetic field at a distance r from the axis, sufficiently far from the plates so we can assume a cylindrical symmetry of the current density.

A possible way to calculate the magnetic field is to use Ampère's law integrated over a circular path Γ of radius r as shown in Fig. 13.1.

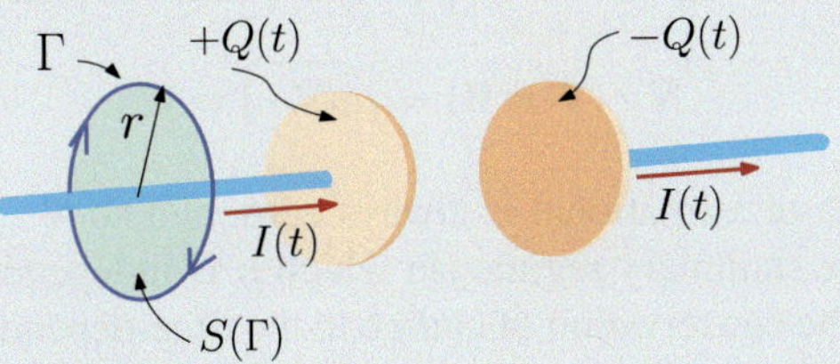

Fig. 13.1 A parallel plates capacitor during a charging process. The planar surface $S(\Gamma)$ is used to apply Ampère's law

We obtain

$$\oint_{\Gamma} \mathbf{B} \cdot d\mathbf{x} = 2\pi r B(r) = \mu_0 \Phi_{S(\Gamma),\mathbf{J}} = \mu_0 \iint_{S(\Gamma)} d\mathbf{S}(\mathbf{x}) \cdot \mathbf{j}(\mathbf{x})$$

where $S(\Gamma)$ is any surface whose bounding curve is Γ. For example, choosing the flat surface coplanar and enclosed by Γ, the flux of $\mathbf{j}$ through $S(\Gamma)$ corresponds to the current I and we obtain the well known result from magnetostatics

$$2\pi r B(r) = \mu_0 I.$$

$$2\pi r B(r) = \mu_0 I$$

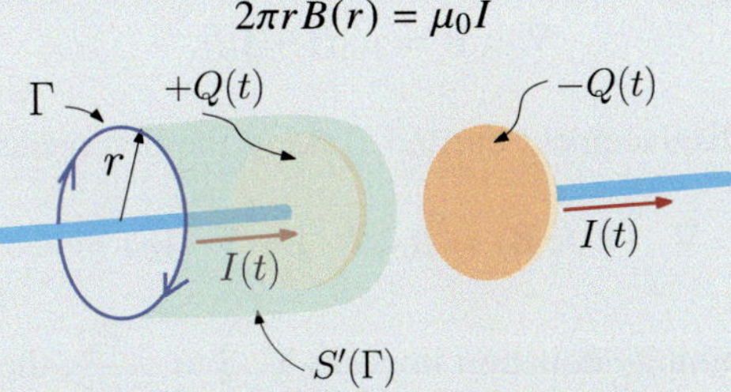

Fig. 13.2 By chosing $S'(\Gamma)$ in Ampère's law, one gets a contradiction

However, we could have chosen instead the surface $S'(\Gamma)$ as shown in Fig. 13.2. In this case, there is no flux of $\mathbf{j}$ through $S'(\Gamma)$ and a contradiction is obtained:

$$2\pi r B(r) = 0$$

the problem arises from the fact that Ampère's law is not compatible with time-varying current densities.

Ampère's law is inconsistent with the conservation law for the electric charge:

$$\frac{\partial \varrho}{\partial t} + \mathbf{\nabla} \cdot \mathbf{j} = 0.$$

Indeed, taking the divergence of Ampère's law in its differential form gives

$$\mathbf{\nabla} \cdot (\mathbf{\nabla} \times \mathbf{B}) = \mu_0 \mathbf{\nabla} \cdot \mathbf{j} = 0$$

since the divergence of a rotational is always zero, and so $\mathbf{\nabla} \cdot \mathbf{j} = 0$. While this is not a problem in the stationary regime (in which ϱ is independent of time), it is in contradiction with the conservation of charge in the time-dependent regime in which

$$\mathbf{\nabla} \cdot \mathbf{j} = -\frac{\partial \varrho}{\partial t} \neq 0.$$

This problem was solved by Maxwell, who added to Ampère's law a second current density term $\mathbf{j}_D$ such that the total current $\mathbf{j}_T = \mathbf{j} + \mathbf{j}_D$, and not $\mathbf{j}$, has zero divergence.

13.2.1 The Displacement Current

Maxwell corrected Ampère's law by adding a term that makes the law consistent with the conservation of charge:

$$\mathbf{\nabla} \times \mathbf{B} = \mu_0(\mathbf{j} + \mathbf{j}_D),$$

where $\mathbf{j}_D$ is called displacement current. Taking the divergence on both sides, we now obtain

$$\mathbf{\nabla} \cdot (\mathbf{\nabla} \times \mathbf{B}) = \mu_0(\mathbf{\nabla} \cdot \mathbf{j} + \mathbf{\nabla} \cdot \mathbf{j}_D) = 0.$$

Since the charge continuity equation imposes $\mathbf{\nabla} \cdot \mathbf{j} = -\dfrac{\partial \varrho}{\partial t}$, then

$$\mathbf{\nabla} \cdot \mathbf{j}_D = \frac{\partial \varrho}{\partial t}.$$

Taking the partial derivative with respect to time in Gauss's law, we find $\nabla \cdot \dfrac{\partial \mathbf{E}}{\partial t} = \dfrac{1}{\epsilon_0} \dfrac{\partial \varrho}{\partial t}$, and we see then that it is possible to choose

$$\mathbf{j}_D = \epsilon_0 \frac{\partial \mathbf{E}}{\partial t}. \tag{13.2}$$

It is this choice that was done by Maxwell (Fig. 13.4) in 1862 to cure the problem of Ampère's law in the time dependent regime and obtain the fourth Maxwell equation. This was a truly brilliant intuition since Maxwell did not know the existence of the electron and did not think in terms of charge conservation when he introduced the displacement current as *one of the chief peculiarities* of his theory. He postulated the set of fundamental equations unifying electric and magnetic phenomena; Maxwell's equations led to the prediction of electromagnetic waves propagating at the speed of light, which, according to A. Zangwill, might be one of the arguments that convinced him of the correctness of the displacement current. It is only after 1888 when Hertz discovered electromagnetic waves behaving as predicted by Maxwell's theory that the rest of the world accepted Maxwell's postulate as correct, that is, the fact that Maxwell's equations are the fundamental laws describing electric and magnetic phenomena in a unified framework and that light is an electromagnetic wave.

Maxwell–Ampère Law

The Maxwell–Ampère law reads

$$\nabla \times \mathbf{B} = \mu_0 \left(\mathbf{j} + \epsilon_0 \frac{\partial \mathbf{E}}{\partial t} \right). \tag{13.3}$$

In the case of fields that do not depend on time, this law reduces to Ampère's law. The introduction of the displacement current $\mathbf{j}_D = \epsilon_0 \frac{\partial \mathbf{E}}{\partial t}$ in the second term represents a major contribution from Maxwell to the theory of electromagnetism, since it allowed him to predict the existence of electromagnetic waves.

Example 13.2—Charging capacitor and Maxwell–Ampère law

Let us consider again the Example 13.1. Assuming for simplicity that the separation between the plates is small enough compared to their lateral dimensions, the electric field $\mathbf{E}$ (and therefore the displacement current $\mathbf{j}_D$) is totally confined between the plates. According to Maxwell–Ampère law, the circulation of the magnetic field over the closed curve Γ equals the flux of $\mu_0(\mathbf{j} + \mathbf{j}_D) = \mu_0\mathbf{j}_T$ over any[2] surface whose bounding curve is Γ.

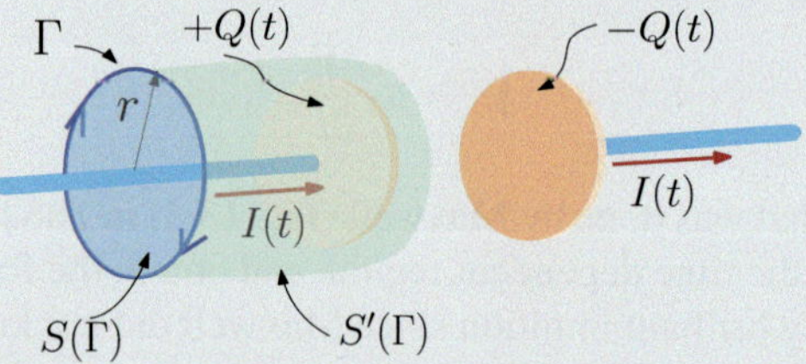

Fig. 13.3 Maxwell–Ampère law gives the same result regardless of the surface chosen

If instead we choose the surface $S(\Gamma)$, as shown in Fig. (13.3), the flux is due to $\mathbf{j}$ and given by $\mu_0\Phi_{S(\Gamma),\mathbf{j}} = \mu_0 I$. If instead we choose the surface $S'(\Gamma)$, the flux is due to $\mathbf{j}_D$ and we have

$$\mu_0\Phi_{S'(\Gamma),\mathbf{j}_D} = \mu_0\epsilon_0\frac{\partial}{\partial t}\Phi_{S'(\Gamma),\mathbf{E}}.$$

The flux of $\mathbf{E}$ over $S'(\Gamma)$ is equal to the flux over the closed surface $S'(\Gamma) - S(\Gamma)$,[3] which by Gauss's law equals $Q(t)/\epsilon_0$ and

$$\mu_0\Phi_{S(\Gamma),\mathbf{j}_D} = \mu_0\frac{\partial Q(t)}{\partial t} = \mu_0 I$$

so that Maxwell–Ampère law gives the same result regardless of the curve that is chosen to calculate the flux of $\mathbf{j} + \mathbf{j}_D$

$$\oint_\Gamma \mathbf{B} \cdot d\mathbf{x} = \mu_0\Phi_{S(\Gamma),\mathbf{j}_T} = \mu_0\Phi_{S'(\Gamma),\mathbf{j}_T} = \mu_0 I.$$

Maxwell's Equations

Maxwell's equations are the following set of coupled equations relating the electric and magnetic fields ($\mathbf{E}$ and $\mathbf{B}$) to their sources (charge density ϱ and current density $\mathbf{j}$):

---

[2] The flux can be taken through *any* surface bounded by Γ because $\nabla \cdot \mathbf{j}_T = 0$.

[3] This is the union of $S'(\Gamma)$ and $-S(\Gamma)$, i.e., $S(\Gamma)$ with the opposite orientation to ensure that the normal to the total surface always points outward.

Fig. 13.4 James Clerk Maxwell (1831–1879), a Scottish physicist who revolutionized science with his unification of electricity, magnetism, and light as aspects of the same phenomenon. His work paved the way for Einstein's theory of relativity

Fig. 13.5 Jean le Rond d'Alembert (1717–1783), a French mathematician, physicist, and philosopher. He co-edited the *Encyclopédie*, a cornerstone of the Enlightenment, with Denis Diderot

$$
\begin{aligned}
\nabla \cdot \mathbf{E} &= \frac{\varrho}{\epsilon_0} &\quad& \text{Maxwell–Gauss} \\[1ex]
\nabla \cdot \mathbf{B} &= 0 &\quad& \text{Maxwell–Thomson} \\[1ex]
\nabla \times \mathbf{E} &= -\frac{\partial \mathbf{B}}{\partial t} &\quad& \text{Maxwell–Faraday} \\[1ex]
\nabla \times \mathbf{B} &= \mu_0 \mathbf{j} + \mu_0 \epsilon_0 \frac{\partial \mathbf{E}}{\partial t} &\quad& \text{Maxwell–Ampère}
\end{aligned}
\tag{13.4}
$$

To Maxwell's equations, we must add the Lorentz force that is exerted on a particle of charge q of velocity $\mathbf{v}$ in an electromagnetic field

$$
\mathbf{F} = q(\mathbf{E} + \mathbf{v} \times \mathbf{B}).
$$

This set of laws gives us a complete description of all classical electromagnetism. Before discussing the possible solutions of Maxwell's equations, a reminder of the wave equation and its general solutions is the following section.

13.3 The d'Alembert Wave Equation

The wave equation in three dimensions corresponds to the following partial differential equation for a scalar field F

$$\nabla^2 F(\mathbf{x}, t) - \frac{1}{c^2} \frac{\partial^2 F(\mathbf{x}, t)}{\partial t^2} = 0.$$

For simplicity, let us first consider the 1D case of a wave propagating along the x direction

$$\frac{\partial^2 F(x, t)}{\partial x^2} - \frac{1}{c^2} \frac{\partial^2 F(x, t)}{\partial t^2} = 0. \tag{13.5}$$

As demonstrated by d'Alembert in 1747 (Fig. 13.5), any function of the form $F(x \pm ct)$, where $X \to F(X)$ is any twice differentiable function of a single variable, is solution to the wave Eq. (13.5).

Indeed, consider the change of variable $X = x \pm ct$. Partial derivatives of X with respect to x or t read $\dfrac{\partial X}{\partial x} = 1$ and $\dfrac{\partial X}{\partial t} = \pm c$. Using the chain rule:

$$\frac{\partial F(X)}{\partial x} = \frac{\partial F(X)}{\partial X} \frac{\partial X}{\partial x} = \frac{\partial F(X)}{\partial X},$$

and so:

$$\frac{\partial^2 F(X)}{\partial x^2} = \frac{\partial^2 F(X)}{\partial X^2}.$$

Similarly, the partial derivatives with respect to t read:

$$\frac{\partial F(X)}{\partial t} = \frac{\partial F(X)}{\partial X} \frac{\partial X}{\partial t} = \pm c \frac{\partial F(X)}{\partial X}$$

and so:

$$\frac{\partial^2 F(X)}{\partial t^2} = \frac{\partial}{\partial t} \left(\pm c \frac{\partial F(X)}{\partial X} \right) = \pm c \frac{\partial^2 F(X)}{\partial X^2} \frac{\partial X}{\partial t} = c^2 \frac{\partial^2 F(X)}{\partial X^2}.$$

We see that $F(X) = F(x \pm ct)$ is solution to d'Alembert's wave equation:

$$\frac{\partial^2 F(X)}{\partial x^2} = \frac{1}{c^2} \frac{\partial^2 F(X)}{\partial t^2}.$$

The latter equation being linear, the superposition principle applies and we conclude that a possible solution is given by

$$F(x, t) = f_+(x - ct) + g_-(x + ct), \tag{13.6}$$

where f_+ and g_- are twice differentiable functions of a single variable.

Fig. 13.6 A solution of the form $f_+(x - ct)$ propagates at speed c toward positive values of x

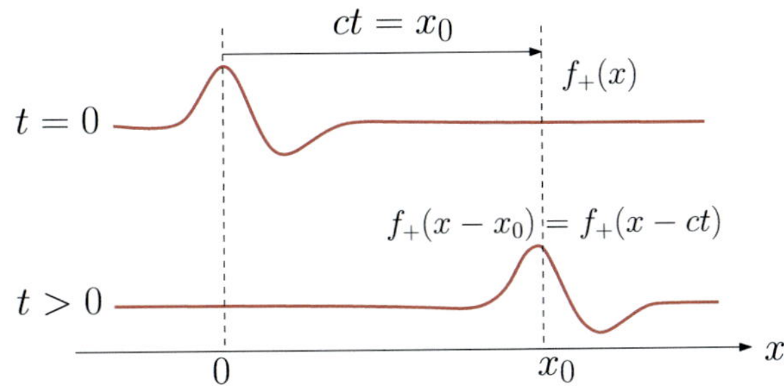

13.3.1 *Physical Interpretation*

Let us consider a solution of the form $f_+(x - ct)$ with $c > 0$, which at $t = 0$ represents a disturbance $f_+(x)$. After a time $t > 0$, the wave corresponds to $f_+(x - x_0)$, which is simply f_+ shifted toward the positive values of x by a quantity $x_0 = ct > 0$, as shown in Fig. 13.6.

We conclude that $f_+(x - ct)$ represents a disturbance that propagates without deformation toward positive values of x at a speed c. It is therefore called a progressive wave.

In the same way, we see that $g_-(x + ct)$ corresponds to a regressive wave, that is, a disturbance that propagates without deformation toward the direction of decreasing values of x. Note finally that the general solution $f_+(x - ct) + g_-(x + ct)$, which is always a superposition of progressive and regressive waves, is not necessarily a traveling wave. It could be, for example, a standing wave.

13.3.2 *Plane Waves*

Consider again the wave equation in three dimensions

$$\nabla^2 F(\mathbf{x}, t) - \frac{1}{c^2} \frac{\partial^2 F(\mathbf{x}, t)}{\partial t^2} = 0.$$

The wave F is said to be a plane wave propagating along the direction of propagation $\mathbf{n}$ if, for every t, the function F has the same value at every point of a plane Π perpendicular to $\mathbf{n}$. For example, for a plane wave propagating in the x-direction, this means that F does not depend on y or z, but only on x. This is represented in Fig. 13.7, F has the same value at P and at P', both belonging to a plane $x = x_0$.

In this case, the definition of a plane wave in three dimensions results in F satisfying the 1D d'Alembert equation:

$$\frac{\partial^2 F}{\partial x^2} - \frac{1}{c^2} \frac{\partial^2 F}{\partial t^2} = 0.$$

Fig. 13.7 A plane wave propagating along the x direction is, at a given time, constant in every plane Π perpendicular to $\mathbf{u}_x$

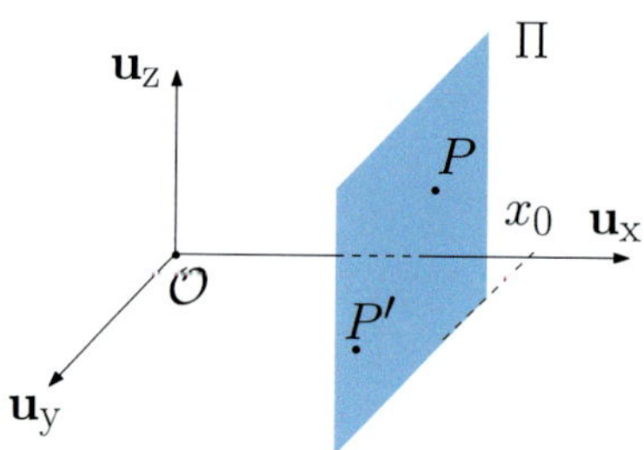

13.3.3 Sinusoidal Plane Waves and Wave Fronts

Now consider a plane wave that propagates along the x-axis. An important solution of d'Alembert's equation corresponds to sinusoidal plane waves. A progressive sinusoidal wave writes

$$f_+(x, t) = \underbrace{A_1}_{\text{amplitude}} \cos(\underbrace{kx - \omega t + \phi_1}_{\text{phase}})$$

and similarly, a regressive sinusoidal wave writes

$$g_-(x, t) = A_2 \cos(kx + \omega t + \phi_2).$$

At a given position $x = x_0$, the wave oscillates in time with a period

$$T = \frac{2\pi}{\omega}$$

whereas at a fixed time $t = t_0$, the wave oscillates in space with a spatial period (also called wavelength)

$$\lambda = \frac{2\pi}{k}.$$

We see then that ω and k correspond to the temporal and spatial frequencies of the wave, respectively. These quantities are not independent. Indeed, injecting f_+ and g_- into d'Alembert's equation leads to the following equation bounding ω and k, also called the dispersion relation

$$\boxed{\omega = kc.} \tag{13.7}$$

A wave front is, at a given instant, a surface of constant phase. For a sinusoidal plane wave propagating along x, this means

$$kx + \phi = \text{constant} \quad \rightarrow \quad x = \text{constant}.$$

Fig. 13.8 A progressive sinusoidal plane wave

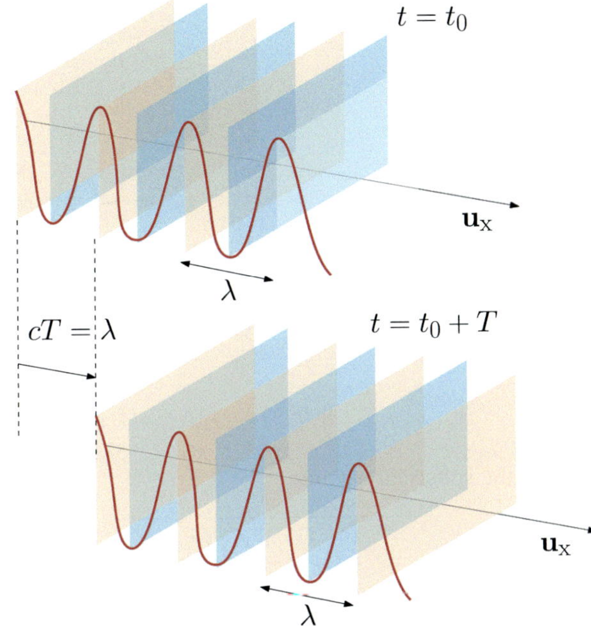

Figure 13.8 represents a progressive sinusoidal wave traveling along the x direction, together with some wave fronts spaced by $\lambda/2$. According to the dispersion relation (13.7), during one temporal period T, the wave has traveled a distance $cT = \lambda$.

13.3.4 Spherical Waves

Consider a wave with spherical symmetry. This would correspond to the case of a wave isotropically emitted from (or focused at) the origin of coordinates. For such a wave, which only depends on r, that is $F = F(r, t)$, the wave equation writes

$$\nabla^2 F(r, t) - \frac{1}{c^2}\frac{\partial^2 F(r, t)}{\partial t^2} = \frac{1}{r}\frac{\partial^2}{\partial r^2}(rF) - \frac{1}{c^2}\frac{\partial^2 F}{\partial t^2} = 0.$$

Writing $\Psi(r, t) = rF(r, t)$, the equation for Ψ reads

$$\frac{\partial^2 \Psi(r, t)}{\partial r^2} - \frac{1}{c^2}\frac{\partial^2 \Psi(r, t)}{\partial t^2} = 0$$

which is the one-dimensional wave d'Alembert equation, whose solution is in general a combination of progressive and regressive waves:

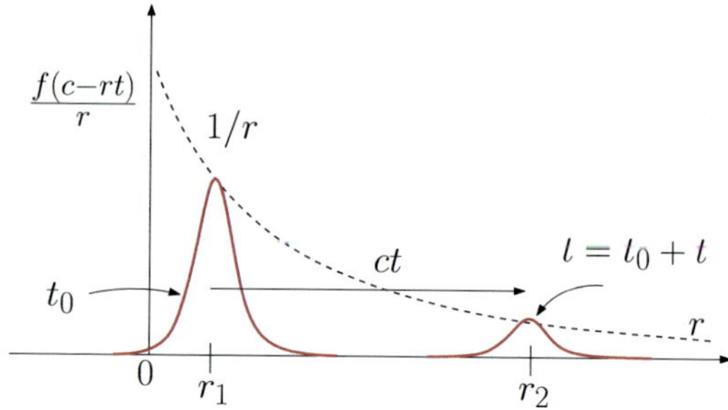

Fig. 13.9 A progressive spherical wave

$$\Psi(r, t) = f(r - ct) + g(r + ct).$$

Finally, the spherical wave is

$$F(r, t) = \left(\frac{f(r - ct)}{r} + \frac{g(r + ct)}{r} \right).$$

Unlike plane waves, the amplitude of spherical waves changes during propagation. In electromagnetism, this reflects the conservation of energy as the wave propagates. Figure 13.9 shows the case of a progressive (along the radial direction) spherical wave of the form $f(r - ct)/r$. The shape of the wave is preserved, although its amplitude decreases as $1/r$ as the wave travels. This typically represents a wave that is emitted from a point source at the origin.

Note that, at a given time, the amplitude of the wave takes the same value at all points such that $r = r_0$, that is, the wave is constant on any sphere centered at $r = 0$. Finally, a solution of the form $g(r + ct)/r$ seems to be emitted from a spherical surface far from the origin and to be focused at $r = 0$. The amplitude of such a wave then increases as r decreases.

If we now choose a sinusoidal spherical wave $f(r - ct) = A\cos(kr - \omega t)$ we have $F(r, t) = \frac{A}{r}\cos(r - ct)$ so that, at any given time, the wave fronts are spheres centered at the origin. At a point P sufficiently far from $r = 0$, these wave fronts can

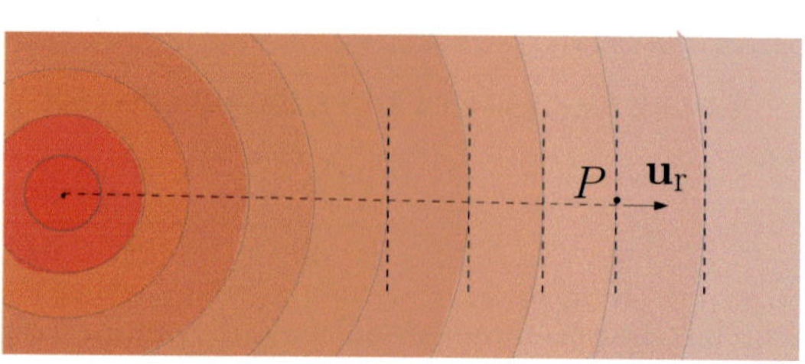

Fig. 13.10 The wavefronts of a spherical wave can be locally approximated by plane waves

be locally approximated by planes perpendicular to the radial direction $\mathbf{u}_r$ so that the wave behaves, locally, like a plane wave propagating along $\mathbf{u}_r$ (Fig. 13.10).

Sinusoidal plane waves can thus be used to describe the local behavior of spherical waves. Near $r = r_0$:

$$F(r_0 + dr, t) = \frac{A\cos(k(r_0 + dr) - \omega t)}{r_0 + dr} \approx \frac{A\cos(kdr - \omega t + kr_0)}{r_0} = A'\cos(kdr - \omega t + \phi)$$

where A' is approximately a constant. Locally, a spherical wave can be approximated by a plane wave.

13.3.5 Complex Representation of Sinusoidal Waves

A sinusoidal wave of the form $u(x, t) = A\cos(kx - \omega t + \phi)$ may be written in a complex representation by using Euler's identity ($e^{ix} = \cos x + i\sin x$) as the real part of the following complex wave

$$\underline{u}(x, t) = \underline{A}e^{i(kx - \omega t)}$$

with $\underline{A} = Ae^{i\phi}$. Indeed

$$u(x, t) = \mathrm{Re}\{\underline{u}(x, t)\} = \mathrm{Re}\{Ae^{i(kx - \omega t + \phi)}\} = A\cos(kx - \omega t + \phi).$$

The complex representation significantly simplifies calculations involving sinusoidal waves, since the result of any linear operation $L(u)$ acting on u may be obtained as $\mathrm{Re}\{L(\underline{u})\}$.

13.3.6 Why Are Sinusoidal Waves so Important?

A sinusoidal wave has an infinite extension in both space and time, and as such, it cannot represent on its own a realistic physical phenomenon. In addition, it may seem arbitrary to give a special importance to sinusoidal plane waves since any twice differentiable function is a solution of d'Alembert's wave equation. But in fact, the restriction to sinusoidal waves is fully justified by the following important result of Fourier analysis: any physical plane wave $f(x, t)$ (bounded in both space and time) may be written as a linear, (infinite) superposition of sinusoidal waves

$$f(x, t) = \int_{\mathbb{R}} A(k)e^{i(kx - \omega(k)t)}\,dk + \int_{\mathbb{R}} B(k)e^{i(kx + \omega(k)t)}\,dk.$$

Sinusoidal plane waves are thus the building blocks of real waves, and due to the linearity of Maxwell's equations, the superposition principle applies, their study is enough to characterize the most general case.

13.4   Electromagnetic Waves in Vacuum

One immediate consequence of Maxwell's equations is the prediction of the existence of electromagnetic waves that propagate in vacuum at the speed of light $c = \dfrac{1}{\sqrt{\mu_0\epsilon_0}} = 3 \times 10^8$ m/s. This is possible because, in the absence of sources, the electric and magnetic fields are coupled according to

$$\underbrace{\nabla \times \mathbf{E} = -\frac{\partial \mathbf{B}}{\partial t}}_{\text{time-varying } \mathbf{B} \text{ accompanies } \mathbf{E}} \qquad \underbrace{\nabla \times \mathbf{B} = \mu_0\epsilon_0 \frac{\partial \mathbf{E}}{\partial t}}_{\text{time-varying } \mathbf{E} \text{ accompanies } \mathbf{B}}$$

which leads to the propagation of the electromagnetic field, $(\mathbf{E}, \mathbf{B})$. This prediction was confirmed experimentally by Hertz in 1888.

13.4.1   Propagation Equation for the Electric Field

The propagation equation is obtained using the following identity, where $\mathbf{F}$ is a vector field:

$$\nabla \times (\nabla \times \mathbf{F}) = \nabla (\nabla \cdot \mathbf{F}) - \nabla^2 \mathbf{F}.$$

For the electric field, the double curl reads

$$\nabla \times (\nabla \times \mathbf{E}) = \nabla \times \left(-\frac{\partial \mathbf{B}}{\partial t} \right) = -\frac{\partial}{\partial t} \nabla \times \mathbf{B},$$

where we have used the Maxwell–Faraday law. Now, using the Maxwell–Ampère law to replace $\nabla \times \mathbf{B}$, we obtain

$$\nabla \times (\nabla \times \mathbf{E}) = -\frac{\partial}{\partial t} \left(\mu_0 \mathbf{j} + \mu_0\epsilon_0 \frac{\partial \mathbf{E}}{\partial t} \right).$$

Thus

$$\nabla \times (\nabla \times \mathbf{E}) = \nabla (\nabla \cdot \mathbf{E}) - \nabla^2 \mathbf{E} = - \left(\mu_0 \frac{\partial}{\partial t} \mathbf{j} + \mu_0\epsilon_0 \frac{\partial^2 \mathbf{E}}{\partial t^2} \right).$$

Finally, using Gauss's law $\nabla \cdot \mathbf{E} = \varrho/\epsilon_0$, we obtain.

$$\boxed{\nabla^2 \mathbf{E} - \mu_0 \epsilon_0 \frac{\partial^2}{\partial t^2} \mathbf{E} = \nabla \frac{\varrho}{\epsilon_0} + \mu_0 \frac{\partial}{\partial t} \mathbf{j}.}$$ (13.8)

Equation (13.8) is the propagation equation for the electric field, establishing a relationship between the spatial and time derivatives of the electric field, the charge density and the variation of the current density. Note that the coupling between the electric and magnetic fields is no longer explicit in this equation. This makes it clear that the sources of the electromagnetic waves are the charges and currents.

13.4.2 Propagation Equation for the Magnetic Field

Using the Ampère-Maxwell law:

$$\nabla \times (\nabla \times \mathbf{B}) = \mu_0 \nabla \times \mathbf{j} + \mu_0 \epsilon_0 \frac{\partial}{\partial t} \nabla \times \mathbf{E} = \nabla (\nabla \cdot \mathbf{B}) - \nabla^2 \mathbf{B}.$$

Since $\nabla \cdot \mathbf{B} = 0$ and $\nabla \times \mathbf{E}$ can be replaced by means of the Maxwell–Faraday law by $-\dfrac{\partial}{\partial t} \mathbf{B}$, we obtain

$$\boxed{\nabla^2 \mathbf{B} - \mu_0 \epsilon_0 \frac{\partial^2}{\partial t^2} \mathbf{B} = -\mu_0 \nabla \times \mathbf{j}}$$ (13.9)

which corresponds to the propagation equation for the magnetic field.

13.4.3 Propagation of Electromagnetic Waves in Vacuum

Next, we consider the behavior of the fields away from the sources that generate them, i.e., in regions where $\varrho = 0$ and $\mathbf{j} = \mathbf{0}$. We see then from (13.8), (13.9) that each component of both $\mathbf{E}$ and $\mathbf{B}$ satisfy the same d'Alembert's wave equation:

$$\nabla^2 \mathbf{E} - \mu_0 \epsilon_0 \frac{\partial^2}{\partial t^2} \mathbf{E} = 0$$ (13.10)

$$\nabla^2 \mathbf{B} - \mu_0 \epsilon_0 \frac{\partial^2}{\partial t^2} \mathbf{B} = 0$$ (13.11)

where $\mu_0 \epsilon_0$ is the inverse of a squared speed, which corresponds to the speed of light

$$c = \frac{1}{\sqrt{\epsilon_0 \mu_0}} \approx 3 \times 10^8 \text{ m/s.}$$

13.4.4 Structure of Electromagnetic Plane Waves in Vacuum

13.4.4.1 Transversality

Consider an electromagnetic plane wave that propagates in vacuum along the x-axis. This means that the fields are of the form $\mathbf{E}(x, t) = \mathbf{E}(x - ct)$ and $\mathbf{B}(x, t) = \mathbf{B}(x - ct)$. Let us define the change of variable $X = x - ct$. Gauss's law in vacuum implies

$$\nabla \cdot \mathbf{E} = \nabla X \cdot \frac{\partial \mathbf{E}}{\partial X} = 0$$

and since $X = x - ct$, $\nabla X = \mathbf{u}_x$, and so

$$\nabla \cdot \mathbf{E} = \mathbf{u}_x \cdot \frac{\partial \mathbf{E}}{\partial X} = \frac{\partial E_x}{\partial X} = 0$$

the longitudinal component of the electric field, parallel to the propagation direction, is thus a constant.

We choose it to be $E_x = 0$, since a constant field does not represent a propagating wave. Similarly, $\nabla \cdot \mathbf{B} = 0$ implies $B_x = 0$. In conclusion, the electric and magnetic fields are transverse: they both belong to a plane Π perpendicular to the propagation axis $\mathbf{n} = \mathbf{u}_x$, as shown in Fig. 13.11.

Hence,

$$\boxed{\mathbf{E} \cdot \mathbf{n} = \mathbf{B} \cdot \mathbf{n} = 0} \tag{13.12}$$

and so

$$\mathbf{E}(x, t) = E_y(x, t)\mathbf{u}_y + E_z(x, t)\mathbf{u}_z,$$
$$\mathbf{B}(x, t) = B_y(x, t)\mathbf{u}_y + B_z(x, t)\mathbf{u}_z.$$

13.4.4.2 E, B and the Propagation Direction n Form a Direct Trihedron

Considering the Maxwell–Faraday equation $\nabla \times \mathbf{E} = -\partial \mathbf{B}/\partial t$ projected along $\mathbf{u}_x$ or $\mathbf{u}_y$:

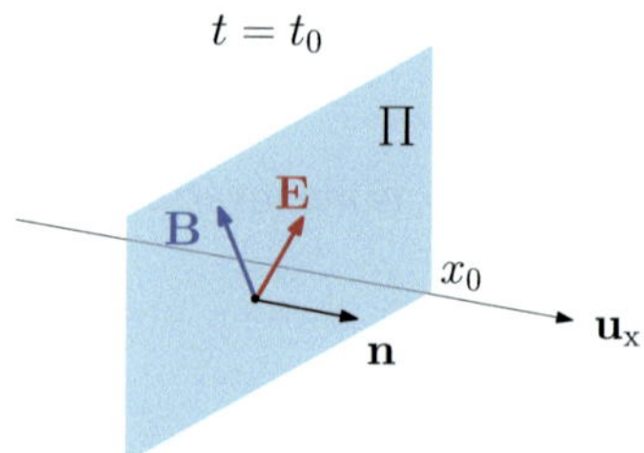

Fig. 13.11 The fields of a plane wave are transverse when propagating in vacuum

$$\frac{\partial E_z(x - ct)}{\partial x} = \frac{\partial B_y(x - ct)}{\partial t}$$
$$\frac{\partial E_y(x - ct)}{\partial x} = -\frac{\partial B_z(x - ct)}{\partial t}.$$

Writing $X = x - ct$

$$\frac{\partial E_z(X)}{\partial X} = -c\frac{\partial B_y(X)}{\partial X},$$
$$\frac{\partial E_y(X)}{\partial X} = c\frac{\partial B_z(X)}{\partial X},$$

and so, excluding constant solutions, we obtain

$$-E_z/c = B_y,$$
$$E_y/c = B_z,$$

which implies that $\mathbf{E}$ and $\mathbf{B}$ are perpendicular to each other, indeed:

$$\mathbf{E} \cdot \mathbf{B} = E_y B_y + E_z B_z = E_y(-E_z/c) + E_z(E_y/c) = 0.$$

Note also that

$$\mathbf{u}_x \times \mathbf{E} = E_y \mathbf{u}_z - E_z \mathbf{u}_y = c\mathbf{B}.$$

More generally, for a wave propagating along the direction $\mathbf{n}$, we have

$$\boxed{\mathbf{B} = \frac{1}{c}\mathbf{n} \times \mathbf{E}.} \tag{13.13}$$

The vectors $(\mathbf{n}, \mathbf{E}, \mathbf{B})$ form therefore a direct trihedron. Note that for a regressive wave, the same result holds after changing $\mathbf{n}$ into $-\mathbf{n}$.

13.4.5 Polarization

By convention, the polarization of the wave specifies the direction of the electric field. For a progressive wave propagating along x, we say that the wave is linearly polarized along the y direction if the electric field is along $\mathbf{u}_y$ at all times:

$$\mathbf{E} = E_y(x - ct)\mathbf{u}_y$$

$$\mathbf{B} = B_z(x - ct)\mathbf{u}_z = \frac{1}{c}E_y(x - ct)\mathbf{u}_z$$

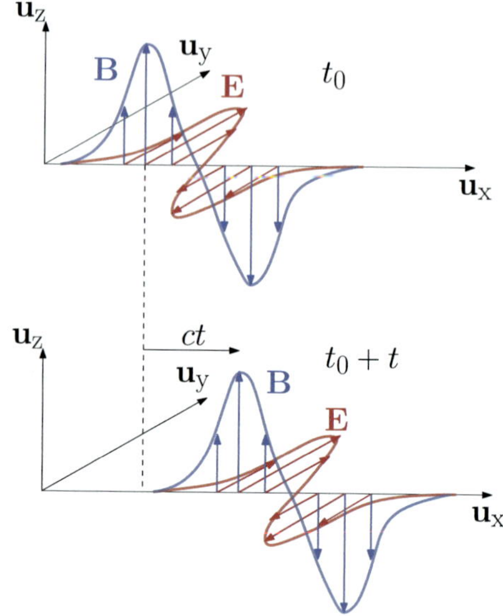

Fig. 13.12 A linearly polarized wave along the y-axis

and such a linearly polarized wave is illustrated in Fig. 13.12.

Note also that any plane wave propagating along x can be written in terms of two independent plane waves according to (13.13):

$$\underbrace{\{E_y, B_z\}}_{\text{linear polarization along } y} \qquad \underbrace{\{E_z, B_y\}}_{\text{linear polarization along } z}$$

Light polarization will be discussed in more detail in Sect. 13.6.

13.4.5.1   Wavelength

As seen in Sect. 13.3.3, a sinusoidal plane wave propagating along the x-axis can be written as

$$\mathbf{E} = \mathbf{E}_0 \cos(kx - \omega t + \phi_0)$$

where $\mathbf{E}_0 \perp \mathbf{u}_x$, and where the spatial and temporal oscillation frequencies of the electromagnetic wave are related by the dispersion relation

$$\omega = kc$$

or, equivalently, the periods are related by

$$\lambda = cT$$

Table 13.1 The electromagnetic spectrum

Name	λ	$f = \omega/2\pi$
Radio waves	10 cm–1 km	$10^5 - 10^{10}$ Hz
Microwaves	1 mm–1 cm	$10^9 - 10^{12}$ Hz
Infrared	0.75 µm–1 mm	$10^{12} - 10^{14}$ Hz
Visible light	400–750 nm	$\sim 5 \times 10^{14}$ Hz
Ultraviolet (UV)	10 nm–400 nm	$10^{15} - 10^{16}$ Hz
X-rays	0.01 nm–10 nm	$10^{16} - 10^{19}$ Hz
γ rays	<10 pm	$>10^{19}$ Hz

where λ is called wavelength. Visible light corresponds to electromagnetic waves in the range of wavelengths 400 nm $< \lambda <$ 800 nm. The wavelengths shorter than 400 nm correspond to ultraviolet radiation, while those longer than 800 nm are called infrared radiation. Table 13.1 shows the wavelength and frequency of the different families of electromagnetic radiation.

13.5 Sinusoidal Plane Waves in 3D

Let us now consider a more general case in which a sinusoidal electromagnetic wave propagates along the unit direction $\mathbf{n}$, and let us choose an orthogonal basis $(\mathbf{u}, \mathbf{v}, \mathbf{n})$ such that $\mathbf{u} \times \mathbf{v} = \mathbf{n}$. The most general sinusoidal wave will have an electric field of the form

$$\mathbf{E} = E_1 \cos(\mathbf{k} \cdot \mathbf{x} - \omega t + \phi_1)\mathbf{u} + E_2 \cos(\mathbf{k} \cdot \mathbf{x} - \omega t + \phi_2)\mathbf{v}$$

where the wave vector is defined as

$$\mathbf{k} = k\mathbf{n} = k_x\mathbf{u}_x + k_y\mathbf{u}_y + k_z\mathbf{u}_z$$

so that its direction coincides with the propagation direction of the wave. At a given instant t, the wave fronts are planes perpendicular to the wave vector $\mathbf{k}$ such that the phase $\mathbf{k} \cdot \mathbf{x} + \phi$ is constant on them. These planes are therefore given by

$$\mathbf{k} \cdot \mathbf{x} = k_x x + k_y y + k_z z = \text{constant}.$$

This is illustrated in Fig. 13.13. In a complex representation, the electric field of such a wave can be rewritten as $\mathbf{E} = \text{Re}\{\underline{\mathbf{E}}\}$ with

$$\underline{\mathbf{E}} = \underline{\mathbf{E}}_0 e^{i(\mathbf{k} \cdot \mathbf{x} - \omega t)}$$

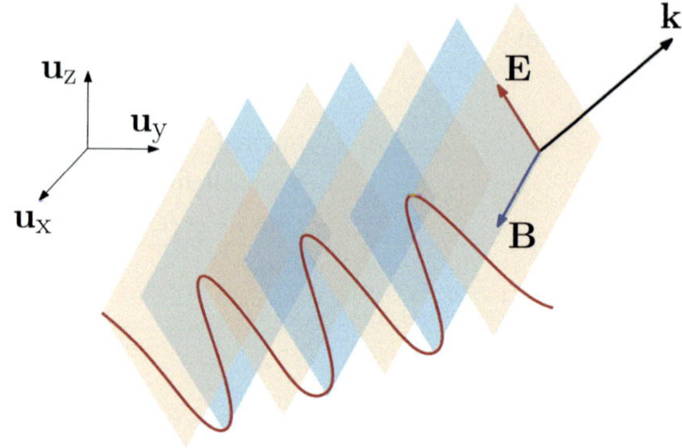

Fig. 13.13 A plane wave propagating along an arbitrary axis

and

$$\underline{\mathbf{E}}_0 = E_1 e^{i\phi_1}\mathbf{u} + E_2 e^{i\phi_2}\mathbf{v}.$$

Injecting this wave in the 3D d'Alembert equation gives the dispersion relation

$$\omega = |\mathbf{k}|c = kc.$$

Finally, if we similarly write the magnetic field as $\mathbf{B} = \mathrm{Re}\{\underline{\mathbf{B}}\} = \mathrm{Re}\{\underline{\mathbf{B}}_0 e^{i(\mathbf{k}\cdot\mathbf{r}-\omega t)}\}$, the four Maxwell equations in vacuum for a sinusoidal plane wave become

$$
\boxed{
\begin{aligned}
\mathbf{k} \cdot \underline{\mathbf{E}} &= 0 && \text{Maxwell--Gauss} \\
\mathbf{k} \cdot \underline{\mathbf{B}} &= 0 && \text{Maxwell--Thomson} \\
\mathbf{k} \times \underline{\mathbf{E}} &= \omega\underline{\mathbf{B}} && \text{Maxwell--Faraday} \\
\mathbf{k} \times \underline{\mathbf{B}} &= -\frac{\omega}{c^2}\underline{\mathbf{E}} && \text{Maxwell--Ampère.}
\end{aligned}
}
\tag{13.14}
$$

Note that the Maxwell–Gauss and Maxwell–Thomson equations for a sinusoidal progressive plane wave implies the transversality of both the electric and magnetic field

$$\mathbf{k} \cdot \underline{\mathbf{E}} = 0 \quad \rightarrow \quad \underline{\mathbf{E}} \perp \mathbf{k}$$

$$\mathbf{k} \cdot \underline{\mathbf{B}} = 0 \quad \rightarrow \quad \underline{\mathbf{B}} \perp \mathbf{k}$$

and since $\mathbf{k} = k\mathbf{n}$, the Maxwell–Faraday law becomes

$$\mathbf{k} \times \underline{\mathbf{E}} = \omega\underline{\mathbf{B}} \quad \rightarrow \quad \underline{\mathbf{B}} = \frac{k\mathbf{n}}{\omega} \times \underline{\mathbf{E}}. \tag{13.15}$$

Using the dispersion relation $k/\omega = 1/c$, we retrieve the previously found relationship between the magnetic and the electric field, now in the particular case of a sinusoidal plane wave

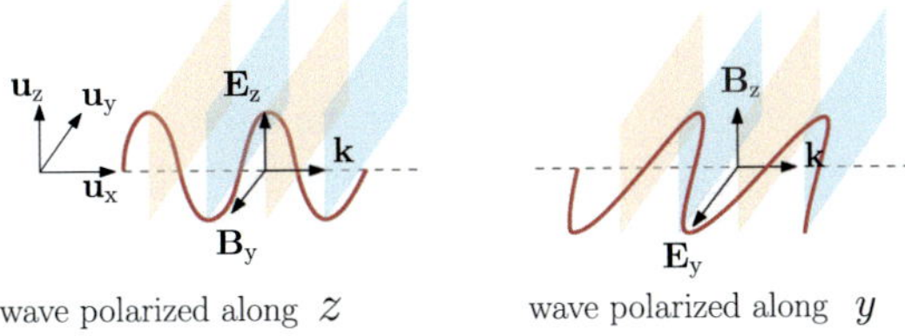

Fig. 13.14 A plane wave propagating along x accepts two independent solutions, polarized along z and y

$$\boxed{\underline{\mathbf{B}} = \frac{1}{c}\mathbf{n} \times \underline{\mathbf{E}}.}$$

(13.16)

13.6 Light Polarization

As seen in Sect. 13.4.5, a sinusoidal plane wave propagating along the x-axis, that is $\mathbf{k} = k\mathbf{u}_x$, accepts two independent solutions:

$$\mathbf{E}_y = E_{0y}\cos(kx - \omega t + \phi_{0y})\mathbf{u}_y,$$

$$\mathbf{E}_z = E_{0z}\cos(kx - \omega t + \phi_{0z})\mathbf{u}_z.$$

These two solutions, shown in Fig. 13.14, form an orthogonal basis of the space of solutions. According to the superposition principle, the general solution for a sinusoidal plane wave propagating along x writes

$$\mathbf{E}(\mathbf{x}, t) = \mathbf{E}_y + \mathbf{E}_z$$

where the amplitudes E_{0y}, E_{0z} and the relative phase $\phi_{0y} - \phi_{0z}$ between them characterize a given polarization state. Using the complex representation:

$$\underline{\mathbf{E}}(\mathbf{x}, t) = \underline{\mathbf{E}}_0 e^{i(\mathbf{k}\cdot\mathbf{x} - \omega t)}$$

the polarization state of the wave is described by

$$\underline{\mathbf{E}}_0 = E_{0y}e^{i\phi_{0y}}\mathbf{u}_y + E_{0z}e^{i\phi_{0z}}\mathbf{u}_z.$$

13.6.1 Linear Polarization

An arbitrary linear polarization consists of a superposition of horizontal and vertical polarization states with no relative phase difference, that is $\phi_{0y} = \phi_{0z} = \phi$. For arbitrary E_{0y} and E_{0z}, it is always possible to find α such that

$$E_{0y} = E_0 \cos\alpha, \qquad E_{0z} = E_0 \sin\alpha \ .$$

This wave is thus characterized by an oblique linear polarization in which the electric field direction forms an angle α with the horizontal direction (Fig. 13.15)

$$\underline{\mathbf{E}} = E_0(\cos\alpha\,\mathbf{u}_y + \sin\alpha\,\mathbf{u}_z)e^{i\phi}e^{i(kx-\omega t)} \quad \rightarrow \quad \mathbf{E}(\mathbf{x},t) = E_0(\cos\alpha\,\mathbf{u}_y + \sin\alpha\,\mathbf{u}_z)\cos(kx - \omega t + \phi).$$

13.6.1.1   Circular Polarization

A circularly polarized wave is such that the direction of the electric field describes a circle in the yz plane. This corresponds to the case $E_{0y} = E_{0z} = E_0$ and $\phi_{0y} - \phi_{0z} = \pm\frac{\pi}{2}$. The right-handed circular polarization is obtained for $\phi_{0y} - \phi_{0z} = \frac{\pi}{2}$, for which the electric field describes a clockwise rotation if looked against the direction of propagation (Fig. 13.16).

$$\underline{\mathbf{E}} = E_0 e^{i\phi_{0y}}(\mathbf{u}_y - i\mathbf{u}_z)e^{i(kx-\omega t)} \quad \rightarrow \quad \mathbf{E}(\mathbf{x},t) = E_0\cos(kx - \omega t + \phi_{0y})\mathbf{u}_y + E_0\sin(kx - \omega t + \phi_{0y})\mathbf{u}_z.$$

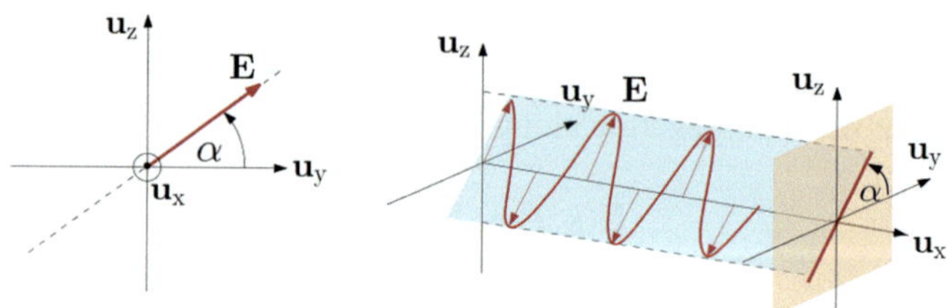

Fig. 13.15 A linearly polarized wave forming an angle of α with respect to the y-axis

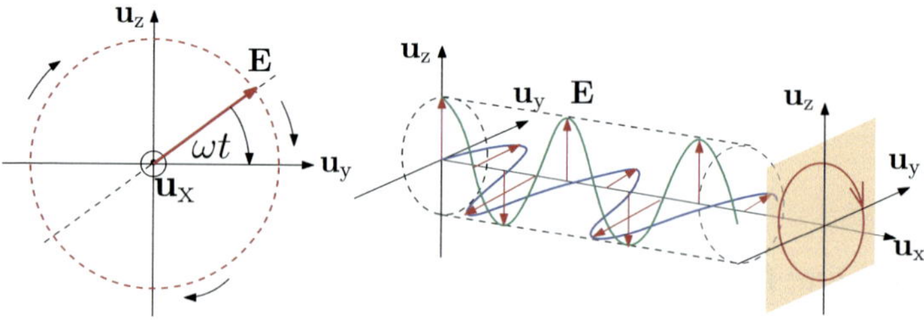

Fig. 13.16 A right-handed circularly polarized wave is a superposition of two linearly polarized waves of same amplitude and relative phase of $\pi/2$

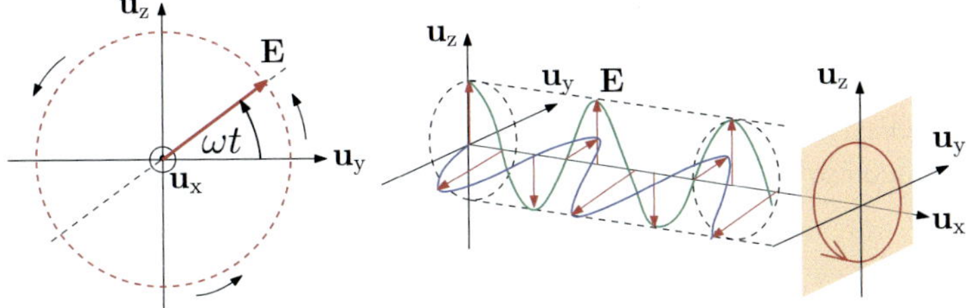

Fig. 13.17 A left-handed circularly polarized wave is a superposition of two linearly polarized waves of same amplitude and relative phase of $-\pi/2$

The left-handed circular polarization is obtained for $\phi_{0y} - \phi_{0z} = -\frac{\pi}{2}$, for which the electric field describes an anti-clockwise rotation if looked against the direction of propagation.

$$\underline{\mathbf{E}_0} = E_0 e^{i\phi_{0y}}(\mathbf{u}_y + i\mathbf{u}_z)e^{i(kz-\omega t)} \quad \rightarrow \quad \mathbf{E}(\mathbf{x}, t) = E_0 \cos(kx - \omega t + \phi_{0y})\mathbf{u}_y - E_0 \sin(kx - \omega t + \phi_{0y})\mathbf{u}_z.$$

This is shown in Fig. 13.17.

13.6.1.2 Elliptical Polarization

The most general case for which E_{0y}, E_{0z}, ϕ_{0y}, and ϕ_{0z} are arbitrary corresponds to the case in which the electric field direction describes an ellipse in the yz plane (Fig. 13.18).

$$\mathbf{E}(\mathbf{x}, t) = E_1 \cos(kx - \omega t + \phi_{0y})\mathbf{u}_y + E_2 \cos(kx - \omega t + \phi_{0z})\mathbf{u}_z = \mathrm{Re}\{(E_1 e^{i\phi_{0y}}\mathbf{u}_y + E_2 e^{i\phi_{0z}}\mathbf{u}_z)e^{i(kx-\omega t)}\}$$

the wave will be right-handed if $\phi_{0y} - \phi_{0z} > 0$ and left-handed otherwise. So far we have seen that any polarization state can be written as a superposition of two orthogonal linear polarization states. However, one can use instead a circular basis given by

Fig. 13.18 Elliptical polarization

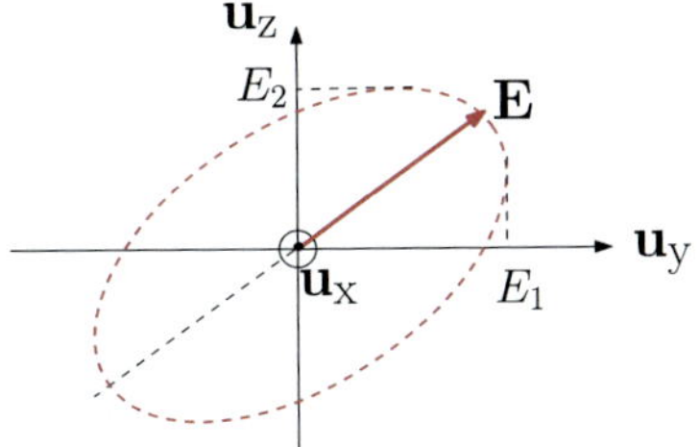

$$\mathbf{u}_r = \frac{\mathbf{u}_y - i\mathbf{u}_z}{\sqrt{2}} \qquad \mathbf{u}_l = \frac{\mathbf{u}_y + i\mathbf{u}_z}{\sqrt{2}}$$

so that any polarization state can be also written as a superposition of a right-handed and left-handed circularly polarized states. Indeed, one can write

$$\mathbf{u}_y = \frac{\mathbf{u}_r + \mathbf{u}_l}{\sqrt{2}} \qquad \mathbf{u}_z = \frac{i(\mathbf{u}_r - \mathbf{u}_l)}{\sqrt{2}}$$

and so:

$$\mathbf{E}(\mathbf{x}, t) = \mathrm{Re}\{(E_r\mathbf{u}_r + E_l\mathbf{u}_l)e^{i(kx-\omega t)}\}$$

with

$$E_r = \frac{E_{0y}e^{i\phi_{0y}} + iE_{0z}e^{i\phi_{0z}}}{\sqrt{2}} \qquad E_l = \frac{E_{0z}e^{i\phi_{0z}} - iE_{0y}e^{i\phi_{0z}}}{\sqrt{2}}$$

both linear and circular basis can be used to describe light polarization.

13.6.1.3 Unpolarized Light

Natural light, which represents most of the light encountered in everyday life such as sunlight or the light emitted from a light bulb, does not have a well defined polarization state. Emission from such sources is made by *wave packets* of light whose polarization varies randomly from one to another. As such, natural light is said to be unpolarized. Polarized light from such sources can be obtained by different methods that will be discussed in Chap. 14.

13.6.2 A Brief History of Light Polarization and Its Applications

The discovery and study of light polarization began in the early 19th century with the work of the French physicist Étienne-Louis Malus in 1808. Malus was the first to apply the term *polarization* to light. He discovered polarization by reflection, a phenomenon in which light reflected off a flat surface, such as glass, under a specific angle, becomes completely polarized. Later, in 1815, the Scottish physicist David Brewster expanded on Malus's experiments.

In the same year, the French physicist Jean-Baptiste Biot (notable for the Biot–Savart law) demonstrated that some liquids, when traversed by a linearly polarized light beam, cause a rotation of the polarization plane. This phenomenon, known as *optical activity*, is a characteristic property of certain chemical compounds, such as sugar, due to their asymmetric (chiral) structure. Instruments called polarimeters, such as the one shown in Fig. 13.19 (left), were used to measure this effect. François

Arago utilized a cyanopolarimeter to quantitatively measure the blueness of the sky, leveraging the partial polarization of light from the sky. This instrument is shown in Fig. 13.19 (right).

In 1816, François Arago and Augustin Fresnel conducted interferometry experiments using polarized light beams. They observed that two light beams with orthogonal polarization states do not interfere with each other (the Fresnel-Arago experiment). This discovery was pivotal, as it provided strong evidence that light behaves as a transverse wave.

In 1845, Michael Faraday conducted an experiment in which a linearly polarized light beam passed through a medium exposed to a magnetic field aligned with the direction of light propagation. He observed that the orientation of the light's linear polarization was altered, demonstrating the influence of a magnetic field on light. This experiment, known as the *Faraday effect*, confirmed a connection between light and electromagnetism—well before Maxwell's equations were established. Today, the Faraday effect is widely used in modern research to investigate the magnetic properties of solids.

In more recent times, the control of light polarization enabled significant technological advancements. In the early 1970s, it led to the development of one of the most widely used electronic components: the liquid crystal display (LCD). This technology also gave rise to 3D glasses, which allow each eye of the viewer to see a different image by utilizing two images projected with light of orthogonal, circular polarization states. Furthermore, light polarization is inherently sensitive to the microstructure of

Fig. 13.19 Left: Instrument to measure polarization rotation by Savart and Biot. Right: Arago's cyanopolarimeter. Pictures taken at the Museum of Ecole Polytechnique (Mus'X)

materials. This property is leveraged in various research fields, such as characterizing the thickness of microstructured electronic components or identifying well-ordered structures in biological tissues, such as collagen fibers. Additionally, polarization is widely used to study the electronic and structural properties of crystalline solids.

13.7   Electromagnetic Energy and Poynting's Theorem

Not only Maxwell's equations contain the law of conservation of charge, but it is also possible to obtain from them an equation reflecting the conservation of electromagnetic energy, called the Poynting theorem. In its integral form, such an equation should read

$$\frac{dU(t)}{dt} = \iiint_\Omega \frac{\partial u_{EM}(\mathbf{x}, t)}{\partial t} d^3x = -W_1 - W_2$$

where U is the total electromagnetic energy contained in an arbitrary volume $\Omega \subseteq \mathbb{R}^3$, and u_{EM} is the density of electromagnetic energy per unit volume (J/m^3). This equation states that the variation of the total energy contained in Ω between t and $t + dt$ may have two origins: an energy exchange with the charges in Ω (at a rate W_1) or a transfer of energy through the frontier $\partial\Omega$ (at a rate W_2). The term W_1 is simply the work done by the Lorentz force on the charges contained in Ω: a fraction of the electromagnetic energy of the wave is thus converted into kinetic energy of the charges. We calculate first the Lorentz force exerted on a volume element d^3x due to the passage of the wave. If the velocity of the charges in d^3x is $\mathbf{v}$, then,

$$d\mathbf{f} = dq(\mathbf{E} + \mathbf{v} \times \mathbf{B}) = \varrho\, d^3x(\mathbf{E} + \mathbf{v} \times \mathbf{B}).$$

Then, the work done by the Lorentz force on Ω is

$$W_1 = \iiint_\Omega d\mathbf{f} \cdot \mathbf{v} = \iiint_\Omega \varrho(\mathbf{E} + \mathbf{v} \times \mathbf{B}) \cdot \mathbf{v}\, d^3x = \iiint_\Omega \varrho\mathbf{v} \cdot \mathbf{E}\, d^3x$$

and since $\varrho\mathbf{v} = \mathbf{j}$ where $\mathbf{j}$ is the current density:

$$W_1 = \iiint_\Omega \mathbf{j} \cdot \mathbf{E}\, d^3x.$$

For the term W_2, we assume that the transfer of energy through the surface $\partial\Omega$ can be expressed as a flux through that surface

$$W_2 = \Phi_{\partial\Omega,\boldsymbol{\Pi}} = \oiint_{\partial\Omega} \boldsymbol{\Pi} \cdot d\mathbf{S} = \iiint_\Omega \nabla \cdot \boldsymbol{\Pi} d^3x$$

where $\boldsymbol{\Pi}$, the Poynting vector, represents a flux of energy associated with the electromagnetic wave, i.e., a density of energy per unit surface and per unit time (W/m^2).

In this way, the conservation equation for the total electromagnetic energy writes

$$\iiint_\Omega \frac{\partial u_{EM}(\mathbf{x}, t)}{\partial t} d^3x = -\iiint_\Omega \mathbf{j} \cdot \mathbf{E}\, d^3x - \iiint_\Omega \nabla \cdot \mathbf{\Pi}\, d^3x$$

or, equivalently, in its local form, the Poynting theorem is

$$\boxed{\frac{\partial u_{EM}}{\partial t} + \nabla \cdot \mathbf{\Pi} = -\mathbf{j} \cdot \mathbf{E}.} \tag{13.17}$$

13.7.1 Identification of $\mathbf{\Pi}$ and u_{EM}

We will now use Maxwell's equations in order to demonstrate (13.17) and identify $\mathbf{\Pi}$ and u_{EM} in terms of the fields $\mathbf{E}$ and $\mathbf{B}$. For this we may start with the Maxwell–Ampère law $\nabla \times \mathbf{B} = \mu_0 \mathbf{j} + \mu_0 \epsilon_0 \dfrac{\partial \mathbf{E}}{\partial t}$ and take the scalar product with $\mathbf{E}/\mu_0$ on both sides of the equation:

$$\frac{1}{\mu_0} \mathbf{E} \cdot (\nabla \times \mathbf{B}) = \mathbf{j} \cdot \mathbf{E} + \epsilon_0 \mathbf{E} \cdot \frac{\partial \mathbf{E}}{\partial t} = \mathbf{j} \cdot \mathbf{E} + \frac{\partial}{\partial t} \left(\frac{\epsilon_0}{2} \mathbf{E}^2 \right). \tag{13.18}$$

Similarly, we use the Maxwell–Faraday law $\nabla \times \mathbf{E} = -\dfrac{\partial \mathbf{B}}{\partial t}$ and we take the scalar product with $\mathbf{B}/\mu_0$:

$$\frac{1}{\mu_0} \mathbf{B} \cdot (\nabla \times \mathbf{E}) = -\frac{1}{\mu_0} \mathbf{B} \cdot \frac{\partial \mathbf{B}}{\partial t} = -\frac{\partial}{\partial t} \left(\frac{1}{2\mu_0} \mathbf{B}^2 \right). \tag{13.19}$$

Subtracting (13.18) from (13.19), we obtain

$$\frac{1}{\mu_0} (\mathbf{E} \cdot (\nabla \times \mathbf{B}) - \mathbf{B} \cdot (\nabla \times \mathbf{E})) = \mathbf{j} \cdot \mathbf{E} + \frac{\partial}{\partial t} \left(\frac{\epsilon_0 \mathbf{E}^2}{2} + \frac{\mathbf{B}^2}{2\mu_0} \right). \tag{13.20}$$

Finally, using the identity $\mathbf{E} \cdot (\nabla \times \mathbf{B}) - \mathbf{B} \cdot (\nabla \times \mathbf{E}) = \nabla \cdot (\mathbf{B} \times \mathbf{E})$, (13.20) is rewritten as

$$\frac{1}{\mu_0} \nabla \cdot (\mathbf{B} \times \mathbf{E}) = \mathbf{j} \cdot \mathbf{E} + \frac{\partial}{\partial t} \left(\frac{\epsilon_0 \mathbf{E}^2}{2} + \frac{\mathbf{B}^2}{2\mu_0} \right)$$

and we recognize an equation of the form of (13.17).

Poynting's Theorem
The conservation of the electromagnetic energy in its local form is given by

Poynting's theorem (Fig. 13.20)

$$\nabla \cdot \left(\frac{\mathbf{E} \times \mathbf{B}}{\mu_0} \right) + \frac{\partial}{\partial t} \left(\frac{\epsilon_0 \mathbf{E}^2}{2} + \frac{\mathbf{B}^2}{2\mu_0} \right) = -\mathbf{j} \cdot \mathbf{E}. \tag{13.21}$$

Or, equivalently,

$$\boxed{\frac{\partial}{\partial t} u_{EM} + \nabla \cdot \mathbf{\Pi} = -\mathbf{j} \cdot \mathbf{E}} \tag{13.22}$$

where $\mathbf{\Pi}$ is the Poynting vector , which gives the direction of electromagnetic energy flow

$$\mathbf{\Pi} = \frac{\mathbf{E} \times \mathbf{B}}{\mu_0} \tag{13.23}$$

and u_{EM} the density of electromagnetic energy

$$u_{EM} = \frac{\epsilon_0 \mathbf{E}^2}{2} + \frac{\mathbf{B}^2}{2\mu_0}. \tag{13.24}$$

We recognize in (13.24) the sum of the electric energy density (3.21) and the magnetic energy density (11.18).

13.7.2   *Energy Flow in a Steady Circuit*

Poynting's theorem states that the energy flow in an electromagnetic system is described by the Poynting vector, which depends on the electric and magnetic fields. How is energy transported in a circuit? To explore this, consider a small section of a current-carrying wire between points A and B, as illustrated in Fig. 13.21. A steady current I flows through the wire, and the fields in close proximity exhibit cylindrical symmetry. Infinitesimally close to the wire's surface, the electric field is determined by the potential difference ΔV across the segment

$$\mathbf{E} = -\nabla V = \frac{V_A - V_B}{\ell} \mathbf{u}_z = \frac{\Delta V}{\ell} \mathbf{u}_z$$

where ℓ the length of the wire section (Fig. 13.21).

Due to the steady current I, a magnetic field $\mathbf{B}$ exists around the wire. At the surface of the wire (radius R), the magnetic field is

$$\mathbf{B} = \frac{\mu_0 I}{2\pi R} \mathbf{u}_\theta.$$

The Poynting vector, representing the energy flow per unit surface, is given by

Fig. 13.20 John Henry Poynting (1852–1914), an English physicist who made significant contributions to electromagnetic theory and the study of radiation

Fig. 13.21 The Poynting vector points radially into a current-carrying wire

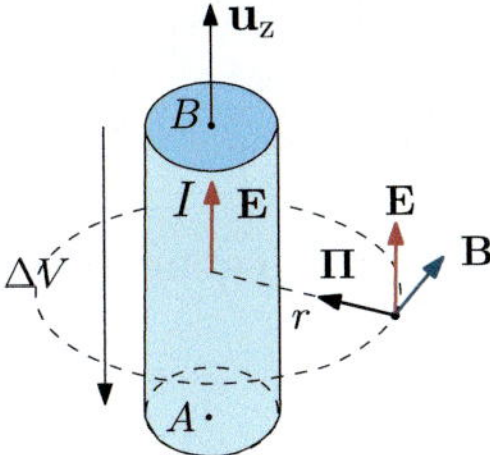

$$\mathbf{\Pi} = \frac{\mathbf{E} \times \mathbf{B}}{\mu_0} = -\frac{I\,\Delta V}{2\pi\,R\ell}\mathbf{u}_r.$$

We find that the Poynting vector points radially into the wire. The total energy leaving the section of wire per unit time is the flux of $\mathbf{\Pi}$ over the lateral surface of the cylindrical wire. The lateral surface has a normal vector $\mathbf{u}_r$ and area $2\pi R\ell$. Integrating the Poynting vector over this surface yields:

$$P = \iint_S \mathbf{\Pi} \cdot d\mathbf{S} = -\frac{2\pi R\ell I\,\Delta V}{2\pi R\ell} = -I\,\Delta V$$

where the minus sign indicates that energy is entering the wire, not leaving it. This is precisely the power dissipated as Joule heating in the wire, confirming that energy flows into the wire through the electromagnetic field at its surface. This energy is generated by the movement of all the charges in the circuit, collectively maintaining the electric and magnetic fields. Thanks to this, the transmission of electrical energy does not need a continuous physical wire to flow from one point to another. Indeed, in between a capacitor, or in the case of a transformer, the continuity of a transmission line is interrupted. This does not prevent the flow of energy.

In Exercise 13.3, a case in which the current varies in time is analyzed for a coaxial cable, consisting of two concentric cylindrical conductors. It is shown that the energy is transported between the cylinders via the electromagnetic field.

13.7.3 Poynting's Vector of a Sinusoidal Plane Wave

Let us apply Poynting's theorem to a sinusoidal, plane electromagnetic wave in vacuum propagating along the x-axis and linearly polarized along y

$$\mathbf{E} = E_0 \cos(\omega t - kx)\mathbf{u}_y$$
$$\mathbf{B} = \frac{1}{c}\mathbf{u}_x \times \mathbf{E} = \frac{1}{c}E_0 \cos(\omega t - kx)\mathbf{u}_z.$$

The electromagnetic energy density associated with such a wave is equally distributed between its electric and magnetic components. According to (13.24)

$$u_{EM} = \frac{\epsilon_0 \mathbf{E}^2}{2} + \frac{\mathbf{B}^2}{2\mu_0} = \frac{\epsilon_0}{2}E_0^2 \cos^2(\omega t - kx) + \frac{E_0^2}{2c^2\mu_0}\cos^2(\omega t - kx) = \epsilon_0 E_0^2 \cos^2(\omega t - kx)$$

and the Poynting vector, given by (13.23) writes

$$\mathbf{\Pi} = \frac{\mathbf{E} \times \mathbf{B}}{\mu_0} = c\epsilon_0 E_0^2 \cos^2(\omega t - kx)\mathbf{u}_x = cu_{EM}\mathbf{u}_x$$

so that the electromagnetic energy is moving at the speed c in the positive x direction. Indeed, the Poynting vector represents the energy passing per unit area and per unit time through a surface element perpendicular to $\mathbf{\Pi}$. Note that both u_{EM} and $\mathbf{\Pi}$ oscillate at the frequency ω, which in the visible range corresponds to a period $T = 2\pi/\omega$ of the order of $T \sim 0.3$ fs, which is too fast to be detected even with our fastest optical detectors. As an example, the eye response time is of the order of 0.1 s, a CCD camera responds typically within 10 ms and a fast photodiode or a photomultiplier may achieve a response of 10 ps which is still much larger than T for visible light. Any light detector thus measures an averaged power over the detector response time $T_{\text{det}} \gg T$. The time average of the electromagnetic energy density and the Poynting vector for a sinusoidal plane wave read

$$\langle u_{EM} \rangle_{T_{\text{det}}} = \epsilon_0 E_0^2 \frac{1}{T_{\text{det}}} \int_0^{T_{\text{det}}} \underbrace{\cos^2(\omega t - kx)dt}_{=1/2(1+\cos(2(\omega t - kx)))}$$

$$= \epsilon_0 E_0^2 \left(\frac{1}{2} + \frac{T}{4\pi T_{\text{det}}} (\sin 2(\omega t - kx))|_0^{T_{\text{det}}} \right) \approx \frac{\epsilon_0 E_0^2}{2}$$

since $T_{\text{det}} \gg T$. Note therefore that for a sinusoidal wave the detector measures an average equal to the average over one light period

$$\langle u_{EM} \rangle_T = \epsilon_0 E_0^2 \frac{1}{T} \underbrace{\int_0^T \cos^2(\omega t - kx)dt}_{=1/2} = \frac{\epsilon_0 E_0^2}{2}$$

and the same applies for the Poynting vector

$$\langle \mathbf{\Pi} \rangle_T = c\epsilon_0 E_0^2 \frac{1}{T} \int_0^T \cos^2(\omega t - kx)dt \ \mathbf{u}_x = \frac{1}{2}c\epsilon_0 E_0^2 \ \mathbf{u}_x = c\langle u_{EM} \rangle_T \mathbf{u}_x.$$

Remarks

1. Since the Poynting vector is not a linear function of the fields, we cannot calculate it by using the complex representation directly, since

$$\mathbf{\Pi} \neq \mathrm{Re}\left\{ \frac{\underline{\mathbf{E}} \times \underline{\mathbf{B}}}{\mu_0} \right\}.$$

However, for the time-averaged energy density and time-averaged Poynting vector, we may write

$$\langle u_{EM} \rangle_T = \frac{1}{2}\mathrm{Re}\left\{ \frac{\epsilon_0}{2}\underline{\mathbf{E}} \cdot \underline{\mathbf{E}}^* + \frac{1}{2\mu_0}\underline{\mathbf{B}} \cdot \underline{\mathbf{B}}^* \right\}$$

and define the complex Poynting vector $\underline{\mathbf{\Pi}} = \frac{\underline{\mathbf{E}} \times \underline{\mathbf{B}}^*}{2\mu_0}$ such that

$$\boxed{\langle \mathbf{\Pi} \rangle_T = \mathrm{Re}\{\underline{\mathbf{\Pi}}\} = \frac{1}{2\mu_0}\mathrm{Re}\left\{ \underline{\mathbf{E}} \times \underline{\mathbf{B}}^* \right\}.} \qquad (13.25)$$

2. The electromagnetic energy density of a plane wave is uniform, so that the plane wave, which is infinite in all directions and in time, carries an infinite electromagnetic energy. Whereas this is not physically acceptable, one must recall that plane waves are useful because they can approximate real waves locally, and because every real wave is a linear superposition of plane waves.

Intensity of an Electromagnetic Wave

The intensity $\mathcal{I}$, also called irradiance, of an electromagnetic wave corresponds to the average power going through a unit surface perpendicular to the propagation axis. It has units of $\mathrm{W/m}^2$ and corresponds to the time-average of the modulus of the Poynting vector over the detector response time T_{det} which, for sinusoidal plane waves and for $T_{\mathrm{det}} \gg T$, coincides with the average over one period T of the wave oscillation

$$\mathcal{I} = \langle |\mathbf{\Pi}| \rangle_T = \left\langle \left\| \frac{\mathbf{E} \times \mathbf{B}}{\mu_0} \right\| \right\rangle_T \qquad (13.26)$$

and for a sinusoidal plane wave, it writes

$$\mathcal{I} = \frac{c\epsilon_0}{2} E_0^2 \tag{13.27}$$

where E_0 is the amplitude of the wave's electric field. Every optical detector of detection surface S, oriented perpendicular to the propagation axis, measures an average power given by $\mathcal{P} = \mathcal{I} S \propto E_0^2$.

13.8   Summary and Essential Formulas

- Maxwell's equations form a complete set of coupled equations relating the electric and magnetic fields to their sources (charge ϱ and current density $\mathbf{j}$):

$$\boxed{\begin{aligned} \boldsymbol{\nabla} \cdot \mathbf{E} &= \frac{\varrho}{\epsilon_0} & &\text{Maxwell–Gauss} \\[2mm] \boldsymbol{\nabla} \cdot \mathbf{B} &= 0 & &\text{Maxwell–Thomson} \\[2mm] \boldsymbol{\nabla} \times \mathbf{E} &= -\frac{\partial \mathbf{B}}{\partial t} & &\text{Maxwell–Faraday} \\[2mm] \boldsymbol{\nabla} \times \mathbf{B} &= \mu_0 \mathbf{j} + \mu_0 \epsilon_0 \frac{\partial \mathbf{E}}{\partial t} & &\text{Maxwell–Ampère} \end{aligned}}$$

which, together with the Lorentz force acting on a point charge q with velocity $\mathbf{v}$

$$\mathbf{F} = q(\mathbf{E} + \mathbf{v} \times \mathbf{B})$$

provide a complete description of all classical electromagnetism.

- In vacuum ($\varrho = 0$, $\mathbf{j} = \mathbf{0}$), the electric and magnetic fields satisfy the same propagation equation

$$\boldsymbol{\nabla}^2 \mathbf{E} - \frac{1}{c^2} \frac{\partial^2}{\partial t^2} \mathbf{E} = 0$$

$$\boldsymbol{\nabla}^2 \mathbf{B} - \frac{1}{c^2} \frac{\partial^2}{\partial t^2} \mathbf{B} = 0$$

where $c = 1/\sqrt{\epsilon_0 \mu_0} \approx 3 \times 10^8$ m/s is the speed of light. Each component of the electric and magnetic fields then satisfy d'Alembert's wave equation

$$\boldsymbol{\nabla}^2 F(\mathbf{x}, t) - \frac{1}{c^2} \frac{\partial^2 F(\mathbf{x}, t)}{\partial t^2} = 0.$$

- A plane wave propagating along x corresponds to a solution of d'Alembert's equation which depends solely on x and t. An important solution corresponds to a sinusoidal plane wave propagating toward positive values of x:

$$f_+(x, t) = A_1 \cos(kx - \omega t + \phi_1)$$

where the spatial (k) and temporal (ω) frequencies are related by the dispersion relation

$$\omega(k) = ck.$$

- For a plane wave propagating along the direction $\mathbf{n}$ in vacuum, the electric and magnetic fields are transverse:

$$\boxed{\mathbf{E} \cdot \mathbf{n} = \mathbf{B} \cdot \mathbf{n} = 0}$$

and perpendicular to each other, so that $\{\mathbf{n}, \mathbf{E}, \mathbf{B}\}$ form a direct trihedral:

$$\boxed{\mathbf{B} = \frac{1}{c}\mathbf{n} \times \mathbf{E}.}$$

- A sinusoidal plane wave propagating along x accepts two independent solutions, linearly polarized along y and z, respectively

$$\mathbf{E}_y = E_{0y} \cos(kx - \omega t + \phi_{0y})\mathbf{u}_y$$

$$\mathbf{E}_z = E_{0z} \cos(kx - \omega t + \phi_{0z})\mathbf{u}_z$$

and the most general polarization state is characterized by the set of parameters $\{E_{0y}, \phi_{0y}, E_{0z}, \phi_{0z}\}$.
- The conservation equation for the electromagnetic energy is given by Poynting's theorem

$$\iiint_\Omega \frac{\partial u_{EM}(\mathbf{x}, t)}{\partial t} d^3x = - \iiint_\Omega \mathbf{j} \cdot \mathbf{E}\, d^3x - \oiint_{\partial\Omega} \mathbf{\Pi} \cdot d\mathbf{S}$$

or, equivalently, in its local form

$$\boxed{\frac{\partial u_{EM}}{\partial t} + \nabla \cdot \mathbf{\Pi} = -\mathbf{j} \cdot \mathbf{E}}$$

where $\mathbf{\Pi}$ is the Poynting vector, which gives the direction of electromagnetic energy flow

$$\mathbf{\Pi} = \frac{\mathbf{E} \times \mathbf{B}}{\mu_0}$$

and u_{EM} the density of electromagnetic energy

$$u_{EM} = \frac{\epsilon_0 \mathbf{E}^2}{2} + \frac{\mathbf{B}^2}{2\mu_0}.$$

- Due to the high frequency of optical waves, every detector of surface S oriented perpendicularly to the direction of propagation of the wave measures a time-averaged power given by $\mathcal{I}S$, where the intensity (also called irradiance) $\mathcal{I}$ (in W/m^2) is

$$\mathcal{I} = \langle |\mathbf{\Pi}| \rangle_T.$$

For a sinusoidal plane wave of amplitude E_0 propagating along x, the time-averaged energy density and the Poynting vector are given by

$$\langle u_{EM} \rangle_T = \frac{\epsilon_0 E_0^2}{2},$$

$$\langle \mathbf{\Pi} \rangle_T = c \langle u_{EM} \rangle_T \mathbf{u}_x = \mathcal{I}\mathbf{u}_x.$$

Problems

13.1 Charged sphere leaking charge radially

Consider a charged sphere located at the origin whose charge is leaking radially in all directions at a rate dQ/dt, so that in spherical coordinates the current density writes $\mathbf{j} = j(r, t)\mathbf{u}_r$.

(a) By using the charge conservation law, determine the current density $\mathbf{j} = \mathbf{j}(r, t)$ as a function of dQ/dt.
(b) Determine the electric field generated by the sphere and the associated displacement current $\mathbf{j}_D = \mathbf{j}_D(r, t)$.
(c) Use the Maxwell–Ampère law to determine the magnetic field everywhere in space. Is this result consistent with the symmetry properties of the total current density?

13.2 Spherical wave

Consider a spherical electromagnetic wave whose electric field writes in spherical coordinates

$$\mathbf{E} = \frac{E_0}{r} \cos(kr - \omega t)\, \mathbf{u}_\theta.$$

(a) Calculate the average electromagnetic energy density, the average Poynting vector and the associated intensity of the wave at any point in space.
(b) What is the total time-averaged electromagnetic energy contained inside a sphere of radius r centered at the origin?
(c) What is the average flux of electromagnetic energy through a spherical surface of radius r centered at the origin? What is the total power emitted by the source of the electromagnetic wave?

13.3 Electromagnetic waves inside a coaxial cable

A coaxial cable consists of two concentric cylindrical conductors: an inner cylinder of radius a and a thin cylindrical metal shell of radius b. We will use the usual notation (r, θ, z) for cylindrical coordinates, where the z-axis is the common axis of the two cylinders. A section of this cable is identified by the z coordinate. A current $I(z, t) = \mathrm{Re}\{\underline{I}(z)e^{-i\omega t}\}$ circulates in the inner conductor, and the current density is assumed to be uniform. The same current circulates as a surface current in the outer cylinder, but in the opposite direction. The symmetries of the current distribution tell us that the electric field $\mathbf{E}$ is radial and the magnetic field $\mathbf{B}$ is orthoradial. In cylindrical coordinates, the fields write $\mathbf{E} = E(r, z, t)\mathbf{u}_r$ and $\mathbf{B} = B(r, z, t)\mathbf{u}_\theta$, respectively.

(a) Calculate $B(r, z, t)$ everywhere as a function of $I(z, t)$ and r by integrating the Maxwell–Ampère equation. Assume that the current density inside the inner cylinder is independent of r.

(b) Link $\dfrac{\partial B}{\partial z}$ to $\dfrac{\partial E}{\partial t}$, and deduce the value of the electric field in the time-dependent regime.

(c) Use the Maxwell–Faraday equation to show that $\underline{I}(z)$ satisfies a differential equation whose solution is

$$\underline{I}(z) = I_0 e^{ikz}$$

and give an expression for k.

(d) Calculate the Poynting vector and the average power transported by the cable.

(e) Consider now the static case $I(z, t) = I_0$ in which a constant potential difference ΔV is applied between the two cylindrical conductors. Calculate the static electric and magnetic fields, the Poynting vector, and the power transported by the cable.

13.4 Incoming radiation from the sun

The irradiance at the top of Earth's atmosphere coming from the Sun is $\mathcal{I} = 1.35 \times 10^3\,\mathrm{Wm}^{-2}$.

(a) Assuming that this radiation behaves locally as a sinusoidal plane wave, what is the magnitude of the electric and magnetic fields of the wave at Earth's atmosphere?

(b) What is the average power emitted by the Sun? The distance between the Earth and the Sun is $R = 1.5 \times 10^{11}$ m.

13.5 Beyond Maxwell's equations—The mass of a photon

Theoretical works in the twentieth century extended Maxwell's equations to allow for the possibility of deviations from the usual laws of electrodynamics. For instance, the simplest generalization of Maxwell's equations (set of Eq. (13.28)) opens up the possibility of deviations from the $1/r^2$ dependence of Coulomb's law and makes Maxwell's equations compatible with massive photons.

$$\nabla \cdot \mathbf{E} = \frac{\varrho}{\varepsilon_0} - \mu^2 V$$

$$\nabla \times \mathbf{E} = -\partial_t \mathbf{B}$$

$$\nabla \cdot \mathbf{B} = 0 \tag{13.28}$$

$$\nabla \times \mathbf{B} = \mu_0 \mathbf{j} + \frac{1}{c^2} \partial_t \mathbf{E} - \mu^2 \mathbf{A}$$

where V and $\mathbf{A}$ are the scalar and vector potential, respectively.

(a) What is the dimension of μ? Show that the generalized Maxwell equations (13.28) preserve the relations between fields and potentials.

(b) What is the consequence of charge conservation? Are the vector and scalar potentials unique?

(c) We are now interested in the propagation of electromagnetic waves: Which propagation equations do the potentials $\mathbf{A}$ and V satisfy?

(d) Demonstrate that $\mathbf{E}$ and $\mathbf{B}$ satisfy similar propagation equations as (13.8) and (13.9).

(e) Consider the propagation of electromagnetic sinusoidal plane waves in vacuum (without free charge or current). Infer the dispersion relation $\omega(k)$ from the propagation equation written in the previous question. Comment.

(f) Show that the dispersion relation is analogous to the dispersion relation for a massive particle of mass m. Express the mass m as a function of μ, c and $\hbar$.
Hint : In special relativity, the energy E and momentum p of a particle of mass m are linked by the relation $E^2 = p^2 c^2 + m^2 c^4$.

(g) A beam of light of finite spatial extension propagates at the so-called group velocity $v_g = \partial \omega / \partial k$ (demonstrated in Chap. 14). Can light propagate at zero velocity? Propose a method to measure the photon mass.

Chapter 14
Electromagnetic Waves in Matter, Polarizers and Wave Plates

Abstract This chapter extends the study of electromagnetic waves from vacuum to *material media*, focusing on how the presence of matter modifies wave propagation. It begins by generalizing *Maxwell's equations in matter*, introducing the auxiliary fields **D** (*electric displacement*) and **H** (*magnetic excitation*) to account for polarization (**P**) and magnetization (**M**) within the medium. *Poynting's theorem* is also generalized to describe energy flow and dissipation in material media. A significant portion of the chapter is dedicated to the *propagation of waves in dielectric media*. *Lorentz model* is introduced to explain the *frequency dependence of permittivity* and, consequently, the *refractive index*, leading to the phenomenon of *dispersion*. This model describes atoms as damped harmonic oscillators, accounting for both absorption (represented by the imaginary part of the complex permittivity and refractive index) and the frequency-dependent speed of light in the medium. Concepts like *phase velocity* and *group velocity* are clearly distinguished, explaining how wave packets deform as they propagate through dispersive media. *Cauchy's law* is presented as an empirical relation for normal dispersion. A significant portion of the chapter is dedicated to the *propagation of waves in dielectric media*. *Lorentz model* is introduced to explain the *frequency dependence of permittivity* and, consequently, the *refractive index*, leading to the phenomenon of *dispersion*. This model describes atoms as damped harmonic oscillators, accounting for both absorption (represented by the imaginary part of the complex permittivity and refractive index) and the frequency-dependent speed of light in the medium. Concepts like *phase velocity* and *group velocity* are clearly distinguished, explaining how wave packets deform as they propagate through dispersive media. *Cauchy's law* is presented as an empirical relation for normal dispersion. The chapter then examines the propagation of *electromagnetic waves in conductors*. Using the *Drude–Lorentz model*, it derives the *complex conductivity* and demonstrates how it leads to a complex wave number, resulting in the exponential decay of wave amplitude within the conductor (the *skin effect*). The relationship between effective permittivity and conductivity is explored, highlighting the unified nature of charge response in both dielectrics and metals. Finally, the chapter provides an overview of practical control of *light polarization* using optical components. *Linear polarizers* are discussed, along with *Malus's Law*, which quantifies the intensity of transmitted light through a polarizer. *Wave plates* (half-wave and quarter-wave plates) are introduced as devices

F. Cadiz and A. Couairon, *Classical Electrodynamics*, Undergraduate Texts in Physics, https://doi.org/10.1007/978-3-031-86785-9_14

that manipulate the phase difference between orthogonal polarization components, enabling transformations between linear, circular, and elliptical polarization states.

Keywords Dispersion · Refractive index · Waveplates · Propagation · Group and phase velocity

14.1 Introduction

In this chapter, we derive Maxwell's equations in matter. We demonstrate that the propagation of waves in a dielectric medium depends on the refractive index, for which we construct a simple Lorentz model to explain its frequency dependence. We define the concepts of phase and group velocities and analyze the effects of dispersion, arising from the wavelength dependence of the refractive index. Next, we discuss the propagation properties of electromagnetic waves in conductors. Finally, we explore how the polarization of light can be controlled using polarizers and waveplates.

14.2 Displacement Current and Maxwell's Equations in Matter

Let us recall that a dielectric medium is characterized by the polarization density vector $\mathbf{P}$ which corresponds to the average dipole moment per unit volume. The polarization charge density writes $\varrho_P = -\nabla \cdot \mathbf{P}$ and must be added to the free charge density ϱ in Gauss's law for the spatially averaged electric field

$$\nabla \cdot \mathbf{E} = \frac{\varrho + \rho_P}{\epsilon_0} = \frac{\varrho}{\epsilon_0} - \frac{1}{\epsilon_0}\nabla \cdot \mathbf{P}$$

and defining the electric displacement field $\mathbf{D} = \epsilon_0\mathbf{E} + \mathbf{P}$, Gauss's law in a dielectric writes

$$\nabla \cdot \mathbf{D} = \varrho$$

Similarly, a magnetized media is characterized by the magnetization vector $\mathbf{M}$ which corresponds to the average magnetic moment per unit volume. The magnetization makes a contribution to the current density via the magnetization current density $\mathbf{j}_M = \nabla \times \mathbf{M}$ that must be added to the free current density $\mathbf{j}$ so that Ampère's law for the average magnetic field writes

$$\nabla \times \mathbf{B} = \mu_0\mathbf{j} + \mu_0\mathbf{j}_M = \mu_0\mathbf{j} + \mu_0\nabla \times \mathbf{M}$$

and defining the the auxiliary magnetic field $\mathbf{H} = \frac{\mathbf{B}}{\mu_0} - \mathbf{M}$, this can be rewritten as:

$$\nabla \times \mathbf{H} = \mathbf{j}$$

We now know that Ampère's law is incomplete and that we must include the displacement current $\mathbf{j}_D = \epsilon_0 \dfrac{\partial \mathbf{E}}{\partial t}$ to the right-hand term:

$$\nabla \times \mathbf{H} = \mathbf{j} + \epsilon_0 \frac{\partial \mathbf{E}}{\partial t} \ .$$

However, this last equation is not yet complete since it does not take into account the polarization charge density $\mathbf{j}_P$ that exists in a dielectric in a time-dependent regime. Indeed, the polarization density $\mathbf{P}$ in a dielectric will depend on time in response to a time-varying electric field. The oscillation of the polarization charges will also contribute to the total current density via the polarization current density $\mathbf{j}_P$ given by[1]

$$\boxed{\mathbf{j}_P = \frac{\partial \mathbf{P}}{\partial t}.} \tag{14.1}$$

Recall that the polarization charge density is $\varrho_P = -\nabla \cdot \mathbf{P}$, and so the conservation law for the polarization charge is verified

$$\frac{\partial \varrho_P}{\partial t} + \nabla \cdot \mathbf{j}_P = -\frac{\partial \nabla \cdot \mathbf{P}}{\partial t} + \nabla \cdot \frac{\partial \mathbf{P}}{\partial t} = 0$$

This current $\mathbf{j}_P$ must be added to the free current density $\mathbf{j}$ and to the displacement current $\mathbf{j}_D$ in Maxwell–Ampère equation so that in a magnetized dielectric medium, the complete equation becomes

$$\nabla \times \mathbf{H} = \mathbf{j} + \underbrace{\frac{\partial \mathbf{P}}{\partial t}}_{\mathbf{j}_P} + \underbrace{\epsilon_0 \frac{\partial \mathbf{E}}{\partial t}}_{\mathbf{j}_D}$$

and recalling that $\mathbf{P} + \epsilon_0 \mathbf{E} = \mathbf{D}$, finally Ampère–Maxwell's equation in matter reads

$$\boxed{\nabla \times \mathbf{H} = \mathbf{j} + \frac{\partial \mathbf{D}}{\partial t}} \tag{14.2}$$

In a dielectric we can therefore define the total displacement current $\mathbf{j}_D = \dfrac{\partial \mathbf{D}}{\partial t}$ which contains contributions from both the actual current $\mathbf{j}_P$ associated with the movement of bound charges, and the displacement current in vacuum $\mathbf{j}_D = \epsilon_0 \partial \mathbf{E}/\partial t$. Note finally that (14.2) is compatible with the conservation law of free charge, since

---

[1] Consider a collection of charges q_i at positions $\mathbf{x}_i$ around a fixed atom. The dipole moment writes $\mathbf{p} = \sum_i q_i \mathbf{x}_i$. We find that $\frac{\partial \mathbf{p}}{\partial t} = \sum_i q_i \mathbf{v}_i = \mathbf{j}_p$, with $\mathbf{j}_p$ the total current density of the atom.

$$\nabla \cdot (\nabla \times \mathbf{H}) = 0 = \nabla \cdot \mathbf{j} + \frac{\partial}{\partial t} \nabla \cdot \mathbf{D}$$

and from Gauss's law $\nabla \cdot \mathbf{D} = \varrho$ and we retrieve

$$\frac{\partial \varrho}{\partial t} + \nabla \cdot \mathbf{j} = 0 \ .$$

Maxwell's Equations in Matter

Maxwell's equations in matter are given by

$$
\boxed{
\begin{aligned}
\nabla \cdot \mathbf{D} &= \varrho && \text{Maxwell–Gauss} \\
\nabla \cdot \mathbf{B} &= 0 && \text{Maxwell–Thomson} \\
\nabla \times \mathbf{E} &= -\frac{\partial \mathbf{B}}{\partial t} && \text{Maxwell–Faraday} \\
\nabla \times \mathbf{H} &= \mathbf{j} + \frac{\partial \mathbf{D}}{\partial t} && \text{Maxwell–Ampère}
\end{aligned}
}
\tag{14.3}
$$

where the auxiliary fields are given by

$$\mathbf{D} = \epsilon_0 \mathbf{E} + \mathbf{P}$$

$$\mathbf{H} = \frac{\mathbf{B}}{\mu_0} - \mathbf{M}$$

The behavior of the different types of materials is thus determined by the so-called constitutive relations

$$\mathbf{P}(\mathbf{E}) \quad \text{and} \quad \mathbf{M}(\mathbf{B})$$

- From now on we will restrict to the propagation in linear, homogeneous and isotropic materials for which $\mathbf{M} = \chi_m \mathbf{H}$ so that $\mathbf{B} = \mu \mathbf{H}$ with μ the magnetic permeability.
- Similarly in linear, homogeneous and isotropic dielectrics we may write:

$$\mathbf{D} = \epsilon \mathbf{E}$$

where the permittivity ϵ is a scalar quantity.
- A conductor is characterized by its conductivity, according to Ohm's law

$$\mathbf{j} = \sigma \mathbf{E}$$

- These constitutive relations where derived in the static case, and assuming their validity in a time-dependent regime would imply an instantaneous response of the medium to the variations of the fields. In reality, the inertia and friction of charges in matter imposes a delay between the cause (for example **E**) and the effects (**D** and **j**). This phenomenom, called dispersion, will be discussed later and is detailed in the Appendix A.10.

14.2.1 Poynting's Theorem in Matter

From Maxwell's equations in matter (14.3) it is possible to generalize Poynting's theorem that was previously demonstrated in vacuum. Taking the scalar product of Ampère–Maxwell's law with **E** on both sides of the equation:

$$\mathbf{E} \cdot (\nabla \times \mathbf{H}) = \mathbf{j} \cdot \mathbf{E} + \mathbf{E} \cdot \frac{\partial \mathbf{D}}{\partial t}. \tag{14.4}$$

Similarly, we use Maxwell–Faraday's law and we take the scalar product with **H**:

$$\mathbf{H} \cdot (\nabla \times \mathbf{E}) = -\mathbf{H} \cdot \frac{\partial \mathbf{B}}{\partial t}. \tag{14.5}$$

Subtracting (13.18) from (13.19), we obtain

$$\underbrace{\mathbf{E} \cdot (\nabla \times \mathbf{H}) - \mathbf{H} \cdot (\nabla \times \mathbf{E})}_{-\nabla \cdot (\mathbf{E} \times \mathbf{H})} = \mathbf{j} \cdot \mathbf{E} + \mathbf{E} \cdot \frac{\partial \mathbf{D}}{\partial t} + \mathbf{H} \cdot \frac{\partial \mathbf{B}}{\partial t} \tag{14.6}$$

and we recognize an equation which is of the form of a local conservation equation

$$\frac{\partial f}{\partial t} + \nabla \cdot (\mathbf{E} \times \mathbf{H}) = -\mathbf{j} \cdot \mathbf{E} \, ,$$

if we remember the expressions for the differential of the density of electric and magnetic free energy

$$df = \underbrace{\mathbf{E} \cdot d\mathbf{D}}_{\text{electric}} + \underbrace{\mathbf{H} \cdot d\mathbf{B}}_{\text{magnetic}} \, .$$

Now for a linear, homogeneous and isotropic medium, the equations of state $\mathbf{D} = \epsilon \mathbf{E}$ and $\mathbf{B} = \mu \mathbf{H}$ allow us to recognize that the free energy density is nothing but the density of electromagnetic energy

$$f = u_{EM} = \epsilon \frac{\mathbf{E}^2}{2} + \frac{\mathbf{B}^2}{2\mu} \, .$$

Integrating $df = \epsilon \mathbf{E} \cdot d\mathbf{E} + \mu \mathbf{H} \cdot d\mathbf{H}$, we find indeed

$$f = \epsilon \frac{\mathbf{E}^2}{2} + \mu \frac{\mathbf{H}^2}{2} = \frac{1}{2}\left(\mathbf{E} \cdot \mathbf{D} + \mathbf{B} \cdot \mathbf{H}\right) ,$$

and we retrieve a local conservation equation similar to that of Eq. (13.17) for the electromagnetic energy in vacuum.

Poynting's Theorem in Matter
Poynting's theorem in matter writes

$$\frac{\partial}{\partial t} u_{EM} + \nabla \cdot \mathbf{\Pi} = -\mathbf{j} \cdot \mathbf{E} \qquad (14.7)$$

where Poynting's vector is given by

$$\mathbf{\Pi} = \mathbf{E} \times \mathbf{H} \qquad (14.8)$$

and the density of electromagnetic energy is

$$u_{EM} = \frac{1}{2}\left(\mathbf{E} \cdot \mathbf{D} + \mathbf{B} \cdot \mathbf{H}\right) . \qquad (14.9)$$

14.3   Propagation of Waves in Dielectric Media

Let us now consider the case of a non-magnetic ($\mu = \mu_0$), linear homogeneous and isotropic dielectric medium in the absence of free charges. Assuming that we may write $\mathbf{D}(\mathbf{x}, t) = \epsilon \mathbf{E}(\mathbf{x}, t)$, it is easy to verify that the propagation equation for the electric and magnetic field becomes[2]

$$\boxed{\begin{aligned} \nabla^2 \mathbf{E} - \epsilon \mu_0 \frac{\partial^2 \mathbf{E}}{\partial t^2} &= 0, \\ \nabla^2 \mathbf{B} - \epsilon \mu_0 \frac{\partial^2 \mathbf{B}}{\partial t^2} &= 0. \end{aligned}} \qquad (14.10)$$

At a first glance, electromagnetic waves propagate in a dielectric just like in vacuum, except with a smaller velocity given by

---

[2] The homogeneity of the medium is crucial here, since otherwise a non-zero polarization charge density $\varrho_P = -\nabla \cdot \mathbf{P} = -\nabla \cdot ((\epsilon - \epsilon_0)\mathbf{E})$ would need to be included in the right-hand side term.

$$v = \frac{1}{\sqrt{\epsilon \mu_0}} = \frac{c}{\sqrt{\epsilon_r}} = \frac{c}{n}$$

where $n = \sqrt{\epsilon_r} > 1$ is the refractive index of the medium and $\epsilon_r = \epsilon/\epsilon_0$ the dielectric constant. All the properties discussed so far for electromagnetic waves in vacuum still hold if we renormalize the speed of light. For example, a plane wave $\underline{\mathbf{E}} = \mathbf{E}_0 e^{i(kx-\omega t)}$ is a solution of Eq. (14.10) with the following dispersion relation

$$k = n\frac{\omega}{c} = \frac{\omega}{v}$$

As we will see, this assumption is justified when dealing with quasi-monochromatic waves (few frequency components) and far from any atomic resonance. The complete picture is, however, slightly more complicated: all charges (free and bound) have inertia and cannot adapt instantaneously to the variations of the oscillating electric field, so that $\mathbf{D}(\mathbf{x}, t) \neq \epsilon \mathbf{E}(\mathbf{x}, t)$. A natural consequence of this is that the permittivity ϵ depends on the frequency ω of the electromagnetic wave, leading to the phenomenom called dispersion. This explains for example the dispersion of sunlight by water droplets which is at the origin of rainbows. Dispersion will be discussed in detail in 14.4.2.

Additionally, a significant fraction of the electromagnetic energy may be absorbed if its frequency ω is close to the resonance frequencies of the atoms or molecules that constitute the medium, or more generally, if it is larger than the frequencies capable of inducing interband transitions in solids. We have all experienced this: the semiconducting materials in our cellphones absorb visible light and this makes possible to take pictures.

Absorption is actually simple to take into account when working with the complex representation of the fields. In the most general case, the response to a complex plane wave $\underline{\mathbf{E}} = \mathbf{E}_0 e^{i(kx-\omega t)}$ will lead to a frequency-dependent, complex permittivity (a explicit model for this frequency dependence will be developed in 14.3.2

$$\underline{\epsilon}(\omega) = \epsilon_R(\omega) + i\epsilon_I(\omega)$$

so that in the absence of free charges and currents, the propagation Eq. (14.10) for a sinusoidal plane wave $\underline{\mathbf{E}} = \underline{\mathbf{E}}_0 e^{i(\underline{k}x-\omega t)}$ gives the following dispersion relation

$$\underline{k}^2 = \frac{\omega^2}{c^2}\frac{1}{\epsilon_0}(\epsilon_R(\omega) + i\epsilon_I(\omega))$$

or, in terms of the complex refractive index $\underline{n} = n_R + in_I = \sqrt{\frac{\epsilon}{\epsilon_0}}$

$$\underline{k} = \frac{\omega}{c}(n_R(\omega) + in_I(\omega)) = k_R + ik_I$$

where the imaginary part of the wave vector's amplitude characterizes the absorption of the electromagnetic wave by matter. Indeed, replacing $\underline{k}$ in the complex plane wave

$\underline{\mathbf{E}} = \underline{\mathbf{E}}_0 e^{i(\underline{k}x - \omega t)}$ and taking the real part, we have

$$\mathbf{E} = \mathbf{E}_0 \cos(k_R x - \omega t) e^{-k_I x} \tag{14.11}$$

so that the amplitude of the wave decays exponentially with a typical decay length $\delta = \frac{1}{k_I}$.

14.3.1   Propagation with Absorption

The magnetic field associated with the electric field (14.11) can be obtained in its complex representation with the help of Eq. (13.15)

$$\underline{\mathbf{B}} = \frac{\underline{k}}{\omega} \mathbf{u}_x \times \underline{\mathbf{E}}_0 e^{i(k_R x - \omega t)} e^{-k_I x}$$

and writing $\underline{k} = |k| e^{-i\phi}$, the real magnetic field reads

$$\mathbf{B} = \frac{|k|}{\omega} \cos(k_R x - \omega t - \phi) e^{-k_I x} (\mathbf{u}_x \times \mathbf{E}_0)$$

so that both fields do not oscillate in phase in the presence of absorption ($\phi \neq 0$ if $k_I \neq 0$). The Poynting vector is given by

$$\boldsymbol{\Pi} = \frac{\mathbf{E} \times \mathbf{B}}{\mu_0} = \frac{|\mathbf{E}_0|^2 |k|}{\mu_0 \omega} \underbrace{\cos(k_R x - \omega t - \phi) \cos(k_R x - \omega t)}_{\frac{1}{2}(\cos(2(k_R x - \omega t) - \phi) + \cos\phi)} e^{-2k_I x} \mathbf{u}_x$$

whose time average reads

$$\langle \boldsymbol{\Pi} \rangle_T = \frac{E_0^2}{2\mu_0 \omega} \underbrace{|k| \cos\phi}_{k_R} e^{-2k_I x} \mathbf{u}_x = \underbrace{\frac{|\mathbf{E}_0|^2 k_R}{2\mu_0 \omega} e^{-2k_I x}}_{I} \mathbf{u}_x \; .$$

Note that the same result may be easily obtained by using the complex Poynting vector $\underline{\boldsymbol{\Pi}}^* = \frac{\underline{\mathbf{E}} \times \underline{\mathbf{B}}^*}{2\mu_0}$, and the result (13.25) still holds for a plane wave in an absorptive medium

$$\underline{\boldsymbol{\Pi}} = \frac{\underline{\mathbf{E}} \times \underline{\mathbf{B}}^*}{2\mu_0} = \frac{1}{2\mu_0 \omega} \underline{\mathbf{E}} \times (\underline{k}^* \mathbf{u}_x \times \underline{\mathbf{E}}) = \frac{\underline{k}^*}{2\mu_0 \omega} (\underline{\mathbf{E}} \cdot \underline{\mathbf{E}}^*) \mathbf{u}_x = \frac{\underline{k}^* |\mathbf{E}_0|^2}{2\mu_0 \omega} e^{-2k_I x} \mathbf{u}_x$$

and

$$\langle \boldsymbol{\Pi} \rangle_T = \mathrm{Re}\{\underline{\boldsymbol{\Pi}}\} = \frac{k_R |\mathbf{E}_0|^2}{2\mu_0 \omega} e^{-2k_I x} \mathbf{u}_x$$

The intensity of the wave, which is proportional to $\langle |\mathbf{E}^2| \rangle$, decays therefore as

$$\mathcal{I}(x) = \frac{\epsilon_0 |\mathbf{E}_0|^2 n_R c}{2} e^{-2k_I x} = I_0 e^{-\alpha x}$$

where $\alpha = 2k_I$ (m^{-1}) is called absorption coefficient. The fraction T of light transmitted through a medium of thickness d follows therefore Beer–Lambert's law

$$T = e^{-\alpha d} \; .$$

In a metal with finite conductivity, a similar law applies, as will be detailed later.

14.3.2   *Lorentz Model for the Classical Atom (1878)*

We will now establish a model for the frequency-dependence of the dielectric constant. Let us recall that in an atom, under the presence of an electric field, the electron cloud orbiting the nucleus will be shifted from its equilibrium position by a quantity $\mathbf{x}$. In consequence an atomic electric dipole $\mathbf{p}$ develops, as shown in Fig. 14.1. Here, we neglect the movement of the nucleus in the electric field due to its much larger mass compared to that of the electron.

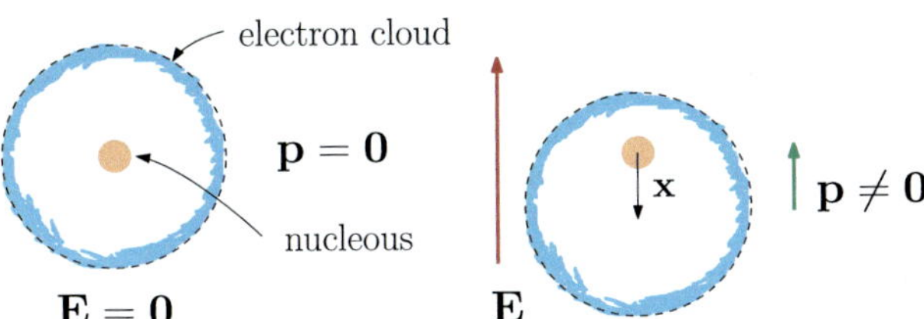

Fig. 14.1 In an atom, a dipole moment appears when the electron cloud is shifted by an external electric field

Fig. 14.2 The atomic dipole will oscillate periodically in the presence of a sinusoidal electromagnetic wave

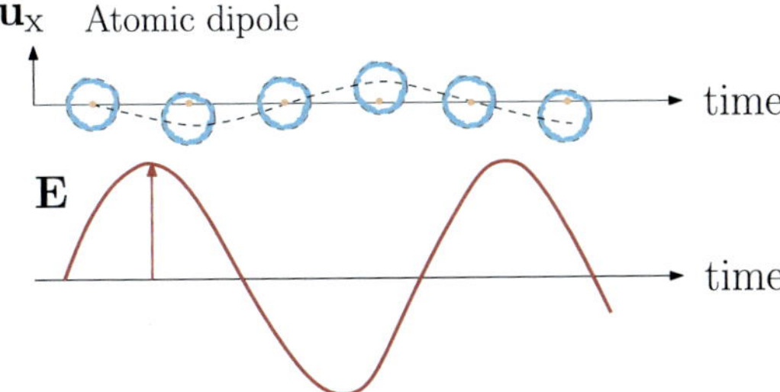

In the presence of an electromagnetic wave, the electric field oscillates periodically, and so does the induced dipole, as shown in Fig. 14.2. We expect that if the frequency ω of the wave is very low, the dipole will instantaneously adapt to the electric field and therefore oscillates in phase with the latter such that $\lim_{\omega \to 0} \epsilon(\omega) = \epsilon$, where ϵ is the static permittivity. In contrast, if the frequency is too high, the inertia of the electron cloud will be such that it will not be able to follow the rapid oscillations of the electric field, and the medium will become transparent. We then expect the permittivity to tend asymptotically to ϵ_0 as $\omega \to 0$, that is $\lim_{\omega \to \infty} \epsilon(\omega) = \epsilon_0$ and waves of very high frequency will propagate as if they were in vacuum.

The Lorentz model take into account two crucial ingredients: firstly, since an oscillating dipole generates in fact an electric current, it creates its own electromagnetic wave. This means that a dipole driven at a frequency ω will loose part of its mechanical energy by emitting radiation at a frequency close to ω. This is called scattering and will be discussed in Chap. 19.

Secondly, we know from experiments that electrons in atoms have discrete energy levels and as such they absorb or emit light at precise resonant frequencies, a fact that arises from the quantum behavior of matter at atomic scales. Lorentz model is a classical model that incorporates these two effects: the movement of an electron around the nucleus is modeled by a damped harmonic oscillator with a natural frequency ω_0. The equation of motion writes

$$m_e \frac{d^2\mathbf{x}}{dt^2} = -e\mathbf{E} - m_e\gamma\frac{d\mathbf{x}}{dt} - m_e\omega_0^2\mathbf{x}$$

where $\mathbf{x}$ is the displacement of the electron with respect to the equilibrium position around the nucleus and γ is a damping factor that accounts for the fact that energy is absorbed from the incident electromagnetic wave and re-emitted by the electron (in an atom γ is typically of the order of 10^8 s^{-1}). Here we have neglected the magnetic component of the Lorentz force since, for a plane electromagnetic wave and for velocities much smaller than the speed of light,

$$|-e\mathbf{v} \times \mathbf{B}| = e|\mathbf{v}||\mathbf{B}| = e\frac{|\mathbf{v}|}{c}|\mathbf{E}| \ll e|\mathbf{E}| \ .$$

Assuming now a sinusoidal plane wave $\underline{\mathbf{E}} = \underline{\mathbf{E}}_0 e^{i(\mathbf{k}\cdot\mathbf{x} - \omega t)}$ with a wavelength much longer than the oscillation amplitude of the atomic dipole, we may consider the field to be spatially uniform at the dipole's position and look for a steady-state solution on the form of a forced oscillator $\underline{\mathbf{x}} = \underline{\mathbf{x}}_0 e^{-i\omega t}$ so we have:

$$-m_e\omega^2\underline{\mathbf{x}} = -e\underline{\mathbf{E}} + i\omega m_e\gamma\underline{\mathbf{x}} - m_e\omega_0^2\underline{\mathbf{x}} \ .$$

Then

$$\underline{\mathbf{x}} = \frac{e}{m_e} \frac{\underline{\mathbf{E}}}{\omega^2 - \omega_0^2 + i\gamma\omega}. \tag{14.12}$$

The squared amplitude of the electron oscillation then reads

$$|\underline{\mathbf{x}_0}|^2 = \frac{e^2}{m_e^2} \frac{|\underline{\mathbf{E}_0}|^2}{(\omega^2 - \omega_0^2) + \gamma^2\omega^2}$$

and exhibits a resonance at $\omega = \omega_0$ for which the amplitude is very large (this would correspond to a transition in quantum mechanics). The width of the resonance is given by γ which, for an isolated atom ($\gamma = 10^8$ s^{-1}) and a visible electromagnetic wave ($\omega_0 \sim 3 \times 10^{15}$) is such that $\gamma/\omega_0 \sim 10^{-8}$. In that case the amplitude as a function of frequency has approximately a Cauchy–Lorentzian shape as shown in Fig. 14.3.

Remarks

- The damping frequency γ is related to the typical time $\tau = 1/\gamma$ that it takes for an oscillator amplitude to decay to zero, and it is related to the energy lost in emitted radiation by the oscillating dipole. In quantum mechanics, τ has a different interpretation: an electron that is initially in an excited state will spontaneously decay to the lowest energy state within a time τ, emitting a wave packet of central frequency ω_0 and of typical duration τ. The frequency of the emitted wave is related to the difference ΔE in energy between the excited and lowest energy state by $\Delta E = \hbar\omega_0$ where $\hbar$ is the reduced Planck constant. This is illustrated in Fig. 14.4.
- This simple model, although representing an atom, can be extended to describe more complex systems such as molecules (whose vibrational modes are quantized as well and therefore exhibit resonant frequencies) or crystals. As such, gas, liquid or solid media can be modeled by an assembly of harmonic oscillators. Typically, the linewidth γ of the resonances in matter is dominated by collisions between the atoms and not by spontaneous emission, as in the case of isolated atoms.

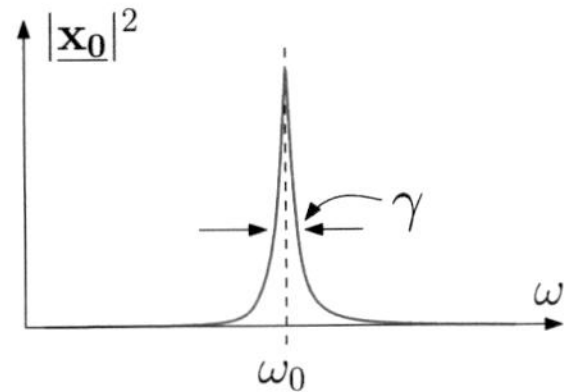

Fig. 14.3 The electron's oscillation amplitude as a function of the incident wave's angular frequency

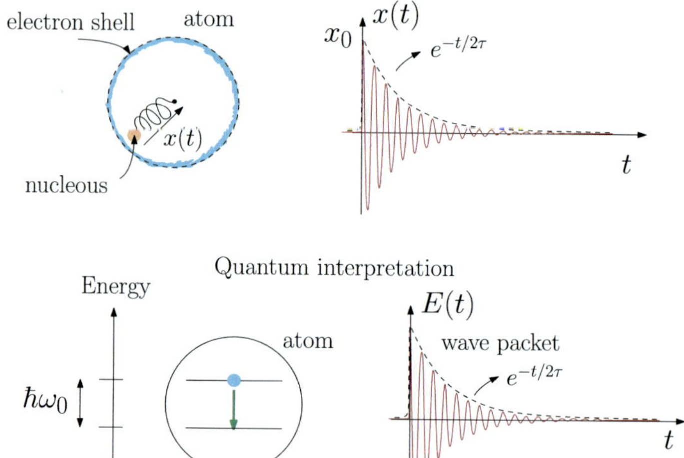

Fig. 14.4 In classical physics, the atom is a damped harmonic oscillator with damping frequency $1/\tau$. In quantum physics, the energy leves are discrete, and a transition between an excited state and the lowest energy state leads to the spontaneous emission of a light wave whose intensity has a duration τ. In both cases, τ is related to the time it takes for the system to relax its energy

14.3.3   The Complex Permittivity

If n is the density of atomic (or molecular) dipoles in the medium, each one having Z electrons and a single resonance ω_0, the polarization vector writes

$$\underline{\mathbf{P}} = n(-Ze\underline{\mathbf{x}}) = \frac{-nZe^2}{m_e}\,\frac{\underline{\mathbf{E}}}{\omega^2 - \omega_0^2 + i\gamma\omega} = (\underline{\epsilon}(\omega) - \epsilon_0)\underline{\mathbf{E}}$$

so that the complex permittivity is given by

$$\underline{\epsilon}(\omega) = \epsilon_0\left(1 + \frac{\omega_p^2}{\omega_0^2 - \omega^2 - i\gamma\omega}\right) \tag{14.13}$$

where $\omega_p = \sqrt{Zne^2/m_e\epsilon_0}$ is called the plasma frequency. Separating explicitly the real and imaginary parts of the permittivity:

$$\underline{\epsilon}(\omega) = \epsilon_0\left(1 + \frac{\omega_p^2(\omega_0^2 - \omega^2)}{(\omega_0^2 - \omega^2)^2 + \gamma^2\omega^2} + i\,\frac{\omega_p^2\gamma\omega}{(\omega_0^2 - \omega^2)^2 + \gamma^2\omega^2}\right). \tag{14.14}$$

Note that in the limit of very high frequencies the medium becomes transparent, $\lim_{\omega \to \infty} \epsilon(\omega) = \epsilon_0$, whereas for very low frequencies we get an expression for the static permittivity

$$\lim_{\omega \to 0} \epsilon(\omega) = \epsilon_0 \left(1 + \frac{\omega_p^2}{\omega_0^2} \right) = \epsilon .$$

14.3.4 Dielectric Heating

Given the sinusoidal plane wave $\underline{\mathbf{E}} = \underline{\mathbf{E}}_0 e^{i(kx - \omega t)}$, the associated displacement vector $\underline{\mathbf{D}} = \underline{\mathbf{D}}_0 e^{i(kx - \omega t)}$ in a linear, homogeneous and isotropic dielectric reads

$$\underline{\mathbf{D}} = \underline{\epsilon}(\omega)\underline{\mathbf{E}} .$$

Thus, the fields $\mathbf{E}$ and $\mathbf{D}$ are in general not in phase since their complex amplitudes satisfy

$$\underline{\mathbf{D}}_0 = \underline{\epsilon}(\omega)\underline{\mathbf{E}}_0 = \left| \underline{\epsilon}(\omega) \right| e^{i\delta} \underline{\mathbf{E}}_0$$

where δ, called the loss angle, corresponds to $\delta = \epsilon_I(\omega)/\epsilon_R(\omega)$. The real notations give (assuming $\underline{\mathbf{E}}_0 \in \mathbb{R}$):

$$\mathbf{E} = \mathbf{E}_0 \cos(kx - \omega t) \quad \text{and} \quad \mathbf{D} = \left| \underline{\epsilon}(\omega) \right| \mathbf{E}_0 \cos(kx - \omega t + \delta) .$$

The work required for the displacement vector to vary from $\mathbf{D}$ to $\mathbf{D} + d\mathbf{D}$ during dt corresponds to the free energy change, hence, per unit volume

$$\delta W = \mathbf{E} \cdot d\mathbf{D} = \mathbf{E} \cdot \frac{d\mathbf{D}}{dt} dt .$$

We obtain the work done over one period T of the electric field by integration. Writing the result as $W = pT$ yields the average power per unit volume, p, delivered to the dielectric medium. The power per unit volume received by the dielectric is

$$p(t) = \mathbf{E} \cdot \frac{d\mathbf{D}}{dt} ,$$

leading to the average power density:

$$p_a = \frac{1}{T} \int_0^T \left| \underline{\epsilon}(\omega) \right| |\mathbf{E}_0|^2 \omega (\sin(kx - \omega t + \delta) \cos(kx - \omega t)) dt = \frac{1}{2} \left| \underline{\epsilon}(\omega) \right| |\mathbf{E}_0|^2 \omega \sin \delta .$$

In order to obtain the same result using complex notations rather than real fields, we have to use the expression for the complex power density:

$$\boxed{\mathcal{P} = \frac{1}{2}\underline{\mathbf{E}} \cdot \left(\frac{d\underline{\mathbf{D}}}{dt}\right)^{*}}$$

where the factor $1/2$ corresponds to the time average of the sinusoidal field over one period. Performing the calculation of the complex power density leads to $(d\underline{\mathbf{D}}/dt)^{*} = (-i\omega\underline{\mathbf{D}})^{*} = i\omega\underline{\mathbf{D}}^{*}$ and $\mathcal{P} = \frac{1}{2}\underline{\mathbf{E}}_0 \cdot i\omega\left|\underline{\epsilon}(\omega)\right| e^{-i\delta}\underline{\mathbf{E}}_0^{*}$. The average power density delivered to the dielectric medium is then the real part of the complex power density:

$$p = \mathrm{Re}\{\mathcal{P}\} = \frac{1}{2}\left|\underline{\epsilon}(\omega)\right| |\mathbf{E}_0|^2 \omega \sin\delta = \frac{1}{2}|\mathbf{E}_0|^2 \omega\epsilon_I(\omega) \ .$$

In conclusion, the imaginary part of the dielectric constant of a dielectric is linked to the power absorbed by the medium. After each cycle, the dielectric medium comes back to its initial electric state. The average power delivered to the dielectric leads to either an increase of its temperature if it is thermally isolated, or to heat radiation for an isothermal dielectric medium.

Heating of the dielectric medium by this process is efficient if the frequency is chosen close to the resonance frequency, that is, the best operation conditions correspond to a frequency close to the frequency for which $\epsilon_I(\omega)$ is maximum. This corresponds typically to the microwave frequency range $5\,\mathrm{MHz} \leq \omega \leq 40\,\mathrm{MHz}$, and electric fields up to $250\,\mathrm{kV/m}$.

Heating with the microwave oven is based on this process: such an oven cooks food because the water molecules inside it absorb the microwave radiation and thereby heat up and heat the surrounding food. The electric field delivered by the microwave oven causes water molecules to vibrate, which generates intermolecular friction between the molecules of the food. The increased friction between the molecules results in heat. Microwave radiation will similarly heat up skin and other body parts. The radiation is harmful mostly to the parts of the body that cannot conduct the heat away very effectively—the eyes especially. If they were able to escape from the microwave oven, microwaves could affect your tissue. This is why modern microwave ovens are designed to allow essentially no leakage of microwaves. For instance, a metallic grid covers the transparent door window, acting as a Faraday cage preventing radiation leakage. The only time for concern would be if the door is broken or damaged, in which case the oven should not be used.

14.3.5 The Complex Refractive Index

Finally, the complex refraction index $\underline{n}$, which fully characterizes the propagation of a wave in a medium, is complex whenever $\underline{\epsilon}(\omega)$ has a non-negligible imaginary part (i.e., close to resonant frequencies). In this case it writes

$$\underline{n}(\omega) = \sqrt{\frac{\epsilon(\omega)}{\epsilon_0}} = n_R(\omega) + in_I(\omega) \, .$$

For a diluted medium, one has $\omega_p \ll \omega_0$ and we can use the following expansion

$$\sqrt{\epsilon(\omega)/\epsilon_0} = \left(1 + \frac{\omega_p^2}{\omega_0^2 - \omega^2 - i\gamma\omega}\right)^{1/2} \approx 1 + \frac{1}{2}\frac{\omega_p^2}{\omega_0^2 - \omega^2 - i\gamma\omega}$$

and so

$$\underline{n}(\omega) \approx 1 + \underbrace{\frac{1}{2}\frac{\omega_p^2(\omega_0^2 - \omega^2)}{(\omega_0^2 - \omega^2)^2 + \gamma^2\omega^2}}_{n_R} + i\underbrace{\frac{1}{2}\frac{\omega_p^2\gamma\omega}{(\omega_0^2 - \omega^2)^2 + \gamma^2\omega^2}}_{n_I} \, . \qquad (14.15)$$

A simple schematics for the real and imaginary parts of the refractive index is shown in Fig. 14.5.

Remarks

- Away from any resonance, $n_I \ll n_R$ so that absorption may be neglected and the refractive index may be considered as a pure real number, $n \approx n_R$.
- The very large change of index of refraction n_R close to a resonance will cause a large dispersion (see 14.4.2): the consequence is that different wavelengths travel at very different speeds near the resonance.
- Note that for frequencies slightly above a very strong resonance the real part of the refractive index could be less than one. This does not mean, however, than an electromagnetic wave centered around this frequency will travel faster than c. In reality, any physical wave is represented by a wave packet, that is a superposition of plane waves. The speed at which such a wave packet moves is not given by $\omega/k(\omega)$ but instead by $\partial\omega/\partial k$, also called group velocity.

Fig. 14.5 Behaviour of the real and imaginary part of the refractive index

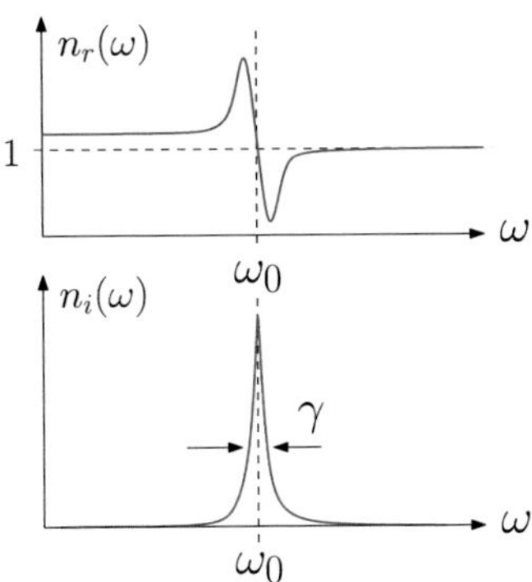

- Finally, in real materials there are several types of resonances distributed in the whole electromagnetic spectrum (vibrations fall in the infrared, electronic resonances in atoms typically in the UV or visible range, etc.). A more realistic model including more than one resonant frequency writes

$$\underline{n}^2(\omega) = 1 + w_p^2 \sum_j \frac{f_j}{w_{0j}^2 - w^2 - i\gamma_j w} \tag{14.16}$$

where f_j, called the oscillator strength, represent the relative weight of the resonance at w_j which can be properly determined within a quantum mechanics framework. When the materials are not diluted, one may refine the model by considering that each atom or molecule will respond to the local field which, for isotropic mediums or cubic crystals, $\mathbf{E}_{\text{local}} = \mathbf{E} + \frac{\mathbf{P}}{3\epsilon_0}$. In that case, (14.16) must be replaced by the generalization of Clausius–Mossotti equation (6.17):

$$3\frac{\underline{n}^2(\omega) - 1}{\underline{n}^2(\omega) + 2} = w_p^2 \sum_j \frac{f_j}{w_{0j}^2 - w^2 - i\gamma_j w} . \tag{14.17}$$

14.4 Propagation in Dielectrics

The propagation equation for a sinusoidal plane wave $\underline{\mathbf{E}} = \underline{\mathbf{E}}_0 e^{i(\mathbf{k}\cdot\mathbf{x} - \omega t)}$ in an homogeneous, linear and isotropic dielectric is given by

$$\nabla^2 \underline{\mathbf{E}} - \frac{\epsilon(\omega)}{\epsilon_0 c^2} \frac{\partial^2 \underline{\mathbf{E}}}{\partial t^2} = \nabla^2 \underline{\mathbf{E}} - \frac{n(\omega)^2}{c^2} \frac{\partial^2 \underline{\mathbf{E}}}{\partial t^2} = 0 .$$

Writing $\mathbf{k} = \underline{k}\mathbf{n}$, with $\mathbf{n}$ the direction of propagation, yields the dispersion relation

$$\boxed{\underline{k} = \underline{n}(\omega)\frac{\omega}{c} = k_R + ik_I} \tag{14.18}$$

with $k_R(\omega) = n_R(\omega)\frac{\omega}{c}$ and $k_I = n_I(\omega)\frac{\omega}{c} > 0$. An electromagnetic wave in the dielectric propagates along x according to

$$\boxed{\underline{\mathbf{E}}(x, t) = \underline{\mathbf{E}}_0 \underbrace{e^{i(k_R x - \omega t)}}_{\text{plane wave, speed } c/n_R} \underbrace{e^{-k_I x}}_{\text{attenuation due to absorption}} .} \tag{14.19}$$

14.4.1 Dispersion, Cauchy's Law

Typical dielectrics used in optics (glass, quartz, plastic, silica) have an index of refraction that is quite transparent (no imaginary part) in the visible range and exhibits a so-called *normal dispersion*, that is an index of refraction that increases at shorter wavelengths (higher frequencies), as shown in Fig. 14.6.

Indeed, away from a resonance ($\omega_0^2 - \omega^2 \gg \gamma\omega$), where there is no absorption, (14.15) can be approximated by

$$\underline{n}(\omega) \approx 1 + \frac{1}{2}\frac{\omega_p^2}{(\omega_0^2 - \omega^2)} \approx 1 + \frac{1}{2}\frac{\omega_p^2}{\omega_0^2}\left(1 + \frac{\omega^2}{\omega_0^2}\right) = n_R(\omega) \tag{14.20}$$

and, in terms of the wavelength in vacuum $\lambda = 2\pi c/\omega$, we obtain Cauchy's law, named after Augustin-Louis Cauchy (Fig. 14.7):

$$n(\lambda) \approx A + \frac{B}{\lambda^2} \tag{14.21}$$

which explains the dispersion of light by a prism, as will be shown in Chap. 17. Normal dispersion implies $n(\text{blue}) > n(\text{red})$ so that blue light is more deviated than red.

14.4.2 Dispersion and Propagation of Wavepackets

We have seen that, quite generally, the refractive index of a medium depends on frequency. This is at the origin of dispersion: a light pulse having different frequency components will be deformed during propagation, since different wavelengths travel at different speeds in matter. From Lorentz model we see that below a resonance $n_R(\omega)$ increases with ω, a regime called normal dispersion and that represents the typical behavior of transparent materials used in optics. This is illustrated in Fig. 14.8.

Fig. 14.6 Refraction index of a typical dielectric used in optics

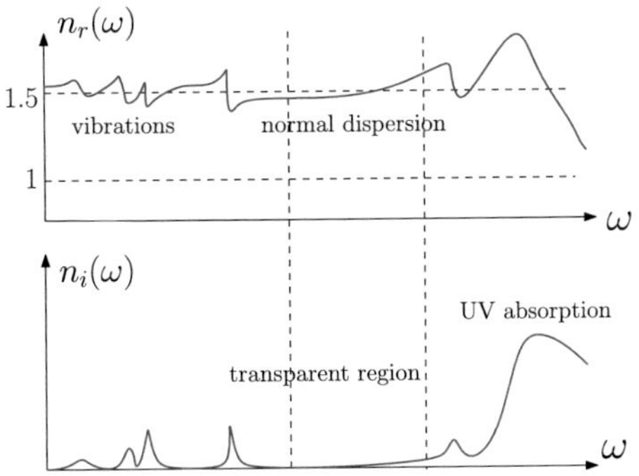

Fig. 14.7 Augustin-Louis Cauchy (1789–1857), a French mathematician and physicist. He made foundational contributions to analysis, algebra, group theory, complex functions, and elasticity. His work laid the groundwork for modern mathematics

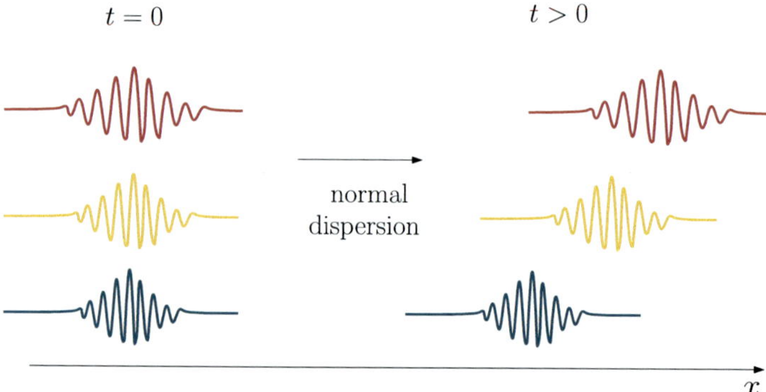

Fig. 14.8 Under normal dispersion, larger wavelengths travel faster than smaller wavelengths

In contrast, above a resonance we have that $n_R(\omega)$ decreases with ω, a regime called anomalous dispersion.

Let us consider first the case in which the wave packet is made of plane waves having wave vectors of magnitude very close to a central value k_0. In this case, the variations of $n(k)$ and therefore $\omega(k)$ are not too large, and we can write up to first order in $k - k_0$:

$$\omega(k) \approx \omega(k_0) + \left(\frac{\partial \omega}{\partial k}\right)_{k_0} (k - k_0) = \omega_0 + v_g (k - k_0) \tag{14.22}$$

where $\omega_0 = \omega(k_0)$ is the central frequency of the wave packet and $v_g = \left(\frac{\partial \omega}{\partial k}\right)_{k_0}$ is called the group velocity. This linear approximation corresponds to the case of a constant refractive index n. It is possible to show that the wave packet propagates without deformation with a velocity equal to the group velocity v_g.

Phase and Group Velocity

The phase velocity v_ϕ is defined as

$$v_\phi = \frac{\omega}{k(\omega)} = \frac{c}{n(\omega)} \tag{14.23}$$

and the group velocity v_g as

$$v_g = \frac{\partial \omega}{\partial k}. \tag{14.24}$$

For a wave packet composed of plane waves centered at frequency ω_0, the phase velocity v_ϕ represents the speed of the crests of the wave, whereas the group velocity v_g is the speed of the envelope, as shown in Fig. 14.9.

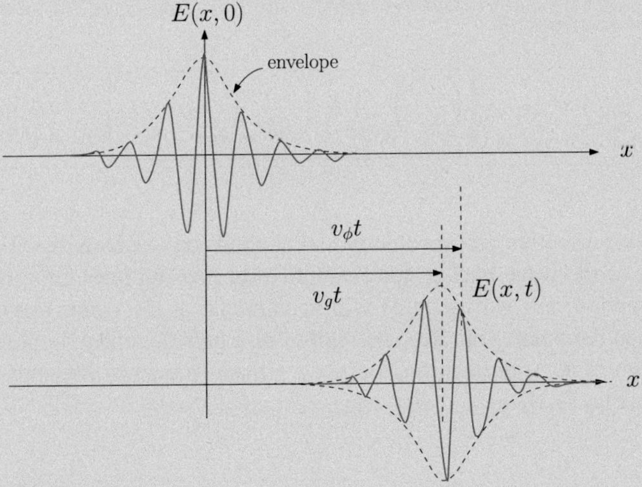

Fig. 14.9 The group velocity is the velocity at which the envelope and intensity of a wavepacket propagates

Note that the velocity of the envelope is the velocity at which the intensity travels. Writing $\omega = ck/n$, we have

$$v_g = \frac{c}{n} + ck\frac{\partial}{\partial k}\left(\frac{1}{n}\right) = \underbrace{\frac{c}{n}}_{v_\phi} + ck\underbrace{\frac{\partial}{\partial \omega}\left(\frac{1}{n}\right)}_{=-\frac{\omega}{n}\frac{\partial n}{\partial \omega}}\underbrace{\frac{\partial \omega}{\partial k}}_{v_g}$$

and so the phase and group velocities are related by

$$v_g = \frac{v_\phi}{1 + \frac{\omega}{n} \frac{\partial n}{\partial \omega}} = \frac{c/n}{1 + \frac{\omega}{n} \frac{\partial n}{\partial \omega}}. \tag{14.25}$$

When the dispersion relation $\omega(k)$ is not linear, which is the case for electromagnetic waves in matter, the linear approximation 14.22 will cease to be valid after the wave-packet has propagated some time. Eventually, the dispersion of the optical index will lead to a deformation of the wave packet. The next term in the expansion would lead to

$$\omega(k) = \omega_0 + \left(\frac{\partial \omega}{\partial k}\right)_{k_0} (k - k_0) + \frac{1}{2} \left(\frac{\partial^2 \omega}{\partial k^2}\right)_{k_0} (k - k_0)^2 + \dots$$

which can be rewritten as

$$\omega(k) \approx \omega_0 + \underbrace{\left\{\left(\frac{\partial \omega}{\partial k}\right)_{k_0} + \frac{1}{2}\left(\frac{\partial^2 \omega}{\partial k^2}\right)_{k_0}(k - k_0)\right\}}_{v_g^{\text{eff}}}(k - k_0)$$

where v_g^{eff} is an effective group velocity that depends on wavelength. This leads to a deformation of the wave packet: for example after passing through a material with normal dispersion, red wavelengths will accumulate at the front, leaving the blue components at the back. This stretched light pulse gets therefore longer in time and can be though of as a monochromatic wave whose frequency depends on time, as illustrated in Fig. 14.10.

14.4.3   An Application of Dispersion that Led to a Nobel Prize

Intense and ultrashort laser pulses, with a duration in the fs range, are difficult to implement since their extremely high peak intensity may damage optical components and specially amplifier crystals. An elegant method to solve this issue is to first stretch

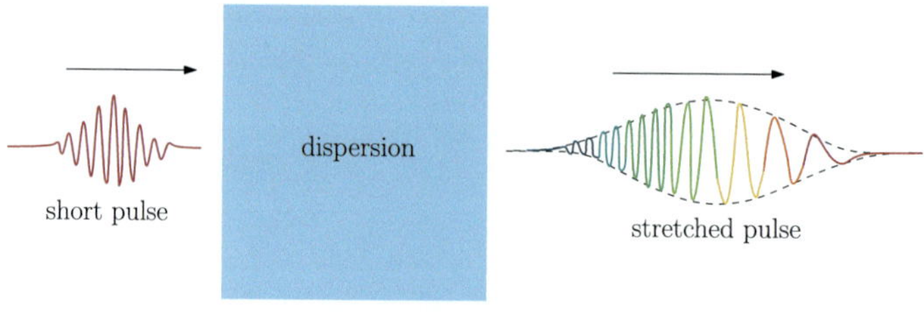

Fig. 14.10 Dispersion leads to the deformation of a wavepacket during propagation

the pulse thanks to a dispersive element, as shown in Fig. 14.10, before going into the amplifier. The pulse is thus stretched from the fs to the ns range, such that high frequencies passes first than the lower frequencies through the amplifier (such a pulse is called chirped). The intensity is therefore spread over a longer time and the peak power is reduced by several orders of magnitude, excluding highly non linear effects and possible damage during the amplification stage. Finally, the pulse is compressed back by inverting the initial dispersion. The physicists who implemented this *chirped pulse amplification* were awarded with the Physics Nobel prize in 2018.

14.5 Electromagnetic Waves in Conductors

Consider the situation in which an electromagnetic wave interacts with a medium in the presence of free charges. This the case of waves in a metal or in a plasma (partially ionized gas), in both cases a density $n \neq 0$ of free electrons is present and a current density $\mathbf{j}$ will appear in response to the electric field of the wave according to Ohm's law. Assuming that the medium preserves its neutrality,[3] $\varrho = 0$ and the equation of propagation (13.8) for the electric field in a conductor becomes

$$\nabla^2 \mathbf{E} - \frac{1}{c^2}\frac{\partial^2}{\partial t^2}\mathbf{E} = \mu_0 \frac{\partial \mathbf{j}}{\partial t}. \tag{14.26}$$

If we assume a sinusoidal plane electromagnetic wave of the form

$$\mathbf{E}(x, t) = \mathrm{Re}\left\{\underline{\mathbf{E}}_0 e^{i(\underline{k}x - \omega t)}\right\} = \mathrm{Re}\left\{\underline{\mathbf{E}}\right\},$$

the current density in its complex representation is given by Ohm's law

$$\underline{\mathbf{j}} = \underline{\sigma}(\omega)\underline{\mathbf{E}}$$

where the conductivity $\underline{\sigma}$ is a function of the wave frequency and in the general case, a complex number. Injecting $\underline{\mathbf{E}}_0 e^{i(\underline{k}x - \omega t)}$ into Eq. (14.26) gives

$$\left(-\underline{k}^2 + \frac{\omega^2}{c^2}\right)\underline{\mathbf{E}} = -\mu_0\underline{\sigma}(\omega)i\omega\underline{\mathbf{E}}$$

and we obtain the following dispersion relation

$$\boxed{\underline{k}^2 = \frac{\omega^2}{c^2} + i\omega\mu_0\underline{\sigma}(\omega).} \tag{14.27}$$

---

[3] That is, the density n of positive ions is the same as that of electrons and the particle distribution is homogeneous.

Again, this dispersion relation leads to a complex wave number $\underline{k} = k_R + ik_I$ so that

$$\underline{\mathbf{E}}(x, t) = \underline{\mathbf{E}}_0 e^{i(k_R x - \omega t)} e^{-ik_I x} .$$

In conclusion, waves in metals and dielectrics take the same general form. From this we see that it is possible to characterize a conductor by an effective permittivity $\underline{\epsilon}$, or equivalently, a complex refractive index $\underline{n} = \sqrt{\underline{\epsilon}/\epsilon_0}$ such that $\underline{k} = \frac{\omega}{c}\underline{n}$ and so

$$\underline{\epsilon}(\omega) = \epsilon_0 \underline{n}^2(\omega) = \epsilon_0 \left(1 - i\frac{\underline{\sigma}(\omega)}{\epsilon_0 \omega} \right) = \epsilon_R(\omega) + i\epsilon_I(\omega) .$$

Conversely, in a dielectric an oscillating polarization vector $\mathbf{P}$ gives rise to a current density $\mathbf{j}_P = \dfrac{\partial \mathbf{P}}{\partial t} = -i\omega(\underline{\epsilon}(\omega) - \epsilon_0)\mathbf{E}$ and so an effective conductivity may be defined as

$$\boxed{\underline{\sigma}(\omega) = -i\omega(\underline{\epsilon}(\omega) - \epsilon_0) .}$$

The fact that we can relate a conductivity to a permittivity in the time-dependent regime is not surprising since, in reality, the distinction between bound and free electrons is somehow arbitrary in the presence of a sinusoidal electromagnetic wave. Both free and bound charges will oscillate around their mean positions generating an oscillating current in both cases. Calculating a conductivity $\underline{\sigma}$ or a permittivity $\underline{\epsilon}$ is therefore only a matter of convention. This, of course, does not mean that metals and dielectrics behave in the same way. A main difference between them comes from the frequency-dependence of the permittivity/conductivity.

14.5.1　Losses in Conductors

According to Poynting's theorem, an electromagnetic wave losses its energy due to the power dissipated in the medium. For a sinusoidal wave, the mean power dissipated per unit volume reads

$$\langle \mathcal{P} \rangle_T = \frac{1}{2}\mathrm{Re}\{\underline{\mathbf{j}} \cdot \underline{\mathbf{E}}^*\} .$$

Using the complex conductivity $\underline{\sigma}$, we find

$$\langle \mathcal{P} \rangle_T = \frac{1}{2}\mathrm{Re}\{\underline{\sigma}(\omega)\}|\underline{\mathbf{E}}|^2 .$$

If instead we use the effective permittivity, $\underline{\mathbf{j}} = -i\omega(\underline{\epsilon}(\omega) - \epsilon_0)\underline{\mathbf{E}}$ and we find:

$$\langle \mathcal{P} \rangle_T = \frac{\omega}{2}\mathrm{Re}\{-i(\underline{\epsilon}(\omega) - \epsilon_0)\}|\underline{\mathbf{E}}|^2 = \frac{\omega}{2}\mathrm{Im}\{\underline{\epsilon}(\omega)\}|\underline{\mathbf{E}}|^2. \tag{14.28}$$

14.5.2 Conductivity in a Time-Dependent Regime: The Drude–Lorentz Model

Now we will derive the equation of motion of an electron of charge $-e$, mass m_e, oscillating around the position x_0 (we assume that the amplitude of the oscillations is negligible with respect to the wavelength of the wave, and so we treat the fields as spatially constants around x_0):

$$m_e \frac{d\mathbf{v}(t)}{dt} = -e\left(\mathbf{E}(x_0, t) + \mathbf{v} \times \mathbf{B}(x_0, t)\right) - \frac{m_e}{\tau}\mathbf{v}$$

where $\mathbf{v}$ is the velocity of the electron and τ is the time between collisions. Assuming that the speed of the electrons is much smaller than c we can neglect the magnetic component of the Lorentz force, and writing $\mathbf{v}(t) = \mathrm{Re}\left\{\underline{\mathbf{v}}_0 e^{-i\omega t}\right\} = \mathrm{Re}\left\{\underline{\mathbf{v}}\right\}$, we get

$$-i\omega m_e \underline{\mathbf{v}} = -e\underline{\mathbf{E}} - \frac{m_e}{\tau}\underline{\mathbf{v}}$$

and so

$$\underline{\mathbf{v}} = \underline{\mathbf{v}}_0 e^{-i\omega t} = \frac{e/m_e}{i\omega - 1/\tau}\underbrace{\underline{\mathbf{E}}_0 e^{i(kx - \omega t)}}_{\underline{\mathbf{E}}} \ .$$

Finally, the current density associated with this oscillatory motion of the electrons is

$$\underline{\mathbf{j}} = -ne\underline{\mathbf{v}} = \frac{ne^2/m_e}{1/\tau - i\omega}\underline{\mathbf{E}}$$

where we have neglected the contribution of the ions, since they are several thousand times heavier than the electrons. Note that from this, the frequency-dependent conductivity reads

$$\boxed{\underline{\sigma}(\omega) = \frac{n^2 e\tau/m_e}{1 - i\omega\tau} = \frac{\sigma_0}{1 - i\omega\tau}} \tag{14.29}$$

where σ_0 is the usual static conductivity. Separating real and imaginary parts and replacing $\underline{\sigma}$ in the dispersion relation Eq. (14.27) yields

$$\boxed{\underline{k}^2 = \frac{\omega^2}{c^2}\left(1 - \frac{\omega_p^2 \tau^2}{1 + \omega^2 \tau^2} + i\frac{\omega_p^2 \tau}{\omega(1 + \omega^2 \tau^2)}\right)} \tag{14.30}$$

where the quantity ω_p corresponds to the plasma frequency previously defined

$$\omega_p = \sqrt{\frac{ne^2}{m_e \epsilon_0}}. \tag{14.31}$$

14.5.3   *The Effective Refractive Index of Metals*

From Eq. (14.30), we see that it is possible to identify a complex permittivity for a metal (or a plasma), and therefore a complex refractive index. The refractive index of a conductor then reads

$$\underline{n}^2(\omega) = 1 - \frac{\omega_p^2}{\omega^2 + i\omega/\tau} = \frac{\underline{\epsilon}(\omega)}{\epsilon_0}. \tag{14.32}$$

In a metal, one has typically $1/\tau \sim 10^{14}$ s^{-1} so that in the visible range $\omega^2 \gg \omega/\tau$ and the refractive index becomes

$$\underline{n}^2(\omega) \approx 1 - \frac{\omega_p^2}{\omega^2} \ .$$

Since the plasma frequency of metals is typically in the UV region, in the visible range $\omega_p^2/\omega^2 \gg 1$ and metals have a purely imaginary refractive index

$$\underline{n}(\omega) \approx i\frac{\omega_p}{\omega}$$

so visible light waves do not propagate at all in metals. As we will see in Chap. 15, these waves are completely reflected by metals and this is how metallic mirrors work. In reality, in real metals we should also include the contribution from bound electrons, so that

$$\underline{n}^2(\omega) = 1 - \frac{\omega_p^2}{\omega^2 + i\omega/\tau} + \omega_p'^2 \sum_k \frac{f_k}{\omega_{0k}^2 - \omega^2 - i\gamma_k\omega} \ .$$

For some metals, some of these resonances may fall in the visible part of the spectrum (blue in the case of gold, red for copper), which then give them a particular color. If no absorption peaks are present in the visible range, a metal will have simply a silver color.

14.5.3.1   Low Frequency Regime ($\omega \ll 1/\tau$)

Consider now the low frequency regime, that is, $\omega\tau \ll 1$. The dynamics of the electrons is thus dominated by collisions, and the dispersion relation (14.30) becomes

$$\underline{k}^2 \approx \frac{\omega^2}{c^2}\left(1 - \omega_p^2\tau^2 + i\frac{\omega_p^2\tau}{\omega}\right) \ .$$

Two distinct cases arise. If $\omega_p^2 \tau^2 \ll 1$, which is verified for a poor conductor, we can further write

$$\underline{k}^2 \approx \frac{\omega^2}{c^2} \left(1 + i \frac{\omega_p^2 \tau}{\omega} \right) = \frac{\omega^2}{c^2} \left(1 + i \frac{\sigma_0}{\epsilon_0 \omega} \right) .$$

By writing the wave number as $k = k_R + i k_I$, we have $k^2 = k_R^2 - k_I^2 + 2 i k_R k_I$. Solving for k_R and k_I gives

$$\underline{k} = \frac{\omega}{c} \left(\sqrt{\frac{\sqrt{\epsilon_0^2 \omega^2 + \sigma_0^2} + \epsilon_0 \omega}{2 \epsilon_0 \omega}} + i \sqrt{\frac{\sqrt{\epsilon_0^2 \omega^2 + \sigma_0^2} - \epsilon_0 \omega}{2 \epsilon_0 \omega}} \right) .$$

The skin depth then reads

$$\delta = \frac{1}{k_I} = \frac{c}{\omega} \sqrt{\frac{2 \epsilon_0 \omega}{\sqrt{\epsilon_0^2 \omega^2 + \sigma_0^2} - \epsilon_0 \omega}} ,$$

which in the limit $\sigma_0^2 \gg \epsilon_0^2 \omega^2$ becomes

$$\delta \approx \sqrt{\frac{2 \epsilon_0 c^2}{\sigma_0 \omega}} = \sqrt{\frac{2}{\sigma_0 \mu_0 \omega}} .$$

In the other case where $\omega_p^2 \tau^2 \gg 1$, which is typically the case in metals, the dispersion relation (14.30) is approximated as

$$\underline{k}^2 \approx \frac{\omega^2}{c^2} \left(-\omega_p^2 \tau^2 + i \frac{\omega_p^2 \tau}{\omega} \right) \approx i \frac{\omega^2}{c^2} \frac{\omega_p^2 \tau}{\omega}$$

and so,

$$\underline{k} = \frac{\omega}{c} \sqrt{\frac{\omega_p^2 \tau}{\omega}} \left(\frac{1 + i}{\sqrt{2}} \right) .$$

The skin depth then reads

$$\delta = \frac{1}{k_I} = \sqrt{\frac{2 c^2}{\omega_p^2 \omega \tau}} .$$

Recalling that the static conductivity of a metal reads $\sigma_0 = \epsilon_0 \omega_p^2 \tau$, one can rewrite the skin depth as

$$\delta = \sqrt{\frac{2}{\mu_0 \omega \sigma_0}} \,.$$

We recover the same low-frequency limit for the skin depth as in the case $\omega_p^2 \tau^2 \ll 1$. In the following we consider the case of intermediate and high frequencies for metals, that is, we assume that we have $\omega \tau \gg 1$.

14.5.3.2   Intermediate Frequency Regime ($1/\tau \ll \omega \ll \omega_p$)

For intermediate frequencies such that $1/\tau \ll \omega \ll \omega_p$ (in the case of a good conductor, this corresponds to $\omega \sim 10^{15}$ s^{-1}, which falls in the visible range), the imaginary part in the dispersion relation (14.30) can be neglected, so that

$$\underline{k}^2 = \frac{\omega^2}{c^2} \left(1 - \frac{\omega_p^2}{\omega^2} \right) \,.$$

For frequencies below the plasma frequency ($\omega < \omega_p$) we obtain a pure imaginary wave number,

$$\underline{k} = i \frac{1}{c} \sqrt{\omega_p^2 - \omega^2} \,.$$

An incident electromagnetic wave is thus completely reflected at the interface and the skin depth is given by

$$\delta = \frac{1}{k_I} = \frac{c}{\sqrt{\omega_p^2 - \omega^2}}$$

14.5.3.3   High Frequency Regime ($1/\tau \ll \omega_p \ll \omega$)

Finally, in the high frequency regime where $\omega > \omega_p$, one obtains a real wave vector

$$k \approx \frac{\omega}{c} \left(\sqrt{1 - \frac{\omega_p^2}{\omega^2}} \right) \,.$$

The medium allows the propagation of electromagnetic waves without attenuation and the refractive index becomes an affine function of the electron density n_e, identified by the small-ω_p/ω expansion of k,

$$k \approx \frac{\omega}{c} \left(1 - \frac{1}{2} \frac{\omega_p^2}{\omega^2} \right) = \frac{\omega}{c} \left(1 - \underbrace{\frac{1}{2} \frac{e^2}{m_e \epsilon_0 \omega^2} n_e}_{n} \right) \,. \tag{14.33}$$

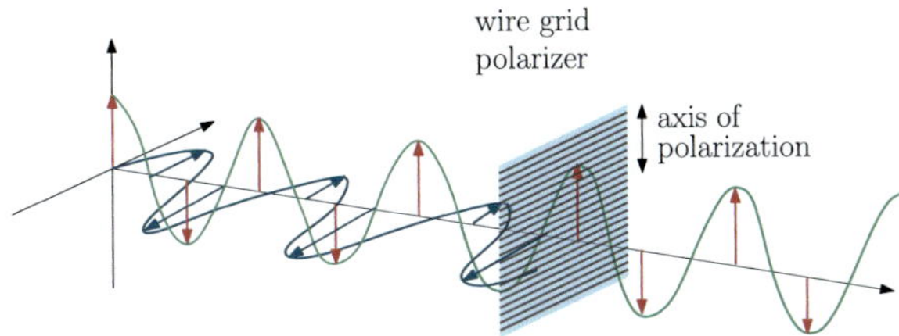

Fig. 14.11 When passing through a wire grid polarizer, the field becomes linearly polarized along the axis perpendicular to the wires

14.6 Controlling Light Polarization: Polarizers and Wave Plates

14.6.1 Linear Polarizers and Malus's Law

A linear polarizer is any anisotropic material that only lets through the light polarized along a specific axis.

An example of a linear polarizer is the so-called wire grid polarizer. It consists of an array of parallel metallic wires spaced by a distance $a \ll \lambda$, on top of a transparent substrate. Electrons in the wire may easily oscillate only along the direction of the wires, so that an electric field parallel to the wire will be highly reflected. In contrast, the grid polarizer will transmit radiation with an electric field vector perpendicular to the wires, which is not coupled to the electrons. The transmitted light will thus be linearly polarized along the axis of polarization of the polarizer, as shown in Fig. 14.11.

There are many other kinds of linear polarizers that rely on the fact that reflection of light off a surface depends, under certain conditions, on its polarization. The legend states that in 1808, Malus (Fig. 14.12) discovered that sunlight could be polarized after reflection from the glasses of Luxembourg Palace in Paris. Polarization-dependent reflection off a dielectric surface will be discussed in detail in Chap. 15. Finally, polarization-dependent polarization in some crystals may be also used to split an incoming beam into two orthogonal polarization states. These so-called beam-splitting polarizers are interesting because they split light into two beams with perpendicular polarization states, in contrast to absorptive polarizers who will only produce a single output beam.

Finally, there are also absorptive polarizers which exhibit a preferential absorption of light polarized along a particular direction. An example are polarizers made of polaroid films consisting of an array of aligned polymers (long chain molecules). Light whose polarization is along the crystal orientation is absorbed by the molecules, thus at the exit of the polarizer light is linearly polarized along the axis perpendicular to the chains. The same principle applies for polarizers made of elliptically shaped nanoparticles embedded in glass.

Fig. 14.12 Étienne-Louis Malus (1775–1812) was a French physicist and engineer, best known for his discovery of the polarization of light by reflection. His work laid the foundation for the study of polarization in optics

14.6.2　Malus's Law (1808)

Suppose that a linearly polarized wave of intensity $\mathcal{I}_0$ is incident on a linear polarizer whose polarization axis forms an angle θ with respect the polarization axis of the incident wave, supposed to be along z as shown in Fig. 14.13.

According to Malus's law, the transmitted wave has a intensity $\mathcal{I}'$ given by

$$\boxed{\mathcal{I}' = \mathcal{I}_0 \cos^2 \theta.} \tag{14.34}$$

Indeed, the incident electric field writes

$$\mathbf{E}(x, t) = E_0 \cos(kx - \omega t + \phi)\mathbf{u}_z \ .$$

Let $\mathbf{v}$ be the axis of polarization of the polarizer, that can be written as

$$\mathbf{v} = \cos\theta\mathbf{u}_z + \sin\theta\mathbf{u}_y \ .$$

Only the component of the electric field parallel to $\mathbf{u}$ will be transmitted by the polarizer. The output wave is therefore given by

$$\mathbf{E}'(x, t) = (\mathbf{E}(x, t) \cdot \mathbf{v})\,\mathbf{v} = E(x, t)(\mathbf{u}_z \cdot \mathbf{v})\mathbf{v}$$
$$= \underbrace{E_0 \cos\theta}_{E_0'} \cos(kx - \omega t + \phi')\mathbf{v}$$

where the phase ϕ' could be eventually different from ϕ due to propagation through the thickness of the polarizer. Finally, since for a plane wave the intensity is proportional to the squared amplitude of the electric field, we find

$$\frac{\mathcal{I}'}{\mathcal{I}} = \frac{|E_0'|^2}{|E_0|^2} = \cos^2 \theta$$

which demonstrates Malus's law.

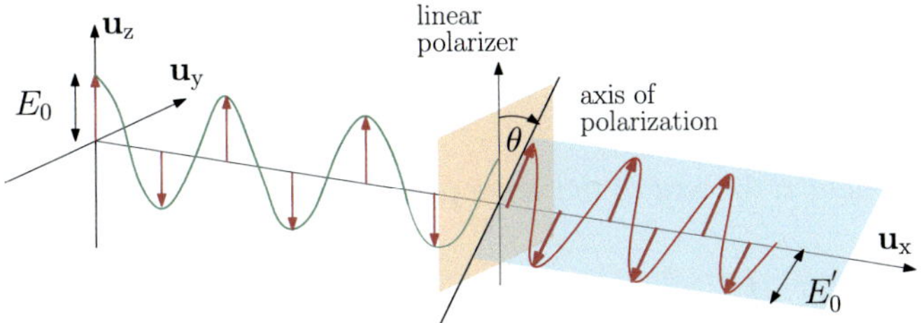

Fig. 14.13 A linear polarizer oriented at an angle θ with respect to the incident light's polarization will transmit a fraction $\cos^2\theta$ of the incident intensity

14.6.3 Waveplates

A wave plate is a transparent plate made of a birefringent material such as quartz or calcite, in which the index of refraction depends on the direction of the field in space. A simple case consists of materials having two main axis perpendicular to each other: a fast axis (f), for example along $\mathbf{u}_y$, for which the refractive index n_f is smaller than the index n_s along the slow axis (s) along $\mathbf{u}_z$. An illustration is shown in Fig. 14.14. Waves polarized along the slow and fast axis do not propagate at the same speed, so that at the exit of the wave plate the wave with polarization along the slow axis is retarded with respect to the wave polarized along the fast axis. As will be shown below, this allows to manipulate the polarization of an electromagnetic wave.

Consider an arbitrary polarization state before the wave plate, that is

$$\underline{\mathbf{E}}_{\mathrm{in}} = \left(E_1 e^{i\phi_{0y}}\mathbf{u}_y + E_2 e^{i\phi_{0z}}\mathbf{u}_z\right) e^{i(kx-\omega t)}$$

so that at the entrance plane of the wave plate, supposed here to be $x = 0$, we have

$$\underline{\mathbf{E}}_{\mathrm{in}}(0, t) = \left(E_1 e^{i\phi_{0y}}\mathbf{u}_y + E_2 e^{i\phi_{0z}}\mathbf{u}_z\right) e^{-i\omega t} \ .$$

Inside the wave plate ($0 < z < e$), the wave number along the slow and fast axis is $k_z = n_z k$ and $k_y = n_y k$, respectively. The wave inside the wave plate then writes

$$\underline{\mathbf{E}}_{\mathrm{inside}}(x, t) = \left(E_1 e^{i(\phi_{0y}+kn_y x)}\mathbf{u}_y + E_2 e^{i(\phi_{0z}+kn_z x)}\mathbf{u}_z\right) e^{-i\omega t} \ .$$

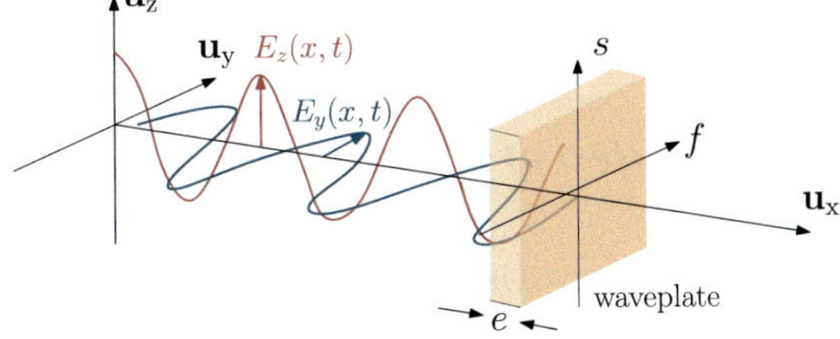

Fig. 14.14 A birefringent material characterized by two principal axes can be used to change the state of the light polarization

The field at the exit of the wave plate ($z = e$) becomes

$$\mathbf{E}_{\text{inside}}(e, t) = \left(E_1 e^{i(\phi_{0y}+kn_ye)}\mathbf{u}_y + E_2 e^{i(\phi_{0z}+kn_ze)}\mathbf{u}_z\right) e^{-i\omega t}$$

so finally, the field at the exit of the wave plate ($z > e$) reads

$$\mathbf{E}_{\text{out}}(x, t) = \left(E_1 \underbrace{e^{i(\phi_{0y}+kn_ye)}}_{e^{i\phi'_{0y}}}\mathbf{u}_y + E_2 \underbrace{e^{i(\phi_{0z}+kn_ze)}}_{e^{i\phi'_{0z}}}\mathbf{u}_z\right) e^{i(k(x-e)-\omega t)} .$$

In conclusion, the phase difference between the two components of the field is modified by the passage through the wave plate and becomes

$$\Delta\phi' = \phi'_{0z} - \phi'_{0y} = k(n_z - n_y)e + \phi_{0z} - \phi_{0y} = k\Delta n e + \Delta\phi$$

with $\Delta n = n_z - n_y = n_s - n_f$. Two cases of great importance are the half-wave plate and quarter-wave plate, for which the phase-shift acquired after the passage through the wave plate is (modulo 2π)

$$\boxed{\underbrace{k\Delta n e = \pi}_{\text{half-wave plate}} \qquad \underbrace{k\Delta n e = \frac{\pi}{2}}_{\text{quarter-wave plate}} .} \qquad (14.35)$$

14.6.4 *Effect of a Half-Wave Plate on Linear Polarization*

Suppose now a linearly polarized wave ($\Delta\phi = 0$) that forms and angle α with the y-axis. That is $E_1 = E_0 \cos\alpha$ and $E_2 = E_0 \sin\alpha$. At the output of the half-wave plate the electric field writes

$$\mathbf{E}_{\text{out}}(x, t) = \left(E_1\mathbf{u}_y + E_2 e^{i\pi}\mathbf{u}_z\right) e^{i(k(x-e)-\omega t+\phi')}$$

$$\mathbf{E}_{\text{out}}(x, t) = \left(E_1\mathbf{u}_y - E_2\mathbf{u}_z\right) e^{i(k(x-e)-\omega t+\phi')}$$

with $\phi' = k(n_y - 1)e$. The component along $\mathbf{u}_z$ therefore changes sign at the output of the wave plate, so that the linear polarization is rotated by 2α. This is shown in Fig. 14.15.

14.6.5 *Effect of a Quarter-Wave Plate on Linear Polarization*

Let us assume now a linearly polarized wave ($\Delta\phi = 0$) that forms and angle $\pi/4$ with the y-axis. That is $E_1 = E_0/\sqrt{2}$ and $E_2 = E_0/\sqrt{2}$. At the output of the half-wave

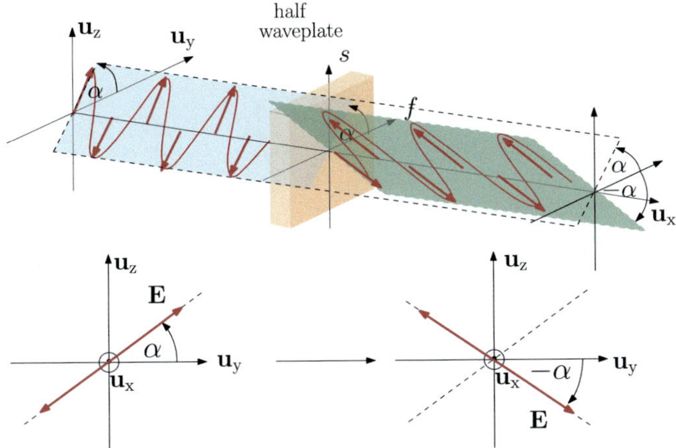

Fig. 14.15 A half-waveplate rotates the plane of polarization of an incident linearly polarized wave

plate the electric field writes

$$\underline{\mathbf{E}}_{\text{out}}(x, t) = \frac{E_0}{\sqrt{2}} \left(\mathbf{u}_y + e^{i\pi/2} \mathbf{u}_z \right) e^{i(k(x-e)-\omega t + \phi')}$$

$$\underline{\mathbf{E}}_{\text{out}}(x, t) = \frac{E_0}{\sqrt{2}} \left(\mathbf{u}_y + i\mathbf{u}_z \right) e^{i(k(x-e)-\omega t + \phi')}$$

with $\phi' = k(n_y - 1)e$. The output wave is a right-handed circular polarization state, as shown. Quarter-wave plates allow the transformation of linear polarization into circular polarization, and vice versa. If the incident angle is changed from $\pi/4$ to $3\pi/4$, the helicity at the output passes from left-handed to right-handed (Fig. 14.16).

14.7 Summary and Essential Formulas

- Maxwell's equations in matter may be rewritten as

$$
\begin{array}{ll}
\nabla \cdot \mathbf{D} = \varrho & \text{Maxwell–Gauss} \\[4pt]
\nabla \cdot \mathbf{B} = 0 & \text{Maxwell–Thomson} \\[4pt]
\nabla \times \mathbf{E} = -\dfrac{\partial \mathbf{B}}{\partial t} & \text{Maxwell–Faraday} \\[8pt]
\nabla \times \mathbf{H} = \mathbf{j} + \dfrac{\partial \mathbf{D}}{\partial t} & \text{Maxwell–Ampère}
\end{array}
$$

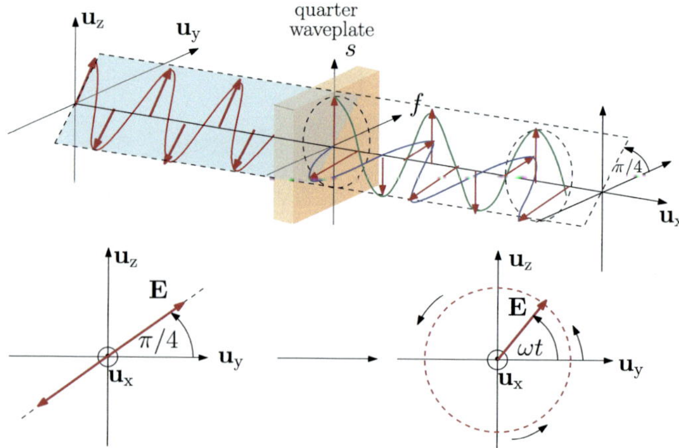

Fig. 14.16 A quarter-waveplate can be used to convert a linearly polarized wave into a circularly polarized one

where the auxiliary fields are given by

$$\mathbf{D} = \epsilon_0 \mathbf{E} + \mathbf{P} ,$$

$$\mathbf{H} = \frac{\mathbf{B}}{\mu_0} - \mathbf{M} .$$

The behavior of the different types of materials is thus determined by the constitutive relations.

The Poynting vector in matter writes

$$\mathbf{\Pi} = \mathbf{E} \times \mathbf{H}$$

and the density of electromagnetic energy

$$u_{EM} = \frac{1}{2} \left(\mathbf{E} \cdot \mathbf{D} + \mathbf{B} \cdot \mathbf{H} \right) .$$

- Due to the inertia of charges in matter, the response to an oscillating electric field is generally not instantaneous. This leads to a frequency-dependence of the permittivity, which is at the origin of dispersion. Lorentz model predicts this dependence by modeling an atom as a damped harmonic oscillator with natural frequency ω_0 and damping constant γ. The polarization density $\underline{\mathbf{P}}$ in the presence of a sinusoidal plane wave $\underline{\mathbf{E}} = \underline{\mathbf{E}}_0 e^{i(kx - \omega t)}$ writes

$$\underline{\mathbf{P}} = (\underline{\epsilon}(\omega) - \epsilon_0)\underline{\mathbf{E}}$$

where the complex permittivity is given by

$$\underline{\epsilon}(\omega) = \epsilon_0 \left(1 + \frac{\omega_p^2}{\omega_0^2 - \omega^2 - i\gamma\omega} \right)$$

with ω_p the plasma frequency which depends on the density of the medium and fundamental constants. The wave vector is thus complex and the dispersion relation becomes

$$\underline{k} = \frac{\omega}{c}\underline{n}(\omega) = k_R + ik_I$$

where $\underline{n}(\omega) = \sqrt{\underline{\epsilon}(\omega)/\epsilon_0}$ is the refractive index. The imaginary part of the wave number k_I is related to the absorption of the electromagnetic energy as the wave propagates in the medium. For a wave propagating along x, we have

$$\underline{\mathbf{E}}(\mathbf{x}, t) = \underline{\mathbf{E}}_0 e^{i(k_R x - \omega t)} e^{-k_I x} \ .$$

- The phase velocity v_ϕ is defined as

$$v_\phi = \frac{\omega}{k(\omega)} = \frac{c}{n(\omega)}$$

and the group velocity v_g as

$$v_g = \frac{\partial \omega}{\partial k} \ .$$

For a wave packet composed of sinusoidal plane waves centered at frequency ω_0, the phase velocity v_ϕ represents the speed of the crests of the wave, whereas the group velocity v_g is the speed of the envelope. Both are related by

$$\boxed{v_g = \frac{v_\phi}{1 + \frac{\omega}{n}\frac{\partial n}{\partial \omega}} = \frac{c/n}{1 + \frac{\omega}{n}\frac{\partial n}{\partial \omega}}} \ .$$

- In a conductor, the response to an oscillating electromagnetic wave is characterized by the complex conductivity which, in the Drude–Lorentz model writes

$$\underline{\sigma}(\omega) = \frac{\sigma_0}{1 - i\omega\tau}$$

where σ_0 is the static conductivity and τ the time between two collisions. Just like in a dielectric, this leads to a complex wave vector and so it is possible to characterize a metal by an effective complex permittivity $\underline{\epsilon}$. Conversely, a dielectric can be characterized by an effective conductivity $\underline{\sigma}$. They are both related by

$$\underline{\epsilon}(\omega) = \epsilon_0 \left(1 - i \frac{\sigma(\omega)}{\epsilon_0 \omega} \right) .$$

14.1 Electromagnetic waves in copper and in the ionosphere

Consider the case of copper, whose free electron density is $n^{Cu} = 8.5 \times 10^{28}$ m^{-3} and for which the mean time between collisions is $\tau^{Cu} \sim 10^{-14}$ s.

(a) What is the skin depth for an incident electromagnetic wave in the visible range ($\omega \sim 4 \times 10^{15}$ s^{-1}) and for a radio wave ($\omega \sim 6 \times 10^{6}$ s^{-1})?

(b) Compare with the case of the ionosphere, which is a plasma of low density $n^{ion} \sim 10^{12}$ m^{-3} with conductivity $\sigma_0^{ion} = 1$ Sm^{-1}.

14.2 Circularly polarized light incident on a linear polarizer

A circularly polarized sinusoidal wave of intensity $\mathcal{I}_0$ is incident on a linear polarizer. What is the transmitted intensity $\mathcal{I}'$ in terms of $\mathcal{I}_0$?

14.3 Skin effect in a metallic plate

A homogeneous metallic plate of conductivity σ, of permittivity ϵ_0 and permeability μ_0, of thickness $2a$, limited by the planes $x = -a$ and $x = a$, is placed in vacuum. This metal plate, of infinite dimensions in the y- and z- directions, is subject to harmonic currents of high frequency $f = \omega/2\pi$, of volume density.

vacuum   metal $\odot$ j $2a$ vacuum

$$\mathbf{j}(x, t) = j_0(x) \cos \omega t \, \mathbf{u}_y .$$

Displacement currents in the metal are negligible compared to conduction currents.

(a) Write Maxwell's equations inside the metal plate and expand them as eight scalar equations which relate the partial derivatives of the components of the electric and magnetic fields.

(b) We will look for a harmonic solution in the form: $\underline{B_z} = \mathcal{B}(x)e^{i\omega t}$, with $\mathcal{B}(x)$ a complex amplitude. Establish the second order differential equation governing $\mathcal{B}(x)$. Use the notation $\delta = \sqrt{2/\mu_0 \sigma \omega}$ and the relation $i = \frac{1}{2}(1+i)^2$ to simplify the result.

(c) Express in complex notation the magnetic and electric fields, $\underline{\mathbf{B}}(x, t)$ and $\underline{\mathbf{E}}(x, t)$, inside the plate. *Hint:* Exploit the symmetries of the system to express integration constants as functions of B_S, the complex amplitude of the magnetic field $\mathbf{B}$ at $x = a$. Write the final expressions of $\mathbf{B}(x, t)$ and $\mathbf{E}(x, t)$ as functions of B_S

(d) Show that the current density is uniform at very low frequency; express it in terms of B_S, a and μ_0.

(e) Express the modulus $j_0(x)$ of the amplitude of the current density as a function of the parameter $u = x/\delta$ and its value $j_0(0)$ at the center of the plate.

(f) We now assume the plate to be very thick ($a \gg \delta$). In these conditions, express the current density vector $\mathbf{j}(x, t)$, in the vicinity of each of the free surfaces $x = \pm a$ of the plate.

14.4 Phase velocity, group velocity, and energy velocity in an absorption band

Consider a homogeneous, isotropic, gaseous dielectric with a refractive index n given by

$$n = 1 + \frac{1}{2}\omega_p^2 \frac{\omega_0^2 - \omega^2}{(\omega_0^2 - \omega^2)^2 + \gamma^2\omega^2} \, .$$

This equation is a simplified form of a more general dispersion relation. It describes the behavior of the refractive index near a resonance frequency. Here ω is the angular frequency of a monochromatic plane wave propagating in the medium, ω_p is the plasma frequency ($\omega_p = 5.67 \times 10^8$ rad s^{-1}), ω_0 is the resonance frequency ($\omega_0 = 3 \times 10^{15}$ rad s^{-1}) and γ is the resonance linewidth ($\gamma = 9 \times 10^9$ rad s^{-1}).

(a) Find expressions for the phase velocity, v_ϕ, and group velocity, v_g, at $\omega = \omega_0$.
(b) Calculate the numerical values of v_ϕ and v_g at $\omega = \omega_0$.
(c) Interpret the result for v_g, particularly in the context of anomalous dispersion.
(d) Find expressions for the time-averaged Poynting vector $\langle \mathbf{\Pi} \rangle$ and time-averaged energy density $\langle u \rangle$ in terms of n and the electric field amplitude E_0.
(e) Deduce the expression for the energy velocity v_e.

14.5 Interaction of a Linearly Polarized Plane Monochromatic Wave with an Atomic Vapor

We are examining the interaction of a linearly polarized plane monochromatic wave with a dilute atomic vapor. Consider a simplified model of an atom within a vapor: Each atom has a single mobile electron that moves relative to the rest of the atom. The nucleus and the other electrons are considered to be fixed and do not move significantly. The electron experiences a restoring force $\mathbf{f} = -m\omega_0^2\mathbf{r}$ that is proportional to its displacement $\mathbf{r}$ from the center. Here $m = 9 \times 10^{-31}$ kg is the electron mass, and $\omega_0 = 1 \times 10^{15}$ rad s^{-1} is the natural angular frequency of the electron oscillation. The electron also experiences a damping force $\mathbf{f}' = -m\gamma\mathbf{v}$ due to interactions with the rest of the atom, where $\mathbf{v}$ is the electron velocity and $\gamma = 5 \times 10^9$ rad s^{-1}.

(a) 1. Derive the equation of motion for the electron under the influence of the electric field and damping force.
    2. Show that the magnetic force is negligible compared to the electric force as long as the electron velocity is much smaller than the velocity of light.
    3. Using complex notations ($\underline{\mathbf{E}} = \mathbf{E}_0 \exp i(\omega t - kz)\mathbf{x}$, find the steady-state solution for the electron motion.

(b) 1. Find the polarization of the medium $\mathbf{P}$ in the steady state regime in terms of $\Omega^2 = Ne^2/m\epsilon_0$, with N the vapor density.
    2. Derive the wave equation for the electric field in the medium

$$\Delta\mathbf{E} - \epsilon_0\mu_0\frac{\partial^2\mathbf{E}}{\partial t^2} = \mu_0\frac{\partial^2\mathbf{P}}{\partial t^2} - \frac{1}{\epsilon_0}\nabla(\nabla\cdot\mathbf{P}) \, ,$$

where $\mathbf{P}$ is the polarization of the medium.

(c) 1. Express the refractive index in terms of its real and imaginary parts.
   2. Interpret the physical meaning of the real and imaginary parts.
   3. Plot the real and imaginary parts of the refractive index as functions of frequency and analyze their behavior.

14.6 Laser Modes

This problem explores the propagation of laser light in a medium with a complex refractive index. The medium is modeled as a gas with atoms that can be excited to a higher energy level.

(a) Consider a linearly polarized plane monochromatic wave propagating along the z-axis in a non-magnetic medium with a complex relative permittivity $\epsilon_r = \epsilon' - i\epsilon''$ such that $|\epsilon' - 1| \ll 1$, and $|\epsilon''| \ll 1$.

We seek a solution to Maxwell's equations in the form of a plane wave

$$\underline{\mathbf{E}} = E_0 \exp i(\omega t - kz)\mathbf{u}_x \qquad \underline{\mathbf{B}} = B_0 \exp i(\omega t - kz)\mathbf{u}_y \ .$$

1. Determine the relationship between the angular frequency ω and the wave number k.
2. Calculate the Poynting vector $\mathbf{\Pi}$ and its time-averaged value $\langle \mathbf{\Pi} \rangle$. Interpret the physical meaning of $\langle \mathbf{\Pi} \rangle$ based on the sign of ϵ''.

(b) The medium consists of a gas where each molecule has two energy levels: a lower energy level E_1 and a higher energy level E_2. The number of molecules in each state is denoted by N_1 and N_2, respectively.

When the angular frequency of the wave, ω, is close to the resonant frequency of the atomic transition between the two energy levels, $\omega_0 = \frac{E_2 - E_1}{\hbar}$, where $\hbar$ is the reduced Planck constant, quantum mechanics predicts the following expression for the relative permittivity:

$$\epsilon_r = 1 + \frac{e^2}{2\,m\epsilon_0\omega_0}(N_1 - N_2)f\,\frac{1}{\omega_0 - \omega + i\Gamma} \ .$$

Here e is the electron charge, m is the electron mass, ϵ_0 is the permittivity of free space, f is a dimensionless oscillator strength and Γ is the linewidth parameter ($\Gamma/\omega_0 \ll 1$).

This expression shows how the properties of the medium, such as its ability to polarize and absorb light, are influenced by the population difference between the energy levels, the frequency of the light, and the atomic properties.

1. In a gas laser, the populations N_1 and N_2 are maintained constant, with $N_2 > N_1$. What happens to the wave as it propagates?
2. The laser cavity is a 1-meter-long tube terminated at both ends by plane mirrors. One mirror is perfectly reflective ($r = -1$), while the other has a reflectivity of $R = 99\%$ ($r' = -\sqrt{R}$). The mirrors are aligned perpendicular to the z-axis.

For an empty laser cavity, determine the angular frequencies ω_p corresponding to the formation of standing waves. Explain why these waves are in fact damped.

3. In the presence of a gas in the cavity, what is the minimum population inversion $(N_2 - N_1)_{\min} = N_0$ required for a wave of angular frequency ω to propagate without damping?

4. For a population inversion of $N_2 - N_1 = 2N_0$, determine the range of angular frequencies for which undamped oscillations (or *locking*) can occur.

5. Within the locking range, only frequencies corresponding to standing waves are selected. Determine the number of undamped standing wave modes.

For numerical application, use: $\omega_0 = 3 \times 10^{15}$ rad s^{-1}, $\Gamma/\omega_0 = 3 \times 10^{-6}$.

14.7 Self-focusing of a laser beam

Consider an infinite, homogeneous, isotropic, non-magnetic, charge-free, and current-free dielectric medium. Under the influence of an electric field $\mathbf{E}$, the dielectric acquires a polarization $\mathbf{P} = \alpha \mathbf{E}$, where α is the polarizability of the dielectric. For very intense fields, α becomes a function of the field amplitude and can be approximated as $\alpha = \epsilon_0(\chi + \beta|E|^2)$ where χ and β are constants, with $\beta|E|^2$ much smaller than χ ($\beta > 0$).

The electric field, polarized along the z-axis, is that of a monochromatic plane wave emitted by a laser and propagating along the x-axis. In complex notation, it can be written as

$$\underline{\mathbf{E}} = E(y)\exp i(\omega t - kx)\mathbf{u}_z \ .$$

(a) Write down Maxwell's equations in the dielectric medium. Give the expressions for the magnetic field $\mathbf{B}$ and the electric displacement $\mathbf{D}$ of the wave. Establish the second-order differential equation, then the first-order differential equation satisfied by $E(y)$. Seek solutions such that both $E(y)$ and its derivative $dE(y)/dy$ vanish as y tends to infinity.

(b) Under what conditions on k^2 does the last equation make physical sense? Show that the previous solution can be expressed as

$$E(y) = \frac{E_0}{\cosh(y/y_0)} \ .$$

Specify the expressions for E_0 and y_0.

(c) Calculate the time-averaged Poynting vector. Determine the average power per unit length carried by the wave along the z-axis as a function of the initial electric field amplitude E_0, the characteristic length y_0, and the wave number k.
Given $P = 3 \times 10^{10}$ W m^{-1}, $\chi = 0.5$, $\beta = 2 \times 10^{-22}$ uSI, and a vacuum wavelength of the laser light $\lambda_0 = 0.694$ μm, calculate E_0, y_0 and k.
Analyze the spatial distribution of the average power density as a function of the radial distance y from the beam axis. Discuss the implications of this distribution for the focusing of the laser beam.

Chapter 15
The Laws of Reflection and Refraction

Abstract This chapter explores the fundamental phenomena of *reflection and refraction* of electromagnetic waves at interfaces between different media. It begins by deriving the *continuity equations* for the normal components of the electric displacement (**D**) and magnetic field (**B**), and the tangential components of the electric field (**E**) and magnetic excitation (**H**), directly from Maxwell's equations. These continuity conditions are crucial for understanding how fields behave at boundaries. The core of the chapter lies in the derivation of the *laws of reflection and refraction* (Snell–Descartes law) as a direct consequence of these continuity conditions. It demonstrates that the frequency of the wave remains unchanged upon reflection and refraction, and that the incident, reflected, and transmitted wave vectors lie in the same plane. The special case of *total internal reflection* is discussed, highlighting its importance in applications like optical fibers. The chapter then introduces *Fresnel's reflection and transmission coefficients*, which quantify the amplitudes of reflected and transmitted electric fields for both s-polarized (perpendicular) and p-polarized (parallel) incident waves. It shows how these coefficients depend on the angles of incidence and the refractive indices of the media. The concept of *Brewster's angle* is presented, where p-polarized light experiences no reflection, leading to the phenomenon of *polarization by reflection*. Furthermore, the chapter defines *reflectance* and *transmittance* as the fractions of incident energy flux that are reflected and transmitted, respectively, and establishes their relationship to Fresnel's coefficients and the *conservation of energy flux* at the interface. The unique case of *reflection by metals* is also analyzed. Finally, the chapter introduces *birefringence*, where the refractive index depends on light polarization, and presents the principles of the *Fabry–Pérot resonator*, an optical cavity that utilizes multiple reflections to achieve wavelength-selective transmission, with applications in spectroscopy and lasers.

Keywords Reflection · Refraction · Fresnel's coefficients · Snell–Descartes law

F. Cadiz and A. Couairon, *Classical Electrodynamics*, Undergraduate Texts in Physics,
https://doi.org/10.1007/978-3-031-86785-9_15

15.1　Introduction

The phenomena of reflection and refraction of light have been known since ancient times. When light encounters the interface between two media, it splits into a reflected beam and a refracted beam, with the latter bending relative to the initial propagation direction. When one of the interfaces is metallic, most of the light is reflected, and if the surface is sufficiently smooth, it acts as a mirror. Refraction, on the other hand, occurs in transparent materials such as glass, where tailoring the interface profile is used to create lenses and optical systems—a topic explored in detail in Chap. 17.

Since refraction is generally a wavelength-dependent phenomenon, it gives rise to effects such as rainbows and the dispersion of sunlight by a prism. The exploitation of the laws of reflection and refraction has led to the development of sophisticated systems with significant applications in modern optics, such as dielectric mirrors with ultra-high reflectivity, interferential filters, anti-reflection coatings, and resonant optical cavities for spectroscopy and laser fabrication.

In this chapter, we will derive the laws of reflection and refraction as consequences of the continuity conditions for electromagnetic fields, as dictated by Maxwell's equations. We will also present Fresnel's formulas for the reflection and transmission coefficients and discuss the conservation of energy flux at an interface.

15.2　Continuity Equations

In this section we will derive from Maxwell's equations the continuity conditions for the electromagnetic fields at an interface separating two media. Of course, no abrupt interface exists at the atomic scale, but it is a convenient way of dealing with the averaged electromagnetic fields in matter. Here, we suppose that the fields are averaged over a typical length scale l such that $a \ll l \ll \lambda$, where $a \sim 0.1\,\mathrm{nm}$ is the typical size of an atom and λ is the wavelength of the electromagnetic wave. When this is possible, matter can be modeled as a continuous medium with refractive index n. Note however that when $\lambda \sim a$, there is no intermediate length scale l and this averaging process is no longer possible. In this regime, one cannot describe the interaction of waves with matter in terms of reflection and refraction. This is the case, for example, of X-rays which are known to be diffracted when incident on crystals. Diffraction will be discussed in Chap. 18.

15.2.1　Continuity Equations for the Normal Components
*　　　　of* $\mathbf{B}$ *and* $\mathbf{D}$ *at an Interface*

The Maxwell–Gauss and Maxwell–Thomson equations for the averaged fields, in their integral form, are given by:

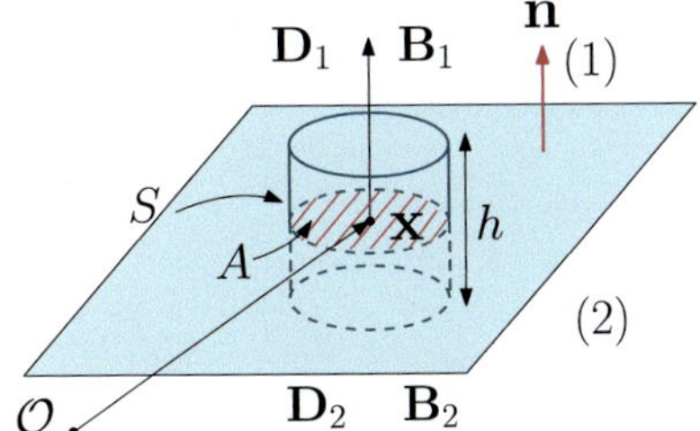

Fig. 15.1 An infinitesimal cylinder of axis parallel to the normal at the interface between two media

$$\oiint_S \mathbf{D} \cdot \mathbf{n}\, dS = Q(S) \qquad \oiint_S \mathbf{B} \cdot \mathbf{n}\, dS = 0$$

Consider now an interface separating two media 1 and 2, and let $\mathbf{n}$ be the normal to the interface at point $\mathbf{x}$, directed from 2 to 1. As shown in Fig. 15.1, we consider an infinitesimal, closed cylindrical surface of height h and transverse area A.

When $h \to 0$, the flux integral over S reduces to the flux through the caps just above and just below the interface

$$\oiint_S \mathbf{D} \cdot \mathbf{n}\, dS = A(\mathbf{D}_2(\mathbf{x}) - \mathbf{D}_1(\mathbf{x})) \cdot \mathbf{n},$$

$$\oiint_S \mathbf{B} \cdot \mathbf{n}\, dS = A(\mathbf{B}_2(\mathbf{x}) - \mathbf{B}_1(\mathbf{x})) \cdot \mathbf{n} .$$

The zero-divergence of the magnetic field ensures the continuity of its component normal to the interface,

$$\boxed{\mathbf{n} \cdot (\mathbf{B}_2 - \mathbf{B}_1)\,|_{\mathbf{x}} = 0,} \tag{15.1}$$

whereas for the displacement vector $\mathbf{D}$, the flux through S may differ from zero if the interface is charged with a surface density σ, in which case the charge $Q(S)$ enclosed by the cylinder is $A\sigma$ and we conclude that the normal component of $\mathbf{D}$ presents a discontinuity whenever $\sigma \neq 0$,

$$\boxed{\mathbf{n} \cdot (\mathbf{D}_2 - \mathbf{D}_1)\,|_{\mathbf{x}} = \sigma(\mathbf{x}).} \tag{15.2}$$

15.2.2 Continuity Equations for the Tangential Components of E and H at an Interface

We recall the Maxwell–Faraday and Maxwell–Ampère laws for the averaged fields in their integral form

$$\oint_\Gamma \mathbf{E} \cdot d\mathbf{l} = -\frac{d}{dt} \iint_{S(\Gamma)} \mathbf{B} \cdot \mathbf{n}\, dS$$

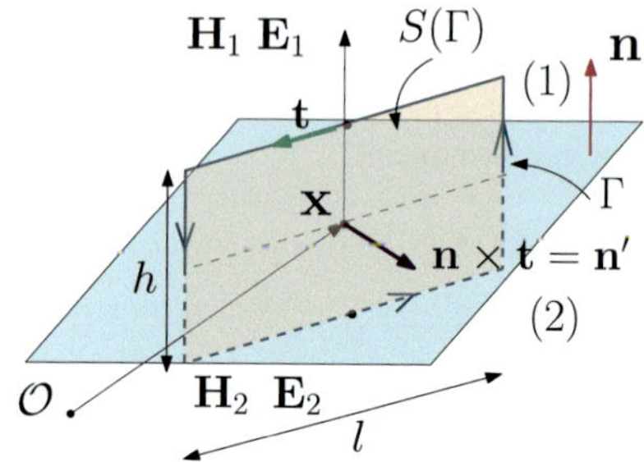

Fig. 15.2 An infinitesimal closed circuit at the interface between two media

$$\oint_{\Gamma} \mathbf{H} \cdot d\mathbf{l} = \iint_{S(\Gamma)} \mathbf{j} \cdot \mathbf{n}\, dS + \frac{d}{dt} \iint_{S(\Gamma)} \mathbf{D} \cdot \mathbf{n}\, dS$$

We can choose for Γ an infinitesimal, closed rectangular path centered at $\mathbf{x}$, of height h and length l and let $S(\Gamma)$ be the planar surface enclosed by Γ, whose normal $\mathbf{n}'$ is tangent to the interface, as shown in Fig. 15.2.

When taking the limit $h \to 0$, the circulation over Γ reduces to the integrals along the horizontal segments parallel to the surface, so that

$$\oint_{\Gamma} \mathbf{E} \cdot d\mathbf{l} = l\mathbf{t} \cdot (\mathbf{E}_1(\mathbf{x}) - \mathbf{E}_2(\mathbf{x}))$$

$$\oint_{\Gamma} \mathbf{H} \cdot d\mathbf{l} = l\mathbf{t} \cdot (\mathbf{H}_1(\mathbf{x}) - \mathbf{H}_2(\mathbf{x}))$$

where $\mathbf{t}$ is a unit vector tangent to the interface along the horizontal path just above it. The flux integrals of $\mathbf{B}$ and $\mathbf{D}$ tend to zero as $h \to 0$. In consequence, the tangential component $\mathbf{E}^{\parallel}$ of the electric field is continuous at any point $\mathbf{x}$ of the interface:

$$\boxed{\mathbf{E}_2^{\parallel}(\mathbf{x}) - \mathbf{E}_1^{\parallel}(\mathbf{x}) = 0.} \qquad (15.3)$$

For the magnetic intensity $\mathbf{H}$, however, a non-zero flux of the current density over $S(\Gamma)$ is possible if a current is confined at the interface, in which case $\iint_{S(\Gamma)} \mathbf{j} \cdot \mathbf{n}\, dS = l\mathbf{J}_s \cdot \mathbf{n}'$ where $\mathbf{J}_S$ is the current density (in A/m) at the interface. Since $\mathbf{t} = \mathbf{n}' \times \mathbf{n}$ and using $(\mathbf{n}' \times \mathbf{n}) \cdot (\mathbf{H}_1 - \mathbf{H}_2) = \mathbf{n}' \cdot (\mathbf{n} \times (\mathbf{H}_1 - \mathbf{H}_2))$ we conclude that

$$\boxed{\mathbf{n} \times (\mathbf{H}_1 - \mathbf{H}_2)\,|_{\mathbf{x}} = \mathbf{J}_S.} \qquad (15.4)$$

15.3 Reflection and Refraction of an Electromagnetic Wave at an Interface

When an incident beam of light arrives at a flat interface between two isotropic media, it gives birth to a reflected and a transmitted beam. Here, we will show that the existence of these waves results from the continuity conditions of the electromagnetic fields at the interface. When the interface is not flat, one can always apply the laws of reflection and refraction locally, at every point of the interface. Recall that a sinusoidal electromagnetic plane wave in matter can be written in its complex representation as

$$\underline{\mathbf{E}} = \underline{\mathbf{E}}_0 e^{i(\mathbf{k}\cdot\mathbf{x}-\omega t)}$$

where the wave vector writes $\mathbf{k} = \underline{k}\mathbf{n}$, with $\mathbf{n}$ a unit vector representing the direction of propagation of the wave and $\underline{k}$, in the general case, a complex number related to the refractive index $\underline{n}$ by the dispersion relation

$$\underline{k}(\omega) = \frac{\omega}{c}\underline{n}(\omega) = \frac{\omega}{c}\left(n_R(\omega) + i n_I(\omega)\right)$$

Let us now consider the situation in which a sinusoidal plane wave encounters an interface between two media, represented by the half-spaces $I : \{x < x_0\}$ and $II : \{x > x_0\}$. Let xy be the plane containing both the normal to the interface and the direction of propagation of the incident wave. This is illustrated in Fig. 15.3. At $x = x_0^-$ (region I), the total electric field is the superposition of an incident wave ($\underline{\mathbf{E}}$) and a wave reflected off the interface ($\underline{\mathbf{E}}'$). The incident wave, whose direction

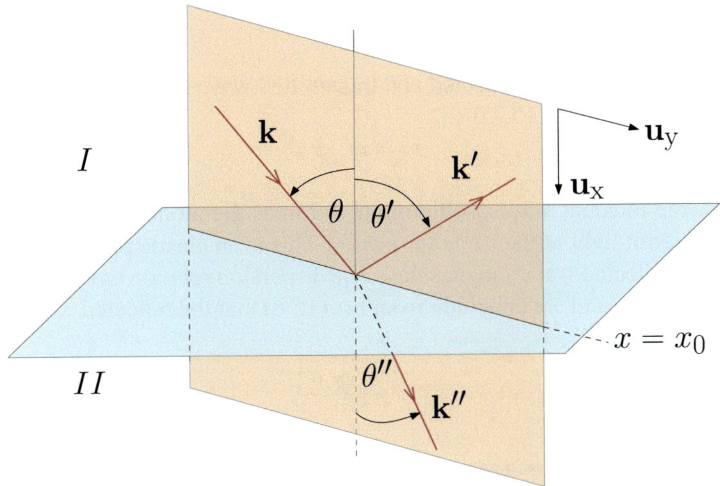

Fig. 15.3 An incident plane wave at an interface $x = x_0$ produces a reflected and a transmitted wave

of propagation forms an angle θ with respect to the normal to the interface is given at $x = x_0^-$ by

$$\underline{\mathbf{E}}(x_0^-, y, t) = \underline{\mathbf{E}}_0 e^{i(\underline{k}\cos\theta x_0 + \underline{k}\sin\theta y - \omega t + \phi)}$$

and the reflected wave writes

$$\underline{\mathbf{E}}'(x_0^-, y, t) = \underline{\mathbf{E}}_0' e^{i(-\underline{k}'\cos\theta' x_0 + \underline{k}'\sin\theta' y - \omega' t + \phi')}$$

Finally, in medium II at $x = x_0^+$ there will be a transmitted wave

$$\underline{\mathbf{E}}''(x_0^+, y, t) = \underline{\mathbf{E}}_0'' e^{i(\underline{k}''\cos\theta'' x_0 + \underline{k}''\sin\theta'' y - \omega'' t + \phi'')}$$

The total electric field $\underline{\mathbf{E}} + \underline{\mathbf{E}}'$ in region I and the field $\underline{\mathbf{E}}''$ in region two are related by the continuity conditions Eqs. (15.2) and (15.3). In particular, the tangential component of the electric field is continuous at $x = x_0$ for arbitrary y and t. This is only possible if the argument of the complex exponentials are the same for all the three waves.[1] At $t = 0$ and $y = 0$, this writes

$$\underline{k}\cos\theta x_0 + \phi = -\underline{k}'\cos\theta' x_0 + \phi' = \underline{k}''\cos\theta'' x_0 + \phi''. \tag{15.5}$$

At $t = 0$ and y arbitrary, and using (15.5)

$$\underline{k}\sin\theta y = \underline{k}'\sin\theta' y = \underline{k}''\sin\theta'' y. \tag{15.6}$$

Finally, at $y = 0$, t arbitrary and using (15.5) and (15.6)

$$\omega t = \omega' t' = \omega'' t''. \tag{15.7}$$

From these conditions we conclude that:

1. The frequency ω of the reflected and transmitted waves is the same as that of the incident one (see Eq. (15.7))
$$\omega = \omega' = \omega'' .$$

   Indeed, the incident wave oscillating at ω forces the atomic dipoles to oscillate and to re-emit light at the same frequency. This is an elastic process.
2. Since the reflected wave satisfies the same dispersion relation as the incident one, we have $\underline{k} = \underline{k}'$ and we conclude from Eq. (15.6) that the reflected wave forms an angle
$$\boxed{\theta' = \theta} \tag{15.8}$$

---

[1] Note that we have already assumed that the reflected and transmitted wave vectors lie both in the plane of incidence. If this was not the case, it would be impossible to fulfill the continuity of the tangential component of $\mathbf{E}$ for every point at the interface.

with respect to the normal, and so the reflected wave vector is the symmetric of
the incident wave vector with respect to the interface.
From Eq. (15.6) we obtain the Snell–Descartes law

$$\boxed{\underline{k}\sin\theta = \underline{k}''\sin\theta''}$$

(15.9)

which can be rewritten in terms of the refractive index as discussed below.

15.3.1   Reflection and Refraction Laws

Imposing the continuity of the tangential component of the electric field at an interface
requires the existence of a reflected an transmitted waves whose wave vectors lie in
the same plane and given by the following reflection and refraction laws:

- **Reflection law**: the reflected wave vector is the symmetric of the incident wave
  vector with respect to the interface Π. They both form the same angle θ_1 with
  respect to the normal to the interface, as shown in Fig. 15.4.
  Specular reflection is the term used to distinguish the reflection from a very flat
  surface (such as a mirror) from the diffuse reflection observed on irregular surfaces
  (most of everyday objects). Specular reflection allows us to form images of objects
  since all the points on the surface reflect light in the same direction, whereas for

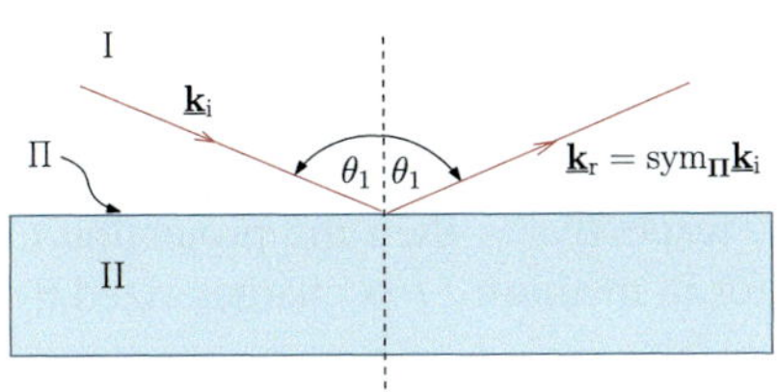

Fig. 15.4 The reflection law

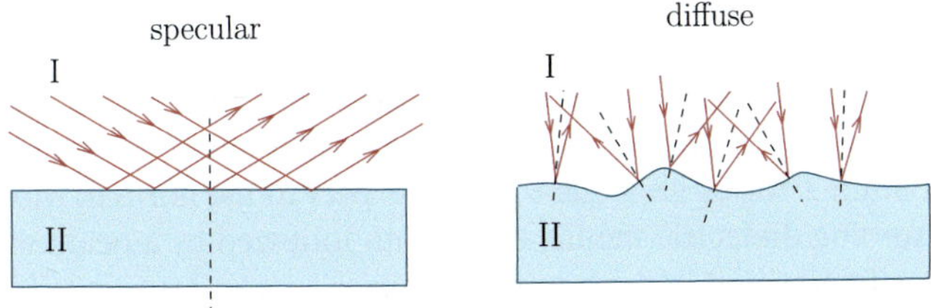

Fig. 15.5 Specular reflection produced by a flat mirror (left), and diffuse reflection produce by a
rough interface (right)

Fig. 15.6 Left: Willebrord Snell (1580–1626), a Dutch mathematician and physicist. His work laid the foundation for modern geometrical optics. Right: René Descartes (1596–1650), French philosopher, mathematician, and scientist. He developed the wave-based explanation of rainbows and made foundational contributions to mathematics, philosophy, and mechanics

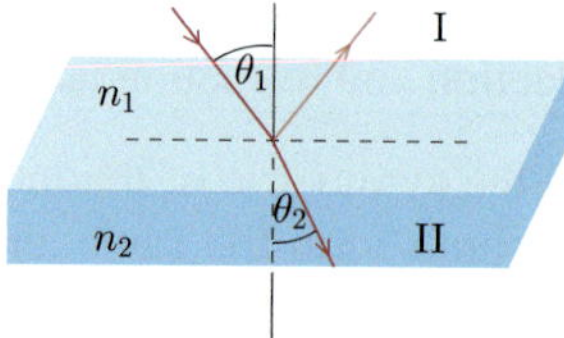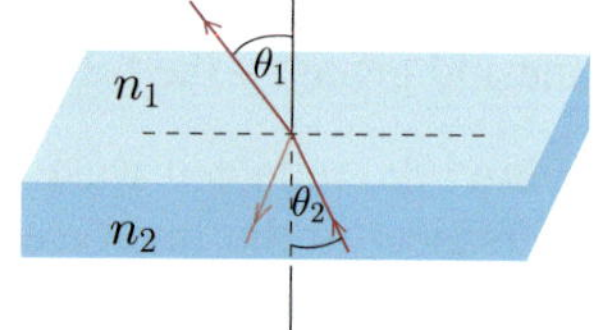

Fig. 15.7 Snell–Descartes law for refraction

diffuse reflection light is scattered in all directions since the normal to the interface changes rapidly from one point to the other (Fig. 15.5). It is this latter type of reflection that allows us to observe the world around us.

- **Snell–Descartes law**: (Figure 15.7) If a monochromatic plane wave coming from medium I is incident on the interface separating the media I and II, of refractive index $\underline{n}_1$ and $\underline{n}_2$, respectively, then the projection along the normal of the transmitted wave vector in medium II is characterized by an angle θ_2 given by

$$\boxed{\underline{n}_1 \sin \theta_1 = \underline{n}_2 \sin \theta_2.}$$

(15.10)

For non-absorbing dielectric media, $\underline{n}_1 = n_1 \in \mathbb{R}$ and $\underline{n}_2 = n_2 \in \mathbb{R}$ and so θ_2 represents the angle that the refracted beam forms with the normal to the interface, as shown in Fig. 15.6.

One consequence of the law of refraction is the principle of backward propagation of light: if light is incident from medium II at an angle θ_2, then the refracted wave in medium I forms an angle θ_1 with respect to the normal. More generally, for non-absorbing dielectric media, every path followed by a beam of light in one direction can be traveled in the opposite sense.

15.3.2 The Case of Total Reflection

In dielectrics far from any absorption resonance, the refractive index is real, and Snell–Descartes law becomes

$$\sin \theta_2 = \frac{n_1}{n_2} \sin \theta_1 \ .$$

A particular case arises whenever the first medium has a refractive index larger than that of medium 2, i.e., when $n_1 > n_2$. In that case, there is no real solution for θ_2 if

$$\frac{n_1}{n_2} \sin \theta_1 \geq 1$$

which is the case when the incident angle exceeds a critical value θ_1^* given by

$$\theta_1^* = \sin^{-1}\left(\frac{n_2}{n_1}\right) \ .$$

In this case, θ_2 is a complex number such that

$$\cos \theta_2 = i \underbrace{\sqrt{\frac{n_1^2}{n_2^2} \sin^2 \theta_1 - 1}}_{>0} \ .$$

The transmitted wave adopts the form of an evanescent wave propagating along the surface (y direction) and decaying exponentially as a function of x inside region II (see Fig. 15.8):

$$\underline{\mathbf{E}}''(x, y, t) = \underline{\mathbf{E}}_0'' e^{i\left(\frac{\omega}{c} n_1 \sin \theta_1 y - \omega t + \phi''\right)} e^{-\frac{\omega}{c}\sqrt{n_1^2 \sin^2 \theta_1 - n_2^2}\, x} \ .$$

This case corresponds to a total reflection of the incident beam.

Example 15.1—Totally reflecting prism
Consider a right-angle prism made of a dielectric material of refractive index n, whose transverse section is an isosceles triangle. The critical angle of incidence θ_i^* above which a beam inside the prism will be totally reflected at the prism-air interface is given by

$$\theta_i^* = \sin^{-1}\left(\frac{1}{n}\right) \ .$$

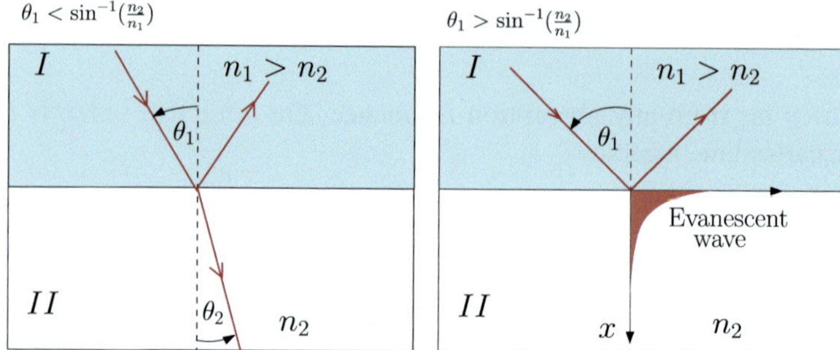

Fig. 15.8 Under total internal reflection, shown here on the right, there can be no transmitted wave propagating along x

A beam entering the prism at zero angle of incidence will encounter the internal prism-air interface forming an angle $\theta_i = \pi/4$ with respect to the normal to the interface, as shown in Fig. 15.9. In order to have total reflection, the critical incidence angle θ_i^* must be then smaller than $\pi/4$

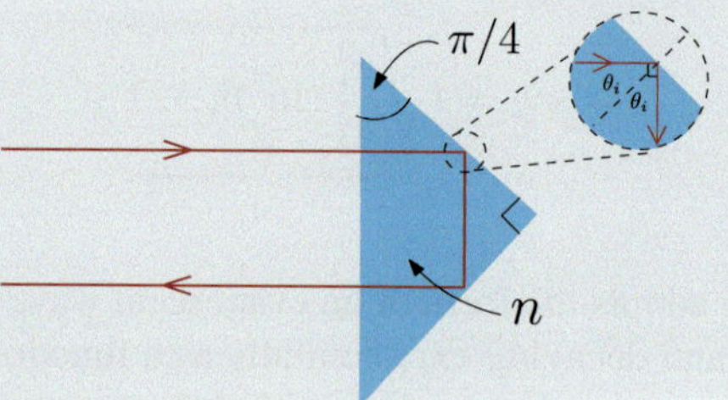

Fig. 15.9 Total internal reflection on a prism

$$\theta_i^* = \sin^{-1}\left(\frac{1}{n}\right) \le \pi/4 \ .$$

that is

$$\frac{1}{n} \le \sin\left(\frac{\pi}{4}\right) = \frac{1}{\sqrt{2}} \ .$$

The prism must me made of a dielectric with refractive index such that

$$n \ge \sqrt{2} \approx 1.41$$

which is the case of glass, whose refractive index is typically close to $n = 1.5$–1.6.

Example 15.2—The optical fiber

Modern day communications rely almost exclusively on optical fibers to transmit information in the form of electromagnetic waves of wavelength $\lambda \sim 1.5\ \mu\text{m}$ over very long distances. An optical fiber is made of a cylindrical dielectric medium of refractive index n_1, called the core, in which light propagates. The core is surrounded by a second, concentric cylinder of refractive index n_2. The indexes are chosen such that $n_2 < n_1$ so that light can be confined in the core by assuring total reflection at the core-cladding interface, as shown in Fig. 15.10.

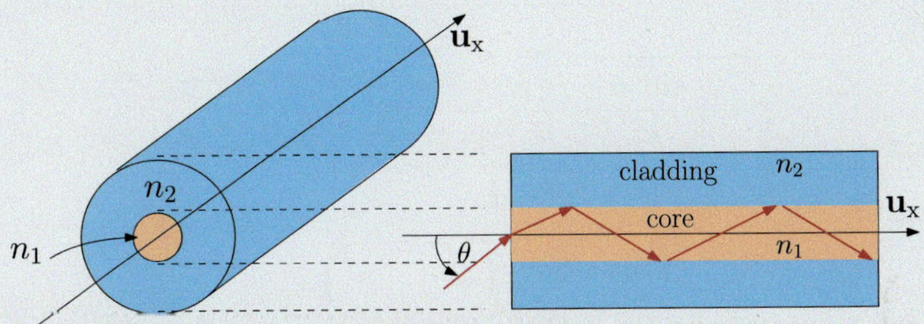

Fig. 15.10 Total internal reflection inside an optical fiber

If the wave enters the core forming an angle θ with respect to the axis of the fiber (and therefore with respect to the air-core interface), then the transmitted wave forms an angle α with the axis given by the Snell–Descartes law (15.10):

$$\sin \theta = n_1 \sin \alpha \ .$$

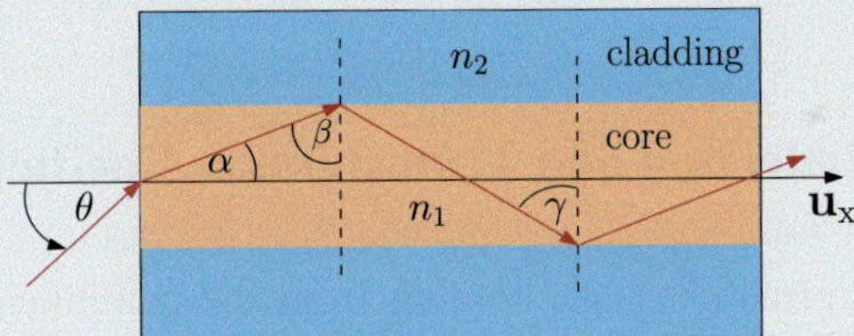

Fig. 15.11 If $\beta > \sin^{-1}(n_2/n_1)$, light will propagate inside the core of the fiber without escaping it

The wave will encounter the core-cladding interface forming an angle $\beta = \pi/2 - \alpha$ with respect to the normal. As shown in Fig. 15.11, the reflection law then assures that $\gamma = \beta$ so that all the subsequent reflections at the core-cladding interface occur at the same incident angle. Light will be confined inside the core if all these reflections are total, that is, if β is larger than the critical angle $\beta^* = \sin^{-1} \dfrac{n_2}{n_1}$, and so

$$\beta = \frac{\pi}{2} - \alpha \geq \sin^{-1} \frac{n_2}{n_1}$$

and in terms of θ

$$\frac{\pi}{2} - \sin^{-1}\left(\frac{\sin\theta}{n_1}\right) \geq \sin^{-1}\frac{n_2}{n_1} \,.$$

Equivalently,

$$\sin\left(\frac{\pi}{2} - \sin^{-1}\left(\frac{\sin\theta}{n_1}\right)\right) = \underbrace{\cos\left(\sin^{-1}\left(\frac{\sin\theta}{n_1}\right)\right)}_{\sqrt{1-\sin^2(\sin^{-1}(\sin\theta/n_1))}} \geq \frac{n_2}{n_1} \,.$$

Finally,

$$\sqrt{1 - \frac{\sin^2\theta}{n_1^2}} \geq \frac{n_2}{n_1} \quad \rightarrow \quad \sin\theta \leq \sqrt{n_1^2 - n_2^2} \,,$$

leading to

$$\theta \leq \sin^{-1}\left(\sqrt{n_1^2 - n_2^2}\right) \,.$$

In practice, no optical fiber is perfect; there will always be some absorption in the material, scattering due to impurities, and light escaping due to fiber bending, to name a few examples. Since the 1990s, optical amplifiers placed every few tens of kilometers across oceans have been used to compensate for losses in the optical fibers.

15.4 Fresnel Reflection and Transmission Coefficients

The reflection and refraction laws, coming from the continuity of the tangential component of the electric field at an interface, give us the propagation direction of the reflected and transmitted waves at an interface. They do not tell us, however, what fraction of the incident intensity is reflected and what fraction is transmitted. We know for example that a metallic mirror will reflect most of the incident light, whereas a transparent dielectric, such as a glass window, reflects only a small percentage of the incident intensity; most of it is transmitted through the material. In order to predict the partition of the energy flux between reflected and transmitted waves, we need to impose the other continuity conditions for the electromagnetic fields.

For a given polarization of the electric field, let us then write the incident, reflected and transmitted waves at an interface $x = x_0$ as follows

$$\text{incident wave } \underline{\mathbf{E}}^i(x_0^-, y, t) = \underline{\mathbf{E}}_0^i e^{i(\mathbf{k}_i \cdot \mathbf{x} - \omega t + \phi_i)}$$

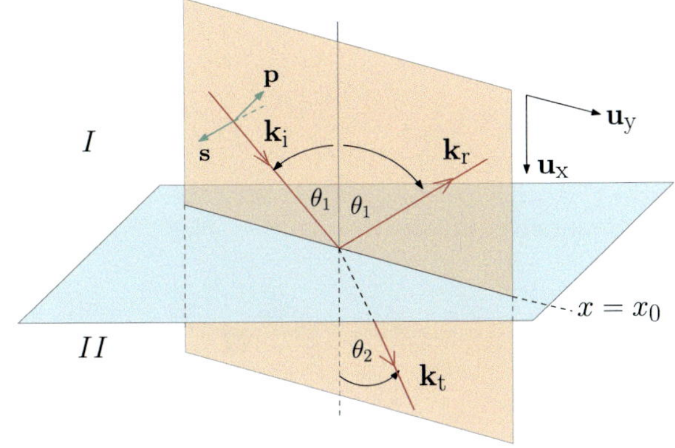

Fig. 15.12 The polarization at the interface can be decomposed into the s and p polarized components

$$\text{reflected wave } \underline{\mathbf{E}}^r(x_0^-, y, t) = \underline{\mathbf{E}}_0^r e^{i(\mathbf{k}_r \cdot \mathbf{x} - \omega t + \phi_r)}$$

$$\text{transmited wave } \underline{\mathbf{E}}^t(x_0^+, y, t) = \underline{\mathbf{E}}_0^t e^{i(\mathbf{k}_t \cdot \mathbf{x} - \omega t + \phi_t)}$$

where we know from the continuity of the tangential component of the electric field at the boundary $x = x_0$ that the arguments of the complex exponentials are the same for every y and t. This imply in particular the laws of reflection and refraction for the angles that the wave vectors form with the normal to the interface.

If we now impose the continuity conditions (15.1), (15.2) and (15.4) at the interface we may compute the amplitude of the reflected and transmitted waves in terms of the incident one, which will allow us to define a reflection r and transmission coefficient t, also called Fresnel coefficients, as follows

$$\underline{E}_0^r = \underline{r}\,\underline{E}_0^i \tag{15.11}$$

$$\underline{E}_0^t = \underline{t}\,\underline{E}_0^i. \tag{15.12}$$

Remarkably, these coefficients depend on the polarization of the incident wave, which in the most general case can be decomposed as a linear superposition of an s polarized wave[2] (electric field perpendicular to the plane of incidence) and a p-polarized wave (electric field in the plane of incidence). These two polarization directions are shown in Fig. 15.12.

The general reflection and transmission coefficients will be therefore a linear combination of the ones for s and p polarization that we separately calculate below.

15.4.1 Fresnel Coefficients for Perpendicular S-Polarization

A s-polarized wave is such that its polarization is perpendicular to the plane of incidence, and so its electric field is tangential to the interface. In Fig. 15.13, this

---

[2] From the German word *senkrecht*.

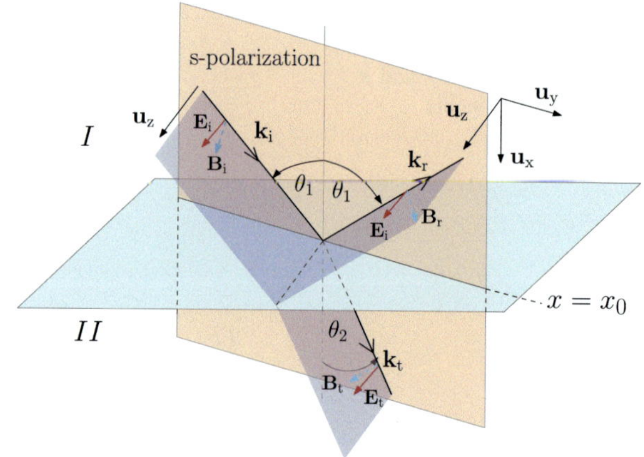

Fig. 15.13 Case of a s-polarized incident wave

corresponds to an electric field along z, so that at the interface, the amplitude of the electric field writes

$$\text{incident wave } \underline{\mathbf{E}}_0^i = \underline{E}_0^i \mathbf{u}_z$$
$$\text{reflected wave } \underline{\mathbf{E}}_0^r = \underline{E}_0^r \mathbf{u}_z$$
$$\text{transmited wave } \underline{\mathbf{E}}_0^t = \underline{E}_0^t \mathbf{u}_z$$

and since according to (15.3), the tangential component of the electric field is continuous, we have

$$\underline{E}_0^i + \underline{E}_0^r = \underline{E}_0^t \tag{15.13}$$

on the other hand, the incident, reflected and transmitted magnetic fields can be obtained from (13.15), which yields

$$\underline{\mathbf{B}}_0^i = \underbrace{\frac{n_1}{c}}_{k/\omega} (\cos\theta_1 \mathbf{u}_x + \sin\theta_1 \mathbf{u}_y) \times \underline{\mathbf{E}}_0^i = \frac{n_1}{c} \underline{E}_0^i (\sin\theta_1 \mathbf{u}_x - \cos\theta_1 \mathbf{u}_y)$$

and similarly

$$\underline{\mathbf{B}}_0^r = \frac{n_1}{c} \underline{E}_0^r (\sin\theta_1 \mathbf{u}_x + \cos\theta_1 \mathbf{u}_y)$$

$$\underline{\mathbf{B}}_0^t = \frac{n_2}{c} \underline{E}_0^t (\sin\theta_2 \mathbf{u}_x - \cos\theta_2 \mathbf{u}_y) \ .$$

It is easy to check that imposing the continuity condition (15.1) for the normal component of the magnetic field gives the same condition as (15.13). On the other hand, (15.4) imposes the continuity of the tangential component (along y) of $\mathbf{H}$, since the media are non-magnetic and we suppose that no current flows at the interface:

$$- \underline{E}_0 \, \underline{n}_1 \cos\theta_1 + \underline{E}_0^r \underline{n}_1 \cos\theta_1 = -\underline{E}_0^t \underline{n}_2 \cos\theta_2. \qquad (15.14)$$

Combining (15.13) and (15.14) allows us to find the Fresnel coefficients giving the reflected and transmitted amplitudes in terms of the incident one for a s-polarized wave:

$$\boxed{\begin{aligned} \underline{r}_s &= \frac{\underline{E}_0^r}{\underline{E}_0} = \frac{\underline{n}_1 \cos\theta_1 - \underline{n}_2 \cos\theta_2}{\underline{n}_1 \cos\theta_1 + \underline{n}_2 \cos\theta_2} \\[2mm] \underline{t}_s &= \frac{\underline{E}_0^t}{\underline{E}_0} = \frac{2\underline{n}_1 \cos\theta_1}{\underline{n}_1 \cos\theta_1 + \underline{n}_2 \cos\theta_2} \end{aligned}} \qquad (15.15)$$

which are, in general, complex numbers.

15.4.2 Fresnel Coefficients for Parallel P-Polarization

A p-polarized wave is such that its polarization is in the plane of incidence, and so its magnetic field is tangential to the interface, as shown in Fig. 15.14.

The amplitude of the electric field writes

$$\text{incident wave } \mathbf{E}_0^i = \underline{E}_0^i(-\sin\theta_1 \mathbf{u}_x + \cos\theta_1 \mathbf{u}_y)$$
$$\text{reflected wave } \mathbf{E}_0^r = \underline{E}_0^r(-\sin\theta_1 \mathbf{u}_x - \cos\theta_1 \mathbf{u}_y)$$
$$\text{transmited wave } \mathbf{E}_0^t = \underline{E}_0^t(-\sin\theta_2 \mathbf{u}_x + \cos\theta_2 \mathbf{u}_y)$$

and since according to (15.3), the tangential component of the electric field (along y) is continuous, we have

$$\underline{E}_0^i \cos\theta_1 - \underline{E}_0^r \cos\theta_1 = \underline{E}_0^t \cos\theta_2. \qquad (15.16)$$

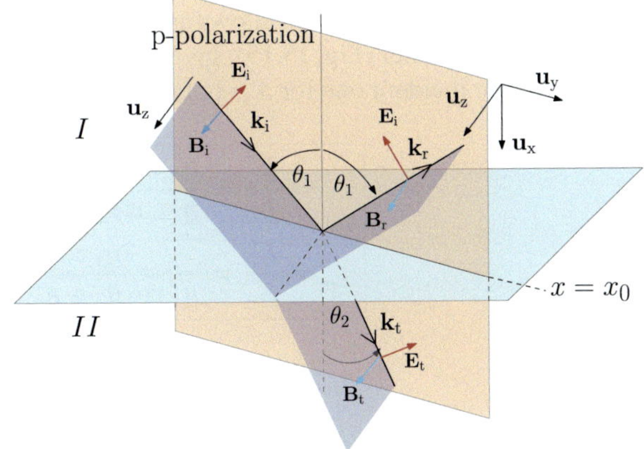

Fig. 15.14 Case of a p-polarized incident wave

Fig. 15.15 Augustin Jean Fresnel (1788–1827) was a French physicist. He made groundbreaking contributions to the wave theory of light by developing the theory of diffraction and polarization. He invented the Fresnel lens, widely used in lighthouses for its ability to focus light efficiently

On the other hand, the incident, reflected and transmitted magnetic fields can be obtained from (13.15), yielding

$$\mathbf{\underline{B}}_0^i = \underbrace{\frac{n_1}{c}}_{k/\omega} (\cos\theta_1\mathbf{u}_x + \sin\theta_1\mathbf{u}_y) \times \mathbf{\underline{E}}_0^i = \frac{n_1}{c}\underline{E}_0^i\mathbf{u}_z$$

and similarly

$$\mathbf{\underline{B}}_0^r = \frac{n_1}{c}\underline{E}_0^r\mathbf{u}_z$$

$$\mathbf{\underline{B}}_0^t = \frac{n_2}{c}\underline{E}_0^t\mathbf{u}_z \ .$$

Assuming a current-free interface, we can impose the continuity condition (15.4) of the tangential component (along z) of the magnetic field:

$$\underline{n}_1\,(\underline{E}_0 + \underline{E}_0^r) = \underline{n}_2\,\underline{E}_0^t. \tag{15.17}$$

Combining (15.16) and (15.17) allows us to find the Fresnel coefficients, named after Augustin Jean Fresnel (Fig. 15.15), giving the reflected and transmitted amplitudes in terms of the incident one for a p-polarized wave:

$$\boxed{\begin{aligned}\underline{r}_p &= \frac{\underline{E}_0^r}{\underline{E}_0} = \frac{\underline{n}_2\cos\theta_1 - \underline{n}_1\cos\theta_2}{\underline{n}_1\cos\theta_2 + \underline{n}_2\cos\theta_1}\\[2mm]\underline{t}_p &= \frac{\underline{E}_0^t}{\underline{E}_0} = \frac{2\underline{n}_1\cos\theta_1}{\underline{n}_1\cos\theta_2 + \underline{n}_2\cos\theta_1}\end{aligned}} \tag{15.18}$$

15.4.3 Brewster's Angle and Polarization by Reflection

Consider the case of two dielectric media far from any resonance so that $\underline{n}_1 = n_1 \in \mathbb{R}$ and $\underline{n}_2 = n_2 \in \mathbb{R}$. Remarkably, the reflection coefficient for a p-polarized wave vanish at a specific incident angle θ_B, called the Brewster angle, such that $r_p = 0$ in (15.18):

$$n_2 \cos \theta_B = n_1 \cos \theta_t^B = n_1 \sqrt{1 - \sin^2 \theta_t^B}$$

$$= n_1 \sqrt{1 - \left(\frac{n_1}{n_2}\right)^2 \sin^2 \theta_B}$$

and so

$$\left(\frac{n_2}{n_1}\right)^2 \cos^2 \theta_B = 1 - \left(\frac{n_1}{n_2}\right)^2 \sin^2 \theta_B \ .$$

Finally since $1 = \sin^2 \theta_B + \cos^2 \theta_B$

$$\frac{n_2^2 - n_1^2}{n_1^2} \cos^2 \theta_B = \frac{n_2^2 - n_1^2}{n_2^2} \sin^2 \theta_B$$

$$\boxed{\theta_B = \tan^{-1}\left(\frac{n_2}{n_1}\right) .} \tag{15.19}$$

In addition, when the incident angle equals the Brewster angle, the transmitted angle is such that

$$\sin \theta_t^B = \frac{n_1}{n_2} \underbrace{\sin(\tan^{-1}(\frac{n_2}{n_1}))}_{\theta_B} = \frac{n_1}{n_2} \frac{n_2/n_1}{\sqrt{1 + (n_2/n_1)^2}} = \frac{n_1}{\sqrt{n_1^2 + n_2^2}} = \cos \theta_B$$

and we conclude that $\theta_t^B = \frac{\pi}{2} - \theta_B$.

Note that no such angle exists in the case of s polarization. Indeed, imposing $r_s = 0$ in (15.15) would lead to $\tan^2 \theta_B < 0$. Figure 15.16 shows the general incident angle dependence of the reflection coefficient at an interface separating two dielectrics, the reflection vanishes at $\theta_i = \theta_B$ for p polarized light. Note that for $n_1 < n_2$ and in the limit of grazing incidence ($\theta_i \to \pi/2$), the reflection coefficient tends to 1. Thus, any dielectric interface becomes a good mirror at grazing incidence. In the case $n_1 > n_2$, there is total reflection for $\theta > \theta_C = \sin^{-1}(\frac{n_2}{n_1}) > \theta_B$.

The reflection of light at a dielectric interface is another way to obtain polarized light, alternative to the use of transmission polarizers discussed in Chap. 14. Indeed, when unpolarized light hits a dielectric interface at the Brewster angle, the reflected wave is completely s polarized, as shown in Fig. 15.17. It is said that this phenomenon

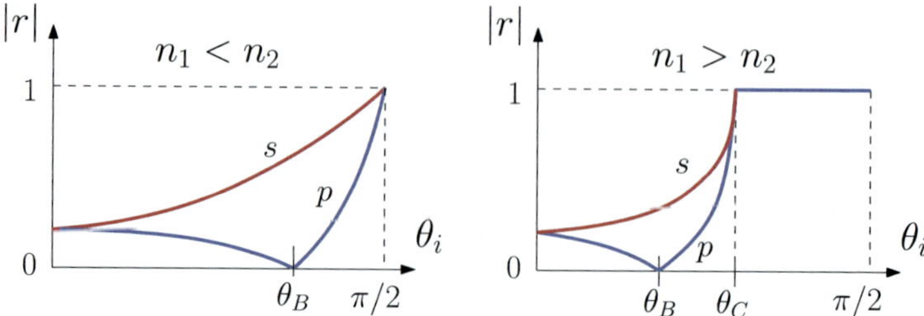

Fig. 15.16 Reflection coefficient as a function of the incident angles for $n_1 < n_2$ (left) and $n_1 > n_2$ (right)

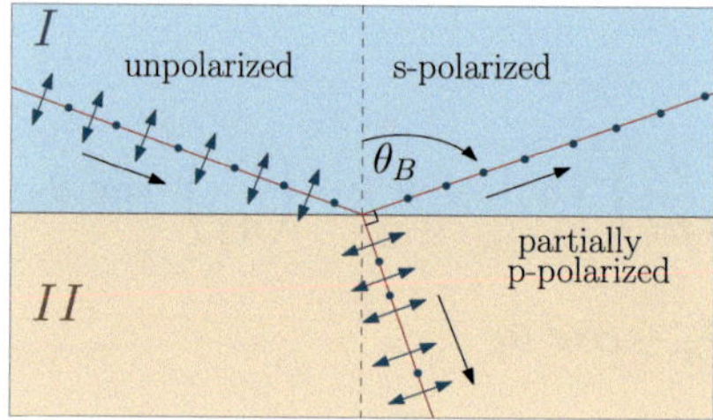

Fig. 15.17 Unpolarized light becomes s polarized after reflection at the Brewster angle at a dielectric interface

Fig. 15.18 David Brewster (1781–1868) was a Scottish physicist, inventor, and writer. He made significant advancements in understanding double refraction and the properties of crystals, and invented the kaleidoscope

of polarization by reflection was observed by Malus after sunlight was reflected by a glass window at the Brewster angle (Fig. 15.18).

In some laser cavities, a window tilted at the Brewster angle with respect to the optical axis (also called Brewster window) is used to control the polarization of the output laser beam. Indeed, a s polarized beam will lose intensity every time it encounters the Brewster window in the cavity, and so the amplified light emission, which makes several round trips after exiting the laser tube, will be exclusively p-polarized. This is the case of the helium–neon laser cavity shown in Fig. 15.19.

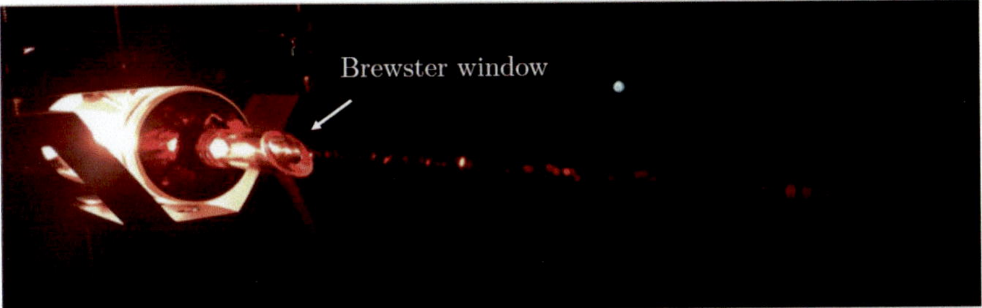

Fig. 15.19 A Helium–Neon laser

15.4.4 Fresnel Coefficients for an Arbitrary Polarization

In the most general case, the incident electric field amplitude can be always written as a linear superposition of s and p polarized fields

$$\underline{\mathbf{E}}_i = \alpha_1 \underline{\mathbf{E}}_s + \alpha_2 \underline{\mathbf{E}}_p \quad \alpha_1, \alpha_2 \in \mathbb{C}$$

for which the reflected and transmitted fields are simply

$$\underline{\mathbf{E}}_r = \alpha_1 \underline{r}_s \underline{\mathbf{E}}_s + \alpha_2 \underline{r}_p \underline{\mathbf{E}}_p$$

$$\underline{\mathbf{E}}_t = \alpha_1 \underline{t}_s \underline{\mathbf{E}}_s + \alpha_2 \underline{t}_p \underline{\mathbf{E}}_p .$$

15.4.5 Normal Incidence

Note that for the particular case of normal incidence, Snell–Descartes law gives $\theta_2 = \theta_1 = 0$, so that the reflection and transmission coefficients are equal for both s and p polarization:

$$\boxed{\begin{aligned} \underline{r}_p &= \frac{E_0^r}{\underline{E}_0} = \frac{\underline{n}_2 - \underline{n}_1}{\underline{n}_1 + \underline{n}_2} = -\underline{r}_s, \\[2mm] \underline{t}_p &= \frac{E_0^t}{\underline{E}_0} = \frac{2\underline{n}_1}{\underline{n}_1 + \underline{n}_2} = \underline{t}_p. \end{aligned}}$$

$$(15.20)$$

15.5 Reflectance and Transmittance

The Fresnel coefficients allows us to determine the amplitude of the reflected and transmitted electric fields at an interface relative to the incident amplitude. If we are interested in the flow of energy, the relevant parameters are the reflectance R and transmittance T defined as the fraction of the electromagnetic energy flux that is reflected and transmitted by the interface, respectively. For this, we can calculate the incident, reflected and transmitted complex Poynting vectors

$$\text{incident wave } \underline{\mathbf{\Pi}}_i = \frac{\mathbf{E}^i \times \mathbf{B}^{i*}}{2\mu_0} = \frac{|\underline{E}_0^i|^2}{2\mu_0\omega}\mathbf{k}_i^*$$

$$\text{reflected wave } \underline{\mathbf{\Pi}}_r = \frac{\mathbf{E}^r \times \mathbf{B}^{r*}}{2\mu_0} = |\underline{r}|^2\frac{|\underline{E}_0^i|^2}{2\mu_0\omega}\mathbf{k}_r^*$$

$$\text{transmited wave } \underline{\mathbf{\Pi}}_t = \frac{\mathbf{E}^t \times \mathbf{B}^{t*}}{2\mu_0} = |\underline{t}|^2\frac{|\underline{E}_0^i|^2}{2\mu_0\omega}\mathbf{k}_t^*$$

and recalling that the average power propagating along the normal to the interface (irradiance) is given by the projection of $\langle\mathbf{\Pi}\rangle_T = \mathrm{Re}\{\underline{\mathbf{\Pi}}\}$ along the normal (here $\mathbf{u}_x$), we find

$$\text{incident } \mathcal{I}^i = \mathrm{Re}\{\underline{\mathbf{\Pi}}_i\} \cdot \mathbf{u}_x = \frac{|\underline{E}_0^i|^2}{2\mu_0\omega}\mathrm{Re}\{\underline{k}_1^* \cos\theta_1^*\}$$

$$\text{reflected } \mathcal{I}^r = \mathrm{Re}\{\underline{\mathbf{\Pi}}_r\} \cdot (-\mathbf{u}_x) = |\underline{r}|^2\frac{|\underline{E}_0^i|^2}{2\mu_0\omega}\mathrm{Re}\{\underline{k}_1^* \cos\theta_1^*\}$$

$$\text{transmited } \mathcal{I}^t = \mathrm{Re}\{\underline{\mathbf{\Pi}}_t\} \cdot \mathbf{u}_x = |\underline{t}|^2\frac{|\underline{E}_0^i|^2}{2\mu_0\omega}\mathrm{Re}\{\underline{k}_2^* \cos\theta_2^*\}$$

and since $\underline{k}_1 = \frac{\omega}{c}\underline{n}_1$, $\underline{k}_2 = \frac{\omega}{c}\underline{n}_2$, we find the reflectance and transmittance of an interface in terms of the Fresnel coefficients

$$\boxed{\begin{aligned} R &= \frac{\mathcal{I}^r}{\mathcal{I}^i} = |\underline{r}|^2, \\[2mm] T &= \frac{\mathcal{I}^t}{\mathcal{I}^i} = |\underline{t}|^2\frac{\mathrm{Re}\{\underline{n}_2 \cos\theta_2\}}{\mathrm{Re}\{\underline{n}_1 \cos\theta_1\}}. \end{aligned}} \tag{15.21}$$

15.5.1 *Conservation of Energy Flux at the Interface*

Note that the reflectance and transmittance of an interface are not independent. This reflects the fact that energy must be conserved at the interface. From Eq. (15.21) we find, for a s-polarized wave:

$$\boxed{R + T = 1 + 2\frac{\mathrm{Im}\{\underline{n}_1 \cos\theta_1\}\mathrm{Im}\{\underline{r}_s\}}{\mathrm{Re}\{\underline{n}_1 \cos\theta_1\}}.} \tag{15.22}$$

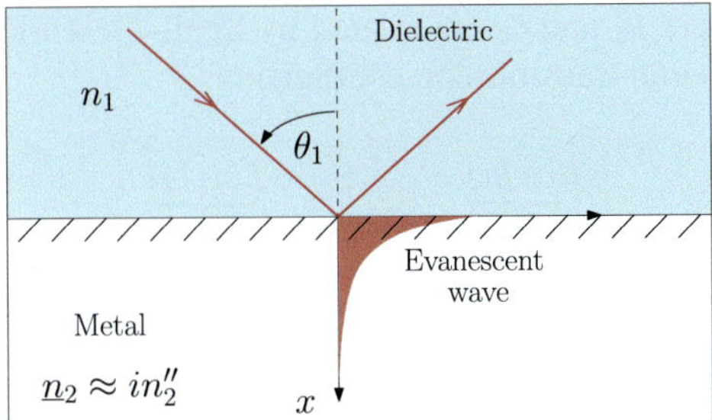

Fig. 15.20 A metal in the visible range reflects all of the incident electromagnetic energy

Remarks

- We see that R+T=1 in the absence of absorption.

- In the case of total reflection ($n_1 > n_2 \in \mathbb{R}$ and $\theta_1 > \sin^{-1}(\frac{n_2}{n_1})$), for which $\cos\theta_2$ is imaginary we find $T = 0$. Conservation of the energy flux dictates that $R = 1$, all the incident energy is reflected back by the interface.

15.5.2 Reflection by Metals

A metal in the visible range is such that the refractive index is essentially a pure imaginary number, $\underline{n}_2 \approx i n_2'' = i\sqrt{(\omega_p/\omega)^2 - 1}$. If an electromagnetic plane wave is incident from a dielectric medium without absorption, $\underline{n}_1 = n_1 \in \mathbb{R}$. From the Fresnel equations (15.15) and (15.18), we see that, regardless of the incident polarization, the wave is completely reflected by the metallic surface as illustrated in Fig. 15.20.

$$\underbrace{\left|\frac{n_1\cos\theta_1 - i n_2''\cos\theta_2}{n_1\cos\theta_1 + i n_2''\cos\theta_2}\right|^2}_{R_s} = 1 = \underbrace{\left|\frac{i n_2''\cos\theta_1 - n_1\cos\theta_2}{n_1\cos\theta_2 + i n_2''\cos\theta_1}\right|^2}_{R_p}$$

and we conclude from (15.22) that $T = 0$. The interface dielectric-metal then constitutes a mirror. In reality, an evanescent wave decays exponentially inside the metal within a typical characteristic length given by the skin depth $\delta = \frac{c}{\omega n_2''}$. In the idealization of a perfect conductor, for which the skin depth δ tends to zero, the electric and magnetic fields are strictly zero inside the metal.

15.5.3 The Stokes Relations

Consider an electromagnetic wave incident at the boundary separating two materials of refractive indexes $\underline{n}_1$ and $\underline{n}_2$, respectively. Depending on whether the wave is incident from region 1 with an angle of θ_1, or whether the wave comes from region

2 at an angle θ_2, where θ_1 and θ_2 are related by Snell–Descartes law, we will have two sets of reflection and transmission coefficients

$$\underbrace{\underline{r}_{12}, \underline{t}_{12}}_{\text{wave incoming from 1}} \qquad \underbrace{\underline{r}_{21}, \underline{t}_{21}}_{\text{wave incoming from 2}}$$

as shown in Fig. 15.21.

The Stokes relations establish a link between these coefficients:

$$\boxed{\underline{r}_{12} = -\underline{r}_{21} \qquad \underline{t}_{12}\underline{t}_{21} = 1 - \underline{r}_{12}^2 = 1 - \underline{r}_{21}^2.} \qquad (15.23)$$

The first relation states that the amplitude of the reflected wave is the same in both cases, but reflection from one side of the interface has a phase shift of π with respect to the reflection from the other side. The second relation in the case of real reflection and transmission coefficients is equivalent to the conservation of energy, since in that case

$$t_{12}t_{21} = \frac{4n_1 \cos\theta_1 n_2 \cos\theta_2}{(n_1 \cos\theta_1 + n_2 \cos\theta_2)^2} = T_{12} = T_{21} = T = 1 - R$$

with $R = \underline{r}_{12}^2 = \underline{r}_{21}^2$. The Stokes relations can be easily demonstrated algebraically from the Fresnel coefficients (15.15) and (15.18). Alternatively, a simple argument based on time-reversal symmetry may be employed in cases where there is no absorption. Indeed, consider first the situation in which a wave of unit amplitude is incident from medium 1, generating a reflected wave of amplitude $\underline{r}_{12}$ and a transmitted wave of amplitude $\underline{t}_{12}$ in medium 2. The system being reversible, the situation in which two waves of amplitude $\underline{t}_{12}$ and $\underline{r}_{12}$, coming from medium 2 and 1 respectively, meet at the interface should result in a wave propagating only in medium 1 of unit amplitude, as shown in Fig. 15.22.

However, both waves generate independently reflected and refracted waves in mediums 1 and 2, and so the situation stated above is the superposition of the two processes shown in Fig. 15.23 and we conclude

$$1 = \underline{t}_{12}\underline{t}_{21} + \underline{r}_{12}^2$$

$$0 = \underline{t}_{12}\left(\underline{r}_{21} + \underline{r}_{12}\right)$$

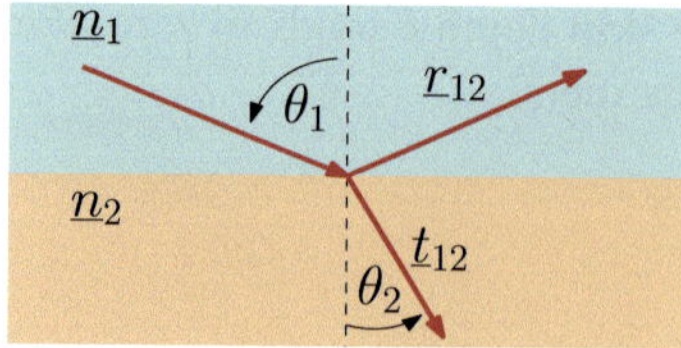

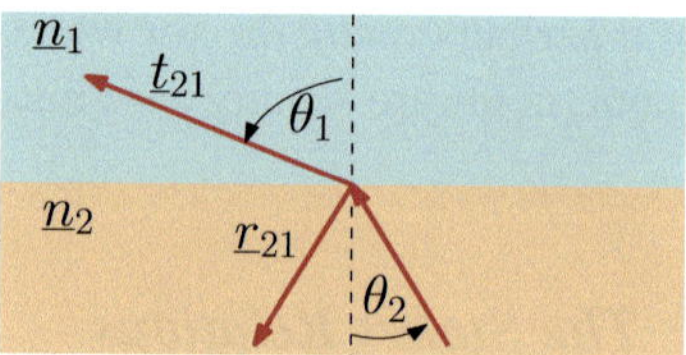

Fig. 15.21 A wave coming from medium 1 at angle θ_1 (left), and a wave coming from medium 2 at angle θ_2 (right), with θ_1 and θ_2 related by Snell–Descartes law

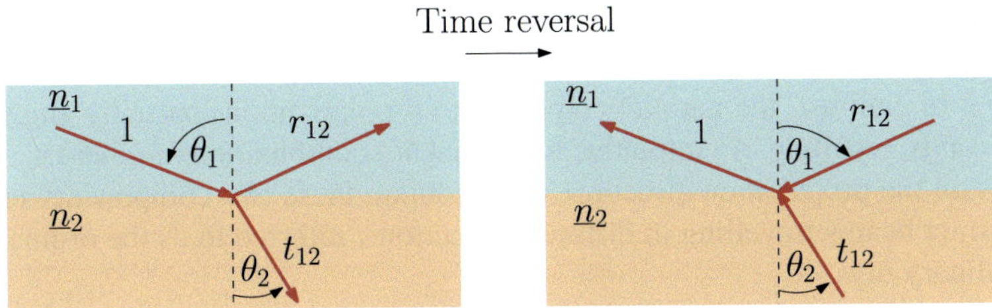

Fig. 15.22 The two processes shown in the figure are related by time reversal

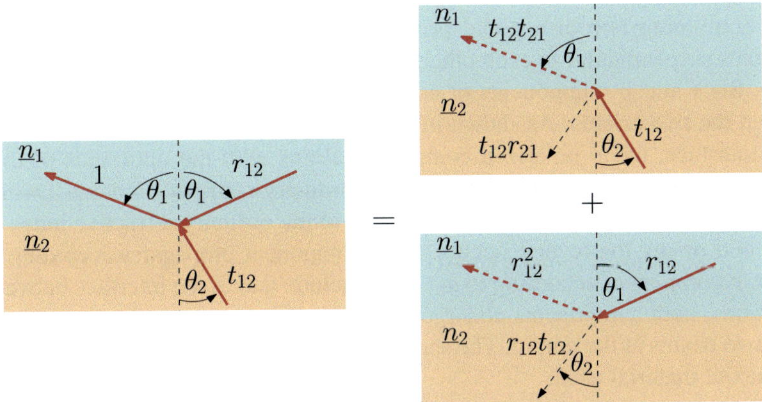

Fig. 15.23 Each incident wave will produce a refracted and a transmitted wave

from which we obtain the Stoke relations.

15.6 Birefringence

Certain anisotropic materials exhibit a refractive index that depends on both the polarization and the propagation direction of light. One such material is calcite, which demonstrates birefringence, also known as double refraction—a phenomenon that causes objects viewed through a clear piece of calcite to appear doubled. This effect was first explained by Christiaan Huygens in 1690.

Calcite is an example of a uniaxial anisotropic material, representing the simplest case of birefringence. In these materials, the optical properties are invariant under rotation about a specific axis called the optic axis. A light wave propagating parallel to the optic axis has its electric field necessarily perpendicular to this axis, resulting in a refractive index of n_o, regardless of the polarization of the incident light. Conversely, when the electric field is aligned parallel to the optic axis, the refractive index is n_e. As illustrated in Fig. 15.24, any incident electric field can be decomposed into two components: one perpendicular and the other parallel to the optic axis. The perpendicular component (s-polarization) is refracted according to the ordinary refractive index n_o. In contrast, the parallel component (p-polarization) partially aligns with the optic axis, resulting in a refractive index that is a combination of n_o and n_e, which depends on the propagation direction. At the ouput, these two components separate into distinct beams travelling in different directions, referred to as the ordinary and extraordinary rays.

An object viewed through calcite appears doubled due to this phenomenon. When a linear polarizer is placed between the calcite and the observer, the two images are shown to be linearly polarized along orthogonal directions, as demonstrated in Fig. 15.25.

By combining two orthogonal prisms made of a uniaxial material, with their optic axes perpendicular to each other, one can construct a Wollaston prism. In this device, the s and p components of an incident beam are refracted at the interface between the two prisms. As shown in Fig. 15.26, after normal incidence at the air-prism interface, the p-polarized component aligns with the optic axis of the first prism and propagates according to the extraordinary refractive index n_e, while the s-polarized component propagates according to the ordinary refractive index n_o. In the second prism, the reverse occurs: the s component propagates according to n_e, and the p component according to n_o. We conclude that at the interface between the two prisms, both components undergo opposite refraction, resulting in a separation of the two beams at the output. The separation angle depends on the wavelength and the uniaxial material used.

15.7   The Fabry–Pérot Resonator (1899)

The Fabry–Pérot (Fig. 15.27) resonator is an optical cavity made of two partially-reflecting surfaces separated by a distance e. These can be for example two highly reflective mirrors deposited onto a transparent substrate, where anti-reflection coatings on the back sides minimize the effect of parasite reflections on the glass-air interface. This is illustrated in Fig. 15.28. In some cases the cavity can be simply a

Fig. 15.24 Double refraction in a uniaxial biregringent material

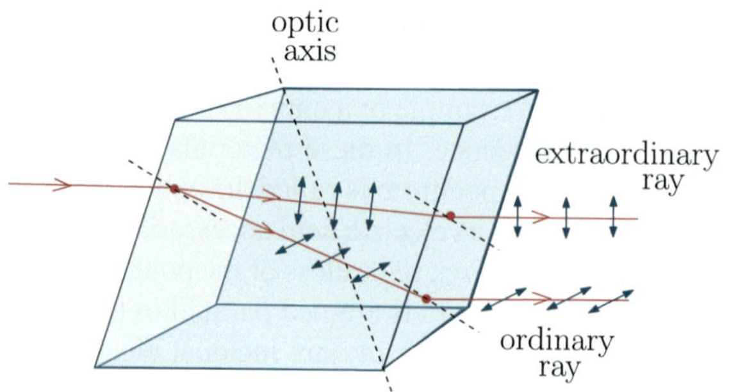

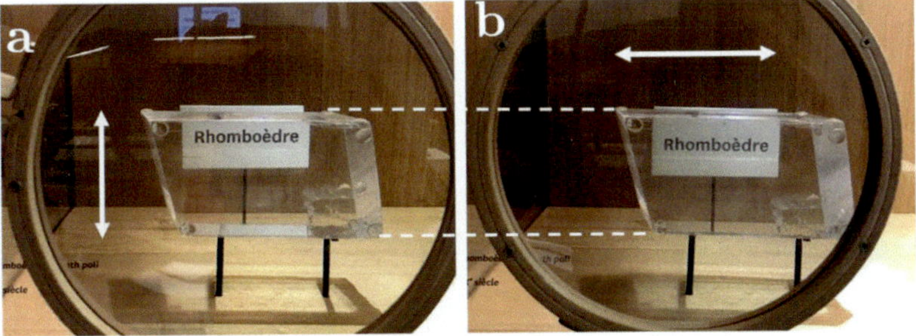

Fig. 15.25 A linear polarizer is placed between the birefringent material and the observer, with the polarization axis vertical (**a**) or horizontal (**b**). Pictures taken at the Museum of Ecole Polytechnique (Mus'X)

Fig. 15.26 A Wollastom prism is constructed by gluing together two right-angle prisms of a uniaxial birefringent material

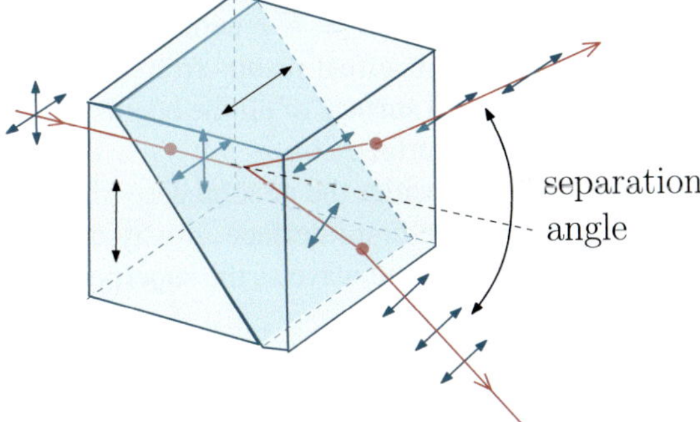

thin dielectric plate whose refractive index provides a high reflection at the interface with air.

Fig. 15.27 Left: Maurice Paul Auguste Charles Fabry (1867–1945), a French physicist who made significant contributions to optics and atmospheric science. He played a key role in discovering the ozone layer. Right: Alfred Pérot (1863–1925), a French physicist. His work in optics and spectroscopy helped lay the foundation for advancements in precision measurements and interferometric techniques

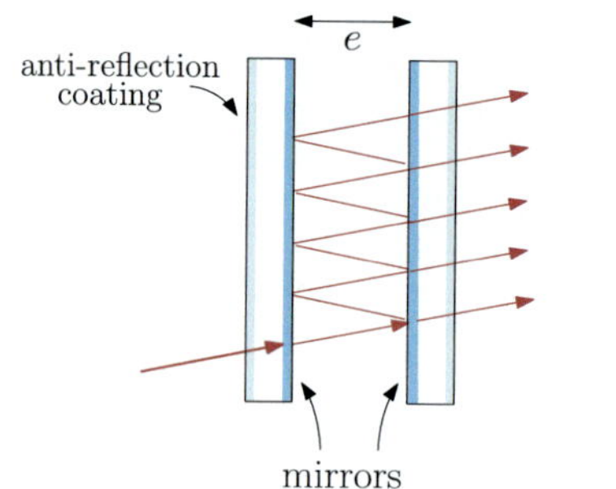

Fig. 15.28 A Fabry–Pérot optical cavity. Picture taken at the Museum of Ecole Polytechnique (Mus'X)

Only light at resonance with the cavity can pass through, since the transmittance is close to 1 whenever $n\pi = \frac{2\pi}{\lambda}e$ with $n \in \mathbb{N}$, where λ is the wavelength of the wave inside the cavity. This effect results from the constructive (destructive) interference at the exit (entrance) surface of all the beams that were subjected to none or multiple reflections at the interfaces.

To show this, consider the general case of an incoming wave incident at an angle θ_i with respect to the first interface. A way of calculating the transmittance consists in considering the output wave as the superposition of the wave $\underline{E}_0$ that went directly

Fig. 15.29 The field at the output can be seen as the superposition of an infinite number of fiels, each produced by a increasing number of round trips inside the cavity

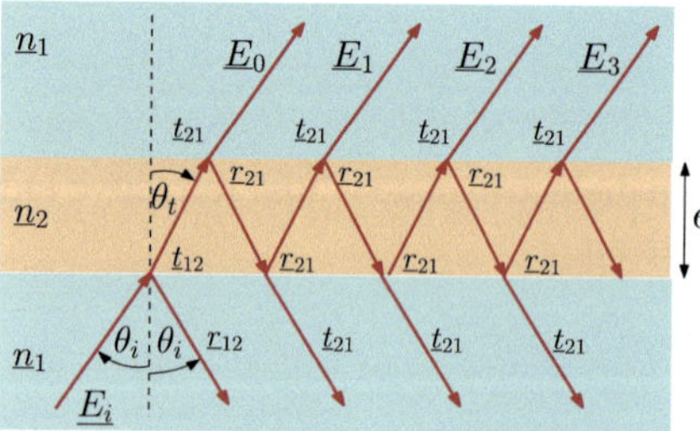

Fig. 15.30 For a given wavelength, a diverging source will produce a series of rings at the output of the cavity, for which the transmission is maximum

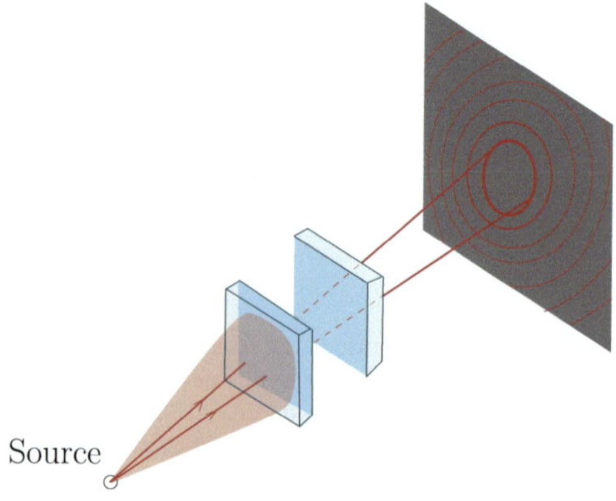

through the two interfaces and the infinite number of waves $(\underline{E}_m)_{m\in\mathbb{N}^+}$, where $\underline{E}_m$ is the wave that performed m round trips inside the cavity, see Fig. 15.29. We have

$$\underline{E}_m = \underline{E}_{m-1}(r_{21})^2 e^{i\,\overbrace{2n_2 ke\cos\theta_t}^{\Delta\phi}} \qquad n \in \mathbb{N}^+$$

and so $\underline{E}_m = \underline{E}_0(r_{21})^{2m}e^{im\Delta\phi}$ $n \in \mathbb{N}^+$ with $\underline{E}_0 = \underline{E}_i t_{12}t_{21}e^{i\Delta\phi/2}$. The total electric field amplitude at the exit of the cavity is then given by the series:

$$\underline{E}_t = \sum_{m=0}^{\infty} \underline{E}_m = \underline{E}_i t_{12}t_{21}e^{i\Delta\phi/2}\left(\sum_{m=0}^{\infty} r_{21}^2 e^{im\Delta\phi}\right) = \underline{E}_i e^{i\Delta\phi/2}\frac{t_{12}t_{21}}{1 - r_{21}^2 e^{i\Delta\phi}} = \underline{E}_i e^{i\Delta\phi/2}\frac{(1-R)^2}{1-Re^{i\Delta\phi}}$$

where we have used the Stokes relations (15.23). The transmittance can be written in terms of the finesse $\mathcal{F} = 2\sqrt{R}/(1-R)$ as

$$\boxed{T = \left|\frac{\underline{E}_t}{\underline{E}_i}\right|^2 = \frac{1}{1 + \mathcal{F}^2 \sin^2(\frac{\Delta\phi}{2})}.} \qquad (15.24)$$

We see that only certain wavelengths in resonance with the Fabry–Pérot cavity are transmitted. Note also that, for a given wavelength, there will be a modulation of the transmittance with the incident angle since $\Delta\phi \propto \cos\theta_t$. In consequence, if a light beam having a certain angular spread, for example if a point-like source is placed at a finite distance to the Fabry–Pérot cavity, then all the rays incoming at the same angle with respect to the normal to the mirrors will have the same transmittance. When a screen is placed behind the Fabry–Pérot, a series of alternating bright and dark rings will appear as shown in (Fig. 15.30): the bright ones correspond to those rays impinging at the front surface at an angle such that the transmittance is one.

Fig. 15.31 For large values of the finesse, the transmission exhibit sharp peaks as a function of the incident wavelength

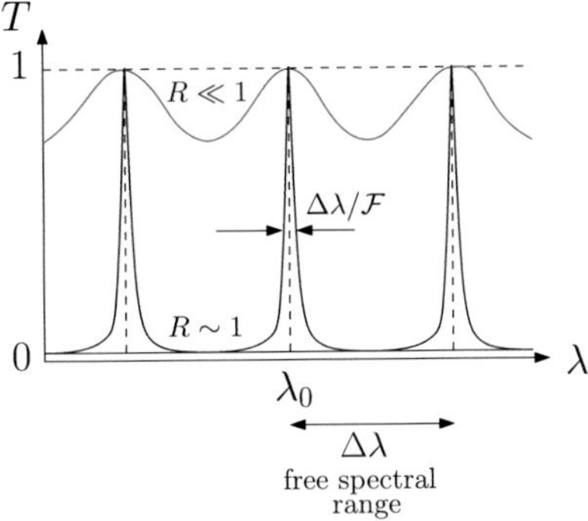

15.7.1   Wavelength Selection of a Fabry–Pérot

Suppose that for some $p \in \mathbb{N}$, $\frac{\Delta\phi}{2} = p\pi = n_2 k e \cos\theta_t$. Let $\lambda_0 = 2\pi/k$ be the wavelength in resonance with this maximum of transmittance, so that

$$\lambda_0 = \frac{2n_2 e}{p} \cos\theta_t \;.$$

Then, the separation $\Delta\lambda$ between two consecutive resonance peaks of transmittance around λ_0, also called the free spectral range, writes

$$\Delta\lambda = 2n_2 e \cos\theta_t \left(\frac{1}{p} - \frac{1}{p+1}\right) = 2n_2 e \cos\theta_t \frac{1}{p(p+1)} = \frac{\lambda_0}{p+1} = \frac{\lambda_0^2}{2n_2 e \cos\theta_t + \lambda_0}$$

and since typically $\lambda_0 \ll 2n_e e \cos\theta_t$, the free spectral range reads

$$\boxed{\Delta\lambda \approx \frac{\lambda_0^2}{2n_2 e \cos\theta_t}.}$$

$$(15.25)$$

The resonances are narrower for large values of the finesse (i.e., high reflectance R of the two surfaces) as shown in Fig. 15.31. Indeed, according to (15.24), the full width at half maximum of a transmission peak around $\Delta\phi/2 = p\pi = 2\pi n_2 e/\lambda_0 \cos\theta_t$ is determined by a change $\delta\phi$ in the phase which, for narrow peaks, is such that $\delta\phi \ll 2\pi$ and

$$\mathcal{F}^2 \underbrace{\sin^2\left(p\pi + \frac{\delta\phi}{2}\right)}_{1/2(1-\cos(2p\pi+\delta\phi))} = 1 \;.$$

For a small $\delta\phi$ we have $\cos(2p\pi + \delta\phi) = \cos\delta\phi \approx 1 - (\delta\phi)^2/2$ and we obtain

$$\frac{\delta\phi^2}{4} = \frac{1}{\mathcal{F}^2} \quad \rightarrow \quad \delta\phi = \frac{2}{\mathcal{F}}.$$

$$(15.26)$$

In terms of wavelength, this corresponds to a change in $\delta\lambda$ that we can find by writing

$$\Delta\phi = \frac{4\pi n_2 e \cos\theta_t}{\lambda_0} \quad \rightarrow \quad \delta\phi = -\frac{4\pi n_2 e \cos\theta_t}{\lambda_0^2}\delta\lambda = -\frac{2}{\Delta\lambda}\delta\lambda$$

and we conclude

$$\boxed{\mathcal{F} = \frac{\Delta\lambda}{\delta\lambda}.}$$

$$(15.27)$$

The Fabry–Pérot resonator has several applications, and we briefly discuss some of them below.

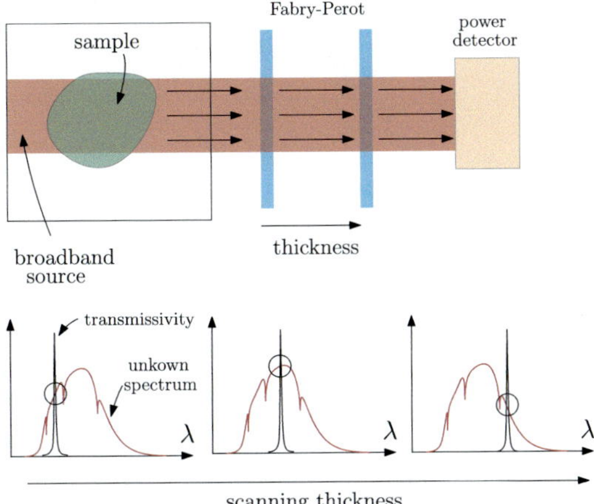

Fig. 15.32 A Fabry–Pérot is typically used as an spectral filter for spectroscopy

- *Spectroscopy*: the very narrow wavelength range that is transmitted by a Fabry–Pérot resonator can be used to analyze the light emitted or absorbed by a sample. By scanning the cavity length and measuring the output intensity, one can reconstruct the emission or absorption spectrum of a given sample provided that the cavity has an appropriate free spectral range and finesse, as illustrated in Fig. 15.32. The higher the finesse of the resonator, the higher the spectral resolution with which one can reconstruct the spectrum of a sample.

  Such a system can be also used as an optical filter. For example, the distance between the two mirrors can be tuned to isolate the emission stemming from a precise atomic resonance and to filter other neighboring lines. These are called interferential filters.

- Lasers are typically made of a Fabry–Pérot cavity[3] inside which a medium with optical gain (gaz, solid or liquid) is placed, as shown in Fig. 15.33.

  The two mirrors of the cavity are highly reflective, allowing for the light to make several round trips before escaping the cavity through one of the two mirrors which has a smaller reflectivity. At each round trip, the light is amplified by the gain medium by a process called stimulated emission, predicted by Einstein roughly 45 years before the invention of the first laser in 1960. In steady state, the gain in a round trip compensate exactly the losses (absorption, transmission through the mirrors, fluoresence). The weak losses of light at the exit mirror constitute the actual laser beam that gets out of the laser cavity.

- The Fabry–Pérot resonator is often called an etalon when the length e of the cavity is fixed. With a well-chosen finesse and free-spectral range, it can select a precise wavelength, and this is used for example inside laser cavities in order to select a single mode out of the multiple possible modes that can eventually lase, giving the laser a more stable frequency output.

---

[3] Although in most lasers, the mirrors are concave in order to minimize diffraction losses.

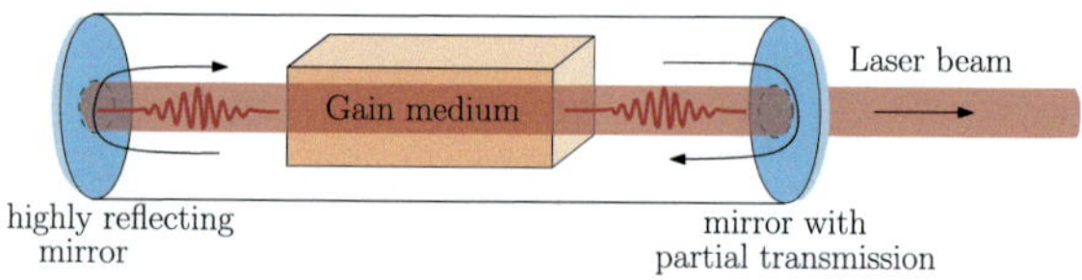

Fig. 15.33 A laser can be constructed by placing a medium with optical gain inside a Fabry–Pérot cavity

- In gravitational wave detection, a Michelson interferometer is employed in which a laser beam is split in two beams that are recombined after traveling different paths. The interference pattern is extremely sensitive to tiny changes to the length of the paths of the two split beams, which can be modified for example by the passage of a gravitational wave. In both arms there is a Fabry–Pérot cavity used to store the light for almost a millisecond as it bounces back and forth between the mirrors, thus increasing the interaction time between the gravitational wave and light inside the cavity. This is schematically shown in Fig. 15.34.

15.8 Summary and Essential Formulas

- When an electromagnetic wave is incident on an interface separating two distinct media, it splits into a reflected and a transmitted (refracted) wave. If $\mathbf{u}_x$ is the normal to the interface, and $\underline{\mathbf{k}}_i$ the wave vector of an incident plane wave, then the plane of incidence is defined as the plane containing both $\mathbf{u}_x$ and $\mathrm{Re}\{\underline{\mathbf{k}}_i\}$.

By imposing the continuity of the tangential component of the electric field at the interface, one concludes that the real parts of the reflected $\mathbf{k}_r$ and transmitted $\mathbf{k}_t$ wave vectors lie also in the plane of incidence. Moreover, by writing $\underline{\mathbf{k}}_i \cdot \mathbf{u}_x =$

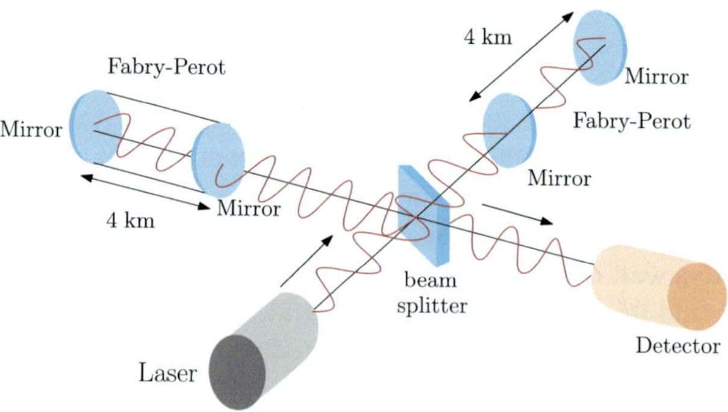

Fig. 15.34 An interferometer used to detect gravitational waves

$\underline{k}_1 \cos \theta_i$, $\mathbf{k}_r \cdot \mathbf{u}_x = -\underline{k}_1 \cos \theta_r$, and $\mathbf{k}_t \cdot \mathbf{u}_x = \underline{k}_2 \cos \theta_t$, the continuity condition gives both the reflection law

$$\boxed{\theta_i = \theta_r}$$

and Snell–Descartes law

$$\boxed{\underline{n}_1 \sin \theta_i = \underline{n}_2 \sin \theta_t}$$

where $\underline{n}_1$ is the refractive index of the medium from which the wave is incident and $\underline{n}_2$ is the refractive index of the medium on which light is refracted.

- When $n_1 > n_2 \in \mathbb{R}$, there is total reflection for angles of incidence above a critical value

$$\theta_i^* = \sin^{-1}\left(\frac{n_2}{n_1}\right)$$

in this case the transmitted wave only propagates along the interface whereas its amplitude decays exponentially in the direction perpendicular to the interface (evanescent wave). An optical fiber operates according to this principle; light is confined inside the fiber core since the fields are totally reflected at its boundary.

- If $\underline{E}_0^i$ is the incident electric field amplitude, the Fresnel reflection $\underline{r}$ and transmission $\underline{t}$ coefficients are used to express the reflected ($\underline{E}^r$) and transmitted ($\underline{E}^t$) electric field amplitude at an interface according to

$$\underline{E}^r = \underline{r}\, \underline{E}_0^i \, ,$$

$$\underline{E}^t = \underline{t}\, \underline{E}_0^i \, .$$

For a s-polarized wave (electric field perpendicular to the plane of incidence), the Fresnel coefficients are

$$\boxed{\begin{aligned}
\underline{r}_s &= \frac{\underline{E}_0^r}{\underline{E}_0} = \frac{\underline{n}_1 \cos \theta_1 - \underline{n}_2 \cos \theta_2}{\underline{n}_1 \cos \theta_1 + \underline{n}_2 \cos \theta_2}, \\
\underline{t}_s &= \frac{\underline{E}_0^t}{\underline{E}_0} = \frac{2\underline{n}_1 \cos \theta_1}{\underline{n}_1 \cos \theta_1 + \underline{n}_2 \cos \theta_2},
\end{aligned}}$$

whereas for a p-polarized wave (electric field in the plane of incidence), they are

$$\boxed{\begin{aligned}
\underline{r}_p &= \frac{\underline{E}_0^r}{\underline{E}_0} = \frac{\underline{n}_2 \cos \theta_1 - \underline{n}_1 \cos \theta_2}{\underline{n}_1 \cos \theta_2 + \underline{n}_2 \cos \theta_1}, \\
\underline{t}_p &= \frac{\underline{E}_0^t}{\underline{E}_0} = \frac{2\underline{n}_1 \cos \theta_1}{\underline{n}_1 \cos \theta_2 + \underline{n}_2 \cos \theta_1}.
\end{aligned}}$$

At normal incidence, we have

$$\underline{r}_p = -\underline{r}_s = \frac{\underline{n}_2 - \underline{n}_1}{\underline{n}_2 + \underline{n}_1}$$

and

$$\underline{t}_s = \underline{t}_p = \frac{2\underline{n}_1}{\underline{n}_1 + \underline{n}_2} \, .$$

- For a p-polarized wave, there exists a particular incident angle, called the Brewster angle θ_B for which there is no reflection at the interface ($\underline{r}_p = 0$). If an initially unpolarized beam of light strikes an interface at Brewster's angle, then the reflected wave is perfectly s-polarized and the transmitted wave partially p-polarized. Such an interface acts as a linear polarizer. Brewster windows are used in laser cavities which transmit the p-polarized component of the electric field but introduce losses for the s-polarized component. The laser beam at the output is therefore p-polarized.
- The reflectance R and transmittance T represent the fraction of the incident electromagnetic energy flux that is reflected and transmitted by the interface, respectively. They are related to the Fresnel reflection and transmission coefficients by

$$\boxed{\begin{aligned} R &= \frac{\mathcal{I}^r}{\mathcal{I}^i} = |\underline{r}|^2, \\[2mm] T &= \frac{\mathcal{I}^t}{\mathcal{I}^i} = |\underline{t}|^2 \frac{\mathrm{Re}\{\underline{n}_2 \cos\theta_2\}}{\mathrm{Re}\{\underline{n}_1 \cos\theta_1\}} \, . \end{aligned}}$$

Conservation of the energy flux at an interface in the absence of absorption yields

$$R + T = 1 \, .$$

Problems

15.1 The anti-reflection coating

Glass, one of the most employed materials in optics, has a refractive index of $n_{\mathrm{glass}} \sim 1.5$.

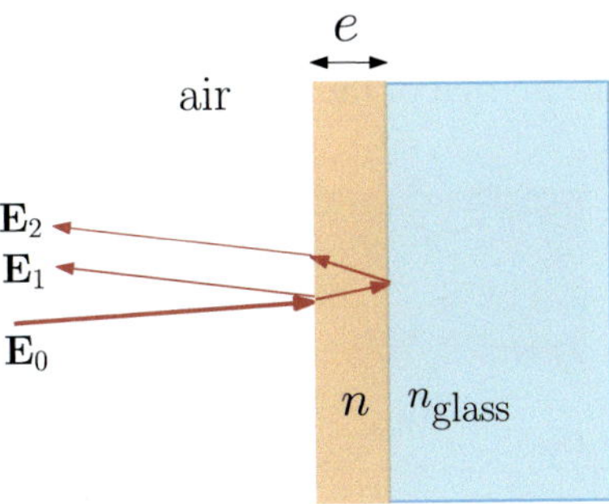

(a) What is the reflectance $R_{\text{air-glass}}$ under normal incidence of the air-glass interface? What fraction of a light beam intensity is transmitted through a glass plate surrounded by air? You may neglect multiple reflections inside the plate (Why?)

(b) Modern objectives are made of several glass components in order to correct optical aberrations. What is the fraction of transmitted light of an objective having 5 glass components (and therefore 10 air-glass interfaces)?

In order to minimize reflections, one deposits on top of glass a thin dielectric layer of thickness e and refractive index n, with $1 < n < n_{\text{glass}}$. The goal of this part is to determine the thickness e and the refractive index n that will minimize the reflectance under normal incidence. Consider an incident field of amplitude $\underline{E}_0$, and we will focus on the two beams $\underline{E}_1$ and $\underline{E}_2$ reflected by the air-dielectric and dielectric-glass interfaces, respectively. Multiple reflections inside the dielectric layer will be neglected.

(c) Write the amplitude of the fields $\underline{E}_1$ and $\underline{E}_2$ in terms of $\underline{E}_0, e, n$ and the reflection coefficients $r_{\text{air-diel}}$ and $r_{\text{diel-glass}}$ for an incident light of wavelength λ.

(d) What are the possible values of e that guarantees that the two reflected fields $\underline{E}_1$ and $\underline{E}_2$ are out of phase? What is the minimum thickness e of dielectric that must be deposited?

(e) Assuming that the fields $\underline{E}_1$ and $\underline{E}_2$ are out of phase, what is the required refractive index n for these fields to interfere destructively?

(f) Anti-reflection coatings are typically obtained by depositing magnesium fluoride $n = 1.38$ on top of glass. Determine the minimum thickness of MgF_2 that should be deposited and the reflectance of a glass-air interface in the visible range ($\lambda = 500\,\text{nm}$) for a glass treated with such a coating.

15.2 The prism spectrometer

Since the refractive index depends on wavelength, refraction can be used in order to separate the different wavelengths present in a beam of light. Isaac Newton demonstrated in this way that white light bent through a prism contained all the visible colors, and William Herschel discovered in 1800 that a thermometer placed beyond the red part of the refracted beam was heated by an invisible light now called infrared. Consider a prism made of a glass of index $n(\lambda)$, of rectangular base and rectangular sides. A monochromatic beam of wavelength λ is incident from air ($n \approx 1$) at an angle i with respect to the normal of the entrance surface, as shown below.

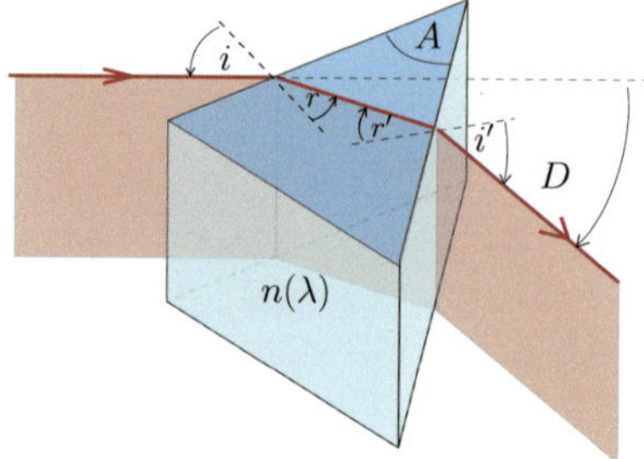

(a) From Snell–Descartes law and the geometry of the prism, find an expression for $\sin i'$ in terms of i, A and $n(\lambda)$.

(b) Show that the beam is deviated by a total angle D given by

$$D = i + i' - A .$$

How does the deviation angle D evolve with the refractive index n?

(c) According to Cauchy's law, the refractive index of a transparent dielectric far and below an absorption resonance is given by

$$n(\lambda) = A + \frac{B}{\lambda^2} ,$$

with $B > 0$. Which color of the visible spectrum is more deviated by the prism?

(d) Show that the deviation angle D has a minimum D_m when $i = i' = i_m$, and determine the value of the incidence angle i_m that produces this minimum deviation. Finally, obtain an expression of D_m in terms of $n(\lambda)$ and A.

15.3 Deviation of sunlight by a water droplet

Consider a spherical water droplet of refractive index n, and an incident monochromatic wave incident at an angle i with respect to the normal to the interface water-air.

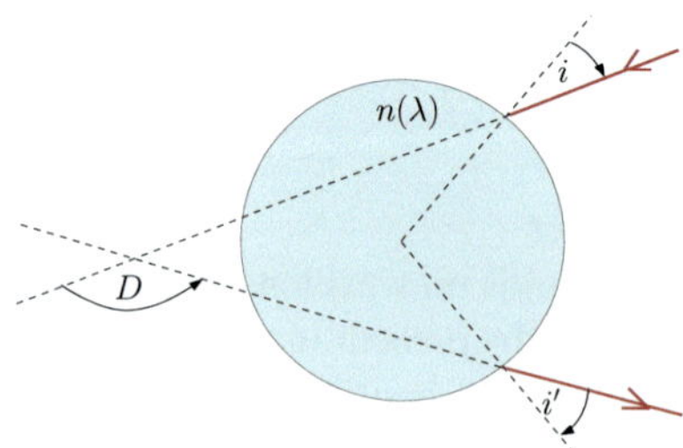

(a) Consider that the incoming beam is reflected once inside the droplet. Determine the angle i' at which the beam exits the droplet and the deviation angle D as a function of i and n.

(b) Determine the minimum deviation angle D_m and the value of $i = i_m$ for which the deviation angle reaches its minimum. What is the value of D_m for a water droplet ($n = 1.33$)?

(c) In reality the refractive index decreases with increasing wavelength in the visible range. Which part of the visible spectrum will be more deviated by the droplet?

15.4 The optical fiber

An optical fiber is made of a cylindrical dielectric medium of refractive index n_1 called the core and in which light propagates. The core ($r < a$) is surrounded by a second, concentric cylinder of refractive index n_2. The indexes are chosen such that $n_2 < n_1$ so that light can be confined in the core by ensuring total reflections at the core-cladding interface.

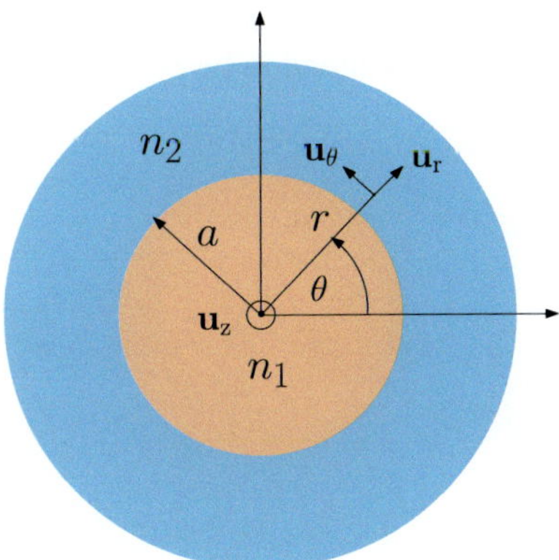

(a) Suppose a monochromatic wave propagating along the z-axis in region 1, which in cylindrical coordinates writes

$$\underline{\mathbf{E}}_i(r, \theta, z, t) = E_i(r, \theta)e^{i(kz - \omega t)}\mathbf{u} \,,$$

where $\mathbf{u} \perp \mathbf{u}_z$ is the wave polarization.

What is the equation that $E_i(r, \theta)$ must satisfy if medium 1 has a constant refractive index n_i?

(b) We write down the solution under the form $E_i(r, \theta) = \mathcal{E}_i(r)e^{\pm il\theta}$ where $l \in \mathbb{N}$. Show that $\mathcal{E}_i$ is therefore a solution of

$$r^2 \frac{d^2 \mathcal{E}_i}{dr^2} + r \frac{d\mathcal{E}_i}{dr} + \left(\frac{r^2}{a^2} q_i(r)^2 - l^2 \right) \mathcal{E}_i = 0$$

and determine $q_i(r)$.

(c) The solutions of the equation

$$x^2\frac{d^2y}{dx^2} + x\frac{dy}{dx} + \left(q^2\frac{x^2}{a^2} - l^2\right)y = 0$$

are called Bessel functions. Here, we will only consider Bessel functions that do not diverge at $x = 0$ or when $x \to \infty$. If $q^2 > 0$, the solution writes in terms of the Bessel function of the first kind, $y(x) = J_l(qx/a)$, which oscillates while decaying slowly to zero as x increases. If $q^2 < 0$, the solution writes in terms of the Bessel function of the second kind, $y(x) = K_l(hx/a)$ where $h^2 = -q^2 > 0$, which decays exponentially as x increases. What are the possible values of k for which the wave is confined in the core of the fiber? Show that this is equivalent to the condition found in Example 15.2 with the Snell–Descartes law for the maximum angle of entrance in the fiber.

15.5 Reflection of a circularly polarized wave on a metallic conductor

A circularly polarized plane wave propagates along the x-axis towards the surface of a conductor located at $x = 0$, assumed to be perfect. The incident electric field is expressed as

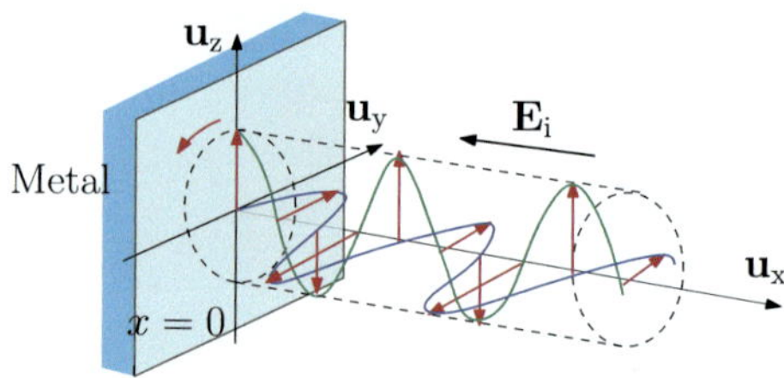

$$\mathbf{E}_i = E_0(\cos(\omega t + kx)\mathbf{u}_y + \sin(\omega t + kx)\mathbf{u}_z) \ .$$

(a) Obtain the incident magnetic field $\mathbf{B}_i$ associated with $\mathbf{E}_i$. Calculate the incident Poynting vector.

(b) What happens to the wave on the conductor?

(c) Calculate the reflected electric and magnetic fields ($\mathbf{E}_r$ and $\mathbf{B}_r$).

(d) Calculate the total field in the region $x > 0$ and the total Poynting vector.

(e) What is the magnetic field at $x = 0^+$? what can you conclude about the presence of charges and/or currents at the interface?

15.6 Electromagnetic cavity

Consider an electromagnetic wave confined in a cavity made of two perfect conductors as shown in the figure below. The region between the two conductors is delimited by the planes located at $x = 0$ and $x = L$. The electric field of the wave writes

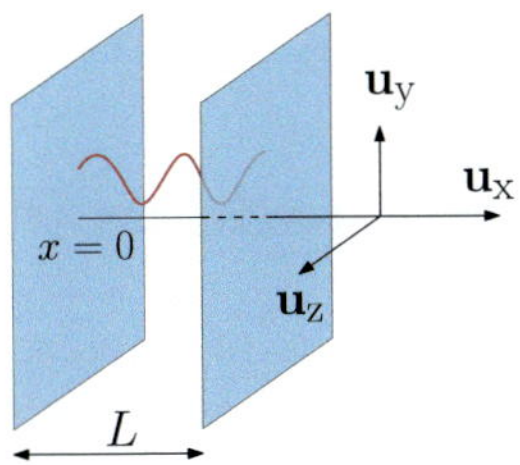

$$\mathbf{E}(x, t) = \underline{E}_1 e^{i(kx - \omega t)} \mathbf{u}_y + \underline{E}_2 e^{i(-kx - \omega t)} \mathbf{u}_y \ .$$

(a) What is the electric field at $x = 0$ and $x = L$? Deduce $\underline{E}_2$ as a function of $\underline{E}_1$ and the possible values k_n that the wavenumber k can take, with $n \in \mathbb{N}^+$.

(b) Determine the relationship between ω and k_n using d'Alembert's wave equation.

(c) Consider a particular mode, that is a particular solution with $\omega = \omega_n$ and $k = k_n$. Write the real electric field $\mathbf{E}_n(x, t)$. How does it evolve in space and time? Does it propagate?

(d) Determine the positions x_p of the nodes, that is, the points for which the electric field is zero for all times t, and give the distance between two successive nodes.

(e) Determine the magnetic field $\mathbf{B}_n(x, t)$ associated with this mode, and determine the magnetic field nodes x'_p.

(f) What is the general form of a one-dimensional (varying with respect to x only) electromagnetic wave in the cavity?

15.7 Guided wave

A wave with electric field $\mathbf{E} = E_0 \sin \dfrac{\pi x}{L} \cos(ky - \omega t)\mathbf{u}_z$ propagates in vacuum between two infinite metallic plates located at $x = 0$ and $x = L$, as shown below.

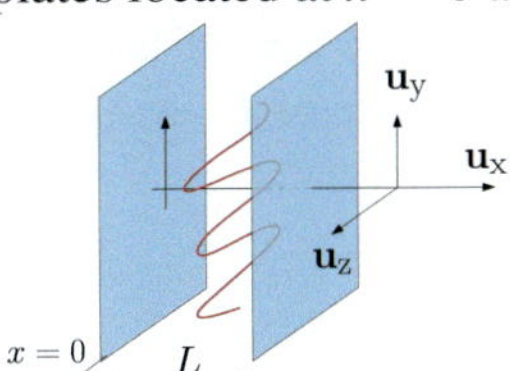

(a) What is the propagation direction? Express the wave as a superposition of two plane waves.

(b) What is the value of the electric field on the metallic plates?

(c) Find the conditions under which this wave is a solution of the propagation equation. What are the possibles values of the frequency ω that allow for a propagating wave inside the guide?

(d) Express the magnetic field associated with $\mathbf{E}$.

(e) Calculate the average (in both space and time) volume energy density and the average Poynting vector.

(f) What is the propagation velocity $\mathbf{v}_e$ for the electromagnetic energy?

15.8 Reflexion and transmission coefficients of a thin dielectric plate: the Fabry–Pérot resonator

A plane wave $\mathbf{E}_i = E_0 \cos(\omega t - n_1 k x)\mathbf{u}_z$ of frequency ω and wavenumber $k = \omega/c$ propagates in air (index n_1) towards an infinite dielectric plate of thickness e and refractive index n_2.

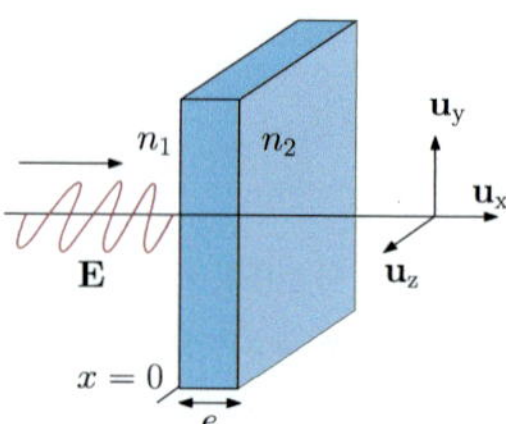

(a) What happens to the wave when it reaches the entrance ($x = 0$) and exit ($x = e$) surfaces of the dielectric plate?

(b) Express the electric and magnetic fields on each side of the plate as a function of $\underline{r} = \underline{E}_r(x = e)/\underline{E}_i(x = 0)$ and $\underline{t} = \underline{E}_t(x = e)/\underline{E}_i(x = 0)$, where $\underline{E}_r$ and $\underline{E}_t$ are the (complex) amplitudes of the reflected and transmitted waves, respectively. Write also the general form for the electric and magnetic fields inside the dielectric plate.

(c) Impose the continuity of the total electric and magnetic fields at each surface and find expressions for the complex reflection $\underline{r}$ and transmission coefficient $\underline{t}$ in terms of Fresnel's coefficients $t_{12}, t_{21}, r_{12} = -r_{21}$ at each interface.

(d) Calculate the transmittance $T = \left| \dfrac{\underline{E}_t}{\underline{E}_i} \right|^2 = |\underline{t}|^2$ of the plate. What are the maximum and minimum values of the transmission T as a function of $\Delta\phi = 2kn_2 e$? Make a sketch of the dependence of T on $\Delta\phi$ for different values of $R = r_{12}^2$.

Chapter 16
Interferences and Coherence

Abstract This chapter introduces the phenomenon of *light interference*, where the superposition of two or more waves results in a redistribution of intensity, leading to patterns of constructive and destructive interference. It begins by establishing the fundamental principles of interference between scalar waves, highlighting that for optical waves, interference patterns are only observable if the waves have the same frequency and a stable relative phase, due to the rapid oscillations of light and the slow response of detectors. The concept of *coherence* is central to the chapter, distinguishing between *temporal coherence* and *spatial coherence*. Temporal coherence quantifies the time duration over which a light source maintains a stable phase, defined by the *coherence time* and *coherence length*. The *autocorrelation function* and *degree of coherence* are introduced as mathematical tools to characterize temporal coherence. The *Wiener–Khintchine theorem* establishes a crucial link between the temporal coherence of a source and its *spectral power density*, demonstrating that a narrower spectral width corresponds to higher temporal coherence. The chapter then explores various *interferometers*, devices designed to split and recombine light waves to produce interference patterns. These include *wavefront splitting interferometers* (like Young's double-slit experiment, Fresnel biprism, and Lloyd's mirror) and *amplitude splitting interferometers* (such as the Michelson and Mach–Zehnder interferometers, and thin films). The historical significance of interferometers in pivotal experiments, including the Michelson–Morley experiment and gravitational wave detection, is emphasized. Finally, the chapter addresses *spatial coherence*, which describes the phase correlation between different points on an extended light source. The *Van Cittert–Zernike formula* is introduced to relate the visibility of interference fringes to the spatial intensity distribution of the source, leading to the definition of the *coherence angle*. This concept is particularly relevant in applications like stellar interferometry, where it allows for the measurement of the angular size of distant stars.

Keywords Interferences · Interferometers · Coherence · Coherence time · Coherence length

16.1   Introduction

Interference is one of the most remarkable phenomena in the behavior of light. When two waves overlap in space, the resulting intensity can exceed the simple sum of the separate intensities of the waves, or it can become negligible, despite the presence of two incoming beams of light. As observed by Francesco Maria Grimaldi in the 17th century, an illuminated object may become obscure when additional light is added to that already present.

In optics, observing interference is often challenging due to a technical limitation: any optical detector averages the wave's intensity over millions of light cycles. As a result, if the waves are not perfectly monochromatic, the interference effects are blurred by this averaging process and may become undetectable if the light beams have *too many* different frequency components. The invention of the laser has overcome this limitation, making it easy to observe interference with simple optical setups. The control of light interference lies at the core of numerous applications across fields such as metrology, spectroscopy, astrophysics, holography, and more.

In the first part of this chapter, we establish the fundamental principles of interference between light waves and introduce the most commonly used interferometers along with their applications. In the second part, we show that the visibility of interference patterns can be directly related to the temporal coherence of a point source, which, in turn, depends on its spectral composition. Finally, we discuss the effect of the size of the light source, introducing the concept of spatial coherence, which is particularly relevant in applications like stellar interferometry.

16.2   Interferences Between Two Scalar Waves

Before stating the problem of interference between light waves, consider the simple case of the overlap at a point x_0 of two real scalar waves $\psi_1(x_0, t)$ and $\psi_2(x_0, t)$. The total intensity at point x_0 is proportional to the square of the total amplitude $\psi_T(x_0, t) = \psi_1(x_0, t) + \psi_2(x_0, t)$ and so, dropping for simplicity the proportionality constant

$$\mathcal{I}(x_0, t) = (\psi_1(x_0, t) + \psi_2(x_0, t))^2 = \psi_1^2(x_0, t) + \psi_2^2(x_0, t) + \underbrace{2\psi_1(x_0, t)\psi_2(x_0, t)}_{\text{interferences}} .$$

We see that whenever the term $\psi_1(x_0, t)\psi_2(x_0, t)$ is different from zero, the intensity is not simply the sum of the separate intensities of the individual waves. In the case of two monochromatic waves oscillating according to $\psi_1(t) = A_1 \cos(\omega_1 t + \phi_1)$ and $\psi_2(t) = A_2 \cos(\omega_2 t + \phi_2)$, the total intensity reads

$$\mathcal{I}(t) = A_1^2 \cos^2(\omega_1 t + \phi_1) + A_2^2 \cos^2(\omega_2 t + \phi_2) + 2A_1 A_2 \cos(\omega_1 t + \phi_1) \cos(\omega_2 t + \phi_2)$$

and using the identity $\cos a \cos b = \frac{1}{2}(\cos(a+b) + \cos(a-b))$, the interference term can be written as

$$
\begin{aligned}
2\psi_1(x_0, t)\psi_2(x_0, t) &= A_1 A_2 \cos((\omega_1 + \omega_2)t + \phi_1 + \phi_2) \\
&\quad + A_1 A_2 \cos((\omega_1 - \omega_2)t + \phi_1 - \phi_2).
\end{aligned}
\tag{16.1}
$$

We conclude that it consists of two terms oscillating in time at the frequencies $\omega_1 + \omega_2$ and $\omega_1 - \omega_2$, respectively.

16.2.1 Interferences in Optics and Time Average

Optical waves have oscillation periods of the order of $T \sim 0.3\,\mathrm{fs}$ that are simply too fast to be resolved. Indeed our fastest optical detectors have typical response times of $T_{\mathrm{det}} \sim 0.1\,\mathrm{ns} \sim 10^6\,T$, so all that we can detect with light waves is the averaged intensity over (at least) millions of light cycles

$$
\langle \mathcal{I}(t) \rangle_{T_{\mathrm{det}}} = \frac{1}{T_{\mathrm{det}}} \int_0^{T_{\mathrm{det}}} \mathcal{I}(t).dt \ .
$$

In the following we will retain that, for a wave oscillating sinusoidally as $\cos(\omega t + \phi)$, with $T = 2\pi/\omega \ll T_{\mathrm{det}}$, we have

$$
\langle \cos(\omega t + \phi) \rangle_{T_{\mathrm{det}}} \approx 0
$$

and

$$
\langle \cos^2(\omega t + \phi) \rangle_{T_{\mathrm{det}}} \approx \frac{1}{2} \ .
$$

In particular, the time average of the term oscillating at $\omega_1 + \omega_2$ in Eq. (16.1) is always zero. Then

$$
\langle \mathcal{I}(t) \rangle_{T_{\mathrm{det}}} = \frac{A_1^2 + A_2^2}{2} + \underbrace{A_1 A_2 \langle \cos((\omega_1 - \omega_2)t + \phi_1 - \phi_2) \rangle_{T_{\mathrm{det}}}}_{\text{interferences}} \ .
\tag{16.2}
$$

We conclude that the interference between two optical waves is detectable only if the term $\cos((\omega_1 - \omega_2)t + \phi_1 - \phi_2)$ does not oscillate in time. That is, if:

1. both waves oscillate at the exact same frequency $\omega = \omega_1 = \omega_2$
2. the relative phase $\Delta\phi = \phi_1 - \phi_2$ does not fluctuate in time. In this case we say that the two waves are mutually coherent.

If these conditions are fulfilled, then the averaged intensity reads

$$\langle \mathcal{I}(t) \rangle_{T_{\text{det}}} = \frac{A_1^2 + A_2^2}{2} + \underbrace{A_1 A_2 \cos(\Delta\phi)}_{\text{interferences}} \ .$$

We conclude that if two scalar light waves have different frequencies or if they are not mutually coherent, then the intensity is simple the sum of their separate intensities

$$\langle \mathcal{I}(t) \rangle_{T_{\text{det}}} = \frac{A_1^2 + A_2^2}{2} \ .$$

This explains the difficulty in observing interferences with optical waves.

16.3   Interferences Between Two Monochromatic Light Waves

Let us consider two point-like sources located at $\mathbf{x}_1$ and $\mathbf{x}_2$, each one emitting a monochromatic spherical wave of frequency ω and whose electric field writes

$$\mathbf{E}_1(\mathbf{x}, t) = \mathbf{E}_{01}(r_1) \cos(kr_1 - \omega t + \phi_1)$$
$$\mathbf{E}_2(\mathbf{x}, t) = \mathbf{E}_{02}(r_2) \cos(kr_2 - \omega t + \phi_2)$$

where $r_i = |\mathbf{x} - \mathbf{x}_i|$ and we assume, for simplicity, a linear polarization for both fields. Note that the dependence of $\mathbf{E}_{0i}$ on r_i is two-fold: both the amplitude and the direction of the electric field may vary with r_i since $\mathbf{E}$ must be always transverse to the radial direction. The vector character of the electric field is a complication with respect to the case of scalar waves. However, in most cases we can forget about the vector nature of the fields as will be discussed below. Experimentally, one can produce these two point-like sources by letting light pass through two small apertures, as Thomas Young did in 1801 in his famous *double slit* experiment which was key for supporting the theory that light was a wave (this was long before Maxwell's equations). This is shown in Fig. 16.1.

The fact that the fied transmitted by two apertures is equivalent to that generated by two spherical waves emitted from two point-like sources will be admitted here, but it can be demonstrated via the Huygens–Fresnel principle that will be discussed in Chap. 18. The averaged intensity $\mathcal{I}$ at a point x on a screen is proportional to the time average of $|\mathbf{E}(x)|^2$, and therefore writes

$$\boxed{\begin{aligned} \mathcal{I}(x) &= A \langle |\mathbf{E}_1(x) + \mathbf{E}_2(x)|^2 \rangle_{T_{\text{det}}} \\ &= \underbrace{A \langle |\mathbf{E}_1(x)|^2 \rangle_{T_{\text{det}}}}_{\mathcal{I}_1(x)} + \underbrace{A \langle |\mathbf{E}_2(x)|^2 \rangle_{T_{\text{det}}}}_{\mathcal{I}_2(x)} + \underbrace{2A \langle |\mathbf{E}_1(x) \cdot \mathbf{E}_2(x)| \rangle_{T_{\text{det}}}}_{\mathcal{I}_{12}(x)} \end{aligned}} \qquad (16.3)$$

where A is a proportionality constant. According to (16.2), the interference term reads

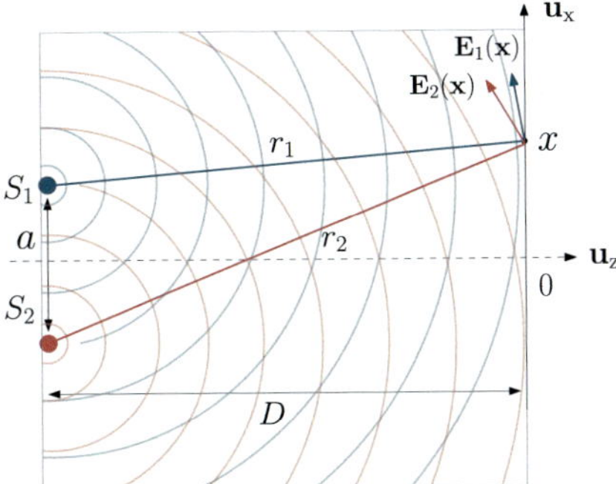

Fig. 16.1 The light from two point sources interfere at a point x on the screen

$$\mathcal{I}_{12}(x) = A\langle \mathbf{E}_{01}(r_1) \cdot \mathbf{E}_{02}(r_2) \cos(k(r_1 - r_2) + \underbrace{\varphi_1 - \varphi_2}_{\Delta\varphi}))\rangle_{T_{\text{det}}} \,.$$

We see that one cannot observe interferences if the two waves are polarized into orthogonal states $(\mathbf{E}_1(M) \cdot \mathbf{E}_2(M) = 0)$, as demonstrated by Fresnel and Arago. Assuming that the polarization does not change in time, we can write

$$\mathcal{I}_{12}(x) = A\mathbf{E}_{01}(r_1) \cdot \mathbf{E}_{02}(r_2)\langle\cos(k(r_1 - r_2) + \Delta\varphi)\rangle_{T_{\text{det}}} \,.$$

Finally, if the phase difference $\Delta\varphi = \varphi_1 - \varphi_2$ is stationnary, that is if the two sources are mutually coherent, we finally have:

$$\boxed{\mathcal{I}_{12}(\mathbf{x}) = A\mathbf{E}_{01}(r_1) \cdot \mathbf{E}_{02}(r_2)\cos(\Delta\phi(x))} \tag{16.4}$$

where $\Delta\phi(x) = k(r_1 - r_2) + \Delta\varphi = \Delta L + \Delta\varphi$ is the total phase difference between the two interfering waves. It has therefore two contributions, one caused by the phase of the waves at their moment of emission, and the second one caused by a difference in the so called optical path length ΔL travelled by the two waves

$$\Delta L = k(r_1 - r_2) = \frac{2\pi}{\lambda}(r_1 - r_2) = \frac{2\pi}{\lambda}dl$$

which increases during propagation, and it is responsible for an x dependence of the phase difference since $dl = r_1 - r_2 = dl(x)$. Along the x-axis on the screen, the interference term then oscillates periodically.

16.3.1 The Scalar Approximation and the Fresnel Formula

If the screen is placed far enough from the sources $(D \gg a)$, we may use the small angle approximation $(\theta \approx x/D \ll 1)$ so that $\mathbf{E}_{01}(r_1) \cdot \mathbf{E}_{02}(r_2) \approx E_{01}(r_1)E_{02}(r_2)$ and

Fig. 16.2 Interference pattern observed in the screen

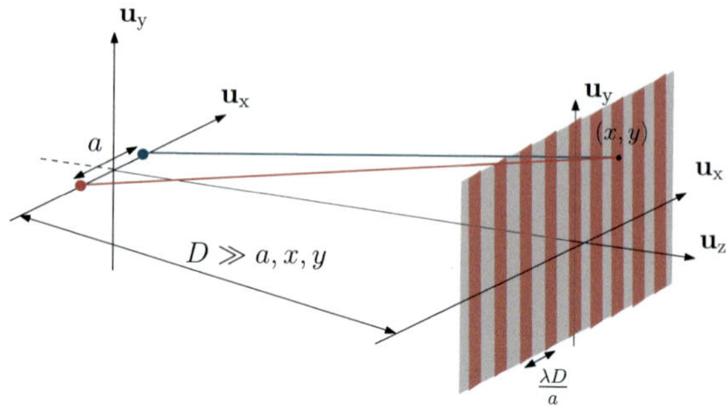

we can forget about the vector nature of the electric field (indeed, the component of both electric fields in the plane containing the sources and x are approximately collinear in the small angle approximation, as illustrated in the figure below). In addition, we can consider than the amplitudes $E_{01}(r_1)$ and $E_{01}(r_2)$ are approximately constant on any point on the screen ($r_1 \approx r_2 \approx D$) and so we write

$$A\mathbf{E}_{01}(r_1) \cdot \mathbf{E}_{02}(r_2) \approx E_{01}(r_1)E_{02}(r_2) = 2\sqrt{\mathcal{I}_1 \mathcal{I}_2} \ .$$

Finally, we obtain the Fresnel formula for the interference between two waves at a point $\mathbf{x}$:

$$\boxed{\mathcal{I}(\mathbf{x}) = \mathcal{I}_1 + \mathcal{I}_2 + 2\sqrt{\mathcal{I}_1 \mathcal{I}_2} \cos \Delta\phi(\mathbf{x})} \tag{16.5}$$

with

$$\Delta\phi(\mathbf{x}) = \Delta L + \Delta\varphi = \frac{2\pi}{\lambda} dl + (\varphi_1 - \varphi_2) \ .$$

The path difference dl depends only on the x coordinate on the screen, so that the interference pattern is invariant under translations along the y-axis. It consists of a series of bright and dark vertical fringes as shown in Fig. 16.2.

Up to first order in x/D, we have

$$(r_1 - r_2) = \frac{ax}{D} \sim a \sin\theta \ ,$$

as shown in Fig. 16.3, and so

$$\Delta\phi(x) = \frac{2\pi a x}{\lambda D} + (\varphi_1 - \varphi_2) \ .$$

On the screen, we observe a periodic modulation of the measured intensity as a function of x, reaching a maximum value of $\mathcal{I}_{\max} = \mathcal{I}_1 + \mathcal{I}_2 + 2\sqrt{\mathcal{I}_1 \mathcal{I}_2}$ for $\Delta\phi = 2m\pi$ with $m \in \mathbb{N}$ (constructive interference) and a minimum value of $\mathcal{I}_{\min} = \mathcal{I}_1 + \mathcal{I}_2 - 2\sqrt{\mathcal{I}_1 \mathcal{I}_2}$ for $\Delta\phi = (2m + 1)\pi$ with $m \in \mathbb{N}$ (destructive interference). This is illustrated in Fig. 16.4. The periodicity on the screen is therefore given by $\Delta x = \frac{\lambda D}{a}$.

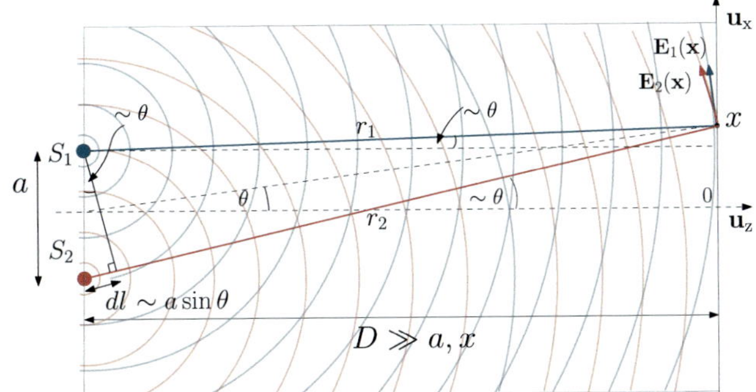

Fig. 16.3 The optical path difference between the two waves arriving at x is $dl \sim a \sin \theta$

Fig. 16.4 Variation of the intensity as a function of the horizontal position x onthe screen

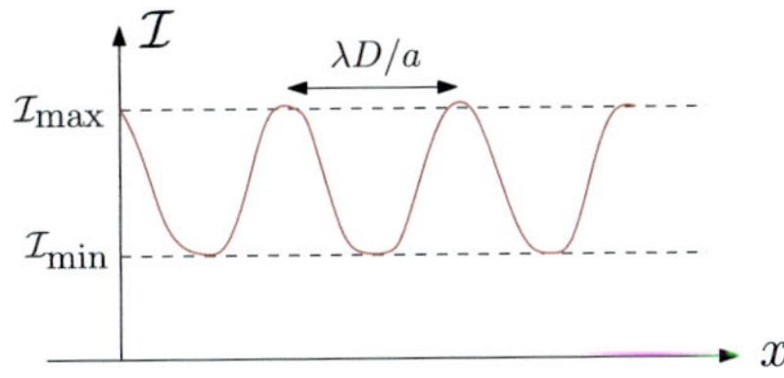

This provides a relatively easy way of determining the wavelength of light by simply measuring the interferences period when a and D are known.

The contrast of the interference pattern is defined as

$$C = \frac{\mathcal{I}_{\max} - \mathcal{I}_{\min}}{\mathcal{I}_{\max} + \mathcal{I}_{\min}} = \frac{2\sqrt{\mathcal{I}_1 \mathcal{I}_2}}{\mathcal{I}_1 + \mathcal{I}_2} \tag{16.6}$$

and it quantifies the visibility of the fringes on the screen. The contrast is maximum $(C = 1)$ when the two waves arriving at the screen have the same amplitude, in which case $\mathcal{I}_1 = \mathcal{I}_2 = \mathcal{I}_0$ and the Fresnel formula becomes

$$\mathcal{I}(x) = 2\mathcal{I}_0 \left(1 + \cos \Delta\phi(x)\right). \tag{16.7}$$

It shows that whenever the waves add up constructively, the maximum intensity may reach twice the sum of the separate intensities of each source, i.e., $4\mathcal{I}_0$.

In the case where the two sources are located along the z-axis, the path difference dl depends only on the distance r with respect to the axis on the screen, as shown in Fig. 16.5. We have in this case

$$dl = a \cos \theta \approx a \left(1 - \frac{\theta^2}{2}\right) = a \left(1 - \frac{r^2}{2D^2}\right)$$

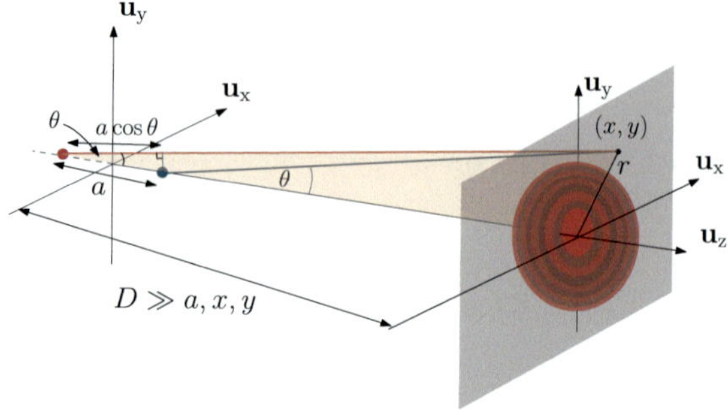

Fig. 16.5 When the point sources are along the z-axis, a series of concentric rings is observed in the screen

and one observes circular fringes on the screen, where the radius of the bright fringe of order m (that is, the radius for which $\Delta\phi = 2m\pi$) is given by

$$r_m = D\sqrt{2\left(1 - m\frac{\lambda}{a}\right)} .$$

Conditions for the observation of interferences

In summary, optical waves generated by two sources will generate an interference pattern if the following conditions are fullfilled:

1. The two waves have the same frequency ω.
2. Their polarization do not vary in time or in an uncorrelated way, and they are not orthogonally polarized.
3. They are mutually coherent, that is, their relative phase does not vary in time.

The third condition tell us that two independent sources, even if they are perfectly monochromatic with the same frequency, will not produce any measurable interference if their relative phase $\Delta\varphi$ varies randomly in timescales much shorter than the detector's response time. Since the light of any source (excepting the case of the laser) is the superposition of waves emitted by a large number of atoms or electrons, the relative phase of two independent sources jumps in time in an unpredictable way. To observe interferences with light, one can instead start with light emitted from a single source, split the wave into two equal and synchronous ($\varphi_1 = \varphi_2$) parts who are recombined later on. Any device capable of doing this is called an interferometer, and the phase difference in this case is only due to the optical path difference ΔL.

16.4 Temporal Coherence

As mentioned earlier, trying to observe interferences between optical waves generated from two independent sources results generally in failure. This is because most of light, either coming from the Sun, an electroluminescent diode (LED) or a discharge lamp, originates from a fundamental and probabilistic process at the atomic scale called spontaneous emission. Light emitted from such objects is therefore the superposition of wavepackets coming from a large number of atoms (or electrons) radiating independently at different frequencies, amplitudes and phases. In consequence, two independent sources emit waves whose relative phase varies randomly in time and without any correlation. We say that they are mutually incoherent.

To get rid of this issue one uses a single source and an interferometer so that the measured intensity at a given point x on the screen is given by the superposition between the field $\underline{E}(t)$ with a retarded version of itself, $\underline{E}(t - \tau_x)$. Naturally, there exists a limit on the maximum retardation $\tau_x = dl/c$ compatible with the observation of interferences.

One defines the coherence time τ_c of a source as the typical timescale over which the phase emitted by the source is stable (a more precise definition will be given in 16.11). This allows to define a coherence length $l_c = c\tau_c$ which is simply the distance travelled by the wave during the coherence time. In order to observe interferences, the phase difference between the two beams overlapping at a point x on the screen should be constant in time. Since the lower beam in Fig. 16.6 travels a larger distance than the upper one, this will be true if the time delay $\tau = dl/c$ between the two is smaller than the coherence time t_c (or, equivalently, if the path lenth difference dl is smaller than the coherence length l_c).

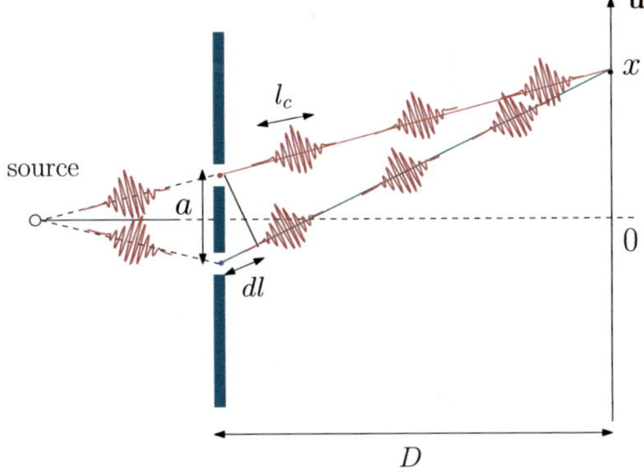

Fig. 16.6 An interferometer can be use to make the supeposition between the field emitted by a single source and a retarded version of itself

16.4.1   Autocorrelation Function and Degree of Coherence of a Source

In an interferometer, the time delay between the two interfering waves depends on the position x on the screen, and so the total intensity at point x is given by

$$\mathcal{I}(x) = A\langle|\underline{E}(t) + \underline{E}(t - \tau_x)|^2\rangle_{T_{\text{det}}}$$
$$= A\langle|\underline{E}(t)|^2\rangle_{T_{\text{det}}} + A\langle|\underline{E}(t - \tau_x)|^2\rangle_{T_{\text{det}}} + 2A\mathrm{Re}\{\langle\underline{E}^*(t)\,\underline{E}(t - \tau_x)\rangle_{T_{\text{det}}}\}\,.$$

If the source is stationnary (that is, if all of the average properties are constant in time), then $A\langle|\underline{E}(t)|^2\rangle_{T_{\text{det}}} = A\langle|\underline{E}(t - \tau_x)|^2\rangle_{T_{\text{det}}} = \mathcal{I}_0$. In the limit of very large T_{det}, the intensity can be rewritten as

$$\mathcal{I}(x) = 2\left(\mathcal{I}_0 + \mathrm{Re}\{G(\tau_x)\}\right)$$

where G is called the autocorrelation function of order 1, defined as

$$G(\tau) = \lim_{T \to \infty} \frac{A}{T} \int_{-T/2}^{T/2} \underline{E}^*(t)\,\underline{E}(t - \tau)dt. \qquad (16.8)$$

Finally, noticing that $G(0) = \mathcal{I}_0$, we can rewrite the intensity as

$$\boxed{\mathcal{I}(x) = 2\mathcal{I}_0\left(1 + \frac{\mathrm{Re}\{G(\tau_x)\}}{G(0)}\right) = 2\mathcal{I}_0\left(1 + \mathrm{Re}\{g(\tau_x)\}\right)} \qquad (16.9)$$

where $g(\tau) = G(\tau)/G(0)$ is the normalized autocorrelation function, also called degree of coherence of the source.

Degree of temporal coherence and coherence time
The degree of temporal coherence of a light source of intensity $\mathcal{I}_0$ is the function g defined by

$$g(\tau) = \lim_{T \to \infty} \frac{A}{T\mathcal{I}_0} \int_{-T/2}^{T/2} \underline{E}^*(t)\,\underline{E}(t - \tau)dt\,. \qquad (16.10)$$

For a given delay τ, g is a measure of the correlation between the field at instants t and $t - \tau$. One has $0 \le |g(\tau)| \le 1$, and $g(0) = 1$. We can define the coherence time τ_c as the typical width of the function g around $\tau = 0$

$$\tau_c^2 = \frac{\int_{-\infty}^{+\infty} t^2 \, g(\tau)|^2 d\tau}{\int_{-\infty}^{+\infty} |g(\tau)|^2} \qquad (16.11)$$

and it represents the memory time of the fluctuations. Interferences are therefore visible for delay times τ such that $\tau \leq \tau_c$.

Note that if $\underline{E}(t)$ and $\underline{E}(t - \tau)$ with $\tau \neq 0$ are two independent variables, that is, with no correlation, then

$$\langle \underline{E}^*(t)\underline{E}^*(t - \tau)\rangle_{T_{\text{det}}} = \langle \underline{E}^*(t)\rangle_{T_{\text{det}}} \langle \underline{E}^*(t - \tau)\rangle_{T_{\text{det}}} = 0 \ .$$

We obtain $g = 0$ and we say that the two sources are incoherent. The measured intensity produced by the overlap of two incoherent sources is just simply the sum of their separate intensities.

Example 16.1—Degree of coherence of a monochromatic wave
Consider the case of a perfectly monochromatic source whose electric field's amplitude at a fixed position writes

$$\underline{E}(t) = \underline{E}_0 e^{-i\omega_0 t} \ .$$

Since this has a well defined constant phase, we expect it to have a degree of coherence such that $|g| = 1$. Indeed $\underline{E}(t) = \underline{E}_0 e^{-i\omega_0 t}$ and so

$$\underline{E}(t - \tau) = \underline{E}_0 e^{-i\omega_0 t} e^{i\omega_0 \tau} = \underline{E}(t) e^{i\omega\tau} \ .$$

The degree of coherence reads

$$g(\tau) = \frac{\langle \underline{E}^*(t)\underline{E}(t - \tau)\rangle_{T_{\text{det}}}}{\langle |\underline{E}(t)|^2\rangle_{T_{\text{det}}}} = \frac{\langle e^{i\omega_0 \tau}|\underline{E}(t)|^2\rangle_{T_{\text{det}}}}{\langle |\underline{E}(t)|^2\rangle_{T_{\text{det}}}} = e^{i\omega_0 \tau} \ .$$

We see that $|g(\tau)| = 1$ and so the coherence time τ_c is infinite. According to Eq. (16.9), if such source is placed at the entrance of an interferometer such as the one depicted in Fig. 16.6, the measured intensity for a path length difference dl is given by

$$I(dl) = 2I_0 \left(1 + \text{Re}\{g(dl/c)\}\right)$$

so that

$$I(dl) = 2I_0 \left(1 + \cos(\frac{\omega_0}{c}dl)\right) = 2I_0 \left(1 + \cos\left(\frac{2\pi dl}{\lambda}\right)\right) \ .$$

the intensity varies sinusoidally between $4\mathcal{I}_0$ and 0 with a periodicity in dl of λ. The contrast is $C = 1$.

16.5   Interferometers

An interferometer is any optical device capable of producing an overlap of a wave with a retarded version of itself. Depending on the way of spliting the initial lightwave, they are classified into two main categories: wavefront and amplitude splitting interferometers.

16.5.1   *Wavefront Splitting Interferometers*

Young's double slit apparatus is an example of a wavefront splitting interferometer, since it selects two beams coming from two different places of the wavefront reaching the double slit. The double slit diffracts the beam and act like two synchronized point sources emitting spherical waves that interfere on a screen. Other examples of wavefront splitting interferometers are the Fresnel biprism and Lloyd's mirror. They are shown in Fig. 16.7. Fresnel biprism is treated in Exercise 16.1, according

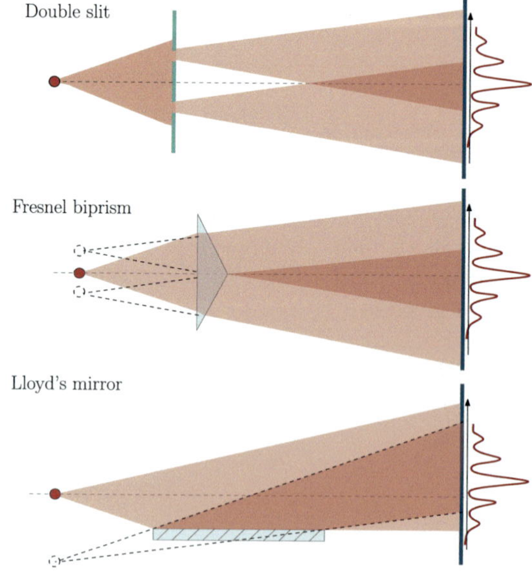

Fig. 16.7 Different examples of wavefront splitting interferometers

to the laws of refraction the wavefront is bended differently (upward or downward) by the two half of the prism, so that at any point in the screen there is interference of two waves that appear to come from two synchronous virtual sources and that have travelled different distances.

In the case of Lloyd's mirror, a part of the beam strikes a mirror at a grazing angle, and so in the screen one observes the interference between the light that travelled directly from the source to the screen and some reflected light off the mirror, appearing to come from a virtual source, syncrhonized with the original one.

16.5.2 *Amplitude Splitting Interferometers*

An amplitude splitting interferometer uses a partially transparent reflector, also called a beam splitter, to divide the amplitude of the incident wave into separate beams of equal intensity generating thus two synchronous waves than be recomined later. One of the most emblematic examples is the so-called Michelson interferometer, in which a semi-reflecting plate splits the beam into two arms of equal intensity that travel different path lengths before striking two flat mirrors forcing the beams to recombine and interfere on a screen (see Fig. 16.8).

Some crucial experiments in the history of physics have been performed with a Michelson interferometer. It is the case of the famous Michelson–Morley experiment that, in 1887, failed to prove the existence of aether, an hypotetical medium filling the whole space and on which light was thought to propagate. This negative result of the Michelson–Morley experiment resulted in Einstein's special theory of relativity.

Another famous use of a Michelson interferometer was the first direct detection, in 2015, of a gravitational wave produced by the merging of two black holes by the observatories LIGO and VIRGO. During the passage of a gravitational wave through earth, extremely small but anisotropic changes in the length of the two arms (of several

Fig. 16.8 The Michelson interferometer

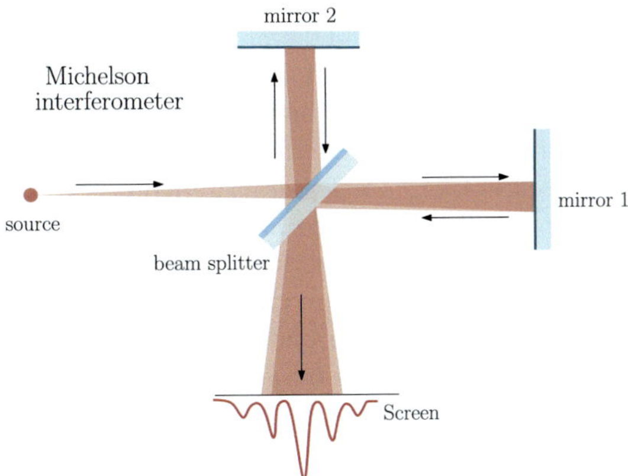

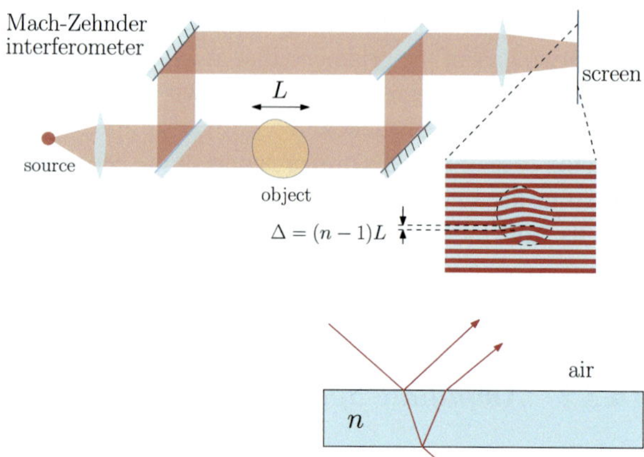

Fig. 16.9 The Mach–Zehnder interferometer

Fig. 16.10 A transparent, thin plate produces interference between the first reflected beam and the one that made one round trip inside the plate

kilometers long) of a Michelson interferometer are produced and detected by looking at the modification of the interference pattern of a laser. These gravitational wave detectors can detect changes in the optical path that are 10^4 times smaller than the diameter of a proton, equating to roughly 10^{-19} m.

A similar interferometer is the Mach–Zehnder shown in Fig. 16.9, in which two collimated beams are separated by a first beamsplitter and later recombined by a second one (instead of the same one as in the case of a Michelson interferometer). Each of the arms is traversed only once, so this interferometer is useful to determine the phase shift introduced by a sample on one of the two arms. Indeed, a transparent object (a gaz, a low density plasma, a flame or a transparent plate) in one of the two arms will change the phase shift between the interfering waves by an amount $\Delta = (n-1)L$, where n is the refractive index and L the characteristic diameter of the object being studied. On a screen, the interference fringes in the zone were the waves went through the object will be shifted with respect to the fringes produced in the absence of the object. The measurement of this shift provides an estimation of the refractive index, and therefore, of the object's density according to (14.33).

Another amplitude splitting interferometer consists on a transparent, thin plate of refractive index n. When an incident beam arrives at the first interface, it is partially transmitted and partially reflected, according to the laws of reflection and refraction discussed in Chap. 15. The transmitted beam will encounter a second interface and part of it will be reflected back toward the source, as shown in Fig. 16.10. If the refractive index is not too large, the light waves reflected by the upper and lower boundaries of the thin film interfere with one another, enhancing or cancelling each other depending on their optical path difference.

In reality, there is an infinite sequence of reflected beams, although usually it is enough to consider the first two reflected beams since they are much more intense than the subsequent ones. The interference by a thin film is exploited in antireflection coatings deposited on glasses and more generally onto optical components for

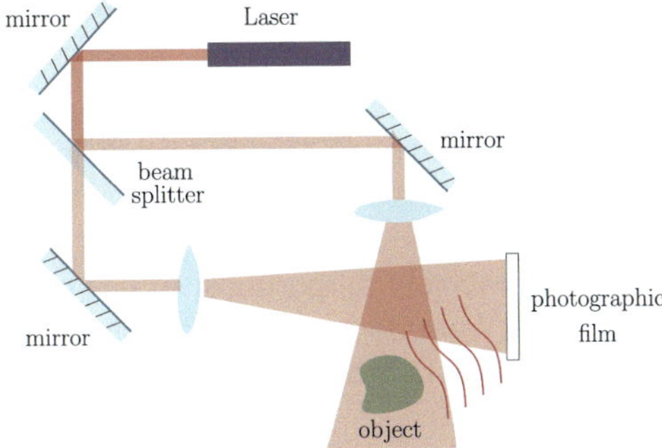

Fig. 16.11 To record hologram of an object, one exposes a film to the interference pattern generated by a reference beam and a beam reflected by the object

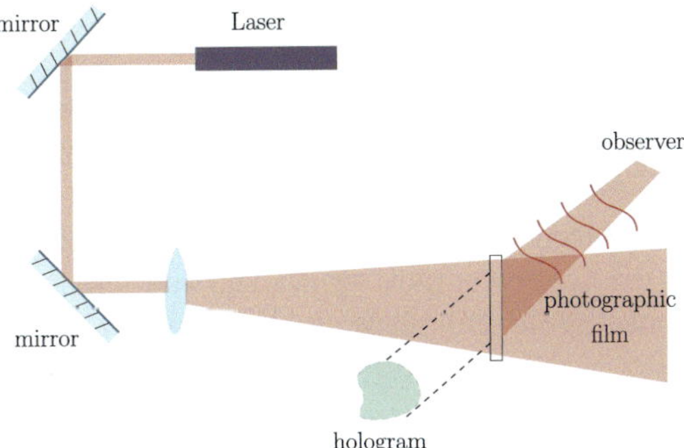

Fig. 16.12 The hologram diffracts light in such a way that it recreates the field of the original object

applications in which one searches to minimize the loss of intensity due to reflections. The thickness and refractive index is chosen so as to minimize reflected light by tuning the destructive interference at a particular wavelength. This type of interference is also the reason why soap bubbles show colorful patterns, that are due to the inhomogeneity of its thickness and the wavelength dependence of the refractive index.

Another famous application of interferences is holography. In contrast to photography which records the intensity of a light beam, holography also records its phase, an essential key to provide the impression of a real object in three dimensions. To create an hologram, one must first record on a photosensitive film the interference pattern between a reference beam and a second beam diffused by an object, whose amplitude describes the reflectivity of the object and the phase describes the surface appearance and its depth. This is illustrated in Fig. 16.11. Lasers are the preferred light source for holography due to their high temporal and spatial coherence. The laser beam is first split into two synchronous beams by a beamsplitter.

The recorded interference pattern acts as a diffraction grating so that when lit with a laser identical to the one used to record the hologram, the interference pattern recorded on the film diffracts the light into an accurate reproduction of the original light field. As illustrated in Fig. 16.12, the view of this diffracted beam from different angles give the observer the impression that the object is really present.

16.6   Coherence Time and Spectral Density of a Source

In nature there are no perfectly monochromatic sources. Indeed, every emission process is defined by a finite lifetime of an excited state, so there is always a frequency dispersion. Any source emits light within a certain frequency range $\Delta\omega$ around a central frequency ω_0. As we will show in this section, the coherence time τ_c of a source is inversely proportional to its spectral width $\Delta\omega$. An intuitive way to see this comes from noticing that the phase difference ΔL introduced by an optical path difference dl between two waves depends on wavelength, or equivalently, on frequency, since

$$\Delta L = \frac{2\pi dl}{\lambda} = \frac{\omega}{c} dl .$$

At a given point on a screen, one observes the sum of the interference patterns generated by each frequency contained in the source (see Exercise 16.2). These patterns are slightly shifted with respect to each other, indeed the phase difference introduced by the optical path will differ between the different frequencies contained in the field by at most $\frac{\Delta\omega}{c} dl$. The interferences will still be visible if this phase shift is negligible with respect to 2π, that is if

$$\frac{\Delta\omega}{c} dl \ll 2\pi \quad \rightarrow \quad dl \ll c\frac{2\pi}{\Delta\omega} .$$

On the other hand, we know that interferences are clearly visible when dl is much smaller than the coherence length l_c, and so we expect a relationship between $\Delta\omega$ and τ_c such that

$$c\frac{2\pi}{\Delta\omega} \sim c\tau_c = l_c$$

We will see that this is actually the case provided that we define properly the spectral width $\Delta\omega$ of a random electric field. For this, we now define the spectral power density.

16.6.1   The Spectral Power Density

The Fourier transform of a function f with respect to time provides an information about its frequency content, and so it would seem natural to determine the spectral content of a source by calculating the electric field's Fourier transform. However, this is not possible for a stationnary electric field $\underline{E}$ since the latter is not characterized by an integrable function. Nevertheless, the average intensity of such a wave is finite over the time response T of a detector

$$\mathcal{I}_T = \frac{A}{T} \int_{-T/2}^{T/2} |\underline{E}(t)|^2 dt = \frac{A}{T} \int_{\mathbb{R}} |\underline{E}_T(t)|^2 dt$$

where $\underline{E}_T = \underline{E}(t) H_T(t)$, with $H_T(t)$ equal to one for $|t| \leq T/2$ and 0 otherwise. This truncated version of the electric field is integrable and admits a Fourier transform $\mathcal{F}\{\underline{E}_T\}(\omega)$. The Parseval–Plancherel theorem (A.61) then states that the intensity $\mathcal{I}_T$ of the field averaged over an interval T can be written as

$$\mathcal{I}_T = \frac{A}{T} \int_{\mathbb{R}} |\underline{E}_T(t)|^2 dt = \frac{2\pi A}{T} \int_{\mathbb{R}} \left| \mathcal{F}\{\underline{E}_T\}(\omega) \right|^2 d\omega. \tag{16.12}$$

The spectral power density

We define the spectral power density $\mathcal{S}$, sometimes simply called the spectrum, such that $\mathcal{S}(\omega)d\omega$ represents the fraction of the wave's intensity contained in the frequency interval $[\omega, \omega + d\omega]$. More precisely

$$\mathcal{I}_0 = \lim_{T \to \infty} \frac{A}{T} \int_{-T/2}^{T/2} |\underline{E}(t)|^2 dt = \int_{\mathbb{R}} \mathcal{S}(\omega)d\omega \tag{16.13}$$

and so comparing with Eq. (16.12) we obtain

$$\mathcal{S}(\omega) = \lim_{T \to \infty} \frac{2\pi A}{T} \left| \mathcal{F}\{\underline{E}_T\}(\omega) \right|^2. \tag{16.14}$$

The spectral width $\Delta\omega$ can be defined as the typical extension of the spectral density around its central frequency ω_0

$$\Delta\omega^2 = \frac{\int_{-\infty}^{+\infty} (\omega - \omega_0)^2 |\mathcal{S}(\omega)|^2 d\omega}{\int_{-\infty}^{+\infty} |\mathcal{S}(\omega)|^2} \tag{16.15}$$

where ω_0 writes

$$\omega_0 = \frac{\int_{-\infty}^{+\infty} \omega |\mathcal{S}(\omega)|^2 d\omega}{\int_{-\infty}^{+\infty} |\mathcal{S}(\omega)|^2}.$$

16.6.2   The Wiener–Khintchine Theorem

In order to demonstrate that the coherence time τ_c and the spectral width $\Delta\omega$ of a source are related, let us calculate the Fourier transform of the autocorrelation function

$$\mathcal{F}\{G\}(\omega) = \frac{1}{2\pi}\int_{\mathbb{R}} G(\tau)e^{-i\omega\tau}\,d\tau = \lim_{T\to\infty}\frac{A}{2\pi T}\int_{\mathbb{R}}\int_{-T/2}^{T/2}\underline{E}^*(t)\,\underline{E}(t-\tau)e^{-i\omega\tau}\,dt\,d\tau$$

$$= \lim_{T\to\infty}\frac{A}{2\pi T}\int_{-T/2}^{T/2}\underline{E}(u)e^{i\omega u}\,du\int_{-T/2}^{T/2}\underline{E}^*(t)\,e^{-i\omega t}\,dt$$

$$= \lim_{T\to\infty}\frac{A}{2\pi T}(2\pi)^2|\mathcal{F}\{\underline{E}_T\}(\omega)|^2 = \mathcal{S}(\omega)$$

where $\mathcal{S}$ is the power spectral density. Note that we have use the fact that $G(\tau) = G^*(-\tau)$ so that $\mathcal{F}\{\underline{E}_T\}(\omega) = \mathcal{F}^*\{\underline{E}_T\}(-\omega)$. The autocorrelation function G and the spectral density $\mathcal{S}$ are thus Fourier transforms of each other. This result, called the Wiener–Khintchine theorem, stablishes the relation between the temporal coherence of a source and its spectrum. It follows that if $\Delta\omega$ is the width of the spectral density defined by (16.15), and if τ_c is the width of G as defined by (16.11), then[1]

$$\tau_c \sim \frac{2\pi}{\Delta\omega}$$

or, in terms of the coherence length and the central wavelength λ_0

$$l_c \sim \frac{\lambda_0^2}{\Delta\lambda}\;.$$

16.6.3   Fourier-Transform Spectroscopy

By analyzing the variation of the interference pattern as a function of the time-delay between the two interferring waves, one obtains what is called an interferogram

$$\mathcal{I}(\tau) = 2\mathcal{I}_0\left(1 + |g(\tau)|\cos\phi(\tau)\right)$$

where we have written $g(\tau) = |g(\tau)|e^{i\phi(\tau)}$. A Michelson interferometer in which one of the two path lengths is varied by using a translating mirror provides an efficient way of measuring the interferogram of a source. The time-delay between the two interfering beams is simply $\tau = dl/c$, where dl is the length difference between the two arms (see Fig. 16.13).

---

[1] Here we suppose that $G(\tau)$ varies slowly inside $|\tau| < \tau_c$. Rigourously, one has the inequality $\tau_c \geq \frac{2\pi c}{\Delta\omega}$.

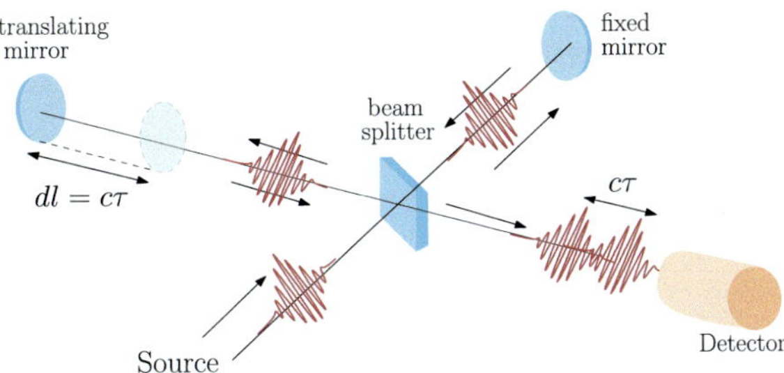

Fig. 16.13 A Michelson interferometer can be used to measure the interferogram associated with a light source

The interferogram contains all the information about the statistical properties of the source: the width is given by the width of $|g(\tau)|$ and provides a measurement of the coherence length, whereas the Fourier transform of the interferogram allows to determine the spectrum of the source. Indeed, we can also write

$$\mathcal{I}(\tau) = 2\left(\mathcal{I}_0 + \mathrm{Re}\left\{G(\tau)\right\}\right)$$

and according to Wiener–Khintchine theorem, G can be written as the inverse Fourier transform of the spectum $\mathcal{S}$

$$\mathcal{I}(\tau) = 2\left(\mathcal{I}_0 + \mathrm{Re}\{\int_{\mathbb{R}} \mathcal{S}(\omega)e^{i\omega\tau}d\omega\}\right).$$

Since $\mathcal{S}(\omega) \in \mathbb{R}$ and $\int_{\mathbb{R}} \mathcal{S}(\omega)d\omega = G(0) = \mathcal{I}_0$ we can rewrite

$$\boxed{\mathcal{I}(\tau) = 2\int_{\mathbb{R}} \mathcal{S}(\omega)\left(1 + \cos\omega\tau\right)d\omega\,.} \tag{16.16}$$

Equation (16.16) is a generalization of (16.7) and of the formula (A.94) obtained for the case of a doublet in Exercise 16.2. Since waves of different frequencies are mutually incoherent, this formula states that the resulting inteferogram is a weighted superposition of the interferograms generated independently by each of the frequency components of the source.

16.7   Coherence Length of a Quasi Monochromatic Light Source

A light source is constituted by a large number of atoms, each one emitting radiation by spontaneous emission in a probabilistic process and over a finite period of time. A simple model consists in considering the field emitted by the source as a sequence of wavepackets such that the phase of the resulting electric field varies randomly over a

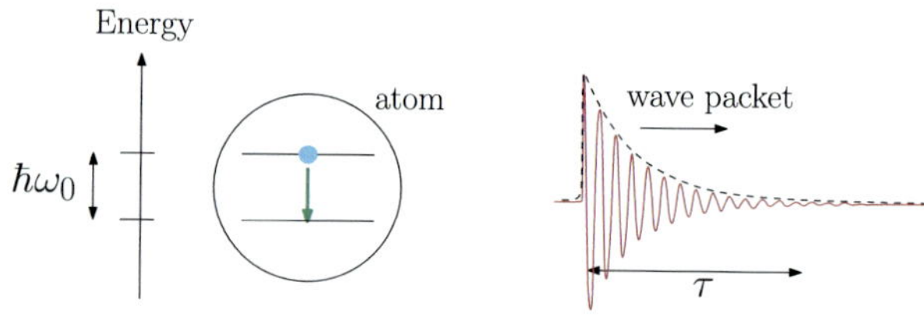

Fig. 16.14 An atom emits a wavepacket of typical duration τ when it relaxes its energy

typical timescale τ_0, that we will show in the following to be equal to the coherence time.

As shown in Fig. 16.14, an atom that is initially in an excited state can spontaneously relax to a lower energy state by emitting, at an instant t_i, a wavepacket whose projection along an arbitrary axis writes

$$\underline{E}_i(t) = \underline{E}_{0i}\, f(t - t_i)e^{-i\omega_0(t-t_i)} = \underline{E}_{0i}\, f(t - t_i)e^{-i\omega_0 t}e^{i\phi_{0i} t}$$

where $\omega_0 = \Delta E/\hbar$ is the central frequency of the wave determined by the energy difference ΔE between the atom's initial and final state, f is an enveloppe of typical length τ related to the lifetime of the excited state, and $\phi_{0i} = \omega_0 t_i$.

For an ensemble of atoms, the emission times t_i, and therefore the phase ϕ_{0i}, are random. For simplicity, we consider that all the wavepackets have the same amplitude $\underline{E}_0$, so that

$$\underline{E}(t) = \sum_i \underline{E}_0\, f(t - t_i)e^{-i\omega_0 t}e^{i\phi_{0i}}$$

$$= \underbrace{\left(\sum_i \underline{E}_0\, f(t - t_i)e^{i\phi_{0i}}\right)}_{\underline{A}e^{i\phi(t)}} e^{-i\omega_0 t}. \tag{16.17}$$

The resulting field has a fluctuating amplitude $\underline{A}$ and phase ϕ. Figure 16.15a shows the real part of the electric field obtained with gaussian wavepackets (for simplicity) of width τ. The arrival times, indicated by the blue dots, are randomly chosen at a small rate such that the average interval between two wavepackets is $\Delta t = 10\tau$. In this case the resulting field fluctuates strongly because the wavepackets are well isolated in time. In contrast, when the interval between two emissions is smaller than τ, as shown in Fig. 16.15b for $\Delta t = \tau/4$, the fluctuations of the field (16.17) are largely dominated by the sudden jumps in the phase $\phi(t)$, indicated by the vertical arrows. The typical duration of each individual wavepacket, determined by τ, sets the timescale for each packet's influence in the sum (16.17). As a result, phase changes tend to occur over a timescale comparable to τ. We see that for a large ensemble of atoms, the relative fluctuations of the amplitude may be neglected with respect to the phase jumps, so we can consider $\underline{A}$ as a constant in (16.17). The time interval

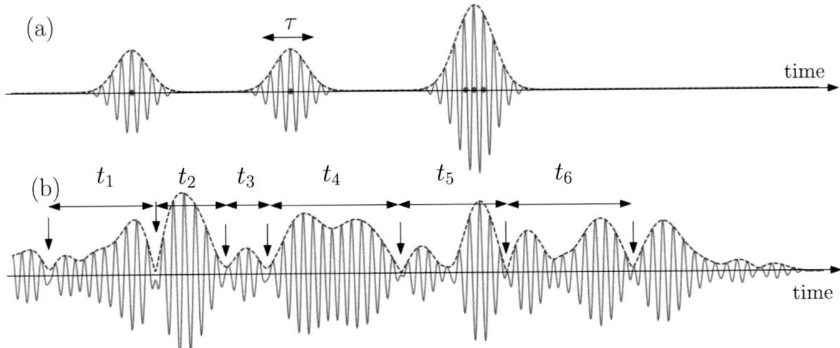

Fig. 16.15 Electric field resulting from the superposition of gaussian wavepackets at low (**a**) and large (**b**) rate of emission events

between two phase jumps is the random variable that will limit the coherence time of the source.

Let $t = 0$ be the time immediately after a phase jump has occured. Assuming that the probability dP that the phase $\phi(t)$ changes between t and $t + dt$ is independent of t, we may write

$$dP = \frac{dt}{\tau_0}$$

and let $\mathbb{P}(t)$ be the probability that the phase has not changed after a time t, then

$$\mathbb{P}(t + dt) = \mathbb{P}(t)\left(1 - \frac{dt}{\tau_0}\right)$$

and for an infinitesimal dt, we conclude that the probability of survival of the phase follows an exponential law

$$\frac{d\mathbb{P}(t)}{dt} = -\frac{\mathbb{P}(t)}{\tau_0} \quad \rightarrow \quad \mathbb{P}(t) = e^{-t/\tau_0} \; .$$

The probability that the phase changes between t and $t + dt$ is therefore $\mathbb{P}(t)\frac{dt}{\tau_0}$, and so the average lifetime of the phase reads

$$\langle t \rangle = \int_0^{+\infty} t\mathbb{P}(t)\frac{dt}{\tau_0} = \int_0^{+\infty} te^{-t/\tau_0} = \tau_0 \; .$$

Let us now calculate the degree of coherence of this source. This reads

$$g(\tau) = \frac{\langle \underline{E}^*(t)\underline{E}(t - \tau)\rangle_{T_{\text{det}}}}{\langle |\underline{E}(t)|^2\rangle_{T_{\text{det}}}} = |\underline{A}|^2 e^{-i\omega_0\tau} \frac{\langle e^{-i\phi(t)} e^{i\phi(t-\tau)}\rangle_{T_{\text{det}}}}{|\underline{A}|^2} \; .$$

Fig. 16.16 Interferogram of
a quasi monochromatic
source

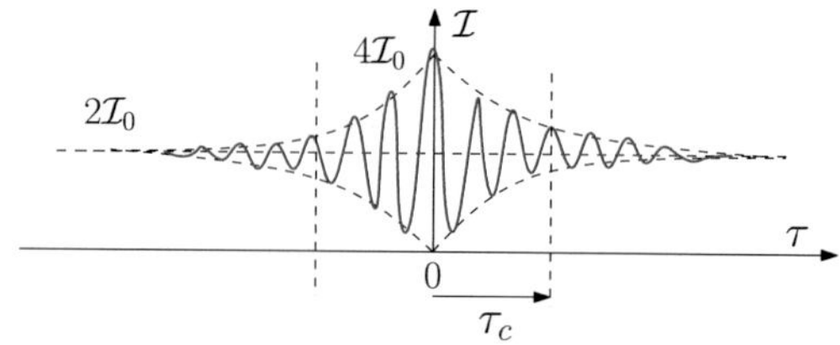

Let us suppose first that $\tau > 0$. The value of $\langle e^{-i\phi(t)} e^{i\phi(t-\tau)} \rangle_{T_{\mathrm{det}}}$ depends on wether the phase has been constant or not during $t - \tau$ and t. We have $\langle e^{-i\phi(t)} e^{i\phi(t-\tau)} \rangle_{T_{\mathrm{det}}} = 1$ with probability $e^{-\tau/t_0}$ and $\langle e^{-i\phi(t)} e^{i\phi(t-\tau)} \rangle_{T_{\mathrm{det}}} \approx 0$ with probability $1 - e^{-\tau/\tau_0}$. It results that

$$g(\tau) = \langle e^{-i\phi(t)} e^{i\phi(t-\tau)} \rangle_{T_{\mathrm{det}}} = e^{-\tau/t_0} e^{-i\omega_0 \tau} \; .$$

For $\tau < 0$, $g(\tau) = e^{+\tau/\tau_0} e^{-i\omega_0 \tau}$ and so the degree of coherence is written

$$g(\tau) = e^{-|\tau|/\tau_0} e^{-i\omega_0 \tau} \; .$$

The visibility of the interferences coming from such a source decays exponentially as a function of the delay time τ between the two beams, with a decay constant of τ_0:

$$\mathcal{I}(\tau) = 2\mathcal{I}_0 \left(1 + e^{-|\tau|/\tau_0} \cos(\omega_0 \tau) \right)$$

as shown in Fig. 16.16. We know that this decay is related to a loss of coherence, which is lost after a typical delay time τ_c given by (16.11)

$$\tau_c = \sqrt{\frac{\int_{\mathbb{R}} \tau^2 |g(\tau)|^2 d\tau}{\int_{\mathbb{R}} |g(\tau)|^2 d\tau}} = \sqrt{\frac{\int_0^\infty \tau^2 e^{-2\tau/\tau_0} d\tau}{\int_0^\infty e^{-2\tau/\tau_0} d\tau}} = \frac{\tau_0}{\sqrt{2}} \; .$$

The coherence time is therefore determined by the average lifetime τ_0 of the phase, which is of the order of the lifetime τ of the wavepackets emitted by the source.

Finally, let us calculate the spectral content of the source. It is given by the spectral density

$$\mathcal{S} = \mathcal{F}\{G\}(\omega) = \frac{\mathcal{I}_0}{2\pi} \int_{\mathbb{R}} e^{-|\tau|/\tau_0} e^{-i\omega_0 \tau} e^{-i\omega\tau} d\tau = \frac{\mathcal{I}_0}{2\pi} \frac{2/\tau_0}{(\omega - \omega_0)^2 + 1/\tau_0^2} \; .$$

The spectrum $\mathcal{S}(\omega)$ (Fig. 16.17) is a Cauchy–Lorentzian centered at $\omega = \omega_0$ with a full width at half maximum of $2/\tau_0$. The spectral width, according to (16.15), equals

Fig. 16.17 Spectrum of the quasi-monochromatic source

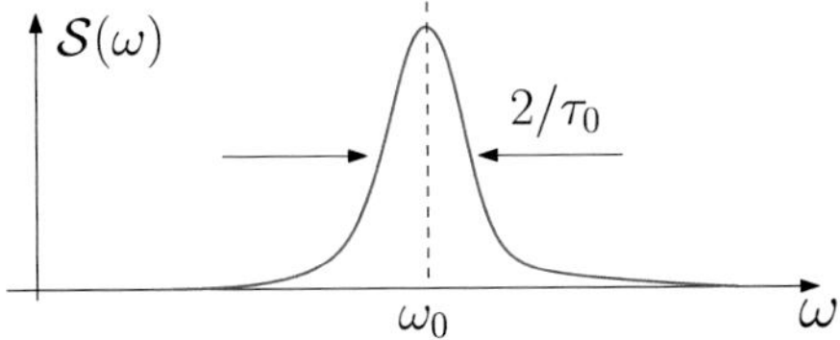

$$\Delta\omega = \sqrt{\frac{\int_{\mathbb{R}}(\omega - \omega_0)^2 |\mathcal{S}(\omega)|^2 d\omega}{\int_{\mathbb{R}} |\mathcal{S}(\omega)|^2 d\omega}} = \frac{1}{\tau_0} \; .$$

We see that the coherence length and the spectral width are such that

$$\tau_c \Delta\omega = \frac{1}{\sqrt{2}} \sim 1 \; .$$

Note that in Sect. 14.3.5 we have seen that the imaginary part of the refractive index (related to the absorption coefficient) exhibits resonances centered at the atomic transitions. For a single transition of frequency ω_0, the linewidth of the transition was found to be $\gamma = 1/\tau$ with τ the typical time it takes for the atom's dipole to decay to zero. This is related to the excited state lifetime and therefore to the size of the emitted wavepacket. We conclude that both in absorption and emission of light, the linewdith of an atomic transition is inversely proportional to its excited state lifetime τ.

16.7.1 The Particular Case of a Laser

In a laser, light-matter interaction is not dominated by spontaneous emission, but by the so-called stimulated emission predicted by Albert Einstein in the very beginning of the 19th century. In this regime a perfect synchronization occurs between the wavepackets emitted, thus creating a collective coherent radiation. In a laser, emission is thus highly controlled in terms of polarization, phase, and spectral purity. The coherence length of a multimode laser is of several tens of cm, whereas that of a monomode laser (almost a single frequency emission) can easily reach hundreds of meters. As a comparison, the coherence length of a spectral lamp (mercury, sodium) is of the order of 0.5 mm, and the coherence length of sunlight only 1 μm. Table 16.1 shows the spectral and coherent properties of typical light sources.

Table 16.1 Coherence time and coherence length of different light sources

Source	λ_0	$\Delta\lambda$	τ_c	$l_c = c\tau_c$
Visible sunlight	0.6 μm	0.4 μm	3 fs	900 nm
Blue LED	450 nm	50 nm	13 fs	4 μm
Sodium lamp	589 nm	0.6 nm	2 ps	580 μm
Multi mode He-Ne laser	633 nm	0.002 nm	0.67 ns	20 cm
Mono mode He-Ne laser	633 nm	1.4 fm	0.9 μs	290 m

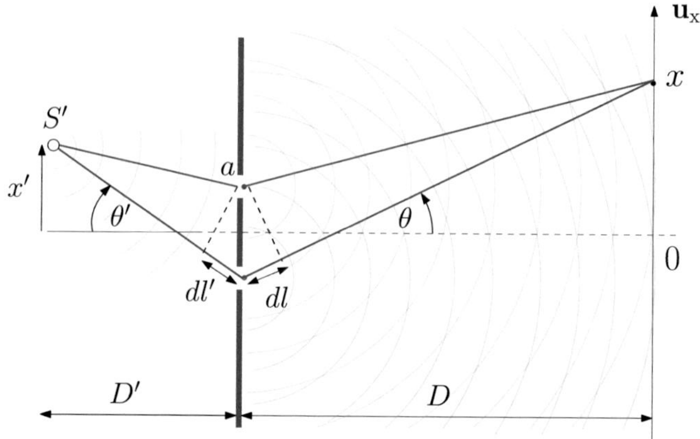

Fig. 16.18 A point source S' off axis of a double slit

16.8 Spatial Coherence

So far, we have dealt with the temporal coherence of point-like sources. Any real source has, however, a certain size which can play a role in the visibility of interferences. Two points on the source will emit light waves which are generally mutually incoherent (an exception is the laser). To understand the effect that the size of a source has on the interference pattern that they produce, let us consider a double slit interferometer and a point source S' that is off-axis by a quantity x', as shown in Fig. 16.18.

In a point x on the screen, the total path difference between the overlapping fields will be given by $dl + dl'$, where $dl \approx ax/D$ and where dl' is the additional path difference due to the fact that the source is no longer equidistant to the two apertures. Assuming that the source is far enough from the double slit (small angle approximation), we have

$$dl' = a \sin \theta' \approx \frac{ax'}{D'} .$$

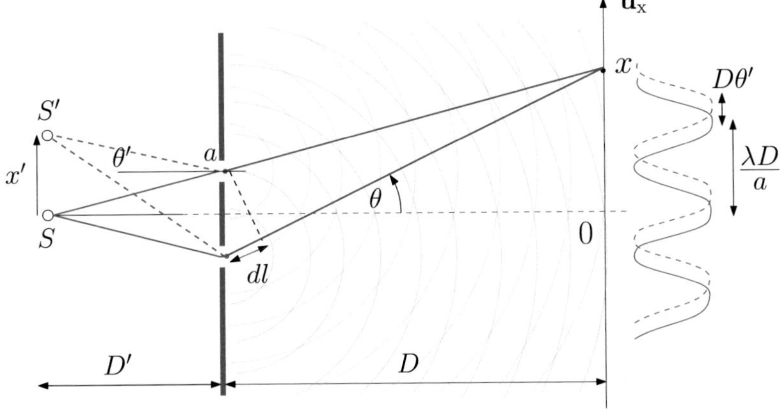

Fig. 16.19 The interference pattern of two mutually incoherence sources is the sum of the patterns generated individually by each source

The interference pattern on the screen is therefore

$$I'(x) = 2I_0 \left(1 + \cos\left(\frac{2\pi}{\lambda} \left(\frac{ax}{D} + \frac{ax'}{D'} \right) \right) \right).$$

We see that the constructive interference is obtained for x_n such that

$$\frac{ax_n}{\lambda D} + \frac{ax'}{\lambda D'} = n \in \mathbb{N}$$

that is

$$x_n = \frac{\lambda D}{a} n - \frac{x'D}{D'} = \frac{\lambda D}{a} n - D\theta'.$$

The interfence pattern is then shifted in the screen by a quantity $D\theta' = x'D/D'$ with respect to the pattern produced by a point source equidistant to both slits. This allows us to understand why the finite size of a source can lead to a complete dissapereance of the interference pattern, even if each point of the source is temporally coherent. If we now consider the case in which the two point sources S (equidistant to the slit) and S' are present, as shown in Fig. 16.19, the interference pattern on the screen will be simply the sum of the interference patterns that each source generates separately, since they are mutually incoherent. We see that for the interferences to be visible, the shift between the two patterns must be much smaller than the distance between two consecutive fringes

$$D\theta' = \frac{x'D}{D'} \ll \frac{\lambda D}{a}$$

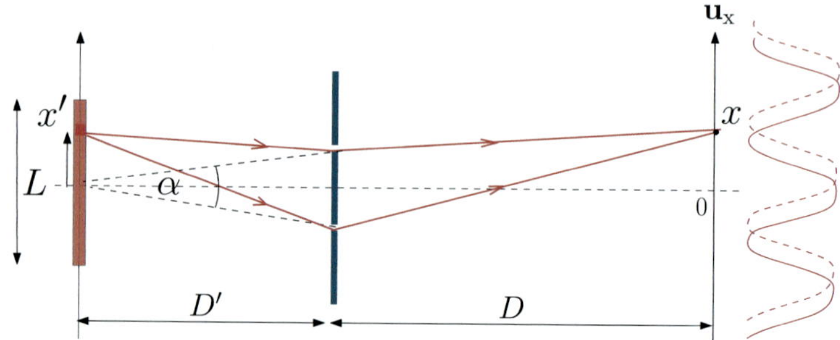

Fig. 16.20 Extended source in front of a double slit

and so the angle θ' must be small enough compared to λ/a. Let us generalize this condition for an extended source.

16.8.1 Interferences with an Extended Source

Consider now a source distributed over a typical length L along the x-axis, as shown in Fig. 16.20, and such that $I(x')dx'$ denotes the intensity emitted by all the points in the source between x' and $x' + dx$. Let $I_0 = \int_{\mathbb{R}} I(x')dx'$ be the total intensity of the source, and let's calculate the interference pattern observed on a screen far from the double slit. Assuming for simplicity a degree of temporal coherence close to 1 (monochromatic source), the infinitesimal element dx' around x' produces an interference pattern on the screen given by

$$dI(x) = 2I(x')dx' \left(1 + \cos\left(\frac{2\pi a}{\lambda}\left(\frac{x}{D} + \frac{x'}{D'}\right)\right)\right).$$

The different points of the source being mutually incoherent, the total intensity on the screen will be the superposition of the intensities generated by each infinitesimal element on the source, that is

$$I(x) = 2 \int_{\mathbb{R}} I(x') \left(1 + \cos\left(\frac{2\pi a}{\lambda}\left(\frac{x}{D} + \frac{x'}{D'}\right)\right)\right) dx'$$

$$= 2I_0 \left(1 + \frac{1}{I_0}\mathrm{Re}\left\{\int_{\mathbb{R}} I(x')e^{-i\frac{2\pi a x'}{\lambda D'}} e^{\frac{-i2\pi a x}{\lambda D}} dx'\right\}\right).$$

We obtain the Van Cittert–Zernike formula

$$\mathcal{I}(x) = 2I_0 \left(1 + \frac{1}{I_0}\mathrm{Re}\left\{ e^{-\frac{2\pi a x}{\lambda D}} \underbrace{\int_{\mathbb{R}} I(x')e^{-i\frac{2\pi a x'}{\lambda D'}}\,dx'}_{2\pi\mathcal{F}\{I\}(\frac{2\pi a}{\lambda D'})} \right\} \right). \qquad (16.18)$$

The interference pattern on the screen is that of a point source at $x' = 0$ with a contrast given by the Fourier transform of the source's spatial intensity distribution

$$C\left(\frac{a}{D'}\right) = 2\pi\mathcal{F}\{I\}\left(\frac{2\pi a}{\lambda D'}\right).$$

Remarks

- The visibility of the fringes does not depend on the position of the screen, but only on $a/D' \sim \alpha$, where α is the angle at which the double slit is seen from the source. Interestingly, from the loss of contrast one may infere the properties on the brightness distribution of the source, a method widely employed in astronomy.
- According to Fourier uncertainty principle, if L is the typical size of I, then its Fourier transform $\mathcal{F}\{I\}(u)$ has a typical width Δu of the order of $\Delta u \sim 2\pi/L$. The contrast C of the fringes takes therefore non negligible values for

$$\frac{2\pi a}{\lambda D'} \lesssim \frac{2\pi}{L}$$

that is

$$\frac{a}{D'} = \alpha \leq \frac{\lambda}{L}.$$

This allows us to define a coherence angle

$$\boxed{\theta_c = \frac{\lambda}{L}.} \qquad (16.19)$$

In summary, the interferences of a spatially incoherent source will be visible if the angle α at which the double slit is seen from the source is smaller than coherence angle λ/L. This is illustrated in Fig. 16.21.

From another point of view, the light produced by a spatially incoherent source can produce interferences in the Young apparatus only if the double slit is small enough. This provides a method to measure the size of distant stars when looked from the earth. Using a giant Young apparatus with variable interslit distance a, the loss of contrast of the fringes gives access to the (angular) size of the source. For example, an arrangement of telescopes can be combined into an astronomical interferometer, improving the angular resolution with respect to that of a single telescope. This is

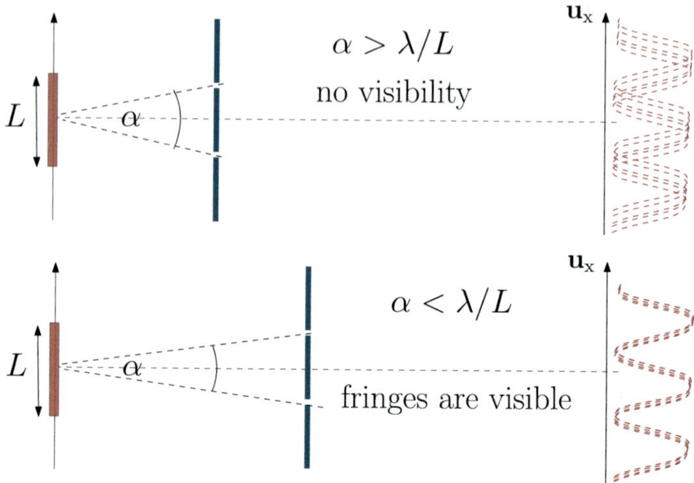

Fig. 16.21 An extended monochromatic source will produce visible interference if α is smaller than λ/L

the case of the Very Large Telescope Inteferometer (VLTI) or the Atacama Large Millimiter/submillimiter Array (ALMA), both located in the north of Chile.

16.9   Summary and Essential Formulas

- Interference can be summarized as the phenomenom by which the measured intensity I in a region where two waves overlap is not simply the sum of their separate intensities. It can be greater than their sum (constructive interference) or negligible (destructive interference).
- Interference between two waves always exist. However, due to the much faster oscillations of optical waves with respect to the time response of an optical detector, one measures an average intensity I over a very large number of cycles. In consequence, the interference term between two optical waves is different from zero if both have the same frequency, their polarizations are not orthogonal, and if the phase difference $\Delta\phi$ does not vary in time. When this conditions are fullfilled, the intensity is given by Fresnel formula

$$I = I_1 + I_2 + 2\sqrt{I_1 I_2}\cos\Delta\phi .$$

- The condition on a constant relative phase $\Delta\phi$ is generally imposible to achieve when the two waves come from independent sources, since their relative phase will vary randomly without correlation in timescales much shorter than the response of any optical detector. We say that in this case the waves are mutually incoherent and no interferences are observed.

- In contrast, if the two waves are mutually coherent, that is, if their relative phase is stationnary, interferences may be observed. This can be achieved, for example, by spliting a single wave originating from a single source into two parts that are later recombined once again after exploring different paths. Even if the phase of the field emitted by the source fluctuates in an unpredictable way, the relative phase between these two waves can be stationnary.
- The degree of coherence g of a source is defined as

$$g(\tau) = \lim_{T \to \infty} \frac{A}{T \mathcal{I}_0} \int_{-T/2}^{T/2} \underline{E}^*(t)\, \underline{E}(t - \tau)dt$$

and it measures the correlation between the field at instants t and $t - \tau$. One has $0 \leq |g(\tau)| \leq 1$, and $g(0) = 1$. The coherence time τ_c is defined as the typical width of the function g around $\tau = 0$ and it represents the memory time of the fluctuations. Interferences are therefore visible for delay times τ such that $\tau \leq \tau_c$. The coherence time defines a coherence length by $l_c = c\tau_c$.

- The spectral density $\mathcal{S}$, or simply the spectrum of a source of intensity $\mathcal{I}_0$ indicates the repartition of the intensity into the frequency components of the source. Therefore

$$\mathcal{I}_0 = \lim_{T \to \infty} \frac{A}{T} \int_{-T/2}^{T/2} |\underline{E}(t)|^2 dt = \int_{\mathbb{R}} \mathcal{S}(\omega)d\omega$$

and it corresponds to the Fourier transform of the correlation function $G(\tau) = \mathcal{I}_0 g(\tau)$ (Wiener–Khintchine theorem)

$$\mathcal{S}(\omega) = \mathcal{F}\{G\}(\omega) \,.$$

The spectral width $\Delta\omega$ of a source corresponds to the width of $\mathcal{S}$ around its central frequency.

- The coherence time τ_c of a source is inversely proportional to its spectral width $\Delta\omega$

$$\tau_c \Delta\omega \sim 1$$

explaining the high coherence of lasers, whose emissions are much narrower than conventional sources such as low pressure vapour lamps or natural light coming from the Sun.

Problems

16.1 The Fresnel biprism

A Fresnel biprism is used to produce interferences. Suppose that a monochromatic wave of wavevector $\mathbf{k} = k\mathbf{u}_x$ is shone at the vertical side ($x = 0$) of a biprism of angle α and refractive index n, as shown below.

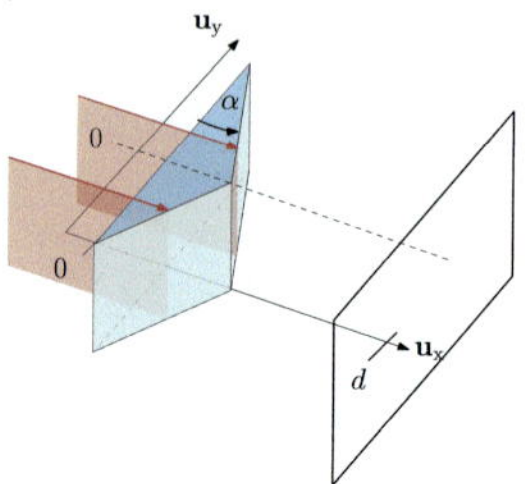

(a) What is the wavevector inside the prism and the wavevector of transmitted light coming out of each part of the biprism?

(b) Suppose that the two outcoming waves have all the same amplitude and phase, so that they can be written as $\underline{\mathbf{E}}_1 = \underline{\mathbf{E}}_0 e^{i(\mathbf{k}_1 \cdot \mathbf{x} - \omega t)}$ and $\underline{\mathbf{E}}_2 = \underline{\mathbf{E}}_0 e^{i(\mathbf{k}_2 \cdot \mathbf{x} - \omega t)}$, where $\mathbf{k}_1$ and $\mathbf{k}_2$ are the wavevectors found in the previous question. What is the real total electric field at a point $P = (x = d, y, z)$ on the screen?

(c) How does the averaged intensity I vary as a function of the position on the screen? What phenomenon is observed?

16.2 Degree of coherence of a doublet

Suppose that the source emits two monochromatic waves of equal amplitude and slightly different frequencies ω_1 and ω_2, also called a doublet. The field of the source is split into two equal parts and then recombined.

(a) Assuming that the detector's response is such that $T_{\text{det}} \gg \frac{2\pi}{|\omega_1 - \omega_2|}$, determine the degree of coherence of this source and the variation of the intensity resulting from interferences as a function of the path difference dl, the mean frequency $\omega_0 = (\omega_1 + \omega_2)/2$ and the frequency difference $\Delta\omega = \omega_1 - \omega_2$. How does the contrast of the interference pattern vary with dl?

(b) The doublet is placed in front of a double slit of size a. Make an sketch of the interference pattern produced as a function of the position x on the screen placed at distance D from the double slit and parallel to the line connecting both slits. Indicate the periodicity of the relevant features in terms of $\Delta\lambda = \lambda_1 - \lambda_2 \ll \lambda_1$ and $\lambda_0 = (\lambda_1 + \lambda_2)/2 \sim \lambda_1$, where λ_1 and λ_2 are the two wavelengths of the source.

16.3 A stellar interferometer

In 1920, Michelson and Pease used an apparatus equivalent to Young's double slit to make the first ever measurement of the diameter of a star. In that occasion they looked at Betelgeuse, a reddish star in the constellation of Orion. he distance of Betelgeuse from earth is estimated to be around $D' = 600$ lightyears by a measurement of the apparent change of its position caused by the movement of the earth around the Sun (parallax). Given that the interferences dissapeared for a distance between the two aperture mirrors of $a \sim 3\,\mathrm{m}$, estimate the diameter L of this star. Compare with the Sun's diameter $L_{\mathrm{Sun}} = 1.39 \times 10^9\,\mathrm{m}$.

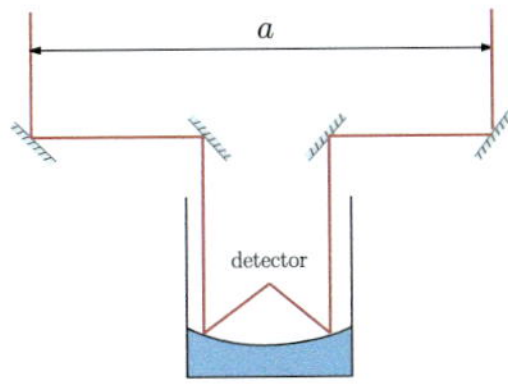

16.4 Interferences with sunlight

Sunlight enters a dark room through two small apertures separated by a distance $a = 2\,\mathrm{mm}$ and the resulting superposition of beams is observed on a screen at distance $D = 2\,\mathrm{m}$ from the two apertures.

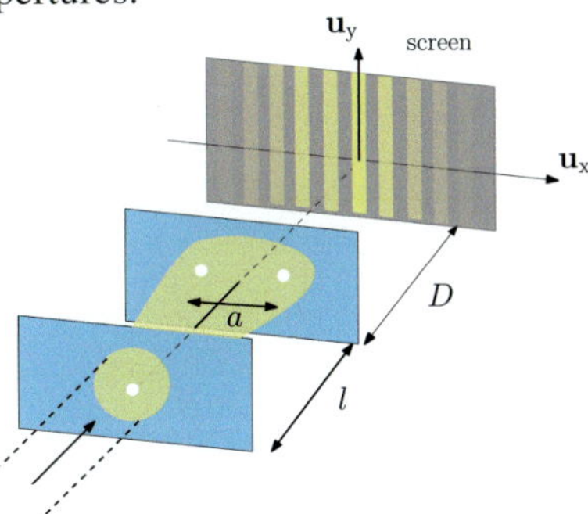

(a) Estimate the coherence angle θ_c of the Sun (radius $R = 6.96 \times 10^8\,\mathrm{m}$) in the visible range $\lambda_0 \sim 500\,\mathrm{nm}$. Considering that the distance between the earth and the Sun is $D' = 1.5 \times 10^{11}\,\mathrm{m}$, is the separation a between the holes consistent with the observation of an interferce pattern on the screen?

(b) Instead of letting the sunlight to enter directly the interferometer, one can put a dark screen with a small, 1 mm pinhole at the entrance, selecting only a fraction of the rays coming from the Sun to enter the interferometer. At what distance l from the two apertures should this pinhole be at so that one can observe interferences for $a = 2\,\mathrm{mm}$?

(c) What would be the distance Δx in the screen between two consecutive bright (or dark) fringes? Assuming that the pinhole is centered with respect to the two apertures, what is the optical path difference between the two interferring beams at the position of the first dark fringe? How does it compare with the coherence length $l_c \sim 0.9$ μm of sunlight in the visible range ? Conclude about the maximum number of fringes that are visible.

Chapter 17
Geometrical Optics

Abstract This chapter introduces *geometrical optics*, an approximation of light propagation valid in the *short wavelength limit*, where light is described as an ensemble of independent *optical rays*. It establishes that these rays propagate in straight lines in homogeneous media and follow curved trajectories in inhomogeneous media, always perpendicular to the *wavefronts*. The fundamental *optical ray equation* is derived, showing that rays bend towards regions of higher refractive index, a generalization that includes *Snell-Descartes law* for interfaces. A cornerstone of geometrical optics, *Fermat's principle of least time*, is presented, stating that light travels along paths that make the travel time stationary. This principle is used to derive the laws of reflection and refraction and to explain the fundamental behavior of optical components. The chapter then establishes the properties of *optical systems* and *image formation*. Key definitions are introduced, including *real and virtual objects and images*, *stigmatism* (the ability of a system to form point-like images), and the concept of *centered optical systems* with an *optical axis*. The *paraxial approximation* is emphasized as a simplification that allows for perfect stigmatism and aplanatism (forming plane images of plane objects) for rays close to the optical axis. Crucial *cardinal points* of optical systems are defined: *focal points*, *principal points*, and *nodal points*. Their relationships are established through *Newton's relation* and the *lens equation*, which relate object and image positions to focal lengths. The *transverse magnification* and *longitudinal magnification* are introduced to quantify image size and orientation. The chapter applies these principles to *spherical lenses* and *spherical mirrors*, deriving their imaging properties and demonstrating the concept of *geometrical aberrations* for non-paraxial rays. It also discusses *aspheric lenses* and *Fresnel lenses* as solutions for aberration correction and compact design, respectively. The *human eye* is analyzed as a complex optical system. Finally, *ray tracing* is presented as a practical graphical method for determining image formation, and the *association of multiple optical systems* is explored, culminating in *Gullstrand's formula* for combined focal lengths.

Keywords Geometrical optics · Refractive index · Optical rays · Fermat's principle · Optical systems · Lenses · Mirrors · Image formation

17.1 Introduction

By the end of the 19th century, the wave nature of light was firmly established, with experimental observations of diffraction (Grimaldi, 1665) and interference (Young, 1801) theoretically supported by Maxwell's equations (1861). However, it had been known since antiquity that many properties of light interacting with obstacles, such as lenses, mirrors, or prisms, could be explained within the framework of geometrical optics—the earliest theory of optics. This theory was developed over centuries through the contributions of Euclid, Snell, Descartes, Fermat, and others. Remarkably, the laws of reflection and refraction, derived in Chap. 15 as consequences of the continuity of electromagnetic fields at an interface, were already known long before Maxwell's equations were formulated.

Geometrical optics describes light as an ensemble of independent rays propagating in straight lines through homogeneous media and following curved trajectories when the refractive index of the medium changes. These paths can be determined using simple geometric principles, which lend the theory its name.

In the first part of this chapter, we will show that geometrical optics is a valid approximation in the limit of very small wavelengths—much smaller than the dimensions of the obstacles encountered by light. As long as the properties of the medium change gradually compared to the wavelength, the geometrical optics approximation holds. However, this description breaks down near the edges of objects, where abrupt changes in the refractive index cause diffraction, a phenomenon addressed in Chap. 18. Additionally, geometrical optics does not account for effects related to interference or polarization.

In the second part, we will explore the fundamental properties of optical systems and derive the equations governing the behavior of lenses and mirrors on optical rays. The chapter concludes with a discussion of image formation, introducing for this a practical graphical method called ray tracing.

17.2 The Short Wavelength Limit and Geometrical Optics

Consider a monochromatic wave of the form

$$\underline{\mathbf{E}} = \underline{\mathbf{E}}_0(\mathbf{x})e^{i(\phi(\mathbf{x})-\omega t)}$$

$$\underline{\mathbf{B}} = \underline{\mathbf{B}}_0(\mathbf{x})e^{i(\phi(\mathbf{x})-\omega t)}$$

We recall that a wave front is a surface such that $\phi(\mathbf{x}) = \text{constant}$, and we note that the particular case of a plane wave with a wave vector $\mathbf{k}$ corresponds to $\phi(\mathbf{x}) = \mathbf{k} \cdot \mathbf{x}$ and $\underline{\mathbf{E}}_0(\mathbf{x}) = \underline{\mathbf{E}}_0$.

From now on, we will consider the limit where the spatial derivatives of the amplitudes $\underline{\mathbf{E}}_0$ and $\underline{\mathbf{B}}_0$ are much smaller than the spatial derivatives of the phase ϕ. We will also consider that the refractive index n varies over length scales much larger than the wavelength. This limit, also called the short-wavelength limit, cannot explain the phenomenon of diffraction which manifests itself when, for example, the refractive index presents discontinuities at the edges of a lens aperture or a diaphragm.

In the short-wavelength limit, Maxwell's equations in the absence of free charges and currents are written as:

$$
\begin{aligned}
\nabla \cdot \mathbf{D} &= 0 & &\rightarrow & \nabla\phi \cdot \underline{\mathbf{E}} &= 0 \\
\nabla \cdot \mathbf{B} &= 0 & &\rightarrow & \nabla\phi \cdot \underline{\mathbf{B}} &= 0 \\
\nabla \times \mathbf{E} &= -\frac{\partial \mathbf{B}}{\partial t} & &\rightarrow & i\nabla\phi \times \underline{\mathbf{E}} &= i\omega\underline{\mathbf{B}} \\
\nabla \times \mathbf{B} &= \mu_0 \frac{\partial \mathbf{D}}{\partial t} & &\rightarrow & i\nabla\phi \times \underline{\mathbf{B}} &= -i\frac{\omega}{c^2}n^2(\omega)\underline{\mathbf{E}}
\end{aligned}
\tag{17.1}
$$

From the first three equations we see that, at every point,

$$
\underline{\mathbf{E}} \perp \nabla\phi(\mathbf{x}), \qquad \underline{\mathbf{B}} \perp \nabla\phi(\mathbf{x}), \qquad \underline{\mathbf{B}} = \frac{\nabla\phi}{\omega} \times \underline{\mathbf{E}}
$$

so that $(\nabla\phi, \mathbf{E}, \mathbf{B})$ form a direct trihedral. Moreover, replacing $\underline{\mathbf{B}}$ into Maxwell–Ampère equation yields

$$
\underbrace{\nabla\phi \times \left(\nabla\phi \times \underline{\mathbf{E}}\right)}_{-|\nabla\phi(\mathbf{x})|^2\underline{\mathbf{E}}+\nabla\phi(\nabla\phi\cdot\underline{\mathbf{E}})} = -\frac{\omega^2 n^2}{c^2}\underline{\mathbf{E}} \ .
$$

Since $\nabla\phi \cdot \underline{\mathbf{E}} = 0$, we obtain the Eikonal equation

$$
\boxed{|\nabla\phi(\mathbf{x})|^2 = \frac{n(\mathbf{x})^2\omega^2}{c^2}.}
\tag{17.2}
$$

We see that locally, the structure of the wave is similar to that of a plane wave with a local wave vector $\mathbf{k} = \nabla\phi$ given by

$$
\mathbf{k}(\mathbf{x}) = \nabla\phi(\mathbf{x}) = \frac{n(\mathbf{x})\omega}{c}\mathbf{t}(\mathbf{x})
\tag{17.3}
$$

where the local direction of propagation $\mathbf{t}(\mathbf{x})$ is perpendicular to the wave fronts since $\mathbf{k}(\mathbf{x}) = \nabla\phi(\mathbf{x})$ (a result known as Malus–Dupin theorem).

17.2.1 Optical Rays

In the short-wavelength limit, the time-averaged Poynting vector reads

$$
\langle \mathbf{\Pi} \rangle_T = \mathrm{Re}\left\{\frac{\mathbf{E} \times \mathbf{B}^*}{2\mu_0}\right\} = \frac{1}{2\mu_0\omega}|\mathbf{E}_0(\mathbf{x})|^2\mathbf{k}(\mathbf{x}) = \frac{n(\mathbf{x})}{2\mu_0 c}|\mathbf{E}_0(\mathbf{x})|^2\mathbf{t}(\mathbf{x}) \ .
$$

On the other hand, the time-averaged electromagnetic energy density reads

$$\langle u_{EM}\rangle_T = \frac{1}{2}\langle \mathbf{E}\cdot\mathbf{D} + \mathbf{B}\cdot\mathbf{H}\rangle_T = \frac{\epsilon_0 n^2(\mathbf{x})}{2}|\mathbf{E}_0(\mathbf{x})|^2$$

and we obtain

$$\boxed{\langle \boldsymbol{\Pi}\rangle_T = \frac{c}{n}\langle u_{EM}\rangle_T \mathbf{t}} \tag{17.4}$$

and so the electromagnetic energy propagates in the direction $\mathbf{t}$ of the local wave vector $\mathbf{k}$ at a speed c/n. This is why in geometrical optics light is described by an ensemble of independent and incoherent optical rays, which are defined as the field lines of $\mathbf{k}(\mathbf{x}) = \boldsymbol{\nabla}\phi(\mathbf{x})$, always perpendicular to the wave fronts and pointing along the unit direction $\mathbf{t} = \dfrac{\boldsymbol{\nabla}\phi}{|\boldsymbol{\nabla}\phi|}$. They coincide with the field lines of the Poynting vector $\langle \boldsymbol{\Pi}\rangle_T$ in the short wavelength limit.

17.2.2   Spherical and Plane Waves in Geometrical Optics

Consider a spherical wave emitted by (or converging to) a single point S. In geometrical optics, such a wave is represented by straight optical rays intersecting at the point S, or equivalently, by wave fronts that are spheres centered at S. The wave is said to be divergent when the rays depart from S and convergent when they are directed toward S. This is shown in Fig. 17.1.

On the other hand, a plane wave is described by straight optical rays parallel to each other or, equivalently, by wave fronts that are parallel planes. A plane wave is a good approximation for the wave created by distant sources in regions whose dimensions are much smaller than the distance to the source, as shown in Fig. 17.2.

Since optical rays are perpendicular to the wave fronts, any transparent object that modifies the phase of a light wave will bend the optical rays. For example, a thin piece of glass with varying thickness can be used to transform plane wave fronts into

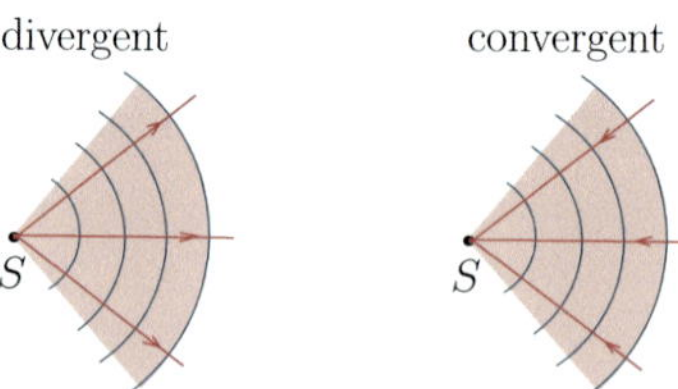

Fig. 17.1   Representation of a spherical wave

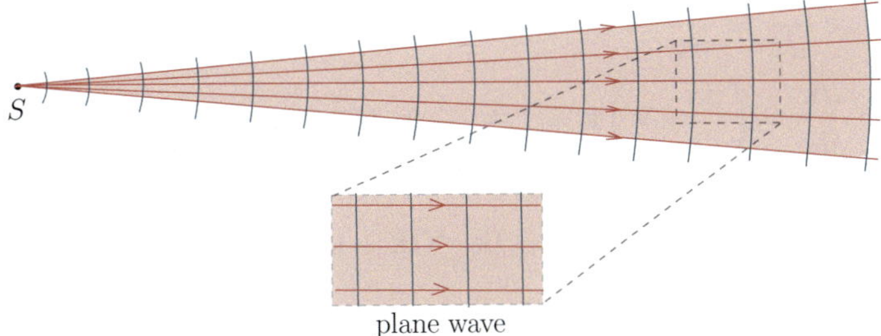

Fig. 17.2 Locally, a spherical wave looks like a plane wave

Fig. 17.3 A convergent lens transforms a plane wave into a converging spherical wave

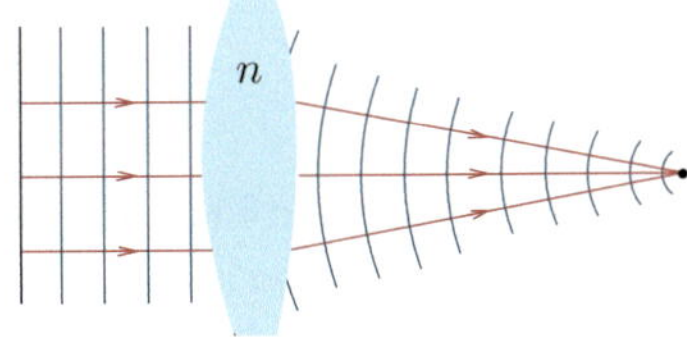

spherical wave fronts, focusing the incoming, parallel light rays into a single point. This is the case of a convergent lens, illustrated in Fig. 17.3.

17.3 The Optical Ray Equation

We now focus on a field line of the field $\mathbf{k}(\mathbf{x})$. The goal is to find an equation for the trajectory $\mathbf{x}(s)$ of an optical ray along that field line. Here s is the arc length parameter of the curve, which gives a parametrization such that $|d\mathbf{x}/ds| = 1$. The arc length s represents, at every point, the total distance travelled along the curve:

$$\int_A^B |d\mathbf{x}| = \int_{s_A}^{s_B} \left| \frac{d\mathbf{x}}{ds} \right| ds = \int_{s_A}^{s_B} ds = s_B - s_A \ .$$

As shown in Fig. 17.4, the tangent $d\mathbf{x}/ds$ to that curve at point $\mathbf{x}$ is equal to the unit direction $\mathbf{t}$ of the local wave vector, which according to (17.3) writes:

$$\frac{d\mathbf{x}}{ds} = \mathbf{t} = \frac{c}{\omega n(s)} \mathbf{k}(s) \ .$$

Fig. 17.4 The trajectory of an optical ray is a field line of the field $\mathbf{k}(\mathbf{x}) = |\mathbf{k}(\mathbf{x})|\mathbf{t}$

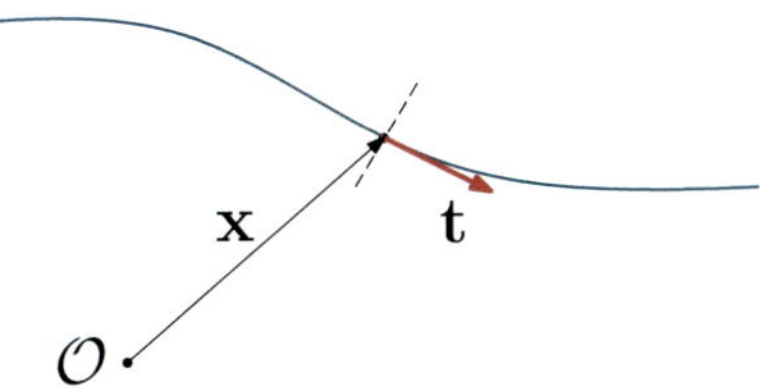

Now, since $\dfrac{dk_i}{ds} = \dfrac{d\mathbf{x}}{ds} \cdot \nabla k_i$ we have

$$\frac{d\mathbf{k}}{ds} = \left(\frac{d\mathbf{x}}{ds} \cdot \nabla\right)\mathbf{k} = \left(\frac{c}{\omega n(s)}\mathbf{k} \cdot \nabla\right)\mathbf{k} = \frac{c}{\omega n(s)}(\mathbf{k} \cdot \nabla)\mathbf{k}$$

and using the identity $\dfrac{1}{2}\nabla(\mathbf{k} \cdot \mathbf{k}) = (\mathbf{k} \cdot \nabla)\mathbf{k} + \mathbf{k} \times (\nabla \times \mathbf{k})$, with $\nabla \times \mathbf{k} = \nabla \times (\nabla\phi) = 0$, we can finally write

$$\frac{d\mathbf{k}}{ds} = \frac{c}{2\omega n(s)}\nabla(\mathbf{k} \cdot \mathbf{k}) = \frac{c}{2\omega n(s)}\nabla \underbrace{|\mathbf{k}^2|}_{n^2(s)\omega^2/c^2}$$

$$= \frac{\omega}{2n(s)c}\nabla(n^2(s)) = \frac{\omega}{c}\nabla n(s) \ .$$

We conclude that in an inhomogeneous dielectric medium, the local wave vector along a ray points toward the gradient of the refractive index. In other words, rays are bent whenever there is an optical index contrast, and we can rewrite the ray equation as follows

$$\boxed{\frac{c}{\omega}\frac{d\mathbf{k}}{ds} = \frac{d}{ds}(n(s)\mathbf{t}) = \nabla n(s).}$$

(17.5)

17.3.1 Propagation in a Homogeneous Medium

The simplest case is when light travels in a homogeneous medium so that $\nabla n = \mathbf{0}$. It follows that the local wave vector $\mathbf{k}$ is constant along the trajectory of a ray, which is given by

$$\frac{d}{ds}(n\mathbf{t}) = n\frac{d}{ds}\mathbf{t} = n\frac{d^2\mathbf{x}}{ds^2} = 0$$

whose general solution is a straight line

$$\mathbf{x} = \mathbf{a}s + \mathbf{b} \ .$$

The idea of light traveling in straight lines was already present in the works of Euclid (300 BC), who postulated that a discrete set of straight rays, called visual rays, diverge infinitely from our eyes allowing us to observe objects.

The fact that light in homogeneous media travels in straight lines can be shown by first *isolating* one ray by placing a small aperture A in front of a light source.[1] Then, a second aperture C is placed so that the beam goes through both A and C

---

[1] Of course, trying to better isolate a ray by reducing the size of the aperture will eventually result in failure since, as the size of the aperture becomes comparable to the wavelength of light, the short-wavelength limit ceases to be valid and diffraction can no longer be neglected.

simultaneously. An intermediate aperture B placed in between A and C allows us to show that light will pass through C only if B belongs to the straight line joining A and C. This is shown in Fig. 17.5.

17.3.2 Generalization of the Snell–Descartes Law

Let us find the general solution of the ray Eq. (17.5) in the presence of a gradient of the refractive index along the z-axis:

$$\frac{d}{ds}(n\mathbf{t}) = \frac{\partial n}{\partial z}\mathbf{u}_z \, .$$

Taking the vector product with $\mathbf{u}_z$ on both sides of the equation:

$$\mathbf{u}_z \times \frac{d}{ds}(n\mathbf{t}) = \frac{d}{ds}(n\mathbf{u}_z \times \mathbf{t}) = 0 \, .$$

We conclude that $n\mathbf{u}_z \times \mathbf{t}$ is a constant vector along the ray, implying that the trajectory lies in a plane. Moreover, since $|\mathbf{u}_z \times \mathbf{t}| = \sin\theta$, with θ the angle that the ray makes with respect to the z-axis, we find:

$$\boxed{\frac{d}{ds}(n\sin\theta) = 0 \quad \rightarrow \quad n\sin\theta = \text{constant}.} \tag{17.6}$$

This remains valid in the case of a jump discontinuity of the refractive index at an interface (which corresponds to the Snell–Descartes law). Indeed, consider the

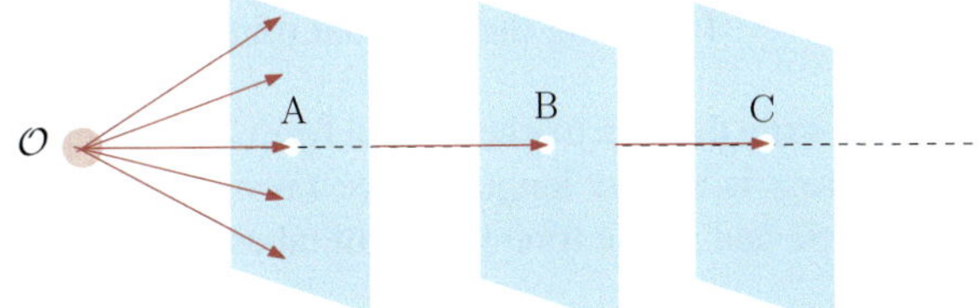

Fig. 17.5 If a light ray passes through A and C, it will also pass through B if the latter belongs to the straight line going from A to C

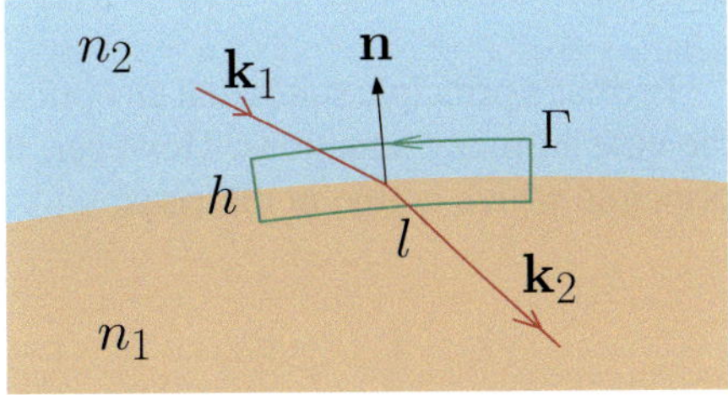

Fig. 17.6 Snell–Descartes law can be retrieved from the optical ray equation

interface separating two media with optical index n_1 and n_2, respectively, and let us calculate the circulation of $\mathbf{k}$ along the closed rectangular curve Γ shown in Fig. 17.6. According to the Stokes theorem (A.20)

$$\oint_\Gamma \mathbf{k} \cdot d\mathbf{l} = \iint_{S(\Gamma)} (\nabla \times \mathbf{k}) \cdot \mathbf{d}S$$

but $\nabla \times \mathbf{k} = \nabla \times (\nabla \phi) = 0$, and so $\oint_\Gamma \mathbf{k} \cdot d\mathbf{l} = 0$. By taking the limit $h \to 0$ this circulation reduces to

$$\oint_\Gamma \mathbf{k} \cdot d\mathbf{l} = l \left(\mathbf{k}_2^{\|} - \mathbf{k}_1^{\|} \right) = 0$$

and so the tangential component of $\mathbf{k} = n\mathbf{t}$ must be continuous at the interface. If the ray forms an angle θ_1 (respectively θ_2) with respect to the normal to the interface in region 1 (respectively in region 2), the continuity of the tangential component of $\mathbf{k}$ writes

$$n_1 \sin \theta_1 = n_2 \sin \theta_2$$

and we retrieve the Snell–Descartes law (15.10).

17.4 Fermat's Principle (1657)

Long before Maxwell's equations, Fermat (Fig. 17.7) postulated the famous principle of least time, which can be stated as:

$$\text{(17.7)} \qquad \begin{array}{l} \text{Light propagates from one point to another by} \\ \text{following the path that minimizes the time of travel.} \end{array}$$

Let us write this in a mathematical form. An optical ray is parametrized by a curve $\mathbf{x}(s)$ where s is the arc length. Since rays travel at a speed $c/n = ds/dt$, we have $n/c\,ds = dt$ and so the time that light takes to go from A to B can be written as

$$\frac{1}{c} \int_A^B n(s)ds .$$

Fermat's principle states that an optical path minimizes this integral. This is indeed the case in many situations. However, there is a more general version of Fermat's principle which is true in all cases:

Fig. 17.7 Pierre de Fermat (1607–1665), a French mathematician and physicist. His work profoundly influenced the development of mathematical analysis and physics

Light propagates from one point to another by following the trajectory Γ for which the travel time is stationary with respect to any small variation of the path Γ.
(17.8)

This means that infinitesimal variations around the optical path lead to no first-order variation in the travel time:

$$\delta\left(\int_A^B n(s)\,ds\right) = 0. \tag{17.9}$$

The path that light takes does not always represent a local minimum of the travel time. For example, the path could correspond to a saddle (or inflection) point where small variations do not change the travel time. The conclusion is that an optical ray path is surrounded by close paths for which the time of travel is almost the same. Fermat's principle is a particular case of a principle that is omnipresent in modern physics: Hamilton's principle of stationary action. It was first proposed by Maupertuis in 1741 and formalized by Euler and Lagrange a decade later.

In order to demonstrate Fermat's principle, one must consider an arbitrary curve $\mathbf{x}(t)$ that joins two points A and B. The time of travel $T[\mathbf{x}(t)]$ along this curve is the integral

$$T[\mathbf{x}(t)] = \frac{1}{c}\int_A^B n(\mathbf{x})|d\mathbf{x}| = \frac{1}{c}\int_{t_A}^{t_B} n(\mathbf{x}(t))\left|\frac{d\mathbf{x}}{dt}\right|dt$$
$$= \frac{1}{c}\int_{t_A}^{t_B} n(x(t), y(t), z(t))\sqrt{(\dot{x})^2 + (\dot{y})^2 + (\dot{z})^2}\,dt. \tag{17.10}$$

Among all the curves such that $\mathbf{x}(t_A) = A$ and $\mathbf{x}(t_B) = B$, the time T is stationary if $\mathbf{x}(t)$ is a solution of Euler–Lagrange's equations, which here writes:

$$\frac{d}{dx_i}\left(n(x(t), y(t), z(t))\sqrt{(\dot{x})^2 + (\dot{y})^2 + (\dot{z})^2}\right) = \frac{d}{dt}\left(\frac{d}{d\dot{x}_i}n(x(t), y(t), z(t))\sqrt{(\dot{x})^2 + (\dot{y})^2 + (\dot{z})^2}\right)$$

for $x_i = x, y, z$. This equation gives, for the x_i component:

$$\frac{dn(\mathbf{x}(t))}{dx_i}\sqrt{(\dot{x})^2 + (\dot{y})^2 + (\dot{z})^2} = \frac{d}{dt}\left(\frac{n(x(t), y(t), z(t))}{\sqrt{(\dot{x}_i)^2 + (\dot{y})^2 + (\dot{z})^2}}\frac{dx_i}{dt}\right).$$

If now t is an arc length parameter, $\left|\frac{d\mathbf{x}}{dt}\right| = \sqrt{(\dot{x}_i)^2 + (\dot{y})^2 + (\dot{z})^2} = 1$ and so

$$\frac{dn(\mathbf{x}(t))}{dx_i} = \frac{d}{dt}\left(n(\mathbf{x})\frac{dx_i}{dt}\right).$$

We recognize the x_i component of the optical ray Eq. (17.5).

The optical path length

Since c is a constant, the time of travel between A and B along a path Γ is proportional to the optical path length, noted (AB):

$$\boxed{(AB) = \int_{\Gamma} n(s)ds.} \tag{17.11}$$

According to Fermat's principle, (AB) is stationary along the ray trajectory.

Remark

We see that in a homogeneous medium, the optical path along a ray is simply equal to the length of the ray multiplied by the refractive index. For instance, in Fig. 17.8 we have

$$(AD) = (AB) + (BC) + (CD) = n_1|AB| + n_2|BC| + n_3|CD|.$$

Fig. 17.8 A ray trajectory is a straight line in each region where the refractive index is constant

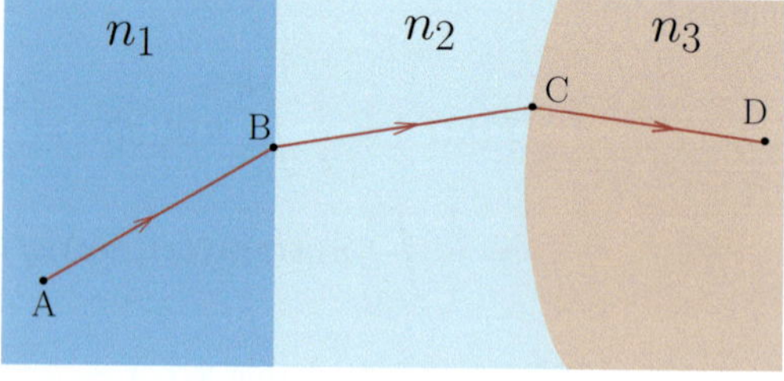

Example 17.1—Fermat's principle and the law of reflection

To go from A to B in a homogeneous medium, light will follow the straight line which gives the global minimum of the optical path. In the presence of a flat reflecting surface, there is also a second possibility illustrated in Fig. 17.9. This trajectory, according to Fermat, represents a local extremum of the optical path length.

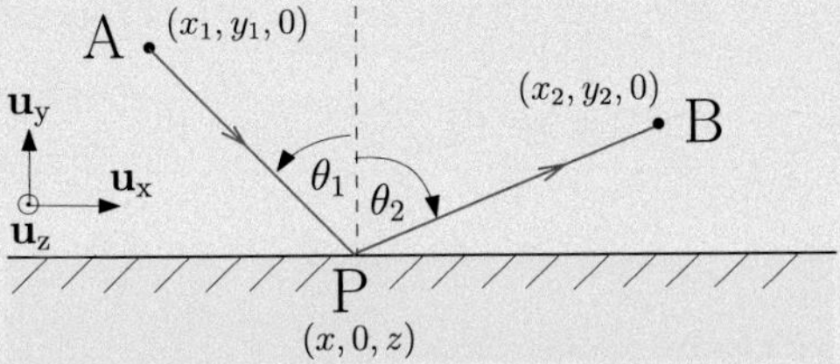

Fig. 17.9 A ray passing through B can reach B after reflection by a flat mirror

Let $P = (x, 0, z)$ be the point at which the ray hits the mirror. Since the medium in which it propagates is homogeneous, the paths AP and PB are both straight lines. As such, the time t_{AB} that it takes to go from A to B is simply

$$t_{AB} = n\frac{AP + PB}{c}$$

where n is the refractive index of the medium.

Replacing AP and PB in terms of x and z:

$$t_{AB}(x, z) = \frac{n}{c}\sqrt{(x - x_1)^2 + y_1^2 + z^2} + \frac{n}{c}\sqrt{(x - x_2)^2 + y_2^2 + z^2}\,.$$

and the position of P must be such that the travel time t_{AB} is stationary. Therefore the partial derivatives with respect to x and z must vanish

$$\frac{\partial t_{AB}(x, z)}{\partial z} = \frac{n}{c}\left(\frac{z}{AP} + \frac{z}{PB}\right) = 0$$

and so $z = 0$, which means that A, P and B belong to the same incidence plane $z = 0$. On the other hand

$$\frac{\partial t_{AB}(x, z)}{\partial x} = \frac{n}{c}\left(\frac{x - x_1}{AP} + \frac{x - x_2}{PB}\right) = 0$$

and so

$$\frac{x - x_1}{AP} = \frac{x_2 - x}{PB}$$

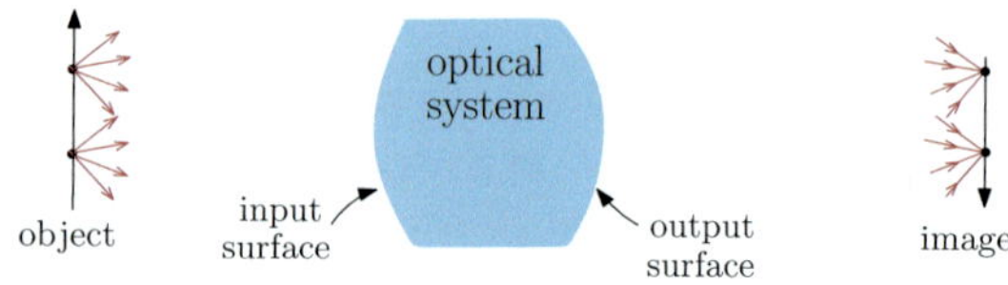

Fig. 17.10 By convention, light always enter an optical system from left to right

which implies $x_1 < x < x_2$ and we retrieve the law of reflection

$$\sin \theta_1 = \sin \theta_2 \quad \rightarrow \quad \theta_1 = \theta_2 .$$

17.5   Optical Systems and Image Formation Definitions

In geometrical optics, an object is any source of light (direct or indirect) and it can always be thought as a superposition of point sources, each one emitting spherical wave fronts. An optical system, generally constituted of transparent or reflecting elements, allows us to observe objects by forming images after multiple reflections and refractions of the optical rays coming from an object. Any arbitrary optical system is limited by an input and output surface, and we will consider by convention that light travels from left to right when entering the system, as shown in Fig. 17.10.

In the following of this chapter we will derive the main laws describing optical systems. Since we have defined a preferred convention for the direction of propagation of light rays (from left to right in Fig. 17.10), it is very useful to define a distance between two points A and B not by its euclidean distance (always positive) but instead by a number that has a sign indicating the direction in which the displacement occurs.

Algebraic measure

We denote by $\overline{AB}$ the algebraic measure of the distance between points A and B. The magnitude of $\overline{AB}$ corresponds to the distance between A and B, while its sign indicates the direction of displacement along a defined axis. Specifically, $\overline{AB}$ is positive for a displacement from A to B and negative for a displacement from B to A. Throughout this chapter, if A and B lie along the horizontal axis, then $\overline{AB}$ is positive when B is to the right of A (indicating a rightward displacement). If A and B lie along the vertical axis, then $\overline{AB}$ is positive when B is above A (indicating an upward displacement).

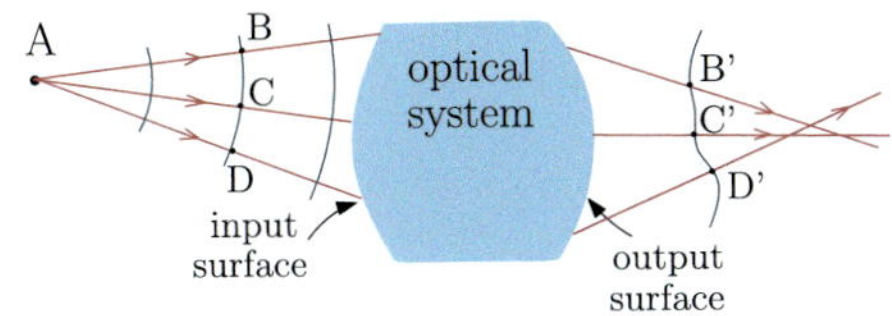

Fig. 17.11 The optical path lengths (BB'), (CC') and (DD') along the rays emitted by A are all equal

Fig. 17.12 Pierre Charles François Dupin (1784–1873), a French mathematician and engineer. He made contributions to differential geometry and cartography

17.5.1 *Geometrical Wave Fronts and Malus–Dupin Theorem*

Let A be a point source of light. A geometrical wave front with respect to A is defined as a surface of constant optical path length from point A, corresponding as well to a surface of constant phase. In Fig. 17.11, the points B, C, and D belong to the same wave front. If they are all in the same medium as A, this wave front simply corresponds to a constant path length, and so it is a sphere centered at A. At the exit of an optical system, however, the geometrical wave fronts are not necessarily spheres.

Since B', C', and D' belong to the same wave front at the output of the system, along the rays shown in Fig. 17.11 the optical path length (BB') equals (CC') and also (DD').

$$(BB') = (CC') = (DD'). \tag{17.12}$$

The theorem of Malus (1808) clarified by Dupin (1822, Fig. 17.12) states that the rays are always perpendicular to the geometrical wave fronts. We know this from (17.3) and the fact that the optical rays are the field lines of **k** in the short wavelength limit.

17.5.2 *Stigmatism*

Suppose that a point object A is a source of optical rays that enter an optical system S. If at the exit of the system all the outgoing rays (or their extensions) intersect at a single point A' as shown in Fig. 17.13, we say that A' is the image of A by S, and

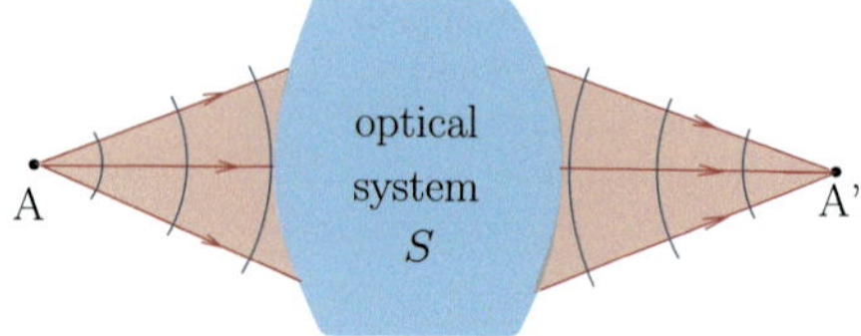

Fig. 17.13 A system with perfect stigmatism for the pair (A, A')

that the optical system S is perfectly stigmatic for the so-called stigmatic pair (A, A'). According to the principle of backward propagation of light, a point source at A' will be imaged at A by the same optical system. This is why A and A' are also said to be conjugate points and we write

$$\text{A} \xrightarrow{S} \text{A}'$$

Truly speaking, there is no such thing as perfect stigmatism: no optical system is able to form point-like images because of diffraction (this is discussed in Sect. 18.5.4). In this chapter, however, we will restrict ourselves to the framework of geometrical optics so that, at least in principle, perfect stigmatism is possible. Still, the most common optical systems only exhibit an approximate stigmatism, meaning that the image of a point A by S is a point only for a certain class of rays coming from A. When the stigmatism is approximate and not perfect, we say that the optical system exhibits geometrical aberrations.

An example of a system with perfect stigmatism is the flat mirror. Indeed, according to the reflection law, every ray emitted by a point source A seems to come from a single point A' after reflection by the mirror, with a conjugation relation given by

$$\text{A}' = \text{sym}_\Pi \text{A}$$

where Π is the plane corresponding to the mirror surface. That is, A' is simply the symmetric of A with respect to the plane of the mirror. This is illustrated in Fig. 17.14 (left).

An example of a system with approximate stigmatism is the plane interface between two dielectric media, also called plane diopter, shown in Fig. 17.14 (right). The rays coming from a point source A and refracted by the diopter seem to come from a single point image A' only for those rays forming small angles with respect to the normal to the interface. Indeed, consider the ray coming from A arriving at P with an incidence angle i with respect to the normal. Snell–Descartes law gives $n_1 \sin i = n_2 \sin r$ and

$$\tan r = \frac{\overline{\text{OP}}}{\overline{\text{A'O}}}, \quad \tan i = \frac{\overline{\text{OP}}}{\overline{\text{AO}}}$$

so the refracted ray seems to come from a point A', at a distance with respect to O given by

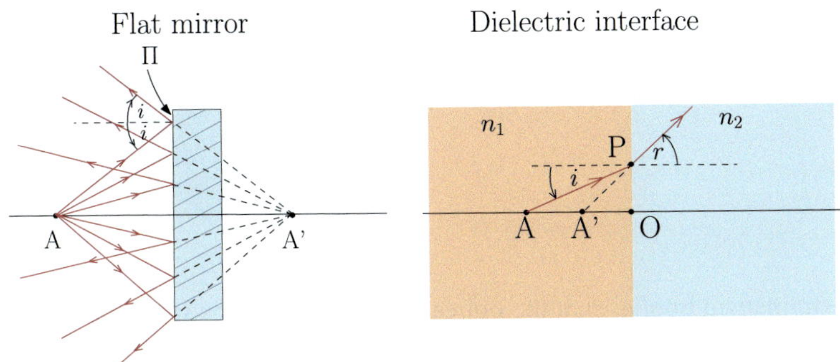

Fig. 17.14 A flat mirror (left) exhibits perfect stigmatism. A dielectric interface (right) only approximates stigmatism for rays close to the axis

$$\overline{\text{A'O}} = \overline{\text{AO}}\frac{\tan i}{\tan r} = \overline{\text{AO}}\frac{n_2}{n_1}\sqrt{\frac{1 - \frac{n_1^2}{n_2^2}\sin^2 i}{\cos^2 i}} \ .$$

The distance depends on the incidence angle i of the ray at the interface, except for rays very close to normal incidence for which the angles are small, $n_1 i \approx n_2 r$ and $\tan i / \tan r \approx n_2/n_1$, in which case all the refracted rays seem to come from a single point A' located at

$$\overline{\text{A'O}} \approx \overline{\text{AO}}\frac{n_2}{n_1} \ .$$

This optical system has therefore only approximate stigmatism, which is generally the case for most systems such as lenses and curved mirrors. Since every detector has a finite spatial resolution anyways (limited for example by the size of a pixel in a CCD, or the cone cells in the retina, or the grains in photographic films), an approximate stigmatism is sometimes more than enough to provide good images of objects.

17.5.3 Stigmatism and Fermat's Principle

Let S be an optical system and (A, A') a stigmatic pair, such that A' is the image of A by S. According to Fermat's principle, a ray going from A to A' follows a trajectory with stationary optical path. Since all the rays stemming from A reach A', it follows that the optical path must be the same for all these rays (Fig. 17.15).

This can also be thought as a particular consequence of the Malus–Dupin theorem. Since A and A' are point-like wave fronts with respect to A, it follows that the optical path length is the same for all rays going from a point source A and its image A'.

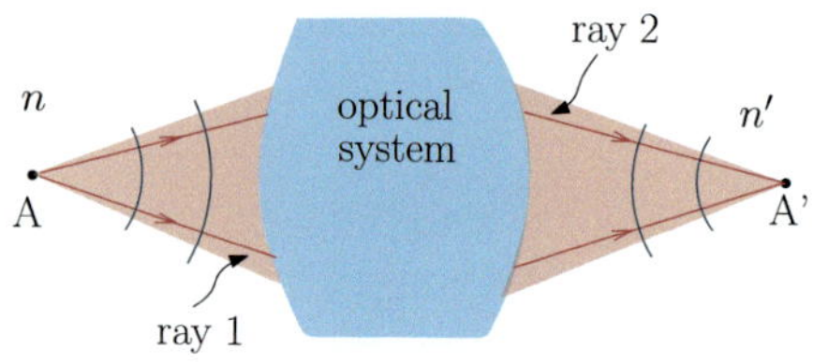

Fig. 17.15 All the rays going from A to A' have the same optical path length

Perfect stigmatism implies that the optical system transforms spherical wave fronts into spherical wave fronts.

17.5.4   Real Versus Virtual Objects and Images

Objects and images can be real or virtual. A point source A located before the input surface and from which rays enter the optical system S is called a real object for S. If the rays emerging from the system converge to A' after the output surface, we say that the image A' of A by S is real. On the other hand, if the rays emerging from the system S seem to come from a point A' before the output surface, one obtains a virtual image. This is illustrated in Fig. 17.16. The rays coming from such a virtual image will be focused by our eyes (which constitute an optical system on their own) and produce a real image on the retina, thus they will be perceived by the brain as if there was a real object in A'.

Now if a beam is converging at the entrance of an optical system S, such that the point of convergence A of the rays is located after the input surface, then A is a virtual object for the system S. Once again, depending on whether the outcoming rays converge to a point A' after the output surface, or seem to come from a point A' before the output surface, the image A' is real or virtual, respectively (Fig. 17.17).

Remarks

- Since light is assumed to be propagating from left to right before entering the system, in many cases the terms *before* and *after* can be replaced by *at the left of* and *at the right of* in the above definitions. This is the case for lenses, for example. However, this will not work in the case where the optical system is a mirror, for

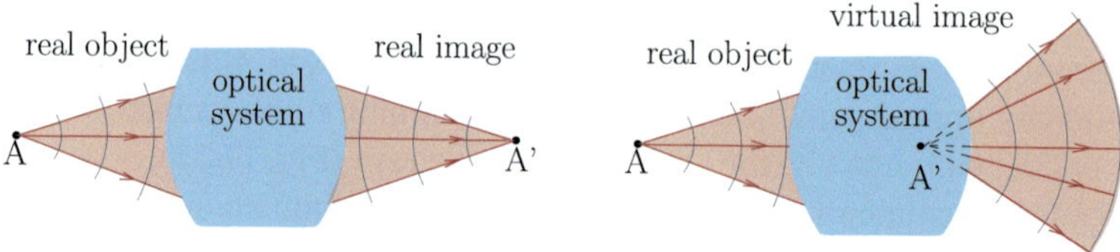

Fig. 17.16 An object may produce a real image (left) or a virtual image (right)

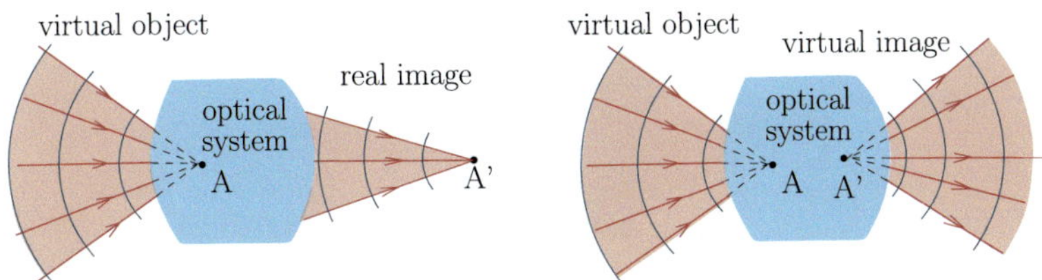

Fig. 17.17 Virtual object and real image (left). Virtual object and virtual image (right)

Fig. 17.18 When the image is virtual, one may still define AA' as (AB)–(A'B)

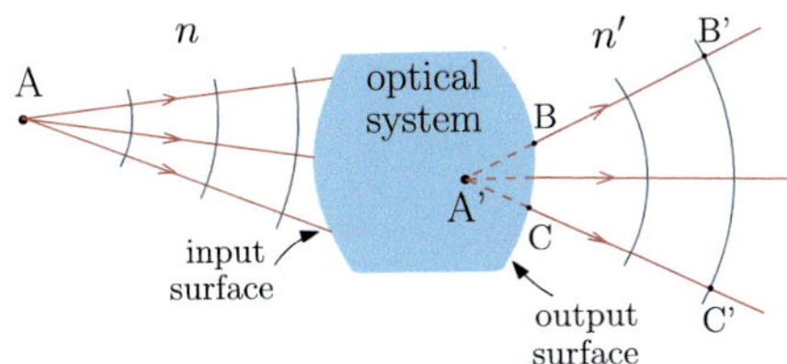

which the input and the output surface is the mirror itself: real objects and images are at the left of the mirror, whereas virtual objects and images are at the right of the mirror.

- When A' is a virtual image, no ray coming from A will actually pass through A', so that strictly speaking there is no optical path going from A to A' along a ray. Nevertheless, we can extend Fermat's theorem by considering two rays coming from A and intersecting a wave front at the exit of the optical system at points B' and C', respectively (Fig. 17.18). We have

$$(AB') = (AB) + (BB') = (AC) + (CC') = (AC'). \tag{17.13}$$

On the other hand, since both B' and C' belong to a spherical wave front centered at A', we have

$$(A'B') = (A'B) + (BB') = (A'C) + (CC') = (A'C') .$$

Note that here, the optical paths starting from A' should be computed as if a real ray started from A' in a medium of refractive index n', even if A' belongs to a region with a different index. Replacing (BB') and (CC') in (17.13), we find

$$(AB) - (A'B) = (AC) - (A'C) .$$

For a real object A and its virtual image A', we can therefore define the optical path (AA') = AB–A'B such that the prolongation A'B contributes with a negative sign. This allows us to conclude that the optical path (AA'), defined in this way, is the same for all rays, as in the case where both the object and its image are real.

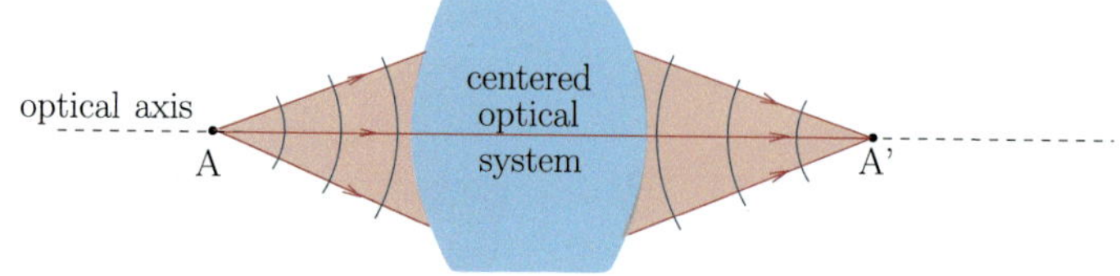

Fig. 17.19 For a centered (and stigmatic) optical system and a point object A on its axis, the image A' will be on the axis as well

This can be further generalized in the same way for the case where objects are virtual.

17.5.5   Centered Optical Systems and Optical Axis

An optical system S is said to be centered if it has a rotational symmetry along a line, called the optical axis. This is usually the case for lenses for which the axis passes through the center of curvature of each surface. As a consequence of the rotational symmetry, a ray along the optical axis passing through the system remain in this axis. We conclude that in particular, for a stigmatic pair (A, A'), if the point object A is located in the optical axis, so will be its conjugate image A', as shown in Fig. 17.19.

Rotational symmetry simplifies the analysis of optical systems. Indeed, for such systems it is enough to analyze the behavior of the rays belonging to a single transverse plane containing the optical axis, called a meridian plane. From now on, all the optical systems will be considered to be centered.

17.5.6   Aperture and Field of View

The object (image) aperture angle α (α') corresponds to the angle of the most inclined optical ray that enters (exits) the system S for a point object placed in the optical axis, as shown in Fig. 17.20. The aperture controls the transmitted flux trough the system.

The field of view along a direction is the maximum size of an object along that direction that can be imaged by the optical system. The field of view is limited by the size of the detector used to record the image, the larger the detector, the larger the field of view (Fig. 17.21). For objects located at infinity, one defines an angular field of view as the maximum angle under which an object can be seen by the system (Fig. 17.22).

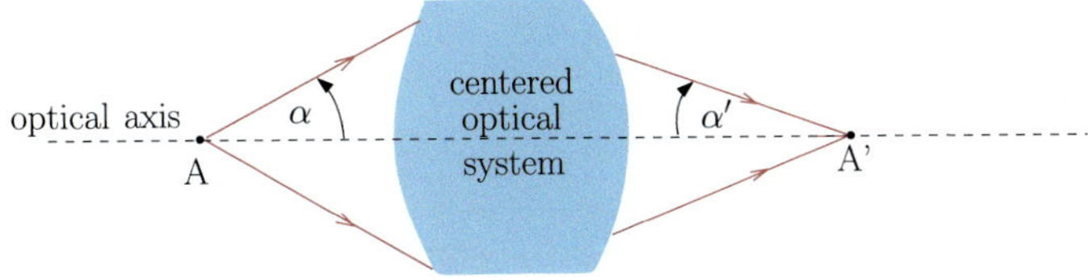

Fig. 17.20 The aperture angles of an optical system are defined by the most inclined optical rays that can enter and leave the system

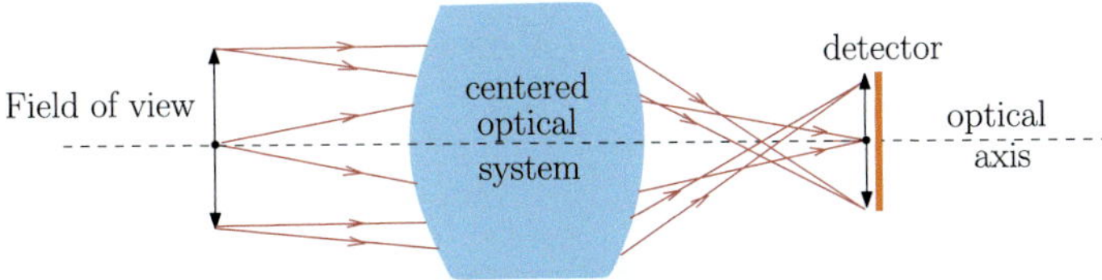

Fig. 17.21 The field of view is the maximum size of an object that can be imaged on the detector

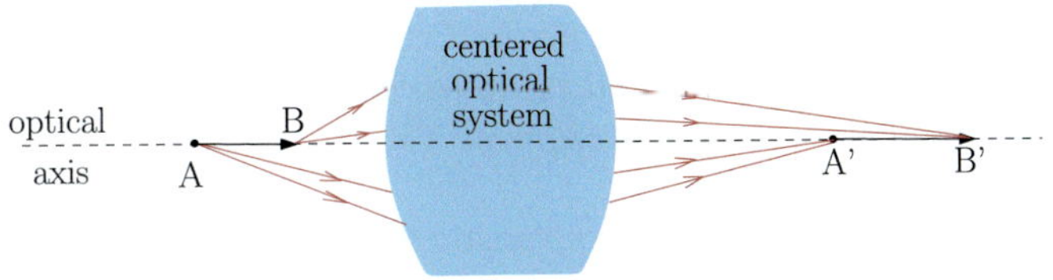

Fig. 17.22 The longitudinal magnification is defined for every centered optical system

17.5.7 *The Paraxial Approximation*

Systems having large apertures and fields of view are potentially accompanied by a degradation in image quality due to geometrical aberrations. From now on, we will assume the validity of the paraxial (or Gauss) approximation: we will consider centered optical systems having perfect stigmatism for rays forming small angles with respect to the optical axis and that remain close to the axis throughout the system.

17.5.8 *Longitudinal Magnification*

Consider a segment AB along the optical axis of a centered system S. Given the rotational symmetry along the axis, every incident ray in the axis must remain in the axis. The image A'B' of the segment is therefore in the axis as well and the longitudinal magnification M_l is defined as

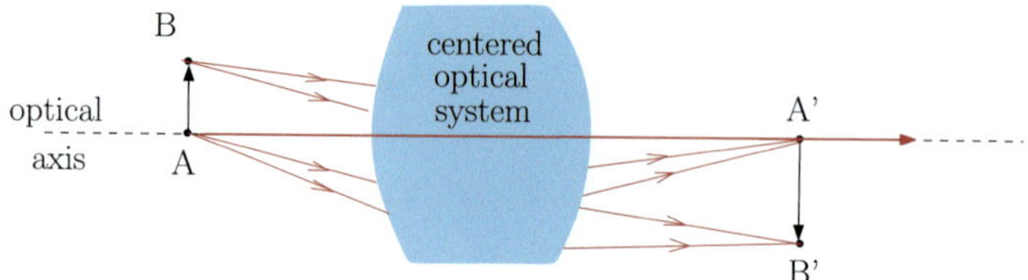

Fig. 17.23 An aplanatic system

$$M_l = \frac{\overline{A'B'}}{\overline{AB}} \ .$$

17.5.9   *Aplanatic Systems and Transverse Magnification*

An optical system S is said to be aplanatic if every object in a plane perpendicular to
the optical axis defines a plane image perpendicular to the optical axis. A centered
optical system is therefore aplanatic if the image of a straight segment AB perpendicular to the optical axis is a straight segment A'B' perpendicular to the optical axis
as well (Fig. 17.23).

The transverse magnification of an aplanatic optical system S is given by

$$\boxed{M = \frac{\overline{A'B'}}{\overline{AB}}.}\tag{17.14}$$

For $|M| > 1$, the image is said to be magnified, otherwise the image is diminished.
For $M > 0$ the image is said to be upright whereas for $M < 0$ the image is inverted.

17.5.10   *Focal Points and Focal Planes*

For every stigmatic and centered optical system S, two points play a very important
role: these are the object and image focal points. The image focal point F' is the
image of a point A_∞ located at infinity in the optical axis, that is

$$A_\infty \xrightarrow{S} F'$$

whereas a point object located at the object focal point F is imaged into a point A'_∞
at infinity on the optical axis

$$F \xrightarrow{S} A'_\infty$$

as illustrated in Fig. 17.24.

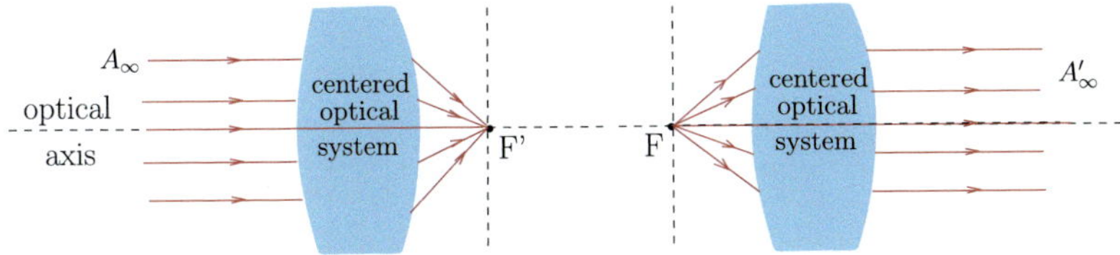

Fig. 17.24 Definition of image and object focal points and planes

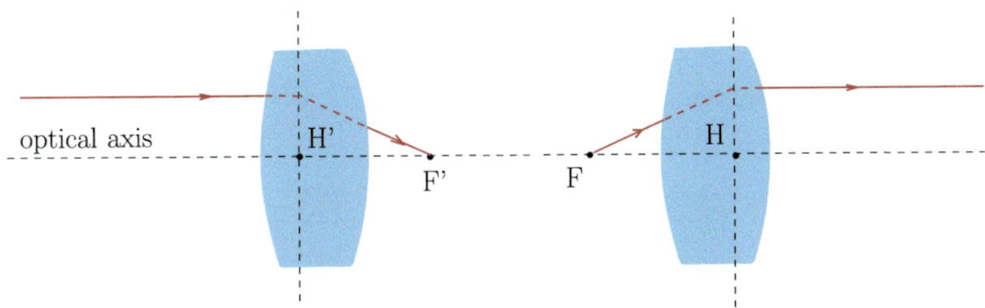

Fig. 17.25 Definition of principal points and principal planes

The object (image) focal plane is the plane perpendicular to the optical axis containing F' (F). They are represented by the dashed lines in Fig. 17.24.

17.5.11   Principal Planes and Principal Points

The image principal plane is the plane perpendicular to the optical axis containing the intersection points between the prolongations of each incident ray parallel to the axis with its corresponding image ray, which therefore passes (or seem to pass) trough the image focal point F'. The image principal point H' is the intersection of the image principal plane with the optical axis. These are shown in Fig. 17.25 (left).

Similarly, the object principal plane H is the plane perpendicular to the optical axis containing the intersection points of all the rays passing through the object focal point F and their corresponding image rays, which are therefore parallel the optical axis. The object principal point H is the intersection of the object principal plane with the optical axis.

The principal planes are conjugated: every point object in the object principal plane has a unique image in the image principal plane. The associated transverse magnification is equal to 1. Finally, we admit that any ray (or its extensions) cuts the principal planes at the same height.

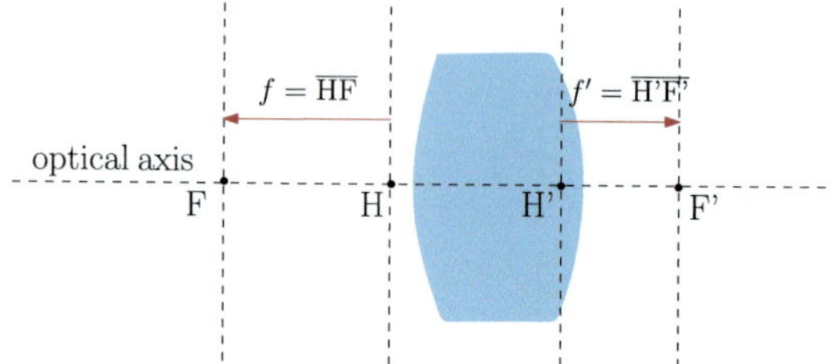

Fig. 17.26 Definition of the focal distances

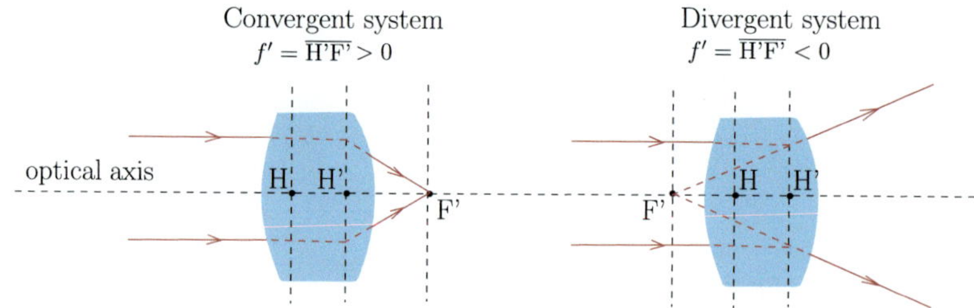

Fig. 17.27 For a convergent system F' is at the right of H'. For a divergent system, F' is at the left of H'

17.5.12 *Focal Distances*

The principal planes allows us to define the object focal distance f and the image focal distance f' as

$$f = \overline{\mathrm{HF}} \qquad f' = \overline{\mathrm{H'F'}}$$

As shown in Fig. 17.26. Note that with this definition, if F is located at the left of H, then $f < 0$.

A system with a positive image focal distance $f' > 0$ is said to be convergent, whereas a system with negative image focal distance $f' < 0$ is called divergent. These two cases are illustrated in Fig. 17.27. A system for which the focal points are located at infinity is said to be afocal.

Remark

If the object and image regions have the same optical index, which is the case for example for lenses in air, then $f = \overline{\mathrm{HF}} = -f' = -\overline{\mathrm{H'F'}}$. More generally, one has $f/n = -f'/n'$, where n (n') is the optical index of the object (image) region. This is the case, for example, of a human eye whose inside region has a different refractive index than air.

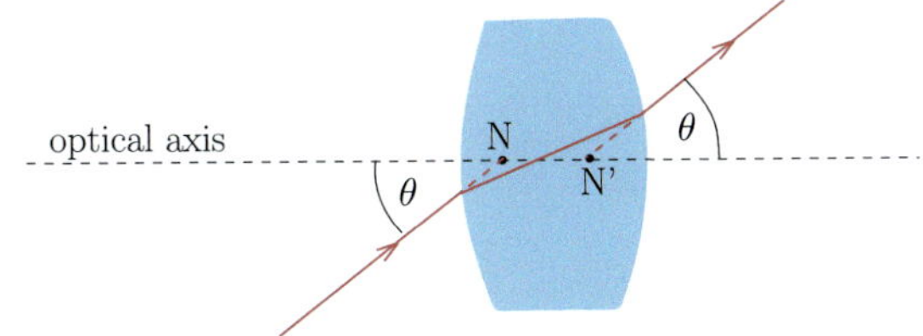

Fig. 17.28 Definition of the nodal points

17.5.13 Nodal Points

There are two particular points N and N' in the optical axis, called nodal points, such that every ray aimed at N will appear to have come from N' after leaving the optical system, forming the same angle with respect to the axis at the input and output of the system, as shown in Fig. 17.28.

Is it possible to show that $\overline{HH'} = \overline{NN'}$ and that

$$\overline{HN} = \overline{H'N'} = f + f' \,.$$

In particular, if the optical index is the same at the object and image regions of the optical system, $f = -f'$ and we obtain that the nodal points N and N' correspond to the principal points H and H', respectively.

17.5.14 The Cardinal Points

The cardinal points of a centered optical system S are three pair of points: the focal points F and F', the principal points H and H', and the nodal points N and N'. All the basic imaging properties of the system such as the magnification, orientation and location of images are completely determined by the locations of the cardinal points.

17.5.15 Afocal Systems and Angular Magnification

In the particular case of an afocal system, the image of an object at infinity is located at infinity. Two points A and B of an object at infinity are thus characterized by the angle that their rays form with the optical axis. If A is on the axis, all the rays coming from A enter the system horizontally and leave it horizontally as well, since the image A' of A should also lie in the optical axis at infinity. If all the rays coming from point B form an angle α with the axis, we say that the object AB is viewed under an angle α (see Fig. 17.29).

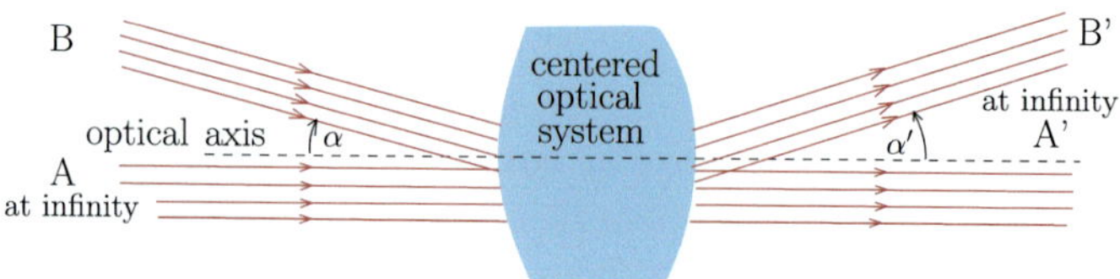

Fig. 17.29   Definition of the nodal points

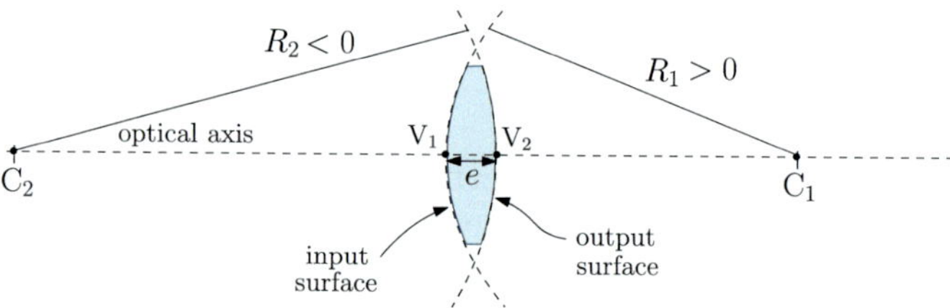

Fig. 17.30   In a spherical lens, the input and output surfaces are sections of spheres

The image B' of B being located at infinity, all the rays coming from B leave the optical system forming an angle α' with respect to the axis. The image A'B' is then viewed under an angle α', and the angular magnification is defined by

$$\gamma = \frac{\alpha'}{\alpha}. \tag{17.15}$$

This can be the case of a distant object, like a star or the moon, imaged with an afocal telescope which operates by producing an angular magnification, creating an image of the star at infinity. Our eyes focus the rays at the output of the telescope into the retina, the angular magnification is thus finally transformed into a transverse magnification by the eye so that the object appears larger when looked through such an optical system.

17.6   Spherical Lenses

A spherical lens is a system formed by a transparent medium of index n_L shaped such that the input and output surfaces are sections of spheres. This is a centered optical system where the optical axis corresponds to the line passing through the center of curvature of both surfaces. As we will show in this section, spherical lenses do not present perfect stigmatism. An approximate stigmatism is achieved, however, in the paraxial approximation for rays close to the optical axis and with small inclinations. The case of a convergent spherical lens of thickness e is shown in Fig. 17.30.

The vertex V of each surface corresponds to its intersection with the optical axis, and we adopt the following sign convention for the radius of curvature: it is positive

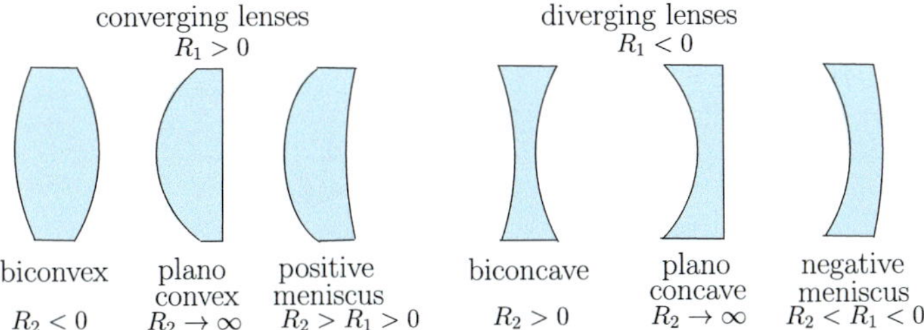

Fig. 17.31 Different types of spherical lenses

if the center of curvature is located at the right of the vertex, and negative otherwise. We can write in algebraic measure $R = \overline{VC}$. We see that the input surface is convex if $R_1 = \overline{V_1 C_1} > 0$ and concave for $R_1 = \overline{V_1 C_1} < 0$. Similarly, the output surface is convex if $R_2 = \overline{V_2 C_2} < 0$ and concave for $R_2 = \overline{V_2 C_2} > 0$. Figure 17.31 shows the most common geometries of converging and divergent lenses, each surface can be either convex, concave, or planar (infinite radius of curvature). Convex-concave (meniscus) lenses can be either converging or diverging, depending on the relative curvatures of the two surfaces

In the following, we will derive the main properties of spherical lenses by using Snell–Descartes's law in the paraxial approximation.

17.6.1 Principal Planes and Focal Distances of a Lens

Consider, for simplicity, a bi-convex converging lens, made of a transparent dielectric of refractive index n_L shaped such that the input and output surfaces are spheres with radius of curvature $R_1 > 0$ and $R_2 < 0$, respectively. As shown in Fig. 17.32, we consider an incident ray coming from a point object A_∞, located at infinity and in a medium of index n. This ray arrives parallel to the optical axis and at a distance h from it. After the lens, this ray passes through the image focal point F' which belongs to a region of index n'. Let us calculate the position of the image principal plane H' and the image focal length $f' = \overline{H'F'}$ (also called back focal length) by using Snell–Descartes's law.

Let us first write Snell–Descartes's law for the refraction at both interfaces. In the small angle approximation, that is, for paraxial rays which stay very close to the optical axis ($i_1, r_1, r_2, i_2 \ll 1$), we can write

$$n \sin i_1 = n_L \sin r_1 \quad \rightarrow \quad n i_1 \approx n_L r_1 \,,$$

$$n' \sin i_2 = n_L \sin r_2 \quad \rightarrow \quad n' i_2 \approx n_L r_2 \,.$$

On the other hand, summing the angles of the triangle $C_1 P_1 B_1$ gives

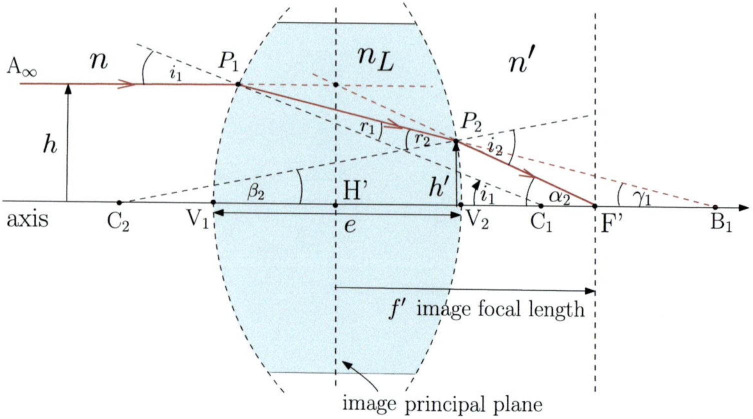

Fig. 17.32 Trayectory of an incoming ray parallel to the axis of a spherical lens

$$r_1 + \pi - i_1 + \gamma_1 = \pi \quad \rightarrow r_1 = i_1 - \gamma_1$$

and those of the triangle $C_2 P_2 B_1$

$$\beta_2 + \gamma_1 + \pi - r_2 = \pi \quad \rightarrow r_2 = \beta_2 + \gamma_1 \ .$$

Moreover, since $i_2 = \alpha_2 + \beta_2$, Snell–Descartes law can be rewritten as

$$n i_1 = n_L (i_1 - \gamma_1) \ ,$$

$$n'(\alpha_2 + \beta_2) = n_L (\beta_2 + \gamma_1) \ .$$

Eliminating γ_1 yields

$$\alpha_2 = \frac{n_L - n'}{n'}\beta_2 + \frac{n_L - n}{n'}i_1. \tag{17.16}$$

For small angles $\alpha_2 = h/f'$, $i_1 \approx h/R_1$ and $\beta_2 \approx -h'/R_2$. Replacing in (17.16)

$$\frac{h}{f'} = \frac{h(n_L - n)}{n' R_1} - \frac{(n_L - n')h'}{n' R_2} \ .$$

In addition $i_1 - r_1 = i_1(1 - n/n_L) = h/R_1(n_L - n)/n_L \approx (h - h')/e$ so that $h' = h - eh(n_L - n)/(R_1 n_L)$. From this, we obtain the image focal length f' (or back focal length) of the lens

$$\boxed{\frac{1}{f'} = \frac{n_L - n}{n' R_1} - \frac{(n_L - n')}{n' R_2} + \frac{e(n_L - n)(n_L - n')}{R_1 R_2 n' n_L} \ .} \tag{17.17}$$

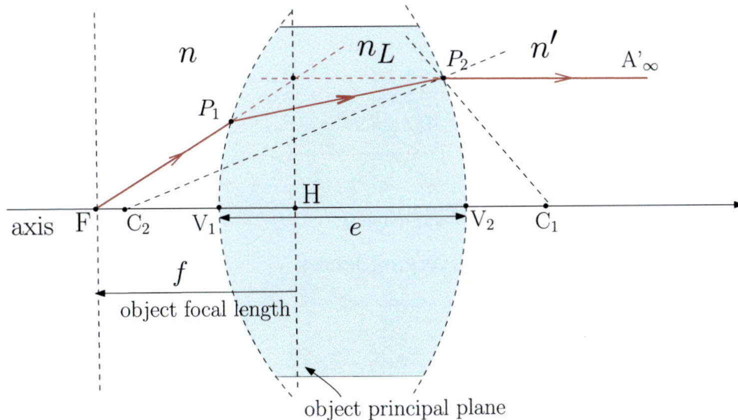

Fig. 17.33 Trayectory of an incoming ray passing through the object focal point

We see that the result does not depend on h, which confirms that the spherical lens is stigmatic in the paraxial approximation. The object focal plane H' is therefore well defined and its location can be found by writing

$$\alpha_2 = \frac{h}{f'} = \frac{h'}{\overline{V_2F'}} = \frac{h'}{\overline{H'F'} - \overline{H'V_2}} = \frac{h'}{f' - \overline{H'V_2}}$$

and so

$$\overline{H'V_2} = \frac{e(n_L - n)f'}{R_1 n_L} \ .$$

In order to find the object focal length and object principal plane, one proceeds in a similar way (Fig. 17.33). An incident ray passing through the object focal point F will exit the lens horizontally. According to the principle of backward propagation of light, a horizontal ray coming from right to left will pass through F after the lens. The situation is therefore identical to the case already discussed and all the previous results apply after exchanging R_1 with R_2 and n with n'.

It results from (17.17) that the object focal length (also called front focal length) is simply $f = \overline{HF} = -\frac{n}{n'} f'$ and that

$$\overline{HV_1} = -\frac{e(n_L - n')f}{R_2 n_L} \ .$$

These results are general and also apply when the input and/or the output surfaces are concave by following the sign convention stated above for the radius of curvature. Note finally than in most cases, $n = n' = 1$, and we find remarkably that $f' = -f$. Figure 17.34 shows the principal points of the different types of spherical lenses in air.

converging lenses

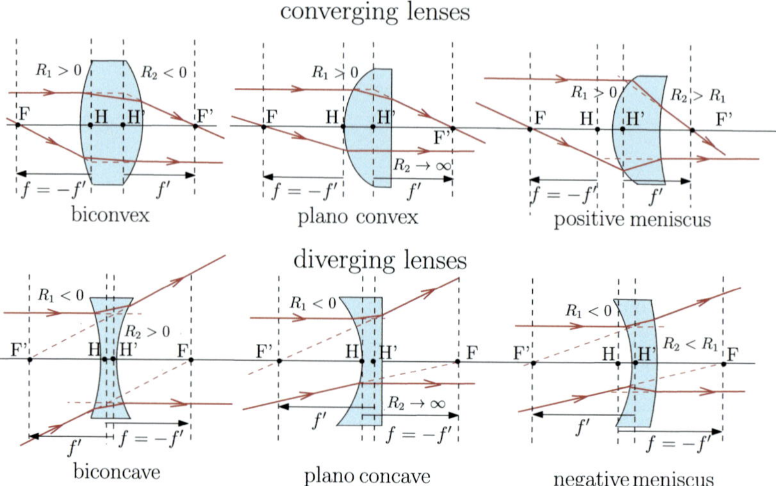

Fig. 17.34 Different type of spherical lenses and their principal points

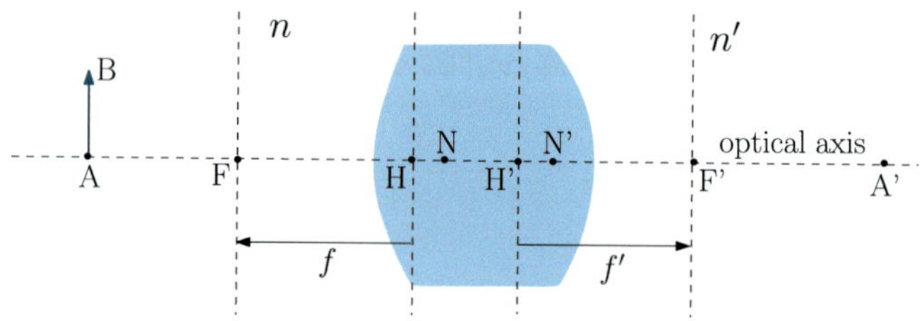

Fig. 17.35 An object AB placed in front of a lens L

17.6.2 *Imaging Properties of a Lens*

Consider an object AB perpendicular to the optical axis of a lens L, where A lies on the axis. The lens is characterized by its cardinal points (F, F'), (H, H'), and (N, N'). Let A'B' be the image of AB by the lens, that is $A \xrightarrow{L} A'$ and $B \xrightarrow{L} B'$ (Fig. 17.35).

We will assume aplanetism and perfect stigmatism between conjugate points in the paraxial approximation. In that case all the rays coming from B will eventually converge (or seem to converge) to a single point B'. The lens being aplanatic, the image A' of A is simply the orthogonal projection of B' along the optical axis. We consider in particular the two rays coming from B shown in Fig. 17.36.

First, the ray entering the lens horizontally will pass through the image focal point F', the intersection of the prolongations of the ray before and after the lens occurs at the principal plane H'. Secondly, the ray coming from B and passing through the object focal point F will exit the lens horizontally, and the prolongations of the ray before and after the lens intersect at the principal plane H. Let α and α' be the angle

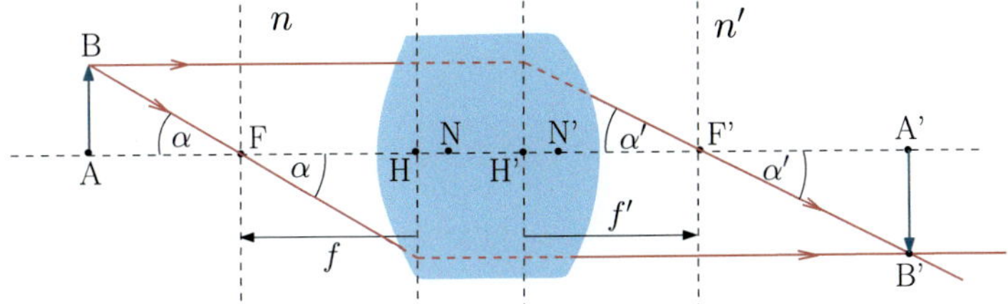

Fig. 17.36 The point B is imaged by the lens into point B'

that this second ray forms with the optical axis before and after the lens, respectively. We have, in the paraxial approximation

$$\alpha = \frac{\overline{AB}}{\overline{AF}} = \frac{\overline{A'B'}}{f}$$

and

$$\alpha' = -\frac{\overline{A'B'}}{\overline{F'A'}} = \frac{\overline{AB}}{f'} \ .$$

From the first relation we obtain $\overline{AB} = \overline{AF}\,\overline{A'B'}/f$, and replacing in the second one

$$-\frac{\overline{A'B'}}{\overline{F'A'}} = \frac{\overline{AF}\,\overline{A'B'}}{f'f} \ .$$

We obtain Newton's relation

$$\boxed{ff' = \overline{FA}\,\overline{F'A'}.} \qquad (17.18)$$

Now, by writing $\overline{FA} = \overline{HA} - f$ and $\overline{F'A'} = \overline{H'A'} - f'$ in Newton's relation we get

$$(\overline{HA} - f)(\overline{H'A'} - f') = ff'$$

or equivalently $ff'(\frac{\overline{HA}}{f} - 1)(\frac{\overline{H'A'}}{f'} - 1) = ff'$ and so

$$1 - \frac{\overline{H'A'}}{f'} - \frac{\overline{HA}}{f} + \frac{\overline{HA}}{f}\frac{\overline{H'A'}}{f'} = 1 \ .$$

Finally

$$\frac{\overline{H'A'}}{f'} + \frac{\overline{HA}}{f} = \frac{\overline{HA}}{f}\frac{\overline{H'A'}}{f'}$$

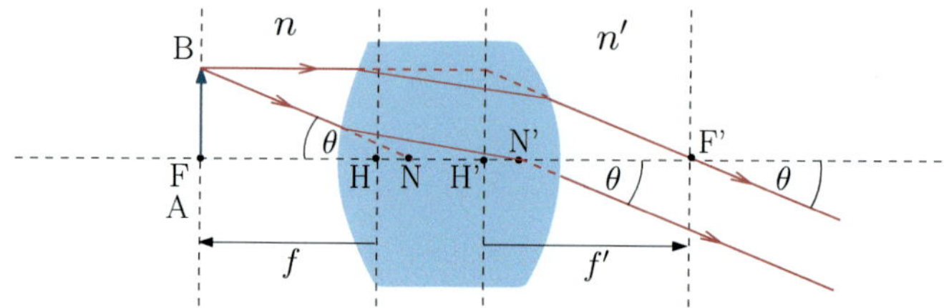

Fig. 17.37 The point B is now at the object focal plane, and so its image is located at infinity

which demonstrates, after dividing both sides by $\frac{\overline{HA}}{f}\frac{\overline{H'A'}}{f'}$, the lens Eq. (17.19)

$$\boxed{\frac{f}{\overline{HA}} + \frac{f'}{\overline{H'A'}} = 1.}$$

(17.19)

Let us now consider the ray coming from B and directed toward the nodal point N, which forms an angle θ with respect to the lens axis, as shown in Fig. 17.37. At the output of the lens, the ray seems to come from N' and it forms the same angle θ with respect to the optical axis. If the object AB is located at the object focal point F, we have $\overline{HA} = f$ and from the lens Eq. (17.19) we conclude that the image A'B' of AB is located at infinity.

All the rays coming from B and leaving the lens are thus parallel, since they never intersect. In particular, the horizontal ray passing through the image focal point F' should form the same angle θ with respect to the axis. In the paraxial approximation

$$\theta = \frac{\overline{AB}}{\overline{FN}} = \frac{\overline{AB}}{\overline{H'F'}} = \frac{\overline{AB}}{f'}$$

and so

$$\overline{FN} = f' = \underbrace{\overline{FH}}_{-f} + \overline{HN}$$

so that

$$\overline{HN} = f + f' .$$

Considering now an object at infinity, whose image is formed in the image focal plane, we conclude in a similar way that

$$\overline{H'N'} = \overline{HN} = f + f' .$$

Finally, let us obtain an expression for the transverse magnification. Since all the rays coming from B should intersect at B' after the lens, it is true in particular for the

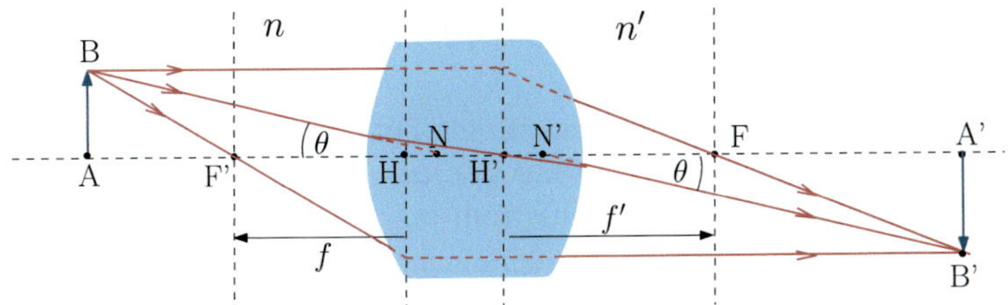

Fig. 17.38 All the rays depicted must converge toward the same point after the lens

ray coming from B and directed toward N, which forms an angle $\theta = \overline{AB}/\overline{AN}$ with respect to the axis in the paraxial approximation, as shown in Fig. 17.38. At the exit of the lens, the ray should appear to come from N' and pass through B', forming the same angle θ with respect to the axis.

We can therefore write

$$\theta = \frac{\overline{AB}}{\overline{AN}} = \frac{-\overline{A'B'}}{\overline{N'A'}}$$

but $\overline{N'A'} = \overline{H'A'} - \overline{H'N'} = \overline{H'A'} - (f + f')$ and, similarly, $\overline{AN} = \overline{AH} + \overline{HN} = -\overline{HA} + f + f'$, so we conclude that

$$M = \frac{\overline{A'B'}}{\overline{AB}} = \frac{\overline{N'A'}}{\overline{NA}} = \frac{\overline{H'A'} - (f + f')}{\overline{HA} - (f + f')} \ .$$

Now, since $f = -\dfrac{n}{n'} f'$, we have $f + f' = f'(1 - \dfrac{n}{n'})$. Moreover, replacing f in terms of f' in the lens Eq. (17.19) yields

$$\frac{1}{f'} = \frac{1}{\overline{H'A'}} - \frac{n/n'}{\overline{HA}} = \frac{\overline{HA} - \dfrac{n}{n'}\overline{H'A'}}{\overline{H'A'}\ \overline{HA}}$$

so that the transverse magnification becomes

$$M = \frac{\overline{H'A'} - f'\left(1 - \dfrac{n}{n'}\right)}{\overline{HA} - f'\left(1 - \dfrac{n}{n'}\right)} = \frac{n\overline{H'A'}}{n'\overline{HA}}$$

$$\boxed{M = \frac{\overline{A'B'}}{\overline{AB}} = \frac{n\overline{H'A'}}{n'\overline{HA}}\ .} \qquad (17.20)$$

17.6.3 Lenses and Fermat's Principle

Another way to see the effect of a lens on the optical rays is by a modification of
the incoming wave fronts by the introduction of a position-dependent phase shift.
Indeed, if all the rays entering the lens from point A intersect at point A', then it
must be that the optical path length is stationary, in fact the same, for all the rays
connecting A and A'. If this were not the case, it would contradict Fermat's principle.

In the case of a convergent lens, since rays entering further from the axis travel a
longer path in vacuum, the only way to achieve a stationary optical path length for
all the rays is to compensate by adding a longer optical path length for those rays
closer to the optical axis. A converging lens is therefore thicker at the center as shown
in Fig. 17.39. For a diverging lens, for which A' is a virtual image, one must have
$(AP)-(A'P) = $ constant for all rays, so that a divergent lens is thicker at the edges.

17.6.4 The Human Eye

A human eye, shown in Fig. 17.40 is constituted by a spherical cavity made of a
transparent medium, called vitreous humor ($n_{\mathrm{vh}} = 1.337$), inside which light enters
trough a circular diaphragm, called the iris. The iris controls the size of the eye
aperture, called the pupil. The lens, together with the cornea ($n_{\mathrm{c}} = 1.376$) and the
aqueous humor ($n_{\mathrm{ah}} = 1.336$), refracts the incoming beams of light to form images
on the retina, which detects light and sends messages to the brain through the optic
nerve. A normal eye which does not require corrective lenses when relaxed is called
emmetropic and is such that an object at infinity is exactly focused on the retina,
creating a sharp image of distant objects. The principal planes, nodal and cardinal
points for the human eye are also illustrated in Fig. 17.40.

For an emmetropic eye at rest, the image focal length is $f' = 22.2$ mm and
the object focal length $f = -16.7$ mm. We have $\overline{NN'} = \overline{HH'} = 0.3$ mm, and
$\overline{HN} = \overline{H'N'} = f + f' = 5.5$ mm. This justifies a simplified model for the eye by

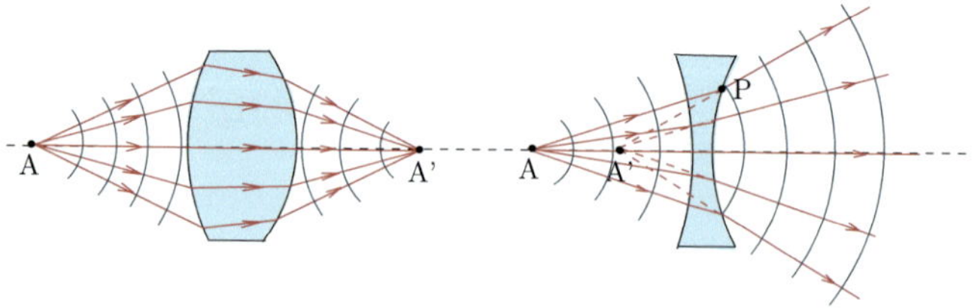

Fig. 17.39 According to Fermat's principle, a convergent lens must be thicker at its center, whereas
the opposite is true for a diverging lens

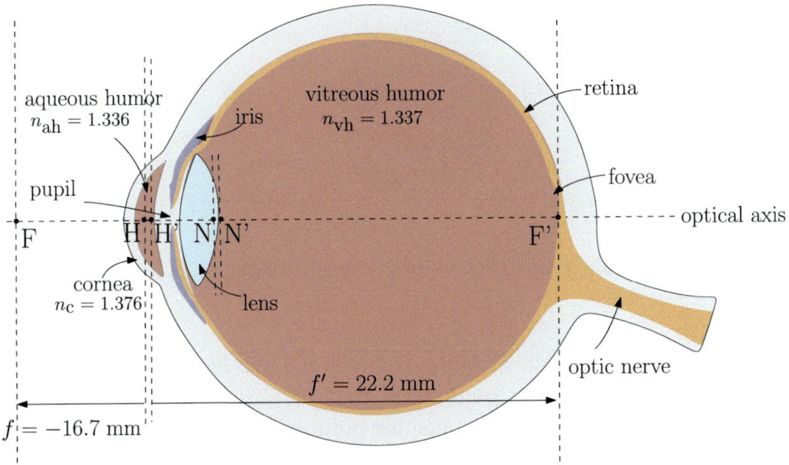

Fig. 17.40 The human eye as a single optical system

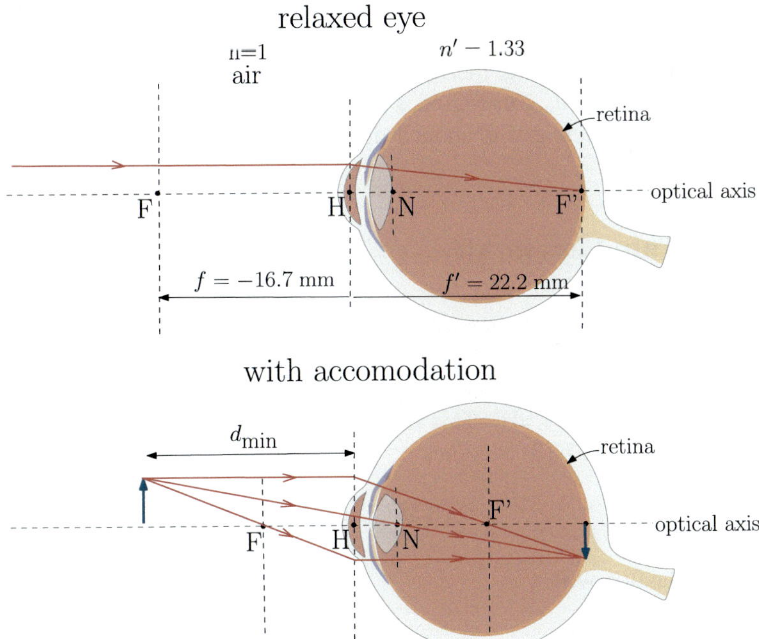

Fig. 17.41 The eye lens can change its curvature in order to form, at the retina, images of objects at different distances

considering $H \approx H'$, $N \approx N'$ so that one principal point and one nodal point are enough to analyze the formation of images by the eye.

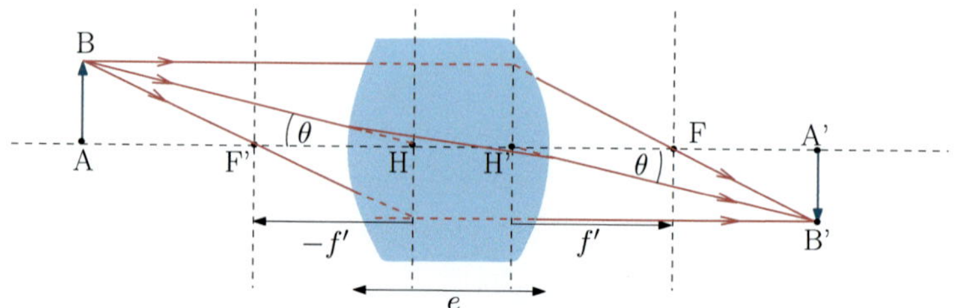

Fig. 17.42 For a lens in air, $f = -f'$ and only 4 cardinals points need to be specified

In order to observe objects at finite distance, the eye lens can accommodate by increasing the curvature of its outer wall, thus increasing its refracting capabilities, as illustrated in Fig. 17.41. The nearest point of distinct vision with maximum accommodation is called the *punctum proximum*, which for a young adult is typically $d_{\min} \sim 25$ cm. Using the lens Eq. (17.19), we see that if $\overline{\text{HA}} = -25$ cm, and $\overline{\text{HA'}} = 2.22$ cm, the new image focal length of the accommodated eye should be $f' = 20.4$ mm. Defining the dioptre of a lens as the reciprocal of its focal length in meters, this represents a change of about 4 dioptres between a relaxed and a contracted eye lens. Due to aging, there is an irreversible loss of the ability of the eye to accommodate so that $d_{\min}$ increases, a condition called presbyopia. By the age of 70, accommodation decreases to about 1 dioptre.

17.7 The Thin Lens in Air

In most cases, lenses are surrounded by air, so that the object and image regions have an index of $n = n' \approx 1$. We see that in this case $f = -f'$ and the nodal points coincide with the intersections of the principal planes with the optical axis (see 17.6.2), that is N=H and N'=H'. This is illustrated in Fig. 17.42

Newton's relation (17.18) becomes

$$\overline{\text{FA}} \; \overline{\text{F'A'}} = -f'^2$$

and the lens equation writes

$$\frac{1}{\overline{\text{H'A'}}} - \frac{1}{\overline{\text{HA}}} = \frac{1}{f'} \; .$$

The transverse magnification is simply $M = \dfrac{\overline{\text{H'A'}}}{\overline{\text{HA}}}$.

Additionally, in many situations we can make the assumption of a thin lens, that is, we consider that the thickness e of the lens is much smaller than the radii of curvature R_1 and R_2 of the input and output surfaces of the lens. In this case, both

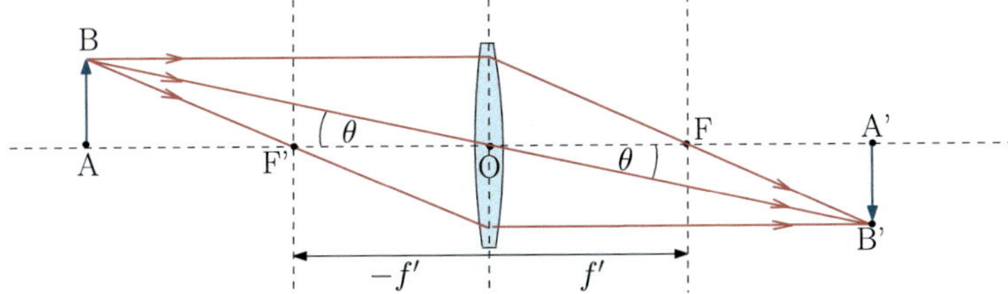

Fig. 17.43 Three cardinal points are required for a thin lens in air

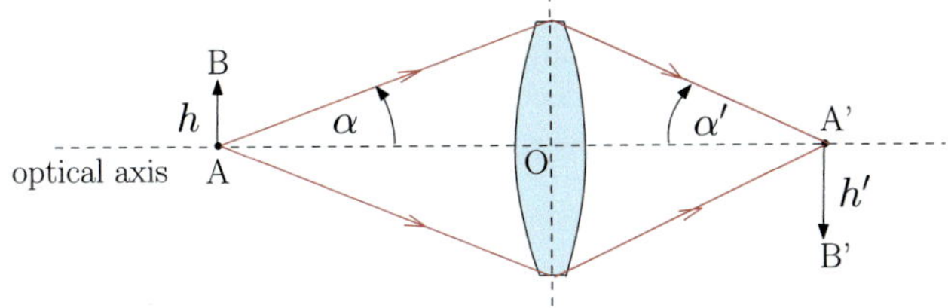

Fig. 17.44 An object AB and its image A'B' by a thin lens

principal points coincide, that is H=H'=O, where O is simply called the center of the lens (Fig. 17.43), and the image focal length (17.17) for a thin lens in air becomes

$$\frac{1}{f'} = (n_L - 1)\left(\frac{1}{R_1} - \frac{1}{R_2}\right).$$

(17.21)

The thins lens equation, also known as Decartes's relation is

$$\frac{1}{\overline{OA'}} - \frac{1}{\overline{OA}} = \frac{1}{f'}.$$

(17.22)

17.7.1 The Lagrange–Helmholtz Invariant

Consider an object AB and its image A'B' by a thin lens, as shown in Fig. 17.44. Let α (resp. α') be the object (image) angular aperture of the lens.

Let D be the diameter of the lens. In the paraxial approximation, we have $\tan \alpha \approx \alpha$ and $\tan \alpha' \approx \alpha'$ so that

$$\frac{D}{2} = \alpha \overline{AO} = \alpha' \overline{OA'} .$$

On the other hand, the transverse magnification M of the lens reads

Fig. 17.45 Joseph-Louis Lagrange (1736–1813): an Italian-French mathematician and physicist. He made monumental contributions to mechanics, calculus, differential equations and number theory. He formulated Lagrangian mechanics, a reformulation of classical mechanics that remains fundamental in physics

$$M = \frac{\overline{A'B'}}{\overline{AB}} = \frac{\overline{OA'}}{\overline{OA}} = -\frac{h'}{h}$$

and so

$$\alpha\overline{OA'}\,\frac{h}{h'} = \alpha'\overline{OA'}$$

which demonstrates that

$$\alpha h = \alpha'h' \ .$$

This result, due to Joseph-Louis Lagrange (Fig. 17.45), can be generalized to the case in which the image region has an index n and the image region an index n'

$$n\alpha h = n'\alpha'h' \ .$$

17.7.2   Beyond Spherical Lenses

So far we have restricted the analysis of spherical lenses to paraxial rays which are close to and form a small angle with the optical axis. In general, however, a ray refracted by a spherical lens will intersect the optical axis at a point which depends on both the angle of incidence and the distance with respect to the axis at which the ray enters the lens. Not all the rays emitted by a point image A will intersect at A', as shown in Fig. 17.46. The spherical lens presents only approximate stigmatism.

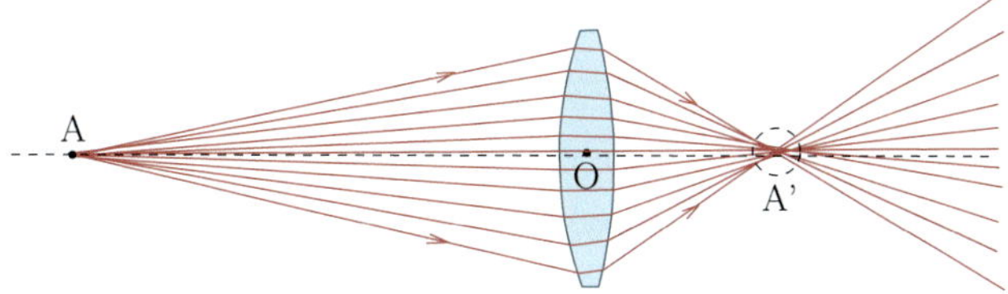

Fig. 17.46 A spherical lens suffers from geometrical aberrations. The rays refracted far from the axis do not intersect the axis at the same point

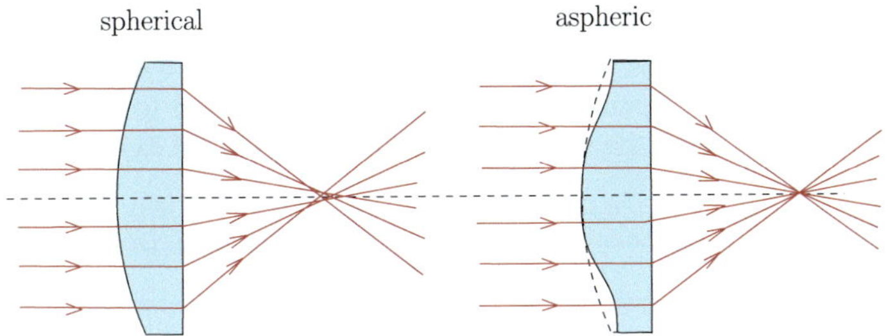
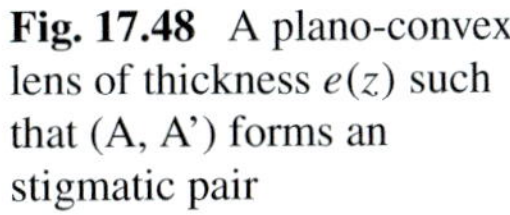

Fig. 17.47 An aspheric lens has perfect stigmatism in the small angle approximation

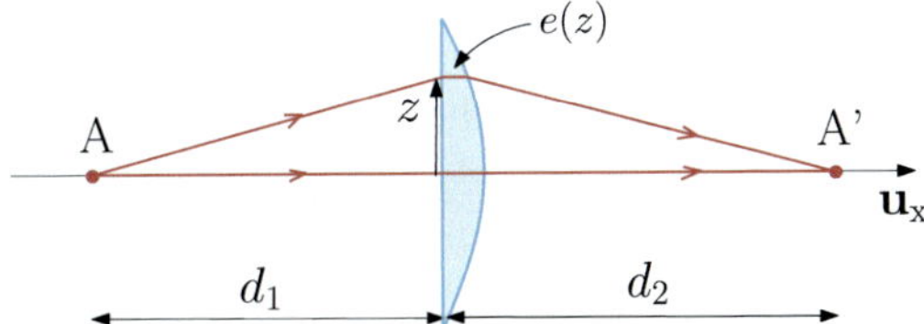

Fig. 17.48 A plano-convex lens of thickness $e(z)$ such that (A, A') forms an stigmatic pair

A more complex surface profile (parabolic, hyperbolic, elliptical) may be used in order to get rid of these aberrations, in which case we talk about aspheric lens (see Fig. 17.47). While these lenses form images with less aberrations than standard, simple spherical lenses, they are more difficult and expensive to produce. Below, we will show that Fermat's principle predicts that a parabolic lens ensures perfect stigmatism in the small angle approximation for point objects in the optical axis.

Consider for simplicity a thin plano-convex lens of index n and two conjugate points A and A' in the optical axis, located at distances d_1 and d_2 from the lens, respectively, as shown in Fig. 17.48. The thickness $e(z)$ of the lens, where z is the distance to the optical axis, is chosen such that every ray stemming from point A reaches point A', that is, the lens has perfect stigmatism for the pair (A, A'). To determine $e(z)$, let us focus on the ray of Fig. 17.48 coming from A and entering the lens at a height z with respect to the optical axis.

The optical path length (AA') is given, in the small angle approximation by

$$(\text{AA'}) = \underbrace{\sqrt{d_1^2 + z^2} + \sqrt{(d_2 - e(z))^2 + z^2}}_{\text{air}} + \underbrace{ne(z)}_{\text{lens}} \; .$$

We further consider z and e small with respect to d_1 and d_2, so that $\sqrt{d_1^2 + z^2} \approx d_1(1 + \frac{z^2}{2d_1^2})$ and $\sqrt{(d_2 - e(z))^2 + z^2} \approx d_2(1 + \frac{z^2 - 2e(z)d_2}{2d_2^2})$ so

$$(\text{AA'}) \approx d_1 + d_2 + \frac{z^2}{2}\left(\frac{1}{d_1} + \frac{1}{d_2}\right) + (n - 1)e(z) \; .$$

Writing $1/f' = 1/d_1 + 1/d_2$ and considering that (A, A') is a stigmatic pair if the optical path (AA') is independent on z (and therefore equal to the path $(\text{AA'}) = d_1 + d_2 + (n - 1)e_0$ along the ray passing through the optical axis)

$$(\text{AA'}) = d_1 + d_2 + \frac{z^2}{2f'} + (n - 1)e(z) = d_1 + d_2 + (n - 1)e_0 \; .$$

We find that the lens must have a parabolic profile

$$e(z) = e_0 - \frac{z^2}{2(n - 1)f'}. \tag{17.23}$$

If we approximate the parabolic surface by a sphere of radius R close to the axis, then $e(z) \approx e_0 - \frac{z^2}{2R}$ and so comparing with (17.23) we retrieve the result $R = (n - 1)f'$ relating R with the focal length f' for a spherical thin lens.

17.7.3 The Fresnel Lens

A different kind of lens is the so-called Fresnel lens, which acts as a conventional lens while having the advantage of being much thinner, thus reducing significantly the amount of mass and volume required for large apertures lenses. The first application of a Fresnel lenses was to replace metallic reflectors in order to produce bright collimated beams in lighthouses in the beginning of the 19th century.

As illustrated in Fig. 17.49, the idea is to divide the lens into a set of concentric annular steps. In each section, the thickness is reduced with respect to that of an equivalent simple lens, by removing all the bulk glass which does not contribute to refraction, indeed glass increases the amount of weight and absorption within the system. The substantial reduction in thickness is obtained at the expense of a poorer image quality when compared to a conventional, single curvature lens. With a Fresnel lens, however, it is possible to achieve a short focal length and a large aperture while keeping the lens light. They can be used to efficiently collimate diverging sources of light or to collect large beams of light and focused them on a small area, for example

Fig. 17.49 A Fresnel lens

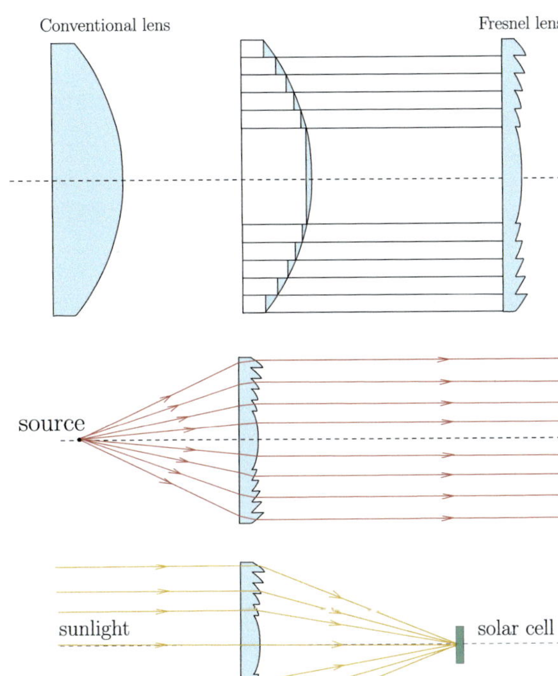

Fig. 17.50 A Fresnel lens can have large apertures while being light

sunlight can be focused onto a solar cell with a Fresnel lens in order to increase the collection efficiency. This is shown in Fig. 17.50.

17.8 Spherical Mirrors

A spherical mirror is made of a highly reflecting spherical surface with center of curvature C and whose axis of symmetry defines the optical axis. If O is the vertex of the mirror surface, then we can distinguish concave ($\overline{OC} < 0$) from convex ($\overline{OC} > 0$) mirrors by the relative position of C with respect to O, as shown in Fig. 17.51. It can be shown that for rays close to the optical axis (paraxial approximation), a spherical mirror focus an incident beam of rays parallel to the axis into the object focal point F' located at $\overline{OF'} = \overline{OC}/2$, the image of a point object at infinity is therefore real (virtual) for the case of a concave (convex) mirror. Due to the principle of backward propagation of light, a point source at the image focal point F' will produce an ensemble of beams parallel to the optical axis after reflection off the mirror. We conclude that F = F'. The object and image principal planes are equal and intersect the axis at H = H' = O. In the paraxial approximation, the principal planes correspond to the mirror surface itself, which is approximately flat very close to the vertex.

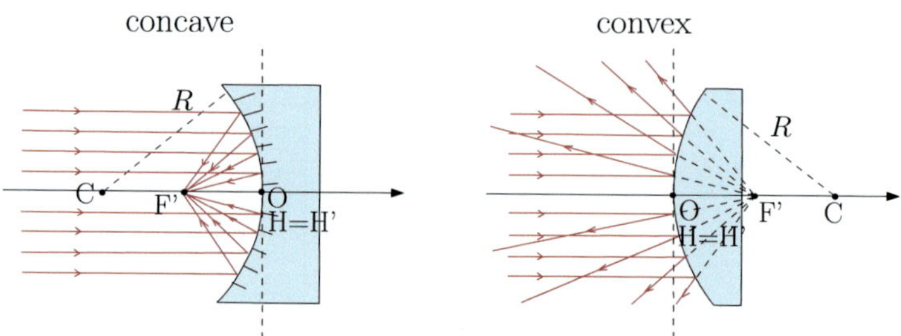

Fig. 17.51 Spherical mirrors can be concave (left) or convex (right)

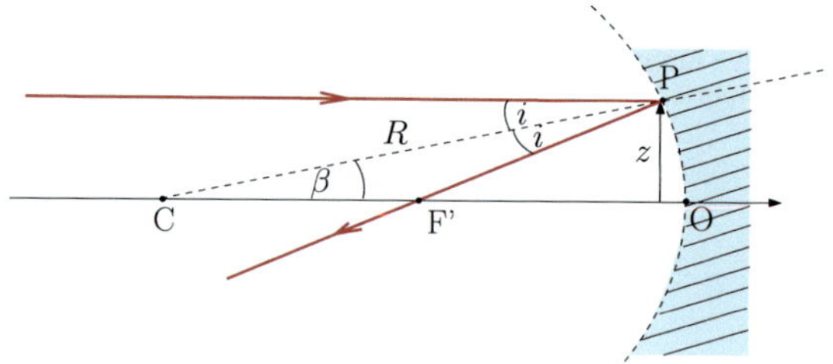

Fig. 17.52 A ray parallel to the optical axis is reflected by the mirror and passes through the image focal point F'

The focal point F of a spherical mirror is located at $\overline{OF} = \overline{OF'} = f' = \frac{\overline{OC}}{2} = R$, that is, at a distance R from the mirror, where R is the radius of curvature of the reflecting surface (negative for a concave mirror, positive for a convex one). Let us show this for the case of the concave mirror. A ray parallel to the optical axis hits the mirror at point P at an incidence angle i (see Fig. 17.52). The law of reflection tells us that the reflected ray is the symmetrical of the incident ray with respect to the normal to the surface, and intersects the optical axis at the focal point F'. We have, for paraxial rays

$$\sin \beta = \frac{z}{R} \approx \beta$$

and since $\beta = i$, the triangle CF'P is isosceles, and it follows that

$$\cos \beta = \cos i = \frac{-R/2}{\overline{CF'}} = \sqrt{1 - \sin^2 i} = \sqrt{1 - \frac{z^2}{R^2}} \, .$$

We find

$$\overline{CF'} = \frac{-R/2}{\sqrt{1 - \dfrac{z^2}{R^2}}} \, .$$

We conclude that, in the paraxial approximation ($z \ll R$), F' is found at the middle point of $\overline{CO} = -R$, and that the spherical mirror has approximate stigmatism for the

Fig. 17.53 Image formation
by a concave mirror

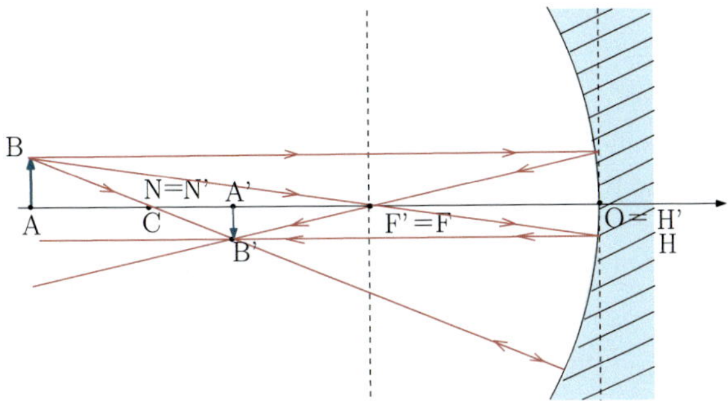

pairs (A$_\infty$, F') and (F, A$_\infty$). A spherical mirror has perfect stigmatism for just one
point, located at the center of curvature C (real object in the case of a concave mirror
and virtual for a convex mirror), whose conjugate is C itself. Indeed, any ray coming
from C will be reflected by the mirror on itself, since the incidence angle is zero. The
center of curvature also coincides with the nodal points C $=$ N $=$ N' of the mirror.

17.8.1 The Mirror Equation

Consider an object AB, where A is in the optical axis of a mirror of center of curvature
C. Let us find the position of the image A'B' of AB in the paraxial approximation.
As shown in Fig. 17.53, a horizontal ray coming from B will pass through the image
focal point F', whereas the ray coming from B and passing through the object focal
point F $=$ F' will leave the mirror horizontally. These two rays intersect at a single
point B', corresponding to the conjugate image of B by the mirror. A' is therefore the
orthogonal projection of B' onto the optical axis. One can check that the ray coming
from B and passing through the center of curvature, which coincides with the nodal
points, will be reflected on itself and intersect B' as well.

For the rays passing through F we conclude that

$$\frac{\overline{AB}}{\overline{AF'}} = -\frac{\overline{A'B'}}{\overline{F'O}}$$

and

$$\frac{\overline{AB}}{\overline{F'O}} = -\frac{\overline{A'B'}}{\overline{AF'}}$$

which proves Newton's relation (17.18) for the case of a mirror ($f = f'$)

$$\overline{F'A}\,\overline{F'A'} = \overline{OF'}^2 = f'^2 \;.$$

Since $\overline{F'A} = \overline{OA} - f'$ and $\overline{F'A'} = \overline{OA'} - f'$ we obtain the mirror equation

$$\boxed{\frac{1}{\overline{OA}} + \frac{1}{\overline{OA'}} = \frac{1}{\overline{OF'}} = \frac{1}{f'}}$$ (17.24)

and the transverse magnification reads

$$M = \frac{\overline{A'B'}}{\overline{AB}} = -\frac{\overline{OA'}}{\overline{OA}} \, .$$

17.8.2   *Beyond Spherical Mirrors*

Perfect stigmatism is fulfilled for an elliptical mirror and the stigmatic pair (A_1, A_2), where A_1 and A_2 are the foci of the ellipse, as shown in Fig. 17.54 for the case of a concave elliptical mirror.

Indeed, for such a mirror, any ray passing through A_1 and A_2 after hitting the mirror at point P is such that

$$|A_1 P| + |A_2 P| = (A_1 A_2) = \text{constant}$$

so that the optical path length is the same for all the rays connecting both foci. As shown in Fig. 17.55, this type of mirror is used in some laser cavities where an amplifying crystal is optically excited by a discharge lamp. In order to increase the efficiency of the system, the rod-like crystal is placed in one of the foci of an elliptical mirror, and the discharge lamp in the other one.

Fig. 17.54 An elliptical mirror is stigmatic for the pair of foci (A_1, A_2)

Fig. 17.55 An elliptical mirror can be used to optimize the coupling between a discharge lamp and an amplifying crystal rod

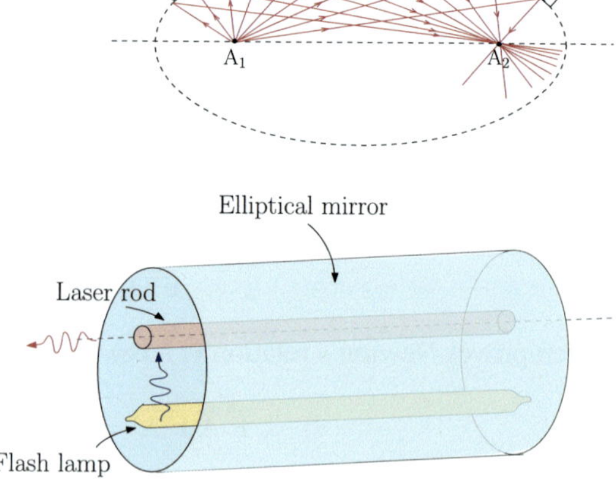

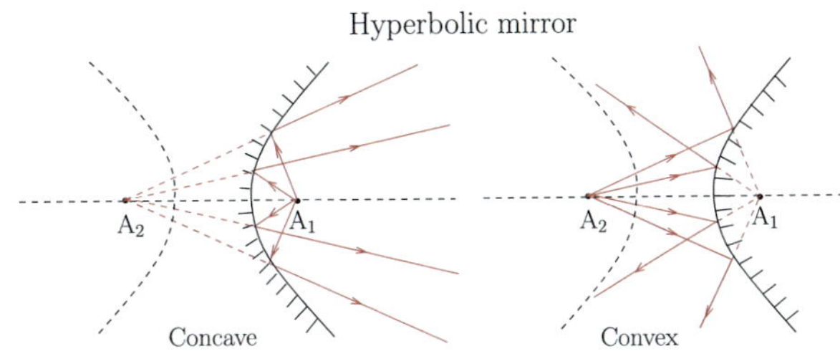

Fig. 17.56 Hyperbolic mirrors present perfect stigmatism for the pair of foci

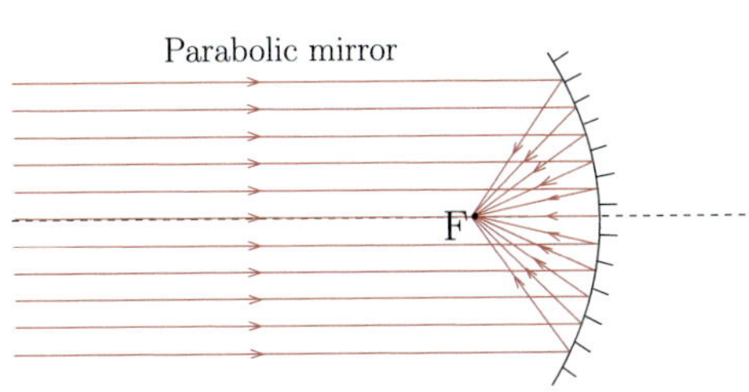

Fig. 17.57 Parabolic mirrors are perfectly stigmatic for a point at infinity and its focal point

When the surface of the mirror is a hyperboloid of revolution with foci A_1 and A_2, one also obtains perfect stigmatism for the pair (A_1, A_2). For any ray coming from one of the foci, striking the mirror at P, and seeming to come from the other foci, we have

$$|\overline{A_1 P}| - |\overline{A_2 P}| = \text{constant}$$

which ensures that all the rays after reflection seem to come from a virtual image, the optical path length of the different rays will be the same at all points of a sphere centered on that image. This is illustrated in Fig. 17.56.

Finally, parabolic mirrors are perfectly stigmatic for a point object at infinity and the focus of the parabola, the latter coincides in this case with the focal point of the system. This is shown in Fig. 17.57. These mirrors can be found for example in telescopes, where the light coming from a star at infinity is collected and focused by such a mirror onto its focal point, or in solar electric generators, where a concave parabolic reflector concentrates sunlight onto a pipe positioned at its focal point. The fluid inside the pump is therefore heated and ultimately this energy is transformed in electricity by first producing steam to power a generator, converting mechanical energy into electrical energy via electromagnetic induction.

17.9   Ray Tracing and Image Formation with Lenses and Mirrors

Ray tracing is a method consisting in drawing the path of different rays when entering and leaving an optical system S according to the laws of reflection and refraction.

For thin lenses and mirrors in the paraxial approximation, some particular rays obey simple geometrical rules, allowing us to easily understand the formation of images by following a general procedure listed below. We present as an example the images of a (real) vertical object formed by a concave mirror and a convergent lens.

- If the object AB is real, it should be placed at the left of the optical system as shown in Fig. 17.58. If it is virtual, at the right of the input surface so that the rays entering the system seem to converge toward the object.
- We assume, in the paraxial approximation, that the system is aplanatic and has perfect stigmatism. To find the image of a vertical object AB, where A is on the optical axis, it is enough to find B', the image of B. Indeed, the image A' of A is on the axis and simply corresponds to the orthogonal projection of B' along the axis.
- The system having perfect stigmatism, we know that all rays entering the system coming from B should meet at B' after leaving the system. To find the position of the image B' of point B, it is therefore enough to find the point at which two particular rays coming from B intersect. For this, the following rules can be applied:

  1. A horizontal ray entering the system will either pass through the image focal point F' (real image), or either seem to come from F' (virtual image). This is shown in Fig. 17.59.
  2. As shown in Fig. 17.60, a ray passing through the object focal point F will exit the system horizontally
  3. A ray passing through the center of the optical system will not be deviated (thin lens) or will be reflected symmetrically with respect to the axis (mirror). This is shown in Fig. 17.61.
  4. For the particular case of a mirror, a ray passing through its center of curvature will be reflected on itself.

- All the above rays should meet at one point B' corresponding to the image of B. The image of a vertical segment AB is therefore given by the orthogonal projection of B' along the optical axis. In the example of the convergent thin lens or the concave mirror, the image of a vertical segment positioned at the left of the object focal point is real and inverted, as shown in Fig. 17.62.

For more complex systems composed of several lenses and mirrors, ray tracing can be applied separately for each optical element in order to find the behavior of the whole system.

Fig. 17.58 A real object is always placed at the left of the optical system

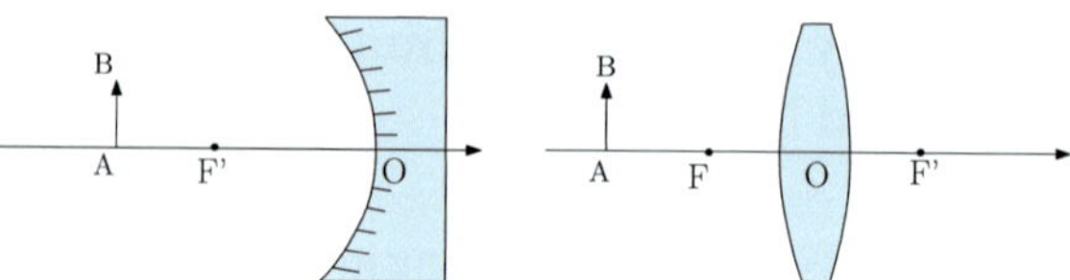

17.10   Association of Two Centered Optical Systems

Let us now consider the association of two centered optical systems S_1 and S_2, aligned such that both present rotational symmetry along a common optical axis. We will show in this section that this composed optical system can be seen as a single system S for which the positions of its cardinal points F', F, H', and H can be determined in terms of those of the individual systems S_1 and S_2 and their separation e. Let us suppose, in the most general case, that the refractive index is n in the object region of the first system, n' in the image region of the second system, and n_{int} in between the two systems, as shown in Fig. 17.63.

In order to determine the image principal point H' and the image focal length f' of the composed system, we need to find the position F' at which a horizontal ray intersects the optical axis. As shown in Fig. 17.64, the prolongation of such a ray will seem to converge to the object focal point F_1' after leaving the first system S_1. This prolongation and that of the ray after leaving the second system will both encounter the principal planes of S_2 at the same height (this is a general property of the principal planes). Finally, the ray passes through the axis at a point F' corresponding by definition to the image focal point of the composed system. The principal point H' is determined by the intersection of the prolongation of the incident horizontal ray and the prolongation of the ray passing through F' after leaving both systems.

Using the lens Eq. (17.19) for the conjugate points F_1' and F' by the second lens, we have

$$\frac{f_2'}{\overline{H_2'F'}} + \frac{f_2}{\overline{H_2F_1'}} = 1 \ .$$

Replacing $f_2 = -n_{\text{int}} f_2'/n'$ and $\overline{H_2F_1'} = \overline{H_1'F_1'} - \overline{H_1'H_2} = f_1' - e$, we have

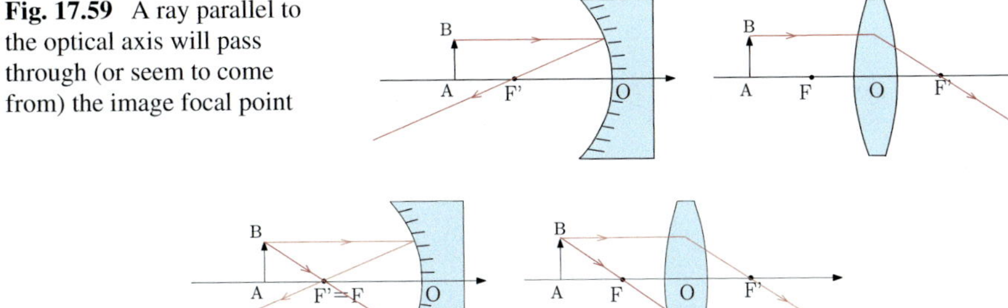

Fig. 17.59   A ray parallel to the optical axis will pass through (or seem to come from) the image focal point

Fig. 17.60   A ray passing through the object focal point will exit the system parallel to the optical axis

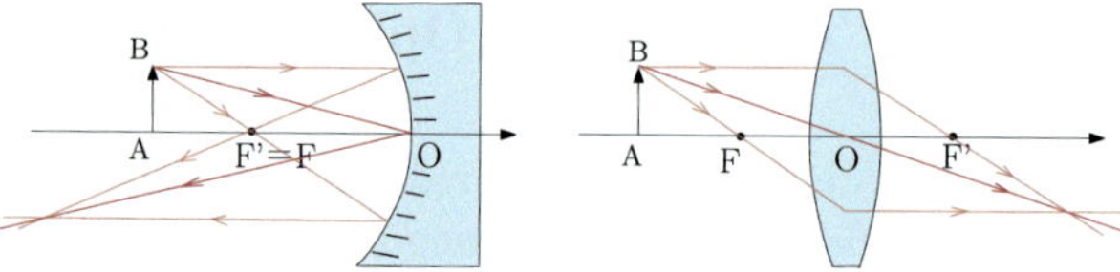

Fig. 17.61 A ray passing through the center of the optical system

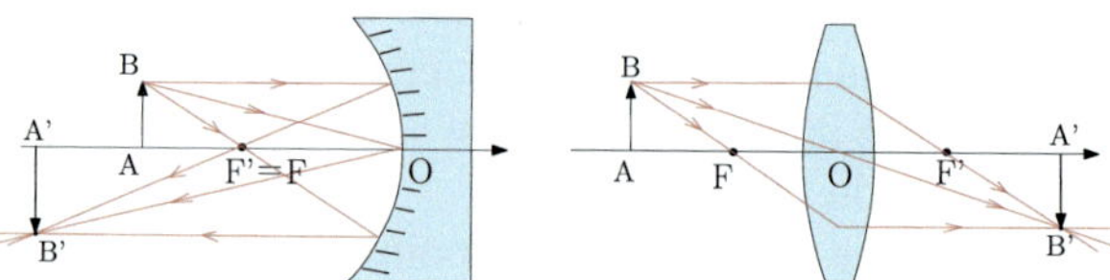

Fig. 17.62 Ray tracing showing the image formation for a convergent lens (left) and a concave mirror (right)

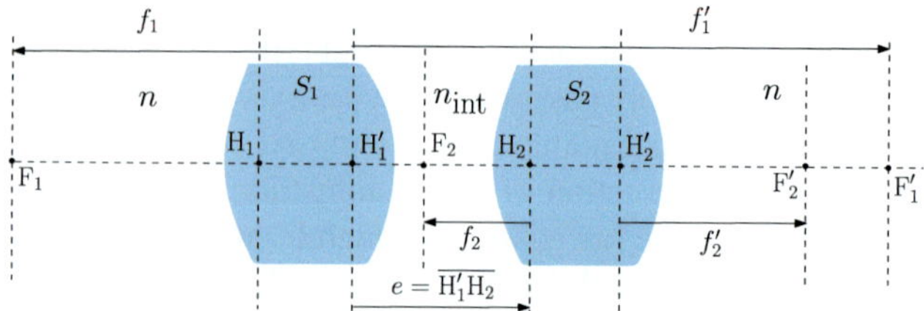

Fig. 17.63 Association of two optical systems with a common optical axis

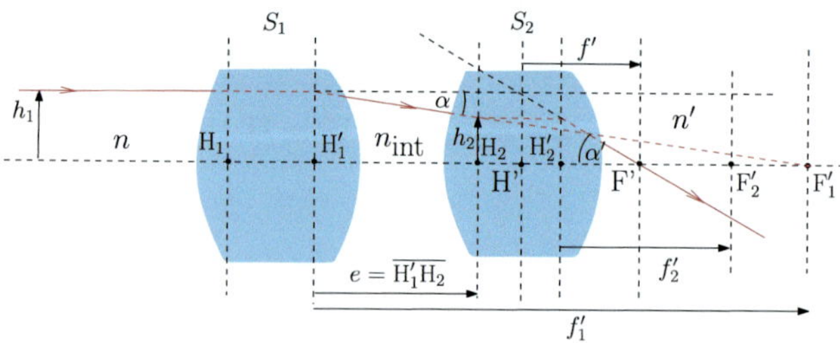

Fig. 17.64 Trajectory of an incoming ray parallel to the optical axis

$$\frac{f_2'}{\overline{H_2'F'}} - \frac{n_{\text{int}}f_2'}{n'(f_1' - e)} = 1 \ .$$

On the other hand, $\alpha' = h_1/\overline{H'F'} = h_1/f' = h_2/\overline{H_2'F'}$ and $\alpha = h_1/f_1' = (h_1 - h_2)/e$ so that $h_1/h_2 = f_1'/(f_1' - e)$. We finally obtain

$$\frac{f_2'f_1'}{f'(f_1' - e)} - \frac{n_{\text{int}}f_2'}{n'(f_1' - e)} = 1$$

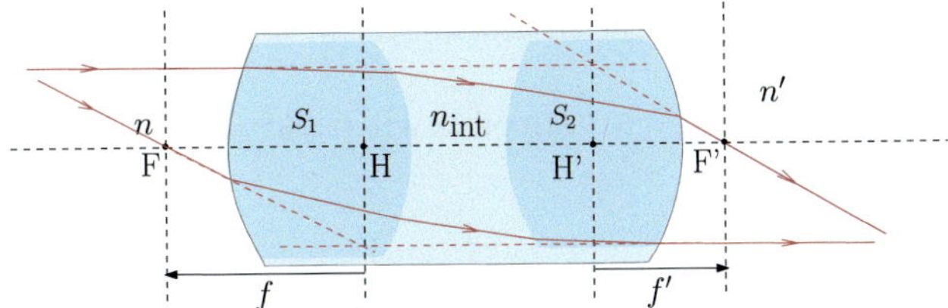

Fig. 17.65 The object focal length of the whole system is simply $f = -nf'/n'$

Fig. 17.66 A lens can be though as the asssembly of two different optical systems

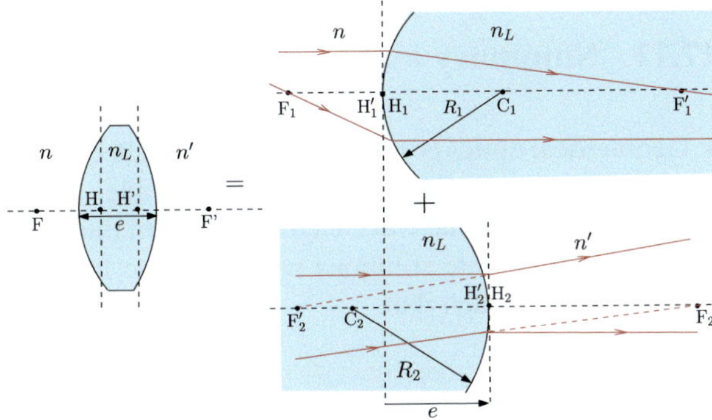

which demonstrates Gullstrand's formula giving the image focal length f' of the whole system in terms of f_1', f_2' and e:

$$\boxed{\frac{n'}{f'} = \frac{n_{\text{int}}}{f_1'} + \frac{n'}{f_2'} - \frac{en'}{f_1' f_2'}.}$$

(17.25)

The object focal length f can be obtained by simply exchanging n with n', f_1' with f_2 and f_2' with f_1 in the above formula. Since $f_1 = -n/n_{\text{int}} f_1'$ and $f_2 = -n_{\text{int}}/n' f_2'$, one obtains simply $n/f = -n'/f'$. The rays passing through F and F' are shown in Fig. 17.65.

Remarks

- The formula (17.19) giving the focal length of a spherical lens can be considered as a particular case of (17.25). Indeed, such a lens can be thought as being composed of two optical systems separated by a distance e: the input and output spherical surfaces and a medium of refractive index n_L in between (Fig. 17.66). Each surface has an image focal length given by $f_1' = R_1 n_L/(n_L - n)$ and $f_2' = -R_2 n'/(n_L - n')$, respectively.

- For two systems surrounded everywhere by the same medium ($n = n' = n_{\text{int}}$), one has simply $f = -f'$ with

$$\frac{1}{f'} = \frac{1}{f_1'} + \frac{1}{f_2'} - \frac{e}{f_1' f_2'}.$$

From this formula, we see that an afocal system is achieved ($f' \to \infty$) if $e = f_1' + f_2'$.

- Two thin lenses mounted next to each other will behave as a single lens of effective focal length f' given by

$$\frac{1}{f'} = \frac{1}{f_1'} + \frac{1}{f_2'} .$$

17.11 Summary and Essential Formulas

- Geometrical optics corresponds to the so-called short wavelength limit, valid when the length scales over which the refractive index and the field amplitudes change are much larger than the wavelength of light. In this framework, Maxwell's equations admit approximate solutions which behave locally as plane waves with a local wave vector $\mathbf{k}$ such that

$$|\mathbf{k}(\mathbf{x})|^2 = \frac{n^2(\mathbf{x})\omega^2}{c^2} .$$

In this limit, light is described by an ensemble of independent optical rays, corresponding to the field lines of the vector field $\mathbf{k}$, which coincide with the field lines of the Poynting vector.

- The local wave vector can be therefore written as $\mathbf{k} = \frac{n\omega}{c}\mathbf{t}$, where $\mathbf{t}$ is a unit vector, and obeys the ray equation:

$$\boxed{\frac{c}{\omega}\frac{d\mathbf{k}}{ds} = \frac{d}{ds}(n\mathbf{t}) = \nabla n}$$

where s is the path length along a flow line of the field $\mathbf{k}$. This equation, which is a generalization of Snell–Descartes law, tells us that optical rays follow curved paths whenever the refractive index n of the propagation medium changes.

- Long before Maxwell's equations, Fermat postulated in 1657 the famous principle of least time, according to which light travels between two points by minimizing the travel time. The optical path length from A to B along the trajectory Γ is defined as

$$(AB) = \int_\Gamma nds .$$

Fermat's principle, under its more general form, states that light follows the trajectory for which the optical path length is stationary.

- An optical system S is any superposition of transparent or reflecting elements. The system S is said to be perfectly stigmatic for the pair of points (A, A') if all the rays (or their prolongations) coming from A intersect at A' when leaving the system. We say then that A' is the conjugate image of A by S and we write

$$A \xrightarrow{S} A'$$

Since all the rays stemming from A reach A', Fermat's principle implies that the optical path must be the same for all the rays going from A to A'.

- An optical system S is said to be centered if it has a rotational symmetry along a line, called the optical axis. In particular, if A belongs to the optical axis and if (A, A') is a stigmatic pair of S, then A' lies on its optical axis.
- The paraxial (or Gauss) approximation consists in considering optical rays close to the axis of a centered system and forming small angles with respect to it. Within this approximation, optical systems composed of lenses and mirrors are stigmatic and aplanatic (the image of an object contained in a plane perpendicular to the optical axis defines a plane image, also perpendicular to the optical axis).
- The transverse magnification of an aplanatic optical system S is given by

$$M = \frac{\overline{A'B'}}{\overline{AB}}$$

  where AB is a straight line perpendicular to the optical axis and $A \xrightarrow{S} A'$, $B \xrightarrow{S} B'$. For $|M| > 1$, the image is said to be magnified, otherwise the image is diminished. For $M > 0$ the image is said to be upright whereas for $M < 0$ the image is inverted.
- For a centered and stigmatic system S, the image focal point F' is the image of a point A_∞ located at infinity in the optical axis, that is

$$A_\infty \xrightarrow{S} F'$$

  whereas the object focal point F is imaged into a point A'_∞ at infinity on the optical axis

$$F \xrightarrow{S} A'_\infty$$

  The object (image) focal plane is the plane perpendicular to the optical axis containing F (F').
- The object principal plane is the plane perpendicular to the optical axis containing the intersection points of all the rays passing through the object focal point F and their corresponding image rays, which are parallel to the optical axis. It intersects the axis at the object principal point H. Similarly, the image principal plane is the plane perpendicular to the optical axis containing the intersection points between the prolongations of each incident ray parallel to the axis with its corresponding image ray, which passes (or seem to pass) through the image focal point F'. It intersects the axis at the image principal point H'. The object focal distance f and the image focal distance f' are defined as

$$f = \overline{\text{HF}}, \qquad f' = \overline{\text{HF'}} ,$$

For a lens, both are related by $f/n = -f'/n'$, where n (n') is the optical index of the medium in which the object (image) is at. In particular, if $n = n'$, one has

$$f = -f' ,$$

- There are two particular points N and N', called the nodal points, such that every ray aimed at one of them will appear to have come from the other one after leaving the optical system, forming always the same angle with respect to the axis. One has $\overline{\text{HH'}} = \overline{\text{NN'}}$ and

$$\overline{\text{HN}} = \overline{\text{H'N'}} = f + f' ,$$

If $n = n'$, the nodal points N and N' correspond to the principal points H and H', respectively.
- The cardinal points are three pairs of points: the focal points, the principal points, and the nodal points. All the imaging properties of a system are completely determined by the locations of the cardinal points.
- If A is the location of an object on the optical axis of a centered system, and A' is the location of its image, the Newton relation states

$$\boxed{\overline{\text{FA}}\ \overline{\text{F'A'}} = ff'}$$

or, equivalently, if both are measured with respect to the principal points, one obtains the relation

$$\boxed{\frac{f}{\overline{\text{HA}}} + \frac{f'}{\overline{\text{H'A'}}} = 1 ,}$$

and the transverse magnification reads

$$M = \frac{n\overline{\text{H'A'}}}{n'\overline{\text{HA}}} .$$

- The composition of two centered and aligned systems S_1 and S_2, separated by a distance $e = \overline{\text{H}'_1 \text{H}_2}$, can be seen as a single optical system having an image focal length f' given by Gullstrand's formula

$$\boxed{\frac{n'}{f'} = \frac{n_{\text{int}}}{f'_1} + \frac{n'}{f'_2} - \frac{en'}{f'_1 f'_2}}$$

where n_{int} is the optical index of the medium between S_1 and S_2, and n (resp. n') that of the object (resp. image) region of S.

Problems

17.1 Fermat's principle and Snell–Descartes's law

Consider two points A and B that belong to two media of refractive index n_1 ($y > 0$) and n_2 ($y < 0$), respectively. Based on Fermat's principle, determine the trajectory of the ray that goes from A to B and retrieve Snell–Descartes's law.

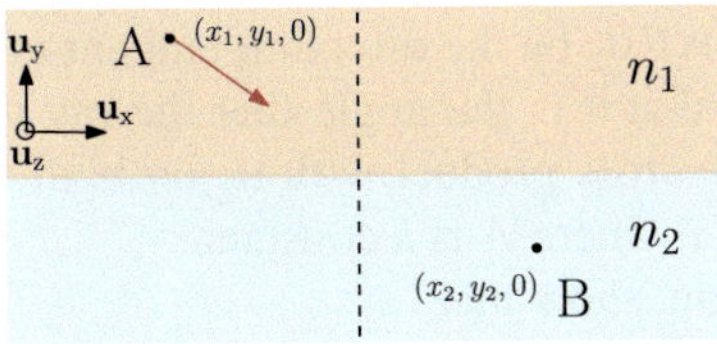

17.2 Deviation of light by the atmosphere

Suppose that the atmosphere has a refractive index with spherical symmetry, so that $n = n(r)$ in spherical coordinates with the origin O at the center of the Earth.

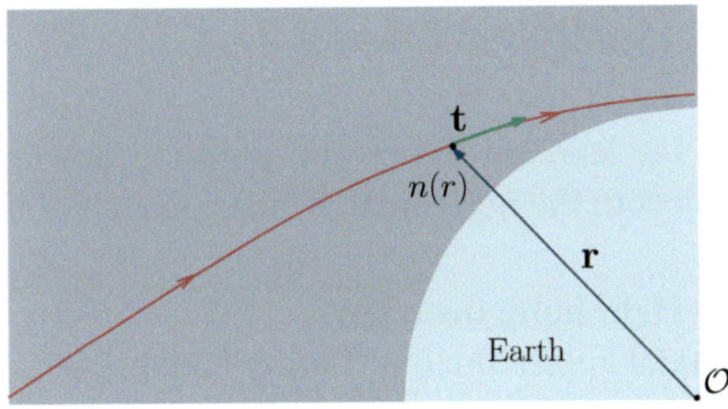

(a) By taking the vector product with $\mathbf{r}$ on both sides of the ray equation, show that $\mathbf{r} \times n\mathbf{t}$ is a constant vector along a ray trajectory.

(b) Let ϕ be the angle formed by $\mathbf{r}$ and $\mathbf{t}$. Show that $nr \sin \phi$ is a constant along the trajectory.

(c) Suppose that Earth's atmosphere has a refractive index of $n(R) = n_0$ at Earth's surface and $n(R + h) = 1$, where R is Earth's radius and $h \ll R$ is the atmosphere thickness. Suppose that a light ray coming from a star makes an angle of θ_i with respect to the horizon when it arrives at the Earth. At what apparent angle θ_0 is the star observed from Earth's surface due to the presence of the atmosphere?

(d) By writing $\theta_0 = \theta_i + \Delta\theta$, find an expression for θ_0 assuming $\Delta\theta \ll 1$.

17.3 The mirage effect

Consider the propagation of a light ray in the vertical plane Oxz where, due to a vertical gradient of the air temperature, the refractive index varies as a function of z according to the law.

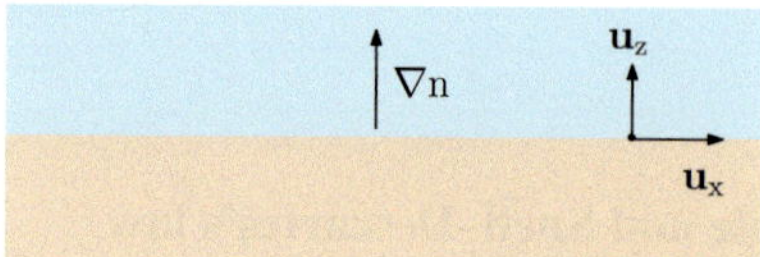

$$z = an^2 + b, \qquad a > 0 .$$

(a) Let $\mathbf{t} = \sin\theta\,\mathbf{u}_x - \cos\theta\,\mathbf{u}_z$ be the direction tangent to the ray trajectory $\mathbf{x}(s) = x(s)\mathbf{u}_x + z(s)\mathbf{u}_z$, so that θ is the angle that the ray makes with respect to the z-axis. By taking the cross product with $\mathbf{u}_z$ on both sides of the ray equation, show that $n\sin\theta = A$, where A is a constant.

(b) Using the ray equation, show that

$$\frac{d^2z}{ds^2} = \frac{\sin^2\theta}{n}\frac{dn}{dz}, \qquad \frac{d^2x}{ds^2} = \frac{\sin\theta\cos\theta}{n}\frac{dn}{dz} .$$

(c) Show that the ray trajectory is defined by the parabola

$$\frac{d^2z}{dx^2} = \frac{1}{2aA^2} .$$

(d) Suppose that the ray starts at the origin, pointing downward and forming an angle θ_0 with respect to the horizontal. Make a sketch of the ray trajectory.

17.4 The Lagrange–Helmholtz invariant

Show that, in the paraxial approximation for a plane diopter separating two homogeneous media of refractive index n and n', the Lagrange–Helmholtz invariant is generalized to

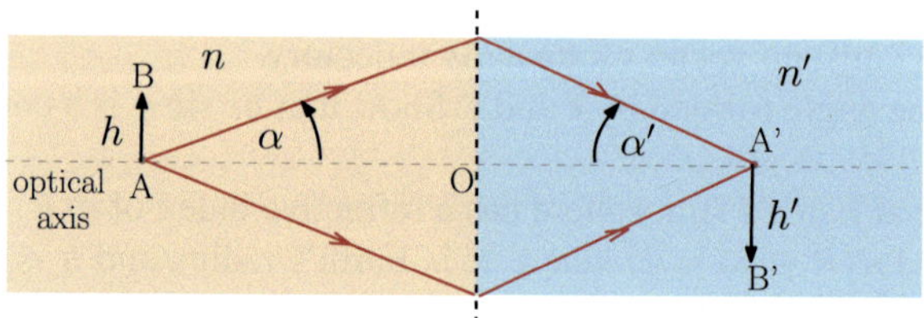

$$n'h'\alpha' = nh\alpha .$$

17.5 Ray tracing with spherical mirrors

Using ray tracing, determine the position of the image of a real, vertical object AB for each of the different cases shown in the figure below. Conclude about the magnification (magnified, diminished), orientation (inverted, upright) and real/virtual character of the image for each case.

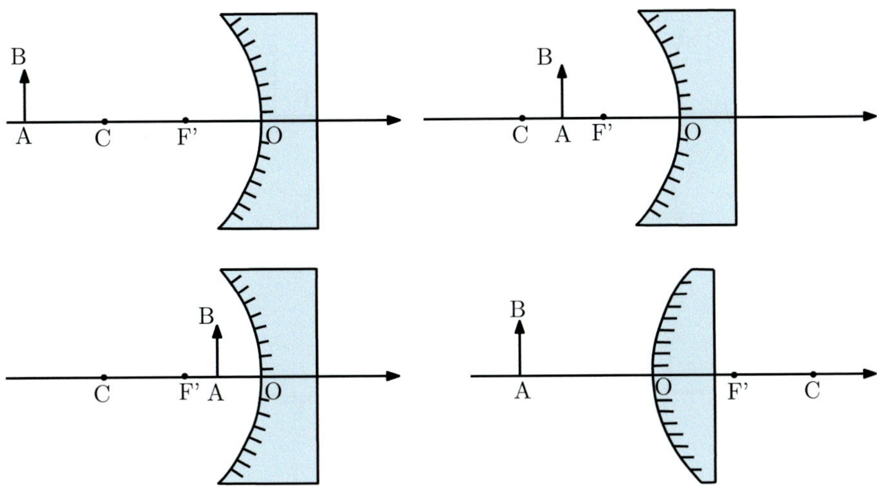

17.6 Image formation with a convergent thin lens

Using ray tracing, determine the image of a vertical object AB for each of the different cases shown in the figure below. Conclude about the magnification (magnified, diminished), orientation (inverted, upright) and real/virtual character of the image.

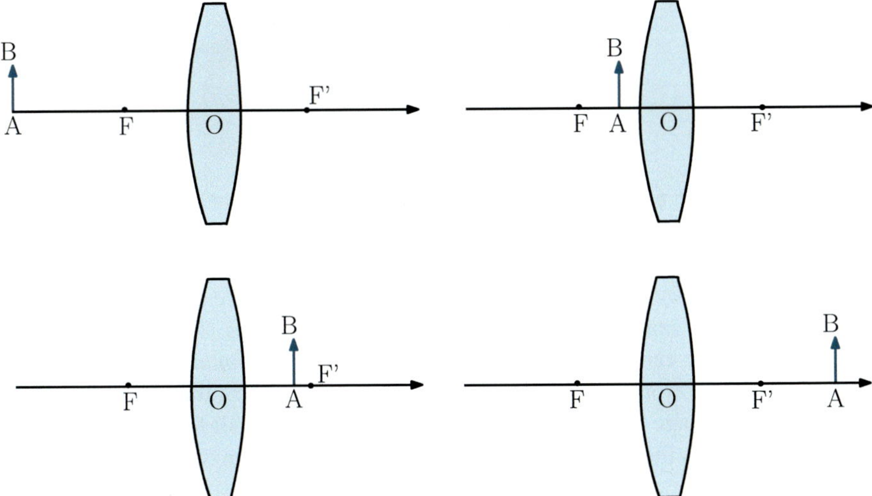

17.7 Image formation with a divergent thin lens

Using ray tracing, determine the image of a vertical object AB for each of the different cases shown in the figure below, and conclude about the magnification (magnified, diminished), orientation (inverted, upright) and real/virtual character of the image.

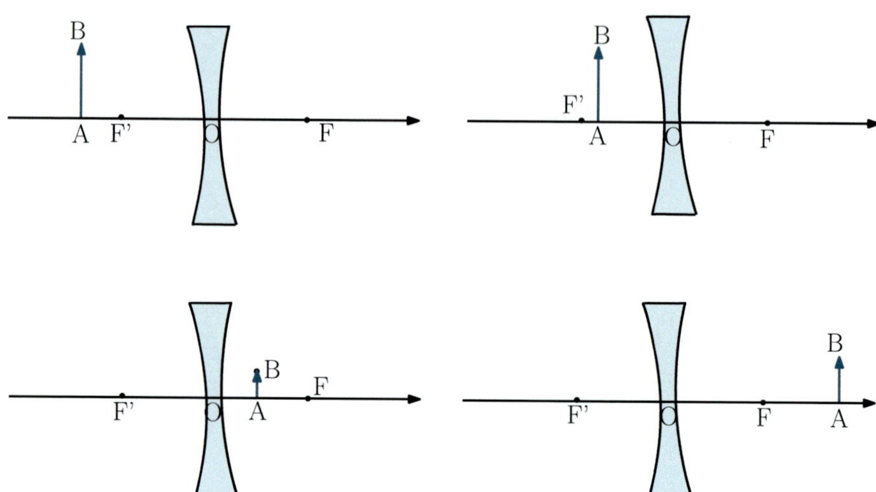

17.8 The human eye and its resolution

The human eye can be modeled by a thin lens, whose principal plane will be considered to be located at the entrance of the eye, intersecting the optical axis at point O, as shown in the figure below. The retina is located at $d_0 = 2.2$ cm from O.

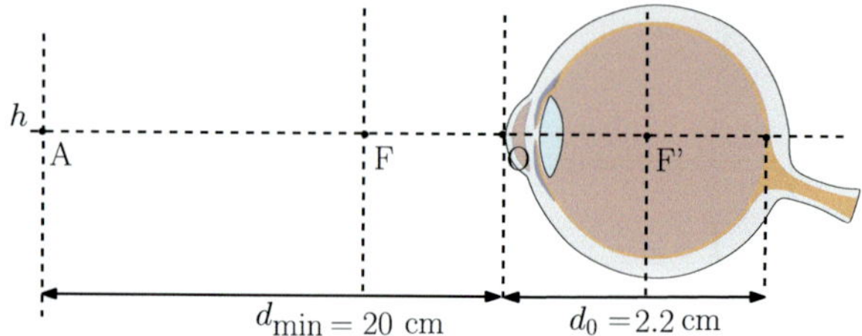

(a) The eye lens, by changing its shape, changes the effective focal distance of the eye so that it can focus on objects at different distances. The closest point on which an eye can focus at is called the *ponctum proximum*, and is located at $d_{\min} \approx 20$ cm from the eye for a young adult. By modeling the eye as a thin lens, what is its image focal length in this situation? For a small feature of transverse size h, what is the maximum angle θ under which such an object can be seen by the eye?

(b) The resolution of the retina is limited by the size and separation of the cells it contains. The cell density is higher at the fovea, at the center of the retina, so that a normal eye is capable of resolving two points separated by an angle of $\theta_{\min} = 3 \times 10^{-4}$ rad. To what distance in the retina this angle corresponds to? Based on this, what is the smallest feature that a normal eye can resolve?

17.9 The magnifying glass

A magnifying glass is made of a convergent lens of short focal length $f' > 0$ and allows us to observe objects under a larger angle than with the naked eye. Increasing the angle that the object subtends from the eye is responsible for an apparent increase in the object size, since a larger image on the retina will be formed by the eye lens. Suppose that a vertical object of height $h_0 = \overline{AB}$, where A is on the optical axis, is placed between F, the object focal plane of the lens, and its center O, at a distance $d = \overline{AO} < f'$. The lens is held at a distance $l = \overline{OE}$ from the eye.

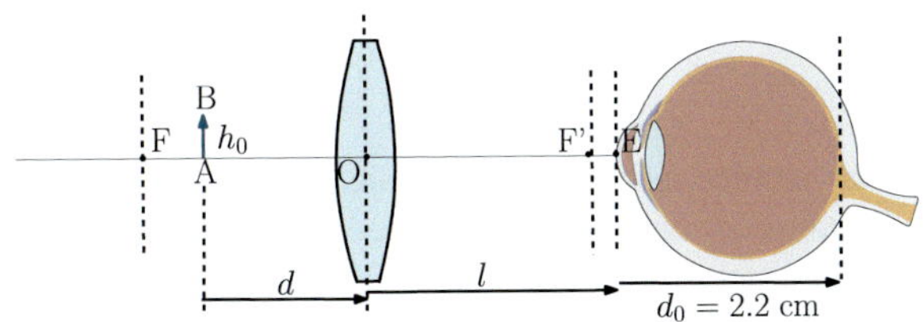

(a) Suppose that the *punctum proximum* is located at $d_{\min} = 25$ cm from the eyes of the observer. Determine the maximum angle $\theta_{\max}$ at which the naked eye (i.e., without the magnifying glass) can observe the object AB, and the maximum physical size h_0' of the object image on the retina.

(b) Determine the position of the image of AB by the lens and make a sketch of it based on ray tracing. Is this image real or virtual, upward or inverted? What is the angle θ that this image subtends from the eye?

(c) The angular magnification γ of the magnifying glass is defined as $\gamma = \theta/\theta_{\max}$. Determine γ as a function of f', l, d and $d_{\min}$.

(d) Typically, one holds the magnifying glass very close to the eye, so that $l \approx 0$. What is the minimum and maximum distance d between the object and the lens that will produce a sharp image on the retina? Conclude about the minimum and maximum angular magnification that can be achieved with the magnifying glass. Evaluate numerically for $f' = 5$ cm and $d_{\min} = 25$ cm.

17.10 The microscope

The simplest microscope is made of two converging lenses: an objective lens of short focal length $f_1' > 0$ and an eyepiece, also named ocular, of longer focal length $f_2' > 0$. When placed in front of the eye, the combination of these two lenses produces a magnified image of an object sufficiently far away from the eye of the observer, who can thus see the object by forming a real image on the retina by accommodating the eye lens. The distance $L = \overline{O_1 O_2} > f_1' + f_2'$ between the two lenses is fixed, and let $d_o = \overline{AO_1} > f_1'$ be the distance between the object h_0, and the objective.

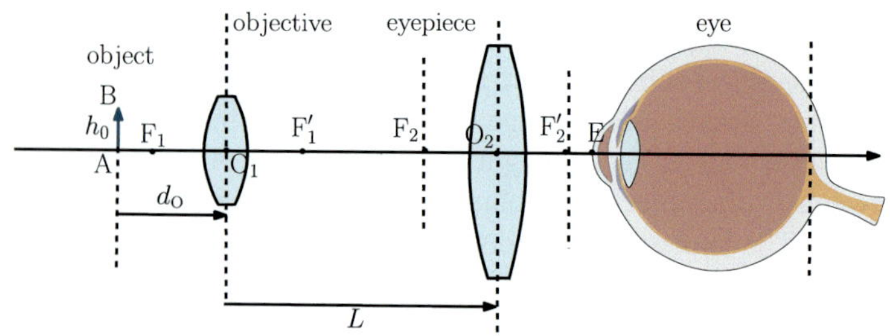

(a) Determine the distance d_i between the objective and the image h_i it creates of h_0, and the distance d_i' between the eyepiece and the image h_i' it creates of h_i. Using ray tracing, determine if the latter is real or virtual, upright or inverted, and magnified or reduced.

(b) Deduce the overall angular magnification of the microscope when the eye is relaxed, i.e., without accommodation. Where should the object placed at with respect to the objective?

17.11 Projecting an image on a screen

Consider a real object AB placed at a distance D from a screen. We want to project a real image of this object on the screen by placing a thin lens of focal length f' in between, as shown below.

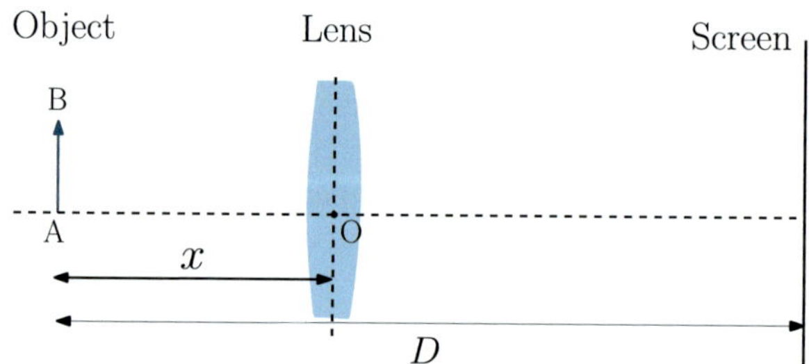

(a) Which type of lens (convergent or divergent) is capable of producing a real image on the screen?

(b) Let $x > 0$ be the distance between the object and the lens. Show that, in order to project an image of the object on the screen, one can always find a position $f' < x < D$ for the lens provided that $D \geq 4f'$, where D is the distance between the object and the screen.

(c) Consider the case $D = 4f'$. Where should the lens be placed? Determine the magnification of the image on the screen, and check this result by using ray tracing.

(d) For $D > 4f'$, where should the lens be placed in order to obtain a magnified ($|M| > 1$) projection of the object on the screen?

17.12 Longitudinal magnification of a thin lens

Consider a thin lens centered at O, and a small object placed at A with a finite extension $\overline{AB}$ along the optical axis, much smaller than $|\overline{OA}|$. Show that in this case the longitudinal magnification M_l of the lens is related to its transverse magnification M.

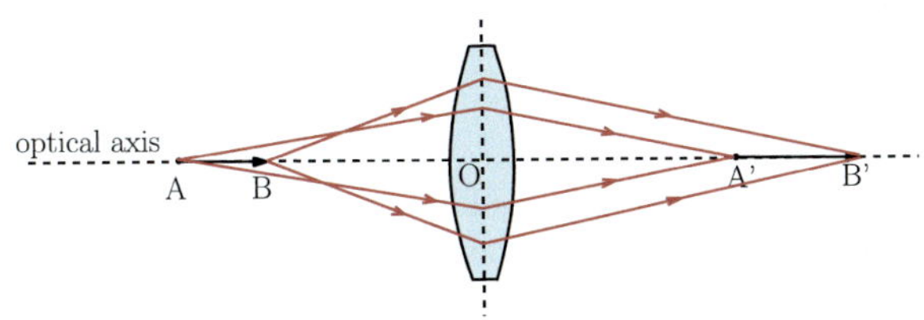

$$M_l = \frac{\overline{A'B'}}{\overline{AB}} = M^2 = \left(\frac{\overline{OA'}}{\overline{OA}}\right)^2 .$$

17.13 The field of depth of an objective lens

An objective made of a thin lens of focal length f' and diameter D is used to image objects in the plane perpendicular to the optical axis and passing through point A. The detector is positioned at the image plane containing A' and perpendicular to the optical axis, where A and A' are conjugated by the lens. The field of depth $\Delta_{\max}$ is the interval inside which an object can be moved with respect to A along the optical axis so that its image on the detector is still sharp enough. Determine $\Delta_{\max}$ by considering that the image of a point on the detector is said to be sharp as long as its diameter is smaller than ϕ_0.[2] Show that in the limit $\overline{AO} \gg f'$.

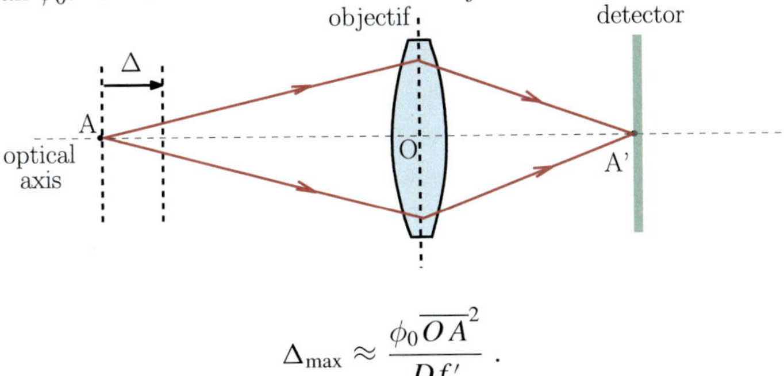

$$\Delta_{\max} \approx \frac{\phi_0 \overline{OA}^2}{Df'} .$$

17.14 The Galilean telescope

The Galilean telescope is a type of refracting telescope: an afocal system that uses two lenses to produce a magnified image of distant objects after passing through the eyes of an observer. It is made of an objective lens, which is a convergent lens of focal length $f_1' > 0$ followed by the eyepiece, which is a divergent lens of focal length $f_2' < 0$, so that the image focal point F_1' of the objective coincides with the object focal point F_2 of the eyepiece, as shown below.

---

[2] The diameter ϕ_0 can be either limited by diffraction or by the finite size of a pixel in the detector.

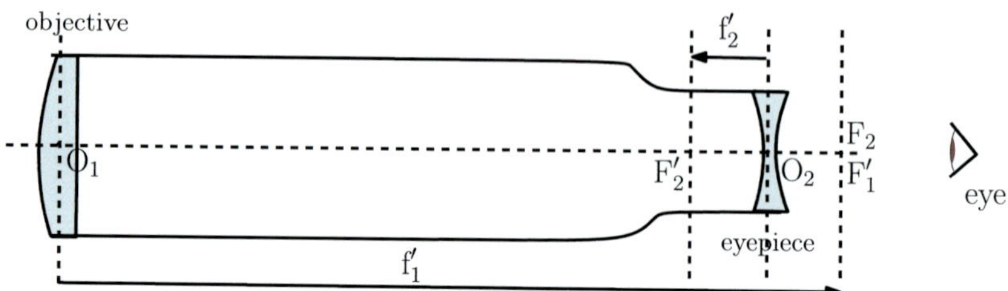

(a) Consider an object at infinity looked by the telescope under an angle α. Determine the angular magnification γ of the telescope in terms of f_1' and f_2'.

(b) In 1609, Galileo built a refracting telescope by the association of a plano-convex objective lens of focal length $f_1' = 98$ cm and a plano-concave eyepiece of focal length $f_2' = -5$ cm. Among his findings, he was able to observe the four largest moons of Jupiter, the first objects found to orbit a body that was neither Earth nor the Sun. Io, the largest moon, has a diameter of $\phi = 3600$ km and it is at a distance $D = 628.3 \times 10^6$ km from the Earth. Under which angle is this moon looked at with the naked eye and with Galileo's telescope?

(c) When an object is located at infinity, the focal length of a relaxed eye is about 2.2 cm. What is the physical size of Io's image on the retina when looked with the naked eye and under the telescope? Given that cone cells in the region of the retina with highest resolution (fovea) are spaced at around 2.5 μm, conclude about the possibility to resolve Io with the naked eye and with Galileo's telescope.

(d) By moving the eyepiece relative to the objective, it is possible to observe objects at finite distances without accommodating the eye. What should be the necessary displacement of the eyepiece in order look at an object A located at a distance $\overline{O_1 A} = -50$ m from the objective?

(e) The eye lens can accommodate in order to form on the retina the image of an object located at finite distance. The minimum distance between the object and the eye depends on age: a person in their forties can observe, in average, objects at distances larger than 20 cm. Considering that the eye is always placed at $d = 15$ mm from the eyepiece and in the conditions of the previous question, what is the closest object that can be observed with the telescope?

17.15 The achromatic doublet

Lenses made of glass suffer from chromatic aberration: since the optical index is a function of wavelength, different colors are focused at different points. A possible solution to correct this chromatism consists in cementing together two optical components, usually made of a convergent, low-index lens and a divergent, high-index lens. This composed system is called an achromatic doublet.

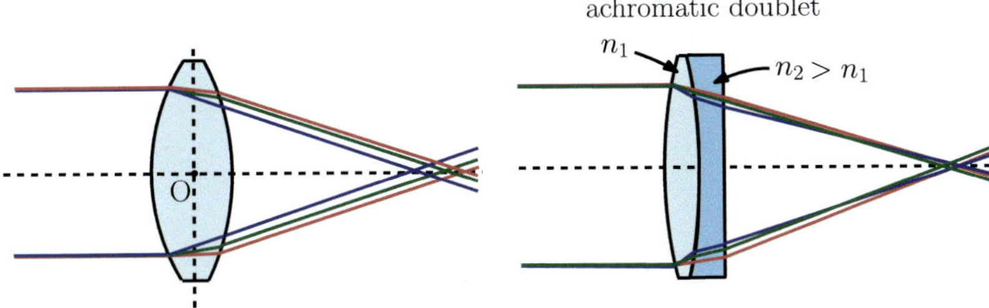

(a) Consider a spherical, biconvex convergent lens of thickness $e_1 = 5$ mm made of lanthanum crown glass ($n_1 = 1.651$ for yellow light). The radius of curvature of the input and output surfaces are $R_1 = 117.1$ mm and $R_2 = -142.1$ mm, respectively. What is the focal length f_1' of this lens when placed in air? Justify the validity of the thin-lens approximation and give an estimate of the error induced in f_1' by this approximation.

(b) Due to dispersion, the refractive index of crown glass differs between blue and red wavelengths by $\Delta n_1 = n_1(\text{blue}) - n_1(\text{red}) = 0.01165$. What is the difference in focal length between blue and red light for the lens of the previous question?

(c) To correct for the dispersion of the lanthanum crown glass, a second, plano-concave lens made of high-index glass ($n_2 = 1.728$ for yellow light) is glued to the first lens to form an achromat doublet. Its concave side has the same radius of curvature R_2 than the output surface of the first lens, and its thickness is $e_2 = 3$ mm. What is the resulting focal length f' of the achromatic doublet for yellow light?

(d) The dispersion of the high-index glass is such that $\Delta n_2 = n_2(\text{blue}) - n_2(\text{red}) = 0.0255$. What is the difference in focal length between blue and red light for the doublet? Compare to the case where a single lens is used.

Chapter 18
Diffraction

Abstract This chapter discusses the phenomenon of *diffraction*, which describes the spreading of light waves as they propagate through apertures or around obstacles, leading to intensity patterns that deviate from the predictions of geometrical optics. It begins by explaining how light propagation in vacuum can be understood as an *expansion in plane waves*, where the transverse spatial frequencies determine the angular divergence of the beam. The *uncertainty principle* is invoked to show that a finite beam size inherently leads to a minimum angular divergence. The core of the chapter introduces the *Huygens–Fresnel principle*, which states that every point on a wavefront can be considered a source of secondary spherical wavelets, and the total field at any point is the superposition of these wavelets. This principle is mathematically formalized through the *Huygens–Fresnel integral*, providing a method to calculate diffraction patterns. The counter-intuitive *Fresnel-Arago-Poisson spot* (a bright spot in the center of a shadow cast by an opaque disk) is presented as a compelling experimental validation of the wave theory of light, alongside *Babinet's theorem*, which relates the diffraction patterns of complementary obstacles. The chapter then introduces the *Fresnel (paraxial) approximation*, a simplification of the Huygens–Fresnel integral valid for small angles and distances much larger than the wavelength. This approximation leads to the *paraxial wave equation*, a fundamental equation for describing beam propagation in optics. The behavior of *Gaussian beams*, a common solution to this equation that describes laser beams, is analyzed in detail, including their waist, Rayleigh length, and divergence. A significant section explores *diffraction by thin lenses*, demonstrating how lenses perform a *Fourier transform* of the incident light field at their focal plane, a concept central to *Fourier optics*. This principle is applied to *spatial filtering*, where masks in the Fourier plane can selectively block or transmit spatial frequencies, enabling image processing techniques like edge enhancement. The chapter also discusses *image formation and the diffraction limit*, showing that the image of a point source formed by a lens is an *Airy pattern*, which defines the fundamental limit of resolution for optical imaging systems (the *Rayleigh criterion* and *numerical aperture*). Finally, the chapter distinguishes between *Fresnel diffraction (near-field)* and *Fraunhofer diffraction (far-field)*, providing analytical solutions for common geometries like single and double slits. It concludes by demonstrating how *diffraction gratings* utilize the principles of interference and diffraction

F. Cadiz and A. Couairon, *Classical Electrodynamics*, Undergraduate Texts in Physics, https://doi.org/10.1007/978-3-031-86785-9_18

to function as *spectrometers*, spatially separating different wavelengths of light and enabling high-resolution spectral analysis, characterized by their *resolving power*.

Keywords Diffraction · Huygens–Fresnel principle · Fraunhoffer diffraction · Fourier optics · Spectrometers · Diffraction limit

18.1 Introduction

Electromagnetic waves are often modeled as sinusoidal plane waves, where the electric field's amplitude is constant across every plane perpendicular to the propagation direction. These waves are solutions to d'Alembert's wave equation, maintaining their homogeneous spatial structure as they propagate. In reality, however, every physical wave has a finite transverse extent, either due to the limited size of the source itself or the presence of obstacles or apertures. Regardless of the cause, we will show in this chapter that the spatial field distribution evolves during propagation as a result of a wave phenomenon known as diffraction. Due to diffraction, the observed intensity pattern beyond an obstacle can be far more intricate than the sharp shadow predicted by geometrical optics, as illustrated in Fig. 18.1.

We will demonstrate how diffraction imposes a minimum divergence on a light beam and introduce the famous Huygens–Fresnel principle. Using this principle, we show that far from an obstacle and near the propagation axis, the diffracted intensity can be determined by calculating Fourier transforms. This analysis is then applied to understand the resolution limit of imaging systems, also known as the diffraction limit. Finally, we discuss how diffraction principles are utilized in constructing optical spectrometers, essential tools in modern science.

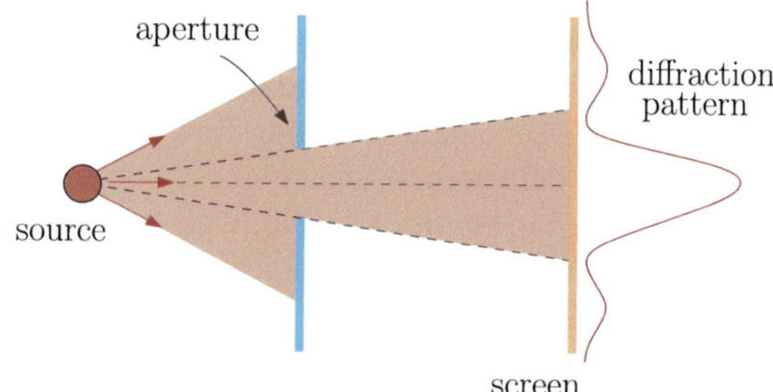

Fig. 18.1 The field's intensity after an obstacle is not simply a clearly defined shadow, but rather a complex diffraction pattern

18.2 Propagation of Light in Vacuum Expansion in Plane Waves

Consider a monochromatic electromagnetic wave that propagates in vacuum along the z direction, and let us suppose that the electric field $\mathbf{E}(x, y, z = 0, t)$ is known at the plane $z = 0$. The goal is to obtain an expression for the electric field at an arbitrary plane $z = z_0$ after propagation. To do this, we recall that according to Fourier's theorem (see Appendix A.8), any integrable function can be expressed as a superposition of sinusoidal plane waves of the form

$$\underline{\mathbf{E}}e^{i(\mathbf{k}\cdot\mathbf{x}-\omega t)} = \underline{\mathbf{E}}e^{i(k_x x + k_y y + k_z z - \omega t)} \ .$$

Imposing the above plane wave to be a solution of d'Alembert's propagation equation in vacuum

$$\nabla^2 \underline{\mathbf{E}} - \frac{1}{c^2}\frac{\partial^2 \mathbf{E}}{\partial t^2} = 0$$

leads to the well-known dispersion relation between the frequency ω and the modulus of the wavevector

$$k^2 = k_x^2 + k_y^2 + k_z^2 = \frac{\omega^2}{c^2}$$

Since we are interested in the evolution of a wave propagating along the z-axis ($k_z > 0$), it is convenient to express k_z in terms of k_x, k_y and ω

$$\boxed{k_z = +\sqrt{\frac{\omega^2}{c^2} - (k_x^2 + k_y^2)}.} \tag{18.1}$$

From this, we see that a plane wave of wavevector $\mathbf{k} = (k_x, k_y, k_z)$ will propagate along z only if $k_z \in \mathbb{R}$, that is, if the transverse component of $\mathbf{k}$ is small enough, $\omega^2/c^2 > k_x^2 + k_y^2$.

In contrast, for those waves such that $\omega^2/c^2 < k_x^2 + k_y^2$, k_z is a pure imaginary number and we obtain either an evanescent wave ($k_z = i\sqrt{k_x^2 + k_y^2 - \omega^2/c}$) or an exponentially increasing wave ($k_z = -i\sqrt{k_x^2 + k_y^2 - \omega^2/c}$) along the direction of increasing values of z. We disregard the latter solution since it diverges at infinity, and so the most general solution of the wave equation for a monochromatic wave for $z \geq 0$ can be written as $\underline{\mathbf{E}}(\mathbf{x}, t) = \underline{E}_x(\mathbf{x}, t)\mathbf{u}_x + \underline{E}_y(\mathbf{x}, t)\mathbf{u}_y + \underline{E}_z(\mathbf{x}, t)\mathbf{u}_z$ with

$$\underline{E}_i(x, y, z, t) = \int_{\mathbb{R}}\int_{\mathbb{R}} \underline{A}_i(k_x, k_y)e^{i(k_x x + k_y y)}e^{i(k_z(k_x, k_y) - \omega t)}\,dk_x dk_y$$

with k_z given by (18.2) and where $A_i(k_x, k_y)$ is a complex coefficient that defines the amplitude and the phase of the plane wave of wavevector $\mathbf{k} = (k_x, k_y, k_z)$ in the superposition. This coefficient is strongly related to the Fourier transform of the field at $z = 0$. Indeed, if we write $\underline{E}_i(x, y, z = 0, t) = \underline{E}_{0i}(x, y)e^{-i\omega t}$, then we see that

$$\underline{E}_i(x, y, z = 0, t) = \underline{E}_{0i}(x, y)e^{-i\omega t} = \int_{\mathbb{R}} \int_{\mathbb{R}} A_i(k_x, k_y)e^{i(k_x x + k_y y)}e^{-i\omega t} dk_x dk_y$$

and we conclude that $A_i = \mathcal{F}\{\underline{E}_{0i}\} = \hat{\underline{E}}_{0i}$ is the 2D Fourier transform of $\underline{E}_{0i}(x, y)$. In the following, we will adopt the scalar approximation in which we suppose that the different wavevectors composing $\mathbf{E}$ form small angles with the z-axis (see Sect. 16.3.1) allowing us to drop the index i of the electric field. We may write finally

$$\boxed{\underline{E}(x, y, z, t) = \int_{\mathbb{R}} \int_{\mathbb{R}} \hat{\underline{E}}_0(k_x, k_y)e^{i(k_x x + k_y y)}e^{i(k_z(k_x, k_y) - \omega t)} dk_x dk_y} \qquad (18.2)$$

where

$$k_z(k_x, k_y) = \sqrt{\frac{\omega^2}{c^2} - (k_x^2 + k_y^2)}\,.$$

Remark: Propagation is a Spatial Filter

From Eq. (18.2), we conclude that every electromagnetic wave is a superposition of propagating and evanescent plane waves. As light propagates along the z-axis, the contribution of the evanescent waves decreases exponentially and eventually the *far-field* contains only those waves with small values of $k_x^2 + k_y^2$. Moreover, we know that the amplitude of the plane wave with $k_\perp = (k_x, k_y)$ in the superposition integral (18.2) is given by the Fourier transform $\hat{\underline{E}}_0$ of the field's amplitude $\underline{E}_0$ at $z = 0$ evaluated at $\mathbf{k} = (k_x, k_y)$. It represents therefore the amplitude of the electric field component that oscillates in the plane $z = 0$ at the spatial frequencies $f_x = k_x/2\pi$ and $f_y = k_y/2\pi$ along x and y, respectively.

Thus, the amplitude of the waves with $k_x^2 + k_y^2 = 4\pi^2(f_x^2 + f_y^2) \gg (\omega/c)^2 = 4\pi^2/\lambda^2$ correspond to the features of $\underline{E}_0$ that vary in space faster than λ. All the information carried by these waves is lost after propagating a distance $z = -i/k_z \sim 1/\sqrt{k_x^2 + k_y^2} < \lambda/2\pi$. Propagation is therefore a low-pass spatial filter, only those features larger than λ may reach a far detector, which explains why the spatial resolution is limited to λ for any optical system working in the far field. This resolution limit will be discussed in Sect. 18.5.4.

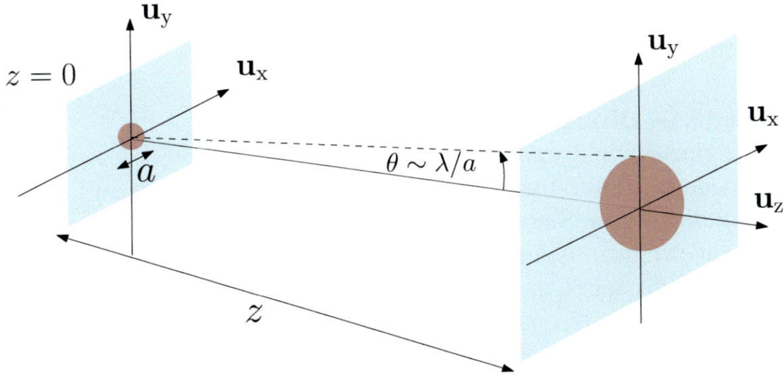

Fig. 18.2 The minimum divergence angle θ is inversely proportional to the size a of the beam at $z = 0$

18.2.1 Diffraction and Angular Divergence

The uncertainty principle is a general property of the Fourier transform that gives us already an idea on how does a light beam evolve while it propagates in vacuum. If Δx is the characteristic extension of $\underline{E}_0(x, y)$ along the x direction, then $\underline{\hat{E}}_0(k_x, k_y)$ is non-negligible for $|k_x| \leq k_{x\max}$ with

$$\Delta x k_{x\max} \geq 2\pi \ .$$

Since these waves move away from the z-axis as they propagate, this allows us to estimate the angular divergence of a light beam. Indeed, the maximum angle θ that the wavevector will form with respect to the z-axis in the plane xz is such that

$$\sin\theta = \frac{k_{x\max}}{k} \geq \frac{\lambda}{\Delta x}$$

and for small angles, we find

$$\sin\theta \approx \theta \geq \frac{\lambda}{\Delta x} \ .$$

The minimum possible divergence allowed by diffraction is therefore $\theta_{\min} = \lambda/\Delta x$, and it is achieved when both the amplitude and the phase of the field $\underline{E}_0$ vary slowly inside Δx (in which case $\Delta x\, k_{x\max} \approx 2\pi$). Figure 18.2 shows the case of a uniform circular beam of diameter a, for which the angular divergence will be $\theta \sim \lambda/a$ in both x and y.

In conclusion, the more the electric field is confined in space at $z = 0$, the larger should be the transverse wavevector values contained in the superposition (18.2), leading to an increase of the lateral size of the field during propagation. When the angular divergence of a beam of size a (at $z = 0$) satisfy $\theta \sim \lambda/a$, we say that the angular divergence is limited by diffraction.

> **Example 18.1—Divergence of a laser beam**
>
> Since the electric field's amplitude and phase vary slowly across the fundamental mode of a laser, the angular divergence θ of a laser beam is typically limited by diffraction. Since the diameter of the beam is usually much larger than the wavelength, θ is small. This is why laser beams keep roughly the same size over large distances, in which case they are said to be collimated. Consider the case of a Helium–Neon laser, which has a typical output diameter of $a \sim 1$ mm and a wavelength of $\lambda = 633$ nm. The angular divergence of the beam will be
>
> $$\theta = \frac{\lambda}{a} \sim 0.6 \text{ mrad}.$$
>
> After propagating $D = 10$ m, the spot size will be of the order of $a' = D\theta \sim 5$ mm. It is a relatively moderate increase in size given the long distance traveled.

18.2.2 Diffraction by an Obstacle

We now consider the case in which a wave encounters a thin object placed at $z = 0$, characterized by a complex transmission factor $\underline{t} = |\underline{t}|e^{i\phi}$, such that $|\underline{t}| \le 1$ represents the fraction of the light's amplitude transmitted by the obstacle, and ϕ takes into account an eventual phase shift generated by the passage through the object. This is illustrated in Fig. 18.3.

We want to find the electric field's amplitude at any point (x, y) of the plane $z = D$. With respect to the case of free propagation (i.e., without an obstacle), all we need to do is to replace the amplitude of the incident wave $\underline{E}_0$ at $z = 0$ by $\underline{E}_0\underline{t}$ and consider the free propagation of the latter field after the obstacle. We say that the field is diffracted by the obstacle, although we highlight the fact that diffraction is due to the wave nature of light and does not require the presence of an obstacle to manifest itself. Equation (18.2) in the presence of an obstacle then becomes

$$\underline{E}(x, y, z, t) = \int_{\mathbb{R}} \int_{\mathbb{R}} \widehat{\underline{E}_0\,\underline{t}_0}(k_x, k_y)e^{i(k_x x + k_y y)}e^{i(k_z(k_x, k_y) - \omega t)}\,dk_x dk_y. \qquad (18.3)$$

This can be written differently by exploiting some properties of the Fourier transform. Indeed, Eq. (18.3) tell us that

$$\widehat{\underline{E_0}\,\underline{t_0}}(k_x, k_y)e^{ik_z(k_x,k_y)z}e^{-i\omega t} = \mathcal{F}\{E(x, y, z, t)\} \tag{18.4}$$

so that, using the convolution property (A.60) of the Fourier transform, the electric field at any value of z can be computed as the following 2D convolution (over the x and y variables only):

$$\boxed{\underline{E}(x, y, z, t) = \frac{e^{-i\omega t}}{(2\pi)^2}(\underline{t_0}\underline{E_0} * h)(x, y, z, t)} \tag{18.5}$$

where $h(x, y, z)$ is the inverse Fourier transform of $\hat{h}(k_x, k_y) = e^{ik_z(k_x,k_y)z}$, and therefore given by the Fourier inversion theorem (A.57)

$$h(x, y, z) = \mathcal{F}^{-1}\{e^{ik_z(k_x,k_y)z}\}(x, y, z) \tag{18.6}$$
$$= \int_{\mathbb{R}}\int_{\mathbb{R}} e^{ik_z(k_x,k_y)z}e^{i(k_x x + k_y y)}dk_x dk_y.$$

The problem of finding the electric field at any position after the obstacle is therefore reduced to calculating a Fourier transform and performing the convolution (18.5). This can be done, for example, numerically. However, in some cases, an analytical expression for the electric field may be obtained after some approximations that will be treated in the subsequent sections of this chapter.

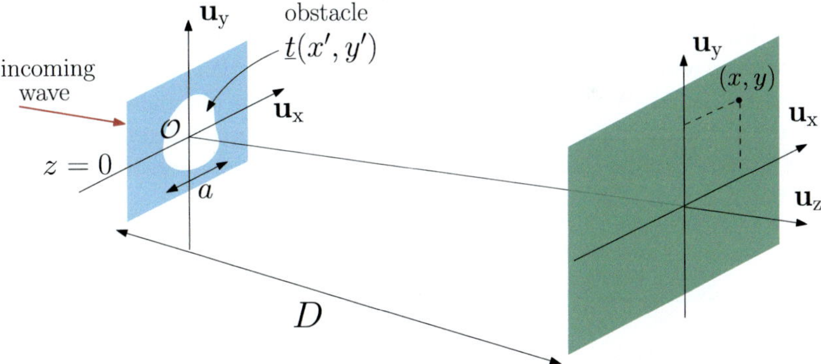

Fig. 18.3 An incident wave faces an obstacle at $z = 0$ defined by its complex transmission factor $\underline{t}(x', y')$

18.2.3 Babinet's Theorem

Before going into the calculation of the diffraction generated by different obstacles, let us show a general theorem relating the diffraction patterns of complementary objects. For this, suppose an object of transmission coefficient $\underline{t}$, and let $\underline{t}' = 1 - \underline{t}$ be its complement, i.e., an object that is transparent at any point where the other is opaque, and vice versa (see Fig. 18.4).

Babinet's theorem, named after Jacques Babinet (Fig. 18.5), states that the diffracted intensities from these two objects are the same at any point where the undisturbed beam would have not reached (that is, outside the geometrical image of the source produced in the absence of an obstacle).

Indeed, let $\underline{E}_0$ be the incident wave at $z = 0$ and let E and E' be the diffracted field generated by the object $\underline{t}$ and $\underline{t}'$, respectively. Then, according to (18.5) we have

$$\underline{E}(\mathbf{x}, t) + \underline{E}'(\mathbf{x}, t) = \frac{e^{-i\omega t}}{(2\pi)^2} \left\{ \underline{t}\underline{E}_0 * h + \underline{t}'\underline{E}_0 * h \right\} (x, y, z, t) .$$

Since $\underline{t}' = 1 - \underline{t}$, and due to the linearity of the convolution product we find

$$\underline{E}(\mathbf{x}, t) + \underline{E}'(\mathbf{x}, t) = \frac{e^{-i\omega t}}{(2\pi)^2} \left\{ \underline{E}_0 * h \right\} (x, y, z, t) = \underline{E}_s(\mathbf{x}, t). \tag{18.7}$$

The sum of the diffraction patterns caused by $\underline{t}$ and $\underline{t}'$ must produce the pattern $\underline{E}_s$ of the unobstructed beam. Where the latter quantity is zero, $E_s = 0$, the diffraction patterns caused by $\underline{t}$ and $\underline{t}'$ must be opposite in phase, but equal in amplitude ($\mathbf{E}(\mathbf{x}, t) = -\mathbf{E}'(\mathbf{x}, t)$). Their intensities are therefore equal.

Figure 18.6a shows the calculated diffracted intensity at a distance $z = 0.4a$ from a slit of width a and height $5a$, when illuminated by a Gaussian beam whose diameter is of the order of $2a$. When the same beam illuminates the complementary mask of the slit, the diffraction pattern is that of Fig. 18.6b. If one sums both electric fields that produce (a) and (b), the total intensity is exactly that of the unobstructed Gaussian beam, shown in Fig. 18.6c.

Fig. 18.4 Two obstacles are complementary if their transmission factors $\underline{t}$ and $\underline{t}'$ are related by $\underline{t}' = 1 - \underline{t}$

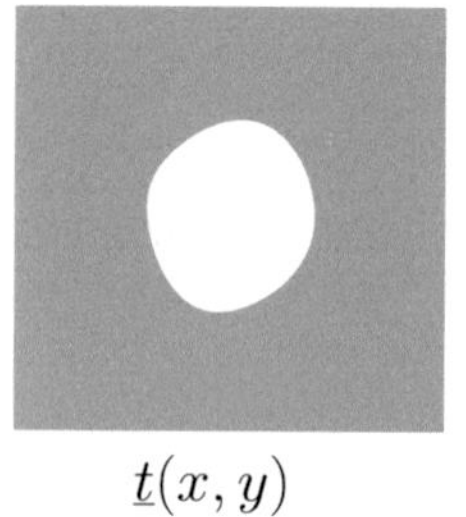
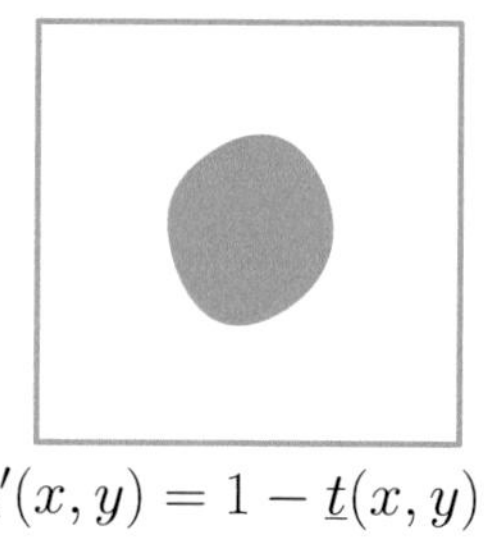

$$\underline{t}(x, y) \qquad\qquad \underline{t}'(x, y) = 1 - \underline{t}(x, y)$$

Fig. 18.5 Jacques Babinet (1794–1872), French engineer and physicist. Among his numerous contributions, we can cite the fabrication of the goniometer and Babinet's compensator, an instrument used to introduce a variable phase retardation between two components of the electric field

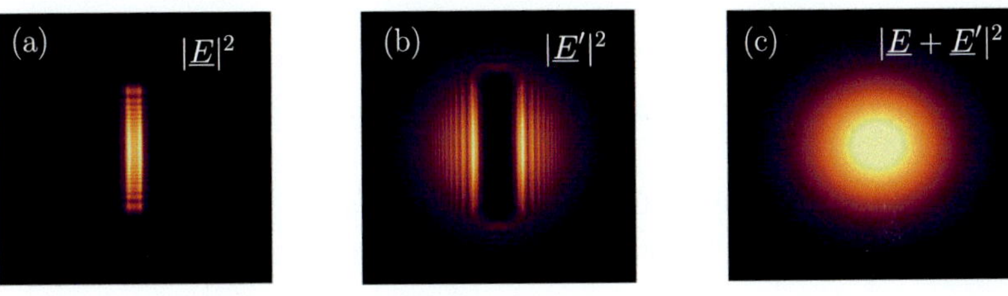

Fig. 18.6 An illustration of Babinet's theorem

18.3 The Huygens–Fresnel Principle

The Huygens–Fresnel principle, named after Christian Huygens (Fig. 18.8) and Augustin Jean Fresnel (Fig. 15.15), tells us that the diffraction by an object can be seen as the result of a superposition of fictive, secondary sources radiating spherical waves in all directions. To demonstrate this principle, we will use an important result known as the Weyl expansion, which allows to write a spherical wave as a superposition of plane waves. For $z > 0$ this reads

$$\frac{e^{ikr}}{r} = \frac{i}{2\pi} \int_{\mathbb{R}} \int_{\mathbb{R}} \frac{e^{i(k_x x + k_y y + k_z(k_x,k_y)z)}}{k_z} dk_x dk_y$$

where k_z is given by (18.2). Differentiating under the integral sign, we see that

$$\frac{\partial}{\partial z} \frac{e^{ikr}}{r} = -\frac{1}{2\pi} \int_{\mathbb{R}} \int_{\mathbb{R}} e^{i(k_x x + k_y y)} e^{ik_z(k_x,k_y)z} dk_x dk_y$$

We recognize in the right-hand side the function h given by Eq. (18.7) , and so

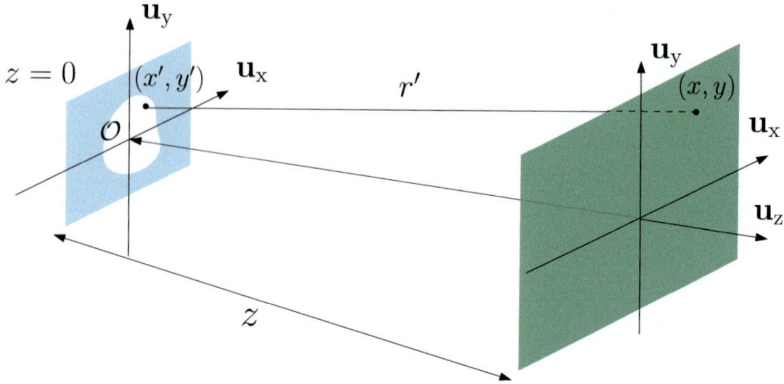

Fig. 18.7 According to the Huygens–Fresnel integral (18.8), each point (x', y') in the plane $z = 0$ can be thought as a fictive source of a spherical wave proportional to the incident field at that point

$$- 2\pi \frac{\partial}{\partial z} \frac{e^{ikr}}{r} = h(x, y, z) \, .$$

Equation (18.5) then tells us that the electric field at (x, y, z, t) can be written in terms of a 2D convolution between $\underline{t}\,\underline{E}_0(x, y, 0, t)$ and $\underline{h}(x, y, z) = -2\pi \frac{\partial}{\partial z} \frac{e^{ikr}}{r}$, that is

$$\underline{E}(x, y, z, t) = -\frac{e^{-i\omega t}}{2\pi} \left(\underline{E}_0 \underline{t}_0 * \frac{\partial}{\partial z} \frac{e^{ikr}}{r} \right) \, .$$

Moreover, since

$$\frac{\partial}{\partial z} \frac{e^{ikr}}{r} = \frac{e^{ik|\mathbf{x}|}}{|\mathbf{x}|} \left(ik \frac{z}{|\mathbf{x}|} - \frac{z}{|\mathbf{x}|^2} \right)$$

we obtain the Rayleigh-Sommerfeld formula

$$\underline{E}(x, y, z, t) = \frac{e^{-i\omega t}}{2\pi} \int_{\mathbb{R}} \int_{\mathbb{R}} \underline{E}_0(x', y')\underline{t}(x', y')\, e^{ik|\mathbf{x}-\mathbf{x}'|} \frac{z}{|\mathbf{x} - \mathbf{x}'|^2} \left(\frac{1}{|\mathbf{x} - \mathbf{x}'|} - ik \right) dx'dy'$$

and in the limit $z \gg \lambda$, we can neglect $1/|\mathbf{x} - \mathbf{x}'|$ with respect to k and we obtain the Huygens–Fresnel integral which justifies the Huygens–Fresnel principle.

$$\underline{E}(x, y, z, t) = -\frac{e^{-i\omega t}}{\lambda} \int_{\mathbb{R}} \int_{\mathbb{R}} \underline{E}_0(x', y')\underline{t}(x', y') \frac{e^{ikr'}}{r'} \frac{z}{r'} dx'dy' \qquad (18.8)$$

with $r' = \sqrt{(x - x')^2 + (y - y')^2 + z^2}$ (see Fig. 18.7).

Diffraction by an obstacle can be interpreted, according to the Huygens–Fresnel's principle, as follows: every unobstructed point (x', y') at $z = 0$ is a source of an spherical wave whose amplitude is proportional to the incident wave $\underline{t}(x', y')\underline{E}(x', y')$ at that point. The diffracted field at point (x, y) is thus obtained by calculating the field radiated by all these secondary, fictive sources placed at $z = 0$.

Example 18.2—Young's interference experiment
The intensity pattern produced far from two small apertures can be interpreted, according to the Huygens–Fresnel principle, as the interference between two spherical waves emitted by two point sources located at $x = \pm a/2$ in the plane $z = 0$, as shown in Fig. 18.9.

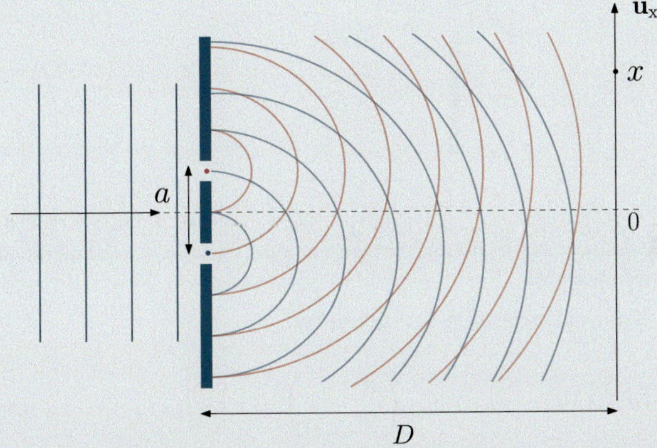

Fig. 18.9 The wave diffracted by two small apertures is equivalent to the superposition of two sherical waves emitted by each point

Suppose that a plane wave of wavelength λ is incident at $z = 0$, so that the field has the same amplitude $\underline{E}_0$ at both apertures. The total electric field $\underline{E}$ at a point x on the screen is given by Eq. (18.8) where the diffracting object corresponds to $\underline{t}$ given by two Dirac distributions of unit amplitude located at $x' = \pm a/2$. Replacing in the Huyguens–Fresnel integral, we have

Fig. 18.8 Christian Huygens (1629–1695), dutch physicist, mathematician, and astronomer with major contributions in optics and mechanics. He invented the pendulum clock

$$\underline{E}(x) = -\frac{i\underline{E}_0 e^{-i\omega t}}{\lambda} \left(\frac{e^{ikr_1}}{r_1} \frac{D}{r_1} + \frac{e^{ikr_2}}{r_2} \frac{D}{r_2} \right) .$$

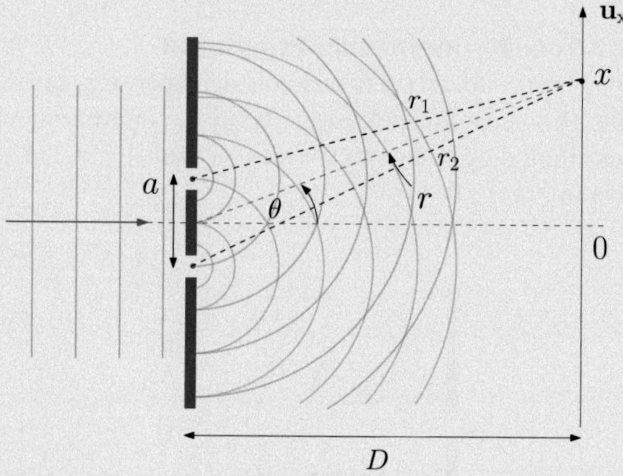

Fig. 18.10 We define r as the distance between a point x on the screen and the middle point between the two apertures

Where the distances r_1 and r_2 are given by

$$r_1^2 = (a/2)^2 + r^2 - ar\cos\left(\frac{\pi}{2} - \theta\right) = r^2 + (a/2)^2 - ar\sin\theta$$

$$r_2^2 = (a/2)^2 + r^2 - ar\cos\left(\frac{\pi}{2} + \theta\right) = r^2 + (a/2)^2 + ar\sin\theta$$

with r the distance betwen the middle point between the two apertures and the point x on the screen, and $\tan\theta = x/D$ (see Fig. 18.10). If $D \gg a, x$, then $\theta \approx x/D \ll 1$ and $r \sim D$. We can therefore expand r_1 and r_2 up to first order in ax/D as:

$$r_1 \approx r - \frac{ax}{2D}$$

$$r_2 \approx r + \frac{ax}{2D} .$$

Neglecting the variation of $1/r$ with x with respect to that of the oscillating phase e^{ikr}, we can write

$$\underline{E}(x) \approx -\frac{i\underline{E}_0}{\lambda r} e^{i(kr_1 - \omega t)} \left(1 + e^{ik(r_2 - r_1)}\right) \approx -\frac{i\underline{E}_0}{\lambda r} e^{i(kr_1 - \omega t)} \left(1 + e^{ikax/D}\right) ,$$

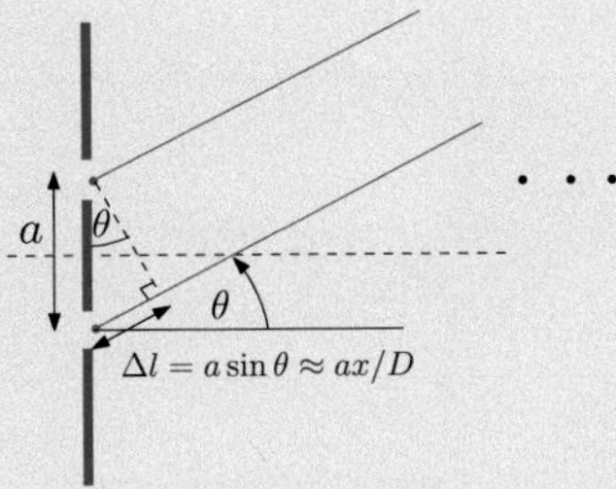

Fig. 18.11 The phase difference between the waves emitted from the two apertures is $a \sin \theta \approx ax/D$

Note that this approximation consists in neglecting the curvature of the wavefronts of the two spherical waves, and considering that the field is the superposition of two plane waves of same amplitude but having travelled difference path lengths from the holes to the point x, therefore with a phase difference of $\Delta \phi = k \Delta l = kax/D$, as shown in Fig. 18.11.

We conclude that the averaged intensity at x, which is proportional to $|\underline{E}|^2$, varies as

$$\mathcal{I} = \mathcal{I}_0 \left|1 + e^{ikax/D}\right|^2 = 4\mathcal{I}_0 \left(1 + \cos\left(\frac{2\pi ax}{\lambda D}\right)\right)$$

where $\mathcal{I}_0$ would correspond to the maximum intensity measured on the screen when only one hole is open. We recognize the formula for the interference between two coherent waves. The intensity on the screen varies periodically between $4\mathcal{I}_0$ (constructive interference) and 0 (destructive interference), with a separation between two consecutive maxima (or minima) given by

$$\delta = \frac{\lambda D}{a} .$$

In reality, the interference pattern has a finite spatial extension in both the x and the y-axis, due to the finite size of each aperture. The diffraction pattern far from two rectangular slits, illustrated in Fig. 18.12, will be calculated later in Sect. 18.6.1.

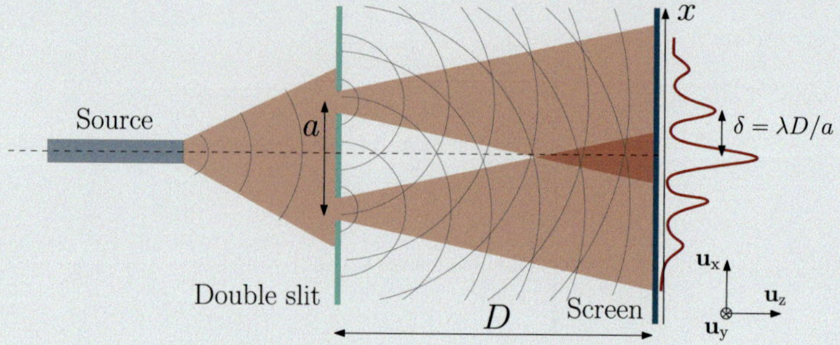

Fig. 18.12 Diffraction by a double slit

18.3.1 The Fresnel-Arago-Poisson Spot

A decade after Young published his results on the double slit experiment there was still an ongoing debate about the nature of light. Eminent scientists such as Siméon Denis Poisson and Pierre-Simon Laplace still believed in the corpuscular theory of light, as proposed by Isaac Newton and René Descartes. Augustin Fresnel submitted his wave theory of light to the Académie des Sciences in 1818, and Poisson realized that Fresnel's theory predicted a very counterintuitive result: when an opaque circular object is placed in front of a light beam, a bright spot appears at the very center of its shadow, in contradiction with the uniform shadow predicted by the corpuscular theory. This spot, usually called Arago's spot is schematically shown in Fig. 18.13. In Exercise 18.1 it is shown that the field diffracted by a circular aperture of radius a is given, in the axis, by Eq. (A.98)

$$\underline{E}_{\text{aperture}}(z, t) \approx \underline{E}_0 e^{i(kz - \omega t)} - \frac{z\underline{E}_0}{\sqrt{z^2 + a^2}} e^{ik(\sqrt{a^2 + z^2} - \omega t)} \ .$$

If $\underline{t}_{\text{aperture}}$ is the transmission factor of a circular aperture of radius a, and $\underline{t}_{\text{disk}}$ that of an opaque disk of the same radius, both centered around the z-axis, then

$$\underline{t}_{\text{aperture}} + \underline{t}_{\text{disk}} = 1 \ .$$

From Babinet's theorem (18.7), the field diffracted by an aperture and the field diffracted by a disk are complementary, and their sum corresponds simply to the incident plane wave, that is

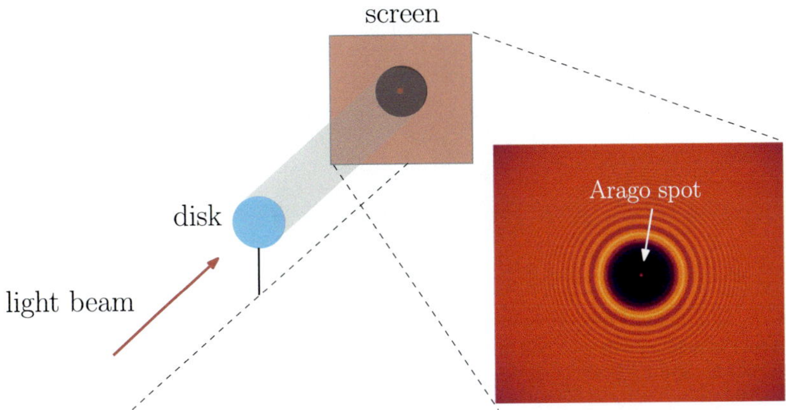

Fig. 18.13 The Arago's spot appears at the center of the shadow of an opaque disk

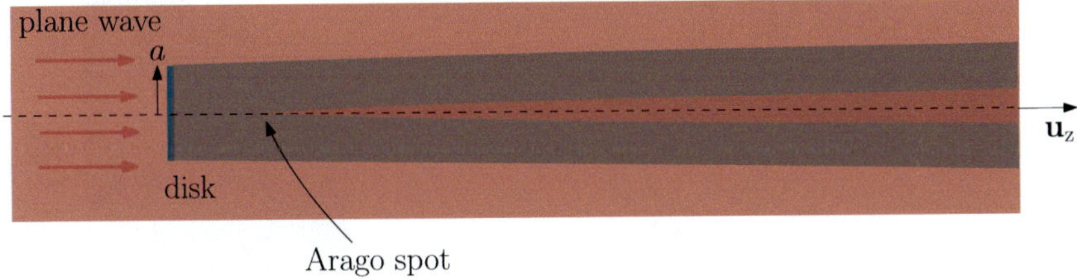

Fig. 18.14 The Arago spot is clearly visible as soon as the distance with respect to the disk is larger than a

$$\underline{E}_{\text{aperture}}(z, t) + \underline{E}_{\text{disk}}(z, t) = \underline{E}_0 e^{i(kz-\omega t)} \ .$$

We conclude that the field in the axis diffracted by an opaque disk is

$$\underline{E}_{\text{disk}}(z, t) = \frac{z\underline{E}_0}{\sqrt{z^2 + a^2}} e^{ik(\sqrt{a^2+z^2}-\omega t)} \ ,$$

so that the intensity $\mathcal{I} \propto |\underline{E}_{\text{disk}}|^2$ at the center of the axis varies according to

$$\mathcal{I}(z) = \frac{\mathcal{I}_0 z^2}{z^2 + a^2} \ .$$

The center is never dark (except very close to the disk) and can be as bright as if the obstacle were not present. This is shown in Fig. 18.14.

Since in most of everyday experiences there is no visible bright spot at the center of a shadow, Poisson believed that this was a strong enough argument to show that Fresnel's theory was wrong. Fortunately, François Arago (Fig. 18.15) decided to actually perform the experiment. The bright spot at the center of the disk's shadow

Fig. 18.15 Dominique François Arago (1786–1853), French physicist and astronomer. Appart from supporting Fresnel's theory of light, he measured the refractive index of air, explained the twinkling of starts, co-discovered Eddy currents, to cite a few contributions

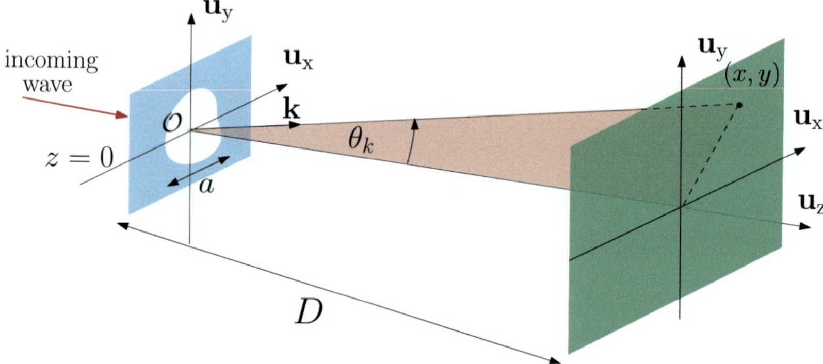

Fig. 18.16 In the paraxial approximation, we consider $\theta_k \ll 1$

was indeed observed, and it was a huge sucess of Fresnel's wave theory. A calculation of the diffraction pattern of the disk close to the axis is the subject of Exercise 18.2.

18.4 The Fresnel Paraxial Approximation

Fresnel's approximation, also called small-angle or paraxial approximation, plays an important role in simplyfing the Huygens–Fresnel diffraction integral (18.8). We will consider that the plane waves composing the electric field $\underline{E}_0$ make a small angle with respect to the z-axis. That is, we consider those wavevectors $\mathbf{k} = (k_x, k_y, k_z)$ for which $k_x^2 + k_y^2 \ll k^2 = 4\pi^2/\lambda^2$ so that the angle θ_k they make with the z-axis, as illustrated in Fig. 18.16, satisfies:

$$\sin \theta_k = \frac{\sqrt{k_x^2 + k_y^2}}{k} \ll 1 \ .$$

Note that this supposes that the distance D at which the diffraction pattern is observed is such that we can neglect the contribution of the evanescent waves in Eq. (18.2) which are known to decay within a typical distance of the order of the wavelength λ of the wave. We can write

$$\hat{h}(k_x, k_y) = e^{ik_z(k_x,k_y)z} = e^{iz\sqrt{k^2-k_x^2-k_y^2}} = e^{ikz\sqrt{1-\sin^2\theta_k}}.$$

In the small-angle approximation, up to first order in $\sin^2\theta$

$$\sqrt{1-\sin^2\theta_k} \approx 1 - \frac{1}{2}\sin^2\theta_k = 1 - \frac{k_x^2+k_y^2}{2k^2}$$

and so we obtain in the Fresnel approximation

$$\hat{h}(k_x, k_y) = e^{ik_z(k_x,k_y)z} \approx e^{ikz}e^{-iz\frac{k_x^2+k_y^2}{2k}}. \tag{18.9}$$

We can calculate its inverse Fourier transform h by recalling that the Fourier transform of a Gaussian is also a Gaussian. Indeed:

$$\mathcal{F}\{e^{-\alpha x^2}\}(k_x) = \frac{1}{\sqrt{4\pi\alpha}}e^{-\frac{k_x^2}{4\alpha}}$$

and so in two dimensions

$$\mathcal{F}\{4\pi\alpha e^{-\alpha(x^2+y^2)}\}(k_x, k_y) = e^{-\frac{(k_x^2+k_y^2)}{4\alpha}}$$

and identifying $\alpha = k/2iz$ in Eq. (18.9), we conclude that in the Fresnel approximation

$$h(x, y) = e^{ikz}\frac{2\pi k}{iz}e^{-\frac{k(x^2+y^2)}{2iz}}$$

and by replacing in Eq. (18.5) we obtain the Fresnel diffraction integral:

$$\underline{E}(x, y, z, t) = \frac{k}{2\pi iz}e^{i(kz-\omega t)}\int_{\mathbb{R}}\int_{\mathbb{R}} \underline{E}_0(x', y')\underline{t}(x', y')e^{i\frac{k}{2z}((x-x')^2+(y-y')^2)}dx'dy'.$$

$$\tag{18.10}$$

Validity of Fresnel's approximation

- The small angle approximation corresponds to the case where the electric field's spatial extensions, Δx and Δy at $z = 0$ are much larger than the wavelength λ. of light. Indeed, due to the properties of the Fourier transform we know that

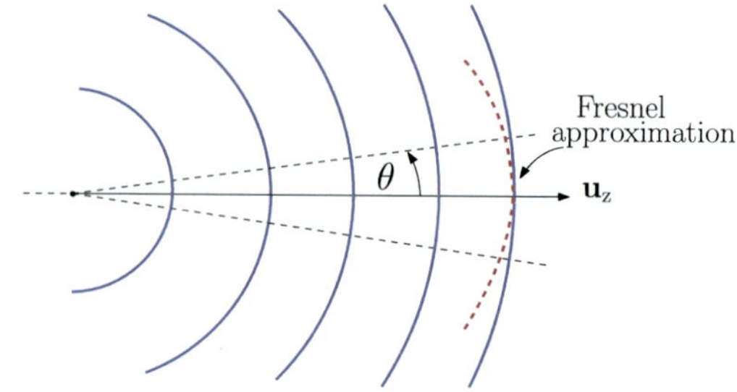

Fig. 18.17 In Fresnel paraxial approximation, the spherical wavefronts are approximated by parabolic surfaces. This is valid near the axis

$$\Delta x\,\Delta k \geq 2\pi \quad \text{and} \quad \Delta y\,\Delta k \geq 2\pi,$$

and Fresnel's approximation requires $k_{x\mathrm{max}}, k_{y\mathrm{max}} \ll k = 2\pi/\lambda$, which imposes

$$\Delta x, \Delta y \gg \lambda .$$

We conclude that the typical size $a = \sqrt{\Delta x^2 + \Delta y^2}$ of an aperture at $z = 0$ should satisfy

$$a \gg \lambda .$$

- The Fresnel approximation is equivalent to replace the spherical wavefronts in the Huygens–Fresnel integral Eq. (18.8) by parabolic surfaces centered around the origin, as shown in Fig. 18.17. Indeed

$$\frac{e^{ikr}}{r}\frac{z}{r} = \frac{e^{ikz\sqrt{1+\frac{x^2+y^2}{z^2}}}}{z\sqrt{1+\frac{x^2+y^2}{z^2}}}\frac{z}{z\sqrt{1+\frac{x^2+y^2}{z^2}}} = \frac{e^{ikz\sqrt{1+\frac{x^2+y^2}{z^2}}}}{z(1+\frac{x^2+y^2}{z^2})} \approx \frac{e^{ikz(1+\frac{x^2+y^2}{2z^2})}}{z} .$$

18.4.1 Paraxial Optics

The Fresnel approximation is a good way to deal with waves such that the electric field has a principal direction of propagation along z and it is not very narrow in the transverse direction (otherwise the minimum divergence angle λ/a becomes too large). Another way to treat diffraction in this limit is by means of the paraxial wave equation. In order to derive this equation, first consider a monochromatic wave propagating along the z-axis in an homogeneous medium of refractive index n. In the scalar approximation, we may write

$$\underline{E}(x, y, z, t) = \mathcal{E}(x, y, z)e^{i(kz-\omega t)}$$

with $k = n\omega/c$. Injecting into d'Alembert's wave equation yields

$$\nabla^2 \underline{E} + \frac{n^2}{c^2}\frac{\partial^2 \underline{E}}{\partial t^2} = 0 \iff \nabla^2 \mathcal{E} + 2ik\frac{\partial \mathcal{E}}{\partial z} \underbrace{-k^2\mathcal{E} + \frac{n^2\omega^2}{c^2}\mathcal{E}}_{=0} = 0$$

and so

$$\nabla^2 \mathcal{E}(x, y, z) + 2ik\frac{\partial \mathcal{E}(x, y, z)}{\partial z} = 0 .$$

If the field describes a light beam propagating essentially in the z-axis, we assume a slowly varying amplitude along this direction, that is

$$\left|\frac{\partial^2 \mathcal{E}}{\partial z^2}\right|^2 \ll \left|k\frac{\partial \mathcal{E}}{\partial z}\right|$$

which allows us to write the wave equation in the paraxial approximation

$$\boxed{\frac{\partial^2 \mathcal{E}}{\partial x^2} + \frac{\partial^2 \mathcal{E}}{\partial y^2} + 2ik\frac{\partial \mathcal{E}(x, y, z)}{\partial z} = 0.}\tag{18.11}$$

Let us now show that the Fresnel integral (18.10) is a solution of the wave equation (18.11). Indeed, let us consider the general solution $\mathcal{E}(x, y, z)e^{i(kz-\omega t)}$ for a wave freely propagating along the z-axis given by Fresnel diffraction

$$\mathcal{E}(x, y, z) = \frac{k}{2\pi iz}\int_{\mathbb{R}}\int_{\mathbb{R}} \underline{E}_0(x', y')e^{i\frac{k}{2z}((x-x')^2+(y-y')^2)} .$$

We have $\frac{\partial \mathcal{E}}{\partial x} = i\frac{\mathcal{E}k}{z}(x - x')$ and so

$$\frac{\partial^2 \mathcal{E}}{\partial x^2} = i\frac{\mathcal{E}k}{z} + i\frac{k}{z}(x - x')\frac{\partial \mathcal{E}}{\partial x} = i\frac{\mathcal{E}k}{z} - \frac{k^2}{z^2}(x - x')^2\mathcal{E}$$

and similarly

$$\frac{\partial^2 \mathcal{E}}{\partial y^2} = i\frac{\mathcal{E}k}{z} - \frac{k^2}{z^2}(y - y')^2\mathcal{E} .$$

On the other hand, $\frac{\partial \mathcal{E}}{\partial z} = -\frac{1}{z}\mathcal{E} - \frac{ik}{2z^2}((x - x')^2 + (y - y')^2)\mathcal{E}$ and we conclude that

$$\frac{\partial^2 \mathcal{E}}{\partial x^2} + \frac{\partial^2 \mathcal{E}}{\partial y^2} = \left(\frac{2ik}{z} - \frac{k^2}{z^2}((x - x')^2 + (y - y')^2)\right)\mathcal{E} = -2ik\frac{\partial \mathcal{E}}{\partial z}$$

which proves that a field propagating according to the Fresnel diffraction integral is a solution of the paraxial wave equation (18.11).

Example 18.3—Gaussian beams

Laser cavities with cylindrical symmetry often exhibit a field distribution well described by a Gaussian mode, which is a particular solution of (18.11) as demonstrated in Exercise 18.3. In such a mode the electric field writes $\underline{E}(x, y, z, t) = \mathcal{E}(x, y, z)e^{ik(z-\omega t)}$ with

$$\mathcal{E}(x, y, z) \propto \left(\frac{1}{R(z)} + i\frac{\lambda}{\pi w^2(z)}\right) e^{ik\frac{x^2+y^2}{2R(z)}} e^{-\frac{x^2+y^2}{w(z)^2}} \tag{18.12}$$

where

$$R(z) = (z - z_0) + \frac{z_R^2}{(z - z_0)} \tag{18.13}$$

and

$$w(z) = w_0\sqrt{1 + \left(\frac{z - z_0}{z_R}\right)^2}. \tag{18.14}$$

Here $z_R = \pi w_0^2/\lambda$ is called the Raleygh length. The intensity $\mathcal{I}(x, y, z) \propto |\mathcal{E}(x, y, z)|^2$ is a Gaussian:

$$\mathcal{I}(x, y, z) = \frac{2\mathcal{I}_0}{\pi w^2(z)} e^{-2\frac{x^2+y^2}{w^2(z)}}. \tag{18.15}$$

We see from (18.12) that $R(z)$ represents the radius of curvature of the wavefront at position z, whereas $w(z)$ defines the characteristic distance over which the field's intensity decays as a function of the distance $x^2 + y^2$ perpendicular to the propagation axis.

- From Eq. (18.14) we see that in the plane $z = z_0$, also called the beam waist, the beam radius is minimum and equal to w_0. Close to the waist $|z - z_0| \ll z_R$, the beam radius increases quadratically (although very weakly) as a function of $z - z_0$:

$$w(z) \approx w_0 \left(1 + \frac{(z - z_0)^2}{2z_R^2}\right)$$

The beam is therefore considered to be approximately collimated in the region $|z - z_0| \leq z_R$. On the other hand, in the far field limit $|z - z_0| \gg z_R$, the beam radius increases linearly along the axis

$$w(z) \approx w_0 \frac{|z - z_0|}{z_R}$$

this linear increase of the beam size with distance allows to define a divergence angle θ given by

$$\theta \approx \frac{w(z)}{|z - z_0|} = \frac{w_0}{z_R} = \frac{\lambda}{\pi w_0} \, .$$

We conclude that a Gaussian mode broadens with the minimum divergence angle $\theta \sim \lambda/w_0$ imposed by diffraction.

- In what concerns the radius of curvature, from Eq. (18.13) we see that it is infinite at the waist where the wavefront is flat, passes through a minimum at $z \pm z_R$ and rises again towards infinity as z is further increased. In the far field limit, it increases as

$$R(z) \approx z - z_0$$

so that the wavefronts are approximately those of a point source located at the waist.

These results are summarized in Fig. 18.19 which shows the evolution of a Gaussian beam as a function of the propagation distance z with $z_0 = 0$ (Fig. 18.18).

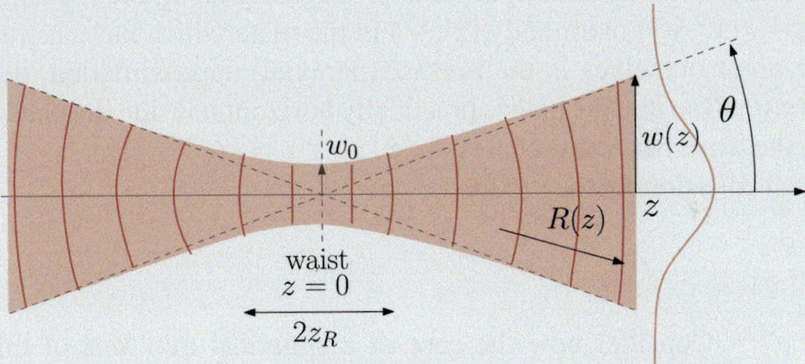

Fig. 18.18 The Gaussian mode is a solution of the paraxial wave equation

18.5 Diffraction by a Thin Lens

A thin lens, or more generally, a thin transparent plate of refractive index n is an example of an object having a complex transmission factor $\underline{t}$. Indeed, the phase acquired by the field after passing through a thin plate at point (x', y') will depend on its thickness $e(x', y')$. Suppose that the thin optical element is contained between the planes $z = 0$ and $z = e_0$, as shown in Fig. 18.19.

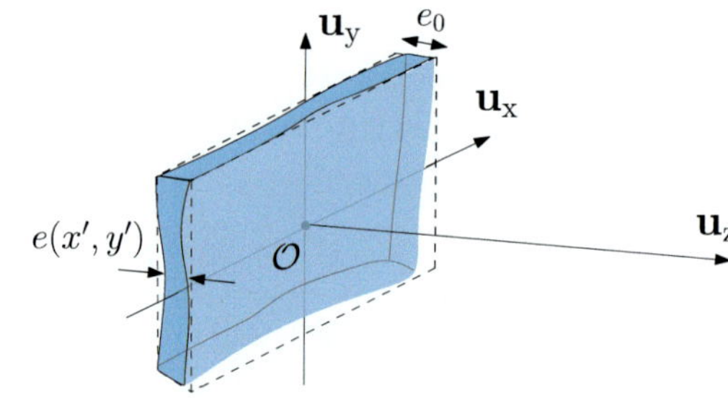

Fig. 18.19 A thin lens of arbitrary shape is considered to be contained between $z = 0$ and $z = e_0$

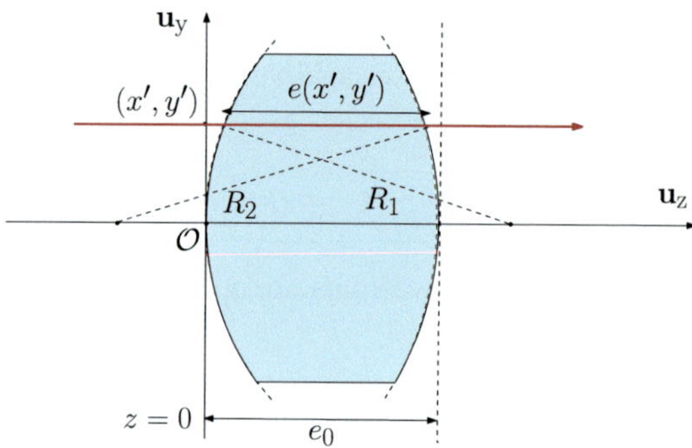

Fig. 18.20 A thin spherical lens

At point (x', y'), an incident light wave propagating along the z-axis will travel a distance $(e_0 - e(x', y'))$ in air and $e(x', y')$ in the plate which has a refractive index n. Here, we place ourselves in the Fresnel (paraxial) approximation, the angles of incidence are small so that the path is practically horizontal inside the plate. At the exit of the plate, the wave has acquired a phase $\phi(x', y') = k(e_0 - e(x', y') + ne(x', y'))$ so that the transmission factor $\underline{t}$ reads

$$\underline{t}(x', y') = e^{i\phi(x',y')} = \underline{t}_0 e^{ik(n-1)e(x',y')}$$

where $\underline{t}_0 = e^{ike_0}$. Consider now the case of a spherical thin lens of thickness e_0, shown in Fig. 18.20. At point (x', y'), the thickness of the lens is given by

$$e(x', y') = e_0 - \left(R_1 - \sqrt{R_1^2 - (x'^2 + y'^2)}\right) - \left(-R_2 - \sqrt{R_2^2 - (x'^2 + y'^2)}\right).$$

Here, we keep the sign convention for the radii of curvature as defined in Chap. 17. In the paraxial approximation, $x^2 + y^2 \ll |R_1|, |R_2|$ so that

$$e(x', y') \approx e_0 - \frac{x'^2 + y'^2}{2R_1} + \frac{x'^2 + y'^2}{2R_2}.$$

The transmission factor of a thin spherical lens is therefore

$$\underline{t}(x', y') = \underline{t}_0 e^{-ik(n-1)\left(\frac{1}{R_2} - \frac{1}{R_2}\right)\frac{x'^2 + y'^2}{2}} .$$

Recalling that the image focal length f' of a thin spherical lens is given by $1/f' = (n - 1)(1/R_1 - 1/R_2)$, we finally obtain the transmission factor of a thin lens in the Fresnel approximation

$$\boxed{\underline{t}(x', y') = \underline{t}_0 e^{-ik\frac{x'^2 + y'^2}{2f'}} = \underline{t}_0 e^{-i\pi\frac{x'^2 + y'^2}{\lambda f'}} .}$$

(18.16)

18.5.1 Focusing of Gaussian Beams

A Gaussian beam propagates according to

$$\underline{E}(x, y, z, t) = \frac{E_0 w_0}{w(z)} e^{i(kz - \tan^{-1}(\frac{z - z_0}{z_R}) - \omega t)} e^{ik\frac{x^2 + y^2}{2R(z)}} e^{-\frac{(x^2 + y^2)}{w(z)^2}}$$

(18.17)

The radius of curvature R and the beam radius w vary along the propagation direction according to (18.13) and (18.14)

$$R(z) = (z - z_0) + \frac{z_R^2}{z - z_0}$$

$$w(z) = w_0 \sqrt{1 + \left(\frac{z - z_0}{z_R}\right)^2}$$

where $z_R = \pi w_0^2/\lambda$ is the Raleygh length.

We will show in this section that Gaussian beams preserve their Gaussian character after passing through a perfect thin lens. Let $z = z_0$ and $z = z_0'$ be the position of the beam waist before and after a converging lens of focal length f', as shown in Fig. 18.21. In the following, we will find an expression for the Raleygh length z_R' and the beam radius w' after the lens as a function of $d > 0$, f', and w_0.

At $z = z_0 + d$, inmediately after the lens, the electric field is of the form $\underline{E}(x, y, z_0 + d, t)\underline{t}_0 e^{-i\pi\frac{x^2 + y^2}{\lambda f'}}$ where $\underline{t}_0$ is a constant and f' is the image focal distance of the lens. We conclude that only the radius of curvature of the beam is modified when passing through the lens, according to

$$\frac{1}{R(z_0 + d)} - \frac{1}{f'} = \frac{1}{R'(z_0 + d)}$$

and replacing R and R' by Eq. (18.13)

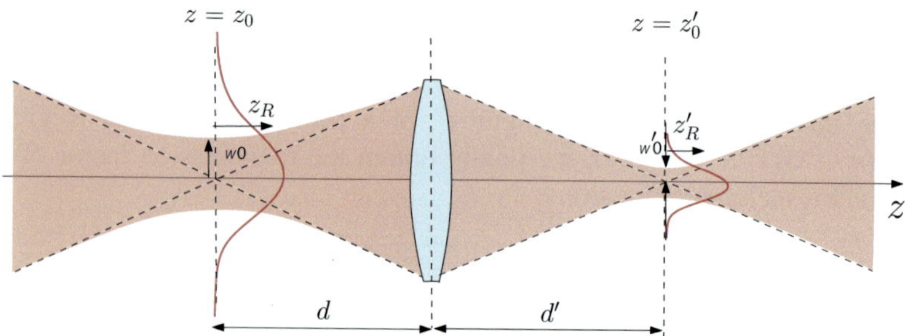

Fig. 18.21 A Gaussian beam remains Gaussian after passing through a thin lens

$$\frac{1}{d + \frac{z_R^2}{d}} - \frac{1}{f'} = -\frac{1}{d' + \frac{z_R'^2}{d'}}. \tag{18.18}$$

Additionally, since the beam radius is the same just before and just after the lens, we can also write $w(z_0 + d) = w'(z_0 + d)$, that is to say

$$w_0 \sqrt{1 + \left(\frac{d}{z_R}\right)^2} = w_0' \sqrt{1 + \left(\frac{d'}{z_R'}\right)^2} \tag{18.19}$$

and injecting $z_R'/z_R = w_0'^2/w_0^2$ into Eq. (18.19), we find the following relation

$$\frac{z_R^2 + d^2}{z_R} = \frac{z_R'^2 + d'^2}{z_R'},$$

from which we can write d' as

$$d' = \sqrt{\frac{z_R'}{z_R}(z_R^2 + d^2) - z_R'^2}.$$

Replacing d' in Eq. (18.18) we obtain the Rayleigh length z_R' in the image region, and therefore the transverse magnification M which reads:

$$\boxed{M = \frac{w_0'}{w_0} = \sqrt{\frac{z_R'}{z_R}} = \frac{f'}{\sqrt{(f' - d)^2 + z_R^2}}.} \tag{18.20}$$

Finally, the distance between the the lens and the waist planes satisfy a relationship similar (but not identical) to the Descartes relation of geometrical optics

$$\boxed{\frac{1}{d'} + \frac{1}{d + \frac{z_R^2}{(d-f')}} = \frac{1}{f'}.}$$

(18.21)

We conclude from (18.20) that to focus a Gaussian beam into a small spot, we need a large incident beam width (large Rayleigh length) and a short focal length f'.

Remark The Descartes relation is retrieved in the limit $z_R^2 \ll d|d - f'|$, in which case

$$\frac{1}{d'} + \frac{1}{d} = \frac{1}{f'}$$

and the magnification is that obtained in geometrical optics in the limit $z_R \ll |d - f'|$

$$M = \frac{f'}{f' - d}.$$

We now focus on two particular cases:

1. The input waist coincides with the lens, $d = 0$ (Fig. 18.22). In this case, we have

$$w_0' = \frac{w_0 f'}{\sqrt{f'^2 + z_R^2}} = \frac{\lambda f'/\pi w_0}{\sqrt{1 + \left(\frac{\lambda f'}{\pi w_0^2}\right)^2}}$$

and

$$d' = \frac{f' z_R^2}{z_R^2 + f'^2} = \frac{f'}{1 + \left(\frac{\lambda f'}{\pi w_0^2}\right)^2}.$$

   If, in addition, $f' \ll z_R$, then $d' \approx f'$. This corresponds to the case where the lens is well within the Rayleigh length of the input beam, the incoming wavefronts are therefore approximately planar. The image plane corresponds to the image focal plane, as one expects for a collimated input beam. Note that an almost collimated beam at the entrance of the lens is also obtained in the case of a very distant input waist with $d \gg f', z_R$, since in this case one obtains $d' \approx f'$ as well. In practice, however, the angular divergence would make the beam much larger than the aperture of the lens.

2. The input waist is at the object focal plane of the lens, so that $d = f'$ (Fig. 18.23). In this case, we have

$$w_0' = \frac{w_0 f'}{z_R} = \frac{\lambda f'}{\pi w_0}$$

   and the output waist is at the image focal plane

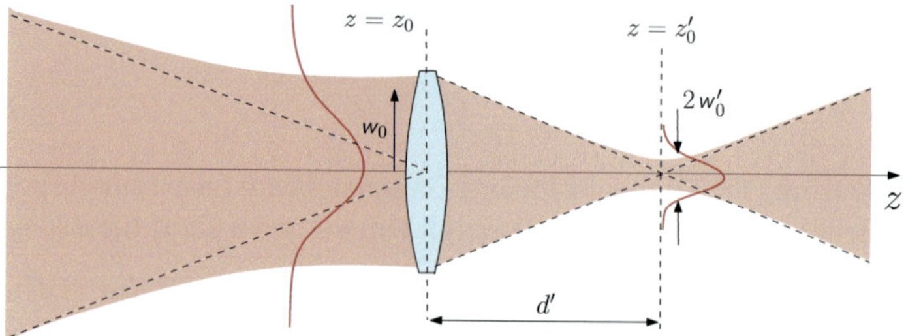

Fig. 18.22 A Gaussian beam whose waist coincides with the plane of the lens is focused at a distance $d' = f'/(1 + (\lambda f'/\pi w_0^2)^2)$ after the lens

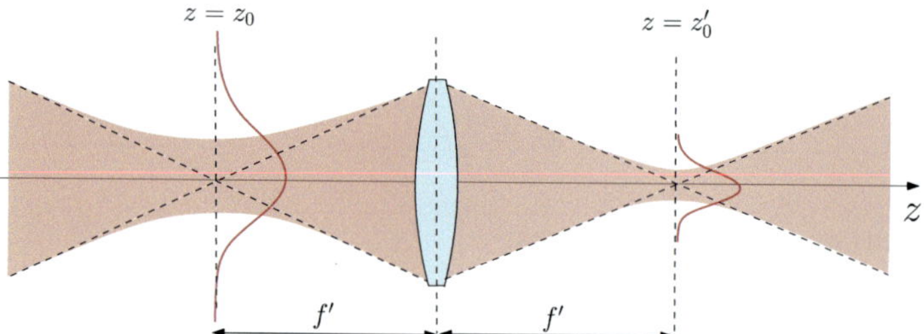

Fig. 18.23 A Gaussian beam whose waist is at $d = f'$ from the lens is focused at the image focal plane

$$d' = f' \ .$$

18.5.2 Lenses and Fourier Optics

In this section, we will give a flavor of Fourier optics, named after Joseph Fourier (Fig. 18.24). We will show that diffraction by a thin lens performs a Fourier transform of the incoming field distribution. Suppose that a field $\underline{E}_0(x', y')$ is incident on a thin lens placed at $z = 0$. Inmediately after the lens, the field is given by $\underline{E}_0(x', y')e^{-i\omega t}\underline{t}_0 e^{-i\pi \frac{x'^2+y'^2}{\lambda f'}}$ where f' is the image focal length of the lens. The diffraction pattern at a distance z along the axis of the lens is given, in the Fresnel approximation, by

$$\underline{E}(x, y, z, t) = -\frac{i\underline{t}_0}{\lambda z}e^{i(\frac{2\pi}{\lambda}z - \omega t)} \int_{\mathbb{R}} \int_{\mathbb{R}} \underline{E}_0(x', y')e^{-i\frac{\pi(x'^2+y'^2)}{\lambda f'}} e^{i\frac{\pi}{\lambda z}((x-x')^2+(y-y')^2)}dx'dy' \ .$$

Fig. 18.24 Joseph Fourier (1768–1830), a French mathematician and physicist. He contributed with foundational tools for mathematics, physics, signal processing and engineering. Fourier made significant contributions to the study of heat conduction, formulating the heat equation

In particular, at the image focal plane of the lens ($z = f'$) we find

$$\underline{E}(x, y, f', t) = -\frac{it_0}{\lambda f'}e^{i(2\frac{\pi}{\lambda}f'-\omega t)} \int_{\mathbb{R}} \int_{\mathbb{R}} \underline{E}_0(x', y')e^{-i\frac{\pi(x'^2+y'^2)}{\lambda f'}} e^{i\frac{\pi}{\lambda f'}((x-x')^2+(y-y')^2)} dx'dy'$$

$$= -\frac{it_0}{\lambda f'}e^{i(\frac{2\pi}{\lambda}f'-\omega t)}e^{i\frac{\pi}{\lambda f'}(x^2+y^2)} \int_{\mathbb{R}} \int_{\mathbb{R}} \underline{E}_0(x', y')e^{-i\frac{2\pi}{\lambda f'}(xx'+yy')} dx'dy'$$

$$= -\frac{4\pi^2 it_0}{\lambda f'}e^{i(\frac{2\pi}{\lambda}f'-\omega t)}e^{i\frac{\pi}{\lambda f'}(x^2+y^2)} \mathcal{F}\{\underline{E}_0\}\left(\frac{2\pi x}{\lambda f'}, \frac{2\pi y}{\lambda f'}\right)$$

We obtain the remarkable result

$$\boxed{\underline{E}(x, y, f', t) = -\frac{4\pi^2 t_0 i}{\lambda f'}e^{i(\frac{2\pi}{\lambda}f'+\frac{\pi}{\lambda f'}(x^2+y^2)-\omega t)}\mathcal{F}\{\underline{E}_0\}\left(\frac{2\pi x}{\lambda f'}, \frac{2\pi y}{\lambda f'}\right)} \qquad (18.22)$$

and so the field distribution at the focal plane is, within a position-dependent phase and a constant multiplicative factor, the Fourier transform of the field just before the lens. It follows that the intensity at a point (x, y) in the image focal plane of the lens is proportional to the squared absolute value of the Fourier transform of the wave at the input plane $z = 0$ of the lens, evaluated at the spatial frequencies $(x/\lambda f', y/\lambda f')$. This is why the image focal plane is also called the Fourier plane.

$$\boxed{\mathcal{I}(x, y, z = f') \propto \left|\mathcal{F}\{\underline{E}_0\}\left(\frac{2\pi x}{\lambda f'}, \frac{2\pi y}{\lambda f'}\right)\right|^2.} \qquad (18.23)$$

Fig. 18.25 Fourier optics correctly describes the magnification of -1 of the optical system formed by two identical lenses separated by $2f'$

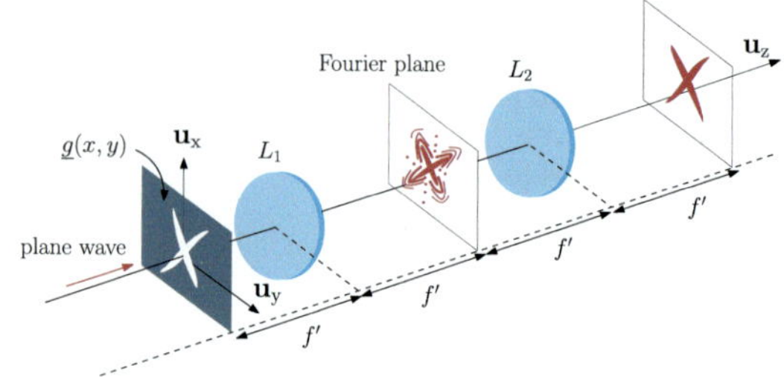

18.5.3　Spatial Filtering

Even though the intensity of the field (18.22) is given by the squared modulus of the Fourier transform of the field at the input plane of the lens ($z = 0$), the amplitude of the field at $z = f'$ is not simply given by the Fourier transform of the field at $z = 0$ due to the phase term in Eq. (18.22) which depends on (x, y).

If we now consider instead $\underline{E}_{-f'}$, the field distribution in the object focal plane ($z = -f'$), we have according to (18.4):

$$\mathcal{F}\{\underline{E}_0\}\left(\frac{2\pi x}{\lambda f'}, \frac{2\pi y}{\lambda f'}\right) = \mathcal{F}\{\underline{E}_{-f'}\}\left(\frac{2\pi x}{\lambda f'}, \frac{2\pi y}{\lambda f'}\right)\hat{h}\left(\frac{2\pi x}{\lambda f'}, \frac{2\pi y}{\lambda f'}\right)$$

$$= \mathcal{F}\{\underline{E}_{-f'}\}\left(\frac{2\pi x}{\lambda f'}, \frac{2\pi y}{\lambda f'}\right) e^{i\left(\frac{2\pi}{\lambda} f'\right)} e^{-i\frac{\pi}{\lambda f'}(x^2+y^2)}$$

where we have used the Fresnel approximation (18.9) of $\hat{h}$. Finally, replacing in (18.22) we obtain

$$\boxed{\underline{E}(x, y, f', t) = -\frac{4\pi^2 t_0 i}{\lambda f'} e^{i\left(\frac{4\pi}{\lambda} f' - \omega t\right)} \mathcal{F}\{\underline{E}_{-f'}\}\left(\frac{2\pi x}{\lambda f'}, \frac{2\pi y}{\lambda f'}\right).} \qquad (18.24)$$

We conclude that the electric field at the image focal plane of the lens is, within a multiplicative factor only, the Fourier transform of $\underline{E}_{-f'}$, the electric field at the object focal point of the lens. This is a key result of Fourier optics, since it allows to access and manipulate the Fourier transform of a field distibution by simply using lenses.

One important application is the spatial filtering of an object by the association of two lenses. Consider a plane wave that shines on an object characterized by a transmission factor $g(x, y)$, placed at the object focal plane of the first lens L_1. In the Fourier plane, located at the image focal plane of L_1, one obtains a field distribution given by the Fourier transform of $\underline{t}$. This is illustrated in Fig. 18.25.

At this position, each point in the plane perpendicular to the axis corresponds to a given spatial frequency in the Fourier decomposition of $\underline{g}$. If a second lens L_2

Fig. 18.26 An aperture in the Fourier plane is equivalent to a low-pass filter, since only the low frequencies of the spatial distribution of the field are transmitted by the aperture

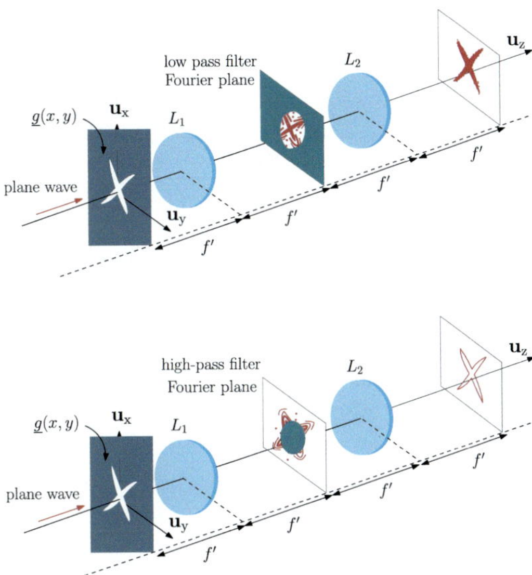

Fig. 18.27 An opaque disk in the Fourier plane is equivalent to a high-pass filter, since only the high frequencies of the spatial distribution of the field will pass through

(supposed here to be identical to L_1) is positioned so that the Fourier plane of L_1 coincides with the object focal plane of L_2, one obtains at point (x, y) in the image focal plane of L_2 a complex amplitude proportional to that of $\underline{g}(-x, -y)$, and the object's intensity distribution is thus reconstructed, except that the image is inversed in x and y. This system applies twice the Fourier transform, which according to Fourier inversion theorem (A.57), should give the original field but evaluated in $(-x, -y)$. This is consistent with the known result of geometrical optics which says that this system has a transverse magnification of -1.

By now placing a mask in the Fourier plane one can change the frequency content of the field. For example, a low-pass filter can be obtained by placing a circular aperture in the Fourier plane as shown in Fig. 18.26. If the radius of the aperture is a, this filter eliminates spatial frequencies higher than $a/\lambda f'$. The resulting image is smoother due to the suppression of rapid spatial changes in the intensity.

A high-pass filter is achieved by placing instead a mask which is transparent except for an opaque central disk, as shown in Fig. 18.27. When placed in the Fourier plane, it blocks the low spatial frequencies of the original object and so one obtains an enhancement of the edges of the image, where the rapid variations in intensity occur. The homogeneous illumination and slow noise are supressed from the image.

Figure 18.28 shows an example of a simple optical system capable of highlighting the edges of a human hair. The specimen is placed onto a transparent glass slide and illuminated uniformly with a plane wave. A lens performs the Fourier transform of the transmission factor of the object, and by placing a small needle in the center of the Fourier plane, only the high spatial frequencies are observed on the image plane.

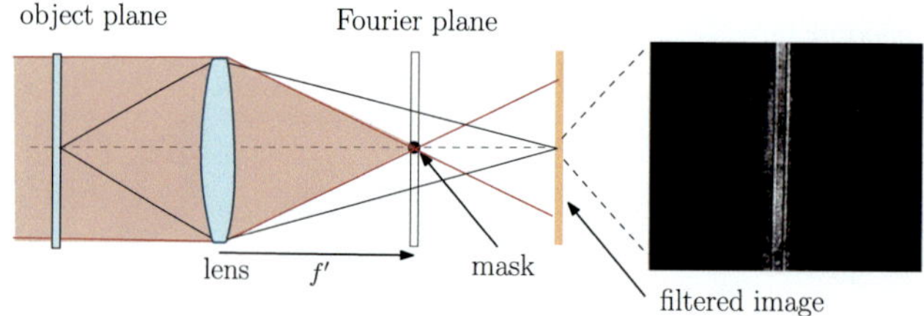

Fig. 18.28 High-pass filtering of an almost transparent object. The image on the right shows the observed image of a human hair after,

Fig. 18.29 The image of a point source by a thin lens will have a finite size due to diffraction

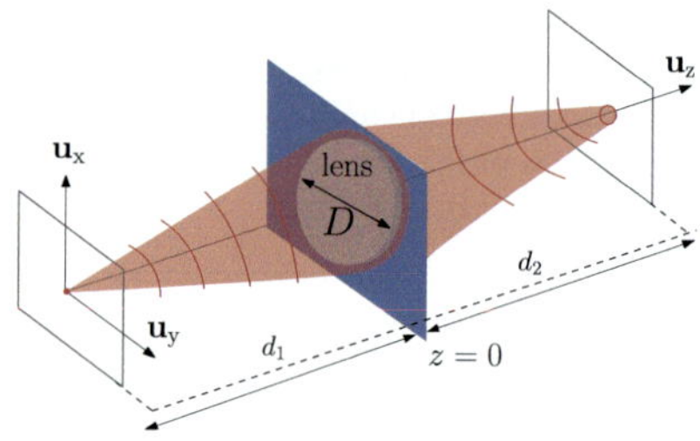

18.5.4 *Image Formation and the Diffraction Limit*

Let us consider the case of an incident plane wave of amplitude $\underline{E}_0$. The intensity at point (x, y) in the object focal plane of a thin lens of infinite extension would be proportional to the Fourier transform of a constant, and so a Dirac at the origin $\mathcal{I}(x, y) = \mathcal{I}_0 \delta(x, y)$, meaning that all the intensity is concentrated at a single point on the optical axis. We retrieve the perfect stigmatism between a point object at infinity and its image by a thin lens in the paraxial approximation. Of course, in reality, neither the lens nor the field can have infinite extension. This will limit the spatial resolution of every optical system. Let us calculate, according to Fresnel's diffraction, what is the image of a point source located at a distance d_1 from a convergent thin lens of aperture D, as shown in Fig. 18.29.

The field incident on the aperture plane (just before the lens) at a point (x', y') corresponds to the paraxial approximation of a spherical wave emitted by a point at a distance d_1, that is:

$$\underline{E}_i(x', y', t) \approx \frac{\underline{E}_0 e^{i(kd_1 - \omega t)}}{d_1} e^{ik\frac{x'^2 + y'^2}{2d_1}}$$

and inmediately after the lens, the field can be written as

$$\underline{E}_0(x', y') = \underline{A} e^{i\pi \frac{x'^2 + y'^2}{\lambda d_1}} \underline{t}(x', y') e^{-i\pi \frac{x'^2 + y'^2}{\lambda f'}}$$

where $\underline{A}$ is a constant and $\underline{t}$ is the transmission factor of the lens pupil. The image is formed, according to Descartes's relation, at a distance d_2 from the lens such that $1/d_1 + 1/d_2 = 1/f'$. We can then write

$$\underline{E}_0(x', y') = \underline{A}\, \underline{t}(x', y')e^{-i\pi\frac{x'^2+y'^2}{\lambda d_2}}$$

The field $\underline{E}_0(x, y, d_2)$ at a point (x, y) in the image plane $(z = d_2)$ is thus given in the Fresnel approximation by:

$$\underline{E}_0(x, y, d_2) = \underline{A}' \int_{\mathbb{R}}\int_{\mathbb{R}} \underline{t}(x', y')\underbrace{e^{-i\pi\frac{x'^2+y'^2}{\lambda d_2}}\, e^{i\frac{\pi}{\lambda d_2}((x-x')^2+(y-y')^2)}}_{e^{\frac{i\pi}{\lambda d_2}(x^2+y^2-2xx'-2yy')}}dx'dy'$$

$$= \underline{A}'e^{i\frac{\pi}{\lambda d_2}(x^2+y^2)}\int_{\mathbb{R}}\int_{\mathbb{R}} \underline{t}(x', y')e^{-i\frac{2\pi x}{\lambda d_2}x'}\, e^{-i\frac{2\pi y}{\lambda d_2}y'}dx'dy'$$

$$= 4\pi^2\underline{A}'e^{i\frac{\pi}{\lambda d_2}(x^2+y^2)}\mathcal{F}\{\underline{t}\}\left(\frac{2\pi x}{\lambda d_2}, \frac{2\pi y}{\lambda d_2}\right).$$

The intensity on the image plane is therefore proportional to the squared modulus of the Fourier transform of the pupil function

$$\underline{I}(x, y, d_2) \propto |\underline{E}_0(x, y, d_2)|^2 \propto \left|\mathcal{F}\{\underline{t}\}\left(\frac{2\pi x}{\lambda d_2}, \frac{2\pi y}{\lambda d_2}\right)\right|^2.$$

For a circular aperture of radius $D/2$, $\underline{t}(x', y') = 1$ if $x'^2 + y'^2 \leq \frac{D^2}{4}$ and zero elsewhere. Then

$$\mathcal{F}\{\underline{t}\}(u, v) = \frac{1}{4\pi^2}\int_{x'^2+y'^2\leq\frac{D}{2}} e^{-i(ux'+vy')}dx'dy'.$$

In cylindrical coordinates, $x' = r'\cos\theta'$, $y' = r'\sin\theta'$ and $dx'dy' = r'dr'd\theta'$:

$$\mathcal{F}\{\underline{t}\}(u, v) = \frac{1}{4\pi^2}\int_0^{2\pi}\int_0^{\frac{D}{2}} e^{-ir'(u\cos\theta'+v\sin\theta')}r'dr'd\theta'.$$

Similarly, by using cylindrical coordinates in the frequency space, $u = r\cos\theta$, $v = r\sin\theta$ (see Fig. 18.30)

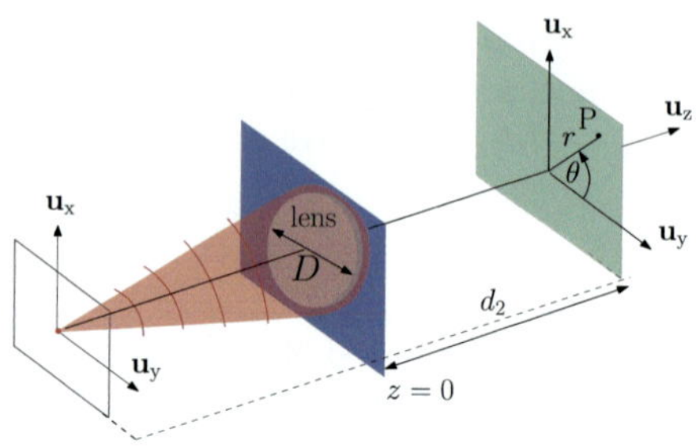

Fig. 18.30 Cylindrical coordinates used for the image plane

$$\mathcal{F}\{\underline{t}\}(r,\theta) = \frac{1}{4\pi^2} \int_0^{2\pi} \int_0^{\frac{D}{2}} e^{-ir'r(\cos\theta'\cos\theta + \sin\theta'\sin\theta)} r'\,dr'\,d\theta'$$

$$= \frac{1}{4\pi^2} \int_0^{2\pi} \int_0^{\frac{D}{2}} e^{-ir'r\cos(\theta'-\theta)} r'\,dr'\,d\theta' \ .$$

Here we need to introduce the Bessel function of the first kind J_1, defined by the identity

$$\int_0^x x' J_0(x')\,dx' = x J_1(x)$$

where $J_0(x) = \frac{1}{\pi} \int_0^\pi e^{ix\cos\phi}\,d\phi$. Since $J_0(x) = J_0(-x)$ we obtain

$$\mathcal{F}\{\underline{t}\}(r,\theta) = \frac{1}{2\pi} \int_0^{D/2} J_0(rr')r'\,dr' = \frac{1}{2\pi r^2} \int_0^{Dr/2} J_0(x)x\,dx = \frac{D}{4\pi r} J_1\left(\frac{Dr}{2}\right) \ .$$

Recall that the intensity in the image plane writes $\mathcal{I}(x,y) \propto |\mathcal{F}\{\underline{t}\}(\frac{2\pi x}{\lambda d_2}, \frac{2\pi y}{\lambda d_2})|^2$ so that

$$\mathcal{I}(r,\theta) = \mathcal{I}_0 \frac{\lambda d_2}{\pi D r^2} J_1^2\left(\frac{\pi D r}{\lambda d_2}\right) = \mathcal{I}_0 \frac{J_1^2\left(\frac{\pi D r}{\lambda d_2}\right)}{\frac{\pi D r^2}{\lambda d_2}} \ .$$

The intensity in the image plane consists of a bright central spot surrounded by a series of concentric rings of decreasing amplitude as the radius of the ring increases. The center of the pattern is almost 60 times brighter than the first ring. This figure is called an Airy pattern and it is illustrated in Fig. 18.31.

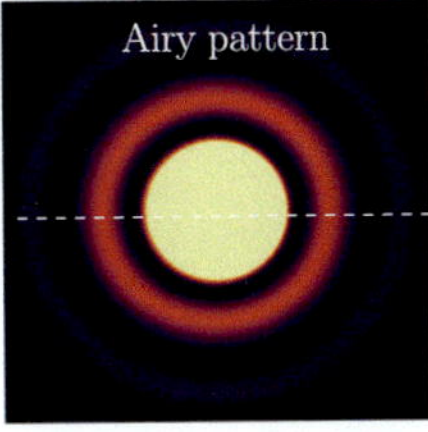

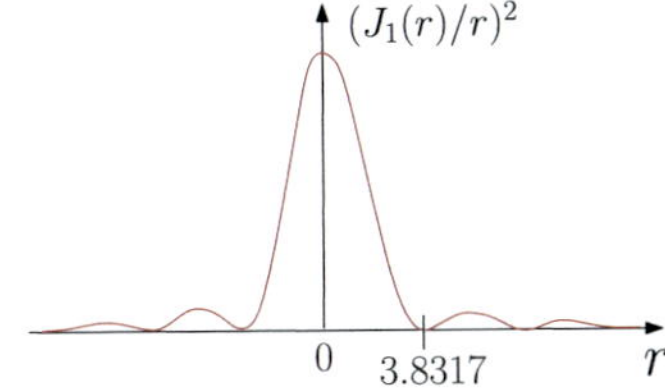

Fig. 18.31 The intensity of the image of a point source by a thin lens is an Airy pattern, as shown on the left. Note that the colormap is saturated at the central spot to make the rings visible. The rings reflect the behaviour of the function $(J_1(r)/r)^2$, shown on the right

Fig. 18.32 Two identical Airy disks separated by $r^*/2$ (left), r^* (middle) and $2r^*$ (right), where r^* is the radius of the central disk of one Airy disk

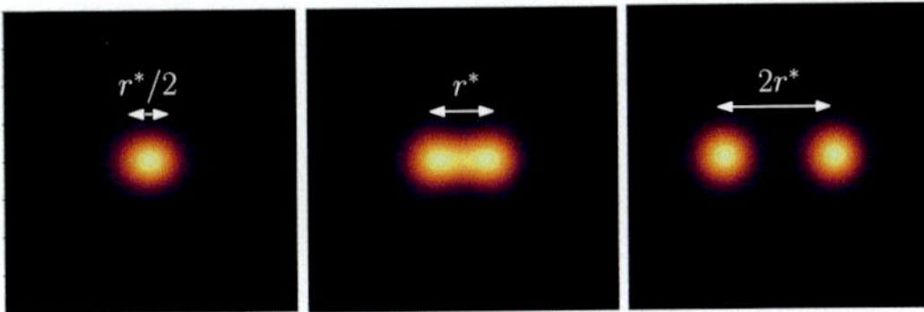

The Airy disk, corresponding to the central spot delimited by the first black ring has a radius r^* such that

$$J_1\left(\frac{\pi D r^*}{\lambda d_2}\right) = 0 \quad \rightarrow \quad \frac{\pi D r^*}{\lambda d_2} = 3.8317$$

and so

$$r^* \approx 1.22\frac{\lambda d_2}{D}. \tag{18.25}$$

This describes the best-focused spot, said to be diffraction-limited, that a perfect lens with a circular aperture can make.

The Rayleigh criterion: limits of resolution

Since the image of a point source by a thin lens gives origin to an Airy disk of radius $r^* > 0$ on the image plane, diffraction imposes a limit for the resolution of an optical system. The Rayleigh criterion defines the minimum resolvable detail on an image. Two point sources are said to be well resolved when the center of one Airy disk corresponds to the first minimum of the second one. Their separation is therefore r^*. As shown in Fig. 18.32, a smaller separation makes the two point sources impossible to distinguish.

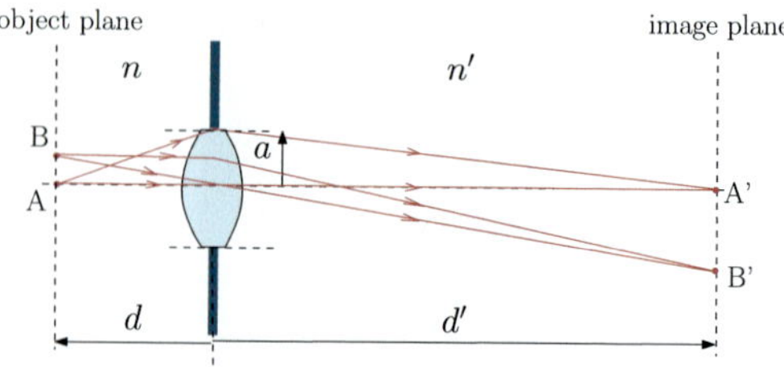

Fig. 18.33 A microscope images two point sources A and B into A' and B'

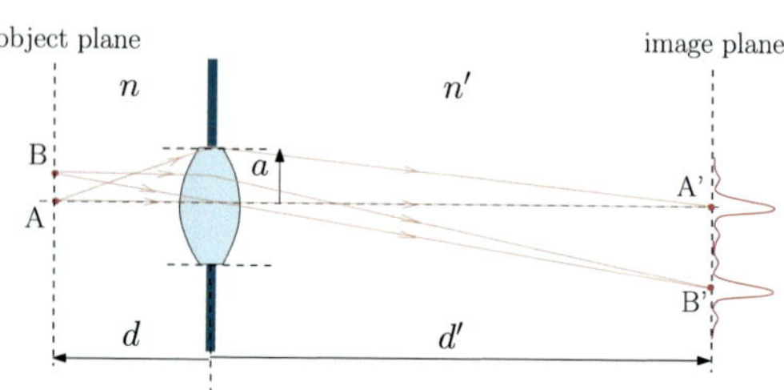

Fig. 18.34 The point sources A and B produce two Airy disks in in the image plane

18.5.5 Diffraction Limit and Numerical Aperture

Consider a microscope objective made of a convergent lens having a circular aperture of radius a. Let n and n' be the refractive index of the object and image region, respectively. The object and image planes are at a distance d and d', respectively, from the center of the lens, as shown in Fig. 18.33.

Consider two point sources at the two extremities of a vertical segment AB in the object plane. According to geometrical optics, in the image plane the intensity is concentrated in two points, corresponding to the conjugates A' and B' of A and B, respectively. If the two point sources are mutually incoherent, the resulting intensity in the image plane is simply the sum of the intensities generated separately by each diffracted wave, therefore two Airy spots. Since the aperture of the lens is a circle of radius a, and since the wavelength in the image region is λ/n', the intensity is the sum of two Airy functions of radius $r^* = 1.22\frac{\lambda d'}{2n'a}$ centered at A' and B', as schematically shown in Fig. 18.34.

The spatial resolution of the microscope corresponds to the smallest spatial separation Δl between two point objects that the microscope is able to resolve. The two points A and B will be clearly resolved if the separation $A'B'$ is larger than the radius of each Airy disk, that is if

$$A'B' > r^* = 1.22\frac{\lambda d'}{2n'a} .$$

If α and α' are the object and image apertures of the objective, respectively (see Fig. 18.35), we have the Lagrange–Helmholtz invariance discussed in Sect. 17.7.1

$$n\Delta l\alpha = n'\Delta l'\alpha'$$

Fig. 18.35 The angles α and α' are the object and image apertures, respectively

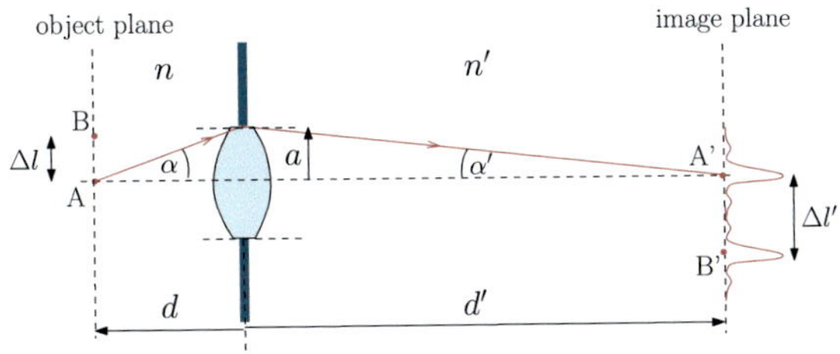

so that the smallest separation AB that can be resolved is such that $\Delta l' = r^*$, that is

$$\Delta l = \frac{n'\alpha'}{n\alpha}\Delta l' = \frac{n'\alpha'}{n\alpha}1.22\frac{\lambda d'}{2n'a}$$

and since in the paraxial approximation $\alpha' = a/d'$,

$$\boxed{\Delta l = 1.22\frac{\lambda d}{2n\alpha} = 1.22\frac{\lambda}{2\text{NA}}}$$

where $\text{NA} = n\sin\alpha \approx n\alpha$ is the numerical aperture of the objective. Since NA is of the order of 1, we see that the diffraction limit of a microscope is of the order of the wavelength used for imaging.

18.5.6 Diffraction by a Single Slit

Here we will introduce a distinction between two regimes for the diffraction of light by an object: Fresnel diffraction (close to the object) and Fraunhofer diffraction (far from the object).

To illustrate the transition between these two regimes, let us calculate the diffraction pattern obtained when a uniform plane wave $\underline{E}_0 e^{i(kz-\omega t)}$ is incident on a rectangular slit of length b and width a, centered around the origin at $z = 0$, shown in

Fig. 18.36 A rectangular slit of width a and height b in the $z = 0$ plane

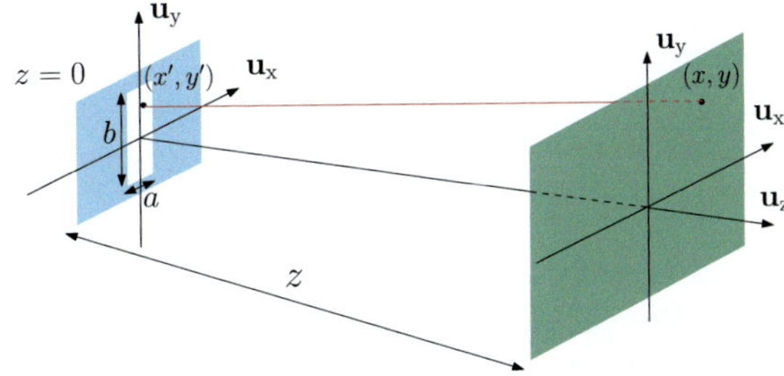

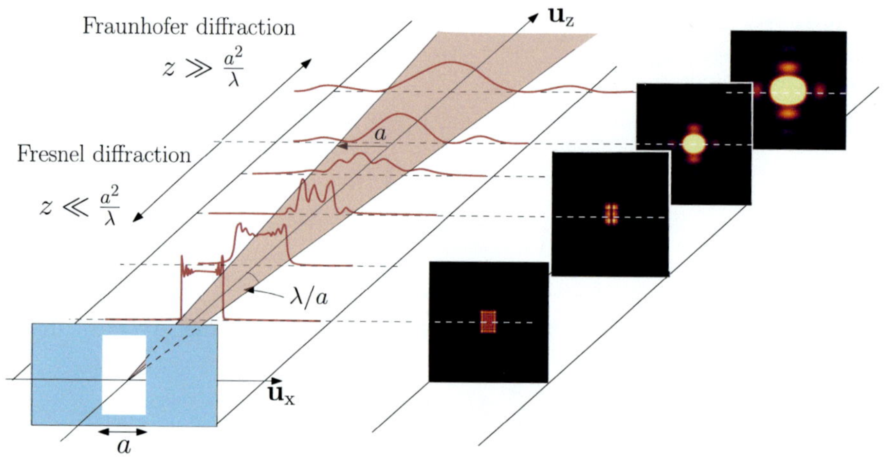

Fig. 18.37 Diffration pattern as a function of the distance to the slit

Fig. 18.36. At a point (x, y) on a screen placed at a distance z from the slit, Fresnel integral (18.10) gives

$$E(x, y, z, t) = \frac{E_0 k}{2\pi i z} e^{i(kz-\omega t)} \int\limits_{-\frac{b}{2}}^{\frac{b}{2}} \int\limits_{-\frac{a}{2}}^{\frac{a}{2}} e^{i\frac{k}{2z}((x-x')^2+(y-y')^2)} dx' dy' \qquad (18.26)$$

and by changing variables according to $u' = \sqrt{\frac{k}{2z}}(x - x')$ and $v' = \sqrt{\frac{k}{2z}}(y - y')$ we obtain

$$E(x, y, z, t) = \frac{E_0}{\pi i} e^{i(kz-\omega t)} \int\limits_{y_1}^{y_2} e^{iv^2} dv \int\limits_{x_1}^{x_2} e^{iu^2} du \qquad (18.27)$$

where $\quad x_1 = (x - \frac{a}{2})\sqrt{\frac{k}{2z}}, \qquad x_2 = (x + \frac{a}{2})\sqrt{\frac{k}{2z}}, \qquad y_1 = (y - \frac{b}{2})\sqrt{\frac{k}{2z}} \quad$ and $y_2 = (y + \frac{b}{2})\sqrt{\frac{k}{2z}}$, and by the definition of the Fresnel integrals $C(x) = \int_0^x \cos t^2 dt$ and $S(x) = \int_0^x \sin t^2 dt$, the field can be rewritten as

$$E(x, y, z, t) = \frac{E_0}{\pi i} e^{i(kz-\omega t)} \left(C(x_2) - C(x_1) + i(S(x_2) - S(x_1))\right)$$
$$\times \left(C(y_2) - C(y_1) + i(S(y_2) - S(y_1))\right) .$$

These Fresnel integrals may be evaluated numerically. Figure 18.37 shows the diffracted intensity by a slit of width a as well as cuts along the x-axis.

Very close to the slit, in the so-called Fresnel regime (also called near-field regime), the diffraction pattern ressembles the geometrical image of the slit, and develops fast oscillations as z increases. There is a strong evolution in the diffraction pattern shape until one reaches a distance far enough, in the so-called far-field or Fraunhofer regime, beyond which the diffraction figure simply streches in the direction perpendicular to the slit . The change in regime occurs when the size $z\theta$ of the figure given by the minimum angle of divergence $\theta = \lambda/a$ becomes comparable to the size a of the slit, that is, for

$$z\frac{\lambda}{a} = a \quad \rightarrow \quad N_F = \frac{a^2}{\lambda z} = 1$$

where N_F is the Fresnel number. Fresnel diffraction is observed for large values of N_F, that is for $N_F \gg 1$, whereas Fraunhofer diffraction occurs for $N_F \ll 1$.

18.6 Fraunhofer Diffraction: The Far-Field Limit

The Fraunhofer diffraction, named after Joseph von Fraunhofer (Fig. 18.39), is also called far-field diffraction and corresponds to the case in which the diffracted wave is observed at long distances $z = D \gg a^2/\lambda \gg a$ and in the paraxial approximation, that is, for small values of the Fresnel number $N_F = \frac{a^2}{\lambda z}$. This is illustrated in Fig. 18.38.

In this limit, the phase term inside the integral in Eq. (18.10) reads

$$\frac{ik}{2D}\left((x - x')^2 + (y - y')^2\right) = \frac{ik}{2D}(x^2 + y^2 - 2(xx' + 2yy') + x'^2 + y'^2) \, .$$

In the far-field limit and within the paraxial approximation $x'^2 + y'^2 < a^2 \ll \lambda D$, we can neglect the last two terms on the right-hand side and write

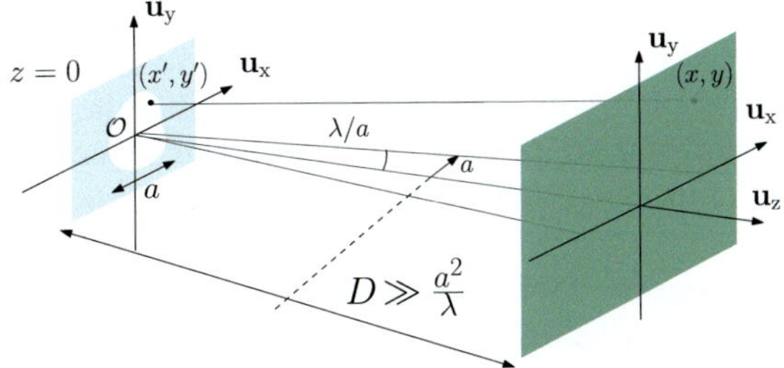

Fig. 18.38 In the far-field limit the distance D is much larger than $a^2/\lambda \gg a$

Fig. 18.39 Joseph von Fraunhofer (1787–1826), a German physicist and optical scientist renowned for his pioneering work in spectroscopy and optical engineering. He discovered the Fraunhofer lines, dark absorption lines in the solar spectrum, which provided crucial insights into the composition of stars

$$\frac{ik}{2D}\left((x-x')^2 + (y-y')^2\right) \approx \frac{ik(x^2 + y^2 - 2(xx' + yy'))}{2D}.$$

Equation (18.10) becomes

$$\underline{E}(x, y, z, t) = \frac{k}{2\pi i D} e^{i(kD-\omega t)} e^{i\frac{k(x^2+y^2)}{2D}} \underbrace{\int_{\mathbb{R}}\int_{\mathbb{R}} \underline{E}_0(x', y')\underline{t}(x', y') e^{-i\frac{k(x'x+y'y)}{D}} dx'dy'}_{4\pi^2 \mathcal{F}\{\underline{E}_0\underline{t}\}(\frac{kx}{D}, \frac{ky}{D})}$$

$$(18.28)$$

which gives the following remarkable result in the far-field approximation $(D\lambda \gg a^2)$, also known as Fraunhofer diffraction

$$\boxed{\underline{E}(x, y, z, t) = \frac{2\pi k}{iD} e^{i(kD-\omega t)} e^{i\frac{k(x^2+y^2)}{2D}} \mathcal{F}\{\underline{E}_0\underline{t}\}\left(\frac{kx}{D}, \frac{ky}{D}\right).} \qquad (18.29)$$

The far field is the Fourier transform of the field at the diffracting object multiplied by a phase factor. The intensity $\mathcal{I} \propto |\mathbf{E}|^2$ at a distance D from the object is therefore proportional to the modulus squared of the Fourier transform of $\underline{E}_0\underline{t}$

$$\mathcal{I}(z = D)(x, y) = A\left|\mathcal{F}\{\underline{E}_0\underline{t}\}\left(\frac{kx}{D}, \frac{ky}{D}\right)\right|^2 = A\left|\mathcal{F}\{\underline{E}_0\underline{t}\}\left(\frac{2\pi x}{\lambda D}, \frac{2\pi y}{\lambda D}\right)\right|^2.$$

$$(18.30)$$

Comparing with Eq. (18.23), we see that the effect of a thin lens is to bring the Fraunhofer limit closer to the diffracting object, from $D \gg a^2/\lambda$ to the focal length f' of the lens, which can be smaller than a^2/λ.

Example 18.4—Far-field diffraction by a rectangular slit
Consider a rectangular slit of width a and height b, illuminated by a uniform field of amplitude $\underline{E}_0$. We assume here that $a, b \gg \lambda$ and that $D \gg a^2/\lambda$ so

that the diffraction pattern at the screen, at distance D from the slit, is well described by the Fraunhofer approximation. Equation (18.30) tells us that the intensity at the screen is proportional to $|\mathcal{F}\{\underline{E}_0 t\}(\frac{kx}{D}, \frac{ky}{D})|^2$, where $\mathcal{F}(\underline{E}_0 t)$ is the Fourier transform of the field profile at $z = 0$. For the rectangular slit, $\underline{t}(x', y') = 1$ for any (x, y) inside the interval $[-a/2, a/2] \times [-b/2, b/2]$ and zero elsewhere. For a uniform incident field $\underline{E}_0(x', y') = \underline{E}_0$, and so

$$\mathcal{F}(\underline{E}_0 t)(u, v) = \frac{E_0}{4\pi^2} \int_{-a/2}^{a/2} \int_{-b/2}^{b/2} e^{-i(ux' + vy')} dx' dy'$$

$$= \frac{E_0}{4\pi^2} \underbrace{\int_{-a/2}^{a/2} e^{-iux'} dx'}_{\frac{2}{ua} \sin(\frac{ua}{2})} \underbrace{\int_{-b/2}^{b/2} e^{-ivy'} dy'}_{\frac{2}{vb} \sin(\frac{vb}{2})} .$$

Recalling the sinus cardinal function defined by $\text{sinc}(x) = \sin x / x$ for $x \neq 0$ and $\text{sinc}(0) = 1$, we can write

$$\mathcal{F}(\underline{E}_0 t)(u, v) = \frac{E_0 ab}{4\pi^2} \text{sinc}\left(\frac{ua}{2}\right) \text{sinc}\left(\frac{vb}{2}\right)$$

and the intensity on the screen has the form $I(x, y) \propto |\mathcal{F}(\underline{E}_0 t)(kx/D, ky/D)|^2$ and so

$$I(x, y) = I_0 \, \text{sinc}^2\left(\frac{kax}{2D}\right) \text{sinc}^2\left(\frac{kyb}{2D}\right)$$

or, in terms of the light's wavelength

$$I(x, y) = I_0 \, \text{sinc}^2\left(\frac{\pi a x}{\lambda D}\right) \text{sinc}^2\left(\frac{\pi b y}{\lambda D}\right) .$$

When $b \gg a$, which corresponds to the case of a very narrow slit aligned along the y-axis, the intensity on the screen decays much faster in the y direction than in the x direction, because the divergence angles in both directions are such that $\theta_y = \lambda/b \ll \theta_x = \lambda/a$. Along the x-axis, the intensity is maximum at $x = 0$ and presents several local maxima with decreasing amplitude as the distance to the axis increases. The first zeros of intensity are found at $x = \pm\lambda D/a$, and so the central spot has a typical size of $\Delta x = 2\lambda D/a$. A schematic of the diffraction pattern produced by a narrow slit is shown in Fig. 18.40.

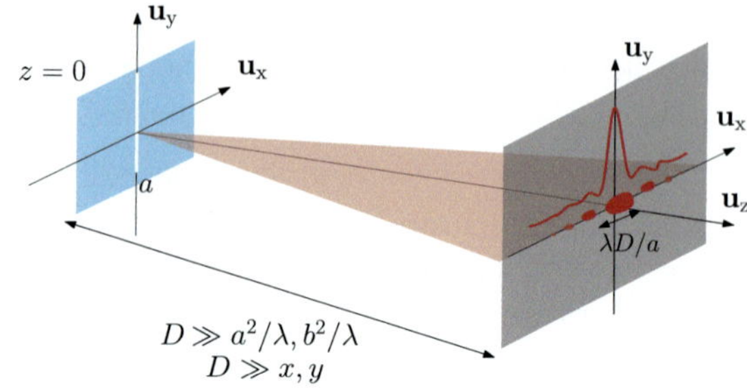

Fig. 18.40 Diffraction pattern by a narrow slit

Example 18.5—Far-field diffraction by a circular aperture

Consider a screen at distance D from a circular aperture of radius a illuminated by a uniform field $\underline{E}_0$, as shown in Fig. 18.41.

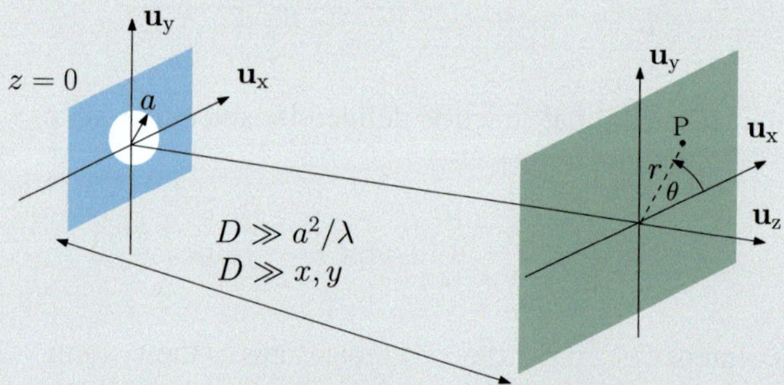

Fig. 18.41 A screen far from a circular aperture

In order to obtain the intensity on the screen, we need to calculate the Fourier transform of the field profile $\underline{E}_0\underline{t}$ at $z = 0$. For the circular aperture, $\underline{t}(x', y') = 1$ if $x'^2 + y'^2 \leq a^2$ and zero elsewhere. For a uniform incident field $\underline{E}_0(x', y') = \underline{E}_0$. The same calculation as in Sect. 18.5.4 yields

$$\mathcal{F}(\underline{E}_0\underline{t})(r, \theta) = \frac{\underline{E}_0}{2\pi r^2} \int\limits_0^{ar} J_0(x)x\,dx = \frac{a\underline{E}_0}{2\pi r} J_1(ar)$$

and since the field at point (x, y) in the screen is given by $\mathcal{F}(\underline{E}_0\underline{t})(kx/D, ky/D)$, the intensity $\mathcal{I}(x, y) \propto |\mathcal{F}(\underline{E}_0\underline{t})(kx/D, ky/D)|^2$ can be written

$$\mathcal{I}(r, \theta) = I_0 \frac{D}{kar^2} J_1^2\left(\frac{kar}{D}\right) = I_0 \frac{J_1^2\left(\frac{2\pi ar}{\lambda D}\right)}{\frac{2\pi ar^2}{\lambda D}}.$$

The observed diffraction figure consists of an Airy disk surrounded by concentric rings, as shown in Fig. 18.42. The radius r_0 of the disk is

$$r_0 \approx 1.22 \frac{\lambda D}{2a} \; .$$

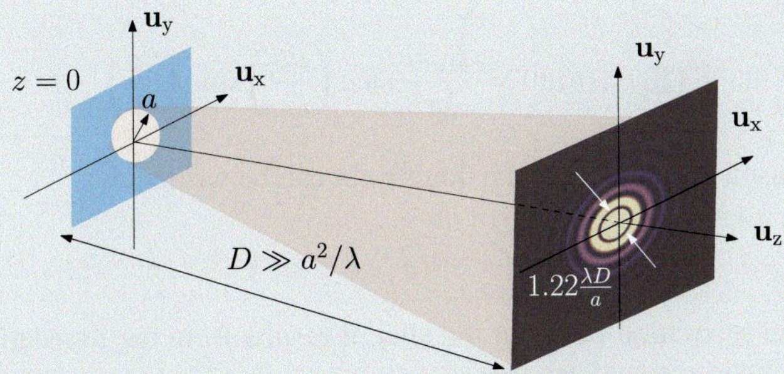

Fig. 18.42 The diffraction pattern by a circular aperture is an Airy disk in the far-field limit

18.6.1   Far Field Diffraction from a Double Slit

In Example 18.3 it was shown that, in the framework of the Huygens–Fresnel principle, the diffraction pattern of two point-like apertures corresponds to the interference between two spherical waves and that in the small-angle approximation, one obtains a periodic sucession of bright and dark fringes. In reality, due to the finite size of the slits, the interference pattern is modulated by an envelope that decays away from the optical axis. Let us show this by calculating the Fraunhofer diffraction of two narrow slits separated by a distance a, shown in Fig. 18.43. Let b and c be the width and the height of each slit, respectively.

Fig. 18.43 Two narrow slits separated by a distance a

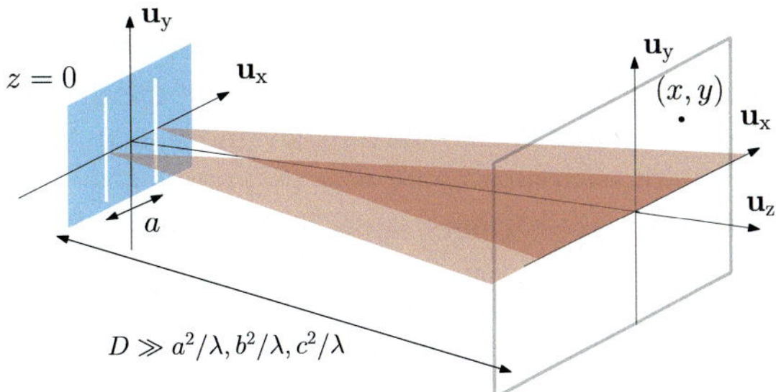

Let us assume that an homogeneous field $\underline{E}_0$ is incident on both slits. In the Fraunhofer limit, the field at any point on the screen (x, y) is proportional to the Fourier transform of the field's distribution $\underline{E}_0\underline{t}$ at $z = 0$ evaluated at $(kx/D, ky/D)$, where $\underline{t}$ is the transmission factor of the double slit.

Using the properties of the Fourier transform, this can be easily calculated by considering first the case of a single slit $\underline{t}_0$ of width b and height c centered around the origin. The Fourier transform of such a field distribution was calculated in Example 18.4:

$$\mathcal{F}\{\underline{E}_0\underline{t}_0\}(u, v) = \frac{E_0 bc}{4\pi^2}\operatorname{sinc}\left(\frac{ub}{2}\right)\operatorname{sinc}\left(\frac{vc}{2}\right).$$

Since the transmission factor of the double slit can be written as

$$\underline{t}(x', y') = \underline{t}_0(x' - a/2, y') + \underline{t}_0(x' + a/2, y')$$

where a is the separation between the slits, it results from the translation property (A.58) of the Fourier transform that

$$\mathcal{F}\{\underline{E}_0\underline{t}\}(u, v) = e^{-iu\frac{a}{2}}\mathcal{F}\{\underline{E}_0\underline{t}_0\}(u, v) + e^{iu\frac{a}{2}}\mathcal{F}\{\underline{E}_0\underline{t}_0\}(u, v)$$
$$= 2\mathcal{F}\{\underline{E}_0\underline{t}_0\}(u, v)\cos\left(\frac{ua}{2}\right)$$

and replacing $\mathcal{F}\{\underline{E}_0\underline{t}_0\}$

$$\mathcal{F}\{\underline{E}_0\underline{t}\}(u, v) = \frac{E_0 ab}{4\pi^2}2\cos\left(\frac{ua}{2}\right)\operatorname{sinc}\left(\frac{ub}{2}\right)\operatorname{sinc}\left(\frac{vc}{2}\right).$$

Finally, the intensity at a point (x, y) on the screen is given, in the far-field limit, by (18.30)

$$\mathcal{I}(x, y) \propto \left|\mathcal{F}\{\underline{E}_0\underline{t}\}\left(\frac{kx}{D}, \frac{ky}{D}\right)\right|^2 = 4\mathcal{I}_0\cos^2\left(\frac{\pi xa}{\lambda D}\right)\operatorname{sinc}^2\left(\frac{\pi xb}{\lambda D}\right)\operatorname{sinc}^2\left(\frac{\pi yc}{\lambda D}\right)$$

where $\mathcal{I}_0$ is the maximum intensity that would be obtained when a single slit is present. For narrow slits, $b \ll c$ so that the diffraction pattern on the screen is concentrated in the horizontal axis around $y = 0$. When $b \ll a$, the variation of the $\cos^2$ term with x is much faster than that of the sinc^2, and so the observed pattern consists of interferences fringes separated by $\Delta x = \frac{\lambda D}{a}$ modulated by a sinc^2 envelope of typical size $\frac{\lambda D}{b} \gg \Delta x$, as shown in Fig. 18.44.

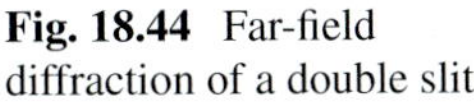

Fig. 18.44 Far-field diffraction of a double slit

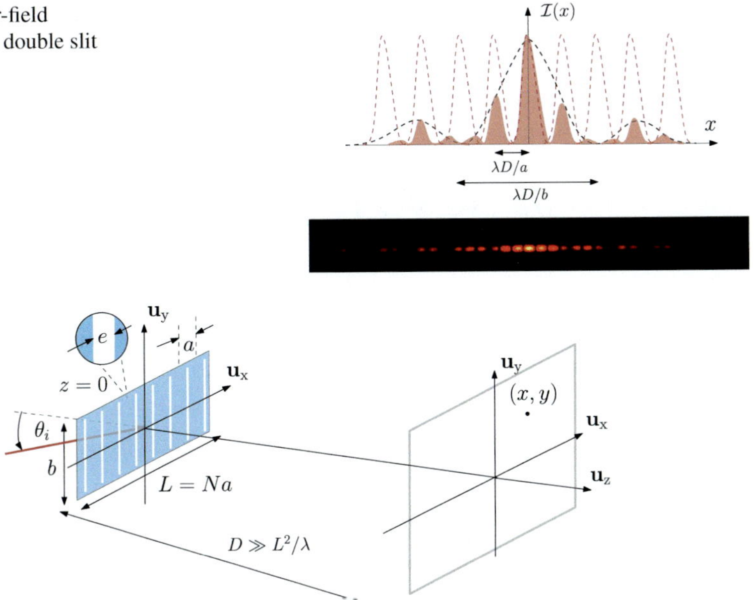

Fig. 18.45 A diffraction grating illuminated by a plane wave at incidence angle θ_i

18.7 Diffraction Spectrometers

In this section we will see how diffraction by a periodic pattern allows for the spatial separation of the spectral components of the incident field. This principle is widely used for the fabrication of spectrometers.

18.7.1 The Diffraction Grating

A very common instrument in optical spectroscopy is based on the diffraction by a grating consisting of N narrow slits (each of width w and height b) equally spaced, of spacing $a \gg w$. Let us calculate the diffracted pattern in the far field when the slit is illuminated with a plane wave forming an angle θ_i with respect to the normal of the grating. This is illustrated in Fig. 18.45.

By exploiting the modulation and translation properties of the Fourier transform, we can determine the diffraction of the grating by determining first the diffraction pattern of a single slit $\underline{t}_0$ illuminated by a plane wave $\underline{E}(x', y', t) = \underline{E}_0 e^{-i\omega t}$ at normal incidence. For a single slit centered around the origin, we know from Example 18.4 that

$$\mathcal{F}\{\underline{E}_0\underline{t}_0\}(u, v) = \frac{\underline{E}_0 e b}{4\pi^2}\,\mathrm{sinc}\left(\frac{bv}{2}\right)\,\mathrm{sinc}\left(\frac{eu}{2}\right).$$

If we now have N slits distributed symmetrically with respect to the origin, the transmission factor, assuming N odd, reads

$$\underline{t}(x', y') = \sum_{p=0}^{N-1} \underline{t}_0\left(x' + \frac{(N-1)a}{2} - pa,\, y'\right)$$

and the far field pattern is given by its Fourier transform which can be easily obtained by linearity and the translation property (A.58) which gives $\mathcal{F}\{\underline{E}_0\underline{t}(x' - x_0)\}(u, v) = e^{-iux_0}\mathcal{F}\{\underline{E}_0\underline{t}(x')\}(u, v)$ so that

$$\mathcal{F}\{\underline{E}_0\underline{t}\}(u, v) = \sum_{p=0}^{N-1} \mathcal{F}\{\underline{E}_0\underline{t}_0(x', y')\}(u, v)e^{iu(\frac{(N-1)a}{2} - pa)}$$

$$= e^{iu\frac{(N-1)a}{2}}\mathcal{F}\{\underline{E}_0\underline{t}_0(x', y')\}(u, v)\underbrace{\sum_{p=0}^{N-1} e^{-ipau}}_{\frac{1-e^{-iNua}}{1-e^{-iua}}}$$

$$= e^{iu\frac{(N-1)a}{2}}\frac{\underline{E}_0 e b}{4\pi^2}\,\mathrm{sinc}\left(\frac{bv}{2}\right)\,\mathrm{sinc}\left(\frac{eu}{2}\right)\frac{1 - e^{-iNua}}{1 - e^{-iua}}.$$

Finally, if at $z = 0$ the incoming electric field has a complex amplitude $\underline{E}(x', y', t) = \underline{E}_0 E^{ik\sin\theta_i x'}e^{-i\omega t}$, the modulation property of the Fourier transform (A.59) gives $\mathcal{F}\{\underline{E}_0 e^{ik\sin\theta_i x'}\underline{t}\}(u, v) = \mathcal{F}\{\underline{E}_0\underline{t}\}(u - k\sin\theta_i, v)$. The intensity on the screen $\mathcal{I}(x, y) \propto |\mathcal{F}\{\underline{E}_0 e^{ik\sin\theta_i}\underline{t}\}(\frac{kx}{D}, \frac{ky}{D})|^2$ is given by

$$\mathcal{I}(x, y) = N^2\mathcal{I}_0\mathrm{sinc}^2\left(\frac{\pi b y}{\lambda D}\right)\underbrace{\mathrm{sinc}^2\left(\frac{\pi e}{\lambda}\left(\frac{x}{D} - \sin\theta_i\right)\right)}_{f(x)}\underbrace{\left(\frac{\sin(\frac{N\pi a}{\lambda}(x/D - \sin\theta_i))}{N\sin(\frac{\pi a}{\lambda}(x/D - \sin\theta_i))}\right)^2}_{g(x)}.$$

$$(18.31)$$

The diffraction pattern is strongly concentrated along the horizontal axis when the vertical dimension b of the slits is much larger than w. The x dependence of the diffracted intensity has many features and reflects all the transverse physical dimensions of the grating (L, a, e). As illustrated in Fig. 18.46, the slowly varying envelope f is a geometrical factor due to the diffraction by a single slit and it tell us that the intensity is centered on the screen at a position x_0 such that $x_0/D = \sin\theta_i$, i.e., in the direction given by the incident wave. Recalling that in the small angle approximation $x/D \approx \sin\theta$, this corresponds to $\theta_0 = \theta_i$ where θ_0 is the angle that the center of the diffraction pattern makes with respect to the x-axis.

In terms of $\sin\theta = x/D$ at $y = 0$ we can write

$$I(x,0) = N^2 I_0 \underbrace{\operatorname{sinc}^2\left(\frac{\pi e(\sin\theta - \sin\theta_i)}{\lambda}\right)}_{f(\sin\theta)} \underbrace{\left(\frac{\sin(\frac{N\pi a}{\lambda}(\sin\theta - \sin\theta_i))}{N\sin(\frac{\pi a}{\lambda}(\sin\theta - \sin\theta_i))}\right)^2}_{g(\sin\theta)}. \qquad (18.32)$$

The function g, which reflects simultaneously the diffraction by the whole aperture $L = Na$ of the slit and the superposition of the N waves diffracted by the slits, presents maxima and mimima with two typical lengthscales $1/a$ and $1/Na$. The maxima are obtained when the denominator goes to zero and this corresponds to a set of diffraction angles θ_n given by the so-called grating equation:

$$\sin\theta_n - \sin\theta_i = n\frac{\lambda}{a} \quad n \in \mathbb{Z}. \qquad (18.33)$$

The intensity decays very fast away from θ_n, also called the nth diffraction order, due to the rapid oscillations of the numerator of g. Indeed the width of each maximum scales as $1/Na$. A sketch of the general shape of g is shown in Fig. 18.47, for $N \sim 10a$.

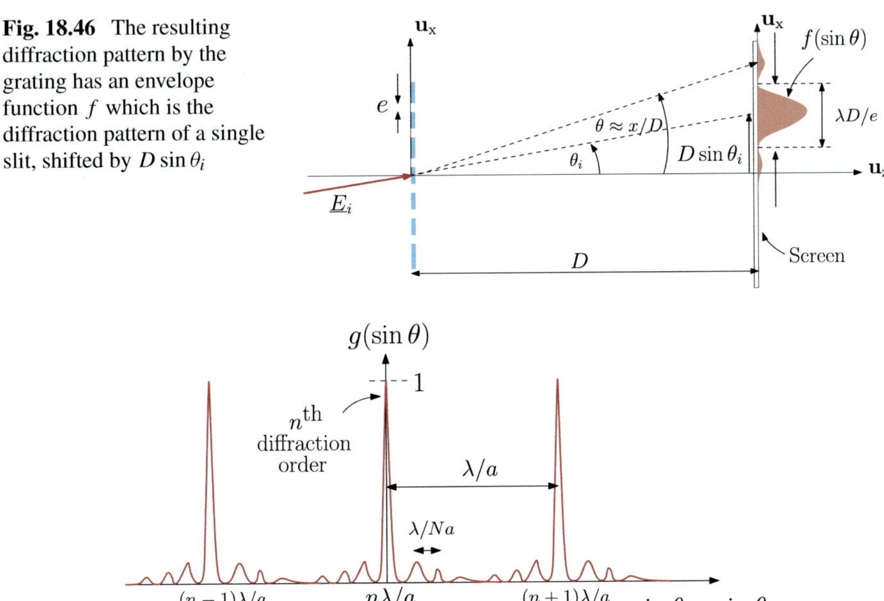

Fig. 18.46 The resulting diffraction pattern by the grating has an envelope function f which is the diffraction pattern of a single slit, shifted by $D\sin\theta_i$

Fig. 18.47 The function g presents equally spaced diffraction peaks

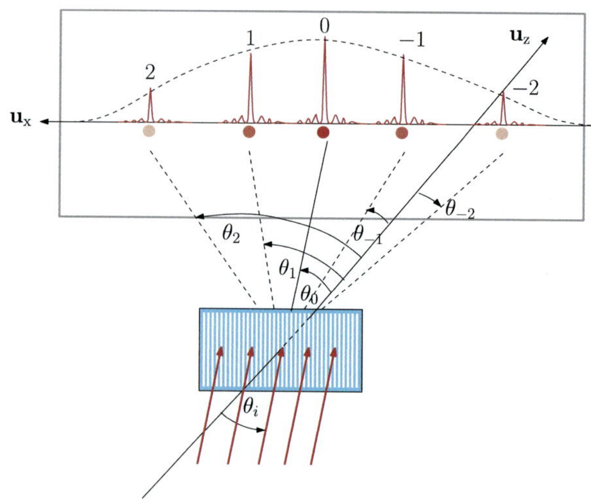

Fig. 18.48 Diffraction pattern produced by a grating

Remarks

- The maximum intensity on the screen reaches $N^2 \mathcal{I}_0$, where $\mathcal{I}_0$ is the maximum intensity that would be obtained with a single slit. This is a strong enhancement due to constructive interference with respect to the simple sum of their separate intensities, that would have given $N \mathcal{I}_0$.
- The diffraction pattern is therefore characterized by the presence of a series of bright spots on the screen, corresponding to the different diffraction orders. This is illustrated in Fig. 18.48.
- The grating equation (18.33) can be easily obtained by using a simple argument: the waves coming out of the grating at an angle θ should all have the same phase in order to interfere constructively. The total phase difference between two consecutive beams diffracted at an angle θ is given by

$$\Delta\phi = \frac{2\pi}{\lambda}(\Delta l_2 - \Delta l_1)$$

as can be seen in Fig. 18.49, and

$$\Delta l_1 = a \sin\theta_i \qquad \Delta l_2 = a \sin\theta \ .$$

All the waves leaving at an angle θ will interfere constructively if

$$\Delta\phi = 2n\pi = \frac{2a\pi}{\lambda}(\sin\theta - \sin\theta_i) \quad n \in \mathbb{Z} \ .$$

- Sometimes a more convenient geometry consists in diffracting light by a reflection grating. This can be obtained for example by patterning parallel grooves in a mirror surface as shown in Fig. 18.50. If the source is placed in front of such a

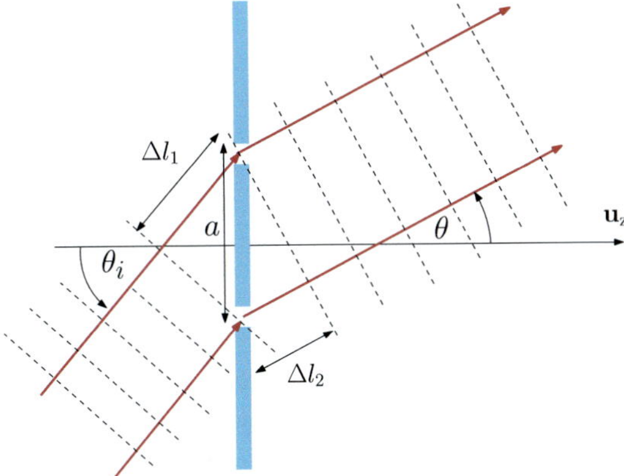

Fig. 18.49 For two diffracted beams to interfere constructively, the phase difference must be a multiple of 2π

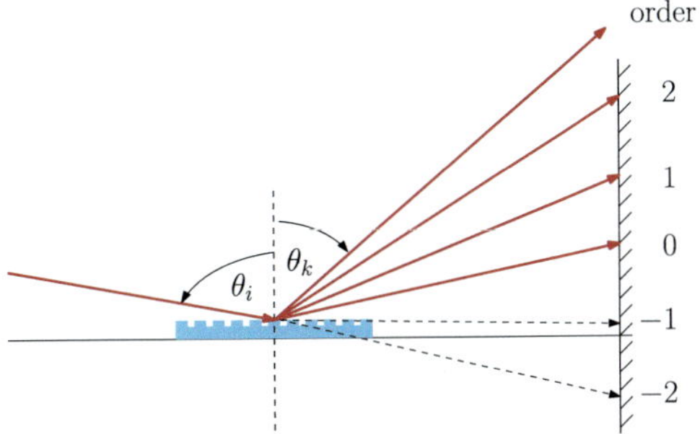

Fig. 18.50 A diffraction grating in reflection configuration

grating, the observed diffraction pattern would be the same as the one obtained by a transmission grating. In this case the grating equation reads

$$\sin\theta_i - \sin\theta_n = n\frac{\lambda}{a} \quad n \in \mathbb{Z}$$

18.7.2 The Diffraction Grating as a Spectrometer

Diffraction gratings are excellent tools for spectral analysis since, as shown by the grating equation (18.33), the n-order diffraction is observed at a position x_n on the screen given by

$$\sin\theta_n = \frac{x_n}{D} = \sin\theta_i + n\frac{\lambda}{a}$$

and so $x_n = x_n(\lambda)$, the position depends on the wavelength of light. A polychromatic beam is then split into its different wavelength components as illustrated in

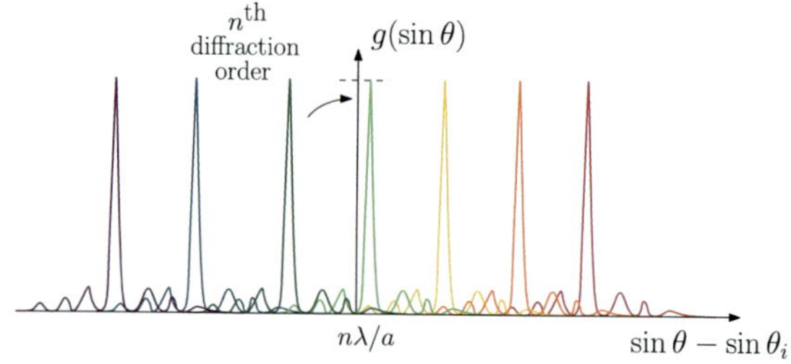

Fig. 18.51 The position of the diffracted peak at order n is a function of wavelength

Fig. 18.51 and the spectral information of a source can be therefore analyzed by placing a detector and calibrating the correspondance between x_n and λ. Here the position on the screen varies approximately linearly with λ instead of $1/\lambda^2$ as in the case of a prism (Cauchy's law for the dispersion of glass).

Let $\theta_n(\lambda_0)$ be the direction of a n-order diffraction for the wavelength λ_0. Then, the dispersive power $\gamma(\lambda, n)$ around λ_0 is defined as

$$\gamma(\lambda, n) = \left(\frac{d\theta_n(\lambda)}{\lambda}\right)_{\lambda_0}$$

and quantifies the angular separation between two neighbouring wavelengths. From the grating equation, we obtain

$$\cos\theta_n \frac{d\theta_n(\lambda)}{\lambda} = \frac{n}{a}$$

and so

$$\gamma(\lambda, n) = \frac{n}{a\cos\theta_n(\lambda_0)} \; .$$

We see that for small angles, $\gamma(\lambda, n) \approx n/a$, and so the dispersion increases with the diffraction order. Naturally, it vanishes for the zero order: all wavelengths are merged together in the transmitted beam (or in the reflected beam, if a reflection grating is used). It is therefore desirable to disperse the light at a high diffraction order. However, the diffraction pattern is multiplied by an envelope due to the finite size of a single groove, and so higher diffraction orders are less intense. A compromise must be found between high dispersion power and high enough intensity, and typically one uses the first or second diffraction order.

There is, however, a clever way to redistribute the energy onto the different diffraction orders. The energy that goes to the zero order, for example, is not useful to resolve the spectrum of a source. Blazed gratings are made of a series of reflecting stripes with a sawtooth-shaped cross section, forming a step structure, as shown in Fig. 18.52. The direction in which maximum efficiency is achieved is called the blaze angle α and depends on the wavelength and the diffraction order. It is possible to concentrate the maximum of energy in a non-zero order, for which there is dispersion, obtaining thus a brighter spectrum with respect to a classic grating.

Another application of the diffraction grating is to select a particular wavelength of a broadband source, in a configuration named monochromator. An schematic of a monochromator based on a diffraction grating is shown in Fig. 18.53. The light coming from a source is focused into the entrance slit, placed at the object focal plane of a concave mirror which collimates the beam and illuminates a reflection diffraction grating. The diffraction off the grating is then focused by a second mirror into the output slit, which selects one particular wavelength. By turning the grating, one can finely tune the output wavelength. One can use this setup to produce, for example, a tunable monochromatic source from a broad incandescent lamp. Of course, this same setup can be used as well as a spectrometer. Indeed, by placing after the output slit a photodetector, one can record the intensity as a function of wavelength to reconstruct the spectrum of the source. Alternatively, by replacing the output slit by a linear array such as a CCD, one can record the spectrum of the source in a single shot.

Fig. 18.52 A blazed grating

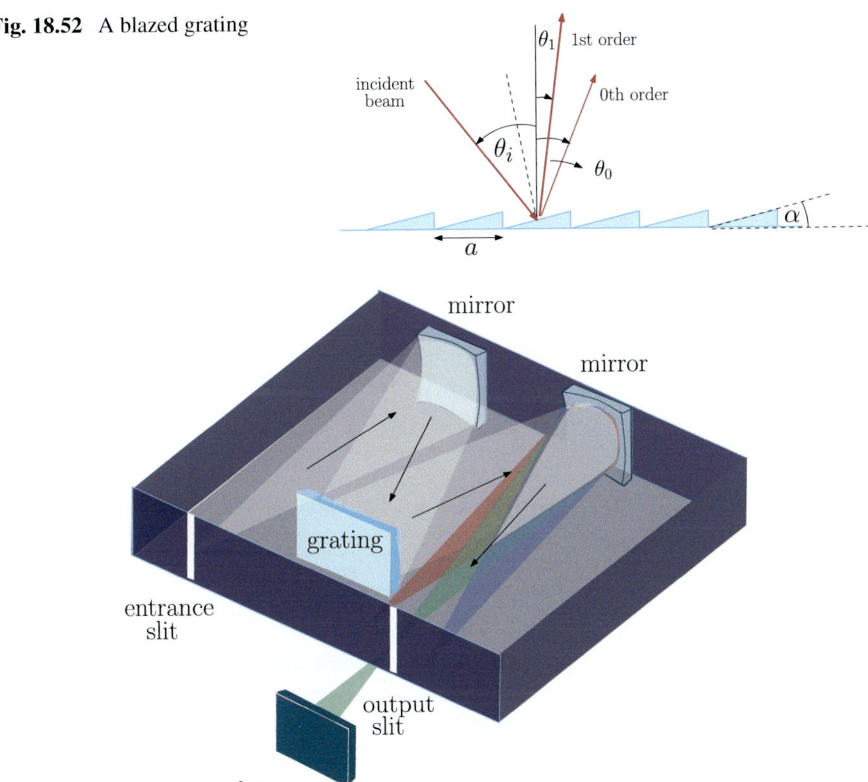

Fig. 18.53 A monochromator

Let us now discuss the minimum wavelength separation that can be resolved with a grating spectrometer. Since the thickness of a diffraction order is $\delta\theta \approx \frac{\lambda}{Na}$, a wavelength separation of $\delta\lambda$ can be resolved if their angular separation $d\theta = \gamma\delta\lambda = n\delta\lambda/a$ is smaller than $\delta\theta$, and so if

$$\frac{n\delta\lambda}{a} < \frac{\lambda}{Na}$$

$$\delta\lambda < \frac{\lambda}{Nn} \; .$$

The resolving power of a grating with N slits in the n diffraction order is

$$\mathcal{R} = \frac{\lambda}{\delta\lambda} = nN \; .$$

The larger the number of slits, the greater the resolution. Typical gratings contain thousands to tens of thousands of grooves. The resolution in wavelength in the visible range can easily reach $\sim 0.5\,\text{Å}$, which corresponds to a resolution in energy of the photons detected of a few hundreds of μeV.

18.8 Summary and Essential Formulas

- The transverse distribution of a light beam evolves during propagation due to a wave phenomenom called diffraction. For a scalar monochromatic wave $\underline{E}$ propagating along the z-axis, the most general solution of the wave-equation reads

$$\underline{E}(x, y, z, t) = \int_{\mathbb{R}} \int_{\mathbb{R}} \hat{E}_0(k_x, k_y) e^{i(k_x x + k_y y)} e^{i(k_z(k_x, k_y)z - i\omega t)} dk_x dk_y$$

where $\hat{E}_0$ is the 2D Fourier transform of the field spatial distribution at $z = 0$, and

$$k_z(k_x, k_y) = \sqrt{\frac{\omega^2}{c^2} - (k_x^2 + k_y^2)} \; .$$

The field is therefore a superposition of propagating ($k_z \in \mathbb{R}$) and evanescent ($\text{Im}\{k_z\} > 0$) waves. The contribution of evanescent waves is negligible at distances from $z = 0$ which are larger than the wavelength λ. Since the wave of transverse wavevector (k_x, k_y) carries away the variations of the field with spatial frequencies $(k_x/2\pi, k_y/2\pi)$, it follows that at distances larger than λ, all the features of the electric field varying faster than the wavelength are lost.

- The uncertainty principle dictates a minimum angular divergence for any beam of finite size. If Δx is the characteristic extension of $\underline{E}_0(x, y)$ along the x direction, then the plane waves composing $\underline{E}_0$ have a maximal component $k_{x\max}$ of the wavevector along x such that

$$\Delta x k_{x\max} \geq 2\pi \; .$$

These waves move away from the z-axis, the minimum possible divergence angle of a beam is therefore $\theta_{\min} = \lambda/\Delta x$, and it is achieved when both the amplitude and the phase of the field $\underline{E}_0$ vary slowly inside Δx.

- Diffraction by a thin obstacle placed at $z = 0$ can be treated as a free propagation problem by simply replacing the field $\underline{E}_0$ at $z = 0$ by $\underline{E}_0 \, \underline{t}$, where $\underline{t}$ is the transmission factor of the obstacle. By exploiting the convolution property (A.60) of the Fourier transform, the electric field at any value of z can be computed as the following 2D convolution (over the x and y variables only):

$$\boxed{\underline{E}(x, y, z, t) = \frac{e^{-i\omega t}}{(2\pi)^2}(\underline{t}_0\underline{E}_0 * \underline{h})(x, y, z, t)}$$

where $h(x, y, z)$ is the inverse Fourier transform of $\hat{h}(k_x, k_y) = e^{ik_z(k_x,k_y)z}$, and therefore given by Fourier inversion theorem (A.57)

$$h(x, y, z) = \mathcal{F}^{-1}\{e^{ik_z(k_x,k_y)z}\}(x, y) = \int_{\mathbb{R}} e^{ik_z(k_x,k_y)z} e^{i(k_x x + k_y y)} dk_x dk_y \; .$$

The problem of finding the electric field at any position after the obstacle is therefore reduced to calculating a Fourier transform and performing the convolution (18.5).

- Using Weyl's expansion it is possible to show that

$$h(x, y, z) = -2\pi \frac{\partial}{\partial z} \frac{e^{ikr}}{r}$$

with $r = \sqrt{x^2 + y^2 + z^2}$. In the limit $z \gg \lambda$, the diffracted field is given by the Huygens–Fresnel integral

$$\boxed{\underline{E}(x, y, z, t) = -\frac{e^{-i\omega t}}{\lambda} \int_{\mathbb{R}} \int_{\mathbb{R}} \underline{E}_0(x', y')\underline{t}(x', y') \frac{e^{ikr'}}{r'} \frac{z}{r'} dx' dy'}$$

with $r' = \sqrt{(x - x')^2 + (y - y')^2 + z^2}$.

This justifies the Huygens–Fresnel principle: every unobstructed point (x', y') at $z = 0$ is a source of an spherical wave whose amplitude is proportional to the incident wave $\underline{t}(x', y')\underline{E}(x', y')$ at that point. The diffracted field at point (x, y)

is thus obtained by calculating the field radiated by all these secondary, fictive
sources placed at $z = 0$.

- The Fresnel approximation, or paraxial approximation, consists in considering the
  diffraction pattern close to the axis so that the small-angle approximation can be
  employed. Spherical wavefronts are approximated by paraboloids, so that

$$h(x, y) = e^{ikz} \frac{2\pi k}{iz} e^{-\frac{k(x^2+y^2)}{2iz}}$$

and we obtain the Fresnel diffraction integral:

$$\underline{E}(x, y, z, t) = \frac{k}{2\pi i z} e^{i(kz-\omega t)} \int_{\mathbb{R}} \int_{\mathbb{R}} \underline{E}_0(x', y')\underline{t}(x', y')e^{i\frac{k}{2z}((x-x')^2+(y-y')^2)}dx'dy' \; .$$

- The Fresnel integral provides a solution of the paraxial wave equation. For a wave
  of the form $\underline{E}(x, y, z, t) = \mathcal{E}(x, y, z)e^{ik(z-\omega t)}$, this equation reads

$$\frac{\partial^2 \mathcal{E}}{\partial x^2} + \frac{\partial^2 \mathcal{E}}{\partial y^2} + 2ik\frac{\partial \mathcal{E}(x, y, z)}{\partial z} = 0$$

and describes a light beam propagating essentially along the z-axis, with a slowly
varying amplitude along this direction.

- The transmission factor of a thin spherical lens in Fresnel's approximation writes

$$\underline{t}(x', y') = \underline{t}_0 e^{-ik\frac{x'^2+y'^2}{2f'}} = \underline{t}_0 e^{-i\pi\frac{x'^2+y'^2}{\lambda f'}} \tag{18.34}$$

where f' is the focal length and $\underline{t}_0$ is a constant factor.

- A Gaussian beam propagates accroding to

$$\underline{E}(x, y, z, t) = \frac{E_0 w_0}{w(z)} e^{i(kz-\tan^{-1}(\frac{z-z_0}{z_R})-\omega t)} e^{ik\frac{x^2+y^2}{2R(z)}} e^{-\frac{(x^2+y^2)}{w(z)^2}}$$

and it satisfies Fresnel diffraction integral (18.10). The radius of curvature R and
the beam radius w vary along the propagation direction as follows

$$R(z) = (z - z_0) + \frac{z_R^2}{z - z_0}$$

$$w(z) = w_0\sqrt{1 + \left(\frac{z - z_0}{z_R}\right)^2}$$

where $z = z_0$ is the position of the beam waist, the plane at which the beam radius
is minimum and equal to w_0, and $z_R = \pi w_0^2/\lambda$ is the Raleygh length. Gaussian

beams preserve their Gaussian character after passing through a perfect thin lens. Let $z = d$ and $z = d'$ be the position of the beam waist before and after a converging lens of focal length f', then

$$\frac{1}{d'} + \frac{1}{d + \frac{z_R^2}{(d-f')}} = \frac{1}{f'} \tag{18.35}$$

and the transverse magnification reads

$$M = \frac{w_0'}{w_0} = \sqrt{\frac{z_R'}{z_R}} = \frac{f'}{\sqrt{(f'-d)^2 + z_R^2}}. \tag{18.36}$$

- The intensity at a point (x, y) in the image focal plane of a lens is proportional to the squared absolute value of the Fourier transform of the wave at the input plane $z = 0$ of the lens, evaluated at the spatial frequencies $(x/\lambda f', y/\lambda f')$. This is why the image focal plane of a lens is also called the Fourier plane.

$$\mathcal{I}(x, y, z = f') \propto \left| \mathcal{F}\{\underline{E}_0\}\left(\frac{2\pi x}{\lambda f'}, \frac{2\pi y}{\lambda f'}\right) \right|^2.$$

- The intensity on the image plane of a lens illuminated by point-source is proportional to the squared modulus of the Fourier transform of its pupil function $\underline{t}$

$$\mathcal{I}(x, y, d_2) \propto |\underline{E}_0(x, y, d_2)|^2 \propto \left| \mathcal{F}\{\underline{t}\}\left(\frac{2\pi x}{\lambda d_2}, \frac{2\pi y}{\lambda d_2}\right) \right|^2$$

which provides a limit for the resolution of every imaging system.
- The Fraunhofer diffraction, also called far-field diffraction, corresponds to the case in which the diffracted wave is observed at long distances $z = D \gg a^2/\lambda \gg a$ and in the paraxial approximation, that is, for small values of the Fresnel number $N_F = \frac{a^2}{\lambda z}$. In this case, the diffracted field is given by

$$\underline{E}(x, y, z, t) = \frac{2\pi k}{iD} e^{i(kD - \omega t)} e^{i \frac{k(x^2 + y^2)}{2D}} \mathcal{F}\{\underline{E}_0 t\}\left(\frac{kx}{D}, \frac{ky}{D}\right).$$

The far field is the Fourier transform of the field at the diffracting object multiplied by a phase factor. The intensity $\mathcal{I} \propto |\mathbf{E}|^2$ at a distance D from the object is therefore proportional to the modulus squared of the Fourier transform of $\underline{E}_0 t$

$$I(z = D)(x, y) = A\left|\mathcal{F}\{\underline{E}_0 t\}\left(\frac{kx}{D}, \frac{ky}{D}\right)\right|^2 = A\left|\mathcal{F}\{\underline{E}_0 t\}\left(\frac{2\pi x}{\lambda D}, \frac{2\pi y}{\lambda D}\right)\right|^2.$$

When the condition $D \gg a^2/\lambda$ of Fraunhofer diffraction is too restrictive, one can still observe the far-field diffraction by placing a lens after the obstacle and observing the resulting intensity at the lens image focal plane.

Problems

18.1 Diffraction Along the Axis of a Circular Aperture

Consider a circular aperture of radius a in the plane $z = 0$ illuminated by a uniform plane wave of amplitude $\underline{E}_0$. The goal is to determine the diffracted intensity at any point in the axis at a distance z from the aperture.

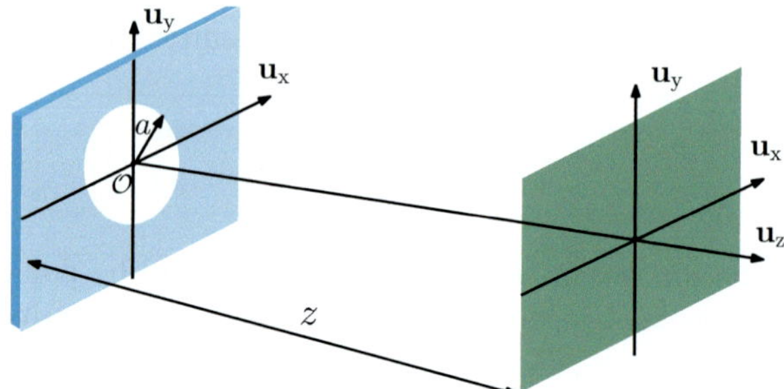

(a) Using the Huygens–Fresnel principle (18.8), show that the electric field in the axis at a distance z from the disk is given by

$$\underline{E}(z, t) = -\frac{i2\pi z \underline{E}_0}{\lambda} \int_z^{\sqrt{a^2+z^2}} \frac{e^{ikv}}{v}\,dv$$

(b) Find an approximate expression for the above electric field by using the Sommerfeld lemma, which establishes that

$$\int_a^b f(x)e^{ikx} \approx \left[\frac{f(x)}{ik}e^{ikx}\right]_a^b$$

when f is a slowly varying function (justified in the limit in which the diffracting object is much larger than the wavelength λ). Show that the resulting field can be interpreted as the interference between the incident plane wave and a wave diffracted by the borders of the aperture.

(c) Find the intensity I along the axis. Show that far from the aperture and in the limit $z \gg a^2/\lambda$, the intensity decays as $1/z^2$ according to

$$I(z) \approx I_0 \frac{k^2 a^4}{4z^2}$$

where $I(0) = I_0$ is the intensity of the incident plane wave. Interpret this decrease in intensity with distance by means of the minimum divergence $\theta = \lambda/a$ of a beam of size a due to diffraction.

(d) Now study the behaviour of the intensity on the axis close to the aperture, and show that the intensity presents several minima for particular values of z given by

$$\sqrt{z^2 + a^2} - z = p\lambda \quad p \in \mathbb{N}^* .$$

18.2 The Poisson-Arago Spot

Consider a disk of radius $a \gg \lambda$ centered at the origin in the plane $z = 0$ and illuminated by a plane wave $\mathbf{E}_0 e^{i(kz-\omega t)}$. The goal is to calculate the diffracted intensity at a point P near the center of a screen placed at a distance z from the disk. Due to the rotational symmetry of the problem, we use the cylindrical coordinates.

(a) Using the Huygens–Fresnel principle, write an expression for the electric field at a point $P = (\rho, \theta)$ on the screen as an integral over the coordinates (ρ', θ') in the plane $z = 0$.

(b) At distances ρ from the axis much smaller than the radius a of the disk, show by a proper change of variables that the electric field can be written approximately as

$$\underline{E}(\rho, \theta, z, t) = -\frac{i\underline{E}_0 z e^{-i\omega t}}{\lambda} \int_0^{2\pi} \int_{R(\rho',\theta',z)}^{\infty} \frac{e^{ikr}}{r} dr d\theta'$$

with

$$R(\rho', \theta', z) \approx \sqrt{a^2 + z^2} + \frac{\rho^2}{2\sqrt{z^2 + a^2}} - \frac{\rho a \cos(\theta - \theta')}{\sqrt{z^2 + a^2}} .$$

(c) Now use the Sommerfeld lemma $\int_a^b f(x) e^{ikx} dx \approx [\frac{f(x)}{ik} e^{ikx}]_a^b$ for a slowly varying function f to show that the intensity of the Arago spot close to the axis writes

$$I(\rho, \theta, z) = \frac{I_0 z^2}{a^2 + z^2} J_0^2 \left(\frac{2\pi \rho a}{\lambda \sqrt{z^2 + a^2}} \right)$$

where $J_0(x) = \frac{1}{2\pi} \int_0^{2\pi} e^{ix \cos\phi} d\phi$ is a Bessel function of the first kind, whose graph is shown schematically in below.

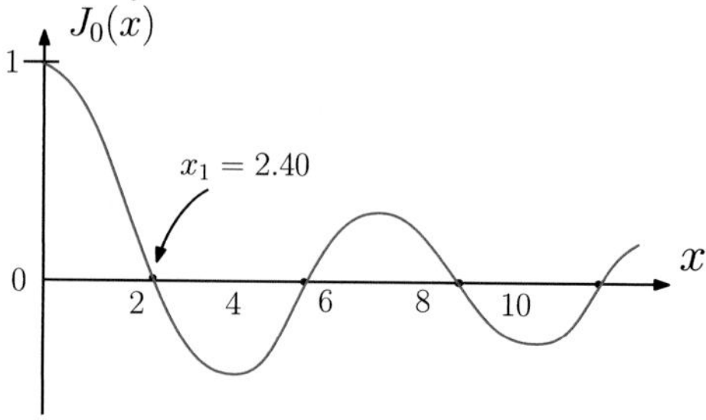

Based on this, sketch the wave intensity in a plane perpendicular to the z-axis.

(d) For which values of z is the Arago spot visible? Evaluate numerically for a disk of radius $a = 50$ μm and a red laser of wavelength $\lambda \sim 650$ nm.

18.3 The Fundamental Gaussian Mode

Gaussian modes play an important role in optics since they represent the electric field distribution at the output of most laser cavities with cylindrical symmetry. For such lasers, the fundamental mode corresponds to a solution of the paraxial equation (18.11) of the form

$$\mathcal{E}(x, y, z) = A(z)e^{ik\frac{x^2+y^2}{2q(z)}}$$

(a) Inject the proposed solution into the paraxial wave equation (18.11) and find the differential equation relating A and q.

(b) Integrate this equation and show that $q(z) = q_0 + (z - z_0)$ and $A(z) = \frac{A_0 q_0}{q(z)}$.

(c) We now consider the case of a complex q which we write as

$$q(z) = (z - z_0) - iz_R$$

where $z_R = \frac{\pi w_0^2}{\lambda}$ is called the Raleygh length. Defining the beam radius $w = w(z)$ and the radius of curvature $R(z)$ by

$$\frac{1}{q(z)} = \frac{1}{R(z)} + i\frac{\lambda}{\pi w(z)^2} \; ,$$

show that the electric field writes

$$\mathcal{E}(x, y, z) = \frac{1}{q(z)}e^{ik\frac{x^2+y^2}{2R(z)}} e^{-\frac{x^2+y^2}{w(z)^2}}$$

and that

$$R(z) = (z - z_0) + \frac{z_R^2}{(z - z_0)}$$

$$w(z) = w_0\sqrt{1 + \left(\frac{z - z_0}{z_R^2}\right)^2}$$

(d) Sketch the field intensity in a plane perpendicular to the z-axis. At which distance with respect to the axis the intensity decrease by a factor of $1/e^2$ with respect to its value on axis?

(e) Sketch the field intensity in a plane perpendicular to the z-axis.

18.4 The Descartes Relation

Consider a point source located on the z-axis at a distance d_1 from a thin lens of focal length f', placed at $z = 0$ such that its optical axis coincides with the z-axis. The source emits a spherical wave whose electric field has the form $\underline{E}_i(\mathbf{x}, t) = \underline{E}_0 \frac{e^{i(k|\mathbf{x}-\mathbf{x}_s|-\omega t)}}{|\mathbf{x}-\mathbf{x}_s|}$, where $\mathbf{x}_s = -d_1\mathbf{u}_z$ is the position of the source.

(a) Write an expression for the electric field $\underline{E}_0$ in the plane $z = 0$ immediately after the lens, at any point (x', y') such that $x'^2 + y'^2 \ll d_1^2, f'^2$ (paraxial approximation).

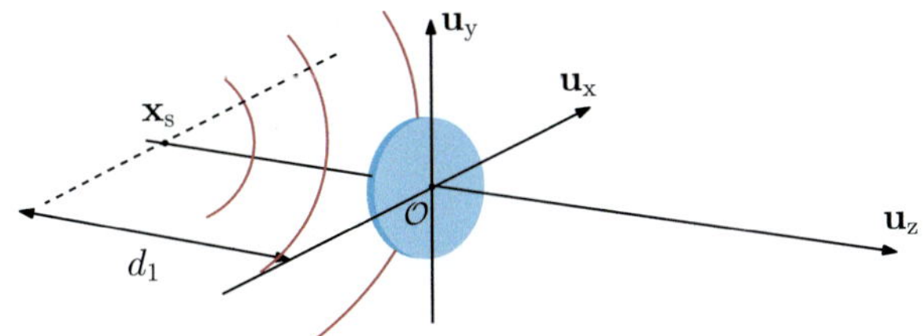

(b) Show that the field immediately after the lens is that of a point source located on the axis at a position $\mathbf{x}_i = d_2\mathbf{u}_z$, where d_2 is given by the Descartes relation

$$\frac{1}{d_1} + \frac{1}{d_2} = \frac{1}{f'}\,.$$

18.5 The Human Eye

The pupil diameter of a normal human eye can vary between 2 mm under bright light to 8 mm in the dark. What is the size of a point-source (very distant source of light) in the retina in the visible range ($\lambda = 500$ nm)? Recall that an emmetropic eye has a focal length $f' = 2.2$ cm and that the refractive index of the vitreous humor is $n' = 1.33$. Given that the minimum distance between cone cells in the fovea is around 2.5 μm, is the resolution of a human eye limited by diffraction?

18.6 Divergence of a Gaussian Beam

Consider a laser beam whose spatial profile in the waist plane $z = 0$ is given by a Gaussian in cylindrical coordinates:

$$\underline{E}(r, z = 0, t) = \underline{E}_0 e^{-i\omega t} e^{-\frac{r^2}{w_0^2}}$$

where w_0 represents the radius for which the intensity decreases by a factor of $1/e^2$ with respect to the center of the spot.

(a) In the case of a red He-Ne laser, $\lambda = 632.8$ nm and $w_0 \sim 1$ mm. Beyond which distance z^* from the laser tube can we approximate the field distribution by the Fraunhofer diffraction figure?

(b) What is the spatial profile of the beam as a function of the distance z from the laser tube in the far field approximation? How does the spatial extent of the laser intensity evolves with z?

(c) What is the angular divergence of the laser beam? Use the numerical values for the red He-Ne laser stated above to determine its angular divergence in mrads.

Chapter 19
Electromagnetic Radiation

Abstract This chapter provides a comprehensive treatment of *electromagnetic radiation*, focusing on how time-varying charge and current distributions generate fields that can propagate independently through space, carrying energy to infinity. It begins by deriving the *general solutions to Maxwell's equations* in the presence of time-dependent sources, introducing the concept of *retarded potentials* (Liénard–Wiechert potentials) to account for the finite speed of light and the causal relationship between sources and fields. The chapter then analyzes the *electromagnetic field generated by a point charge*, demonstrating that only an *accelerating charge* produces radiation fields that decay as $1/r$ and transport energy to large distances. It distinguishes between *near fields* (which behave like static fields at short distances) and *far fields* (*radiation fields*), emphasizing that the latter are responsible for electromagnetic radiation. Generalizing from a point charge, the chapter develops expressions for the *radiation fields* produced by arbitrary time-varying charge and current distributions confined within a finite volume. It introduces the *Poynting vector* to quantify the *radiated power* and its *angular distribution*, defining the *radiation pattern* as a key characteristic of emitting systems like antennas. A significant portion is dedicated to *electric dipole radiation*, which represents the dominant term in the multipole expansion for sources much smaller than the wavelength of the emitted radiation. The scalar and vector potentials for electric dipole radiation are derived, leading to the expressions for the electric and magnetic radiated fields, which exhibit the structure of a quasi-plane wave propagating radially. *Larmor's formula* for the total power radiated by an accelerating charge is derived, and its implications for classical atomic models (e.g., the instability of the classical hydrogen atom) are discussed. A significant portion is dedicated to *electric dipole radiation*, which represents the dominant term in the multipole expansion for sources much smaller than the wavelength of the emitted radiation. The scalar and vector potentials for electric dipole radiation are derived, leading to the expressions for the electric and radiated magnetic fields, which exhibit the structure of a quasi-plane wave propagating radially. *Larmor's formula* for the total power radiated by an accelerating charge is derived, and its implications for classical atomic models (e.g., the instability of the classical hydrogen atom) are discussed. Finally, the chapter explores the phenomenon of *scattering of electromagnetic waves* by matter. It details *Thomson scattering* (scattering by free electrons) and extends the analysis to *scattering from atoms and molecules*,

© The Author(s), under exclusive license to Springer Nature Switzerland AG 2025

F. Cadiz and A. Couairon, *Classical Electrodynamics*, Undergraduate Texts in Physics,
https://doi.org/10.1007/978-3-031-86785-9_19

incorporating Lorentz model to account for the binding forces on electrons. This leads to the derivation of the *scattering cross section*, which quantifies the effective area for scattering, and explains phenomena like *Rayleigh scattering* (responsible for the blue color of the sky) and *resonant scattering*.

Keywords Radiation · Larmor's formula · Scattering · Retarded potentials

19.1   Introduction

In the previous chapters, we have studied solutions to Maxwell's equations in the form of plane waves propagating in regions that are either empty of conduction charges (vacuum, linear homogeneous and isotropic dielectrics) or may contain free charge and current densities (metals). We discussed propagation, reflection and refraction of the electromagnetic field at an interface between two media, dispersion and absorption. However, we only considered waves propagating away from their emission region and did not investigate how these waves detach from their sources before propagating.

In this chapter, we explore the emission of electromagnetic radiation. Specifically, among all the fields generated by a charge or current distribution, we focus on those capable of detaching from their source and propagating independently. We begin by examining the fields produced by a point charge, demonstrating that only an accelerating charge generates an electromagnetic field that can propagate to infinity. This analysis is then extended to the case of an arbitrary time-varying charge distribution, which radiates energy in the form of electromagnetic fields.

Electromagnetic radiation is the fundamental physical process enabling signal transmission through space between two antennas, one serving as an emitter and the other as a receiver. While numerous types of antennas have been developed since the twentieth century, the physical principles governing their operation remain the same, relying on the emission and absorption of waves by moving charges.

These same principles also explain how an incident wave is scattered by an atom or molecule, producing a scattered field emitted by charges set in motion by the incident field. We will explore how these concepts help us understand natural phenomena such as the color of the sky and scattering phenomena in general.

19.2   General Solution to Maxwell's Equations

To start with, we will establish the general solution to Maxwell's equations in vacuum, with time dependent current or charge density. This will generalize what we already know from electrostatics and magnetostatics, and will show that electromagnetic radiation exhibits retardation with respect to its source due to the finite propagation velocity of electromagnetic waves.

We have already established in Chap. 13 that the propagation equations for the fields read

$$\nabla^2 \mathbf{E}(\mathbf{x}, t) - \epsilon\mu \frac{\partial^2 \mathbf{E}(\mathbf{x}, t)}{\partial t^2} = \frac{\nabla \varrho(\mathbf{x}, t)}{\epsilon} + \mu \frac{\partial \mathbf{j}(\mathbf{x}, t)}{\partial t} \tag{19.1}$$

for the electric field and

$$\nabla^2 \mathbf{B}(\mathbf{x}, t) - \epsilon\mu \frac{\partial^2 \mathbf{B}(\mathbf{x}, t)}{\partial t^2} = -\mu \nabla \times \mathbf{j}(\mathbf{x}, t) , \tag{19.2}$$

for the magnetic field. These equations, derived from Maxwell's equations, involve the six unknown components of the electric and magnetic fields and the source terms $\varrho(\mathbf{x}, t)$ and $\mathbf{j}(\mathbf{x}, t)$. However, only four components of the fields are independent since Maxwell's equations involve 8 scalar equations for 6 field components, meaning that two scalar equations can be viewed as constraints, leaving only 4 independent field-components. To simplify the search for general solutions to Eqs. (19.1) and (19.2) that also satisfy the original Maxwell equations,[1] we introduced the scalar and vector potentials, $V(\mathbf{x}, t)$ and $\mathbf{A}(\mathbf{x}, t)$, which indeed amounts to finding 4 scalar quantities. The potentials allow us to determine all field components from

$$\mathbf{B}(\mathbf{x}, t) = \nabla \times \mathbf{A}(\mathbf{x}, t),$$

and

$$\mathbf{E}(\mathbf{x}, t) = -\frac{\partial \mathbf{A}(\mathbf{x}, t)}{\partial t} - \nabla V(\mathbf{x}, t).$$

In order to find the equation of propagation for the potentials, Maxwell–Ampère equation can be rewritten using the potentials instead of the fields:

$$\nabla \times (\nabla \times \mathbf{A}) = \mu \mathbf{j} - \epsilon\mu \frac{\partial}{\partial t}\left(\nabla V + \frac{\partial \mathbf{A}}{\partial t}\right).$$

Then use the double curl identity (A.4) $\nabla \times (\nabla \times \mathbf{A}) = \nabla(\nabla \cdot \mathbf{A}) - \nabla^2 \mathbf{A}$ and group the terms that make up d'Alembert's equation on the left-hand side. On the right-hand side, two terms appear in the form of a gradient, after commuting spatial and temporal differentiation:

$$\nabla^2 \mathbf{A} - \epsilon\mu \frac{\partial^2 \mathbf{A}}{\partial t^2} = -\mu \mathbf{j} + \nabla\left(\epsilon\mu \frac{\partial V}{\partial t} + \nabla \cdot \mathbf{A}\right).$$

Now remember that the potentials V and $\mathbf{A}$ are not uniquely defined. It is always possible to add the gradient of a scalar quantity $\phi(\mathbf{x}, t)$ to the vector potential and subtract

---

[1] Solving directly Eqs. (19.1) and (19.2) does not guarantee that $\nabla \cdot \mathbf{B} = 0$ and $\nabla \cdot (\epsilon\mathbf{E}) = \varrho$, even if we used these equations to derive the wave equations (19.1) and (19.2).

the time-derivative of this quantity from the scalar potential to define exactly the same electric and magnetic fields. This means that an additional condition is required to uniquely define the potentials. This is called a *Gauge condition*. A convenient choice is the *Lorenz gauge* (Fig. 19.1), which reads

$$\epsilon\mu\frac{\partial V}{\partial t} + \mathbf{\nabla}\cdot\mathbf{A} = 0.$$

With this condition, the vector potential satisfies the propagation equation

$$\boxed{\mathbf{\nabla}^2\mathbf{A} - \epsilon\mu\frac{\partial^2\mathbf{A}}{\partial t^2} = -\mu\mathbf{j}\,.}\tag{19.3}$$

The scalar potential also satisfies a propagation equation that can be obtained from Maxwell–Gauss equation

$$\mathbf{\nabla}\cdot\mathbf{E} = \frac{\varrho}{\epsilon} \quad\rightarrow\quad \mathbf{\nabla}\cdot\left(-\mathbf{\nabla}V - \frac{\partial\mathbf{A}}{\partial t}\right) = \frac{\varrho}{\epsilon} \quad\rightarrow\quad \mathbf{\nabla}^2 V + \mathbf{\nabla}\cdot\frac{\partial\mathbf{A}}{\partial t} = -\frac{\varrho}{\epsilon}.$$

By differentiating the Lorenz gauge with respect to time, we find

$$\epsilon\mu\frac{\partial^2 V}{\partial t^2} = -\frac{\partial\mathbf{\nabla}\cdot\mathbf{A}}{\partial t} = -\mathbf{\nabla}\cdot\frac{\partial\mathbf{A}}{\partial t},$$

which allows us to write the propagation equation for the scalar potential as d'Alembert's equation

$$\boxed{\mathbf{\nabla}^2 V - \epsilon\mu\frac{\partial^2 V}{\partial t^2} = -\frac{\varrho}{\epsilon}\,.}\tag{19.4}$$

Both potentials satisfy an inhomogeneous wave equation, the solution of which guarantees that the electric and magnetic fields are solutions to the original Maxwell equations. In the static case for which the sources do not depend on time $\varrho = \varrho(\mathbf{x})$, $\mathbf{j} = \mathbf{j}(\mathbf{x})$, the Lorenz gauge is equivalent to the Coulomb gauge ($\mathbf{\nabla}\cdot\mathbf{A} = 0$) and the well-known Poisson equation is retrieved for both V and $\mathbf{A}$

$$\mathbf{\nabla}^2 V(\mathbf{x}) = -\frac{\varrho(\mathbf{x})}{\varepsilon},$$

$$\mathbf{\nabla}^2\mathbf{A}(\mathbf{x}) = -\mu\mathbf{j}(\mathbf{x}),$$

which admit the static solutions

$$V(\mathbf{x}) = \frac{1}{4\pi\epsilon}\iiint_{\mathbb{R}^3}\frac{\varrho(\mathbf{x}')}{|\mathbf{x} - \mathbf{x}'|}d^3x',$$

Fig. 19.1 Ludvig Lorenz (1829–1891), a Danish physicist and mathematician. He made major contributions to electromagnetism and optics. He developed the Lorenz–Mie theory, which explains light scattering by spherical particles

$$\mathbf{A}(\mathbf{x}) = \frac{\mu}{4\pi} \iiint_{\mathbb{R}^3} \frac{\mathbf{j}(\mathbf{x}')}{|\mathbf{x} - \mathbf{x}'|} d^3 x'.$$

For the following of this chapter, we will assume that the sources are embedded in vacuum so that $\epsilon_0 = \epsilon$ and $\mu = \mu_0$.

19.2.1 Retarded Potentials—Solution of the Inhomogeneous Wave Equation

Since the three components of the vector potential and the scalar potential satisfy similar equations only involving a different source term, we can find the complete electromagnetic field if we find the formal solution to the following scalar inhomogneneous wave equation

$$\left(\nabla^2 - \frac{1}{c^2}\frac{\partial^2}{\partial t^2}\right)\phi(\mathbf{x}, t) = -4\pi f(\mathbf{x}, t) , \tag{19.5}$$

where $\phi(\mathbf{x}, t)$ represents the scalar potential or any component of the vector potential, and $f(\mathbf{x}, t)$ denotes the corresponding function of space and time that plays the role of the source term. In this aim, we define the Green function that corresponds to the solution that decays to zero at infinity when we have a point source at $x = x'$ and $t = t'$:

$$\left(\nabla^2 - \frac{1}{c^2}\frac{\partial^2}{\partial t^2}\right) G(\mathbf{x} - \mathbf{x}', t - t') = -4\pi \delta(t - t')\delta(\mathbf{x} - \mathbf{x}') . \tag{19.6}$$

This is a generalization of Eq. (4.10) for the Green function defined in Sect. 4.5 for Poisson's equation, which is a particular case of (19.5). In terms of Green's function, the general solution for $\phi(\mathbf{x}, t)$ takes the form

$$\phi(\mathbf{x}, t) = \int\limits_{-\infty}^{+\infty} \iiint\limits_{\mathbb{R}^3} G(\mathbf{x} - \mathbf{x}', t - t') f(\mathbf{x}', t') d^3 x' dt'.$$

In the Appendix A.11, it is shown that there are two possible solutions for the Green function:

$$G(\mathbf{x} - \mathbf{x}', t - t') = \frac{\delta(t - t' \mp |\mathbf{x} - \mathbf{x}'|/c)}{|\mathbf{x} - \mathbf{x}'|},$$

The retarded Green function is obtained by taking the solution with the minus sign

$$G_-(\mathbf{x} - \mathbf{x}', t - t') = \frac{\delta(t - t' - |\mathbf{x} - \mathbf{x}'|/c)}{|\mathbf{x} - \mathbf{x}'|}.$$

The advanced Green function is obtained with the plus sign

$$G_+(\mathbf{x} - \mathbf{x}', t - t') = \frac{\delta(t - t' + |\mathbf{x} - \mathbf{x}'|/c)}{|\mathbf{x} - \mathbf{x}'|}.$$

Mathematically, both are solutions to the wave equation. Physically, note that $\frac{|\mathbf{x} - \mathbf{x}'|}{c} = \frac{r}{c}$ represents the time it takes for a signal to propagate at speed c over a distance $r = |\mathbf{x} - \mathbf{x}'|$. Thus the retarded Green function represents the impulse response at point $\mathbf{x}$ and time t to a source located at $\mathbf{x}'$ in the state it had at the retarded time $t' = t - r/c$, that is, the signal at an observation point $\mathbf{x}$ and time t follows the emission by a source a distance r away from $\mathbf{x}$ after a delay corresponding to the signal propagation from the source to the observation point. This solution is in agreement with our intuition that cause precedes effect.

Similarly, the advanced Green function represents the impulse response observed at point $\mathbf{x}$ and time t to a source located at $\mathbf{x}'$, a distance r away from the observer, in its future state at time $t' = t + r/c$. Although it is a perfectly valid solution from the mathematical point of view, the advanced solution is not causal.

The retarded Green function is causal, and it is the only physically acceptable solution, therefore:

$$\phi(t, \mathbf{x}) = \int\limits_{-\infty}^{+\infty} \iiint\limits_{\mathbb{R}^3} \frac{\delta(t - t' - |\mathbf{x} - \mathbf{x}'|/c)}{|\mathbf{x} - \mathbf{x}'|} f(t', \mathbf{x}') d^3 x' dt'$$

$$= \iiint\limits_{\mathbb{R}^3} \frac{f(t - |\mathbf{x} - \mathbf{x}'|/c, \mathbf{x}')}{|\mathbf{x} - \mathbf{x}'|} d^3 x'.$$

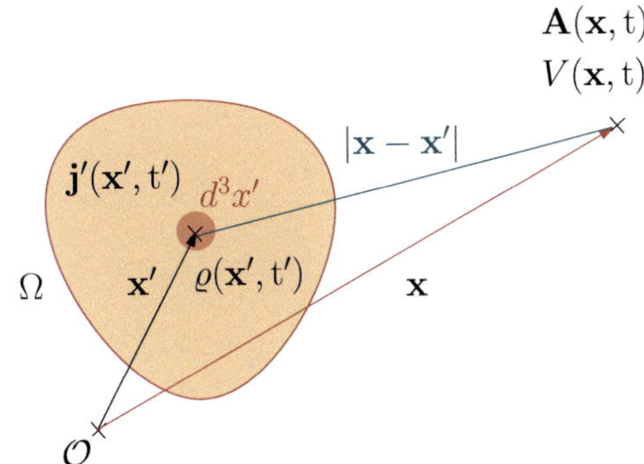

Fig. 19.2 The influence of the charge and current at point $\mathbf{x}'$ takes a time $|\mathbf{x} - \mathbf{x}'|/c$ to influence the values of the potentials (and fields) at position x. Electrodynamics propagates at speed c

The retarded potentials

The (causal) solution for the propagation equation for the potentials is given by

$$V(\mathbf{x}, t) = \frac{1}{4\pi \epsilon_0} \iiint_{\mathbb{R}^3} \frac{\varrho(\mathbf{x}', t - |\mathbf{x} - \mathbf{x}'|/c)}{|\mathbf{x} - \mathbf{x}'|} d^3 x' , \tag{19.7}$$

$$\mathbf{A}(\mathbf{x}, t) = \frac{\mu_0}{4\pi} \iiint_{\mathbb{R}^3} \frac{\mathbf{j}(\mathbf{x}', t - |\mathbf{x} - \mathbf{x}'|/c)}{|\mathbf{x} - \mathbf{x}'|} d^3 x' . \tag{19.8}$$

In these expressions, the delay $|\mathbf{x} - \mathbf{x}'|/c$ that appears in the time dependency corresponds to the time it takes light to travel the distance that separates the point of emission $\mathbf{x}'$ from the observation point $\mathbf{x}$, see Fig. 19.2. This solution to Maxwell's equations makes it obvious that electrodynamics is spreading at speed c. This is why, for example, when we observe a star located several light-years from the Earth, this star may have already disappeared. We also see that in Eqs. (19.7) and (19.8), the potentials decay as $1/r$, which is necessary to satisfy the conservation of energy, as will be shown later.

19.3 Electromagnetic Field Generated by a Point Charge

Consider a point charge q moving along a trajectory $\mathbf{r}(t)$. The charge density at instant t is given by $\varrho(\mathbf{x}, t) = q\delta(\mathbf{x} - \mathbf{r}(t))$. To the motion of the charge there is also an associated current density $\mathbf{j}(\mathbf{x}, t) = q\mathbf{v}(t)\delta(\mathbf{x} - \mathbf{r}(t))$ with $\mathbf{v} = \frac{d\mathbf{r}}{dt}$ the velocity

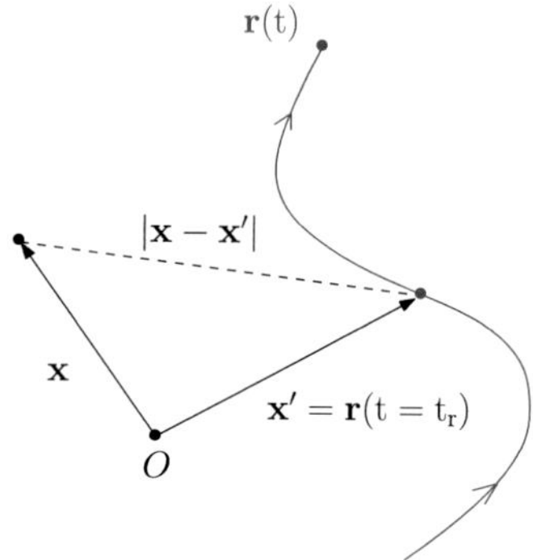

Fig. 19.3 The potentials at $\mathbf{x}$ at instant t depend on the position of the charge at time $t_r = t - |\mathbf{x} - \mathbf{r}(t_r)|/c$

of the charge at time t. Defining $t_r = t - |\mathbf{x} - \mathbf{x}'|/c$, and replacing into (19.7) and (19.8) we obtain the potentials

$$V(\mathbf{x}, t) = \frac{q}{4\pi \epsilon_0} \iiint_{\mathbb{R}^3} \frac{\delta\left(\mathbf{x}' - \mathbf{r}(t_r)\right)}{|\mathbf{x} - \mathbf{x}'|} d^3 x',$$

$$\mathbf{A}(\mathbf{x}, t) = \frac{q}{4\pi \epsilon_0 c^2} \iiint_{\mathbb{R}^3} \frac{\mathbf{v}(t_r)\delta\left(\mathbf{x}' - \mathbf{r}(t_r)\right)}{|\mathbf{x} - \mathbf{x}'|} d^3 x'.$$

Here, we must be careful when *integrating* the Dirac distribution because t_r is a function of $\mathbf{x}'$. We use the following property of the Dirac distribution

$$\delta(\mathbf{g}(\mathbf{x}')) = \sum_{\mathbf{x}_0} \frac{\delta(\mathbf{x}' - \mathbf{x}_0)}{|\det J(\mathbf{x}_0)|} \tag{19.9}$$

where the sum is made over all the isolated roots $\mathbf{x}_0$ of $\mathbf{g}$ and J is the Jacobian matrix of $\mathbf{g}$ with respect to $\mathbf{x}'$. Here, we consider $\mathbf{g}(\mathbf{x}') = \mathbf{x}' - \mathbf{r}(t_r) = \mathbf{x}' - \mathbf{r}(t - |\mathbf{x} - \mathbf{x}'|/c)$. This has an isolated root given by the condition

$$\mathbf{g}(\mathbf{x}') = \mathbf{0} \quad \rightarrow \quad \mathbf{x}' = \mathbf{r}(t - |\mathbf{x} - \mathbf{x}'|/c).$$

This condition tell us that $\mathbf{x}'$ must be a point along the past trajectory of the particle. At instant t, $\mathbf{x}'$ is the position of the particle at $t - \Delta t$, such that light emitted from $\mathbf{x}'$ reaches $\mathbf{x}$ at time t. This is shown in Fig. 19.3.

Let us calculate the elements of the Jacobian matrix

$$J_{ij} = \frac{\partial g_i}{\partial x'_j} = \frac{\partial (x'_i - r_i(t_r))}{\partial x'_j} = \delta_{ij} - \frac{\partial r_i}{\partial t_r} \frac{\partial t_r}{\partial x'_j}.$$

We have $\frac{\partial r_i}{\partial t_r} = v_i(t_r)$ with v the velocity of the charge, and

$$\frac{\partial t_r}{\partial x'_j} = \frac{1}{c} \frac{x_j - x'_j}{|\mathbf{x} - \mathbf{x}'|}$$

so that, finally

$$J_{ij} = \delta_{ij} - v_i(t_r)\frac{1}{c} \frac{x_j - x'_j}{|\mathbf{x} - \mathbf{x}'|}.$$

The determinant is

$$|\det J(\mathbf{x}_0)| = 1 - \mathbf{v}(t_r) \cdot \frac{\mathbf{x} - \mathbf{x}'}{c|\mathbf{x} - \mathbf{x}'|} > 0$$

and so, according to (19.9)

$$\delta(\mathbf{x}' - \mathbf{r}(t_r)) = \frac{\delta(x' - \mathbf{r}(t_r))}{1 - \frac{\mathbf{v}(t_r) \cdot (\mathbf{x} - \mathbf{r}(t_r))}{c|\mathbf{x} - \mathbf{r}(t_r)|}},$$

which gives the Liénard–Wiechert potentials (Fig. 19.4)

$$V(\mathbf{x}, t) = \frac{q}{4\pi \epsilon_0} \frac{1}{|\mathbf{x} - \mathbf{r}(t_r)| - \frac{\mathbf{v}(t_r)}{c} \cdot (\mathbf{x} - \mathbf{r}(t_r))}, \tag{19.10}$$

$$\mathbf{A}(\mathbf{x}, t) = \frac{q}{4\pi \epsilon_0 c^2} \frac{\mathbf{v}(t_r)}{|\mathbf{x} - \mathbf{r}(t_r)| - \frac{\mathbf{v}(t_r)}{c} \cdot (\mathbf{x} - \mathbf{r}(t_r))}, \tag{19.11}$$

with $t_r = t - \frac{|\mathbf{x}-\mathbf{r}(t_r)|}{c}$. From these potentials we can compute the electromagnetic field of the moving charge by using $\mathbf{E} = -\nabla V - \frac{\partial \mathbf{A}}{\partial t}$ and $\mathbf{B} = \nabla \times \mathbf{A}$. The detailed calculation can be found in A.11.1. The electric field generated by the moving charge writes

$$\mathbf{E}(\mathbf{r}, t) = \mathbf{E}_{\text{near}}(\mathbf{r}, t) + \mathbf{E}_{\text{far}}(\mathbf{r}, t) \tag{19.12}$$

where

$$\boxed{\mathbf{E}_{\text{near}}(\mathbf{r}, t) = \frac{q}{4\pi \epsilon_0 |\mathbf{x} - \mathbf{r}(t_r)|^2} \frac{\left(\mathbf{u}_r(t_r) - \frac{\mathbf{v}(t_r)}{c}\right)\left(1 - \frac{|\mathbf{v}(t_r)|^2}{c^2}\right)}{\left(1 - \frac{\mathbf{v}(t_r)}{c} \cdot \mathbf{u}(t_r)\right)^3}} \tag{19.13}$$

and

$$\mathbf{E}_{\mathrm{far}}(\mathbf{r}, t) = \frac{q}{4\pi\epsilon_0 |\mathbf{x} - \mathbf{r}(t_r)|} \frac{\frac{\mathbf{a}(t_r)}{c^2} \cdot \mathbf{u}_r(t_r)\left(\mathbf{u}_r(t_r) - \frac{\mathbf{v}(t_r)}{c}\right) - \frac{\mathbf{a}(t_r)}{c^2}\left(1 - \frac{\mathbf{v}(t_r)}{c} \cdot \mathbf{u}_r(t_r)\right)}{\left(1 - \frac{\mathbf{v}(t_r)}{c} \cdot \mathbf{u}_r(t_r)\right)^3}, \tag{19.14}$$

where $\mathbf{u}_r(t_r) = \frac{\mathbf{x} - \mathbf{r}(t_r)}{|\mathbf{x} - \mathbf{r}(t_r)|}$ is the unit vector that joins the retarded position to point $\mathbf{x}$, and $\mathbf{a}$ is the acceleration of the charge. By using the identity $\mathbf{u} \times (\mathbf{z} \times \mathbf{w}) = (\mathbf{u} \cdot \mathbf{w})\mathbf{z} - (\mathbf{u} \cdot \mathbf{z})\mathbf{w}$ with $\mathbf{u} = \frac{\mathbf{x} - \mathbf{r}}{|\mathbf{x} - \mathbf{r}|}$, $\mathbf{z} = \frac{\mathbf{x} - \mathbf{r}}{|\mathbf{x} - \mathbf{r}|} - \frac{\mathbf{v}}{c}$, and $\mathbf{w} = \frac{\mathbf{a}}{c^2}$, Eq. (19.14) can be also written as:

$$\mathbf{E}_{\mathrm{far}}(\mathbf{r}, t) = \frac{q}{4\pi\epsilon_0 |\mathbf{x} - \mathbf{r}(t_r)|} \frac{\mathbf{u}_r(t_r) \times \left(\left(\mathbf{u}_r(t_r) - \frac{\mathbf{v}}{c}\right) \times \frac{\mathbf{a}(t_r)}{c^2}\right)}{\left(1 - \frac{\mathbf{v}(t_r)}{c} \cdot \mathbf{u}_r(t_r)\right)^3}. \tag{19.15}$$

Finally, the magnetic field is found according to $\mathbf{B} = \nabla \times \mathbf{A}$. It is found that:

$$\mathbf{B}(\mathbf{r}, t) = \mathbf{B}_{\mathrm{near}}(\mathbf{r}, t) + \mathbf{B}_{\mathrm{far}}(\mathbf{r}, t) \tag{19.16}$$

with

$$\mathbf{B}_{\mathrm{near}}(\mathbf{r}, t) = \frac{q}{4\pi\epsilon_0 c^2 |\mathbf{x} - \mathbf{r}(t_r)|^2} \frac{\mathbf{v}(t_r) \times \mathbf{u}_r(t_r)\left(1 - \frac{|\mathbf{v}(t_r)|^2}{c^2}\right)}{\left(1 - \frac{\mathbf{v}(t_r)}{c} \cdot \mathbf{u}_r(t_r)\right)^3} \tag{19.17}$$

and

$$\mathbf{B}_{\mathrm{far}}(\mathbf{r}, t) = \frac{q}{4\pi\epsilon_0 c^2 |\mathbf{x} - \mathbf{r}(t_r)|} \frac{\mathbf{v}(t_r) \times \mathbf{u}_r(t_r)\left(\frac{\mathbf{a}}{c^2} \cdot \mathbf{u}_r(t_r)\right) - \mathbf{u}_r(t_r) \times \frac{\mathbf{a}(t_r)}{c}\left(1 - \frac{\mathbf{v}(t_r)}{c} \cdot \mathbf{u}_r(t_r)\right)}{\left(1 - \frac{\mathbf{v}(t_r)}{c} \cdot \mathbf{u}_r(t_r)\right)^3}. \tag{19.18}$$

We see that

$$\mathbf{B}_{\mathrm{near}}(\mathbf{r}, t) = \frac{\mathbf{u}_r(t_r)}{c} \times \mathbf{E}_{\mathrm{near}}(\mathbf{r}, t) = \frac{\mathbf{v}(t_r)}{c^2} \times \mathbf{E}_{\mathrm{near}},$$

$$\mathbf{B}_{\mathrm{far}}(\mathbf{r}, t) = \frac{\mathbf{u}_r(t_r)}{c} \times \mathbf{E}_{\mathrm{far}}(\mathbf{r}, t),$$

so

$$\mathbf{B}(\mathbf{r}, t) = \frac{\mathbf{u}_r(t_r)}{c} \times \mathbf{E}(\mathbf{r}, t). \tag{19.19}$$

19.3.1 The Near Fields

Note that $\mathbf{B}_{\text{near}}$ and $\mathbf{E}_{\text{near}}$ are independent on the charge's acceleration and both decay as the inverse of the distance squared, just like for the static fields. In the limit of low velocities compared to the speed of light ($|\mathbf{v}|/c \to 0$), we retrieve the static formulas (1.4) and (8.7). The direction of the electric field (19.13) is that of Fig. 19.5.

$$\frac{\mathbf{x} - \mathbf{r}(t_r)}{|\mathbf{x} - \mathbf{r}(t_r)|} - \frac{\mathbf{v}(t_r)}{c} = \frac{\mathbf{x} - \mathbf{r}(t_r) - \mathbf{v}(t_r)|\mathbf{x} - \mathbf{r}(t_r)|/c}{|\mathbf{x} - \mathbf{r}(t_r)|} = \frac{\mathbf{x} - (\mathbf{r}(t_r) + \mathbf{v}(t_r)(t - t_r))}{|\mathbf{x} - \mathbf{r}(t_r)|}.$$

As illustrated in Fig. 19.5, the near field at time t and position $\mathbf{x}$ is radial from the instantaneous position $\mathbf{r}(t_r) + (t - t_r)\mathbf{v}(t_r)$ that the charge would have at time t if it was moving at constant velocity.

In particular, this means that for a charge moving at constant velocity, the near field updates instantaneously along the radial line that joins the current position of the charge and point $\mathbf{x}$. Indeed, for a constant velocity $\mathbf{v}$, we have $\mathbf{r}(t_r) + (t - t_r)\mathbf{v} = \mathbf{r}(t)$. Equation(19.13) simplifies into:

$$\mathbf{E}_{\text{near}}(\mathbf{r}, t) = \frac{q}{4\pi\epsilon_0|\mathbf{x} - \mathbf{r}(t_r)|^3} \frac{(\mathbf{x} - \mathbf{r}(t))\left(1 - \frac{v^2}{c^2}\right)}{\left(1 - \frac{\mathbf{v}}{c} \cdot \frac{\mathbf{x}-\mathbf{r}(t_r)}{|\mathbf{x}-\mathbf{r}(t_r)|}\right)^3}. \tag{19.20}$$

Moreover, we can write

$$|\mathbf{x} - \mathbf{r}(t)|^2 = |\mathbf{x} - \mathbf{r}(t_r) - \mathbf{v}|\mathbf{x} - \mathbf{r}(t_r)|/c|^2$$
$$= |\mathbf{x} - \mathbf{r}(t_r)|^2\left(1 + \frac{v^2}{c^2}\right) - 2\frac{\mathbf{v}}{c} \cdot (\mathbf{x} - \mathbf{r}(t_r))|\mathbf{x} - \mathbf{r}(t_r)|$$

Fig. 19.4 Left: Alfred-Marie Liénard (1869–1958), a French physicist and engineer best known for his work in electromagnetism. He contributed to the study of oscillatory circuits and dynamical systems. Right: Emil Wiechert (1861–1928), a German physicist and geophysicist who independently derived the Liénard–Wiechert potentials. He also made significant contributions to seismology, including the development of early seismographs

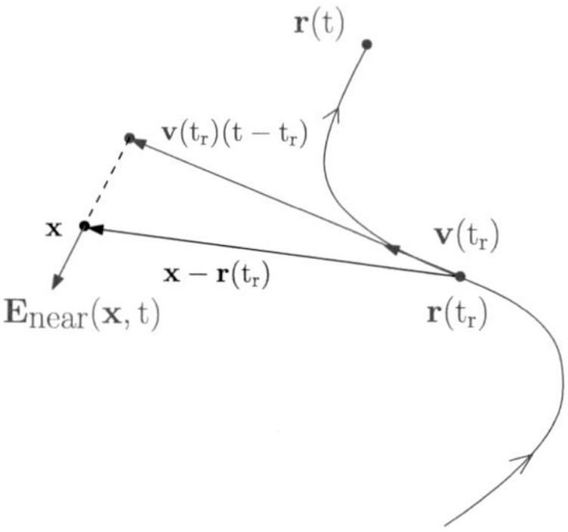

Fig. 19.5 The near field at **x** is along the line that joins **x** with the instantaneous position of the charge if the latter moves at constant velocity

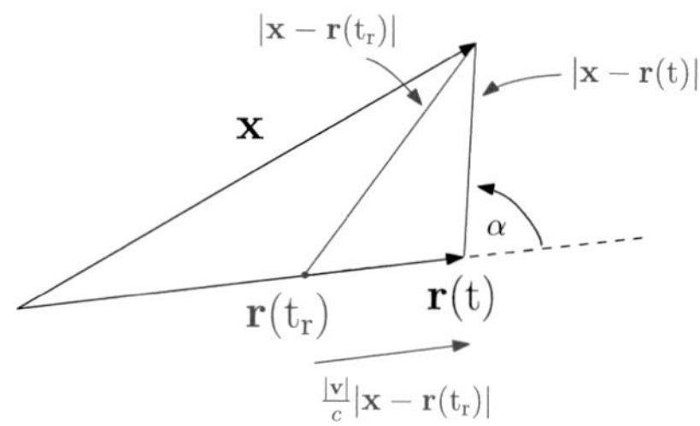

Fig. 19.6 The angle α is the angle between the velocity and $\mathbf{x} - \mathbf{r}(t)$

and using the cosine theorem on the triangle shown in Fig. 19.6,

$$|\mathbf{x} - \mathbf{r}(t_r)|^2 = \frac{v^2}{c^2}|\mathbf{x} - \mathbf{r}(t_r)|^2 + |\mathbf{x} - \mathbf{r}(t)|^2 + 2\frac{v}{c}|\mathbf{x} - \mathbf{r}(t_r)||\mathbf{x} - \mathbf{r}(t)|\cos\alpha ,$$

$$(19.21)$$

we can solve for the denominator in (19.20)

$$|\mathbf{x} - \mathbf{r}(t_r)|\left|1 - \frac{\mathbf{v}}{c}\cdot\frac{\mathbf{x} - \mathbf{r}(t_r)}{|\mathbf{x} - \mathbf{r}(t_r)|}\right| = |\mathbf{x} - \mathbf{r}(t)|\left|\frac{|\mathbf{x} - \mathbf{r}(t_r)|}{|\mathbf{x} - \mathbf{r}(t)|}\left(1 - \frac{v^2}{c^2}\right) - \frac{v}{c}\cos\alpha\right| .$$

From (19.21), $\frac{|\mathbf{x} - \mathbf{r}(t_r)|}{|\mathbf{x} - \mathbf{r}(t)|}(1 - \frac{v^2}{c^2}) = \frac{v}{c}\cos\alpha + \sqrt{(v/c)^2\cos^2\alpha + 1 - (v/c)^2}$ and we get the final result:

$$\mathbf{E}_{\text{near}}(\mathbf{r}, t) = \frac{q}{4\pi\epsilon_0|\mathbf{x} - \mathbf{r}(t)|^3}\frac{(\mathbf{x} - \mathbf{r}(t))\left(1 - \frac{v^2}{c^2}\right)}{\left(1 - \left(\frac{v}{c}\right)^2\sin^2\alpha\right)^{3/2}} .$$

$$(19.22)$$

We see from (19.22) that the field of the moving charge is radial and takes the form of Coulomb's law when $v/c \to 0$, which is valid only in the frame where the charge is at rest. Note that the amplitude of the field is corrected by the factor $(1 - (v/c)^2)/(1 - \left(\frac{v}{c}\right)^2\sin^2\alpha)^{3/2}$, which breaks the spherical symmetry of the field.

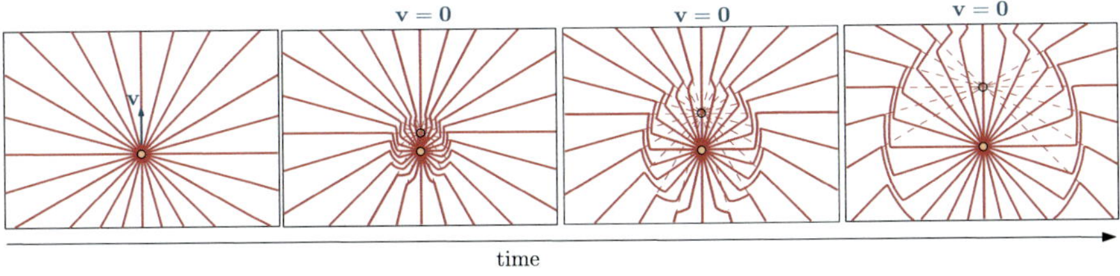

time

Fig. 19.7 An electromagnetic wave is produced by a charge's acceleration

This asymmetry is a relativistic effect that will be further explored in Chap. 20. Indeed, the Liénard–Wiechert fields derived here are fully compatible with special relativity, which is not surprising given that Maxwell's equations form the foundation of its construction.

The fact that the field no longer depends on the retarded time and responds to the instantaneous position of the charge is also consistent with the principle of relativity: in the inertial frame where the charge is at rest, it produces a static Coulomb field. In this frame, it is the observer that moves at constant velocity, and perceives a static electric field whose direction updates instantaneously to remain radial, with the charge at its origin. In summary, there are no propagation effects on the near fields.

19.3.2 The Far Fields

Let us now discuss the fields far from the charge, in which case the fields given by (19.14) and (19.18) dominate significantly with respect to the near fields when the charge's acceleration is non-zero. These fields depend on the position of the charge at the retarded time and are responsible for the emission of electromagnetic radiation. Figure 19.7 shows the electric field lines generated by a point charge initially moving at constant speed. When the charge suddenly stops, the field does not respond instantaneously. After a certain time, the fields update according to the new state of the charge, but only within a finite region of space. Outside this region, the field lines remain those of a charge still moving at constant speed. The frontier between these two regions expands outward at the speed of light as time increases. Such disturbance, created by the rapid deceleration of the charge, propagates through space and carries away energy to infinity. This phenomenom is called radiation.

A charge that continuously oscillates back and forth emits electromagnetic waves persistently, and this is the principle behind an antenna. As Fig. 19.8 shows, the density of electromagnetic radiation (indicated by closer field lines) is maximum in the plane perpendicular to the motion of the charge, and vanishes along the axis of motion. Indeed, Eq. (19.15) shows that the far fields are zero in the direction parallel to the acceleration. These results will be further generalized in the following section.

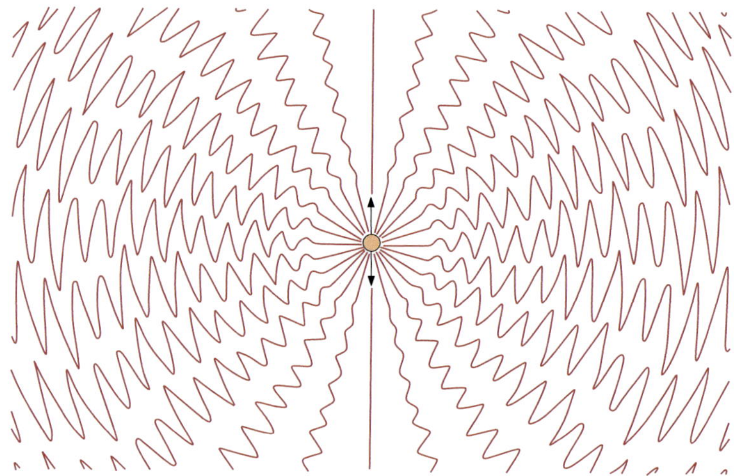

Fig. 19.8 An oscillating charge is a source of electromagnetic waves

19.4 Radiation

The purpose of this section is to generalize the results obtained for a point charge and to obtain expressions for the potentials and fields radiated by an arbitrary charge distribution that we suppose to be confined within a finite volume Ω. We will see that the dipole moment of the distribution plays a fundamental role. While the charge density is time dependent in general, the total charge within Ω is not, in agreement with charge conservation and our assumption that the conduction current at the surface of Ω is zero for the charge to remain localized.

19.4.1 The Scalar Potential

The scalar potential is given by (19.7)

$$V(\mathbf{x}, t) = \frac{1}{4\pi\epsilon_0} \iiint_{\Omega} \frac{\varrho(\mathbf{x}', t - |\mathbf{x} - \mathbf{x}'|/c)}{|\mathbf{x} - \mathbf{x}'|} d^3 x'.$$

Let us first examine what happens near the volume Ω, that is, when $|\mathbf{x} - \mathbf{x}'| \ll cT$, where T denotes a time scale for the variation of the charge density. If we consider a sinusoidal signal of frequency ω, then $T = 2\pi/\omega$ and the condition reads $|\mathbf{x} - \mathbf{x}'| \ll \lambda$, defining a near zone as the region at less than a wavelength from the source Ω.

Using a Taylor expansion to first order, we can write

$$\varrho(\mathbf{x}', t - |\mathbf{x} - \mathbf{x}'|/c) \approx \varrho(\mathbf{x}', t) - \frac{1}{c}\frac{\partial \varrho(\mathbf{x}', t)}{\partial t}|\mathbf{x} - \mathbf{x}'| . \tag{19.23}$$

This expansion leads to

$$V(\mathbf{x}, t) = \frac{1}{4\pi \epsilon_0}\left\{ \iiint_\Omega \frac{\varrho(\mathbf{x}', t)}{|\mathbf{x} - \mathbf{x}'|} d^3 x' - \frac{1}{c}\iiint_\Omega \frac{\partial \varrho(\mathbf{x}', t)}{\partial t} d^3 x' + \cdots \right\} .$$

The second term is null, given the conservation of the total charge in Ω. Finally, the potential near the source is obtained:

$$V_{\text{near}}(\mathbf{x}, t) = \frac{1}{4\pi \epsilon_0}\iiint_\Omega \frac{\varrho(\mathbf{x}', t)}{|\mathbf{x} - \mathbf{x}'|} d^3 x' .$$

We see that in the near zone, the delay that represents the propagation time from the source to the observation point is neglected, and the potential obtained is equal to the static potential.

Now we consider the opposite case, far enough from Ω, i.e., using an origin close to Ω,

$$|\mathbf{x}| \gg |\mathbf{x}'|, \qquad \forall \mathbf{x}' \in \Omega.$$

Writing in spherical coordinates $\mathbf{x} = r\mathbf{u}_r$ leads to the expansion

$$|\mathbf{x} - \mathbf{x}'| = r\sqrt{1 - \frac{2\mathbf{u}_r \cdot \mathbf{x}'}{r} + \frac{|\mathbf{x}'|^2}{r^2}} \approx r\left(1 - \frac{\mathbf{u}_r \cdot \mathbf{x}'}{r} + O\left(\frac{|\mathbf{x}'|^2}{r^2}\right)\right) . \tag{19.24}$$

Then, at first order, this yields the following expression:

$$\boxed{V_{\text{far}}(\mathbf{x}, t) = \frac{1}{4\pi \epsilon_0 r}\iiint_\Omega \varrho\left(\mathbf{x}', t - \frac{r}{c} + \frac{\mathbf{u}_r \cdot \mathbf{x}'}{c}\right) d^3 x' .} \tag{19.25}$$

In the denominator it is sufficient to retain only the first term r of the expansion since only the fields decaying as $1/r$ can propagate energy to infinity, while in the numerator, we retain a correction to the main term. For this reason, the charge density is evaluated at time $t^* = t - r/c + \mathbf{u}_r \cdot \mathbf{x}'/c$, which is simply interpreted as the sum of the retarded time

$$\tau = t - \frac{r}{c}$$

and a correction $\mathbf{u}_r \cdot \mathbf{x}'/c$ that has the order or magnitude of L/c, the time required for a signal to propagate across the source of dimension L at velocity c. This correction is also important as it retains the angular dependence of the fields in the far-zone.[2]

19.4.2   Vector Potential

We can use the same expansions to express the vector potential

$$\mathbf{A}(\mathbf{x}, t) = \frac{\mu_0}{4\pi} \iiint_{\mathbb{R}^3} \frac{\mathbf{j}(\mathbf{x}', t - |\mathbf{x} - \mathbf{x}'|/c)}{|\mathbf{x} - \mathbf{x}'|} d^3 x' \tag{19.26}$$

in the near zone and in the far zone. Without any surprise, an expansion of the current density in Eq. (19.26) similarly to the expansion of the charge density Eq. (19.23) in the near zone $|\mathbf{x} - \mathbf{x}'| \ll \lambda$ leads to

$$\boxed{\mathbf{A}_{\mathrm{near}}(\mathbf{x}, t) = \frac{\mu_0}{4\pi} \iiint_{\mathbb{R}^3} \frac{\mathbf{j}(\mathbf{x}', t)}{|\mathbf{x} - \mathbf{x}'|} d^3 x'} \tag{19.27}$$

which is the static vector potential. We see that the expression known from magnetostatics is a particular case of the general solution. Simply allowing the current density to be time-dependent in this expression for the static vector potential leads to a solution in the time-dependent regime that has a validity range limited to the near-zone.

In the far zone $|\mathbf{x}'| \ll |\mathbf{x}| = r$, the expansion Eq. (19.24) at leading order yields

$$\boxed{\mathbf{A}_{\mathrm{far}}(\mathbf{x}, t) = \frac{\mu_0}{4\pi r} \iiint_{\mathbb{R}^3} \mathbf{j}\left(\mathbf{x}', t - \frac{r}{c} + \frac{\mathbf{u}_r \cdot \mathbf{x}'}{c}\right) d^3 x' \, .} \tag{19.28}$$

19.4.3   Radiation Fields

Radiation fields denote the prevailing terms of the electromagnetic field in the far-zone. As will be seen below, the electromagnetic field in the far-zone decays as $1/r$, in contrast with the $1/r^2$ Coulomb field of a point charge in electrostatics for instance. This has important consequences on the transported electromagnetic energy. Before discussing these consequences, expressions for the radiation fields are required. We will use the potentials in the far-zone Eqs. (19.28) and (19.25) to

---

[2] It is clear that without this correction, the scalar potential would have spherical symmetry.

find these expressions by means of the equations linking potentials and fields:

$$\mathbf{E}(\mathbf{x}, t) = -\frac{\partial \mathbf{A}(\mathbf{x}, t)}{\partial t} - \nabla V(\mathbf{x}, t) , \qquad (19.29)$$

$$\mathbf{B}(\mathbf{x}, t) = \nabla \times \mathbf{A}(\mathbf{x}, t) . \qquad (19.30)$$

We start with the magnetic field, retaining only the prevailing terms. The final result, whose demonstration is proposed as problem 19.2, reads

$$\mathbf{B}_{\mathrm{rad}}(\mathbf{x}, t) = -\frac{\mu_0}{4\pi c r} \mathbf{u}_r \times \frac{\partial}{\partial t} \left(\iiint_\Omega \mathbf{j}\left(\mathbf{x}', t - \frac{r}{c} + \frac{\mathbf{u}_r \cdot \mathbf{x}'}{c}\right) d^3 x' \right) . \qquad (19.31)$$

Proceeding with the radiated electric field and retaining only the dominant terms varying as $1/r$, the final result, proposed as problem 19.3, reads

$$\mathbf{E}_{\mathrm{rad}}(\mathbf{x}, t) = \frac{\mu_0}{4\pi r} \mathbf{u}_r \times \left(\mathbf{u}_r \times \frac{\partial}{\partial t} \iiint_\Omega \mathbf{j}\left(\mathbf{x}', t - \frac{r}{c} + \frac{\mathbf{u}_r \cdot \mathbf{x}'}{c}\right) d^3 x' \right) . \qquad (19.32)$$

Note that $\mathbf{E}_{\mathrm{rad}}$ and $\mathbf{B}_{\mathrm{rad}}$ can be entirely determined from the radiation vector potential

$$\mathbf{A}_{\mathrm{rad}}(\mathbf{x}, t) = \frac{\mu_0}{4\pi r} \iiint_\Omega \mathbf{j}\left(\mathbf{x}', t - \frac{r}{c} + \frac{\mathbf{u}_r \cdot \mathbf{x}'}{c}\right) d^3 x' . \qquad (19.33)$$

The radiation fields indeed read

$$\mathbf{B}_{\mathrm{rad}}(\mathbf{x}, t) = -\frac{\mathbf{u}_r}{c} \times \frac{\partial}{\partial t} \mathbf{A}_{\mathrm{rad}}(\mathbf{x}, t),$$

$$\mathbf{E}_{\mathrm{rad}}(\mathbf{x}, t) = -c \mathbf{u}_r \times \mathbf{B}_{\mathrm{rad}}(\mathbf{x}, t),$$

or equivalently to the last equation, we can write

$$\mathbf{B}_{\mathrm{rad}}(\mathbf{x}, t) = \frac{\mathbf{u}_r}{c} \times \mathbf{E}_{\mathrm{rad}}(\mathbf{x}, t).$$

This last expression confirms the result (19.19) obtained for a point charge. The electromagnetic radiation field thus corresponds to a spherical wave due to the dependence of $\mathbf{E}_{\mathrm{rad}}$ and $\mathbf{B}_{\mathrm{rad}}$ as $1/r$. As will be seen below, this means that the Poynting vector will transport electromagnetic radiation energy undiminished to radial distances arbitrarily far from the source. The facts that $|\mathbf{B}_{\mathrm{rad}}| = |\mathbf{E}_{\mathrm{rad}}|/c$ and $(\mathbf{E}_{\mathrm{rad}}, \mathbf{B}_{\mathrm{rad}}, \mathbf{u}_r)$ forms a right-handed orthogonal triad means that the electromagnetic radiation field has the structure of a quasi-plane wave, propagating along the radial direction.

19.4.4   The Poynting Vector and Radiated Energy

The electromagnetic power that is radiated far from the source is obtained from the flux of the Poynting vector through a spherical surface S of radius r centered on the source. Since $\mathbf{E} = -c\mathbf{u}_r \times \mathbf{B}$, the Poynting vector is radial and reads

$$\mathbf{\Pi}_{\text{rad}}(\mathbf{x}, t) = \frac{1}{\mu_0} \mathbf{E}_{\text{rad}}(\mathbf{x}, t) \times \mathbf{B}_{\text{rad}}(\mathbf{x}, t) = \frac{c}{\mu_0} \left| \mathbf{B}_{\text{rad}}(\mathbf{x}, t) \right|^2 \mathbf{u}_r$$

$$= \frac{1}{\mu_0 c} \left| \mathbf{u}_r \times \frac{\partial \mathbf{A}_{\text{rad}}(\mathbf{x}, t)}{\partial t} \right|^2 \mathbf{u}_r .$$

The power radiated by the distribution of charges is obtained by integrating the Poynting vector on a closed surface containing Ω. Considering a sphere of radius r and using the solid angle element $d\Omega$, defined from the surface element $d\mathbf{S}$ as $d\Omega = d\mathbf{S}.\mathbf{u}_r / r^2 = \sin\theta\, d\theta\, d\varphi$:

$$P = \oiint_S d\mathbf{S}(\mathbf{x}) \cdot \mathbf{\Pi}_{\text{rad}}(\mathbf{x}, t) = \frac{1}{\mu_0 c} \int_0^{4\pi} d\Omega \left| r\mathbf{u}_r \times \frac{\partial \mathbf{A}_{\text{rad}}(\mathbf{x}, t)}{\partial t} \right|^2 .$$

Since an expansion in powers of $1/r$ shows that $\mathbf{A}_{\text{rad}}(\mathbf{x}, t)$ varies as $1/r$ at leading order, we have

$$\lim_{r \to \infty} \left| r\mathbf{u}_r \times \frac{\partial \mathbf{A}_{\text{rad}}(\mathbf{x}, t)}{\partial t} \right|^2 = f(\theta, \phi),$$

that is, the amount of radiated power dP into a differential element of solid angle $d\Omega$, defined as

$$dP = \lim_{r \to +\infty} \mathbf{u}_r \cdot \mathbf{\Pi}(\mathbf{x}, t) r^2 d\Omega,$$

does not depend on r but only on θ and ϕ, the polar and azimuthal angles of spherical coordinates. We can then define a general formula that quantifies the angular distribution of electromagnetic radiation:

$$\frac{dP}{d\Omega} = \frac{1}{c\mu_0} \left| \mathbf{r} \times \frac{\partial \mathbf{A}_{\text{rad}}(\mathbf{x}, t)}{\partial t} \right|^2 . \tag{19.34}$$

Radiative sources and in particular antennas are characterized by their radiation pattern, which informs us about the angular distribution of the radiated electromagnetic power and is entirely determined by Eq. (19.34). An example can be found in the section on antennas Sect. 19.6.

Total radiation power and its angular distribution

The electromagnetic power radiated through a spherical surface of radius r reads

$$P = \frac{1}{\mu_0 c} \int_0^{4\pi} d\Omega \left| \mathbf{r} \times \frac{\partial \mathbf{A}_{\text{rad}}(\mathbf{x}, t)}{\partial t} \right|^2 \tag{19.35}$$

where $d\Omega$ denotes the infinitesimal element of solid angle defined from the surface element $d\mathbf{S}$ for the sphere: $d\Omega = d\mathbf{S} \cdot \mathbf{u}_r / r^2 = \sin\theta \, d\theta \, d\phi$, where θ and ϕ denote the polar and azimuthal angles of spherical coordinates. To calculate the angular distribution of radiation power, it is convenient to introduce the *radiation vector*

$$\boldsymbol{\alpha}(\mathbf{x}, t) = \frac{\partial}{\partial t} \iiint_\Omega \mathbf{j}\left(\mathbf{x}', t - \frac{r}{c} - \frac{\mathbf{u}_r \cdot \mathbf{x}'}{c} \right) d^3 x'$$

such that

$$\frac{\partial \mathbf{A}_{\text{rad}}(\mathbf{x}, t)}{\partial t} = \frac{\mu_0}{4\pi r} \boldsymbol{\alpha}(\mathbf{x}, t).$$

The angular distribution of radiated power is then given by the formula

$$\frac{dP}{d\Omega} = \frac{\mu_0}{16\pi^2 c} \left| \mathbf{u}_r \times \boldsymbol{\alpha}(\mathbf{x}, t) \right|^2 .$$

The total power is obtained by integration over the solid angle:

$$P = \frac{\mu_0}{16\pi^2 c} \int_0^{4\pi} d\Omega \left| \mathbf{u}_r \times \boldsymbol{\alpha}(\mathbf{x}, t) \right|^2 . \tag{19.36}$$

At large distances $r \to \infty$, the radiation vector $\boldsymbol{\alpha}(\mathbf{x}, t)$ and $\dfrac{dP}{d\Omega}$ only depend on the spherical angles θ and ϕ. The characterization of the radiation pattern consists in plotting the time average value for $\left\langle \dfrac{dP}{d\Omega} \right\rangle$ as a function of the polar and azimuthal angles. The pattern exhibits lobes that provide information about the directivity of the emitter. Note finally that the radiated fields are entirely determined by the radiation vector potential. The radiated magnetic field reads

$$\mathbf{B}_{\text{rad}}(\mathbf{x}, t) = -\frac{\mathbf{u}_r}{c} \times \frac{\partial \mathbf{A}_{\text{rad}}(\mathbf{x}, t)}{\partial t} \tag{19.37}$$

and the radiated electric field reads

$$\mathbf{E}_{\text{rad}}(\mathbf{x}, t) = \mathbf{u}_r \times \left(\mathbf{u}_r \times \frac{\partial \mathbf{A}_{\text{rad}}(\mathbf{x}, t)}{\partial t} \right). \tag{19.38}$$

19.5   Electric Dipole Radiation

Even after reducing the general solutions to Maxwell's equations expressed in Eqs. (19.7) and (19.8) to their far-zone limit, that is, to the potentials for electromagnetic radiation fields Eqs. (19.25) and (19.28), it is often a difficult task to go further and find analytic expressions of radiation fields for physically relevant current density distributions. Fortunately, simplifications are possible in the case of current distributions whose dimension L remains small compared to the distance cT traveled by the signal during a characteristic evolution time T of the current source. Note that for a point charge this is a non-relativistic approximation, where we suppose that the characteristic speed is much smaller than the speed of light $v \sim L/T \ll c$. For a harmonic signal, this simply means that the source dimension is smaller than the wavelength of the signal. It is then possible to perform an expansion of the source terms in powers of L/cT. Retaining only the dominant term in such an expansion leads to the laws for electric dipole radiation, valid at distances r from the source satisfying $L \ll \underbrace{cT}_{\sim\lambda} \ll r$.

To illustrate this procedure, consider the radiation vector

$$\boldsymbol{\alpha}(\mathbf{x}, t) = \frac{\partial}{\partial t} \iiint\limits_{\Omega} \mathbf{j}\left(\mathbf{x}', t - \frac{r}{c} + \frac{\mathbf{u}_r \cdot \mathbf{x}'}{c} \right) d^3 x'.$$

The current density under the integral can be expanded as

$$\mathbf{j}\left(\mathbf{x}', t - \frac{r}{c} + \frac{\mathbf{u}_r \cdot \mathbf{x}'}{c} \right) = \mathbf{j}(\mathbf{x}', \tau) + \frac{\mathbf{u}_r \cdot \mathbf{x}'}{c} \frac{\partial}{\partial t} \mathbf{j}(\mathbf{x}', \tau) + \cdots$$

where we introduced the retarded time

$$\tau = t - \frac{r}{c}.$$

Note that the order of magnitude for the second term on the right-hand side is L/cT or L/λ for a harmonic current of wavelength λ, justifying the validity range $L \ll \lambda \ll r$ of this expansion. Integrating over the volume Ω, the radiation vector is expressed as

$$\boldsymbol{\alpha}(\mathbf{x}, t) = \frac{d}{dt}\left(\int_{\Omega} \mathbf{j}(\mathbf{x}', \tau)d^3x'\right) + \frac{\mathbf{u}_r}{c} \cdot \frac{d^2}{dt^2}\left(\int_{\Omega} \mathbf{x}'\mathbf{j}(\mathbf{x}', \tau)d^3x'\right) + \cdots$$

Similarly, the charge density that appears in the radiation potential (see Eq. (19.25)) can be expanded as

$$\varrho\left(\underbrace{\mathbf{x}', t - \frac{r}{c} + \frac{\mathbf{u}_r \cdot \mathbf{x}'}{c}}_{\tau}\right) \approx \varrho\left(\mathbf{x}', \tau\right) + \frac{\mathbf{u}_r \cdot \mathbf{x}'}{c}\frac{\partial}{\partial t}\varrho\left(\mathbf{x}', \tau\right) + \cdots$$

These expansions, after truncation at leading order, will yield the electromagnetic dipole radiation fields.

19.5.1 Scalar Potential for the Electric Dipole Radiation

From the expansion of the charge density, we expand the scalar potential as

$$V(\mathbf{x}, t) \approx \frac{1}{4\pi\epsilon_0 r}\left\{\iiint_{\Omega} \varrho(\mathbf{x}', \tau) + \left(\frac{\mathbf{u}_r \cdot \mathbf{x}'}{c}\right)\frac{\partial}{\partial t}\varrho(\mathbf{x}', \tau)\right\}d^3x'$$

$$= \frac{1}{4\pi\epsilon_0 r}\underbrace{\iiint_{\Omega} \varrho(\mathbf{x}', \tau)d^3x'}_{Q} + \frac{1}{4\pi\epsilon_0 cr}\mathbf{u}_r \cdot \frac{d}{dt}\underbrace{\iiint_{\Omega} \mathbf{x}'\varrho(\mathbf{x}', \tau)d^3x'}_{\mathbf{p}(t-r/c)} + \cdots$$

where $Q = \iiint_{\Omega} \varrho(\mathbf{x}', t)d^3x'$ is the total charge contained in Ω, which is independent of time, and

$$\mathbf{p}(t) \equiv \iiint_{\Omega} \mathbf{x}'\varrho(\mathbf{x}', t)d^3x'$$

is the dipole moment of the charge distribution at time t. Finally

$$V(\mathbf{x}, t) \approx \underbrace{\frac{Q}{4\pi\epsilon_0 r}}_{\text{Static}} + \underbrace{\frac{1}{4\pi\epsilon_0 cr}\mathbf{u}_r \cdot \frac{d\mathbf{p}}{dt}(t - r/c)}_{\text{Radiative}}.$$

The first term corresponds to a static potential, exactly equal to the electrostatic potential of a point charge Q at the origin. The second term exists as long as the total dipole moment of the distribution depends on time (that is, when the charges move within Ω). Calling $\dot{\mathbf{p}}(t) \equiv d\mathbf{p}/dt$, we write the scalar potential for electric dipole radiation

$$\boxed{V_{\mathrm{rad,e}}(\mathbf{x}, t) = \frac{1}{4\pi\epsilon_0 cr}\,\mathbf{u}_r \cdot \dot{\mathbf{p}}(\tau)\, .} \qquad (19.39)$$

Note again that the $1/r$-dependence of the radiative potential differs significantly from the $1/r^2$ dependence of the electrostatic potential generated by a dipole charge distribution.

For example, for a dipole at the origin whose magnitude oscillates harmonically in time:

$$\mathbf{p}(t) = \mathbf{p}_0 \cos \omega t,$$

we obtain:

$$\frac{d}{dt}\mathbf{p}(t) = -\mathbf{p}_0 \omega \sin \omega t$$

and so

$$V_{\mathrm{rad}}(\mathbf{x}, t) = -\frac{1}{4\pi\epsilon}\mathbf{p}_0 \cdot \mathbf{u}_r \left(\frac{\omega}{rc}\right) \sin[\omega(t - r/c)].$$

19.5.2   Current Associated with a Dipole

A variation in time of the dipole moment is associated with a current density (Remember the polarization current $\mathbf{j}_P = \dfrac{\partial \mathbf{P}}{\partial t}$), and therefore, with the generation of a magnetic field. Indeed:

$$\frac{d\mathbf{p}(t)}{dt} = \frac{d}{dt}\iiint\limits_{\Omega} \mathbf{x}'\varrho(\mathbf{x}', t)d^3x' = \iiint\limits_{\Omega} \mathbf{x}'\frac{\partial}{\partial t}\varrho(\mathbf{x}', t)d^3x'.$$

The law of conservation of charge reads $\partial\varrho/\partial t + \nabla \cdot \mathbf{j} = 0$, then

$$\frac{d\mathbf{p}(t)}{dt} = -\iiint\limits_{\Omega} \mathbf{x}' \left(\nabla' \cdot \mathbf{j}(\mathbf{x}', t)\right)d^3x'.$$

Using the vector calculus identity $\nabla \cdot (x\mathbf{j}) = j_x + x\nabla \cdot \mathbf{j}$, we can rewrite the component along the x-axis as

$$\frac{dp_x(t)}{dt} = -\iiint\limits_{\Omega} x' \left(\nabla' \cdot \mathbf{j}(\mathbf{x}', t)\right)d^3x'$$

$$= -\iiint\limits_{\Omega} \left[\nabla' \cdot (x'\mathbf{j}(\mathbf{x}', t)) - j_x(\mathbf{x}', t)\right]d^3x'.$$

The first integral over the volume Ω on the right-hand side can be transformed into a surface integral by Ostrogradsky's theorem and is identically zero since $\mathbf{j} = 0$ at the frontier $\partial\Omega$. Reproducing this calculation with the y- and z-components of $d\mathbf{p}/dt$, we obtain

$$\frac{d\mathbf{p}(t)}{dt} = \iiint_{\Omega} \mathbf{j}(\mathbf{x}', t)\, d^3 x'.$$

19.5.3   Vector Potential for Electric Dipole Radiation

Now we calculate the vector potential associated with the current density in Ω:

$$\mathbf{A}_{\mathrm{rad}}(\mathbf{x}, t) = \frac{\mu_0}{4\pi r} \iiint_{\Omega} \mathbf{j}\left(\mathbf{x}', \underbrace{t - \frac{r}{c}}_{\tau} + \frac{\mathbf{u}_r \cdot \mathbf{x}'}{c}\right) d^3 x' \approx \frac{\mu_0}{4\pi r} \iiint_{\Omega} \mathbf{J}\left(\mathbf{x}', \tau\right) d^3 x'.$$

We see that we can use the relation

$$\iiint_{\Omega} d^3 x'\, \mathbf{j}\left(\mathbf{x}', \tau\right) = \frac{d}{dt}\mathbf{p}(\tau) \equiv \dot{\mathbf{p}}(\tau),$$

so that the radiation vector potential for the electric dipole radiation reads

$$\mathbf{A}_{\mathrm{rad,e}}(\mathbf{x}, t) = \frac{\mu_0}{4\pi r}\dot{\mathbf{p}}(\tau) \tag{19.40}$$

where $\tau = t - r/c$ and, noting that $\mathbf{p}(t)$ is a function of a single variable, $\dot{\mathbf{p}} \equiv \dfrac{d\mathbf{p}}{dt}$. Hence, the potentials Eqs. (19.39) and (19.40) involve $\dot{\mathbf{p}}$ evaluated at the retarded time $t - r/c$.

19.5.4   Fields for Electric Dipole Radiation

The electric dipole radiation corresponds to the leading order term in the expansion for the radiation vector $\boldsymbol{\alpha}(\mathbf{x}, t)$ in powers of L/λ. In this case, using the link between the current density and the dipole moment, we have

$$\boldsymbol{\alpha}(\mathbf{x}, t) = \frac{d}{dt}\frac{d\mathbf{p}(\tau)}{dt} + \cdots \approx \frac{d^2\mathbf{p}(\tau)}{dt^2} = \ddot{\mathbf{p}}(\tau).$$

Fig. 19.9 The fields far from a source of dimension smaller than the wavelength depends on the time-dependence of the total dipole moment of the source, and exhibit the structure of a quasi-plane wave propagating along the radial direction

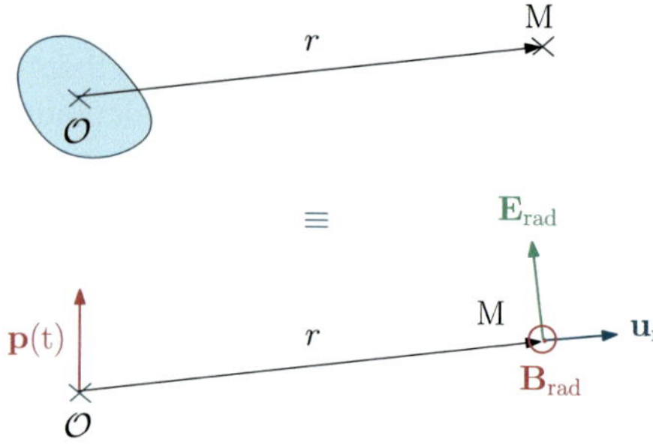

We see that this truncation makes the radiation vector dependent only on time. We can then use Eq. (19.38) to find the radiated electric field:

$$\mathbf{E}_{\mathrm{rad,e}}(\mathbf{x}, t) = \frac{\mu_0}{4\pi r}\mathbf{u}_r \times (\mathbf{u}_r \times \ddot{\mathbf{p}}(t - r/c)) \tag{19.41}$$

and Eq. (19.37) will in turn lead to the radiated magnetic field

$$\mathbf{B}_{\mathrm{rad,e}}(\mathbf{x}, t) = -\frac{\mu_0}{4\pi cr}(\mathbf{u}_r \times \ddot{\mathbf{p}}(t - r/c)) \ . \tag{19.42}$$

Note that $\mathbf{E}_{\mathrm{rad,e}}$ and $\mathbf{B}_{\mathrm{rad,e}}$ still correspond to spherical waves[3] which satisfy

$$\mathbf{B}_{\mathrm{rad,e}}(\mathbf{x}, t) = \frac{\mathbf{u}_r}{c} \times \mathbf{E}_{\mathrm{rad,e}}(\mathbf{x}, t).$$

The electromagnetic radiation field far from the source, when the phase front flattens, keeps the structure of a quasi-plane wave, propagating along the radial direction. This is illustrated in Fig. 19.9.

19.5.5   *Poynting Vector and Radiation Power for Electric Dipole Radiation*

Using the vector radiation $\boldsymbol{\alpha}(\mathbf{x}, t) = \ddot{\mathbf{p}}(\tau)$, the angular distribution of radiated power in the case of electric dipole radiation is easy to evaluate from the general formula Eq. (19.34):

$$\left[\frac{dP}{d\Omega}\right]_e = \frac{\mu_0}{16\pi^2 c}|\mathbf{u}_r \times \boldsymbol{\alpha}(\mathbf{x}, t)|^2 = \frac{\mu_0}{16\pi^2 c}|\mathbf{u}_r \times \ddot{\mathbf{p}}(\tau)|^2.$$

---

[3] The propagation direction is radial, the wavefronts coincide with the surface of a sphere and the amplitude varies as $1/r$.

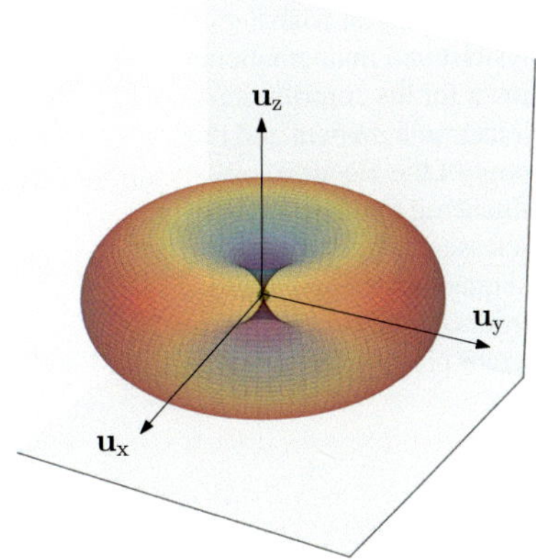

Fig. 19.10 Radiation pattern at a fixed time of an electric dipole oriented along the z-axis

If we chose the axes of the spherical basis of coordinates such that the dipole moment is oriented along the z-axis, we have

$$|\mathbf{u}_r \times \ddot{\mathbf{p}}(t - r/c)|^2 = |\ddot{\mathbf{p}}(t - r/c)|^2 \sin^2 \theta$$

and

$$\left[\frac{dP}{d\Omega}\right]_e = \frac{\mu_0}{16\pi^2 c} |\ddot{\mathbf{p}}(t - r/c)|^2 \sin^2 \theta.$$

The radiation pattern displayed in Fig. 19.10 exhibits an azimuthally symmetric lobe orthogonal to the dipole, that is, characterizing a vertical dipole antenna emitting in all directions but with maximum power directed in a horizontal plane perpendicular to the antenna.

The total radiated power for an electric dipole can be evaluated by integration over the azimuthal and polar angles:

$$P = \frac{\mu_0}{(4\pi)^2 c} |\ddot{\mathbf{p}}(t - r/c)|^2 \underbrace{\int_0^{2\pi} d\varphi}_{2\pi} \underbrace{\int_0^{\pi} d\theta \sin^3 \theta}_{4/3}$$

$$\boxed{P = \frac{\mu_0}{6\pi c} |\ddot{\mathbf{p}}(t - r/c)|^2 = \frac{1}{6\pi \epsilon_0 c^3} |\ddot{\mathbf{p}}(t - r/c)|^2 \,.} \tag{19.43}$$

At any distance r far from the source, the electromagnetic power transported through a spherical surface of radius r only depends on time. This power is transported undiminished at arbitrarily large distances, thanks to the $1/r$ dependence of the radiation fields. Equation (19.43) is known as Larmor formula, named after Joseph Larmor (Fig. 19.11), for the case of a point charge, for which $\mathbf{p} = q\mathbf{r}$ and writes.

Fig. 19.11 Joseph Larmor (1857–1942), an Irish physicist and mathematician known for his contributions to electromagnetism and the theory of the electron, influencing the development of classical electrodynamics and quantum theory. He introduced the concept of the Larmor precession

$$P = \frac{\mu_0 q^2}{6\pi c} \, |\mathbf{a}(t - r/c)|^2 = \frac{q^2}{6\pi \epsilon_0 c^3} \, |\mathbf{a}(t - r/c)|^2 \tag{19.44}$$

with $\mathbf{a}$ the charge's acceleration.

Example 19.1—Cyclotron radiation

When non-relativistic charges are deflected by a magnetic field, they emit the so-called cyclotron radiation. As discussed in Sect. 8.6, a cyclotron is a special type of particle accelerator in which charged particles are accelerated by following a spiral-like trajectory due to the presence of a magnetic field and an alternating electric field. Under the presence of a uniform magnetic field, a charge q describes a circular orbit in the plane perpendicular to the magnetic field, as shown in Fig. 19.12.

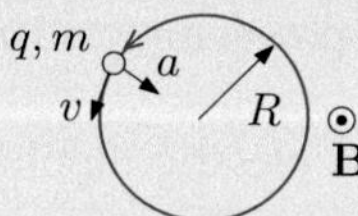

Fig. 19.12 A charge in a cyclotron orbit radiates energy

The acceleration is given by $a = \frac{qvB}{m}$, so that the power radiated by the charge, according to (19.44), is simply

$$P = \frac{\mu_0}{6\pi c} |q\mathbf{a}|^2 = \frac{q^4 v^2 B^2 \mu_0}{6\pi m^2 c}.$$

The frequency of the emitted wave corresponds to the cyclotron frequency $\omega = qB/m$, which is independent on the particle's velocity in the non relativistic case. In a cyclotron accelerator it is necessary to overcome this loss of energy to keep the acceleration.

Example 19.2—Radiation of the hydrogen atom

In a classical model, the electron in hydrogen describes a circular motion of radius $r = a_0$ around the proton, where $a_0 = 0.529$ Å is the Bohr radius. If the proton is assumed to be at rest at the origin, this is equivalent to a rotating dipole moment $\mathbf{p}(t) = -e\mathbf{r}(t)$ where $\mathbf{r}$ is the position of the electron. According to (19.43), the electron then radiates energy in the form of electromagnetic waves, making the atom instable. Let us assume that the electron starts revolving at a distance $r = a_0$ and compute what would be the time τ required for the electron to collapse into the nucleous by releasing energy in the form of radiation.

The acceleration of the electron is given by the Coulomb force divided by the mass of the electron:

$$\mathbf{a}(t) = -\frac{e^2}{4\pi\epsilon_0 m_e r^2(t)}\mathbf{u}_r(t).$$

Thus, the power emitted by the electron is given by

$$P = \left(\frac{\mu_0 e^2}{6\pi c}\right)|\mathbf{a}|^2 = \left(\frac{\mu_0}{6\pi c}\right)\frac{e^6}{(4\pi\epsilon_0)^2\, m_e^2 r(t)^4}.$$

Note that we are assuming that at instant t the orbit desribes a uniform circular motion. To justify this, consider the energy loss ΔE during a period $T = 2\pi r/v$ of revolution around the nucleus, $\Delta E = -PT$. The speed of the electron on its orbit is $v = \sqrt{ar} = e/\sqrt{4\pi\epsilon_0 m_e r}$, so that

$$\Delta E = -\left(\frac{\mu_0}{3c}\right)\frac{e^5}{(4\pi\epsilon_0 m_e r)^{3/2}\, r}.$$

Initially, the total energy of the electron is given by $E = -\dfrac{1}{2}\dfrac{e^2}{4\pi\epsilon_0 r_B} = -13.6$ eV, then during the first revolution:

$$\frac{\Delta E}{E} = \frac{2\mu_0 e^3}{3c}\frac{1}{\sqrt{4\pi\epsilon_0}(m_e r_B)^{3/2}} = 3.24\times 10^{-6}.$$

The energy loss during a period is negligible with respect to the initial energy of the electron, which justifies the use of the equations of uniform circular motion for the precedent calculations. If at instant t the radius of the orbit is $r(t)$, the energy is:

$$E(t) = -\frac{1}{2}\left(\frac{e^2}{4\pi\epsilon_0 r(t)}\right).$$

Thus,

$$\frac{dE(t)}{dt} = \frac{dE(t)}{dr}\frac{dr}{dt} = \left(\frac{e^2}{8\pi\epsilon_0 r^2(t)}\right)\frac{dr}{dt} = -P,$$

leading to

$$\left(\frac{e^2}{8\pi\epsilon_0 r^2(t)}\right)\frac{dr}{dt} = -\left(\frac{\mu_0}{6\pi c}\right)\frac{e^6}{(4\pi\epsilon_0)^2\, m_e^2 r^4(t)}$$

which is simplified into

$$\frac{dr(t)}{dt} = -\frac{4}{3}\frac{e^4}{(4\pi\epsilon_0)^2\, m_e^2 c^3 r^2(t)} = -\frac{4}{3}c\frac{r_c^2}{r^2(t)},$$

where $r_c \equiv \dfrac{e^2}{4\pi\epsilon_0}\dfrac{1}{mc^2}$ is the classical electron radius. Let τ be the time it takes the electron to collapse on the proton. Integrating the equation above by separating variables, we obtain

$$\int_{r_B}^{0} r^2 dr = -\frac{4}{3}c r_c^2 \int_0^{\tau} dt,$$

where τ is the collapse time. We find

$$\frac{1}{3}r_B^3 = \frac{4}{3}r_c^2 c\tau$$

and

$$\tau = \frac{1}{4}\frac{r_B^3}{c r_c^2}.$$

Evaluating the collapse time with $r_c = \alpha^2 r_B$ and $\alpha \approx 1/137$ yields

$$\tau \approx 1.6 \times 10^{-11} \text{ s}.$$

In conclusion, the classic view of the hydrogen atom consisting of an electron spinning around the proton is not consistent with electrodynamics: the latter predicts that a hydrogen atom does not survive more than 16 picoseconds. Of course, we know that the good description is a quantum mechanical one, the electron in its lower energy state is not actually moving, it is in fact in a stationary state. There is no orbital motion and therefore no radiation of energy.

19.5.6 Quasi-Plane Wave Approximation

It is instructive to note that without knowing the general expressions Eqs. (19.32) and (19.31), the radiative fields Eqs. (19.41) and (19.42) can be derived in a simple way from the potentials Eqs. (19.39) and (19.40), using the *quasi-plane wave approximation*, which consists in considering the amplitude of the radiated electric and magnetic fields as quasi uniform, that is, in keeping r constant in the denominator of Eqs. (19.39) and (19.40) when taking the curl. Indeed, starting with Eq. (19.40) and taking $\nabla \times \mathbf{A}_{\mathrm{rad,e}}(\mathbf{x}, t)$, we find

$$\mathbf{B}(\mathbf{x}, t) = \frac{\mu_0}{4\pi r} \nabla \times \dot{\mathbf{p}}(\tau) + \frac{\mu_0}{4\pi} \underbrace{\left(\nabla \frac{1}{r} \right)}_{-1/r^2} \times \dot{\mathbf{p}}(\tau),$$

where the first term prevails over the second. This becomes clear if we remember that $\nabla \times \dot{\mathbf{p}}(\tau)$ involves a derivative with respect to r of a function that depends on $\tau = t - r/c$. Its order of magnitude is then

$$\left| \frac{\partial}{\partial r} \dot{\mathbf{p}}(\tau) \right| = \left| \underbrace{\frac{\partial \tau}{\partial r}}_{-1/c} \underbrace{\ddot{\mathbf{p}}(\tau)}_{\sim \dot{p}/T} \right| = \frac{1}{cT} |\dot{\mathbf{p}}(\tau)| \sim \frac{|\dot{\mathbf{p}}(\tau)|}{\lambda},$$

where $\dot{\mathbf{p}}/T$ is the order of magnitude for $\ddot{\mathbf{p}}$. If we assume a signal dominated by a sinusoidal signal of frequency $\omega = 2\pi/T$, whose wavelength is $\lambda = cT$, the order of magnitude is $\dot{\mathbf{p}}/\lambda$. Hence the ratio of the second to the first term in our expression for the magnetic field scales as λ/r, showing that the first term prevails for $r \gg \lambda$, that is, in the region farther from the source than the wavelength of the signal. The radiated magnetic field then writes

$$\mathbf{B}_{\mathrm{rad,e}}(\mathbf{x}, t) \approx \frac{\mu_0}{4\pi r} \nabla \times \dot{\mathbf{p}}(\tau).$$

For the calculation of $\nabla \times \dot{\mathbf{p}}(\tau)$, we can use spherical coordinates (r, θ, ϕ). A rigorous derivation would require the expression for the curl in spherical coordinates. However, this expression involves terms of higher order varying as $1/r$, which can be neglected with respect to the dominant terms. The result is then the same as that obtained by assuming that $\mathbf{u}_r, \mathbf{u}_\theta, \mathbf{u}_\phi$ is a quasi-Cartesian basis. The radiated magnetic field then reads

$$\mathbf{B}_{\mathrm{rad,e}}(\mathbf{x}, t) = \frac{\mu_0}{4\pi r} \begin{pmatrix} \partial_r \\ 0 \\ 0 \end{pmatrix} \times \begin{pmatrix} p_r'(\tau) \\ p_\theta'(\tau) \\ p_\phi'(\tau) \end{pmatrix} = \frac{\mu_0}{4\pi r} \mathbf{u}_r \partial_r \times \dot{\mathbf{p}}(\tau).$$

Finally differentiating with respect to r using $\partial_r = -(1/c)\partial_\tau$, we find

$$\boxed{\mathbf{B}_{\text{rad,e}}(\mathbf{x}, t) = \frac{\mu_0}{4\pi r}\left(-\frac{\mathbf{u}_r}{c} \times \ddot{\mathbf{p}}(\tau)\right).}$$

The radiated electric field can then be calculated from Maxwell–Ampère equation

$$\nabla \times \mathbf{B} = \frac{1}{c^2}\frac{\partial \mathbf{E}}{\partial t},$$

by performing consistently the same approximation to evaluate the curl

$$\nabla \times \mathbf{B}_{\text{rad,e}} = \frac{\mu_0}{4\pi r}\underbrace{(\mathbf{u}_r \partial_r)}_{-\dfrac{\mathbf{u}_r}{c}\dfrac{\partial}{\partial \tau}} \times \left(-\frac{\mathbf{u}_r}{c} \times \ddot{\mathbf{p}}(\tau)\right) = \frac{1}{c^2}\frac{\partial \mathbf{E}_{\text{rad,e}}}{\partial t}.$$

Finally integrating in time leads to

$$\boxed{\mathbf{E}_{\text{rad,e}}(\mathbf{x}, t) = \frac{\mu_0}{4\pi r}\mathbf{u}_r \times (\mathbf{u}_r \times \ddot{\mathbf{p}}(\tau)) = -\mathbf{u}_r \times c\mathbf{B}_{\text{rad,e}}(\mathbf{x}, t).}$$

We retrieve the structure of a plane wave, in agreement with our assumption at the beginning of this section. Summarizing, the shortest way to derive the electromagnetic radiation fields for electric dipole radiation is to start from the expression for the vector potential Eq. (19.40), take the curl using the quasi-plane wave approximation to find the magnetic field, and finally use Maxwell–Ampère equation to find the electric field from the magnetic field.

19.5.7　Multipole Expansion of the Radiation Fields

Just like for the static case, it is possible to express the far fields as a multipole series, for which the dipole approximation consists in keeping only the first term. We recall the expression for the vector potential (19.33) far from the source:

$$\mathbf{A}_{\text{rad}}(\mathbf{x}, t) = \frac{\mu_0}{4\pi r}\iiint\limits_{\Omega} \mathbf{j}\left(\mathbf{x}', t - \frac{r}{c} + \frac{\mathbf{u}_r \cdot \mathbf{x}'}{c}\right) d^3x' . \tag{19.45}$$

We first compute its (inverse) Fourier transform with respect to time, $\hat{\mathbf{A}}_{\text{rad}}(\mathbf{x}, \omega) = \frac{1}{2\pi}\int_{-\infty}^{+\infty} \mathbf{A}_{\text{rad}}(\mathbf{x}, t)e^{+i\omega t}\, dt$. By linearity of the Fourier transform and the translation property (A.58), we obtain

$$\hat{\mathbf{A}}_{\text{rad}}(\mathbf{x}, \omega) = \frac{1}{2\pi}\frac{\mu_0}{4\pi r}\iiint\limits_{\Omega} e^{i\frac{\omega r}{c} - i\frac{\omega \mathbf{u}_r \cdot \mathbf{x}'}{c}} \hat{\mathbf{j}}(\mathbf{x}', \omega)d^3x' . \tag{19.46}$$

Defining $k(\omega) = \omega/c$, this rewrites as

$$\hat{\mathbf{A}}_{\text{rad}}(\mathbf{x}, \omega) = \frac{\mu_0}{4\pi r} e^{ikr} \underbrace{\frac{1}{2\pi} \iiint_{\Omega} \hat{\mathbf{j}}(\mathbf{x}', \omega) e^{-ik\mathbf{u}_r \cdot \mathbf{x}'} d^3 x'}_{\hat{\mathbf{j}}(\mathbf{k}, \omega)} \tag{19.47}$$

where $\hat{j}(\mathbf{k}, \omega)$ is the Fourier transform of the current density with respect to time and space, evaluated at $\mathbf{k} = k\mathbf{u}_r$. We obtain then the following result for the vector potential in the frequency domain:

$$\boxed{\hat{\mathbf{A}}_{\text{rad}}(\mathbf{x}, \omega) = \frac{\mu_0}{4\pi r} e^{ikr} \hat{\mathbf{j}}(\mathbf{k}, \omega)} \tag{19.48}$$

which gives, in the temporal domain

$$\mathbf{A}_{\text{rad}}(\mathbf{x}, t) = \frac{1}{2\pi} \int_{\mathbb{R}} \hat{\mathbf{A}}_{\text{rad}}(\mathbf{x}, \omega) e^{-i\omega t} d\omega = \frac{\mu_0}{4\pi} \frac{1}{2\pi} \int_{\mathbb{R}} \frac{e^{i(\frac{\omega}{c} r - \omega t)}}{r} \hat{j}(\mathbf{k}, \omega) d\omega \ . \tag{19.49}$$

We recognize in (19.49) a superposition where the spherical waves of frequency ω and wavevector $\mathbf{k} = k\mathbf{u}_r = \frac{\omega}{c}\mathbf{u}_r$. Each wave has an amplitude given by the Fourier transform of the source $\hat{\mathbf{j}}(\mathbf{k}, \omega)$.

The multipole expansion involves truncating in (19.47) the series expansion of the exponential $e^{-i\mathbf{k}\cdot\mathbf{x}'}$, given by

$$\sum_{n=0}^{\infty} \frac{(-i\mathbf{k} \cdot \mathbf{x}')^n}{n!} \ ,$$

which is justified when $|\mathbf{k}||\mathbf{x}'| = \frac{2\pi}{\lambda}|\mathbf{x}'| \ll 1$, meaning that the size of the source is smaller than the wavelength. Keeping the first term leads to the dipole approximation

$$\hat{\mathbf{A}}_{\text{dipole}}(\mathbf{x}, \omega) = \frac{\mu_0}{4\pi r} e^{i\frac{\omega r}{c}} \iiint_{\Omega} \hat{\mathbf{j}}(\mathbf{x}', \omega) d^3 x' \tag{19.50}$$

which, in the time domain, leads to the known result

$$\mathbf{A}_{\text{dipole}}(\mathbf{x}, t) = \frac{1}{2\pi} \int_{\mathbb{R}} \hat{\mathbf{A}}_{\text{dipole}}(\mathbf{x}, \omega) e^{-i\omega t} d\omega = \frac{\mu_0}{4\pi r} \iiint_{\Omega} \mathbf{j}\left(\mathbf{x}', t - \frac{r}{c}\right) d^3 x' \ . \tag{19.51}$$

The next term in the expansion is the quadripolar term

$$\hat{\mathbf{A}}_{\text{quadripole}}(\mathbf{x}, \omega) = \frac{-ik\mu_0}{4\pi r} e^{i\frac{\omega}{c} r} \iiint_{\Omega} \hat{\mathbf{j}}(\mathbf{x}', \omega)(\mathbf{u}_r \cdot \mathbf{x}') d^3 x' \ . \tag{19.52}$$

Example 19.3—Radiation of a loop of current

Let us suppose a circular loop Γ of current $I(t)$ located in the xy plane, as shown in Fig. 19.13.

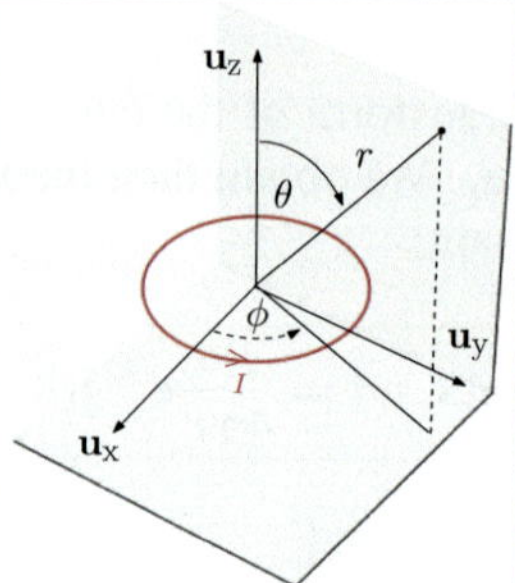

Fig. 19.13 A circular current loop

The dipolar term, given by (19.51), writes

$$\mathbf{A}_{\text{dipole}}(\mathbf{x}, t) = \frac{\mu_0 I\left(t - \frac{r}{c}\right)}{4\pi r} \oint_\Gamma d\mathbf{l} = \mathbf{0}$$

which is not surprising since the net dipole moment of the loop is zero. For the quadripolar term, we have

$$\hat{\mathbf{A}}_{\text{quadripole}}(\mathbf{x}, \omega) = \frac{-i\omega\mu_0}{4\pi cr} e^{i\frac{\omega}{c}r} \oint_\Gamma I(\omega) d\mathbf{l}(\mathbf{x}')(\mathbf{u}_r \cdot \mathbf{x}')$$

and using Kelvin Formula (A.18)

$$\hat{\mathbf{A}}_{\text{quadripole}}(\mathbf{x}, \omega) = \frac{i\mu_0\omega}{4\pi cr} e^{i\frac{\omega}{c}r} I(\omega) \iint_{S(\Gamma)} \nabla'(\mathbf{u}_r \cdot \mathbf{x}') \times d\mathbf{S}.$$

By using (A.8), one can show that $\nabla'(\mathbf{u}_r \cdot \mathbf{x}') = \mathbf{u}_r$

$$\hat{\mathbf{A}}_{\text{quadripole}}(\mathbf{x}, \omega) = \frac{i\mu_0\omega}{4\pi cr} e^{i\frac{\omega}{c}r} I(\omega)\mathbf{u}_r \times \iint_{S(\Gamma)} d\mathbf{S} = \frac{i\mu_0\omega}{4\pi cr} e^{i\frac{\omega}{c}r} \mathbf{u}_r \times I(\omega)\mathbf{S}$$

where $\mathbf{S}$ is the surface vector associated with the contour Γ. Defining the ω-frequency component of the dipole moment $\mathbf{m}(\omega) = I(\omega)\mathbf{S}$, we get finally

$$\hat{\mathbf{A}}_{\text{quadripole}}(\mathbf{x}, \omega) = \frac{\mu_0\omega}{4\pi cr} e^{i\frac{\omega}{c}r} i\mathbf{u}_r \times \mathbf{m}(\omega)$$

from which we can calculate the frequency components of the radiated fields

$$\hat{\mathbf{B}}_{\text{rad}}(\mathbf{x}, \omega) = \nabla \times \hat{\mathbf{A}}_{\text{rad}}(\mathbf{x}, \omega) = \frac{\mu_0 \omega^2}{4\pi c^2 r} e^{i\frac{\omega}{c}r} i\mathbf{u}_r \times (i\mathbf{u}_r \times \mathbf{m}(\omega))$$

$$= \frac{\mu_0 \omega^2}{4\pi c^2 r} e^{i\frac{\omega}{c}r} (\mathbf{m} - (\mathbf{u}_r \cdot \mathbf{m})\mathbf{u}_r)$$

and since the loop of current is neutral, $\mathbf{E} = -\frac{\partial \mathbf{A}}{\partial t}$ so that

$$\hat{\mathbf{E}}_{\text{rad}}(\mathbf{x}, \omega) = i\omega\hat{\mathbf{A}}_{\text{rad}}(\mathbf{x}, \omega) = \frac{\mu_0}{4\pi r} e^{i\frac{\omega}{c}r} \frac{\omega^2}{c} \mathbf{m}(\omega) \times \mathbf{u}_r \ . \tag{19.53}$$

If the current in the loop varies as $I(t) = I_0 \cos \omega_0 t$, then $\mathbf{m}(\omega) = \mathbf{m}_0 \pi (\delta(\omega + \omega_0) + \delta(\omega - \omega_0))$ with $\mathbf{m}_0 = I_0 \mathbf{S}$, which gives

$$\mathbf{E}_{\text{rad}}(\mathbf{x}, t) = \frac{\mu_0}{4\pi r} \frac{\omega_0^2}{c} \cos\left(\frac{\omega_0}{c}(r - ct)\right) \mathbf{m}_0 \times \mathbf{u}_r \ . \tag{19.54}$$

$$\mathbf{B}_{\text{rad}}(\mathbf{x}, t) = \frac{\mu_0 \omega_0^2}{4\pi r c^2} \cos\left(\frac{\omega_0}{c}(r - ct)\right) (\mathbf{m}_0 - (\mathbf{u}_r \cdot \mathbf{m}_0)\mathbf{u}_r) = \frac{\mathbf{u}_r}{c} \times \mathbf{E}_{\text{rad}}(\mathbf{x}, t) \tag{19.55}$$

The Poynting vector reads

$$\mathbf{\Pi}_{\text{rad}}(\mathbf{x}, t) = \frac{\mathbf{E}_{\text{rad}} \times \mathbf{B}_{\text{rad}}(\mathbf{x}, t)}{\mu_0} = \frac{|\mathbf{E}_{\text{rad}}(\mathbf{x}, t)|^2}{\mu_0 c} \mathbf{u}_r$$

$$= \frac{\mu_0 \omega_0^4}{(4\pi)^2 r^2 c^3} |\mathbf{m}_0|^2 \sin^2 \theta \cos^2\left(\frac{\omega_0}{c}(r - ct)\right) \mathbf{u}_r$$

where θ is the angle between the position vector and $\mathbf{m}_0$. The total radiated power is therefore

$$P = \frac{\mu_0 \omega_0^4}{(4\pi)^2 c^3} |\mathbf{m}_0|^2 \cos^2\left(\frac{\omega_0}{c}(r - ct)\right) \underbrace{\int_0^{2\pi} \int_0^{\pi} \sin^3 \theta \, d\theta \, d\phi}_{8\pi/3}$$

and the time average is

$$\langle P \rangle = \frac{\mu_0 \omega_0^4}{12\pi c^3} |\mathbf{m}_0|^2 .$$

19.6 Thin-Wire Antennas

Thin-wire antennas constitute prototypical radiating systems. Incident waves on a thin wire conductor induce a time dependent current in the wire, which in turn drives a voltage across a load resistance. The same principles govern emission of electromagnetic waves, that is, a voltage applied to the terminals of a a thin wire conductor drives a time dependent current that will radiate an electromagnetic wave.

Consider a straight, thin-wire antenna of length $2d$ driven at its center by a sinusoidal input voltage of frequency ω. This is illustrated in Fig. 19.14. We will consider complex notations and write the current induced in the conductor by this voltage as $I(z, t) = \mathrm{Re}\,[\underline{I}(z, t)]$ and

$$\underline{I}(z, t) = I_0 \sin k(d - |z|)e^{-i\omega t} \quad \text{for} \quad -d \leq z \leq d,$$

where $k = \omega/c$.

We assume that the radius of the wire is infinitesimally small so that we can calculate the radiation vector potential from the retarded potential formula as follows

$$\underline{\mathbf{A}}_{\mathrm{rad}} = \frac{\mu_0}{4\pi} \iiint_{\Omega} \frac{\mathbf{j}(\mathbf{x}', t - |\mathbf{x} - \mathbf{x}'|/c)}{|\mathbf{x} - \mathbf{x}'|} d^3 x'$$

$$\approx \frac{\mu_0 \mathbf{u}_z}{4\pi r} \int_{-d}^{d} \underline{I}(z', t - r/c + \mathbf{u}_r \cdot (z'\mathbf{u}_z)/c)\,dz'.$$

Since we wish to characterize the antenna with its radiation pattern, we retained the signal travel time across the source, that is $\mathbf{u}_r \cdot \mathbf{x}'/c$, as a correction to the retarded time. Now using a basis of spherical coordinates, the current at retarded time reads

$$\underline{I}(z', t - r/c + \mathbf{u}_r \cdot (z'\mathbf{u}_z)/c) = I_0 \sin k(d - |z'|)e^{-ikrz'\cos\theta}e^{ikr}e^{-i\omega t}$$

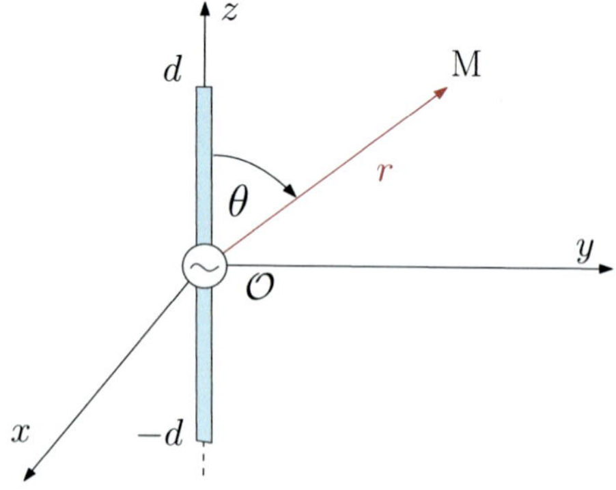

Fig. 19.14 A time-dependent current on a thin wire radiates electromagnetic energy

and the radiated vector potential,

$$\mathbf{A}_{\text{rad}}(\mathbf{x}, t) = \frac{\mu_0 I_0 \mathbf{u}_z}{4\pi} \frac{e^{ikr-i\omega t}}{r} 2 \int_0^d \cos(k \cos\theta\, z') \sin[k(d - z')]dz'.$$

The integral can be calculated using first a trigonometric formula.[4] We find

$$\mathbf{A}_{\text{rad}}(\mathbf{x}, t) = \frac{\mu_0 I_0}{2\pi} \frac{e^{ikr-i\omega t}}{kr} \frac{[\cos(kd \cos\theta) - \cos(kd)]}{\sin^2\theta} \mathbf{u}_z.$$

We aim at finding the radiation pattern of the antenna, which informs us about the angular distribution of the radiated electromagnetic power. We know that the angular distribution of radiated power is given by time average of the quantity

$$\frac{dP}{d\Omega} = \frac{1}{c\mu_0} \left| r\mathbf{u}_r \times \frac{\partial \mathbf{A}_{\text{rad}}(\mathbf{x}, t)}{\partial t} \right|^2.$$

Introducing the expression of $\mathbf{A}_{\text{rad}}(\mathbf{x}, t)$ for the thin-wire antenna, we calculate the angular distribution of radiated power:

$$\left\langle \frac{dP}{d\Omega} \right\rangle = \frac{1}{2c\mu_0} |\mathbf{r}|^2 |-i\omega\mathbf{A}_{\text{rad}}|^2 \sin^2\theta = \frac{1}{2} \frac{\mu_0 c I_0^2}{(2\pi)^2} \frac{[\cos(kd \cos\theta) - \cos(kd)]^2}{\sin^2\theta},$$

where the multiplicative factor $1/2$ takes care of the time average procedure for the harmonic signal.[5]

Figure 19.15 shows the radiation pattern obtained for different lengths of the antenna: $2d = (2n - 1)\lambda/2$ with $n = 1, 2, 3$. The vertical axis is the z-axis, coinciding with the axis of the antenna. There is revolution symmetry around the z-axis (no dependence upon ϕ). The patterns show that there is a single lobe pointing in all horizontal directions when the length of the antenna is equal to $\lambda/2$. Increasing the length of the antenna at fixed wavelength[6] shows that the number of lobes is equal to the number of half-wavelengths fitting the antenna. and that the main lobe becomes narrower and closer to the axis of the antenna.

---

[4] $\sin\alpha \cos\beta = (1/2)[\sin(\alpha + \beta) + \sin(\alpha - \beta)]$.

[5] $\langle \cos^2(\omega t - kr) \rangle = 1/2$.

[6] This is perfectly fine as we did not use the expansion in powers of L/λ in this section.

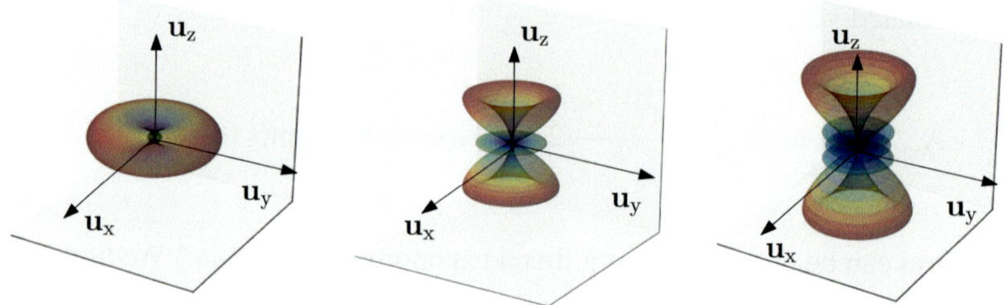

Fig. 19.15 Radiation pattern for an antena of length $\lambda/2$ (left), $3\lambda/2$ (middle) and $5\lambda/2$ (right)

19.7   Scattering of Electromagnetic Waves

An incident electromagnetic wave on a medium induces a motion of charges. These charges are secondary sources of radiation, thus generating a scattered wave, which can have the same frequency as that of the incident wave (elastic diffusion), but it can also have a different frequency (in which case diffusion is then said to be inelastic).

19.7.1   Scattering by an Electron: Thomson Scattering

Thomson scattering denotes the phenomenon of scattering occurring to an electromagnetic wave interacting with a free electron. Consider a free electron (charge $q = -e$, mass m) that undergoes the action of a monochromatic electromagnetic wave of the form

$$\mathbf{E}(\mathbf{r}, t) = \mathrm{Re}\left\{\underline{\mathbf{E}}e^{i(\mathbf{k}\cdot\mathbf{r}-\omega t)}\right\}.$$

Under the action of this field, the electron will oscillate at a frequency $f = \omega/2\pi$, thus behaving like an oscillating dipole that radiates at the same frequency as that of the incident field, thus creating an elastic diffusion field. If $\mathbf{r}$ represents the position of the electron, we have:

$$m_e\frac{d^2\mathbf{r}}{dt^2} = q[\mathbf{E}(\mathbf{r}, t) + \mathbf{v} \times \mathbf{B}(\mathbf{r}, t)].$$

If the speed of the electron is much less than the speed of light c, we can disregard the magnetic force since, for an electromagnetic wave, $|\mathbf{B}| = |\mathbf{E}|/c$. In addition, if the wavelength of the incident field is much greater than the oscillation amplitude of the electron, we can neglect the spatial variation of $\mathbf{E}(\mathbf{r}, t)$ and write $\mathbf{E}(\mathbf{r}, t) = \mathrm{Re}\left\{\underline{\mathbf{E}}e^{-i\omega t}\right\}$. Looking for a stationary solution of the form:

$$\mathbf{r}(t) = \mathrm{Re}\left\{\underline{\mathbf{r}}_0 e^{-i\omega t}\right\},$$

We have:

$$\mathrm{Re}\left\{-m_e\omega^2\underline{\mathbf{r}}_0 e^{-i\omega t}\right\} = \mathrm{Re}\left\{q\underline{E}e^{-i\omega t}\right\}.$$

Then, the amplitude of the electron motion is given by:

$$\underline{\mathbf{r}}_0 = -\frac{q}{m_e\omega^2}\underline{\mathbf{E}} \tag{19.56}$$

The acceleration of an electron that oscillates under the influence of an electromagnetic wave at frequency ω is then given by:

$$\mathbf{a} = \frac{q}{m_e}\mathbf{E}\cos(\omega t)$$

and the power radiated throughout space is obtained using (19.43) for a dipole $\mathbf{p}(t) = q\mathbf{r}(t)$:

$$P = \frac{\mu_0 q^2}{6\pi c}\frac{q^2|\mathbf{E}|^2}{m_e^2}\cos^2(\omega t),$$

the time average of which is:

$$\langle P\rangle = \frac{\mu_0 q^2}{6\pi c}\frac{|\mathbf{E}|^2}{2m_e^2},$$

which can be rewritten as

$$\langle P\rangle = \frac{q^2}{4\pi\epsilon_0}\frac{|\mathbf{E}|^2}{3m_e^2 c^3} = \frac{4\pi}{3}r_c^2\epsilon_0 c|E|^2, \tag{19.57}$$

where $r_c = q^2/(4\pi\epsilon_0 m_e c^2)$ is the classic electron radius (see Example 3.5). In expression Eq. (19.57), we recognize the time average of the Poynting vector of the incident plane wave $\epsilon_0 c|E|^2/2$. The diffusion power can therefore be written as the product of an energy flux per unit area (modulus of the incident Poynting vector) multiplied by a quantity that is homogeneous to a surface, called **cross section**.

Scattering cross section

The scattering cross section σ corresponds to the surface such that, exposed to the incident flux of an electromagnetic wave, intercepts a flux of energy equal to the flux of the radiated field, which is the long distance part of the scattered electromagnetic wave. The effective cross section σ_e of a free electron, called the Thomson effective cross section, is obtained from Eq. (19.57) and is given by:

$$\sigma_e = \frac{8\pi}{3}r_c^2 \approx 2.66\pi r_c^2.$$

An electron is thus seen by an electromagnetic wave as a circular surface of radius similar to the classical radius $r_c \sim 10^{-15}$ m.

19.7.2 Scattering from an Atom or a Molecule

Atoms are made up of positive charges (protons), whose mass is three orders of magnitude greater than that of the electron. This means that the square of the acceleration, and then the radiated power, is 10^6 times less for a proton than the power radiated by the electrons. It is then reasonable to consider only the energy radiated by the electrons of an atom when it interacts with an electromagnetic wave. We can take into account the fact that the electrons are bound to the atomic nucleus by an elastic force. We therefore anticipate a strong dependence of scattering upon the frequency of the incident wave, due to the presence of a resonance frequency ω_0 corresponding to the natural frequency of oscillation of the electron around the nucleus. We will use the Lorentz model that treats the electron-atom system as a damped harmonic oscillator (see Eq. (14.12)). The equation of motion of the electron reads

$$m\frac{d^2\mathbf{r}}{dt^2} = q\mathbf{E} - m\gamma\frac{d\mathbf{r}}{dt} - m\omega_0^2\mathbf{r}$$

where $\mathbf{r}$ is the displacement of the electron with respect to the equilibrium position. Again, assuming a monochromatic incident wave with a much longer wavelength than the oscillation amplitude of the electrons, we look for a solution of the form $\mathbf{r} = \mathrm{Re}\left\{\mathbf{r}_0 e^{-i\omega t}\right\}$ and we have:

$$\mathrm{Re}\left\{-m\omega^2\underline{\mathbf{r}}_0 e^{-i\omega t}\right\} = \mathrm{Re}\left\{q\underline{\mathbf{E}}e^{-i\omega t} + i\omega m\gamma\underline{\mathbf{r}}_0 e^{-i\omega t} - m\omega^2\underline{\mathbf{r}}_0 e^{-i\omega t}\right\}.$$

We retrieve Eq. (14.12)

$$\underline{\mathbf{r}}_0 = -\frac{q}{m}\frac{\underline{\mathbf{E}}}{\omega^2 - \omega_0^2 + i\gamma\omega}\,. \tag{19.58}$$

Since the radiation power is proportional to the square of the acceleration, and as in a harmonic motion, the latter is proportional to the motion amplitude $\mathbf{r}_0$, we have:

$$\frac{\sigma}{\sigma_e} = \frac{|\underline{\mathbf{r}}_0|^2}{|\underline{\mathbf{r}}_0^{\mathrm{free}}|^2},$$

where $\underline{\mathbf{r}}_0^{\mathrm{free}}$ corresponds to the result found previously for a free electron—See Eq. (19.56). We have:

$$\frac{\sigma}{\sigma_e} = \frac{\omega^4}{(\omega^2 - \omega_0^2)^2 + \gamma^2\omega^2}$$

so that the cross section of an electron that is elastically bound to an atom reads

$$\sigma(\omega) = \sigma_e \frac{\omega^4}{(\omega^2 - \omega_0^2)^2 + \gamma^2 \omega^2} \cdot \qquad (19.59)$$

19.7.2.1 Thomson Scattering $\omega \gg \omega_0, \gamma$

In the case of high frequencies such that $\omega \gg \omega_0, \gamma$, we retrieve Thomson scattering since $\sigma(\omega) \approx \sigma_e$. When the electromagnetic wave has a frequency much larger than vibration the frequencies ω_0 of electrons in atoms, electrons behave as free electrons. Diffraction by X-rays by electrons in a crystal is typically well described by Thomson scattering.

19.7.2.2 Resonant Scattering

We see from Eq. (19.59) that the interaction between the electromagnetic wave and an electron bound to an atom is particularly strong when the frequency of the wave is in resonance with the vibration frequency of the electron in the atom, ω_0. From Eq. (19.59), we see that in the limit $\omega \approx \omega_0$

$$\sigma(\omega) = \sigma_e \frac{\omega^4}{[(\omega - \omega_0)(\omega + \omega_0)]^2 + \gamma^2 \omega^2}$$

$$\approx \sigma_e \frac{\omega_0^4}{[(\omega - \omega_0)(2\omega_0)]^2 + \gamma^2 \omega_0^2} = \sigma_e \frac{\omega_0^2}{4(\omega - \omega_0)^2 + \gamma^2} \cdot$$

The dependence of $\sigma(\omega)$ on ω close to resonance is approximately a Lorentzian of width $1/\gamma$. The maximum value of $\sigma(\omega)$ is obtained for $\omega = \omega_0$:

$$\sigma(\omega_0) = \sigma_e \left(\frac{\omega_0}{\gamma} \right)^2 \cdot$$

In the visible domain, $\omega_0 \sim 10^{15}$ s^{-1}, and $\gamma = 2/3\omega_0\tau \sim 10^8$ s^{-1}, so that the cross section can be many orders of magnitude larger than that of the free electron, $\sigma(\omega_0) \sim 10^{14} \times \sigma_e$.

19.7.2.3 Rayleigh Scattering $\omega \ll \gamma \ll \omega_0$

If $\omega \ll \gamma \ll \omega_0$, we are in the case of Rayleigh scattering. The scattering cross section in this case becomes

$$\sigma(\omega) = \sigma_e \frac{\omega^4}{\omega_0^4}.$$

The scaling as ω^4 shows that scattering becomes inefficient at low frequencies, toward the far-infrared domain. Rayleigh scattering is responsible for the blue color of the sky. The ω^4 scaling means that blue light, with a frequency roughly twice that of red light, is scattered 16 times more when white light from the Sun interacts with nitrogen and oxygen molecules in the atmosphere. As a result, blue light is scattered much more effectively, creating the appearance of a blue sky as it seems to come from every direction.

In contrast, the Sun appears yellow because the direct sunlight we perceive has lost a significant portion of its blue component due to scattering. This effect is even more pronounced at sunset, when the thickness of the atmosphere through which sunlight passes is at its maximum. In this case, blue light is scattered even more efficiently compared to red light, causing the Sun to appear redder.

19.8   Summary and Essential Formulas

- From Maxwell's equations one can derive the propagation equations for the potentials, which in the Lorenz gauge ($\frac{1}{c^2}\frac{\partial V}{\partial t} + \nabla \cdot \mathbf{A} = 0$) read:

$$\nabla^2 \mathbf{A} - \frac{1}{c^2}\frac{\partial^2 \mathbf{A}}{\partial t^2} = -\mu \mathbf{j},$$

$$\nabla^2 V - \frac{1}{c^2}\frac{\partial^2 V}{\partial t^2} = -\frac{\varrho}{\epsilon}.$$

The solution to these equations is given by

$$V(\mathbf{x}, t) = \frac{1}{4\pi\epsilon_0} \iiint_{\mathbb{R}^3} \frac{\varrho(\mathbf{x}', t - |\mathbf{x} - \mathbf{x}'|/c)}{|\mathbf{x} - \mathbf{x}'|} d^3 x',$$

$$\mathbf{A}(\mathbf{x}, t) = \frac{\mu_0}{4\pi} \iiint_{\mathbb{R}^3} \frac{\mathbf{j}(\mathbf{x}', t - |\mathbf{x} - \mathbf{x}'|/c)}{|\mathbf{x} - \mathbf{x}'|} d^3 x'.$$

The value of the potentials at position x and time t depends on the charge and current densities at position $\mathbf{x}'$ in the past, at a time $t - |\mathbf{x} - \mathbf{x}'|/c$, which indicates that electrodynamics propagates at speed c.

- Electromagnetic radiation corresponds to the electromagnetic fields that transport energy arbitrarily far away from their source without attenuation of the total energy observed at distance r. The radiation field that has the structure of a spherical wave with amplitudes of the electric and magnetic fields varying as $1/r$ and becoming quasi-plane waves propagating along $\mathbf{u}_r$, far from the source, with $\mathbf{E}_{rad}$, $\mathbf{B}_{rad}$ and $\mathbf{u}_r$ forming a right-handed orthogonal triad. The radiated fields are entirely determined by the radiation vector potential

$$\mathbf{A}_{rad}(\mathbf{x}, t) = \frac{\mu_0}{4\pi r} \iiint_\Omega \mathbf{j}(\mathbf{x}', t^*) d^3 x' \tag{19.60}$$

with

$$t^* = t - \frac{r}{c} + \frac{\mathbf{u}_r \cdot \mathbf{x}'}{c}.$$

The radiated magnetic field reads

$$\boxed{\mathbf{B}_{rad}(\mathbf{x}, t) = -\frac{\mathbf{u}_r}{c} \times \frac{\partial \mathbf{A}_{rad}(\mathbf{x}, t)}{\partial t}} \tag{19.61}$$

and the radiated electric field reads

$$\boxed{\mathbf{E}_{rad}(\mathbf{x}, t) = \mathbf{u}_r \times \left(\mathbf{u}_r \times \frac{\partial \mathbf{A}_{rad}(\mathbf{x}, t)}{\partial t} \right).} \tag{19.62}$$

They satisfy

$$\mathbf{B}_{rad}(\mathbf{x}, t) = \frac{\mathbf{u}_r}{c} \times \mathbf{E}_{rad}(\mathbf{x}, t).$$

The amount of radiated power dP into a differential element of solid angle $d\Omega$ is:

$$\frac{dP}{d\Omega} = \frac{1}{c\mu_0} \left| \mathbf{r} \times \frac{\partial \mathbf{A}_{rad}(\mathbf{x}, t)}{\partial t} \right|^2. \tag{19.63}$$

- In the case of charge distributions whose maximal dimension L is small compared to the wavelength of the emitted radiation, the radiation fields write as:

$$\mathbf{E}_{rad,e}(\mathbf{x}, t) = \frac{\mu_0}{4\pi r} \mathbf{u}_r \times (\mathbf{u}_r \times \ddot{\mathbf{p}}(t - r/c))$$

and

$$\mathbf{B}_{rad,e}(\mathbf{x}, t) = -\frac{\mu_0}{4\pi c r} (\mathbf{u}_r \times \ddot{\mathbf{p}}(t - r/c))$$

where $\mathbf{p}$ is the dipole moment of the distribution. The radiation electromagnetic field far from the source, when the phase front flattens, has a structure of a quasi-plane wave propagating along the radial direction.

The angular distribution of radiated power in the case of an electric dipole oriented along z is

$$\left[\frac{dP}{d\Omega}\right]_e = \frac{\mu_0}{16\pi^2 c}|\ddot{\mathbf{p}}(t - r/c)|^2 \sin^2\theta.$$

The total radiated power for an electric dipole is evaluated by integration over the azimuthal and polar angles:

$$P = \frac{\mu_0}{6\pi c}|\ddot{\mathbf{p}}(t - r/c)|^2 = \frac{1}{6\pi\epsilon_0 c^3}|\ddot{\mathbf{p}}(t - r/c)|^2.$$

- Scattering occurs whenever a system oscillates in the presence of an incident electromagnetic wave. If the scattering is elastic, the system radiates a wave of same frequency as the incident one. The scattering cross section σ of a system corresponds to the surface such that, exposed to the incident flux of an electromagnetic wave, intercepts a flux of energy equal to the flux of the radiated field by the scattering process. The cross section of an atom or molecule of resonant frequency ω_0 and damping factor γ reads

$$\sigma(\omega) = \sigma_e \frac{\omega^4}{(\omega^2 - \omega_0^2)^2 + \gamma^2\omega^2}$$

where $\sigma_e = \frac{8\pi r_c^2}{3}$ is the scattering cross section of a free electron, with r_c the classical electron radius.

Problems

19.1 The Lorenz gauge

Starting from arbitrary potentials $\mathbf{A}$ and V, show that it is always possible to perform a gauge transformation $\{\mathbf{A}, V\} \to \{\mathbf{A}' = \mathbf{A} + \nabla\phi, V' = V - \frac{\partial\phi}{\partial t}\}$ that leaves the electric and magnetic fields unchanged, and such that $\frac{1}{c^2}\frac{\partial V'}{\partial t} + \nabla \cdot \mathbf{A}'$. The potentials $\mathbf{A}'$ and V' satisfy therefore the Lorenz gauge condition. What is the equation that ϕ must satisfy?

19.2 Derivation of the radiated magnetic field from the vector potential

Use Eq. (19.33) together with Eq. (19.30) to derive the expression Eq. (19.31) for the radiated magnetic field corresponding to the leading order of $\mathbf{B}(\mathbf{x}, t)$ varying as $1/r$ in the far-zone.

19.3 Derivation of the radiated electric field from the potentials

Use Eqs. (19.25) and (19.33) together with Eq. (19.29) to find the expression (19.32) for the radiative electric field.

19.4 Energy radiated by a charged particle

A particle of charge q falls radially on a repulsive potential $V(r) = \dfrac{\alpha}{r}$ with $\alpha > 0$. The particle at infinity has a radial velocity v_0. The particle approaches the origin up to a return point $r_{\min}$ and then it goes back to infinity. Determine the total energy radiated by the particle during the entire journey.

19.5 Electron in a gravitational field

An electron is released from rest and falls under the influence of gravity. What portion of the potential energy is radiated during the first centimeter?

19.6 Charge suspended by a spring

Suppose that a charge of mass m and charge q is suspended on a spring with constant K. If the system is set to oscillate, how long must it take for the system to lose half of its initial mechanical energy?

19.7 Cyclotron radiation

Consider a charge q of mass m and initial velocity v_0 that describes a circular orbit in a uniform magnetic field B, perpendicular to the movement of the charge. Show that the radius R of the orbit evolves according to $R(t) = R_0 e^{-t/\tau}$. Calculate the total energy radiated by the particle.

19.8 Radiation power and polarization of a rotating dipole

An electric dipole moment of constant modulus p_0 rotates in the plane yz at a constant rotation frequency ω.

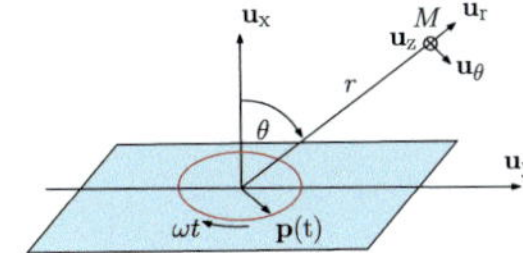

(a) Determine, at a point M on the plane xy the components of the magnetic field $\mathbf{B}(\mathbf{r}, t)$ and of the electric field $\mathbf{E}(\mathbf{r}, t)$ on the right-handed basis $\mathbf{u}_r$, $\mathbf{u}_\theta$, $\mathbf{u}_z$, with $r = OM$ and $\theta = (\widehat{\mathbf{Ox}, \mathbf{OM}})$.

(b) Express the averaged value of the magnetic field squared $\langle \mathbf{B}^2 \rangle$ over time.

(c) What is the direction of radiation? Calculate the average radiated power $\langle d\mathcal{P} \rangle / d\Omega$ per unit of solid angle in the different directions θ with respect to the x-axis.

(d) Show that the power $\langle \mathcal{P} \rangle$ radiated by the rotating dipole is: $\mathcal{P} = K\omega^4$. Express the proportionality coefficient K as a function of p_0 and the constants μ_0 and c.

(e) In which directions θ is the radiation from the rotating dipole linearly polarized? circularly polarized? elliptically polarized?

Chapter 20
Special Relativity and Covariant Electrodynamics

Abstract This concluding chapter integrates the principles of *special relativity* with classical electrodynamics, leading to a *covariant formulation of Maxwell's equations* that is explicitly invariant under Lorentz transformations. It begins by introducing Einstein's postulates of special relativity, emphasizing the relativity principle (laws of physics are the same in all inertial frames) and the constancy of the speed of light in vacuum for all observers. Experimental evidence refuting the concept of aether and confirming these postulates (e.g., Michelson–Morley, Bertozzi's experiment) is reviewed. The chapter then presents *relativistic kinematics*, demonstrating the profound consequences of Einstein's postulates on space and time. Key concepts like the *relative nature of simultaneity, time dilation*, and *length contraction* are derived from the *Lorentz transformation*, which replaces the Galilean transformation of Newtonian mechanics. The *relativistic addition of velocities* is also presented, showing how velocities combine at high speeds. The invariance of the *spacetime interval* is established, leading to the classification of events into time-like, null (light-like), and space-like separations, and the introduction of the *Minkowski spacetime diagram* and the *light cone*. Building on this kinematic foundation, the chapter introduces *relativistic dynamics*, defining *four-vectors* (e.g., four-position, four-velocity, four-momentum, four-acceleration) in Minkowski space. The *Lorentz invariance of mass* is postulated, leading to the definitions of *relativistic momentum* ($\mathbf{p} = \gamma m\mathbf{v}$) and *energy* ($E = \gamma mc^2$), and the famous mass-energy equivalence ($E = mc^2$). The relativistic equation of motion for a charged particle in an electromagnetic field is derived, extending Newton's law to the relativistic regime. Finally, the chapter applies these relativistic principles to electromagnetic quantities. It demonstrates that the *four-current* ($\mathbf{J}_4 = (\mathbf{j}, ic\varrho)$) and *four-potential* ($\mathbf{A}_4 = (\mathbf{A}, iV/c)$) are four-vectors, ensuring the covariance of the continuity equation and the Lorenz gauge condition. The *transformation laws for electric and magnetic fields* between inertial frames are derived, showing their intermixing. The *relativistic Larmor formula* for the power radiated by an accelerating point charge is presented, highlighting the phenomenon of *synchrotron radiation* and *relativistic beaming*, where radiation is strongly concentrated in the direction of the particle motion. The chapter concludes with the *covariant formulation of Maxwell's equations* using the *electromagnetic field-strength tensor* and its dual, and the *stress-energy tensor*, explicitly

F. Cadiz and A. Couairon, *Classical Electrodynamics*, Undergraduate Texts in Physics, https://doi.org/10.1007/978-3-031-86785-9_20

demonstrating their invariance under Lorentz transformations and unifying classical electrodynamics within the framework of special relativity.

Keywords Special relativity · Four-vectors · Minkowski space · Covariant electrodynamics

20.1 Introduction

This concluding chapter is dedicated to the covariant formulation of Maxwell's equations—a formulation that is explicitly invariant under any Lorentz transformation, which describes the relationship between two inertial frames in the framework of special relativity. Before presenting this formulation, we introduce Einstein's theory of special relativity, proposed in 1905, and highlight the pivotal role Maxwell's equations played in its development.

To establish the groundwork for the covariant formulation, we introduce Minkowski four-dimensional spacetime and the associated four-vectors. These concepts are essential for expressing physical laws in a form that is consistent with special relativity. As an illustration of four-vectors and their applications, we provide a brief overview of the dynamics of relativistic particles, offering a practical introduction to Einstein's theory.

We also explore how electric and magnetic fields transform between inertial frames and generalize Larmor's formula to account for the radiation emitted by a relativistic point charge. Finally, we demonstrate how Einstein's principle of relativity naturally extends to Maxwell's equations, which can be written as a single equation having the same form in all Galilean reference frames.

20.2 The Relativity Principle

Until the nineteenth century, Galilean relativity treated all observers as equivalent no matter how fast they were moving. Galileo's relativity principle states that the laws of physics remain the same in all inertial reference frames (also called Galilean frames). In order to underline what special relativity has changed, we briefly review useful definitions and assumptions of Newtonian mechanics.

20.2.1 *SpaceTime Framework in Newtonian Mechanics*

A reference frame $\mathcal{R}$ denotes an oriented system of coordinates, for instance Cartesian coordinates, equipped with rulers and clocks to measure position and time. The spacetime framework of Newtonian mechanics then relies on the following assumptions and properties:

- An *event* is a localized phenomenon in time and space. It is given by its spacetime coordinates (x, y, z, t) in a reference frame $\mathcal{R}$.
- The *physical space* is assumed to be the Euclidian space with three dimensions. It is homogeneous and isotropic, which means that no modification of physical properties can occur by translation or rotation of coordinate axes.
  In an Euclidian space, the distance l between two fixed points $\mathbf{x_1} \equiv (x_1, y_1, z_1)$ and $\mathbf{x_2} \equiv (x_2, y_2, z_2)$ is obtained by the Euclidian distance

$$l^2 = \Delta x^2 + \Delta y^2 + \Delta z^2 = (x_2 - x_1)^2 + (y_2 - y_1)^2 + (z_2 - z_1)^2 \, .$$

  The distance l is invariant: it is the same for all observers. If Alice is at rest and holds a ruler of length l, Bob who is looking at Alice while riding his bike at velocity $\mathbf{v}$, will see a ruler of the same length.
- *Time* is defined by a reproducible phenomenon, that is, a clock, and represented by a parameter t. Time is absolute, meaning that it is independent of the reference frame. The duration Δt between two events is an invariant. Imagine that Alice throws a ball at Bob. With their respective clocks, Alice and Bob will measure the same duration for the ball motion from the instant Alice releases the ball to the instant Bob catches it while still in his bike (Alice's clock is at rest and Bob's clock is in motion). An event, for instance Bob catching the ball, takes place at the same time in different reference frames.

20.2.2 Galilean Transformation

In Newtonian mechanics, *Galilean* (or *inertial*) frames play a special role. Galilean frames are defined as the frames where objects move at constant velocity if they are not acted upon by external forces. The Galilean transformation is one of the most important concepts of Newtonian mechanics as it allows us to transform the spacetime coordinates of an event in a Galilean frame $\mathcal{R}$ into the spacetime coordinates of the same event viewed from another Galilean frame $\mathcal{R}'$.

For instance, assume two Galilean frames $\mathcal{R}(0, x, y, z)$ and $\mathcal{R}'(0', x', y', z')$ such that $\mathcal{R}'$ moves at constant velocity $\mathbf{v}$ with respect to $\mathcal{R}'$ ($\mathbf{v} \equiv \mathbf{v}(\mathcal{R}/\mathcal{R}')$), as illustrated

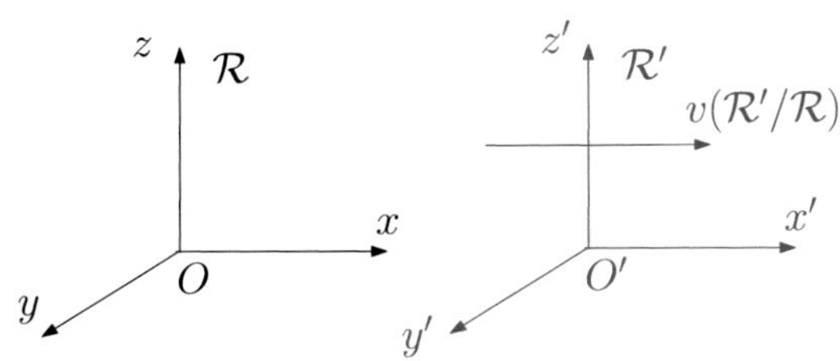

Fig. 20.1 The Galilean frame $\mathcal{R}'$ moves at uniform velocity $v(\mathcal{R}'/\mathcal{R})$ with respect to the frame $\mathcal{R}$

in Fig. 20.1. The coordinate axes can always be chosen in such a way that the x-axis is parallel to the x'-axis and to $\mathbf{v}$. The Galilean transformation from $\mathcal{R}$ to $\mathcal{R}'$ reads

$$
\begin{cases}
x' = x - vt \\
y' = y \\
z' = z \\
t' = t
\end{cases}
\tag{20.1}
$$

and the inverse transformation reads

$$
\begin{cases}
x = x' + vt' \\
y = y' \\
z = z' \\
t = t'
\end{cases}
$$

The inverse transformation is simply obtained by changing the sign of the velocity, since from the point of view of an observer at rest in $\mathcal{R}'$, it is the reference frame $\mathcal{R}$ that is moving at velocity $-\mathbf{v}$ with respect to $\mathcal{R}'$.

Let us discuss the implications of the Galilean transformation on velocities. Suppose Alice, observer in $\mathcal{R}$, throws a ball along the x-axis. Alice observes the ball moving at velocity $\mathbf{v}_{\text{ball}}(t) = \dfrac{dx(t)}{dt}\mathbf{u}_x$. Bob, observer from his bike in $\mathcal{R}'$ will see the ball moving at velocity $\mathbf{v}'_{\text{ball}} = \dfrac{dx'(t')}{dt'}\mathbf{u}'_x$. Differentiating the coordinates of the ball with respect to time in each frame leads to the velocities observed by each observer:

$$
\frac{dx'(t')}{dt'} = \frac{dx'}{dt}\underbrace{\frac{dt}{dt'}}_{1} = \frac{d(x(t) - vt)}{dt} = \frac{dx(t)}{dt} - v
$$

Thus,

$$
v'_{\text{ball}} = v_{\text{ball}} - vx \ .
$$

This is the composition law for velocities. Differentiating again with respect to time yields accelerations:

$$
a'(t') = \frac{d}{dt'}\mathbf{v}'_{\text{ball}} = \underbrace{\frac{dt}{dt'}}_{1}\left(\frac{d}{dt}\mathbf{v}_{\text{ball}} - \underbrace{\frac{dv}{dt}}_{0}\right) = \frac{d}{dt}\mathbf{v}_{\text{ball}} = a(t) \ .
$$

Hence Alice and Bob observe the same acceleration of the ball. Assuming that the force $\mathbf{F}$ is the same in all Galilean frames, they conclude that Newton's law

$$
\mathbf{F} = m\mathbf{a}
$$

is the same in both reference frames.

> **Galileo's relativity principle**
>
> - The laws of mechanics are the same in all Galilean reference frames. In other words, Galileo's transformation leaves invariant Newton's law.
> - These laws are said to be covariant with respect to a change of Galilean reference frame.

It is reasonable to ask ourselves: can we extend this relativity principle to the laws of electrodynamics, while preserving the same spacetime framework?

20.2.3 *Electromagnetism and Relativity Principle*

Maxwell's theory, published in 1854, established that at a given instant and at each point of space, properties of space are modified by a *charge distribution in motion*. The charge distribution is responsible for the generation of an electromagnetic field, characterized by an electric field $\mathbf{E}$ and a magnetic field $\mathbf{B}$. A charge q undergoes the action of the electromagnetic field via the Lorentz force

$$\mathbf{F} = q(\mathbf{E} + \mathbf{v} \times \mathbf{B}) \, .$$

The electromagnetic field, $\mathbf{E}$ and $\mathbf{B}$, can be calculated from Maxwell's equations. These equations led to the prediction that electromagnetic fields propagate in vacuum at velocity $v = (\epsilon_0 \mu_0)^{-1/2}$, where μ_0 and ϵ_0 characterize the electric and magnetic properties of vacuum. Up to measurement uncertainties, it was shown that $v \sim c$, that is, electromagnetic waves propagate at the speed of light. This led to the conclusion that light is an electromagnetic wave.

Fizeau was the first to succeed in a terrestrial measurement of the speed of light in 1849, sending a light beam along a 17 km round trip path across the outskirts of Paris. At the light source, the exiting beam was chopped by a rotating toothed wheel as shown in Fig. 20.2; the measured rotational rate of the wheel at which the beam, upon its return, was eclipsed by the toothed rim was used to determine the travel time of the beam. Fizeau reported a light speed that differs by only about 5 percent from the currently accepted value.

Hertz produced radio waves for the first time in 1888 and confirmed Maxwell's prediction that they propagate at the same velocity as light, and undergo similar phenomena of reflection, refraction and polarization. If the velocity of electromagnetic waves predicted by Maxwell's equations were only valid in a specific reference frame, this frame would have an absolute nature, that is, Maxwell's equations would not be valid in any reference frame. It was clear by then that experiments on light

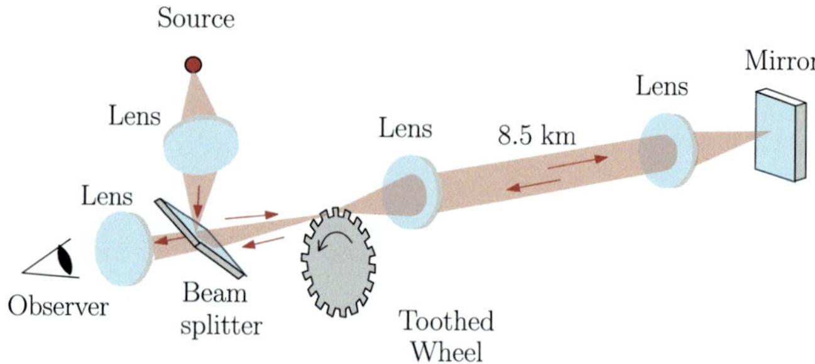

Fig. 20.2 Fizeau's experiment for the measurement of the speed of light

waves would allow to single out this specific frame. Two point of views were then debated to conciliate Maxwell's theory and the concept of absolute reference frame originating from classical Newtonian mechanics.

- Scientists in the nineteenth century have assumed that Maxwell's equations needed modifications and concluded that a medium was necessary for the propagation of electromagnetic waves and light, and that the speed of light with respect to that medium was constant, just like the propagation of sound in air. This medium was named *aether* and was seen as an invisible, massless, jelly like medium filling all space and fixed with respect to an absolute reference frame identified as the Copernic reference frame. Aether had other exotic properties like its high rigidity, allowing light to travel so fast in it while allowing Earth or objects to move in it without any drag. Needless to say that no consensus was reached regarding aether.
- There was a second point of view: Maxwell's equations are equally valid in *all* Galilean reference frames, that is, they obey the relativity principle. As a result, the velocity of electromagnetic waves is unique: it is the same in all reference frames. The main problem is that this principle is not compatible with the laws of Newtonian mechanics since velocity is frame-dependent. If we accept this principle, then,

> The spacetime framework of physics is not that of classical kinematics.

Experiments have proven this second point of view to be valid.

20.3 Experimental Results

Here we review a couple of important experiments that allowed scientists to refute the concept of aether on the one hand, and to confirm the validity of the second point of view on the other hand.

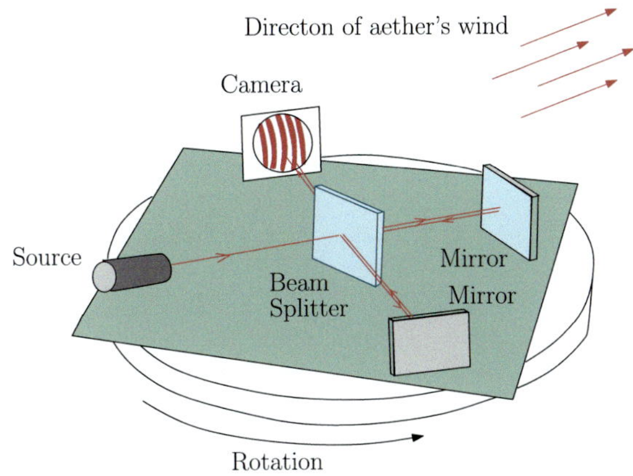

Fig. 20.3 Michelson and Morley's interferometric setup

20.3.1 *Michelson and Morley Experiment*

Michelson and Morley attempted to detect the existence of aether in a famous experiment in 1887. Their goal was to measure the orbital velocity of the Earth by optical means. Assuming the Earth is moving through the aether, the speed of light in the direction of Earth's motion should be lower than it is in a perpendicular direction. Michelson and Morley thus built an experiment to measure these speeds in order to detect Earth's absolute velocity relative to the aether.

The principle of their experiment, illustrated in Fig. 20.3, is the following: A light source sends light into the two arms of an interferometer, thanks to a glass plate that partly transmits and partly reflects the beam. In both arms, the light beams are reflected by mirrors and recombine after hitting the glass plate again. This is the Michelson interferometer previously discussed in Sect. 16.5.2. The recombination leads to interference fringes that depend on the length difference between the two arms. Importantly, the whole interferometer is mounted on a block that floats in an annular trough of mercury, allowing for a rotation of the device with very little friction.

The basic assumption to understand the principle of the experiment is that the Earth constitute a frame $\mathcal{R}'_{\text{Earth}}$ that moves in aether at the absolute velocity $\mathbf{v} \equiv \mathbf{v}_{\text{Earth}/\mathcal{R}_{\text{abs}}}$. The light velocity $\mathbf{c}'_{\mathcal{R}'_{\text{Earth}}}$ in the Earth's frame is then different from that in aether $\mathbf{c} = \mathbf{c}_{\mathcal{R}_{\text{abs}}}$, satisfying the Galilean composition of velocities

$$\mathbf{c}_{\mathcal{R}_{\text{abs}}} = \mathbf{c}'_{\mathcal{R}'_{\text{Earth}}} + \mathbf{v}_{\text{Earth}/\mathcal{R}_{\text{abs}}} \, .$$

Now, if one of the arms of the interferometer is parallel to the velocity of Earth in aether, there will be a difference of phase between the two arms, that is, the time required by light to make the round trip from the glass plate to the mirror and back to the glass plate will differ for the longitudinal ray parallel to $\mathbf{v}$ and for the perpendicular ray. As shown in Fig. 20.4, for the longitudinal arm the light indeed travels toward

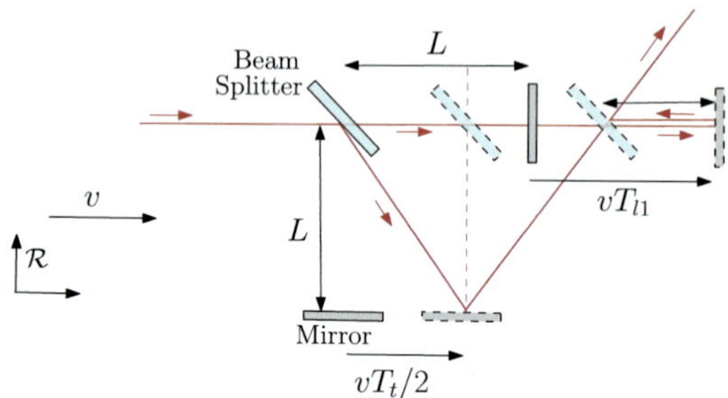

Fig. 20.4 Light's trajectories as seen from the aether's frame

the mirror at velocity c during a time T_{l1}, over a distance equal to the sum of the length of the arm and the distance traveled by the mirror, hence $L + vT_{l1} = cT_{l1}$, leading to

$$T_{l1} = \frac{L}{c - v}$$

where $v = |\mathbf{v}|$. On its way back, the light will travel a shorter distance to hit the glass plate, $L - vT_{l2} = cT_{l2}$, leading to a travel time

$$T_{l2} = \frac{L}{c + v} \ .$$

The total duration of the light journey in the longitudinal arm is then

$$T_l = T_{l1} + T_{l2} = \frac{2Lc}{c^2 - v^2} \ .$$

In the perpendicular arm, the light travels at c along twice the length of the hypothenuse of a right angle triangle, hence the travel time satisfies

$$(cT_t/2)^2 = L^2 + (vT_t/2)^2 \qquad \rightarrow \qquad T_t = \frac{2L}{\sqrt{c^2 - v^2}} \ .$$

The time difference between the longitudinal and the transverse arms is, at leading order in v/c:

$$(T_l - T_t) \sim \frac{L}{c} \frac{v^2}{c^2}$$

and is thus responsible for a phase difference $\Delta\phi = 2\pi c(T_l - T_t)/\lambda$ where λ is the wavelength of the light source. For the orbital velocity of the Earth, $v \sim 3 \times 10^4$ m/s, arms of length $L = 10$ m and a wavelength of 600 nm, we find

$$\Delta\phi \sim 2\pi \frac{L}{\lambda} \frac{v^2}{c^2} = 2 \times 3 \frac{10}{600 \times 10^{-9}} \frac{9 \times 10^8}{9 \times 10^{16}} = 1 \text{ rad} \ ,$$

which is large enough to be visible.

Now, it is not guaranteed that the two arms of the interferometer have exactly the same length, thus leading to a fringe pattern even if $v = 0$. However, to disambiguate the fringes due to the imperfect setup from those due to the motion of Earth in aether, it is sufficient to rotate the interferometer by $90°$, thus exchanging the roles of the arms, so as to observe a change in the fringe pattern corresponding to the Earth motion. In spite of numerous experiments performed at different times of the day and year, the result was always negative. Michelson and Morley could not observe a change in the fringe pattern and concluded that if there is a motion of Earth with respect to aether, the velocity must be extremely small. An absolute motion cannot be identified using electromagnetic waves.

20.3.2 *Measurement of the Velocity of Light*

Measurement of the velocity of light produced by a moving source were performed by Alväger et al. in 1964, at the CERN Proton Synchrotron. They executed a time of flight measurement to test the Newtonian momentum relations for light, being valid in the so-called emission theory. In this experiment, gamma rays were produced in the decay of 6-GeV pions π^0 traveling at $0.99975\,c$. If the Newtonian momentum $p = mv$ were valid, those gamma rays should have traveled at superluminal speeds. However, testing a Newtonian relation in the form $c' = v + c$, they found no difference with a source at rest:

$$\text{Accelerated pion: } c'_{\text{Lab}} = (2.9977 \pm 4 \times 10^{-4}) \times 10^8 \text{m/s}$$

$$\text{Source at rest: } c_{\text{Lab}} = (2.99792458 \pm 1.2) \times 10^8 \text{m/s}$$

Hence, they gave an upper limit of $\Delta v/c \sim 10^{-5}$.

They concluded that $c \simeq c'$: the speed of light is independent of the reference frame. This result is not compatible with Newtonian mechanics.

20.3.3 *Bertozzi's Experiment (1964)*

Bertozzi's experiment was performed in 1964. Special relativity (and general relativity) had already been successfully tested by earlier experiments but the relativistic dynamics of particles had been tested only indirectly. The development of particle accelerators in the twentieth century allowed measurements with increasing accuracy to be performed to test the predictions of relativity. Bertozzi's experiment is the first exrperiment leading to a direct measurement of the kinetic energy of accelerated particles.

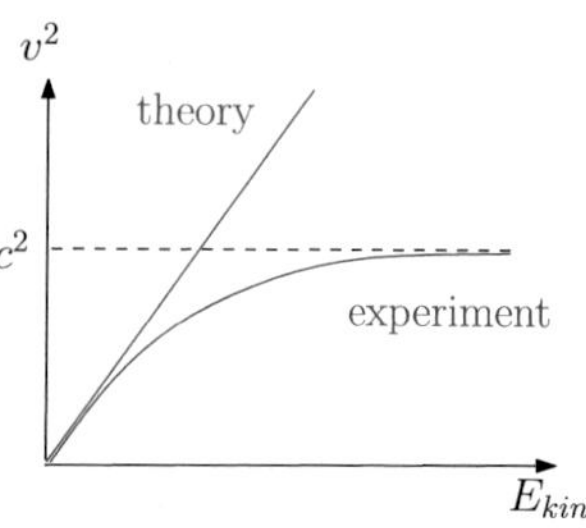

Fig. 20.5 Bertozzi's experiment. The theory here refers to Newtonian mechanics

Using an electron accelerator facility, electrons were accelerated by very high electric fields and their energy was measured. The experiment showed an upper velocity limit for accelerated particles. Newtonian mechanics simply predicts that the kinetic energy of electrons can increase up to any value, linearly with v^2. In contrast, Bertozzi's experiments demonstrated a saturation, as shown in Fig. 20.5. The velocity of the particle could not exceed the velocity of light.

In conclusion, all experiments aimed at identifying an absolute reference frame have consistently produce negative results, indicating that such a frame does not exist. The speed of light in vacuum is independent of the direction of propagation, affirming the fundamental isotropy of space. Moreover, the speed of light is invariant across all reference frames. Maxwell's equations are fully consistent with the relativity principle. It is Newtonian mechanics that must be revised, not Maxwell's equations.

20.4 Einstein's Special Relativity Principle (1905)

In 1905, Einstein (Fig. 20.6) published the theory of special relativity, which follows from the enunciation of the following relativity principle.

Relativity principle

- The fundamental laws of physics are covariant with respect to a change of Galilean reference frame.
- Light propagates in vacuum isotropically, with a speed c that is the same in every Galilean reference frame, given by[1]

$$c = 299\,792\,458\,\mathrm{m\,s^{-1}}$$

- A velocity limit exists for the propagation of a signal, corresponding to the velocity of light in vacuum.

---

[1] Nowadays, the speed of light is not measured aymore. It has a fixed value which in fact allows to define the meter from the definition of the second and the numerical value of c.

Fig. 20.6 Albert Einstein (1879–1955), a German-born physicist who revolutionized science with his theory of general relativity, describing gravity as the curvature of spacetime and fundamentally changing our understanding of the universe. He also made major contributions to quantum mechanics and statistical mechanics

A presentation of special relativity will be detailed in the next sections. We note that a requirement of the new theory is the equivalence principle, stating that Newtonian mechanics must be retrieved as a particular case of special relativity.

Equivalence principle
If $v \ll c$, the laws of Newtonian mechanics must be recovered.

For the purpose of showing that Maxwell's theory is fully compatible with the relativity principle, special relativity is all we need. However, special relativity is not the end of the story. In 1916, Einstein published another fundamental paper in the history of physics, presenting the theory of general relativity, which extends the theory of special relativity to the case of non Galilean reference frames. General relativity goes beyond the scope of this book.

20.5 Relativistic Kinematics

Einstein's postulates imply that our Universe is different from the Newtonian Universe that appears intuitive to our everyday experience. Velocities of material objects cannot be infinite and Newton's concept of absolute time is no longer valid. Similarly, the simultaneity of events for all observers does not hold. The kinematics characterizing this Universe, where infinite velocity is impossible, is presented in this section.

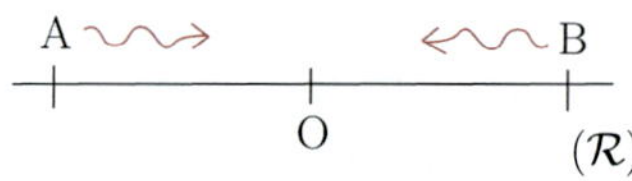

Fig. 20.7 Alice and Bob emit a light signal toward the middle point, where Oliver is located

20.5.1   Relative Nature of Time

Two events are simultaneous if they happen at the same instant. Therefore, we need clocks to help us measure simultaneity. Since the two events we wish to measure may happen at different locations, we also need to synchronize the clocks used for the measurement.

Suppose Alice and Bob are at fixed positions, at rest, in a reference frame $\mathcal{R}$. An observer, Oliver, stands exactly in the middle of Alice and Bob. Alice and Bob emit an electromagnetic signal toward Oliver, as shown in Fig. 20.7. We know that these signals travel at c, the speed of light.[2] If Oliver receives the signals from Alice and Bob at the same instant, he observes that the two events (emission of signal S_A by Alice and S_B by Bob) are simultaneous. It is then possible for Alice and Bob to synchronize their clocks.

20.5.1.1   Synchronization of Fixed Clocks in a Reference Frame $\mathcal{R}$

To observe simultaneous events in a given reference frame is therefore possible but it requires to synchronize fixed clocks. Synchronization can be achieved in the following way: If the propagation velocity of the signal is unknown, it is possible to first measure it. To do so, Alice first emits a signal (at $t = 0$) toward Bob, who holds a mirror, a distance a away from her. She will receive the signal at $t = 2a/c$ and knowing the distance, she can infer the propagation velocity of the signal in $\mathcal{R}$ (Fig. 20.8).

Then Alice emits emits another signal at t_0 in Bob's direction. Bob will receive the signal at $t_0 + \dfrac{a}{c}$ and can thus synchronize his clock with that of Alice. Fixed clocks in a reference frame $\mathcal{R}$ can be synchronized.

Fig. 20.8 Alice measures the velocity of the signal

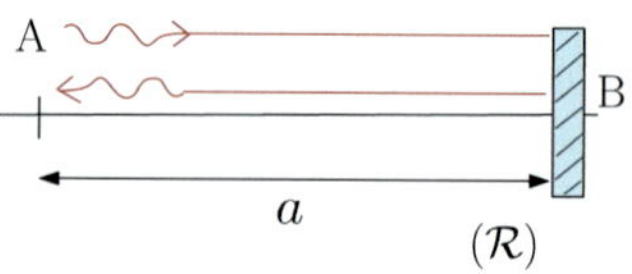

---

[2] If Alice and Bob use a slower communication means, for instance pigeons rather than radio waves, they will simply follow the same procedure, adding a preliminary measurement of the velocity c of the signal.

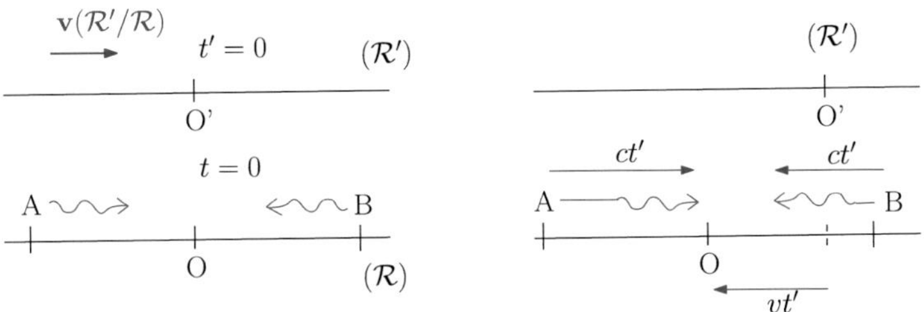

Fig. 20.9 An observer at O' moves at velocity **v** with respect to A, O and B

20.5.1.2   Relativity (Non Absolute Nature) of Simultaneous Events

Now consider another observer in a different reference frame, Oscar, at O' in the frame $\mathcal{R}'$ moving at velocity **v** with respect to $\mathcal{R}$. These frames are one-dimensional with x-axis parallel to the velocity **v**. Oscar and Oliver stand at O' and O, respectively, and synchronize their clocks when they pass in front of each other. At this instant $(O' \equiv O)$, $t = t' = 0$. This is illustrated in Fig. 20.9.

At the same instant, Alice and Bob who are at rest in $\mathcal{R}$ and have clocks that are synchronized with that of Oliver, emit their light signals S_A (from A), and S_B (from B). Of course, nothing changed from the point of view of Oliver in reference frame $\mathcal{R}$. He receives the signals at the same instant. However, Oscar in $\mathcal{R}'$ observes that Oliver does not receive the signals at the same time.

From Oscar's point of view, the signal from Alice takes a shorter time to reach Oliver since Oliver is moving toward the signal emitted by Alice. Since light travels at c in all reference frames, in reference $\mathcal{R}'$ the signal from Alice takes a time t'_A to reach Oliver, with $ct'_A = (a/2) - vt'_A$, that is $t'_A = (a/2)/(c + v)$. Similarly, in Oscar's frame Oliver is moving away from the signal emitted by Bob, that will take a longr time t'_B to reach him, given by $ct'_B = (a/2) + vt'_B$, so $t'_B = (a/2)/(c - v)$. Oscar therefore concludes that the signals are out of synchronization by an amount

$$\Delta T' = \frac{a/2}{c - v} - \frac{a/2}{c + v} = \frac{a}{c}\,\frac{v/c}{(1 - v^2/c^2)} \; .$$

This is a very small quantity if $v \ll c$. Nevertheless, Oscar concludes that the events (reception of signals S_A and S_B by Oliver) are not simultaneous, while Oliver receives the signals at the same time. When v approaches c, the delay observed by Oscar can become arbitrarily large. Simultaneity depends on the reference frame, and so time is relative to the observer.

Consequently, the postulate of the absolute nature of time in Newtonian mechanics turns out to be an approximation of the kinematics of our Universe which has a limit velocity; it breaks down when the relative velocity between frames approaches the speed of light. The invariance of the velocity limit in the Universe means that the

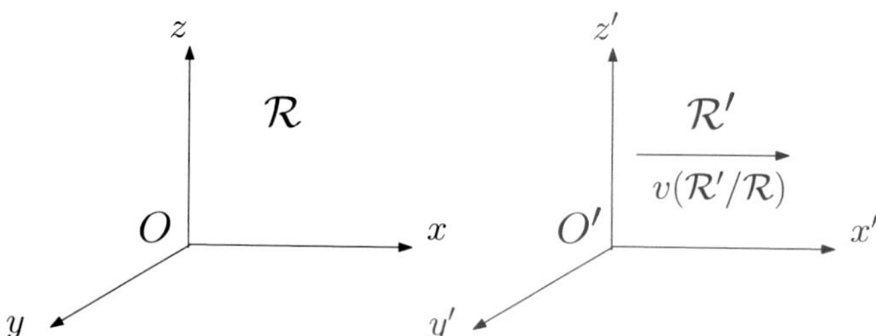

Fig. 20.10 Two Galilean frames in relative motion at constant velocity

velocity of an electromagnetic signal is the same in $\mathcal{R}'$ as in $\mathcal{R}$: $\mathbf{c}_{S_B/\mathcal{R}} = \mathbf{c}_{S_B/\mathcal{R}'}$, which implies that the spacetime framework need to be reconstructed.

20.5.2 Lorentz Special Transformation

Consider an event of spacetime coordinates (x, y, z, t) in a Galilean frame $\mathcal{R}$. In another Galilean frame $\mathcal{R}'$, the same event has spacetime coordinates (x', y', z', t'). The Lorentz transformation, derived by Hendrik Lorentz and Henri Poincaré (Fig. 20.11), is the kinematic transformation that replaces the Galilean transformation of Newtonian mechanics. It links the spacetime coordinates of an event in $\mathcal{R}$ to those in $\mathcal{R}'$, and it is demonstrated from homogeneity and isotropy of space, invariance of c and the relativity principle.

Suppose two Galilean frames $\mathcal{R}$ and $\mathcal{R}'$ equipped with rulers and clocks. All clocks of a given frame are synchronized. The frame $\mathcal{R}'$ moves at velocity $\mathbf{v} = \mathbf{v}_{\mathcal{R}'/\mathcal{R}}$ with respect to $\mathcal{R}$, as shown in Fig. 20.10.

The origins of $\mathcal{R}$ and $\mathcal{R}'$ are chosen so as to coincide when the clocks in $\mathcal{R}$ mark $t = 0$. The clock attached to $\mathcal{R}'$ placed at O' marks $t' = 0$ when O' is at O, that is, the origin events coincide in $\mathcal{R}$ and $\mathcal{R}'$. We chose the axes of coordinates and their orientation in $\mathcal{R}$ and $\mathcal{R}'$ such that $\mathbf{v}_{\mathcal{R}'/\mathcal{R}} \parallel Ox \parallel O'x'$, and in the transverse direction, $Oy \parallel O'y'$, and $Oz \parallel O'z'$. Before writing down the Lorentz transformation, we define standard notations in special relativity:

- The β factor represents a dimensionless velocity of the frame in motion, normalized by the speed of light. It is therefore smaller than one.

$$\beta = \frac{v}{c} \leq 1 \,.$$

- The Lorentz factor

$$\gamma = \frac{1}{\sqrt{1 - \dfrac{v^2}{c^2}}} = \frac{1}{\sqrt{1 - \beta^2}} \geq 1$$

Fig. 20.11 Left: Hendrik Lorentz (1853–1928), a Dutch physicist who made foundational contributions to electromagnetism and the electron theory of matter. Right: Henri Poincaré (1854–1912), a French mathematician, physicist, and philosopher of science. He made critical contributions to the development of special relativity. He introduced the concept of spacetime symmetry and formalized the mathematics of the Lorentz transformations

can take all values from 1 to infinity.

The Lorentz transformation and its inverse then reads

$$
x' = \frac{1}{\sqrt{1 - v^2/c^2}}(x - vt)
$$
$$
y' = y
$$
$$
z' = z
$$
$$
t' = \frac{1}{\sqrt{1 - v^2/c^2}}\left(t - \frac{vx}{c^2}\right)
$$

$$
x = \frac{1}{\sqrt{1 - v^2/c^2}}(x' + vt')
$$
$$
y = y'
$$
$$
z = z'
$$
$$
t = \frac{1}{\sqrt{1 - v^2/c^2}}\left(t' + \frac{vx'}{c^2}\right)
$$

or in terms of the β and Lorentz (γ) factors:

$$
x' = \gamma(x - \beta ct)
$$
$$
y' = y
$$
$$
z' = z
$$
$$
ct' = \gamma(ct - \beta x)
$$

$$
x = \gamma(x' + \beta ct')
$$
$$
y = y'
$$
$$
z = z'
$$
$$
ct' = \gamma(ct + \beta x')
$$

The Lorentz transformation ensures that the equivalence principle is satisfied, which means that if the velocity of $\mathcal{R}'$ is much smaller than the velocity of light, the Lorentz transformation tends to the Galilean transformation (20.1)

$$v \ll c \quad \rightarrow \quad \gamma \sim 1 \quad \rightarrow \quad \begin{cases} x' & \sim x - vt, \\ t' & \sim t. \end{cases}$$

Hence, this new kinematics is compatible with Newtonian mechanics at low speeds.

20.5.3   Invariance of SpaceTime Intervals

The Lorentz transformation introduces a coupling between space and time. This has several consequences, among which a change in the invariant quantities, that is the quantities that take the same numerical value in all Galilean frames. For instance, the speed of light is an invariant quantity by Einstein's postulate. A second invariant quantity is the charge of a particle. The spacetime interval constitutes a third important invariant quantity, as it allows us to distinguish past events from future events and thus, causes from effects. If we consider two events, which have spacetime coordinates in $\mathcal{R}$ and $\mathcal{R}'$:

$$\text{In } \mathcal{R}: \quad \text{Event 1 } (x_1, y_1, z_1, t_1); \quad \text{Event 2 } (x_2, y_2, z_2, t_2),$$
$$\text{In } \mathcal{R}': \quad \text{Event 1 } (x'_1, y'_1, z'_1, t'_1); \quad \text{Event 2 } (x'_2, y'_2, z'_2, t'_2),$$

the square of the space time interval between the two events is defined as

$$\text{In } \mathcal{R}: \quad \Delta s^2 = \Delta x^2 + \Delta y^2 + \Delta z^2 - c^2 \Delta t^2$$
$$\text{In } \mathcal{R}': \quad \Delta s'^2 = \Delta x'^2 + \Delta y'^2 + \Delta z'^2 - c^2 \Delta t'^2.$$

We will show that the spacetime interval is invariant:

$$\boxed{(\Delta s)^2 = (\Delta s')^2 = (\Delta l)^2 - c^2 (\Delta t)^2.}$$

Expanding the space time interval in $\mathcal{R}'$ and replacing the event coordinates in $\mathcal{R}'$ by those in $\mathcal{R}$ using the Lorentz transformation and the fact that $\gamma^2(1 - \beta^2) = 1$, we have

$$\begin{aligned}
\Delta s'^2 &= (x'_2 - x'_1)^2 + (y'_2 - y'_1)^2 + (z'_2 - z'_1)^2 - (ct'_2 - ct'_1)^2 \\
&= \gamma^2 (x_2 - vt_2 - x_1 + vt_1)^2 + (y_2 - y_1)^2 + (z_2 - z_1)^2 \\
&\quad - \gamma^2 (ct_2 - \beta x_2 - ct_1 + \beta x_1)^2 \\
&= \gamma^2 (x_2 - x_1)^2 \left(1 - \beta^2\right) - \gamma^2 (t_2 - t_1)^2 c^2 \left(1 - \beta^2\right) + \Delta y^2 + \Delta z^2 \\
&= (x_2 - x_1)^2 - c^2 (t_2 - t_1)^2 + \Delta y^2 + \Delta z^2 \\
&= \Delta s^2.
\end{aligned}$$

Therefore, we find

$$\Delta s = \pm \Delta s'$$

and the limit at low velocity allows us to choose the positive sign:

$$\text{if } v \to 0, \quad \Delta s' \to \Delta s \quad \Rightarrow \quad \Delta s = \Delta s' .$$

The spacetime interval is an invariant.

20.5.4 Time Dilation

20.5.4.1 Proper Time

The proper time denotes the interval between two events taking place at the same point of a Galilean reference frame. It is therefore measured by a clock attached to the Galilean frame. This frame is called the proper reference frame for these two events.

For instance, consider two events taking place in a Galilean reference frame $\mathcal{R}'$ moving at velocity v with respect to the frame $\mathcal{R}$: Event 1 corresponds to the emission of an electromagnetic signal from O' in the direction of the y-axis, and event 2 corresponds to the reception of the signal at O' after it is reflected by a mirror attached to $\mathcal{R}'$, a distance a away from O'. This is illustrated in Fig. 20.12. The two events take place at the same point O' of $\mathcal{R}'$, hence, the interval between the two events is the proper time interval in $\mathcal{R}'$.

Now let O_1 be the point in $\mathcal{R}$ that coincides with O' at event 1, when the signal is emitted. Similarly, O_2 denotes the point in $\mathcal{R}$ that coincides with O' at event 2, when the signal is received. If we use a clock of $\mathcal{R}$ to measure the time interval Δt between the two events viewed from $\mathcal{R}$, we find

$$\Delta t = 2\frac{O_1 M}{c} \quad \to \quad O_1 M = \frac{c \Delta t}{2} ,$$

while the interval between the two events viewed from the proper frame $\mathcal{R}'$ is

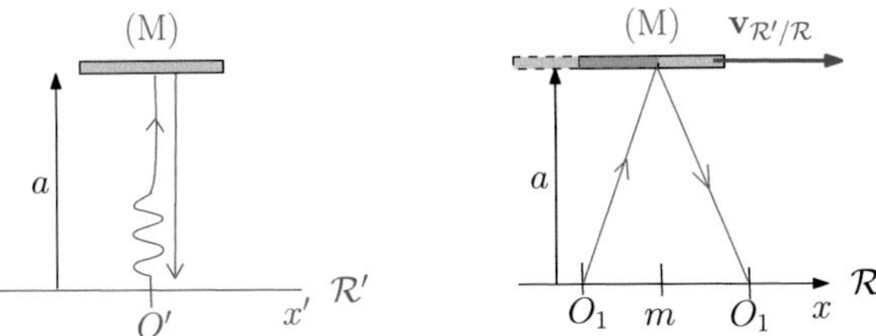

Fig. 20.12 Two events taking the same place in $\mathcal{R}'$ will take place at different positions in another frame $\mathcal{R}$. The proper time interval of these two events is by definition the time measured in $\mathcal{R}'$

$$\Delta t' = \frac{2a}{c} \rightarrow a = \frac{c\Delta t'}{2} \, .$$

In the triangle $O_1 m M$, we can use the Pythagorean theorem to link the time intervals measured by the two clocks in $\mathcal{R}$ and in $\mathcal{R}'$:

$$(O_1 M)^2 = a^2 + (O_1 m)^2, \quad \text{with} \quad O_1 m = \frac{v\Delta t}{2} \, .$$

Replacing all distances by their expressions as functions of the time intervals, we have

$$\frac{\Delta t^2 c^2}{4} = \frac{(\Delta t')^2 c^2}{4} + \frac{v^2 \Delta t^2}{4} \, .$$

Thus,

$$\Delta t^2 (c^2 - v^2) = c^2 \Delta t'^2 \quad \rightarrow \quad \Delta t = \frac{\Delta t'}{\sqrt{1 - \dfrac{v^2}{c^2}}} \, .$$

We find

$$\boxed{\Delta t = \gamma \Delta t'}$$

and since $\gamma \geq 1$, the proper time is smaller than the time interval measured in $\mathcal{R}$: $\Delta t' \leq \Delta t$. The observer in the laboratory $\mathcal{R}$ reports a longer elapsed time than the observer in the moving frame $\mathcal{R}'$.

20.5.4.2 Time Dilation Equation

The expression for the elapsed time between two events observed by an observer who is not in the proper reference frame is called the time dilation equation. It can be obtained directly from the Lorentz transformation between the Galilean frames $\mathcal{R}$ and $\mathcal{R}'$, moving at v with respect to $\mathcal{R}$. Indeed if two events are localized at the same point in the Galilean frame $\mathcal{R}'$, they occur at times t_1' and t_2', respectively, and at the same position $x_1' = x_2'$, that is $\Delta x' = 0$. In the reference frame of the observer $\mathcal{R}$, we have from the Lorentz transformation

$$t = \gamma \left(t' + \frac{vx'}{c^2} \right)$$

The elapsed time for the observer thus reads

$$\Delta t = \gamma \left(\Delta t' + \frac{v\Delta x'}{c^2} \right) \quad \rightarrow \quad \boxed{\Delta t = \gamma \Delta t' \, .}$$

The observer in $\mathcal{R}$ reports a longer elapsed time than the observer in the moving frame $\mathcal{R}'$. Similarly, if two events occur in $\mathcal{R}$ at the same point $x_1 = x_2$, then $\Delta x = 0$, and using the Lorentz transformation, we find, following similar arguments,

$$t' = \gamma \left(t - \frac{vx}{c^2} \right) \quad \rightarrow \quad \boxed{\Delta t' = \gamma \, \Delta t}$$

The observer in the moving frame $\mathcal{R}'$ observes a longer elapsed time than the observer in $\mathcal{R}$. The situation seems to be perfectly symmetric. From the point of view of $\mathcal{R}'$, it is the reference frame $\mathcal{R}$ that is moving at velocity $-v$, and so now $\mathcal{R}$ is the proper reference frame for the two events. This means that extreme care is required when applying the time dilation equation. It is important to appreciate the role of the proper reference frame in which two events occur at the same position.

Time dilation is a real phenomenon that has been experimentally verified through the radioactive decay of muons, which are elementary particles produced, for example, by cosmic rays in the atmosphere. These particles have a finite lifetime of 2.2×10^{-6} s corresponding to the characteristic exponential decay time of an initial muon population. This decay process effectively acts as a proper clock, measuring the proper time for the muons. In 1977, experiments conducted at CERN showed that the decay time measured in the Earth's reference frame depended on the mean-velocity of muons accelerated to relativistic speeds, where $\gamma \sim 30$. The measured lifetime was found to be in excellent agreement with the prediction of the time-dilation equation, with a precision of approximately 10^{-4}.

20.5.4.3 Proper Time and Proper Reference Frame for a Particle

Even if it is accelerated, at a fixed time t a particle can be considered as being at rest in a Galilean reference frame $\mathcal{R}_0$ having the particle velocity. This frame is called the tangent reference frame as its velocity with respect to the laboratory frame coincides with that of the particle at time t. Let $M(x, y, z)$ be the position and $\mathbf{v}(M/\mathcal{R})$ the velocity of the particle at time t in the Galilean reference frame $\mathcal{R}$ (the laboratory frame). From t to $t + dt$, the trajectory of the particle in $\mathcal{R}$ is described by $d\mathbf{M} = \mathbf{v}dt$. The space time interval between $M(t)$ and $M(t + dt)$ writes

$$(ds)^2 = (dx)^2 + (dy)^2 + (dz)^2 - c^2(dt)^2 \; .$$

Since the space time interval is an invariant through a Lorentz transformation, it can also be calculated in the tangent Galilean frame $\mathcal{R}_0$:

$$\boxed{(ds)^2 = -c^2(dt_0)^2 \quad \text{by definition of the proper time} \, ,}$$

which defines the *proper time* for the particle as a measure of the motion of the particle along its spacetime trajectory. Using the invariance of the spacetime interval, we can write

$$v^2 (dt)^2 - c^2 (dt)^2 = -c^2 (dt_0)^2 \; ,$$

that is

$$(dt)^2 = \frac{c^2}{c^2 - v^2} (dt_0)^2 \quad \Rightarrow \quad dt = \frac{dt_0}{\sqrt{1 - v^2/c^2}}$$

We can then define a Lorentz factor of the particle, $\gamma(v) = \dfrac{1}{\sqrt{1 - v^2(t)/c^2}}$, using its instantaneous velocity $v(t)$ in order to find a relation between time in the laboratory and the proper time of the particle

$$\boxed{dt = \gamma \, dt_0 \; .}$$

20.5.5   Length Contraction

20.5.5.1   Proper Length

The *proper length* of an object denotes its length in the Galilean reference frame where it is at rest.

   Consider an object $A'B'$ at rest in the reference frame $\mathcal{R}'$, moving at velocity v with respect to $\mathcal{R}$ as shown in Fig. 20.13. An observer in $\mathcal{R}'$ measures with a ruler the length $\Delta x'$ of the the object $A'B'$. $\Delta x'$ is the proper length of the object. An observer in the laboratory frame $\mathcal{R}$ observes the coincidence of the object ends with fixed points of $\mathcal{R}$ at the same instant t and measures the distance between these fixed points with a ruler at rest in $\mathcal{R}$. The observer in the laboratory frame reports a shorter length than the observer in the moving frame. He concludes that moving objects contract in the direction of motion.

20.5.5.2   Length Contraction Equation

The length measured by the observer in $\mathcal{R}$ and the proper length in $\mathcal{R}'$ are linked by the Lorentz transformation:

$$x' = \gamma(x - vt) \; ,$$

Fig. 20.13 The length of an object depends on the reference frame

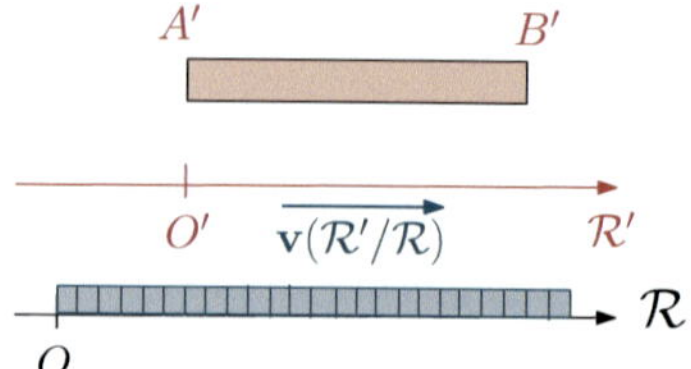

that can be applied to the positions of the ends of the object, measured at the same time in $\mathcal{R}$, that is, $\Delta t = 0$.

$$\Delta x' = \gamma (\Delta x - v \Delta t) \,.$$

We finally find

$$\boxed{\Delta x' = \gamma \, \Delta x \,,}$$

Since $\gamma \geq 1$, we have $\Delta x < \Delta x'$ which confirms that the observer in the laboratory reports a shorter length than the observer in the moving frame. Length contraction is a purely kinematic effect, it does not result from any compression force that would occur due to the motion of the object. Rather, it is again a manifestation of the non-absolute nature of time in a Universe without infinite velocity.

20.5.6 *Relativistic Addition of Velocities*

Suppose Alice is onboard a rocket, flying at a relativistic speed equal to $0.9\,c$, toward Bob who is waiting for her, sitting on a remote star (assumed at rest). She sends him a message using radio waves. What is the speed of the message, relative to Bob's star? If Alice and Bob were living in a Newtonian world, the message would travel at $0.9\,c + c = 1.9c$, which is obviously wrong since c, the speed of light, is the highest possible speed for the radio-wave signal to travel. The composition law for velocities must be adapted to be compatible with the Lorentz transformation.

Consider the velocity $\mathbf{v}(M/\mathcal{R})$ of a material point M in the Galilean frame $\mathcal{R}$. By definition of a velocity, it is given by differentiating the position with respect to time, which yields the three components

$$v_x = \frac{dx}{dt}, \quad v_y = \frac{dy}{dt}, \quad v_z = \frac{dz}{dt} \,.$$

Now consider a Galilean reference frame $\mathcal{R}'$, moving at velocity $\mathbf{u}(\mathcal{R}'/\mathcal{R})$ with respect to $\mathcal{R}$. Without loss of generality, we can assume that the axes of $\mathcal{R}$ and $\mathcal{R}'$ are chosen such that the x-axis is parallel to the x'-axis and to $\mathbf{u}$. The transverse axes are also parallel, that is, $Oy \parallel O'y'$ and $Oz \parallel O'z'$. In the frame $\mathcal{R}'$, the velocity of the material point is $\mathbf{v}'(M/\mathcal{R}')$, defined by

$$v'_{x'} = \frac{dx'}{dt'}, \quad v'_{y'} = \frac{dy'}{dt'}, \quad v'_{z'} = \frac{dz'}{dt'} \,.$$

Now we can use the Lorentz transformation to express the velocity components in $\mathcal{R}'$ as functions of those in $\mathcal{R}$. For the component along the x-axis, we find

$$dx' = \gamma (dx - u\,dt)$$

$$dt' = \gamma \left(dt - \frac{u\,dx}{c^2} \right)$$

leading to

$$v'_{x'} = \frac{dx - u\,dt}{dt - \dfrac{u\,dx}{c^2}} \quad \rightarrow \quad \boxed{\; v'_{x'} = \frac{v_x - u}{1 - \dfrac{u v_x}{c^2}} \;} \; .$$

We see that the numerator $v_x - u$ is just what we would expect in a Newtonian world but the denominator brings a correction. Going back to the question on the velocity of Alice's message to Bob, we have $u = 0.9\,c$, $v'_{x'} = c$ and the message velocity is obtained as

$$v_x = \frac{v'_{x'} + u}{1 + \dfrac{u v'_{x'}}{c^2}} = \frac{c + 0.9\,c}{1 + \dfrac{0.9\,c\,c}{c^2}} = c$$

As it should, the radio wave carrying Alice's message travels at c in Bob's reference frame. For the velocity components in the transverse directions, we find

$$\begin{cases} dy' = dy, \\ dz' = dz \end{cases} \quad \rightarrow \quad \begin{cases} \dfrac{dy'}{dt'} = \dfrac{dy}{\gamma \left(dt - \dfrac{u\,dx}{c^2} \right)} \\[4mm] \dfrac{dz'}{dz'} = \dfrac{dz}{\gamma \left(dt - \dfrac{u\,dx}{c^2} \right)} \end{cases}$$

leading to

$$\boxed{\; v'_{y'} = \frac{v_y}{\gamma \left(1 - \dfrac{u v_x}{c^2} \right)}, \qquad v'_{z'} = \frac{v_z}{\gamma \left(1 - \dfrac{u v_x}{c^2} \right)} \;} \; .$$

We see that the velocity components transverse to the direction of motion of the frame $\mathcal{R}'$ are also modified. This is not the case in a Newtonian Universe where only the longitudinal velocity component is modified. Again, this is a kinematic effect due to the relative nature of time.

We also note that the equivalence principle is verified, that is, the composition rule for velocities known in Newtonian mechanics is retrieved for $u \ll c$:

$$v'_{x'} \rightarrow v_x - u, \quad v'_{y'} \rightarrow v_y, \quad v'_{x'} \rightarrow v_z \; .$$

For an electromagnetic signal (satisfying $v_x^2 + v_y^2 + v_z^2 = c^2$), it is easy to verify that

$$v'^2_{x'} + v'^2_{y'} + v'^2_{z'} = c^2 \; .$$

Indeed, squaring the velocity components and replacing $1/\gamma^2$ by $(1 - u^2/c^2)$, we find

$$
\begin{aligned}
v'^2_{x'} + v'^2_{y'} + v'^2_{z'} &= \frac{(v_x - u)^2}{\left(1 - \frac{uv_x}{c^2}\right)^2} + \frac{v_y^2}{\gamma^2\left(1 - \frac{uv_x}{c^2}\right)^2} + \frac{v_z^2}{\gamma^2\left(1 - \frac{uv_x}{c^2}\right)^2} \\
&= \frac{(v_x^2 - 2uv_x + u^2) + (v_y^2 + v_z^2)(1 - u^2/c^2)}{\left(1 - \frac{uv_x}{c^2}\right)^2} \\
&= \frac{v_x^2 - 2uv_x + u^2 + (c^2 - v_x^2) - u^2(1 - v_x^2/c^2)}{\left(1 - 2uv_x/c^2 + u^2v_x^2/c^4\right)} = c^2 \ .
\end{aligned}
$$

20.5.6.1 Simultaneity and Localization

It results from the Lorentz transformation that space intervals and time intervals in $\mathcal{R}$ and in $\mathcal{R}'$ are linked by the relations

$$
\Delta x' = \gamma(\Delta x - u\Delta t)
$$
$$
\Delta t' = \gamma(\Delta t - u\Delta x/c^2)
$$

where u denotes the velocity of $\mathcal{R}'$ with respect to $\mathcal{R}$. Thus, if two events are localized at the same position in $\mathcal{R}$ ($\Delta x = 0$), an observer of $\mathcal{R}'$ does not observe them at the same position as

$$
\Delta x' = -\gamma u\Delta t \neq 0 \ .
$$

Similarly, if two events are simultaneous in $\mathcal{R}$ ($\Delta t = 0$), an observer of $\mathcal{R}'$ observes them at different times

$$
\Delta t' = -\gamma u\Delta x/c^2 \neq 0 \ .
$$

20.5.7 SpaceTime Interval Between Two Events

We have seen that the spacetime interval between two events is an invariant. This means that it is invariant to translations and rotations in space, invariant to translations in time, and invariant to a Lorentz transformation. Hence , the interval

$$
(\Delta s)^2 = \underbrace{(\Delta x)^2 + (\Delta y)^2 + (\Delta z)^2}_{d^2} - (c^2\Delta t)^2
$$

takes the same value in all Galilean reference frames. Since $(\Delta s)^2$ can take different signs, this leads to a classification of the intervals between two events, depending whether the spatial distance d is smaller or larger than the distance an electromagnetic signal can cover during the time interval Δt. We thus have three classes:

$$
\begin{aligned}
(\Delta s)^2 &< 0 \quad \text{time-like separation} \qquad & d < c\Delta t \\
(\Delta s)^2 &= 0 \quad \text{null separation} \qquad & d = c\Delta t \\
(\Delta s)^2 &> 0 \quad \text{space-like separation} \qquad & d > c\Delta t
\end{aligned}
$$

which receives the following interpretation:

- Events with a null separation $d = c\Delta t$ can be connected by a signal traveling at the speed of light.
- Two events with a space-like separation are separated by a greater distance than the distance an electromagnetic signal can cover during Δt, that is $|c\Delta t/\Delta x| < 1$. It is then always possible to perform a Lorentz transformation to a Galilean frame $\mathcal{R}'$ where the events are simultaneous ($\Delta t' = 0$) with

$$
(\Delta s')^2 = (\Delta x')^2 - (c\Delta t')^2 = (\Delta x')^2 \ .
$$

  This transformation is characterized by its normalized velocity $\beta = |c\Delta t/\Delta x|$.
- For events separated by a time-like interval, the distance in space is smaller than the distance an electromagnetic signal can cover during Δt. It is possible to perform a Lorentz transformation so that the events are localized at the same position ($\Delta x' = 0$) with

$$
(\Delta s')^2 = (\Delta x')^2 - (c\Delta t')^2 = -c^2(\Delta t')^2 \ .
$$

  This transformation is characterized by its normalized velocity $\beta = |\Delta x/c\Delta t|$.

This classification leads to the spacetime Minkowski diagram shown in Fig. 20.14, which is a representation of the interval classes in the 4-dimensional spacetime, using a single variable for two spatial variables. For instance, using $\rho^2 = x^2 + y^2$, we obtain two cones for the class of null separation, of equation $\rho^2 + z^2 = c^2 t^2$, which is called the light cone. The light cone separates the class of space-like intervals (outside the light cone, in the direction of space axes) from the class of time-like intervals (inside the light cone, in the direction of time).

Consider an event (observer) occupying the origin of coordinates, $t = 0$, $z = 0$, $\rho = 0$. The plane that comprises spatial coordinates is the hyperplane of the present. Any event, say T, inside the light cone lies either inside the future light cone and will occur after the event at O in all Galilean frames, or inside the past light cone and occurred before the event at O in all Galilean frames. The time-like interval between the events O and T make it possible for a cause and effect relationship to exist between them, that is, for a signal of velocity smaller than the light velocity to travel between the two events.

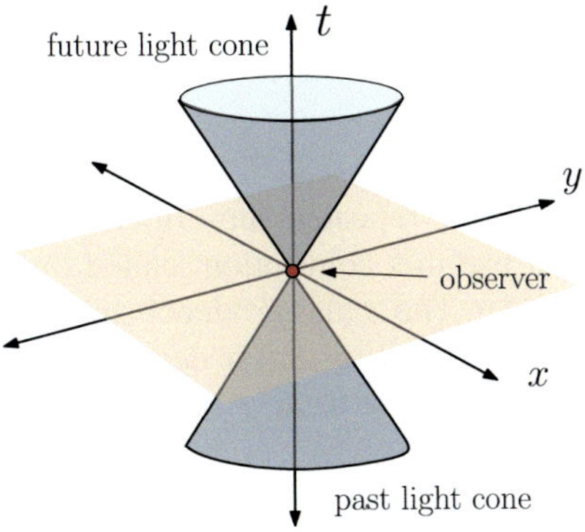

Fig. 20.14 Only events inside the light cones can have a cause-effect relantionship with the event O

future light cone

t

y

observer

x

past light cone

Fig. 20.15 Hermann Minkowski (1864–1909), a German mathematician and physicist. His work provides the mathematical framework for both special and general relativity

In contrast, for a space-like interval, that is, an event S lying outside the light cone, a cause and effect relationship between O and S is not possible. The time interval between these events may have different signs for different observers. No signal propagating at a velocity smaller than c can travel between O and S. It is impossible to transform these events to the same point in space as would be required to compare their clocks, if we wanted to check a cause and effect relationship.

20.6 Relativistic Dynamics

Einstein's postulate of relativity states that the laws of physics have the same form in every Galilean frame. This invariance can be formalized using a suited mathematical framework which was introduced by Minkowski (Fig. 20.15) and is presented in this section.

20.6.1 Euclidian Space

In the usual *Euclidian space*, vectors are defined by their three spatial components, for instance, $\mathbf{a} = (a_1, a_2, a_3)$, and $\mathbf{b} = (b_1, b_2, b_3)$. We will call them *three-vectors*. The scalar product between the three-vectors $\mathbf{a}$ and $\mathbf{b}$ is defined by

$$\mathbf{a} \cdot \mathbf{b} \equiv \sum_{k=1}^{3} a_k b_k = a_k b_k \ ,$$

where the repeated subscripts k means that a summation over k is performed. We will use this convention, called the Einstein convention, in the remaining of this chapter. The scalar product is invariant to translations and rotations of the system of coordinates. If $\mathbf{a}'$ and $\mathbf{b}'$ denotes the new components of $\mathbf{a}$ and $\mathbf{b}$ after a rotation or a translation, we have

$$\mathbf{a} \cdot \mathbf{b} = a_k b_k = a'_k b'_k = \mathbf{a}' \cdot \mathbf{b}'$$

The norm $\sqrt{\mathbf{a} \cdot \mathbf{a}}$ is also invariant to translations or rotations of the coordinate system in an Euclidian space, this property being a special case of the invariance of the scalar product.

20.6.2   *Minkowski's Space and Four Vectors*

For special relativity, Minkowski's space plays the role of the Euclidian space for Newtonian mechanics. The spacetime framework is naturally described by four coordinates and thus by *four-vectors*

$$\mathbf{a} = (a_1, a_2, a_3, a_4) \ .$$

A prototypical four-vector in special relativity is the spacetime coordinate vector

$$\mathbf{r} = (x, y, z, ict) \ ,$$

where the fourth component is purely imaginary. In Minkowski's space, the scalar product between $\mathbf{a}$ and $\mathbf{b} = (b_1, b_2, b_3, b_4)$ is defined in the same way as in an Euclidian space, just with one more component, and according to the usage, component subscripts are written with a greek letter, which is repeated to indicate a sum over the subscript values. The scalar product is invariant to translations, rotations and Lorentz transformations from a Galilean frame to another, that is,

$$\mathbf{a} \cdot \mathbf{b} = a_\mu b_\mu = a'_\mu b'_\mu = \mathbf{a}' \cdot \mathbf{b}' \ .$$

Still in analogy with the Euclidian space, the norm $\sqrt{\mathbf{a} \cdot \mathbf{a}}$ is an invariant as it is a particular case of an invariant scalar product. Now we see that applied to spacetime coordinates, this yields the invariant spacetime interval

$$(\Delta s)^2 = \Delta \mathbf{r} \cdot \Delta \mathbf{r} = (\Delta x)^2 + (\Delta y)^2 + (\Delta z)^2 - c^2 (\Delta t)^2 \ ,$$

as the imaginary i in the fourth component gives the minus sign in the calculation of the spacetime interval as a dot product.

20.6.3 Lorentz Transformation in 4-Vector Notation

Four-vectors are particularly useful in writing the Lorentz transformation in a compact form of a matrix-vector product, using the convention of repeated subscripts for a sum. The Lorentz transformation between the Galilean frames $\mathcal{R}$ and $\mathcal{R}'$, and its inverse indeed read

$$r'_\mu = L_{\mu\nu} r_\nu, \quad \text{and} \quad r_\mu = L^{-1}_{\mu\nu} r'_\nu ,$$

where $L_{\mu\nu}$ and $L^{-1}_{\mu\nu}$ denotes the 4×4 Lorentz matrix and its inverse:

$$L = \begin{bmatrix} \gamma & 0 & 0 & i\beta\gamma \\ 0 & 1 & 0 & 0 \\ 0 & 0 & 1 & 0 \\ -i\beta\gamma & 0 & 0 & \gamma \end{bmatrix} \qquad L^{-1} = \begin{bmatrix} \gamma & 0 & 0 & -i\beta\gamma \\ 0 & 1 & 0 & 0 \\ 0 & 0 & 1 & 0 \\ i\beta\gamma & 0 & 0 & \gamma \end{bmatrix}$$

Here it is assumed, as usual in this chapter, that the motion of $\mathcal{R}'$ is along $(Ox') \parallel (Ox)$ at velocity βc with respect to $\mathcal{R}$.

The expression for the inverse of the Lorentz matrix shows that L is an orthogonal matrix: $L^T = L^{-1}$. The determinant of the Lorentz matrix is one: $|L| = |L^{-1}| = 1$. Now, we can apply a more general Lorentz transformation to an arbitrary four vector $\mathbf{a}$ whose space-components are a_1, a_2, a_3 and whose time-component is a_4. In fact, the very definition of a four-vector of Minkowski space is precisely that it is transformed exactly like the spacetime coordinate $\mathbf{r}$:

$$\boxed{a'_\mu = L_{\mu\nu} a_\nu .}$$

20.6.4 General Lorentz Transformation

So far, we considered the Lorentz transformation between two Galilean frames $\mathcal{R}$ and $\mathcal{R}'$ where coordinate axes in $\mathcal{R}$ and $\mathcal{R}'$ were chosen parallel and orientation of the Ox and $O'x'$ axes parallel to $\mathbf{v}$, the velocity of $\mathcal{R}'$ with respect to $\mathcal{R}$. We might need to express the Lorentz transformation between two Galilean frames in a more general configuration, that is, in the case where $\mathbf{v}$ is not aligned with any axis of the frames, or in the case where the Cartesian axes of the two frames are not aligned. In the latter case, it is possible to perform first a rotation of axes using an Euler angle transformation to align the axes. We then end up with aligned Cartesian axes between

Fig. 20.16 A Lorentz boost

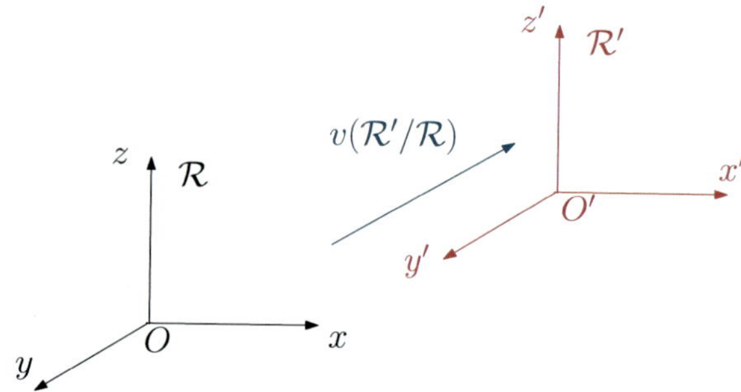

the two frames and with an arbitrary orientation of the vector quantity

$$\boldsymbol{\beta} \equiv \mathbf{v}/c \, ,$$

which represents the dimensionless velocity of $\mathcal{R}'$ with respect to $\mathcal{R}$. The frame $\mathcal{R}'$ is often called the *boosted frame*, and this situation is represented in Fig. 20.16.

Then, decompose any vector $\mathbf{r}$ into its components $\mathbf{r}_\parallel$ and $\mathbf{r}_\perp$ that are parallel and perpendicular to $\boldsymbol{\beta}$, respectively. The Lorentz transformation and its inverse takes simple expressions using this decomposition:

$$\begin{cases} \mathbf{r}'_\perp = \mathbf{r}_\perp \\ \mathbf{r}'_\parallel = \gamma(\mathbf{r}_\parallel - \boldsymbol{\beta} ct) \\ ct' = \gamma(ct - \boldsymbol{\beta} \cdot \mathbf{r}_\parallel) \end{cases} \qquad \begin{cases} \mathbf{r}_\perp = \mathbf{r}'_\perp \\ \mathbf{r}_\parallel = \gamma(\mathbf{r}'_\parallel + \boldsymbol{\beta} ct') \\ ct = \gamma(ct' + \boldsymbol{\beta} \cdot \mathbf{r}'_\parallel) \end{cases} \qquad (20.2)$$

And, for a general four-vector $\mathcal{A} = (\mathbf{a}, a_4)$

$$\begin{cases} \mathbf{a}'_\perp = \mathbf{a}_\perp \\ \mathbf{a}'_\parallel = \gamma(\mathbf{a}_\parallel + i\boldsymbol{\beta} a_4) \\ a'_4 = \gamma(a_4 - i\boldsymbol{\beta} \cdot \mathbf{a}_\parallel) \end{cases} \qquad \begin{cases} \mathbf{a}_\perp = \mathbf{a}'_\perp \\ \mathbf{a}_\parallel = \gamma(\mathbf{a}'_\parallel - i\boldsymbol{\beta} a'_4) \\ a_4 = \gamma(a'_4 + i\boldsymbol{\beta} \cdot \mathbf{a}'_\parallel) \end{cases} \qquad (20.3)$$

Note that these expressions assume aligned axes between the two frames. The most general Lorentz transformation requires applying first the Euler angle transformation to align the axes. One can choose the order of the rotation of axes and Lorentz's transformation, but the $\boldsymbol{\beta}$ vector and the Euler angles must be adapted as they depend on the chosen order.

20.6.5   Four-Velocity

Let us introduce another example of four-vector, the four-velocity, that will be useful in the context of dynamics of relativistic particles. In an Euclidian space, the three-velocity is obtained by differentiating in time the position three-vector. Since we wish to obtain a four-velocity $\mathbf{U}$ that represents a relativistic version of the velocity of a particle, we will differentiate the four-position with respect to the proper time of the particle, that is, we take into account the Lorentz factor and write

$$\mathbf{U} = \underbrace{\frac{d\mathbf{r}}{dt_0}}_{\text{Proper time}} = \gamma(v)\frac{d}{dt}(x, y, z, ict) = \gamma(v)\left(\frac{dx}{dt}, \frac{dy}{dt}, \frac{dz}{dt}, ic\right) = \gamma(v)(\mathbf{v}, ic)$$

$$= (U_1, U_2, U_3, U_4) \ .$$

This tells us that $\mathbf{U}$ is a time-like four-vector, as the squared norm of $\mathbf{U}$ is equal to a negative invariant quantity:

$$\mathbf{U} \cdot \mathbf{U} = (U_1^2 + U_2^2 + U_3^3 + U_4^4) = \gamma^2(v^2 - c^2) = -c^2 \ .$$

It is therefore always possible to find a Galilean frame in which the three-velocity of the particle $\mathbf{v}$ is instantaneously zero.

20.6.6   *Four-Acceleration*

Differentiating the four-velocity with respect to the proper time, we obtain the four-acceleration

$$\mathcal{A} = \frac{d\mathbf{U}}{dt_0} = \gamma(v)\frac{d}{dt}\gamma(v)(\mathbf{v}, ic) = (\mathcal{A}_1, \mathcal{A}_2, \mathcal{A}_3, \mathcal{A}_4) \ .$$

After differentiation, we find

$$\mathcal{A} = (\underbrace{\gamma^2\mathbf{a} + \gamma^4\mathbf{v}(\mathbf{v} \cdot \mathbf{a})/c^2}_{\mathcal{A}_1,\mathcal{A}_2,\mathcal{A}_3}, \underbrace{i\gamma^4(\mathbf{v} \cdot \mathbf{a})/c}_{\mathcal{A}_4}) \ ,$$

where $\mathbf{a} = d\mathbf{v}/dt$ denotes the acceleration three-vector. We obtain the following invariant quantity

$$\mathcal{A} \cdot \mathcal{A} = \gamma^4\mathbf{a}^2 + \gamma^6\frac{(\mathbf{a} \cdot \mathbf{v})^2}{c^2}$$

and in terms of the parallel ($\mathbf{a}_\parallel$) and perpendicular ($\mathbf{a}_\perp$) components of the acceleration

$$\mathcal{A} \cdot \mathcal{A} = \gamma^4\left(\mathbf{a}_\perp^2 + \gamma^2\mathbf{a}_\parallel^2\right) \ . \tag{20.4}$$

20.6.7 *Relativistic Momentum and Energy*

A scalar quantity $\mathcal{E}$, the energy of the system, can be associated with any physical system. This quantity is a state function in the sense that its variation only depends on the initial and final states of the system. If the system is isolated, its energy is constant. We are looking for an expression of the energy that is compatible with the relativity principle. Energy satisfies a conservation law, and a law for the conservation of momentum will be found as a consequence of the conservation of energy. We will build another four-vector, the four-momentum $\mathbf{P}$, and identify from its expression the relativistic energy and momentum that are compatible with the relativity principle. It is only after this identification that we will know that components of $\mathbf{P}$ indeed correspond to a momentum or an energy. The four-momentum will play an important role in the dynamics of relativistic particles as it will allow us to extend Newton's law to the relativistic regime. We define $\mathbf{P}$ by multiplying the four-velocity by a mass m. We postulate:

$$\boxed{\text{The mass } m \text{ is a Lorentz invariant quantity}\, .}$$

That is, invariant to a Lorentz transformation. The four-momentum then reads

$$\mathbf{P} = m\mathbf{U} = m\gamma\,(\mathbf{v}, ic)\, .$$

Writing $\mathbf{P}$ as

$$\mathbf{P} = (\mathbf{p},\, P_4)\, ,$$

we recognize in the first three components the three-momentum

$$\mathbf{p} = m(U_1, U_2, U_3) \quad \rightarrow \quad \boxed{\mathbf{p} = \gamma m\mathbf{v}\, ,}$$

after performing a Taylor expansion for $v \ll c$:

$$\mathbf{p} = \gamma m\mathbf{v} = m\mathbf{v}\left[1 + \frac{1}{2}\frac{v^2}{c^2} + \frac{3}{8}\frac{v^4}{c^4} + \cdots\right]$$

which shows that $\mathbf{p}$ reduces to its Newtonian expression for small particle velocities with respect to the light velocity. We write the fourth component as $P_4 = i\mathcal{E}/c$, where

$$\mathcal{E}/c = -imU_4 = \gamma mc \quad \rightarrow \quad \boxed{\mathcal{E} = \gamma mc^2\, .}$$

Hence, we can interpret the fourth component of the four-momentum vector as the ratio of an energy by the speed of light. Indeed, performing a Taylor expansion of $\mathcal{E}$ for $v \ll c$, we find

$$\mathcal{E} = \frac{mc^2}{\sqrt{1 - v^2/c^2}} = \underbrace{mc^2}_{\text{Rest energy}} + \underbrace{\frac{1}{2}mv^2}_{\text{Classical } E_{\text{kin}}} + \frac{3}{8}m\frac{v^4}{c^2} + \cdots$$

where, by neglecting the higher-order terms of the expansion, we recognize the sum of a rest energy, when $v = 0$, and the classical kinetic energy $mv^2/2$. The four-momentum is thus a four-vector that generalizes to the relativistic regime the classical momentum and total energy of a particle. The rest energy corresponds to the famous Einstein formula

$$\mathcal{E} = mc^2 \, ,$$

which tells us that the mass of a particle is equivalent to its energy.

A consequence of this equivalence is that the mass defect of a nucleus, defined as the difference between the mass of a nucleus and the sum of the masses of the nucleons within the nucleus, represents the energy binding the nucleus together.

$$\Delta m = m_{\text{nucleus}} - \sum_k m_{k,\text{nucleons}} < 0 \, .$$

The negative sign of the mass defect means that the nucleus can liberate the corresponding amount of energy, $(\Delta m)c^2$ if the nucleons are separated. This is the basis of the nuclear energy technology by fission of heavy nuclei (Uranium).

A second consequence is that in relativistic dynamics, mass is not conservative as it can be exchanged with energy. Mass is however invariant to a Lorentz transformation.

20.6.7.1 Kinetic Energy

The kinetic energy T for a relativistic particle can be defined by expressing the total energy $\mathcal{E} = \gamma mc^2$ as the sum of the rest energy and the relativistic kinetic energy. Hence,

$$\boxed{T = \mathcal{E} - mc^2 = (\gamma - 1)mc^2 \, .}$$

20.6.8 Relation Between Momentum and Energy

The expressions for the relativistic momentum and energy allow us to calculate the squared norm of the four-momentum:

$$\mathbf{P} \cdot \mathbf{P} = \mathbf{p}^2 - \frac{\mathcal{E}^2}{c^2} = (\gamma m\mathbf{v})^2 - \frac{(\gamma mc^2)^2}{c^2}$$
$$= -m^2 c^2 \, .$$

This is in keeping with the postulate that the mass is a Lorentz invariant. From this invariance, we infer the relation between the relativistic momentum and energy:

$$\boxed{\mathcal{E} = \sqrt{p^2c^2 + m^2c^4} = \gamma mc^2}$$

Another useful relation is obtained from the ratio of the momentum to total energy, which yields the velocity of a relativistic particle

$$\mathbf{v} = \frac{c^2\mathbf{p}}{\mathcal{E}} \ .$$

These relations apply the particular case of photons, which are massless particles. Indeed for $m = 0$, we find

$$\boxed{\mathcal{E} = pc} \quad \text{and} \quad \boxed{\mathbf{v} = c\mathbf{p}/p \ ,}$$

which shows that photons travel at the speed of light ($|\mathbf{v}| = c$).

20.6.9   Relativistic Particle Dynamics

Relativistic particle dynamics

The motion of a particle of velocity $\mathbf{v}$ is characterized by the Lorentz factor $\gamma = \dfrac{1}{\sqrt{1 - v^2/c^2}}$. From the relativistic momentum $\mathbf{p} = \gamma m\mathbf{v}$, the equation of motion for a particle with charge q and mass m in an electromagnetic field is

$$\frac{d\mathbf{p}}{dt} = \frac{d}{dt}[\gamma m\mathbf{v}] = q(\mathbf{E} + \mathbf{v} \times \mathbf{B}) \ .$$

More generally,

$$\frac{d\mathbf{p}}{dt} = \mathbf{F} \ ,$$

where $\mathbf{F}$ denotes the force applied to the particle. Note that $\mathbf{F} = \gamma m\mathbf{a} + m\mathbf{v}\frac{d\gamma}{dt}$ and only in the non-relativistic limit $\mathbf{F} = m\mathbf{a}$. This law constitutes a postulate that extends Newton's law to the relativistic regime. Its validity was confirmed by experiments.

20.6.10 Conservation of Power

The rate of change of the kinetic energy of a particle reads

$$\frac{dT}{dt} = \frac{d}{dt}[\mathcal{E} - mc^2] = \frac{d\mathcal{E}}{dt} \ .$$

From the relation between $\mathcal{E}$ and $\mathbf{p}$,

$$\mathcal{E}^2 = p^2 c^2 + m^2 c^4 \ ,$$

we obtain by differentiation

$$\mathcal{E}\frac{d\mathcal{E}}{dt} = c^2 \mathbf{p} \cdot \frac{d\mathbf{p}}{dt} \ .$$

Replacing $c^2\mathbf{p}$ by $\mathcal{E}\mathbf{v}$

$$\frac{d\mathcal{E}}{dt} = \mathbf{v} \cdot \frac{d\mathbf{p}}{dt} = \mathbf{v} \cdot \mathbf{F} \ ,$$

we retrieve the kinetic energy theorem: the time rate of change of the kinetic energy of a particle moving at velocity $\mathbf{v}$ is equal to the time rate of change of the work done on the particle by the force $\mathbf{F}$, that is, the power of force $\mathbf{F}$.

$$\boxed{\frac{dT}{dt} = \mathbf{F} \cdot \mathbf{v} \ .}$$

This law for conservation of power is the same as in classical mechanics provided the kinetic energy is replaced by its relativistic counterpart.

20.6.10.1 Four-Force

Defining the four-force by

$$\mathcal{F} = \gamma\left(\mathbf{F}, i\frac{\mathbf{v} \cdot \mathbf{F}}{c}\right)$$

it is found that

$$\mathcal{F} = \frac{d}{dt_0}\mathbf{P} = m\mathcal{A} \tag{20.5}$$

The spatial part of (20.5) yields the relativistic equation of motion

$$\frac{d\mathbf{p}}{dt} = \mathbf{F}$$

whereas its temporal component is the kinetic energy theorem:

$$\mathbf{v} \cdot \mathbf{F} = \frac{d}{dt}\mathcal{E} \, .$$

20.7   Electromagnetic Quantities

In this section we will revisit the electromagnetic quantities such as the fields, potentials, charge and current density that allowed us to formulate Maxwell's equations. Our objective is to discover how these quantities are transformed via the Lorentz transformation. Since Maxwell's equations are expressed using vector calculus operators such as the gradient, the divergence, the curl, etc., our first task is to perform a Lorentz transformation for derivatives with respect to space and time.

Consider the Galilean reference frame $\mathcal{R}$ and another Galilean frame $\mathcal{R}'$ moving at velocity $\mathbf{v}$ with respect to $\mathcal{R}$. Assume the Cartesian axes of $\mathcal{R}$ and $\mathcal{R}'$ are aligned but the velocity $\mathbf{v}$ is arbitrary. We can define the boost $\boldsymbol{\beta}$:

$$\boldsymbol{\beta} = \frac{\mathbf{v}}{c} \, .$$

Writing $\nabla \equiv \dfrac{\partial}{\partial \mathbf{r}}$, we use the chain rule and obtain the following derivatives

$$\frac{\partial}{\partial \mathbf{r}'_\parallel} = \underbrace{\frac{\partial \mathbf{r}_\parallel}{\partial \mathbf{r}'_\parallel} \cdot \frac{\partial}{\partial \mathbf{r}_\parallel}}_{\gamma} + \underbrace{\frac{\partial \mathbf{r}_\perp}{\partial \mathbf{r}'_\parallel} \cdot \frac{\partial}{\partial \mathbf{r}_\perp}}_{0} + \underbrace{\frac{\partial t}{\partial \mathbf{r}'_\parallel}\frac{\partial}{\partial t}}_{\gamma\beta/c} = \gamma \left(\frac{\partial}{\partial \mathbf{r}_\parallel} + \frac{\boldsymbol{\beta}}{c}\frac{\partial}{\partial t} \right)$$

$$\frac{\partial}{\partial \mathbf{r}'_\perp} = \frac{\partial}{\partial \mathbf{r}_\perp}$$

$$\frac{\partial}{\partial t'} = \underbrace{\frac{\partial \mathbf{r}_\parallel}{\partial t'} \cdot \frac{\partial}{\partial \mathbf{r}_\parallel}}_{\gamma c\beta} + \underbrace{\frac{\partial t}{\partial t'}\frac{\partial}{\partial t}}_{\gamma} = \gamma \left(c\boldsymbol{\beta} \cdot \frac{\partial}{\partial \mathbf{r}_\parallel} + \frac{\partial}{\partial t} \right)$$

Here we have used the expression of Lorentz transformation (20.2) to recognize the Lorentz factor γ and the $\boldsymbol{\beta}$ vector. Identifying the parallel and perpendicular components of the gradient operator

$$\nabla \equiv \frac{\partial}{\partial \mathbf{r}}, \quad \nabla_\parallel \equiv \frac{\partial}{\partial \mathbf{r}_\parallel}, \quad \nabla_\perp \equiv \frac{\partial}{\partial \mathbf{r}_\perp}, \quad \nabla'_\parallel \equiv \frac{\partial}{\partial \mathbf{r}'_\parallel}, \quad \nabla'_\perp \equiv \frac{\partial}{\partial \mathbf{r}'_\perp} \, ,$$

the transformation equations for partial derivatives take the form of a Lorentz transformation for the four gradient ∇_4:

$$\boxed{\begin{aligned}
\mathbf{\nabla}'_{\parallel} &= \gamma\left(\mathbf{\nabla}_{\parallel} + i\beta\,\frac{\partial}{\partial(ict)}\right)\\[4pt]
\mathbf{\nabla}'_{\perp} &= \mathbf{\nabla}_{\perp}\\[4pt]
\frac{\partial}{\partial(ict')} &= \gamma\left(\frac{\partial}{\partial(ict)} - i\beta\mathbf{\nabla}_{\parallel}\right)
\end{aligned}}
\qquad
\boxed{\mathbf{\nabla}_4 = \left(\mathbf{\nabla},\,\frac{\partial}{\partial(ict)}\right).}$$

The norm of a four-vector is an invariant, thus the norm or the four-gradient is a covariant operator, invariant to a Lorentz transformation:

$$\mathbf{\nabla}_4 \cdot \mathbf{\nabla}_4 = \mathbf{\nabla}^2 - \frac{1}{c^2}\frac{\partial^2}{\partial t^2} = \mathbf{\nabla}'^2 - \frac{1}{c^2}\frac{\partial^2}{\partial t'^2}\ .$$

We see that the wave propagation operator (the d'Alembertian) is invariant to Lorentz transformations, that is, waves propagate at the speed of light in all Galilean frames, and we are now ready to check this invariance for propagation equations after expressing the transformation laws for the potentials and fields.

20.7.1 Continuity Equation

Einstein's relativity principle states that the laws of electrodynamics are valid in every Galilean frame. This must be the case for the conservation of charge that we have written in the form of the continuity equation

$$\mathbf{\nabla}\cdot\mathbf{j} + \frac{\partial\varrho}{\partial t} = 0. \tag{20.6}$$

The goal of this section is to show that (20.6) is indeed invariant to a Lorentz transformation. This equation can be rewritten using the four-gradient defined in the previous section and the possible four-vector[3]

$$\boldsymbol{J} = (\mathbf{j}, ic\varrho). \tag{20.7}$$

The continuity equation then reads

$$\mathbf{\nabla}_4 \cdot \boldsymbol{J} = 0, \tag{20.8}$$

where the operator $\mathbf{\nabla}_4\cdot$ denotes the four-divergence.

Now reading the resulting equation as the dot product of the four-gradient, which we have shown to be a four-vector, with $\boldsymbol{J}$ defined in (20.7), we conclude that (20.8) is invariant to a Lorentz transformation only if the four-current $\boldsymbol{J} = (\mathbf{j}, ic\varrho)$ is indeed a four-vector, that is, if it is invariant to a Lorentz transformation.

---

[3] This is a definition. We do not know yet if it is a four-vector.

In order to check that $\boldsymbol{j}$ is a four-vector, consider a charge element $dq' = \varrho' dx'dy'dz'$, at rest in the Galilean frame $\mathcal{R}'$. In the frame $\mathcal{R}$, this charge element is expressed as $dq = \varrho dxdydz$, where the charge densities ϱ and ϱ' correspond to the densities viewed from their respective frame. If we assume that the direction of motion of $\mathcal{R}'$ with respect to $\mathcal{R}$ is along the x-axis (parallel to the x'-axis), there is length contraction and therefore, $dx = dx'/\gamma$. The electric charge is an invariant, which means $dq = dq'$. Thus, the charge densities in $\mathcal{R}$ and in $\mathcal{R}'$ must satisfy

$$dq = \varrho dxdydz = \varrho\frac{dx'}{\gamma}dy'dz' = \frac{\varrho}{\gamma}dx'dy'dz' = \varrho'dx'dy'dz' = dq' \,,$$

which implies density dilation

$$\varrho = \gamma\varrho' \,.$$

Finally, to be a four-vector, the four-current must transform according to the Lorentz transformation

$$\mathbf{j}'_{\parallel} = \gamma\,(\mathbf{j}_{\parallel} - \varrho\mathbf{v}) \,,$$
$$\mathbf{j}'_{\perp} = \mathbf{j}_{\perp} \,,$$
$$\varrho' = \gamma\left(\varrho - \frac{jv}{c^2}\right) \,.$$

From the definition of the current $\mathbf{j} = \varrho\mathbf{v}$, we can check that the equations for the parallel and perpendicular components are satisfied ($\mathbf{j}'_{\perp} = \mathbf{j}_{\perp} = 0$ and $\mathbf{j}'_{\parallel} = 0$ since the charge is at rest in $\mathcal{R}'$). For the last equation

$$\varrho' = \gamma(1 - \beta^2)\varrho = \frac{\varrho}{\gamma} \,,$$

as found above. In conclusion $\boldsymbol{j}$ is a four-vector and the continuity equation is frame invariant and reads

$$\boxed{\nabla_4 \cdot \boldsymbol{j} = 0 \quad \text{with} \quad \boldsymbol{j} = (\mathbf{j}, ic\varrho) \,.}$$

In any other Galilean frame $\mathcal{R}'$, the continuity equation reads

$$\nabla'_4 \cdot \boldsymbol{j}' = 0 \quad \text{with} \quad \boldsymbol{j}' = (\mathbf{j}', ic\varrho') \,.$$

20.7.2 Lorenz Gauge Potentials

In this section, we will show that the Lorenz gauge condition

$$\nabla \cdot \mathbf{A} + \frac{1}{c^2}\frac{\partial V}{\partial t} = 0$$

is invariant to a Lorentz transformation.[4] We first note that if we define the four-potential as

$$\mathbf{A}_4 = \left(\mathbf{A}, i\frac{V}{c}\right),$$

then the Lorenz gauge can be rewritten in the form of a four-divergence

$$\mathbf{\nabla}_4 \cdot \mathbf{A}_4 = 0,$$

or in other words, a dot product between the four-gradient and the four-potential $\mathbf{A}_4$. If we show that $\mathbf{A}_4$ is indeed a four-vector, that is, if $\mathbf{A}_4$ is transformed according to (20.2) under a Lorentz transformation between two Galilean frames, then the Lorenz gauge condition will be indeed invariant to a Lorentz transformation.

To check that $\mathbf{A}_4$ is a four-vector, we recall that the potentials in the Lorenz gauge satisfy the inhomogeneous d'Alembert wave equations

$$\left[\nabla^2 - \frac{1}{c^2}\frac{\partial^2}{\partial t^2}\right] V = -\frac{\varrho}{\epsilon_0} \quad \text{and} \quad \left[\nabla^2 - \frac{1}{c^2}\frac{\partial^2}{\partial t^2}\right] \mathbf{A} = -\mu_0 \mathbf{j}.$$

Thus, it is a straightforward step to conclude that the four-potential satisfies the inhomogeneous d'Alembert wave equation where the four-current is the source term:

$$\left[\nabla^2 - \frac{1}{c^2}\frac{\partial^2}{\partial t^2}\right]\left(\mathbf{A}, i\frac{V}{c}\right) = -\mu_0(\mathbf{j}, ic\varrho). \tag{20.9}$$

Now the d'Alembertian operator $\left[\nabla^2 - \dfrac{1}{c^2}\dfrac{\partial^2}{\partial t^2}\right]$ is a Lorentz invariant operator: A Lorentz transformation from $\mathcal{R}$ to $\mathcal{R}'$ leads to

$$\left[\nabla^2 - \frac{1}{c^2}\frac{\partial^2}{\partial t^2}\right] \quad \rightarrow \quad \left[\nabla'^2 - \frac{1}{c^2}\frac{\partial^2}{\partial t'^2}\right].$$

The four-current $\jmath$ is a four-vector, which implies that the four-potential $\mathbf{A}_4$ must be a four-vector as well, otherwise the transformation properties of left-hand-side and right-hand-side of Eq. (20.9) will differ.

In conclusion, the Lorenz gauge condition takes the same form in all Galilean frames.

---

[4] Incidentally, this section makes it clear that Ludvig Lorenz, a Dane physicist who gave his name to the gauge, is not the same physicist as Hendrik Lorentz (from the Netherlands), who gave his name to the force and to the transformation.

20.8　Field Transformation Laws

The objective of this section is to express the electric and magnetic fields $\mathbf{E}'(\mathbf{r}', t')$ and $\mathbf{B}'(\mathbf{r}', t')$ viewed by an observer of a Galilean frame $\mathcal{R}'$ moving at velocity $\mathbf{v} = c\boldsymbol{\beta}$ with respect to $\mathcal{R}$, as a function of the fields $\mathbf{E}(\mathbf{r}, t)$ and $\mathbf{B}(\mathbf{r}, t)$ in $\mathcal{R}$. For this task, we will exploit the definition of the fields in terms of the potentials

$$\mathbf{B} = \nabla \times \mathbf{A}, \qquad \mathbf{E} = -\nabla V - \frac{\partial \mathbf{A}}{\partial t} ,$$

in conjunction with the transformation rules for the four-gradient ∇_4 and the four-potential $\mathbf{A}_4$. Starting with the magnetic field, we perform a decomposition into the parallel and perpendicular components to $\mathbf{v}$:

$$\begin{aligned}
\mathbf{B}' = \nabla' \times \mathbf{A}' &= (\nabla'_\parallel + \nabla'_\perp) \times (\mathbf{A}'_\parallel + \mathbf{A}'_\perp) \\
&= \nabla'_\parallel \times \mathbf{A}'_\parallel + \nabla'_\perp \times \mathbf{A}'_\perp + \nabla'_\parallel \times \mathbf{A}'_\perp + \nabla'_\perp \times \mathbf{A}'_\parallel .
\end{aligned}$$

The first term vanishes $\nabla'_\parallel \times \mathbf{A}'_\parallel = \mathbf{0}$ and the remaining terms allow us to identify the parallel and perpendicular components of the field

$$\begin{aligned}
\mathbf{B}'_\parallel &= (\nabla' \times \mathbf{A}')_\parallel = \nabla'_\perp \times \mathbf{A}'_\perp , \\
\mathbf{B}'_\perp &= (\nabla' \times \mathbf{A}')_\perp = \nabla'_\parallel \times \mathbf{A}'_\perp + \nabla'_\perp \times \mathbf{A}'_\parallel .
\end{aligned}$$

Both three-vectors $\nabla'_\perp$ and $\mathbf{A}'_\perp$ are perpendicular components of the space-part of a four-vector, hence, they are invariant under the Lorentz transformation and so the first equation tells us that the parallel component of the magnetic field is invariant:

$$\mathbf{B}'_\parallel = \mathbf{B}_\parallel .$$

For the perpendicular components of the field, we substitute the prime quantities using the Lorentz transformation:

$$\begin{aligned}
\mathbf{B}'_\perp &= \gamma \left(\nabla_\parallel + \frac{\mathbf{v}}{c^2} \frac{\partial}{\partial t} \times \mathbf{A}_\perp \right) + \nabla_\perp \times \gamma \left(\mathbf{A}_\parallel - \frac{\mathbf{v}}{c^2} V \right) \\
&= \gamma \left(\nabla_\parallel \times \mathbf{A}_\perp + \nabla_\perp \times \mathbf{A}_\parallel \right) + \frac{\gamma}{c^2} \left(\mathbf{v} \times \frac{\partial \mathbf{A}_\perp}{\partial t} - \nabla_\perp \times (\mathbf{v} V) \right) .
\end{aligned}$$

We recognize $\mathbf{B}_\perp$ in the first two terms of the right-hand-side, i.e., the perpendicular component of $\nabla \times \mathbf{A}$ if we perform a similar decomposition as for $\nabla' \times \mathbf{A}'$, and the fact that $\mathbf{v}$ is a constant vector leads to

$$\mathbf{B}'_\perp = \gamma \mathbf{B}_\perp - \frac{\gamma \mathbf{v}}{c^2} \times \left(-\frac{\partial \mathbf{A}_\perp}{\partial t} - \nabla_\perp V \right) .$$

Finally, from the definition of $\mathbf{E}$ as a function of the potentials, we can write

$$\mathbf{B}'_{\perp} = \gamma \left(\mathbf{B}_{\perp} - \frac{\mathbf{v}}{c^2} \times \mathbf{E}_{\perp} \right) = \gamma \left(\mathbf{B} - \frac{\mathbf{v}}{c^2} \times \mathbf{E} \right)_{\perp} ,$$

or equivalently,

$$c\mathbf{B}'_{\perp} = \gamma (c\mathbf{B} - \boldsymbol{\beta} \times \mathbf{E})_{\perp} .$$

For the electric field, we start from the parallel component

$$\mathbf{E}'_{\parallel} = -\boldsymbol{\nabla}'_{\parallel} V' - \frac{\partial \mathbf{A}'_{\parallel}}{\partial t'}$$

and we use the inverse Lorentz transformation to substitute all prime quantities:

$$\mathbf{E}'_{\parallel} = -\gamma \left(\boldsymbol{\nabla}_{\parallel} + \frac{\mathbf{v}}{c^2} \frac{\partial}{\partial t} \right) \gamma (V - \mathbf{v} \cdot \mathbf{A}) - \gamma \left(\frac{\partial}{\partial t} + \mathbf{v} \cdot \boldsymbol{\nabla} \right) \gamma \left(\mathbf{A}_{\parallel} - \frac{\mathbf{v}}{c^2} V \right) .$$

Expanding the right-hand-side, we find

$$\mathbf{E}'_{\parallel} = \gamma^2 \left(-\boldsymbol{\nabla}_{\parallel} V + \cancel{\boldsymbol{\nabla}_{\parallel}(\mathbf{v} \cdot \mathbf{A})} - \cancel{\frac{\mathbf{v}}{c^2} \frac{\partial V}{\partial t}} + \frac{\mathbf{v}}{c^2} \frac{\partial (\mathbf{v} \cdot \mathbf{A})}{\partial t} - \frac{\partial \mathbf{A}_{\parallel}}{\partial t} + \cancel{\frac{\mathbf{v}}{c^2} \frac{\partial V}{\partial t}} \right.$$
$$\left. - \cancel{(\mathbf{v} \cdot \boldsymbol{\nabla})\mathbf{A}_{\parallel}} + \frac{\mathbf{v}}{c^2} (\mathbf{v} \cdot \boldsymbol{\nabla}) V \right)$$

Four terms cancel, including $\boldsymbol{\nabla}_{\parallel}(\mathbf{v} \cdot \mathbf{A})$ and $-(\mathbf{v} \cdot \boldsymbol{\nabla})\mathbf{A}_{\parallel}$, since $\mathbf{v}$ is a constant vector. The remaining terms can then be rewritten as

$$\mathbf{E}'_{\parallel} = -\gamma^2 \left(1 - \frac{v^2}{c^2} \right) \left(\boldsymbol{\nabla}_{\parallel} V + \frac{\partial \mathbf{A}_{\parallel}}{\partial t} \right) = \left(\boldsymbol{\nabla}_{\parallel} V + \frac{\partial \mathbf{A}_{\parallel}}{\partial t} \right) = \mathbf{E}_{\parallel} .$$

Finally, for the perpendicular component of the electric field, we have

$$\mathbf{E}'_{\perp} = -\boldsymbol{\nabla}'_{\perp} V' - \frac{\partial \mathbf{A}'_{\perp}}{\partial t'} .$$

Again we substitute the prime quantities using the Lorentz transformation:

$$\mathbf{E}'_{\perp} = -\boldsymbol{\nabla}_{\perp} \gamma (V - \mathbf{v} \cdot \mathbf{A}) - \gamma \left(\frac{\partial}{\partial t} + \mathbf{v} \cdot \boldsymbol{\nabla} \right) \mathbf{A}_{\perp}$$
$$= -\gamma \left(\boldsymbol{\nabla}_{\perp} V + \frac{\partial \mathbf{A}_{\perp}}{\partial t} \right) + \gamma \boldsymbol{\nabla}_{\perp} (\mathbf{v} \cdot \mathbf{A}) - \gamma (\mathbf{v} \cdot \boldsymbol{\nabla}) \mathbf{A}_{\perp}$$

The first two terms on the right-hand-side yield $\mathbf{E}_{\perp}$ and we can use $\boldsymbol{\nabla} = \boldsymbol{\nabla}_{\parallel} + \boldsymbol{\nabla}_{\perp}$ in the third term:

$$\mathbf{E}'_\perp = \gamma \mathbf{E}_\perp + \gamma \nabla(\mathbf{v} \cdot \mathbf{A}) - \gamma \nabla_\|(\mathbf{v} \cdot \mathbf{A}) - \gamma(\mathbf{v} \cdot \nabla)\mathbf{A}_\perp .$$

For two three-vector fields $\mathbf{a}$ and $\mathbf{b}$, we have the identity (A.8)

$$\nabla(\mathbf{a} \cdot \mathbf{b}) = (\mathbf{a} \cdot \nabla)\mathbf{b} + (\mathbf{b} \cdot \nabla)\mathbf{a} + \mathbf{a} \times (\nabla \times \mathbf{b}) + \mathbf{b} \times (\nabla \times \mathbf{a}) .$$

Applying it to $\mathbf{a} = \mathbf{v}$ and $\mathbf{b} = \mathbf{A}$ and the fact that $\mathbf{v}$ is a constant vector reduces the right-hand side to two terms

$$\nabla(\mathbf{v} \cdot \mathbf{A}) = (\mathbf{v} \cdot \nabla)\mathbf{A} + \mathbf{v} \times (\nabla \times \mathbf{A}) .$$

We know that $\nabla \times \mathbf{A} = \mathbf{B} = \mathbf{B}_\| + \mathbf{B}_\perp$. Taking the cross product with $\mathbf{v}$ yields

$$\mathbf{v} \times (\nabla \times \mathbf{A}) = \mathbf{v} \times \mathbf{B}_\perp .$$

We then replace $\nabla(\mathbf{v} \cdot \mathbf{A})$ into the expression for $\mathbf{E}'_\perp$:

$$\mathbf{E}'_\perp = \gamma\left(\mathbf{E}_\perp + \mathbf{v} \times \mathbf{B}_\perp + (\mathbf{v} \cdot \nabla)\mathbf{A} - \nabla_\|(\mathbf{v} \cdot \mathbf{A}) - (\mathbf{v} \cdot \nabla)\mathbf{A}_\perp\right) .$$

The last three terms in the latter equation cancel out since

$$\begin{aligned}
(\mathbf{v} \cdot \nabla)\mathbf{A} &= (\mathbf{v} \cdot \nabla)\mathbf{A}_\| + (\mathbf{v} \cdot \nabla)\mathbf{A}_\perp \\
&= (\mathbf{v} \cdot \nabla_\|)\mathbf{A}_\| + (\mathbf{v} \cdot \nabla)\mathbf{A}_\perp \\
&= \nabla_\|(\mathbf{v} \cdot \mathbf{A}_\|) + (\mathbf{v} \cdot \nabla)\mathbf{A}_\perp \\
&= \nabla_\|(\mathbf{v} \cdot \mathbf{A}) + (\mathbf{v} \cdot \nabla)\mathbf{A}_\perp .
\end{aligned}$$

Hence, we see that the transformation for the perpendicular component of the field reads

$$\mathbf{E}'_\perp = \gamma\left(\mathbf{E}_\perp + \mathbf{v} \times \mathbf{B}_\perp\right) ,$$

or equivalently

$$\mathbf{E}'_\perp = \gamma\left(\mathbf{E}_\perp + \boldsymbol{\beta} \times c\mathbf{B}_\perp\right) .$$

Finally, the transformation law for the electromagnetic field and its inverse read

$$\begin{cases}
\mathbf{E}'_\| = \mathbf{E}_\| \\
\mathbf{E}'_\perp = \gamma(\mathbf{E} + \boldsymbol{\beta} \times c\mathbf{B})_\perp \\
\mathbf{B}'_\| = \mathbf{B}_\| \\
c\mathbf{B}'_\perp = \gamma(c\mathbf{B} - \boldsymbol{\beta} \times \mathbf{E})_\perp
\end{cases}
\qquad
\begin{cases}
\mathbf{E}_\| = \mathbf{E}'_\| \\
\mathbf{E}_\perp = \gamma(\mathbf{E}' - \boldsymbol{\beta} \times c\mathbf{B}')_\perp \\
\mathbf{B}'_\| = \mathbf{B}_\| \\
c\mathbf{B}_\perp = \gamma(c\mathbf{B}' + \boldsymbol{\beta} \times \mathbf{E}')_\perp
\end{cases}
\qquad (20.10)$$

This transformation law does not take the form of the transformation law established for a four-vector. We note from these expressions that the electric or magnetic nature of a field are intrinsically observer dependent. In $\mathcal{R}'$, the electric field is a combination of the electric and magnetic fields in $\mathcal{R}$ and so does the magnetic field in $\mathcal{R}'$. Observers in different Galilean frames will reach different conclusion regarding the electric or magnetic origin of an electromagnetic phenomenon.

Nevertheless, Eq. (20.10) constitute the transformation rule that we need to apply, in conjunction with the Lorentz transformation for the four-gradient, in order to conclude on the invariance of Maxwell's equations under a Lorentz transformation.

20.8.1 Fields of a Charge Moving at Constant Velocity

Consider a charge q moving with velocity $\mathbf{v}$ with respect to a Galilean frame $\mathcal{R}$. Let $\mathcal{R}'$ be the frame in which the charge is at rest at the origin, so that the fields in $\mathcal{R}'$ at position $\mathbf{r}'$ write;

$$\mathbf{E}'(\mathbf{r}') = \frac{q}{4\pi\epsilon_0}\frac{\mathbf{r}'}{|\mathbf{r}'|^3}, \qquad\qquad \mathbf{B}'(\mathbf{r}') = \mathbf{0} \ .$$

Supposing that both frames are aligned and that their origins coincide at $t = 0$, applying (20.10), the fields in $\mathcal{R}$ are found:

$$\mathbf{E}_\parallel = \mathbf{E}'_\parallel$$
$$\mathbf{E}_\perp = \gamma(\mathbf{E}' - \boldsymbol{\beta} \times c\mathbf{B}')_\perp$$
$$\mathbf{B}_\parallel = \mathbf{B}'_\parallel = \mathbf{0}$$
$$c\mathbf{B}_\perp = \gamma(c\mathbf{B}' + \boldsymbol{\beta} \times \mathbf{E}')_\perp$$

For the electric field, we have

$$\mathbf{E} = \mathbf{E}_\parallel + \mathbf{E}_\perp = \frac{q}{4\pi\epsilon_0}\frac{\mathbf{r}'_\parallel + \gamma\mathbf{r}'_\perp}{|\mathbf{r}'|^3} \ .$$

According to (20.2), $\mathbf{r}'_\parallel + \gamma\mathbf{r}'_\perp = \gamma(\mathbf{r}_\parallel - \mathbf{v}t) + \gamma\mathbf{r}_\perp = \gamma(\mathbf{r} - \mathbf{v}t)$, and

$$\begin{aligned}
|\mathbf{r}'|^2 &= |\gamma(\mathbf{r}_\parallel - \mathbf{v}t) + \mathbf{r}_\perp|^2 = |\gamma(\mathbf{r} - \mathbf{v}t) + (1 - \gamma)\mathbf{r}_\perp|^2 \\
&= \gamma^2|\mathbf{r} - \mathbf{v}t|^2 + (1 - \gamma)^2|\mathbf{r}_\perp|^2 + 2(1 - \gamma)\gamma\underbrace{(\mathbf{r} - \mathbf{v}t)\cdot\mathbf{r}_\perp}_{|\mathbf{r}_\perp|^2} \\
&= \gamma^2|\mathbf{r} - \mathbf{v}t|^2 + (1 - \gamma^2)|\mathbf{r}_\perp|^2 \ .
\end{aligned}$$

Defining α as the angle shown in Fig. 20.17, we finally find

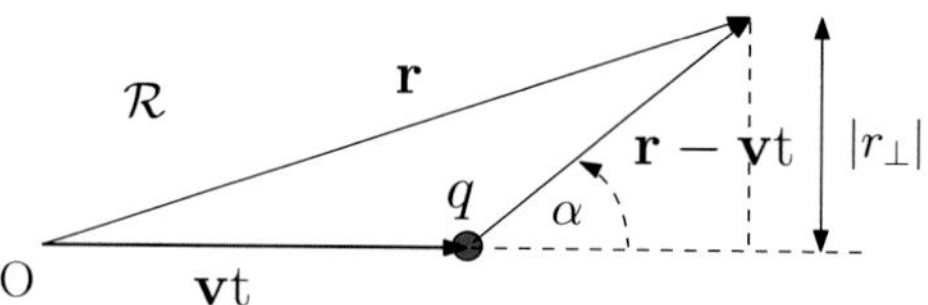

Fig. 20.17 At time t the charge is at position $\mathbf{v}t$ with respect to the origin of $\mathcal{R}$

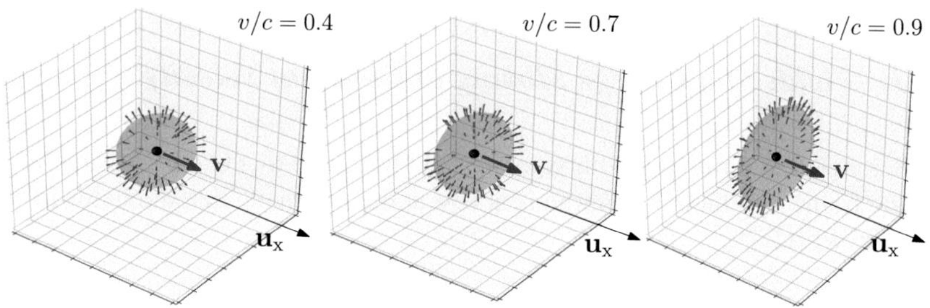

Fig. 20.18 Electric field at constant magnitude for different charge velocities

$$|\mathbf{r}'|^2 = |\mathbf{r} - \mathbf{v}t|^2 \left(\gamma^2 + (1 - \gamma^2)\sin^2 \alpha\right) = \gamma^2 |\mathbf{r} - \mathbf{v}t|^2 (1 - (v/c)^2 \sin^2 \alpha) \ .$$

We find the electric field in $\mathcal{R}$ that is the relativistic version of Coulomb's law, which is retrieved in the limit of small velocities ($v/c \ll 1, \gamma \approx 1$).

$$\mathbf{E}(\mathbf{r}) = \frac{q}{4\pi\epsilon_0} \frac{\mathbf{r} - \mathbf{v}t}{|\mathbf{r} - \mathbf{v}t|^3} \frac{1}{\gamma^2(v)(1 - \frac{v^2}{c^2}\sin^2 \alpha)^{3/2}} \ .$$

Note that this result was obtained as well with the Liénard–Wiechert potentials, see Eq. (19.22).

The electric field remains a radial field in all reference frames, but the spherical symmetry that it has in $\mathcal{R}'$ is absent in $\mathcal{R}$. This is significant when v is comparable to c and is a consequence of length contraction along the direction parallel to the velocity. Figure 20.18 shows, for different values of v/c, the electric field in $\mathcal{R}$ at points where its magnitude its constant, belonging to an isosurface $|\mathbf{E}(\mathbf{r})| = C$ that is also shown. This surface is a sphere at $v/c \to 0$, but gets significantly compressed along the direction of motion of the charge, taking an oblate ellipsoidal shape. The field gets much stronger in directions perpendicular to the axis of motion.

Similarly, for the magnetic field

$$\mathbf{B} = \mathbf{B}_\perp = \frac{\gamma}{c^2}\mathbf{v} \times \mathbf{E}' = \frac{q\mu_0}{4\pi}\frac{\mathbf{v} \times \gamma r'_\perp}{|\mathbf{r}'|^3} = \frac{q\mu_0}{4\pi}\frac{\mathbf{v} \times \gamma r_\perp}{|\mathbf{r}'|^3}$$

from which we retrieve the relativistic version of Biot–Savart law

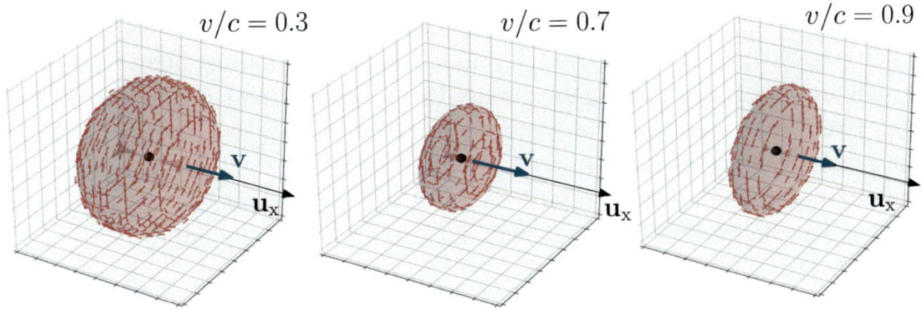

Fig. 20.19 Magnetic field at constant magnitude for different charge velocities

$$\mathbf{B(r)} = \frac{\mu_0}{4\pi} \frac{q\mathbf{v} \times (\mathbf{r} - \mathbf{v}t)}{|\mathbf{r} - \mathbf{v}t|^3} \frac{1}{\gamma^2(v)(1 - \frac{v^2}{c^2}\sin^2\alpha)^{3/2}} = \frac{1}{c^2}\mathbf{v} \times \mathbf{E(r)} \ .$$

The magnetic field has no-component along the charge velocity and the field lines are concentric circles. The field at surfaces of constant field magnitude is shown in Fig. 20.19 for different values of v/c.

Example 20.1—A wire of current
Consider a neutral wire carrying a current I due to electrons moving along the $-\mathbf{u}_z$ direction in the laboratory frame $\mathcal{R}$. In this frame, there is only a magnetic field, given in cylindrical coordinates by

$$\mathbf{B(x)} = \frac{\mu_0 I}{2\pi r}\mathbf{u}_\theta$$

If $\mathbf{v} = -v\mathbf{u}_z$ is the drift velocity of the electrons with respect to the wire, one may think that the magnetic field should be zero in the frame $\mathcal{R}'$ in which the electrons are at rest. However, this is not the case because the wire being neutral in $\mathcal{R}$ means that the wire is necessarily composed of a system of electrons of linear charge density $\lambda_e < 0$ that is exactly cancelled by a the density $\lambda_i = \lambda_0 = -\lambda_e$ of positive ions in the wire. These ions are at rest in $\mathcal{R}$, but no longer in $\mathcal{R}'$, where the current will be due exclusively to the ions. Since the fields in $\mathcal{R}$ are perpendicular to $\mathbf{v}$, the relevant coordinates which are (r, θ) do not change under the Lorentz transformation. A simple application of (20.10) then yields:

$$\mathbf{B'(x')} = \frac{\gamma\mu_0 I}{2\pi r}\mathbf{u}_\theta \tag{20.11}$$

$$\mathbf{E'(x')} = \frac{\lambda_0}{2\pi\epsilon_0 r}\left(\frac{\gamma^2 - 1}{\gamma}\right)\mathbf{u}_r = \gamma\frac{v^2}{c^2}\frac{\lambda_0}{2\pi\epsilon_0 r}\mathbf{u}_r \tag{20.12}$$

where we have used the relation $I = \lambda_0 v$. In $\mathcal{R}'$ the current is due to the ions, which move at speed v along the z-axis. This gives a current $I' = \gamma \lambda_0 v$, where $\gamma \lambda_0$ is the increased positive charge density due to length contraction along the z-axis. Indeed, if Δx is the distance between two ions along the wire in the laboratory frame, in $\mathcal{R}'$ these ions move and their distance is reduced by a factor of γ, increasing the ion density and therefore the current to $I' = \gamma I$ in (20.11). This is illustrated in Fig. 20.20.

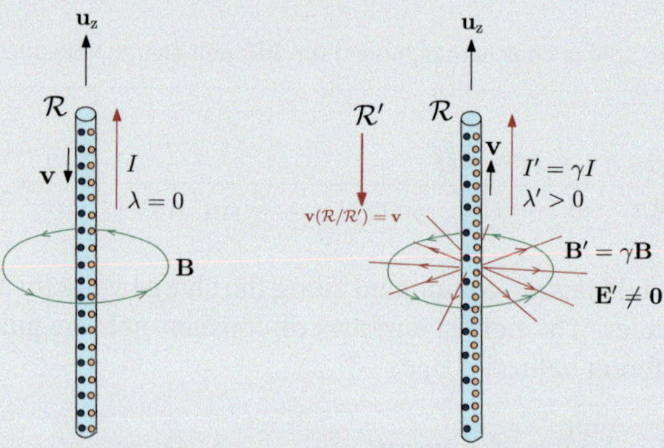

Fig. 20.20 A neutral wire carrying an electronic current in the laboratory frame acquires a positive charge density in the frame $\mathcal{R}'$ at which the electrons are at rest and the ions move. An electric field appears in $\mathcal{R}'$

The electric field that appears in $\mathcal{R}'$ is also a consequence of length contraction. In $\mathcal{R}'$, the electron density is decreased by γ with respect to $\mathcal{R}$, the wire acquires a net charge density

$$\lambda' = \gamma \lambda_i + \frac{1}{\gamma} \lambda_e = \left(\frac{\gamma^2 - 1}{\gamma} \right) \lambda_0$$

which explains the electric field expression given by (20.12).

20.8.2   Plane Waves

Since electromagnetic waves propagate at the speed of light in vacuum, it is instructive to perform a Lorentz transformation to a plane wave and check the transformation rules for their characteristics. Consider a monochromatic plane wave propagating in vacuum at velocity $c = \omega/k$ in the Galilean reference frame $\mathcal{R}$. The electromagnetic fields in $\mathcal{R}$ read, in complex representation

$$\mathbf{E}(\mathbf{r}, t) = \mathbf{E}_0 \exp[i(\mathbf{k} \cdot \mathbf{r} - \omega t)], \qquad c\mathbf{B} = \frac{\mathbf{k}}{k} \times \mathbf{E} \,.$$

Since the electric and magnetic fields in $\mathcal{R}$ are solutions to d'Alembert's wave equation, and the d'Alembertian is invariant to a Lorentz transformation, a plane wave has exactly the same form if it is observed from another Galilean frame $\mathcal{R}'$ moving at velocity $\mathbf{v}$ with respect to $\mathcal{R}$. Hence, we can write

$$\mathbf{E}'(\mathbf{r}', t') = \mathbf{E}'_0 \exp[i(\mathbf{k}' \cdot \mathbf{r}' - \omega' t')], \qquad c\mathbf{B}' = \frac{\mathbf{k}'}{k'} \times \mathbf{E}' .$$

To relate the frequency and wavenumber of the wave in $\mathcal{R}'$ to their counterparts in $\mathcal{R}$, we can perform a direct transformation of the expression of the wave in $\mathcal{R}$. First we relate the component of $\mathbf{E}'_0$ to $\mathbf{E}_0$ and $\mathbf{B}_0$ components using field transformation laws. Once this step is done, we have

$$\mathbf{E}'(\mathbf{r}, t) = \mathbf{E}'_0 \exp[i(\mathbf{k} \cdot \mathbf{r} - \omega t)] .$$

Then we apply the Lorentz transformation $(\mathbf{r}, ict) \rightarrow (\mathbf{r}', ict')$ to eliminate the variables $\mathbf{r}$ and t in favor of $\mathbf{r}'$ and t'. This yields

$$\mathbf{E}'(\mathbf{r}, t) = \mathbf{E}'_0 \exp[i(\mathbf{k} \cdot \{\mathbf{r}'_\perp + \gamma(\mathbf{r}'_\parallel + \boldsymbol{\beta} t')\} - \omega\gamma(t' + \boldsymbol{\beta} \cdot \mathbf{r}'_\parallel/c))] .$$

Finally, we compare the two expressions for the wave in $\mathcal{R}'$. We find

$$\mathbf{k}'_\parallel = \gamma(\mathbf{k}_\parallel - \boldsymbol{\beta}\omega/c), \qquad \mathbf{k}'_\perp = \mathbf{k}_\perp, \qquad \omega' = \gamma(\omega - \mathbf{v} \cdot \mathbf{k}_\parallel). \tag{20.13}$$

The last identity in (20.13) is the relativistic Doppler effect. If the plane wave propagates parallel to the velocity, then $\mathbf{v} \cdot \mathbf{k}_\parallel = \frac{v\omega}{c}$ and

$$\frac{\omega'}{\omega} = \gamma\left(1 - \frac{v}{c}\right) = \frac{\sqrt{1 - \frac{v}{c}}}{\sqrt{1 + \frac{v}{c}}} \tag{20.14}$$

and, contrary to the classic Doppler effect which is only due to a relative motion between source and the receptor of the wave, Eq. (20.14) includes time dilation effects.

From (20.13) it is seen that the wavevector and frequency is thus transformed by a Lorentz transformation law, allowing us to define the four-vector

$$\mathbf{K} = (\mathbf{k}, i\omega/c) .$$

From the general property that the scalar product between two four-vectors is an invariant, we find that the phase of a plane wave

$$\phi = \mathbf{K} \cdot \mathbf{R} = \mathbf{k} \cdot \mathbf{r} - \omega t$$

is an invariant: $\phi = \phi' = \mathbf{K}' \cdot \mathbf{R}' = \mathbf{k}' \cdot \mathbf{r}' - \omega' t'$. From the property that the norm of a four-vector is an invariant, we find

$$\mathbf{K} \cdot \mathbf{K} = \mathbf{k}^2 - \frac{\omega^2}{c^2} = 0$$

that is, the dispersion relation of a plane wave in vacuum, $\omega = kc$, takes the same form in all Galilean frames.

20.9 Relativistic Larmor Formula for a Point Charge

In Sect. 19.3, the Liénard–Wiechert fields generated by a point charge were determined. From these, we can calculate the radiated power at time t as the flux of the Poynting vector over a sphere centered around the charge. Suppose that in the frame $\mathcal{R}$ the trajectory of the charge is given by $\mathbf{r}(t)$, and that at time t the velocity is $\mathbf{v}(t)$. Now consider the frame $\mathcal{R}'$ such that at time t the particle is at rest. In that frame, the Liénard–Wiechert fields (19.13), (19.14), (19.17), (19.18) close to the charge simplify to:

$$\mathbf{E}'(\mathbf{x}', t') \approx \frac{q\mathbf{u}'_r(t)}{4\pi\epsilon_0|\mathbf{x}' - \mathbf{r}'(t')|^2} + \frac{q}{4\pi\epsilon_0 c^2|\mathbf{x}' - \mathbf{r}'(t')|}\left(\mathbf{a}'(t') \cdot \mathbf{u}'_r(t')\mathbf{u}'_r(t') - \mathbf{a}'(t')\right)$$

$$(20.15)$$

$$\mathbf{B}'(\mathbf{x}', t') \approx \frac{q}{4\pi\epsilon_0 c^2|\mathbf{x}' - \mathbf{r}'(t')|} \frac{\mathbf{a}'(t')}{c} \times \mathbf{u}'_r(t') \qquad (20.16)$$

with $\mathbf{u}'_r(t') = (\mathbf{x}' - \mathbf{r}'(t'))/|\mathbf{x}' - \mathbf{r}'(t')|$. Here we have neglected retardation effects ($t'_r \approx t'$), which is true infinitely close to the charge. The Poynting vector is $\mathbf{\Pi}' = \frac{1}{\mu_0}\mathbf{E}' \times \mathbf{B}'$. Ignoring the term that decays as the inverse of the distance cubed, which comes from the near field part of the electric field, we obtain the radiative part of the Poynting vector

$$\mathbf{\Pi}'_{\text{rad}}(\mathbf{x}', t') = \frac{q^2\mu_0}{(4\pi)^2 c|\mathbf{x}' - \mathbf{r}'(t')|^2}\left(|\mathbf{a}'(t')|^2 - \left(\mathbf{a}'(t') \cdot \mathbf{u}'_r(t')\right)^2\right)\mathbf{u}'_r(t') .$$

The emitted power P' can be obtained by integrating the flux of the Poynting vector over a sphere of radius R' centered on the charge. The result is, of course, independent on R' which justifies the approximation of the Liénard–Wiechert fields to the case without retardation. The normal to the sphere is $\mathbf{u}_r = \mathbf{u}'_r(t') = (\mathbf{x}' - \mathbf{r}'(t'))/|\mathbf{x}' - \mathbf{r}'(t')|$ and choosing $\mathbf{a}'$ along the z-axis of the spherical coordinates,

$$P' = \frac{\mu_0 q^2}{(4\pi)^2 c} \int_0^{2\pi} \int_0^{\pi} \left(|\mathbf{a}'(t')|^2 - (\mathbf{a}'(t') \cdot \mathbf{u}_r)^2 \right) \mathbf{u}_r \cdot \mathbf{u}_r \sin\theta \, d\theta \, d\phi$$

$$= \frac{\mu_0 q^2}{(4\pi)^2 c} |\mathbf{a}'(t')|^2 \underbrace{\int_0^{2\pi} \int_0^{\pi} \left(1 - \cos^2\theta\right)^2 \sin\theta \, d\theta \, d\phi}_{=8\pi/3}$$

$$= \frac{\mu_0 q^2}{6\pi c} |\mathbf{a}'(t')|^2 \ .$$

We have retrieved Larmor's formula (19.44) which is valid in the instantaneous reference frame of the particle, but not necessarily in $\mathcal{R}$ since we can no longer use the dipole approximation if the charge moves at relativistic speeds with respect to $\mathcal{R}$. To find the relativistic version of Larmor's formula, we now apply a Lorentz transformation from reference $\mathcal{R}'$ (in which the particle is at rest, $\mathbf{v}' = \mathbf{0}$ and $\gamma'(\mathbf{v}') = 1$) and the laboratory frame $\mathcal{R}$, we obtain

$$\gamma^2(v)\mathbf{a}_\perp = \gamma^2(v')\mathbf{a}'_\perp = \mathbf{a}'_\perp$$

$$\gamma^4(v)\mathbf{a}_\parallel = \gamma(v)\gamma^2(v')\mathbf{a}'_\parallel = \gamma(v)\mathbf{a}'_\parallel \ .$$

We then find

$$|\mathbf{a}'(t')|^2 = \gamma^4 \mathbf{a}_\perp^2 + \gamma^6 \mathbf{a}_\parallel^2$$

which, according to (20.4), is a Lorentz invariant quantity. Therefore, the same is true for the power radiated by a point charge and we obtain the relativistic version of Larmor's formula in the laboratory frame

$$\boxed{P = P' = \frac{\mu_0 q^2}{6\pi c} \gamma^6 \left(\mathbf{a}_\parallel^2 + \frac{1}{\gamma^2} \mathbf{a}_\perp^2 \right) = \frac{\mu_0 q^2}{6\pi c} \gamma^6 \left(|\mathbf{a}|^2 - \frac{v^2}{c^2} \mathbf{a}_\perp^2 \right)} \tag{20.17}$$

Since $m\mathcal{A} = \gamma \frac{d\mathbf{P}}{dt}$, Eq. (20.17) can be rewritten in terms of the particle momentum (not to be confused here witht he dipole moment). We have $\gamma m \mathbf{a}_\perp = \left(\frac{d\mathbf{p}}{dt}\right)_\perp$ and $\gamma^3 m \mathbf{a}_\parallel = \left(\frac{d\mathbf{p}}{dt}\right)_\parallel$, leading to

$$P = \frac{\mu_0 q^2}{6\pi c m^2} \left(\left(\frac{d\mathbf{p}}{dt}\right)_\parallel^2 + \gamma^2 \left(\frac{d\mathbf{p}}{dt}\right)_\perp^2 \right) . \tag{20.18}$$

20.9.1 *Synchrotron Radiation and Relativistic Beaming*

The power emitted for a change in momentum perpendicular to the velocity is γ^2 times stronger than the power associated with a change in momentum parallel to the velocity. This phenomenom in exploited in synchroton accelerators, where relativistic electrons are deflected by a magnetic field. For circular motion $\mathbf{a}_{\parallel} = \mathbf{0}$ and the radiated power is given by

$$P = \frac{\mu_0 q^2}{6\pi c}\gamma^4 a^2 = \frac{\mu_0 q^2}{6\pi c R^2}\gamma^4 v^4$$

with R the radius of the circular trajectory. At relativistic velocities, this radiation becomes highly anisotropic. As illustrated in Fig. 20.21, an electron emits intense radiation predominantly in the direction of its motion, an effect known as *beaming*. This arises from the denominator in Eq. (19.14), where the fields are significantly enhanced when $\mathbf{v}/c \cdot \mathbf{u}_r$ approaches unity. Such an enhancement only occurs when the velocity of the charge approaches the speed of light.

In the laboratory frame $\mathcal{R}$, a stationary detector observes periodic bursts of intense electromagnetic radiation whenever the electron velocity, evaluated at the retarded time, points toward the detector. These bursts, or light beams, are widely used for research purposes. Their short duration implies a broad frequency spectrum, making them suitable for irradiating matter to study its properties.

Let us demonstrate this beaming effect from the fields radiated by a point charge. At a given instant, let us introduce a coordinate system such that the circular trajectory of an electron lies in the xz plane. Suppose $\mathbf{r}(t_r)$ to be at the origin, the retarded velocity $\mathbf{v}(t_r)$ along $\mathbf{u}_z$, and the acceleration $\mathbf{a}(t_r)$ along $\mathbf{u}_x$, and let us use the spherical coordinates as shown in Fig. 20.22.

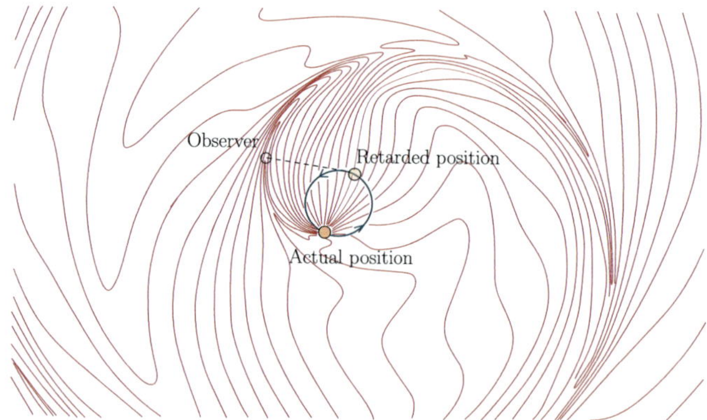

Fig. 20.21 Electric field lines of a charge describing a circular motion at a speed $v = 0.9\,c$

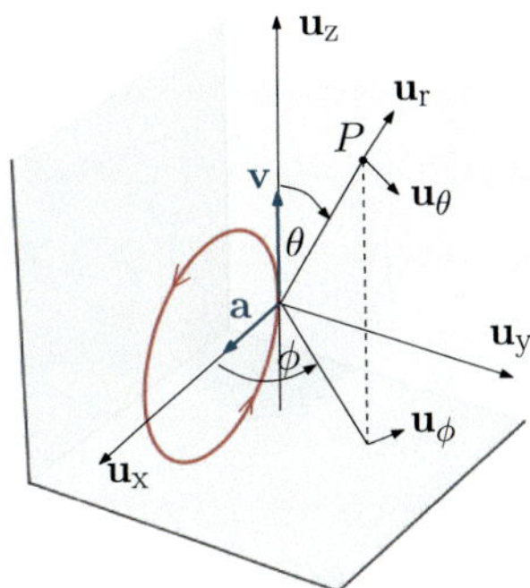

Fig. 20.22 Coordinate system used to express the angular dependence of the Poynting vector

The Poynting vector writes

$$\mathbf{\Pi} = \frac{\mathbf{E} \times \left(\frac{\mathbf{u}_r}{c} \times \mathbf{E}\right)}{\mu_0} = \frac{|\mathbf{E}|^2}{\mu_0 c}\mathbf{u}_r = \frac{q^2 c}{(4\pi)^2 \epsilon_0 r^2} \frac{\left|\mathbf{u}_r \times \left((\mathbf{u}_r - \frac{\mathbf{v}}{c}) \times \frac{\mathbf{a}(t_r)}{c^2}\right)\right|^2 \mathbf{u}_r}{\left(1 - \frac{\mathbf{v}(t_r)}{c} \cdot \mathbf{u}_r\right)^6}$$

$$\mathbf{\Pi} = \frac{q^2 a(t_r)^2}{(4\pi)^2 \epsilon_0 c^3 r^2} \frac{\left(1 - \frac{v}{c}\cos\theta\right)^2 - \left(1 - \frac{v^2}{c^2}\right)\cos^2\phi\sin^2\theta}{\left(1 - \frac{v}{c}\cos\theta\right)^6}\mathbf{u}_r \ .$$

The angular distribution of radiated power writes

$$\frac{dP}{d\Omega} = r^2 \mathbf{u}_r \cdot \mathbf{\Pi} = \frac{\mu_0 q^2 a(t_r)^2}{16\pi^2 c} \frac{1}{\left(1 - \frac{v}{c}\cos\theta\right)^4}\left(1 - \left(1 - \frac{v^2}{c^2}\right)\frac{\cos^2\phi\sin^2\theta}{\left(1 - \frac{v}{c}\cos\theta\right)^2}\right). \tag{20.19}$$

Figure 20.23 shows isosurfaces defined by Eq. (20.19). In the non-relativistic limit, we retrieve the emission pattern of a dipole, indeed in this limit $dP/d\Omega \propto \cos^2\phi\sin^2 \propto \mathbf{a}\cdot\mathbf{u}_r^2$. For relativistic velocities, the radiated power is maximum for $\theta = 0$ and so strongly concentrates along the direction of propagation of the particle.

In the small angle approximation, we may expand (20.19) to leading order in θ

$$\frac{dP}{d\Omega} \approx \frac{\mu_0 q^2 a(t_r)^2}{16\pi^2 c} \frac{1}{\left(1 - \frac{v}{c} + \frac{v}{2c}\theta^2\right)^6}\left(\left(1 - \frac{v}{c}\right)^2 + \left(1 - \frac{v}{c}\right)\frac{v}{c}\theta^2\right.$$

$$\left. - \left(1 - \frac{v^2}{c^2}\right)\cos^2\phi\,\theta^2\right). \tag{20.20}$$

In the ultra relativistic regime $(v/c = \beta \approx 1)$, we can write $(1 - \beta^2) = \gamma^{-2} \approx 2(1 - \beta)$ so that, in terms of the γ factor

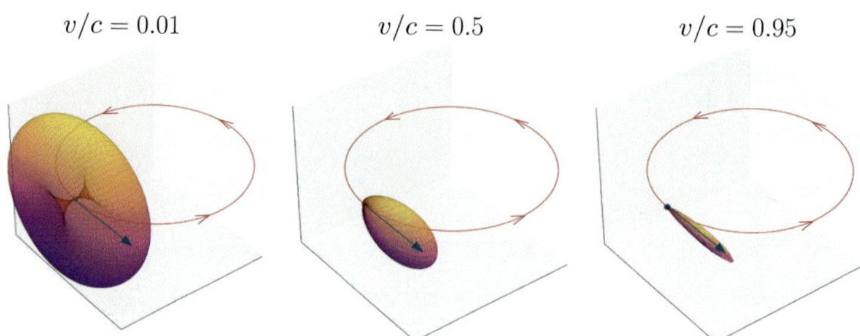

Fig. 20.23 Relativistic beaming

$$\frac{dP}{d\Omega} = \frac{\mu_0 q^2 a(t_r)^2}{16\pi^2 c}(2\gamma^2)^4 \frac{1 + 2\gamma^2\theta^2 - -4\gamma^2\theta^2 \cos^2\phi}{(1 + \gamma^2\theta^2)^6}. \tag{20.21}$$

The denominator decreases very fast with γ, limiting the angular spreading of the radiation pattern to a cone which makes an angle of $\pm 1/\gamma$ with respect to the particle propagation direction.

20.10   Covariant Formulation of Maxwell's Equations

In order to show that Maxwell's equations take the same form in every Galilean frame, one should use the transformation laws established in previous sections for the partial derivatives, the sources (charge density and current), the fields and the potentials. This is far from obvious even if a cumbersome step by step calculation will lead to the conclusion. There is a more elegant way to demonstrate this covariance, which takes advantage of writing of Maxwell's equation using four-tensors, making their covariance fully explicit. The objective of the present section is to present Maxwell's equations in this manifestly covariant form. This requires the introduction of four-tensors, which are multidimensional arrays obeying a transformation rule under a Lorentz transformation of spacetime coordinates.

20.10.1   Cartesian Tensors

In the Euclidian space, Cartesian tensors are multidimensional arrays defined by their behavior under orthogonal transformations. We start with a short reminder on orthogonal transformations: Consider two sets of orthogonal Cartesian unit vectors $\mathbf{u}_1, \mathbf{u}_2, \mathbf{u}_3$ and $\mathbf{u}'_1, \mathbf{u}'_2, \mathbf{u}'_3$. Since each set constitutes a complete basis of the three-dimensional space, we have

$$\mathbf{u}'_i = A_{ij}\mathbf{u}_j \ ,$$

and the matrix A is an orthogonal matrix, that is, its transpose is equal to its inverse. The matrix A thus describes an orthogonal transformation of coordinates: it corresponds to a rotation around an axis, a reflection (mirror symmetry $(\mathbf{u}_1, \mathbf{u}_2, \mathbf{u}_3) \rightarrow (-\mathbf{u}_1, \mathbf{u}_2, \mathbf{u}_3))$ or a combination of these options. For instance an inversion $(\mathbf{u}_1, \mathbf{u}_2, \mathbf{u}_3) \rightarrow (-\mathbf{u}_1, -\mathbf{u}_2, -\mathbf{u}_3)$ is a combination of reflections.

A Cartesian tensor of rank 0 is a scalar $s(\mathbf{r})$ that is invariant under a rotation of coordinates:

$$s'(\mathbf{r}') = s(\mathbf{r}) \ .$$

A Cartesian tensor of rank 1 is a vector $\mathbf{V}$ whose three components $V_i, i = 1, 2, 3,$ transform under rotations like the three components of $\mathbf{r}$:

$$V'_i(\mathbf{r}') = A_{ij} V_j(\mathbf{r}) \ .$$

A Cartesian tensor of rank 2 is a nine component quantity T (a 3×3 matrix) whose components T_{km} transform under rotations by the rule

$$T'_{ij}(\mathbf{r}') = A_{ik} A_{jm} T_{km}(\mathbf{r}) \ .$$

An example that we have already encountered is the electric permittivity in the case of an anisotropic dielectric medium.

20.10.2 *Lorentz Tensors*

Lorentz tensors are the counterparts of Cartesian tensors in Minkowski space: they are multidimensional arrays obeying a transformation rule under a Lorentz transformation of coordinates between two Galilean frames. This section is devoted to a short presentation of this transformation rule.

A Lorentz tensor of rank 0 is simply a scalar quantity. For instance the velocity of light, which is invariant to a change of Galilean reference frame: $c' = c$. The scalar product of two four-vectors is another example of Lorentz tensor of rank zero.

A Lorentz tensor of rank 1 is a vector whose components are transformed according to the Lorentz transformation

$$a'_\mu = L_{\mu\nu} a_\nu \ .$$

In other words, it is a four-vector. We have already seen several examples of four-vectors, for instance, the four-current $\jmath$ or the four-potential $\mathbf{A}_4$.

A Lorentz tensor of rank 2 is a 4×4 matrix $T_{\mu\nu}$ whose 16 components are transformed according to the transformation rule

$$T'_{\mu\nu} = L_{\mu\alpha} L_{\nu\beta} T_{\alpha\beta}. \tag{20.22}$$

20.10.3 *Homogeneous Maxwell's Equations*

Maxwell–Thomson and Maxwell–Faraday equations allowed us to define the electric and magnetic fields from the scalar and vector potentials

$$\mathbf{B} = \nabla \times \mathbf{A} \quad \text{and} \quad \mathbf{E} = -\nabla V - \frac{\partial \mathbf{A}}{\partial t}.$$

Using Cartesian axes (x, y, z) and the time-coordinate ict to build four-vectors with subscripts $1 \equiv x, 2 \equiv y, 3 \equiv z$, and $4 \equiv ict$, we can rewrite these field definitions as

$$\begin{cases} B_x = \partial_y A_z - \partial_z A_y = \partial_2 A_3 - \partial_3 A_2 \, , \\ B_y = \partial_z A_x - \partial_x A_z = \partial_3 A_1 - \partial_1 A_3 \, , \\ B_z = \partial_x A_y - \partial_y A_x = \partial_1 A_2 - \partial_2 A_1 \, , \end{cases}$$

for the magnetic field and

$$\begin{cases} i E_x/c = \partial_{(ict)} A_x - \partial_x (i V/c) = \partial_4 A_1 - \partial_1 A_4 \, , \\ i E_y/c = \partial_{(ict)} A_y - \partial_y (i V/c) = \partial_4 A_2 - \partial_2 A_4 \, , \\ i E_z/c = \partial_{(ict)} A_z - \partial_z (i V/c) = \partial_4 A_3 - \partial_3 A_4 \, , \end{cases}$$

for the electric field, where A_ν denote the four components of the four-potential

$$(A_1, A_2, A_3, A_4) = (A_x, A_y, A_z, i V/c) \, .$$

We see that we obtain the six components of a generalized curl (in the 4D Minskowski space), which defines the second-rank electromagnetic field-strength tensor

$$\boxed{F_{\mu\nu} = \partial_\mu A_\nu - \partial_\nu A_\mu.} \tag{20.23}$$

In terms of the original field components, the electromagnetic field strength tensor $F_{\mu\nu}$ takes the form of the 4×4 matrix

$$[\mathbf{F}] = \begin{bmatrix} 0 & B_z & -B_y & -i E_x/c \\ -B_z & 0 & B_x & -i E_y/c \\ B_y & -B_x & 0 & -i E_z/c \\ i E_x/c & i E_y/c & i E_z/c & 0 \end{bmatrix}$$

As a generalized curl, $[\mathbf{F}]$ is an antisymmetric second-rank Lorentz tensor. This means that $[\mathbf{F}]$ has only six independent components, as explicitly shown in its matrix expression. In addition, **[F]** follows the field transformation laws (20.22). This can be easily shown by expressing the components of **[F]** as functions of the four-potential components using the definition (20.23):

$$L_{\mu\alpha}L_{\nu\beta}F_{\alpha\beta} = L_{\mu\alpha}L_{\nu\beta}\left(\partial_\alpha A_\beta - \partial_\beta A_\alpha\right) .$$

Since the components of the Lorentz matrix L do not depend on coordinates, they commute with the partial derivatives with respect to spacetime coordinates:

$$L_{\mu\alpha}L_{\nu\beta}F_{\alpha\beta} = L_{\mu\alpha}\partial_\alpha L_{\nu\beta}A_\beta - L_{\mu\alpha}L_{\nu\beta}\partial_\beta A_\alpha$$

For the first term on the right-hand side, we recognize the transformation rules for the four-gradient and for the four-potential:

$$L_{\mu\alpha}\partial_\alpha L_{\nu\beta}A_\beta = \partial'_\mu A'_\nu .$$

For the second term on the right-hand side, we first substitute the central term $L_{\nu\beta}\partial_\beta$ using the transformation rule for the four-gradient, then we commute the resulting components of the four gradient in $\mathcal{R}'$ with the components of the remaining Lorentz matrix, and we finally transform the four-vector, which writes

$$L_{\mu\alpha}L_{\nu\beta}\partial_\beta A_\alpha = L_{\mu\alpha}\partial'_\nu A_\alpha = \partial'_\nu L_{\mu\alpha}A_\alpha = \partial'_\nu A'_\mu .$$

In result,
$$L_{\mu\alpha}L_{\nu\beta}F_{\alpha\beta} = \partial'_\mu A'_\nu - \partial'_\nu A'_\mu = F'_{\mu\nu} .$$

The field strength tensor follows the transformation rule (20.23). It is thus a Lorentz invariant.

20.10.4 Inhomogeneous Maxwell's Equations

It is straightforward to write the two inhomogeneous Maxwell equations in terms of the electromagnetic field strength tensor, using the four-current $(\mathbf{j}, ic\varrho)$ as a source term:

$$\boxed{\partial_\nu F_{\mu\nu} = \mu_0 j_\mu.} \tag{20.24}$$

On the left-hand side, the repeated ν subscript indicates a sum over the second index of the field tensor $F_{\mu\nu}$, which amounts to taking the divergence of a second-rank tensor. The result is a first-rank tensor, that is, a four-vector.

To see how Maxwell's equations are encoded in (20.24), it is instructive to expand if for the different values of μ. We retrieve Maxwell–Gauss equation from the $\mu = 4$ component

$$\partial_\nu F_{4\nu} = \mu_0 j_4 \quad \text{and} \quad j_4 = ic\varrho \, .$$

Writing it out, we find

$$\frac{\partial}{\partial x}\left(\frac{i E_x}{c}\right) + \frac{\partial}{\partial y}\left(\frac{i E_y}{c}\right) + \frac{\partial}{\partial z}\left(\frac{i E_z}{c}\right) + 0 = i\mu_0 c\varrho \, ,$$

which, after replacing μ_0 by $1/\epsilon_0 c^2$ is identified to Maxwell–Gauss equation

$$\mathbf{\nabla} \cdot \mathbf{E} = \frac{\varrho}{\epsilon_0} \, .$$

The three other components for $\mu = 1, 2, 3$ are the x-, y-, and z-components of Maxwell–Ampère equation. For instance, $\mu = 1$ yields

$$\partial_\nu F_{1\nu} = \mu_0 j_x \, ,$$

that we expand as

$$0 + \frac{\partial}{\partial y} B_z - \frac{\partial}{\partial z} B_y + \frac{\partial}{\partial ict}\left(\frac{-i E_x}{c}\right) = \mu_0 j_x$$

and we recognize the x-component of Maxwell–Ampère equation

$$[\mathbf{\nabla} \times \mathbf{B}]_x - \frac{1}{c^2}\frac{\partial E_x}{\partial t} = \mu_0 j_x$$

Similarly, $\mu = 2, 3$ would give the y- and z-components of Maxwell–Ampère equation.

Equation (20.24) is therefore a compact way of writing the Maxwell's equations with source terms (four scalar equations). It has the important advantage of showing that the Maxwell's equations with source terms are manifestly covariant. Indeed the left-hand side is a four-vector obtained by the contraction of the four-gradient with the field strength tensor, which are both covariant under a Lorentz transformation. The right-hand side is the four current which is covariant. Hence, Eq. (20.24) is manifestly covariant under a Lorentz transformation.

The two homogeneous Maxwell's equations can also be written in terms of the field tensor $F_{\mu\nu}$. This is visible from the third-rank tensor

$$\partial_\lambda F_{\mu\nu} + \partial_\mu F_{\nu\lambda} + \partial_\nu F_{\lambda\mu} = 0, \tag{20.25}$$

where we see that there is no repeated subscript but a cyclic permutation of the indices from one term to the next. Since the field strength tensor is antisymmetric, we have $F_{\mu\mu} = 0$ and $F_{\lambda\mu} = -F_{\mu\lambda}$, so the left-hand side of (20.25) is zero if any two indices are equal.

Now for $\lambda = 1$, $\mu = 2$, $\nu = 3$, Eq. (20.25) yields

$$0 = \partial_1 F_{23} + \partial_2 F_{31} + \partial_3 F_{12} \, ,$$

that is, Maxwell–Thomson equation

$$\frac{\partial B_x}{\partial x} + \frac{\partial B_y}{\partial y} + \frac{\partial B_z}{\partial z} = \nabla \cdot \mathbf{B} = 0 \, .$$

The three components of Maxwell–Faraday equation are obtained by the three remaining triads of indices. For instance $\lambda = 4$, $\mu = 1$, $\nu = 2$ yields the z-component of Maxwell–Faraday equation:

$$\begin{aligned}
0 &= \partial_4 F_{12} + \partial_1 F_{24} + \partial_2 F_{41} \\
&= \frac{\partial B_z}{\partial ict} + \frac{\partial}{\partial x}\left(\frac{-iE_y}{c}\right) + \frac{\partial}{\partial y}\left(\frac{iE_x}{c}\right) \\
&= -i\left[\frac{\partial \mathbf{B}}{\partial t} + \nabla \times \mathbf{E}\right]_z \, .
\end{aligned}$$

The other choices of indices would lead to the x- and y-components of Maxwell–Faraday equation.

20.10.5 Dual Tensor

There is an alternative way to write the homogeneous Maxwell's equations. The field strength tensor is actually not the only way to build a second-rank tensor from which Maxwell's equations can be retrieved. The field transformation laws are indeed invariant to the transformation $\mathbf{B} \to -\mathbf{E}/c$ and $\mathbf{E}/c \to \mathbf{B}$. This is called the duality transformation. Applying the duality transformation to the field strength tensor provides another second-rank tensor $G_{\mu\nu}$, which reads in matrix form

$$[\mathbf{G}] = \begin{bmatrix}
0 & -E_z/c & E_y/c & -iB_x \\
E_z/c & 0 & -E_x/c & -iB_y \\
-E_y/c & E_x/c & 0 & -iB_z \\
iB_x & iB_y & iB_z & 0
\end{bmatrix}$$

The homogeneous Maxwell's equations

$$\begin{cases}
\nabla \cdot \mathbf{B} = 0 \\[2mm]
\nabla \times \mathbf{E} = -\dfrac{\partial \mathbf{B}}{\partial t}
\end{cases}$$

can then be written using the dual field-strength tensor $G_{\mu\nu}$ as

$$\partial_\nu G_{\mu\nu} = 0 \,,$$

which is a manifestly covariant equation under a Lorentz transformation, and so are the homogeneous Maxwell's equations.

20.11   Conservation Laws

To complete this chapter on covariant electrodynamics, note that conservation laws can be obtained from the symmetric, second rank, electromagnetic stress-energy tensor:

$$\Theta_{\mu\sigma} = \frac{1}{\mu_0} \left[F_{\mu\nu} F_{\sigma\nu} - \frac{1}{4}\delta_{\mu\sigma} F_{\alpha\beta} F_{\alpha\beta} \right] = \Theta_{\sigma\mu} \,,$$

where $\delta_{\mu\sigma}$ denotes the Kronecker delta symbol. In order to relate the electromagnetic stress-energy tensor to physical quantities previously introduced in conservation laws, it is useful to give the expression for its 16 components in matrix form:

$$[\Theta] = \begin{bmatrix} -T_{xx} & -T_{xy} & -T_{xz} & i\Pi_x/c \\ -T_{yx} & -T_{yy} & -T_{yz} & i\Pi_y/c \\ -T_{zx} & -T_{zy} & -T_{zz} & i\Pi_z/c \\ i\Pi_x/c & i\Pi_y/c & i\Pi_z/c & -u_{\text{em}} \end{bmatrix} \,,$$

where the space-space components are the negative of the components of the Maxwell stress tensor

$$T_{ij} = \epsilon_0 \left[E_i E_j + c^2 B_i B_j - \frac{1}{2}\delta_{ij}(E^2 + c^2 B^2) \right]$$

and the remaining components involve the spatial components of the Poynting vector:

$$\mathbf{\Pi} = \frac{1}{\mu_0} \mathbf{E} \times \mathbf{B}$$

and the electromagnetic energy density

$$u_{\text{em}} = \frac{\epsilon_0}{2}(\mathbf{E}^2 + c^2 \mathbf{B}^2) \,.$$

A manifestly covariant conservation equation under Lorentz transformation writes

$$\partial_\sigma \Theta_{\sigma\mu} = -j_\nu F_{\mu\nu} = -f_\mu, \tag{20.26}$$

where the source term f_μ denotes the components of a force density four-vector defined as

$$f_\mu = (\varrho \mathbf{E} + \mathbf{j} \times \mathbf{B}, i\mathbf{j} \cdot \mathbf{E}/c) \ .$$

A careful expansion of (20.26) shows that its time component ($\mu = 4$) is just the expression for the Poynting theorem

$$-\mathbf{j} \cdot \mathbf{E} = \nabla \cdot \mathbf{\Pi} + \frac{\partial u_{\mathrm{em}}}{\partial t} \ .$$

The space components of (20.26) also correspond to the three components of the momentum conservation equation, which reads

$$\frac{1}{c^2} \frac{\partial \mathbf{\Pi}}{\partial t} = \nabla \cdot [T] - \mathbf{f} \ .$$

20.12 Summary and Essential Formulas

- The Lorentz transformation is the transformation law between the Galilean frames $\mathcal{R}$ and $\mathcal{R}'$, moving at velocity $\mathbf{v}$ with respect to $\mathcal{R}$. Using Cartesian axes with $(Ox) \parallel (O'x') \parallel \mathbf{v}, (Oy) \parallel (O'y')$ and $(Oz) \parallel (O'z')$, the Lorentz transformation reads

$$\begin{bmatrix} x' \\ y' \\ z' \\ ict' \end{bmatrix} = \begin{bmatrix} \gamma & 0 & 0 & i\beta\gamma \\ 0 & 1 & 0 & 0 \\ 0 & 0 & 1 & 0 \\ -i\beta\gamma & 0 & 0 & \gamma \end{bmatrix} \begin{bmatrix} x \\ y \\ z \\ ict \end{bmatrix}$$

where $\beta = \dfrac{v}{c} \leq 1$ and $\gamma = \dfrac{1}{\sqrt{1-\beta^2}} \geq 1$.

- A four-vector is a vector of the four-dimensional Minkowski space whose coordinates transform according to the Lorentz transformation between two Galilean frames.
- The four-momentum $\mathbf{P} = (\mathbf{p}, i\mathcal{E}/c)$ transforms according to Lorentz transformation.
- For a particle of mass m and velocity v, the relativistic momentum and energy read

$$\text{Momentum} \quad \mathbf{p} = \gamma m \mathbf{v}$$

$$\text{Energy} \quad \mathcal{E} = \gamma m c^2 = \sqrt{p^2 c^2 + m^2 c^4}$$

where $\gamma = (1 - v^2/c^2)^{-1/2}$ is the Lorentz factor associated with the particle.
- The dynamics of a relativistic particle of mass m and charge q that is subject to an electromagnetic field $\mathbf{E}, \mathbf{B}$, satisfies the equation of motion

$$\frac{d\mathbf{p}}{dt} = \frac{d(\gamma m\mathbf{v})}{dt} = q(\mathbf{E} + \mathbf{v} \times \mathbf{B}) .$$

- The four-current $\mathbf{J} = (\mathbf{j}, ic\varrho)$, and the four-potential $\mathbf{A}_4 = (\mathbf{A}, iV/c)$ are four-vectors.
- Field strength tensor: $\boxed{F_{\mu\nu} = \partial_\mu A_\nu - \partial_\nu A_\mu}$ and its dual $G_{\mu\nu}$ are expressed in matrix form as functions of the field components:

$$[\mathbf{F}] = \begin{bmatrix} 0 & B_z & -B_y & -i\dfrac{E_x}{c} \\ -B_z & 0 & B_x & -i\dfrac{E_y}{c} \\ B_y & -B_x & 0 & -i\dfrac{E_z}{c} \\ i\dfrac{E_x}{c} & i\dfrac{E_y}{c} & i\dfrac{E_z}{c} & 0 \end{bmatrix}, \quad [\mathbf{G}] = \begin{bmatrix} 0 & -\dfrac{E_z}{c} & \dfrac{E_y}{c} & -iB_x \\ \dfrac{E_z}{c} & 0 & -\dfrac{E_x}{c} & -iB_y \\ -\dfrac{E_y}{c} & \dfrac{E_x}{c} & 0 & -iB_z \\ iB_x & iB_y & iB_z & 0 \end{bmatrix}$$

- The covariant formulation of Maxwell's equations read

$$\boxed{\begin{aligned} \partial_\nu F_{\mu\nu} &= \mu_0 j_\mu, \\ \partial_\nu G_{\mu\nu} &= 0, \end{aligned}}$$

and means that Maxwell's equations take the same form in all Galilean reference frames.

Problems

20.1 Global Positioning System and Relativity

The Positioning System (GPS) relies on data sent by satellites that fly at an altitude of 20000 km at a speed of 14000 km/h with respect to the surface of the Earth. Each satellite carries an extremely precise atomic clock ticking at 10.23 MHz, and periodically sends signals to the receivers in GPS devices, which calculate their position by measuring the time it took the signal to reach them, and thus determining their relative distance to the satellites. Assuming that the clocks on the satellites and the receiver are synchronized at $t = t_0$, what is the error in the position calculated after 12 h if relativistic effects are not taken into account?

20.2 Acceleration to Relativistic Speeds

An electron, initially at rest, is subjected to a constant accelerating force $\mathbf{F} = F_0 \mathbf{u}_x$. Determine the energy and velocity of the electron at time t. What is the speed of the electron when its energy reaches 10 MeV?

20.3 Duration of a Synchrotron Pulse

An ultrarelativistic charge describes a circular path of radius R in the presence of a homogeneous magnetic field. Show that an observer at rest will receive a periodic series of electromagnetic pulses of duration $\Delta T \approx \frac{R}{c\gamma^3}$. How does this compare with the time interval T between two consecutive pulses?

20.4 Relativistic Composition of Velocities

Let $\mathbf{v}$ and $\mathbf{v}'$ be the velocity of a particle in the aligned Galilean reference frames $\mathcal{R}$ and $\mathcal{R}'$, respectively. If $\mathbf{u}$ is the velocity of $\mathcal{R}'$ with respect to $\mathcal{R}$, show that

$$\mathbf{v} = \frac{1}{1 + \frac{\mathbf{u} \cdot \mathbf{v}_\parallel}{c^2}} \left(\frac{\mathbf{v}'}{\gamma_u} + \mathbf{u} + \left(1 - \frac{1}{\gamma_u} \right) \frac{(\mathbf{u} \cdot \mathbf{v}')}{u^2} \mathbf{u} \right)$$

with γ_u the Lorentz factor associated with the boost velocity $\mathbf{u}$.

20.5 Magnetic Interaction and Relativity

Consider a Galilean frame $\mathcal{R}'$ where a point charge q_2 moves with velocity $\mathbf{v}'$ and a point charge q_1 is at rest. In this frame, the force acting on q_2 is purely of electrostatic nature, given by Coulomb's law

$$\mathbf{F}' = \frac{q_1 q_2 (\mathbf{r}_2' - \mathbf{r}_1')}{4\pi \epsilon_0 |\mathbf{r}_2' - \mathbf{r}_1'|^3} \, .$$

Using the Lorentz transformation for aligned frames, show that in a reference frame $\mathcal{R}$ that moves with velocity $-\mathbf{u}$ with respect to $\mathcal{R}'$, the force writes

$$\mathbf{F} = q\mathbf{E} + q\mathbf{v} \times \mathbf{B}$$

with $\mathbf{v}$ the velocity of q_2 in $\mathcal{R}$. Give an expression for the fields $\mathbf{E}$ and $\mathbf{B}$. Comment.

Appendix A

A.1   Mathematical Toolbox for Classical Electrodynamics

Introduction

This section presents mathematical theorems and relations that are used throughout the course. The goal of this section is to serve as reminders and as a place to find the relevant formulas. Proofs are not given in the mathematical sense but some details are provided for relations of vector calculus.

A.1.1   Coordinate Systems

The following coordinate systems are used throughout this book.

A.1.1.1   Cartesian Coordinates

As shown in Fig. A.1, the Cartesian basis is the orthonormal set $\{\mathbf{u}_x, \mathbf{u}_y, \mathbf{u}_z\}$. Any vector $\mathbf{x}$ is fully determined by its coordinates (x, y, z)

$$\mathbf{x} = x\mathbf{u}_x + y\mathbf{u}_y + z\mathbf{u}_z \ .$$

A line element reads

$$d\mathbf{x} = dx\mathbf{u}_x + dy\mathbf{u}_y + dz\mathbf{u}_z$$

and corresponds to the infinitesimal displacement between the point of coordinates (x, y, z) and the point $(x + dx, y + dy, z + dz)$. A surface element parallel to the xy-plane reads

$$dS = dxdy \ ,$$

F. Cadiz and A. Couairon, *Classical Electrodynamics*, Undergraduate Texts in Physics,
https://doi.org/10.1007/978-3-031-86785-9

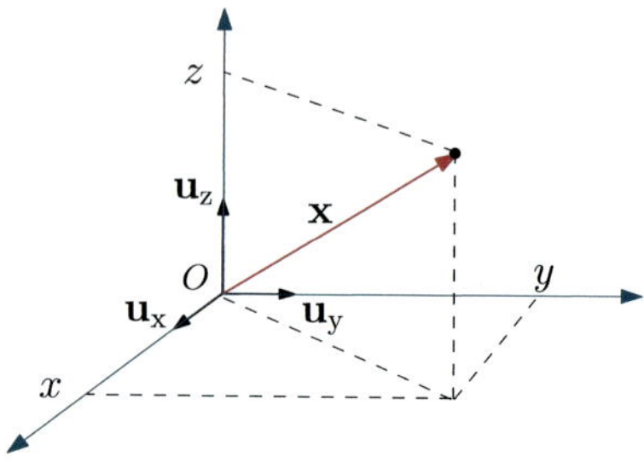

Fig. A.1 A Cartesian coordinate system

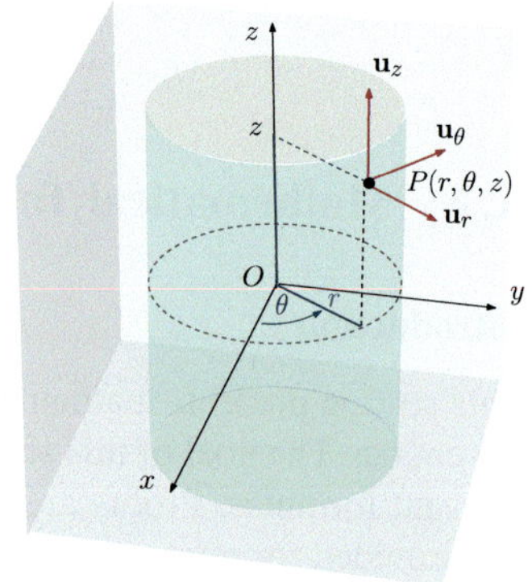

Fig. A.2 Cylindrical coordinates

and permutations can be used for surfaces that are parallel to the xz- or yz-plane. The volume element reads

$$d^3x = dx\,dy\,dz \ .$$

Note that it is common to use the same symbol to refer to both a Cartesian coordinate and the corresponding Cartesian axis. For example, depending on the context, x may refer to the axis defined by $\mathbf{u}_x$, or to the x-component of a vector.

A.1.1.2 Cylindrical Coordinates

As shown in Fig. A.2, a point P is determined by the following cylindrical coordinates:

- The distance $r \in [0, +\infty)$ between P and the z-axis
- The azimuthal angle $\theta \in [0, 2\pi)$ between the x-axis and the line connecting the origin and the point's projection in the xy plane
- The vertical position $z \in \mathbb{R}$ of the point along the z-axis

Fig. A.3 Infinitesimal
displacement in cylindrical
coordinates

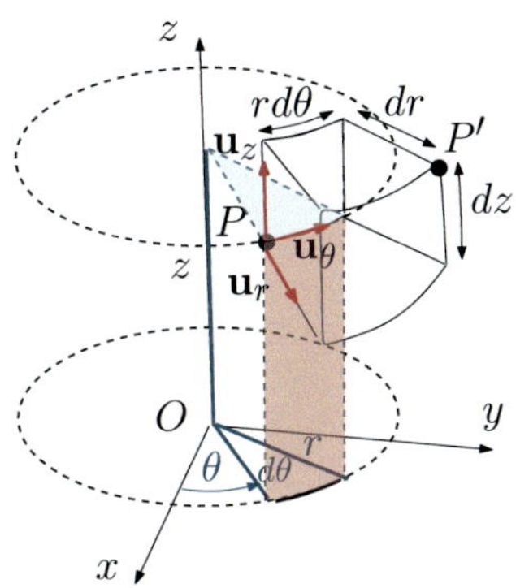

This set of coordinates defines the local orthonormal basis $\{\mathbf{u}_r, \mathbf{u}_\theta, \mathbf{u}_z\}$, with $\mathbf{u}_\theta = \mathbf{u}_z \times \mathbf{u}_r$. This basis is local because it depends on the position in space, unlike the fixed basis vectors in Cartesian coordinates. Indeed, $\mathbf{u}_r$ and $\mathbf{u}_\theta$ change direction depending on the azimuthal angle θ. In terms of the Cartesian basis, we have

$$\mathbf{u}_r = \cos\theta\,\mathbf{u}_x + \sin\theta\,\mathbf{u}_y$$

$$\mathbf{u}_\theta = -\sin\theta\,\mathbf{u}_x + \cos\theta\,\mathbf{u}_y \; .$$

Any point is written in cylindrical coordinates as $\mathbf{x} = r\mathbf{u}_r + z\mathbf{u}_z$. Note that all points with the same value of r lie on the lateral surface of a cylinder with radius r and axis along the z-axis.

A line element reads

$$d\mathbf{x} = dr\,\mathbf{u}_r + rd\theta\,\mathbf{u}_\theta + dz\,\mathbf{u}_z$$

and corresponds to an infinitesimal displacement between a point P of coordinates (r, θ, z) and a point P' of coordinates $(r + dr, \theta + d\theta, z + dz)$, as shown in Fig. A.3. It can be calculated as

$$d\mathbf{x} = \underbrace{\frac{\partial\mathbf{x}}{\partial r}}_{\mathbf{u}_r} dr + \underbrace{\frac{\partial\mathbf{x}}{\partial \theta}}_{r\frac{d\mathbf{u}_r}{d\theta}=r\mathbf{u}_\theta} d\theta + \underbrace{\frac{\partial\mathbf{x}}{\partial z}}_{\mathbf{u}_z} dz \; .$$

A surface element writes as a product of the relevant components of the line element. For instance,

$$dS = rdrd\theta \qquad \text{for a surface orthogonal to the } z\text{-axis}$$

The volume element reads

$$d^3x = rdrd\theta dz \; .$$

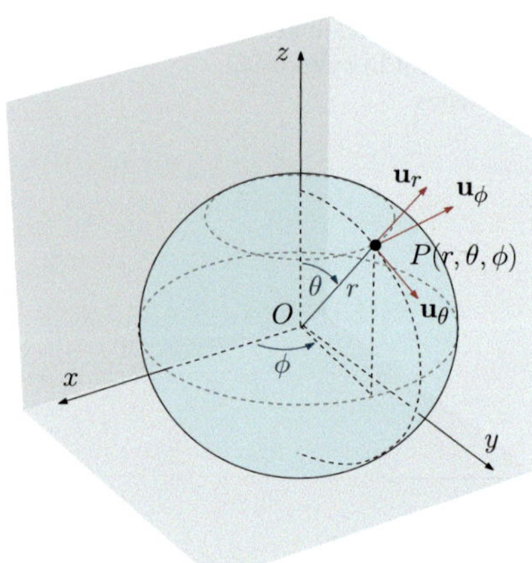

A.1.1.3 Spherical Coordinates

As shown in Fig. A.4, the spherical coordinates of a point P are

- The distance $r \in [0, +\infty)$ from the origin to the point
- The azimuthal angle $\phi \in [0, 2\pi)$ between the x-axis and the projection of the point onto the xy plane
- The polar angle $\theta \in [0, \pi]$ between the z-axis and the line connecting the origin to the point

These define the local orthonormal basis $\{\mathbf{u}_r, \mathbf{u}_\theta, \mathbf{u}_\phi\}$, with $\mathbf{u}_\phi = \mathbf{u}_r \times \mathbf{u}_\theta$ and $\mathbf{u}_\theta = \mathbf{u}_\phi \times \mathbf{u}_r$. In terms of the Cartesian basis,

$$\mathbf{u}_r = \sin\theta \cos\phi \,\mathbf{u}_x + \sin\theta \sin\phi \,\mathbf{u}_y + \cos\theta \,\mathbf{u}_z$$

$$\mathbf{u}_\theta = \cos\theta \cos\phi \,\mathbf{u}_x + \cos\theta \sin\phi \,\mathbf{u}_y - \sin\theta \,\mathbf{u}_z$$

$$\mathbf{u}_\phi = -\sin\phi \,\mathbf{u}_x + \cos\phi \,\mathbf{u}_y$$

and any point writes $\mathbf{r} = r\mathbf{u}_r$. All the points that have the same coordinate r belong to the surface of a sphere of radius r centered at the origin.

Figure A.5 shows an infinitesimal displacement from a point P of coordinates (r, θ, ϕ) to a point P' of coordinates $(r + dr, \theta + d\theta, \phi + d\phi)$. The infinitesimal line element $d\mathbf{r} = \mathbf{OP}' - \mathbf{OP}$ reads

$$d\mathbf{r} = \frac{\partial \mathbf{r}}{\partial r}dr + \frac{\partial \mathbf{r}}{\partial \theta}d\theta + \frac{\partial \mathbf{r}}{\partial \phi}d\phi$$

$$d\mathbf{r} = dr\mathbf{u}_r + rd\theta\mathbf{u}_\theta + r\sin\theta\mathbf{u}_\phi \,.$$

Fig. A.5 An infinitesimal displacement in spherical coordinates

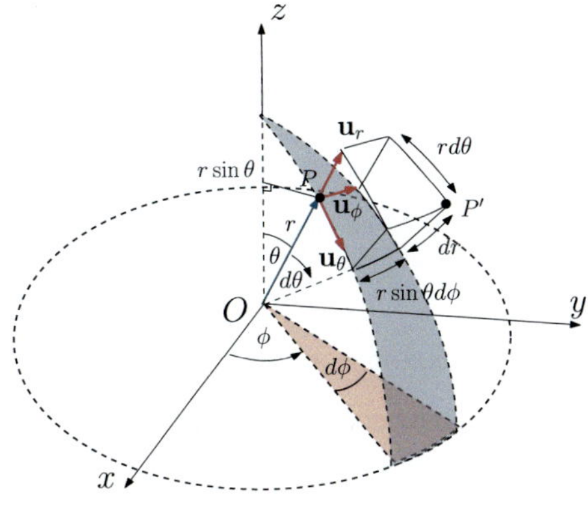

A surface element writes as a product of two components of the line element,

$$dS = r^2 \sin\theta d\theta d\phi$$

for a spherical surface element spanned by the θ and ϕ coordinates. The volume element reads

$$d^3 r = r^2 \sin\theta dr d\theta d\phi \,.$$

A comment on notation

Throughout this book, we will use two different notations for the vector position. At times, it will be written as $\mathbf{x}$, while at other times it will appear as $\mathbf{r}$. The latter notation is particularly convenient when working in spherical coordinates, where $\mathbf{r} = r\mathbf{u}_r$. On the other hand, $\mathbf{x}$ is typically used when expressing equations in their most general form, as a reminder to the reader that the equation is independent of the coordinate system being used. Of course, at the end, it is only a matter of notation and nothing prevents us from writing general expressions with $\mathbf{r}$ as well.

A.1.2   *Vector Differential Operators*

Gradient operator

Let V be a scalar field. The gradient operator, noted ∇, acts on a scalar field and results in a vector field with the following Cartesian representation

$$\nabla V = \frac{\partial V}{\partial x}\mathbf{u}_x + \frac{\partial V}{\partial y}\mathbf{u}_y + \frac{\partial V}{\partial z}\mathbf{u}_z. \qquad \text{(gradient of } V\text{)}$$

where (x, y, z) denote the Cartesian coordinates of the position vector $\mathbf{x}$. A coordinate-free definition of the gradient is the following:

$$V(\mathbf{x} + d\mathbf{x}) = V(\mathbf{x}) + d\mathbf{x} \cdot \nabla V(\mathbf{x})$$

where $d\mathbf{x}$ is an infinitesimal displacement.

As an example, let us calculate the gradient of the scalar field $V(\mathbf{r}) = \dfrac{1}{r}$ where r is the distance with respect to the origin. Writing $r = \sqrt{x^2 + y^2 + z^2}$, we can use chain's rule to differentiate

$$\frac{\partial}{\partial x} \frac{1}{r} = -\frac{1}{2} \frac{2x}{(x^2 + y^2 + z^2)^{3/2}} = -\frac{x}{r^3} \, .$$

Similarly, we have $\partial V / \partial y = -y/r^3$ and $\partial V / \partial z = -z/r^3$, thus,

$$\nabla V = -\frac{x\mathbf{u}_x + y\mathbf{u}_y + z\mathbf{u}_z}{(x^2 + y^2 + z^2)^{3/2}} = -\frac{\mathbf{r}}{r^3}. \tag{A.1}$$

The divergence operator

Let $\mathbf{A}$ be a vector field with the Cartesian components

$$\mathbf{A}(\mathbf{x}) = A_x(\mathbf{x})\mathbf{u}_x + A_y(\mathbf{x})\mathbf{u}_y + A_z(\mathbf{x})\mathbf{u}_z$$

The divergence operator $\nabla \cdot$ acts on a vector and results in a scalar field with the following Cartesian representation

$$\nabla \cdot \mathbf{A} = \frac{\partial A_x}{\partial x} + \frac{\partial A_y}{\partial y} + \frac{\partial A_z}{\partial z}. \qquad \text{(divergence of } \mathbf{A}\text{)}$$

We see that, in Cartesian coordinates, applying the divergence operator amounts to calculating a dot product between column vectors

$$\nabla \cdot \mathbf{A} = \begin{vmatrix} \frac{\partial}{\partial x} \\ \frac{\partial}{\partial y} \\ \frac{\partial}{\partial z} \end{vmatrix} \cdot \begin{vmatrix} A_x \\ A_y \\ A_z \end{vmatrix} = \frac{\partial A_x}{\partial x} + \frac{\partial A_y}{\partial y} + \frac{\partial A_z}{\partial z} \, .$$

This justifies the notation $\nabla \cdot$ used for the divergence. As an example, let us calculate the divergence of the vector field $\mathbf{x}$. Applying the definition in Cartesian coordinates, we find

$$\mathbf{\nabla} \cdot \mathbf{x} = \underbrace{\frac{\partial x}{\partial x}}_{1} + \underbrace{\frac{\partial y}{\partial y}}_{1} + \underbrace{\frac{\partial z}{\partial z}}_{1} = 3 \, .$$

The divergence of $\mathbf{x}$ is therefore a constant scalar field.

The curl operator

The curl operator $\mathbf{\nabla} \times$ acts on a vector and results in a vector field with the Cartesian representation

$$\mathbf{\nabla} \times \mathbf{A} = \left(\frac{\partial A_z}{\partial y} - \frac{\partial A_y}{\partial z} \right) \mathbf{u}_x + \left(\frac{\partial A_x}{\partial z} - \frac{\partial A_z}{\partial x} \right) \mathbf{u}_y + \left(\frac{\partial A_y}{\partial x} - \frac{\partial A_x}{\partial y} \right) \mathbf{u}_z. \quad \text{(curl of } \mathbf{A}\text{)}$$

In Cartesian coordinates, applying the curl operator amounts to calculating a cross product using column vectors

$$\mathbf{\nabla} \times \mathbf{A} = \begin{vmatrix} \frac{\partial}{\partial x} \\ \frac{\partial}{\partial y} \\ \frac{\partial}{\partial z} \end{vmatrix} \times \begin{vmatrix} A_x \\ A_y \\ A_z \end{vmatrix} = \begin{vmatrix} \frac{\partial A_z}{\partial y} - \frac{\partial A_y}{\partial z} \\ \frac{\partial A_x}{\partial z} - \frac{\partial A_z}{\partial x} \\ \frac{\partial A_y}{\partial x} - \frac{\partial A_x}{\partial y} \end{vmatrix} \, .$$

We conclude that, just as with the divergence, the notation $\mathbf{\nabla} \times$ allows us to easily obtain the expression for the curl in Cartesian coordinates.

The Laplacian operator

The Laplacian of a scalar field V yields a scalar field defined as the divergence of the gradient of V. Its Cartesian representation reads

$$\nabla^2 V = \mathbf{\nabla} \cdot (\mathbf{\nabla} V) = \frac{\partial^2 V}{\partial x^2} + \frac{\partial^2 V}{\partial y^2} + \frac{\partial^2 V}{\partial z^2}. \qquad \text{(Laplacian of } V\text{)}$$

The Laplacian of a vector field $\mathbf{A}$ is a vector field with the Cartesian representation

$$\nabla^2 \mathbf{A} = \nabla^2 A_x \mathbf{u}_x + \nabla^2 A_y \mathbf{u}_y + \nabla^2 A_z \mathbf{u}_z. \qquad \text{(Laplacian of } \mathbf{A}\text{)}$$

In many books, particularly in mathematical literature, the Laplacian is noted Δ. However, in this book we will only use the notation ∇^2 which helps us to easily remember that the Laplacian is the divergence of a gradient, $\nabla^2 = \nabla \cdot \nabla$.

The A *dot nabla* operator

For any vector field $\mathbf{A}$, the $\mathbf{A}$ *dot nabla* operator is defined as the dot product of vector $\mathbf{A}$ with the gradient operator. In Cartesian representation, this reads

$$(\mathbf{A} \cdot \nabla) = A_x \frac{\partial}{\partial x} + A_y \frac{\partial}{\partial y} + A_z \frac{\partial}{\partial z} \ .$$

The order matters! Do not switch $\mathbf{A}$ and ∇ in the dot product. Like the Laplacian, the $\mathbf{A}$ *dot nabla* operator can be applied to scalar- as well as vector fields. In the latter case, it applies component-wise.

For example, let us apply $(\mathbf{A} \cdot \nabla)$ to the position vector $\mathbf{r}$. The components of the position vector are x, y, and z. For the first component,

$$(\mathbf{A} \cdot \nabla)x = A_x \frac{\partial}{\partial x} x = A_x \ .$$

For the second and third components, we find similarly $(\mathbf{A} \cdot \nabla)y = A_y$ and $(\mathbf{A} \cdot \nabla)z = A_z$. Finally,

$$(\mathbf{A} \cdot \nabla)\mathbf{r} = \mathbf{A} \ .$$

A.1.2.1   Variation of a Field

The increment dV of a scalar field $V(\mathbf{r}, t)$ along a line element

$$d\mathbf{r} = dx\,\mathbf{u}_x + dy\,\mathbf{u}_y + dz\,\mathbf{u}_z \ ,$$

during a time dt is given by

$$dV = \frac{\partial V}{\partial x} dx + \frac{\partial V}{\partial y} dy + \frac{\partial V}{\partial z} dz + \frac{\partial V}{\partial t} dt$$

which rewrites as

$$dV = d\mathbf{r} \cdot (\nabla V) + \frac{\partial V}{\partial t} dt = (d\mathbf{r} \cdot \nabla)V + \frac{\partial V}{\partial t} dt \ .$$

Applying the latter result component-wise to a vector field $\mathbf{A}(\mathbf{r}, t)$, we find

$$dA = (d\mathbf{r} \cdot \nabla)\mathbf{A} + \frac{\partial \mathbf{A}}{\partial t} dt \ .$$

The first term on the right-hand side stands for the non-uniform character of $\mathbf{A}$ while the second term stands for its non-stationary character.

A.1.3 Vector Differential Operators in Other Coordinates

So far, we have seen the expression for the differential operators (gradient, divergence, curl, and Laplacian) in Cartesian coordinates. In a more general case, when using a set of coordinates (u, v, w) so that $\mathbf{x} = \mathbf{x}(u, v, w)$, we can write the infinitesimal line element as

$$d\mathbf{x} = \frac{\partial \mathbf{x}}{\partial u} du + \frac{\partial \mathbf{x}}{\partial v} dv + \frac{\partial \mathbf{x}}{\partial w} dw.$$

We will suppose that (u, v, w) defines an orthogonal coordinate system, that is, the set of unit vectors $\{\mathbf{u}_u, \mathbf{u}_v, \mathbf{u}_w\}$, where

$$\mathbf{u}_u = \frac{\frac{\partial \mathbf{x}}{\partial u}}{\left|\frac{\partial \mathbf{x}}{\partial u}\right|}, \quad \mathbf{u}_v = \frac{\frac{\partial \mathbf{x}}{\partial v}}{\left|\frac{\partial \mathbf{x}}{\partial v}\right|}, \quad \mathbf{u}_w = \frac{\frac{\partial \mathbf{x}}{\partial w}}{\left|\frac{\partial \mathbf{x}}{\partial w}\right|}$$

forms a right-handed orthogonal trihedron ($\mathbf{u}_u \times \mathbf{u}_v = \mathbf{u}_w$). Now we can use the general definition of the gradient, so that the infinitesimal change dV of a scalar field V between points $\mathbf{x}$ and $\mathbf{x} + d\mathbf{x}$ is

$$dV = \frac{\partial V}{\partial u} du + \frac{\partial V}{\partial v} dv + \frac{\partial V}{\partial w} dw = \nabla V \cdot d\mathbf{x}.$$

From this equation we can obtain the expression of the gradient operator in the basis $\{\mathbf{u}_u, \mathbf{u}_v, \mathbf{u}_w\}$:

$$\nabla = \frac{1}{\left|\frac{\partial \mathbf{x}}{\partial u}\right|}\mathbf{u}_u \frac{\partial}{\partial u} + \frac{1}{\left|\frac{\partial \mathbf{x}}{\partial v}\right|}\mathbf{u}_v \frac{\partial}{\partial v} + \frac{1}{\left|\frac{\partial \mathbf{x}}{\partial w}\right|}\mathbf{u}_w \frac{\partial}{\partial w},$$

where it is important to place the unit vectors at the left of the partial derivatives, since in general they will depend upon the coordinates (u, v, w). Once the gradient operator is known in cylindrical or spherical coordinates, it is easy to compute the divergence, curl, and Laplacian. One must not forget that the unit vectors should also be differentiated with respect to the coordinates.

A.1.3.1 Vector Differential Operators in Cylindrical Coordinates

Let $V(\mathbf{x}) \equiv V(r, \theta, z)$ be a scalar field, and

$$\mathbf{A}(\mathbf{x}) = A_r(r, \theta, z)\mathbf{u}_r + A_\theta(r, \theta, z)\mathbf{u}_\theta + A_z(r, \theta, z)\mathbf{u}_z,$$

a vector field, expressed in cylindrical coordinates. Then

$$\nabla V = \frac{\partial V}{\partial r}\mathbf{u}_r + \frac{1}{r}\frac{\partial V}{\partial \theta}\mathbf{u}_\theta + \frac{\partial V}{\partial z}\mathbf{u}_z$$

$$\nabla \cdot \mathbf{A} = \frac{1}{r}\frac{\partial(r A_r)}{\partial r} + \frac{1}{r}\frac{\partial A_\theta}{\partial \theta} + \frac{\partial A_z}{\partial z}$$

$$\nabla \times \mathbf{A} = \left[\frac{1}{r}\frac{\partial A_z}{\partial \theta} - \frac{\partial A_\theta}{\partial z}\right]\mathbf{u}_r + \left[\frac{\partial A_r}{\partial z} - \frac{\partial A_z}{\partial r}\right]\mathbf{u}_\theta + \frac{1}{r}\left[\frac{\partial(r A_\theta)}{\partial r} - \frac{\partial A_r}{\partial \theta}\right]\mathbf{u}_z.$$

$$\nabla^2 V = \frac{1}{r}\frac{\partial}{\partial r}\left[r\frac{\partial V}{\partial r}\right] + \frac{1}{r^2}\frac{\partial^2 V}{\partial \theta^2} + \frac{\partial^2 V}{\partial z^2}.$$

A.1.4 Vector Differential Operators in Spherical Coordinates

Let $V(\mathbf{r}) \equiv V(r, \theta, \phi)$ be a scalar field, and

$$\mathbf{A}(\mathbf{r}) = A_r(r, \theta, \phi)\mathbf{u}_r + A_\theta(r, \theta, \phi)\mathbf{u}_\theta + A_\phi(r, \theta, \phi)\mathbf{u}_\phi,$$

a vector field, expressed in spherical coordinates. Then

$$\nabla V = \frac{\partial V}{\partial r}\mathbf{u}_r + \frac{1}{r}\frac{\partial V}{\partial \theta}\mathbf{u}_\theta + \frac{1}{r \sin \theta}\frac{\partial V}{\partial \phi}\mathbf{u}_\phi$$

$$\nabla \cdot \mathbf{A} = \frac{1}{r^2}\frac{\partial(r^2 A_r)}{\partial r} + \frac{1}{r \sin \theta}\frac{\partial(\sin \theta A_\theta)}{\partial \theta} + \frac{1}{r \sin \theta}\frac{\partial A_\phi}{\partial \phi}.$$

$$\nabla \times \mathbf{A} = \frac{1}{r \sin \theta}\left[\frac{\partial(\sin \theta A_\phi)}{\partial \theta} - \frac{\partial A_\theta}{\partial \phi}\right]\mathbf{u}_r + \frac{1}{r}\left[\frac{1}{\sin \theta}\frac{\partial A_r}{\partial \phi} - \frac{\partial(r A_\phi)}{\partial r}\right]\mathbf{u}_\theta$$
$$+ \frac{1}{r}\left[\frac{\partial(r A_\theta)}{\partial r} - \frac{\partial A_r}{\partial \theta}\right]\mathbf{u}_\phi.$$

$$\nabla^2 V = \frac{1}{r^2}\frac{\partial}{\partial r}\left[r^2\frac{\partial V}{\partial r}\right] + \frac{1}{r^2\sin\theta}\frac{\partial}{\partial\theta}\left[\sin\theta\frac{\partial V}{\partial\theta}\right] + \frac{1}{r^2\sin^2\theta}\frac{\partial^2 V}{\partial\phi^2}.$$

A.1.5 Vector Calculus Identities

In this section, V and U are scalar fields, $\mathbf{A}$ and $\mathbf{B}$ are vector fields.

- The divergence of the curl of a vector field is equal to zero

$$\nabla \cdot (\nabla \times \mathbf{A}) = 0. \tag{A.2}$$

- The curl of the gradient of a scalar field is equal to zero

$$\nabla \times (\nabla V) = \mathbf{0}. \tag{A.3}$$

- A useful identity for the physics of electromagnetic waves:

$$\nabla^2\mathbf{A} = \nabla(\nabla \cdot \mathbf{A}) - \nabla \times (\nabla \times \mathbf{A}). \tag{A.4}$$

The next three identities look similar to the derivative of a product.

- Gradient of a product
$$\nabla(U\,V) = U\,\nabla V + V\,\nabla U. \tag{A.5}$$

- Curl of the product of a scalar field and a vector field

$$\nabla \times (U\mathbf{A}) = U\,\nabla \times \mathbf{A} + \nabla U \times \mathbf{A}. \tag{A.6}$$

- Divergence of a product of scalar field by vector field

$$\nabla \cdot (U\mathbf{A}) = U\,\nabla \cdot \mathbf{A} + \nabla U \cdot \mathbf{A}. \tag{A.7}$$

The next 3 identities (A.8), (A.9) and (A.10) look *different* from the derivative of a product!

- Gradient of a dot product between vector fields

$$\nabla(\mathbf{A}\cdot\mathbf{B}) = (\mathbf{A}\cdot\nabla)\mathbf{B} + \mathbf{A}\times(\nabla\times\mathbf{B}) + (\mathbf{B}\cdot\nabla)\mathbf{A} + \mathbf{B}\times(\nabla\times\mathbf{A}). \tag{A.8}$$

This identity can be used to calculate the electric field for a dipole of (uniform) dipole moment $\mathbf{p}$: $\nabla(\mathbf{p}\cdot\mathbf{r}) = (\mathbf{p}\cdot\nabla)\mathbf{r} = \mathbf{p}, \quad (\nabla\times\mathbf{r} = 0)$.
- Curl of a cross product between vector fields

$$\nabla \times (\mathbf{A}\times\mathbf{B}) = (\nabla\cdot\mathbf{B})\mathbf{A} + (\mathbf{B}\cdot\nabla)\,\mathbf{A} - (\nabla\cdot\mathbf{A})\mathbf{B} - (\mathbf{A}\cdot\nabla)\,\mathbf{B}. \tag{A.9}$$

This identity can be used to calculate the magnetic field for a dipole of (uniform) dipole moment **m**:

$$\nabla \times \left(\mathbf{m} \times \frac{\mathbf{r}}{r^3}\right) = -(\mathbf{m} \cdot \nabla)\frac{\mathbf{r}}{r^3} = -\nabla\left(\frac{\mathbf{m} \cdot \mathbf{r}}{r^3}\right) \qquad \left(\nabla \cdot \frac{\mathbf{r}}{r^3} = 0 \text{ for } \mathbf{r} \neq 0\right)$$

- Divergence of a cross product between vector fields

$$\nabla \cdot (\mathbf{A} \times \mathbf{B}) = -\mathbf{A} \cdot \nabla \times \mathbf{B} + \mathbf{B} \cdot \nabla \times \mathbf{A}. \tag{A.10}$$

A.1.6 Some Important Results of Vector Calculus

In the following we will require the notions of oriented surfaces, oriented contours, flux, and circulation.

A.1.6.1 Orientable Surfaces

We say that S is an orientable surface if it is possible to choose a direction for the normal vector that remains continuous across the entire surface without any ambiguities. Figure A.6 shows the example of an oriented surface, for which one can choose the orientation of the normal to point either upwards or downwards.

A famous example of a non-orientable surface is the so-called Möbius strip, which can be made by attaching the ends of a strip with a half-twist. As shown in Fig. A.7, one can try to define the normal at every point on the closed loop, pointing upwards at the starting point. After one turn, however, the normal ends up pointing downwards. It is not possible to define a continuously varying normal vector at each point on the Möbius strip.

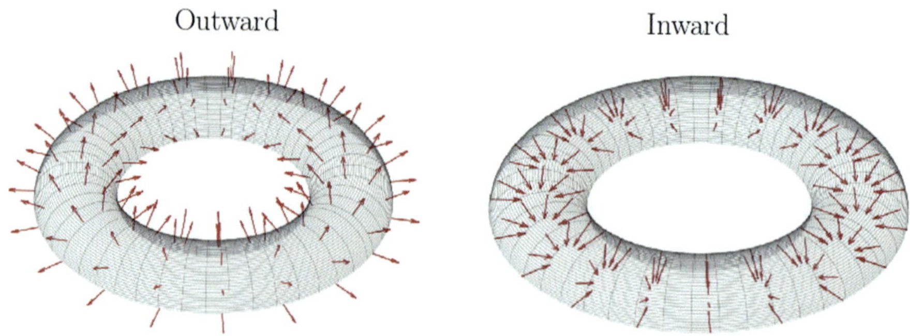

Fig. A.6 An oriented surface with the two possible orientations of the normal vector

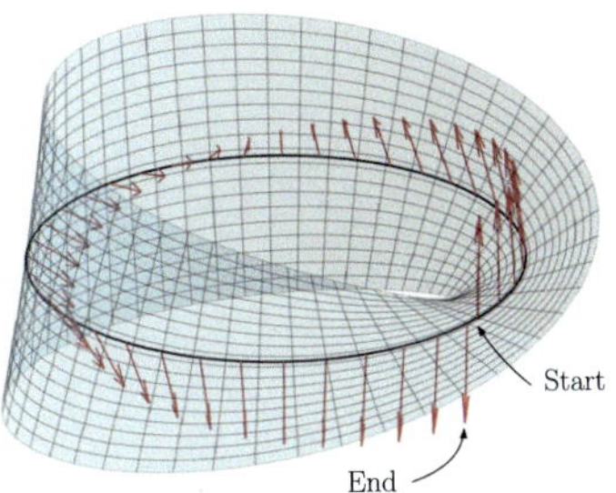

Fig. A.7 A non-orientable surface

A.1.6.2 Oriented Contours and Line Integrals

A contour is a closed curve, also called loop. Let $S(\Gamma)$ be the oriented surface delimited by the contour Γ. The orientation of Γ is given by the orientation of the surface S according to the right-hand rule: the direction of positive circulation of the contour is dictated by the normal $\mathbf{n}$ to the surface. The right hand's fingers circulate along Γ, and then the thumb is directed along $\mathbf{n}$. This is shown in Fig. A.8.

A.1.6.3 Flux and Circulation

Given an oriented surface S, the flux of a vector field $\mathbf{F}$ across S is given by the integral

$$\iint_S \mathbf{F} \cdot d\mathbf{S}. \tag{A.11}$$

The flux is positive when the field is primarily aligned with the normal to the surface, and it is negative when aligned in the opposite direction. In other words, the

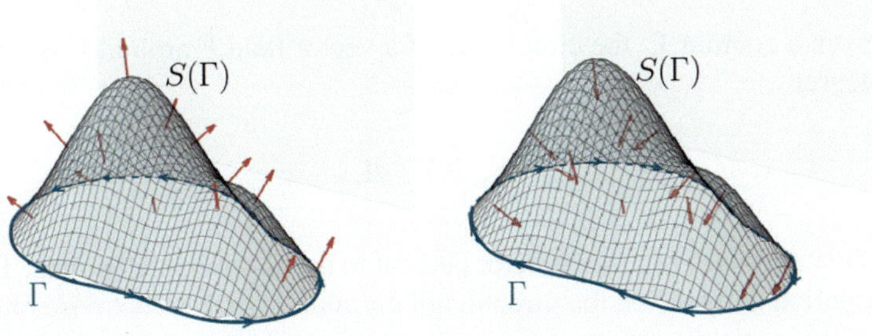

Fig. A.8 An oriented surface uniquely defines the direction of circulation of its contour according to the right-hand rule. The dashed portion indicates the hidden part of the contour

Fig. A.9 Flux of a field **F** through an oriented surface parametrized by $\mathbf{x}(u, v)$

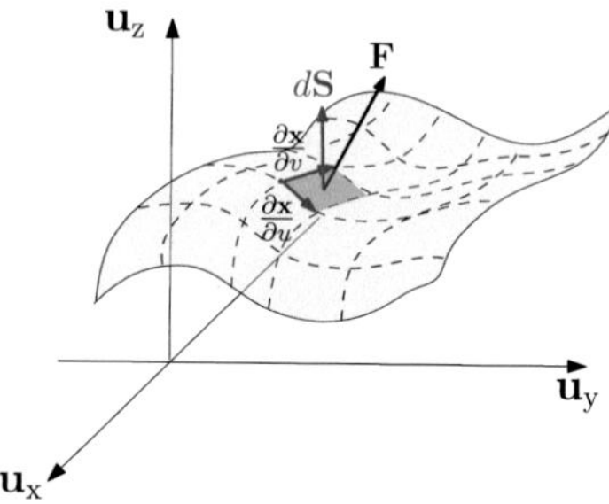

sign of the flux depends on the orientation of the surface. If the surface is parametrized by the two variables (u, v) such that every $\mathbf{x} \in S$ is written as

$$\mathbf{x}(u, v) = (x(u, v), y(u, v), z(u, v)) \quad (u, v) \in \mathcal{D}.$$

Then, the surface vector element writes $d\mathbf{S} = \frac{\partial \mathbf{x}}{\partial u} \times \frac{\partial \mathbf{x}}{\partial v} du dv$, as shown in Fig. A.9. The flux of a vector field is then calculated as

$$\iint_S \mathbf{F} \cdot d\mathbf{S} = \iint_{\mathcal{D}} \mathbf{F}(\mathbf{x}(u, v)) \cdot \left(\frac{\partial \mathbf{x}}{\partial u} \times \frac{\partial \mathbf{x}}{\partial v} \right) du dv \qquad (A.12)$$

and it can be shown that its value is independent of the parametrization chosen for the surface.

The scalar surface element is $dS = |d\mathbf{S}| = \left| \frac{\partial \mathbf{x}}{\partial u} \times \frac{\partial \mathbf{x}}{\partial v} \right| du dv$ and it appears in the surface integral of scalar fields. For example the integral of a scalar field V over a surface is written

$$\iint_S V dS = \iint_{\mathcal{D}} V(\mathbf{x}(u, v)) \left| \frac{\partial \mathbf{x}}{\partial u} \times \frac{\partial \mathbf{x}}{\partial v} \right| du dv. \qquad (A.13)$$

Now, given a contour Γ, the circulation of a vector field $\mathbf{F}$ around Γ is given by the path integral

$$\oint_\Gamma \mathbf{F} \cdot d\mathbf{l}, \qquad (A.14)$$

where $d\mathbf{l} = dl \mathbf{t}$, with $\mathbf{t}$ the unit vector tangent to the contour at all points. The sign of the circulation depends on the direction of circulation of Γ (clockwise or anticlockwise). If the parametrization of the curve is $\mathbf{x}(t) = (x(t), y(t), z(t))$ for $t \in [a, b]$, then $d\mathbf{l} = d\mathbf{x} = (x'(t), y'(t), z'(t)) dt$ and

$$\int_{\Gamma} \mathbf{F} \cdot dl = \sum_{i=x,y,z} \int_{a}^{b} F_i(\mathbf{x}(t)) \frac{dx_i(t)}{dt} dt.$$

The following results are all particular cases of a general formula by George Stokes (1819–1903).

The divergence theorem (Green–Ostrogradsky formula)

Let S be a closed, positively oriented surface defining an interior volume Ω such that $\partial\Omega = S$, then:

$$\boxed{\iiint_{\Omega} \nabla \cdot \mathbf{F} d^3r = \oiint_{\partial\Omega} \mathbf{F} \cdot d\mathbf{S}} \tag{A.15}$$

that is, the flux of a field across a closed surface equals the integral of its divergence over the volume enclosed by this surface. Note that S is positively oriented if it has an outward normal vector, meaning that $d\mathbf{S}$ points outwards from Ω.

The differential form of Gauss's law (2.3) demonstrated in Chap. 2 is obtained by using the divergence theorem (A.15). For this theorem to hold, it is often assumed that the field $\mathbf{F}$ is continuously differentiable within Ω. However, in electrodynamics, we frequently encounter singularities. For example, the field of a point charge at the origin is not defined at the charge's location, and its divergence, in the sense of distributions, corresponds to a Dirac delta centered at the origin. A generalization of Gauss's law for such cases is discussed in Sect. A.2.2.4.

The gradient formula

For a scalar field V and a volume Ω bounded by the closed, positively oriented surface $\partial\Omega$,

$$\oiint_{\partial\Omega} V(\mathbf{r}) d\mathbf{S}(\mathbf{r}) = \iiint_{\Omega} \nabla V(\mathbf{r}) d^3r. \tag{A.16}$$

Let us demonstrate the gradient formula component wise. First project the formula on a unit vector $\mathbf{u}$ in a Cartesian basis:

$$\mathbf{u} \cdot \oiint_{\partial\Omega} V(\mathbf{r}) d\mathbf{S}(\mathbf{r}) = \oiint_{\partial\Omega} V(\mathbf{r}) \mathbf{u} \cdot d\mathbf{S}(\mathbf{r}).$$

Apply Ostrogradsky's theorem (A.15) to the vector $V\mathbf{u}$:

$$\oiint_{\partial\Omega} V(\mathbf{r})\mathbf{u} \cdot d\mathbf{S}(\mathbf{r}) = \iiint_{\Omega} \nabla \cdot (V\mathbf{u})d^3r.$$

Since $\nabla \cdot (V\mathbf{u}) = \mathbf{u} \cdot \nabla V + V\nabla \cdot \mathbf{u}$ (Eq. (A.7)) and $\nabla \cdot \mathbf{u} = 0$, we find

$$\iiint_{\Omega} \nabla \cdot (V\mathbf{u})d^3r = \iiint_{\Omega} \mathbf{u} \cdot \nabla V d^3r = \mathbf{u} \cdot \iiint_{\Omega} \nabla V d^3r.$$

We have shown

$$\mathbf{u} \cdot \oiint_{\partial\Omega} V(\mathbf{r}, t)d\mathbf{S}(\mathbf{r}) = \mathbf{u} \cdot \iiint_{\Omega} \nabla V d^3r.$$

Since this is valid for all three axes of the Cartesian basis, the gradient formula is demonstrated.

Stokes theorem
Let Γ be the contour of an oriented surface $S(\Gamma)$, with the direction of circulation of Γ given by the orientation of the normal to $S(\Gamma)$ according to the right-hand rule. Then:

$$\boxed{\iint_{S} \nabla \times \mathbf{F} \cdot d\mathbf{S} = \oint_{\Gamma} \mathbf{F} \cdot d\mathbf{l}} \tag{A.17}$$

that is, the flux of the curl of a field $\mathbf{F}$ across a surface equals the circulation of $\mathbf{F}$ over the contour of that surface.

Kelvin formula
For a scalar field V and an open surface $S(\Gamma)$ bounded by a closed curve Γ,

$$\oint_{\Gamma} V d\mathbf{l} = -\iint_{S(\Gamma)} \nabla V \times d\mathbf{S}. \tag{A.18}$$

Let us demonstrate Kelvin's formula component-wise. Taking the dot product of the left-hand-side with a unit vector $\mathbf{u}$ of a Cartesian basis, and applying Stokes's Theorem A.20, we have

$$\mathbf{u} \cdot \oint_{\Gamma} V d\mathbf{r} = \oint_{\Gamma} V\mathbf{u} \cdot d\mathbf{r} = \iint_{S(\Gamma)} (\nabla \times V\mathbf{u}) \cdot d\mathbf{S}.$$

Since $\mathbf{u}$ is uniform, $\boldsymbol{\nabla} \times V\mathbf{u} = \boldsymbol{\nabla} V \times \mathbf{u}$ (we have used (A.6)). Replacing the integral and using a circular permutation in the triple product, we find

$$\iint_{S(\Gamma)} (\boldsymbol{\nabla} \times V\mathbf{u}) \cdot d\mathbf{S} = \iint_{S(\Gamma)} (\boldsymbol{\nabla} V \times \mathbf{u}) \cdot d\mathbf{S} = - \iint_{S(\Gamma)} (\boldsymbol{\nabla} V \times d\mathbf{S}) \cdot \mathbf{u}.$$

Since this calculation is similar for all unit vectors of a Cartesian basis, Kelvin's formula is demonstrated component-wise.

Curl formula
For a vector field $\mathbf{A}$ and a volume Ω enclosed by the positively oriented surface $\partial\Omega$:

$$\oiint_{\partial\Omega} \mathbf{A} \times d\mathbf{S} = - \iiint_{\Omega} (\boldsymbol{\nabla} \times \mathbf{A}) d^3r. \tag{A.19}$$

Again, this identity is shown component-wise. Take first the dot product with a unit vector $\mathbf{u}$, which can be pushed under the integral, and then apply a circular permutation.

$$\mathbf{u} \cdot \oiint_{\partial\Omega} \mathbf{A} \times d\mathbf{S} = \oiint_{\partial\Omega} \mathbf{u} \cdot (\mathbf{A} \times d\mathbf{S}) = \oiint_{\partial\Omega} (\mathbf{u} \times \mathbf{A}) \cdot d\mathbf{S}.$$

Apply Ostrogradsky's theorem (A.15):

$$\oiint_{\partial\Omega} (\mathbf{u} \times \mathbf{A}) \cdot d\mathbf{S} = \iiint_{\Omega} \boldsymbol{\nabla} \cdot (\mathbf{u} \times \mathbf{A}) d^3r.$$

From vector calculus identity (A.10), we have

$$\boldsymbol{\nabla} \cdot (\mathbf{u} \times \mathbf{A}) = -\mathbf{u} \cdot \boldsymbol{\nabla} \times \mathbf{A} + \mathbf{A} \times \underbrace{(\boldsymbol{\nabla} \times \mathbf{u})}_{\mathbf{0}}.$$

Thus

$$\iiint_{\Omega} \boldsymbol{\nabla} \cdot (\mathbf{u} \times \mathbf{A}) d^3r = \iiint_{\Omega} -\mathbf{u} \cdot (\boldsymbol{\nabla} \times \mathbf{A}) d^3r = -\mathbf{u} \cdot \iiint_{\Omega} (\boldsymbol{\nabla} \times \mathbf{A}) d^3r.$$

We have shown for any unit vector

$$\mathbf{u} \cdot \oiint_{\partial\Omega} \mathbf{A} \times d\mathbf{S} = -\mathbf{u} \cdot \iiint_{\Omega} (\boldsymbol{\nabla} \times \mathbf{A}) d^3r,$$

which completes the demonstration of the curl formula.

Another important result is the following. Let γ be a curve that goes from A to B, and ϕ is a scalar field. Then:

$$\int_{\gamma} \nabla\phi \cdot d\mathbf{l} = \phi(B) - \phi(A). \tag{A.20}$$

A.1.7 *Existence of Scalar and Vector Potentials*

As we observed in (A.3), the curl of a gradient is always zero. Now, if the curl of a field is zero, can we conclude that the field is the gradient of a scalar field?

Similarly, from (A.2), we see that the divergence of a curl is always zero. If the divergence of a field is zero, can we deduce that the field is the curl of some vector field?

The answer to both questions is yes provided the fields are defined everywhere, this is an important result from Poincaré.

Existence of the scalar potential

Let $\mathbf{F}$ be a vector field defined over the whole space $\mathbb{R}^3$, and suppose it verifies

$$\nabla \times \mathbf{F} = \mathbf{0}$$

everywhere. Then, there exists a function ϕ, called the scalar potential, such that

$$\mathbf{F} = \nabla\phi.$$

Existence of the vector potential

Let $\mathbf{F}$ be a vector field defined over the whole space $\mathbb{R}^3$, such that its divergence is zero everywhere

$$\nabla \cdot \mathbf{F} = 0.$$

Then, there exists a potential vector $\mathbf{A}$ such that

$$\mathbf{F} = \nabla \times \mathbf{A}.$$

A.2   Distributions and Point Charges

Let us first restrict ourselves to one dimension. A distribution is a linear application over the space $\mathcal{D}$ of *test* functions, that is, functions which are infinitely differentiable and are non-zero inside a bounded interval. A distribution T acts on any $\varphi \in \mathcal{D}$ and yields a complex number noted (T, φ):

$$T : \mathcal{D} \mapsto \mathbb{C} \tag{A.21}$$

$$\varphi \mapsto (T, \varphi). \tag{A.22}$$

Distributions are natural generalizations of functions, since any function f that is locally integrable defines a regular distribution via the integral:

$$(f, \varphi) = \int_{\mathbb{R}^3} f(x)\varphi(x)dx,$$

where φ is a test function. However, the converse is not true: not all distributions correspond to locally integrable functions. Such distributions are called singular. The Dirac distribution is an example of a singular distribution, since it is not defined by an integral, but by the following relation:

$$\boxed{(\delta, \varphi) = \varphi(0) \in \mathbb{C} \quad \varphi \in \mathcal{D}.} \tag{A.23}$$

It does not make any sense, for example, trying to evaluate δ at a particular value of x. Despite this, the practical notation $\delta(x)$ is used to designate the Dirac distribution centered at $x = 0$, whereas $\delta(x - x_0)$ denotes $\delta(x_0)$, the Dirac centered at $x = x_0$, which is defined as

$$(\delta_{x_0}, \varphi) = \varphi(x_0) \in \mathbb{C} \quad \varphi \in \mathcal{D}.$$

A.2.1   Functions that Tend to the Dirac Distribution

Consider the following sequence of functions:

$$\delta_n(x) = \begin{cases} \dfrac{n}{2} & |x| < 1/n, \\ 0 & |x| > 1/n. \end{cases} \tag{A.24}$$

For all $n \in \mathbb{N}$ the area under the curve of δ_n is equal to 1. As n increases, the function δ_n describes a rectangle that is increasingly concentrated around the origin, as illustrated in Fig. A.10.

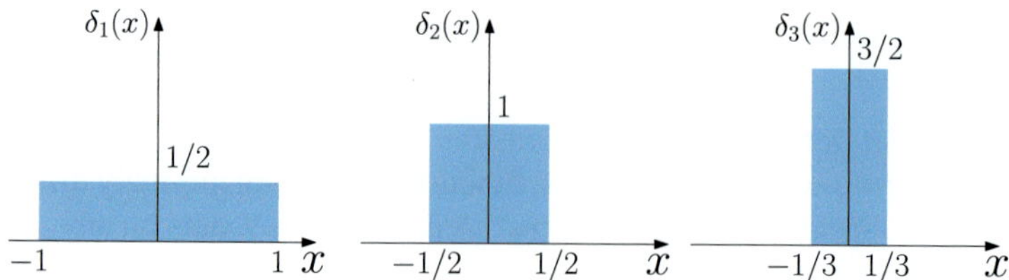

Fig. A.10 First three functions given by the sequence δ_n defined by (A.24)

This sequence tends to zero for all $x \neq 0$, that is, it converges to the null function at all points except for $x = 0$. In consequence $\int_{\mathbb{R}} \lim_{n \to \infty} \delta_n(x)dx = 0$. However, for any test function $\varphi \in \mathcal{D}$, one has

$$\int_{\mathbb{R}} \varphi(x)\delta_n(x)dx = \frac{n}{2} \int_{-1/n}^{1/n} \varphi(x)dx = \varphi(\theta_n/n)$$

with $-1 \leq \theta_n \leq 1$ (theorem of intermediate values). Taking the limit when $n \to \infty$

$$\lim_{n \to \infty} \int_{\mathbb{R}} \varphi(x)\delta_n(x)dx = \lim_{n \to \infty} \varphi(\theta_n/n) = \varphi(0) = (\delta, \varphi).$$

The sequence of functions δ_n has therefore the desired property (A.23) of the Dirac δ as $n \to \infty$. Indeed it defines a sequence of regular distributions δ_n defined by

$$(\delta_n, \varphi) = \int_{\mathbb{R}} \delta_n(x)\varphi(x)dx,$$

we see that the sequence $(\delta_n)_{n \in \mathbb{N}}$ converges to δ in the sense of distributions, since for every test function φ:

$$\lim_{n \to \infty} (\delta_n, \varphi) = (\delta, \varphi) = \varphi(0).$$

There are in fact many ways to construct sequences that converge to δ, for example, a sequence of Gaussian functions of the form:

$$h_n(x) = \frac{n}{\sqrt{\pi}}e^{-n^2 x^2}$$

converges to δ as well, since $\lim_{n \to \infty} \int_{\mathbb{R}} h_n(x)\varphi(x)dx = \varphi(0)$ for every test function φ.

Note

Throughout the text, we will sometimes use the following abuse of notation

$$\int_{\mathbb{R}} \delta(x)\varphi(x)dx = \varphi(0) \tag{A.25}$$

even if this integral is not really defined, since δ is not a function. Every time we write this, we can think of it as a limit of integrals $\int_{\mathbb{R}} \delta(x)\varphi(x)dx \equiv \lim_{n\to\infty} \int_{\mathbb{R}} h_n(x)\varphi(x)\, dx = \varphi(0)$, where h_n is any sequence of functions that converges to δ in the sense stated above.

A.2.2 *Distributions in $\mathbb{R}^3$ and Their Derivatives*

One can extend the above definitions by now considering $\mathcal{D}$ as the space of test functions defined over $\mathbb{R}^3$. The Dirac distribution in this case writes

$$(\delta, \varphi) = \varphi(\mathbf{0}) \quad \varphi \in \mathcal{D}.$$

and a regular distribution defined by the locally integrable function f is defined by

$$(f, \varphi) = \iiint_{\mathbb{R}^3} f(\mathbf{x})\varphi(\mathbf{x})d^3x \quad \varphi \in \mathcal{D}.$$

The partial derivative with respect to x_i of a distribution T is defined by

$$\left(\frac{\partial T}{\partial x_i}, \varphi \right) = - \left(T, \frac{\partial \varphi}{\partial x_i} \right) \tag{A.26}$$

which is a generalization of an integration by parts for the case of regular distributions. Every distribution is infinitely differentiable, which is very convenient. In the same way, one can define the Laplacian of a distribution by

$$\left(\nabla^2 T, \varphi \right) = - \left(T, \nabla^2\varphi \right). \tag{A.27}$$

As an example, let us calculate the gradient of the Dirac distribution. One of the components is given, according to (A.26), by

$$\left(\frac{\partial}{\partial x_i}\delta, \varphi \right) = - \left(\delta, \frac{\partial}{\partial x_i}\varphi \right) = - \frac{\partial \varphi}{\partial x_i}(\mathbf{0}).$$

And so we conclude that $\nabla\delta = \sum_{i=x,y,z} \frac{\partial \delta}{\partial x_i}\mathbf{u}_i$ is given by

$$(\nabla\delta, \varphi) = -\nabla\varphi(\mathbf{0}). \tag{A.28}$$

A.2.2.1 The Divergence of $\mathbf{x}/|\mathbf{x}|^3$

The field generated by a point charge, proportional to $\mathbf{x}/|\mathbf{x}|^3$, has a singularity at the origin where its divergence is not defined. Despite this, this field can be considered to be locally integrable because in spherical coordinates the volume element cancels out the singularity. We can therefore consider the regular distribution defined as

$$\left(\frac{\mathbf{x}}{|\mathbf{x}|^3}, \varphi\right) = \iiint_{\mathbb{R}^3} \varphi(\mathbf{x}) \frac{\mathbf{x}}{|\mathbf{x}|^3} d^3x \quad \varphi \in \mathcal{D}$$

and calculate its divergence according to (A.26). For a field distribution $\mathbf{F}$:

$$(\nabla \cdot \mathbf{F}, \varphi) = \sum_{i=x,y,z} \left(\frac{\partial F_i}{\partial x_i}, \varphi\right) = -\sum_{i=x,y,z} \left(F_i, \frac{\partial \varphi}{\partial_i}\right)$$

so that

$$\left(\nabla \cdot \frac{\mathbf{x}}{|\mathbf{x}|^3}, \varphi\right) = -\sum_{i=x,y,z} \left(\frac{x_i}{|\mathbf{x}|^3}, \frac{\partial \varphi}{\partial_i}\right) = -\iiint_{\mathbb{R}^3} \frac{\mathbf{x}}{|\mathbf{x}|^3} \cdot \nabla\varphi(\mathbf{x}) d^3x.$$

In spherical coordinates

$$\left(\nabla \cdot \frac{\mathbf{x}}{|\mathbf{x}|^3}, \varphi\right) = -\int_0^{2\pi} \int_0^{\pi} \int_0^{\infty} \frac{\mathbf{u}_r}{r^2} \cdot \nabla\varphi r^2 dr \sin\theta d\theta d\phi \tag{A.29}$$

$$= -\int_0^{2\pi} \int_0^{\pi} \underbrace{\int_0^{\infty} \frac{\partial \varphi}{\partial r} dr}_{\varphi(\mathbf{0})} \sin\theta d\theta d\phi = 4\pi\varphi(\mathbf{0}) = 4\pi(\delta, \varphi) \tag{A.30}$$

since φ is zero at infinity. We end up with the following result, valid in the sense of distributions

$$\boxed{\nabla \cdot \frac{\mathbf{x}}{|\mathbf{x}|^3} = 4\pi\delta(\mathbf{x}).} \tag{A.31}$$

Multiplying Eq. (A.31) by $1/(4\pi\epsilon_0)$ yields the differential form of Gauss law (2.3) for a point charge distribution $\varrho(\mathbf{r}) = q\delta(\mathbf{x})$. Note also that Eq. (3.1) holds for the distribution $1/|\mathbf{x}|$ defined as

$$\left(\frac{1}{|\mathbf{x}|}, \varphi\right) = \iiint_{\mathbb{R}^3} \frac{\varphi(\mathbf{x})}{|\mathbf{x}|} d^3x \quad \varphi \in \mathcal{D}.$$

So that in the sense of distributions

$$\nabla \frac{1}{|\mathbf{x}|} = -\frac{\mathbf{x}}{|\mathbf{x}|^3}$$

and one concludes that, for the distribution $1/|\mathbf{x}|$,

$$\nabla^2 \frac{1}{|\mathbf{x}|} = \nabla \cdot \nabla \frac{1}{|\mathbf{x}|} = -4\pi \delta(\mathbf{x}) \qquad (A.32)$$

which is identical to Eq. (1.10).

A.2.2.2 Convolution Between Distributions

The convolution product between two locally integrable functions f and g, noted $f * g$, is given by

$$f * g(\mathbf{x}) = \iiint_{\mathbb{R}^3} f(\mathbf{x}')g(\mathbf{x} - \mathbf{x}')d^3x' \qquad (A.33)$$

and one has $f * g = g * f$. Convolutions appear in electrostatics very often, simply look at Eq. (1.5) or (3.2). We can define the regular distribution $f * g$ by

$$(f * g, \varphi) = \iiint_{\mathbb{R}^3} f * g(\mathbf{x})\varphi(\mathbf{x})d^3x = \iiint_{\mathbb{R}^3} \left(\iiint_{\mathbb{R}^3} f(\mathbf{x}')g(\mathbf{x} - \mathbf{x}')d^3x' \right) \varphi(\mathbf{x})d^3x$$

$$= \iiint_{\mathbb{R}^3} \iiint_{\mathbb{R}^3} f(\mathbf{x}')g(\mathbf{u})\varphi(\mathbf{u} + \mathbf{x}')d^3u\, d^3x' = (g(\mathbf{x}), f(\mathbf{x}')\varphi(\mathbf{x} + \mathbf{x}'))$$

where we have used the change of variables $\mathbf{u} = \mathbf{x} - \mathbf{x}'$. The generalization for the convolution $[T * S]$ between two distributions T and S (not necessarily regular) is the following

$$([T * S], \varphi) = \big(T(\mathbf{x}), \big(S(\mathbf{x}'), \varphi(\mathbf{x} + \mathbf{x}')\big)\big) \quad \varphi \in \mathcal{D}. \qquad (A.34)$$

If $T * S$ is well defined, it can be shown that $T * S = S * T$. We will use the following properties:

- The Dirac distribution acts as unity for the convolution product

$$\delta * T = T * \delta = T$$

  which is very easy to demonstrate from the definition (A.34).
- To calculate the derivative of a convolution, it is enough to take the derivative of one of the factors:

$$\frac{\partial}{\partial x_i}[T * S] = \left[\frac{\partial T}{\partial x_i} * S\right] = \left[T * \frac{\partial S}{\partial x_i}\right] \qquad (A.35)$$

and this formula still works for the derivative of order n with respect to x_i.

A.2.2.3 Electric Field and Potential as Distributions

When a charge density ϱ is a function, the absolute potential can be calculated by the integral

$$V(\mathbf{x}) = \frac{1}{4\pi\epsilon_0} \iiint_{\mathbb{R}^3} \frac{\varrho(\mathbf{x}')}{|\mathbf{x} - \mathbf{x}'|} d^3 x' = \frac{1}{4\pi\epsilon_0} \varrho * \frac{1}{|\mathbf{x}|}.$$

Here $*$ stands for the convolution product. The way to generalize this formula is to consider V as a distribution, given by the convolution between the distribution ϱ and the distribution $1/|\mathbf{x}|$:

$$V = \frac{1}{4\pi\epsilon_0} \left[\varrho * \frac{1}{|\mathbf{x}|} \right]. \tag{A.36}$$

From this, the distribution associated with the electric field writes, according to (A.35)

$$\mathbf{E} = -\nabla V = -\frac{1}{4\pi\epsilon_0} \left[\varrho * \nabla \frac{1}{|\mathbf{x}|} \right]$$

and since $\nabla \frac{1}{|\mathbf{x}|} = -\mathbf{x}/|\mathbf{x}|^3$

$$\mathbf{E} = \frac{1}{4\pi\epsilon_0} \left[\varrho * \frac{\mathbf{x}}{|\mathbf{x}|^3} \right] \tag{A.37}$$

which is a generalization of the Coulomb integral (1.5) which is indeed a convolution.

A.2.2.4 Gauss's Law for Distributions

Here we will generalize the differential version of Gauss's law (2.3) to the case of distributions. According to (A.37) and using (A.31):

$$\nabla \cdot \mathbf{E} = \frac{1}{4\pi\epsilon_0} \left[\varrho * \underbrace{\nabla \cdot \frac{\mathbf{x}}{|\mathbf{x}|^3}}_{4\pi\delta} \right] = \frac{1}{\epsilon_0} [\rho * \delta] = \frac{\varrho}{\epsilon_0}.$$

Note that similarly we can retrieve Poisson's equation in the framework of distributions

$$\nabla^2 V = \frac{1}{4\pi\epsilon_0}\left[\varrho * \underbrace{\nabla^2 \frac{1}{|\mathbf{x}|}}_{-4\pi\delta}\right] = -\frac{\varrho}{\epsilon_0}$$

where we have used (A.32). All the fundamental equations of electrodynamics are valid in the sense of distributions.

Note that in physics literature the divergence theorem is often applied to fields with singularities. This is because even though the theorem applies to continuously differentiable fields, it is possible to generalize it in the framework of distributions. For example, for a point charge at the origin, $\mathbf{E}(\mathbf{r}) = \frac{q}{4\pi\epsilon_0}\frac{\mathbf{r}}{|r|^3}$ and $\varrho(\mathbf{r}) = q\delta(\mathbf{r})$. Let Ω be the volume enclosed by a sphere of radius R centered around the origin. Then

$$\oiint_{\partial\Omega} \mathbf{E}\cdot d\mathbf{S} = \frac{q}{\epsilon_0}$$

and since $\nabla\cdot\frac{\mathbf{r}}{|r|^3} = 4\pi\delta(\mathbf{r})$ in the sense of distributions, an abuse of notation allows one to write

$$\iiint_\Omega \nabla\cdot\mathbf{E}d^3r = \frac{q}{\epsilon_0}\iiint_\Omega \delta(\mathbf{r})d^3r = \frac{q}{\epsilon_0}$$

By comparing with the flux integral above, we see that the divergence theorem *works* in the framework of distributions despite the singularity.

A.2.2.5   Charge Density and Potential of an Infinitesimal Dipole

An application of (A.28) helps us to identify the distribution of an infinitesimal dipole $\mathbf{p}$ at the origin. For a dipole consisting of two-point charges $\pm q$ separated by a along the unit direction $\mathbf{d}$, the charge density is written as two displaced Diracs

$$\varrho = q\delta(\mathbf{x} - \frac{a}{2}\mathbf{d}) - q\delta(\mathbf{x} + \frac{a}{2}\mathbf{d}).$$

Let's see how this distribution acts on a test function φ

$$(\varrho, \varphi) = q(\delta(\mathbf{x} - \frac{a}{2}\mathbf{d}), \varphi) - q(\delta(\mathbf{x} + \frac{a}{2}\mathbf{d}), \varphi) = q\left(\varphi(\frac{a}{2}\mathbf{d}) - \varphi(-\frac{a}{2}\mathbf{d})\right).$$

In the limit $a \to 0$, we have

$$\lim_{a\to 0}\left(\varphi(\frac{a}{2}\mathbf{d}) - \varphi(-\frac{a}{2}\mathbf{d})\right) = \nabla\varphi(\mathbf{0})\cdot\mathbf{d}.$$

Comparing with (A.28), we conclude that

$$\lim_{a \to 0} \varrho = -q\mathbf{d} \cdot \nabla \delta = -\mathbf{p} \cdot \nabla \delta.$$

This allows one to immediately retrieve the potential (3.8) far from a dipole by using (A.36):

$$V = -\frac{\mathbf{p}}{4\pi \epsilon_0} \cdot \left[\nabla \delta * \frac{1}{|\mathbf{x}|} \right] = -\frac{\mathbf{p}}{4\pi \epsilon_0} \cdot \left[\delta * \nabla \frac{1}{|\mathbf{x}|} \right]$$

$$V = -\frac{\mathbf{p}}{4\pi \epsilon_0} \cdot \left[\delta * \frac{\mathbf{x}}{|\mathbf{x}|^3} \right]$$

and since $\delta * T = \delta$ for any distribution T, we retrieve Eq. (3.8), here written as a distribution

$$V = \frac{1}{4\pi \epsilon_0} \mathbf{p} \cdot \frac{\mathbf{x}}{|\mathbf{x}|^3}.$$

A.3   From Discrete to Continuous Charge Distributions

Since any piece of matter is composed of elementary particles, which have no physical dimension, the charge density of any object will always be a collection of Dirac distributions of the form $\varrho(\mathbf{r}) = \sum_i q_i \delta(\mathbf{r} - \mathbf{r}_i)$, where the sum is made over all the charged elementary particles that are present. In practice, it would be impossible to track all the positions of the individual particles of a macroscopic object. But even if we could, quantum mechanics actually prevents us to define a precise trajectory for elementary particles. Instead, we could define the charge density in terms of the mean positions of the particles, for example in a hydrogen atom the electron is, on average, at a distance of about 0.05 nm with respect to the proton. In a solid, the distance between neighboring atoms is, in average, of the order of $a = 0.1$ nm. The charge density in matter thus varies very rapidly, as illustrated in Fig. A.11. Compared to a macroscopic scale, this is still closer to a sum of Dirac distributions than to a continuous charge density.

The case of continuous charge distributions seen in Chap. 1 consists in spatially averaging both the charge density and the resulting electric field. The reason for doing so is that, in many cases, we are not interested in these rapid variations. This is the case, for example, when dealing with macroscopic objects, for which we only care about variations of the charge density and of the electric field over a length scale $d \gg a$. For this, we define the macroscopic charge density $\langle \varrho \rangle$ and the macroscopic electric field $\langle \mathbf{E} \rangle$ at every point $\mathbf{x}$ as the average over a volume $V(\mathbf{x}) \sim d^3$ centered around $\mathbf{x}$

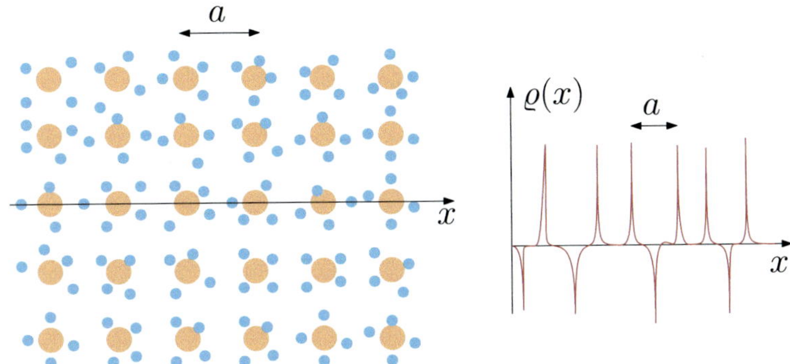

Fig. A.11 Left: Electrons and nuclei in a crystal lattice. Right: charge density across the x-axis that corresponds to one crystallographic axis

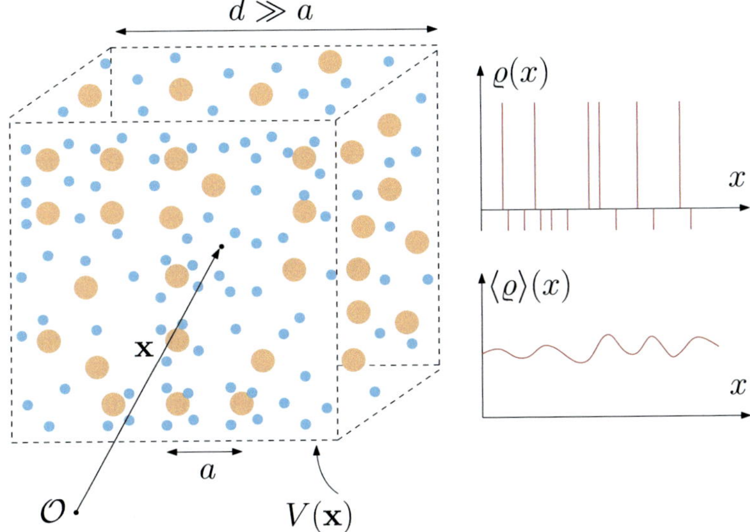

Fig. A.12 Averaging the charge density over a volume V of typical dimensions $d \gg a$ provides a smooth charge density without the fast jumps at the atomic scale

$$\langle \varrho \rangle (\mathbf{x}) = \frac{1}{V} \iiint_{V(\mathbf{x})} \varrho(\mathbf{x}')d^3x'$$

$$\langle \mathbf{E} \rangle (\mathbf{x}) = \frac{1}{V} \iiint_{V(\mathbf{x})} \mathbf{E}(\mathbf{x}')d^3x'$$

they are therefore smooth functions of $\mathbf{x}$ provided that $d \gg a$. This is illustrated for the charge density in Fig. A.12.

This averaging may be written in a more convenient way if we consider the function φ defined as

$$\varphi(\mathbf{x}) = \begin{cases} \dfrac{1}{V} & \text{if} \quad \mathbf{x} \in V(\mathbf{0}) \\ 0 & \text{otherwise} \end{cases}$$

then $\varphi(\mathbf{x} - \mathbf{x}')$ equals to $1/V$ in a volume V centered at $\mathbf{x}'$ and 0 everywhere else, and we can write

$$\langle \varrho \rangle (\mathbf{x}) = \iiint_{\mathbb{R}^3} \varrho(\mathbf{x}')\varphi(\mathbf{x} - \mathbf{x}')d^3x' = \varrho * \varphi \tag{A.38}$$

$$\langle \mathbf{E} \rangle (\mathbf{x}) = \iiint_{\mathbb{R}^3} \mathbf{E}(\mathbf{x}')\varphi(\mathbf{x} - \mathbf{x}')d^3x' = \mathbf{E} * \varphi \tag{A.39}$$

where $*$ stands for the convolution product. More generally, the spatial average over the scale d of any quantity f can be written as

$$\boxed{\langle f(\mathbf{x}) \rangle = \iiint_{\mathbb{R}^3} f(\mathbf{x}')\varphi(\mathbf{x} - \mathbf{x}')d^3x' = f * \varphi .} \tag{A.40}$$

For example, a point charge q located at $\mathbf{x}_i$ defines a charge distribution $\varrho(\mathbf{x}) = q\delta(\mathbf{x} - \mathbf{x}_i)$. The associated averaged charge distribution then writes

$$\langle \varrho \rangle (\mathbf{x}) = \iiint_{\mathbb{R}^3} q\delta(\mathbf{x}' - \mathbf{x}_i)\varphi(\mathbf{x} - \mathbf{x}')d^3x' = q\varphi(\mathbf{x} - \mathbf{x}_i)$$

which represents a charge q uniformly distributed over a volume V centered around $\mathbf{x}_i$. A smoother charge density can be achieved by choosing $\varphi(\mathbf{x})$ to be a normalized Gaussian function centered at $\mathbf{x}$ and of typical extension V. For a system of N charges q_i located at positions $\mathbf{x}_i$, with $i = 1, 2, \ldots, N$, the average charge density then reads

$$\langle \varrho \rangle (\mathbf{x}) = \sum_{i=1}^{N} q_i\varphi(\mathbf{x} - \mathbf{x}_i).$$

Consider now a volume Ω representing a macroscopic object, and let us consider one particular atom whose center of mass is at position $\mathbf{x}'$. It is composed of N charges q_i located at $\mathbf{x}_i$ with respect to $\mathbf{x}'$, with $i = 1, 2, \ldots, N$, as illustrated in Fig. A.13.

The microscopic charge density associated with this atom is $\varrho_{\text{at}}(\mathbf{x}) = \sum_{i=1}^{N} q_i\delta(\mathbf{x} - \mathbf{x}' - \mathbf{x}_i)$. The average charge density then reads

$$\langle \varrho_{\text{at}} \rangle (\mathbf{x}) = \sum_{i=1}^{N} q_i\varphi(\mathbf{x} - \mathbf{x}' - \mathbf{x}_i)$$

Fig. A.13 An atom at position $\mathbf{x}'$ composed of elementary charges

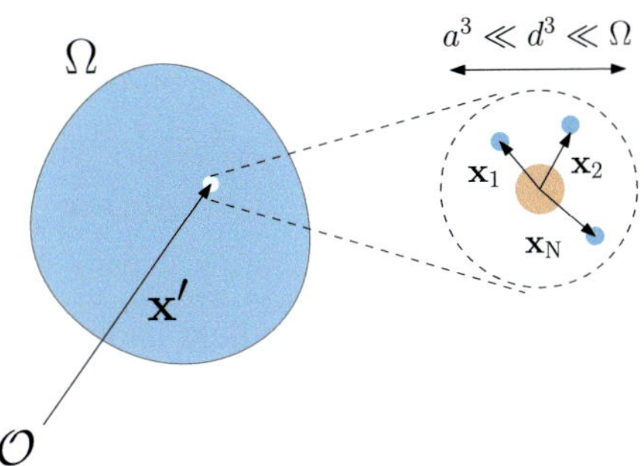

where φ is a function centered around the origin and of characteristic extension d along each axis. Now, if $d \gg a$, where $a \sim 0.05\,\text{nm}$ is the typical size of an atom, the function φ is slowly varying in the above sum, so that to the lowest order, $\varphi(\mathbf{x} - \mathbf{x}' - \mathbf{x}_i) \approx \varphi(\mathbf{x} - \mathbf{x}')$ and

$$\langle \varrho_{\text{at}}(\mathbf{x}) \rangle \approx \sum_{i=1}^{N} q_i \varphi(\mathbf{x} - \mathbf{x}') = Q_{\text{at}} \varphi(\mathbf{x} - \mathbf{x}'),$$

where $Q_{\text{at}} = \sum_{i=1}^{N} q_i$ is simply the total charge contained by the atom. On the other hand, the electric field at position $\mathbf{x}$ generated by this atom is given by

$$\mathbf{E}_{\text{at}}(\mathbf{x}) = \frac{1}{4\pi\epsilon_0} \sum_{i=1}^{N} \frac{q_i (\mathbf{x} - \mathbf{x}' - \mathbf{x}_i)}{|\mathbf{x} - \mathbf{x}' - \mathbf{x}_i|^3}$$

and by neglecting $|\mathbf{x}' - \mathbf{x}_i|$ with respect to $|\mathbf{x} - \mathbf{x}'|$, we obtain the electric field of a point charge Q_{at} at $\mathbf{x}'$

$$\mathbf{E}_{\text{at}}(\mathbf{x}) \approx \frac{Q_{\text{at}}}{4\pi\epsilon_0} \frac{(\mathbf{x} - \mathbf{x}')}{|\mathbf{x} - \mathbf{x}'|^3}$$

and the average electric field writes

$$\langle \mathbf{E}_{\text{at}}(\mathbf{x}) \rangle = \iiint_{\mathbb{R}^3} \mathbf{E}_{\text{at}}(\mathbf{u})\varphi(\mathbf{x} - \mathbf{u})d^3u \approx \frac{1}{4\pi\epsilon_0} \iiint_{\mathbb{R}^3} \frac{Q_{\text{at}}\varphi(\mathbf{x} - \mathbf{u})(\mathbf{u} - \mathbf{x}')}{|\mathbf{u} - \mathbf{x}'|^3}d^3u$$

by changing variables $\mathbf{v}' = \mathbf{x}' + \mathbf{x} - \mathbf{u}$

$$\langle \mathbf{E}_{\text{at}}(\mathbf{x}) \rangle \approx \frac{1}{4\pi\epsilon_0} \iiint_{\mathbb{R}^3} \frac{Q_{\text{at}}\varphi(\mathbf{v}' - \mathbf{x}')(\mathbf{x} - \mathbf{v}')}{|\mathbf{x} - \mathbf{v}'|^3}d^3v' = \frac{1}{4\pi\epsilon_0} \iiint_{\mathbb{R}^3} \frac{\langle \varrho_{\text{at}} \rangle(\mathbf{v}')(\mathbf{x} - \mathbf{v}')}{|\mathbf{x} - \mathbf{v}'|^3}d^3v'$$

so that the averaged electric field created by the atom is given by a Coulomb integral where the charge density ϱ_{at} is replaced by the averaged charge density of the atom.

Finally, since the volume Ω is a collection of M atoms located at $\mathbf{x}'_k$ with $k = 1, 2, \ldots M$, one defines the continuous charge distribution ρ at every point of Ω as

$$\varrho(\mathbf{x}) = \sum_{k=1}^{M} Q_k \varphi(\mathbf{x} - \mathbf{x}'_k)$$

and the average electric field is obtained by calculating the Coulomb integral with the above-average charge distribution.

A.4 Invariances and Curie's Principle

Here it is shown that the invariances of the charge distribution ϱ translate into the same invariances for the electric potential as well as for the electric field. We first write Coulomb's integral for the potential:

$$V(\mathbf{x}) = \frac{1}{4\pi\epsilon_0} \iiint \frac{\varrho(\mathbf{x}')}{|\mathbf{x} - \mathbf{x}'|} d^3 x'.$$

Consider a transformation T that leaves invariant the charge distribution, i.e., if $\mathbf{X} = T(\mathbf{x})$, then $\varrho(\mathbf{X}) = \varrho(\mathbf{x})$. In addition, we assume that T is an isometry: it is linear and preserves distances and volumes: $\det T = |T| = 1$.

Let's show that $V(\mathbf{X}) = V(\mathbf{x})$ for any $\mathbf{x}$.

$$V(\mathbf{X}) = \frac{1}{4\pi\epsilon_0} \iiint \frac{\varrho(\mathbf{x}')}{|\mathbf{X} - \mathbf{x}'|} d^3 x'.$$

Let's perform a change of variable $\mathbf{x}' = T(\mathbf{x}'')$. The Jacobian of the transformation is

$$\frac{\partial(x', y', z')}{\partial(x'', y'', z'')} = \begin{vmatrix} \frac{\partial x'}{\partial x''} & \frac{\partial x'}{\partial y''} & \frac{\partial x'}{\partial z''} \\ \frac{\partial y'}{\partial x''} & \frac{\partial y'}{\partial y''} & \frac{\partial y'}{\partial z''} \\ \frac{\partial z'}{\partial x''} & \frac{\partial z'}{\partial y''} & \frac{\partial z'}{\partial z''} \end{vmatrix} = \det T = 1$$

The potential at $\mathbf{X}$ then reads

$$V(\mathbf{X}) = \frac{1}{4\pi\epsilon_0} \iiint \frac{\varrho(T(\mathbf{x}''))}{|T(\mathbf{x}) - T(\mathbf{x}'')|} \frac{\partial(x', y', z')}{\partial(x'', y'', z'')} d^3 x''.$$

Since $\varrho(T(\mathbf{x}'')) = \varrho(\mathbf{x}'')$ and the distance between $T(\mathbf{x})$ and $T(\mathbf{x}'')$ is the same as the distance between $\mathbf{x}$ and $\mathbf{x}''$, we can write $|T(\mathbf{x}) - T(\mathbf{x}'')| = |\mathbf{x} - \mathbf{x}''|$. Hence,

$$V(\mathbf{X}) = \frac{1}{4\pi\epsilon_0} \iiint \frac{\varrho(\mathbf{x}'')}{|\mathbf{x} - \mathbf{x}''|} d^3 x'' = V(\mathbf{x}).$$

For the electric field, an additional argument is needed as in the Coulomb integral, the vector $T(\mathbf{x}) - T(\mathbf{x}'')$ appears, which has the same modulus as $\mathbf{x} - \mathbf{x}''$ but not necessarily the same orientation. So it must be transformed using the linearity of T:

$$\varrho(T(\mathbf{x}''))\frac{T(\mathbf{x}) - T(\mathbf{x}'')}{|T(\mathbf{x}) - T(\mathbf{x}'')|^3} = \varrho(T(\mathbf{x}''))\frac{T(\mathbf{x} - \mathbf{x}'')}{|T(\mathbf{x} - \mathbf{x}'')|^3} = \varrho(\mathbf{x}'')\frac{T(\mathbf{x} - \mathbf{x}'')}{|\mathbf{x} - \mathbf{x}''|^3},$$

hence

$$\varrho(T(\mathbf{x}''))\frac{T(\mathbf{x}) - T(\mathbf{x}'')}{|T(\mathbf{x}) - T(\mathbf{x}'')|^3} = T\left(\frac{\varrho(\mathbf{x}'')(\mathbf{x} - \mathbf{x}'')}{|\mathbf{x} - \mathbf{x}''|^3}\right).$$

Finally, using again the linearity of T, we can swap the integral and T:

$$\begin{aligned}
\mathbf{E}(\mathbf{X}) &= \frac{1}{4\pi\epsilon_0} \iiint T\left(\frac{\varrho(\mathbf{x}'')(\mathbf{x} - \mathbf{x}'')}{|\mathbf{x} - \mathbf{x}''|^3}\right) d^3 x'' \\
&= T\left(\frac{1}{4\pi\epsilon_0} \iiint \frac{\varrho(\mathbf{x}'')(\mathbf{x} - \mathbf{x}'')}{|\mathbf{x} - \mathbf{x}''|^3} d^3 x''\right) = T(\mathbf{E}(\mathbf{x})).
\end{aligned}$$

In conclusion, if ϱ is invariant under the transformation T ($\varrho(T\mathbf{x}) = \varrho(\mathbf{x})$), then

$$V(T\mathbf{x}) = V(\mathbf{x})$$

$$\mathbf{E}(T\mathbf{x}) = T\mathbf{E}(\mathbf{x})$$

which demonstrates Curie's principle.

A.5   Solution of Poisson's Equation in a Finite Volume

Let us consider a region in space $\Omega \subseteq \mathbb{R}^3$. It is possible to write the potential as an integral over Ω and over its boundary $\partial\Omega$. This makes use of Green's theorem and Green's identity that will be demonstrated below.

A.5.1   Green's Theorem

Let us write a vector field $\mathbf{A}$ under the form

$$\mathbf{A} = \phi\nabla\psi$$

where ϕ and ψ are two scalar fields. Then according to (A.7)

$$\nabla \cdot \mathbf{A} = \nabla \cdot (\phi \nabla \psi) = \nabla \phi \cdot \nabla \psi + \phi \nabla^2 \psi$$

and using Green–Ostrogradsky theorem (A.15), we obtain *Green's first identity*

$$\iiint_\Omega \underbrace{\left(\nabla \phi \cdot \nabla \psi + \phi \nabla^2 \psi \right)}_{\nabla \cdot \mathbf{A}} d^3x = \oiint_{\partial\Omega} \underbrace{\phi \nabla \psi}_{\mathbf{A}} \cdot d\mathbf{S} \tag{A.41}$$

Now, we can exchange the roles of ψ and ϕ to obtain an equivalent identity

$$\iiint_\Omega \left(\nabla \phi \cdot \nabla \psi + \psi \nabla^2 \phi \right) d^3x = \oiint_{\partial\Omega} \psi \nabla \phi \cdot d\mathbf{S}$$

and subtracting the latter equation from (A.41), we obtain *Green's theorem*

$$\iiint_\Omega (\phi \nabla^2 \psi - \psi \nabla^2 \phi) d^3x = \oiint_{\partial\Omega} (\phi \nabla \psi - \psi \nabla \phi) \cdot d\mathbf{S} \tag{A.42}$$

Poisson's integral equation

If V satisfies Poisson's equation in Ω, then, for $\mathbf{x} \in \Omega$:

$$V(\mathbf{x}) = \frac{1}{4\pi\epsilon_0} \iiint_\Omega \frac{\rho(\mathbf{x}')d^3x'}{|\mathbf{x} - \mathbf{x}'|} - \frac{1}{4\pi} \oiint_{\partial\Omega} \frac{d\mathbf{S}(\mathbf{x}') \cdot \mathbf{E}(\mathbf{x}')}{|\mathbf{x} - \mathbf{x}'|}$$

$$+ \frac{1}{4\pi} \oiint_{\partial\Omega} d\mathbf{S}(\mathbf{x}') \cdot \frac{\mathbf{x}' - \mathbf{x}}{|\mathbf{x}' - \mathbf{x}|^3} V(\mathbf{x}') \tag{A.43}$$

In consequence, the potential at any point $\mathbf{x} \in \Omega$ can be written in terms of the charge density in Ω and the values of the potential and the electric field at the boundary $\partial\Omega$. All the information about the exterior of Ω is therefore contained by V and $\mathbf{E}$ in $\partial\Omega$.

Proof We use Green's theorem (A.42) with $\phi(\mathbf{x}') = V(\mathbf{x}')$ and $\psi(\mathbf{x}') = 1/|\mathbf{x} - \mathbf{x}'|$. Recalling the fundamental identity (A.32) in the sense of distributions

$$\nabla'^2 \psi(\mathbf{x}') = -4\pi \delta(\mathbf{x} - \mathbf{x}')$$

and that $\nabla^2 V = \varrho/\epsilon_0$, then

$$\iiint_\Omega \left\{ V(\mathbf{x}')\nabla'^2\psi(\mathbf{x}') - \psi(\mathbf{x}')\nabla'^2 V(\mathbf{x}') \right\} d^3x' = -4\pi \iiint_\Omega \left\{ V(\mathbf{x}')\delta(\mathbf{x} - \mathbf{x}') - \frac{\varrho(\mathbf{x}')}{4\pi\epsilon_0 \, |\, \mathbf{x} - \mathbf{x}' \,|} \right\} d^3x'$$

$$= \oiint_{\partial\Omega} \left\{ V(\mathbf{x}')\nabla'\frac{1}{|\mathbf{x} - \mathbf{x}'|} \cdot d\mathbf{S} - \frac{1}{|\mathbf{x} - \mathbf{x}'|}\nabla' V(\mathbf{x}') \cdot d\mathbf{S} \right\}$$

Considering $\mathbf{x} \in \Omega$ and that $\nabla'\frac{1}{|\mathbf{x}-\mathbf{x}'|} = \frac{\mathbf{x}-\mathbf{x}'}{|\mathbf{x}-\mathbf{x}'|^3}$, one obtains Poisson's integral equation.

Corollary By taking $\Omega = \mathbb{R}^3$, and for an integrable charge density ($\iiint_{\mathbb{R}^3} |\varrho(\mathbf{x}')| d^3x' < \infty$), we obtain the known result (3.2):

$$V(\mathbf{x}) = \frac{1}{4\pi\epsilon_0} \iiint_{\mathbb{R}^3} \frac{\varrho(\mathbf{x}')}{|\mathbf{x} - \mathbf{x}'|} d^3x'.$$

A.5.2 Uniqueness of the Solution to the Poisson–Dirichlet Problem

The Poisson–Dirichlet problem reads

$$\begin{cases} \mathbf{x} \in \Omega : \ \nabla^2 V(\mathbf{x}) = -\dfrac{\varrho(\mathbf{x})}{\epsilon_0} \\ \mathbf{x} \in \partial\Omega : V(\mathbf{x}) = V_D(\mathbf{x}) \end{cases}$$

To show that the solution is unique, let $V_1(x)$ and $V_2(x)$ be two solutions of this problem. Then

$$\psi(\mathbf{x}) = V_2(\mathbf{x}) - V_1(\mathbf{x})$$

is a solution to Laplace's equation with homogeneous Dirichlet's boundary conditions, that is

$$\begin{cases} \mathbf{x} \in \Omega : \ \nabla^2\psi(\mathbf{x}) = 0 \\ \mathbf{x} \in \partial\Omega : \psi(\mathbf{x}) = V_D(\mathbf{x}) \end{cases}$$

and from Green's first identity (A.41)

$$\iiint_\Omega \{\underbrace{\nabla\psi \cdot \nabla\psi}_{|\nabla\psi|^2} + \psi \underbrace{\nabla^2\psi}_{=0}\} d^3x = \oiint_{\partial\Omega} \psi\nabla\psi \cdot \mathbf{n}\, dS = 0$$

but since $|\nabla\psi| \geq 0$, we conclude that

$$\nabla\psi(\mathbf{x}) = 0 \quad \text{for } \mathbf{x} \in \Omega.$$

Then $\psi(\mathbf{x})$ is constant for $\mathbf{x} \in \Omega$, and since $\psi = 0$ on $\partial\Omega$, the value of this constant is zero. Finally

$$\psi(\mathbf{x}) = V_2(\mathbf{x}) - V_1(\mathbf{x}) = 0 \quad \text{for } \mathbf{x} \in \Omega$$

and $V_1 = V_2$ in Ω, which demonstrates that the solution of the Poisson–Dirichlet problem is unique.

A.5.3 *Uniqueness of the Solution to the Poisson–Neumann Problem*

The Poisson–Neumann problem reads

$$\begin{cases} \mathbf{x} \in \Omega: \quad \nabla^2 V(\mathbf{x}) = -\dfrac{\varrho(\mathbf{x})}{\epsilon_0} \\ \mathbf{x} \in \partial\Omega: \nabla V(\mathbf{x}) \cdot \mathbf{n}(\mathbf{x}) = -E_N(\mathbf{x}) \end{cases}$$

Suppose that $V_1(\mathbf{x})$ and $V_2(\mathbf{x})$ are two solutions of this problem, and define

$$\psi(\mathbf{x}) = V_2(\mathbf{x}) - V_1(\mathbf{x}).$$

It follows that $\psi(\mathbf{x})$ satisfies Laplace's equation with a homogeneous Neumann boundary condition

$$\begin{cases} \mathbf{x} \in \Omega: \quad \nabla^2 \psi(\mathbf{x}) = 0, \\ \mathbf{x} \in \partial\Omega: \nabla \psi(\mathbf{x}) \cdot \mathbf{n}(\mathbf{x}) = 0. \end{cases}$$

Green's first identity (A.41) gives

$$\iiint_\Omega |\nabla\psi|^2 d^3x = \oiint_{\partial\Omega} \psi \underbrace{\nabla\psi \cdot \mathbf{n}}_{=0} \, dS = 0$$

and we conclude $\nabla\psi = \mathbf{0}$, so $\psi = V_1 - V_2$ is constant in Ω. The two solutions V_1 and V_2 are equal up to an arbitrary constant which does not modify the resulting electric field. In this sense, the solution to the Poisson–Neumann problem is also unique.

A.6 Properties of the Legendre Polynomials

The main properties of Legendre polynomials that are useful in electrostatics are listed below.

1. For $l \in \mathbb{N}_0$, the Legendre polynomial P_l is a polynomial of order l, which is a solution of Legendre's differential equation

$$\frac{d}{dx}\left\{(1-x^2)\frac{dP_l(x)}{dx}\right\} + l(l+1)P_l(x) = 0 \tag{A.44}$$

for $|x| \leq 1$. P_l can be explicitly computed by Rodrigues formula:

$$\boxed{P_l(x) = \frac{1}{2^l l!}\frac{d^l}{dx^l}[(x^2-1)^l]} \tag{A.45}$$

with $P_l(1) = 1$ and $P_l(-1) = (-1)^l$ for every $l \in \mathbb{N}_0$. The first Legendre polynomials are

$$P_0(x) = 1$$
$$P_1(x) = x$$
$$P_2(x) = \frac{1}{2}\left(3x^2 - 1\right)$$

and satisfy the recursive formula

$$(l+1)P_{l+1}(x) = (2l+1)x P_l(x) - l P_{l-1}(x) \tag{A.46}$$

2. The Legendre polynomials are such that

$$P_l(-x) = (-1)^l P_l(x)$$

   and so if l is even (resp. odd), P_l is an even function (resp. odd function).
3. One has the identity

$$P_l(x) = \frac{1}{2l+1}\frac{d}{dx}[P_{l+1}(x) - P_{l-1}(x)]. \tag{A.47}$$

4. The set of Legendre polynomials $\{P_l\}_{l\in\mathbb{N}_0}$ is orthogonal in $[-1, 1]$. This means:

$$\int_{-1}^{1} P_{l'}(x)P_l(x)dx = \begin{cases} 0 & \text{if } l \neq l' \\ \dfrac{2}{(2l+1)} & \text{if } l = l' \end{cases} \tag{A.48}$$

5. The set of Legendre polynomials $\{P_l\}_{l\in\mathbb{N}_0}$ is a complete orthogonal set in $[-1, 1]$. In consequence, if $f : [-1, 1] \mapsto \mathbb{C}$ is such that $\int_{[-1,1]} |f(x)|^2 dx < \infty$, then it admits an expansion in Legendre polynomials

$$f(x) = \sum_{l=0}^{\infty} A_l P_l(x) \tag{A.49}$$

where the coefficients A_l are given by

$$A_l = \frac{2l+1}{2} \int_{-1}^{1} P_l(x) f(x)\, dx \tag{A.50}$$

6. We will admit the following result

$$\frac{1}{\sqrt{1 - 2xr + r^2}} = \sum_{l=0}^{\infty} P_l(x) r^l \quad r < 1 \tag{A.51}$$

which allows, in electrostatics, to expand $1/|\mathbf{x} - \mathbf{x}'|$ in terms of Legendre polynomials. Indeed, considering that in spherical coordinates $|\mathbf{x}| = r$, $|\mathbf{x}'| = r'$ and $\mathbf{x} \cdot \mathbf{x}' = rr' \cos\theta$, then:

$$\frac{1}{|\mathbf{x} - \mathbf{x}'|} = \frac{1}{\sqrt{r^2 + r'^2 - 2rr' \cos\theta}}.$$

Assuming, without loss of generality, that $r > r'$:

$$\frac{1}{|\mathbf{x} - \mathbf{x}'|} = \frac{1}{r\sqrt{1 + (r'/r)^2 - 2(r'/r) \cos\theta}}$$

and comparing with the identity (A.51), we obtain

$$\boxed{\frac{1}{|\mathbf{x} - \mathbf{x}'|} = \sum_{l=0}^{\infty} P_l(\cos\theta) \frac{r'^l}{r^{l+1}} \quad \text{for } r > r'.} \tag{A.52}$$

A.7   Electric Field of a Penning Trap

In this section we will derive the potential and the field generated by the quadripole configuration discussed in Sect. 8.7 of Chap. 8. Let us use cylindrical coordinates and place ourselves close to the center, as shown in Fig. A.14.

The potential generated by the ring at point $\mathbf{r} = r\mathbf{u}_r + z\mathbf{u}_z$ is given by

$$V(\mathbf{r})_{\text{ring}} = \frac{\lambda a}{4\pi\epsilon_0} \int_0^{2\pi} \frac{d\theta}{|z\mathbf{u}_z + r\mathbf{u}_r - a\mathbf{u}'_r|} \tag{A.53}$$

where θ is the angle between $\mathbf{u}_r$ and $\mathbf{u}'_r$. We can write the denominator inside the integral as

$$\frac{1}{|z\mathbf{u}_z + r\mathbf{u}_r - a\mathbf{u}'_r|} = \left(z^2 + r^2 + a^2 - 2ar\cos\theta\right)^{-1/2} = \frac{1}{a}\left(1 + \frac{z^2 + r^2}{a^2} - \frac{2r}{a}\cos\theta\right)^{-1/2}$$

Fig. A.14 Quadrupole configuration used in a Penning trap

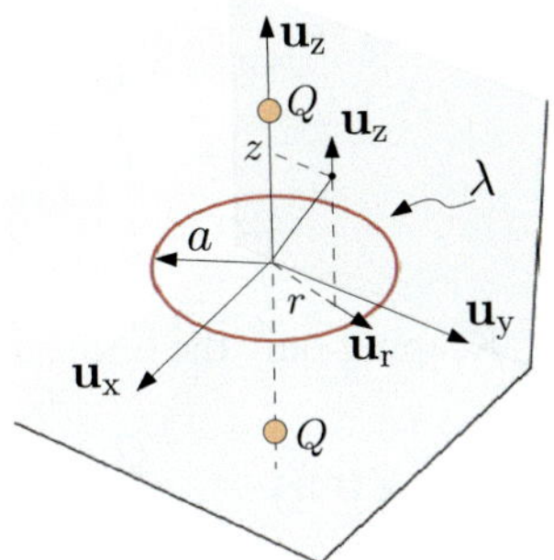

so that

$$V(\mathbf{r})_{\text{ring}} = \frac{\lambda}{4\pi\epsilon_0} \int_0^{2\pi} \left(1 + \frac{z^2 + r^2}{a^2} - \frac{2r}{a}\cos\theta\right)^{-1/2} d\theta$$

We now place ourselves close to the origin so that $r/a \ll 1$, $z/a \ll 1$ and use the following expansion $(1 + x)^{-1/2} = 1 - \frac{1}{2}x + \frac{3}{8}x^2 - \frac{5}{16}x^3 + \frac{35}{128}x^4 + \cdots$ and keep only the terms up to order four.

$$\left(1 + \frac{z^2 + r^2}{a^2} - \frac{2r}{a}\cos\theta\right)^{-1/2} \approx 1 - \frac{z^2 + r^2}{2a^2} + \frac{r}{a}\cos\theta$$
$$+ \frac{3}{8}\left(\frac{(z^2 + r^2)^2}{a^4} + \frac{4r^2\cos^2\theta}{a^2} - \frac{4r(z^2 + r^2)\cos\theta}{a}\right)$$
$$- \frac{5}{16}\left(\frac{12z^2r^2\cos^2\theta}{a^4} + \frac{12r^4\cos^2\theta}{a^4} - \frac{8r^3\cos^3\theta}{a^3}\right)$$
$$+ \frac{35}{128}\left(\frac{16r^4\cos^4\theta}{a^4}\right)$$

In the integral (A.53), the terms proportional to $\cos\theta$ and $\cos^3\theta$ vanish, and since $\int_0^{2\pi}\cos^2\theta\, d\theta = \pi$ and $\int_0^{2\pi}\cos^4\theta\, d\theta = \frac{3\pi}{4}$

$$V(\mathbf{r})_{\text{ring}} \approx \frac{\lambda}{2\epsilon_0}\left(1 - \frac{z^2}{2a^2} + \frac{r^2}{4a^2} - \frac{9}{8}\frac{r^2z^2}{a^4} + \frac{3}{8}\frac{z^4}{a^4} + \frac{9}{64}\frac{r^4}{a^4}\right) \tag{A.54}$$

We now proceed to the potential due to the two-point charges

$$V(\mathbf{r})_Q = \frac{Q}{4\pi\epsilon_0}\left(\frac{1}{\sqrt{(z-a)^2+r^2}} + \frac{1}{\sqrt{(z+a)^2+r^2}}\right)$$

$$= \frac{Q}{4\pi\epsilon_0 a}\left(\left(1+\frac{r^2+z^2}{a^2}-\frac{2z}{a}\right)^{-1/2} + \left(1+\frac{r^2+z^2}{a^2}+\frac{2z}{a}\right)^{-1/2}\right)$$

Keeping only the terms up to the fourth order

$$V(\mathbf{r})_Q \approx \frac{Q}{4\pi\epsilon_0 a}\left(2 - \frac{r^2}{a^2} + \frac{2z^2}{a^2} - 6\frac{z^2 r^2}{a^4} + 2\frac{z^4}{a^4} + \frac{3}{4}\frac{r^4}{a^4}\right) \qquad (A.55)$$

By comparing (A.54) and (A.55), we see that if $Q = -\frac{3\pi a\lambda}{8}$ the terms proportional to $z^2 r^2$, r^4 and r^4 cancel each other out. The total potential becomes, up to a constant

$$V(r,z) = \frac{7Q}{12\pi\epsilon_0 a}\left(\frac{2z^2}{a^2} - \frac{r^2}{a^2}\right)$$

Which gives an electric field

$$\mathbf{E} = -\nabla V = -\alpha z\mathbf{u}_z + \frac{\alpha r}{2}\mathbf{u}_r$$

with $\alpha = \frac{7Q}{3\pi\epsilon_0 a^3}$.

A.8 Properties of the Fourier Transform

The Fourier transform $\mathcal{F}$ is a linear application defined over the set of integrable functions. If f is integrable, then its Fourier transform $\mathcal{F}\{f\}$ is the function $\hat{f}$ defined for every $k \in \mathbb{R}$ by

$$\mathcal{F}\{f\}(k) = \hat{f}(k) = \frac{1}{2\pi}\int_{\mathbb{R}} f(x)e^{-ikx}\,dx \qquad (A.56)$$

This definition (A.56) is very restrictive in practice since it requires f to be integrable, indeed this is necessary for the integral in (A.56) to be well defined

$$|\hat{f}(k)| \le \frac{1}{2\pi}\int_{\mathbb{R}}|f(x)|\,dx < \infty$$

This definition can be extended to square-integrable functions and even to distributions. Indeed, for a distribution T one defines its Fourier transform by

$$(\mathcal{F}\{T\}, \varphi) = -(T, \hat{\varphi})$$

where φ is a test function that is infinitely differentiable and such that φ and all of its derivatives decay faster than any power at infinity. This restricts the distributions to a relatively broad subset called tempered distributions. Below we list the properties of the Fourier transform that are used in this book. Some of these properties have special requirements for their validity, but in the framework of tempered distributions they are always valid, although some precautions must be taken with the convolution properties, since the convolution between two distributions is not always well defined.

- Inverse

$$f(x) = \int_{\mathbb{R}} e^{+ikx} \mathcal{F}\{f\}(k)dk \tag{A.57}$$

Or, equivalently

$$2\pi \mathcal{F}\{\mathcal{F}\{f\}\}(u) = f(-u)$$

- Translation

$$\mathcal{F}\{f(x-a)\} = e^{-iak}\mathcal{F}(k) \tag{A.58}$$

- Modulation

$$\mathcal{F}\left\{e^{iax}f(x)\right\} = \mathcal{F}\{f\}(k-a) \tag{A.59}$$

- The Fourier transform exchanges multiplication and differentiation

$$\mathcal{F}\left\{\frac{df}{dx}\right\}(k) = ik\hat{f}(k)$$

$$\mathcal{F}\{-ixf\}(k) = \frac{d\hat{f}(k)}{dk}$$

- The Fourier transforms exchanges convolution and multiplication

$$\mathcal{F}\{f * g\}(k) = 2\pi\hat{g}(k)\hat{f}(k) \tag{A.60}$$

$$\mathcal{F}\{fg\}(k) = \hat{f} * \hat{g}(k)$$

- Plancherel–Parseval

$$2\pi \int_{\mathbb{R}} \hat{f}(k)\hat{g}(k)dk = \int_{\mathbb{R}} f(x)^* g(x)dx \tag{A.61}$$

- Uncertainty principle
  The width of f is given by $\Delta x = \sqrt{\langle x^2 \rangle - \langle x \rangle^2}$, where

$$\langle x^n \rangle = \frac{\int_{\mathbb{R}} x^n |f(x)|^2 dx}{\int_{\mathbb{R}} |f(x)|^2 dx}$$

Similarly, the width Δk of its Fourier transform is given by $\Delta k = \sqrt{\langle k^2 \rangle - \langle k \rangle^2}$, where

$$\langle k^n \rangle = \frac{\int_{\mathbb{R}} k^n |f(k)|^2 dk}{\int_{\mathbb{R}} |\hat{f}(k)|^2 dk}$$

Then, one has

$$\Delta x \, \Delta k \geq \frac{1}{2} \tag{A.62}$$

which means that the broader f is, the narrower its Fourier transform tends to be, and vice versa.
- The generalization of (A.56) to functions defined over $\mathbb{R}^3$ is

$$\mathcal{F}\{f\}(\mathbf{k}) = \hat{f}(\mathbf{k}) = \frac{1}{(2\pi)^3} \int_{\mathbb{R}^3} f(\mathbf{x}) e^{-i\mathbf{k}\cdot\mathbf{x}} d^3 x. \tag{A.63}$$

A.9 Reminders on Energy and Potentials in Thermodynamics

The goal of this section is to serve as reminders on the concepts of thermodynamics. In particular we remind the definition and physical interpretation of the (Helmholtz) free energy and free enthalpy (also called Gibbs free energy).

A.9.0.1 First Principle

In a mechanical conservative system, the total energy is shared between kinetic and potential energy. Equilibrium states are generally found by looking for minima of the potential energy. For thermodynamical systems, the equivalent principle of energy conservation is the first principle, which states that in a transformation from initial state i to final state f, the variation of energy of the system is equal to the energy received in the form of work W performed by external forces acting on the system and in the form of heat Q exchanged with the environment

$$\Delta U = U(f) - U(i) = W + Q.$$

The energy function U is a state function, that depends only on the state of the system, and therefore, its variation only depends on the initial and final states but not on the path followed between.

A.9.0.2 Second Principle

The second principle states that if the system is in thermal contact with the environment at a constant temperature T, there exists another state function, the entropy S, such that the exchanged heat during the transformation from i to f satisfies

$$Q = \int_i^f \delta Q \leq T(S(f) - S(i)). \tag{A.64}$$

This means that there is an upper limit to the amount of heat the system can receive from the environment, and this limit determines the entropy of the system.

A.9.0.3 The Free Energy

Introducing inequality (A.64) into the first principle leads to

$$U(f) - U(i) - T(S(f) - S(i)) \leq W,$$

which means that the amount of work that the system can receive from external forces during the transformation is an upper limit for the variation of the function $U - TS$. This function is another state function, the free energy $F = U - TS$, also called the Helmholtz free energy, which satisfies

$$F(f) - F(i) \leq W. \tag{A.65}$$

Now if the transformation is reversible, the equality sign holds in (A.65), that is, in a reversible transformation between two states which have a temperature equal to that of the environment, the work of external forces received by the system is equal to the variation of free energy of the system. In other words, for reversible transformations with heat exchange with the environment at temperature T, the free energy in a thermodynamical system plays a role analogous to that of energy for mechanical systems. In addition, for systems that do not exchange work with the environment $(W = 0)$,[1] the free energy cannot increase and consequently, reaches a minimum when the system is in a state of stable equilibrium, similarly to a dynamically iso-

---

[1] This is the case, for example, for a gas undergoing an isochore transformation, that is, a gas of uniform pressure enclosed in a container of fixed volume.

lated mechanical system that exchanges no work from the environment and that has equilibrium states corresponding to minima of its potential energy.

A.9.0.4 Natural Variables of State Functions

The free energy plays the role of a potential in thermodynamics. As with any potential, it is used to find minima, and thus equilibrium states of the system. In this aim, it should be expressed as a function of its natural variables, which denote the independent thermodynamical state variables describing the system. Since thermodynamical systems are usually governed by an equation of state, that is, a relation between the thermodynamical variables, the number of independent variables is reduced accordingly. For example, an ideal gas satisfies the law of ideal gases $pv \propto T$ connecting the pressure p to the temperature T and volume v. Hence, a set of two independent variables, e.g., (v, T) or (p, T), is sufficient to define the state of a system and the dependence of state functions.

Although the choice of independent variables is arbitrary, it is convenient to use a set of independent variables suited to the system transformation under investigation. The set (v, T) is convenient for studying isochore and isothermal transformations. The set (p, T) is convenient for studying isobar and isotherm transformations. For a gas, the free energy $F(v, T)$ is appropriately expressed as a function of the volume v and temperature T. For this reason, it is also called the potential for isotherm and isochore transformations. v and T are called the *natural variables* of the free energy.

Applied to infinitesimal reversible transformations, the principles of thermodynamics allow us to express differential forms of state functions like the free energy. The state function can then be expressed as a function of its natural variables by integration, using the fact that mathematically, a differential form for a state function corresponds to an exact differential.

A.9.0.5 Exact Differentials

Consider a differential form dz of two independent variables x and y:

$$dz = P(x, y)dx + Q(x, y)dy$$

This differential form is exact (or perfect) if it is the differential of a function of x and y. If that is the case, P and Q must satisfy

$$\frac{\partial P(x, y)}{\partial y} = \frac{\partial Q(x, y)}{\partial x} \tag{A.66}$$

and vice versa. Equation (A.66) is the Schwartz property. When it is satisfied, it is possible to integrate the differential form and find a function of x and y that satisfy this equation. In physics, such functions are called state functions. For example, for an ideal gas, the differential form for the free energy reads

$$dF = -pdv - SdT,$$

and the equation of state for the pressure $p(v, T) = nRT/v$ and energy $U(T) = C_v T$ (it is a property of ideal gases that U is independent of v), allows for finding the entropy $S(v, T)$ and in turn the free energy $F(v, T)$.

A.9.0.6   The Free Enthalpy

The free energy is not the only thermodynamic potential that can be defined. Taking the example of a gas, the free energy $F(v, T)$ is useful to find equilibrium states for isothermal and isochore transformations at constant volume and temperature. If in contrast a transformation is isothermal and isobaric, that is, if it occurs at constant temperature T and pressure p, with a change of volume from $v(i)$ to $v(f)$, the work of external forces is that of pressure forces $-p\Delta v$, leading to the inequality

$$W = -p(v(f) - v(i)) \geq F(f) - F(i)$$

Defining a new state function G, the free enthalpy, or Gibbs free energy,

$$G = F + pv = U - TS + pv,$$

we see that
$$G(f) \leq G(i)$$

In other words the free enthalpy G cannot increase for isothermal and isobaric transformations. Therefore it plays the role of a thermodynamic potential at constant pressure. For these transformations, minima of the free enthalpy are stable equilibrium states.

Differentiating the definition of the free enthalpy, we find

$$dG = dF + pdv + vdp$$

and introducing $dF = -pdv - SdT$ leads to

$$dG = vdp - SdT$$

which shows that the natural variables for the free enthalpy $G(p, T)$ are the pressure p and temperature T, that is, $G(p, T)$ is the thermodynamic potential for isobar and isothermal transformations.

A.10   Dispersion, Temporal Response and Causality

A.10.1   Dispersion

A medium is dispersive if its response depends on frequency. The cause for dispersion is quite general, it comes from the non-instantaneous response of the medium to an external excitation. That is the case for the polarization $\mathbf{P}$ of a dielectric medium under the influence of an electromagnetic wave (or the conduction current in the case of a conductor). Whereas in the static regime, one may write $\mathbf{P} = \epsilon_0 \chi \mathbf{E}$, where χ is the susceptibility, in the time-dependent regime we generally have

$$\mathbf{P}(\mathbf{x}, t) \neq \epsilon_0 \chi \mathbf{E}(\mathbf{x}, t),$$

since friction and inertia are responsible for a retarded response of the medium polarization to the varying electric field. The polarization vector at instant t depends not only on the value of the electric field at t but also on the history of the electric field, that is, on the previous values of the field at times preceding t. For a linear, homogeneous, and isotropic medium, the polarization generally reads

$$\mathbf{P}(\mathbf{x}, t) = \epsilon_0 \int_{-\infty}^{\infty} \chi(t') \mathbf{E}(\mathbf{x}, t - t') dt' \tag{A.67}$$

we recognize in Eq. (A.67) a convolution product which makes dealing with Maxwell's equations difficult a priori. On the other hand, $\mathbf{P}$ can always be written as a superposition of sinusoidal waves due to Fourier's theorem

$$\mathbf{P}(\mathbf{x}, t) = \int_{-\infty}^{\infty} \underline{\mathbf{P}}(\mathbf{x}, \omega) e^{i\omega t} dt$$

where $\underline{\mathbf{P}}(\mathbf{x}, \omega)$ is the Fourier transform of $\mathbf{P}$ with respect to time. Equation (A.67) and the property (A.60) of the Fourier transform then implies that

$$\underline{\mathbf{P}}(\mathbf{x}, \omega) = \epsilon_0 \underline{\chi}(\omega) \underline{\mathbf{E}}(\mathbf{x}, \omega),$$

where $\underline{\chi}(\omega)$ is the Fourier transform of χ. This last equality and the linearity of Maxwell's equations justifies working with monochromatic plane waves: the response of the medium to a sinusoidal plane wave $\underline{\mathbf{E}} = \underline{\mathbf{E}}(\omega) e^{i(\underline{k}x - \omega t)}$ simply reads

$$\underline{\mathbf{P}} = \left(\underline{\epsilon}(\omega) - \epsilon_0\right) \underline{\mathbf{E}} = \epsilon_0 \underline{\chi}(\omega) \underline{\mathbf{E}}.$$

A.10.2   Causality

This section presents general properties of the susceptibility that are a consequence of the causality principle. Namely, relations called Kramers–Kronig relations are established between the real and imaginary parts of the complex susceptibility. Complex analysis, integration in the complex plane, and mathematical notions about distributions constitute prerequisites to follow the detailed derivation of Kramers–Kronig relations.

Causality imposes that in Eq. (A.67), $\chi(t) = 0$ if $t < 0$. Otherwise, the polarization at time t would depend on future values of the electric field, violating the causality principle that states that any effect cannot precede its cause. We will now show that causality introduces a constraint linking the real and imaginary parts of the complex susceptibility $\underline{\chi}(\omega)$ (and therefore, of the complex permittivity). Fourier's inversion formula together with $\chi = 0$ for $t < 0$ implies

$$\underline{\chi}(\omega) = \frac{1}{2\pi} \int_0^\infty \chi(t) e^{+i\omega t} dt.$$

To calculate the latter integral, we may use an extension of the complex susceptibility $\underline{\chi}$ to the complex frequency plane, that is, we consider $\omega = \omega' + i\omega''$, and note that for $\mathrm{Im}\{\omega\} = \omega'' > 0$, we have

$$|\underline{\chi}(\omega)| < \frac{1}{2\pi} \int_0^\infty |\chi(t) e^{+i\omega t}| dt < \frac{1}{2\pi} \int_0^\infty |\chi(t)| dt,$$

so that $\underline{\chi}(\omega)$ extended to the complex plane is bounded for every ω such that $\mathrm{Im}\{\omega\} > 0$ and therefore has no poles in this half plane. The function $\underline{\chi}(\omega)$ is therefore analytic. We now define the function $f(\omega) = \underline{\chi}(\omega)/(\omega - \omega_0)$ where ω_0 is real. This function has a pole in ω_0 and we can apply Cauchy's residues theorem by integrating f over the closed curve shown in Fig. A.15.

The integral over the half-circle tends to zero when $R \rightarrow \infty$ provided that $\lim_{|\omega| \to \infty} |\underline{\chi}(\omega)| = 0$, which is the case since all materials become transparent at high frequencies due to the inertia of charged particles. Cauchy's theorem then gives

$$i\pi \underline{\chi}(\omega_0) + P \int \frac{\underline{\chi}(\omega')}{\omega' - \omega_0} d\omega' = 2i\pi \underline{\chi}(\omega_0)$$

Fig. A.15 Close curve used
to apply Cauchy's theorem

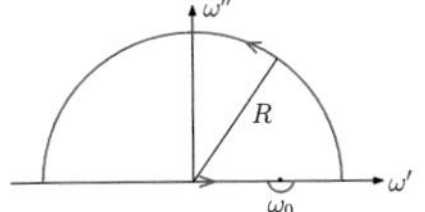

were P stands for the principal value of the integral. We obtain

$$\underline{\chi}(\omega_0) = \frac{1}{i\pi} P \int_{-\infty}^{\infty} \frac{\underline{\chi}(\omega')}{\omega' - \omega_0} d\omega'$$

and by taking the real and imaginary part and by writing $\underline{\chi}(\omega) = \underline{\epsilon}(\omega)/\epsilon_0 - 1$, we obtain the Kramers–Kronig relations,

$$\boxed{\begin{aligned} \frac{\epsilon_R(\omega_0)}{\epsilon_0} &= 1 + \frac{1}{\pi} P \int_{-\infty}^{\infty} \frac{\epsilon_I(\omega')/\epsilon_0}{\omega' - \omega_0} d\omega', \\ \frac{\epsilon_I(\omega_0)}{\epsilon_0} &= -\frac{1}{\pi} P \int_{-\infty}^{\infty} \frac{\epsilon_R(\omega')/\epsilon_0 - 1}{\omega' - \omega_0} d\omega'. \end{aligned}}$$

(A.68)

Considering finally that $\epsilon(t)$ is real, we have $\underline{\epsilon}(-\omega) = \underline{\epsilon}^*(\omega)$ and

$$\frac{\epsilon_R(\omega_0)}{\epsilon_0} = 1 + \frac{2}{\pi} P \int_{-\infty}^{\infty} \frac{\omega' \epsilon_I(\omega')/\epsilon_0}{\omega'^2 - \omega_0^2} d\omega'. \tag{A.69}$$

This relation allows us, for example, to determine the real part of the permittivity from absorption measurements, which provide access to the imaginary part of the permittivity.

Remark The Kramers–Kronig relations may be more easily obtained by using some well-known results of Fourier analysis. Indeed, since causality forces the function χ to be zero for $t < 0$, this means that χ can be written as

$$\chi(t) = \chi(t)\Theta(t),$$

where Θ is the Heaviside function, that is,

$$\Theta(t) = \begin{cases} 1 & \text{if } t > 0, \\ 0 & \text{if } t < 0. \end{cases}$$

It follows that

$$\underline{\chi}(\omega) = (\underline{\Theta} * \underline{\chi})(\omega),$$

where $*$ stands for convolution. The Fourier transform of the Heaviside (in the framework of distributions) reads

$$\underline{\Theta}(\omega) = \frac{1}{2\pi i} \text{p.v} \frac{1}{\omega} + \frac{1}{2}\delta(\omega),$$

where p.v stands for Cauchy's principal value. Finally,

$$\chi(\omega) = \frac{1}{2}\chi(\omega) - \frac{i}{2\pi}P\int_{-\infty}^{\infty}\frac{\chi(\omega')}{\omega - \omega'}d\omega',$$

which leads to the Kramers–Kronig relations.

A.11 Green's Function for the Retarded Potentials

The propagation equation for the Green function is

$$\left(\nabla^2 - \frac{1}{c^2}\frac{\partial^2}{\partial t^2}\right)G(\mathbf{x} - \mathbf{x}', t - t') = -4\pi\delta(t - t')\delta(\mathbf{x} - \mathbf{x}') \qquad (A.70)$$

This equation can be solved in the Fourier domain. Indeed, if $G(\mathbf{x} - \mathbf{x}', t - t')$ is the Fourier transform with respect to $t - t'$ of $\hat{G}(\mathbf{x} - \mathbf{x}', \omega)$

$$G(\mathbf{x} - \mathbf{x}', t - t') = \frac{1}{2\pi}\int_{-\infty}^{+\infty}\hat{G}(\mathbf{x} - \mathbf{x}', \omega)e^{-i\omega(t-t')}d\omega,$$

applying the d'Alembert operator $\nabla^2 - \dfrac{1}{c^2}\dfrac{\partial^2}{\partial t^2}$ to $G(\mathbf{x} - \mathbf{x}', t - t')$ and commuting it with the integral with respect to ω, we obtain

$$\left(\nabla^2 - \frac{1}{c^2}\frac{\partial^2}{\partial t^2}\right)\left(\frac{1}{2\pi}\int_{-\infty}^{+\infty}\hat{G}(\mathbf{x} - \mathbf{x}', \omega)e^{-i\omega(t-t')}d\omega\right)$$

$$= \frac{1}{2\pi}\int_{-\infty}^{+\infty}\left(\nabla^2 - \frac{1}{c^2}\frac{\partial^2}{\partial t^2}\right)\hat{G}(\mathbf{x} - \mathbf{x}', \omega)e^{-i\omega(t-t')}d\omega.$$

Using Eq. (19.6), we find

$$\frac{1}{2\pi}\int_{-\infty}^{+\infty}\left(-\frac{1}{c^2}\frac{\partial^2}{\partial t^2} + \nabla^2\right)\hat{G}(\mathbf{x} - \mathbf{x}', \omega)e^{-i\omega(t-t')}d\omega$$

$$= -4\pi\underbrace{\frac{1}{2\pi}\int_{-\infty}^{+\infty}e^{-i\omega(t-t')}\delta(\mathbf{x} - \mathbf{x}')d\omega,}_{=\delta(\mathbf{x}-\mathbf{x}')\delta(t-t')}$$

that is,

$$\frac{1}{2\pi}\int_{-\infty}^{+\infty}d\omega\left\{\left(\frac{\omega^2}{c^2} + \nabla^2\right)\hat{G}(\mathbf{x} - \mathbf{x}', \omega) + 4\pi\delta(\mathbf{x} - \mathbf{x}')\right\}e^{-i\omega(t-t')} = 0,$$

leading to the equation satisfied by the (inverse) Fourier transform of the Green function

$$\left(\nabla^2 + k^2\right) \hat{G}(\mathbf{x} - \mathbf{x}', \omega) = -4\pi \delta(\mathbf{x} - \mathbf{x}'),$$

where we introduced the (real) variable

$$k \equiv \frac{\omega}{c},$$

which will include the frequency dependence of the Green function. Note the relative similarity of the result with the Laplace equation, where the source is a point charge located at $\mathbf{x}'$. The solution to this problem must have spherical symmetry with respect to $\mathbf{x}'$, we will then look for a solution of the form

$$\hat{G}(\mathbf{x} - \mathbf{x}', \omega) = \hat{G}(|\mathbf{x} - \mathbf{x}'|, \omega).$$

In spherical coordinates, we obtain

$$\nabla^2 \hat{G}(r, \omega) = \frac{1}{r} \frac{d^2}{dr^2} r \hat{G}(r, \omega),$$

and so

$$\frac{1}{r} \frac{d^2}{dr^2} r \hat{G}(r, \omega) + k^2 \hat{G}(r, \omega) = -4\pi \delta(r)$$

The function $r\hat{G}(r, \omega)$ satisfies a second-order ordinary differential equation which is that for a harmonic oscillator with pseudo-time r and pseudo-frequency $k = \omega/c$ for all $r \neq 0$. Hence the solution is a linear superposition of the two components

$$\hat{G}_{\pm}(r, \omega) = \frac{g_{\pm}}{r} \exp(\pm ikr).$$

The constants $g_{\pm}$ are found from the boundary condition at $r = 0$, which is obtained by integrating the original equation

$$\left(\nabla^2 + k^2\right) \hat{G}(r, \omega) = -4\pi \delta(r).$$

over the volume Ω bounded by a sphere of radius r. We find

$$\iiint_{\Omega} \nabla \cdot \nabla \hat{G}(r, \omega) \, d^3x = -4\pi \iiint_{\Omega} \delta(r) d^3x - k^2 \iiint_{\Omega} \hat{G}(r, \omega) \, d^3x,$$

that is, after using Ostrogradsy's theorem to transform the left-hand side into a surface integral and using the property $\iiint_{\Omega} \delta(r) d^3x = 1$

$$(\nabla \hat{G}(r, \omega)) \cdot \mathbf{u}_r 4\pi r^2 = -4\pi - k^2 \int_0^r \hat{G}(r, \omega) 4\pi r^2 dr.$$

For the left-hand side:

$$(\nabla \hat{G}(r, \omega)) \cdot \mathbf{u}_r 4\pi r^2 = 4\pi r^2 \frac{\partial \hat{G}}{\partial r} = 4\pi g_\pm e^{\pm ikr} \left(\pm ikr - 1 \right).$$

For the last term on the right-hand side, integrating by parts, we find

$$- k^2 \int_0^r \hat{G}(r, \omega) 4\pi r^2 dr = -k^2 4\pi g_\pm \int_0^r e^{\pm ikr} r \, dr = 4\pi g_\pm \left(\pm ikr e^{\pm ikr} - e^{\pm ikr} + 1 \right).$$

Identifying both sides, we finally find

$$g_\pm = 1.$$

$$\hat{G}(r, \omega) = \frac{1}{r} e^{\pm ikr}.$$

Therefore

$$\hat{G}(\mathbf{x} - \mathbf{x}', \omega) = \frac{1}{|\mathbf{x} - \mathbf{x}'|} e^{\pm i(\omega/c)|\mathbf{x}-\mathbf{x}'|}$$

The inverse Fourier transform yields

$$G(\mathbf{x} - \mathbf{x}', t - t') = \frac{1}{2\pi} \int_{-\infty}^{+\infty} \hat{G}(\mathbf{x} - \mathbf{x}', \omega) e^{-i\omega(t-t')} d\omega = \frac{1}{2\pi} \int_{-\infty}^{+\infty} \frac{e^{\pm i(\omega/c)|\mathbf{x}-\mathbf{x}'|}}{|\mathbf{x} - \mathbf{x}'|} e^{-i\omega(t-t')} d\omega$$

$$= \frac{1}{2\pi |\mathbf{x} - \mathbf{x}'|} \int_{-\infty}^{+\infty} e^{-i\omega(t-t' \mp |\mathbf{x}-\mathbf{x}'|/c)} d\omega = \frac{\delta(t - t' \mp |\mathbf{x} - \mathbf{x}'|/c)}{|\mathbf{x} - \mathbf{x}'|}.$$

A.11.1 Liénard–Wiechert Fields Generated by a Moving Point Charge

In Chap. 19, the retarded potentials for a point charge were determined to be

$$V(\mathbf{x}, t) = \frac{q}{4\pi \epsilon_0} \frac{1}{|\mathbf{x} - \mathbf{r}(t_r)| - \frac{\mathbf{v}(t_r)}{c} \cdot (\mathbf{x} - \mathbf{r}(t_r))} \tag{A.71}$$

$$\mathbf{A}(\mathbf{x}, t) = \frac{q}{4\pi \epsilon_0 c^2} \frac{\mathbf{v}(t_r)}{|\mathbf{x} - \mathbf{r}(t_r)| - \frac{\mathbf{v}(t_r)}{c} \cdot (\mathbf{x} - \mathbf{r}(t_r))} \tag{A.72}$$

with $t_r = t - \frac{|\mathbf{x} - \mathbf{r}(t_r)|}{c}$. Let us now compute the electromagnetic field generated by this charge at instant t. For the electric field we use $\mathbf{E} = -\nabla V - \frac{\partial \mathbf{A}}{\partial t}$. For this, we calculate

$$\nabla |\mathbf{x} - \mathbf{r}(t_r)| = \nabla \sqrt{(x - r_x(t_r))^2 + (y - r_y(t_r))^2 + (z - r_z(t_r))^2}$$

let us compute the j component of this gradient

$$\frac{\partial |\mathbf{x} - \mathbf{r}(t_r)|}{\partial x_j} = \frac{\sum_{i=x,y,z}(x_i - r_i(t_r))\frac{\partial}{\partial x_j}(x_i - r_i(t_r))}{|\mathbf{x} - \mathbf{r}(t_r)|} \tag{A.73}$$

now, $\frac{\partial}{\partial x_j}(x_i - r_i(t_r)) = \delta_{ij} - \frac{\partial r_i(t_r)}{\partial t_r}\frac{\partial t_r}{\partial x_j}$. Since $\frac{\partial r_i(t_r)}{\partial t_r} = v_i(t_r)$ and $t_r = t - |\mathbf{x} - \mathbf{r}(t_r)|/c$, we find

$$\frac{\partial}{\partial x_j}(x_i - r_i(t_r)) = \delta_{ij} + \frac{v_i(t_r)}{c}\frac{\partial}{\partial x_j}|\mathbf{x} - \mathbf{r}(t_r)|$$

replacing into (A.73) gives

$$\frac{\partial |\mathbf{x} - \mathbf{r}(t_r)|}{\partial x_j} = \frac{\sum_{i=x,y,z}(x_i - r_i(t_r))(\delta_{ij} + \frac{v_i(t_r)}{c}\frac{\partial}{\partial x_j}|\mathbf{x} - \mathbf{r}(t_r)|)}{|\mathbf{x} - \mathbf{r}(t_r)|}$$

$$= \frac{x_j - r_j(t_r)}{|\mathbf{x} - \mathbf{r}(t_r)|} + \frac{\mathbf{v}(t_r) \cdot (\mathbf{x} - \mathbf{r}(t_r))}{c|\mathbf{x} - \mathbf{r}(t_r)|}\frac{\partial}{\partial x_j}|\mathbf{x} - \mathbf{r}(t_r)|$$

We can solve for $\frac{\partial}{\partial x_j}|\mathbf{x} - \mathbf{r}(t_r)|$ and obtain

$$\frac{\partial |\mathbf{x} - \mathbf{r}(t_r)|}{\partial x_j} = \frac{\frac{x_j - r_j(t_r)}{|\mathbf{x} - \mathbf{r}(t_r)|}}{1 - \frac{\mathbf{v}(t_r)}{c} \cdot \frac{\mathbf{x} - \mathbf{r}(t_r)}{|\mathbf{x} - \mathbf{r}(t_r)|}}$$

which gives the gradient

$$\nabla |\mathbf{x} - \mathbf{r}(t_r)| = \frac{\frac{\mathbf{x} - \mathbf{r}(t_r)}{|\mathbf{x} - \mathbf{r}(t_r)|}}{1 - \frac{\mathbf{v}(t_r)}{c} \cdot \frac{\mathbf{x} - \mathbf{r}(t_r)}{|\mathbf{x} - \mathbf{r}(t_r)|}}$$

Similarly, we need to calculate $\nabla(\mathbf{v}(t_r) \cdot (\mathbf{x} - \mathbf{r}(t_r)))$. For this, let us consider the j component

$$\frac{\partial}{\partial j}(\mathbf{v}(t_r) \cdot (\mathbf{x} - \mathbf{r}(t_r))) = \sum_{i=x,y,z} \frac{\partial v_i(t_r)}{\partial x_j}(x_i - r_i(t_r)) + v_i(t_r)\frac{\partial}{\partial x_j}(x_i - r_i(t_r))$$

$$= \sum_{i=x,y,z} \frac{\partial v_i(t_r)}{\partial t_r}\frac{\partial t_r}{\partial x_j}(x_i - r_i(t_r)) + \sum_{i=x,y,z} v_i(t_r)\left(\delta_{ij} - v_i(t_r)\frac{\partial t_r}{\partial x_j}\right)$$

since $\partial t_r/\partial x_j = -\frac{1}{c}\frac{\partial}{\partial x_j}|\mathbf{x} - \mathbf{r}(t_r)|$ and $\partial v_i/\partial t_r = a_i(t_r)$ with $\mathbf{a}$ the acceleration of the charge, we have

$$\frac{\partial}{\partial j}(\mathbf{v}(t_r) \cdot (\mathbf{x} - \mathbf{r}(t_r))) = v_j(t_r) + \frac{\partial}{\partial x_j}|\mathbf{x} - \mathbf{r}(t_r)|\left(\frac{|\mathbf{v}(t_r)|^2}{c} - \mathbf{a}(t_r) \cdot \frac{\mathbf{x} - \mathbf{r}(t_r)}{c}\right)$$

$$= v_j(t_r) + \frac{\frac{\mathbf{x}-\mathbf{r}(t_r)}{|\mathbf{x}-\mathbf{r}(t_r)|}}{1 - \frac{\mathbf{v}(t_r)}{c} \cdot \frac{\mathbf{x}-\mathbf{r}(t_r)}{|\mathbf{x}-\mathbf{r}(t_r)|}}\left(\frac{|\mathbf{v}(t_r)|^2}{c} - \mathbf{a}(t_r) \cdot \frac{\mathbf{x} - \mathbf{r}(t_r)}{c}\right)$$

We then have found

$$\frac{1}{c}\nabla(\mathbf{v}(t_r) \cdot (\mathbf{x} - \mathbf{r}(t_r))) = \frac{\mathbf{v}(t_r)}{c} + \left(\frac{|\mathbf{v}(t_r)|^2}{c^2} - \mathbf{a}(t_r) \cdot \frac{\mathbf{x} - \mathbf{r}(t_r)}{c^2}\right)\frac{\frac{\mathbf{x}-\mathbf{r}(t_r)}{|\mathbf{x}-\mathbf{r}(t_r)|}}{1 - \frac{\mathbf{v}(t_r)}{c} \cdot \frac{\mathbf{x}-\mathbf{r}(t_r)}{|\mathbf{x}-\mathbf{r}(t_r)|}}$$

From this, we have the expression for $-\nabla V$ by applying the chain rule

$$-\nabla V(\mathbf{x}) = \frac{q}{4\pi\epsilon_0}\frac{\nabla\left(|\mathbf{x} - \mathbf{r}(t_r)| - \frac{1}{c}\nabla(\mathbf{v}(t_r) \cdot (\mathbf{x} - \mathbf{r}(t_r)))\right)}{\left(|\mathbf{x} - \mathbf{r}(t_r)| - \frac{\mathbf{v}(t_r)}{c} \cdot (\mathbf{x} - \mathbf{r}(t_r))\right)^2}$$

Replacing

$$-\nabla V(\mathbf{x}) = \frac{q}{4\pi\epsilon_0|\mathbf{x} - \mathbf{r}(t_r)|^2}$$

$$\times \frac{\frac{\mathbf{x}-\mathbf{r}(t_r)}{|\mathbf{x}-\mathbf{r}(t_r)|}\left(1 - \frac{|\mathbf{v}(t_r)|^2}{c^2} + \mathbf{a}(t_r) \cdot \frac{\mathbf{x}-\mathbf{r}(t_r)}{c^2}\right) - \frac{\mathbf{v}(t_r)}{c}\left(1 - \frac{\mathbf{v}(t_r)}{c} \cdot \frac{\mathbf{x}-\mathbf{r}(t_r)}{|\mathbf{x}-\mathbf{r}(t_r)|}\right)}{\left(1 - \frac{\mathbf{v}(t_r)}{c} \cdot \frac{\mathbf{x}-\mathbf{r}(t_r)}{|\mathbf{x}-\mathbf{r}(t_r)|}\right)^3}$$

$$\tag{A.74}$$

For the calculation of $\partial\mathbf{A}/\partial t$, we have

$$-\frac{\partial\mathbf{A}}{\partial t} = -\frac{q}{4\pi\epsilon_0 c^2}\left(\frac{\frac{\partial\mathbf{v}(t_r)}{\partial t}}{|\mathbf{x} - \mathbf{r}(t_r)| - \frac{\mathbf{v}(t_r)}{c} \cdot (\mathbf{x} - \mathbf{r}(t_r))}\right.$$

$$\left. - \frac{\mathbf{v}(t_r)\frac{\partial}{\partial t}\left(|\mathbf{x} - \mathbf{r}(t_r)| - \frac{\mathbf{v}(t_r)}{c} \cdot (\mathbf{x} - \mathbf{r}(t_r))\right)}{\left(|\mathbf{x} - \mathbf{r}(t_r)| - \frac{\mathbf{v}(t_r)}{c} \cdot (\mathbf{x} - \mathbf{r}(t_r))\right)^2}\right)$$

Here, we need to evaluate $\frac{\partial\mathbf{v}(t_r)}{\partial t} = \frac{\partial\mathbf{v}(t_r)}{\partial t_r}\frac{\partial t_r}{\partial t} = \mathbf{a}(t_r)\frac{\partial t_r}{\partial t}$ with

$$\frac{\partial t_r}{\partial t} = 1 - \frac{1}{c}\frac{\partial |\mathbf{x} - \mathbf{r}(t_r)|}{\partial t} = 1 + \frac{1}{c}\frac{\sum_{i=x,y,z}(x_i - r_i(t_r))\frac{\partial r_i(t_r)}{\partial t}}{|\mathbf{x} - \mathbf{r}(t_r)|}$$

$$= 1 + \frac{1}{c}\frac{\sum_{i=x,y,z}(x_i - r_i(t_r))v_i(t_r)\frac{\partial t_r}{\partial t}}{|\mathbf{x} - \mathbf{r}(t_r)|} = 1 + \frac{\mathbf{v}(t_r)\cdot(\mathbf{x} - \mathbf{r}(t_r))}{c|\mathbf{x} - \mathbf{r}(t_r)|}\frac{\partial t_r}{\partial t}$$

From this, we obtain

$$\frac{\partial t_r}{\partial t} = \frac{1}{\left(1 - \frac{\mathbf{v}}{c}\cdot\frac{\mathbf{x}-\mathbf{r}(t_r)}{|\mathbf{x}-\mathbf{r}(t_r)|}\right)}$$

so that

$$\frac{\partial |\mathbf{x} - \mathbf{r}(t_r)|}{\partial t} = \frac{-\sum_{i=x,y,z}(x_i - r_i(t_r))\frac{\partial r_i}{\partial t}}{|\mathbf{x} - \mathbf{r}(t_r)|}$$

$$= \frac{-\sum_{i=x,y,z}(x_i - r_i(t_r))v_i(t_r)\frac{\partial t_r}{\partial t}}{|\mathbf{x} - \mathbf{r}(t_r)|} = -\frac{\mathbf{v}\cdot(\mathbf{x} - \mathbf{r}(t_r))}{|\mathbf{x} - \mathbf{r}(t_r)|}\frac{\partial t_r}{\partial t}$$

$$= -\mathbf{v}\cdot\frac{\frac{\mathbf{x}-\mathbf{r}(t_r)}{|\mathbf{x}-\mathbf{r}(t_r)|}}{1 - \frac{\mathbf{v}}{c}\cdot\frac{\mathbf{x}-\mathbf{r}(t_r)}{|\mathbf{x}-\mathbf{r}(t_r)|}}$$

$$\frac{1}{c}\frac{\partial \mathbf{v}(t_r)\cdot(\mathbf{x} - \mathbf{r}(t_r))}{\partial t} = \frac{1}{c}\sum_{i=x,y,z}\frac{\partial}{\partial t}v_i(t_r)(x_i - r_i(t_r))$$

$$= \frac{1}{c}\sum_{i=x,y,z}\frac{\partial v_i(t_r)}{\partial t}(x_i - r_i(t_r)) - \frac{1}{c}\sum_{i=x,y,z}v_i(t_r)\frac{\partial r_i(t_r)}{\partial t}$$

$$= \frac{1}{c}\sum_{i=x,y,z}a_i(t_r)\frac{\partial t_r}{\partial t}(x_i - r_i(t_r)) - \frac{1}{c}\sum_{i=x,y,z}v_i^2(t_r)\frac{\partial t_r}{\partial t}$$

$$= \left(\frac{\mathbf{a}(t_r)}{c}(\mathbf{x} - \mathbf{r}(t_r)) - \frac{|\mathbf{v}(t_r)|^2}{c}\right)\frac{1}{1 - \frac{\mathbf{v}}{c}\cdot\frac{\mathbf{x}-\mathbf{r}(t_r)}{|\mathbf{x}-\mathbf{r}(t_r)|}}$$

We finally obtain the term $-\frac{\partial \mathbf{A}}{\partial t}$

$$-\frac{\partial \mathbf{A}(\mathbf{r}, t)}{\partial t} = \frac{q}{4\pi\epsilon_0|\mathbf{x} - \mathbf{r}(t_r)|^2}$$

$$\times \frac{\frac{\mathbf{v}}{c}\cdot\left(-\frac{\mathbf{v}}{c}\cdot\frac{\mathbf{x}-\mathbf{r}(t_r)}{|\mathbf{x}-\mathbf{r}(t_r)|} + \frac{|\mathbf{v}(t_r)|^2}{c^2} - \frac{\mathbf{a}}{c^2}\cdot(\mathbf{x} - \mathbf{r}(t_r))\right) - |\mathbf{x} - \mathbf{r}(t_r)|\frac{\mathbf{a}(t_r)}{c^2}\left(1 - \frac{\mathbf{v}}{c}\cdot\frac{\mathbf{x}-\mathbf{r}(t_r)}{|\mathbf{x}-\mathbf{r}(t_r)|}\right)}{\left(1 - \frac{\mathbf{v}(t_r)}{c}\cdot\frac{\mathbf{x}-\mathbf{r}(t_r)}{|\mathbf{x}-\mathbf{r}(t_r)|}\right)^3}$$

$$\tag{A.75}$$

Taking the sum of (A.74) and (A.75) gives the electric field generated by the moving charge, which writes

$$\mathbf{E}(\mathbf{r}, t) = \mathbf{E}_{\text{near}}(\mathbf{r}, t) + \mathbf{E}_{\text{far}}(\mathbf{r}, t) \tag{A.76}$$

where

$$\mathbf{E}_{\text{near}}(\mathbf{r}, t) = \frac{q}{4\pi\epsilon_0 |\mathbf{x} - \mathbf{r}(t_r)|^2} \frac{\left(\frac{\mathbf{x} - \mathbf{r}(t_r)}{|\mathbf{x} - \mathbf{r}(t_r)|} - \frac{\mathbf{v}(t_r)}{c} \right) \left(1 - \frac{|\mathbf{v}(t_r)|^2}{c^2} \right)}{\left(1 - \frac{\mathbf{v}(t_r)}{c} \cdot \frac{\mathbf{x} - \mathbf{r}(t_r)}{|\mathbf{x} - \mathbf{r}(t_r)|} \right)^3} \tag{A.77}$$

$$\mathbf{E}_{\text{far}}(\mathbf{r}, t) = \frac{q}{4\pi\epsilon_0 |\mathbf{x} - \mathbf{r}(t_r)|} \frac{\frac{\mathbf{a}(t_r)}{c^2} \cdot \frac{\mathbf{x} - \mathbf{r}(t_r)}{|\mathbf{x} - \mathbf{r}(t_r)|} \left(\frac{\mathbf{x} - \mathbf{r}(t_r)}{|\mathbf{x} - \mathbf{r}(t_r)|} - \frac{\mathbf{v}(t_r)}{c} \right) - \frac{\mathbf{a}(t_r)}{c^2} \left(1 - \frac{\mathbf{v}(t_r)}{c} \cdot \frac{\mathbf{x} - \mathbf{r}(t_r)}{|\mathbf{x} - \mathbf{r}(t_r)|} \right)}{\left(1 - \frac{\mathbf{v}(t_r)}{c} \cdot \frac{\mathbf{x} - \mathbf{r}(t_r)}{|\mathbf{x} - \mathbf{r}(t_r)|} \right)^3}$$

$$\tag{A.78}$$

Finally, the magnetic field is found according to $\mathbf{B} = \nabla \times \mathbf{A}$. Using (A.6), we can write

$$\mathbf{B}(\mathbf{r}, t) = \frac{q}{4\pi\epsilon_0 c^2} \left(\frac{\nabla \times \mathbf{v}(t_r)}{|\mathbf{x} - \mathbf{r}(t_r) - \frac{\mathbf{v}}{c} \cdot (\mathbf{x} - \mathbf{r}(t_r))|} \right.$$
$$\left. - \frac{\nabla (|\mathbf{x} - \mathbf{r}(t_r)| - \frac{\mathbf{v}}{c} \cdot (\mathbf{x} - \mathbf{r}(t_r)))}{\left(|\mathbf{x} - \mathbf{r}(t_r) - \frac{\mathbf{v}}{c} \cdot (\mathbf{x} - \mathbf{r}(t_r))| \right)^2} \times \mathbf{v}(t_r) \right)$$

For $\nabla \times \mathbf{v}(t_r)$, let us first calculate

$$\frac{\partial}{\partial x_j} v_i(t_r) = \frac{\partial v_i(t_r)}{\partial t_r} \frac{\partial t_r}{\partial x_j} = a_i(t_r) \frac{\partial t_r}{\partial x_j}$$

and

$$\frac{\partial t_r}{\partial x_j} = -\frac{1}{c} \frac{\partial}{\partial x_j} |\mathbf{x} - \mathbf{r}(t_r)| = -\frac{\frac{x_j - r_j(t_r)}{c|\mathbf{x} - \mathbf{r}(t_r)|}}{1 - \frac{\mathbf{v}(t_r)}{c} \frac{\mathbf{x} - \mathbf{r}(t_r)}{|\mathbf{x} - \mathbf{r}(t_r)|}}$$

so that:

$$\frac{\partial v_i(t_r)}{\partial x_j} - \frac{\partial v_j(t_r)}{\partial x_i} = \frac{\frac{(x_i - r_i(t_r))}{c|\mathbf{x} - \mathbf{r}(t_r)|} a_j - \frac{(x_j - r_j(t_r))}{c|\mathbf{x} - \mathbf{r}(t_r)|} a_i}{1 - \frac{\mathbf{v}(t_r)}{c} \frac{\mathbf{x} - \mathbf{r}(t_r)}{|\mathbf{x} - \mathbf{r}(t_r)|}}$$

which gives finally

$$\nabla \times \mathbf{v}(t_r) = \frac{\mathbf{a}(t_r) \times \frac{\mathbf{x} - \mathbf{r}(t_r)}{c|\mathbf{x} - \mathbf{r}(t_r)|}}{1 - \frac{\mathbf{v}(t_r)}{c} \frac{\mathbf{x} - \mathbf{r}(t_r)}{|\mathbf{x} - \mathbf{r}(t_r)|}}$$

We find for the magnetic field:

$$\mathbf{B}(\mathbf{r}, t) = \mathbf{B}_{\mathrm{near}}(\mathbf{r}, t) + \mathbf{B}_{\mathrm{far}}(\mathbf{r}, t) \tag{A.79}$$

with

$$\mathbf{B}_{\mathrm{near}}(\mathbf{r}, t) = \frac{q}{4\pi\epsilon_0 c^2 |\mathbf{x} - \mathbf{r}(t_r)|^2} \frac{\mathbf{v}(t_r) \times \frac{\mathbf{x}-\mathbf{r}(t_r)}{|\mathbf{x}-\mathbf{r}(t_r)|}\left(1 - \frac{|\mathbf{v}(t_r)|^2}{c^2}\right)}{\left(1 - \frac{\mathbf{v}(t_r)}{c} \cdot \frac{\mathbf{x}-\mathbf{r}(t_r)}{|\mathbf{x}-\mathbf{r}(t_r)|}\right)^3} \tag{A.80}$$

$$\mathbf{B}_{\mathrm{far}}(\mathbf{r}, t) = \frac{q}{4\pi\epsilon_0 c^2 |\mathbf{x} - \mathbf{r}(t_r)|^2}$$
$$\times \frac{\mathbf{v}(t_r) \times \frac{\mathbf{x}-\mathbf{r}(t_r)}{|\mathbf{x}-\mathbf{r}(t_r)|}\frac{\mathbf{a}}{c^2} \cdot (\mathbf{x} - \mathbf{r}(t_r)) - (\mathbf{x} - \mathbf{r}(t_r)) \times \frac{\mathbf{a}(t_r)}{c}\left(1 - \frac{\mathbf{v}(t_r)}{c} \cdot \frac{\mathbf{x}-\mathbf{r}(t_r)}{|\mathbf{x}-\mathbf{r}(t_r)|}\right)}{\left(1 - \frac{\mathbf{v}(t_r)}{c} \cdot \frac{\mathbf{x}-\mathbf{r}(t_r)}{|\mathbf{x}-\mathbf{r}(t_r)|}\right)^3} \tag{A.81}$$

Solutions

Problems of Chap. 1

1.1 Operation of an electroscope

Due to the symmetry of the problem, it is sufficient to analyze the forces for one of the charges, say the charge on the left.

There are three forces acting on the charge, as shown in Fig. A.16 the repulsive electrostatic force $\mathbf{F}_e = -F_e\mathbf{u}_x$ (charges of the same sign), the tension of the string $\mathbf{T}$, and the gravitational force $m\mathbf{g} = -mg\mathbf{u}_y$. From Newton's second law for the charge, we have

$$T \sin\theta - F_e = ma_x,$$
$$T \cos\theta - mg = ma_y.$$

As the system is in equilibrium, $a_x = a_y = 0$, and the first equation leads to

$$T \sin\theta = F_e = \frac{q^2}{4\pi\epsilon_0 r^2},$$

where r is the separation between the two charges, which can be written as a function of θ as $r = 2l \sin\theta$. Then, we obtain

$$q^2 = 4\pi\epsilon_0 r^2 T \sin\theta = 16l^2\pi\epsilon_0 T \sin^3\theta.$$

Or equivalently,

$$q = \pm 4l \sin\theta \sqrt{\pi\epsilon_0 T \sin\theta}.$$

From the balance of forces projected on the vertical axis, we obtain:

$$T = \frac{mg}{\cos\theta},$$

F. Cadiz and A. Couairon, *Classical Electrodynamics*, Undergraduate Texts in Physics, https://doi.org/10.1007/978-3-031-86785-9

Fig. A.16 Forces acting on
the left charge

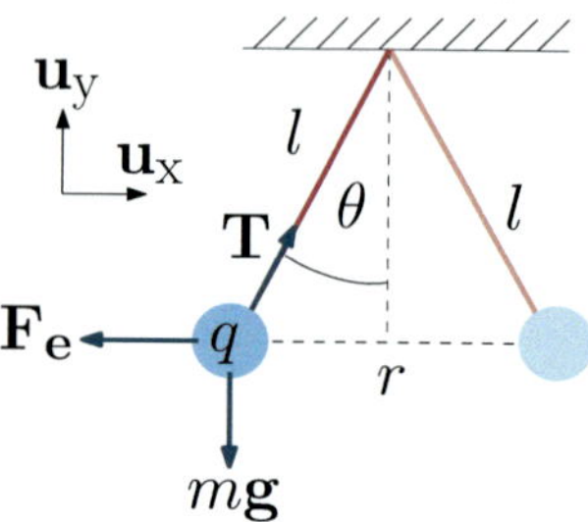

which can be introduced in the expression for the charge. Finally:

$$|q| = 4l \sin\theta \sqrt{\pi \epsilon_0 mg \tan\theta}.$$

Note that it is impossible to determine the sign of the charge knowing only θ.
By evaluating numerically the minimum charge that can be detected for $l = 0.01\,\mathrm{m}$,
$m = 0.5\,\mathrm{kg}$, and $\theta = 1\,\mathrm{deg}$, we find

$$|q_{\min}| \sim 10^{-9} C.$$

1.2 Superposition of forces

This problem can be solved by using the superposition principle. Consider first the
force exerted on q by two charges located at opposite corners, as shown in Fig. A.17.
The components of both forces along the vertical axis add up, while in the plane
perpendicular to $\mathbf{u}_z$, they cancel out.

The electric force on the charge q due to these two charges will then be

$$\mathbf{F}'_q = 2\frac{qQ}{4\pi\epsilon_0 r^2}\sin\theta\,\mathbf{u}_z,$$

where $r^2 = h^2 + L^2/2$ and $\sin\theta = h/r$, so that

Fig. A.17 Forces acting on
the left charge

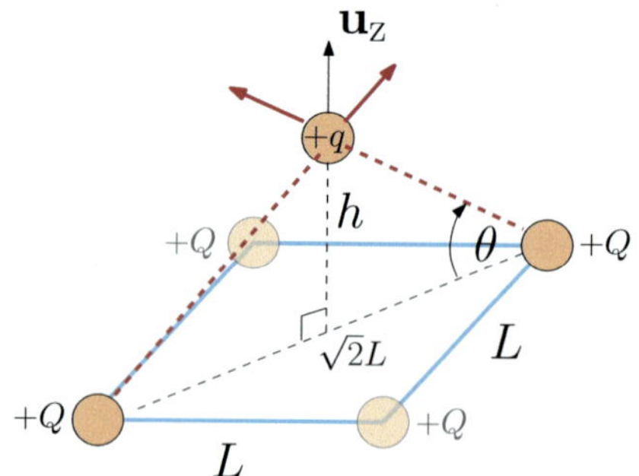

$$\mathbf{F}'_q = \frac{qQh}{2\pi\epsilon_0 r^3}\mathbf{u}_z.$$

The same result holds for the two other charges, and the total force will be

$$\mathbf{F}_q = 2\mathbf{F}'_q = \frac{qQh}{\pi\epsilon_0 r^3}\mathbf{u}_z.$$

This force must be counteracted by the weight of the charge q. The balance of forces will occur when h is such that $|\mathbf{F}_q| = mg$, that is

$$qQ = \frac{mg\pi\epsilon_0 r^3}{h}, \quad \text{with } r = \sqrt{h^2 + L^2/2}.$$

1.3 Field of a quadripole

The electric field at P can be obtained using the superposition principle. The field generated at P by the charge $2Q$ is

$$\mathbf{E}_1 = \frac{2Q}{4\pi\epsilon_0 r^2}\mathbf{u}_y,$$

where the y-axis has been defined as the vertical axis.

From the symmetry of the charge distribution, the superposition of the fields $\mathbf{E}_2 + \mathbf{E}_3$ will generate a resulting field with a component along $\mathbf{u}_y$ only (see Fig. A.18), given by

$$\mathbf{E}_2 + \mathbf{E}_3 = -2\frac{Q}{4\pi\epsilon_0}\frac{1}{r^2 + d^2}\cos\theta\,\mathbf{u}_y,$$

where θ is the angle that $\mathbf{E}_2$ (or $\mathbf{E}_3$) forms with the y-axis. We have $\cos\theta = r/\sqrt{r^2 + d^2}$, and then

$$\mathbf{E}_2 + \mathbf{E}_3 = -\frac{Q}{2\pi\epsilon_0 r^2}\cos^3\theta\,\mathbf{u}_y.$$

Thus, the superposition of the three fields will give a total field equal to

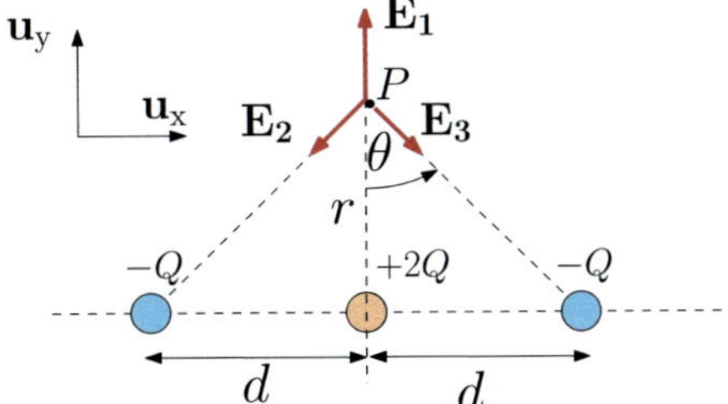

Fig. A.18 Forces acting on the left charge

$$\mathbf{E}(P) = \frac{Q}{2\pi\epsilon_0 r^2}\left(1 - \cos^3\theta\right)\mathbf{u}_y.$$

To analyze the limit $r \gg d$, we use a small-d/r Taylor expansion of $\theta = \mathrm{atan}(d/r) \sim d/r$ and a small-θ Taylor expansion of $\cos^3\theta \sim (1 - \theta^2/2)^3 \sim 1 - 3\theta^2/2$, and we rewrite the electric field as

$$\mathbf{E}(P) \simeq \frac{Q}{2\pi\epsilon_0 r^2}\left(1 - 1 + \frac{3}{2}\frac{d^2}{r^2}\right)\mathbf{u}_y = \frac{3Qd^2}{4\pi\epsilon_0 r^4}\mathbf{u}_y.$$

Note that when the distance is large compared to the separation between the charges ($r \gg d$), this charge distribution generates a field that decays as $1/r^4$ (faster than the field of a point charge and that of a dipole), due to the screening effect (the two charges $-Q$ screen the charge $+2Q$).

1.4 Uniformly charged disk

Consider a surface element on the disk at the position $\mathbf{x}' = r\mathbf{u}_r$ in cylindrical coordinates. The associated infinitesimal area is $ds' = r\,dr\,d\theta$, and the infinitesimal charge element is then $dq = \sigma r\,dr\,d\theta$ (see Fig. A.19). The electric field generated by this surface element at $\mathbf{x} = z\mathbf{u}_z$ is given by

$$d\mathbf{E}(z\mathbf{u}_z) = \frac{dq}{4\pi\epsilon_0}\frac{\mathbf{x} - \mathbf{x}'}{|\mathbf{x} - \mathbf{x}'|^3} = \frac{\sigma}{4\pi\epsilon_0}\frac{r\,dr\,d\theta(z\mathbf{u}_z - r\mathbf{u}_r)}{(z^2 + r^2)^{3/2}}.$$

Thus, the total electric field is obtained by adding up the contributions of all the surface elements on the disk:

$$\mathbf{E}(z\mathbf{u}_z) = \frac{\sigma}{4\pi\epsilon_0}\mathbf{u}_z\int_0^{2\pi} d\theta\int_0^{R} dr\,\frac{r\,z}{(z^2 + r^2)^{3/2}},$$

where the integral along $\mathbf{u}_r$ cancels out, as can be shown by calculating it directly or using symmetry arguments for field elements generated by symmetric charge elements with respect to the disk center. The resulting electric field will only have a component along $\mathbf{u}_z$. Let $\tan\alpha = r/z$, then $dr = z\,d\alpha/\cos^2\alpha$ and $\cos\alpha = z/(z^2 + r^2)^{1/2}$, $\sin\alpha = r/(z^2 + r^2)^{1/2}$, so that $zr\,dr/(z^2 + r^2)^{3/2} = \sin\alpha\,d\alpha$. With this,

Fig. A.19 A uniformly charged disk with charge density σ

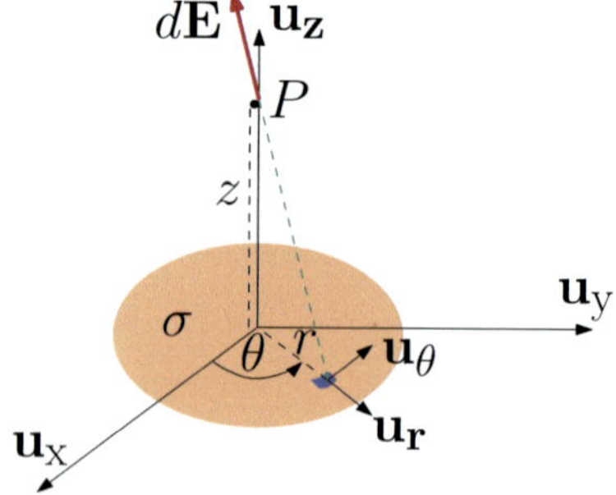

$$\mathbf{E}(z\mathbf{u}_z) = \frac{\sigma}{4\pi\epsilon_0}\mathbf{u}_z 2\pi \int_0^{\alpha_R} \sin\alpha \, d\alpha,$$

where the factor 2π results from integrating over θ between 0 and 2π. Finally,

$$\mathbf{E}(z\mathbf{u}_z) = \frac{\sigma}{2\epsilon_0}(1 - \cos\alpha_R)\mathbf{u}_z,$$

where $\cos\alpha_R = z/(z^2 + R^2)^{1/2}$. Note that if $R \to \infty$ (equivalently, $z \to 0$, or $\alpha_R \to \pi/2$, $\cos\alpha_R \to 0$), the field generated by an infinite plane is obtained

$$\mathbf{E}_{R\to\infty}(z\mathbf{u}_z) = \frac{\sigma}{2\epsilon_0}\mathbf{u}_z.$$

This field does not depend on the distance z to the plane. Moreover, in this case the axis $\mathbf{u}_z$ is equivalent to any axis perpendicular to the plane. That is, the magnitude of the electric field of an infinite plane is uniform throughout the space.

1.5 Superposition of two known distributions

(a) A natural way to solve this problem is to use the superposition principle. That is, consider the system as the superposition of an infinite plane and a disk of radius R of surface density $-\sigma$. We already saw in Exercise 1.4 that the field generated by a disk of radius R and density $-\sigma$ is given by

$$\mathbf{E}_1(z\mathbf{u}_z) = \frac{-\sigma}{2\epsilon_0}\left(1 - \frac{z}{\sqrt{R^2 + z^2}}\right)\mathbf{u}_z.$$

The field generated by the infinite plane of density σ was also obtained by considering a disk of infinite radius:

$$\mathbf{E}_1(z\mathbf{u}_z) = \frac{\sigma}{2\epsilon_0}\mathbf{u}_z.$$

The total electric field will be the superposition of $\mathbf{E}_1$ and $\mathbf{E}_2$

$$\mathbf{E}(z\mathbf{u}_z) = \frac{\sigma}{2\epsilon_0}\mathbf{u}_z + \frac{\sigma}{2\epsilon_0}\left(\frac{z}{\sqrt{R^2 + z^2}} - 1\right)\mathbf{u}_z = \frac{\sigma}{2\epsilon_0}\frac{z}{\sqrt{R^2 + z^2}}\mathbf{u}_z.$$

(b) To calculate the force on the charge line, we first take a differential element of length dz (distance z from the plane, see Fig. A.20). The force on this element is given by

$$d\mathbf{F} = dq\mathbf{E}(z) = \lambda dz\mathbf{E}(z) = \frac{\sigma z}{2\epsilon_0}\frac{\lambda}{\sqrt{R^2 + z^2}}dz\mathbf{u}_z.$$

The total force on the charge line is obtained by integration

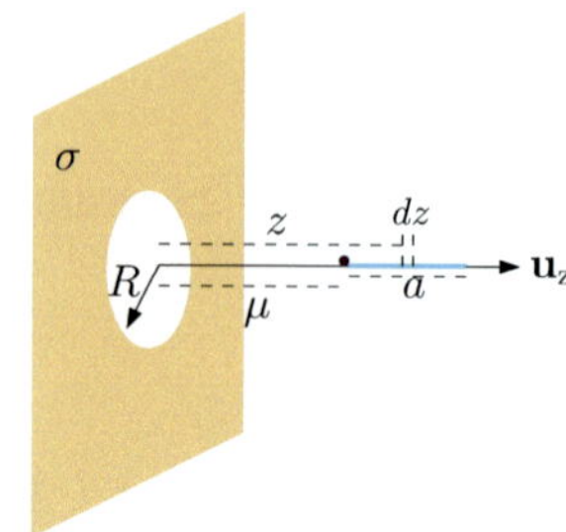

Fig. A.20 An infinitesimal length element dz at distance z from the plane

$$\mathbf{F} = \int_{\mu}^{\mu+a} dz \left(\frac{\sigma z}{2\epsilon_0} \frac{\lambda}{\sqrt{R^2 + z^2}} \right) \mathbf{u}_z.$$

Let $u = R^2 + z^2$, whereby $du = 2z\,dz$. Finally

$$\mathbf{F} = \frac{\lambda \sigma}{4\epsilon_0} \int_{R^2+\mu^2}^{R^2+(\mu+a)^2} \frac{du}{\sqrt{u}} \mathbf{u}_z = \frac{\lambda \sigma}{2\epsilon_0} \left(\sqrt{R^2 + (\mu + a)^2} - \sqrt{R^2 + \mu^2} \right) \mathbf{u}_z.$$

1.6 Field on the axis of a charged line

An infinitesimal element dl of the charged line, located at distance l from the left end, carries a charge element λdl, as shown in Fig. A.21.

The electric field at point P due to this charge element reads

$$d\mathbf{E}(P) = \frac{1}{4\pi \epsilon_0} \frac{\lambda dl}{(a + x - l)^2} \mathbf{u}_x.$$

The total electric field is obtained by integrating over the entire length of the charged line and reads

$$\mathbf{E}(P) = \frac{1}{4\pi \epsilon_0} \int_0^a \frac{\lambda dl}{(a + x - l)^2} \mathbf{u}_x.$$

Let $z = a + x - l$. With this, $dz = -dl$ and the field is

$$\mathbf{E}(P) = \frac{1}{4\pi \epsilon_0} \int_{a+x}^x -\frac{\lambda dz}{z^2} \mathbf{u}_x = \frac{Q}{4\pi \epsilon_0 x (a + x)} \mathbf{u}_x,$$

with $Q = \lambda a$ the total charge of the line. Note that if $x \gg a$, at the lowest order in a/x

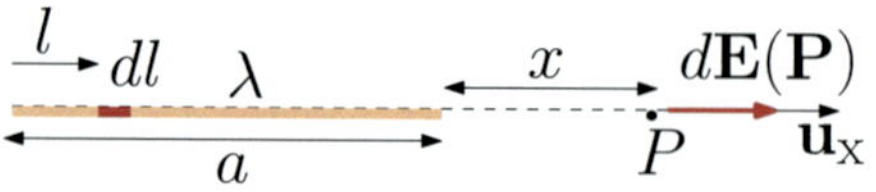

Fig. A.21 Electric field at P generated by an infinitesimal charge element dl along the charged line

$$\mathbf{E}(P) = \frac{Q}{4\pi\epsilon_0 x^2}\mathbf{u}_x.$$

Far from the charged line, the field matches the Coulomb field of a point charge.

1.7 Off-axis field of a charged line

(a) An infinitesimal element of length dl at position $\mathbf{x}'$ carries a charge $dq = \lambda dl$. Let $\mathbf{x}$ denote the position of point P. According to Coulomb's law, this charge element generates an electric field at P that is given by

$$d\mathbf{E}(P) = \frac{1}{4\pi\epsilon_0}\frac{dq\,(\mathbf{x} - \mathbf{x}')}{|\mathbf{x} - \mathbf{x}'|^3}.$$

As shown in Fig. A.22, we have $\mathbf{x} = r\mathbf{u}_r$, where $\mathbf{u}_r$ is the unit radial vector in cylindrical coordinates that points from the origin to point P. For an infinitesimal charge element at distance l from the origin, $\mathbf{x}' = l\mathbf{u}_z$, then:

$$\mathbf{x} - \mathbf{x}' = r\mathbf{u}_r - l\mathbf{u}_z, \quad |\mathbf{x} - \mathbf{x}'| = \sqrt{r^2 + l^2}.$$

Using the superposition principle, the total field at P will be the sum of all infinitesimal contributions of the distribution:

$$\mathbf{E}(P) = \frac{\lambda}{4\pi\epsilon_0}\left(\int_{-l_2}^{l_1} \frac{(r\mathbf{u}_r - l\mathbf{u}_z)dl}{(r^2 + l^2)^{3/2}}\right).$$

The integral is solved with the trigonometric substitution $l = r\tan\alpha$ (see Fig. A.23), so $dl = rd\alpha/\cos^2\alpha$ and $\cos\alpha = r/(r^2 + l^2)^{1/2}$, $\sin\alpha = l/(r^2 + l^2)^{1/2}$,

$$\mathbf{E}(P) = \frac{\lambda}{4\pi\epsilon_0 r}\int_{-\alpha_2}^{\alpha_1}(\cos\alpha\,\mathbf{u}_r - \sin\alpha\,\mathbf{u}_z)d\alpha.$$

With this, the total field at P is

Fig. A.22 Electric field at P generated by an infinitesimal charge element dl along the charged line

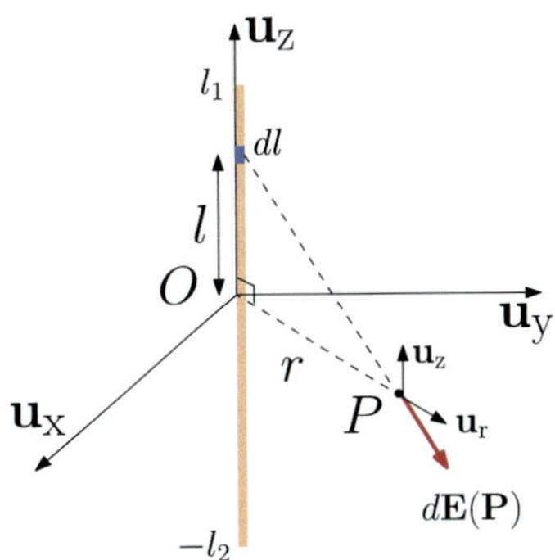

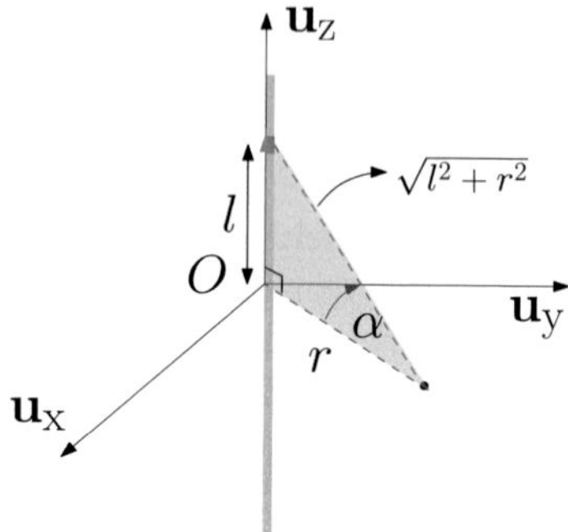

Fig. A.23 The angle α is defined by the relation $l = r \tan \alpha$

$$\mathbf{E}(P) = \frac{\lambda}{4\pi\epsilon_0 r}\left([\sin\alpha]_{-\alpha_2}^{\alpha_1}\mathbf{u}_r + [\cos\alpha]_{-\alpha_2}^{\alpha_1}\mathbf{u}_z\right),$$

with $\sin\alpha_i = l_i/(r^2 + l_i^2)^{1/2}$ and $\cos\alpha_i = r/(r^2 + l_i^2)^{1/2}$.

(b) In the limit when $r \gg l_1$ and $r \gg l_2$, we have (at second order in $1/r$):

$$\sin\alpha_i \sim \frac{l_i}{r}, \quad \cos\alpha_i \sim 1 - \frac{l_i^2}{2r^2}, \quad i = 1, 2.$$

Then the field is

$$\mathbf{E}(P) \simeq \frac{\lambda}{4\pi\epsilon_0 r}\frac{l_1 + l_2}{r}\mathbf{u}_r,$$

but $l = l_1 + l_2$ corresponds to the total length of the distribution and the total charge is $Q = \lambda l$, then the field is rewritten

$$\mathbf{E}(P) \simeq \frac{Q}{4\pi\epsilon_0 r^2}\mathbf{u}_r,$$

which is the Coulomb field of a point charge Q.

(c) Now, in the limit when $l_1 \gg r$ and $l_2 \gg r$, we have

$$\sin\alpha_i \sim 1 - \frac{r^2}{2l_i^2} \sim 1, \quad \cos\alpha_i \sim \frac{r}{l_i} \ll 1, \quad i = 1, 2,$$

and the field behaves as

$$\mathbf{E}(P) \simeq \frac{\lambda}{2\pi\epsilon_0 r}\mathbf{u}_r,$$

which is the electric field generated by an infinite charged wire.

1.8 Force between a charged ribbon and a charged line

We can choose to calculate the force exerted by the line charge distribution on the ribbon charge distribution, or vice versa. We use the first option: we need to obtain

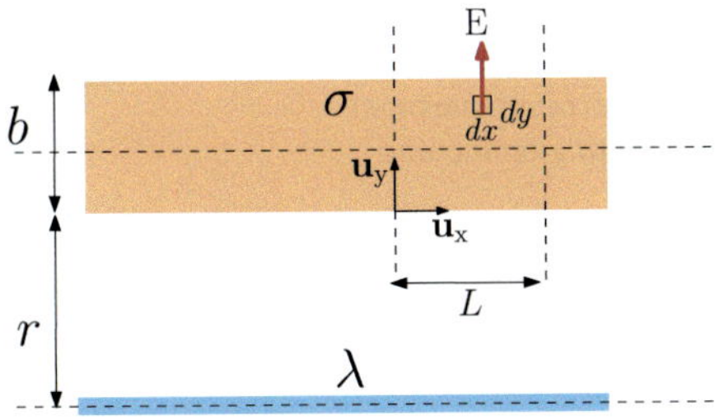

Fig. A.24 An infinitesimal surface element $dxdy$ used to calculate an infinitesimal force acting on the ribbon

the electric field generated by a very long line charge distribution, as obtained in Problem 1.7:

$$\mathbf{E}(\mathbf{r}) = \frac{\lambda}{2\pi\epsilon_0 d}\mathbf{u}_r,$$

where d is the distance to the wire. Now, in the plane that contains both distributions, we define the directions $\mathbf{u}_x$, $\mathbf{u}_y$ as indicated in Fig. A.24.

The electric field generated by the line charge distribution at a point (x, y) has the form:

$$\mathbf{E}(\mathbf{x}) = \frac{\lambda}{2\pi\epsilon_0 (r+y)}\mathbf{u}_y.$$

Then the force exerted on a surface element of infinitesimal charge $dq = \sigma dxdy$ at (x, y) is given by

$$d\mathbf{F}(x, y) = dxdy\sigma\frac{\lambda}{2\pi\epsilon_0 (r+y)}\mathbf{u}_y.$$

The force exerted on a portion of length L of the flat distribution is

$$\mathbf{F} = \int_0^L dx \int_0^b \frac{\sigma\lambda dy}{2\pi\epsilon_0 (r+y)}\mathbf{u}_y .$$

Integrating over x first and then over y, we obtain:

$$\mathbf{F} = \frac{\sigma\lambda L}{2\pi\epsilon_0}\int_0^b \frac{dy}{(r+y)}\mathbf{u}_y = \frac{\sigma\lambda L}{2\pi\epsilon_0}[\ln(r+y)]_0^b \,\mathbf{u}_y = \frac{\sigma\lambda L}{2\pi\epsilon_0}\ln\frac{r+b}{r}\mathbf{u}_y$$

and the force per unit length is

$$\frac{\mathbf{F}}{L} = \frac{\sigma\lambda}{2\pi\epsilon_0}\ln\frac{r+b}{r}\mathbf{u}_y.$$

Alternative solution

We could also have calculated the force exerted by the two-dimensional distribution on the charge line. For this, consider a charge element $dq = \lambda dx$ located at position

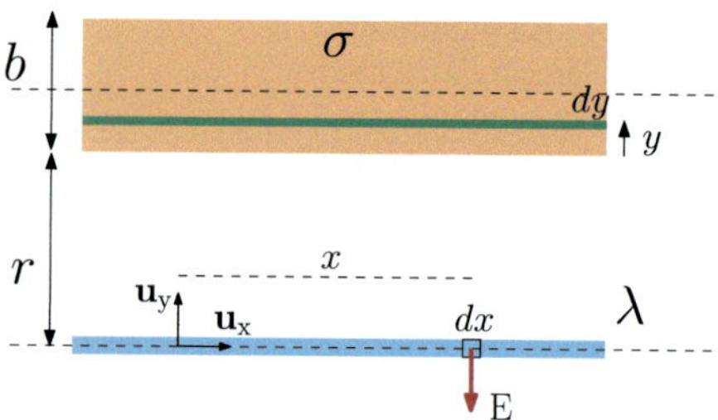

Fig. A.25 An infinitesimal length element dx used to calculate an infinitesimal force acting on the line

x on the line, as shown in Fig. A.25, and calculate the force exerted on it by a very thin strip of the plane of width dy, as shown in the figure.

This strip can be seen as an infinite line of charge density σdy. Then, the field $\mathbf{E}(x)$ that is generated at the position of the charge element is given by

$$d\mathbf{E}(x) = -\frac{\sigma dy}{2\pi\epsilon_0(r+y)}\mathbf{u}_y$$

and the total field will be that generated by the superposition of all the stripes on the plane:

$$\mathbf{E}(x) = -\frac{\sigma}{2\pi\epsilon_0}\int_0^b \frac{dy}{(r+y)}\mathbf{u}_y = -\frac{\sigma}{2\pi\epsilon_0}\ln\frac{r+b}{r}\mathbf{u}_y.$$

Then, the force on dq is

$$d\mathbf{F}(x) = \lambda dx \mathbf{E}(x) = -\frac{\lambda dx \sigma}{2\pi\epsilon_0}\ln\frac{r+b}{r}\mathbf{u}_y.$$

Since the force is independent of x (constant on the charge line), the force on a segment of the long line L is, simply, $\mathbf{F} = -\dfrac{\lambda L \sigma}{2\pi\epsilon_0}\ln\dfrac{r+b}{r}\mathbf{u}_y.$

From here, we get immediately the force per unit length that is exerted on the line:

$$\frac{\mathbf{F}}{L} = \frac{-\sigma\lambda}{2\pi\epsilon_0}\ln\frac{r+b}{r}\mathbf{u}_y.$$

1.9 Asymmetric rod

We will first obtain the electric field generated by the upper segment of total charge q. We define the axes as shown in Fig. A.26, where the origin is $\mathbf{x} = \mathbf{0}$. The vector $\mathbf{x}' = R\mathbf{u}_r$, with $0 < \theta < \pi/2$, designates the position of an infinitesimal arc element, of length $dl = Rd\theta$, whose charge is

$$dq = \lambda dl = \lambda R d\theta = \frac{2q}{\pi}d\theta.$$

This charge element generates an electric field at $\mathbf{x} = \mathbf{0}$ given by

Fig. A.26 Axes and definitions for this problem

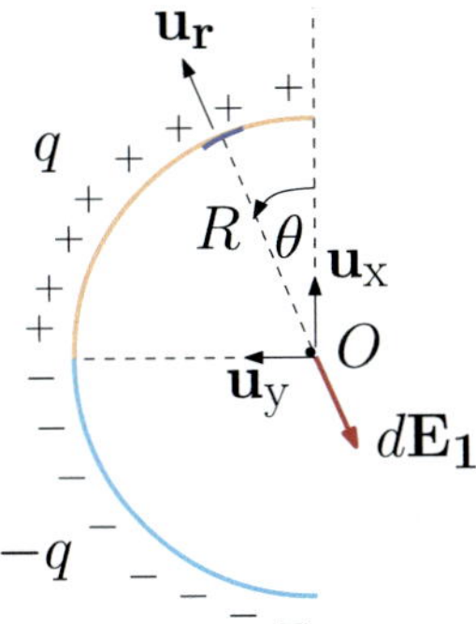

Fig. A.27 Superposition of the fields generated by the upper and bottom segments

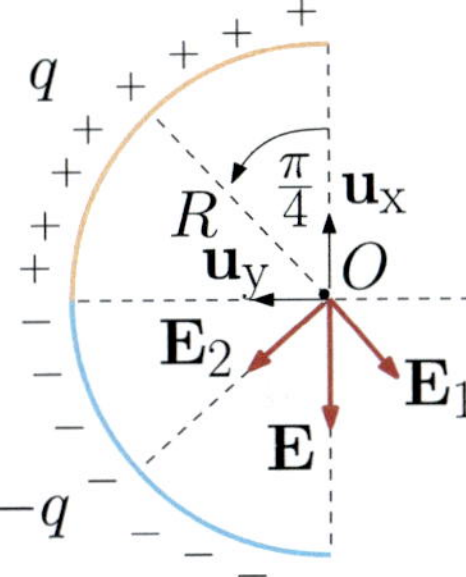

$$dE_1(O) = \frac{1}{4\pi\epsilon_0} \frac{dq}{|\mathbf{x} - \mathbf{x}'|^3}(\mathbf{x} - \mathbf{x}') = -\frac{2q\,d\theta}{4\pi^2\epsilon_0 R^2}\mathbf{u}_r.$$

In this way, the total electric field due to the upper part is obtained using the superposition principle

$$\mathbf{E}_1(O) = -\frac{2q}{4\pi^2\epsilon_0 R^2}\int_0^{\pi/2} d\theta\,\mathbf{u}_r = -\frac{q}{2\pi^2\epsilon_0 R^2}\int_0^{\pi/2} d\theta\,(\cos\theta\,\mathbf{u}_x + \sin\theta\,\mathbf{u}_y)$$

$$= -\frac{q}{2\pi^2\epsilon_0 R^2}(\mathbf{u}_x + \mathbf{u}_y),$$

which forms an angle of $\pi/4$ with the horizontal axis. The field generated by the segment carrying the charge $-q$ can be obtained using a simple symmetry argument: the field $\mathbf{E}_2$ generated at the origin would be identical to $\mathbf{E}_1$ except for the inversion of the sign of the component along $\mathbf{u}_x$ and of the charge (see Fig. A.27). That is to say

$$\mathbf{E}_2(O) = \frac{q}{2\pi^2\epsilon_0 R^2}(\mathbf{u}_y - \mathbf{u}_x).$$

Adding both fields:

$$\mathbf{E}(O) = -\frac{q}{\pi^2\epsilon_0 R^2}\mathbf{u}_x.$$

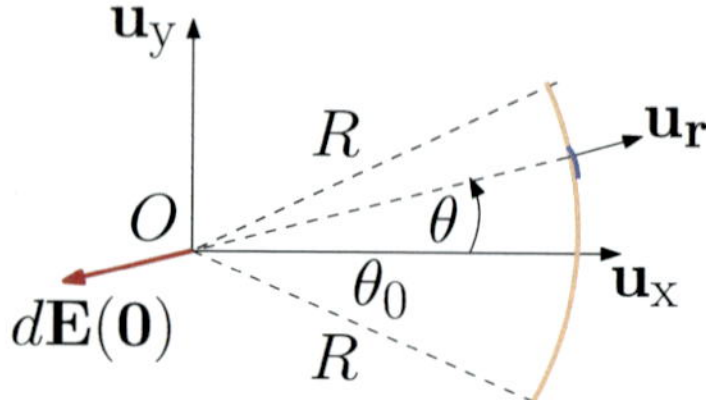

Fig. A.28 Field at **0** generated by an infinitesimal arc element

1.10 Circle arc

Consider an infinitesimal arc element of length $dl = Rd\theta$, which forms an angle θ with respect to the x-axis (see Fig. A.28).

The amount of charge carried by this element is $dq = \lambda dl = \lambda R d\theta$. Its contribution to the electric field at O is

$$d\mathbf{E}(O) = -\frac{1}{4\pi\epsilon_0}\frac{dq}{R^2}\mathbf{u}_r = -\frac{\lambda d\theta}{4\pi\epsilon_0 R}\mathbf{u}_r.$$

Integrating θ between $-\theta_0$ and θ_0 and noticing that $\mathbf{u}_r = \frac{d\mathbf{u}_\theta}{d\theta}$, we obtain

$$\mathbf{E}(O) = \frac{\lambda}{4\pi\epsilon_0 R}\int_{-\theta_0}^{\theta_0}\frac{d\mathbf{u}_\theta}{d\theta}d\theta = \frac{\lambda}{4\pi\epsilon_0 R}(\mathbf{u}_{\theta_0} - \mathbf{u}_{-\theta_0}) = -\frac{\lambda\sin\theta_0}{2\pi\epsilon_0 R}\mathbf{u}_x.$$

We see that the electric field has a component along the x-axis only, which agrees with the symmetry of the problem. If we take the limit when $\theta_0 \to \pi$, the arc becomes a circular ring. Since $\sin\pi \to 0$, the above equation implies that the electric field in the center of a uniformly charged ring is zero, which is to be expected by symmetry arguments. On the other hand, for very small angles, $\sin\theta_0 \sim \theta_0$, we recover the case of a point charge.

$$\mathbf{E}(O) \simeq -\frac{1}{4\pi\epsilon_0}\frac{2\lambda\theta_0}{R}\mathbf{u}_x = -\frac{1}{4\pi\epsilon_0}\frac{Q}{R^2}\mathbf{u}_x,$$

where the total charge of the arc is $Q = \lambda l = \lambda(2R\theta_0)$.

1.11 Charged hemisphere

We take as origin the center of curvature of the hemisphere. From this, we can determine the contribution of an infinitesimal charge element on the surface to the electric field on the axis $\mathbf{u}_z$ passing through the center of curvature (see Fig. A.29). By symmetry, it can be anticipated that the total field will only have a component along $\mathbf{u}_z$. An infinitesimal surface element on the sphere is given by

$$ds = a^2 \sin\theta d\theta d\phi$$

and as the charge is evenly distributed, the surface density is simply $\sigma = Q/2\pi a^2$. Then

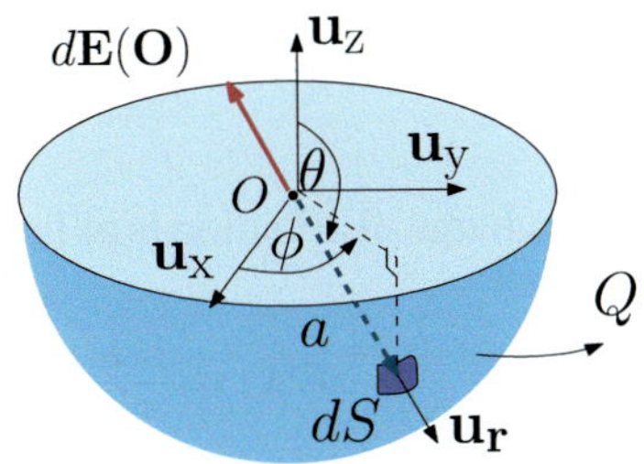

Fig. A.29 Infinitesimal electric field at **O** generated by a surface element dS on the hemisphere

$$d\mathbf{E}(\mathbf{x}) = \frac{Q}{8\pi^2 a^2 \epsilon_0} \frac{\mathbf{x} - a\mathbf{u}_r}{|\mathbf{x} - a\mathbf{u}_r|^3} a^2 \sin\theta d\theta d\phi.$$

With this, we can determine the total field by integration over the entire surface

$$\mathbf{E}(\mathbf{x}) = \frac{Q}{8\pi^2 \epsilon_0} \int_0^{2\pi} d\phi \int_{\pi/2}^{\pi} \sin\theta d\theta \frac{\mathbf{x} - a\mathbf{u}_r}{|\mathbf{x} - a\mathbf{u}_r|^3}.$$

By evaluating the integral at $\mathbf{x} = \mathbf{0}$, we obtain the field at the center of curvature

$$\mathbf{E}(\mathbf{0}) = \frac{Q}{8\pi^2 \epsilon_0} \int_0^{2\pi} d\phi \int_{\pi/2}^{\pi} \frac{-a\mathbf{u}_r}{a^3} \sin\theta d\theta.$$

Since $\mathbf{u}_r = \cos\theta\, \mathbf{u}_z + \sin\theta \cos\phi \mathbf{u}_x + \sin\theta \sin\phi \mathbf{u}_y$, the components along $\mathbf{u}_x$ and $\mathbf{u}_y$ vanish after integration over ϕ. To avoid the calculation, we can use symmetry arguments relying on the symmetry of the charge distribution, and conclude that only the component along $\mathbf{u}_z$ is non-zero (same argument used as for the charged ring). With this,

$$\mathbf{E}(\mathbf{0}) = -\frac{Q}{4\pi \epsilon_0 a^2} \int_{\pi/2}^{\pi} \cos\theta \sin\theta d\theta\, \mathbf{u}_z.$$

Finally, using $s = \sin\theta$ and $\int_{\pi/2}^{\pi} \cos\theta \sin\theta d\theta = \int_1^0 s ds = -1/2$, the field in the center of curvature is

$$\mathbf{E}(\mathbf{0}) = \frac{Q}{8\pi \epsilon_0 a^2}\, \mathbf{u}_z.$$

1.12 Discrete distribution as a volume density

To see that the charge distribution can be written in this way, we must verify that the electric field it generates is the correct one. Indeed, using the general formula (1.5)

$$\mathbf{E}(\mathbf{x}) = \frac{1}{4\pi \epsilon_0} \iiint_{\mathbb{R}^3} \rho(\mathbf{x}') \frac{\mathbf{x} - \mathbf{x}'}{|\mathbf{x} - \mathbf{x}'|^3} d^3 x'$$

$$\mathbf{E}(\mathbf{x}) = \frac{1}{4\pi\epsilon_0} \iiint_{\mathbb{R}^3} \left[\sum_{i=1}^{N} q_i \delta(\mathbf{x}' - \mathbf{x}_i) \right] \frac{\mathbf{x} - \mathbf{x}'}{|\mathbf{x} - \mathbf{x}'|^3} d^3 x'.$$

If the charge distribution is of finite extent we can exchange the integral with the sum

$$\mathbf{E}(\mathbf{x}) = \frac{1}{4\pi\epsilon_0} \sum_{i=1}^{N} q_i \iiint_{\mathbb{R}^3} \delta(\mathbf{x}' - \mathbf{x}_i) \frac{\mathbf{x} - \mathbf{x}'}{|\mathbf{x} - \mathbf{x}'|^3} d^3 x'.$$

The volume integral involving the Dirac delta is calculated by extracting the value of the integrand at $\mathbf{x}' = \mathbf{x}_i$

$$\mathbf{E}(\mathbf{x}) = \frac{1}{4\pi\epsilon_0} \sum_{i=1}^{N} q_i \frac{\mathbf{x} - \mathbf{x}_i}{|\mathbf{x} - \mathbf{x}_i|^3},$$

which is exactly the field generated by the discrete charge distribution, according to the superposition principle.

Problems of Chap. 2

2.1 Charge density in the atmosphere
(a) To obtain the charge density ρ, we can use the differential form of Gauss's law (2.3):

$$\nabla \cdot \mathbf{E} = \alpha a e^{-\alpha z} = \frac{\partial E(z)}{\partial z} = \rho(z)/\epsilon_0$$

so that the charge density in the atmosphere decreases exponentially with height

$$\rho(z) = \epsilon_0 \alpha a e^{-\alpha z}.$$

(b) The total charge contained in a vertical column of section S is

$$Q_S = S \int_0^{\infty} \epsilon_0 \alpha a e^{-\alpha z} dz = S\epsilon_0 a$$

and the total charge Q_T in the atmosphere is obtained by considering S as the Earth surface, that is $S = 4\pi R_T^2$, with $R_T = 6400$ km the Earth radius.

$$Q_T = 4\pi \epsilon_0 R_T^2 a = 4.55 \times 10^5 \text{ C}.$$

2.2 Electric flux through a square surface
Let us first try to calculate the flux explicitly, that is, we evaluate the integral

Fig. A.30 Electric flux over an infinitesimal surface element

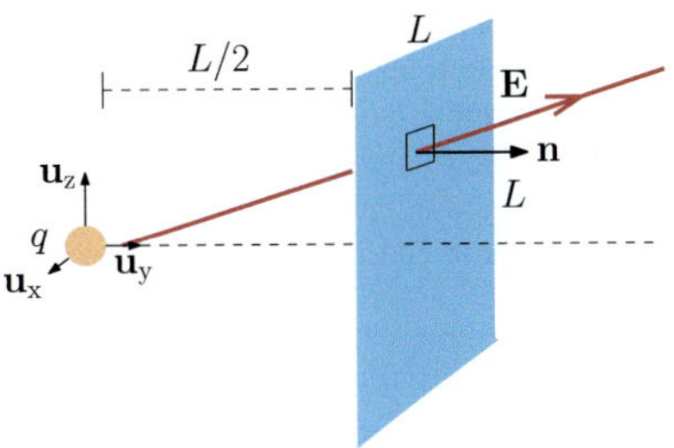

$$\Phi_{\Sigma,\mathbf{E}} = \iint_{\Sigma} dS(\mathbf{x}')\mathbf{n}(\mathbf{x}') \cdot \mathbf{E}(\mathbf{x}').$$

Using a Cartesian coordinate system whose origin coincides with the point charge, the flux through a surface element dS at the point $\mathbf{x}' = (x', d = L/2, z') \in \Sigma$ (see Fig. A.30) is given by

$$dS(\mathbf{x}')\mathbf{n}(\mathbf{x}') \cdot \mathbf{E}(\mathbf{x}') = dx'dz'\mathbf{u}_y \cdot \frac{q(x'\mathbf{u}_x + d\mathbf{u}_y + z'\mathbf{u}_z)}{4\pi\epsilon_0(d^2 + x'^2 + z'^2)^{3/2}}$$

and so,

$$\Phi_{\Sigma,\mathbf{E}} = \frac{qd}{4\pi\epsilon_0} \int_{-d}^{d} dx' \int_{-d}^{d} dz' \frac{1}{(d^2 + x'^2 + z'^2)^{3/2}}.$$

Now let us consider

$$I = \int_{-d}^{d} \frac{dz'}{(d^2 + x'^2 + z'^2)^{3/2}} = \frac{1}{(d^2 + x'^2)^{3/2}} \int_{-d}^{d} \frac{dz'}{(1 + \frac{z'^2}{d^2+x'^2})^{3/2}}$$

$$= \frac{1}{d^2 + x'^2} \int_{-\frac{d}{\sqrt{d^2+x'^2}}}^{\frac{d}{\sqrt{d^2+x'^2}}} \frac{du}{(1 + u^2)^{3/2}} = \frac{2d}{(d^2 + x'^2)(2d^2 + x'^2)^{1/2}},$$

where we have used $\int \frac{du}{(1+u^2)^{3/2}} = \frac{u}{\sqrt{1+u^2}} + \text{Cst}$. We then have

$$\Phi_{\Sigma,\mathbf{E}} = \frac{2qd^2}{4\pi\epsilon_0} \int_{-d}^{d} \frac{dx'}{(d^2 + x'^2)(2d^2 + x'^2)^{1/2}} = \frac{2q}{4\pi\epsilon_0} \int_{-1}^{1} \frac{du}{(1 + u^2)\sqrt{2 + u^2}}$$

and using the substitution $u = \sqrt{2}\tan\theta$, one can show that $\int \frac{du}{(1+u^2)\sqrt{2+u^2}} = \arctan\left(\frac{u}{\sqrt{2+u^2}}\right) + \text{Cst}$ and finally

$$\Phi_{\Sigma,\mathbf{E}} = \frac{2q}{4\pi\epsilon_0} 2\arctan\left(\frac{1}{\sqrt{3}}\right) = \frac{q}{\pi\epsilon_0}\frac{\pi}{6} = \frac{q}{6\epsilon_0}.$$

Fig. A.31 The flux is the same over each face of a cube centered at the charge position. The flux is the same over each face of a cube centered at the charge position

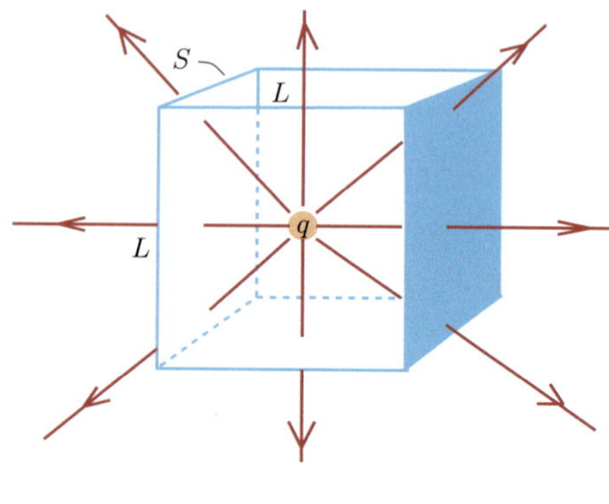

Fig. A.32 Cylindrical coordinates used to determine the electric field of a charged wire

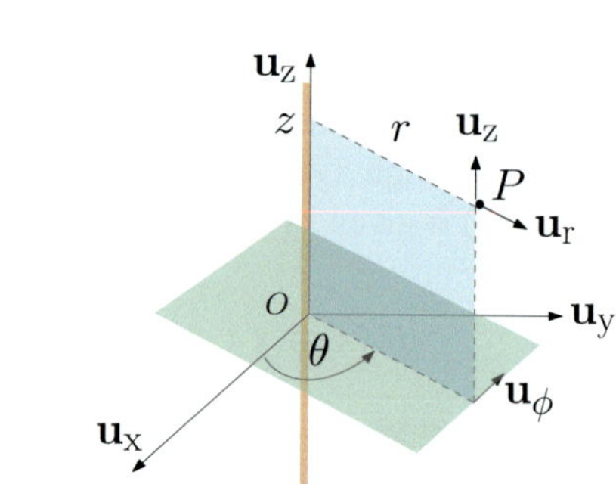

Instead of this very long calculation, we can use the Gauss theorem instead. Consider the closed surface S of a cube where Σ is one of the faces (Fig. A.31).

The flux through S, by the Gauss theorem, is simply $\Phi_{S,\mathbf{E}} = q/\epsilon_0$. Since the point charge is located in the center of the cube, the flux is identical through each of the six faces of S. If Σ is one of them, we then find

$$\Phi_{\Sigma,\mathbf{E}} = \iint_\Sigma dS(\mathbf{x}')\mathbf{n}(\mathbf{x}') \cdot \mathbf{E}(\mathbf{x}') = \frac{1}{6} \oiint_S dS(\mathbf{x}')\mathbf{n}(\mathbf{x}') \cdot \mathbf{E}(\mathbf{x}') = \frac{q}{6\epsilon_0}.$$

2.3 Field of an infinite charged line

Here it is convenient to choose cylindrical coordinates with the origin located at some point of the charged line, as shown in Fig. A.32. By Curie's principle, invariance of the charge distribution to translations along z and rotations around the z-axis results in an independence of the electric field upon θ and z, $\mathbf{E}(r, \theta, z) = \mathbf{E}(r)$.

For any point P, one can find the two symmetry planes containing P shown in Fig. A.32. The orientation of the electric field at P is then along the intersection of the two planes, that is the radial direction $\mathbf{u}_r$. Therefore $\mathbf{E}(r, \theta, z) = E(r)\mathbf{u}_r$.

Due to this cylindrical symmetry, we take as a Gauss surface a cylinder of radius r and length l and whose axis of symmetry coincides with that of the line distribution of charge, see Fig. A.33. Due to the radial field orientation, there is no flux through the cylinder caps, and the electric field on the lateral surface S_{lat} is always normal to the surface and of constant magnitude, then

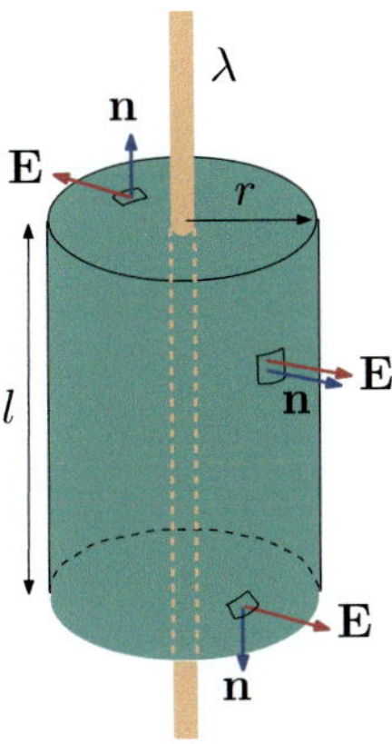

Fig. A.33 Cylindrical surface used to calculate the electric flux

$$\oiint_{S} \mathbf{E} \cdot \mathbf{n}\, dS = \iint_{S_{\mathrm{lat}}} \mathbf{u}_r \cdot E(r)\mathbf{u}_r dS = E(r) \iint_{S_{\mathrm{lat}}} dS = E(r)2\pi rl.$$

The charge enclosed by this cylinder is $Q(S) = \lambda l$, then

$$E(r)2\pi rl = \frac{\lambda l}{\epsilon_0}.$$

Finally,

$$\mathbf{E}(r) = \frac{\lambda}{2\pi \epsilon_0 r}\mathbf{u}_r.$$

The same result was obtained in Problem 1.7.

2.4 A coaxial cable

We use a system of cylindrical coordinates (r, θ, z) with origin at the center of the cable. The cylindrical symmetry of the charge distribution implies an invariance of the electric field on the θ and z coordinates. In addition, and similar to the case of the infinitely long line of charge, for any point P one can find two planes of symmetry containing P (one perpendicular to Oz and another containing both $\mathbf{u}_r$ and $\mathbf{u}_z$) as shown in Fig. A.34. Curie's principle then translates into an orientation of the electric field along the intersection of these two planes, $\mathbf{u}_r$, and therefore $\mathbf{E}(r, \theta, z) = E(r)\mathbf{u}_r$.

We distinguish three regions, $I : \{b < r\}$, $II : \{a < r < b\}$ and $III : \{r \le a\}$. To calculate the electric field, we use Gauss's law, with a cylindrical surface (radius r, height h). The case $r > b$ is shown in Fig. A.35.

Through the basis surfaces, there will be a zero flux since $\mathbf{E} \cdot \mathbf{n} = \pm \mathbf{E} \cdot \mathbf{k} = 0$ on these surfaces. The flux integral is reduced to the integral on the lateral surface of the cylinder:

$$\oiint_{S} dS(\mathbf{x}')\mathbf{n}(\mathbf{x}') \cdot \mathbf{E}(\mathbf{x}') = \iint_{\mathrm{lateral}} dS(\mathbf{x}')\mathbf{n}(\mathbf{x}') \cdot \mathbf{E}(\mathbf{x}').$$

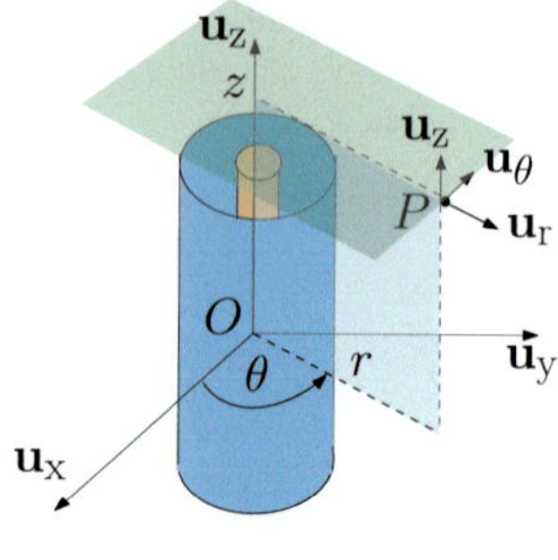

Fig. A.34 For any point P one can find two planes of symmetry whose intersection is the radial direction

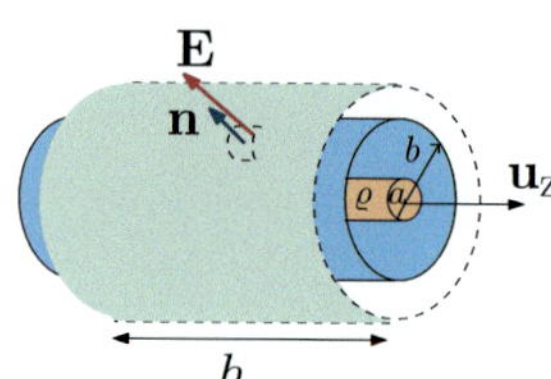

Fig. A.35 Cylindrical surface used to calculate the electric flux

For all $\mathbf{x}'$ on the lateral surface, $dS(\mathbf{x}')\mathbf{n}(\mathbf{x}') = r\,d\phi\,dz\,\mathbf{u}_r(\phi)$, and $\mathbf{E}(\mathbf{x}') = E(r)\mathbf{u}_r(\phi)$, so that

$$\iint_{\text{lateral}} dS(\mathbf{x}')\mathbf{n}(\mathbf{x}') \cdot \mathbf{E}(\mathbf{x}') = r E(r) \int_0^{2\pi} d\phi \int_0^h dz = 2\pi r h E(r).$$

Note that this result is general and true for all r. With this information,

$$\oiint_S d\mathbf{S}(\mathbf{x}') \cdot \mathbf{E}(\mathbf{x}') = E(r)2\pi r h,$$

but since the entire cable is neutral, the charge enclosed by S is zero, $Q(S) = 0$. Then, by Gauss's law, the electric field is zero outside of the cable:

$$E(r)2\pi r h = 0 \rightarrow E(r) = 0 \quad r > b.$$

Note that

$$Q(S) = \pi a^2 h \varrho + \sigma 2\pi b h = 0,$$

then the relationship between ρ and σ is as follows:

$$\sigma = -\frac{\varrho a^2}{2b}.$$

In region $II : \{a < r < b\}$, we again use a cylinder S of radius r and height h as a closed Gauss surface. We know that

$$\oiint_S d\mathbf{S}(\mathbf{x}') \cdot \mathbf{E}(\mathbf{x}') = E(r)2\pi rh = \frac{Q(S)}{\epsilon_0}$$

and the enclosed charge is, this time, that contained in the volume of the inner solid cylinder that is inside S:

$$Q(S) = \varrho\pi a^2 h,$$

so, by Gauss's theorem,

$$E(r)2\pi rh = \frac{\varrho\pi a^2 h}{\epsilon_0},$$

which yields

$$\mathbf{E}(r) = \frac{a^2 \varrho}{2\epsilon_0 r}\mathbf{u}_r \quad \text{if} \quad a \le r \le b.$$

Finally, for region $III : \{r \le a\}$, a cylinder of radius r and height h is again chosen as a Gauss surface. The flux through the Gauss surface reads

$$\oiint_S d\mathbf{S}(\mathbf{x}') \cdot \mathbf{E}(\mathbf{x}') = E(r)2\pi rh.$$

The volume bounded by the Gauss surface holds a charge

$$Q_S = \varrho\pi r^2 h,$$

so,

$$E(r)2\pi rh = \frac{\varrho\pi r^2 h}{\epsilon_0}$$

and

$$\mathbf{E}(\mathbf{x}) = \frac{\varrho r}{2\epsilon_0}\mathbf{u}_r.$$

In summary:

$$\mathbf{E}(\mathbf{x}) = \begin{cases} \dfrac{\varrho r}{2\epsilon_0} & \text{if } 0 \le r \le a, \\[2mm] \dfrac{\varrho a^2}{2\epsilon_0 r} & \text{if } a \le r < b, \\[2mm] 0 & \text{if } b < r. \end{cases}$$

Note the discontinuity when crossing the charged surface $r = b$,

$$\lim_{r \to b^+} E(r) - \lim_{r \to b^-} E(r) = -\frac{a^2 \varrho}{2\epsilon_0 b} = \frac{\sigma}{\epsilon_0}.$$

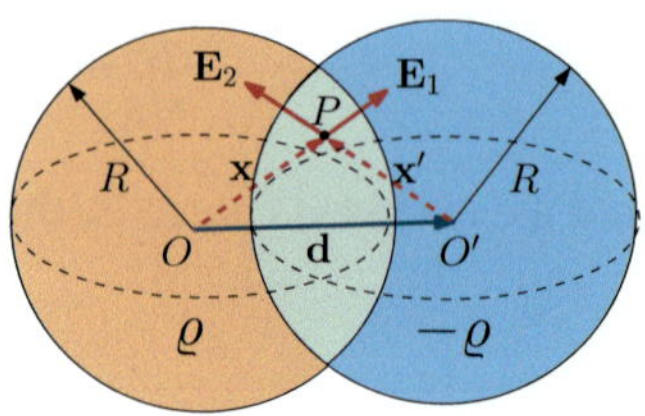

Fig. A.36 The field at point P can be calculated as the superposition of the fields created by two uniformly charged spheres

2.5 Superposition of spheres

This problem can be solved by superposition and by remembering the expression for the electric field generated by a homogeneous spherical distribution of charge of density ϱ:

$$
\mathbf{E}(\mathbf{x}) =
\begin{cases}
\dfrac{\varrho r}{3\epsilon_0}\mathbf{u}_r & \text{if } r \le R, \\[2ex]
\dfrac{\varrho R^2}{3\epsilon_0 r^2}\mathbf{u}_r & \text{if } r > R,
\end{cases}
$$

where r is the distance to the center of the sphere. Now, let P be an arbitrary point within the region of the intersection between the two spheres (see Fig. A.36). The field at P produced by the positive charge distribution is

$$
\mathbf{E}_1(P) = \frac{\varrho}{3\epsilon_0}\mathbf{x}
$$

and that of the negative charge distribution

$$
\mathbf{E}_2(P) = \frac{-\varrho}{3\epsilon_0}\mathbf{x}' \,,
$$

where $\mathbf{x} = \mathbf{OP}$ and $\mathbf{x}' = \mathbf{O'P}$ are the vectors defined in Fig. A.36. By the superposition principle, the electric field at P is given by

$$
\mathbf{E}(P) = \mathbf{E}_1 + \mathbf{E}_2 = \frac{\varrho}{3\epsilon_0}\mathbf{x} - \frac{\varrho}{3\epsilon_0}\mathbf{x}'.
$$

But $\mathbf{x} - \mathbf{x}' = \mathbf{OP} - \mathbf{O'P} = \mathbf{OO'} = \mathbf{d}$, so finally,

$$
\mathbf{E}(P) = \frac{\varrho}{3\epsilon_0}\mathbf{x} - \frac{\varrho}{3\epsilon_0}(\mathbf{x} - \mathbf{d}) = \frac{\varrho}{3\epsilon_0}\mathbf{d}.
$$

P was chosen arbitrarily, then, for every point within the intersection zone, the result is the same, and the electric field is uniform in that region.

Problems of Chap. 3

3.1 Potential difference between two charged planes

(a) The electric field generated by the two planes can be obtained by the superposition principle. If the origin is chosen as a point on the left plane, the field generated by the plane of density σ is (see Example 6.3)

$$\mathbf{E}_1(\mathbf{x}) = \begin{cases} \frac{\sigma}{2\epsilon_0}\mathbf{u}_x & \text{if } x > 0, \\ -\frac{\sigma}{2\epsilon_0}\mathbf{u}_x & \text{if } x < 0 \end{cases}.$$

Similarly, for the plane with charge density $-\sigma$

$$\mathbf{E}_2(\mathbf{x}) = \begin{cases} -\frac{\sigma}{2\epsilon_0}\mathbf{u}_x & \text{if } x > d \\ \frac{\sigma}{2\epsilon_0}\mathbf{u}_x & \text{if } x < d \end{cases}.$$

The total electric field is obtained from the superposition of both fields and so:

$$\mathbf{E}(\mathbf{x}) = \begin{cases} \frac{\sigma}{\epsilon_0}\mathbf{u}_x & \text{if } 0 < x < d, \\ 0 & \text{otherwise} \end{cases}.$$

The superposition gives a null electric field in the region outside the planes, and a uniform field in between the planes. Since the charge q is in equilibrium, along the x-axis we have

$$- T \sin\theta + F_e = 0,$$

with $F_e = q|\mathbf{E}|$ the modulus of the electrostatic force acting on the charge. Along the y-axis:

$$T \cos\theta - mg = 0,$$

so that

$$- mg \tan\theta + q|\mathbf{E}| = 0$$

and we find

$$\mathbf{E} = \frac{mg \tan\theta}{q}\mathbf{u}_x = \frac{\sigma}{\epsilon_0}\mathbf{u}_x.$$

The surface charge density is then given by

$$\sigma = \frac{mg\epsilon_0 \tan\theta}{q}.$$

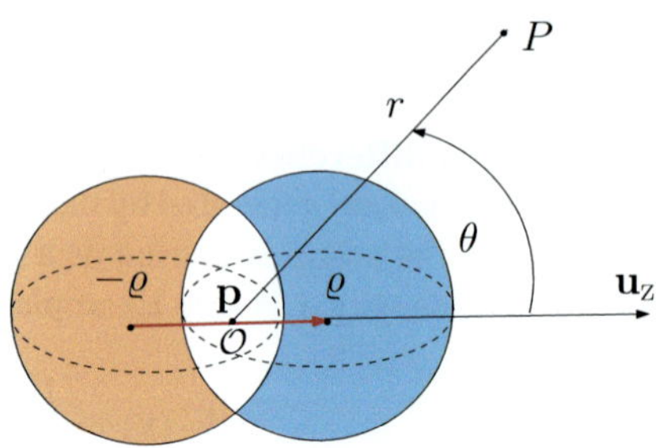

Fig. A.37 The field outside the spheres is that of an electric dipole

(b) The potential difference between the planes can be obtained by integrating the electric field over any path joining both planes. Choosing an horizontal path such that $d\mathbf{x} = \mathbf{u}_x$:

$$V(0) - V(d) = \int_0^d \mathbf{E} \cdot d\mathbf{x} = \int_0^d \frac{mg \tan\theta}{q} dx = \frac{mg \tan\theta d}{q} .$$

Remark Note that in this example, we have supposed that the charge density is uniform in both planes, even in the presence of the charge q in between. This is only an approximation, valid when the electric field generated by q can be neglected with respect to the field of the planes. In general, the charge distribution on the surface of a conductor is modified in the presence of other charges nearby.

3.2 Superposition of two spheres

Since the electric field outside a uniformly charged sphere of total charge Q is the same as the field generated by a point charge Q at the center of the sphere, this system is equivalent to the superposition of the two fields generated by point charges $Q_1 = -\frac{4}{3}\pi R^3 \varrho$ and $Q_2 = \frac{4}{3}\pi R^3 \varrho$ located at O_1 and O_2, respectively. The superposition of two spheres is, therefore, equivalent to an electric dipole moment

$$\mathbf{p} = \frac{4}{3}\pi R^3 \varrho \mathbf{d} .$$

At any point P at distance $r \gg |\mathbf{d}|$ from the origin, the potential is that of a electric dipole (see Fig. A.37):

$$V(P) = \frac{|\mathbf{p}| \cos\theta}{4\pi\epsilon_0 r^2} = \frac{\pi R^3 \varrho d \cos\theta}{3\pi\epsilon_0 r^2}$$

and the electric field

$$\mathbf{E}(P) = \frac{|\mathbf{p}|}{4\pi\epsilon_0 r^3}(2\cos\theta \mathbf{u}_r + \sin\theta \mathbf{u}_\theta) ,$$

whose modulus is

$$|\mathbf{E}(P)| = \frac{|\mathbf{p}|}{4\pi\epsilon_0 r^3}\sqrt{4\cos^2\theta + \sin^2\theta} = \frac{\pi R^3 \varrho d}{3\pi\epsilon_0 r^3}\sqrt{3\cos^2\theta + 1}\,.$$

3.3 Linear particle accelerator

(a) Assuming a uniform electric field between the gates g_k and g_{k+1} of the form $\mathbf{E}_k = E_k \mathbf{u}_x$, we have

$$V(g_{k+1}) - V(g_k) = -\int_{g_k}^{g_{k+1}} E_k dx = -E_k(g_{k+1} - g_k) = -E_k l_{k+1}$$

so that

$$\mathbf{E}_k = \frac{V(g_k) - V(g_{k+1})}{l_{k+1}}\mathbf{u}_x\,.$$

If k is even, then $g_k = 0$ and $g_{k+1} = +V$ since the electron enters the region between these two gates at $t = kT$ which is a multiple of $2T$. The electric field $\mathbf{E}_k$ is therefore

$$\mathbf{E}_k = -\frac{V}{l_{k+1}}\mathbf{u}_x\,.$$

For k odd, $g_k = -V$ and $g_{k+1} = 0$ since the electron enters the region between the gates at $t = kT$ which is of the form $t = (2n+1)T$ with $n \in \mathbb{N}$. We conclude that, for all k

$$\mathbf{E}_k = -\frac{V}{l_{k+1}}\mathbf{u}_x$$

so that the electrons are constantly accelerated in the $\mathbf{u}_x$ direction.

(b) The work W_k done by the electric force $-e\mathbf{E}$ while the electron passes from g_k to g_{k+1} is simply

$$W_k = \int_{g_k}^{g_{k+1}} (-e\mathbf{E}_k) \cdot d\mathbf{x} = -e\left(V(g_k) - V(g_{k+1})\right) = eV\,.$$

If the electron starts from rest at $t = 0$, then the kinetic energy when it passes g_k is simply

$$K_k = keV = \frac{1}{2}mv_k^2$$

and the velocity

$$v_k = \sqrt{\frac{2\,keV}{m}}\,.$$

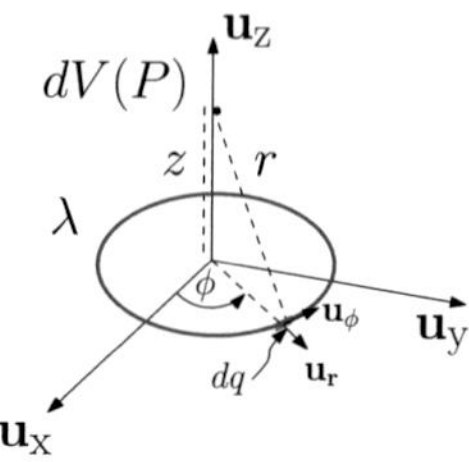

Fig. A.38 An infinitesimal charge element generates an infinitesimal potential dV at P

(c) At large k, the velocity is roughly constant between two consecutive gates and approximately equal to

$$v = v_k = \sqrt{\frac{2\,k e V}{m}}$$

so that

$$l_{k+1} = vT = \sqrt{\frac{2\,k e V}{m}}\, T \,.$$

3.4 Potential in the axis of a charged ring

Let us consider an infinitesimal line element on the ring $dl = R d\phi$. It posses a differential charge $dq = \lambda dl = d\phi R \lambda$ (see Fig. A.38). Its contribution to the potential at P is

$$dV(P) = \frac{1}{4\pi \epsilon_0} \frac{dq}{r} = \frac{1}{4\pi \epsilon_0} \frac{d\phi R \lambda}{\sqrt{R^2 + z^2}} \,.$$

Therefore, the potential due to the whole ring is the following superposition:

$$V(P) = \int_0^{2\pi} \frac{d\phi}{4\pi \epsilon_0} \frac{R\lambda}{\sqrt{R^2 + z^2}} = \frac{1}{4\pi \epsilon_0} \frac{2\pi R \lambda}{\sqrt{R^2 + z^2}} \,.$$

Noting that $Q = 2\pi R \lambda$ is the total charge of the ring, then:

$$V(P) = \frac{1}{4\pi \epsilon_0} \frac{Q}{\sqrt{R^2 + z^2}} \,.$$

We see that in the limit $z \gg R$, at zero order in R/z one obtains the potential generated by a point charge Q at the origin

$$V(P) \approx \frac{1}{4\pi \epsilon_0} \frac{Q}{|z|} \,.$$

Note that the electric field at P can be obtained from the potential as

Fig. A.39 Hollow charged disk of inner radius a and outer radius b

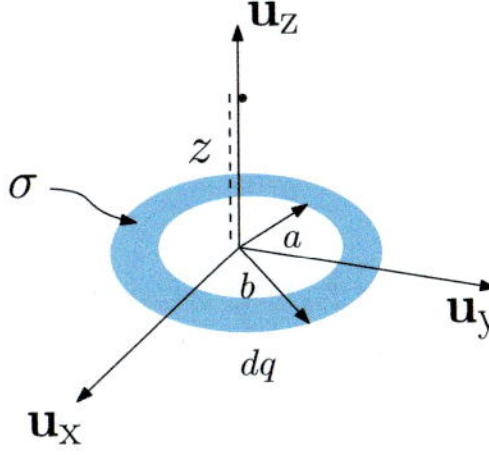

$$\mathbf{E}(P) = -\nabla V(P) = \frac{1}{4\pi\epsilon_0} \frac{zQ}{(R^2 + z^2)^{3/2}} \mathbf{u}_z \,.$$

To obtain the potential of a hollow disk, such as the one shown in Fig. A.39, one can consider the latter as the superposition of infinitely thin rings. Let us consider a section of the disk of radius r such that $a < r < b$ with an infinitesimal width dr, the potential generated at P by this section is that of a charged ring:

$$dV(P) = \frac{1}{4\pi\epsilon_0} \frac{dQ}{\sqrt{r^2 + z^2}}$$

with $dQ = \sigma 2\pi r dr$ the charge contained in that infinitesimal ring. Replacing:

$$dV(P) = \frac{2\pi\sigma}{4\pi\epsilon_0} \frac{r dr}{\sqrt{r^2 + z^2}}$$

and the potential of the hollow disk is

$$V(P) = \frac{\sigma}{2\epsilon_0} \int_a^b \frac{r dr}{\sqrt{r^2 + z^2}} \,.$$

Consider the substitution $u = r^2 + z^2$, so $du = 2r dr$ which gives

$$\int \frac{dr\, r}{\sqrt{r^2 + z^2}} = \frac{1}{2} \int \frac{du}{u^{1/2}} = \sqrt{u} = \sqrt{r^2 + z^2}$$

and:

$$V(P) = \frac{\sigma}{2\epsilon_0} \left(\sqrt{b^2 + z^2} - \sqrt{a^2 + z^2} \right).$$

Finally, the case of an infinite charged plane is obtained by taking the limit $a = 0, b \to \infty$. We see that the absolute potential of an infinite plane diverges, as a consequence of the fact that the charge of the plane is not integrable. One can, however, for a finite disk of radius b, add a constant to the potential such that $V(0) = 0$:

$$V(P) = \frac{\sigma}{2\epsilon_0} \left(\sqrt{b^2 + z^2} - |z| - b \right).$$

in which case the limit $b \to \infty$ for an infinite plane is well defined

$$\lim_{b \to \infty} V(P) = -\frac{\sigma}{2\epsilon_0}|z|\,.$$

3.5 Potential at a point outside the axis of a charged ring

The absolute potential in any point $\mathbf{x} \in \mathbb{R}^3$ is defined by the integral over the ring Γ:

$$V(\mathbf{x}) = \frac{1}{4\pi\epsilon_0} \oint_\Gamma dl \frac{\lambda(\mathbf{x}')}{|\mathbf{x} - \mathbf{x}'|}\,.$$

If the linear density in the ring is constant,

$$\lambda = \frac{Q}{2\pi R}\,.$$

Moreover, given the invariance of the charge distribution by rotation around the z-axis, the potential is of the form $V(\mathbf{x}) = V(r, z)$ in cylindrical coordinates. We can choose to calculate V in the plane $\theta = 0$ (xz) so that at any point $\mathbf{x} = x\mathbf{u}_x + z\mathbf{u}_z$. A point in the ring is characterized by $\mathbf{x}' = R\cos\theta\mathbf{u}_x + R\sin\theta\mathbf{u}_y$, so that:

$$V(\mathbf{x}) = \frac{Q}{8\pi^2\epsilon_0 R} \int_0^{2\pi} \frac{d\theta}{|(x - R\cos\theta)\mathbf{u}_x - R\sin\theta\mathbf{u}_y + z\mathbf{u}_z|}$$

$$= \frac{Q}{8\pi^2\epsilon_0 R} \int_0^{2\pi} \frac{d\theta}{\sqrt{(R\cos\theta - x)^2 + R^2\sin^2\theta + z^2}}\,.$$

Now, let us consider the integral:

$$I = \int_0^{2\pi} \frac{d\theta}{\sqrt{(R\cos\theta - x)^2 + R^2\sin^2\theta + z^2}} = \int_0^{2\pi} \frac{d\theta}{\sqrt{R^2 + x^2 + z^2 - 2xR\cos\theta}}$$

$$= \int_0^{2\pi} \frac{d\theta}{\sqrt{2xR}\sqrt{\frac{R^2+x^2+z^2}{2xR} - \cos\theta}}\,.$$

Let $b = \frac{R^2+x^2+z^2}{2xR}$, so:

$$I = \frac{1}{\sqrt{2xR}} \int_0^{2\pi} \frac{d\theta}{\sqrt{b - \cos\theta}} = \frac{2}{\sqrt{2xR}} \int_0^{\pi} d\theta \frac{1}{\sqrt{b - \cos\theta}} = \sqrt{\frac{2}{xR}}\sqrt{2\lambda}K(\lambda)\,.$$

Finally, the potential in terms of a complete elliptical integral of the first kind writes

$$V(x, z) = \frac{Q}{4\pi^2\epsilon_0 R}\sqrt{\frac{\lambda}{xR}}K(\lambda)\,,$$

with

$$\lambda = \frac{2}{1 + \frac{R^2 + x^2 + z^2}{2xR}} \; .$$

3.6 Ion between two cylindrical electrodes

Let us first find the velocity v at which ions enter the region between the cylinders. Since they are accelerated from rest at a potential V_0 toward a region with potential $V = 0$ (ground), conservation of the total mechanical energy gives

$$q V_0 = \frac{1}{2}mv^2 \;\rightarrow\; v^2 = \frac{2q V_0}{m} \; .$$

Now, since in between the two half-cylinders the trajectory is semi-circular, it follows that the electric field must be of the form $\mathbf{E} = E(r)\mathbf{u}_r$ in cylindrical coordinates with the origin at the common center of curvature of both electrodes. This approximation comes from neglecting edge effects so that both the θ dependence of the electric field and its $\mathbf{u}_\theta$ component are neglected. The equilibrium of forces in the radial direction then writes

$$q\mathbf{E} = q E(r)\mathbf{u}_r = -m\frac{v^2}{r}\mathbf{u}_r$$

with m the mass of an ion. Replacing v, the electric field for $a < r < b$ is given by

$$\mathbf{E}(r) = -\frac{mv^2}{qr}\mathbf{u}_r = -\frac{2V_0}{r}\mathbf{u}_r \; .$$

The potential difference between the two half-cylinders is then:

$$\Delta V = V(b) - V(a) = -\int_a^b \mathbf{E}(\mathbf{x}) \cdot d\mathbf{x}$$

and choosing a radial path:

$$\Delta V = \int_a^b \frac{2V_0}{r}dr = 2V_0 \ln\left(\frac{b}{a}\right) \; .$$

3.7 Infinite plane and point charge

(a) To obtain the potential in the x-axis we calculate first the total electric field. For the infinite plane, the electric field for $x > 0$ is

$$\mathbf{E}_1(x) = \frac{\sigma}{2\epsilon_0}\mathbf{u}_x$$

and the superposition principle states that, for $0 < x < a$

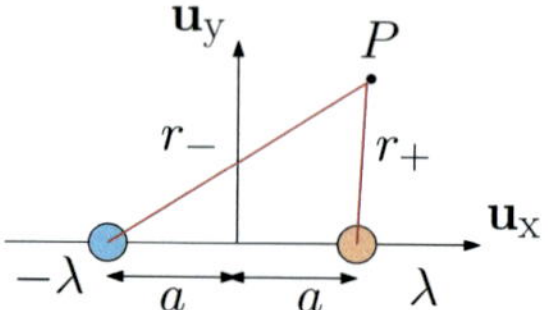

Fig. A.40 The point P is defined by the distances r_+ and r_- with respect to the wires

$$\mathbf{E}(x) = \mathbf{E}_1(x) + \frac{q}{4\pi\epsilon_0(a-x)^2}\mathbf{u}_x = \left(\frac{\sigma}{2\epsilon_0} + \frac{q}{4\pi\epsilon_0(a-x)^2}\right)\mathbf{u}_x \ .$$

The potential at position x with respect to the potential at $x = 0$ is given by

$$V(x) - V(0) = -\int_0^x \mathbf{E}(\mathbf{x}) \cdot d\mathbf{x} \ .$$

Choosing a path along the x-axis, $d\mathbf{x} = dx\,\mathbf{u}_x$ and

$$V(x) - V(0) = -\int_0^x E(x)dx = -\frac{\sigma x}{2\epsilon_0} + \frac{q}{4\pi\epsilon_0}\left(\frac{1}{a} - \frac{1}{(a-x)}\right) \ .$$

(b) If a particle of mass m and charge $-e$ is placed in $x = a/2$ at rest, it will feel a force along $-\mathbf{u}_x$ and it will eventually reach the plane at $x = 0$ with a speed v. Conservation of energy gives

$$-eV(a/2) = -eV(0) + \frac{1}{2}mv^2$$

and we find

$$v^2 = \frac{2e}{m}(V(0) - V(a/2)) = \frac{2e}{m}\left(\frac{-q}{4\pi\epsilon_0 a} + \frac{2q}{4\pi\epsilon_0 a} + \frac{\sigma}{4\epsilon_0}a\right) = \frac{2e}{m}\left(\frac{q}{4\pi\epsilon_0 a} + \frac{\sigma a}{4\epsilon_0}\right) \ .$$

3.8 Superposition of infinite wires

(a) The charge distribution being invariant under a translation along the z-axis, the potential is of the form $V(P) = V(x, y)$. Thus, we can restrict the calculation to any plane parallel to $z = 0$.

By the superposition principle, the potential at point P will be the superposition of the potentials generated by each wire separately (see Fig. A.40). In Problem 1.7, the electric field of an infinitely long wire is calculated. From this one can deduce that

$$V_1(P) = -\frac{\lambda}{2\pi\epsilon_0}\ln r_+ + C_1$$

and the potential due to the wire with density $-\lambda$ is

$$V_2(P) = \frac{\lambda}{2\pi \epsilon_0} \ln r_- + C_2 .$$

By using the potential at the origin as a reference, we have

$$V(P) - V(0) = \frac{\lambda}{2\pi \epsilon_0} (\ln r_- - \ln r_+)$$

and

$$r_+ = \sqrt{(x - a)^2 + y^2} \qquad r_- = \sqrt{(x + a)^2 + y^2}$$

so that

$$V(P) - V(0) = \frac{\lambda}{2\pi \epsilon_0} \ln \left(\frac{\sqrt{(x + a)^2 + y^2}}{\sqrt{(x - a)^2 + y^2}} \right) = \frac{\lambda}{4\pi \epsilon_0} \ln \left(\frac{(x + a)^2 + y^2}{(x - a)^2 + y^2} \right) .$$

(b) The equipotential surfaces are given by

$$\frac{(x + a)^2 + y^2}{(x - a)^2 + y^2} = k ,$$

where k is a constant. Equivalently,

$$(x + a)^2 + y^2 = k((x - a)^2 + y^2) ,$$

$$x^2(k - 1) + y^2(k - 1) + a^2(k - 1) - 2ax(k + 1) = 0 .$$

Dividing by $k - 1$:

$$x^2 + y^2 + a^2 - 2ax \frac{k + 1}{k - 1} = 0 ,$$

which is the equation of a circle with center $x_k = a\frac{k+1}{k-1}$ and radius $R_k = \frac{2a\sqrt{k}}{|k-1|}$. Since the potential is invariant under translation along the z-axis, we conclude that the equipotential surfaces are cylinders parallel to the z-axis of radius R_k and center x_k. Figure A.41 shows the equipotentials and field lines in a plane perpendicular to the $\mathbf{u}_z$ axis.

3.9 Force between a dipole and a charged line

(a) The field generated by an infinitely long line of charge is given by

$$\mathbf{E}(r) = \frac{\lambda}{2\pi \epsilon_0 r} \mathbf{u}_r .$$

The force on the dipole is

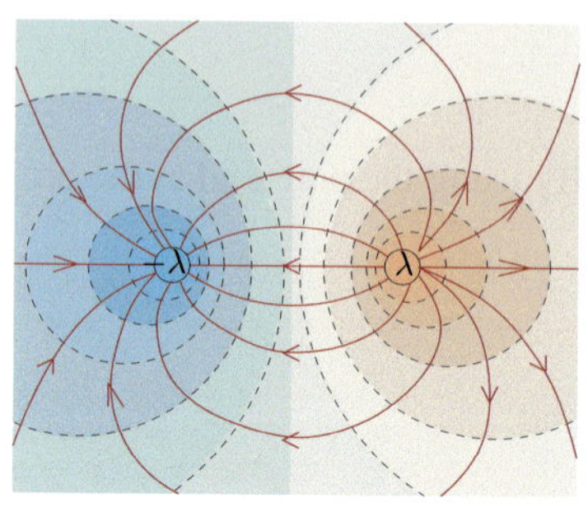

Fig. A.41 Electric field lines and equipotentials created by the two infinite wires

$$\mathbf{F} = q\mathbf{E}(r + d/2) - q\mathbf{E}(r - d/2) = \frac{q\lambda}{2\pi\epsilon_0}\left(\frac{1}{r + d/2} - \frac{1}{r - d/2}\right)\mathbf{u}_r$$

$$= \frac{q\lambda}{2\pi\epsilon_0 r}\left(\left(1 + \frac{d}{2r}\right)^{-1} - \left(1 - \frac{d}{2r}\right)^{-1}\right)\mathbf{u}_r.$$

By using the first-order expansion $(1 + x)^n \approx 1 + nx$ for $x \ll 1$, the force can be written as

$$\mathbf{F} \approx \frac{q\lambda}{2\pi\epsilon_0 r}\left(1 - \frac{d}{2r} - \left(1 + \frac{d}{2r}\right)\right)\mathbf{u}_r = \frac{q\lambda}{2\pi\epsilon_0 r}\left(-\frac{d}{r}\right)\mathbf{u}_r = \frac{-p\lambda}{2\pi\epsilon_0 r^2}\mathbf{u}_r,$$

where $p = qd$ is the magnitude of the dipole moment.

(b) Note that if $l \ll r$, we expect the force to be given by

$$\mathbf{F}(\mathbf{x}) = (\mathbf{p} \cdot \nabla)\,\mathbf{E}(\mathbf{x}).$$

In cylindrical coordinates, and given the symmetry of the electric field,

$$(\mathbf{p} \cdot \nabla)\,\mathbf{E}(r) = \left(p\mathbf{u}_r \cdot \mathbf{u}_r\frac{\partial}{\partial r}\right)\mathbf{E}(r) = p\frac{\partial}{\partial r}\left(\frac{\lambda}{2\pi\epsilon_0 r}\right)\mathbf{u}_r = -p\frac{\lambda}{2\pi\epsilon_0 r^2}\mathbf{u}_r,$$

which is indeed the result found previously.

3.10 Dipole–dipole interaction

(a) The potential generated by $\mathbf{p}$ is $V(\mathbf{r}) = \dfrac{1}{4\pi\epsilon_0}\dfrac{\mathbf{p} \cdot \mathbf{r}}{r^3}$.

and so

$$\mathbf{E} = -\frac{1}{4\pi\epsilon_0}\left(-3\frac{(\mathbf{p} \cdot \mathbf{r})\mathbf{r}}{r^5} + \frac{1}{r^3}\nabla\mathbf{p} \cdot \mathbf{r}\right), \quad \text{thus,} \quad \mathbf{E}(r) = \frac{1}{4\pi\epsilon_0 r^5}\left(3(\mathbf{p} \cdot \mathbf{r})\mathbf{r} - r^2\mathbf{p}\right).$$

(b) The potential energy is simply

$$U = -\mathbf{p}' \cdot \mathbf{E}(r) = -\frac{1}{4\pi\epsilon_0 r^5}\left(3(\mathbf{p} \cdot \mathbf{r})(\mathbf{p}' \cdot \mathbf{r}) - r^2(\mathbf{p} \cdot \mathbf{p}')\right).$$

(c) As $\mathbf{p}'$ does not depend on $\mathbf{E}(\mathbf{r})$, we can differentiate the potential energy:

$$
\begin{aligned}
\mathbf{F}_{\mathbf{p}\to\mathbf{p}'} &= -\nabla U \\
&= \frac{1}{4\pi\epsilon_0 r^5}\left(3(\mathbf{p}\cdot\mathbf{r})\nabla(\mathbf{p}'\cdot\mathbf{r}) + 3(\mathbf{p}'\cdot\mathbf{r})\nabla(\mathbf{p}\cdot\mathbf{r}) - (\mathbf{p}\cdot\mathbf{p}')\nabla(r^2)\right) \\
&\quad - 5\frac{\mathbf{r}}{4\pi\epsilon_0 r^7}\left(3(\mathbf{p}\cdot\mathbf{r})(\mathbf{p}'\cdot\mathbf{r}) - r^2(\mathbf{p}\cdot\mathbf{p}')\right) \\
&= \frac{1}{4\pi\epsilon_0 r^5}\left(3(\mathbf{p}\cdot\mathbf{r})\mathbf{p}' + 3(\mathbf{p}'\cdot\mathbf{r})\mathbf{p} - 2(\mathbf{p}\cdot\mathbf{p}')\mathbf{r} - 15(\mathbf{p}\cdot\mathbf{r})(\mathbf{p}'\cdot\mathbf{r})\frac{\mathbf{r}}{r^2} + 5(\mathbf{p}\cdot\mathbf{p}')\mathbf{r}\right).
\end{aligned}
$$

Finally,

$$
\mathbf{F}_{\mathbf{p}\to\mathbf{p}'} = \frac{1}{4\pi\epsilon_0 r^5}\left(3(\mathbf{p}\cdot\mathbf{r})\mathbf{p}' + 3(\mathbf{p}'\cdot\mathbf{r})\mathbf{p} + 3(\mathbf{p}\cdot\mathbf{p}')\mathbf{r} - 15(\mathbf{p}\cdot\mathbf{r})(\mathbf{p}'\cdot\mathbf{r})\frac{\mathbf{r}}{r^2}\right).
$$

(d) Yes there is a force acting on $\mathbf{p}$. Newton's third law tells us that the force acting on $\mathbf{p}$ is $\mathbf{F}_{\mathbf{p}'\to\mathbf{p}} = -\mathbf{F}_{\mathbf{p}\to\mathbf{p}'}$. This can also be seen by remarking the symmetry $\mathbf{p} \to \mathbf{p}'$ in the expression found at question (b), thus $\mathbf{F}_{\mathbf{p}'\to\mathbf{p}}$ is obtained by changing $\mathbf{r}$ into $-\mathbf{r}$, i.e., $\mathbf{F}_{\mathbf{p}'\to\mathbf{p}} = -\mathbf{F}_{\mathbf{p}\to\mathbf{p}'}$.

Problems of Chap. 4

4.1 Conductive sphere and sparks

If the sphere has a charge Q, it generates an electric field given by $\mathbf{E} = Q/(4\pi\epsilon_0 r^2)\mathbf{u}_r$ at a distance r from its center. For a spark to form between this sphere and a grounded sphere at a distance d from it, the electric field must be larger than the breakdown field of air ($E = 3 \times 10^6$ V m^{-1} at $d = 1$ m. Thus:

$$
\frac{|\,Q_{max}\,|}{4\pi\epsilon_0 d^2} = 3 \times 10^6 \text{ V m}^{-1},
$$

which gives

$$
|\,Q_{max}\,| \sim 300 \ \mu\text{C}.
$$

The potential at the surface of the sphere is given by

$$
V(R) = \int_R^d \mathbf{E}\cdot d\mathbf{x} = \int_R^d E(r)dr = \frac{Q}{4\pi\epsilon_0}\left(\frac{1}{R} - \frac{1}{d}\right),
$$

where a path has been chosen such that $d\mathbf{x} = dr\mathbf{u}_r$ with $\mathbf{u}_r$ the radial direction joining the centers of both spheres. Then, the maximum potential at which the sphere can be connected before generating a spark is

$$V_{max} = \frac{|Q_{max}|}{4\pi\epsilon_0}\left(\frac{1}{R} - \frac{1}{R+d}\right) = 6.3 \times 10^6 \text{ V}.$$

4.2 Screening length in semiconductors

(a) The potential must satisfy Poisson's equation $\nabla^2 V = -\frac{\varrho}{\epsilon_0}$, where the charge density for $r \neq 0$ writes, in spherical coordinates

$$\varrho(r) = -en(r) + ep(r) = e(p(r) - n(r)) = en_0\left(e^{eV(r)/k_BT} - e^{eV(r)/k_BT}\right),$$

where we assume a spherical symmetry so that the potential depends only on the r coordinate in spherical coordinates. Then the differential equation for the potential writes

$$\nabla^2 V(r) = \frac{\partial^2 V(r)}{\partial r^2} + \frac{2}{r}\frac{\partial V(r)}{\partial r} = -\frac{en_0}{\epsilon_0}\left(e^{-eV(r)/k_BT} - e^{eV(r)/k_BT}\right).$$

(b) Since $k_BT \gg eV$, we have up to first order in eV/k_BT

$$e^{\frac{-eV(r)}{k_BT}} - e^{\frac{eV(r)}{k_BT}} \approx 1 - \frac{eV(r)}{k_BT} - 1 - \frac{eV(r)}{k_BT} = -2\frac{eV(r)}{k_BT}$$

and the equation for the potential becomes a linear equation of the form

$$\frac{\partial^2 V(r)}{\partial r^2} + \frac{2}{r}\frac{\partial V(r)}{\partial r} = \frac{2en_0}{\epsilon_0}\frac{eV(r)}{k_BT}.$$

(c) Let us introduce the change of variable $rV(r) = U(r)$. Then, we have

$$\frac{\partial V(r)}{\partial r} = -\frac{1}{r^2}U(r) + \frac{1}{r}\frac{\partial U(r)}{\partial r},$$

$$\frac{\partial^2 V(r)}{\partial r^2} = \frac{2}{r^3}U(r) - \frac{2}{r^2}\frac{\partial U(r)}{\partial r} + \frac{1}{r}\frac{\partial^2 U(r)}{\partial r^2},$$

and so

$$\frac{\partial^2 V(r)}{\partial r^2} + \frac{2}{r}\frac{\partial V(r)}{\partial r} = \frac{1}{r}\frac{\partial^2 U(r)}{\partial r^2} = \frac{2e^2n_0}{k_BT\epsilon_0}\frac{U(r)}{r},$$

and the function U satisfies the equation

$$\frac{\partial^2 U(r)}{\partial r^2} - \frac{2e^2n_0}{k_BT\epsilon_0}U(r) = 0,$$

which admits solutions of the form $U(r) = Ae^{-r/\lambda_D} + Be^{r/\lambda_D}$, where we identify the following length scale $\lambda_D = \sqrt{\frac{k_BT\epsilon_0}{2e^2n_0}}$, also called the Debye length. The solution for which the potential generated by the ion vanishes at infinity reads

$$V(r) = \frac{V_0}{r} e^{-r/\lambda_D}$$

so that the electric field of the positive ion decays exponentially as a function of distance due to the screening effect of the negative charges surrounding it. Only those charges within a distance of the order of the Debye length will feel the electric field of the ion.

4.3 Charge density of the hydrogen atom

The potential is a solution of Poisson's equation $\nabla^2 V = -\frac{\varrho}{\epsilon_0}$, from which one can determine the charge density ϱ if the potential is known. Let us define $\psi(r) = \frac{1}{r}$ and $\chi(r) = r V(r)$, so that $V(r) = \psi(r)\chi(r)$. Then

$$\nabla^2 \underbrace{(\psi\chi)}_{V} = \nabla \cdot (\nabla\psi\chi) = \nabla \cdot (\psi\nabla\chi + \chi\nabla\psi) = \psi\nabla^2\chi + \chi\nabla^2\psi + 2\nabla\psi \cdot \nabla\chi$$

but $\nabla^2\psi = -4\pi\delta(r)$ and $\nabla\psi = -\frac{1}{r^2}\mathbf{u}_r$, so that:

$$\nabla^2 V(r) = \frac{1}{r} \underbrace{\nabla^2\chi(r)}_{\frac{1}{r}\frac{d^2}{dr^2}(r\chi(r))} - \underbrace{4\pi\delta(r)\chi(r)}_{4\pi\delta(r)\chi(0)} - \frac{2}{r^2} \underbrace{\mathbf{u}_r \cdot \nabla\chi(r)}_{\frac{d\chi(r)}{dr}}$$

and since

$$\frac{d^2}{dr^2}(r\chi(r)) = \frac{d}{dr}\left(\chi(r) + r\frac{d\chi(r)}{dr}\right) = 2\frac{d\chi(r)}{dr} + r\frac{d^2\chi(r)}{dr^2}$$

we obtain

$$\nabla^2 V(r) = \frac{1}{r}\frac{d^2\chi(r)}{dr^2} - 4\pi\delta(r)\underbrace{\chi(0)}_{\frac{q}{4\pi\epsilon_0}} = \frac{q\alpha^3}{8\pi\epsilon_0}e^{-\alpha r} - \delta(r)\frac{q}{\epsilon_0} = -\frac{\varrho(r)}{\epsilon_0} .$$

Finally, the charge density is given by

$$\varrho(r) = q\delta(r) - \frac{q\alpha^3}{8\pi}e^{-\alpha r} .$$

The term $q\delta(r)$ represents a proton at the origin, whereas $-\frac{q\alpha^3}{8\pi}e^{-\alpha r}$ is the averaged density representing the electron surrounding the proton. Note that integrated over the whole space, this second term has a total charge of $-q$, and the system is globally neutral.

4.4 Charge between cylinders

The charge distribution is invariant under rotations along the axis of the cylinders. In cylindrical coordinates this translates into an invariance of the potential upon the θ coordinate. In addition, if we neglect edge effects (or, equivalently, we consider the cylinders to be infinitely long), we have also an invariance on z, so that $V = V(r)$. Poisson's equation for $a < r < b$ then writes

$$\nabla^2 V(r) = \frac{1}{r}\frac{d}{dr}\left(r\frac{dV(r)}{dr}\right) = -\frac{\varrho}{\epsilon_0}\,.$$

Integrating once with respect to r, we obtain

$$\frac{dV(r)}{dr} = -\frac{r\varrho}{2\epsilon_0} + \frac{A}{r}$$

and integrating a second time with respect to r

$$V(r) = -\frac{r^2\varrho}{4\epsilon_0} + A\ln r + B$$

where the constants A and B can be determined with the proper boundary conditions. Since the inner cylinder is grounded, we must have $V(a) = 0$

$$V(a) = -\frac{a^2\varrho}{4\epsilon_0} + A\ln a + B = 0\,.$$

The outer cylinder is grounded as well, so that

$$V(b) = -\frac{b^2\varrho}{4\epsilon_0} + A\ln b + B = 0\,.$$

Taking the difference of the latter two equations

$$0 = A\ln\left(\frac{b}{a}\right) - \frac{(b^2 - a^2)\varrho}{4\epsilon_0}$$

which gives the constant A

$$A = \frac{(b^2 - a^2)\varrho}{4\epsilon_0\ln(b/a)}$$

and therefore

$$B = \frac{a^2\varrho}{4\epsilon_0} - \ln a\,\frac{(b^2 - a^2)\varrho}{4\epsilon_0\ln(b/a)} = \frac{\varrho(a^2\ln b - b^2\ln a)}{4\epsilon_0\ln(b/a)}$$

and the potential, for $a < r < b$, is given by

$$V(r) = -\frac{r^2 \varrho}{4\epsilon_0} + \frac{(b^2 - a^2)\varrho}{4\epsilon_0 \ln(b/a)} \ln r + \frac{\varrho(a^2 \ln b - b^2 \ln a)}{4\epsilon_0 \ln(b/a)} \, .$$

From this, the electric field in the region $a < r < b$ is obtained

$$\mathbf{E}(r) = -\frac{dV(r)}{dr}\mathbf{u}_r = \left(\frac{r\varrho}{2\epsilon_0} - \frac{(b^2 - a^2)\varrho}{4\epsilon_0 r \ln(b/a)} \right) \mathbf{u}_r$$

and by using Coulomb's theorem, the charge densities at $r = a$ and $r = b$ as well. For $r = a$ we have

$$\mathbf{E}(a) = \frac{\sigma_a}{\epsilon_0}\mathbf{u}_r = \frac{a\varrho}{2\epsilon_0} - \frac{(b^2 - a^2)\varrho}{4\epsilon_0 a \ln(b/a)}\mathbf{u}_r$$

and so

$$\sigma_a = \frac{a\varrho}{2} - \frac{(b^2 - a^2)\varrho}{4a \ln(b/a)}$$

For $r = b$, we have

$$\mathbf{E}(b) = \frac{\sigma_b}{\epsilon_0}(-\mathbf{u}_r) = \frac{b\varrho}{2\epsilon_0} - \frac{(b^2 - a^2)\varrho}{4\epsilon_0 b \ln(b/a)}\mathbf{u}_r$$

and we obtain

$$\sigma_b = -\frac{b\varrho}{2} + \frac{(b^2 - a^2)\varrho}{4b \ln(b/a)}$$

If the cylinders have a total length L, the total charge contained in the region between them is

$$Q = \pi(b^2 - a^2)L\varrho$$

The charge contained in $r = a$ is

$$Q_a = 2\pi a L \sigma_a = \pi a^2 L \varrho - \frac{2Q}{4 \ln(b/a)}$$

and in $r = b$

$$Q_b = 2\pi b L \sigma_b = -\pi b^2 L \varrho + \frac{2Q}{4 \ln(b/a)}$$

Note that $Q_b + Q_a = -Q$.

4.5 System of concentric conductors

(a) Although the spherical shell is neutral, a charge separation within the shell must occur so that it cancels the electric field at any point inside it. Since the sphere and

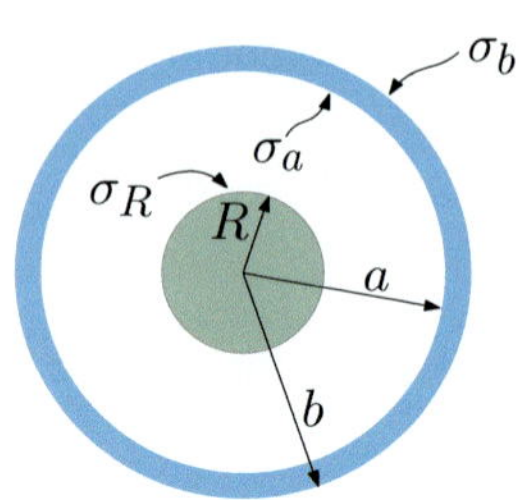

Fig. A.42 Charge densities on the three surfaces of the problem

the shell are both conductors, all their charge must be distributed on their surface. We then define the following charge densities: σ_R for the metallic sphere, σ_a and σ_b for the inner ($r = a$) and outer ($r = b$) surfaces of the spherical shell, respectively. These are shown in Fig. A.42. Finally we note that only homogeneous charge densities will preserve the spherical symmetry of the problem so that the electric field writes $\mathbf{E} = E(r)\mathbf{u}_r$.

The sphere having a total charge Q distributed uniformly on its surface, we have

$$\sigma_R = \frac{Q}{4\pi R^2}.$$

Since the electric field is zero inside the shell, $\mathbf{E}(\mathbf{x}) = \mathbf{0}$ for $a < r < b$. Using a spherical surface S of radius r, with $a < r < b$, Gauss's law gives

$$\oiint_S d\mathbf{S}(\mathbf{x}') \cdot \mathbf{E}(\mathbf{x}') = \frac{Q(S)}{\epsilon_0} = 0,$$

where the charge $Q(S)$ contained in S is

$$Q(S) = Q + \sigma_a 4\pi a^2 = 0.$$

The total charge induced on the inner surface of the shell is thus $-Q$, and the charge density is given by

$$\sigma_a = -\frac{Q}{4\pi a^2}.$$

Since the shell is electrically neutral, its total charge is zero

$$\underbrace{\sigma_a 4\pi a^2}_{-Q} + \sigma_b 4\pi b^2 = 0.$$

This means that the charge in the outer surface must be $+Q$ and so,

$$\sigma_b = \frac{Q}{4\pi b^2}.$$

(b) To determine the absolute potential at the center of the sphere we can first determine the electric field everywhere in space. For this, we use once again Gauss's law with a closed spherical surface S of radius r, such that

$$\oiint_S d\mathbf{S}(\mathbf{x}') \cdot \mathbf{E}(\mathbf{x}') = 4\pi r^2 E(r) = \frac{Q(S)}{\epsilon_0}.$$

It follows that the electric field, at every point outside the conductors ($R < r < a$ or $r > b$), is identical to that of a point charge Q at the origin $\mathbf{E} = \dfrac{Q}{4\pi\epsilon_0 r^2}\mathbf{u}_r$, and zero otherwise. The absolute potential at the origin can be obtained as

$$V(0) = \int_0^\infty \mathbf{E}(\mathbf{x}') \cdot d\mathbf{x}'$$

and by choosing a radial path, $d\mathbf{x}' = dr\mathbf{u}_r$, we obtain

$$V(0) = \int_0^\infty E(r)dr = \int_0^R \underbrace{E(r)\,dr}_{0} + \int_R^a \underbrace{E(r)\,dr}_{\frac{Q}{4\pi\epsilon_0 r^2}} + \int_a^b \underbrace{E(r)\,dr}_{0} + \int_b^\infty \underbrace{E(r)\,dr}_{\frac{Q}{4\pi\epsilon_0 r^2}}$$

$$= \frac{Q}{4\pi\epsilon_0}\left(\frac{1}{b} + \frac{1}{R} - \frac{1}{a}\right).$$

A much faster way of obtaining this result is to recall that the potential anywhere inside a charged metallic sphere of charge Q' and radius R' is simply $V = \frac{Q'}{4\pi\epsilon_0 R'}$. Here, we have the superposition of three charged spheres, one with charge Q and radius R, a second one with charge $-Q$ and radius a, and a third one with charge Q and radius b. So we can write directly

$$V(0) = \frac{Q}{4\pi\epsilon_0 R} - \frac{Q}{4\pi\epsilon_0 a} + \frac{Q}{4\pi\epsilon_0 b}.$$

(c) Now the outer surface $r = b$ of the shell is grounded. Nothing changes for the charge in the inner surface, since the induced charge $-Q$ should still be there in order to cancel the electric field inside the shell, as shown in Fig. A.43. However, the potential at $r = b$ must now be zero. This means that for any path connecting $r = b$ and infinity, we must have

$$V(b) = \int_b^\infty \mathbf{E}(\mathbf{x}') \cdot d\mathbf{x}' = 0,$$

which can only be true if the electric field itself is zero everywhere outside the shell, $\mathbf{E} = \mathbf{0}$ for $r > b$. This implies, by Gauss's law, that the total charge inside any closed surface surrounding the shell is zero. We conclude that the outer shell is no longer

Fig. A.43 The shell is now
grounded

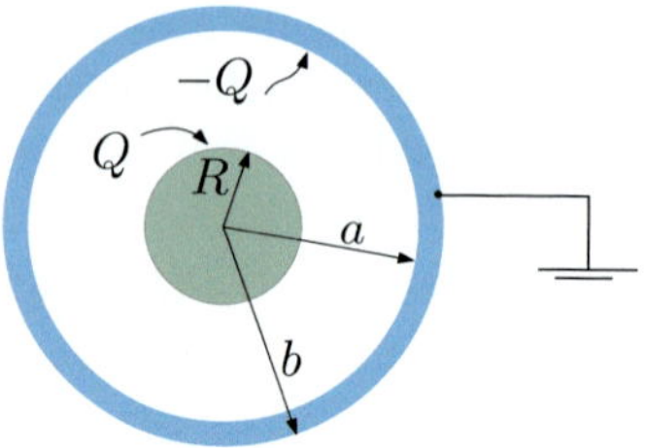

electrically neutral, it must have a total charge $-Q$. This means that the charge $+Q$
that was initially induced in the outer surface is evacuated by connecting the shell to
the ground. Therefore

$$\sigma_b = 0.$$

4.6 Three conductive spheres

(a) Due to the spherical symmetry of the system, the electric field must be of the
form $\mathbf{E}(\mathbf{x}) = E(r)\mathbf{u}_r$ in spherical coordinates. By using Gauss's law with a closed
spherical surface S of radius r we obtain

$$\oiint_S \mathbf{E}(\mathbf{x}') \cdot \mathbf{n}(\mathbf{x}')dS(\mathbf{x}') = 4\pi r^2 E(r) = Q(S),$$

where the charge $Q(S)$ inside S is given by

$$Q(S) = \begin{cases} 0 & \text{if } r < b, \\ -Q & \text{if } b < r < c, \\ 0 & \text{if } r > c, \end{cases}$$

so that the electric field reads

$$\mathbf{E}(r) = \begin{cases} \dfrac{-Q}{4\pi\epsilon_0 r^2}\mathbf{u}_r & \text{if } b < r < c, \\ \mathbf{0} & \text{otherwise.} \end{cases}$$

The absolute potential in the outer shell is

$$V(c) = \int_c^\infty \mathbf{E}(\mathbf{x}) \cdot d\mathbf{x}' = 0$$

since for any path going from $r = c$ to infinity contained in the region $r > c$, $\mathbf{E} = \mathbf{0}$.
For the potential in the sphere of radius b, we can choose a radial path from b to
infinity so that

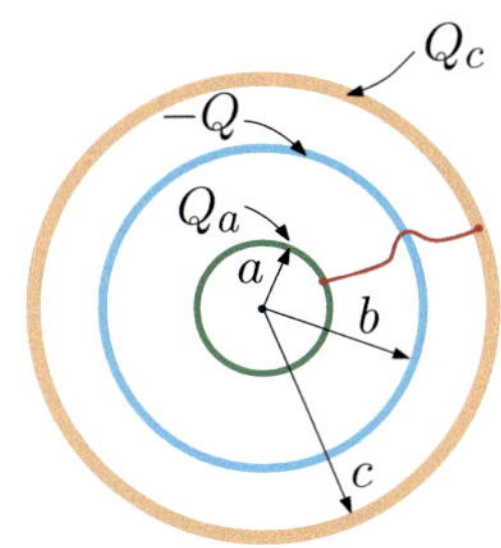

Fig. A.44 The inner and outer shell are now connected

$$V(b) = \int_b^\infty E(r)\mathbf{u}_r \cdot \mathbf{u}_r dr = \int_b^c \frac{-Q}{4\pi\epsilon_0 r^2} dr = \frac{Q}{4\pi\epsilon_0}\left(\frac{1}{c} - \frac{1}{b}\right)$$

and since the electric field is zero for $r < b$, the potential is constant and equal to $V(b)$ at any point such that $r < b$. In particular, for $r = a$,

$$V(a) = V(b) = \frac{Q}{4\pi\epsilon_0}\left(\frac{1}{c} - \frac{1}{b}\right).$$

(b) Now the shells $r = a$ and $r = c$ are at the same potential, as illustrated in Fig. A.44. For this to happen, the charge $+Q$ that was initially at $r = c$ is now distributed over the two conductors. We must have

$$Q_a + Q_c = Q.$$

Ignoring any effect of the thin wire connecting both surfaces, we assume that the radial symmetry of the electric field is preserved. Therefore, Gauss's law over a closed spherical surface S of radius r such that $a < r < b$ gives

$$\oiint_S \mathbf{E}(\mathbf{x}') \cdot \mathbf{n}(\mathbf{x}') dS(\mathbf{x}') = E(r)4\pi r^2 = \frac{Q_a}{\epsilon_0}.$$

Thus,

$$\mathbf{E}(r) = \frac{Q_a}{4\pi\epsilon_0 r^2}\mathbf{u}_r, \quad \text{for } a < r < b,$$

whereas for $b < r < c$, Gauss's law writes

$$E(r)4\pi r^2 = \frac{-Q + Q_a}{\epsilon_0}$$

and so,

$$\mathbf{E}(r) = \frac{Q_a - Q}{4\pi\epsilon_0 r^2}\mathbf{u}_r, \quad \text{for } b < r < c.$$

Since the total charge in the system continues to be zero, the electric field is, again, null for $r > c$. This means that the potential at $r = c$ is zero as well and so is the potential at $r = a$. Therefore

$$V(a) = \int_a^\infty E(r)dr = \int_a^b E(r)dr + \int_b^c E(r)dr + \underbrace{\int_c^\infty E(r)\,dr}_{=0}$$

$$= \int_a^b \frac{Q_a}{4\pi\epsilon_0 r^2}dr + \int_b^c \frac{Q_a - Q}{4\pi\epsilon_0 r^2}dr = \frac{Q_a}{4\pi\epsilon_0}\left(\frac{b-a}{ab}\right) + \frac{Q_a - Q}{4\pi\epsilon_0}\left(\frac{c-b}{bc}\right) = 0.$$

from this, we obtain

$$Q_a\left(\frac{b-a}{ab} + \frac{c-b}{bc}\right) = Q\left(\frac{c-b}{bc}\right)$$

and the charge Q_a is given by

$$Q_a = Q\frac{a(c-b)}{b(c-a)}$$

and

$$Q_c = Q - Q_a = \frac{c(b-a)}{b(c-a)}.$$

Finally, for $r = b$, we have

$$V(b) = \int_b^\infty E(r)dr = \int_b^c E(r)dr = \int_b^c \frac{Q_a - Q}{4\pi\epsilon_0 r^2}dr = \frac{Q_a - Q}{4\pi\epsilon_0}\left(\frac{c-b}{bc}\right)$$

and by replacing $Q_a - Q = -Q_c$, we find

$$V(b) = \frac{Q}{4\pi\epsilon_0}\frac{c(a-b)}{b(c-a)}\left(\frac{c-b}{bc}\right).$$

4.7 Cavities in a conductor

(a) Let us first consider the case of a single point charge q at the center of a spherical cavity of radius r inside a conductor. Let Q_r be the induced charge at the surface of the cavity. Now, consider an arbitrary closed surface S contained in the conductor and surrounding the spherical cavity (see Fig. A.45). The total charge contained inside S is $q + Q_r$, and since the electric field is zero at any point inside the conductor, Gauss's law writes

$$\oiint_S \mathbf{E}(\mathbf{x}')\mathbf{n}(\mathbf{x}')dS(\mathbf{x}') = \frac{q + Q_r}{\epsilon_0} = 0.$$

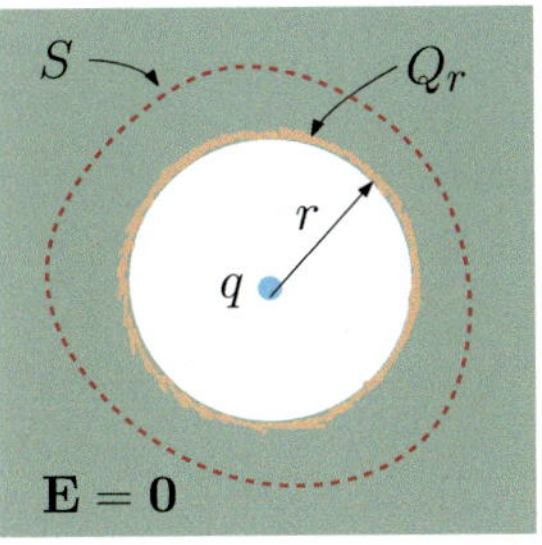

Fig. A.45 A gauss surface S fully within the conductor is surrounding the charge and the cavity

Hence, at the interface between the conductor and the cavity, a charge equal to $Q_r = -q$ is induced so as to cancel the electric field of the point charge q anywhere outside the cavity. The system is therefore equivalent to a point charge q at the center of a charged sphere with surface charge Q_r. For the potential to be constant at the surface of the cavity, it follows that the charge Q_r must be homogeneously distributed, thus with a surface density

$$\sigma_r = \frac{Q_r}{4\pi r^2} = -\frac{q}{4\pi r^2}.$$

For any point outside the cavity, the superposition of q and Q_r has therefore no influence whatsoever, and when looked at from outside, it is as if the cavity were not existing. Now, if there are N cavities inside a conductor, they can be treated separately as we did for a single cavity. The charge densities on the surfaces of the cavities a and b are therefore given by

$$\sigma_a = \frac{-q_a}{4\pi a^2} \quad \text{and} \quad \sigma_b = -\frac{q_b}{4\pi b^2}.$$

Since the conductive sphere is neutral, the total charge induced on the outer surface of radius R will be $Q_R = -Q_a - Q_b = q_a + q_b$ and for the potential to be constant at the surface $r = R$, this charge should be distributed homogeneously. The charge density at this surface is therefore

$$\sigma_R = \frac{q_a + q_b}{4\pi R^2}.$$

(b) In the region outside the sphere, the system is equivalent to a charged sphere of radius R with total charge Q_R uniformly distributed on its surface. This is due to the fact that each cavity and its respective point charge generate no electric field outside them. Then, the electric field for $r > R$ is given by

$$\mathbf{E}(r) = \frac{q_a + q_b}{4\pi\epsilon_0 r^2}\mathbf{u}_r, \quad r > R.$$

Fig. A.46 Electric field
lines. The field is zero
anywhere inside the
conductor

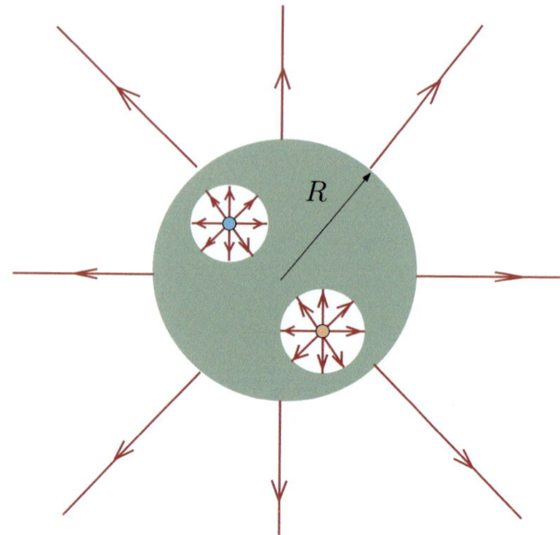

(c) The charge on the surface of each cavity does not contribute to the electric field
inside the cavity. This follows from Gauss's law applied to any surface included in
the cavity. Therefore, the field inside each cavity is simply the field generated by the
point charge at its center. Thus,

$$\mathbf{E}_a = \frac{q_a}{4\pi\epsilon_0 r^2}\mathbf{u}_r^{(a)} \text{ for cavity } a \text{ and } \mathbf{E}_b = \frac{q_b}{4\pi\epsilon_0 r^2}\mathbf{u}_r^{(b)} \text{ for cavity } b,$$

with $\mathbf{u}_r^{(a)}$ (resp. $\mathbf{u}_r^{(b)}$) the radial direction with respect to the origin at q_a (resp. q_b).
The electric field lines are schematically shown in Fig. A.46.

4.8 Charge on Earth's surface

(a) The electric field is non-zero in the atmosphere since the latter is not in electrostatic
equilibrium. Indeed, as seen in Problem 2.1, there is a non-zero charge density in the
atmosphere because of its constant ionization due to the incident cosmic and solar
radiation. This charge density is responsible for an exponentially decaying electric
field for $R_E < r < h$

$$\mathbf{E}(\mathbf{x}) = E_0 e^{-r/H}\mathbf{u}_r.$$

The potential difference between the ionosphere and the ground can be obtained as

$$\Delta V = -\int_0^h \mathbf{E}\cdot d\mathbf{l} = -E_0\int_0^h e^{-r/H}dr = E_0 H(e^{-h/H} - 1),$$

where a radial path has been used. The numerical evaluation of the magnitude of
the electric field at Earth's surface for $H = 3.5\,\text{km}$, $h = 100\,\text{km}$, and $\Delta V = 350\,\text{kV}$
yields

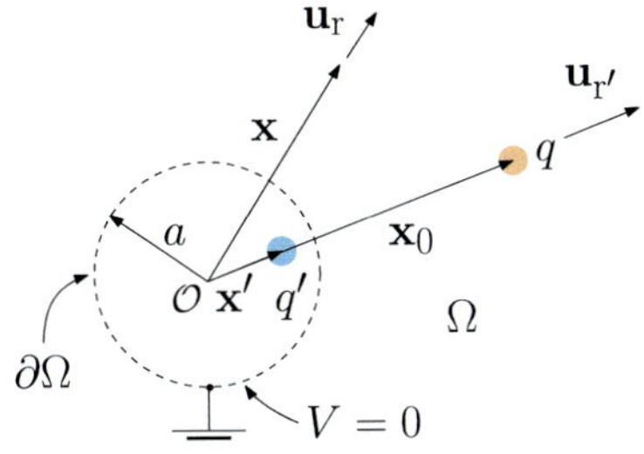

Fig. A.47 The grounded sphere is replaced by an image charge q' along the axis that join O and the charge q.

$$E_0 = -\frac{\Delta V}{H(1 - e^{-\frac{h}{H}})} \approx -\frac{\Delta V}{H} = -100 \,\text{V}\,\text{m}^{-1}.$$

(b) Since the electric field at the surface of a conductor reads, according to Coulomb's theorem, $\mathbf{E} = \dfrac{\sigma}{\epsilon_0}\mathbf{n}$, where $\mathbf{n}$ is the normal to the surface, we see that

$$\mathbf{E}(0) = E_0\mathbf{u}_r = \frac{\sigma}{\epsilon_0}\mathbf{u}_r.$$

The ground is therefore negatively charged, since $E_0 < 0$. The total charge on Earth's surface is

$$Q_E = 4\pi R_E^2 \sigma = 4\pi \epsilon_0 R_E^2 E_0.$$

Given Earth's radius $R_E = 6400\,\text{km}$, we obtain a total charge of

$$Q_E = -4.55 \times 10^5 \,\text{C}.$$

4.9 **Point charge in the presence of a grounded conductive sphere**
We need to solve the following Poisson–Dirichlet problem in the volume Ω outside the sphere:

$$\begin{cases} \mathbf{x} \in \Omega : \quad \nabla^2 V(\mathbf{x}) = -\dfrac{q}{\epsilon_0}\delta(\mathbf{x} - \mathbf{x}_0), \\ \mathbf{x} \in \partial\Omega : V(\mathbf{x}) = 0. \end{cases}$$

We can solve this problem by using the method of images. In spherical coordinates, we write the position of the charge q as $\mathbf{x}_0 = d\mathbf{u}_{r'}$. Let us place an image charge q' in the volume inside the sphere at position $\mathbf{x}' = d'\mathbf{u}_{r'}$, as shown in Fig. A.47.

The potential at any point $\mathbf{x} = r\mathbf{u}_r \in \Omega$ is therefore

$$V(\mathbf{x}) = \frac{1}{4\pi\epsilon_0}\frac{q}{|r\mathbf{u}_r - d\mathbf{u}_{r'}|} + \frac{1}{4\pi\epsilon_0}\frac{q'}{|r\mathbf{u}_r - d'\mathbf{u}_{r'}|},$$

which satisfies $\nabla^2 V(\mathbf{x}) = -\dfrac{q}{\epsilon_0}\delta(\mathbf{x} - \mathbf{x}_0)$ in Ω. To find q' and d' we impose Dirichlet's boundary condition at the surface of the sphere

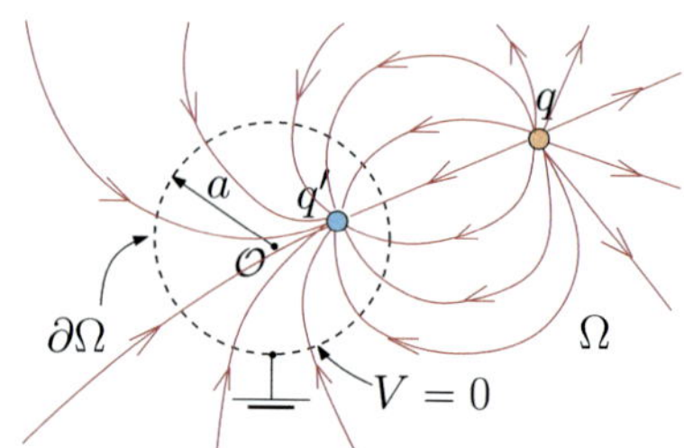

Fig. A.48 Electric field lines created by the two charges q and q'

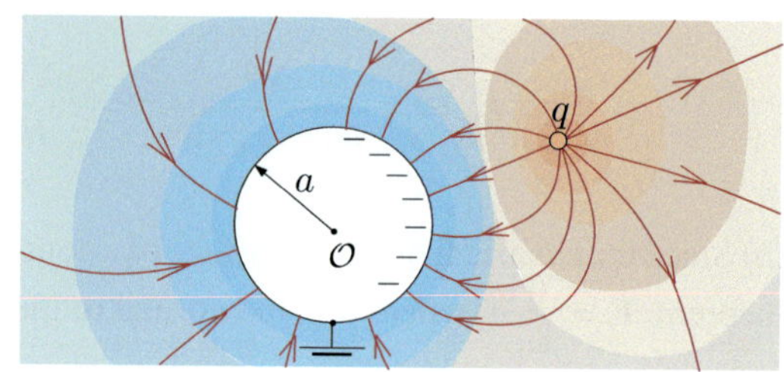

Fig. A.49 Electric field lines and equipotentials created by the charge q and the grounded sphere

$$
V(r = a) = \frac{1}{4\pi\epsilon_0} \left\{ \frac{q/a}{\left| \mathbf{u}_r - \dfrac{d}{a}\mathbf{u}_{r'} \right|} + \frac{q'/d'}{\left| \mathbf{u}_{r'} - \dfrac{a}{d'}\mathbf{u}_r \right|} \right\}
$$

$$
= \frac{1}{4\pi\epsilon_0} \left\{ \frac{\dfrac{q}{a}\left| \mathbf{u}_{r'} - \dfrac{a}{d'}\mathbf{u}_r \right| + \dfrac{q'}{d'}\left| \mathbf{u}_r - \dfrac{d}{a}\mathbf{u}_{r'} \right|}{\left| \mathbf{u}_r - \dfrac{d}{a}\mathbf{u}_{r'} \right| \left| \mathbf{u}_{r'} - \dfrac{a}{d'}\mathbf{u}_r \right|} \right\} = 0
$$

So, we can take $d' = \dfrac{a^2}{d}$ and $q' = -\dfrac{a}{d}q$. Introducing the notation $\alpha = \dfrac{a}{d}$, we write $d' = a\alpha$ and $q' = -\alpha q$. The potential in Ω is therefore given by

$$
V(\mathbf{x}) = \frac{q}{4\pi\epsilon_0} \left\{ \frac{1}{|\mathbf{x} - d\mathbf{u}_{r'}|} - \frac{\alpha}{|\mathbf{x} - a\alpha\mathbf{u}_{r'}|} \right\}.
$$

The electric field lines generated by these two charges are shown in Fig. A.48. It is seen that the electric field is normal to the surface of the sphere.

The electric field lines and equipotentials of the original problem are thus identical to the ones of the two-point charges outside the sphere and are shown in Fig. A.49. The induced charge density on the sphere can be obtained by Coulomb's theorem:

$$
\mathbf{x} \in \partial\Omega : \sigma(\mathbf{x}) = \epsilon_0 \mathbf{E}(\mathbf{x}) \cdot \mathbf{n}(\mathbf{x})|_{\mathbf{x}=a\mathbf{u}_r}
$$

where the electric field $\mathbf{E} = -\nabla V$ reads

$$\mathbf{E}(\mathbf{x}) = \frac{q}{4\pi\epsilon_0}\left\{\frac{r\mathbf{u}_r - d\mathbf{u}_{r'}}{|r\mathbf{u}_r - d\mathbf{u}_{r'}|^3} - \alpha\frac{r\mathbf{u}_r - a\alpha\mathbf{u}_{r'}}{|r\mathbf{u}_r - a\alpha\mathbf{u}_{r'}|^3}\right\}.$$

At the surface of the sphere, $r = a$ and $\mathbf{n} = \mathbf{u}_r$, and so

$$\sigma(\mathbf{x}) = \frac{q}{4\pi}\left\{\frac{a - d\mathbf{u}_{r'}\cdot\mathbf{u}_r}{|a\mathbf{u}_r - d\mathbf{u}_{r'}|^3} - \alpha\frac{a - a\alpha\mathbf{u}_{r'}\cdot\mathbf{u}_r}{|a\mathbf{u}_r - a\alpha\mathbf{u}_{r'}|^3}\right\}.$$

If we choose the orientation of the Cartesian directions such that $\mathbf{u}_{r'} = \mathbf{u}_z$, then $\mathbf{u}_r \cdot \mathbf{u}_{r'} = \cos\theta$, and

$$|a\mathbf{u}_r - d\mathbf{u}_{r'}|^3 = (a^2 + d^2 - 2ad\cos\theta)^{3/2}$$
$$= d^3\left(1 + \alpha^2 - 2\alpha\cos\theta\right)^{3/2},$$
$$|a\mathbf{u}_r - a\alpha\mathbf{u}_{r'}|^3 = a^3\left(1 + \alpha^2 - 2\alpha\cos\theta\right)^{3/2}.$$

The charge density on the surface of the sphere can then be rewritten as

$$\sigma(\mathbf{x}) = \frac{q}{4\pi\left(1 + \alpha^2 - 2\alpha\cos\theta\right)^{3/2}} \times \left\{\frac{\alpha - \cos\theta}{d^2} - \alpha\frac{1 - \alpha\cos\theta}{a^2}\right\}$$
$$= \frac{q}{4\pi ad}\frac{\alpha^2 - 1}{\left(1 + \alpha^2 - 2\alpha\cos\theta\right)^{3/2}}$$

with $\alpha = a/d$. The induced charge density has a maximum at $\theta = 0$, which corresponds to the point on the sphere that is the closest to the charge q. The total charge Q induced on the sphere is

$$Q = \oiint_{\partial\Omega} dS(\mathbf{x})\sigma(\mathbf{x})$$
$$= a^2\int_0^{2\pi}d\phi\int_0^{\pi}\sin\theta\,\sigma(\cos\theta)\,d\theta = 2\pi a^2\int_{-1}^{1}\sigma(\mu)d\mu$$
$$= -\frac{q}{2}\alpha\left(1 - \alpha^2\right)\int_{-1}^{1}\frac{d\mu}{(A - B\mu)^{3/2}}$$

with $A = 1 + \alpha^2$, $B = 2\alpha$. Moreover, using the change of variable $v = A - B\mu$,

$$\int_{-1}^{1} (A - B\mu)^{-3/2} d\mu = -\frac{1}{B} \int_{A+B}^{A-B} v^{-3/2} dv$$

$$= \frac{2}{B} \left\{ \frac{(A+B)^{1/2} - (A-B)^{1/2}}{(A^2 - B^2)^{1/2}} \right\} = \frac{2}{1-\alpha^2}.$$

Finally, the total charge induced on the sphere equals the image charge q'

$$Q = \oiint_{\partial\Omega} dS(\mathbf{x})\sigma(\mathbf{x}) = -q\alpha = q'.$$

The force $\mathbf{F}_q$ that the sphere exerts on the point charge q can be calculated by considering instead the force that the image charge q' exerts on q

$$\mathbf{F}_q = \frac{qq'}{4\pi\epsilon_0} \frac{\mathbf{x}' - \mathbf{x}''}{|\mathbf{x}' - \mathbf{x}''|^3} = -\frac{q^2\alpha}{4\pi\epsilon_0} \frac{d - a\alpha}{|d - a\alpha|^3} \mathbf{u}_{r'}$$

$$= -\frac{q^2}{a^2 4\pi\epsilon_0} \alpha^3 \frac{\mathbf{u}_{r'}}{\left(1 - \alpha^2\right)^2}.$$

Another way to obtain this result is to consider the electrostatic pressure (4.4), $P_e = \dfrac{\sigma^2}{2\epsilon_0}$, such that the force $d\mathbf{F}_e$ acting on a differential surface element dS on the sphere is

$$d\mathbf{F}(\mathbf{x}) = P_e(\mathbf{x})dS(\mathbf{x})\mathbf{n}(\mathbf{x}) = \frac{\sigma^2(\mathbf{x})dS(\mathbf{x})}{2\epsilon_0} \mathbf{u}_r(\mathbf{x})$$

and the total force on the sphere becomes

$$\mathbf{F} = \frac{1}{2\epsilon_0} \oiint_{\partial\Omega} dS(\mathbf{x})\sigma^2(\mathbf{x})\mathbf{u}_r(\mathbf{x})$$

and after an extensive calculation, one can show that

$$\mathbf{F}_e = \frac{q^2}{4\pi\epsilon_0 a^2} \alpha^3 \left\{1 - \alpha^2\right\}^{-2} \mathbf{u}_{r'} = -\mathbf{F}_q.$$

4.10 Dipole inside a spherical shell

Outside the sphere, the potential is solution to Laplace's equation with the boundary condition $V(R, \theta) = 0$, which gives simply

$$V(r, \theta) = 0, \quad r \geq R.$$

In the region $r < R$, the potential is the superposition of the potential far from a dipole and a solution of Laplace's equation invariant to rotations around the z-axis, which then writes in terms of Legendre polynomials

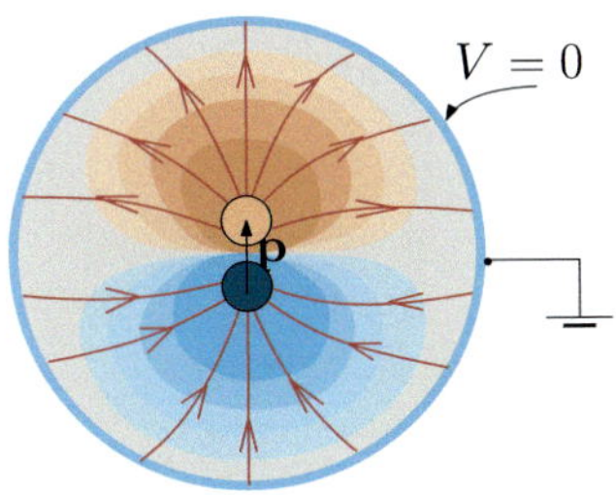

Fig. A.50 Electric field lines and equipotentials created by the dipole and surrounding grounded sphere

$$V(r, \theta) = \frac{p \cos \theta}{4\pi \epsilon_0 r^2} + \sum_{l=0}^{\infty} (A_l r^l + B_l r^{-(l+1)}) P_l(\cos \theta) \quad r \leq R.$$

For the potential to be finite at $r = 0$, $B_l = 0$ for every l. If, in addition, we impose the Dirichlet boundary condition $V(r = R, \theta) = 0$, we obtain

$$V(R, \theta) = \frac{p \cos \theta}{4\pi \epsilon_0 R^2} + \sum_{l=0}^{\infty} A_l R^l P_l(\cos \theta) = 0,$$

that is

$$\sum_{l=0}^{\infty} A_l R^l P_l(\cos \theta) = -\frac{p \cos \theta}{4\pi \epsilon_0 R^2}$$

and following the orthogonality of the Legendre polynomials, we conclude that $A_l = 0$ for $l \neq 1$, and

$$A_1 = -\frac{p \cos \theta}{4\pi \epsilon_0 R^3}.$$

Finally:

$$V(r, \theta) = \frac{p \cos \theta}{4\pi \epsilon_0} \left(\frac{1}{r^2} - \frac{r}{R^3} \right), \quad r \leq R.$$

The electric field lines and equipotential surfaces are shown in Fig. A.50. The charge density $\sigma(\theta)$ induced on the shell is given by Coulomb's theorem (4.1):

$$\sigma(\theta) = \epsilon_0 \mathbf{E}(r, \theta) \cdot \underbrace{\mathbf{n}}_{-\mathbf{u}_r} = \epsilon_0 \left. \frac{\partial V(r, \theta)}{\partial r} \right|_{r=R}.$$

This relation yields

$$\sigma(\theta) = -\frac{3p \cos \theta}{4\pi R^3}.$$

The total charge induced in the upper hemisphere is

$$Q = \int_0^{\pi/2} 2\pi R^2 \sigma(\theta) \sin\theta \, d\theta$$

$$= -\frac{3p}{2R} \int_0^1 x \, dx = -\frac{3p}{4R}$$

and since $p = 2qa$:

$$Q = -\frac{3}{2} q \frac{a}{R}.$$

4.12 A corollary of the mean value theorem

In vacuum, the potential satisfies Laplace's equation $\nabla^2 V(\mathbf{x}) = 0$. Using a Green function $G(\mathbf{x}, \mathbf{x}') = 1/|\mathbf{x} - \mathbf{x}'|$, we have according to (4.11)

$$V(\mathbf{x}) = -\frac{1}{4\pi} \oiint_{\partial\Omega} \left[d\mathbf{S}(\mathbf{x}') \cdot \nabla' \frac{1}{|\mathbf{x} - \mathbf{x}'|} \right] V(\mathbf{x}') + \frac{1}{4\pi} \oiint_{\partial\Omega} \frac{d\mathbf{S}(\mathbf{x}') \cdot \nabla' V(\mathbf{x}')}{|\mathbf{x} - \mathbf{x}'|}$$

$$= -\frac{1}{4\pi} \oiint_{\partial\Omega} \frac{d\mathbf{S}(\mathbf{x}') \cdot \mathbf{E}(\mathbf{x}')}{|\mathbf{x} - \mathbf{x}'|} + \frac{1}{4\pi} \oiint_{\partial\Omega} d\mathbf{S}(\mathbf{x}') \cdot \underbrace{\frac{\mathbf{x}' - \mathbf{x}}{|\mathbf{x}' - \mathbf{x}|^3}}_{= d S(\mathbf{x}')/a^2} V(\mathbf{x}')$$

where $a = |\mathbf{x} - \mathbf{x}'|$ if one chooses $\partial\Omega$ as a sphere of radius a centered at $\mathbf{x}$. Now, using the mean value theorem

$$V(\mathbf{x}) = \frac{1}{4\pi a^2} \oiint_{\partial\Omega} dS(\mathbf{x}') V(\mathbf{x}'),$$

then, we conclude that

$$\oiint_{\partial\Omega} \frac{d\mathbf{S}(\mathbf{x}') \cdot \mathbf{E}(\mathbf{x}')}{|\mathbf{x} - \mathbf{x}'|} = 0$$

and, in addition, Green–Ostrogradsky theorem (A.15) gives

$$\oiint_{\partial\Omega} d\mathbf{S}(\mathbf{x}') \cdot \frac{\mathbf{E}(\mathbf{x}')}{|\mathbf{x} - \mathbf{x}'|} = \iiint_{\Omega} \nabla' \cdot \frac{\mathbf{E}(\mathbf{x}')}{|\mathbf{x} - \mathbf{x}'|} d^3 x'$$

$$= \iiint_{\Omega} \left\{ \frac{\nabla' \cdot \mathbf{E}(\mathbf{x}')}{|\mathbf{x} - \mathbf{x}'|} + \mathbf{E}(\mathbf{x}') \cdot \nabla' \frac{1}{|\mathbf{x} - \mathbf{x}'|} \right\} d^3 x'$$

but in vacuum $\nabla \cdot \mathbf{E} = 0$, and so

$$\oiint_{\partial\Omega} d\mathbf{S}(\mathbf{x}') \cdot \frac{\mathbf{E}(\mathbf{x}')}{|\mathbf{x} - \mathbf{x}'|} = \iiint_{\Omega} \mathbf{E}(\mathbf{x}') \cdot \frac{\mathbf{x} - \mathbf{x}'}{|\mathbf{x} - \mathbf{x}'|^3} d^3 x'$$

$$= -\iiint_{\Omega} \mathbf{E}(\mathbf{x}') \cdot \frac{\mathbf{x}' - \mathbf{x}}{|\mathbf{x}' - \mathbf{x}|^3} d^3 x' = 0.$$

4.12 Electrostatics with massive photons

(a) Given the spherical symmetry of the charge distribution, the modified Poisson's equation writes in spherical coordinates as

$$\frac{1}{r}\frac{d^2}{dr^2}(rV(r)) - \mu^2 V(r) = -\frac{q}{\epsilon_0}\delta(\mathbf{r}) .$$

For $\mathbf{r} \neq 0$, we must then have

$$\frac{1}{r}\frac{d^2}{dr^2}(rV(r)) - \mu^2 V(r) = 0 \quad \rightarrow \quad \frac{d^2}{dr^2}(rV(r)) - \mu^2 rV(r) = 0 .$$

From this we see that the function $U(r) = rV(r)$ satisfies

$$\frac{d^2}{dr^2}U(r) - \mu^2 U(r) = 0$$

and the general solution is

$$U(r) = Ae^{-\mu r} + Be^{+\mu r} .$$

The potential is then given by

$$V(r) = \frac{U(r)}{r} = \frac{A}{r}e^{-\mu r} + \frac{B}{r}e^{+\mu r} .$$

Knowing that the potential must be finite at infinity, $B = 0$ and since for $\mu = 0$ we must retrieve the well-known result $V(r) = \frac{q}{4\pi\epsilon_0 r} = A$, we conclude that

$$V(r) = \frac{q}{4\pi\epsilon_0 r}e^{-\mu r} .$$

This predicts a potential that behaves like the Coulomb potential at short scales ($r \ll \mu^{-1}$) but decays exponentially for $r \gg \mu^{-1}$.

(b) The electric field is given as usual by

$$\mathbf{E}(r) = -\nabla V = \frac{q}{4\pi\epsilon_0}e^{-\mu r}\left(\frac{1}{r^2} + \frac{\mu}{r}\right)\mathbf{u}_r .$$

(c) For an arbitrary charge distribution in a volume Ω one may use the superposition principle so that

$$V(\mathbf{r}) = \frac{1}{4\pi\epsilon_0}\iiint_\Omega \frac{\varrho(\mathbf{x}')e^{-\mu|\mathbf{x}-\mathbf{x}'|}}{|\mathbf{x}-\mathbf{x}'|}d^3x'$$

and

$$\mathbf{E}(\mathbf{r}) = \frac{1}{4\pi\epsilon_0} \iiint_\Omega \varrho(\mathbf{x}')e^{-\mu|\mathbf{x}-\mathbf{x}'|} \left(\frac{\mathbf{x} - \mathbf{x}'}{|\mathbf{x} - \mathbf{x}'|^3} + \mu \frac{\mathbf{x} - \mathbf{x}'}{|\mathbf{x} - \mathbf{x}'|^2} \right) d^3x'.$$

(d) Since the sphere is uniformly charged, the charge density at its surface is $Q/(4\pi a^2)$ and the potential at any point $\mathbf{r}$ can be written as

$$V(\mathbf{r}) = \frac{Q}{16\pi^2 a^2 \epsilon_0} \int_0^{2\pi} \int_0^{\pi} \frac{e^{-\mu|\mathbf{r}-a\mathbf{u}'_r|}}{|\mathbf{r} - a\mathbf{u}'_r|} a^2 \sin\theta' d\theta' d\phi'.$$

Since the charge distribution is invariant under rotations, $V(\mathbf{r}) = V(r)$ so that we can choose $\mathbf{r} = r\mathbf{u}_z$ without loss of generality. In this case $|\mathbf{r} - a\mathbf{u}'_r| = \sqrt{r^2 + a^2 - 2ar\cos\theta'}$ and

$$V(r) = \frac{Q}{8\pi\epsilon_0} \int_0^{\pi} \frac{e^{-\mu\sqrt{r^2+a^2-2ar\cos\theta'}}}{\sqrt{r^2 + a^2 - 2ar\cos\theta'}} \sin\theta' d\theta'.$$

Using the integration variable s such that $s^2 = a^2 + r^2 - 2ar\cos\theta'$, one has $2sds = 2ar\sin\theta' d\theta'$ and since $r \leq a$

$$V(r) = \frac{Q}{8\pi ar\epsilon_0} \int_{a-r}^{a+r} \frac{e^{-\mu s}}{s} s\, ds = \frac{Q}{8\pi a\mu r\epsilon_0} \left(e^{-\mu(a-r)} - e^{-\mu(a+r)} \right).$$

(e) From the previous result we see that the potential is no longer uniform inside the shell, the electric field is not zero which is in contradiction with Gauss's law which is valid for $\mu = 0$. This suggests that a non-zero photon mass should produce a measurable electric field inside the shell. Another option would be to measure a non-zero potential difference between the surface and the center of the shell. Note that close to the center, $\mu r \ll 1$, and one may write

$$V(r) \approx \frac{Q}{8\pi\epsilon_0\mu ar} e^{-\mu a} (1 + \mu r - 1 + \mu r) = \frac{Q}{4\pi\epsilon_0 a} e^{-\mu a}$$

and assuming a very small photon mass so $\mu a \ll 1$

$$V(a) - V(0) \approx \frac{Q}{4\pi\epsilon_0} \mu .$$

Experimentally, the upper limit for the photon mass is of the order of 1×10^{-50} kg.

Problems of Chap. 5

5.1 Spherical capacitor

Let us assume that the internal sphere has a total charge Q, while the outer sphere $-Q$, as shown in Fig. A.51. The spherical symmetry of this charge distribution translates into an electric field of the form $\mathbf{E}(\mathbf{x}) = E(r)\mathbf{u}_r$ in spherical coordinates, with the origin at the center of the spheres.

Considering a closed spherical surface S of radius r such that $a < r < b$, then

$$\oiint_S \mathbf{E}(\mathbf{x}') \cdot \underbrace{\mathbf{n}(\mathbf{x}')}_{\mathbf{u}_r}\, dS(\mathbf{x}') = E(r)4\pi r^2 = \frac{Q(S)}{\epsilon_0}$$

where the charge $Q(S)$ inside S is simply Q. The electric field then reads

$$\mathbf{E}(\mathbf{x}) = \frac{Q}{4\pi\epsilon_0 r^2}\mathbf{u}_r.$$

The potential difference between the spheres is therefore

$$\Delta V = V(b) - V(a) = -\int_a^b \mathbf{E}\cdot d\mathbf{l}$$

and choosing a radial path

$$\Delta V = -\frac{Q}{4\pi\epsilon_0}\int_a^b \frac{dr}{r^2} = \frac{Q}{4\pi\epsilon_0}\left(\frac{b-a}{ab}\right).$$

The capacitance is

$$C = \frac{Q}{\Delta V} = 4\pi\epsilon_0\left(\frac{ab}{b-a}\right).$$

5.2 Capacitance between Earth's surface and the ionosphere

Based on the results of the mentioned exercise, the capacity of the system ground-ionosphere is

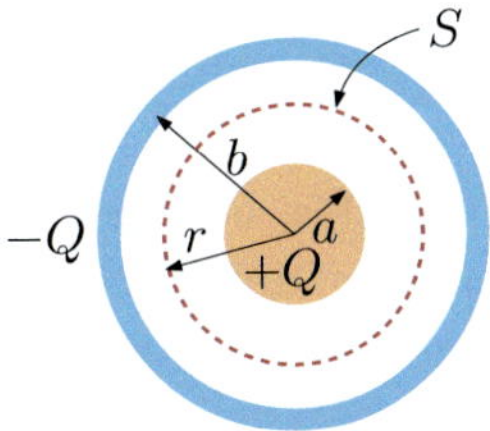

Fig. A.51 The capacitor is supposed to carry a charge Q

$$C = \frac{|Q|}{\Delta V} = 1.3\,\text{F}.$$

Note that if all the charge present in the atmosphere were confined exclusively in the ionosphere (i.e., the case of an insulating atmosphere between the ground and the ionosphere), the capacitance would be that of a spherical capacitor (see Exercise 5.1), resulting in a much smaller value of the capacitance

$$C = 4\pi \epsilon_0 R_T (1 + R_T/h) = 0.046\,\text{F}.$$

5.3 Capacitance of three concentric spheres

In Exercise 5.1 the capacitance C_0 of a spherical capacitor was determined. Assuming that the sphere of radius a has a charge Q on its surface, while the sphere of radius b has a charge $-Q$, we obtain

$$C_0 = \frac{Q}{\Delta V} = 4\pi \epsilon_0 \left(\frac{ab}{b - a} \right).$$

The introduction of a spherical shell between the spheres does not change the symmetry and invariance properties of the charge distribution, and so the electric field maintains its radial dependence $\mathbf{E}(\mathbf{x}) = E(r)\mathbf{u}_r$. Using Gauss's law over a closed spherical surface S of radius r, we get

$$\oiint_S \mathbf{E}(\mathbf{x}') \cdot \mathbf{n}(\mathbf{x}')dS(\mathbf{x}') = E(r)4\pi r^2 = \frac{Q(S)}{\epsilon_0}.$$

For $a < r < c$, the charge inside S is simply the charge on the sphere of radius a, which we assume to be $Q(S) = +Q$. Then

$$\mathbf{E}(r) = \frac{Q}{4\pi \epsilon_0 r^2}\mathbf{u}_r \quad \text{for } a < r < c$$

whereas for $c < r < d$, $\mathbf{E}(r) = \mathbf{0}$ since it corresponds to the region inside a conductor. This happens because an induced charge $-Q$ appears on the inner surface of the shell, at $r = c$. The shell being neutral, an induced charge $+Q$ is distributed on its outer surface at $r = d$. The total charge inside a closed spherical surface of radius r with $d < r < b$ is therefore $+Q$, and we obtain for the electric field the same result as for $a < r < c$:

$$\mathbf{E}(r) = \frac{Q}{4\pi \epsilon_0 r^2}\mathbf{u}_r \quad \text{for } d < r < b.$$

From this, we see that the effect of the shell inside the capacitor is to reduce the potential difference between the spheres $r = a$ and $r = b$, since in the region $b < r < c$ the electric field vanishes. In consequence, the capacitance of the system increases. Indeed, by choosing a radial path from $r = a$ to $r = b$

Fig. A.52 Coordinate system chosen to determine the potential

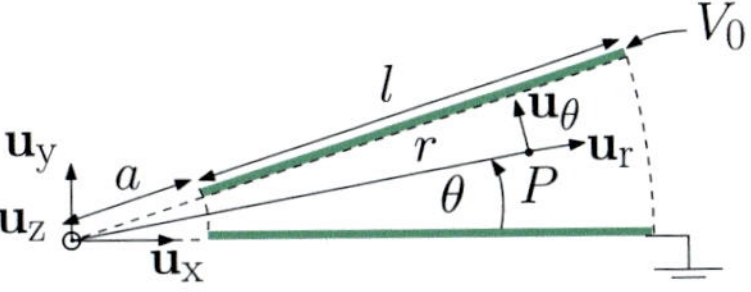

$$
\begin{aligned}
\Delta V &= V(a) - V(b) = \int_a^b \mathbf{E}(\mathbf{x}) \cdot d\mathbf{l}(\mathbf{x}) = \int_a^b E(r)\,dr \\
&= \int_a^c E(r)\,dr + \underbrace{\int_c^d E(r)\,dr}_{0} + \int_d^b E(r)\,dr \\
&= \int_a^c dr\,\frac{Q}{4\pi\epsilon_0 r^2} + \int_d^b dr\,\frac{Q}{4\pi\epsilon_0 r^2} \\
&= \left(\frac{1}{a} - \frac{1}{b} + \underbrace{\frac{1}{d} - \frac{1}{c}}_{<0} \right) \frac{Q}{4\pi\epsilon_0} ,
\end{aligned}
$$

giving finally

$$
C = \frac{Q}{\Delta V} > C_0.
$$

5.4 Plates at an angle

(a) Let us choose a system of polar coordinates with the origin at the intersection point of the plates, as shown in Fig. A.52. The system being invariant to a translation along the z-axis, the potential will be independent of the z coordinate.

In the region between the plates, the potential satisfies Laplace's equation

$$
\nabla^2 V(r, \theta) = 0.
$$

Expanding the Laplacian operator, Laplace's equation reads

$$
\frac{1}{r}\frac{\partial}{\partial r}\left(r\frac{\partial}{\partial r} V(r, \theta) \right) + \frac{1}{r^2}\frac{\partial^2 V(r, \theta)}{\partial \theta^2} = 0.
$$

Looking for a solution of the form $V(r, \theta) = R(r)\Theta(\theta)$, we obtain

$$
\frac{r^2}{R(r)}\frac{d^2 R(r)}{dr^2} + \frac{r}{R(r)}\frac{dR(r)}{dr} + \frac{1}{\Theta(\theta)}\frac{d^2\Theta(\theta)}{d\theta^2} = 0.
$$

The boundary conditions are given by $V(r, 0) = 0$ and $V(r, \theta_0) = V_0$ for $a \le r \le a + l$. If we neglect edge effects, we can consider these conditions to remain valid for any $r \ge 0$, in which case the solution for the potential is obtained by taking $R(r) = C$

where C is a constant, so that $\dfrac{dR(r)}{dr} = 0$ and therefore

$$\frac{1}{\Theta(\theta)} \frac{d^2 \Theta(\theta)}{d\theta^2} = 0 \quad \rightarrow \quad \Theta(\theta) = A\theta + B.$$

Finally, the boundary conditions give $A = V_0/\theta_0$ and $B = 0$

$$V(r, \theta) = V_0 \frac{\theta}{\theta_0}$$

and the electric field is simply

$$\mathbf{E} = -\nabla V = -\frac{1}{r} \frac{dV(\theta)}{d\theta} \mathbf{u}_\theta = -\frac{1}{r} \frac{V_0}{\theta_0} \mathbf{u}_\theta.$$

(b) The charge Q on the upper plate can be obtained by Coulomb's theorem

$$\mathbf{E} \cdot \mathbf{n}|_{\theta=\theta_0} = -\frac{1}{r} \frac{V_0}{\theta_0} \mathbf{u}_\theta \cdot (-\mathbf{u}_\theta) = \frac{\sigma(r, \theta_0)}{\epsilon_0}$$

from which the charge density is

$$\sigma(r, \theta_0) = \frac{\epsilon_0}{r} \frac{V_0}{\theta_0},$$

the charge Q is

$$Q = L \int_a^{a+l} \sigma(r, \theta_0) dr = L \int_a^{a+l} \frac{\epsilon_0}{r} \frac{V_0}{\theta_0} dr$$
$$= L\epsilon_0 \frac{V_0}{\theta_0} \ln\left(\frac{a+l}{a}\right)$$

and the capacitance is given by

$$C = \frac{Q}{V_0} = \frac{\epsilon_0 L}{\theta_0} \ln\left(1 + \frac{l}{a}\right).$$

5.5 Conic capacitor

The charge distribution being invariant to rotations around the z-axis, Curie's principle results in an invariance of the potential upon the ϕ coordinate, so that $V = V(r, \theta)$ in spherical coordinates. Let Ω be the region in between the cones ($\theta_1 < \theta < \theta_2$). The potential satisfies Laplace's equation in Ω

$$\nabla^2 V(r, \theta) = 0$$

with boundary conditions $V(\theta_1) = 0$ and $V(\theta_2) = V_0$ which are independent of the coordinate r. It results that the potential depends solely on θ and

$$\nabla^2 V(\theta) = \frac{1}{r^2 \sin\theta} \frac{\partial}{\partial\theta} \left(\sin\theta \frac{\partial V}{\partial\theta} \right) = 0$$

and so

$$\sin\theta \frac{\partial V(\theta)}{\partial\theta} = C_1 \quad \rightarrow \quad \frac{\partial V(\theta)}{\partial\theta} = \frac{C_1}{\sin\theta}.$$

The solution is obtained by integrating with respect to θ

$$V(\theta) = C_1 \int \frac{d\theta}{\sin\theta} + C_2 = C_1 \ln\left(\tan\frac{\theta}{2} \right) + C_2.$$

Imposing the boundary conditions

$$V(\theta_1) = C_1 \ln\left(\tan\frac{\theta_1}{2} \right) + C_2 = 0,$$

$$V(\theta_2) = C_1 \ln\left(\tan\frac{\theta_2}{2} \right) + C_2 = V_0,$$

we obtain

$$C_1 = \frac{V_0}{\ln\left(\dfrac{\tan\frac{\theta_2}{2}}{\tan\frac{\theta_1}{2}} \right)}, \quad C_2 = -\frac{V_0 \ln\left(\tan\frac{\theta_1}{2} \right)}{\ln\left(\dfrac{\tan\frac{\theta_2}{2}}{\tan\frac{\theta_1}{2}} \right)}.$$

The electric field is obtained by using $\mathbf{E}(\mathbf{x}) = -\nabla V(\mathbf{x})$. Since V only depends on θ

$$\mathbf{E}(r, \theta) = -\frac{1}{r} \frac{\partial V(\theta)}{\partial\theta} \mathbf{u}_\theta = -\frac{C_1}{r \sin\theta} \mathbf{u}_\theta$$

$$= -\frac{V_0}{\ln\left(\dfrac{\tan\frac{\theta_2}{2}}{\tan\frac{\theta_1}{2}} \right)} \frac{1}{r \sin\theta} \mathbf{u}_\theta.$$

5.6 Two conductive hemispheres

The potential satisfies Laplace's equation at every point such that $r \neq R$. Since the potential does not depend on ϕ, it can be written as an expansion of Legendre polynomials according to (4.9)

$$V(r, \theta) = \sum_{l=0}^{\infty} \left(A_l r^l + B_l r^{-(l+1)} \right) P_l(\cos \theta) \quad r \leq R,$$

$$V(r, \theta) = \sum_{l=0}^{\infty} \left(C_l r^l + D_l r^{-(l+1)} \right) P_l(\cos \theta) \quad r > R.$$

For $r \leq R$, we must have $B_l = 0$ since the potential must be finite at $r = 0$, and imposing the boundary condition on the surface of the sphere

$$V(R, \theta) = V_D(\theta) = \sum_{l=0}^{\infty} A_l R^l P_l(\cos \theta),$$

we recognize the expansion of V_D as a Legendre series of the form (A.49). Then

$$A_l = \frac{2l + 1}{2R^l} \int_0^{\pi} \sin \theta \, P_l(\cos \theta) V_D(\theta) d\theta.$$

For $R > a$, for the potential to remain finite as $r \to \infty$, we must have $C_l = 0$ for every l. Then

$$V(r, \theta) = \sum_{l=0}^{\infty} \frac{D_l}{r^{l+1}} P_l(\cos \theta), \quad r > a.$$

The continuity of the potential at $r = R$ gives

$$V(R, \theta) = \sum_{l=0}^{\infty} \frac{D_l}{R^{l+1}} P_l(\cos \theta) = \sum_{l=0}^{\infty} A_l a^l P_l(\cos \theta)$$

and so $D_l = R^{2l+1} A_l$. The potential then writes

$$V(r, \theta) = \sum_{l=0}^{\infty} A_l r^l P_l(\cos \theta) \quad \text{for } r \leq R,$$

$$V(r, \theta) = \sum_{l=0}^{\infty} R^{2l+1} \frac{A_l}{r^{l+1}} P_l(\cos \theta) \quad \text{for } r \geq R,$$

with

$$A_l = \frac{2l+1}{2R^l} \int_0^\pi \sin\theta \, P_l(\cos\theta) V(R,\theta) \, d\theta$$

$$= \frac{2l+1}{2R^l} V_0 \left(\int_0^{\pi/2} \sin\theta \, P_l(\cos\theta) \, d\theta - \int_{\pi/2}^\pi \sin\theta \, P_l(\cos\theta) \, d\theta \right)$$

$$= V_0 \frac{2l+1}{2R^l} \left(\int_0^1 P_l(x) dx - \int_{-1}^0 P_l(x) dx \right).$$

If l is even, $P_l(x) = P_l(-x)$, whereas if l is odd, $P_l(x) = -P_l(-x)$. It follows that $A_l = 0$ for l even and

$$A_l = V_0 \frac{2l+1}{R^l} \int_0^1 P_l(x) dx \quad \text{for } l \text{ odd}$$

and so

$$V(r,\theta) = \sum_{l=1}^\infty A_{2l-1} r^{2l-1} P_{2l-1}(\cos\theta), \quad r \le R,$$

$$V(r,\theta) = \sum_{l=1}^\infty R^{4l-1} \frac{A_{2l-1}}{r^{2l}} P_{2l-1}(\cos\theta), \quad r \ge R,$$

and using the recursion formula (A.47):

$$P_n(x) = \frac{1}{2n+1} \frac{d}{dx} [P_{n+1}(x) - P_{n-1}(x)],$$

we have

$$C_n = \int_0^1 P_n(x) dx = \frac{1}{2n+1} [P_{n-1}(0) - P_{n+1}(0)], \quad n \ge 1$$

since $P_l(1) = 1$ for every l. Now using the recursion (A.46) ($P_n(0) = -(n-1)/n P_{n-2}(0)$), we see that for every $n \ge 1$, $P_{2n-1}(0) = 0$ and $P_{2n}(0) = (-1)^n (2n-1)!!/(2n)!!$ In consequence $C_n = 0$ for every even n, and

$$C_{2n-1} = \int_0^1 P_{2n-1}(x) dx = -\frac{P_{2n}(0)}{2n-1} = (-1)^{n+1} \frac{(2n-1)!!}{(2n)!!(2n-1)} \quad \text{for } n \ge 1.$$

Finally

$$V(r,\theta) = V_0 \sum_{l=1}^\infty (-1)^{l+1} \frac{(4l-1)(2l-1)!!}{(2l)!!(2l-1)} \times P_{2l-1}(\cos\theta) \begin{cases} \dfrac{r^{2l-1}}{R^{2l-1}} & r \le R, \\[2mm] \dfrac{R^{2l}}{r^{2l}} & r \ge R, \end{cases}$$

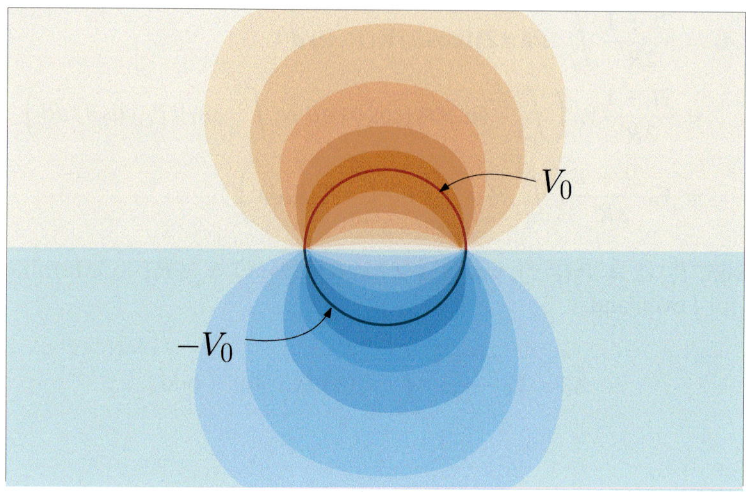

Fig. A.53 Equipotentials created by the pair of hemispheres

and we see that the potential decays as $1/r^2$ at infinity, at leading order, as for an electric dipole. The equipotentials are shown in Fig. A.53

To find the surface charge density, we use the fact that the normal component of the electric field is discontinuous at a charged interface, the discontinuity being related to the surface charge density:

$$\lim_{r \to R^+} \mathbf{E}(r, \theta) \cdot \mathbf{u}_r - \lim_{r \to R^-} \mathbf{E}(r, \theta) \cdot \mathbf{u}_r = \frac{\sigma(R, \theta)}{\epsilon_0}$$

from which we obtain the charge density as

$$
\begin{aligned}
\sigma(R, \theta) &= -\epsilon_0 \left(\left. \frac{\partial V}{\partial r} \right|_{r=R^+} - \left. \frac{\partial V}{\partial r} \right|_{r=R^-} \right) \\
&= \frac{V_0 \epsilon_0}{R} \left(\sum_{l=1}^{\infty} (-1)^{l+1} \frac{(4l-1)(2l-1)!!}{(2l)!!(2l-1)} P_{2l-1}(\cos\theta) \right)
\end{aligned}
$$

and we know that

$$
\begin{aligned}
V(R, \theta) &= V_0 \sum_{l=1}^{\infty} (-1)^{l+1} \frac{(4l-1)(2l-1)!!}{(2l)!!(2l-1)} P_{2l-1}(\cos\theta) , \\
&= \begin{cases} V_0 & \text{if } 0 \leq \theta \leq \pi/2 \\ -V_0 & \text{if } \pi/2 < \theta \leq \pi \end{cases} .
\end{aligned}
$$

It follows that the charge density is homogeneous in each hemisphere

$$\sigma(R, \theta) = \begin{cases} V_0\epsilon_0/R & \text{if } 0 \le \theta \le \pi/2, \\ -V_0\epsilon_0/R & \text{if } \pi/2 < \theta \le \pi . \end{cases}$$

The total charge in the upper hemisphere is

$$Q = 2\pi R^2 \frac{V_0\epsilon_0}{R} = 2\pi R V_0\epsilon_0$$

and the capacity is therefore

$$C = Q/(2V_0) = \epsilon_0\pi R.$$

5.7 Electrostatic force from Helmholtz and Gibbs free energies

The Helmholtz free energy expressed as a function of charges reads

$$F(Q_1, \ldots, Q_N) = \frac{1}{2}\sum_{i=1}^{N}\sum_{j=1}^{N} Q_i\Gamma_{ij}(\mathbf{x})Q_j = \frac{1}{2}(Q)^T(\Gamma(\mathbf{x}))(Q),$$

where $(Q)^T$ denotes the transpose vector of charges $(Q_1, \ldots, Q_N)$ and $(\Gamma(\mathbf{x}))$ the inverse of the capacitance matrix, with coefficients $\Gamma_{ij}(\mathbf{x})$. The electric force is obtained by taking the gradient of Helmholtz's free energy assuming constant charges:

$$\mathbf{F}_e = -\nabla F|_{Q_i} = -\frac{1}{2}(Q)^T(\nabla\Gamma)(Q),$$

where the matrix $(\nabla\Gamma)$ is the matrix with vector coefficients $\nabla\Gamma_{ij}(\mathbf{x})$. The matrix (Γ) is the inverse of the capacitance matrix (C) and therefore satisfies $(C)(\Gamma) = (I)$ where (I) is the identity matrix. By differentiating (taking the gradient) of this relation, we obtain

$$(\nabla C)(\Gamma) + (C)(\nabla\Gamma) = 0,$$

or, multiplying by $(C)^{-1} = (\Gamma)$:

$$(\nabla\Gamma) = -(\Gamma)(\nabla C)(\Gamma).$$

The expression for the electric force then becomes

$$\mathbf{F}_e = \frac{1}{2}(Q)^T(\Gamma)(\nabla C)(\Gamma)(Q).$$

Using the transpose vector of potentials $(V)^T = (V_1, \ldots, V_N)$, the equation of state for the system of conductors reads

$$(V) = (\Gamma)(Q), \quad \text{and transposing,} \quad (V)^T = (Q)^T (\Gamma)^T = (Q)^T (\Gamma),$$

where the far right equal sign is due to the symmetry of (Γ) that implies $(\Gamma)^T = (\Gamma)$. Finally,

$$\mathbf{F}_e = \frac{1}{2}(V)^T (\nabla C)(V) = \frac{1}{2} \sum_{i=1}^{N} \sum_{j=1}^{N} V_i \nabla C_{ij} V_j.$$

This expression coincides with the formula for the electric force derived from the Gibbs free energy:

$$\mathbf{F}_e = -\nabla_{\mathbf{x}} G|_{V_1, \ldots, V_N}, \quad \text{with } G = -\frac{1}{2} \sum_{i=1}^{N} \sum_{j=1}^{N} V_i C_{ij}(\mathbf{x}) V_j.$$

5.8 Separation of capacitor plates

The charge on the isolated plate is invariant. Since the capacitance decreases when the plates are separated, the potential of the isolated plate increases to keep the charge constant. Hence $Q = C_1 V_1 = C_2 V_2$ and $V_2 = \frac{C_1}{C_2} V_1$.

5.9 Energy in a capacitor

The capacitance of the upper and lower capacitors are

$$C_1 = \frac{\epsilon_0 S}{d_1}, \qquad C_2 = \frac{\epsilon_0 S}{d_2},$$

where d_i denotes the distance between the plates for capacitor i. From the figure, we see that

$$d_1 + d_2 = b - a.$$

For the two capacitors in series, the total capacitance C satisfies

$$\frac{1}{C} = \frac{1}{C_1} + \frac{1}{C_2} = \frac{d_1}{\epsilon_0 S} + \frac{d_2}{\epsilon_0 S} = \frac{b - a}{\epsilon_0 S},$$

whose inverse is $C = \epsilon_0 S/(b - a)$. As C is independent of d_1 and d_2, it is independent of the position of the center section. The total energy stored in the capacitor is

$$U = \frac{1}{2} C V_0^2 = \frac{S \epsilon_0 V_0^2}{2(b - a)}.$$

If the center section is removed, the remaining pair of plates separated by a distance b form a capacitor of capacitance $\epsilon_0 S/b$, which stored the energy

$$U' = \frac{S \epsilon_0 V_0^2}{2b},$$

and the energy change reads

$$U - U' = \frac{S\epsilon_0 V_0^2}{2(b-a)} \frac{a}{b}.$$

The series combination of capacitors allows us to store a larger energy than that stored by a single capacitor with the plate separation as the distance between the outside plates.

5.10 Work to separate plates of a charged capacitor

Neglecting edge effects, the capacitance of the parallel-plate capacitor is $C = \epsilon_0 \pi r^2 / d$ and the stored energy is $U = \frac{1}{2}CV^2$. When the capacitor is disconnected, it becomes isolated. The charges on the plates, $Q = \pm CV$, do not vary with the separation. Then, we have

$$V' = \frac{C}{C'} V = \frac{d'}{d} V.$$

The energy stored when separation is d' is

$$U' = \frac{1}{2} C' V'^2 = \frac{1}{2} \frac{Q^2}{C'}.$$

Thus the change of the energy stored in the capacitor is

$$\Delta U = U' - U = \frac{Q^2}{2}\left(\frac{1}{C'} - \frac{1}{C}\right) = \frac{Q^2}{2C}\left(\frac{d'}{d} - 1\right) = \frac{1}{2}CV^2\left(\frac{d'}{d} - 1\right).$$

Therefore, the work done in changing the separation from d to d' is

$$W = \frac{\epsilon_0 \pi r^2 (d' - d) V^2}{2d^2}.$$

5.11 System of three conductors

(a) We note

$$\kappa = \frac{1}{4\pi\epsilon_0}.$$

For sphere A, all the charge is on the surface, at distance r from A. The potential at A due to Q_1 is

$$V_{11} = \kappa \frac{Q_1}{r}.$$

The potential at A due to Q_2 is

$$V_{12} \simeq \kappa \frac{Q_2}{a}.$$

Since $a \gg r$, B is *far* from A and we indeed use the dominant term of the multipolar expansion of the potential. The same argument applies for the potential at A due to Q_3 which reads

$$V_{13} \simeq \kappa \frac{Q_3}{a}.$$

By using the superposition theorem, the potential at A due to the system of charges is

$$V_1 = V_{11} + V_{12} + V_{13} = \kappa \left(\frac{Q_1}{r} + \frac{Q_2}{a} + \frac{Q_3}{a} \right).$$

Thus we can write potentials as functions of charges on each sphere:

$$\begin{pmatrix} V_1 \\ V_2 \\ V_3 \end{pmatrix} = \frac{\kappa}{r} \begin{pmatrix} 1 & x & x \\ x & 1 & x \\ x & x & 1 \end{pmatrix} \begin{pmatrix} Q_1 \\ Q_2 \\ Q_3 \end{pmatrix}$$

where $x = r/a \ll 1$. We write $V_i = \sum_{j=1}^{3} \Gamma_{ij} Q_j$ where $[\Gamma_{ij}]$ denotes the inverse of the capacitance matrix, with $\Gamma_{ii} = \kappa/r$ and $\Gamma_{ij} = \kappa x/r$ for $i \neq j$. In order to find the capacitance matrix, we need to invert the 3×3 matrix

$$[M] = \begin{pmatrix} 1 & x & x \\ x & 1 & x \\ x & x & 1 \end{pmatrix}.$$

The determinant is

$$\det M = 1 - 3x^2 + 2x^3 \simeq 1 - 3x^2$$

and the inverse matrix reads

$$[M^{-1}] = \frac{1}{\det M} \begin{pmatrix} 1 - x^2 & -x + x^2 & -x + x^2 \\ -x + x^2 & 1 - x^2 & -x + x^2 \\ -x + x^2 & -x + x^2 & 1 - x^2 \end{pmatrix} \simeq (1 + 3x^2) \begin{pmatrix} 1 - x^2 & -x + x^2 & -x + x^2 \\ -x + x^2 & 1 - x^2 & -x + x^2 \\ -x + x^2 & -x + x^2 & 1 - x^2 \end{pmatrix}.$$

Finally, at second order in x, we obtain

$$[M^{-1}] = \begin{pmatrix} 1 + 2x^2 & -x + x^2 & -x + x^2 \\ -x + x^2 & 1 + 2x^2 & -x + x^2 \\ -x + x^2 & -x + x^2 & 1 + 2x^2 \end{pmatrix}$$

and

$$C_{ii} = 4\pi \epsilon_0 r (1 + 2x^2), \qquad C_{ij} = -4\pi \epsilon_0 r x (1 - x) \quad \text{for } i \neq j.$$

(b) If we ground sphere A, the potential at A becomes 0, and the charge after the new equilibrium is established is Q_1', while the charges of spheres B and C are unchanged. The Γ_{ij} coefficients are unchanged: $0 = \Gamma_{11}Q_1' + \Gamma_{12}Q_2 + \Gamma_{13}Q_3$. Hence

$$Q_1' = -x(Q_2 + Q_3).$$

The charge is preserved on sphere A after the connection is cut. If B is grounded, the potential at B becomes 0, and the charge becomes Q_2' after the new equilibrium is established, whereas spheres A and C carry charge Q_1' and Q_3. Hence

$$Q_2' = -x(Q_1' + Q_3).$$

This charge is preserved after the connection is cut. Similarly, when C is grounded, its potential becomes $V_3' = 0$ and its charge after the new equilibrium is Q_3':

$$Q_3' = -x(Q_1' + Q_2').$$

We can replace Q_1' in the expressions for Q_2', which yields

$$Q_2' = -xQ_3 + x^2(Q_2 + Q_3),$$

and by introducing now Q_1' and Q_2' in the expression for Q_3', we obtain

$$Q_3' = x^2(Q_2 + 2Q_3).$$

(c) The energy of the system is

$$U = \frac{1}{2}\sum_i Q_i V_i.$$

If the three charges are identical, the system is invariant by rotation of angle $2\pi/3$ and the potentials must also be the same $V_1 = V_2 = V_3 = V$. Hence

$$U = \frac{3}{2}QV.$$

Moreover,

$$V = \kappa Q \left(\frac{1}{r} + \frac{2}{a}\right) = \kappa \frac{Q}{r}(2x + 1).$$

Thus,

$$U_1 = \frac{3\kappa}{2}\frac{Q^2}{r}(2x + 1) \simeq \frac{3\kappa}{2}\frac{Q^2}{r}.$$

After the three operations, the charges on the spheres A, B, and C are Q'_1, Q'_2 and Q'_3, respectively, and their potentials are V'_1, V'_2 and $V'_3 = 0$, satisfying $V'_i = \sum_j \Gamma_{ij} Q'_j$. Thus, since $V'_3 = 0$, the electrostatic energy of the system reads

$$U_2 = \frac{1}{2}(Q'_1 V'_1 + Q'_2 V'_2).$$

Since $\Gamma_{ii} = \dfrac{\kappa}{r}$ and $\Gamma_{ij} = \dfrac{\kappa}{a}$ for $i \neq j$,

$$W_2 = \frac{\kappa}{2r}\left\{ Q'^2_1 + Q'^2_2 + x[2Q'_1 Q'_2 + Q'_3(Q'_1 + Q'_2)] \right\}.$$

Reporting the expressions for Q'_1, Q'_2 and Q'_3, we find

$$Q'_1 = -2xQ, \qquad Q'_2 = Q(-x + 2x^2) \simeq -Qx, \qquad Q'_3 = 3x^2 Q.$$

Hence

$$Q'^2_1 = 4x^2 Q^2, \quad Q'^2_2 = x^2 Q^2(1 - 4x + 4x^2) \simeq x^2 Q^2,$$

$$2Q'_1 Q'_2 = 4Q^2 x^2(1 - 2x) \simeq 4x^2 Q^2, \quad Q'_3(Q'_1 + Q'_2) = 3x^2 Q^2(-3x + 2x^2) \simeq -9x^3 Q^2$$

and finally:
$$U_2 = \frac{\kappa Q^2 x^2}{2r}(5 - 13x^2) \simeq \frac{5\kappa Q^2 x^2}{2r}.$$

(d) $Q = 10^{-4}$ C. $a = 1$ m. $r = 5 \times 10^{-2}$ m

$$C_{ii} = 5.6 \times 10^{-12} \text{ F} = 5.6 \text{ pF}; \qquad C_{ij} = -0.27 \text{ pF};$$

$$Q'_1 = -10^{-5} \text{ C} = -10\,\mu\text{C}; \qquad Q'_2 = -4.5\,\mu\text{C}; \qquad Q'_3 = 0.75\,\mu\text{C};$$

$$\frac{U_2}{U_1} = 0.0037.$$

Problems of Chap. 6

6.1 Dielectric shell

The spherical symmetry and mirror properties of both free and bound charge are such that $\mathbf{D} = D(r)\mathbf{u}_r$ in spherical coordinates (origin at the center of the sphere). By taking an arbitrary spherical surface S of radius $r > 0$ and centered at the position of the free charge q, Gauss's law for dielectrics leads to

$$\oiint_S \mathbf{D}(\mathbf{x}') \cdot d\mathbf{S}(\mathbf{x}') = 4\pi r^2 D(r) = q \; .$$

Hence,

$$\mathbf{D}(r) = \frac{q}{4\pi r^2}\mathbf{u}_r$$

For all $r > 0$. Now, we can write the electric field as $\mathbf{E} = \mathbf{D}/\epsilon$. In the region $r < a$, we obtain

$$\mathbf{E}(r) = \frac{q}{4\pi\epsilon_0 r^2}\mathbf{u}_r \; .$$

for $a < r < b$,

$$\mathbf{E}(r) = \frac{q}{4\pi\epsilon r^2}\mathbf{u}_r,$$

and for $r > b$

$$\mathbf{E}(r) = \frac{q}{4\pi\epsilon_0 r^2}\mathbf{u}_r \; .$$

Note that the field is weakened in the inner region of the dielectric shell. Now, for the bound charges, there is no volume density since

$$\varrho_P(\mathbf{x}) = -\nabla \cdot \mathbf{P}(\mathbf{x}) = -\nabla \cdot (\epsilon - \epsilon_0)\,\mathbf{E}(\mathbf{x}) = 0$$

Let σ_a denote the surface density of bound charges on the inner shell and σ_b the surface density of bound charges on the outer shell. Then

$$\sigma_a = (\epsilon - \epsilon_0)\mathbf{E}(a^+) \cdot (-\mathbf{u}_r) = -\frac{q(\epsilon - \epsilon_0)}{4\pi\epsilon a^2},$$

$$\sigma_b = (\epsilon - \epsilon_0)\mathbf{E}(b^-) \cdot (\mathbf{u}_r) = \frac{q(\epsilon - \epsilon_0)}{4\pi\epsilon b^2}.$$

We can verify that the total polarization charge is zero, since $Q_a = -q(\epsilon - \epsilon_0)/\epsilon = -Q_b$.

6.2 Conductive sphere surrounded by a dielectric shell

(a) To find the electric field in all space we can use Gauss's law in dielectrics with S a closed sphere of radius r (the symmetries of the charge distributions lead to $\mathbf{D} = D(r)\mathbf{u}_r$ in spherical coordinates):

$$\oiint_S \mathbf{D}(\mathbf{x}) \cdot d\mathbf{S}(\mathbf{x}) = 4\pi r^2 D(r) = Q_{\text{in}} \; .$$

For $r < a$, $Q_{in} = 0$ and the field is identically zero (the density of free charge is zero inside a conductor, all of its charge lies on its surface). For $a < r < b$, the charge enclosed by the surface is Q, then

$$4\pi r^2 D(r) = Q \quad \rightarrow \quad D(r) = \frac{Q}{4\pi r^2}\,.$$

We also note that in this region $\mathbf{D}(\mathbf{x}) = \epsilon_0 \mathbf{E}(\mathbf{x})$, hence

$$\mathbf{E}(\mathbf{x}) = \frac{Q}{4\pi \epsilon_0 r^2}\mathbf{u}_r\,.$$

For $b < r < c$, we have again

$$\oiint_S d\mathbf{S}(\mathbf{x}) \cdot \mathbf{D}(\mathbf{x}) = 4\pi r^2 D(r) = Q$$

and

$$\mathbf{D}(\mathbf{x}) = \epsilon \mathbf{E}(\mathbf{x}).$$

Then, using $\epsilon = \epsilon_r \epsilon_0$,

$$\mathbf{E}(\mathbf{x}) = \frac{Q}{4\pi \epsilon_r \epsilon_0 r^2}\mathbf{u}_r\,.$$

Finally, for $r > c$, we obtain

$$\mathbf{E}(\mathbf{x}) = \frac{Q}{4\pi \epsilon_0 r^2}\mathbf{u}_r\,.$$

(b) The relation between the electrostatic field and potential reads

$$V(r) = \int_r^\infty \mathbf{E}(\mathbf{x}) \cdot d\mathbf{x}.$$

Given the radial nature of the electric field, a radial path can be chosen to evaluate the integral ($d\mathbf{x} = \mathbf{u}_r dr$), and then the potential in the conducting sphere is

$$V(a) = \int_a^b E(r)dr + \int_b^c E(r)dr + \int_c^\infty E(r)dr$$

$$= -\frac{Q}{4\pi \epsilon_0 r}\Big|_a^b - \frac{Q}{4\pi \epsilon_r \epsilon_0 r}\Big|_b^c - \frac{Q}{4\pi \epsilon_0 r}\Big|_c^\infty$$

$$= \frac{Q}{4\pi \epsilon_0}\left[\left(\frac{1}{a} - \frac{1}{b}\right) + \frac{1}{\epsilon_r}\left(\frac{1}{b} - \frac{1}{c}\right) + \frac{1}{c}\right]$$

$$= \frac{Q}{4\pi \epsilon_0}\left(\frac{1}{a} + \frac{(b-c)\,(\epsilon_r - 1)}{bc}\frac{1}{\epsilon_r}\right).$$

(c) The energy required to move a charge q from infinity to $r = c$ is

$$U = qV(c)$$

where

$$V(c) = \int_c^\infty E(r)dr = \int_c^\infty \frac{Q}{4\pi\epsilon_0 r^2}dr = \frac{Q}{4\pi\epsilon_0 c}.$$

Therefore

$$U = \frac{qQ}{4\pi\epsilon_0 c}.$$

6.3 Concentric spheres

(a) By connecting the spheres of radius a and c, the charge q_2 that was initially on the external sphere will be now redistributed on both spheres so that they remain at the same potential, then

$$q_2 = q_A + q_C.$$

Furthermore, for the region given by $r < a$ which corresponds to the interior of a conductive sphere, $\mathbf{E}(\mathbf{x}) = \mathbf{0}$. For $a < r < b$ Gauss's law applied to a spherical surface of radius r reads

$$\oiint_{S(r)} \mathbf{D}(\mathbf{x}') \cdot d\mathbf{S}(\mathbf{x}') = 4\pi r^2 D(r) = q_A.$$

Thus,

$$\mathbf{D} = \frac{q_A}{4\pi r^2}\mathbf{u}_r$$

and then,

$$\mathbf{E}(r) = \frac{q_A}{4\pi\epsilon_0\epsilon_r r^2}\mathbf{u}_r.$$

For the region $b < r < c$

$$\oiint_{S(r)} \mathbf{D}(\mathbf{x}') \cdot d\mathbf{S}(\mathbf{x}') = 4\pi r^2 D(r) = (q_A + q_1).$$

Thus,

$$\mathbf{E}(r) = \frac{(q_A + q_1)}{4\pi\epsilon_0 r^2}\mathbf{u}_r.$$

Finally, for $r > c$

$$\oiint_{S(r)} \mathbf{D}(\mathbf{x}') \cdot d\mathbf{S}(\mathbf{x}') = 4\pi r^2 D(r) = q_1 + q_A + q_C.$$

In summary

$$\mathbf{E}(\mathbf{x}) = \begin{cases} \mathbf{0} & \text{if } 0 < r < a, \\ \dfrac{q_A}{4\pi\epsilon_0\epsilon_r r^2}\mathbf{u}_r & \text{if } a < r < b, \\ \dfrac{q_A+q_1}{4\pi\epsilon_0 r^2}\mathbf{u}_r & \text{if } b < r < c, \\ \dfrac{q_A+q_1+q_C}{4\pi\epsilon_0 r^2}\mathbf{u}_r & \text{if } c < r. \end{cases}$$

The potential of the sphere of radius a is

$$V(a) = \int_a^\infty \mathbf{E}\cdot d\mathbf{x}$$

$$= \int_a^b E(r)\,dr + \int_b^c E(r)\,dr + \int_c^\infty E(r)\,dr$$

where we have used a radial path such that $d\mathbf{x} = dr\,\mathbf{u}_r$. The last integral corresponds to

$$\int_c^\infty E(r)\,dr = V(c) = V(a).$$

Hence,

$$V(a) = \frac{q_A}{4\pi\epsilon_r\epsilon_0}\int_a^b \frac{dr}{r^2} + \frac{q_A+q_1}{4\pi\epsilon_0}\int_b^c \frac{dr}{r^2} + V(a),$$

then

$$\frac{q_A}{4\pi\epsilon_r\epsilon_0}\left(\frac{1}{a}-\frac{1}{b}\right) + \frac{q_A+q_1}{4\pi\epsilon_0}\left(\frac{1}{b}-\frac{1}{c}\right) = 0,$$

$$q_A\frac{b-a}{ab} + (q_A+q_1)\frac{c-b}{cb} = 0,$$

$$\frac{q_A}{\epsilon_r}\left(\frac{1}{a}+\frac{1}{b}(\epsilon_r-1)-\frac{\epsilon_r}{c}\right) = q_1\left(\frac{1}{c}-\frac{1}{b}\right).$$

Finally

$$q_A = \frac{\epsilon_r q_1 a(b-c)}{bc + (\epsilon_r-1)ac - \epsilon_r ab},$$

$$q_C = q_2 - q_A = q_2 - \frac{\epsilon_r q_1 a(b-c)}{bc + (\epsilon_r-1)ac - \epsilon_r ab}.$$

(b) On the internal surface of the dielectric, $\sigma_P = \mathbf{P}\cdot\mathbf{n}$, with $\mathbf{P} = (\epsilon-\epsilon_0)\mathbf{E}$, $\mathbf{E} = \dfrac{q_A}{4\pi\epsilon_r\epsilon_0 a^2}\mathbf{u}_r$ and $\mathbf{n} = -\mathbf{u}_r$. Then

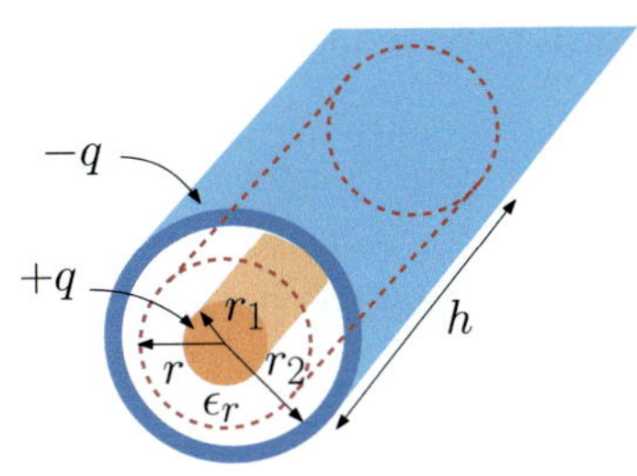

Fig. A.54 Cylindrical surface used to apply Gauss's law

$$\sigma_P|_{r=a} = -\frac{(\epsilon - \epsilon_0)\,q_A}{4\pi\,\epsilon_r\epsilon_0 a^2} = -\frac{(\epsilon_r - 1)\,q_A}{4\pi\,\epsilon_r a^2}.$$

and on the outer surface $\mathbf{E} = \dfrac{q_A}{4\pi\,\epsilon_r\epsilon_0 b^2}\mathbf{u}_r$, $\mathbf{n} = \mathbf{u}_r$ and $\sigma_P = \mathbf{P}\cdot\mathbf{n}$, then

$$\sigma_P|_{r=b} = (\epsilon - \epsilon_0)\,\frac{q_A}{4\pi\,\epsilon_r\epsilon_0 b^2} = \frac{(\epsilon_r - 1)\,q_A}{4\pi\,\epsilon_r b^2}$$

so that the total bound charge at $r = a$ (resp. $r = b$) is $-(\epsilon_r - 1)q_A/\epsilon_r$ (resp. $+(\epsilon_r - 1)q_A/\epsilon_r$).

6.4 Cylindrical capacitor filled with an inhomogeneous dielectric

(a) The cylindrical symmetry of the free charge distribution, together with its mirror symmetries, assures that the displacement vector $\mathbf{D}$ is written in the form $\mathbf{D} = D(r)\mathbf{u}_r$ in cylindrical coordinates. To obtain the electrostatic field between the two conductors, we use Gauss's law for dielectrics, with a cylindrical surface S of radius r, with $r_1 < r < r_2$ and length h, as shown in Fig. A.54

This leads to

$$\oiint_S \mathbf{D}(\mathbf{x}') \cdot d\mathbf{S}(\mathbf{x}') = \oiint_S \mathbf{u}_r \cdot D(r)\mathbf{u}_r\, dS(\mathbf{x}') = q\frac{h}{l}.$$

Thus

$$D(r)2\pi r h = q\frac{h}{l},$$

hence,

$$\mathbf{D}(r) = \frac{q}{2\pi r l}\mathbf{u}_r.$$

Besides $\mathbf{D}$ is related to $\mathbf{E}$ according to

$$\mathbf{D}(\mathbf{x}) = \epsilon\mathbf{E}(\mathbf{x}) = \epsilon_r(r)\epsilon_0\mathbf{E}(r)$$

then

$$\mathbf{E}(r) = \frac{q}{2\pi r l \epsilon_r(r)\epsilon_0}\mathbf{u}_r.$$

For $\mathbf{E}$ to be independent on r, $\epsilon_r(r)$ must be of the form $\epsilon_r(r) = \alpha/r$, so

$$\mathbf{E}(r) = \frac{q}{2\pi \alpha l \epsilon_0} \mathbf{u}_r.$$

(b) Let's calculate the potential difference between the cylindrical shells

$$\int_{r_1}^{r_2} \mathbf{E}(\mathbf{x}) \cdot d\mathbf{x} = V(r_1) - V(r_2) = \Delta V.$$

By choosing a radial path, $d\mathbf{x} = dr\mathbf{u}_r$

$$\int_{r_1}^{r_2} \frac{q}{2\pi \alpha l \epsilon_0} \mathbf{u}_r \cdot dr\mathbf{u}_r = \frac{q}{2\pi \alpha l \epsilon_0} \int_{r_1}^{r_2} dr$$
$$= \frac{q(r_2 - r_1)}{2\pi \alpha l \epsilon_0},$$

and the capacitance of this capacitor is

$$C = \frac{q}{\Delta V} = \frac{2\pi \alpha l \epsilon_0}{r_2 - r_1}.$$

(c) The surface density of bound charges in a surface S is given by

$$\sigma_P(\mathbf{x}) = \mathbf{P}(\mathbf{x}) \cdot \mathbf{n}(\mathbf{x}), \quad \mathbf{x} \in S$$

and for $r_1 < r < r_2$
$$\mathbf{P} = (\epsilon - \epsilon_0)\mathbf{E} = \left(\frac{\alpha}{r} - 1\right)\epsilon_0 \mathbf{E}.$$

On the inner surface $r = r_1$, $\mathbf{n} = -\mathbf{u}_r$, and

$$\mathbf{P}(r_1) = \left(\frac{\alpha}{r_1} - 1\right)\epsilon_0 \mathbf{E} = \left(\frac{\alpha}{r_1} - 1\right)\frac{q}{2\pi \alpha l}\mathbf{u}_r,$$

and so
$$\sigma_1 = \mathbf{P} \cdot \mathbf{n}\big|_{r=r_1} = -\left(\frac{\alpha}{r_1} - 1\right)\frac{q}{2\pi \alpha l}.$$

For the outer surface $r = r_2$, $\mathbf{n} = \mathbf{u}_r$ and

$$\sigma_2 = \mathbf{P} \cdot \mathbf{n}\big|_{r=r_2} = \left(\frac{\alpha}{r_2} - 1\right)\frac{q}{2\pi \alpha l}.$$

(d) The volume density of bound charges is $\varrho_P = -\nabla \cdot \mathbf{P}$, and as $\mathbf{P} = P(r)\mathbf{u}_r$

$$
\begin{aligned}
\varrho_P &= -\nabla \cdot \mathbf{P} = -\frac{1}{r}\frac{\partial}{\partial r}\left(r P(r)\right) \\
&= -\frac{1}{r}\frac{\partial}{\partial r}\left[r\left(\frac{\alpha - r}{r}\right)\frac{q}{2\pi\alpha l}\right] \\
&= -\frac{1}{r}\frac{\partial}{\partial r}\left[\frac{(\alpha - r)q}{2\pi\alpha l}\right] \\
&= \frac{q}{2\pi r\alpha l}
\end{aligned}
$$

(e) The total polarization charge is

$$
Q_p = \iiint_\Omega \varrho_P(\mathbf{x}')d^3x' + \oiint_{S_1} dS(\mathbf{x}')\sigma_1(\mathbf{x}') + \oiint_{S_2} dS(\mathbf{x}')\sigma_2(\mathbf{x}')
$$

where S_1 (resp. S_1) denotes the two cylindrical surfaces of the conductors, of radius $r = r_1$ (resp. $r = r_2$) and length l, and Ω denotes the volume between S_1 and S_2 ($r_1 < r < r_2$). The surface integrals read

$$
\begin{aligned}
\oiint_{S_1} dS(\mathbf{x}')\sigma_1(\mathbf{x}') &= -\left(\frac{\alpha}{r_1} - 1\right)\frac{q}{2\pi\alpha l}\underbrace{\oiint_{S_1} dS(\mathbf{x}')}_{2\pi r_1 l}, \\
&= -(\alpha - r_1)\frac{q}{\alpha}.
\end{aligned}
$$

In the same way

$$
\oiint_{S_2} dS(\mathbf{x}')\sigma_2(\mathbf{x}') = \sigma_2\oiint_{S_1} dS(\mathbf{x}') = (\alpha - r_2)\frac{q}{\alpha}.
$$

Finally, remembering that the volume element in cylindrical coordinates is $d^3x' = r\,dr\,d\theta\,dz$

$$
\begin{aligned}
\iiint_\Omega \varrho_P(\mathbf{x}')d^3x' &= \int_0^{2\pi} d\theta \int_0^l dz \int_{r_1}^{r_2} dr\, r\,\frac{q}{2\pi\alpha l r} \\
&= \frac{2\pi q}{2\pi\alpha}\int_{r_1}^{r_2} dr = \frac{q}{\alpha}(r_2 - r_1).
\end{aligned}
$$

Hence the total bound charge reads

$$
Q_P = \frac{q}{\alpha}(r_2 - r_1) - (\alpha - r_1)\frac{q}{\alpha} + (\alpha - r_2)\frac{q}{\alpha} = 0.
$$

This is due to the fact that the dielectric is electrically neutral.

(f) The energy stored by the capacitor is

$$U = \frac{1}{2}\frac{q^2}{C}$$

where

$$C = \frac{2\pi \alpha l \epsilon_0}{(r_2 - r_1)}.$$

Hence

$$U = \frac{1}{2}\frac{q^2 (r_2 - r_1)}{2\pi \alpha l \epsilon_0}.$$

(g) The initial energy is

$$U_i = \frac{1}{2}\frac{q^2 (r_2 - r_1)}{2\pi \alpha l \epsilon_0}$$

and the final energy reads

$$U_f = \frac{1}{2}\frac{q^2}{C_f}$$

where C_f is the capacity of the empty capacitor. The capacitance of a cylindrical capacitor was determined in Example 5.3

$$C_f = \frac{2\pi l \epsilon_0}{\ln\left(\frac{r_2}{r_1}\right)}$$

and the final energy is then

$$U_f = \frac{1}{2}\frac{q^2}{C_f} = \frac{1}{2}\frac{q^2 \ln\left(\frac{r_2}{r_1}\right)}{2\pi l \epsilon_0}.$$

Finally, the work necessary to remove the dielectric is

$$W = U_f - U_i = \frac{1}{2}\frac{q^2}{2\pi l \epsilon_0}\left(\frac{(r_1 - r_2)}{\alpha} + \ln\left(\frac{r_2}{r_1}\right)\right).$$

6.5 Dielectric sphere in a uniform field

Given the symmetry of the problem, we choose a system of spherical coordinates as shown in Fig. A.55.

Since there are no charges except at $r = a$, the electrostatic potential satisfies Laplace's equation for $r \neq a$

$$\nabla^2 V = 0$$

Fig. A.55 Spherical coordinates with origin centered on the sphere

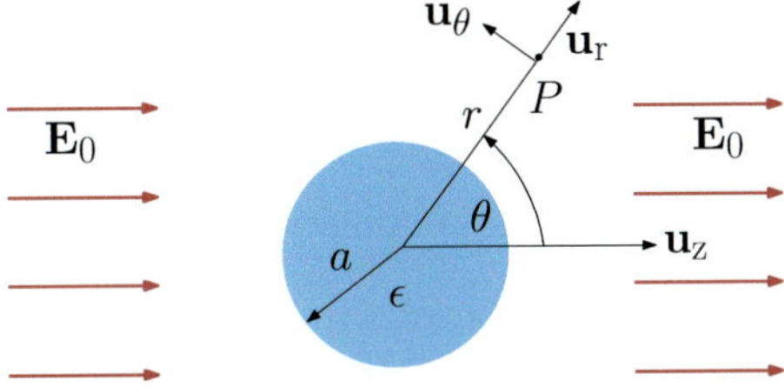

Since there is an invariance of the bound charge distribution on the angle ϕ, the general solution for V takes the form given by Eq. (4.9)

$$V(r, \theta) = \sum_{l=0}^{\infty} A_l r^l P_l(\cos\theta) \quad r < a,$$

$$V(r, \theta) = \sum_{l=0}^{\infty} \left(B_l r^l P_l + C_l r^{-(l+1)} \right) P_l(\cos\theta) \quad r > a.$$

Far from the influence of the dielectric sphere, $\mathbf{E}(\mathbf{x}) = E_0 z$, or, equivalently

$$\lim_{r\to\infty} V(r, \theta) = -E_0 z = -E_0 r \cos\theta,$$

that is to say

$$\lim_{r\to\infty} V(r, \theta) = \sum_{l=0}^{\infty} B_l r^l P_l(\cos\theta) = -E_0 r \cos\theta.$$

From here, it is clear that $B_l = 0$ for $l \neq 1$, and

$$B_1 r \cos\theta = -E_0 r \cos\theta \quad \to \quad B_1 = -E_0.$$

With this result, the potential for $r > a$ takes the form

$$V(r, \theta) = -E_0 r \cos\theta + \sum_{l=0}^{\infty} C_l r^{-(l+1)} P_l(\cos\theta) \quad r > a.$$

Now we will use the boundary conditions on the surface $r = a$. First, the tangential component of the electric field in $r = a$ must be continuous (Eq. 6.21)

$$\mathbf{E}_{\text{int}} \cdot \mathbf{u}_\theta|_{r=a^-} = \mathbf{E}_{\text{ext}} \cdot \mathbf{u}_\theta|_{r=a^+},$$

which reads:

$$-\nabla V(r, \theta) \cdot \mathbf{u}_\theta|_{r=a^-} = -\nabla V(r, \theta) \cdot \mathbf{u}_\theta|_{r=a^+},$$

$$-\frac{1}{r}\frac{\partial V(r,\theta)}{\partial \theta}\bigg|_{r=a^-} = -\frac{1}{r}\frac{\partial V(r,\theta)}{\partial \theta}\bigg|_{r=a^+}.$$

Equivalently, the continuity of V at $r=a$ can be imposed for all θ:

$$\sum_{l=0}^{\infty} A_l a^l P_l(\cos\theta) = -E_0 a\cos\theta + \sum_{l=0}^{\infty} C_l a^{-(l+1)} P_l(\cos\theta),$$

$$\sum_{l=0}^{\infty}\left(A_l a^l - C_l a^{-(l+1)}\right) P_l(\cos\theta) = -E_0 a\cos\theta.$$

From here, we obtain

$$A_1 a - \frac{C_1}{a^2} = -E_0 a,$$

and for $l \neq 1$, $A_l a^l - C_l a^{-(l+1)} = 0$. Also, if there is no free charge density on the dielectric surface, the normal component of the field $\mathbf{D}$ is continuous (Eq. 6.19)

$$\mathbf{D}\cdot\mathbf{u}_r|_{r=a^-} = \mathbf{D}\cdot\mathbf{u}_r|_{r=a^+}$$

$$-\epsilon\nabla V(r,\theta)\cdot\mathbf{u}_r|_{r=a^-} = -\epsilon_0\nabla V(r,\theta)\cdot\mathbf{u}_r|_{r=a^+}$$

$$-\epsilon\frac{\partial V(r,\theta)}{\partial r}\bigg|_{r=a^-} = -\epsilon_0\frac{\partial V(r,\theta)}{\partial r}\bigg|_{r=a^+}$$

$$\epsilon\sum_{l=0}^{\infty} A_l l a^{l-1} P_l(\cos\theta) = -\epsilon_0 E_0\cos\theta - \epsilon_0(l+1)\sum_{l=0}^{\infty} C_l a^{-(l+2)} P_l(\cos\theta),$$

which reads:

$$\sum_{l=0}^{\infty}\left(\left(\frac{\epsilon}{\epsilon_0}\right) A_l l a^{l-1} + (l+1)C_l a^{-(l+2)}\right) P_l(\cos\theta) = -E_0\cos\theta.$$

From here, we obtain

$$\left(\frac{\epsilon}{\epsilon_0}\right) A_1 + 2C_1 a^{-3} = -E_0 \quad \text{(for } l=1\text{)},$$

$$\left(\frac{\epsilon}{\epsilon_0}\right) A_l l a^{l-1} + (l+1)C_l a^{-(l+2)} = 0 \quad \text{if } l \neq 1.$$

In summary, the boundary conditions at $r = a$ impose

$$A_1 a - \frac{C_1}{a^2} = -E_0 a,$$
$$A_l a^l - C_l a^{-(l+1)} = 0, \quad l \neq 1,$$
$$\left(\frac{\epsilon}{\epsilon_0}\right) A_1 + 2C_1 a^{-3} = -E_0,$$
$$\left(\frac{\epsilon}{\epsilon_0}\right) A_l l a^{l-1} + (l+1)C_l a^{-(l+2)} = 0, \quad l \neq 1.$$

From the second equation

$$A_l = C_l \frac{1}{a^{2l+1}}, \quad l \neq 1,$$

which is introduced in the fourth equation

$$\left(\frac{\epsilon}{\epsilon_0}\right) C_l \frac{1}{a^{2l+1}} l a^{l-1} + (l+1)C_l a^{-(l+2)} = 0, \quad l \neq 1,$$

leading to

$$\left(\frac{\epsilon}{\epsilon_0}\right) C_l \left(\frac{l}{a^{l+2}} + \frac{l+1}{a^{l+2}}\right) = 0, \quad l \neq 1.$$

Then

$$C_l = A_l = 0, \quad l \neq 1.$$

For $l = 1$, the first equation yields

$$A_1 = \frac{C_1}{a^3} - E_0.$$

By replacing A_1 in the third equation, we obtain

$$\left(\frac{\epsilon}{\epsilon_0}\right) \left(\frac{C_1}{a^3} - E_0\right) + 2C_1 a^{-3} = -E_0,$$

which is reorganized as

$$\frac{C_1}{a^3} \left(\frac{\epsilon}{\epsilon_0} + 2\right) = E_0 \left(\frac{\epsilon}{\epsilon_0} - 1\right).$$

Finally

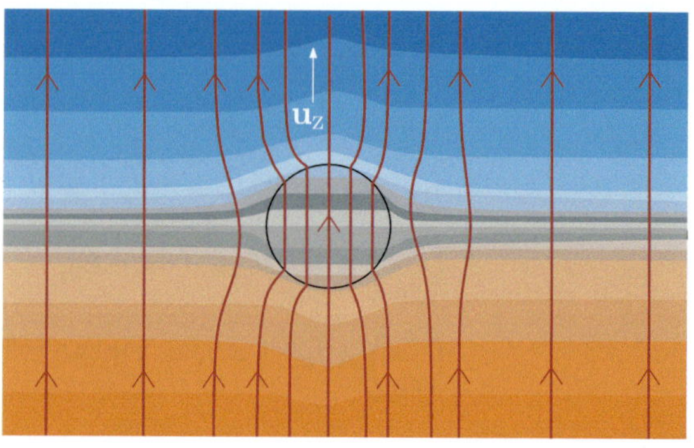

Fig. A.56 Electric field lines and equipotentials near the dielectric sphere

$$C_1 = a^3 E_0 \left(\frac{\epsilon - \epsilon_0}{\epsilon + 2\epsilon_0} \right) = a^3 E_0 \left(\frac{\epsilon_r - 1}{\epsilon_r + 1} \right),$$

$$A_1 = E_0 \left(\frac{\epsilon_r - 1}{\epsilon_r + 2} \right) - E_0 = -E_0 \left(\frac{3}{\epsilon_r + 2} \right)$$

with $\epsilon_r = \epsilon/\epsilon_0$. With this result, we obtain

For $r < a$:

$$V(r, \theta) = -E_0 \left(\frac{3}{\epsilon_r + 2} \right) r \cos\theta = -E_0 \left(\frac{3}{\epsilon_r + 2} \right) z.$$

For $r \geq a$:

$$V(r, \theta) = -E_0 r \cos\theta + E_0 \left(\frac{\epsilon_r - 1}{\epsilon_r + 2} \frac{a^3}{r^2} \cos\theta \right).$$

Note that the potential in the sphere defines a uniform electric field along z

$$\mathbf{E}_{\text{int}} = E_0 \left(\frac{3}{\epsilon_r + 2} \right) \mathbf{u}_z \text{ with } E_{\text{int}} \leq E_0.$$

The polarization is also uniform inside the sphere and reads

$$\mathbf{P} = \epsilon_0 (\epsilon_r - 1) \mathbf{E}_{\text{int}} = \frac{3\epsilon_0 (\epsilon_r - 1)}{\epsilon_r + 2} \mathbf{E}_0.$$

Figure A.56 shows an sketch of the equipotentials and electric field lines for the case $\epsilon/\epsilon_0 = \epsilon_r = 5$, corresponding to a reduction of the electric field inside the sphere by a factor of 0.43.

Note that a dielectric with a very large dielectric constant behaves similarly to a conductor in the sense that for $\epsilon \to \infty$

$$\nabla \cdot \mathbf{E}(\mathbf{x}) = \frac{\varrho}{\epsilon} \to 0$$

and then the field is forced to be zero inside the dielectric.

6.6 Classical model for the Van der Waals force

(a) The field generated by a dipole at any point of space when the dipole is at the origin of coordinates was established in Chap. 3. We can then apply the result (3.10) to the case of a dipole $\mathbf{p}_0$ located at $\mathbf{d}$ and generating a field at the origin ($\mathbf{r} = -\mathbf{d}$):

$$\mathbf{E}_0 = \frac{1}{4\pi \epsilon_0 d^5}(3(\mathbf{p}_0 \cdot \mathbf{d})\mathbf{d} - d^2 \mathbf{p}).$$

(b) The field $\mathbf{E}_0$ is assumed to be uniform in the sphere. Hence, Exercise 6.5 has shown that the sphere will be polarized with uniform polarization $\mathbf{P} = 3\epsilon_0 \dfrac{\epsilon_r - 1}{\epsilon_r + 2}\mathbf{E}_0$, which represents the density of dipole moment in the sphere, i.e., $\mathbf{p}/(\frac{4}{3}\pi R^3)$. The field $\mathbf{E}_0$ will therefore induce a dipole moment in the sphere $\mathbf{p} = 4\pi \epsilon_0 R^3 \dfrac{\epsilon_r - 1}{\epsilon_r + 2}\mathbf{E}_0$. Introducing the expression for $\mathbf{E}_0$, we find

$$\mathbf{p} = \frac{\epsilon_r - 1}{\epsilon_r + 2}\frac{R^3}{d^5}(3(\mathbf{p}_0 \cdot \mathbf{d})\mathbf{d} - d^2 \mathbf{p}_0).$$

(c) The force between two dipoles was calculated in Exercise 3.10. We can thus apply this result:

$$\mathbf{F}_{\mathbf{p} \to \mathbf{p}_0} = \frac{1}{4\pi \epsilon_0 d^5}\left(3(\mathbf{p} \cdot \mathbf{d})\mathbf{p}_0 + 3(\mathbf{p}_0 \cdot \mathbf{d})\mathbf{p} + 3(\mathbf{p} \cdot \mathbf{p}_0)\mathbf{d} - 15(\mathbf{p} \cdot \mathbf{d})(\mathbf{p}_0 \cdot \mathbf{d})\frac{\mathbf{d}}{d^2}\right)$$

Using the expression for $\mathbf{p}$, and the intermediate variable $\alpha = \dfrac{\epsilon_r - 1}{\epsilon_r + 2}\dfrac{R^3}{d^3}$, we calculate

$$(\mathbf{p} \cdot \mathbf{d})\mathbf{p}_0 = \frac{\alpha}{d^2}(3(\mathbf{p}_0 \cdot \mathbf{d})d^2 - d^2(\mathbf{p}_0 \cdot \mathbf{d}))\mathbf{p}_0 = 2\alpha(\mathbf{p}_0 \cdot \mathbf{d})\mathbf{p}_0,$$

$$(\mathbf{p}_0 \cdot \mathbf{d})\mathbf{p} = \frac{\alpha}{d^2}[3(\mathbf{p}_0 \cdot \mathbf{d})^2 \mathbf{d} - d^2(\mathbf{p}_0 \cdot \mathbf{d})\mathbf{p}_0],$$

$$(\mathbf{p} \cdot \mathbf{p}_0)\mathbf{d} = \frac{\alpha}{d^2}[3(\mathbf{p}_0 \cdot \mathbf{d})^2 \mathbf{d} - d^2 p_0^2 \mathbf{d}],$$

$$(\mathbf{p} \cdot \mathbf{d})(\mathbf{p}_0 \cdot \mathbf{d})\frac{\mathbf{d}}{d^2} = 2\alpha(\mathbf{p}_0 \cdot \mathbf{d})^2 \frac{\mathbf{d}}{d^2}.$$

Finally the force reads

$$\mathbf{F}_{\mathbf{p}\to\mathbf{p}_0} = \frac{1}{4\pi\epsilon_0 d^5}\left[6\alpha(\mathbf{p}_0\cdot\mathbf{d})\mathbf{p}_0 + \frac{9\alpha}{d^2}(\mathbf{p}_0\cdot\mathbf{d})^2\mathbf{d} - 3\alpha(\mathbf{p}_0\cdot\mathbf{d})\mathbf{p}_0\right.$$

$$\left. + \frac{9\alpha}{d^2}(\mathbf{p}_0\cdot\mathbf{d})^2\mathbf{d} - 3\alpha\mathbf{p}_0^2\mathbf{d} - 30\alpha(\mathbf{p}_0\cdot\mathbf{d})^2\frac{\mathbf{d}}{d^2}\right],$$

$$= \frac{\alpha}{4\pi\epsilon_0 d^4}\left[\frac{3}{d}(\mathbf{p}_0\cdot\mathbf{d})\mathbf{p}_0 - \frac{12}{d^3}(\mathbf{p}_0\cdot\mathbf{d})^2\mathbf{d} - \frac{3}{d}\mathbf{p}_0^2\mathbf{d}\right],$$

$$\text{i.e.,}\quad \mathbf{F}_{\mathbf{p}\to\mathbf{p}_0} = -\frac{3}{4\pi\epsilon_0}\frac{\epsilon_r - 1}{\epsilon_r + 2}\frac{R^3}{d^7}\left[-(\mathbf{p}_0\cdot\hat{\mathbf{r}})\mathbf{p}_0 + 4(\mathbf{p}_0\cdot\hat{\mathbf{r}})^2\hat{\mathbf{r}} + \mathbf{p}_0^2\hat{\mathbf{r}}\right].$$

(d) To perform the average, let us define a new basis for spherical coordinates with $\hat{\mathbf{r}}$ as the z-axis and θ, the angle between $\mathbf{p}_0$ and $\hat{\mathbf{r}}$. Then $\mathbf{p}_0\cdot\hat{\mathbf{r}} = p_0\cos\theta$ and averaging over the direction of $\mathbf{p}_0$ amounts to averaging over θ and φ:

$$\langle(\mathbf{p}_0\cdot\hat{\mathbf{r}})^2\rangle = \frac{1}{4\pi}\int_0^\pi p_0^2\cos^2\theta\, 2\pi\sin\theta\, d\theta = \frac{p_0^2}{3}.$$

The average of $(\mathbf{p}_0\cdot\hat{\mathbf{r}})\mathbf{p}_0$ is the average of its projection on $\hat{\mathbf{r}}$:

$$\langle(\mathbf{p}_0\cdot\hat{\mathbf{r}})\mathbf{p}_0\rangle = \langle(\mathbf{p}_0\cdot\hat{\mathbf{r}})(\mathbf{p}_0\cdot\hat{\mathbf{r}})\hat{\mathbf{r}}\rangle = \langle(\mathbf{p}_0\cdot\hat{\mathbf{r}})^2\rangle\hat{\mathbf{r}} = \frac{p_0^2}{3}\hat{\mathbf{r}}.$$

Thus, we find $\langle\mathbf{F}_{\mathbf{p}\to\mathbf{p}_0}\rangle = -\frac{3}{4\pi\epsilon_0}\frac{\epsilon_r - 1}{\epsilon_r + 2}\frac{R^3}{d^7}p_0^2\left[-\frac{1}{3}\hat{\mathbf{r}} + \frac{4}{3}\hat{\mathbf{r}} + \hat{\mathbf{r}}\right],$

$$\text{i.e.,}\quad \langle\mathbf{F}_{\mathbf{p}\to\mathbf{p}_0}\rangle = -\frac{6p_0^2}{4\pi\epsilon_0}\frac{\epsilon_r - 1}{\epsilon_r + 2}\frac{R^3}{d^7}\hat{\mathbf{r}}.$$

The average force is a central force falling off as d^{-7} which is the correct behavior for Van der Waals forces between neutral and polarizable molecules (for instance water): The dipole–dipole force varies as d^{-4} and the induced dipole is proportional to the electric field generated by the first dipole, which goes like d^{-3}.

6.7 Point charge and dielectric sphere

(a) If we assume that the charge is on the z-axis, its position is written $\mathbf{x}' = d\mathbf{u}_z$. For the region inside the sphere, the electrostatic potential will be the superposition of the potential generated by the point charge and the potential generated by the dielectric sphere, which is a solution of Laplace's equation assuming no charge inside the sphere. In spherical coordinates, there is a clear invariance of the charge distribution under rotation around the z-axis. Thus, for $r < a$,

$$V(r,\theta) = \sum_{l=0}^{\infty} A_l r^l P_l(\cos\theta) + \frac{1}{4\pi\epsilon_0}\frac{q}{|r\mathbf{u}_r - d\mathbf{u}_z|}$$

While for the outer region $r > a$,

$$V(r, \theta) = \sum_{l=0}^{\infty} B_l r^{-(l+1)} P_l(\cos\theta) + \frac{1}{4\pi\epsilon_0} \frac{q}{|r\mathbf{u}_r - d\mathbf{u}_z|}.$$

Using the expansion of $1/|\mathbf{x} - \mathbf{x}'|$ in Legendre polynomials (A.52)

$$V(r, \theta) = \sum_{l=0}^{\infty} A_l r^l P_l(\cos\theta) + \frac{q}{4\pi\epsilon_0} \sum_{l=0}^{\infty} \frac{r^l}{d^{l+1}} P_l(\cos\theta), \quad r < a,$$

$$V(r, \vartheta) = \sum_{l=0}^{\infty} B_l r^{-(l+1)} P_l(\cos\theta) + \frac{q}{4\pi\epsilon_0} \sum_{l=0}^{\infty} \frac{r^l}{d^{l+1}} P_l(\cos\theta), \quad a < r < d,$$

$$V(r, \theta) = \sum_{l=0}^{\infty} B_l r^{-(l+1)} P_l(\cos\vartheta) + \frac{q}{4\pi\epsilon_0} \sum_{l=0}^{\infty} \frac{d^l}{r^{l+1}} P_l(\cos\theta), \quad r > d.$$

Now the appropriate boundary conditions must be imposed on the surface $r = a$. The tangential component of the electric field must be continuous

$$-\frac{1}{a}\frac{\partial V(r, \theta)}{\partial\theta}\bigg|_{r=a^-} = -\frac{1}{a}\frac{\partial V(r, \theta)}{\partial\theta}\bigg|_{r=a^+}$$

or, equivalently $V(r = a^-, \theta) = V(r = a^+; \theta)$, then

$$\sum_{l=0}^{\infty} A_l a^l P_l(\cos\theta) + \frac{q}{4\pi\epsilon_0} \sum_{l=0}^{\infty} \frac{a^l}{d^{l+1}} P_l(\cos\theta)$$

$$= \sum_{l=0}^{\infty} B_l a^{-(l+1)} P_l(\cos\theta) + \frac{q}{4\pi\epsilon_0} \sum_{l=0}^{\infty} \frac{a^l}{d^{l+1}} P_l(\cos\theta).$$

From here we obtain
$$B_l = A_l a^{2l+1}.$$

Since there is no free charge density in the sphere, the normal component of $\mathbf{D}$ is continuous at $r = a$

$$\epsilon \frac{\partial V(r, \theta)}{\partial r}\bigg|_{r=a^-} = \epsilon_0 \frac{\partial V(r, \theta)}{\partial r}\bigg|_{r=a^+}$$

and recalling that $B_l = A_l a^{2l+1}$

$$\frac{\epsilon}{\epsilon_0}\left\{\sum_{l=0}^{\infty} l A_l a^{l-1} P_l(\cos\theta) + \frac{q}{4\pi\epsilon_0}\sum_{l=0}^{\infty}\frac{l a^{l-1}}{d^{l+1}} P_l(\cos\theta)\right\}$$

$$= \sum_{l=0}^{\infty} -(l+1) A_l a^{(l-1)} P_l(\cos\theta) + \frac{q}{4\pi\epsilon_0}\sum_{l=0}^{\infty}\frac{l a^{l-1}}{d^{l+1}} P_l(\cos\theta)$$

Then, identifying both sides and using $\epsilon_r = \epsilon/\epsilon_0$, we must have

$$\epsilon_r\left(l A_l a^{l-1} + \frac{q}{4\pi\epsilon_0}\frac{l a^{l-1}}{d^{l+1}}\right) = -(l+1) A_l a^{l-1} + \frac{q}{4\pi\epsilon_0}\frac{l a^{l-1}}{d^{l+1}}$$

and

$$A_l\left(\epsilon_r l + l + 1\right) = \frac{q}{4\pi\epsilon_0}\frac{l}{d^{l+1}}\left(1 - \epsilon_r\right).$$

Finally

$$A_l = \frac{q l\,(1 - \epsilon_r)}{4\pi\epsilon_0 d^{l+1}\,(1 + l(\epsilon_r + 1))} = \frac{q\alpha_l(\epsilon_r)}{4\pi\epsilon_0 d^{l+1}},$$

$$B_l = \frac{q l\,(1 - \epsilon_r)\,a^{2l+1}}{4\pi\epsilon_0 d^{l+1}\,(1 + l(\epsilon_r + 1))} = \frac{q a^l \beta_l(\epsilon_r)}{4\pi\epsilon_0},$$

with

$$\alpha_l(\epsilon_r) = \frac{l(1 - \epsilon_r)}{(1 + l(\epsilon_r + 1))},$$

$$\beta_l(\epsilon_r) = \frac{\alpha_l(\epsilon_r)a^{l+1}}{d^{l+1}}.$$

We obtain

For $r < a$,

$$V(r,\theta) = \frac{q}{4\pi\epsilon_0 d}\sum_{l=0}^{\infty}\left(\alpha_l(\epsilon_r) + 1\right)\frac{r^l}{d^l} P_l(\cos\theta).$$

For $a < r < d$,

$$V(r,\theta) = \frac{q}{4\pi\epsilon_0 r}\sum_{l=0}^{\infty}\left(\beta_l(\epsilon_r)\frac{a^l}{r^l} + \frac{r^{l+1}}{d^{l+1}}\right) P_l(\cos\theta).$$

For $r > d$,

$$V(r,\theta) = \frac{q}{4\pi\epsilon_0 r}\sum_{l=0}^{\infty}\left(\beta_l(\epsilon_r)\frac{a^l}{r^l} + \frac{d^l}{r^l}\right) P_l(\cos\theta).$$

Figure A.57 shows a sketch of the field lines and the equipotentials for the case $d \sim 2a$ and $\epsilon_r = 5$.

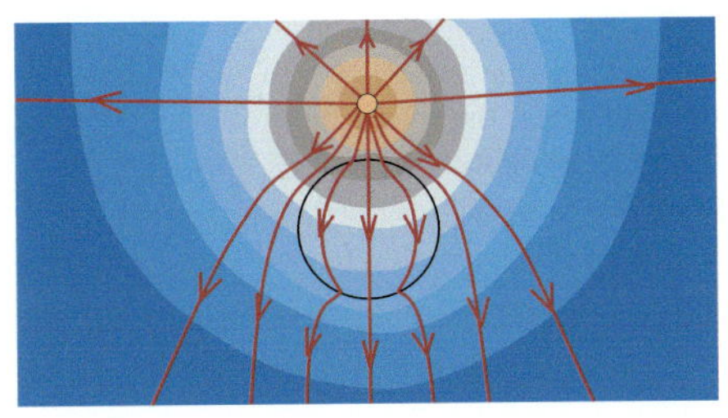

Fig. A.57 Electric field lines and equipotentials generated by a point charge in the presence of a dielectric sphere

(b) The potential within the sphere is

$$V(r, \theta) = \frac{q}{4\pi \epsilon_0 d} \sum_{l=0}^{\infty} (\alpha_l(\epsilon_r) + 1) \frac{r^l}{d^l} P_l(\cos \theta).$$

The electric field is given by

$$\mathbf{E}(r, \theta) = -\frac{\partial V(r, \theta)}{\partial r} \mathbf{u}_r - \frac{1}{r} \frac{\partial V(r, \theta)}{\partial \theta} \mathbf{u}_\theta,$$

where

$$\frac{\partial V(r, \theta)}{\partial r} = \frac{q}{4\pi \epsilon_0 d^2} \sum_{l=0}^{\infty} (\alpha_l(\epsilon_r) + 1) \frac{l r^{l-1}}{d^{l-1}} P_l(\cos \theta),$$

$$\frac{1}{r} \frac{\partial V(r, \theta)}{\partial \theta} = \frac{q}{4\pi \epsilon_0 d^2} \sum_{l=0}^{\infty} (\alpha_l(\epsilon_r) + 1) \frac{r^{l-1}}{d^{l-1}} \frac{\partial [P_l(\cos \theta)]}{\partial \theta}.$$

We then have, for $r/d \ll 1$ (center of the sphere), and using the first-order approximation in r/d:

$$E_r \approx -\frac{q}{4\pi \epsilon_0 d^2} \left\{ \frac{3\cos\theta}{(2 + \epsilon_r)} + \frac{5r(3\cos^2\theta - 1)}{d(3 + 2\epsilon_r)} \right\},$$

$$E_\theta \approx \frac{q}{4\pi \epsilon_0 d^2} \left\{ \frac{3\sin\theta}{(2 + \epsilon_r)} + \frac{15r\cos\theta\sin\theta}{d(3 + 2\epsilon_r)} \right\}.$$

At the lowest order, we find

$$E_r \approx -\frac{3q\cos\theta}{4\pi \epsilon_0 d^2 (2 + \epsilon_r)},$$

$$E_\theta \approx \frac{3q\sin\theta}{4\pi \epsilon_0 d^2 (2 + \epsilon_r)},$$

and

Fig. A.58 Electric field lines and equipotentials in the case where the sphere has a very large dielectric constant ($\epsilon_r = 100$)

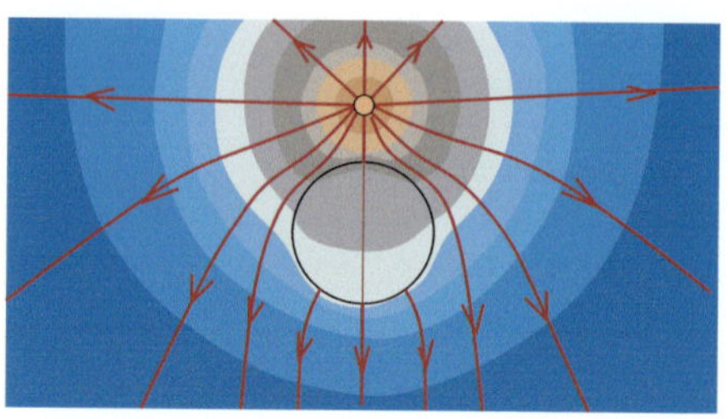

$$|\mathbf{E}| = \sqrt{E_r^2 + E_\theta^2} = \frac{q}{4\pi\epsilon_0 d^2}\frac{3}{2+\epsilon_r}.$$

Note that if $\epsilon \to 1$, $|\mathbf{E}| = \dfrac{q}{4\pi\epsilon_0 d^2}$, which is simply the field generated by the point charge in vacuum. Finally

$$E_z = E_r\cos\theta - E_\theta\sin\theta \approx -\frac{q}{4\pi\epsilon_0 d^2}\frac{3}{(2+\epsilon_r)}$$

and we then deduce that the E_y and E_x components of the field are negligible at the zero order of r/d.

(c) It is immediate that in the limit $\epsilon_r = \epsilon/\epsilon_0 \to \infty$, $\mathbf{E} \to \mathbf{0}$ inside the sphere, as it must occur in a conductor at equilibrium. Figure A.58 shows an sketch of the field lines and the equipotentials for the case $d \sim 2a$ and $\epsilon_r \sim 100$.

6.8 Spherical capacitor with two dielectrics

To find the capacitance, we can assume that the capacitor is charged with a charge Q. In this state, there will be a certain potential difference $\Delta V = V(a) - V(c)$ between $r = a$ and $r = c$. To find ΔV, we need to determine the electric field in the region between the two electrodes $\Omega : \{a \leq r \leq c\}$. To do this, we use Gauss's law for dielectrics with a spherical surface of radius r contained in Ω

$$\oiint_S d\mathbf{S}(\mathbf{x}) \cdot \mathbf{D}(\mathbf{x}) = 4\pi r^2 D(r),$$

where the integration was carried out by considering the clear spherical and mirror symmetries of the charge distribution. By Gauss's law

$$\oiint_S d\mathbf{S}(\mathbf{x}) \cdot \mathbf{D}(\mathbf{x}) = Q,$$

from which it is possible to obtain the electric displacement field in Ω

$$\mathbf{D}(r) = \frac{Q}{4\pi r^2}\mathbf{u}_r, \quad a \leq r \leq c.$$

Now we use $\mathbf{D}(\mathbf{x}) = \epsilon_0 \epsilon_r \mathbf{E}(\mathbf{x})$, where ϵ_r is the dielectric constant. So, for region $\{a \leq r \leq b\}$

$$\epsilon_1 \mathbf{E}(\mathbf{x}) = \frac{Q}{4\pi r^2} \mathbf{u}_r, \qquad a \leq r \leq b,$$

leading to

$$\mathbf{E}(\mathbf{x}) = \frac{Q}{4\pi \epsilon_1 r^2} \mathbf{u}_r, \qquad a \leq r \leq b.$$

And for the region $\{b \leq r \leq c\}$, a similar calculation gives

$$\mathbf{E}(\mathbf{x}) = \frac{Q}{4\pi \epsilon_2 r^2} \mathbf{u}_r, \qquad b \leq r \leq c.$$

With these results, the potential difference reads

$$\Delta V = V(a) - V(c) = \int_a^c \mathbf{E}(\mathbf{x}) \cdot d\mathbf{x}.$$

Taking a radial path of the form $\mathbf{x} = r\mathbf{u}_r$, then $d\mathbf{x} = dr\, \mathbf{u}_r$

$$\Delta V = \int_a^b \frac{dr\, Q}{4\pi \epsilon_1 r^2} + \int_b^c \frac{dr\, Q}{4\pi \epsilon_2 r^2} = \frac{Q}{4\pi} \left[\frac{(b-a)}{\epsilon_1 ab} + \frac{(c-b)}{\epsilon_2 bc} \right].$$

Then

$$C = \frac{Q}{\Delta V} = 4\pi \left(\frac{\epsilon_1 \epsilon_2 abc}{\epsilon_2 c(b-a) + \epsilon_1 a(c-b)} \right).$$

Another way to solve this problem is to note that the capacitor is equivalent to two spherical capacitors in series. The capacitance for a spherical capacitor of inner radius r_1 and outer radius r_2, filled with a dielectric of constant ϵ_r (Example 6.3) is given by

$$C = 4\pi \epsilon_0 \epsilon_r \left(\frac{r_1 r_2}{r_2 - r_1} \right).$$

The equivalent capacity of the system is then

$$\frac{1}{C} = \frac{(b-a)}{4\pi \epsilon_0 \epsilon_{r_1} ab} + \frac{(c-b)}{4\pi \epsilon_0 \epsilon_{r_2} bc} = \frac{\epsilon_{r_2} c(b-a) + \epsilon_{r_1} a(c-b)}{4\pi \epsilon_0 \epsilon_{r_1} \epsilon_{r_2} abc}.$$

Hence

$$C = \frac{4\pi \epsilon_0 \epsilon_{r_1} \epsilon_{r_2} abc}{\epsilon_{r_2} c(b-a) + \epsilon_{r_1} a(c-b)},$$

which is the same result as obtained previously.

6.9 Force on a dielectric bar

The potential energy associated with this configuration is

$$U(x) = \frac{1}{2}\frac{Q^2}{C(x)}.$$

Note that the total capacitance of the capacitor is a function of x. The system can be seen as two parallel capacitors

$$C_{eq} = C_1 + C_2 = \frac{\epsilon_0 a x}{d} + \frac{\epsilon_r \epsilon_0 a (a - x)}{d} = \frac{\epsilon_0 a (x + \epsilon_r (a - x))}{d}.$$

With this result, the stored potential energy reads

$$U(x) = \frac{1}{2}\frac{d Q^2}{\epsilon_0 a (x + \epsilon_r (a - x))},$$

and the force exerted on the dielectric will be

$$\mathbf{F} = -\frac{dU}{dx}\mathbf{u}_x = \frac{Q^2 d}{2\epsilon_0 a}\frac{(1 - \epsilon_r)}{(x + \epsilon_r (a - x))^2}\mathbf{u}_x.$$

Note that $(1 - \epsilon_r) < 0$, that is, the force points toward negative values of x. This means that the dielectric plate is attracted to the inside of the capacitor, so as to minimize the potential energy of the system.

6.10 Capacitor with different dielectrics

The system can be treated as two capacitors in parallel, one formed by the dielectrics of permittivity ϵ_1 and ϵ_2, and the other formed by the dielectric of permittivity ϵ_3 and vacuum. The first capacitor (C_1), can be seen as two capacitors in series, filled with dielectrics 1 and 2, with capacitance

$$\frac{1}{C_1} = \frac{1}{C_1'} + \frac{1}{C_1''},$$

with

$$C_1' = \frac{\epsilon_1 S_1}{d_1}, \qquad C_1'' = \frac{\epsilon_2 S_1}{d_2}.$$

Then,

$$\frac{1}{C_1} = \frac{d_1}{\epsilon_1 S_1} + \frac{d_2}{\epsilon_2 S_1},$$

from which we obtain

$$C_1 = \frac{\epsilon_1 \epsilon_2 S_1}{d_1 \epsilon_2 + d_2 \epsilon_1}.$$

For capacitor C_2, we proceed in a similar way:

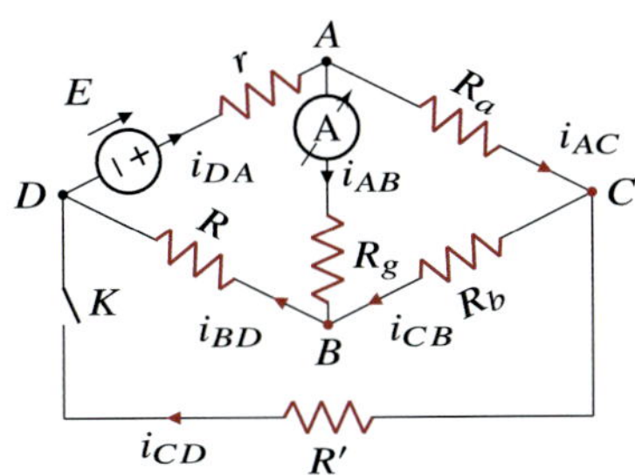

Fig. A.59 Wheatstone bridge

$$\frac{1}{C_2} = \frac{1}{C_2'} + \frac{1}{C_2''},$$

where

$$C_2' = \frac{\epsilon_0 S_2}{d_1}, \qquad C_2'' = \frac{\epsilon_3 S_2}{d_2},$$

and we find

$$C_2 = \frac{\epsilon_3 \epsilon_0 S_2}{d_1 \epsilon_3 + d_2 \epsilon_0}.$$

Finally, since C_1 and C_2 are in parallel, the total capacity will be

$$C = C_1 + C_2 = \frac{\epsilon_1 \epsilon_2 S_1}{d_1 \epsilon_2 + d_2 \epsilon_1} + \frac{\epsilon_3 \epsilon_0 S_2}{d_1 \epsilon_3 + d_2 \epsilon_0}.$$

Problems of Chap. 7

7.1 Measurement of the resistance of a generator with a Wheatstone bridge

We consider first the case when the switch K is open. The currents in each branch are labeled as in Fig. A.59, e.g., i_{DA} in branch DA, etc. In addition, we consider the *loop currents* j_1 and j_2 such that $j_1 = -i_{DA}$ and $j_2 = -i_{AC}$. KCL at nodes D and C give us:

$$i_{CD} = 0, \quad i_{AC} = i_{CB}, \quad \text{and} \quad i_{BD} = i_{DA}.$$

We can apply KVL to the two loops:

$$Ri_{BD} + ri_{DA} + R_g i_{AB} - E = 0, \tag{A.82}$$

$$R_a i_{AC} + R_b i_{CB} - R_g i_{AB} = 0. \tag{A.83}$$

To solve the above system of equations, we first replace the branch currents by their expression as functions of the loop currents: $i_{BD} = i_{DA} = -j_1, i_{AB} = j_2 - j_1,$

$i_{AC} = i_{CB} = -j_2$:

$$\begin{bmatrix} R + r + R_g & -R_g \\ -R_g & R_a + R_b + R_g \end{bmatrix} \begin{bmatrix} j_1 \\ j_2 \end{bmatrix} = \begin{bmatrix} -E \\ 0 \end{bmatrix}.$$

This system admits the solutions

$$j_1 = \frac{1}{\Delta_2} \begin{vmatrix} -E & -R_g \\ 0 & R_a + R_b + R_g \end{vmatrix} \quad \text{and} \quad j_2 = \frac{1}{\Delta_2} \begin{vmatrix} R + r + R_g & -E \\ -R_g & 0 \end{vmatrix},$$

where

$$\Delta_2 = \begin{vmatrix} R + r + R_g & -R_g \\ -R_g & R_a + R_b + R_g \end{vmatrix}.$$

By expanding the determinants, we find

$$i_{AB} = j_2 - j_1 = \frac{(R_a + R_b)E}{\Delta_2},$$

and using the condition $r R_b = R_a R$,

$$\Delta_2 = \frac{(R_a + R_b)[R(R_a + R_b) + R_g(R_b + R)]}{R_b},$$

hence

$$i_{AB} = \frac{E R_b}{R(R_a + R_b) + R_g(R_b + R)},$$

which is independent of R'. Now consider the case when the switch K is closed. We have a third loop current j_3, which is linked to the branch currents by $i_{CB} = j_3 - j_2$ and $i_{BD} = j_3 - j_1$. KVL for the three loops read

$$R i_{BD} + r i_{DA} + R_g i_{AB} - E = 0, \tag{A.84}$$

$$R_a i_{AC} + R_b i_{CB} - R_g i_{AB} = 0, \tag{A.85}$$

$$R_b i_{CB} + R i_{BD} - R' i_{CD} = 0. \tag{A.86}$$

This system can be written in terms of loop currents only:

$$\begin{bmatrix} R + r + R_g & -R_g & -R \\ -R_g & R_a + R_b + R_g & -R_b \\ -R & -R_b & R + R_b + R' \end{bmatrix} \begin{bmatrix} j_1 \\ j_2 \\ j_3 \end{bmatrix} = \begin{bmatrix} -E \\ 0 \\ 0 \end{bmatrix}.$$

Note that the system reads as $[R][j] = [e]$, i.e., it is Ohm's law in matrix form. We find on the diagonal of the $[R]$ matrix the total resistance for each loop and the non-diagonal elements represent the opposite of the resistance that is common to

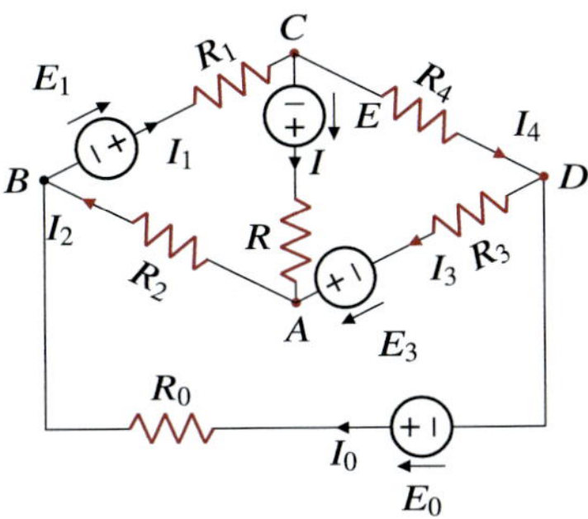

Fig. A.60 Electrical network and definition of the currents

two loops. For example, R is common to loops 1 and 3 and so, the matrix elements $[R]_{13} = -R = [R]_{31}$. On the right-hand side, the column vector $[e]$ represents the sources for each loop.

The solution to the system of equations reads

$$
j_1 = \frac{1}{\Delta_3} \begin{vmatrix} -E & -R_g & -R \\ 0 & R_a + R_b + R_g & -R_b \\ 0 & -R_b & R + R_b + R' \end{vmatrix}, \quad
j_2 = \frac{1}{\Delta_3} \begin{vmatrix} R + r + R_g & -E & -R \\ -R_g & 0 & -R_b \\ -R & 0 & R + R_b + R' \end{vmatrix},
$$

$$
j_3 = \frac{1}{\Delta_3} \begin{vmatrix} R + r + R_g & -R_g & -E \\ -R_g & R_a + R_b + R_g & 0 \\ -R & -R_b & 0 \end{vmatrix}, \quad
\Delta_3 = \begin{vmatrix} R + r + R_g & -R_g & -R \\ -R_g & R_a + R_b + R_g & -R_b \\ -R & -R_b & R + R_b + R' \end{vmatrix}.
$$

Again, $i_{AB} = j_2 - j_1 = \dfrac{E[(R_a + R_b)R' + R_a(R + R_b)]}{\Delta_3}$. At this stage, it looks like i_{AB} depends on R', however, an expansion of Δ_3 shows that

$$
\Delta_3 = \frac{\left[R(R_a + R_b) + R_g(R_b + R) \right]}{R_b} \left[(R + R_b)R_a + R'(R_a + R_b) \right],
$$

hence

$$
i_{AB} = \frac{Eb}{R(R_a + R_b) + R_g(R_b + R)}.
$$

This current is the same as previously and does not depend on R'.

7.2 Electrical network

Let us assign currents to each branch, as indicated in Fig. A.60.

Applying KCL to nodes B, D, and C, we get

$$
I_2 = I_1 - I_0,
$$
$$
I_3 = I_4 - I_0,
$$
$$
I = I_1 - I_4.
$$

Applying KVL to each loop, we obtain

$$\begin{aligned}
\text{loop } BCA \quad & R_1 I_1 + R I + R_2 I_2 = E_1 + E, \\
\text{loop } CDA \quad & R_4 I_4 + R_3 I_3 - R I = E_3 - E, \\
\text{loop } BAD \quad & -R_2 I_2 - R_3 I_3 + R_0 I_0 = E_3 - E_0.
\end{aligned}$$

Substituting I_2, I_3, and I, we can rewrite this system in matrix-vector format as

$$\begin{pmatrix} R_1 + R_2 + R & -R & -R_2 \\ -R & R + R_3 + R_4 & -R_3 \\ -R_2 & -R_3 & R_0 + R_2 + R_3 \end{pmatrix} \begin{pmatrix} I_1 \\ I_4 \\ I_0 \end{pmatrix} = \begin{pmatrix} E_1 + E \\ E_3 - E \\ E_0 - E_3 \end{pmatrix}.$$

This is a linear system of equations $(I_1, I_4, I_0)^T$. Now insert in the system the data given for the resistances (in Ω) and voltage sources (in A):

$$\begin{pmatrix} 27 & -4 & -3 \\ -4 & 28 & -20 \\ -3 & -20 & 25 \end{pmatrix} \begin{pmatrix} I_1 \\ I_4 \\ I_0 \end{pmatrix} = \begin{pmatrix} 8 \\ 6 \\ 22 \end{pmatrix}.$$

Solving this system (using either Gaussian elimination or a numerical solver), we find

$$I_1 = 1\,\text{A}, \qquad I_4 = 2.5\,\text{A}, \qquad I_0 = 3\,\text{A}.$$

KCL yields the currents in other branches:

$$I_2 = -2\,\text{A}, \qquad I_3 = -0.5\,\text{A}, \qquad I = -1.5\,\text{A}.$$

From these values, we find the voltage differences:

$$\begin{aligned}
V_A - V_D &= E_3 - R_3 I_3 = 18\,\text{V}, \\
V_B - V_C &= -E_1 + R_1 I_1 = 14\,\text{V}, \\
V_A - V_C &= E - R I = 8\,\text{V}, \\
V_B - V_D &= E_0 - R_0 I_0 = 24\,\text{V}.
\end{aligned}$$

7.3 Equivalent resistance of a rectangular grid

As shown in Fig. A.61, the network exhibits symmetry about the center of the rectangle $ADA'D'$. his implies that nodes A and A', B and B', etc., are electrically equivalent. As a result, the current distribution must also be symmetric. Consider nodes A and A'. If a current I enters node A, the same current I must exit node A'. Let us denote the current in branch AE as i_1. By symmetry, the current in branch $E'A'$ is also i_1. Applying Kirchhoff's Current Law (KCL) at node A, the current in branch AB must be $I - i_1$. Similarly, the current in branch $A'B'$ is also $I - i_1$. We can continue this symmetry argument to label the currents in the remaining branches:

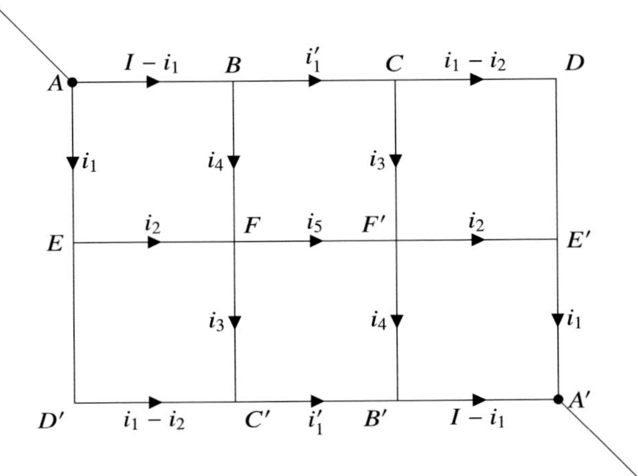

Fig. A.61 Rectangular grid

i_2 in EF; $i_1 - i_2$ in ED' (from KCL at node E), i_3 in FC'; i_4 in BF, i_5 in FF' and i'_1 in $C'B'$.

Applying KCL at nodes B, F, and C', we obtain the following equations:

$$\text{Node } (B) \qquad i_4 = (I - i_1) - i'_1,$$
$$\text{Node } (F) \qquad i_5 = i_2 + i_4 - i_3,$$
$$\text{Node } (C') \qquad i'_1 = i_1 - i_2 - i_3.$$

By systematically applying symmetry and KCL, we have established the current relationships throughout the network.

By applying KVL to the indicated loops, we obtain (clockwise)

$$\text{Loop } (ABFE) \qquad r(I - i_1) + ri_4 + -ri_2 - ri_1 = 0,$$
$$\text{Loop } (EFC'D') \qquad ri_2 + ri_3 - 2r(i_1 - i_2) = 0,$$
$$\text{Loop } (FF'B'C') \qquad ri_5 + ri_4 - ri'_1 - ri_3) = 0.$$

Substituting the expressions for i_4, i_5 and i'_1 obtained from KCL into the KVL equations, we get

$$\text{Loop } (ABFE) \qquad 4i_1 + i_3 = 2I,$$
$$\text{Loop } (EFC'D') \qquad -2i_1 + 3i_2 + i_3 = 0,$$
$$\text{Loop } (FF'B'C') \qquad 5i_1 - 4i_2 + 5i_3 = 2I.$$

Solving this system of linear equations yields

$$i_1 = \frac{32}{69}I, \quad i_2 = \frac{6}{23}, \quad i_3 = \frac{10}{69}.$$

Fig. A.62 Triangular grid

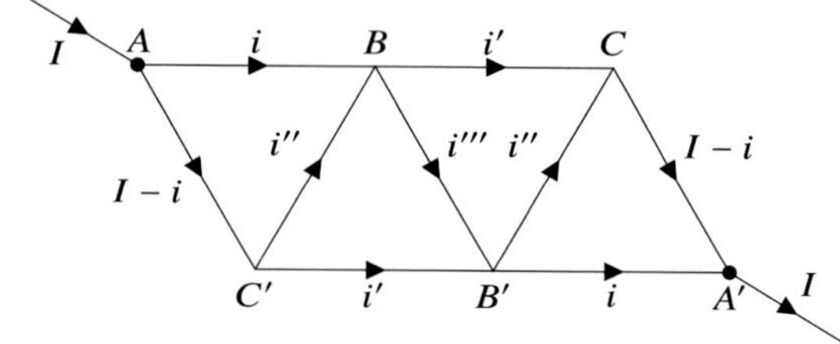

Using KCL, we can then calculate the remaining currents:

$$i_1' = \frac{24}{69}I, \quad i_4 = \frac{13}{69}I, \quad i_5 = \frac{21}{69}I.$$

To find the equivalent resistance between points A and A', we can consider the voltage drop along the path $ABB'A'$:

$$V_A - V_{A'} = 2ri_1 + 2ri_2 + ri_5 = r\left(\frac{2 \times 32}{69} + \frac{2 \times 18}{69} + \frac{21}{69}\right)rI = \frac{121}{69}rI.$$

Comparing this to Ohm's law, $V = IR$, we see that the equivalent resistance R between A and A' is

$$R = \frac{121}{69}r.$$

This relation is Ohm's law: It shows that the network is equivalent to a resistor with resistance between A and A' equal to $R = 121r/69$.

7.4 Equivalent resistance of a triangular grid

If a current I enters the bipole AA' through A, the same current exits through A'. Due to the symmetry of the network (see Fig. A.62), the current i in branch AB is the same as the current in branch $B'A'$. Applying KCL at node A, we get the current $I - i$ in the branch AC', which is the same as in the branch CA' for symmetry reasons. Symmetric branches carry the same currents, as indicated in the figure: Branches $C'B'$ and BC carry the current i', branches $C'B$ and BC' carry the current i'' and branch BB' carries the current i'''.

Applying KCL at nodes C' and B, we get the currents i'' and i''':

$$i'' = I - i - i',$$
$$i''' = i + i'' - i' = I - 2i'.$$

Applying KVL to the first two loops, we obtain

$$\text{Loop } ABC' \qquad ri - ri'' - r(I - i) = 0,$$
$$\text{Loop } BB'C' \qquad ri''' - ri' + ri'' = 0,$$

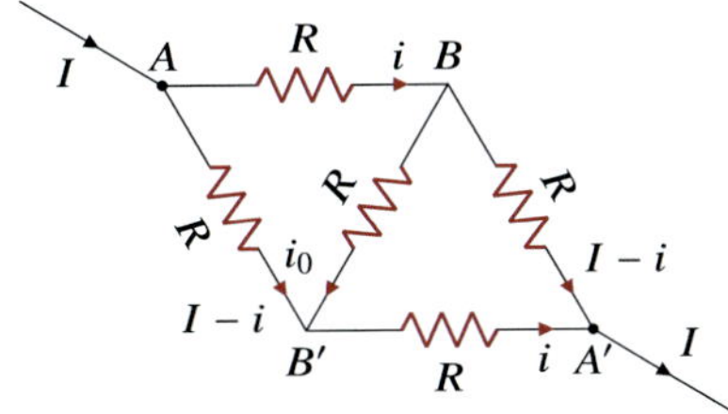

Fig. A.63 Reorganized circuit

Substituting the expressions for i'' and i''', we get

$$\text{Loop } ABC' \qquad 3i + i' = 2I,$$
$$\text{Loop } BB'C' \qquad I + 4i' = 2I,$$

This system of equations admits the solution

$$i = \frac{6}{11}I, \quad i' = \frac{4}{11}I.$$

We then calculate from KCL equations the remaining currents

$$i'' = \frac{1}{11}I, \quad i''' = \frac{3}{11}I.$$

Finally, we can express the voltage difference between A and A', following the path $ABB'A'$:

$$V_A' - V_A = ri + ri''' + ri = \left(\frac{6}{11} + \frac{3}{11} + \frac{6}{11}\right)rI = \frac{15}{11}rI.$$

This relation (Ohm's law) shows that the equivalent resistance between A and A' is

$$R = \frac{15r}{11}.$$

7.5 Equivalent resistance

By exploiting the symmetry of the circuit, which is apparent on the reorganized circuit plotted in Fig. A.63, we can simplify the analysis. Let us denote the current entering node A (and exiting node A') as I. If we label the current in branch AB as i, then due to symmetry, the current in branch $A'B'$ must also be i. Applying Kirchhoff's Current Law at node A, we can determine that the current in branches AB' and BA' is $I - i$.

Let us denote the current in branch BB' as i_0. Applying KCL at node B, we get

$$i_0 = 2i - I.$$

Fig. A.64 Simplified circuit

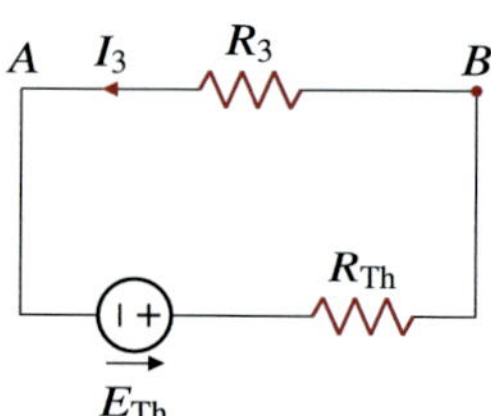

Fig. A.65 Finding the
Thévenin's equivalent
resistance

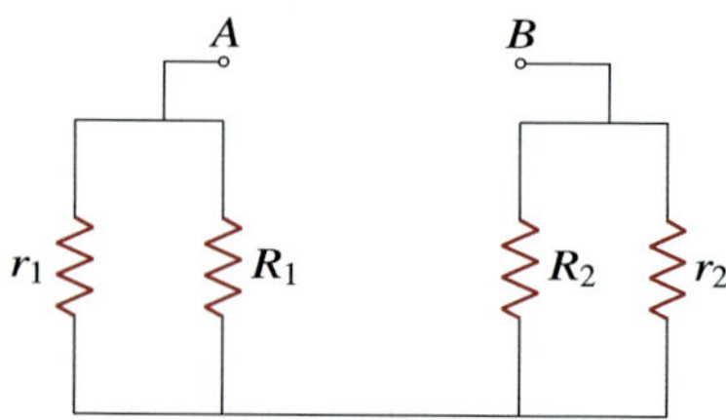

Applying KVL to loop ABB', we obtain

$$Ri + Ri_0 - R(I - i) = 0.$$

Simplifying the second equation, we get

$$i_0 = I - 2i.$$

To satisfy both KCL and KVL, the only possible solution is $i_0 = 0$. This implies
that no current flows through branch BB'. Therefore, this branch can be effectively
removed from the circuit. With branch BB' removed, the circuit simplifies to two
identical branches, each consisting of two resistors R connected in series. These
two branches are effectively in parallel. The equivalent resistance of a single branch
is $2R$. The equivalent resistance of the entire circuit: is $R_{(2R)\|(2R)} = R$. Hence, the
entire network is equivalent to a single resistor of resistance R.

7.6 Application of Thévenin's theorem

The resistor R_3 is connected in series with a Thévenin equivalent circuit, as shown
in Fig. A.64.

The current through R_3 is given by

$$I_3 = \frac{E_{Th}}{R_3 + R_{Th}}.$$

The equivalent resistance, R_{Th}, is the resistance between points A and B when
the circuit is open-circuited and the voltage sources are short-circuited, as shown in
Fig. A.65. In this case, we have a series combination of $(R_1 \parallel r_1)$ and $(R_2 \parallel r_2)$:

Fig. A.66 Finding the Thévenin's voltage

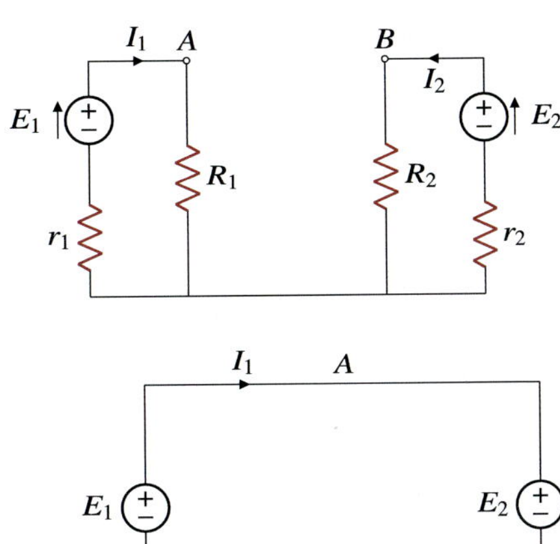

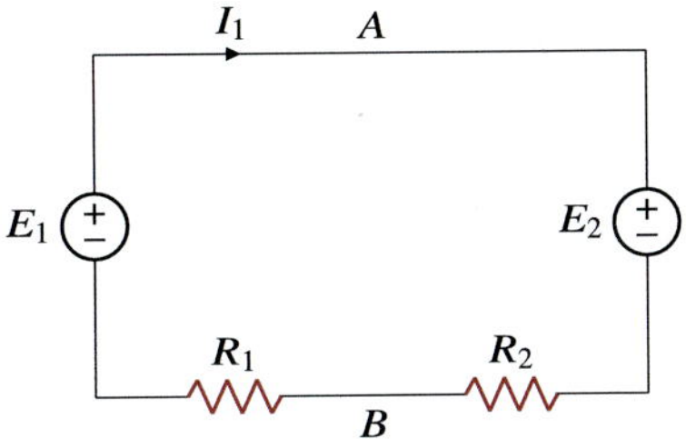

Fig. A.67 Determining Thévenin's equivalent voltage

$$R_{\text{Th}} = \frac{r_1 R_1}{R_1 + r_1} + \frac{r_2 R_2}{R_2 + r_2}.$$

Given the numerical values, we find $R_{\text{Th}} = 2.1\,\Omega$. The Thévenin voltage, E_{Th}, is the open-circuit voltage between points A and B.

Applying KVL to each loop as shown in Fig. A.66, we obtain

$$E_1 = (r_1 + R_1)I_1,$$
$$E_2 = (r_2 + R_2)I_2.$$

Using Ohm's law, the potential difference between points B and A can be expressed as

$$V_B - V_A = R_2 I_2 - R_1 I_1.$$

Substituting the expressions for I_1 and I_2 from the KVL equations, we get

$$E_{\text{Th}} = \frac{R_2}{r_2 + R_2} E_2 - \frac{R_1}{r_1 + R_1} E_1.$$

Given the numerical values, we find $E_{\text{Th}} = 0.3\,\text{V}$. Therefore, the current through R_3 is $I_3 = 0.3/(6 + 2.1) \approx 37\,\text{mA}$.

7.7 Thévenin's theorem versus Norton's theorem

The Thévenin equivalent voltage is calculated by determining the open-circuit voltage across terminals A and B (see Fig. A.67).

Applying Ohm's law and KCL to the loop, we can express the potential difference between A and B:

Fig. A.68 Equivalent circuit

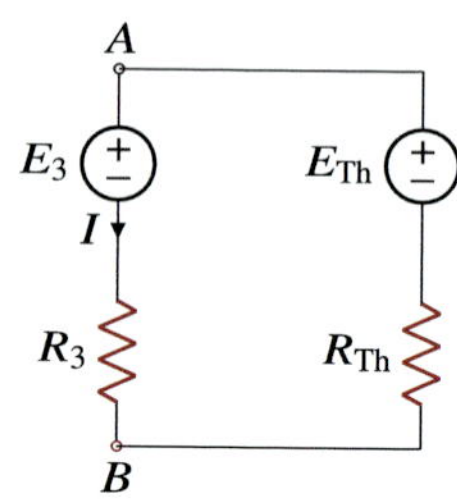

$$V_A - V_B = (E_1 - R_1 I_1) = (E_2 + R_2 I_1).$$

Solving for I_1, we get

$$I_1 = \frac{E_1 - E_2}{R_1 + R_2}.$$

Substituting I_1 into the expression for $V_A - V_B$, we obtain the Thévenin equivalent voltage:

$$E_{Th} = \frac{R_2 E_1 + R_1 E_2}{R_1 + R_2}.$$

The Thévenin equivalent resistance, obtained by short-circuiting the voltage sources, is the parallel combination of R_1 and R_2:

$$R_{Th} = \frac{R_1 R_2}{R_1 + R_2}.$$

Connecting now the branch AB, we obtain the circuit illustrated in Fig. A.68. The current flowing through R_3 is given by

$$I = \frac{E_{Th} - E_3}{R_{Th} + R_3}.$$

Substituting the expressions for E_{Th} and R_{Th}, we get

$$I = \frac{(E_1 R_2 + E_2 R_1) - E_3(R_1 + R_2)}{R_1 R_2 + R_1 R_3 + R_2 R_3}.$$

Using Norton's theorem, the Norton equivalent resistance is identical to the Thévenin equivalent resistance. For convenience, we will represent this equivalent resistance as a conductance, $G_{No} = 1/R_{Th}$. To apply Norton's theorem, we can replace each branch of the circuit with its Norton equivalent, consisting of a current source and a parallel conductance. The Norton current source, η_i, is equal to the short-circuit current of the branch (E_i/R_i). The Norton conductance, G_i, is the reciprocal of the branch resistance ($1/R_i$). This is shown in Fig. A.69.

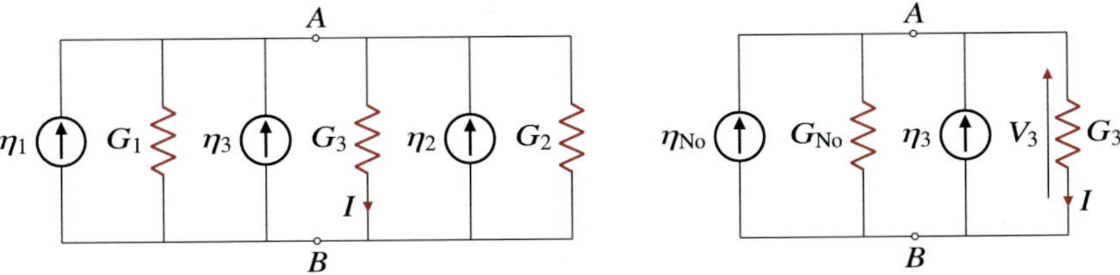

Fig. A.69 Application of Norton's theorem

The Norton equivalent current, η_{No}, and the Norton equivalent conductance, G_{No}, (see figure, right) are given by

$$\eta_{No} = \eta_1 + \eta_2,$$
$$G_{No} = G_1 + G_2.$$

When branch AB (current source η_3 and resistor R_3) is connected, we can apply Ohm's law to the loop containing V_3, R_3 and the Norton equivalent:

$$I = G_3 V_3 - \eta_3,$$
$$I = -G_{No} V_3 + \eta_{No}.$$

Solving these equations for V_3, we get

$$V_3 = \frac{\eta_{No} + \eta_3}{G_{No} + G_3}.$$

Substituting V_3 back into the first equation for I, we find

$$I = \frac{G_3 \eta_{No} - G_{No} \eta_3}{G_{No} + G_3}.$$

Substituting the Norton equivalent current and conductance, we obtain

$$I = \frac{G_3(\eta_1 + \eta_2) - (G_1 + G_2)\eta_3}{G_1 + G_2 + G_3}.$$

Finally, substituting $G_i = 1/R_i$ and $\eta_i = E_i/R_i$, and simplifying, we arrive at the same result obtained using Thévenin's theorem.

7.8 Variable resistor and equivalent generator

(a) In Fig. A.70, the resistance between A and C is xR. Between C and b, it is $(1 - x)R$. When the source is short-circuited, points A and B have the same potential and the circuit between A and C is equivalent to a resistance, R_{Th}:

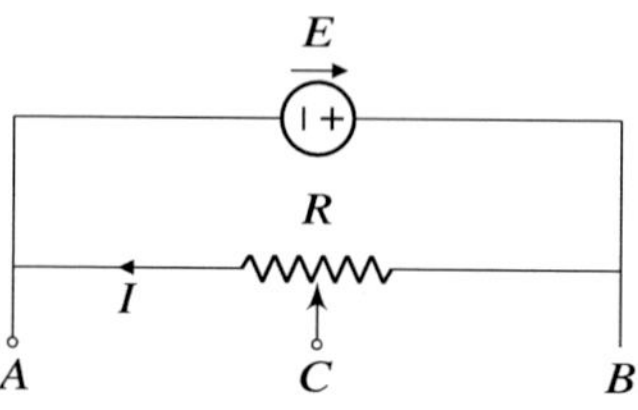

Fig. A.70 Circuit with variable resistor

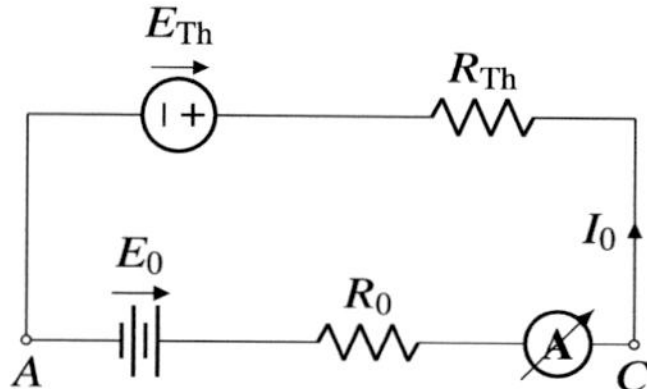

Fig. A.71 Equivalent circuit using Thévenin's theorem

$$R_{\text{Th}} = xR \parallel (1-x)R = \frac{xR(1-x)R}{xR + (1-x)R} = x(1-x)R.$$

The Thévenin voltage, E_{Th}, is the open-circuit voltage between points A and C. Applying Ohm's law between A and C for the two branches of the circuit, we obtain

$$V_C - V_A = E + (1-x)RI = xRI.$$

Solving for the current in the circuit, we get

$$I = \frac{E}{(2x-1)R}.$$

The circuit between A and C can be replaced by a Thévenin equivalent source with Thévenin voltage $E_{\text{Th}} V_C - V_A = xRI$:

$$E_{\text{Th}} = \frac{xE}{2x-1}.$$

(b) Substituting the Thévenin equivalent circuit between A and C and connecting the battery, we obtain the circuit illustrated in Fig. A.71.

The current in the circuit is obtained by applying Ohm's law:

$$I_0 = \frac{E_{\text{Th}} - E_0}{R_{\text{Th}} + R_0},$$

where the resistance of the ammeter was assumed to be negligible. Substituting the expressions for E_{Th} and R_{Th}, we get the current $I_0(x)$:

$$I_0(x) = \frac{E_0 + x(E - 2E_0)}{(2x - 1)(R_0 + x(1 - x)R)}.$$

The current measured by the ammeter is zero if $E_{\text{Th}} = E_0$ or, equivalently,

$$x = x_0 = \frac{E_0}{2E_0 - E}.$$

Problems of Chap. 8

8.1 Mass spectrometer

(a) The speed v with which an ion enters the selector is obtained by energy conservation

$$\frac{1}{2}m(v^2 - v_0^2) = q\Delta V.$$

Since initially, $\mathbf{v} = 0$, we find

$$\mathbf{v} = \sqrt{\frac{2q\Delta V}{m}}\,\mathbf{u}_y.$$

(b) In order for the velocity of a particle to remain constant and exit through the selector, the action of the Lorentz force on it must be null. This is the case if

$$\mathbf{F} = q\,(\mathbf{E} + \mathbf{v} \times \mathbf{B}) = \mathbf{0},$$

where $\mathbf{E} = E\mathbf{u}_x$, $\mathbf{B} = -B\mathbf{u}_z$, and $\mathbf{v} = v\mathbf{u}_y$, that is,

$$qE\mathbf{u}_x - qv\mathbf{u}_y \times B\mathbf{u}_z = 0 \quad \rightarrow \quad qE = qvB,$$

and so,

$$B = \frac{E}{v} = E\sqrt{\frac{m}{2q\Delta V}}.$$

(c) After entering the magnetic field region $\mathbf{B}_0$, the particles describe a circumference of radius r. The magnitude of the magnetic force is qvB_0. Then,

$$qvB_0 = m\frac{v^2}{r},$$

that is,

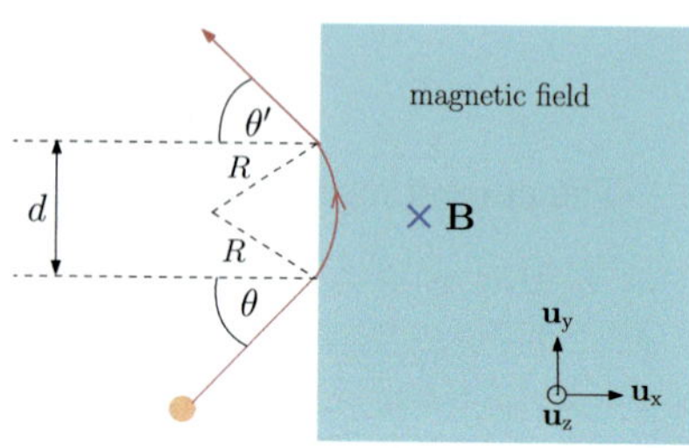

Fig. A.72 Path followed by the proton

$$q B_0 = \frac{\sqrt{2mq\,\Delta V}}{r}.$$

Finally, the mass of the particles can be determined by

$$m = \frac{q B_0^2 r^2}{2\Delta V}.$$

8.2 Proton in a uniform magnetic field

(a) Since the uniform magnetic field is perpendicular to the velocity, the proton will describe a circular path before leaving the magnetic field, as shown in Fig. A.72.

Both at the entrance and at the exit, the speed is tangent to the circular path. By symmetry, we then have $\theta' = \theta$.

(b) We have

$$\frac{d}{2} = R \cos\theta \quad \rightarrow \quad d = 2R \cos\theta.$$

In addition, from the dynamics of a circular motion,

$$q v B = \frac{mv^2}{R}.$$

From here,

$$R = \frac{mv}{qB}$$

and then the distance d is

$$d = \frac{2mv}{qB} \cos\theta.$$

8.3 Electron in a uniform magnetic field

(a) Since the initial velocity of the incident electron is perpendicular to the magnetic field, the trajectory will be a circumference in the Oxy-plane as long as the electron remains in the region between the plates. To analyze how the electron will be deflected, it is enough to analyze what happens with the magnetic force that acts on the electron at the moment $t = 0$ when it enters the region between the plates:

$$\mathbf{F}_L(t = 0) = -e \underbrace{\mathbf{v}(0) \times \mathbf{B}}_{v\mathbf{u}_x \times -B\mathbf{u}_z} = -evB\mathbf{u}_y.$$

That is, the particle is diverted toward the lower plate.

(b) Since the electron hits the right end of the lower plate, the trajectory of the particle is that shown in Fig. A.73. The magnitude of the magnetic force on the electron is constant and given by

$$|\mathbf{F}| = evB.$$

Newton's law for the circular dynamics results in the relation

$$evB = \frac{mv^2}{R},$$

then

$$R = \frac{mv}{eB}.$$

On the other hand, we have

$$l^2 + \left(R - \frac{d}{2}\right)^2 = R^2.$$

Solving for R, we find

$$l^2 + R^2 - Rd + \frac{d^2}{4} = R^2 \quad \rightarrow \quad R = \frac{l^2}{d} + \frac{d}{4},$$

so

$$\frac{mv}{eB} = \frac{l^2}{d} + \frac{d}{4}$$

and the modulus of the electron velocity turns out to be

$$v = \frac{eB}{m}\left(\frac{l^2}{d} + \frac{d}{4}\right).$$

8.4 Charge in a uniform magnetic field

The magnetic force on the charge q is given by the Lorentz force

$$\mathbf{F} = q\mathbf{v} \times \mathbf{B}.$$

Thus, only the component of $\mathbf{B}$ perpendicular to $\mathbf{v}$ will exert a force on the particle. The velocity component parallel to $\mathbf{B}$ is preserved, while the component perpendicular to $\mathbf{B}$ describes a circular path. In this way, the trajectory followed by the particle is a helix. If the projected motion on the plane perpendicular to the magnetic field is observed, the trajectory describes a semicircle (see Fig. A.74). Let R be the radius of the circumference, then we can determine the time it takes to leave as

$$T = \frac{2\pi R}{2v_\perp} = \frac{\pi R}{v_\perp},$$

where $v_\perp$ is the velocity component that is perpendicular to the magnetic field. Now we must determine R. From Newton's second law,

$$\frac{mv_\perp^2}{R} = F,$$

where $F = F = qv_\perp B$ is the magnitude of the magnetic force exerted on the particle. Thus

$$\frac{mv_\perp^2}{R} = qv_\perp B$$

and

$$R = \frac{mv_\perp}{qB}.$$

Finally

$$T = \frac{\pi}{v_\perp}\frac{mv_\perp}{qB} = \frac{\pi m}{qB}.$$

8.5 A proton in a cyclotron accelerator

In a uniform magnetic field, the radius r that describes a charge particle q, mass m

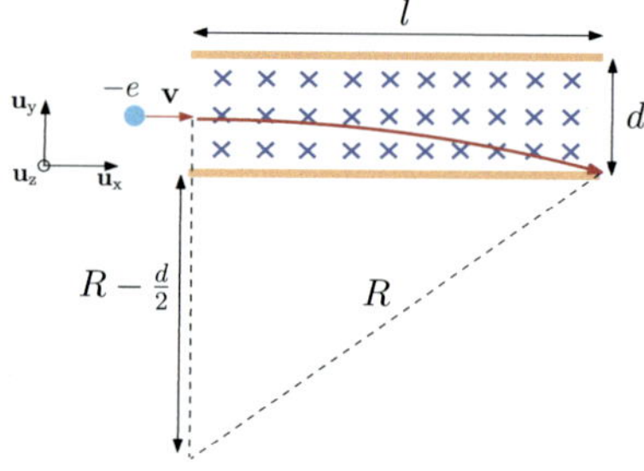

Fig. A.73 Trajectory followed by the electron

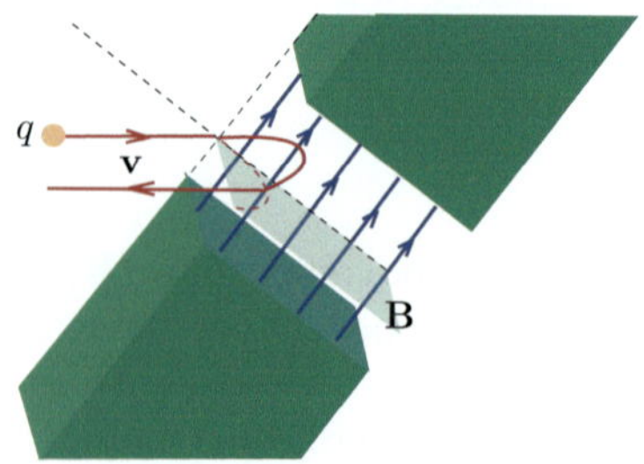

Fig. A.74 The motion projected on the plane perpendicular to **B** is a semicircle in the region of the field

and velocity v is given by

$$r = \frac{mv}{qB}$$

so that the time it takes to describe a semicircle on one of the electrodes is

$$T = \frac{\pi r}{v} = \frac{\pi m}{qB}.$$

The electric field must then oscillate at a frequency $f = 1/T$. Evaluating for a proton, $m_p = 1.67 \times 10^{-27}$ kg, $q_p = 1.6 \times 10^{-19}$ C and for a magnetic field $B = 1$ T:

$$f = 30.5\,\text{MHz}.$$

The maximum speed that a proton can acquire is limited by the maximum possible radius R of the orbit

$$v_{\max} = \frac{qBR}{m},$$

where $R = 2$m is the radius of the cyclotron. Then, the maximum kinetic energy will be

$$E = \frac{1}{2}m_p v_{\max}^2 = 3.06 \times 10^{-11}\,\text{J} = 191\,\text{MeV}.$$

During each rotation, the proton receives a kinetic energy equal to $\Delta E = 2q\Delta V = 100$ keV, then, the required number of turns is

$$N = \frac{191}{100} \times 10^3 = 1910.$$

8.6 The Hall effect

(a) The current density is given by $\mathbf{J} = nq\mathbf{v}$, where $\mathbf{v}$ is the speed of the free charges on the conductor. Since the current flows in the direction $\mathbf{u}_y$, we have $\mathbf{v} = v\mathbf{u}_y$ and then,

$$\mathbf{J} = qnv\mathbf{u}_y.$$

The magnetic field $\mathbf{B}$ exerts an additional force on a charge, given by the magnetic force

$$\mathbf{F}_B = q\mathbf{v} \times \mathbf{B} = qv\mathbf{u}_y \times B\mathbf{u}_z = qvB\mathbf{u}_x.$$

Then, the magnetic force accelerates the charges along the axis $\mathbf{u}_x$ perpendicular to the current. If $q > 0$, the trajectory of the charges will be diverted to the wall $x = a$, while if $q < 0$, the charges will be diverted to the wall $x = 0$. There is then an accumulation of charges on one of the walls and a deficit of charges on the opposite wall.

(b) The electric field $\mathbf{E}_H = -E_H\mathbf{u}_x$ between the plates generates an additional electrostatic force on the free charges, given by

$$\mathbf{F}_H = q\mathbf{E}_H = -qE_H\mathbf{u}_x.$$

Note that if $q > 0$, the plate $x = d$ will be positively charged and $E_H > 0$. If $q < 0$, $E_H < 0$. In all cases, the force associated with the Hall field will be in the direction of $-\mathbf{u}_x$, that is, in the direction opposite to the magnetic force.

(c) For the effect of the magnetic field on the charges to be null in the stationary regime, we must have

$$\mathbf{F}_B + \mathbf{F}_H = 0,$$

that is,

$$qvB\mathbf{u}_x - qE_H\mathbf{u}_x = 0 \quad \rightarrow \quad E_H = vB.$$

In the stationary regime, the charges move along a straight line parallel to the current density $\mathbf{J}$.

(d) The Hall voltage V_H that appears between the side walls is

$$V_H = V(x = 0) - V(x = a) = -aE_h = -avB.$$

Note that if I is the current flowing through the plate, then $I = abJ = abnqv$, so $v = I/(abnq)$ and

$$V_H = -\frac{BI}{bnq} = R_H\frac{BI}{b},$$

where $R_H = -1/(nq)$ denotes the Hall resistance, which depends solely on the density of free charges in the conductor. Thus, if it is desired to determine the magnitude of an unknown magnetic field $\mathbf{B}$, a current I is applied in the plate perpendicular to the magnetic field, the voltage V_H that appears between the side walls is measured and so,

$$B = \frac{V_H b}{I R_H}.$$

8.7 Application of the Hall effect I

(a) The velocity $\mathbf{v}$ of the electrons is related to the current density $\mathbf{J}$ by

$$\mathbf{J} = -ne\mathbf{v}.$$

Then, its magnitude is given by

$$v = \frac{J}{ne} = \frac{I}{abne} = 1.95 \times 10^{-4}\,\mathrm{m\,s^{-1}}.$$

(b) The Hall voltage is written:

$$V_H = R_H\frac{BI}{b} = \frac{BI}{ben} = 7.81\,\mu\mathrm{V}$$

and the Hall field is

$$E_H = V_H/a = 3.9 \times 10^{-4}\,\mathrm{V\,m^{-1}}.$$

8.8 Application of the Hall effect II

We have

$$V_H = R_H \frac{BI}{b}.$$

The Hall field is given by $E_H = V_H/a$, where a is the width of the conductor, and considering that the current density can be written $J = I/(ab)$, where b is the thickness, we obtain

$$E_H = \frac{V_H}{a} = R_H B J$$

and then the Hall resistance R_H is given by

$$R_H = \frac{E_H}{BJ} = 2.5 \times 10^{-10}\,\mathrm{m^3\,C^{-1}} = \frac{1}{en}.$$

With this, we obtain the volume density of electrons in sodium:

$$n = \frac{1}{e R_H} = 2.5 \times 10^{28}\,\mathrm{m^{-3}} = 2.5 \times 10^{22}\,\mathrm{cm^{-3}}.$$

The mass of a sodium atom is given by

$$m_{\mathrm{Na}} = \frac{23\,\mathrm{g}}{6.02 \times 10^{23}} = 3.8206 \times 10^{-23}\,\mathrm{g},$$

and given the mass density $\rho = 0.97\,\mathrm{g\,cm^{-3}}$, the density of sodium atoms is

$$\frac{\rho}{m_{\mathrm{Na}}} = 2.5 \times 10^{22}\,\mathrm{cm^{-3}},$$

so that each sodium atom provides a free electron.

8.9 Field of a current line

Consider a differential line element $d\mathbf{x}' = dx\,\mathbf{u}_x$ at the point $x\mathbf{u}_x$ through which a current I passes in the direction $\mathbf{u}_x$. To calculate the magnetic field at point P, we use the Biot–Savart law. According to our choice of origin, we have $\mathbf{x} = a\mathbf{u}_y$ and $\mathbf{x}' = x\mathbf{u}_x$. So,

$$d\mathbf{B}(\mathbf{x}) = \frac{\mu_0}{4\pi} I \frac{d\mathbf{x}' \times (\mathbf{x} - \mathbf{x}')}{|\mathbf{x} - \mathbf{x}'|^3}.$$

Here, $\mathbf{x} - \mathbf{x}' = a\mathbf{u}_y - x\mathbf{u}_x$ and then $|\mathbf{x} - \mathbf{x}'| = \sqrt{a^2 + x^2}$. With this, the contribution to the field of this differential current element reads

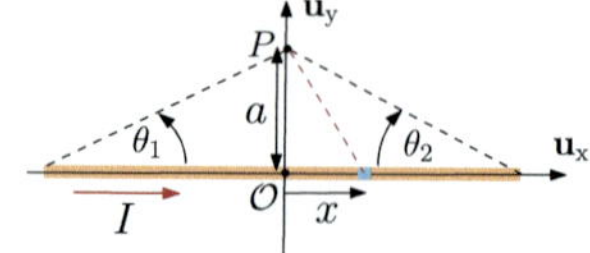

Fig. A.75 The position of the line with respect to P is characterized by the two angles θ_1 and θ_2

$$d\mathbf{B}(\mathbf{x}) = \frac{\mu_0}{4\pi} I \frac{dx\,\mathbf{u}_x \times \left(a\,\mathbf{u}_y - x\,\mathbf{u}_x\right)}{(a^2 + x^2)^{3/2}} = \frac{\mu_0}{4\pi} I \frac{a\,dx\,\mathbf{u}_z}{(a^2 + x^2)^{3/2}}$$

and the total field at P will be

$$\mathbf{B}(P) = \frac{\mu_0}{4\pi} I \int_{-L_1}^{L_2} \frac{a\,dx\,\mathbf{u}_z}{(a^2 + x^2)^{3/2}},$$

where L_1 and L_2 are defined by the angles θ_1 and θ_2, shown in Fig. A.75 and given by

$$\tan \theta_1 = \frac{a}{L_1} \quad \rightarrow \quad -L_1 = a \cot(-\theta_1),$$

$$\tan \theta_2 = \frac{a}{L_2} \quad \rightarrow \quad L_2 = a \cot \theta_2.$$

To solve this integral, we use the following change of variables:

$$x = a \cot \theta \quad \rightarrow \quad dx = -a \csc^2 \theta \, d\theta$$

and the integral becomes

$$\mathbf{B} = \frac{\mu_0}{4\pi} I \int_{-\theta_1}^{\theta_2} \frac{a(-a \csc^2 \theta)\,d\theta\,\mathbf{u}_z}{(a^2 + a^2 \cot^2 \theta)^{3/2}}$$

$$= -\frac{\mu_0}{4\pi a} I \int_{-\theta_1}^{\theta_2} \sin \theta \, d\theta \, \mathbf{u}_z$$

$$= \frac{\mu_0 I}{4\pi a} (\cos \theta_1 + \cos \theta_2) \, \mathbf{u}_z.$$

In the symmetric case, the magnetic field is obtained by imposing $\theta_2 = \theta_1$, and $L_1 = L_2 = L/2$, so

$$\cos \theta_1 = \frac{L}{\sqrt{L^2 + a^2}}$$

and the field is given by

$$\mathbf{B}(P) = \frac{\mu_0 I}{2\pi a} \frac{L}{\sqrt{L^2 + a^2}} \mathbf{u}_z.$$

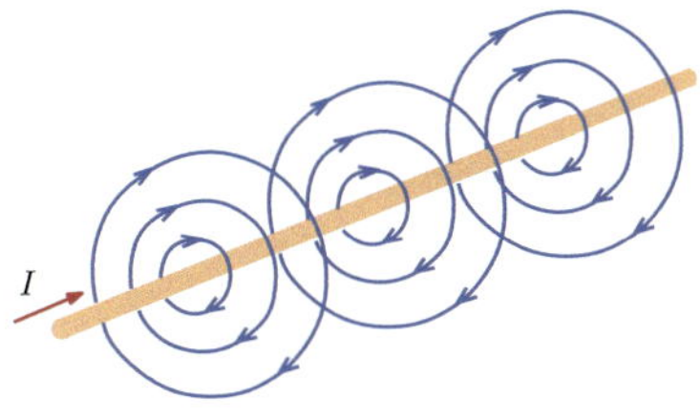

Fig. A.76 Magnetic field lines of an infinitely long wire

In the case of an infinitely long wire ($L \to \infty$), we find

$$\mathbf{B}(P) = \frac{\mu_0 I}{2\pi a}\mathbf{u}_z.$$

Note that in this limit, the system acquires cylindrical symmetry, and the magnetic field lines are concentric circles perpendicular to the wire axis, see Fig. A.76. For $a = 10$ cm, the current needed to generate a field of $B = 2 \times 10^{-5}$ T will be

$$I = \frac{2\pi a B}{\mu_0} = 100\,\text{A}.$$

8.10 Field of a charged disc in rotation

Recall that the magnetic field on the axis of a ring of radius r, carrying a current I is given by

$$\mathbf{B}(z) = \frac{\mu_0 I r^2}{2(z^2 + r^2)^{3/2}}\mathbf{u}_z.$$

We can see the disc as a superposition of rings of current of infinitesimal width dr. In this way, by the superposition principle, the total magnetic field will be the sum of the contributions of these rings. Consider the disc portion between r and $r + dr$ shown in Fig. A.77, with $r \in (0, R)$.

The charge contained in this ring is

$$dq = \sigma 2\pi r dr.$$

The rotation of the disc associates this ring with a current. The charge dq goes one full turn in a period $\Delta t = 2\pi/\omega$, so that the associated current is

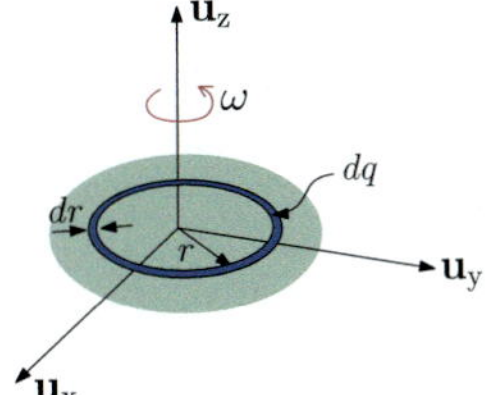

Fig. A.77 The disc can be thought as the superposition of infinitesimal rings of current

$$dI = \frac{dq}{\Delta t} = \frac{\omega\sigma 2\pi r dr}{2\pi} = \omega\sigma r dr.$$

The magnetic field generated by this ring at a distance z on the axis of the disc is

$$d\mathbf{B}(z) = \frac{\mu_0 dI r^2}{2(z^2 + r^2)^{3/2}}\mathbf{u}_z = \frac{dr\mu_0\omega\sigma r^3}{2(z^2 + r^2)^{3/2}}\mathbf{u}_z.$$

Integrating over the entire surface of the disc, we find

$$\mathbf{B}(z) = \int_0^R \frac{dr\mu_0\omega r^3\sigma}{2(z^2 + r^2)^{3/2}}\mathbf{u}_z = \frac{\mu_0\omega\sigma}{2}\int_0^R \frac{r^3 dr}{(z^2 + r^2)^{3/2}}\mathbf{u}_z.$$

Making the change of variable $u^2 = z^2 + r^2$, $2u du = 2r dr$, we obtain

$$\int \frac{r^3 dr}{(z^2 + r^2)^{3/2}} = \int \frac{u(u^2 - z^2)du}{u^3} = \int du\left(1 - \frac{z^2}{u^2}\right) = u + \frac{z^2}{u}.$$

Then,

$$\int \frac{r^3 dr}{(z^2 + r^2)^{3/2}} = \frac{u^2 + z^2}{u} = \frac{2z^2 + r^2}{\sqrt{z^2 + r^2}}.$$

Thus,

$$\mathbf{B}(z) = \frac{\mu_0\omega\sigma}{2}\left[\frac{2z^2 + r^2}{\sqrt{z^2 + r^2}}\right]_0^R \mathbf{u}_z = \frac{\mu_0\omega\sigma}{2}\left(\frac{2z^2 + R^2}{\sqrt{z^2 + R^2}} - 2|z|\right)\mathbf{u}_z.$$

8.11 Field generated by an annular quadrant

To obtain the magnetic field at P, we will use the principle of superposition, noting that the total field at P will be the sum of the contributions from sections 1, 2, 3, 4, and 5 defined in Fig. A.78.

If we set the origin at P, we immediately note that segments 1 and 5 do not contribute to the field since in both cases, the line element $d\mathbf{x}'$ along the curve is parallel to the vector $\mathbf{x} - \mathbf{x}'$, with $\mathbf{x} = \mathbf{0}$. For the segment 2, we first calculate the field $d\mathbf{B}_2$ generated by the current element $I d\mathbf{x}'$ shown in Fig. A.79. For this, we use Biot–Savart law to write

Fig. A.78 The field is obtained by superposition of five different segments

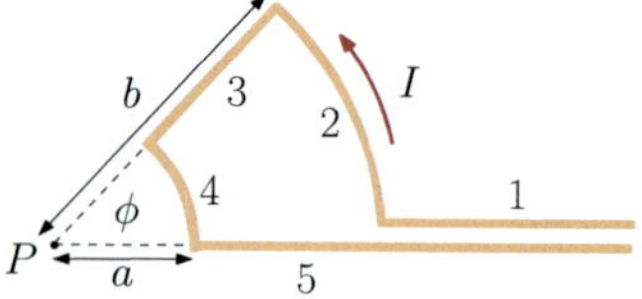

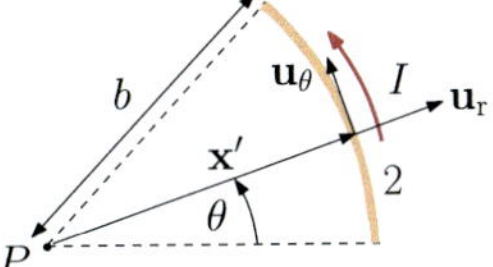

Fig. A.79 Biot–Savart law is applied to calculate the contribution of segment 2 to the field at P

$$d\mathbf{B}_2 = \frac{\mu_0}{4\pi} \frac{I\, d\mathbf{x}' \times (\mathbf{x} - \mathbf{x}')}{|\mathbf{x} - \mathbf{x}'|^3}.$$

Using cylindrical coordinates, $\mathbf{x}' = b\mathbf{u}_r$, and the line element reads $d\mathbf{x}' = b\, d\theta\, \mathbf{u}_\theta$. In addition, $|\mathbf{x} - \mathbf{x}'|^3 = |-\mathbf{x}'|^3 = b^3$, leading to

$$d\mathbf{B}_2 = \frac{\mu_0}{4\pi} I b\, d\theta \frac{\mathbf{u}_\theta \times (-b\mathbf{u}_r)}{b^3} = \frac{\mu_0}{4\pi b} I\, d\theta\, \mathbf{u}_z.$$

Thus, the total field at P due to the outer arc (section 2) is

$$\mathbf{B}_2 = \frac{\mu_0}{4\pi b} I \int_0^\phi d\theta\, \mathbf{u}_z = \frac{\mu_0 I \phi}{4\pi b} \mathbf{u}_z.$$

Now, for the inner arc (section 4) we proceed in the same manner, with the difference that the distance from point P is now a, and that the current goes in the opposite direction to that of the outer arc, then,

$$\mathbf{B}_4 = -\frac{\mu_0 I \phi}{4\pi a} \mathbf{u}_z.$$

Finally, section 3 does not contribute to the field since in this case, the vector $d\mathbf{x}'$ points in the direction opposite to the radial unit vector in cylindrical coordinates, and in turn $\mathbf{x}'$ points in the radial direction. The vector product $d\mathbf{x}' \times (\mathbf{x} - \mathbf{x}')$ is equal to $\mathbf{0}$ all along segment 3. In short, only segments 2 and 4 contribute to the field at P, and the total field will be the sum of both contributions,

$$\mathbf{B} = \mathbf{B}_2 + \mathbf{B}_4 = \frac{\mu_0 I \phi}{4\pi} \left(\frac{1}{b} - \frac{1}{a} \right) \mathbf{u}_z.$$

8.12 Field near a corner

Let P be a point on the plane (x, y) with $x > 0$, $y > 0$. The magnetic field at P will be the superposition of the field generated by the vertical section and that of the horizontal section. Let Γ_1 be the horizontal section. The magnetic field at P is obtained from the Biot–Savart law,

$$\mathbf{B}_1(x, y) = \frac{\mu_0}{4\pi} \int_{\Gamma_1} d\mathbf{x}' I \times \frac{\mathbf{x} - \mathbf{x}'}{|\mathbf{x} - \mathbf{x}'|^3} = \frac{\mu_0 I}{4\pi} \int_0^{\infty} dx' \mathbf{u}_x \times \frac{(x - x')\mathbf{u}_x + y\,\mathbf{u}_y}{\left((x - x')^2 + y^2\right)^{3/2}}$$

$$= \frac{\mu_0 I}{4\pi y^2} \mathbf{u}_z \int_0^{\infty} dx' \frac{1}{\left(1 + (x - x')^2/y^2\right)^{3/2}}.$$

We can now use the substitution $\tan \theta = (x - x')/y$, then $dx' = -y d\theta / \cos^2 \theta$ and

$$(1 + (x - x')^2/y^2)^{-3/2} = \cos^2 \theta.$$

Hence,

$$\int \frac{dx'}{\left(1 + (x - x')^2/y^2\right)^{3/2}} = -y \int d\theta \cos \theta = -y \sin \theta.$$

Now since $\tan \theta = \dfrac{(x - x')}{y}$, we can write $\sin \theta = \dfrac{(x - x')}{\sqrt{y^2 + (x - x')^2}}$, so that

$$\mathbf{B}_1(x, y) = -\frac{\mu_0 I}{4\pi y} \mathbf{u}_z \left(\frac{(x - x')}{\sqrt{y^2 + (x - x')^2}} \right)\Bigg|_0^{\infty} = \frac{\mu_0 I}{4\pi} \left(\frac{1}{y} + \frac{x}{y\sqrt{x^2 + y^2}} \right) \mathbf{u}_z.$$

Now, for the horizontal segment, we simply exchange the roles of x and y, thus,

$$\mathbf{B}_2(x, y) = \frac{\mu_0 I}{4\pi} \left(\frac{1}{x} + \frac{y}{x\sqrt{x^2 + y^2}} \right) \mathbf{u}_z.$$

Finally, using the variable $r = \sqrt{x^2 + y^2}$ and calculating $x/yr + y/xr = r/xy$, we obtain

$$\mathbf{B}(x, y) = \frac{\mu_0 I}{4\pi} \left(\frac{1}{x} + \frac{1}{y} + \frac{r}{xy} \right) \mathbf{u}_z.$$

8.13 Helmholtz coils

We know that the field generated by a circular loop of radius R carrying a current I, at a distance z from its center, is parallel to the axis of symmetry and reads

$$\mathbf{B}(\mathbf{x}) = \frac{\mu_0 I}{2} \frac{R^2}{(R^2 + z^2)^{3/2}} \mathbf{u}_z.$$

Since a coil consists of N circular turns separated by a negligible distance with respect to z, we can use the principle of superposition and obtain the field generated by a coil of N turns

$$\mathbf{B}(\mathbf{x}) = \frac{\mu_0 N I}{2} \frac{R^2}{(R^2 + z^2)^{3/2}} \mathbf{u}_z.$$

Now, consider a point on the $\mathbf{u}_z$ axis, at distance z from the origin, the field generated by the first coil, centered at $z_1 = -l$ is then

$$\mathbf{B}_1(z) = \frac{\mu_0 N I}{2} \frac{R^2}{(R^2 + (z+l)^2)^{3/2}} \mathbf{u}_z$$

and the field generated by the second coil, centered at $z_2 = +l$, reads

$$\mathbf{B}_2(z) = \frac{\mu_0 N I}{2} \frac{R^2}{(R^2 + (z-l)^2)^{3/2}} \mathbf{u}_z.$$

The total magnetic field, by the superposition principle, is $\mathbf{B}(z) = B(z)\mathbf{u}_z$ with

$$B(z) = \frac{\mu_0 N I}{2R} \left(\left[1 + \frac{(z-l)^2}{R^2} \right]^{-\frac{3}{2}} + \left[1 + \frac{(z+l)^2}{R^2} \right]^{-\frac{3}{2}} \right).$$

Note that $B(z)$ is an even function since changing z into $-z$ does not change the field $B(z)$. The first derivative with respect to z at the midpoint $z = 0$ therefore satisfies

$$\left. \frac{dB}{dz} \right|_{z=0} = 0.$$

It can also be shown that if the distance $2l$ between the coils is chosen equal to R (coil radius), we obtain

$$\left. \frac{d^2 B}{dz^2} \right|_{z=0} = \left. \frac{d^3 B}{dz^3} \right|_{z=0} = 0.$$

This configuration known as Helmholtz coils produces a magnetic field approximately uniform around $z = 0$. A Taylor expansion of $B(z)$ around this point reads

$$B(z) = B(0) + \frac{1}{4!} \frac{d^4 B}{dz^4}(0) z^4 + \cdots,$$

that is, the change in the magnitude of the magnetic field at the midpoint for small variations of z is of the order of z^4.

8.13 Force between two rectilinear currents

The magnetic force $\mathbf{F}_1$ exerted by wire 2 on wire 1, defined as in Fig. A.80, can be obtained as

$$\mathbf{F}_1 = I_1 \int_1 d\mathbf{l}_1(\mathbf{x}') \times \mathbf{B}(\mathbf{x}'),$$

where $\mathbf{B}$ is the magnetic field generated by conductor 2.

For an infinitely long rectilinear conductor, the magnetic field lines are concentric circles in the plane perpendicular to the conductor. Thus, as $a \ll l$, we can consider

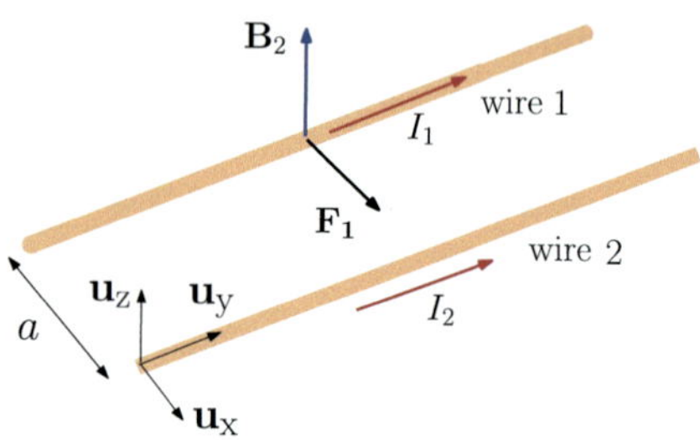

Fig. A.80 Magnetic force $\mathbf{F}_1$ acting on wire 1 due to the field $\mathbf{B}_2$ generated by wire 2

both conductors as infinitely long and at an arbitrary point P (far from the ends) on conductor 1, the magnetic field due to wire conductor 2 is

$$\mathbf{B}_2 = \frac{\mu_0}{2\pi a}\mathbf{u}_z.$$

Since this field is uniform along conductor 1, and this being a rectilinear wire, the force it experiences is given by

$$\mathbf{F}_1 = I_1\mathbf{l} \times \mathbf{B}_2 = I_1 l\mathbf{u}_y \times \left(\frac{\mu_0 I_2}{2\pi a}\right)\mathbf{u}_z = \frac{\mu_0 I_1 I_2 l}{2\pi a}\mathbf{u}_x.$$

This force points toward conductor 2 if $I_1 I_2 > 0$, thus, two parallel wires carrying currents in the same direction will be attracted to each other. On the other hand, if the currents flow in opposite directions, the force will be repulsive.

Problems of Chap. 9

9.1 Force between a rectangular loop and a rectilinear current

To obtain the total force on the loop, we add the forces on sections 1, 2, 3 and 4, shown in Fig. A.81.

We have

$$\mathbf{F}_1 = I_2 \int_0^l d\mathbf{l}_1 \times \mathbf{B}_1,$$

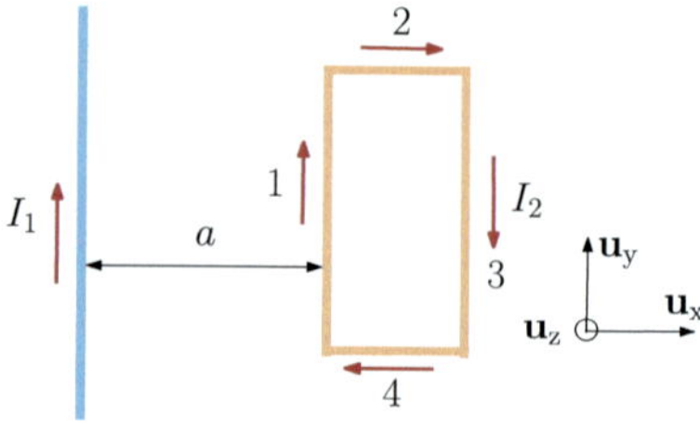

Fig. A.81 The force on the loop can be expressed as the sum of the forces on the 4 segments shown

where l is the length of the rectangular loop. The field in section 1 is uniform and given by

$$\mathbf{B}_1 = -\frac{\mu_0 I_1}{2\pi a}\mathbf{u}_z.$$

Thus, we obtain

$$\mathbf{F}_1 = I_2 \int_0^l dy\mathbf{u}_y \times -\frac{\mu_0 I_1}{2\pi a}\mathbf{u}_z = -\frac{\mu_0 l I_1 I_2}{2\pi a}\mathbf{u}_x.$$

Now, for section 3, the field is obtained by replacing a with $a + b$ and I_2 with $-I_2$ in the above formula. We obtain

$$\mathbf{F}_3 = \frac{\mu_0 l I_1 I_2}{2\pi (a + b)}\mathbf{u}_x.$$

The force on section 2 reads

$$\mathbf{F}_2 = I_2 \int_0^b d\mathbf{l}_2 \times \mathbf{B}_2,$$

where b is the width of the rectangular loop, then,

$$\mathbf{F}_2 = -I_2 \int_0^b dx\mathbf{u}_x \times \frac{\mu_0 I_1}{2\pi (a + x)}\mathbf{u}_z.$$

Note that

$$\mathbf{F}_4 = I_2 \int_b^0 -dx\mathbf{u}_x \times \frac{-\mu_0 I_1}{2\pi (a + x)}\mathbf{u}_z = -\mathbf{F}_2.$$

We find that F_2 and F_4 cancel each other, then, the total force on the loop is

$$\mathbf{F} = \mathbf{F}_1 + \mathbf{F}_3 = \frac{\mu_0 I_1 I_2 l}{2\pi}\left(\frac{1}{(a + b)} - \frac{1}{a}\right)\mathbf{u}_x.$$

This force is attractive.

9.2 Magnetic field of a coaxial cable

Given the cylindrical symmetry of both current distributions, it is appropriate to choose a cylindrical coordinate system with the origin at the center of the internal conductor. The exact same symmetry properties as for the infinitely long wire hold here, so that the magnetic field is of the form $\mathbf{B} = B(r)\mathbf{u}_\theta$. To use Ampère's law, let us define four circular curves of radius r, such that $r < R_1$ (Γ_1), $R_1 < r < R_2$ (Γ_2), $R_2 < r < R_3$ (Γ_3), and $R_3 < r$ (Γ_4). These are shown in Fig. A.82.

For each curve, $\mathbf{B} \cdot d\mathbf{x} = B(r)rd\theta$. Ampère's law writes

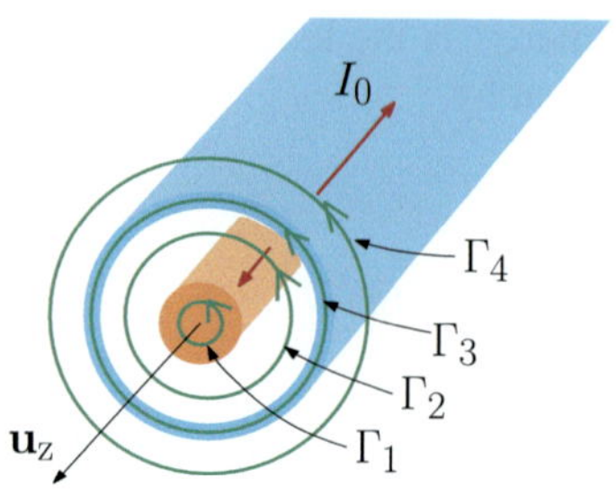

Fig. A.82 Ampère's law applied to four circular loops of different radii

$$\oint_{\Gamma_i} \mathbf{B}(\mathbf{x}) \cdot d\mathbf{x} = 2\pi r B(r) = \mu_0 I_{\Gamma_i}, \quad i = 1, 2, 3, 4.$$

Assuming a uniform current density $\mathbf{J} = J_0 \mathbf{u}_z$ in the internal conductor, one has $I_0 = J_0 \pi R_1^2$ and the current passing through the wire section enclosed by Γ_1 is then given by

$$I_{\Gamma_1} = J_0 \pi r^2 = \frac{I_0 r^2}{R_1^2}.$$

The current enclosed by Γ_2 is simply $I_{\Gamma_2} = I_0$. If the current density in the outer conductor is uniform, it can be written as $\mathbf{J}' = -J_0' \mathbf{u}_z$ with J_0' such that $I_0 = \pi(R_3^2 - R_2^2)J_0'$. The current passing through the section enclosed by Γ_3 is then

$$I_3 = I_0 - J_0' \pi (r^2 - R_2^2) = I_0 \left(1 - \frac{r^2 - R_2^2}{R_3^2 - R_2^2} \right).$$

Finally, the total current enclosed by Γ_4 is zero. One obtains

$$\mathbf{B}(\mathbf{x}) = \begin{cases} \dfrac{\mu_0 I_0 r}{2\pi R_1^2} \mathbf{u}_\theta & \text{if } r \leq R_1, \\[2ex] \dfrac{\mu_0 I_0}{2\pi r} \mathbf{u}_\theta & \text{if } R_1 \leq r \leq R_2, \\[2ex] \dfrac{\mu_0 I_0}{2\pi r} \left(1 - \dfrac{r^2 - R_2^2}{R_3^2 - R_2^2} \right) \mathbf{u}_\theta & \text{if } R_2 \leq r \leq R_3, \\[2ex] \mathbf{0} & \text{if } R_3 \leq r. \end{cases}$$

9.3 Non uniform cylindrical current

The cylindrical symmetry of the current density still holds here since $\mathbf{J}(r, \theta, z) = \mathbf{J}(r)$. The orientation of the magnetic field is constrained to be along $\mathbf{u}_\theta$ due to the mirror symmetries of $\mathbf{J}$ with respect to any plane that contains the axis of the cylinder. Let Γ be a circular path of radius r, with $0 \leq r \leq R$, concentric and perpendicular to the symmetry axis of the conductor. This is illustrated in Fig. A.83.

At every point of Γ, $\mathbf{B}(\mathbf{x}) \cdot d\mathbf{x} = B(r)\mathbf{u}_\theta \cdot r\theta\,\mathbf{u}_\theta$, therefore,

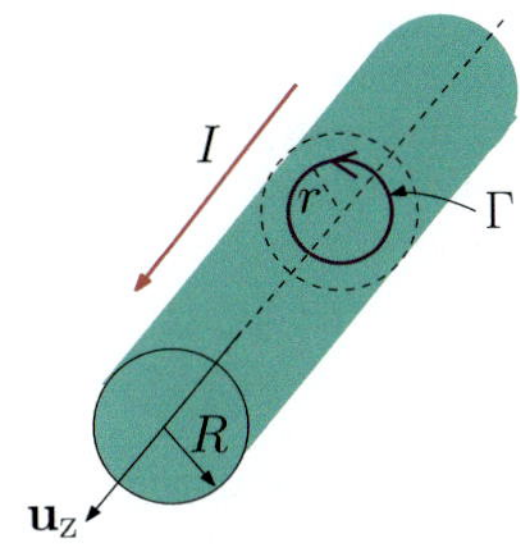

Fig. A.83 Circular loop used to calculate the field inside the conductor

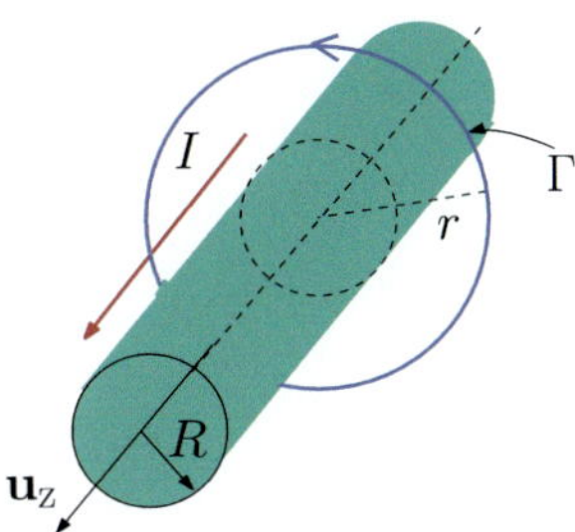

Fig. A.84 Circular loop used to calculate the field outside the conductor

$$\oint_{\Gamma} \mathbf{B}(\mathbf{x}) \cdot d\mathbf{x} = \int_{0}^{2\pi} B(r)r\,d\theta = 2\pi r B(r).$$

The current enclosed by Γ can be calculated as the flux of $\mathbf{J}$ through any surface $S(\Gamma)$ bounded by Γ. The simplest one is the plane cross section (inside the circle Γ) with normal everywhere equal to $\mathbf{u}_z$. In cylindrical coordinates, the surface element reads $dS(\mathbf{x}) = r\,dr\,d\theta$. Then,

$$I_{\Gamma} = \iint_{S(\Gamma)} \mathbf{J}(\mathbf{x}) \cdot \mathbf{n}(\mathbf{x})\,dS(\mathbf{x}) = \int_{0}^{2\pi} d\theta \int_{0}^{r} \mathbf{J}(r) \cdot \mathbf{u}_z r\,dr$$

$$= 2\pi\alpha \int_{0}^{r} r^2\,dr = \frac{2\pi\alpha r^3}{3}.$$

Ampère's law then writes

$$2\pi r B(r) = \mu_0 \frac{2\pi\alpha r^3}{3}.$$

Thus,

$$\mathbf{B}(\mathbf{x}) = \frac{\mu_0 \alpha r^2}{3}\mathbf{u}_\theta.$$

To find the magnetic field outside the conductor, a circle of radius $r \geq R$ is used as Ampère's contour (see Fig. A.84) and Ampère's law gives in this case

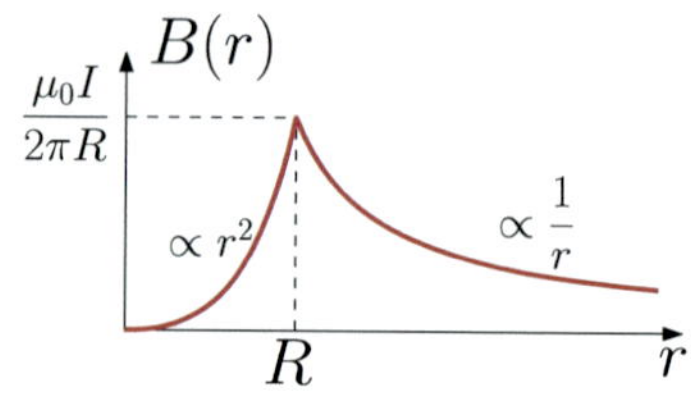

Fig. A.85 Magnetic field's strength as a function of the distance with respect to the center of the conductor

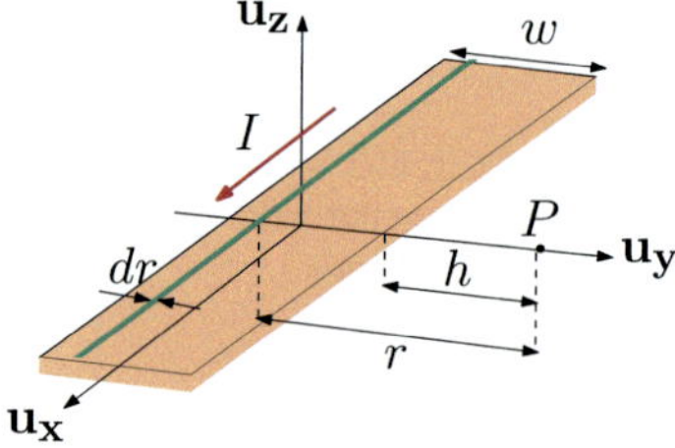

Fig. A.86 The ribbon can be divided into differential line elements

$$\oint_{\Gamma} \mathbf{B}(\mathbf{x}) \cdot d\mathbf{x} = 2\pi r B(r) = \mu_0 I_{\Gamma} = \mu_0 I.$$

This allows us to determine the constant α by applying the latter result at $r = R$:

$$I_{\Gamma}(r = R) = \frac{2\pi \alpha R^3}{3} = I,$$

so that

$$\alpha = \frac{3I}{2\pi R^3}$$

and

$$\mathbf{B}(\mathbf{x}) = \begin{cases} \dfrac{\mu_0 I r^2}{2\pi R^3} \mathbf{u}_\theta & \text{if } r \leq R, \\[2ex] \dfrac{\mu_0 I}{2\pi r} \mathbf{u}_\theta & \text{if } R \leq r. \end{cases}$$

The magnitude of the magnetic field as a function of r is illustrated in Fig. A.85.

9.4 Magnetic field of a plane ribbon

Consider a small section of the conductor parallel to the direction of the current, of width dr and at distance r from P, as shown in Fig. A.86.

The amount of current passing through this section is $dI = I\left(\dfrac{dr}{w}\right)$. The magnetic field at P produced by this long element is the same as that generated by an infinite wire carrying a current dI. It is given by

$$d\mathbf{B}(P) = \frac{\mu_0 dI}{2\pi r} \mathbf{u}_z = \frac{\mu_0 I dr}{2\pi r w} \mathbf{u}_z.$$

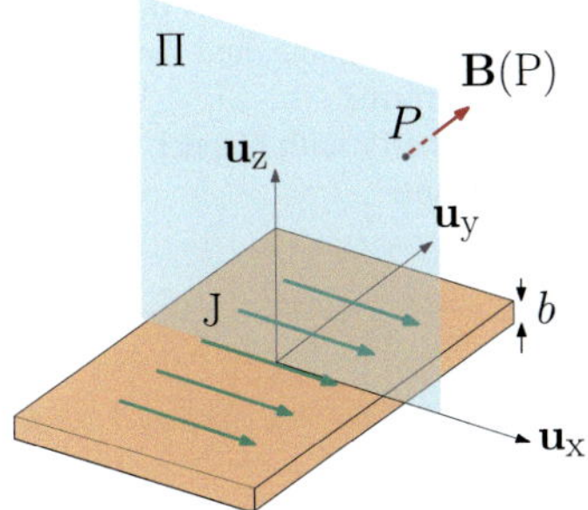

Fig. A.87 The plane Π is a symmetry plane, the field at P is therefore perpendicular to Π

By the superposition principle, the total magnetic field is obtained by integrating over the whole width of the ribbon:

$$\mathbf{B}(P) = \int_h^{h+w} dr\, \frac{\mu_0 I}{2\pi r w}\mathbf{u}_z = \frac{\mu_0 I}{2\pi w}\int_h^{h+w} \frac{dr}{r}\mathbf{u}_z = \frac{\mu_0 I}{2\pi w}\ln\left(\frac{h+w}{h}\right)\mathbf{u}_z.$$

Note that in the limit $w \ll h$,

$$\ln\frac{h+w}{h} = \ln\left(1 + \frac{w}{h}\right) \approx \frac{1}{h}w$$

and one recovers the magnetic field at distance h from an infinitely long and thin wire:

$$\mathbf{B}(P) \approx \frac{\mu_0 I}{2\pi h}\mathbf{u}_z.$$

9.5 Magnetic field generated by a current confined between two planes
Invariance of the current density to translations along the x- or y-axis implies an independence of the magnetic field on x and y, so that in Cartesian coordinates $\mathbf{B}(x, y, z) = \mathbf{B}(z)$. In addition, for any point P, the plane Π containing P and parallel to the Oxz plane is a symmetry plane for the current density, so that the magnetic field at P must be perpendicular to this plane. This is illustrated in Fig. A.87. One concludes that $\mathbf{B}(x, y, z) = B(z)\mathbf{u}_y$.

Moreover, if we choose the origin such that $z = 0$ is in the middle between the two planes, then the plane Oxy is a symmetry plane for the current density. One concludes that $\mathbf{B}$ has mirror anti-symmetry with respect to the Oxy plane, so that the magnetic field changes sign upon reflection on this plane (see Fig. A.88).

To determine the magnitude of the magnetic field, one can use Ampère's law. A convenient closed and oriented curve Γ for which the circulation of $\mathbf{B}$ is simple to calculate is the plane curve shown in Fig. A.89.

For the segments parallel to the z-axis, one has $\mathbf{B}(\mathbf{x}) \cdot d\mathbf{x} = 0$ so that only the segments of width l contribute to the line integral. Recalling the mirror anti-symmetry of $\mathbf{B}$ with respect to the Oxy plane $(B(z) = -B(-z))$, we find

Fig. A.88 The plane $z = 0$ is a symmetry plane, and the field has mirror anti-symmetry with respect to this plane

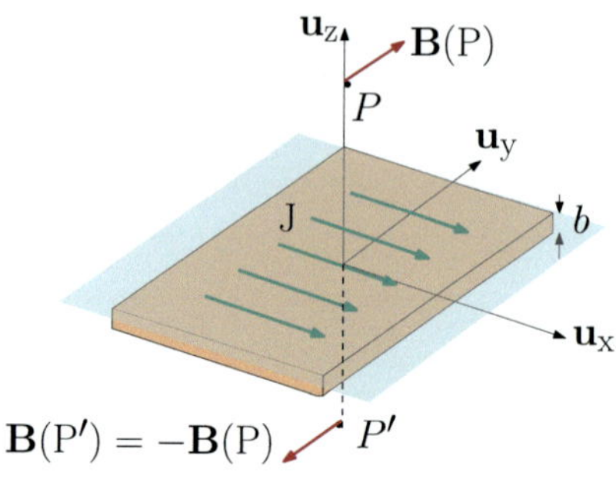

Fig. A.89 The circulation of **B** over the closed loop is determined only by the two horizontal segments at z and $-z$

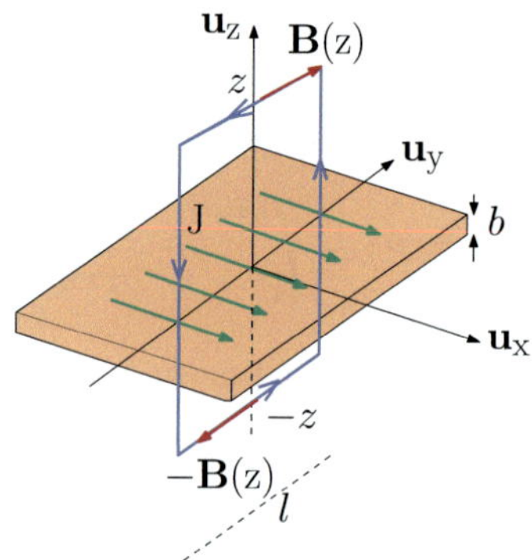

$$\oint_{\Gamma} \mathbf{B}(\mathbf{x}) \cdot d\mathbf{x} = -B(z)l + B(-z)l = -2B(z)l,$$

where we have assumed $z > 0$, and according to Ampère's law, the circulation of **B** is proportional to the current enclosed by Γ, which reads

$$I_{\Gamma} = \iint_{S} \mathbf{J}(\mathbf{x}) \cdot \mathbf{n}(\mathbf{x}) dS(\mathbf{x}),$$

where S is any surface bounded by Γ. Choose S as the plane surface contained in the same plane as the loop Γ, so that the normal is, at every point, $\mathbf{n}(\mathbf{x}) = \mathbf{u}_x$, then,

$$I_{\Gamma} = \int_{-z}^{z} \int_{-l/2}^{l/2} J(z) dy \, dz.$$

Here, two cases must be distinguished, since $J(z) = J_0$ is non-zero only for $|z| \leq b/2$.

$$I_{\Gamma} = \begin{cases} \displaystyle\int_{-z}^{z} \int_{-l/2}^{l/2} J_0 dy \, dz = 2zl J_0 & \text{if } 0 \leq z \leq b/2, \\ \displaystyle\int_{-b/2}^{b/2} \int_{-l/2}^{l/2} J_0 dy \, dz = bl J_0 & \text{if } z \geq b/2. \end{cases}$$

Ampère's law then gives

$$
\mathbf{B}(\mathbf{x}) =
\begin{cases}
-\mu_0 z J_0\, \mathbf{u}_y & \text{if } z \le \dfrac{b}{2}, \\[2mm]
-\dfrac{\mu_0 b J_0}{2}\, \mathbf{u}_y & \text{if } \dfrac{b}{2} \le z.
\end{cases}
$$

For $z < 0$, the magnetic field is found by mirror anti-symmetry of the solution found for $z > 0$.

9.6 Superposition of magnetic fields

The magnetic field of the plane current distribution was obtained in Problem 9.5. In the region above the plane current, the magnetic field reads

$$
\mathbf{B}_1(\mathbf{x}) = -\frac{\mu_0 J_1 b}{2}\mathbf{u}_y.
$$

For the cylinder, one can use Ampère's law. The cylindrical symmetry of its current density and its mirror symmetries constrain the magnetic field to vary as $\mathbf{B}_2(r, \theta, z) = B_2(r)\mathbf{u}_\theta$ in a cylindrical coordinate system with the z-axis being the cylinder axis of symmetry.

The most convenient curve Γ for calculating the circulation of $\mathbf{B}_2$ is then a circle of radius r (here, $r \le R$) for which

$$
\oint_\Gamma \mathbf{B}_2(\mathbf{x}) \cdot d\mathbf{x} = 2\pi r B_2(r).
$$

This is shown in Fig. A.90. On the other hand, the current enclosed by Γ is

$$
\begin{aligned}
I_\Gamma &= \int_0^{2\pi} d\theta \int_0^r J_0 \left(1 - \frac{r}{R}r\right) dr \\
&= J_0 2\pi \left(\frac{r^2}{2} - \frac{r^3}{3R}\right),
\end{aligned}
$$

so that for $r \le R$, Ampère's law gives

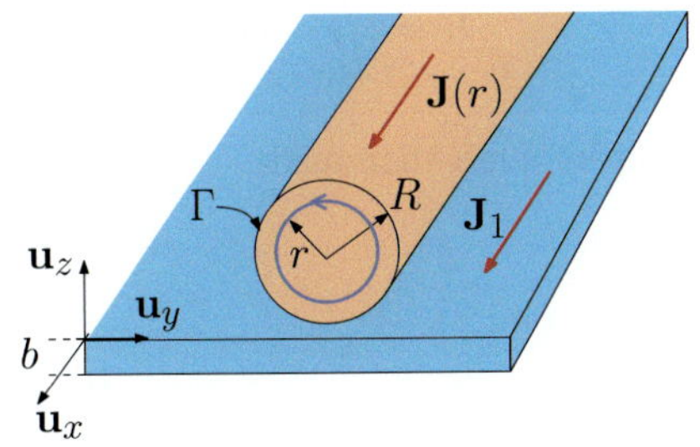

Fig. A.90 Ampère's law is used to calculate the field generated by the cylinder

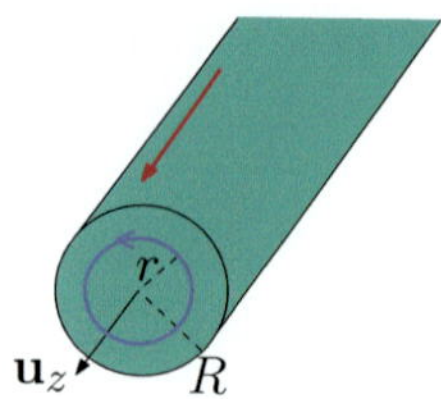

Fig. A.91 Ampère's law used to calculate the uniform current density of the cylinder

$$\mathbf{B}_2(\mathbf{x}) = \mu_0 J_0 \left(\frac{r}{2} - \frac{r^2}{3R} \right) \mathbf{u}_\theta.$$

At point P, the total magnetic field is given by the superposition of $\mathbf{B}_1$ and $\mathbf{B}_2$. At P, $\mathbf{u}_\theta = \mathbf{u}_y$ so that

$$\mathbf{B}(P) = \mu_0 \left(J_0 \left(\frac{R}{4} - \frac{(R/2)^2}{3R} \right) - \frac{bJ_1}{2} \right) \mathbf{u}_y = \frac{\mu_0}{2} \left(\frac{J_0 R}{3} - bJ_1 \right) \mathbf{u}_y.$$

This field is zero if

$$J_0 = \frac{3bJ_1}{R}.$$

9.7 Wire with a cavity

The current distribution is equivalent to the superposition of a cylindrical wire carrying a uniform current density $J_0\mathbf{u}_z$, and a second wire with current density $-J_0\mathbf{u}_z$. The magnetic field inside a cylindrical wire with uniform current density $\mathbf{J} = J_0\mathbf{u}_z$ is easily obtained by means of Ampère's law considering the cylindrical symmetry of the current density and its mirror symmetries. In cylindrical coordinates with the z-axis being the cylinder axis, one has $\mathbf{B}(r, \theta, z) = B(r)\mathbf{u}_\theta$.

Choosing Γ as a circle of radius r, as shown in Fig. A.91, the circulation of $\mathbf{B}$ along Γ reads

$$\oint_\Gamma \mathbf{B}(\mathbf{x}) \cdot d\mathbf{x} = B(r)2\pi r$$

and the current enclosed by Γ can be obtained by calculating the flux of $\mathbf{J}$ through the disk bounded by Γ with normal $\mathbf{n} = \mathbf{u}_z$, which yields

$$I_\Gamma = \iint_{S(\Gamma)} J_0\mathbf{u}_z \cdot \mathbf{u}_z dS(\mathbf{x}) = J_0 \int_0^{2\pi} \int_0^r r\, dr\, d\theta = J_0\pi r^2.$$

Ampère's law then reads

$$B(r)2\pi r = \mu_0 J_0\pi r^2$$

and we find

$$\mathbf{B}(r) = \frac{\mu_0 J_0 r}{2} \mathbf{u}_\theta.$$

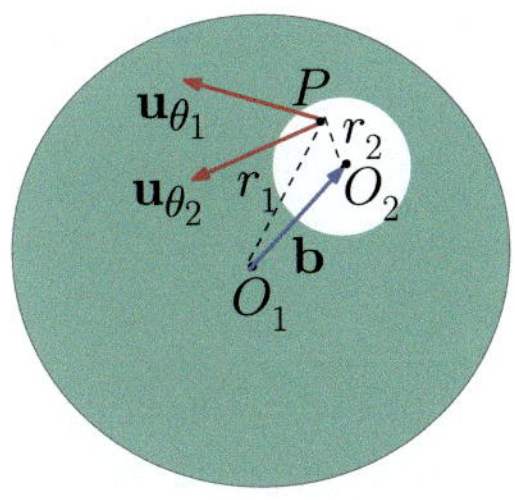

Fig. A.92 The point P is any point inside the cavity

Let now P be an arbitrary point inside the cavity, as shown in Fig. A.92. The superposition of the two magnetic fields yields

$$\mathbf{B} = \frac{\mu_0 J_0 r_1}{2}\mathbf{u}_{\theta_1} - \frac{\mu_0 J_0 r_2}{2}\mathbf{u}_{\theta_2}.$$

Moreover, $\mathbf{u}_{\theta_1} = \mathbf{u}_z \times \mathbf{u}_{r_1}$ and $\mathbf{u}_{\theta_2} = \mathbf{u}_z \times \mathbf{u}_{r_2}$, so that

$$\mathbf{B} = \frac{\mu_0 J_0}{2}\left(r_1\mathbf{u}_z \times \mathbf{u}_{r_1} - r_2\mathbf{u}_z \times \mathbf{u}_{r_2}\right) = \frac{\mu_0 J_0}{2}\mathbf{u}_z \times (\mathbf{r}_1 - \mathbf{r}_2).$$

Whatever the position of point P within the cavity, the vector $\mathbf{r}_1 - \mathbf{r}_2$ is constant and equal to $\mathbf{b} = \mathbf{O_1P} - \mathbf{O_2P} \equiv \mathbf{O_1O_2}$. Thus we obtain a uniform magnetic field inside the cavity,

$$\mathbf{B} = \frac{\mu_0 J_0}{2}\mathbf{u}_z \times \mathbf{b}.$$

9.8 Magnetic force between conductors

(a) The plane current can be seen as a superposition of infinitely long wires of width dx', as shown in Fig. A.93. The infinitesimal segment between x' and $x' + dx'$ carries a current $dI' = I'\dfrac{dx'}{a}$. The magnetic field at a point on the x-axis, a distance $x > a$ from the origin, was obtained previously for an infinitely long and thin wire:

$$d\mathbf{B}_1(x) = -\frac{\mu_0 I'dx'}{2\pi a(x - x')}\mathbf{u}_z.$$

Fig. A.93 The plane current is seen as the superposition of infinitesimal line of currents

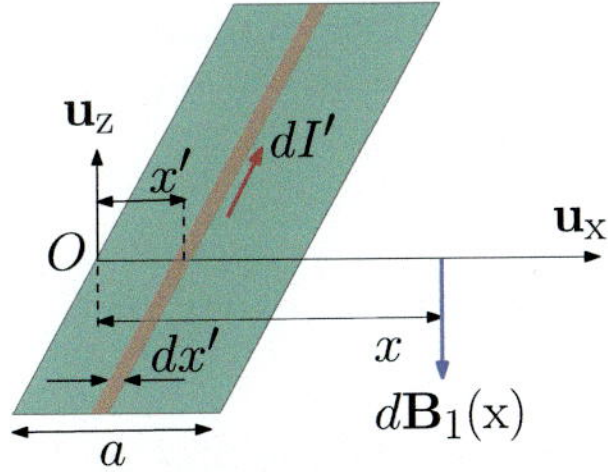

The total magnetic field due to the plane current distribution is, for $x > a$

$$\mathbf{B}_1(x) = -\frac{\mu_0 I'}{2\pi a} \int_0^a \frac{dx'}{x - x'} \mathbf{u}_z = -\frac{\mu_0 I'}{2\pi a} \ln\left(\frac{x}{x - a}\right) \mathbf{u}_z.$$

For the cylinder of radius R, the two regions inside and outside the cylinder must be distinguished. At any point x outside the cylinder, with $a < x < b - R$,

$$\mathbf{B}_2(x) = \frac{\mu_0 I}{2\pi(b - x)} \mathbf{u}_z,$$

whereas for $x > b + R$,

$$\mathbf{B}_2(x) = -\frac{\mu_0 I}{2\pi(x - b)} \mathbf{u}_z.$$

The field inside the cylinder was obtained previously in Exercise 9.7 for a wire carrying a uniform current density J_0. Considering the system of cylindrical coordinates with the cylinder axis as the z-axis and using the current $I = J_0 \pi R^2$, the magnetic field generated by the cylinder reads

$$\mathbf{B}(r) = \mu_0 \frac{Ir}{2\pi R^2} \mathbf{u}_\theta.$$

Finally,

$$\mathbf{B}(x) = \frac{\mu_0 I'}{2\pi a} \left(\ln \frac{x - a}{x} + \frac{I}{I'} f(x) \right) \mathbf{u}_z,$$

where $f(x)$ denotes the dimensionless number

$$f(x) = \begin{cases} \dfrac{a}{(b - x)} & \text{if } a \le x \le b - R, \\[2ex] \dfrac{a(b - x)}{R^2} & \text{if } b - R \le x \le b + R, \\[2ex] -\dfrac{a}{(x - b)} & \text{if } b + R \le x. \end{cases}$$

(b) Let us calculate the force per unit length that the cylinder exerts on the plane current. For this, let us first consider the force per unit length acting on the infinitesimal current element of width dx' at position x' that is shown in Fig. A.94. This force is given by

$$d\mathbf{F}(x') = \frac{I' dx'}{a} \mathbf{u}_y \times B_2(x') \mathbf{u}_z = \frac{I' dx'}{a} \frac{\mu_0 I}{2\pi(b - x)} \mathbf{u}_y \times \mathbf{u}_z = \frac{\mu_0 I I' dx'}{2\pi a(b - x)} \mathbf{u}_x.$$

By superposition, the total force per unit length that acts on the plane current is

Fig. A.94 The force on the plane current is the sum of the infinitesimal forces $d\mathbf{F}$ acting on the infinitesimal line elements

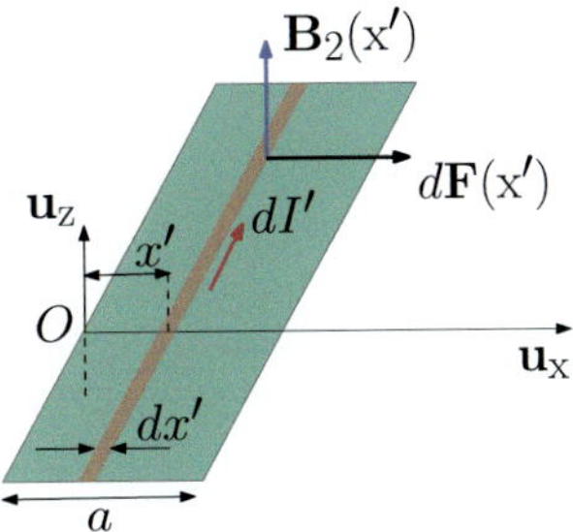

$$\mathbf{F} = \int_0^a \frac{\mu_0 I I'}{2\pi a(b-x)} dx' \mathbf{u}_x = \frac{\mu_0 I I'}{2\pi a} \ln\left(\frac{b}{b-a}\right) \mathbf{u}_x.$$

Problems of Chap. 10

10.1 Magnetic moment of a rotating charged sphere

(a) Since both the charge $-e$ and the mass m_e are uniformly distributed inside the sphere, we have simply

$$\varrho = \frac{-e}{\frac{4\pi}{3}a^3}, \qquad \rho_m = \frac{m_e}{\frac{4\pi}{3}a^3}.$$

(b) Consider at a certain instant a point $\mathbf{x} = r\mathbf{u}_r$ inside the sphere. The infinitesimal volume $d^3x = r^2 dr \sin\theta d\theta d\phi$ around $\mathbf{x} = r\mathbf{u}_r$ describes a circular trajectory of radius $r \sin\theta$ with angular velocity ω. This is shown in Fig. A.95.

This elementary volume contains a mass $dm = \rho_m d^3x$ and moves at velocity $\mathbf{v} = \omega r \sin\theta\, \mathbf{u}_\theta$. Its angular momentum is therefore given by

$$d\mathbf{L}(\mathbf{x}) = \underbrace{\rho_m d^3x}_{dm}\, r\, \mathbf{u}_r \times \underbrace{(\omega r \sin\theta\, \mathbf{u}_\phi)}_{\mathbf{v}} = \rho_m \omega r^4 \sin^2\theta\, dr d\theta d\phi\, (\mathbf{u}_r \times \mathbf{u}_\phi)$$

Fig. A.95 An infinitesimal volume element in the sphere describes a circular trajectory of radius $r \sin\theta$

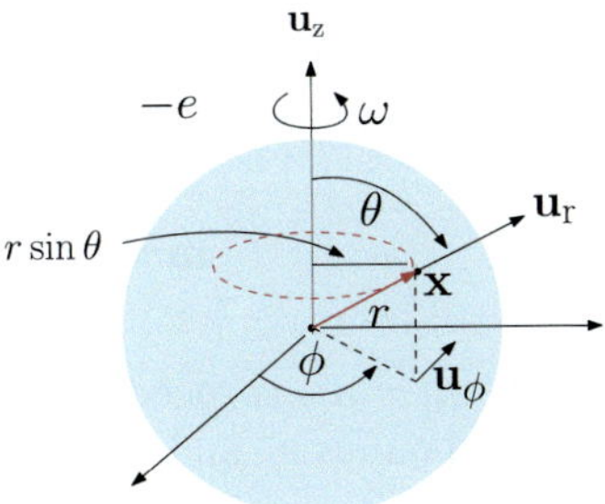

and writing $\mathbf{u}_r \times \mathbf{u}_\phi = -\mathbf{u}_\theta = \sin\theta\, \mathbf{u}_z - \cos\theta(\cos\phi\, \mathbf{u}_x + \sin\phi\, \mathbf{u}_y)$, this can be rewritten as

$$d\mathbf{L}(\mathbf{x}) = \rho_m \omega r^4 \sin^3\theta\, dr d\theta d\phi \mathbf{u}_z - \rho_m \omega r^4 \times \sin^2\theta \cos\theta\, dr d\theta d\phi\, (\cos\phi\, \mathbf{u}_x + \sin\phi\, \mathbf{u}_y).$$

The total angular momentum is obtained by integrating over all the points in the volume Ω of the sphere, after which only the z-component is different from zero

$$\mathbf{L} = \iiint_\Omega d\mathbf{L} = \int_0^{2\pi} d\phi \int_0^\pi \int_0^a \rho_m \omega r^4 \sin^3\theta\, dr d\theta \mathbf{u}_z$$

$$= 2\pi\rho_m\omega \underbrace{\int_0^\pi \sin^3\theta\, d\theta}_{=\frac{4}{3}} \int_0^a r^4 dr \mathbf{u}_z$$

$$= \frac{8\pi\rho_m a^5 \omega}{15}\mathbf{u}_z = \frac{2m_e a^2 \omega}{5}\mathbf{u}_z = \frac{2m_e a^2}{5}\mathbf{\Omega},$$

where $\mathbf{\Omega} = \omega\mathbf{u}_z$. For the magnetic moment, we have

$$\mathbf{m} = \frac{1}{2}\iiint_\Omega \mathbf{x} \times \mathbf{J}(\mathbf{x}) d^3 x,$$

where the current density at point $\mathbf{x}$ is simply $\mathbf{J}(\mathbf{x}) = \varrho(\mathbf{x})\mathbf{v}(\mathbf{x})$ and so

$$\mathbf{m} = \frac{1}{2}\underbrace{\frac{\varrho}{\rho_m}}_{-e/m_e} \underbrace{\iiint_\Omega \rho_m \mathbf{x} \times \mathbf{v}(\mathbf{x}) d^3 x}_{\mathbf{L}} = \frac{-e}{2m_e}\mathbf{L}.$$

We find the following relationship between the magnetic moment and the angular momentum of the sphere

$$\mathbf{m} = g\frac{-e}{2m_e}\mathbf{L} = \frac{-e}{2m_e}\mathbf{L}$$

so that for a classical rotating sphere $g = 1$, which differs from the experimental value $g \approx 2$ for the electron spin. The spin is a purely quantum mechanical angular momentum that does not come from the electron self rotation.

(c) The magnetic moment of a rotating sphere can be written

$$\mathbf{m} = -\frac{e}{2m_e}\mathbf{L} = -\frac{ea^2}{5}\omega\mathbf{u}_z = -\frac{ea}{5}v_{\text{eq}}\mathbf{u}_z,$$

where $v_{\text{eq}} = a\omega$ is speed at the equator of the sphere. Imposing a to be the electron classical radius r_e, and $|\mu| = \mu_B$ yields

$$\frac{er_e}{5}v_{\text{eq}} = \mu_B$$

and so

$$v_{\text{eq}} = \frac{5\mu_B}{er_e} \sim 1.6 \times 10^{11}\ \text{m s}^{-1}$$

that is, about 500 times the speed of light! Not only the classical view of the electron spin as a self-rotation fails to explain the observed g-factor, but it would also require a rotational speed incompatible with special relativity.

10.2 Magnetic spin resonance

(a) The spin $\mathbf{m}$ will feel a torque given by

$$\tau = \mathbf{m} \times (\mathbf{B}_0 + \mathbf{B}_1(t)).$$

Since the spin is initially along $\mathbf{B}_0$, we have, according to (10.5)

$$\frac{d}{dt}\mathbf{m} = \gamma\,[\mathbf{m} \times (\mathbf{B}_0 + \mathbf{B}_1(t))] - \frac{\mathbf{m} - \mu_B \mathbf{u}_z}{T_1}.$$

(b) In the rotating reference frame, we have

$$\mathbf{m} = m'_x \mathbf{u}'_x(t) + m'_y \mathbf{u}'_y(t)$$

and so

$$\frac{d\mathbf{m}}{dt} = \frac{dm'_x}{dt}\mathbf{u}'_x + m'_x\frac{d}{dt}\mathbf{u}'_x + \frac{dm'_y}{dt}\mathbf{u}'_y + m'_y\frac{d}{dt}\mathbf{u}'_y.$$

The basis vectors of the rotating frame satisfy

$$\frac{d}{dt}\mathbf{u}'_x = \omega\mathbf{u}'_y(t)\quad\text{and}\quad \frac{d}{dt}\mathbf{u}'_y = -\omega\mathbf{u}'_x(t).$$

Finally

$$\frac{d\mathbf{m}}{dt} = \underbrace{\frac{dm'_x}{dt}\mathbf{u}'_x + \frac{dm'_y}{dt}\mathbf{u}'_y}_{\frac{d\mathbf{m}'}{dt}} + \underbrace{\omega m'_x \mathbf{u}'_y - \omega m'_y \mathbf{u}'_x}_{\omega\mathbf{u}_z \times \mathbf{m}'}.$$

(c) The equation of motion writes

$$\frac{d}{dt}\mathbf{m} = \frac{d\mathbf{m}'}{dt} + \omega\mathbf{u}_z \times \mathbf{m}' = \underbrace{\gamma\left[\mathbf{m}' \times (B_0\mathbf{u}_z + B_1\mathbf{u}'_x)\right]}_{-\omega_0\mathbf{m}'\times\mathbf{u}_z - \omega_1\mathbf{m}'\times\mathbf{u}'_x} - \frac{\mathbf{m}' - \mu_B\mathbf{u}_z}{T_1}$$

and this can be rewritten as an equation of motion equivalent to that in a static frame

Fig. A.96 In the rotating frame, the spin precesses around an effective, static magnetic field

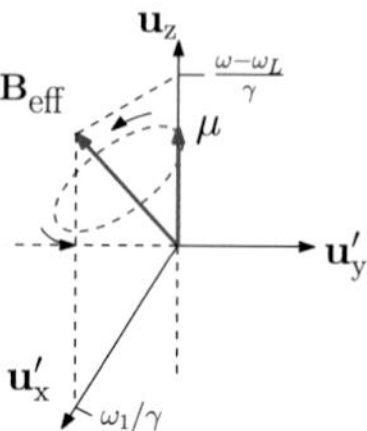

$$\frac{d\mathbf{m}'}{dt} = \mathbf{m}' \times \underbrace{\left((\omega - \omega_0)\mathbf{u}_z - \omega_1\mathbf{u}'_x\right)}_{\gamma\mathbf{B}_{\text{eff}}} - \frac{\mathbf{m}' - \mu_B\mathbf{u}_z}{T_1}.$$

We see that in the rotating frame, and neglecting the relaxation term, the spin precesses around a static effective magnetic field

$$\mathbf{B}_{\text{eff}} = \frac{\omega - \omega_0}{\gamma}\mathbf{u}_z + B_1\mathbf{u}'_x = \frac{\omega - \omega_0}{\gamma}\mathbf{u}_z + \frac{\omega_1}{\gamma}\mathbf{u}'_x.$$

This is illustrated in Fig. A.96.

(d) In the case $\omega_1 \ll |\omega_0 - \omega|$, the effective magnetic field is along the z-axis

$$\mathbf{B}_{\text{eff}} \approx \frac{\omega - \omega_0}{\gamma}\mathbf{u}_z$$

and no torque acts on the spin, the latter stays in its equilibrium position $\mathbf{m} = \mathbf{m}(0) = \mu_B\mathbf{u}_z$. On the other hand, for the case $\omega_1 \gg |\omega_0 - \omega|$ which is verified when $\omega \approx \omega_0$, the effective magnetic field is mostly along $\mathbf{u}'_x$

$$\mathbf{B}_{\text{eff}} \approx B_1\mathbf{u}'_x = \frac{\omega_1}{\gamma}\mathbf{u}'_x$$

and so the static magnetic field $\mathbf{B}_0$ appears to be considerably reduced or even zero at resonance ($\omega = \omega_0$). In the rotating frame the spin precesses around $\mathbf{u}'_x$ at a frequency $\omega_1 = \gamma B_1$, describing a circle in the $y'z$ plane. As such, in resonance the spin can be completely reversed with respect to its initial thermodynamical equilibrium position in the magnetic field $\mathbf{B}_0$. In the laboratory frame, the spin motion is a complex superposition of this precession and a rotation around the z-axis at angular velocity ω.

(e) At resonance, $\mathbf{B}_{\text{eff}} \approx B_1\mathbf{u}'_x$ and the spin precesses in the rotating frame around the $\mathbf{u}'_x$ axis at frequency $\omega_1 = \gamma B_1$. At $t = T_\omega$, the spin has rotated by an angle $\theta = \gamma B_1 T_\omega$ around the $\mathbf{u}'_x$ axis. For a pulse of duration $T_\omega = \dfrac{\pi}{2\gamma B_1}$, $\theta = \dfrac{\pi}{2}$ so that the spin is aligned along the y'-axis just after the pulse.

Fig. A.97 The trajectory describes the return to equilibrium of the spin

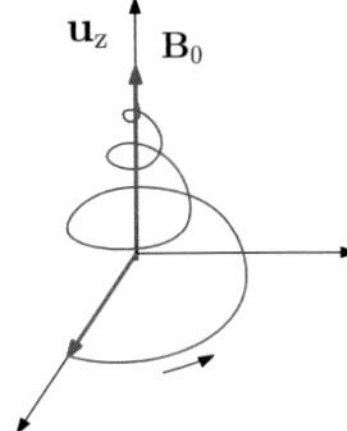

(f) After the pulse, the spin precesses around the z-axis and simultaneously aligns itself with the static field $\mathbf{B}_0$ after a typical time T_1. It describes an helical trajectory as shown in Fig. A.97.

This free spin precession will induce in a neighboring coil an oscillating voltage at frequency ω_0 (thanks to the phenomenon of electromagnetic induction, see Chap. 11) allowing a precise measurement of the Larmor frequency ω_0 of the atoms present in the studied sample, a method called nuclear magnetic resonance (NMR). The atomic electron cloud tends to screen the effective magnetic field seen by the nuclei and as a result, the nuclear spin resonance depends on the electron density distribution in the corresponding molecular orbitals. This *chemical shift* is the reason why NMR is able to probe, at the atomic scale, the chemical structure of molecules.

10.3 Magnetic field of a magnetized sphere

Since there are no free currents, Ampère's law for the magnetic excitation $\mathbf{H}$ writes

$$\nabla \times \mathbf{H} = \nabla \times \left(\frac{\mathbf{B}}{\mu_0} - \mathbf{M} \right) = \mathbf{0}$$

and since $\mathbf{M}$ is uniform inside and outside the sphere, we obtain

$$\nabla \times \mathbf{B} = \mathbf{0}.$$

We conclude that the magnetic field is irrotational (curl-free) and can be written as the gradient of a scalar potential

$$\mathbf{B}(\mathbf{x}) = -\nabla \phi(\mathbf{x}).$$

Since $\nabla \cdot \mathbf{B} = 0$, ϕ satisfies Laplace's equation:

$$\nabla^2 \phi(\mathbf{x}) = 0.$$

In spherical coordinates, the potential must be invariant to any rotation along the z-axis (see Fig. A.98). According to Sect. A.6, the solution can be written as an expansion in the basis of Legendre polynomials. For $r > a$, and imposing $\lim_{r \to \infty} \phi(r, \theta) = 0$, the expansion writes

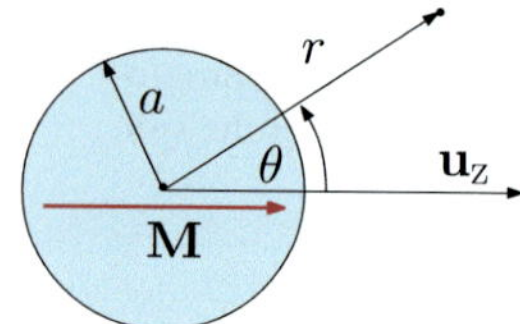

Fig. A.98 Spherical coordinates used to solve Laplace's equation

$$\phi_{\text{ext}}(r,\theta) = \sum_{l=0}^{\infty} \frac{A_l}{r^{l+1}} P_l(\cos\theta).$$

Assuming that the sphere is homogeneous and isotropic, the magnetic field inside the sphere is constant:

$$\mathbf{B}(\mathbf{x}) = B_0 \mathbf{u}_z,$$

so that the magnetic excitation field $\mathbf{H}(\mathbf{x})$ is

$$\mathbf{H}(\mathbf{x}) = \left(\frac{B_0}{\mu_0} - M_0\right)\mathbf{u}_z.$$

The scalar potential for $r < a$ is then

$$\phi_{\text{in}}(\mathbf{x}) = -B_0 z = -B_0 r \cos\theta.$$

The normal component of the magnetic field is continuous at $r = a$, so that (10.15) writes

$$-\nabla\phi_{\text{in}}(\mathbf{x}) \cdot \mathbf{u}_r|_{r=a} = -\nabla\phi_{\text{ext}}(\mathbf{x}) \cdot \mathbf{u}_r|_{r=a},$$

which yields

$$B_0 \cos\theta = \sum_{l=0}^{\infty} \frac{(l+1)A_l}{a^{l+2}} P_l(\cos\theta).$$

In addition, since there are no free currents at the surface of the sphere, (10.16) states that the tangential component of the field $\mathbf{H}(\mathbf{x})$ is continuous at $r = a$:

$$\left(\frac{1}{\mu_0}\mathbf{B}_{\text{in}}(\mathbf{x}) - \mathbf{M}(\mathbf{x})\right) \cdot \mathbf{u}_\theta|_{r=a} = \frac{1}{\mu_0}\mathbf{B}_{\text{ext}}(\mathbf{x}) \cdot \mathbf{u}_\theta|_{r=a},$$

which yields

$$\left(\frac{1}{\mu_0}B_0 - M_0\right)\mathbf{u}_z \cdot \mathbf{u}_\theta|_{r=a} = -\frac{1}{\mu_0}\nabla\phi_{\text{ext}}(\mathbf{x}) \cdot \mathbf{u}_\theta|_{r=a},$$

thus

$$-\left(\frac{1}{\mu_0}B_0 - M_0\right)\sin\theta = -\frac{1}{\mu_0}\sum_{l=0}^{\infty}\frac{A_l}{a^{l+2}}\frac{dP_l(\cos\theta)}{d\theta}.$$

In conclusion, the boundary conditions yield

$$B_0\cos\theta = \sum_{l=0}^{\infty}\frac{(l+1)A_l}{a^{l+2}}P_l(\cos\theta),$$

$$(B_0 - \mu_0 M_0)\sin\theta = \sum_{l=0}^{\infty}\frac{A_l}{a^{l+2}}\frac{dP_l(\cos\theta)}{d\theta}.$$

From the first condition we obtain $A_l = 0,\ \forall l \neq 1$, and

$$B_0 = \frac{2A_1}{a^3}.$$

The second condition gives

$$(B_0 - \mu_0 M_0)\sin\theta = -\frac{A_1}{a^3}\sin\theta.$$

Then

$$A_1 = \frac{a^3}{2}B_0 = a^3(\mu_0 M_0 - B_0),$$

which admits the solution

$$B_0 = \frac{2}{3}\mu_0 M_0,$$

$$A_1 = \frac{a^3\mu_0}{3}M_0.$$

Hence,

$$H_0 = \frac{1}{\mu_0}B_0 - M_0 = -\frac{M_0}{3}.$$

Finally for $r < a$

$$\mathbf{B}(\mathbf{x}) = \frac{2}{3}\mu_0 M_0\mathbf{u}_z = \frac{2\mu_0}{3}\mathbf{M},$$

$$\mathbf{H}(\mathbf{x}) = -\frac{M_0}{3}\mathbf{u}_z = -\frac{1}{3}\mathbf{M}.$$

Note that the magnetic excitation tends to demagnetize the sphere, whose effective permeability is $\mu = (1 + \chi_m)\mu_0 = (1 - 3)\mu_0 = -2\mu_0$. For $r > a$, the scalar potential is given by

$$\phi_{\text{ext}}(r, \theta) = \frac{a^3 \mu_0 M_0}{3 r^2} \cos \theta.$$

The fields for $r > a$ are given by

$$\mathbf{B}(\mathbf{x}) = -\nabla \phi_{\text{ext}}(\mathbf{x}),$$

$$\mathbf{H}(\mathbf{x}) = \frac{1}{\mu_0} \mathbf{B}(\mathbf{x}),$$

and are explicitly expressed as

$$\mathbf{B}(\mathbf{x}) = -\frac{a^3 \mu_0 M_0}{3} \left(\frac{-2}{r^3} \cos \theta \mathbf{u}_r - \frac{\sin \theta}{r^3} \mathbf{u}_\theta \right).$$

Considering that $\mathbf{m} = 4/3 \pi a^3 M_0 \mathbf{u}_z$ is the total magnetic moment of the sphere, we have for $r > a$:

$$\mathbf{B}(\mathbf{x}) = \frac{\mu_0 m}{4 \pi r^3} (2 \cos \theta \, \mathbf{u}_r + \sin \theta \, \mathbf{u}_\theta).$$

Finally as $\mathbf{u}_z = -\sin \theta \, \mathbf{u}_\theta + \cos \theta \, \mathbf{u}_r$, we have

$$\mathbf{B}(\mathbf{x}) = \frac{\mu_0 m}{4 \pi r^3} (3 \cos \theta \, \mathbf{u}_r - \mathbf{u}_z)$$

$$= \frac{\mu_0}{4 \pi r^3} (3(\mathbf{m} \cdot \mathbf{u}_r)\mathbf{u}_r - \mathbf{m}) \quad \text{for} \quad r > a,$$

which corresponds to the magnetic field generated far from a magnetic dipole moment $\mathbf{m}$ at the origin. The magnetic field lines are illustrated in Fig. A.99.

10.5 Magnetic shielding

We distinguish three regions of space given by $r < a$ (region enclosed by the shell), $a < r < b$ (inside the spherical shell), and $r > b$ (outside). In all these regions the magnetic field satisfies

Fig. A.99 Magnetic field lines generated by the magnetized sphere

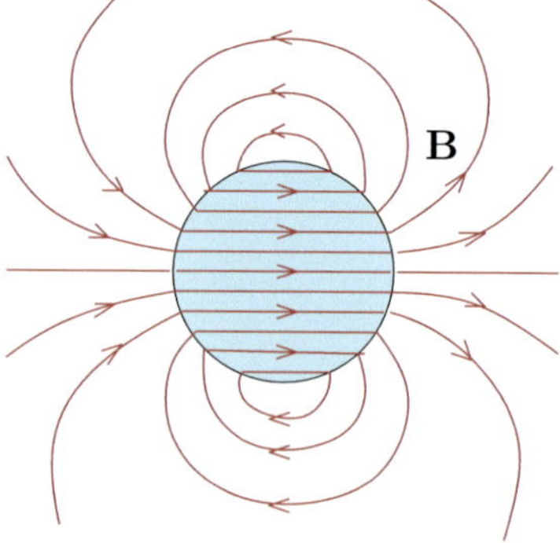

$$\nabla \cdot \mathbf{B}(\mathbf{x}) = 0,$$
$$\nabla \times \mathbf{B}(\mathbf{x}) = 0,$$

so that the magnetic field is irrotational and this problem can be solved by defining a scalar magnetic potential ϕ such that

$$\mathbf{B}(\mathbf{x}) = -\nabla \phi(\mathbf{x}).$$

The condition that the magnetic field is divergence free imposes that the potential satisfies Laplace's equation. The invariance by rotation around the z-axis allows us to expand the potential on the basis of Legendre polynomials. For $r > b$,

$$\phi_3(r, \theta) = -B_0 r \cos \theta + \sum_{l=0}^{\infty} \frac{A_l}{r^{l+1}} P_l(\cos \theta).$$

Note that this matches the condition for the potential at infinity, because for r large enough,

$$\phi_3 \approx -B_0 r \cos \theta = -B_0 z.$$

For $a < r < b$, the solution can take the most general form

$$\phi_2(r, \theta) = \sum_{l=0}^{\infty} \left(B_l r^l + C_l \frac{1}{r^{l+1}} \right) P_l(\cos \theta),$$

while for $r < a$, the condition that the potential is finite when $r \to 0$ translates into

$$\phi_1(r, \theta) = \sum_{l=0}^{\infty} D_l r^l P_l(\cos \theta).$$

Boundary conditions for the fields must be imposed to determine A_l, B_l, C_l and D_l. The normal component of the magnetic field is continuous at $r = a$ and $r = b$, that is

$$\left. \frac{\partial \phi_1(r, \theta)}{\partial r} \right|_{r=a} = \left. \frac{\phi_2(r, \theta)}{\partial r} \right|_{r=a},$$
$$\left. \frac{\partial \phi_2(r, \theta)}{\partial r} \right|_{r=b} = \left. \frac{\phi_3(r, \theta)}{\partial r} \right|_{r=b}.$$

The second condition implies

$$- B_0 \cos \theta - \sum_{l=0}^{\infty} \frac{(l+1)A_l}{b^{l+2}} P_l(\cos \theta) = \sum_{l=0}^{\infty} \left(l B_l a^{l-1} - \frac{(l+1)C_l}{b^{l+2}} \right) P_l(\cos \theta),$$

that is

$$- B_0 \cos \theta = \sum_{l=0}^{\infty} \left(l B_l a^{l-1} + \frac{(l+1)(A_l - C_l)}{b^{l+2}} \right) P_l(\cos \theta).$$

This equation is fulfilled if all the coefficients A_l, B_l, C_l are null for $l \neq 1$. In addition,

$$- B_0 = B_1 + \frac{2(A_1 - C_1)}{b^3}.$$

The first condition also yields

$$D_1 = B_1 - 2 \frac{C_1}{a^3}$$

and $D_l = 0, \forall l \neq 1$. With this, the potential takes the form

$$\phi_1(r, \theta) = D_1 \cos \theta \, r,$$

$$\phi_2(r, \vartheta) = \left(B_1 r + C_1 \frac{1}{r^2} \right) \cos \theta,$$

$$\phi_3(r, \vartheta) = \left(- B_0 r + \frac{A_1}{r^2} \right) \cos \theta.$$

Replacing the two boundary conditions already mentioned

$$\phi_1(r, \theta) = \left(- B_0 + \frac{2(C_1 - A_1)}{b^3} - \frac{2C_1}{a^3} \right) \cos \theta \, r,$$

$$\phi_2(r, \theta) = \left(- B_0 r + r \frac{2(C_1 - A_1)}{b^3} + C_1 \frac{1}{r^2} \right) \cos \theta,$$

$$\phi_3(r, \theta) = \left(- B_0 r + \frac{A_1}{r^2} \right) \cos \theta.$$

In addition, since there are no currents at the interfaces, the tangential component of **H** is continuous at $r = a$ and $r = b$, which means

$$\frac{1}{\mu_0} \left. \frac{\partial \phi_1(r, \theta)}{\partial \theta} \right|_{r=a} = \frac{1}{\mu} \left. \frac{\partial \phi_2(r, \theta)}{\partial \theta} \right|_{r=a}$$

and

$$\frac{1}{\mu} \left. \frac{\partial \phi_2(r, \theta)}{\partial \theta} \right|_{r=b} = \frac{1}{\mu_0} \left. \frac{\partial \phi_3(r, \theta)}{\partial \theta} \right|_{r=b}.$$

The first one is equivalent to

$$\frac{1}{\mu_0}\left(\frac{2C_1}{a^2} + aB_0 + \frac{2a(A_1 - C_1)}{b^3}\right) = \frac{1}{\mu}\left(aB_0 + \frac{2a(A_1 - C_1)}{b^3} - \frac{C_1}{a^2}\right)$$

and the second

$$\frac{1}{\mu}\left(B_0 b + \frac{(2A_1 - 3C_1)}{b^2}\right) = \frac{1}{\mu_0}\left(B_0 b - \frac{A_1}{b^2}\right).$$

It is possible to solve the latter equation to find A_1 as a function of C_1:

$$A_1 = \frac{B_0 b^3 (\mu - \mu_0)}{2\mu_0 + \mu} + \left(\frac{3\mu_0}{2\mu_0 + \mu}\right) C_1.$$

Using the first condition, we obtain

$$C_1 = \frac{3a^3 b^3 B_0 \mu (\mu_0 - \mu)}{b^3(2\mu_0 + \mu)(2\mu + \mu_0) - 2a^3(\mu_0 - \mu)^2},$$

$$A_1 = \frac{B_0 b^3 (b^3 - a^3)(\mu - \mu_0)(2\mu + \mu_0)}{b^3(2\mu_0 + \mu)(2\mu + \mu_0) - 2a^3(\mu_0 - \mu)^2}.$$

Finally

$$\phi_1(r, \theta) = \left(-B_0 + \frac{2(C_1 - A_1)}{b^3} - \frac{2C_1}{a^3}\right)\cos\theta\, r,$$

$$\phi_2(r, \theta) = \left(-B_0 r + r\frac{2(C_1 - A_1)}{b^3} + C_1 \frac{1}{r^2}\right)\cos\theta,$$

$$\phi_3(r, \theta) = \left(-B_0 r + \frac{A_1}{r^2}\right)\cos\theta.$$

Note that the inner potential ϕ_1 represents a uniform magnetic field in the direction $\mathbf{u}_z$. An interesting case occurs in the limit $\mu \gg \mu_0$, that is, when the spherical cavity has a high magnetic permeability. For example, a mu-metal, an alloy that contains mainly Ni, Fe, Cu, and Cr, has a magnetic permeability that can reach $\mu/\mu_0 = 10^5$. In this case it is easy to see that

$$A_1 \approx \frac{B_0(b^3 - a^3)2\mu^2}{2\mu^2 - 2a^3/b^3\mu^2} = B_0 b^3,$$

$$C_1 \approx \frac{-3a^3 b^3 B_0 \mu^2}{2\mu^2 b^3 - 2a^3 \mu^2} = -\frac{3B_0 a^3 b^3}{2(b^3 - a^3)}.$$

Thus

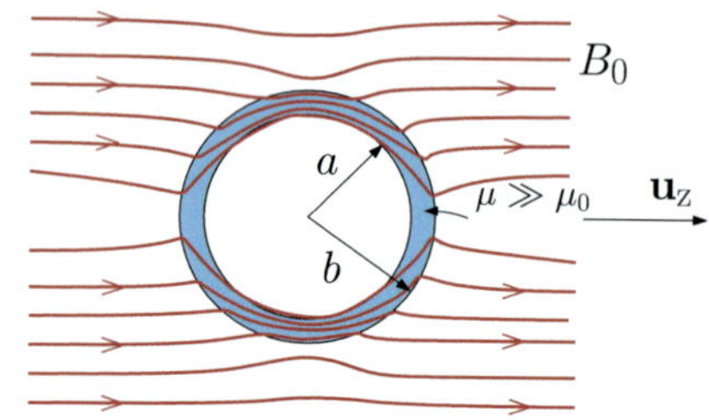

Fig. A.100 Magnetic field lines generated by a uniform field in the presence of a spherical shell of permeability $\mu \gg \mu_0$

$$\phi_1(r,\theta) = 0,$$

$$\phi_2(r,\theta) = -\frac{B_0 r}{2(b^3 - a^3)}\left(4b^3 - a^3 + \frac{3a^3 b^3}{r^3}\right)\cos\theta,$$

$$\phi_3(r,\vartheta) = -B_0 r\left(1 + \frac{b^3}{r^3}\right)\cos\theta.$$

The magnetic field is then null within the cavity. This is a magnetic shield analogous to a Faraday cage for the electric field. A material with high magnetic permeability has the important property of concentrating the field lines within the material, as illustrated in Fig. A.100.

10.5 Field in the presence of a superconducting sphere

(a) The normal component of the magnetic field is continuous when crossing an interface. In this case, we have

$$\mathbf{B}_{\text{in}}(\mathbf{x}) \cdot \mathbf{u}_r|_{r=a} = \mathbf{B}_{\text{ext}}(\mathbf{x}) \cdot \mathbf{u}_r|_{r=a}.$$

Since the field inside the sphere is identically null, it satisfies

$$\mathbf{B}_{\text{ext}}(\mathbf{x}) \cdot \mathbf{u}_r|_{r=a} = 0,$$

so that the magnetic field is tangential to the surface of the conductor.

(b) Since for $r > a$ there are no currents ($\mathbf{J}(\mathbf{x}) = 0$), the magnetic field is irrotational and then there is a scalar magnetic potential ϕ such that

$$\mathbf{B}(\mathbf{x}) = -\nabla\phi(\mathbf{x})$$

which satisfies the Laplace equation. Given the azimuthal symmetry, in spherical coordinates (see Fig. A.101) the solution is expanded in terms of Legendre polynomials

$$\phi(r,\theta) = \sum_{l=0}^{\infty}\left(A_l r^l + B_l r^{-(l+1)}\right) P_l(\cos\theta).$$

For r large enough, the potential must approach that of the external field

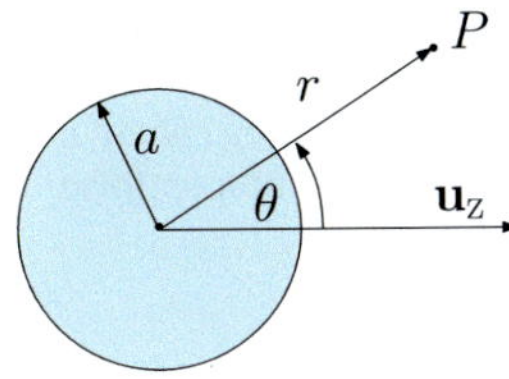

$$\phi(r, \theta) \approx -B_0 z = -B_0 r \cos \theta.$$

From this, it follows that $A_1 = -B_0$ and $A_l = 0, \forall l \neq 1$, hence

$$\phi(r, \theta) = -B_0 r \cos \theta + \sum_{l=0}^{\infty} B_l r^{-(l+1)} P_l(\cos \theta).$$

In addition, the magnetic field at the interface must also be tangential to the sphere, i.e.,

$$\mathbf{B}(\mathbf{x}) \cdot \mathbf{u}_r|_{r=a} = -\nabla \phi(r, \theta) \cdot \mathbf{u}_r|_{r=a}$$
$$= -\frac{\partial \phi(r, \theta)}{\partial r}\bigg|_{r=a} = 0,$$

which yields

$$-B_0 \cos \theta - \sum_{l=0}^{\infty} (l+1) a^{-(l+2)} B_l P_l(\cos \theta) = 0.$$

From this equation, we find $B_l = 0, \forall l \neq 1$, and

$$-B_0 \cos \theta = \frac{2}{a^3} B_1 \cos \theta, \quad \text{i.e.} \quad B_1 = -\frac{a^3 B_0}{2}.$$

Finally the scalar potential is given by

$$\phi(r, \theta) = -B_0 \cos \theta \left(r + \frac{a^3}{2r^2}\right) \quad r > a.$$

(c) We have

$$\mathbf{B}(\mathbf{x}) = -\nabla \phi(\mathbf{x})$$
$$= -\frac{\partial \phi(r, \theta)}{\partial r} \mathbf{u}_r - \frac{1}{r} \frac{\partial \phi(r, \theta)}{\partial \theta} \mathbf{u}_\theta.$$

Then for $r > a$

Fig. A.102 Magnetic field lines produced by a superconducting sphere in the presence of a uniform field

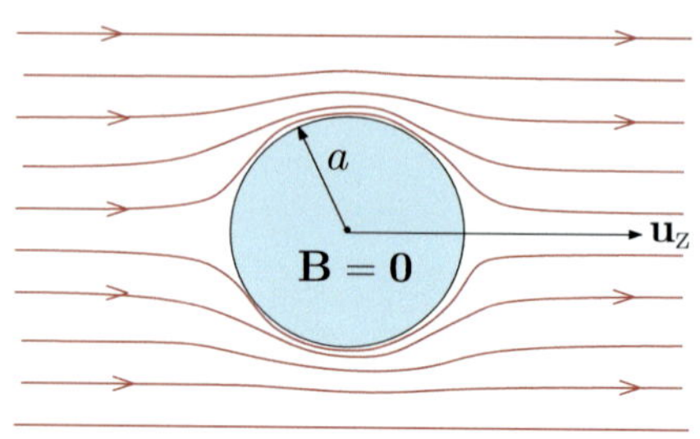

$$\mathbf{B}(r, \theta) = B_0 \cos \theta \left(1 - \frac{a^3}{r^3} \right) \mathbf{u}_r - B_0 \sin \theta \left(1 - \frac{a^3}{2r^3} \right) \mathbf{u}_\theta,$$

which can be rewritten using the fact that $\cos \theta \mathbf{u}_r - \sin \theta \mathbf{u}_\theta = \mathbf{u}_z$, as

$$\mathbf{B}(r, \theta) = B_0 \mathbf{u}_z - \frac{B_0 a^3}{r^3} \left(\cos \theta \mathbf{u}_r - \frac{1}{2} \sin \theta \mathbf{u}_\theta \right)$$

$$= \mathbf{B}_0 - \frac{B_0 a^3}{2r^3} (3 \cos \theta \mathbf{u}_r - \mathbf{u}_z)$$

$$= \mathbf{B}_0 - \frac{3 B_0 a^3}{2r^3} \cos \theta \mathbf{u}_r + \frac{B_0 a^3}{2r^3} \mathbf{u}_z.$$

Defining

$$\mathbf{m} = -\frac{2\pi a^3}{\mu_0} B_0 \mathbf{u}_z,$$

we obtain

$$\mathbf{B}(r, \theta) = \mathbf{B}_0 + \frac{\mu_0}{4\pi} \frac{3m \cos \theta}{r^3} \mathbf{u}_r - \frac{\mu_0}{4\pi} \frac{m}{r^3} \mathbf{u}_z.$$

Finally

$$\mathbf{B}(r, \theta) = \mathbf{B}_0 + \frac{\mu_0}{4\pi r^3} \{ 3(\mathbf{m} \cdot \mathbf{u}_r) \mathbf{u}_r - \mathbf{m} \}.$$

We see that for $r > a$, the superconducting sphere is equivalent to a magnetic dipole $\mathbf{m} = -(2\pi a^3/\mu_0)\mathbf{B}_0$ which opposes the external magnetic field $\mathbf{B}_0$. The field lines are shown in Fig. A.102.

10.6 Curie paramagnetism of localized spins

(a) The energy of a magnetic dipole $\mathbf{m}$ in a magnetic field $\mathbf{B}$ writes $E = -\mathbf{m} \cdot \mathbf{B}$. Therefore, the two possible orientations of a given spin have energies given by

$$E(\mathbf{m} = +\mu_B \mathbf{u}_z) = -\mu_B B_0 = E_1,$$

$$E(\mathbf{m} = -\mu_B \mathbf{u}_z) = +\mu_B B_0 = E_2 > E_1.$$

Fig. A.103 The field favors a particular orientation of the magnetic dipole

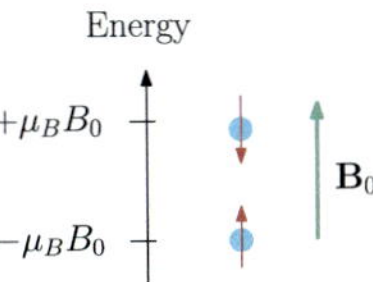

When the spin is parallel to the magnetic field, the energy is lower by $2\mu_B B_0$, and therefore more energetically favorable than when the spin is antiparallel to the field. This is shown in Fig. A.103.

(b) the probability for the spin to occupy the state $\pm\mu_B \mathbf{u}_z$ is given by

$$P_\pm = C e^{\pm W_0}, \quad \text{with } W_0 = \frac{\mu_B B_0}{k_B T},$$

where C is a constant that can be determined by considering that the probability for the spin to occupy either $+\mu_B \mathbf{u}_z$ or $-\mu_B \mathbf{u}_z$ is simply equal to one

$$P_+ + P_- = 1 \quad \rightarrow \quad C = \frac{1}{e^{\frac{\mu_B B_0}{k_B T}} + e^{\frac{-\mu_B B_0}{k_B T}}}$$

$$P_+ + P_- = 1 \quad \rightarrow \quad C = \frac{1}{e^{W_0} + e^{-W_0}}.$$

The average magnetic moment $\langle m \rangle$ along the z-axis of a single spin is therefore

$$\langle m \rangle = +\mu_B P_+ - \mu_B P_- = \mu_B (P_+ - P_-)$$
$$= \mu_B \frac{e^{+W_0} - e^{-W_0}}{e^{W_0} + e^{-W_0}} = \mu_B \tanh (W_0).$$

We see that the spin aligns itself with the magnetic field ($\langle m \rangle \sim \mu_B$) when $W_0 \gg 1$, i.e., when the magnetic interaction energy $\mu_B B_0$ is large compared with the available thermal energy $k_B T$. In contrast, in the regime where $\mu_B B_0 \ll k_B T$, then $\tanh\left(\frac{\mu_B B_0}{k_B T}\right) \approx \frac{\mu_B B_0}{k_B T}$ so that

$$\langle m \rangle \approx \frac{\mu_B^2 B_0}{k_B T} \ll \mu_B.$$

This is illustrated in Fig. A.104.

At room temperature, $k_B T \approx 26 \,\text{meV}$ and for 1 Tesla, $\mu_B B_0 = 56 \,\mu\text{eV}$ so that $\mu_B B_0 / k_B T \sim 10^{-3}$.

(c) If n is the density of spins, then the magnetization $\mathbf{M}$ is given by

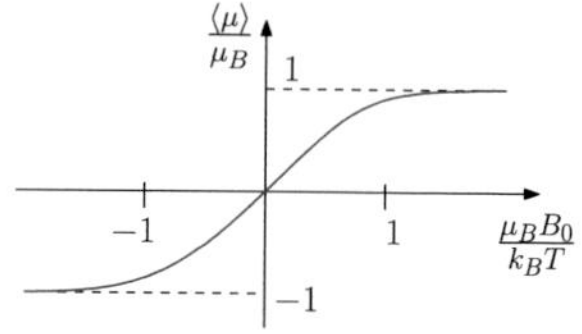

Fig. A.104 Averaged magnetic moment as a function of temperature

$$\mathbf{M} = n\langle m \rangle \mathbf{u}_z = n\mu_B \tanh\left(\frac{\mu_B B_0}{k_B T}\right) \mathbf{u}_z.$$

In the limit $\mu_B B_0 \ll k_B T$

$$\mathbf{M} \approx n\frac{\mu_B^2 B_0}{k_B T}\mathbf{u}_z = n\frac{\mu_B^2}{k_B T}\mathbf{B}$$

and since $\mathbf{M} = \dfrac{\mathbf{B}}{\mu_0}\dfrac{\chi_m}{(1 + \chi_m)}$, for a weak susceptibility, we find

$$\mathbf{M} \approx \frac{\mathbf{B}}{\mu_0}\chi_m = n\frac{\mu_B^2}{k_B T}\mathbf{B},$$

so finally

$$\chi_m = n\mu_0 \frac{\mu_B^2}{k_B T}.$$

For $n = 10^{21}$ cm^{-3} and $k_B T \sim 26$ meV, we find

$$\chi_m \sim 10^{-5}.$$

10.7 Transition between paramagnetism and ferromagnetism

(a) Assuming that the magnetization writes $\mathbf{M} = M\mathbf{u}_z$, a generalization of the magnetization found in Exercise 10.7 gives

$$M = n\langle m \rangle = n\mu_B \tanh\left(\frac{\mu_B(B_0 + \lambda M)}{k_B T}\right).$$

(b) At high temperatures,

$$\tanh\left(\frac{\mu_B(B_0 + \lambda M)}{k_B T}\right) \approx \frac{\mu_B(B_0 + \lambda M)}{k_B T},$$

so we can write

$$M \approx \frac{n\mu_B^2 (B_0 + \lambda M)}{k_B T} = \frac{n\mu_B^2 B_0}{k_B T} + \frac{n\mu_B^2 \lambda M}{k_B T}.$$

Solving for M and considering the limit of a weak susceptibility ($\chi_m \ll 1$)

$$M = \frac{\dfrac{n\mu_B^2 B_0}{k_B T}}{1 - \dfrac{n\mu_B^2 \lambda}{k_B T}} \approx \frac{B_0}{\mu_0} \chi_m,$$

we find the magnetic susceptibility

$$\chi_m = \frac{\dfrac{n\mu_0 \mu_B^2}{k_B T}}{1 - \dfrac{n\mu_B^2 \lambda}{k_B T}} = \frac{\dfrac{n\mu_0 \mu_B^2}{k_B}}{T - \dfrac{n\mu_B^2 \lambda}{k_B}}.$$

We identify the temperature $T_C = \dfrac{n\mu_B^2 \lambda}{k_B}$ and $C = n\mu_0 \mu_B^2 / k_B$ and we see that χ_m follows the Curie–Weiss law:

$$\chi_m = \frac{C}{T - T_C}.$$

Note that for $T \gg T_C$ this reduces to the result obtained in Problem 10.6. As it will be shown below, T_C represents a critical temperature below which another magnetic phase emerges: ferromagnetism.

(c) If $\mathbf{B} = \mathbf{0}$, the magnetization is given by

$$M = n\langle m \rangle = n\mu_B \tanh\left(\frac{\lambda \mu_B M}{k_B T}\right)$$

and this equation needs to be solved in a self-consistent way. We can rewrite it in terms of the critical temperature $T_C = \dfrac{n\mu_B^2 \lambda}{k_B}$ so that

$$\frac{M}{n\mu_B} = \tanh\left(\frac{M/(n\mu_B)}{T/T_C}\right)$$

and we see that, apart from the solution $M = 0$ (which corresponds to the paramagnetic phase), there exists a solution with $M \neq 0$ only if $T < T_C$. The material can be magnetized even in the absence of an external magnetic field. The temperature T_C is called Curie's temperature and marks the transition between paramagnetism and ferromagnetism.

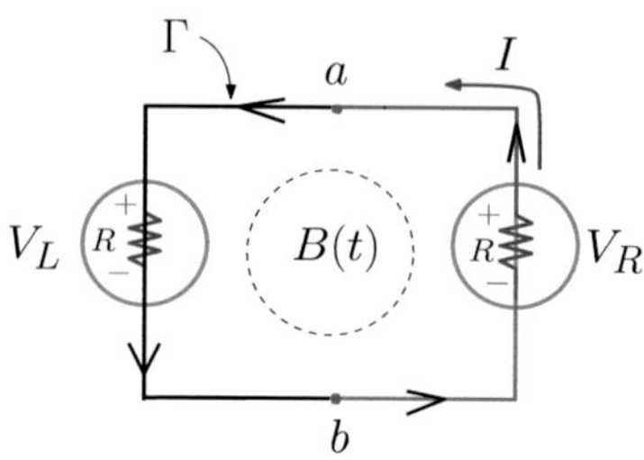

Fig. A.105 Induced current flowing counterclockwise, along the orientation of Γ

Problems of Chap. 11

11.1 Path dependence of the potential difference

(a) Suppose there is an induced current I flowing anticlockwise when looked from above, as shown in Fig. A.105. In the absence of an electromotive force in the voltmeters, we have

$$V_R = -IR = -0.2\,\text{V},$$

so that the induced current reads

$$I = \frac{V_R}{R} = 2 \times 10^{-7}\,\text{A}.$$

Now let us consider the closed path Γ shown in the figure above, oriented as indicated by the arrows. We have, from generalized Ohm's law:

$$\varepsilon = I(R + R) = 2RI = 0.4\,\text{V}.$$

(b) The left voltmeter will simply read

$$V_L = IR = +0.2\,\text{V},$$

which differs from the lecture of the right voltmeter since the electric field is no longer conservative.

Indeed, using generalized Ohm's law once again to the paths connecting e to c and d to f in the anticlockwise direction (see Fig. A.106), we find

$$V_e - V_c = -\varepsilon_{ec} \quad \text{and} \quad V_d - V_f = -\varepsilon_{df}.$$

The sum of the latter two equations gives

$$\underbrace{V_e - V_f}_{V_R} - \underbrace{(V_c - V_d)}_{V_L} = -(\varepsilon_{ec} + \varepsilon_{df}) = -\varepsilon,$$

Fig. A.106 Ohm's law can be applied to the paths ec and df

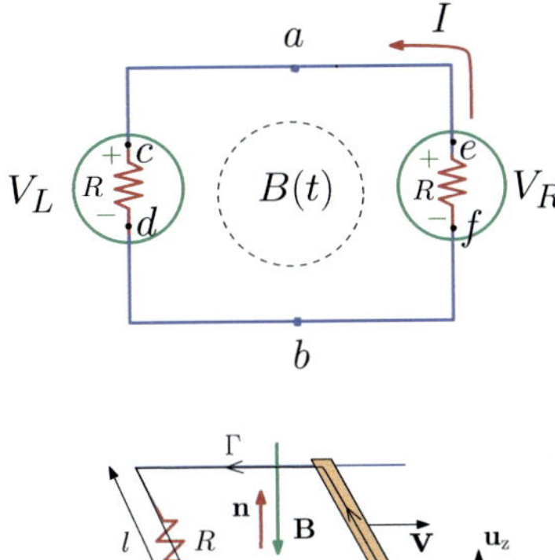

Fig. A.107 The curve Γ is oriented such that the normal to the surface enclosed by it is along z

so that in general the difference between V_L and V_R corresponds to the electromotive force around the circuit. In this particular case,

$$V_L = V_R + \varepsilon = -0.2 + 0.4 = +0.2 \text{ V}.$$

11.2 Mobile rail in a uniform magnetic field

(a) Consider the closed path Γ formed by the moving rail, the resistor, and the wires connecting them. The orientation of the curve is anticlockwise, i.e., such that the normal $\mathbf{n}$ to the planar surface $S(\Gamma)$ enclosed by Γ is along $\mathbf{u}_z$. This is shown in Fig. A.107. If at a given instant the rail is at a distance x from the resistor, the magnetic flux through $S(\Gamma)$ is

$$\Phi_{S(\Gamma),\mathbf{B}} = -\iint_{S(\Gamma)} B\mathbf{u}_z \cdot \mathbf{u}_z dS(\mathbf{x}) = -Blx.$$

Faraday's law states that the electromotive force induced in the circuit is

$$\varepsilon = -\frac{d\Phi_{S(\Gamma),\mathbf{B}}}{dt} = \frac{d}{dt}(Blx) = Bl\frac{dx}{dt} = Blv.$$

One can verify that this electromotive force corresponds to the circulation of the electromotive field along the circuit:

$$\varepsilon = \int_0^l (\mathbf{v} \times \mathbf{B}) \cdot dy\,\mathbf{u}_y = vBl.$$

Only the rail is moving, thus, the other branches of the circuit do not contribute to the emf. Finally, since the magnetic field is static, one can show that the time derivative of the flux is completely due to the term

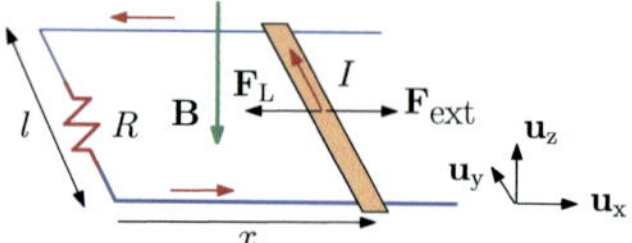

Fig. A.108 An external force must oppose the Laplace force to keep a constant velocity for the rail

$$- (\mathbf{v} \cdot \nabla)\Phi_{S(\Gamma),\mathbf{B}} = -v\frac{\partial}{\partial x}(-Blx) = Blv = -\frac{d}{dt}\Phi_{S(\Gamma),\mathbf{B}}.$$

The induced current, in the direction given by the tangent $d\mathbf{l}$ to the curve Γ (counterclockwise direction seen from above the plane xy) is

$$I = \frac{\varepsilon}{R} = \frac{Blv}{R}.$$

Note that for $v > 0$, the magnetic flux due to the external field is decreasing (increasing toward negative values), so that the induced current generates a positive flux that opposes this change. This is in agreement with Lenz's law.

(b) We see that the Laplace force $\mathbf{F}_L$ acting on the rail generates a force in the direction opposite to $\mathbf{v}$, thus trying to stop the motion of the rail. An external force $\mathbf{F}_{ext}$ is then required to cancel $\mathbf{F}_L$ and keep the bar in a uniform motion. This is illustrated in Fig. A.108. The Laplace force reads

$$\mathbf{F}_L = Il\mathbf{u}_y \times (-B\mathbf{u}_z) = -IlB\mathbf{u}_x = -\frac{B^2l^2v}{R}\mathbf{u}_x,$$

so that

$$\mathbf{F}_{ext} = -\mathbf{F}_L = \frac{B^2l^2v}{R}\mathbf{u}_x.$$

The mechanical power provided by the external force is

$$P_m = \mathbf{F}_{ext} \cdot \mathbf{v} = \frac{B^2l^2v^2}{R},$$

which equals the electric power consumed by the resistance

$$P_e = \frac{\varepsilon^2}{R} = I^2R = \frac{(Blv)^2}{R}.$$

11.3 Rail on an inclined plane surface

(a) Let us suppose that at a certain instant, the bar is at distance s from the top of the plane, as shown in Fig. A.109.

The closed path Γ formed by the rails and the sliding bar encloses a planar surface $S(\Gamma)$ whose normal $\mathbf{n}$ can be written as

Fig. A.109 At a given instant, the circuit is defined by the curve Γ

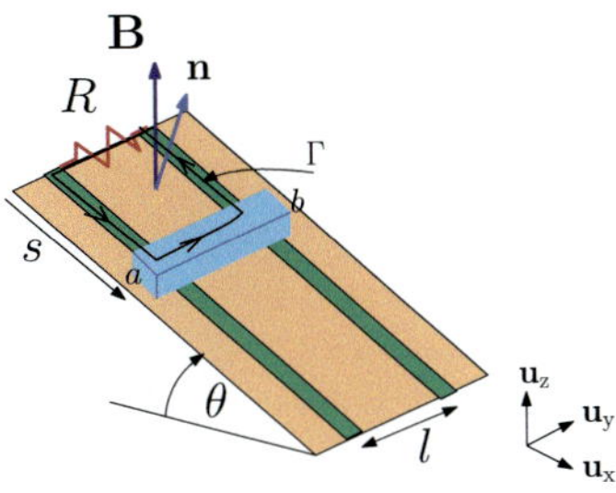

$$\mathbf{n} = \sin\theta\,\mathbf{u}_x + \cos\theta\,\mathbf{u}_z.$$

The magnetic flux through this surface is

$$
\begin{aligned}
\Phi_{S(\Gamma),\mathbf{B}} &= \iint_{S(\Gamma)} \mathbf{B}\,\mathbf{u}_z \cdot \mathbf{n}\,dS(\mathbf{x}) \\
&= \iint_{S(\Gamma)} B\cos\theta\,dS(\mathbf{x}) \\
&= Bsl\cos\theta,
\end{aligned}
$$

so that the electromotive force in the circuit Γ reads

$$
\varepsilon = -\frac{d\Phi_{S(\Gamma),\mathbf{B}}}{dt} = -Bl\cos\theta\,\frac{ds}{dt}
$$

and the induced current is then

$$
I = \frac{\varepsilon}{R} = -\frac{Bl}{R}\cos\theta\,\frac{ds}{dt}.
$$

In agreement with Lenz's law, if the bar falls, $ds/dt > 0$ and the current is in the clockwise direction when the circuit is seen from above (flowing from b to a). The Laplace force on the bar is then

$$
\mathbf{F}_L = Il\mathbf{u}_y \times B\mathbf{u}_z = -\frac{B^2 l^2}{R}\cos\theta\,\frac{ds}{dt}\,\mathbf{u}_x
$$

and indeed opposes the falling of the bar.

(b) The forces acting on the rail are shown in Fig. A.110. Projecting the equation of motion onto the direction parallel to the rails gives

$$
mg\sin\theta - F_L\cos\theta = m\frac{d^2s}{dt^2},
$$

which yields

Fig. A.110 Forces acting on
the rail

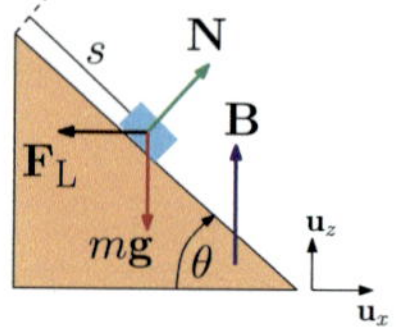

$$mg \sin \theta - \frac{l^2 B^2}{R} \cos^2 \theta \frac{ds}{dt} = m \frac{d^2 s}{dt^2}.$$

Rearranging the terms in the equation, and writing $v = ds/dt$, one obtains a first-order ordinary differential equation for the velocity

$$\frac{dv}{dt} + \frac{l^2 B^2}{mR} \cos^2 \theta \, v - g \sin \theta = 0,$$

which we rewrite as

$$\frac{dv}{dt} + \frac{v}{\tau} - g \sin \theta = 0,$$

with

$$\tau = \frac{mR}{l^2 B^2 \cos^2 \theta}.$$

Integrating this equation we obtain

$$v(t) = \tau g \sin \theta \left(1 - e^{-t/\tau}\right)$$

and the terminal velocity is given by

$$v_\infty = \lim_{t \to \infty} v(t) = \tau g \sin \theta = \frac{mgR \sin \theta}{l^2 B^2 \cos^2 \theta}.$$

(c) The current in the regime of terminal velocity is given by

$$I = -\frac{Bl}{R} \cos \theta \, v_\infty = -\frac{mg \tan \theta}{lB}.$$

(d) The electromotive force is

$$\varepsilon = IR = -\frac{mgR \tan \theta}{lB}$$

and the rate at which energy is dissipated in the resistance is therefore equal to the power delivered by the electromotive force

Fig. A.111 Choice of coordinates used

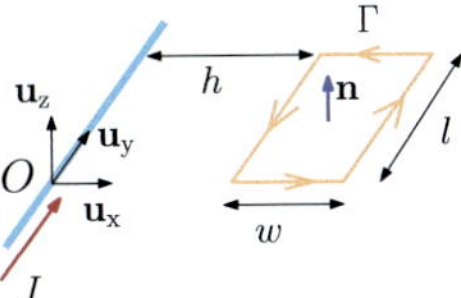

$$P = \varepsilon I = \frac{m^2 g^2 R \tan^2 \theta}{l^2 B^2}.$$

(e) The rate of work provided by gravity is

$$P_g = mg \sin \theta \, v_\infty = \frac{m^2 g^2 R \tan^2 \theta}{l^2 B^2},$$

which is equal to the power dissipated by the circuit.

11.4 Induction by a time-varying magnetic field

(a) We define xy as the plane containing both the wire and the rectangular loop and choose the origin as shown in Fig. A.111.

The magnetic field produced by the wire at any point P in the xy plane is

$$\mathbf{B} = -\frac{\mu_0 I}{2\pi x}\mathbf{u}_z.$$

Consider the closed path Γ which coincides with the rectangular loop. The orientation of the curve is such that the planar surface $S(\Gamma)$ enclosed by Γ has a normal $\mathbf{n} = \mathbf{u}_z$. The magnetic flux through this surface reads

$$\Phi_{S(\Gamma),\mathbf{B}} = -\iint_{S(\Gamma)} \frac{\mu_0 I}{2\pi x}\mathbf{u}_z \cdot \mathbf{u}_z \, dx\, dy = -\frac{\mu_0 I l}{2\pi}\int_h^{h+w}\frac{dx}{x} = -\frac{\mu_0 I l}{2\pi}\ln\left(1 + \frac{w}{h}\right).$$

(b) According to Faraday's induction law

$$\varepsilon = -\frac{d\Phi_{S(\Gamma),\mathbf{B}}}{dt} = \frac{\mu_0 l}{2\pi}\ln\left(1 + \frac{w}{h}\right)\frac{dI}{dt}.$$

Since $dI/dt = b$, we find

$$\varepsilon = \frac{\mu_0 l b}{2\pi}\ln\left(1 + \frac{w}{h}\right)$$

and the induced current is

$$I = \frac{\varepsilon}{R} = \frac{\mu_0 l b}{2\pi R}\ln\left(1 + \frac{w}{h}\right).$$

Fig. A.112 Definition of Γ and its normal **n**

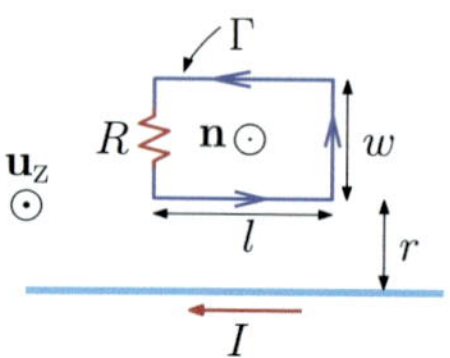

In agreement with Lenz's law, for $b > 0$ ($b < 0$) the current flows in the anticlockwise (clockwise) direction when looked from above, so as to produce a positive (negative) flux through $S(\Gamma)$ which opposes the decrease (increase) of flux due to the current in the wire.

11.5 Relative motion between currents

In Exercise 11.4, we calculated the magnetic flux of the field generated by the infinite wire through the surface enclosed by the loop. If we define the closed path Γ as the one corresponding to the rectangular loop in such a way that the normal to the planar surface $S(\Gamma)$ enclosed by Γ is along $\mathbf{u}_z$ (see Fig. A.112), then,

$$\Phi_{S(\Gamma),\mathbf{B}} = -\frac{\mu_0 I l}{2\pi} \ln\left(1 + \frac{w}{r}\right).$$

Faraday's induction law gives the electromotive force

$$\varepsilon = -\frac{d\Phi_{S(\Gamma),\mathbf{B}}}{dt} = \frac{\mu_0 I l}{2\pi} \frac{d}{dt} \ln\left(1 + \frac{w}{r}\right)$$
$$= -\frac{\mu_0 I l}{2\pi} \left(-\frac{w}{r^2} \frac{1}{1 + w/r}\right) \frac{dr}{dt} = -\frac{\mu_0 I l v}{2\pi} \frac{w}{r(r + w)},$$

with $\dfrac{dr}{dt} = v$, the speed of the rectangular loop. This electromotive force is due to the electromotive field acting on the two horizontal segments for which $(\mathbf{v} \times \mathbf{B}) \cdot d\mathbf{l} \neq 0$. We can check that indeed the time derivative of the flux reads

$$-(\mathbf{v} \cdot \nabla)\Phi_{S(\Gamma),\mathbf{B}} = -v\frac{\partial}{\partial r} \Phi_{S(\Gamma),\mathbf{B}} = -\frac{d}{dt} \Phi_{S(\Gamma),\mathbf{B}}.$$

Now, the induced current is

$$I = \frac{\varepsilon}{R} = -\frac{\mu_0 I l v}{2\pi R} \frac{w}{r(r + w)}.$$

We can check that it is consistent with Lenz's law: If $v > 0$, the flux through the rectangular loop is increasing (toward positive values), and therefore the current flows in the clockwise direction when seen from above, creating a negative flux through $S(\Gamma)$ which tends to oppose the change in the magnetic flux.

11.6 Current induced by gravity

(a) Suppose that the bar is at a distance x from the extremity of the rails. Let Γ be

Fig. A.113 Orientation of Γ with respect to the axes of the problem

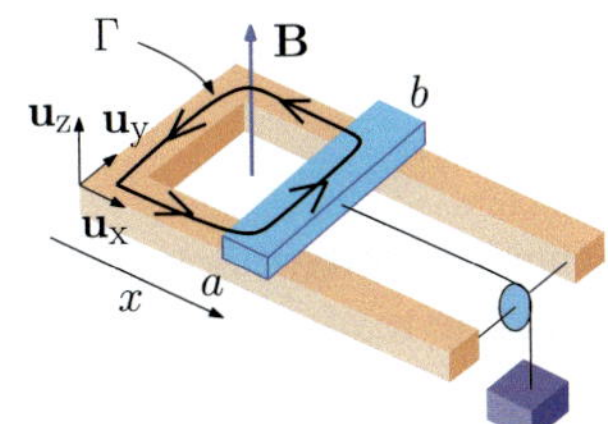

Fig. A.114 Forces acting on the bar and on the mass M

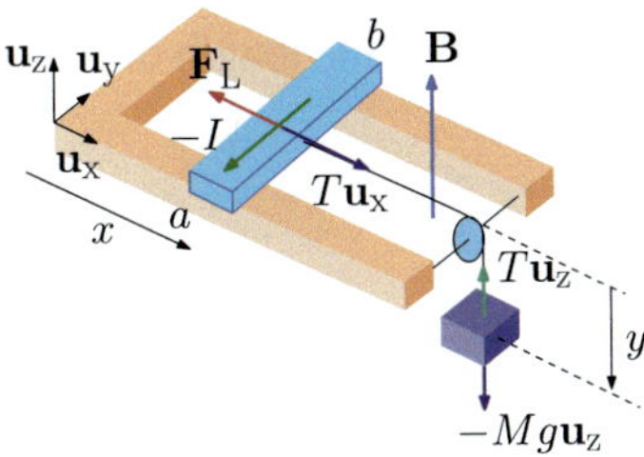

the closed path connecting the rails and the bar, and oriented so the normal to the planar surface $S(\Gamma)$ enclosed by Γ is $\mathbf{n} = \mathbf{u}_z$ as shown in Fig. A.113. The magnetic flux through $S(\Gamma)$ is then

$$\Phi_{S(\Gamma),\mathbf{B}} = \iint_{S(\Gamma)} B\,\mathbf{u}_z \cdot \mathbf{u}_z \, dS(\mathbf{x}) = Bxl$$

and the electromotive force is

$$\varepsilon = -\frac{d\Phi_{S(\Gamma),\mathbf{B}}}{dt} = -Bl\frac{dx}{dt} = -Blv.$$

The time derivative of the flux simply corresponds to the circulation of the electromotive field $\mathbf{v} \times \mathbf{B}$ over Γ, where only the segment ab contributes

$$\varepsilon_{ab} = \int_a^b (\mathbf{v} \times \mathbf{B}) \cdot dy\,\mathbf{u}_y = -vB \int_a^b dy = -vlB.$$

The induced current is given by

$$I = \frac{\varepsilon}{R} = -\frac{Blv}{R}.$$

(b) The bar is subjected to the Laplace force $\mathbf{F}_L$ due to the induced current (note that if $v > 0$, $I < 0$ and this force opposes the movement of the bar, according to Lenz's law) and to the tension of the cable. The same tension acts on the block M, as shown in Fig. A.114.

For the bar, the force balance gives us the equation

$$T\mathbf{u}_x + \underbrace{Il\mathbf{u}_y \times B\mathbf{u}_z}_{\mathbf{F}_L} = (T + IlB)\mathbf{u}_x = m\frac{dx^2}{dt^2}\mathbf{u}_x,$$

whereas for the block of mass M, at distance y from the plane of the rails, the Newton equation reads

$$Mg - T = M\frac{dy^2}{dt^2}.$$

Suppose that the length of the cable does not vary, $dx^2/dt^2 = dy^2/dt^2$, and so

$$Mg - M\frac{dy^2}{dt^2} = Mg - M\frac{dx^2}{dt^2} = T,$$

which gives

$$Mg - \frac{l^2B^2}{R}v = (m + M)\frac{dx^2}{dt^2}.$$

The latter equation can be written as a first-order ordinary differential equation for the velocity of the bar

$$\frac{dv}{dt} + \frac{l^2B^2}{R(M+m)}v - \frac{Mg}{M+m} = 0,$$

rewritten as

$$\frac{dv}{dt} + \frac{v}{\tau} - v_\infty = 0,$$

with

$$\tau = \frac{R(M+m)}{l^2B^2} \quad \text{and} \quad v_\infty = \frac{MgR}{l^2B^2}.$$

Integrating this equation with initial condition $v(0) = 0$ gives the solution

$$v(t) = v_\infty\left(1 - e^{-t/\tau}\right).$$

11.7 Rotation generated by induction

Consider the curve Γ given by $r = b$ and oriented so the normal to the planar surface $S(\Gamma)$ enclosed by Γ is along $\mathbf{u}_z$. At $t = 0$, the magnetic flux through this surface is given by

$$\Phi_{S(\Gamma),\mathbf{B}}(0) = \iint_{S(\Gamma)} B\,\mathbf{u}_z \cdot \mathbf{u}_z dS = B\pi a^2$$

since the magnetic field is zero for $r > a$. Faraday's law states that a circulating electric field $\mathbf{E} = E(r)\mathbf{u}_\theta$ appears whenever the magnetic field varies in time, and it is responsible for the electromotive force acting on the charges at $r = b$

Fig. A.115 The electric force acting on the charges produce a torque τ

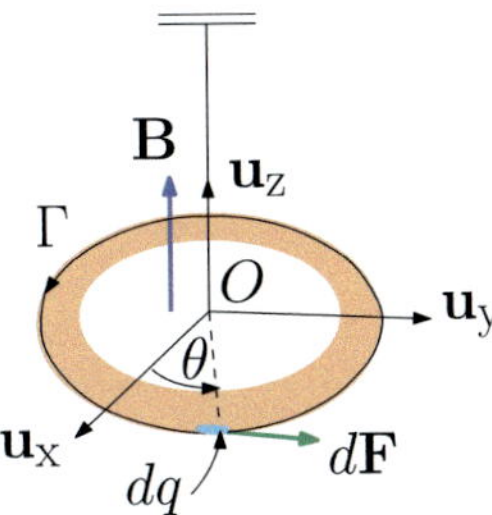

$$\varepsilon = \oint_{\Gamma} \mathbf{E} \cdot d\mathbf{l} = E\, 2\pi b = -\frac{d\Phi_{S(\Gamma),\mathbf{B}}}{dt},$$

so that

$$E = -\frac{1}{2\pi b}\frac{d\Phi_{S(\Gamma),\mathbf{B}}}{dt}.$$

Now, Consider a differential path length $dl = b\,d\theta$ on Γ. The infinitesimal charge contained in this arc is $dq = \lambda dl$ with $\lambda = q/2\pi b$. The force $d\mathbf{F}$ acting on this charge is

$$\mathbf{d}F = dq\, E\mathbf{u}_\theta = \lambda b\, d\theta\, E\, \mathbf{u}_\theta.$$

This force, shown in Fig. A.115, is responsible for a differential torque $d\boldsymbol{\tau}$ acting on the disk

$$d\boldsymbol{\tau} = b\mathbf{u}_r \times d\mathbf{F} = E\lambda b^2 d\theta \mathbf{u}_z = -\frac{\lambda b}{2\pi}\frac{d\Phi_{S(\Gamma),\mathbf{B}}}{dt}d\theta \mathbf{u}_z$$

and the total torque on the disk is

$$\boldsymbol{\tau} = -\int_0^{2\pi}\frac{\lambda b}{2\pi}\frac{d\Phi_{S(\Gamma),\mathbf{B}}}{dt}\, d\theta\, \mathbf{u}_z = -\lambda b\frac{d\Phi_{S(\Gamma),\mathbf{B}}}{dt}\mathbf{u}_z.$$

If $\mathbf{L} = L\mathbf{u}_z$ where $L = \mathcal{J}\omega$ is the angular momentum of the disk, then,

$$\boldsymbol{\tau} = \frac{d\mathbf{L}}{dt},$$

which yields

$$-\lambda b\frac{d\Phi_{S(\Gamma),\mathbf{B}}(t)}{dt} = \frac{dL}{dt}.$$

Integrating the latter equality with respect to time between $t = 0$ and $t = t_f$, we find

$$-\lambda b\left(\Phi_{S(\Gamma),\mathbf{B}}(t_f) - \Phi_{S(\Gamma),\mathbf{B}}(0)\right) = L_f - L_0.$$

If the disk is at rest at $t = 0$, $L_0 = 0$, and if t_f denotes a time after the magnetic field has been switched off, $\Phi_{S(\Gamma),\mathbf{B}}(t_f) = 0$ and so,

Fig. A.116 Definition of the loop Γ

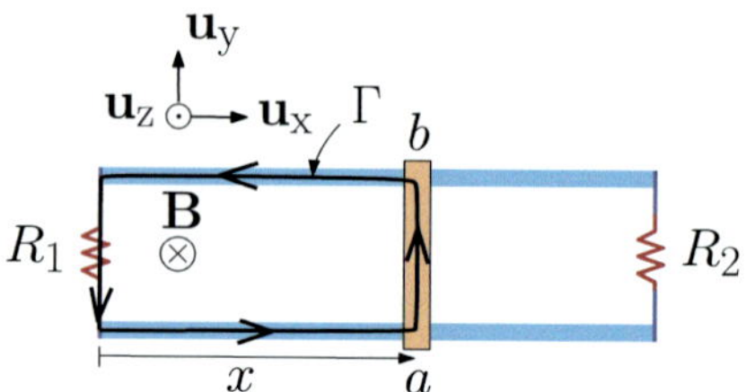

$$\lambda b\pi a^2 B = \mathcal{J}\omega_f,$$

with ω_f the final angular velocity of the disk,

$$\omega_f = \frac{\lambda b\pi a^2 B}{\mathcal{J}} = \frac{qa^2 B}{2\mathcal{J}}.$$

11.8 Mobile bar between two resistors

(a) Let us suppose that the bar is at distance x from the resistor R_1, and consider the closed path Γ oriented so the normal to the surface $S(\Gamma)$ enclosed by Γ is, at every point, $\mathbf{u}_z$. This is shown in Fig. A.116. The magnetic flux through $S(\Gamma)$ is then

$$\Phi_{S(\Gamma),\mathbf{B}} = \iint_{S(\Gamma)} -B\mathbf{u}_z \cdot \mathbf{u}_z dS = -Bxl$$

and the electromotive force is then given by Faraday's law

$$\varepsilon = -\frac{d\Phi_{S(\Gamma),\mathbf{B}}}{dt} = Bl\frac{dx}{dt} = Blv.$$

This electromotive force corresponds to the circulation around Γ of the electromotive field. Only the bar ab with velocity $v\mathbf{u}_x$ gives a non-zero contribution, thus,

$$\varepsilon_{ab} = \int_a^b (v\mathbf{u}_x \times (-B\mathbf{u}_z)) \cdot dy\,\mathbf{u}_y = lvB = \varepsilon.$$

(b) The electromotive force ε generates an induced current through the bar that is distributed between both resistors, as shown in Fig. A.117. We have, by using generalized Ohm's law,

$$I_1 = \frac{\varepsilon}{R_1} = \frac{Blv}{R_1}, \quad \text{and} \quad I_2 = \frac{\varepsilon}{R_2} = \frac{Blv}{R_2}.$$

The total induced current through the bar is then

$$I = I_1 + I_2 = Blv\left(\frac{1}{R_1} + \frac{1}{R_2}\right).$$

Fig. A.117 The induced current is the sum $I_1 + I_2$

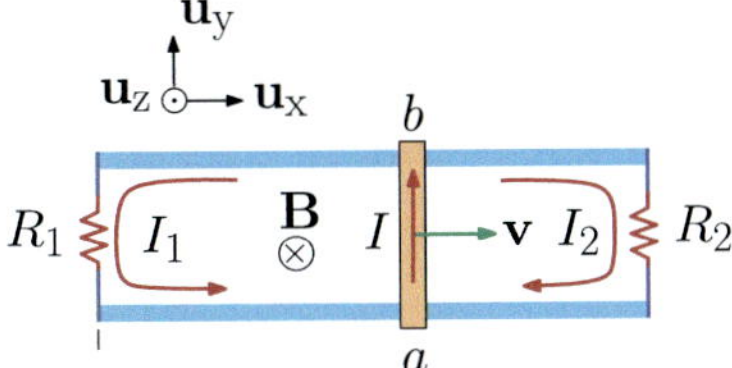

Fig. A.118 Definition of the coordinate system and orientation of Γ

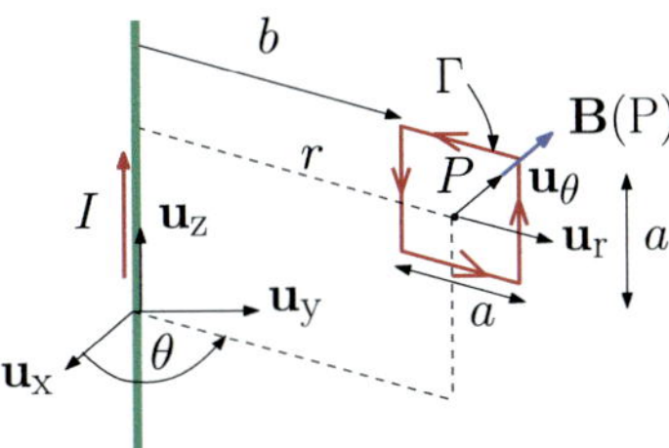

(c) The power dissipated in the resistors is given by

$$P_e = I_1^2 R_1 + I_2^2 R_2 = B^2 l^2 v^2 \left(\frac{1}{R_1} + \frac{1}{R_2} \right).$$

(d) The electric power P_e consumed by the circuit must equal the mechanical power P_m supplied by the external force $\mathbf{F}_{\text{ext}} = F \mathbf{u}_x$, so that

$$P_m = F v = B^2 l^2 v^2 \left(\frac{1}{R_1} + \frac{1}{R_2} \right).$$

Finally,

$$F = B^2 l^2 v \left(\frac{1}{R_1} + \frac{1}{R_2} \right) = I l B.$$

One recognizes in $\mathbf{F}_{\text{ext}}$ a force that exactly cancels the Laplace force $\mathbf{F}_L$ acting on the bar,

$$\mathbf{F}_L = I l \mathbf{u}_y \times \mathbf{B} = -I l B \mathbf{u}_x = -\mathbf{F}_{\text{ext}}.$$

11.9 Induction in a toroid

We choose a system of cylindrical coordinates such that the z-axis corresponds to the axis of the wire. Consider the closed path Γ representing one of the loops of the toroid, oriented so the normal to the planar surface $S(\Gamma)$ enclosed by Γ is, at every point, $\mathbf{n} = -\mathbf{u}_\theta$. This is shown in Fig. A.118. At any point P, the magnetic field generated by the wire is

$$\mathbf{B}(P) = \frac{\mu_0 I}{2\pi r} \mathbf{u}_\theta.$$

The magnetic flux over $S(\Gamma)$ then writes

$$\Phi_{S(\Gamma),\mathbf{B}}(t) = -\iint_{S(\Gamma)} \frac{\mu_0 I}{2\pi r} \mathbf{u}_\theta \cdot \mathbf{u}_\theta \, dS$$
$$= -\frac{\mu_0 I}{2\pi} \int_0^a dz \int_b^{b+a} \frac{dr}{r} = -\frac{\mu_0 I a}{2\pi} \ln\left(\frac{b+a}{b}\right).$$

The system is invariant by rotation around the z-axis, hence the flux will be the same for every loop of the toroid. Thus, the flux through the surface enclosed by all the loops in the toroid is

$$\Phi_T(t) = N\Phi_{S(\Gamma),\mathbf{B}}(t) = -\frac{N\mu_0 I(t)a}{2\pi} \ln\left(1 + \frac{a}{b}\right)$$

and the electromotive force is, according to Faraday's law,

$$\varepsilon(t) = -\frac{d\Phi_T(t)}{dt} = \frac{N\mu_0 \omega a}{2\pi} \ln\left(1 + \frac{a}{b}\right) I_0 \cos \omega t.$$

Note that if $a \ll b$, $\ln(1 + a/b) \sim a/b$ and the flux is approximated by

$$\Phi_T(t) = N\Phi_{S(\Gamma),\mathbf{B}}(t) = -N\frac{\mu_0 I(t)}{2\pi b} a^2,$$

which is simply equal to N times the product of the magnetic field amplitude at the position b of the loop, by the surface of the square loop.

11.10 Solenoid moving in a magnetic field

Let Γ be one loop of the solenoid oriented so the normal to the planar surface $S(\Gamma)$ enclosed by Γ is $\mathbf{u}_z$. The total flux through the solenoid is simply N times the flux through (Γ), and initially given by

$$\Phi(0) = N \iint_{S(\Gamma)} B \, \mathbf{u}_z \cdot \mathbf{u}_z \, dS = NB \iint_{S(\Gamma)} dS = NB\pi \left(\frac{D}{2}\right)^2.$$

While the solenoid is turning, an electromotive force ε appears between its terminals, since the magnetic flux is changing in time. The induced current is

$$I(t) = \frac{\varepsilon}{R} = -\frac{1}{R}\frac{d\Phi(t)}{dt}.$$

If at $t = T$ the solenoid has been turned upside down, then the total charge that flowed through the circuit is

$$Q = \int_0^T I(t)dt = -\frac{1}{R}\int_0^T \frac{d\Phi(t)}{dt} dt = \frac{1}{R}\left(\Phi(0) - \Phi(T)\right),$$

Fig. A.119 The train has entered a distance $x(t)$ in the region of the magnetic field

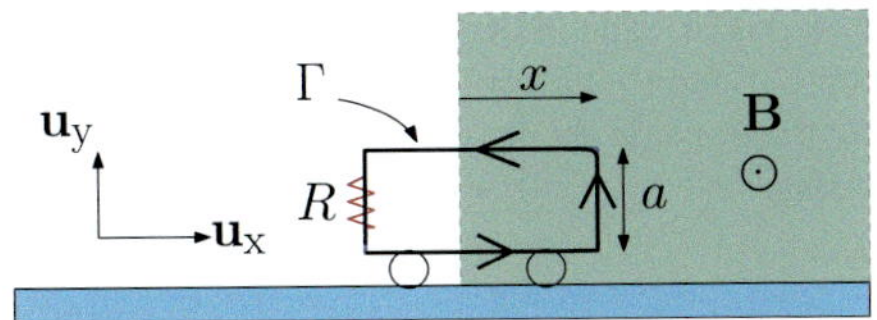

with $\Phi(T) = -\Phi(0)$, so that

$$Q = \frac{2\Phi(0)}{R} = \frac{NB\pi D^2}{2R}$$

and the magnetic field is therefore

$$B = \frac{2QR}{\pi N D^2}.$$

11.11 Magnetic brake

Suppose that at a given instant, the train has entered a distance $x(t)$ into the region of the magnetic field. Let Γ be the closed path corresponding to the loop of the train, oriented so the normal to the planar surface $S(\Gamma)$ enclosed by Γ is $\mathbf{n} = \mathbf{u}_z$, as shown in Fig. A.119.

The magnetic flux through $S(\Gamma)$ is

$$\Phi_{S(\Gamma),\mathbf{B}}(t) = \iint_{S(\Gamma)} B(x)\mathbf{u}_x \cdot \mathbf{u}_x \, dS$$
$$= \begin{cases} aBx(t), & \text{if } x(t) < l \\ aBL & \text{if } x(t) \geq l. \end{cases}$$

As long as $x(t) < l$, the flux is time dependent, hence an electromotive force is induced in Γ and it is given by

$$\varepsilon = -\frac{d\Phi_{S(\Gamma),\mathbf{B}}(t)}{dt} = -aB\frac{dx(t)}{dt} = -aBv.$$

The induced current in the loop is then

$$I = \frac{\varepsilon}{R} = -\frac{aBv}{R}$$

so that, for $v > 0$, it turns clockwise ($I < 0$) when looked from the region $z > 0$, opposing the change in magnetic flux in agreement with Lenz's law.

The net Laplace force $\mathbf{F}_L$ on the train, shown in Fig. A.120, is given by the force acting on the vertical segment inside the magnetic field region (the forces on the two horizontal segments cancel each other out), so that

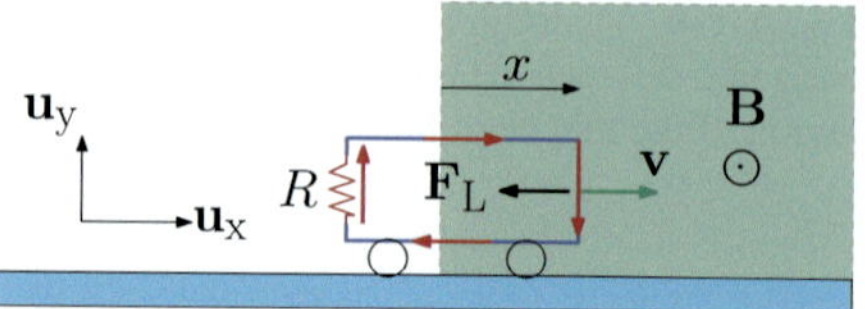

Fig. A.120 The net Laplace force is due to the vertical segment inside the field region

$$\mathbf{F}_L = I a \mathbf{u}_y \times B \mathbf{u}_z = I a B \mathbf{u}_x = -\frac{(aB)^2 v}{R} \mathbf{u}_x.$$

Newton's law projected onto the x-axis then yields

$$m \frac{dv(t)}{dt} = \frac{-(aB)^2}{R} v(t) \quad \text{for} \quad x(t) < l$$

and the solution is given by an exponentially decaying velocity as long as $x(t) < l$,

$$v(t) = v_0 e^{-t/\tau}, \quad t < t^*, \quad \text{with} \quad \tau = \frac{mR}{a^2 B^2},$$

where t^* can be determined by writing

$$l = x(t^*) = v_0 \tau \left(1 - e^{-t^*/\tau}\right), \quad \text{hence,} \quad t^* = -\tau \ln\left(1 - \frac{l}{v_0 \tau}\right).$$

Note that this is valid provided that the initial velocity is high enough $v_0 \tau > \ell$. If it's not the case, the train will stop before entering completely the magnetic field region. Once the train has completely entered the region with the magnetic field, the flux is constant and the electromotive force vanishes. The train then moves at a constant velocity given by

$$v(t^*) = v_0 - \frac{l}{\tau} = v_0 - \frac{la^2 B^2}{mR}.$$

11.12 Magnetic parachute

Define the closed path Γ corresponding to the loop and oriented so the normal to the enclosed surface $S(\Gamma)$ is $\mathbf{u}_x$, as in Fig. A.121. The magnetic flux through $S(\Gamma)$ is then

$$\Phi_{S(\Gamma),\mathbf{B}} = \iint_{S(\Gamma)} B(z) \mathbf{u}_x \cdot \mathbf{u}_x \, dS = \int_0^a dy \int_z^{z+a} (B_0 - \alpha z) \, dz = a^2 B_0 - \frac{a\alpha}{2}(2az + a^2).$$

The electromotive force is given by Faraday's law,

$$\varepsilon = -\frac{d\Phi_{S(\Gamma),\mathbf{B}}}{dt} = -\frac{d\Phi_{S(\Gamma),\mathbf{B}}}{dz}\frac{dz}{dt} = a^2 \alpha v.$$

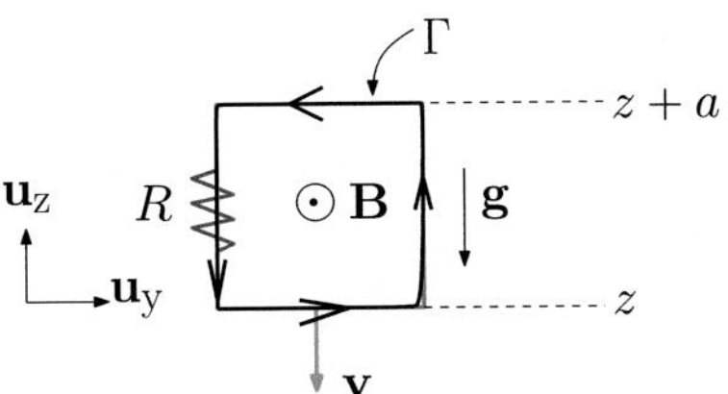

Fig. A.121 Orientation of the circuit Γ

It is due in this case to the electromotive field acting on the two horizontal segments, since for the vertical segments we have $\mathbf{v} \times \mathbf{B} \perp d\mathbf{l}$. Taking the time derivative of the magnetic flux then amounts to applying the $(\mathbf{v} \cdot \nabla)$ operator, which reads

$$- (\mathbf{v} \cdot \nabla)\Phi_{S(\Gamma),\mathbf{B}} = -v\frac{\partial}{\partial z}\Phi_{S(\Gamma),\mathbf{B}} = -\frac{d}{dt}\Phi_{S(\Gamma),\mathbf{B}}.$$

The induced current is

$$I = \frac{\varepsilon}{R} = \frac{a^2 \alpha v}{R},$$

so that for $v > 0$ (falling loop), the current is in the anticlockwise direction when looked at from $x > 0$ and therefore opposes the decrease in magnetic flux through $S(\Gamma)$ (Lenz's law). The net Laplace force $\mathbf{F}_L$ acting on the loop is then the sum of the forces acting on the two horizontal segments (the forces on the two vertical segments cancel each other out). The force on the segment at height $z + a$ is

$$\mathbf{F}_1 = -Ia\mathbf{u}_y \times B(z+a)\mathbf{u}_x = IaB(z+a)\mathbf{u}_z$$

and the force on the segment at height z is

$$\mathbf{F}_2 = Ia\mathbf{u}_y \times B(z)\mathbf{u}_x = -IaB(z)\mathbf{u}_z.$$

The net Laplace force is

$$\mathbf{F}_L = \mathbf{F}_1 + \mathbf{F}_2 = \frac{\alpha^2 a^4 v}{R}\mathbf{u}_z.$$

The equation of motion for the loop becomes

$$m\frac{dv}{dt} = mg - \frac{\alpha^2 a^4 v}{R}$$

and the terminal velocity is reached when $\dfrac{dv}{dt} = 0$, therefore given by

$$v_\infty = R\frac{mg}{\alpha^2 a^4}.$$

Problems of Chap. 12

12.1 Kirchhoff voltage law for a 2-loops RC circuit

We first mark on the circuit the currents i_1 flowing in the $2R$ resistor, i_2 in the capacitor $2C$ and i in the R resistor as shown in Fig. A.122. We use the passive sign convention (currents arrows for i_1 and i_2 are opposite to voltage arrows u_1 and u_2, respectively).

(a) Kirchhoff's Voltage Law (KVL) can be applied to each loop together with Ohm's law for the resistors:

$$\text{Large loop:} \quad u_1 + 2Ri_1 + Ri = E,$$
$$\text{Lower loop:} \quad u_2 + Ri = E,$$
$$\text{Upper loop:} \quad u_1 + 2Ri_1 - u_2 = 0.$$

The third equation is the sum of the first two equations, meaning that in what follows, we will use only two of them (for the lower and upper loops).

When the switch is turned on at $t = 0^+$, the capacitors do not carry any charge and the voltage across their terminals is therefore $u_1(0^+) = u_2(0^+) = 0$. Substituting in the second and third equations, we get $i_1(0^+) = 0$ and $i(0^+) = E/R$. KCL $i = i_1 + i_2$ yields the initial value of the current $i_2(0^+) = i(0^+) = E/R$.

The steady-state values for the currents satisfy $i_1 = dq_1/dt = 0$ and $i_2 = dq_2/dt = 0$ (corresponding to charged capacitors). KCL $i = i_1 + i_2$ yields the steady-state value of the current $i = 0$. Substituting in the second equation, we find $u_2 = E$. Substituting in the third equation, we find $u_1 = u_2 = E$.

(b) The current in a capacity is proportional to the rate of change of the voltage across the capacity. Thus,

$$i_1 = C\frac{du_1}{dt} \quad \text{and} \quad i_2 = 2C\frac{du_2}{dt}.$$

Substituting these relations and the KCL $i = i_1 + i_2$ into the KVLs obtained in question 1, we obtain

Fig. A.122 Definitions of the currents i, i_1 and i_2

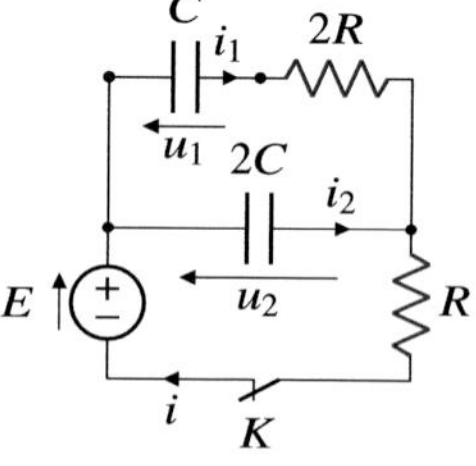

$$u_2 + R\left(C\frac{du_1}{dt} + 2C\frac{du_2}{dt}\right) = E,$$

$$u_1 + 2RC\frac{du_1}{dt} - u_2 = 0.$$

For convenience, we introduce the time constant $\tau = RC$. Calculating u_2 from the second equation and substituting in the third, we obtain a second-order ordinary differential equation (ode) for u_1:

$$4\tau^2 u_1'' + 5\tau u_1' + u_1 = E,$$

where a prime denotes a time derivative. This is a non-homogeneous ode. A particular solution is $u_1 = E$. We will add the general solution to the homogeneous equation. The characteristic polynomial is

$$4\tau^2 r^2 + 5\tau r + 1 = 0, \quad \text{with roots} \quad r_- = -\frac{1}{\tau}, \quad r_+ = -\frac{1}{4\tau}.$$

Therefore the solution is

$$u_1(t) = E + A\exp(r_+ t) + B\exp(r_- t).$$

The integration constants are found from the initial conditions $u_1(0^+) = 0$ and $u_1'(0+) = q_1'(0^+)/2C = i_1(0^+)/2C = 0$. Therefore, A and B satisfy

$$A + B = -E,$$
$$r_+ A + r_- B = 0.$$

The solution is

$$A = \frac{r_-}{r_+ - r_-}E = -\frac{4}{3}E, \qquad B = -\frac{r_+}{r_+ - r_-}E = \frac{1}{3}E.$$

Finally, the solution for $u_1(t)$ reads

$$u_1(t) = E\left(1 - \frac{4}{3}\exp\left(-\frac{t}{4\tau}\right) + \frac{1}{3}\exp\left(-\frac{t}{\tau}\right)\right).$$

As shown in Fig. A.123 for the voltage $u_1(t)$, the circuit response is overdamped.

12.2 Neon lamp

The circuit is shown in Fig. A.124. At $t = 0^+$, the switch is closed. The resistance of the lamp is considered infinite, resulting in zero current through it: $i_2 = 0$. This effectively reduces the circuit to a simple RC loop ($i_1 = i$). Applying Kirchhoff's Voltage Law (KVL) to this loop, we obtain

Fig. A.123 Voltage u_1 as a function of time

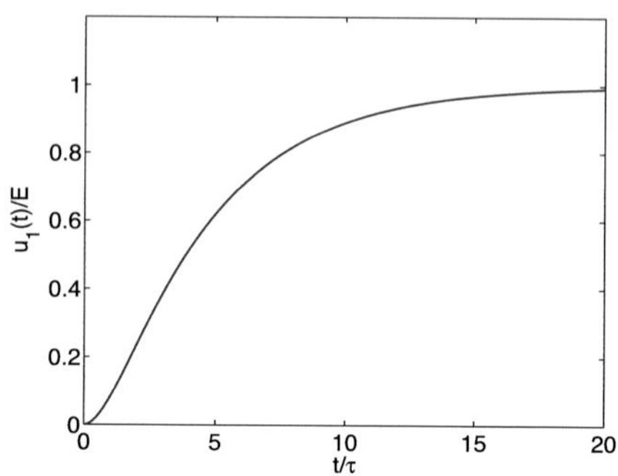

Fig. A.124 Circuit connected to a Neon lamp

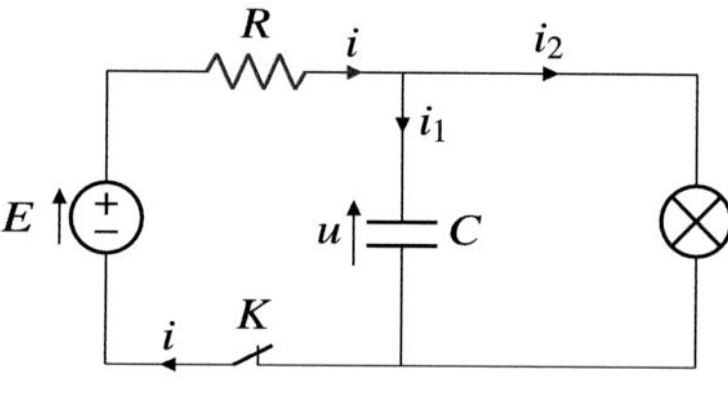

$$Ri + u = E.$$

Substituting the capacitor current-voltage relationship

$$i = C\frac{du}{dt},$$

into the KVL equation yields a first-order linear differential equation

$$RC\frac{du}{dt} + u = E.$$

This is a non-homogeneous ode. Adding a particular solution ($u = E$) to the general solution of the homogeneous equation and determining the integration constant by applying the initial condition $u(0^+) = 0$ (capacitor uncharged), we get

$$u = E(1 - e^{-t/\tau}), \quad \text{where} \quad \tau = RC.$$

The lamp turns on when the voltage across the capacitor, $u(t)$, reaches a threshold voltage, V_i. This occurs at time t_i, which can be determined by solving the equation:

$$V_i = E(1 - e^{-t_i/\tau}).$$

Solving for t_i, we get

$$t_i = \tau \ln(1 - V_a/E)^{-1}.$$

At time $t = t_i^+$, the capacitor is charged to a voltage V_i and its charge is $q = CV_i$. Applying KVL to the two loops, 1-voltage source–resistor–capacitor and 2-voltage source–resistor–lamp, we have

$$\text{(KVL1)} \quad Ri + u = E,$$
$$\text{(KVL2)} \quad Ri + ri_2 = E.$$

Applying Kirchhoff's Current Law (KCL) at the node, we get

$$i = i_1 + i_2.$$

Substituting this relation into (KVL2), we get

$$\text{(KVL2')} \quad Ri_1 + (R + r)i_2 = E.$$

Substituting $i_1 = C\dfrac{du}{dt}$ and $i_2 = \dfrac{u}{r}$ into the (KVL2'), we obtain a first-order linear differential equation for the voltage across the capacitor:

$$RC\frac{du}{dt} + \left(\frac{R}{r} + 1\right)u = E,$$

This can be rewritten as

$$\frac{du}{dt} + \frac{u}{\tau'} = \frac{E}{\tau},$$

where

$$\tau' = \frac{rRC}{R + r}.$$

To find the solution, we add a particular solution to the non-homogeneous equation, $u = E\tau'/\tau = Er/(R + r) \equiv E_\infty$, to the general solution of the homogeneous equation, $u(t) = Ae^{-(t-t_i)/\tau'}$. Applying the initial condition $u(t_i^+) = V_i$, we find the constant $A = V_i - E_\infty$. Therefore, the solution is

$$u(t) = E_\infty + (V_i - E_\infty)\,e^{-(t-t_i)/\tau'}.$$

Depending on the ignition voltage V_i, the behavior of the circuit can be categorized into three cases:

- $E_\infty > V_i$: The voltage $u(t)$ increases from V_i to E_∞ and the lamp remains on.
- $V_e \le E_\infty \le V_i$: The voltage $u(t)$ decreases from V_i toward E_∞. It remains above the extinction threshold since $V_e \le E_\infty$. The lamp remains on.
- $E_\infty \le V_e < V_i$: The voltage $u(t)$ decreases from V_i to V_e, causing the lamp to extinguish before reaching the steady-state value E_∞.

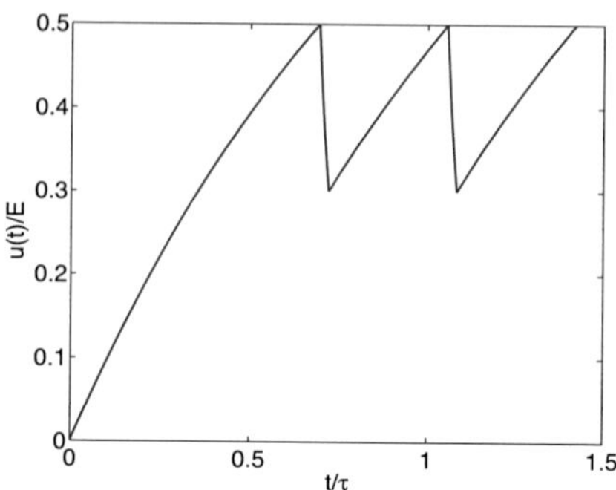

Fig. A.125 Voltage across the lamp as a function of time

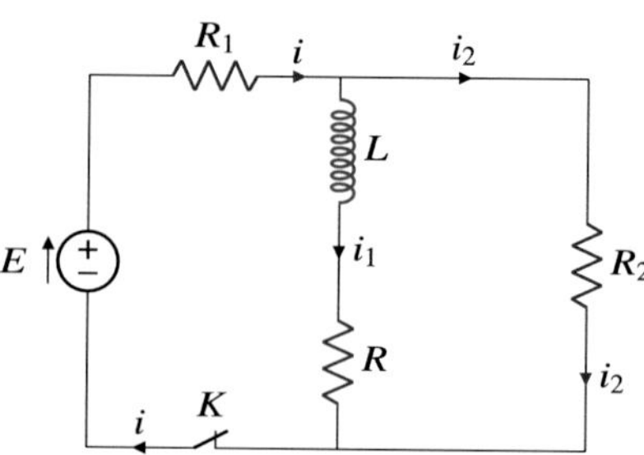

Fig. A.126 Definition of the currents

Assuming the condition $V_e \leq E_\infty \leq V_i$ is fulfilled, the voltage $u(t)$ will reach the extinction voltage V_e at time $t = t_e$. Solving the equation

$$u(t_e) = V_e = E_\infty + (V_a - E_\infty)\, e^{-(t_e - t_a)/\tau'}$$

for t_e, we get

$$t_e - t_a = \tau' \ln\left(\frac{V_i - E_\infty}{V_e - E_\infty}\right),$$

where $\tau' = \frac{rRC}{r+R}$. Once the lamp turns off, its resistance becomes effectively infinite. The circuit returns to a similar state as the initial condition but with an initial voltage across the capacitor of V_e. The voltage will then rise again toward E with a time constant $\tau = RC$. When the voltage reaches the ignition threshold V_i, the lamp turns on again, and the cycle repeats.

As shown in Fig. A.125, if the lamp resistance r is much smaller than the resistor R, then $\tau \gg \tau'$. This implies that the lamp's on-time is significantly shorter than its off-time.

12.3 Kirchhoff's laws in the transient regime

The current i, i_1 and i_2 flow in each branch as indicated in Fig. A.126. Applying KCL to the upper node and KVL to each loop, from left (KVL1) to right (KVL2), we obtain

Fig. A.127 Definition of the currents on the circuit

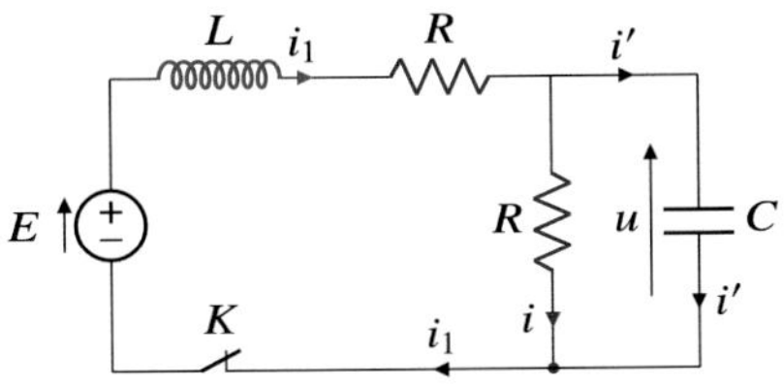

$$\text{(KCL)} \quad i = i_1 + i_2,$$

$$\text{(KVL1)} \quad R_1 i + L\frac{di_1}{dt} + Ri_1 = E,$$

$$\text{(KVL2)} \quad R_1 i + R_2 i_2 = E.$$

Substituting (KCL) into (KVL2), solving for i_2, and using (KCL) to express i, we get

$$i_2 = \frac{E - R_1 i_1}{R_1 + R_2} \quad \text{and} \quad i = \frac{E + R_2 i_1}{R_1 + R_2}.$$

Substituting these equations into (KVL1), we obtain

$$\left(\frac{RR_1 + R_1 R_2 + RR_2}{R_1 + R_2}\right) i_1 + L\frac{di_1}{dt} = \frac{R_2 E}{R_1 + R_2}.$$

This equation is a first-order linear ode of the form

$$\frac{di_1}{dt} + \frac{1}{\tau}i_1 = I_0,$$

where τ and I_0 are a time constant and a current normalization constant, respectively expressed as

$$I_0 = \frac{E R_2}{RR_1 + R_1 R_2 + RR_2} \quad \text{and} \quad \tau = \frac{L(R_1 + R_2)}{RR_1 + R_1 R_2 + RR_2}.$$

Using the initial condition $i(0^+) = 0$, the solution to this equation for the current in the inductor reads

$$i_1(t) = I_0(1 - e^{-t/\tau}).$$

The previously found expression for the current i_2 in R_2 as a function of i_1 allows us to obtain

$$i_2(t) = I_0\left(\frac{R}{R_2} + \frac{R_1}{R_1 + R_2}e^{-t/\tau}\right).$$

The current through R is increasing from 0 to I_0 with a time constant τ. The current through R_2 is decreasing from $E/(R_1 + R_2)$ to $I_0 R/R_2$ with the same time constant.

12.4 **RL and RC circuits in series**

When the switch K is closed, the currents i_1, i', and i flow through the inductor, capacitor and resistor, respectively. They are defined as shown in Fig. A.127. The time constants for the coil-resistor in series and for the resistor-capacitor in parallel are equal, hence

$$\tau = RC = \frac{L}{R}.$$

Applying KCL at the lower node and KVL for the loops on the left (KVL1) and on the right (KVL2) results in the system

$$(\text{KCL}) \quad i_1 = i + i',$$

$$(\text{KVL1}) \quad L\frac{di_1}{dt} + Ri_1 + Ri = E,$$

$$(\text{KVL2}) \quad Ri = u.$$

The voltage-current relationship for the capacitor combined with (KVL2) yields the current

$$i' = C\frac{du}{dt} = RC\frac{di}{dt}.$$

The current i is obtained by rewriting (KVL1) as

$$i = \frac{E}{R} - \tau\frac{di_1}{dt} - i_1.$$

By differentiating this relation with respect to t and substituting in the expression for i', we get

$$i' = -\tau^2\frac{d^2 i_1}{dt^2} - \tau\frac{di_1}{dt}.$$

Substituting the expressions for i and i' into (KCL), we obtain

$$i_1 = i + i' = \frac{E}{R} - \tau\frac{di_1}{dt} - i_1 - \tau^2\frac{d^2 i_1}{dt^2} - \tau\frac{di_1}{dt}.$$

This can be rewritten as a non-homogeneous second ode in i_1, using prime notations for time derivatives:

$$i_1'' + \frac{2}{\tau}i_1' + \frac{2}{\tau^2}i_1 = \frac{2}{\tau^2}I_0,$$

where $I_0 = \frac{E}{2R}$ is the steady-state current (particular solution to the ode). The roots of the characteristic polynomial are

$$r_\pm = -\frac{1}{\tau} \pm \frac{i}{\tau}.$$

The solution is expressed as the sum of the particular solution $i_1 = I_0$ and the general solution to the homogeneous equation, $i_{1,h}(t) = (A\cos vt + B\sin vt)e^{-vt}$, where

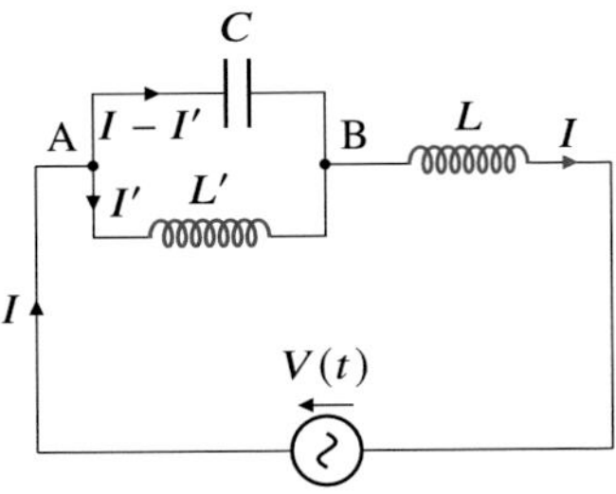

Fig. A.128 LC circuit and definition of the currents I and I'

$\nu = \tau^{-1}$ and the constants A and B are determined by initial conditions $i_1(0) = 0$ (continuity of current through the inductor) and $i_1'(0) = 0$ (continuity of voltage across the coil). The solution is expressed as

$$i_1(t) = I_0 - I_0 e^{-t/\tau} \left(\cos \frac{t}{\tau} - \sin \frac{t}{\tau} \right).$$

This shows that the current through the coil increases with a time constant τ.

12.5 AC Analysis of an LC circuit

As shown in Fig. A.128, let I and I' denote the currents flowing through the inductors L and L', respectively. We will use the same notation for the corresponding complex currents. Applying KCL at node A, the current flowing through the capacitor is $I - I'$. The inductor L' and capacitor C form a parallel combination. Their equivalent impedance, $Z_{L'\|C}$, is given by

$$Z_{L'\|C} = \frac{Z_{L'} Z_C}{Z_{L'} + Z_C} = \frac{j L' \omega}{j \left(\omega(j L' \omega + \frac{1}{jC\omega}) \right)} = \frac{j L' \omega}{1 - L' C \omega^2}.$$

This impedance is in series with the inductor L. Therefore, the total impedance of the circuit is

$$Z_{\text{tot}} = Z_{L'\|C} + Z_L = \frac{j L' \omega}{1 - L' C \omega^2} + j L \omega.$$

The current I through the inductor L can be calculated using Ohm's law:

$$I = \frac{V}{Z_{\text{tot}}} = \frac{V (1 - L' C \omega^2)}{j\omega[L' + L(1 - L' C \omega^2)]}.$$

The voltage difference $V_A - V_B$ can be calculated using either the capacitor branch or the inductor L' branch:

$$V_A - V_B = j L' \omega I' = \frac{1}{jC\omega}(I - I').$$

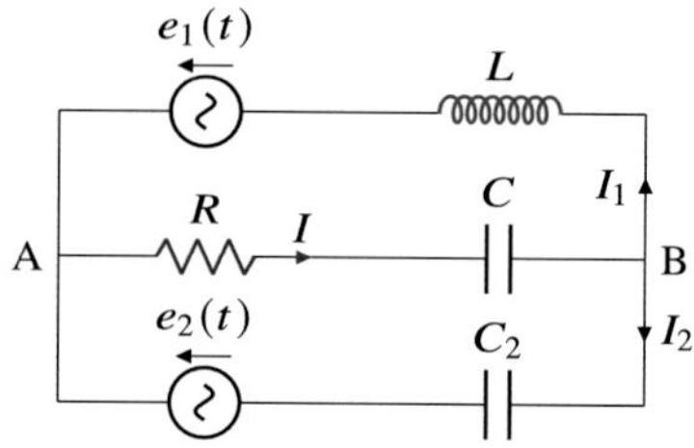

Fig. A.129 RLC circuit and definition of the currents I, I_1 and I_2

Rearranging this equation, we get

$$I = (1 - L'C\omega^2)I'.$$

If $L'C\omega^2 = 1$, then $I = 0$ and $I' = V/(jL'\omega) = jC\omega V$.
If $L'C\omega^2 \neq 1$,

$$I' = \frac{V}{j\omega[L' + L(1 - L'C\omega^2)]}.$$

Using real notations:

$$v = \mathrm{Re}(Ve^{j\omega t}) = V\cos\omega t \quad \text{and} \quad i' = \mathrm{Re}(I'e^{j\omega t}) = |I'|\cos(\omega t - \phi'),$$

where V is the peak voltage of the generator (a positive real quantity). We can determine the phase ϕ' from the complex expression for I',

$$I' = |I'|e^{-j\phi'} = e^{-j\pi/2}\frac{V}{\omega|L' + L(1 - L'C\omega^2)|}.$$

Thus, $\phi' = \pi/2$. If $L'C\omega^2 = 1$, then,

$$I' = |I'|e^{-j\phi'} = e^{-j\pi/2}\frac{V}{L'\omega}.$$

Again, we find that $\phi' = \pi/2$.

12.6 Current in an RLC circuit with two AC generators

Let E_1 and E_2 represent the complex amplitudes of the voltage sources e_1 and e_2, respectively. Similarly, let I, I_1 and I_2 represent the complex amplitudes of the currents as shown in Fig. A.129.

Applying KCL at node B and KVL to each loop yields

$$\begin{aligned}
\text{(KCL)} \quad & I = I_1 + I_2, \\
\text{(KVL1)} \quad & (Z_C + R)I + Z_L I_1 = E_1, \\
\text{(KVL2)} \quad & (Z_C + R)I + Z_{C_2} I_2 = E_2.
\end{aligned}$$

An equation for the current I in the resistance is obtained by $Z_{C_2} \times (\text{KVL1}) + Z_L \times (\text{KVL2})$:

$$(Z_C + R)(Z_{C_2} + Z_L)I = Z_{C_2}E_1 + Z_L E_2 - Z_{C_2}Z_L(I_1 + I_2).$$

Substituting $I_1 + I_2$ (KCL) in this equation and solving in I, we obtain

$$I = \frac{Z_{C_2}E_1 + Z_L E_2}{(Z_C + R)(Z_{C_2} + Z_L) + Z_{C_2}Z_L}.$$

The complex amplitudes E_1 and E_2 are determined by the following relationships:

$$e_1(t) = |E_1|\cos \omega t = \text{Re}\left(E_1 e^{j\omega t}\right),$$
$$e_2(t) = |E_2|\cos \left(\omega t + \frac{\pi}{4}\right) = \text{Re}\left(E_2 e^{j\omega t}\right).$$

Since $e_1(t)$ is chosen as the phase reference, its complex amplitude E_1 is a real number: $E_1 = |E_1|$. For the second voltage source $e_2(t)$, the complex amplitude E_2 is found by comparing the phasor representations:

$$|E_2|e^{j(\omega t + \pi/4)} \equiv E_2 e^{j\omega t}.$$

This implies that $E_2 = |E_2|e^{j\pi/4}$.

We can now substitute the impedances with their expressions in terms of ω into the equation for the current I:

$$I = \frac{\dfrac{|E_1|}{jC_2\omega} + jL\omega|E_2|e^{j\pi/4}}{\left(\dfrac{1}{jC\omega} + R\right)\left(\dfrac{1}{jC_2\omega} + jL\omega\right) + \dfrac{L}{C_2}} = \frac{\left(|E_1| - LC_2\omega^2|E_2|e^{j\pi/4}\right)jC\omega}{1 - L(C + C_2)\omega^2 + jRC\omega(1 - LC_2\omega^2)}.$$

The values for the quantities appearing in this expression are

$L\omega = 2 \times 10^3\,\Omega$, $C\omega = 5 \times 10^{-4}\,\Omega^{-1}$, $C_2\omega = 1 \times 10^{-3}\,\Omega^{-1}$, $R = 1 \times 10^3\,\Omega$,
$LC_2\omega^2 = 2$, $\quad L(C + C_2)\omega^2 = 3$, $\quad RC\omega = 5 \times 10^{-1}$, $\quad L\omega/R = 2$,
$|E_1| = 2\,\text{V}$, $\quad |E_2|e^{j\pi/4} = (1 + j)\,\text{V}$

Hence,

$$I = -\frac{j}{j + 1/4} \times 5 \times 10^{-4}\,\text{A} = \frac{2}{\sqrt{17}}e^{j\phi}\,\text{mA},$$

where the phase of the current, ϕ, defined by the relation

$$i(t) = \text{Re}(I) = I_0\cos(\omega t + \phi),$$

is determined by $e^{j\phi} = (-4 + j)/\sqrt{17}$, rewritten as

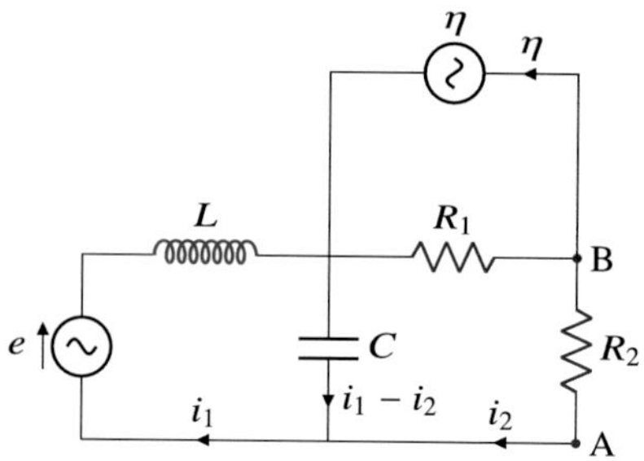

Fig. A.130 Definition of the currents i_1 and i_2

$$\tan\phi = -\frac{1}{4}, \quad \text{and} \quad \cos\phi < 0.$$

Solving in ϕ, we get

$$\phi = 166\,\text{deg} = 2.9\,\text{rad}.$$

The current amplitude is

$$I_0 = |I| = \frac{2}{\sqrt{17}}\,\text{mA} \sim 0.48\,\text{mA}.$$

12.7 Application of Thévenin's or Norton's theorem

We first give a solution using only Kirchhoff's laws.

If i_1 and i_2 are the currents indicated in Fig. A.130, applying KCL yields the current through the capacitor, $i_1 - i_2$ and the current through resistor R_1: $i_2 + \eta$ (KCL at node B). Applying KVL to each loop yields

$$\text{(KVL1)} \quad jL\omega\,i_1 + \frac{1}{jC\omega}(i_1 - i_2) = e,$$

$$\text{(KVL2)} \quad \frac{1}{jC\omega}(i_2 - i_1) + R_1(i_2 + \eta) + R_2 i_2 = 0.$$

Multiplying each equation by $jC\omega$, this system can be rewritten as

$$\begin{pmatrix} 1 - LC\omega^2 & -1 \\ -1 & 1 + j(R_1 + R_2)C\omega \end{pmatrix} \begin{pmatrix} i_1 \\ i_2 \end{pmatrix} = jC\omega \begin{pmatrix} e \\ -\eta R_1 \end{pmatrix}.$$

The current through R_2 reads

$$i_2 = \frac{jC\omega \begin{vmatrix} 1 - LC\omega^2 & e \\ -1 & -R_1\eta \end{vmatrix}}{\begin{vmatrix} 1 - LC\omega^2 & -1 \\ -1 & 1 + j(R_1 + R_2)C\omega \end{vmatrix}} = jC\omega\,\frac{-\eta R_1(1 - LC\omega^2) + e}{(1 - LC\omega^2)(1 + j(R_1 + R_2)C\omega) - 1}.$$

which can be simplified into

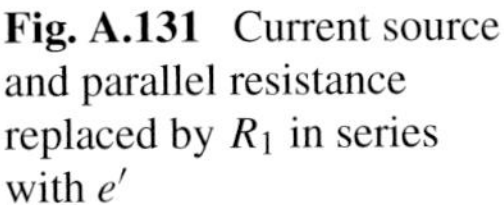
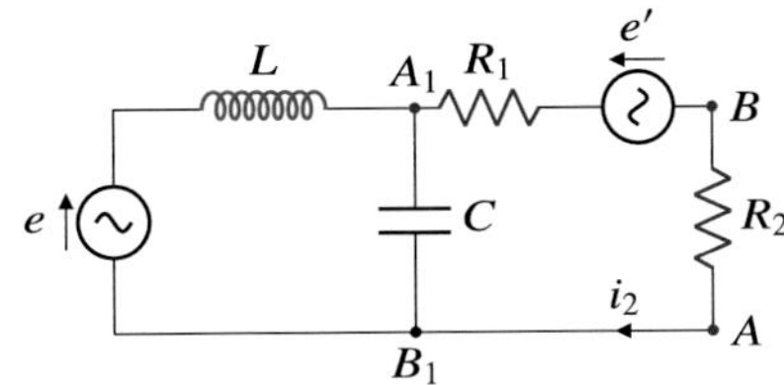

Fig. A.131 Current source and parallel resistance replaced by R_1 in series with e'

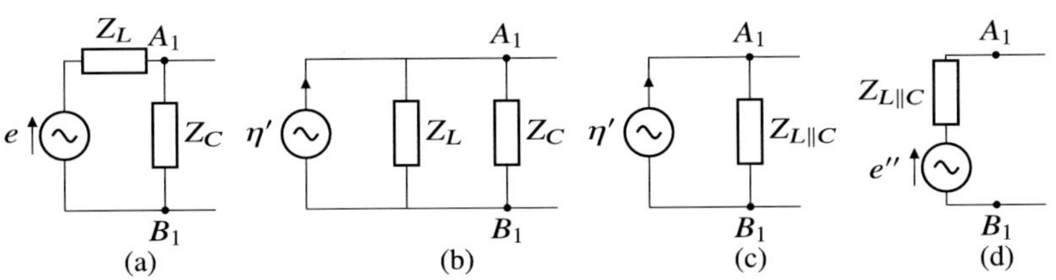

Fig. A.132 **a–d** Successive transformation of the first loop to find the equivalent voltage source between nodes A_1 and B_1. Impedance values: $Z_L = jL\omega$; $Z_C = (jC\omega)^{-1}$. Current source $\eta' = Z_C e$. Voltage source $e'' = Z_{L\|C}\eta'$

$$i_2 = \frac{e - \eta R_1(1 - LC\omega^2)}{(R_1 + R_2)(1 - LC\omega^2) + jL\omega}$$

and the voltage $V_B - V_A$ is obtained using Ohms law

$$V_B - V_A = R_2 i_2 = \frac{R_2[e - \eta R_1(1 - LC\omega^2)]}{(R_1 + R_2)(1 - LC\omega^2) + jL\omega}.$$

We give now a different solution that uses the Thévenin and Norton theorems. We first replace the current source and the parallel resistance R_1 by its equivalent voltage source $e' = R_1\eta$ in series with the resistance R_1 (Thévenin's theorem), as shown in Fig. A.131. Then we will replace the first loop between terminals A_1 and B_1 (including the capacity) with its equivalent source and impedance.

We first open the circuit and consider the loop on the left of nodes A_1 and B_1. Figure A.132a shows this loop with the inductor and capacitor represented by bipoles with their corresponding impedance. We will transform this circuit into an equivalent voltage source by applying Thévenin's and Norton's theorems. In this aim, we apply successive transformations of this circuit from Fig. A.132a to d.

- From Fig. A.132a to b, we replace the voltage generator e in series with the impedance $Z_L = jL\omega$ by a current source (Norton's theorem) delivering the equivalent Norton current

$$\eta' = \frac{e}{Z_L}$$

and the equivalent impedance Z_L is now in parallel.

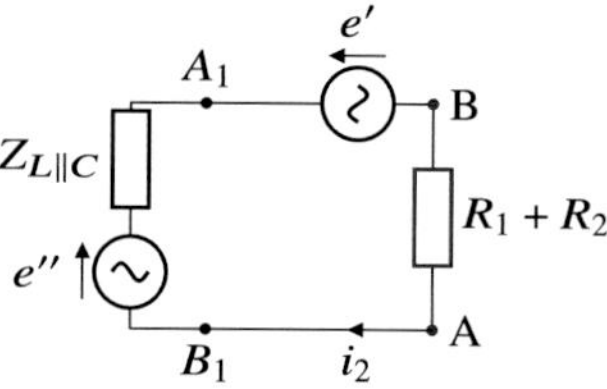

Fig. A.133 Equivalent circuit simplified by using Thévenin and Norton's theorems

- From Fig. A.132b to c, we replace the parallel connection of the impedances Z_C and Z_L by a single equivalent impedance $Z_{L\|C}$ branches carrying the coil and the capacity by a single impedance Z', given by

$$Z_{L\|C}^{-1} = \frac{1}{jL\omega} + jC\omega, \quad \text{i.e.,} \quad Z_{L\|C} = \frac{jL\omega}{1 - LC\omega^2}.$$

- From Fig. A.132c to d, we use Thévenin's theorem and replace the current source η' and the impedance $Z_{L\|C}$ by a voltage source $e'' = Z_{L\|C}\eta'$, with the impedance $Z_{L\|C}$ connected in series. Using the expressions for $Z_{L\|C}$ and η', we find

$$e'' = \frac{e}{1 - LC\omega^2}.$$

We can now connect again the loop on the right of nodes A_1 and B_1. Swapping the resistor R_1 and the voltage source e', this loop is equivalent to the series connection of the voltage source e' with a single resistor $R_1 + R_2$, equivalent to the series connection of R_1 and R_2. This is shown in Fig. A.133.

The current flowing in resistor R_2 flows in the entire loop. Applying KVL to this loop, we get

$$(Z_{L\|C} + R_1 + R_2)i_2 = e'' - e' = 0,$$

Therefore, solving for i_2 and substituting the expressions for e'' and e', we can express the current i_2 in resistor R_2 as

$$i_2 = \frac{e'' - e'}{Z_{L\|C} + R_1 + R_2} = \frac{e - \eta R_1(1 - LC\omega^2)}{(R_1 + R_2)(1 - LC\omega^2) + jL\omega},$$

We retrieve the same expression as previously for the voltage across R_2: $V_B - V_A = R_2 i_2$.

Given the numerical data, we obtain

$$V_B - V_A = 100\frac{(10 - 0.1 \times 50(1 - 150/75))}{150 \times (-1) + j150} = 10\frac{-1 - j}{2} = \frac{10}{\sqrt{2}}e^{-j3\pi/4}.$$

In real notations

$$v_B - v_A = 10\cos(\omega t - 3\pi/4).$$

Fig. A.134 High-pass RC circuit

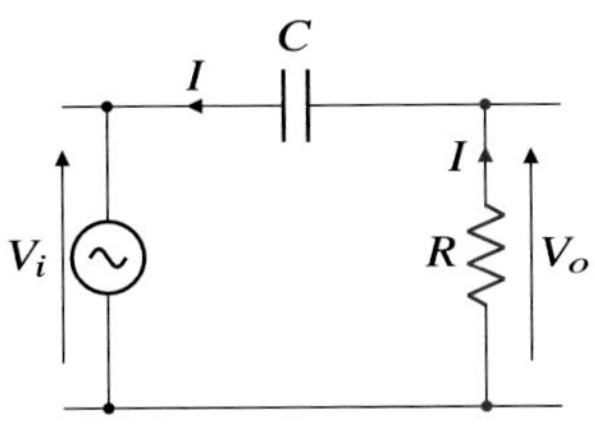

12.8 **High-pass RC filter**

(a) As shown in Fig. A.134, let I, V_i, and V_o represent the complex amplitudes of the current through the capacitor, the input voltage and the output voltage, respectively. Applying Ohm's law, we can express the input and output voltages are expressed as

$$V_i = (R + Z_C)I,$$
$$V_o = RI.$$

The transfer function, defined as the ratio of the output voltage to the input voltage, can be obtained using voltage division:

$$H = \frac{V_o}{V_i} = \frac{R}{R + Z_C} = \frac{1}{1 + 1/(jRC\omega)}.$$

Introducing the notation $\omega_0 = 1/RC$, the transfer function becomes

$$H = \frac{1}{1 - j\omega_0/\omega}.$$

(b) The transfer function can be expressed in polar form as

$$H(j\omega) = G(\omega)\exp(j\phi(\omega)),$$

where the magnitude (gain) $G(\omega)$ and phase $\phi(\omega)$ are given by

$$G(\omega) = \frac{1}{\sqrt{1 + (\omega_0/\omega)^2}}, \quad \phi(\omega) = \arctan(\omega_0/\omega).$$

These functions are plotted in Fig. A.135.

(c) The gain can be rewritten using the variable $x = \omega/\omega_0$:

$$G = \frac{1}{\sqrt{1 + (1/x)^2}} = \frac{x}{\sqrt{1 + x^2}}.$$

The gain in decibels is given by

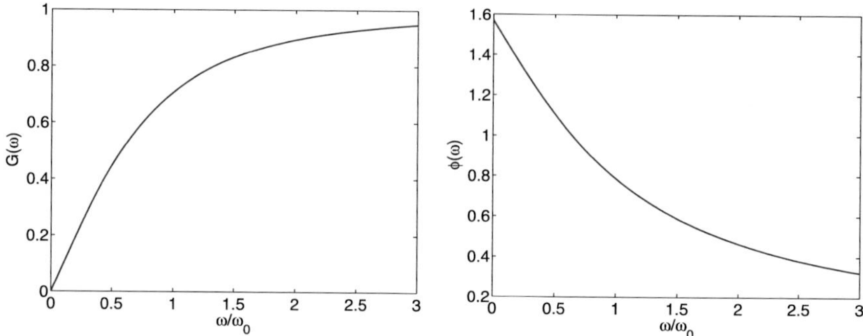

Fig. A.135 Gain (left) and phase (right) of the transfer function

Fig. A.136 Gain in decibels

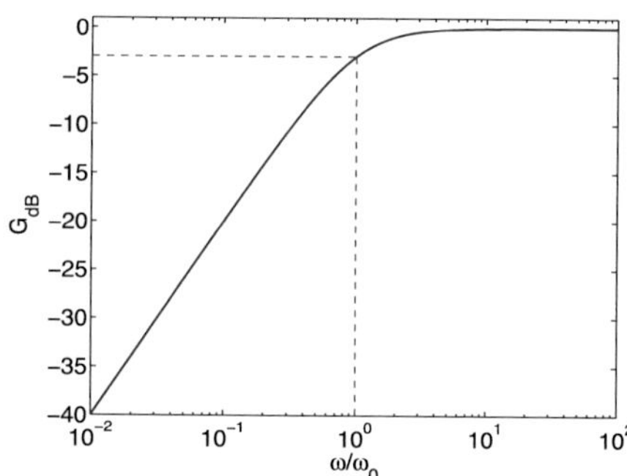

$$G_{\mathrm{dB}} = 20 \log_{10} G = 20 \log_{10} x - 10 \log_{10}(1 + x^2).$$

Consider the asymptotic behavior:

- Low-frequency limit ($x \ll 1$): $G \approx x$ and $G_{\mathrm{dB}} \approx 20 \log x$. This corresponds to a straight line with a slope of $20\,\mathrm{dB/decade}$.
- High-frequency limit ($x \gg 1$): $G \approx 1$, hence $G_{\mathrm{dB}} \approx 0$ dB. This corresponds to a horizontal line.

These two asymptotes intersect at $x = 1$, where $G = 0$ dB (see Fig. A.136). This RC circuit behaves as a high-pass filter. It attenuates low-frequency signals and passes high-frequency signals with minimal attenuation and phase shift.

(d) The cutoff frequency is defined as the frequency at which the gain is reduced by 3 dB ($G_{\mathrm{dB}} = -3$ dB), or equivalently, the magnitude of the transfer function is reduced by a factor of $\sqrt{2}$ ($G(\omega_c) = G_{\max}/\sqrt{2}$):

$$\frac{1}{\sqrt{1 + (1/x)^2}} = \frac{1}{\sqrt{2}}.$$

Solving for x, we get $x = 1$, or $\omega_c = \omega_0 = 1/RC$. With the given numerical values, the cutoff frequency is $f_c = \omega_c/(2\pi) \approx 796$ Hz. For frequencies greater than 796 Hz, the gain is larger than $1/\sqrt{2}$.

(e) To achieve a 10 dB attenuation, the gain must be

$$G_{\text{dB}} = 20 \log_{10} G = -10\,\text{dB},$$

This implies $G = 0.316$ ($\log_{10} G = -0.5$). Solving for x:

$$\frac{1}{\sqrt{1 + x^{-2}}} = 0.316.$$

We find $x \approx 1/3$ or

$$x = \frac{\omega}{\omega_0} = \frac{2\pi f}{RC} \sim \frac{1}{3}, \quad \text{thus,} \quad R = \frac{1}{6\pi f C} = 2.7\,\text{k}\Omega.$$

(f) As illustrated in Fig. A.137, the calculation of the transfer function presented in the response to the first question remains valid, regardless of whether the resistor R' is present or not.

Therefore, the transfer function is the same as before:

$$H = \frac{1}{1 + 1/jRC\omega} = \frac{1}{1 - j\omega_0/\omega}.$$

(g) The complex input impedance $Z_i = V_i/I_i$ can be determined by considering the parallel combination of R' and the series combination of R and Z_C:

$$Z_i = \frac{R'(R + Z_C)}{R' + R + Z_C} = R'\frac{1 + jRC\omega}{1 + j(R + R')C\omega}.$$

The magnitude of the input impedance is

$$|Z_i| = R'\sqrt{\frac{1 + R^2 C^2 \omega^2}{1 + (R + R')^2 C^2 \omega^2}},$$

This magnitude is plotted as a function of ω/ω_0 in Fig. A.138

12.9 Second order filter

The circuit is shown in Fig. A.139.

(a) The current I through the series combination of the inductor and capacitor is the same as the current through the resistor. Using Ohm's law, the input and output voltages are

$$V_i = (Z_L + Z_C + R)I, \quad V_o = RI,$$

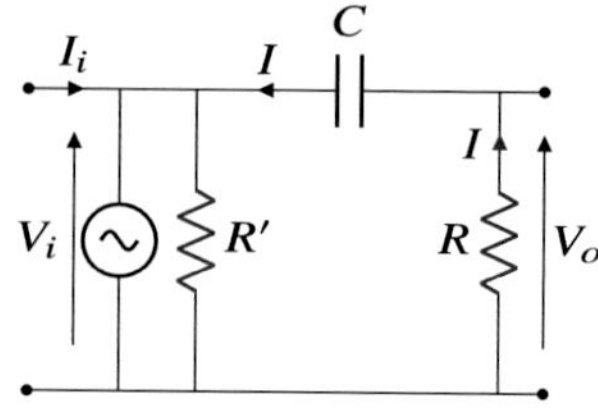

Fig. A.137 A resistor R' is connected in parallel to the input source

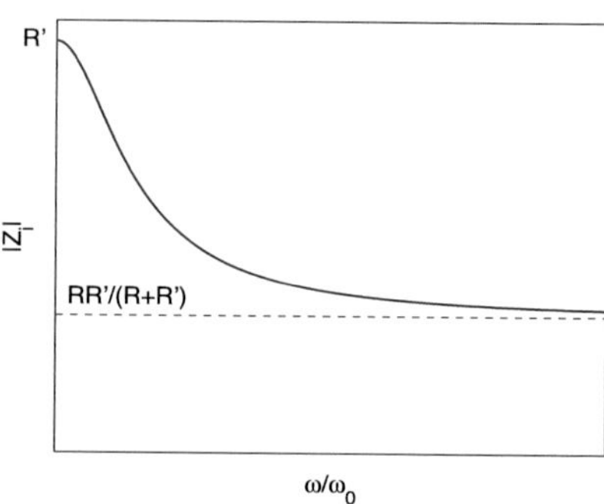

Fig. A.138 Magnitude of the input impedance

Therefore, the transfer function is

$$H(j\omega) = \frac{V_o}{V_i} = \frac{R}{Z_L + Z_C + R} = \frac{R}{R + j\left(L\omega - \dfrac{1}{C\omega}\right)}.$$

The magnitude of the transfer function (gain) is given by

$$G = \frac{R}{\sqrt{R^2 + \left(L\omega - \dfrac{1}{C\omega}\right)^2}}.$$

Introducing the resonant frequency $\omega_0 = 1/\sqrt{LC}$ and the quality factor $Q = \dfrac{L\omega_0}{R} = \dfrac{1}{RC\omega_0}$, we can rewrite the gain as

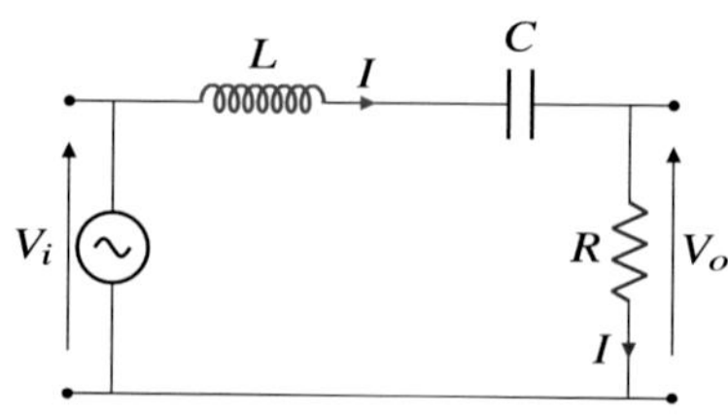

Fig. A.139 Second order filter

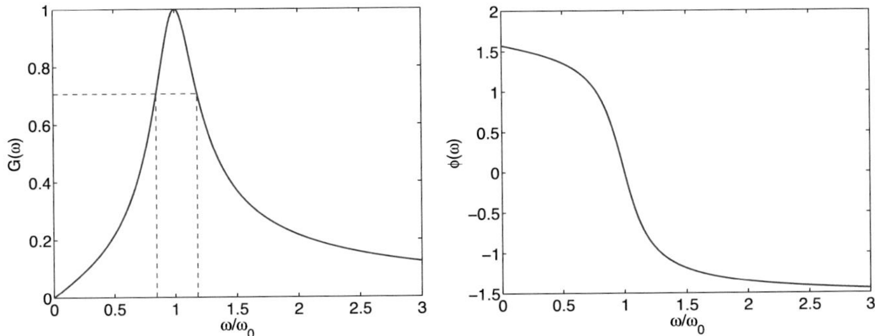

Fig. A.140 Gain (left) and phase (right) of the transfer function

$$G(x) = \frac{1}{1 + Q^2 \left(x - \dfrac{1}{x} \right)^2},$$

where $x = \omega/\omega_0$. The phase of the transfer function is given by

$$\phi(x) = -\arctan\left(\frac{L\omega - \dfrac{1}{C\omega}}{R} \right) = -\arctan\left(Q\left(x - \frac{1}{x} \right) \right).$$

(b) The gain is presented as a function of frequency in Fig. A.140. The dashed lines intersect the gain curve at $G = 1/\sqrt{2}$.

(c) The gain in decibels is expressed as

$$G_{\text{dB}} = 20\log_{10} G = -10\log_{10}\left(1 + Q^2 \left(x - \frac{1}{x} \right)^2 \right).$$

Consider the asymptotic behavior:

- At $x = 1$, $G_{\text{dB}} = 0\,\text{dB}$.
- For $x \ll 1$, $G_{\text{dB}} \approx -20\log_{10} Q + 20\log_{10} x$ (indicating a straight line with a slope of $+20\,\text{dB/decade}$).
- For $x \gg 1$, $G_{\text{dB}} \approx -20\log_{10} Q - 20\log_{10} x$ (indicating a straight line with a slope of $-20\,\text{dB/decade}$).

(d) The gain is maximum (equal to 1) at $\omega = \omega_0$. Both low and high frequencies are attenuated. This is a band-pass filter. Transmitted frequencies are in phase with the input signal.

12.10 T-Π transformation and filter

The two quadrupoles are shown in Fig. A.141.

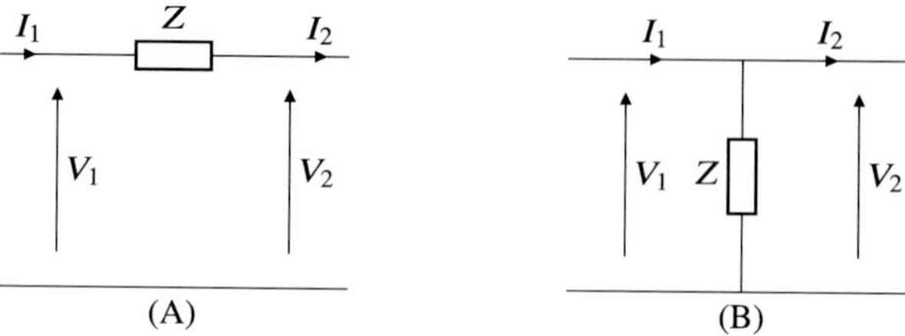

Fig. A.141 Two quadrupoles

(a) For quadrupole (A), it's clear that $I_1 = I_2$. Applying Ohm's law to the impedance, we have $V_1 - V_2 = ZI_1$, or equivalently, $V_2 = V_1 - ZI_1$.

For quadrupole (B), $V_1 = V_2$. Applying Kirchhoff's Current Law (KCL) to the node, we get $I_1 - I_2$ as the current through the impedance. Ohm's law then gives $V_1 = Z(I_1 - I_2)$, or $I_2 = I_1 - V_1/Z$.

Expressing these relationships in matrix form, we obtain

$$M_{(A)} = \begin{pmatrix} 1 & -Z \\ 0 & 1 \end{pmatrix}, \qquad M_{(B)} = \begin{pmatrix} 1 & 0 \\ -Z^{-1} & 1 \end{pmatrix}.$$

(b) Quadrupole (T) is equivalent to connecting three quadrupoles (A), (B), and (A) in series. Therefore, its matrix representation is

$$M_{(T)} = M_{(A)}(Z_2) M_{(B)}(Z_3) M_{(A)}(Z_1)$$
$$= \begin{pmatrix} 1 & -Z_2 \\ 0 & 1 \end{pmatrix} \begin{pmatrix} 1 & 0 \\ -Z_3^{-1} & 1 \end{pmatrix} \begin{pmatrix} 1 & -Z_1 \\ 0 & 1 \end{pmatrix}$$
$$= Z_3^{-1} \begin{pmatrix} Z_3 + Z_2 & -(Z_1 Z_3 + Z_2 Z_3 + Z_1 Z_2) \\ -1 & Z_3 + Z_1 \end{pmatrix}.$$

Similarly, quadrupole (Π) is equivalent to connecting three quadrupoles (B), (A), and (B) in series. Its matrix representation is

$$M_{(\Pi)} = M_{(B)}(Z_1') M_{(A)}(Z_3') M_{(B)}(Z_2')$$
$$= \begin{pmatrix} 1 & 0 \\ -1/Z_1' & 1 \end{pmatrix} \begin{pmatrix} 1 & -Z_3' \\ 0 & 1 \end{pmatrix} \begin{pmatrix} 1 & 0 \\ -1/Z_2' & 1 \end{pmatrix}$$
$$= (Z_1' Z_2')^{-1} \begin{pmatrix} Z_1'(Z_2' + Z_3') & -Z_1' Z_2' Z_3' \\ -(Z_2' + Z_1' + Z_3') & Z_2'(Z_1' + Z_3') \end{pmatrix}.$$

(c) To determine the conditions for equivalence between quadrupoles (T) and (Π), we equate the corresponding elements of their respective matrices:

$$M_{(T)} = M_{(\Pi)}.$$

This leads to the following relations:

$$\frac{Z_2}{Z_3} = \frac{Z_3'}{Z_2'}, \qquad Z_3' = \frac{Z_1 Z_2 + Z_1 Z_3 + Z_2 Z_3}{Z_3},$$

$$\frac{1}{Z_3} = \frac{Z_1' + Z_2' + Z_3'}{Z_1' Z_2'}, \qquad \frac{Z_1}{Z_3} = \frac{Z_3'}{Z_1'}.$$

These equations can be rearranged to express the impedances of (Π) in terms of the impedances of (T), or vice versa:

These relations allow us to express the impedances for the (Π) quadrupole as functions of the impedances for the (T) quadrupole, or vice versa. Thus we obtain two sets of three relations:

- (T) to (Π):

$$Z_1' = \frac{Z_1 Z_3 + Z_2 Z_3 + Z_1 Z_2}{Z_1}, \, Z_2' = \frac{Z_1 Z_3 + Z_2 Z_3 + Z_1 Z_2}{Z_2}, \, Z_3' = \frac{Z_1 Z_3 + Z_2 Z_3 + Z_1 Z_2}{Z_3}.$$

- (Π) to (T):

$$Z_1 = \frac{Z_2' Z_3'}{Z_1' + Z_2' + Z_3'}, \quad Z_2 = \frac{Z_1' Z_3'}{Z_1' + Z_2' + Z_3'}, \quad Z_3 = \frac{Z_1' Z_2'}{Z_1' + Z_2' + Z_3'}.$$

(d) We seek solutions of the form

$$\begin{pmatrix} V_2 \\ I_2 \end{pmatrix} = \lambda \begin{pmatrix} V_1 \\ I_1 \end{pmatrix}, \quad \text{i.e.,} \quad \mathcal{M}_{(T)} \begin{pmatrix} V_1 \\ I_1 \end{pmatrix} = \lambda \begin{pmatrix} V_1 \\ I_1 \end{pmatrix},$$

where $\mathcal{M}_{(T)}$ is the matrix representation of the (T)-type quadrupole. Substituting the impedances $Z_1 = Z_2 = \dfrac{jL\omega}{1 - LC\omega^2}$ (parallel association of a coil and a capacity) and $Z_3 = \dfrac{1}{4jC\omega} + \dfrac{jL\omega}{4}$ (series association), we obtain

$$\begin{pmatrix} 1 + \dfrac{Z_1}{Z_3} - \lambda & -Z_1 \left(2 + \dfrac{Z_1}{Z_3} \right) \\ -Z_3^{-1} & 1 + \dfrac{Z_1}{Z_3} - \lambda \end{pmatrix} \begin{pmatrix} V_1 \\ I_1 \end{pmatrix} = 0.$$

This is an eigenvalue problem, where λ is an eigenvalue and $(V_1, I_1)^T$ is the corresponding eigenvector. To find non-trivial solutions for $(V_1, I_1)^T$, the determinant of the coefficient matrix must be zero:

$$\left(1 + \frac{Z_1}{Z_3} - \lambda \right)^2 - \frac{Z_1}{Z_3} \left(2 + \frac{Z_1}{Z_3} \right) = 0.$$

This leads to a quadratic equation in λ:

$$\lambda^2 - 2\left(1 + \frac{Z_1}{Z_3}\right)\lambda + 1 = 0.$$

The solutions to this equation are

$$\lambda_{\pm} = -\left(1 + \frac{Z_1}{Z_3}\right) \pm \sqrt{\left(1 + \frac{Z_1}{Z_3}\right)^2 - 1},$$

where $Z_1 = (jL\omega) \,\|\, (jC\omega)^{-1}$ and $Z_3 = (j4C\omega)^{-1} + (jL\omega/4)$, with a real negative ratio of the complex impedances given by

$$\frac{Z_1}{Z_3} = -\frac{4LC\omega^2}{(1 - LC\omega^2)^2}.$$

For the signal to be transmitted without attenuation, $|\lambda|$ must be equal to 1. The quadratic equation in λ indicates that $\lambda_+\lambda_- = 1$. Therefore, we must have two complex conjugate roots. This implies that the discriminant must be negative. In contrast, the signal is attenuated if the discriminant is positive:

$$\delta = \left(1 + \frac{Z_1}{Z_3}\right)^2 - 1 > 0,$$

or equivalently: $-2 < Z_1/Z_3 < 0$. Substituting Z_1/Z_3, we obtain

$$\frac{4LC\omega^2}{(1 - LC\omega^2)^2} < 2.$$

This inequality is satisfied for

$$X^2 - 4X + 1 > 0,$$

where $X = LC\omega^2$. Solving this quadratic inequality, we find that the filter attenuates frequencies in the band: $2 - \sqrt{3} < X < 2 + \sqrt{3}$, or equivalently:

$$(2 - \sqrt{3})^{1/2}(LC)^{-1/2} < \omega < (2 + \sqrt{3})^{1/2}(LC)^{-1/2}.$$

Therefore, this is a band-stop filter that blocks frequencies within this band and allows frequencies outside this band to pass.

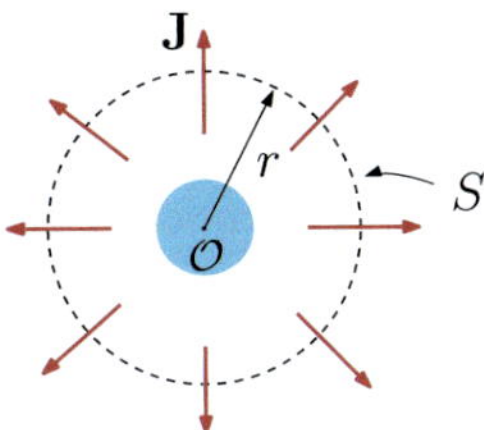

Fig. A.142 A sphere S concentric to the charged sphere

Problems of Chap. 13

13.1 Charged sphere leaking charge radially

(a) Consider a spherical surface S of radius r concentric to the charged sphere, as shown in Fig. A.142.

The charge conservation law states that the current through S equals the rate at which the charge $Q(t)$ contained inside S decreases in time. This reads

$$\oiint_S \underbrace{\mathbf{j}(r,t)}_{J(r,t)\mathbf{u}_r} \cdot \underbrace{\mathbf{n}}_{\mathbf{u}_r}\, dS = 4\pi r^2 j(r,t) = -\frac{dQ(t)}{dt},$$

so that

$$\mathbf{j}(r,t) = -\frac{1}{4\pi r^2}\frac{dQ(t)}{dt}\mathbf{u}_r.$$

(b) To determine the electric field, we can use Gauss's law with a spherical surface of radius r. Due to the spherical symmetry and mirror properties of the charge distribution, we deduce that the electric field writes, in spherical coordinates, as $\mathbf{E} = E(r,t)\,\mathbf{u}_r$ (these symmetry arguments, valid only in the static case, remain valid here since the magnetic field is zero, as will be shown later). Then,

$$\oiint_S \mathbf{E}(r,t) \cdot \mathbf{n}dS = 4\pi r^2 E(r,t) = \frac{Q(S)}{\epsilon_0},$$

where the charge $Q(S)$ contained in S is, simply, $Q(S) = Q(t)$. The electric field is then given by

$$\mathbf{E}(r,t) = \frac{Q(t)}{4\pi\epsilon_0 r^2}\mathbf{u}_r$$

and the displacement current is

$$\mathbf{j}_D = \epsilon_0\frac{\partial\mathbf{E}}{\partial t} = \frac{1}{4\pi r^2}\frac{dQ(t)}{dt}\mathbf{u}_r = -\mathbf{j}.$$

(c) Maxwell–Ampère law reads

$$\nabla \times \mathbf{B} = \mu_0(\mathbf{j} + \mathbf{j}_D) = \mathbf{0}.$$

In addition, since $\nabla \cdot \mathbf{B} = 0$, Helmholtz's theorem ensures that the magnetic field must be identically zero. Since the total current density $\mathbf{j}_T = \mathbf{j} + \mathbf{j}_D$ has spherical symmetry, it results that at any point M there is an infinity of planes containing OM which are planes of symmetry of the current density. The magnetic field is perpendicular to all these planes, which imposes $\mathbf{B}(M) = \mathbf{0}$. Note finally that Ampère's law, valid only in magnetostatics, would have predicted a non-zero magnetic field at M, which is inconsistent with the above symmetry arguments.

13.2 Spherical wave

(a) The magnetic field can be obtained by first using Maxwell–Faraday law,

$$\nabla \times \mathbf{E} = \frac{1}{r}\left(\frac{\partial}{\partial r} r E_\theta\right)\mathbf{u}_\phi = -\frac{\partial \mathbf{B}}{\partial t}.$$

This gives

$$-\frac{E_0}{r}\left(k \sin(kr - \omega t)\right)\mathbf{u}_\phi = -\frac{\partial \mathbf{B}}{\partial t}.$$

Integrating with respect to time and recalling that $k/\omega = c$, we obtain

$$\mathbf{B} = \frac{E_0}{cr}\left(\cos(kr - \omega t)\right)\mathbf{u}_\phi.$$

The average electromagnetic energy density reads

$$\langle u_{EM}\rangle_T = \frac{\epsilon_0}{2}\langle|\mathbf{E}|^2\rangle_T + \frac{1}{2\mu_0}\langle|\mathbf{B}|^2\rangle_T = \frac{\epsilon_0 E_0^2}{r^2}\langle\cos^2(kr - \omega t)\rangle_T = \frac{\epsilon_0 E_0^2}{2r^2}$$

and the time-averaged Poynting vector is simply

$$\langle\mathbf{\Pi}\rangle = \frac{1}{\mu_0}\langle\mathbf{E} \times \mathbf{B}\rangle = \frac{E_0^2}{\mu_0 c r^2}\langle\cos(kr - \omega t)\rangle_T \mathbf{u}_r = \frac{c\epsilon_0}{2}\frac{E_0^2}{r^2}\mathbf{u}_r.$$

Finally, the intensity reads

$$\mathcal{I} = \langle|\mathbf{\Pi}|\rangle = \frac{c\epsilon_0}{2}\frac{E_0^2}{r^2},$$

which decays as $1/r^2$ from the source of the wave, located at the origin.

(b) The time-averaged electromagnetic energy U contained in a sphere of radius r is given by

$$U = \int_0^{2\pi}\int_0^{\pi}\int_0^{r}\langle u_{EM}(r)\rangle_T r^2 dr \sin\theta d\theta d\phi = 2\pi\epsilon_0 E_0^2 \int_0^{r} dr = 2\pi\epsilon_0 r E_0^2.$$

(d) The time-averaged flux of energy through a spherical surface S of radius r is

$$\Phi_{S,\langle\Pi\rangle} = \oiint \langle\Pi\rangle_T \cdot \underbrace{\mathbf{n}}_{\mathbf{u}_r}\, dS = \frac{E_0^2 c\epsilon_0}{2}\int_0^{2\pi} d\varphi \int_0^{\pi} \sin\vartheta\, d\vartheta = \frac{4\pi\epsilon_0 c E_0^2}{2}.$$

We see that it is independent of the sphere radius and corresponds to the total averaged power emitted by the source of the wave.

13.3 Electromagnetic waves inside a coaxial cable
(a) The Stokes theorem states

$$\oint_\Gamma \mathbf{B}\cdot d\mathbf{x} = \iint_{S(\Gamma)} (\nabla\times\mathbf{B})\cdot\mathbf{n}\, dS,$$

where $S(\Gamma)$ is any surface whose bounding curve is Γ. Maxwell–Ampère equation gives $\nabla\times\mathbf{B} = \mu_0\mathbf{j} + \mu_0\epsilon_0\dfrac{\partial\mathbf{E}}{\partial t}$ and so,

$$\oint_\Gamma \mathbf{B}\cdot d\mathbf{x} = \mu_0\iint_{S(\Gamma)} \mathbf{j}\cdot\mathbf{n}\, dS + \mu_0\epsilon_0\frac{\partial}{\partial t}\iint_{S(\Gamma)} \mathbf{E}\cdot\mathbf{n}\, dS.$$

By choosing for Γ a circular path of radius r at constant z such that $d\mathbf{x} = rd\theta\mathbf{u}_\theta$, and $S(\Gamma)$ the planar surface delimited by the circle such that $\mathbf{n} = \mathbf{u}_z$, we have

$$\oint_\Gamma \underbrace{\mathbf{B}\cdot d\mathbf{x}}_{B(r,z,t)rd\theta} = 2\pi r B(r,z,t) = \mu_0\underbrace{\iint_{S(\Gamma)} \mathbf{j}\cdot\mathbf{u}_z dS}_{=I(\Gamma)} + \mu_0\epsilon_0\frac{\partial}{\partial t}\iint_{S(\Gamma)} E(r,z,t)\underbrace{\mathbf{u}_r\cdot\mathbf{u}_z}_{=0} dS,$$

where the circulation on the left-hand side is easily calculated by the product of the magnetic field amplitude, independent of θ, by the circumference of the circle. On the right-hand side, the first term represents the current enclosed by Γ, with density $I(z,t)/(\pi a^2)$ for $r < a$. Thus, the enclosed current reads

$$I(\Gamma) = I(z,t)\begin{cases} \dfrac{r^2}{a^2} & \text{if } r < a, \\ 1 & \text{if } a \le r \le b, \\ 0 & \text{if } r > b. \end{cases}$$

The last term on the right-hand side of the generalized Maxwell–Ampère law is zero due to fact that the radial electric field is tangent to the cable section and does not contribute to the flux. To make the presentation of the solution more compact, let us define the function

$$f(r) = \begin{cases} r/a & \text{if } r < a, \\ a/r & \text{if } a \le r \le b, \\ 0 & \text{if } r > b. \end{cases}$$

The magnetic field obtained from the generalized Maxwell–Ampère law then reads

$$\mathbf{B}(r, z, t) = \frac{\mu_0}{2\pi a} I(z, t) f(r) \mathbf{u}_\theta.$$

(b) Maxwell–Ampère law in its local form reads

$$\nabla \times \mathbf{B} = -\frac{\partial B_\theta}{\partial z} \mathbf{u}_r + \frac{1}{r}\left(\frac{\partial}{\partial r} r B_\theta\right) \mathbf{u}_z = \mu_0 \mathbf{j} + \mu_0 \epsilon_0 \frac{\partial E_r}{\partial t} \mathbf{u}_r.$$

Projecting on the radial direction, we find

$$-\frac{\partial B_\theta}{\partial z} \mathbf{u}_r = \mu_0 \epsilon_0 \frac{\partial E(r, z, t)}{\partial t} \mathbf{u}_r.$$

Since the operators on each side of the equality are linear, the last equation can be seen as the real part of equation

$$\frac{\partial \underline{E}(r, z, t)}{\partial t} = -\frac{f(r)}{2\pi \epsilon_0 a} \frac{d\underline{I}(z)}{dz} e^{-i\omega t},$$

where the real electric field is $E(r, z, t) = \text{Re}\{\underline{E}(r, z, t)\}$. Integrating with respect to time yields

$$\underline{E}(r, z, t) = -\frac{i f(r)}{2\pi \epsilon_0 a \omega} \frac{d\underline{I}(z)}{dz} e^{-i\omega t}.$$

(c) Maxwell–Faraday equation reads

$$\nabla \times \mathbf{E} = \frac{\partial E(r, z, t)}{\partial z} \mathbf{u}_\theta = -\frac{\partial B(r, z, t)}{\partial t} \mathbf{u}_\theta$$

and using the complex representation, we obtain

$$\frac{d^2 \underline{I}(z)}{dz^2} + \epsilon_0 \mu_0 \omega^2 \underline{I}(z) = 0,$$

which admits $\underline{I}(z) = I_0 e^{ikz}$ as a solution if

$$-k^2 + \frac{\omega^2}{c^2} = 0$$

so that the most general solution reads

$$\underline{I}(z) = I_0^+ e^{ikz} + I_0^- e^{-ikz} \text{ with } k = \omega/c.$$

By keeping only the first term, the current represents a wave propagating in the direction of positive z, which reads

$$I(z, t) = \text{Re}\{I_0 e^{ikz} e^{-i\omega t}\} = I_0 \cos(kz - \omega t),$$

with $k = \omega/c$.

(d) Replacing $d\underline{I}(z)/dz = ik\underline{I}(z)$ in the expression for the electric field and taking the real part, we obtain

$$\mathbf{E}(r, z, t) = cB(r, z, t)\mathbf{u}_r,$$

with

$$\mathbf{B}(r, z, t) = \frac{\mu_0 I_0}{2\pi a} f(r) \cos(kz - \omega t)\mathbf{u}_\theta,$$

and the Poynting vector is given by

$$\mathbf{\Pi} = \frac{1}{\mu_0}\mathbf{E} \times \mathbf{B} = \frac{c}{\mu_0}|\mathbf{B}(r, z, t)|^2\mathbf{u}_z,$$

whose time average reads

$$\langle\mathbf{\Pi}\rangle_T = \frac{I_0^2}{8\pi^2 c\epsilon_0 a^2} f^2(r)\mathbf{u}_z,$$

with

$$f^2(r) = \begin{cases} r^2/a^2 & \text{if } r < a, \\ a^2/r^2 & \text{if } a \leq r \leq b, \\ 0 & \text{if } r > b. \end{cases}$$

The average power carried by the cable corresponds to the flux of $\langle\mathbf{\Pi}\rangle_T$ through the circular surface S of radius $r = b$ and normal $\mathbf{n} = \mathbf{u}_z$, which reads

$$\mathcal{P} = \iint_S \underbrace{\langle\mathbf{\Pi}\rangle_T}_{I(r)\mathbf{u}_z} \cdot \mathbf{u}_z dS = 2\pi \int_0^b I(r) r\, dr,$$

with the intensity $I(r) = \langle|\mathbf{\Pi}|\rangle_T = I_0^2 f^2(r)/(8\pi^2 c\epsilon_0 a^2)$. Finally,

$$\mathcal{P} = \frac{I_0^2}{4\pi c\epsilon_0}\left(\int_0^a \frac{r^3}{a^4} dr + \int_a^b \frac{dr}{r}\right) = \frac{I_0^2}{4\pi c\epsilon_0}\left(\frac{1}{4} + \ln\frac{b}{a}\right).$$

The factor $1/4$ is the contribution of the flux of energy inside the inner cylinder, where we have supposed a uniform current density. However, we have already seen that in a time-varying regime, a magnetic field cannot penetrate inside a conductor beyond the skin depth $\delta = \sqrt{\dfrac{2}{\mu_0 \varsigma \omega}}$, where ς is the static conductivity. If the conductivity is high enough, we can consider that the current density is confined very near the surface $r = a$, in which case the fields vanish for $r < a$ and the electromagnetic energy is confined in between $r = a$ and $r = b$, yielding

$$\mathcal{P} = \frac{I_0^2}{4\pi c\epsilon_0} \ln \frac{b}{a}$$

(e) Since the static electric field is non-zero only in the region $a < r < b$, the same is true for the static Poynting vector. Nothing changes for the magnetic field, which in the static case is simply given by Ampère's law:

$$\mathbf{B}(r) = \frac{\mu_0 I_0}{2\pi r} \mathbf{u}_\theta.$$

The static electric field can be obtained by integrating Gauss's law over a cylinder S of radius r and height h, concentric to the conductive cylinders, and with $a < r < b$. The cylindrical symmetry of the charge distribution yields $\mathbf{E} = E(r)\mathbf{u}_r$ and so,

$$\oiint_S \mathbf{E} \cdot \mathbf{n} dS = 2\pi r h E(r) = \frac{\sigma}{\epsilon_0} 2\pi a h,$$

where σ is the charge density in the inner cylinder. Finally,

$$\mathbf{E}(r) = \frac{\sigma a}{\epsilon_0 r} \mathbf{u}_r.$$

The potential difference ΔV can be related to σ since

$$\Delta V = V(a) - V(b) = \int_a^b E(r) dr = \frac{\sigma a}{\epsilon_0} \ln \left(\frac{b}{a}\right)$$

and the electric field can be rewritten as

$$\mathbf{E}(r) = \frac{\Delta V}{r \ln(b/a)} \mathbf{u}_r.$$

The static Poynting vector is

$$\mathbf{\Pi} = \frac{1}{\mu_0} \mathbf{E} \times \mathbf{B} = \frac{\Delta V I_0}{2\pi r^2 \ln(b/a)} \mathbf{u}_z$$

and the power transported by the cable simply equals the power supplied by the battery

$$\mathcal{P} = 2\pi \int_a^b \frac{\Delta V I_0}{2\pi r^2 \ln(b/a)} r\, dr = \Delta V I_0.$$

13.4 Incoming radiation from the sun

(a) Since the distance R between the Earth and the Sun is much larger than both the Earth and the Sun radius, the spherical wave emitted by the Sun can be approximated by a plane wave at the Earth position. Thus, the irradiance can be related to the electric field modulus by Eq. (13.27)

$$I = \frac{\epsilon_0 c E_0^2}{2}$$

from which we obtain

$$E_0 = \sqrt{\frac{2I}{c\epsilon_0}} = 10^3 \text{ Vm}^{-1} = 1\,\text{kVm}^{-1}$$

and the magnetic field amplitude is, simply

$$B_0 = \frac{E_0}{c} = \frac{10^3 \text{ Vm}^{-1}}{3 \times 10^8 \text{ ms}^{-1}} = 3.4\ \mu\text{T},$$

which is an order of magnitude smaller than Earth's magnetic field, which varies between 30 and 60 μT.

(b) Assuming an isotropic emission, the average power emitted by the Sun is simply

$$\mathcal{P} = 4\pi R^2 I = 3.86 \times 10^{26} \text{ W}.$$

13.5 Beyond Maxwell's equations—The mass of a photon

(a) A dimensional analysis of the first equation yield

$$[\mu] = \sqrt{\frac{[\nabla \cdot \mathbf{E}]}{[V]}} = \sqrt{\frac{[V]/[L]^2}{[V]}} = \frac{1}{[L]}.$$

μ is therefore the inverse of a length. The equation $\nabla \cdot \mathbf{B} = 0$ guarantees the existence of a vector potential $\mathbf{A}$ such that $\mathbf{B} = \nabla \times \mathbf{A}$. Maxwell–Faraday equation is unchanged as well, which also means that the field $\mathbf{E} + \frac{\partial \mathbf{A}}{\partial t}$ has zero curl, it derives from a potential V such that

$$\mathbf{E} = -\nabla V - \frac{\partial \mathbf{A}}{\partial t}.$$

(b) Since $\nabla \cdot (\nabla \times \mathbf{B}) = 0$ we obtain

$$\mu_0 \nabla \cdot \mathbf{j} + \frac{1}{c^2} \nabla \cdot \frac{\partial \mathbf{E}}{\partial t} - \mu^2 \nabla \cdot \mathbf{A} = 0$$

and writing $\nabla \cdot \frac{\partial \mathbf{E}}{\partial t} = \frac{\partial}{\partial t} \nabla \cdot \mathbf{E} = \frac{1}{\epsilon_0} \frac{\partial \varrho}{\partial t} - \mu^2 \frac{\partial V}{\partial t}$ this can be rewritten as

$$\underbrace{\nabla \cdot \mathbf{j} + \frac{\partial \varrho}{\partial t}}_{=0} - \frac{\mu^2}{\mu_0 c^2} \frac{\partial V}{\partial t} - \frac{\mu^2}{\mu_0} \nabla \cdot \mathbf{A} = 0 \ .$$

The conservation of charge implies that

$$\nabla \cdot \mathbf{A} + \frac{1}{c^2} \frac{\partial V}{\partial t} = 0 \ .$$

In other words, the potentials must satisfy Lorenz gauge. The potentials are uniquely defined.

(c) Replacing $\mathbf{E} = -\nabla V - \frac{\partial \mathbf{A}}{\partial t}$ into the modified Maxwell–Gauss equation gives

$$- \nabla^2 V - \frac{\partial}{\partial t} \nabla \cdot \mathbf{A} = \frac{\varrho}{\epsilon_0} - \mu^2 V$$

and since $\nabla \cdot \mathbf{A} = -\frac{1}{c^2} \frac{\partial V}{\partial t}$ we find

$$\left(\nabla^2 - \mu^2 \right) V - \frac{1}{c^2} \frac{\partial^2 V}{\partial t^2} = -\frac{\varrho}{\epsilon_0} \ .$$

Similarly, replacing $\mathbf{B} = \nabla \times \mathbf{A}$ into the modified Maxwell–Ampère equation yields

$$\nabla \times \nabla \times \mathbf{A} = \nabla (\nabla \cdot \mathbf{A}) - \nabla^2 \mathbf{A} = \mu_0 \mathbf{j} + \frac{1}{c^2} \frac{\partial \mathbf{E}}{\partial t} - \mu^2 \mathbf{A} \ ,$$

that is

$$\underbrace{\nabla (\nabla \cdot \mathbf{A})}_{=-\frac{1}{c^2} \frac{\partial}{\partial t} \nabla V} - \nabla^2 \mathbf{A} = \mu_0 \mathbf{j} + \frac{1}{c^2} \frac{\partial}{\partial t} \left(-\nabla V - \frac{\partial \mathbf{A}}{\partial t} \right) - \mu^2 \mathbf{A} \ ,$$

$$\left(\nabla^2 - \mu^2 \mathbf{A} \right) \mathbf{A} - \frac{1}{c^2} \frac{\partial^2 \mathbf{A}}{\partial t^2} = -\mu_0 \mathbf{j} \ .$$

(d) Taking the curl of Maxwell–Faraday equation

$$\nabla (\nabla \cdot \mathbf{E}) - \nabla^2 \mathbf{E} = \nabla \times \nabla \times \mathbf{E} = -\frac{\partial}{\partial t} \nabla \times \mathbf{B}$$

and so

$$\nabla\frac{\varrho}{\epsilon_0} - \mu^2\nabla V - \nabla^2\mathbf{E} = -\frac{\mu_0\partial\mathbf{j}}{\partial t} - \frac{1}{c^2}\frac{\partial^2\mathbf{E}}{\partial t^2} + \mu^2\frac{\partial\mathbf{A}}{\partial t},$$

that is

$$\nabla\left(\frac{\varrho}{\epsilon_0}\right) + \mu_0\frac{\partial\mathbf{j}}{\partial t} = \nabla^2\mathbf{E} - \frac{1}{c^2}\frac{\partial^2\mathbf{E}}{\partial t^2} + \mu^2\underbrace{\left(\frac{\partial\mathbf{A}}{\partial t} + \nabla V\right)}_{-\mathbf{E}}.$$

Finally

$$\left(\nabla^2 - \mu^2\right)\mathbf{E} - \frac{1}{c^2}\frac{\partial^2\mathbf{E}}{\partial t^2} = \nabla\left(\frac{\varrho}{\epsilon_0}\right) + \mu_0\frac{\partial\mathbf{j}}{\partial t}. \tag{A.87}$$

Similarly, taking the curl of the modified Maxwell–Ampère equation yields

$$\underbrace{\nabla(\nabla\cdot\mathbf{B})}_{=0} - \nabla^2\mathbf{B} = \nabla\times\nabla\times\mathbf{B} = \mu_0\nabla\times\mathbf{j} - \mu^2\nabla\times\mathbf{A} + \frac{1}{c^2}\frac{\partial}{\partial t}\nabla\times\mathbf{E},$$

$$-\nabla^2\mathbf{B} = \mu_0\nabla\times\mathbf{j} - \mu^2\mathbf{B} - \frac{1}{c^2}\frac{\partial^2\mathbf{B}}{\partial t^2},$$

$$\left(\nabla^2 - \mu^2\right)\mathbf{B} - \frac{1}{c^2}\frac{\partial^2\mathbf{B}}{\partial t^2} = -\mu_0\nabla\times\mathbf{j}.$$

We have obtained the same propagation equations for the fields as in (13.8) and (13.9) after replacing ∇^2 by $\nabla^2 - \mu^2$.

(e) In vacuum, $\varrho = 0$ and $\mathbf{j} = \mathbf{0}$. A sinusoidal plane wave of the form $\underline{\mathbf{E}} = \underline{\mathbf{E}}_0 e^{i(\mathbf{k}\cdot\mathbf{x}-\omega t)}$ is a solution of (A.87) if

$$(i\mathbf{k})^2 - \mu^2 - \frac{1}{c^2}(-i\omega)^2 = 0$$

in other words, if

$$\omega(k) = c\sqrt{k^2 + \mu^2}$$

with $k = |\mathbf{k}|$. This differs from the dispersion relation $\omega(k) = c|k|$ because of the non-zero value of μ.

(f) To a wave of frequency ω one can assign an energy $E = \hbar\omega$, where $\hbar$ is the reduced Planck's constant. We have therefore

$$E^2 = c^2(\hbar k)^2 + c^2(\hbar\mu)^2$$

which is analogous to the energy of a particle of mass m and momentum p, with $p = \hbar k$ and $m = \hbar \mu / c$. This particle is called a photon.

(g) The group velocity writes

$$v_g = \frac{\partial \omega(k)}{\partial k} = \pm \frac{ck}{\sqrt{k^2 + \mu^2}} \, .$$

It is seen that this velocity can never exceed the value of c if we assume a non-zero mass $\mu \neq 0$ for the photon. A non-zero photon mass could be measured experimentally by measuring the difference between the speed of light and the value of c. Note that the velocity of a beam of light can be zero for $k = 0$ if $\mu \neq 0$, which would correspond to a photon at rest.

Problems of Chap. 14

14.1 Electromagnetic waves in copper and in the ionosphere

(a) We evaluate the plasma frequency $\omega_p = \sqrt{\dfrac{ne^2}{m_e \epsilon_0}}$ for copper

$$\omega_p^{\mathrm{Cu}} = 1.6 \times 10^{16} \ \mathrm{s}^{-1}$$

and so, $\omega_p \tau \gg 1$. For copper in the visible range, $\omega \tau^{\mathrm{Cu}} \sim 40$ and $\omega < \omega_p^{\mathrm{Cu}}$, which corresponds to the intermediate frequency regime. The wave number is purely imaginary and the skin depth reads

$$\frac{1}{k_I} = \frac{c}{\sqrt{\omega_p^2 - \omega^2}} = 18 \ \mathrm{nm}.$$

For a radio wave, $\omega \tau^{\mathrm{Cu}} \ll 1$ and therefore

$$\delta = \sqrt{\frac{2}{\mu_0 \omega \sigma_0}} = \sqrt{\frac{2c^2}{\omega_p^2 \omega \tau}} \sim 10 \ \mathrm{km}.$$

(b) For the ionosphere, the plasma frequency is

$$\omega_p^{\mathrm{ion}} = 5.6 \times 10^7 \ \mathrm{s}^{-1},$$

whereas the time between collisions in the atmosphere can be obtained from the static conductivity, since

$$\sigma_0 = \frac{ne^2\tau}{m_e} = \epsilon_0 \omega_p^2 \tau$$

and so,

$$\tau^{\text{ion}} \sim 3.5 \times 10^{-5} \text{ s.}$$

Once again $\omega_p \tau \gg 1$. The radio wave does not propagate in the ionosphere, since $\omega < \omega_p^{\text{ion}}$, and since $\omega \tau^{\text{ion}} \gg 1$

$$\delta = \frac{1}{k_I} = \frac{c}{\sqrt{\omega_p^2 - \omega^2}} = 5.3 \text{ m,}$$

whereas visible light is such that $\omega \gg \omega_p$ and the ionosphere becomes transparent.

14.2 Circularly polarized light incident on a linear polarizer

Suppose, without loss of generality, that the polarizer axis is along the z-axis and that a circularly polarized wave is incident on the polarizer. In the complex representation, the incident field reads

$$\underline{\mathbf{E}}_0 = E_0 e^{i\phi} \frac{(\mathbf{u}_y \pm i\mathbf{u}_z)}{\sqrt{2}} e^{i(kx-\omega t)},$$

where $\pm$ corresponds to left and right-handed polarization, respectively, and the intensity reads

$$\mathcal{I}_0 = \frac{c\epsilon_0}{2} |\underline{\mathbf{E}}_0|^2 = \frac{c\epsilon_0 E_0^2}{2}.$$

Only the component along the z-axis is transmitted by the polarizer, so that at the output the field is

$$\underline{\mathbf{E}}' = \pm i \frac{E_0 e^{i\phi}}{\sqrt{2}} e^{i(kx-\omega t)} \mathbf{u}_z$$

and therefore the transmitted intensity is

$$\mathcal{I}' = \frac{c\epsilon_0}{2} |\underline{\mathbf{E}}'|^2 = \frac{c\epsilon_0}{4} E_0^2 = \frac{\mathcal{I}_0}{2}.$$

regardless of the orientation of the polarizer.

14.3 Skin effect in a metallic plate

(a) The system is invariant to translations along the y- and z-axis. Every component of the electric and magnetic field thus depends on x only: $E_x(x)$, $E_y(x)$, $E_z(x)$, $B_x(x)$, $B_y(x)$, $B_z(x)$. Hence, In Maxwell's equations $\partial/\partial y = 0$ and $\partial/\partial z = 0$. In the metal, the volume charge density is zero, the conduction current density satisfies Ohm's law $\mathbf{j} = \sigma \mathbf{E}$ and the displacement current is neglected with respect to the conduction current.

$$\mathbf{\nabla} \cdot \mathbf{E} = 0, \quad \text{i.e.,} \qquad \partial E_x / \partial x = 0 \quad (1)$$

$$\mathbf{\nabla} \cdot \mathbf{B} = 0, \quad \text{i.e.,} \qquad \partial B_x / \partial x = 0 \quad (2)$$

$$\mathbf{\nabla} \times \mathbf{E} = -\partial_t \mathbf{B}, \quad \text{i.e.,} \quad
\begin{cases}
0 = -\partial B_x / \partial t & (3) \\
-\partial E_z / \partial x = -\partial B_y / \partial t & (4) \\
\partial E_y / \partial x = -\partial B_z / \partial t & (5)
\end{cases}$$

$$\mathbf{\nabla} \times \mathbf{B} = \mu_0 \sigma \mathbf{E}, \quad \text{i.e.,} \quad
\begin{cases}
0 = \mu_0 \sigma E_x & (6) \\
-\partial B_z / \partial x = \mu_0 \sigma E_y & (7) \\
\partial B_y / \partial x = \mu_0 \sigma E_z & (8)
\end{cases}$$

(b) Differentiating Eq. (7) with respect to x and introducing Eq. (5), we find

$$\frac{\partial^2 B_z}{\partial x^2} = -\mu_0 \sigma \frac{\partial E_y}{\partial x} = \mu_0 \sigma \frac{\partial B_z}{\partial t}.$$

Introducing the complex amplitude $\underline{B_z} = \underline{\mathcal{B}} e^{i\omega t}$, we find

$$\frac{\partial^2 \underline{\mathcal{B}}}{\partial x^2} = i\omega \mu_0 \sigma \underline{\mathcal{B}}, \quad \text{with} \quad i = \frac{(1+i)^2}{2} \quad \text{and} \quad \omega \mu_0 \sigma = \frac{2}{\delta^2}.$$

$$\text{Thus,} \quad \frac{\partial^2 \underline{\mathcal{B}}}{\partial x^2} - \left(\frac{1+i}{\delta}\right)^2 \underline{\mathcal{B}} = 0 \quad (9).$$

(c) From Eq. (6), $E_x = 0$. From Eqs (2) and (3), $B_x = \mathrm{Cst} = 0$. Since $\mathbf{j} = \sigma \mathbf{E}$ is directed along the y-axis, $E_z = 0$, thus, from Eq. (8), $B_y = 0$ and

$$\underline{\mathbf{B}} = \underline{B_z} \mathbf{u}_z = \underline{\mathcal{B}}(x) e^{i\omega t} \mathbf{u}_z \quad \text{and} \quad \underline{\mathbf{E}} = \underline{E_y} \mathbf{u}_y.$$

Equation (9) admits the solution

$$\underline{\mathcal{B}}(x) = b_1 \exp\left(\frac{1+i}{\delta} x\right) + b_2 \exp\left(-\frac{1+i}{\delta} x\right).$$

The system is symmetric with respect to the Oyz-plane, thus $\underline{\mathcal{B}}(-x) = -\underline{\mathcal{B}}(x)$, which requires $b_2 = -b_1$. Hence,

$$\underline{\mathcal{B}}(x) = b_1 \left[\exp\left(\frac{1+i}{\delta} x\right) - \exp\left(-\frac{1+i}{\delta} x\right)\right] = 2b_1 \sinh\left[\frac{(1+i)x}{\delta}\right].$$

At $x = a$, $\mathcal{B}(a) = B_S$, fixing $b_1 = \dfrac{B_S}{2\sinh[(1+i)a/\delta]}$ and

$$\underline{\mathbf{B}}(x,t) = B_S \frac{\sinh[(1+i)x/\delta]}{\sinh[(1+i)a/\delta]} e^{i\omega t}\mathbf{u}_z.$$

From Eq. (7), the electric field in the metallic plate reads

$$\underline{\mathbf{E}} = -\frac{1}{\mu_0\sigma}\frac{\partial \underline{B_z}}{\partial x}\mathbf{u}_y,$$

i.e., $$\underline{\mathbf{E}}(x,t) = -\frac{(1+i)B_S}{\mu_0\sigma\delta}\frac{\cosh[(1+i)x/\delta]}{\sinh[(1+i)a/\delta]} e^{i\omega t}\mathbf{u}_y.$$

(d) The complex amplitude of the volume current density reads

$$\underline{\mathbf{j}} = \sigma\underline{\mathbf{E}} = -\frac{(1+i)B_S}{\mu_0\delta}\frac{\cosh[(1+i)x/\delta]}{\sinh[(1+i)a/\delta]} e^{i\omega t}\mathbf{u}_y .$$

At very low frequency ($\omega \to 0$) $\delta \to \infty$ and

$$\cosh[(1+i)x/\delta] \simeq 1 \quad \text{and} \quad \sinh[(1+i)a/\delta] \simeq \frac{(1+i)a}{\delta} .$$

Thus

$$\underline{\mathbf{j}} = -\frac{(1+i)B_S}{\mu_0\delta}\frac{\delta}{(1+i)a}e^{i\omega t}\mathbf{u}_y \quad \text{i.e.} \quad \underline{\mathbf{j}} = -\frac{B_S}{\mu_0 a}e^{i\omega t}\mathbf{u}_y.$$

At low frequency, the current density is thus uniform and of amplitude $\dfrac{B_S}{\mu_0 a}$.

(e) $j_0(x) = |\underline{\mathbf{j}}| = \sqrt{\underline{\mathbf{j}}\cdot\underline{\mathbf{j}}^*}$, thus $j_0(x) = \dfrac{\sqrt{2}B_S}{\mu_0\delta}\left[\dfrac{\cosh(2x/\delta) + \cos(2x/\delta)}{\cosh(2a/\delta) - \cos(2a/\delta)}\right]^{1/2} .$

At $x = 0$, $j_0(0) = \dfrac{\sqrt{2}B_S}{\mu_0\delta}\left[\dfrac{2}{\cosh(2a/\delta) - \cos(2a/\delta)}\right]^{1/2} .$

Finally, with $u = x/\delta$, $j_0(x) = j_0(0)\left[\dfrac{\cosh(2u) + \cos(2u)}{2}\right]^{1/2} .$

The current is located close to the free surfaces of the metallic plate. The ratio $j_0(\pm a)/j_0(0)$ is all the higher as the skin depth δ is small (i.e., as the frequency is high). At low frequency, the current is uniform and equal to $B_S/\mu_0 a$. The higher the frequency, the more the current density is localized close to the free surfaces $x = \pm a$.

(f) For $x \simeq a$ and a very thick plate $a/\delta \gg 1$ and $x/\delta \gg 1$, we have

$$\sinh[(1+i)a/\delta] \simeq \frac{1}{2}\exp((1+i)a/\delta) \quad \text{and} \quad \cosh[(1+i)x/\delta] \simeq \frac{1}{2}\exp((1+i)x/\delta)$$

hence,

$$\underline{\mathbf{j}} = -\frac{(1+i)Bs}{\mu_0\delta}e^{-(a-x)/\delta}e^{i(\omega t-(a-x)/\delta)}\mathbf{u}_y \qquad \text{for } x \simeq a .$$

For $x \sim -a$, we find

$$\underline{\mathbf{j}} = \frac{(1+i)Bs}{\mu_0\delta}e^{-(a+x)/\delta}e^{i(\omega t-(a+x)/\delta)}\mathbf{u}_y \qquad \text{for } x \simeq -a .$$

14.4 Phase velocity, group velocity, and energy velocity in an absorption band

(a) The phase velocity of the wave is $v_\phi = \omega/k = c/n$. For $\omega = \omega_0$, the refractive index is $n(\omega_0) = 1$. Therefore, $v_\phi(\omega_0) = c$.

The group velocity is

$$v_g = \frac{d\omega}{dk} = \left(\frac{dk}{d\omega}\right)^{-1}$$

where $k(\omega) = n(\omega)\omega/c$. Therefore,

$$\frac{dk}{d\omega} = \frac{1}{c}\frac{d(n(\omega)\omega)}{d\omega} = \frac{1}{c}\left(n(\omega) + \omega\frac{dn(\omega)}{d\omega}\right).$$

Differentiating $n(\omega)$, we evaluate the result at ω_0:

$$\frac{dn(\omega)}{d\omega}(\omega_0) = 1 - \omega_p^2\tau^2.$$

Therefore,

$$v_g = \frac{c}{1 - \omega_p^2\tau^2}.$$

Given $\omega_p = 5.67 \times 10^8 \text{ s}^{-1}$ and $\tau^{-1} = 9 \times 10^9 \text{ s}^{-1}$, we obtain

$$v_g = c(1 + 4\ 10^{-3}).$$

(b) The group velocity exceeds the speed of light in vacuum. This seemingly paradoxical result is obtained for a range of frequencies around ω_0 satisfying $dn/d\omega < 0$ (anomalous dispersion region). However, this does not violate the principle of causality. The rapid change in the refractive index near the resonant frequency ω_0 causes different frequency components of the wavepacket to propagate at significantly different phase velocities. This dispersion leads to the rapid spreading of the wavepacket, rendering the concept of group velocity less meaningful in this regime.

(c) Let us denote $\mathbf{E}$ the electric field of the wave, E_0 its peak amplitude and $\mathbf{u}_z$ the unit vector in the direction of propagation. The corresponding magnetic field is

$$\mathbf{B} = \frac{k}{\omega}\mathbf{u}_z \times \mathbf{E} = \frac{n}{c}\mathbf{u}_z \times \mathbf{E}.$$

The Poynting vector is expressed as

$$\mathbf{\Pi} = \frac{1}{\mu_0}\mathbf{E} \times \mathbf{B} = \frac{n}{\mu_0 c}E^2\mathbf{u}_z,$$

where $E = E_0\cos(\omega t + \phi)$. Averaging in time, we get

$$\langle\mathbf{\Pi}\rangle = \frac{n}{\mu_0 c}E_0^2\langle\cos^2(\omega t + \phi)\rangle\mathbf{u}_z = \frac{n}{2\mu_0 c}E_0^2\mathbf{u}_z.$$

The volume density of electromagnetic energy at a point in the medium is

$$u = \epsilon_0\epsilon_r\frac{E^2}{2} + \frac{B^2}{2\mu_0}.$$

Substituting $\epsilon_r = n^2$, and $B^2 = (nE/c)^2$, we obtain

$$u = \epsilon_0 n^2\frac{E^2}{2} + n^2\frac{E^2}{2\mu_0 c^2} = n^2\epsilon_0 E^2,$$

where we used the relation $\epsilon_0\mu_0 c^2 = 1$ to simplify the result. The time average energy density can then be expressed as

$$\langle u\rangle = \frac{1}{2}n^2\epsilon_0 E_0^2.$$

(d) Consider a cylindrical slice of matter of volume Sdz, where S is the surface area of the cylinder basis, dz is the cylinder length, with the cylinder axis parallel to the z-axis. The average electromagnetic energy content of the cylinder is

$$\delta U = \langle u\rangle Sdz.$$

During an interval of time dt, the average electromagnetic energy entering the cylinder through S is equal to product of the averaged Poynting vector flux (average power flux) by the time interval dt:

$$\delta U' = \langle\mathbf{\Pi}\rangle \cdot \mathbf{u}_z Sdt.$$

Assuming the energy propagates at velocity v_e, the energy passing through S between times t and $t + dt$ is contained at t in a cylinder of length $dz = v_e dt$.

Therefore, $\delta U = \delta U'$:

$$\langle u \rangle S v_e dt = \langle \mathbf{\Pi} \rangle \cdot \mathbf{u}_z dt.$$

The energy velocity v_e is the ratio of $\langle \mathbf{\Pi} \rangle$ to $\langle u \rangle$. Substituting $\langle u \rangle$ and $\langle \mathbf{\Pi} \rangle$, we obtain

$$\frac{1}{2} n^2 \epsilon_0 E_0^2 v_e = \frac{n}{2 \mu_0 c} E_0^2.$$

After simplification, we get

$$v_e = \frac{c}{n}.$$

In this particular case, the propagation velocity of the electromagnetic energy is equal to the phase velocity.

14.5 Interaction of a Linearly Polarized Plane Monochromatic Wave with an Atomic Vapor

(a)   1. Newton's second law applied to each electron of the atomic vapor writes:

$$m \frac{d^2 \mathbf{r}}{dt^2} = -m \omega_0^2 \mathbf{r} - m \gamma \mathbf{v} - e(\mathbf{E} + \mathbf{v} \times \mathbf{B}).$$

2. For a plane wave, the magnitude of the magnetic field is related to the magnitude of the electric field by $|\mathbf{B}| = |\mathbf{E}|/c'$, where c' denotes the phase velocity of the wave in the vapor. The order of magnitude of c' is c. As electrons are nonrelativistic, $|\mathbf{v}| \ll c \sim c'$. Therefore $|\mathbf{v} \times \mathbf{B}| \sim (|\mathbf{v}|/c)|\mathbf{E}| \ll |\mathbf{E}|$, i.e., the magnetic force can be neglected in comparison with the electric force. The equation governing the electron motion becomes

$$\frac{d^2 \mathbf{r}}{dt^2} + \gamma \mathbf{v} + \omega_0^2 \mathbf{r} = -\frac{e}{m} \mathbf{E}.$$

3. Using complex notations, the electric field can be expressed as

$$\underline{\mathbf{E}} = \mathbf{E}_0 \exp i(\omega t - kz) \mathbf{u}_x.$$

We look for a steady-state solution for the electron motion in the form

$$\underline{\mathbf{r}} = \underline{\mathbf{r}}_0 \exp i(\omega t - kz).$$

Substituting these expressions in Newton's equation, we get

$$-\omega^2 \underline{\mathbf{r}}_0 + i \gamma \omega \underline{\mathbf{r}}_0 + \omega_0^2 \underline{\mathbf{r}}_0 = -\frac{e}{m} E_0 \mathbf{u}_x.$$

Solving for $\underline{\mathbf{r}}_0$, we obtain

$$\underline{\mathbf{r}}_0 = \frac{-eE_0/m}{\omega_0^2 - \omega^2 + i\gamma\omega}\mathbf{u}_x.$$

The steady-state solution is

$$\underline{\mathbf{r}}(t) = \frac{-eE_0/m}{\omega_0^2 - \omega^2 + i\gamma\omega}\exp i(\omega t - kz)\mathbf{u}_x.$$

(b) 1. Each electron of the atomic vapor forms with the nucleus of the atom a dipole moment $\mathbf{p} = -e\mathbf{r}$. Substituting the expression for $\underline{\mathbf{r}}$ found previously and using the constant $\epsilon_0\Omega^2 = Ne^2/m$, we can write the polarization:

$$\underline{\mathbf{P}} = \epsilon_0\chi\underline{\mathbf{E}},$$

where

$$\chi = \frac{\Omega^2}{\omega_0^2 - \omega^2 + i\gamma\omega}.$$

2. The vapor is a globally neutral, nonconducting medium. Maxwell's equations in matter read

$$\nabla \cdot \mathbf{D} = 0, \qquad\qquad \nabla \cdot \mathbf{B} = 0,$$

$$\nabla \times \mathbf{E} = -\frac{\partial \mathbf{B}}{\partial t}, \qquad\qquad \nabla \times \mathbf{H} = \frac{\partial \mathbf{D}}{\partial t}.$$

The constitutive relations of the medium are

$$\mathbf{B} = \mu_0\mathbf{H}, \quad \mathbf{D} = \epsilon_0\mathbf{E} + \mathbf{P}.$$

Substituting in Maxwell's equations, we obtain

$$\nabla \times \mathbf{B} = \epsilon_0\mu_0\frac{\partial \mathbf{E}}{\partial t} + \mu_0\frac{\partial \mathbf{P}}{\partial t},$$

$$\nabla \cdot \mathbf{D} = \epsilon_0\nabla \cdot \mathbf{E} + \nabla \cdot \mathbf{P} = 0.$$

Taking the curl of Faraday–Maxwell's equation and substituting the expression for $\nabla \times \mathbf{B}$ in the resulting equation, we have

$$\nabla \times (\nabla \times \mathbf{E}) = -\nabla \times \left(\frac{\partial \mathbf{B}}{\partial t}\right) = -\frac{\partial}{\partial t}(\nabla \times \mathbf{B}) = -\epsilon_0\mu_0\frac{\partial^2 \mathbf{E}}{\partial t^2} - \mu_0\frac{\partial^2 \mathbf{P}}{\partial t^2}.$$

The double curl is related to the vectorial Laplacian by

$$\nabla \times (\nabla \times \mathbf{E}) = \nabla(\nabla \cdot \mathbf{E}) - \Delta\mathbf{E}.$$

Therefore, we obtain the propagation equation for the electric field in the atomic vapor with polarization $\mathbf{P}$:

$$\Delta\mathbf{E} - \epsilon_0\mu_0\frac{\partial^2\mathbf{E}}{\partial t^2} = \mu_0\frac{\partial^2\mathbf{P}}{\partial t^2} - \frac{1}{\epsilon_0}\nabla(\nabla\cdot\mathbf{P}).$$

(c) 1. Differentiating the plane wave with respect to t ($\partial/\partial t = i\omega$) and z ($\partial/\partial z = -ik$), we have $\Delta\underline{\mathbf{E}} = -k^2\underline{\mathbf{E}}$, $\partial^2\underline{\mathbf{E}}/\partial t^2 = -\omega^2\underline{\mathbf{E}}$, $\partial^2\underline{\mathbf{P}}/\partial t^2 = -\epsilon_0\omega^2\chi\underline{\mathbf{E}}$ and $\nabla\cdot\underline{\mathbf{P}} = -ik\mathbf{u}_z\cdot\underline{\mathbf{P}} = 0$. Substituting in the propagation equation, we find

$$(-k^2 + \epsilon_0\mu_0\omega^2)\underline{\mathbf{E}} = -\epsilon_0\omega^2\chi\underline{\mathbf{E}}.$$

Now substituting the expression for $\chi(\omega)$ and using the relation $\epsilon_0\mu_0c^2 = 1$, we obtain the dispersion relation

$$k^2 = \frac{\omega^2}{c^2}\left[1 + \frac{\Omega^2}{\omega_0^2 - \omega^2 + i\gamma\omega}\right].$$

The refractive index is $n(\omega) = k(\omega)c/\omega$:

$$n^2(\omega) = 1 + \frac{\Omega^2}{\omega_0^2 - \omega^2 + i\gamma\omega}.$$

Let us denote the real and imaginary parts of the refractive index as

$$n(\omega) = n'(\omega) - in''(\omega).$$

Squaring this relation, we have

$$n'^2(\omega) - n''^2(\omega) - 2in'(\omega)n''(\omega) = 1 + \Omega^2\frac{\omega_0^2 - \omega^2 - i\gamma\omega}{(\omega_0^2 - \omega^2)^2 + \gamma^2\omega^2}.$$

We can now identify the real and imaginary parts, with the simplifying assumption that the imaginary part is negligible in comparison with the real part ($n''^2 \ll n'^2$):

$$n'^2(\omega) = 1 + \Omega^2\frac{\omega_0^2 - \omega^2}{(\omega_0^2 - \omega^2)^2 + \gamma^2\omega^2},$$

$$n'(\omega)n''(\omega) = \frac{\Omega^2}{2}\frac{\gamma\omega}{(\omega_0^2 - \omega^2)^2 + \gamma^2\omega^2}.$$

We choose the positive root for $n'(\omega)$ as it represents a refractive index.

2. The wavenumber can be written as

$$k = n(\omega)\frac{\omega}{c} = \frac{\omega}{c}(n'(\omega) - in''(\omega)).$$

The electric field is the given by

$$\underline{\mathbf{E}} = E_0 \exp i(\omega t - kz)\mathbf{u}_x = E_0 \exp\left(-\frac{\omega}{c}n''(\omega)z\right)\exp i\omega\left(t - n'(\omega)\frac{z}{c}\right)\mathbf{u}_x.$$

This is a plane wave propagating along the z-axis toward increasing z (as $n'(\omega) > 0$) at the phase velocity $c/n'(\omega)$. The real part of the complex valued function $n(\omega)$ therefore represents the real refractive index in the medium. The real exponential term describes the amplitude of the wave, which is attenuating along propagation for $n''(\omega) > 0$. This condition is indeed ensured by the fact that the product $n'(\omega)n''(\omega)$ is positive (result of the last question), therefore $n''(\omega)$ and $n'(\omega)$ have the same sign. The imaginary part of the complex valued function $n(\omega)$ therefore characterizes absorption of the electromagnetic wave by the vapor. Large coefficients $n''(\omega)$ mean a stronger (faster) absorption.

3. Differentiating the function $n'^2(\omega)$, we get

$$\frac{dn'^2(\omega)}{d\omega} = 2n'(\omega)\frac{dn'(\omega)}{d\omega} = \frac{2\omega\left[(\omega_0^2 - \omega^2)^2 - \omega_0^2\gamma^2\right]}{(\omega_0^2 - \omega^2)^2 + \gamma^2\omega^2}.$$

Therefore, the refractive index $n'(\omega)$ has extrema for

$$\omega_\pm^2 = \omega_0^2\left(1 \pm \frac{\gamma}{\omega_0}\right)$$

and $n'(\omega)$ is decreasing as ω is increasing in the range $\omega_- \leq \omega \leq \omega_+$, and increasing with ω out of this range. The extrema are given by

$$n'^2(\omega_\pm) = \left(1 \pm \frac{\Omega^2}{2\gamma\omega_0}\right).$$

Given the data, we have $\gamma/\omega_0 \sim 10^{-6} \ll 1$ and $\Omega^2/\gamma\omega_0 \sim 10^{-6} \ll 1$, therefore,

$$\omega_\pm = \omega_0\left(1 \pm \frac{\gamma}{2\omega_0}\right) \quad \text{and} \quad n'(\omega_\pm) = \left(1 \pm \frac{\Omega^2}{4\gamma\omega_0}\right).$$

Calculating the numerical values, we obtain

$$\Omega^2 = 1.61 \times 10^{24}\,\text{rad}^2\,\text{s}^{-2}\,,$$

$$\Omega = 1.27 \times 10^{12}\,\text{rad}\,\text{s}^{-1}\,,$$

$$\omega_- - \omega_0 \approx -2.5 \times 10^9\,\text{s}^{-1}\,,$$

$$\omega_+ - \omega_0 \approx +2.5 \times 10^9\,\text{s}^{-1}\,,$$

$$n'(\omega_-) \approx 1.08\,,$$

$$n'(\omega_+) \approx 0.92\,.$$

Far from ω_0, the quartic term in ω dominates in the denominator of the squared refractive index which implies $n'(\omega) \approx 1$. The refractive index is illustrated in Fig. A.143a.

As the departure of the refractive index from 1 does not exceed 8%, the absorption coefficient $n''(\omega)$ can by approximated by the function $n'(\omega)n''(\omega)$:

$$n''(\omega) \approx \frac{\Omega^2}{2} \frac{\gamma\omega}{(\omega_0^2 - \omega^2)^2 + \gamma^2\omega^2}.$$

Differentiating with respect to frequency, we get

$$\frac{dn''(\omega)}{d\omega} = \frac{\gamma\Omega^2}{2} \frac{-3\omega^4 + (2\omega_0^2 - \gamma^2)\omega^2 + \omega_0^4}{\left[(\omega_0^2 - \omega^2)^2 + \gamma^2\omega^2\right]^2}.$$

This expression vanishes at frequency

$$\omega_r^2 = \frac{1}{6}\left(2\omega_0^2 - \gamma^2 + \sqrt{(2\omega_0^2 - \gamma^2)^2 + 12\omega_0^2}\right).$$

Given the data, we find $\omega_r \approx \omega_0$ and the maximum absorption coefficient is then given by

$$n''(\omega_r) \approx n''(\omega_0) \approx \frac{\Omega^2}{2\gamma\omega_0}.$$

Numerically, we find $n''(\omega_r) \approx 0.16$. Far from ω_0, the absorption coefficient is rapidly negligible as its denominator is dominated by the quartic power in ω. As $n''(\omega_\pm) \approx n''(\omega_r)/2 = 0.08$, the width of the absorption band where absorption is significant is approximately $\omega_+ - \omega_- = \gamma$. The absorption coefficient is illustrated in Fig. A.143b.

For frequencies far from the absorption band, we have $n''(\omega) \approx 0$ and $n'(\omega) \approx 1$. The atomic vapor behaves as vacuum. The wave propagates at the speed of light c without absorption. For frequencies closer to the absorption band, the refractive index increases with increasing frequencies, or equivalently, with decreasing wavelengths. This is the region of normal dispersion.

Frequencies within the very narrow absorption band correspond to the region of anomalous dispersion: the refractive index and the propagation velocity of

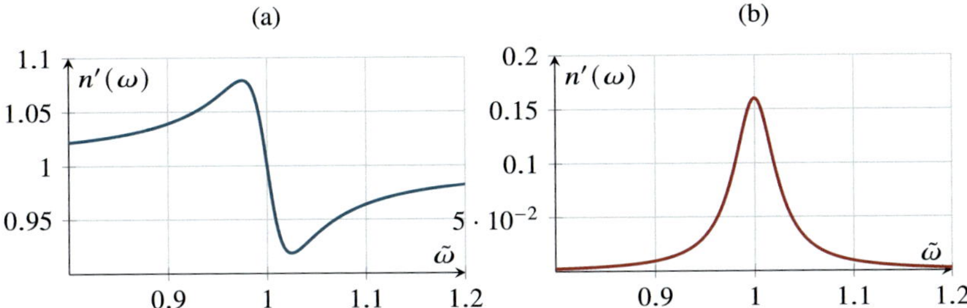

Fig. A.143 **a** Refractive index $n'(\omega)$ and **b** absorption coefficient $n''(\omega)$ as functions of normalized frequency $\tilde{\omega} = \omega/\omega_0$. The $\tilde{\omega}$-axis was stretched by a factor of 10^4 to make the absorption band visible

the wave vary rapidly with frequency. Absorption becomes significant. At the resonant frequency ω_0, the penetration depth is

$$\delta = \frac{c}{\omega_0 n''(\omega_0)} = \frac{2\gamma c}{\Omega^2} \approx 1.6\,\mu\text{m}.$$

A wave of frequency ω_0 cannot penetrate into the vapor.

14.6 Laser Modes

(a)   1. Applying the operators curl, and time differentiation to a plane wave, we can replace in Maxwell–Ampère and Maxwell–Faraday equations $\nabla \times$ by $-ik\mathbf{u}_z \times$ and ∂_t by $i\omega$: The system

$$\nabla \times \mathbf{E} = -\frac{\partial \mathbf{B}}{\partial t}, \qquad\qquad -ik\mathbf{u}_z \times \underline{\mathbf{E}} = -i\omega\underline{\mathbf{B}},$$
$$\text{becomes}$$
$$\nabla \times \mathbf{B} = \frac{1}{c^2}\frac{\partial \mathbf{E}}{\partial t}, \qquad\qquad -ik\mathbf{u}_z \times \underline{\mathbf{B}} = \frac{\epsilon_r}{c^2}i\omega\underline{\mathbf{E}}.$$

Substituting the expressions for $\mathbf{E}$ and $\mathbf{B}$, we get

$$kE_0 = \omega B_0,$$
$$kB_0 = \frac{\epsilon_r}{c^2}\omega E_0.$$

This system can be rewritten as

$$B_0 = \frac{k}{\omega}E_0,$$
$$k^2 = \frac{\epsilon_r \omega^2}{c^2} = (\epsilon' - i\epsilon'')\frac{\omega^2}{c^2}.$$

We look for a complex valued wavenumber k written as

$$k = \frac{\omega}{c}(n' - in''),$$

where n' and n'' satisfy

$$n'^2 - n''^2 - 2in'n'' = \epsilon' - i\epsilon''.$$

Identifying real and imaginary parts, we obtain

$$n'^2 - n''^2 = \epsilon',$$
$$2n'n'' = \epsilon''.$$

The modulus of n^2 is equal to the modulus of ϵ, providing the additional relation

$$n'^2 + n''^2 = \sqrt{\epsilon'^2 + \epsilon''^2}.$$

Solving for n', we get

$$n'^2 = \frac{1}{2}\left(\epsilon' + \sqrt{\epsilon'^2 + \epsilon''^2}\right).$$

Since $|\epsilon''| \ll 1$ and $|\epsilon' - 1| \ll 1$, we know that ϵ' is close to 1 and we can neglect ϵ'' in this expression. Therefore,

$$n' \approx \sqrt{\epsilon'} \quad \text{and} \quad n'' = \frac{\epsilon''}{2}.$$

The dispersion relation is then expressed as

$$k = \frac{\omega}{c}\left(\sqrt{\epsilon'} - i\frac{\epsilon''}{2}\right).$$

2. Substituting the wavenumber into the expressions for the electric and magnetic fields, we obtain

$$\underline{\mathbf{E}} = E_0 \exp\left(-n''\frac{\omega}{c}z\right)\exp i\omega\left(t - n'\frac{z}{c}\right)\mathbf{u}_x$$
$$\underline{\mathbf{B}} = \frac{\mathbf{k}}{\omega} \times \underline{\mathbf{E}} = \frac{E_0}{c}(n' - in'')\exp\left(-n''\frac{\omega}{c}z\right)\exp i\omega\left(t - n'\frac{z}{c}\right)\mathbf{u}_y.$$

Using real notations, we get

$$\mathbf{E} = E_0 \exp\left(-n''\frac{\omega}{c}z\right)\cos\omega\left(t - n'\frac{z}{c}\right)\mathbf{u}_x,$$
$$\mathbf{B} = \frac{E_0}{c}\exp\left(-n''\frac{\omega}{c}z\right)\left[n'\cos\omega\left(t - n'\frac{z}{c}\right) + n''\sin\omega\left(t - n'\frac{z}{c}\right)\right]\mathbf{u}_y.$$

Substituting these expressions in the Poynting vector $\mathbf{\Pi} = \mathbf{E} \times \mathbf{B}/\mu_0$, we obtain

$$\mathbf{\Pi} = \frac{E_0^2}{\mu_0 c} \exp\left(-2n''\frac{\omega}{c}z\right) \left[n' \cos^2 \omega\left(t - n'\frac{z}{c}\right) + \frac{n''}{2} \sin 2\omega\left(t - n'\frac{z}{c}\right)\right]\mathbf{u}_z.$$

The time average Poynting vector is then

$$\langle\mathbf{\Pi}\rangle = \frac{E_0^2}{2\mu_0 c} \exp\left(-\epsilon''\frac{\omega}{c}z\right)\mathbf{u}_z.$$

The flux of the Poynting vector through a plane surface perpendicular to the z-axis is equal to the electromagnetic power crossing this surface. The wave propagates toward increasing z; the exponential term in the expression for the time average Poynting vector shows that the wave is damped along the propagation axis if $\epsilon'' > 0$, or amplified if $\epsilon'' < 0$.

(b)   1. Taking the real and imaginary parts of the relative permittivity,

$$\epsilon_r = 1 + \frac{e^2}{2m\epsilon_0\omega_0}(N_1 - N_2)f\frac{1}{\omega_0 - \omega + i\Gamma},$$

we obtain expressions for ϵ' and ϵ'':

$$\epsilon' = 1 + \frac{e^2}{2m\epsilon_0\omega_0}(N_1 - N_2)f\frac{\omega_0 - \omega}{(\omega_0 - \omega)^2 + \Gamma^2},$$

$$\epsilon'' = \frac{e^2}{2m\epsilon_0\omega_0}(N_1 - N_2)f\frac{\Gamma}{(\omega_0 - \omega)^2 + \Gamma^2}.$$

The imaginary part ϵ'' has the same sign as $N_1 - N_2$. Since $N_2 > N_1$, ϵ'' is negative and the wave is amplified as it propagates.

This amplification is a fundamental property of laser amplifiers. The condition required for amplification, $N_2 > N_1$ is called population inversion. It refers to the fact that in standard conditions corresponding to a gas equilibrium governed by Boltzmann's statistics, the higher energy level is less populated than the lower energy level:

$$N_1 \propto \exp\left(-\frac{E_1}{k_B T}\right) > N_2 \propto \exp\left(-\frac{E_2}{k_B T}\right),$$

for a gas at temperature T, where k_B denotes the Boltzmann constant. The wave is therefore damped as it propagates in standard condition. Amplification only occurs if population inversion is realized, usually by exciting (*pumping*) the molecules to a higher energy state.

2. For an empty cavity, the wave propagates in a vacuum and reflects on the metallic mirrors, which must correspond to nodes for the wave. Standing

waves can therefore be established if the cavity length L corresponds to an integer number n of half-wavelengths:

$$L = p\frac{\lambda_p}{2}.$$

The angular frequencies corresponding to these wavelengths λ_p are therefore

$$\omega_p = \frac{2\pi c}{\lambda_p} = \frac{p\pi c}{L}.$$

If the reflectivity of one of the mirrors is 99%, the wave undergoes damping for any reflection on this mirror, at each round trip in the cavity.

3. In the presence of a gas in the cavity, assuming the wave amplitude is E_0 on the perfect mirror ($z = 0$), its amplitude after a round trip including a reflection on each mirror is

$$E_1 = E_0 r r' \exp\left(-\epsilon''\frac{\omega}{c}2L\right),$$

where $r = 1$ for the perfect mirror, $r' = \sqrt{R}$ for the mirror at $z = L$, and the exponential accounts for the decay (or gain) as the wave propagates in the gas. For the wave to propagate without damping, the relation $E_1 \geq E_0$ must be satisfied, since the gain must at least compensate for the damping of the wave on the mirror:

$$\sqrt{R}\exp\left(-\epsilon''\frac{\omega}{c}2L\right) \geq 1.$$

Solving for ϵ'', we obtain

$$\epsilon'' \leq \frac{c}{2L\omega}\ln R.$$

As $R < 1$, $\ln R < 0$ and this condition implies that ϵ'' must be negative, or equivalently, $N_2 > N_1$. Substituting the expression for ϵ'' found previously, this condition can be written as

$$(N_2 - N_1) \geq N_0 F(\tilde{\omega}),$$

where $\tilde{\omega} = \omega/\omega_0$ and

$$N_0 = \frac{c\Gamma \ln R^{-1}}{Lf}\frac{\epsilon_0 m}{e^2},$$

$$F(\tilde{\omega}) = \frac{\omega_0^2}{\Gamma^2}\left[\tilde{\omega} - 2 + \frac{1}{\tilde{\omega}}\left(1 + \frac{\Gamma^2}{\omega_0^2}\right)\right].$$

Since $\Gamma \ll \omega_0$, we can neglect the ratio Γ^2/ω_0^2 in the function $F(\tilde{\omega})$ to evaluate its behavior: this function behaves as $1/\tilde{\omega}$ at low frequencies and as $\tilde{\omega}$ at high frequencies. It exhibits a minimum at $\tilde{\omega}_m \approx 1 + \dfrac{\Gamma^2}{2\omega_0^2}$ that is equal to $F(\tilde{\omega}_m) \approx$ 1. Therefore, the minimum population inversion to ensure propagation without damping for a range of frequencies is

$$(N_2 - N_1)_{\min} = N_0.$$

For a population inversion of $N_2 - N_1 = 2N_0$, undamped oscillations correspond to the range of frequencies satisfying $F(\tilde{\omega}) \leq 2$. Solving for $\tilde{\omega}$, we obtain locking for

$$1 - \frac{\Gamma}{\omega_0} \lesssim \tilde{\omega} \lesssim 1 + \frac{\Gamma}{\omega_0}.$$

This range of frequencies represents a narrow bandwidth around ω_0.

4. The refractive index is approximately equal to 1. Standing waves selected in the laser cavity must therefore satisfy the same conditions as in the empty cavity:

$$\omega_p = \frac{p\pi c}{L},$$

where p is an integer. Substituting ω_p in the condition for undamped oscillations, we obtain

$$\omega_0 - \Gamma \leq \frac{p\pi c}{L} \leq \omega_0 + \Gamma.$$

Solving for p, we get

$$\frac{L}{\pi c}(\omega_0 - \Gamma) \leq p \leq \frac{L}{\pi c}(\omega_0 + \Gamma).$$

Given the data, we find the numerical values

$$3.18 \times 10^6 - 9.54 \leq p \leq 3.18 \times 10^6 + 9.54.$$

In total, 19 standing waves (laser modes) can be amplified in the laser cavity with angular frequencies around $(3 \times 10^{15} \pm 3 \times 10^{-6})\,\mathrm{rad\,s^{-1}}$.

14.7 Self-focusing of a laser beam

(a) Maxwell's equations for a monochromatic plane wave with angular frequency ω propagating in a charge-free and current-free dielectric medium are

$$\nabla \cdot \underline{\mathbf{D}} = 0,$$

$$\nabla \cdot \underline{\mathbf{B}} = 0,$$

$$\nabla \times \underline{\mathbf{E}} = -i\omega\underline{\mathbf{B}},$$

$$\nabla \times \underline{\mathbf{H}} = i\omega\underline{\mathbf{D}},$$

where $\mathbf{H} = \mathbf{B}/\mu_0$ since the medium is non-magnetic. Substituting the electric field $\underline{\mathbf{E}} = E(y) \exp i(\omega t - kx)\mathbf{u}_z$ into Faraday's law, we obtain

$$B_x = \frac{i}{\omega}\frac{dE(y)}{dy}\exp i(\omega t - kx),$$

$$B_y = -\frac{k}{\omega}E(y)\exp i(\omega t - kx),$$

$$B_z = 0.$$

Substituting these components into the divergence equation for the magnetic field,

$$\nabla \cdot \mathbf{B} = -kB_x + \frac{B_y}{dy} + \frac{B_z}{dz} = 0,$$

confirms the conservation of magnetic flux.

The constitutive relation of the nonlinear medium is given by

$$\mathbf{D} = \epsilon_0 \mathbf{E}(1 + \chi + \beta|E|^2).$$

Substituting the electric field expression, we can write

$$\underline{\mathbf{D}} = \epsilon_0 E(y)(1 + \chi + \beta|E|^2)\exp i(\omega t - kx)\mathbf{u}_z.$$

This can be expressed in the form $\mathbf{D} = \epsilon\mathbf{E}$, where the permittivity ϵ is now a function of the field amplitude:

$$\epsilon = \epsilon_0(1 + \chi + \beta|E(y)|^2).$$

Gauss's law, $\nabla \cdot \mathbf{D} = 0$, is satisfied as $\partial D_z/\partial z = 0$. Additionally, $\nabla \cdot \mathbf{E} = 0$. Taking the curl of Faraday's law and combining it with Maxwell–Ampère law, we obtain

$$\nabla \times (\nabla \times \underline{\mathbf{E}}) = -i\omega\nabla \times \underline{\mathbf{B}} = \mu_0\omega^2\underline{\mathbf{D}}.$$

Since $\nabla \cdot \mathbf{E} = 0$, the curl of the curl simplifies to the Laplacian:

$$\nabla^2\mathbf{E} = -\epsilon\mu_0\omega^2\mathbf{E}.$$

Considering the linear polarization of the electric field along the z-axis, we project this equation onto the z-axis. After calculating the Laplacian, we obtain

$$\frac{\partial^2 E_z}{\partial x^2} + \frac{\partial^2 E_z}{\partial y^2} + \frac{\partial^2 E_z}{\partial z^2} = \left(-k^2 E(y) + \frac{d^2 E(y)}{dy^2}\right)\exp i(\omega t - kx).$$

Substituting the expression for ϵ and simplifying the exponential term, we arrive at a nonlinear second-order differential equation for $E(y)$:

$$\frac{d^2 E(y)}{dy^2} = \left(k^2 - \frac{\omega^2}{c^2}(1+\chi)\right) E(y) - \frac{\omega^2}{c^2}\beta E^3(y),$$

where $c^2 \epsilon_0 \mu_0 = 1$.

Multiplying this equation by $dE(y)/dy$ and integrating with respect to y, assuming appropriate boundary conditions ($E(y)$ and dE/dy vanish as $y \to \infty$), we obtain

$$\left(\frac{dE(y)}{dy}\right)^2 = \left(k^2 - \frac{\omega^2}{c^2}(1+\chi)\right) E^2(y) - \frac{\omega^2}{c^2}\frac{\beta}{2} E^4(y).$$

(b) For a physically meaningful solution, the right-hand side of the differential equation must be non-negative. This condition implies

$$k^2 \geq \frac{\omega^2}{c^2}\left(1 + \chi + \frac{\beta}{2} E^2(y)\right) \quad \text{for all } y.$$

For the maximum electric field amplitude, E_0, this condition becomes

$$k^2 \geq \frac{\omega^2}{c^2}\left(1 + \chi + \frac{\beta}{2} E_0^2\right).$$

We seek a solution of the form

$$E(y) = \frac{E_0}{\cosh(y/y_0)}.$$

Differentiating,

$$\frac{dE(y)}{dy} = \frac{E_0}{y_0}\frac{\tanh(y/y_0)}{\cosh(y/y_0)},$$

and substituting into the differential equation, we obtain

$$\frac{E_0^2}{y_0^2}\frac{\tanh^2(y/y_0)}{\cosh^2(y/y_0)} = \left(k^2 - \frac{\omega^2}{c^2}(1+\chi)\right)\frac{E_0^2}{\cosh^2(y/y_0)} - \frac{\omega^2}{c^2}\frac{\beta}{2}\frac{E_0^4}{\cosh^4(y/y_0)}.$$

Using the identity $\tanh^2(y/y_0) = 1 - \cosh^{-2}(y/y_0)$ and equating the coefficients of the terms $\cosh^{-2}(y/y_0)$ and $\cosh^{-4}(y/y_0)$ to zero in order to satisfy this equation for all y, we obtain the following conditions:

$$E_0^2 = \frac{2}{\beta}\left(\frac{k^2 c^2}{\omega^2} - (1+\chi)\right),$$

$$y_0^2 = \left(k^2 - \frac{\omega^2}{c^2}(1+\chi)\right)^{-1}.$$

Introducing the parameter

$$\kappa^2 = k^2 - \frac{\omega^2}{c^2}(1 + \chi),$$

we can rewrite these conditions as

$$E_0^2 = \frac{2}{\beta}\frac{\kappa^2 c^2}{\omega^2},$$

$$y_0^2 = \kappa^{-2}.$$

(c) The Poynting vector is expressed in real notation as

$$\mathbf{\Pi} = \frac{1}{\mu_0}\mathbf{E} \times \mathbf{B}.$$

Substituting the electric and magnetic fields into this expression, we get

$$\mathbf{E} = E(y)\cos(\omega t - kz)\mathbf{u}_z,$$

$$\mathbf{B} = -\frac{1}{\omega}\left[\frac{dE(y)}{dy}\sin(\omega t - kx)\mathbf{u}_x + kE(y)\cos(\omega t - kx)\mathbf{u}_y\right],$$

$$\mathbf{\Pi} = \frac{k}{\mu_0\omega}E^2(y)\cos^2(\omega t - kz)\mathbf{u}_x - \frac{1}{2\mu_0\omega}\frac{dE(y)}{dy}E(y)\sin 2(\omega t - kx)\mathbf{u}_y.$$

Averaging in time, we obtain

$$\langle\mathbf{\Pi}\rangle = \frac{k}{2\mu_0\omega}E^2(y)\mathbf{u}_x.$$

The average power carried by the wave through a rectangular surface element $d\mathbf{S} = dy\,\Delta z\,\mathbf{u}_x$ is equal to the time average flux of the Poynting vector through the surface:

$$dP = \langle\mathbf{\Pi}\rangle \cdot d\mathbf{S} = \frac{k}{2\mu_0\omega}E^2(y)dy\,\Delta z.$$

For a slice with length $\Delta z = 1$ (unit length), the power per unit length is

$$dP_1 = \frac{kE_0^2}{2\mu_0\omega\cosh^2(y/y_0)}dy.$$

Integrating along the y-axis, we obtain the average power per unit length carried by the wave:

$$\frac{P}{\Delta z} = P_1 = \frac{kE_0^2}{2\mu_0\omega}\int_{-\infty}^{+\infty}\frac{1}{\cosh^2(y/y_0)}dy = \frac{ky_0 E_0^2}{2\mu_0\omega}[\tanh(y/y_0)]_{-\infty}^{+\infty} = \frac{ky_0 E_0^2}{\mu_0\omega}.$$

The quantities k, E_0 and y_0 can be calculated by solving the system of three equations obtained previously:

$$y_0^2 E_0^2 = \frac{2}{\beta k_0^2},$$

$$y_0^{-2} = \kappa^2 = k^2 - 2k_1^2,$$

$$k y_0 E_0^2 = \mu_0 \omega P_1,$$

where $k_0 = \omega/c$, and $k_1^2 = k_0^2(1 + \chi)/\sqrt{2}$.
Given $P_1 = 3 \times 10^{10}\,\mathrm{W\,m^{-1}}$, $\chi = 0.5$, $\beta = 2 \times 10^{-22}$ uSI, and the vacuum wavelength of the laser light $\lambda_0 = 0.694\,\mu\mathrm{m}$, we can calculate

$$k_0 = \omega/c = 2\pi/\lambda_0 = 9.05 \times 10^6\,\mathrm{m^{-1}}, \quad \text{and} \quad k_1 = 9.60 \times 10^6\,\mathrm{m^{-1}}.$$

Introducing the power density, $\Pi_0 = E_0^2/(2\mu_0 c)$, and constants

$$\Pi_1 = \frac{k_0 P_1}{2} = 1.36 \times 10^{17}\,\mathrm{W\,m^{-2}} \quad \text{and} \quad P_0 = \frac{1}{\mu_0 c k_0^2 \beta} = 0.162\,\mathrm{MW},$$

we can rewrite this system as

$$y_0^2 \Pi_0 = P_0,$$

$$y_0^{-2} = k^2 - 2k_1^2$$

$$k y_0 \Pi_0 = \Pi_1.$$

Solving for y_0, we divide the first equation by the third to get $y_0 = k P_0/\Pi_1$. Substituting in the second equation yields a quadratic equation in k^2:

$$k^4 - 2k_1^2 k^2 - (\Pi_1/P_0)^2 = 0 .$$

Solving for k, we obtain

$$k = \left(k_1^2 + \sqrt{k_1^4 + (\Pi_1/P_0)^2} \right)^{1/2} .$$

Numerically, we obtain the values:

$$k = 1.36 \times 10^7\,\mathrm{m^{-1}}, \quad y_0 = 13.2\,\mu\mathrm{m}, \quad I_0 = 6.18 \times 10^{14}\,\mathrm{W\,m^{-2}}$$

and the peak electric field

$$E_0 = 6.82 \times 10^8\,\mathrm{V\,m^{-1}}.$$

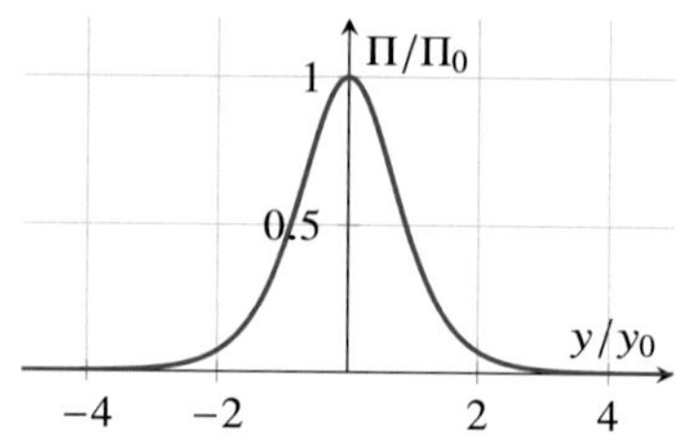

Fig. A.144 Spatial distribution of the average power density for self-focused wave in a nonlinear medium

The average power density is given by

$$\Pi = \frac{dP}{dy} = \frac{\Pi_0}{\cosh^2(y/y_0)}.$$

This equation describes a narrow, intense peak (see Fig. A.144), indicating a localized solution to Maxwell's equations within the nonlinear medium. This localization arises from a phenomenon known as self-focusing. As the wave propagates, it modifies the refractive index of the medium due to the nonlinear term in the permittivity, which is proportional to the beam intensity. Consequently, the refractive index becomes higher along the propagation axis ($y \lesssim y_0$), acting as a self-focusing lens. This effect counteracts the natural tendency of the beam to spread, leading to the formation of a self-guided, localized wave.

Problems of Chap. 15

15.1 The anti-reflection coating
(a) Under normal incidence, the reflectance is given by

$$R_{\text{air-glass}} = \left| \frac{1 - n_{\text{glass}}}{1 + n_{\text{glass}}} \right|^2 = 0.04$$

and $T_{\text{air-glass}} = 1 - R_{\text{air-glass}} = 0.96$. At every air-glass interface, 96% of the incident intensity is transmitted and 4% is reflected back. At the exit of a glass plate, light passes twice through an air-glass interface, and so the fraction of transmitted light is

$$T_{\text{plate}} = (1 - R_{\text{air-glass}})^2 = 92.16\%$$

and so $1 - T_{\text{plate}} = 7.84\%$ of light is reflected back.

Here, we neglect multiple reflections inside the plate, shown in Fig. A.145, which is a reasonable approximation since the reflectance of the air-glass interface is small. We see that the beam that has made two round-trips inside the plate has already an intensity of only $T_{\text{air-glass}}^2 R_{\text{air-glass}}^3 = 0.006\%$ of that of the incident beam, whereas the single reflections by the front and back interfaces are much larger and of similar intensities (4 and 3.7% of the incident intensity, respectively).

(b) Each glass component will transmit approximately $T_{\text{plate}} = 92\%$ of the incident light, and therefore an objective made of 5 components transmits

$$T_{\text{obj}} = T_{\text{plate}}^5 = 0.66$$

that is, only 66% of light is transmitted. This is why it is necessary to treat each surface in order to minimize the loss of intensity due to the multiple reflections.

(c) The field $\underline{E}_1$ reflected by the air-dielectric interface simply writes

$$\underline{E}_1 = \underline{E}_0 r_{\text{air-diel}} \cdot$$

With respect to the field $\underline{E}_1$, $\underline{E}_2$ has traveled an additional distance $2e$ in a medium of index n, so

$$\underline{E}_2 = \underline{E}_0 t_{\text{air-diel}} e^{i\frac{2\pi}{\lambda}2ne} r_{\text{diel-glass}} t_{\text{diel-air}} \cdot$$

(d) Since $1 < n < n_{\text{glass}}$, we see that $r_{\text{air-diel}}$ and $r_{\text{diel-glass}}$ are of same sign. Since the transmission coefficients are positive, the total phase difference between $\underline{E}_1$ and $\underline{E}_2$ is simply

$$\Delta = \frac{4\pi\, e n}{\lambda} \cdot$$

For the two beams to be out of phase, we must have $\Delta = (2m + 1)\pi$ with $m \in \mathbb{N}$, and so

$$e = \frac{(2m + 1)\lambda}{4n} \qquad m \in \mathbb{N}$$

and the minimum value for the thickness is

$$e_{\text{min}} = \frac{\lambda}{4n} \cdot$$

(e) For the fields $\underline{E}_1$ and $\underline{E}_2$ to perfectly cancel each other, both must have equal amplitudes. This is true if

$$r_{\text{air-diel}} = t_{\text{air-diel}} r_{\text{diel-glass}} t_{\text{diel-air}} \cdot$$

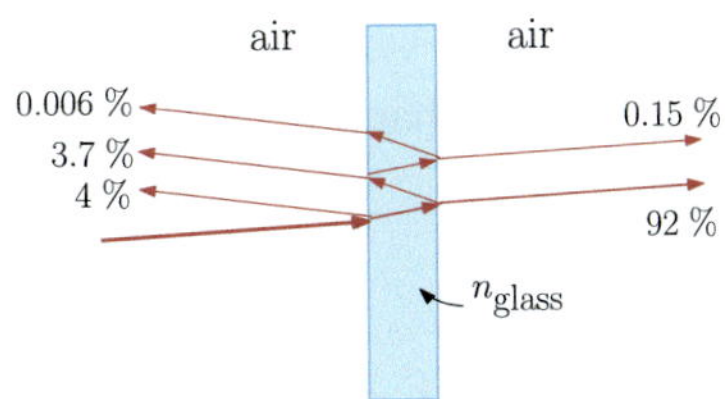

Fig. A.145 Multiple reflections here can be neglected

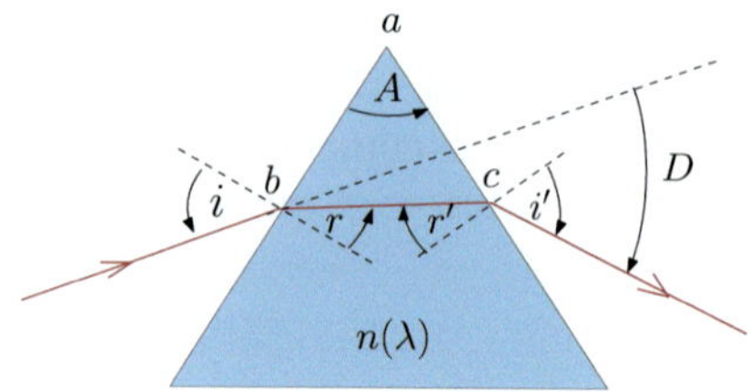

Fig. A.146 Trajectory of a beam refracted by the prism

Replacing the Fresnel coefficients in terms of $n_{\text{air}} \sim 1$, n and n_{glass}, we have

$$\frac{1-n}{1+n} = \frac{2}{1+n} \frac{n - n_{\text{glass}}}{n + n_{\text{glass}}} \frac{2n}{n + n_{\text{glass}}}$$

that is

$$(n + n_{\text{glass}})^2 (1 - n) = 4n (n - n_{\text{glass}}) \ .$$

Solving numerically, we find

$$n \approx 1.2 \ .$$

(f) The minimum thickness of MgF_2 to be deposited is

$$e = \frac{\lambda}{4n} = \frac{500 \times 10^{-9}}{4 \times 1.38} = 90.6 \text{ nm} \ .$$

Let us evaluate the amplitude of the reflected beam

$$\underline{E}_r = \underline{E}_1 + \underline{E}_2 = \underline{E}_0 \left(r_{\text{air-diel}} - t_{\text{air-diel}} r_{\text{diel-glass}} t_{\text{diel-air}} \right) \ .$$

Evaluating for $n = 1.38$ and $n_{\text{glass}} = 1.5$

$$|\underline{E}_r|^2 = |\underline{E}_0|^2 (0.159 - 0.84 \times 0.0416 \times 0.958)^2 = 0.0157 |\underline{E}_0|^2 \ .$$

That is, only 1.6% is reflected, compared to 4% for an air-bare glass interface.

15.2 The prism spectrometer

(a) From Snell–Descartes law at the entrance and exit surfaces we have

$$\sin i = n(\lambda) \sin r \quad \text{and} \quad n(\lambda) \sin r' = \sin i'.$$

On the other hand, adding the angles of the triangle abc shown in Fig. A.146, we find

$$A + \left(\frac{\pi}{2} - r\right) + \left(\frac{\pi}{2} - r'\right) = \pi$$

and so, $A = r + r'$. Replacing $r' = A - r$ into Snell–Descartes law,

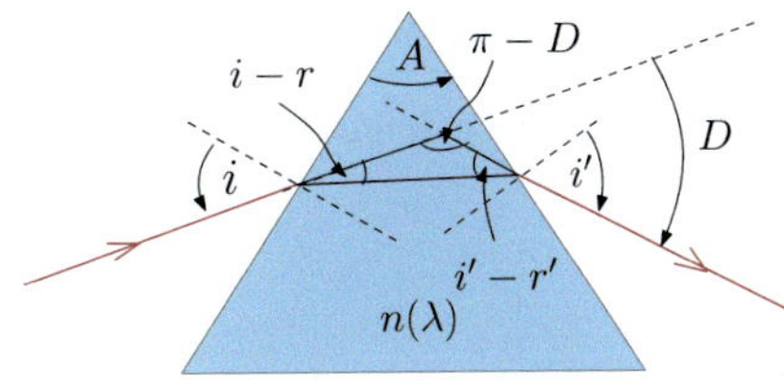

Fig. A.147 The sum of the angles of the highlighted triangle is π

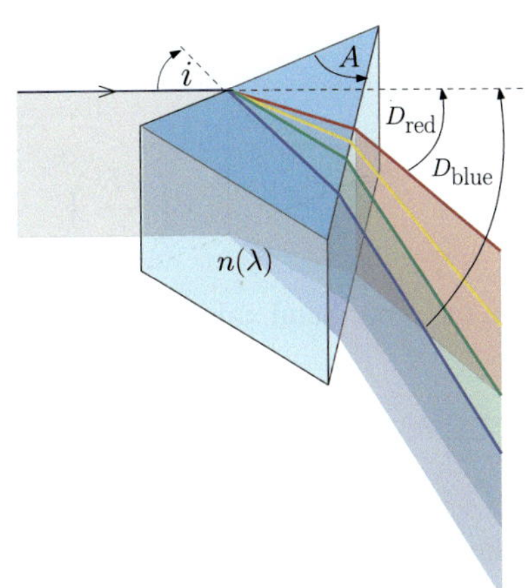

Fig. A.148 Under normal dispersion, blue light is more deviated than red

$$\sin i' = n(\lambda) \sin \left(A - \sin^{-1} \left(\frac{\sin i}{n(\lambda)} \right) \right). \tag{A.88}$$

(b) Summing the angles of the triangle highlighted in Fig. A.147, we obtain

$$(i - r) + (i' - r') + \pi - D = \pi$$

and so,

$$i + i' - (r + r') = i + i' - A = D.$$

D depends on the refractive index since $i' = i'(n)$. Replacing i', we have

$$D = i - A + \sin^{-1} \left(n(\lambda) \sin(A - \sin^{-1}(\frac{\sin i}{n(\lambda)}) \right).$$

We see that if n increases, so does the deviation angle.

(c) Since $B > 0$ (normal dispersion), the refractive index of shorter wavelengths (blue) is higher than for longer wavelengths (red). Blue is more deviated than red, as shown in Fig. A.148.

(d) The angle $i_{\min}$ that minimizes the deviation angle is such that $(\frac{dD}{di})_{i=i_{\mathrm{m}}} = 0$. Since $D = i + i' - A$, we have

Fig. A.149 Trajectory of a
light beam after one
reflection inside the droplet

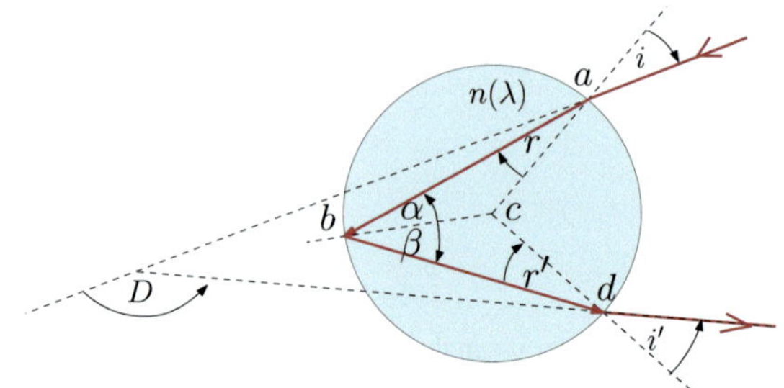

$$\left(\frac{dD}{di}\right)_{i=i_{\mathrm{m}}} = 1 + \left(\frac{di'}{di}\right)_{i=i_{\mathrm{m}}} = 0.$$

On the other hand, taking the derivative with respect to i at both sides of (A.88), we
find

$$\cos i' \frac{di'}{di} = -\frac{\cos(A - \sin^{-1}(\frac{\sin i}{n(\lambda)}))}{\cos(\sin^{-1}(\frac{\sin i}{n(\lambda)}))} \cos i,$$

where we have used $\frac{d}{dx}(\sin^{-1}(x)) = 1/\sqrt{1 - x^2} = 1/\cos(\sin^{-1} x)$. Evaluating at
$i = i_m$, we have $(di'/di)_{i=i_m} = -1$ and $i' = i = i_m$, so

$$1 = \frac{\cos(A - \sin^{-1}(\frac{\sin i_m}{n(\lambda)}))}{\cos(\sin^{-1}(\frac{\sin i_m}{n(\lambda)}))}$$

and from this we see that

$$A - \sin^{-1}(\frac{\sin i_m}{n(\lambda)}) = \sin^{-1}\left(\frac{\sin i_m}{n(\lambda)}\right).$$

The deviation angle is minimum for

$$i = i' = \sin^{-1}\left(n(\lambda)\sin\left(\frac{A}{2}\right)\right).$$

Finally, since $D = i + i' - A$, we have $D_m = 2i_m - A$ and so

$$D_m(\lambda) = 2\sin^{-1}\left(n(\lambda)\sin\left(\frac{A}{2}\right)\right) - A.$$

15.3 Deviation of sunlight by a water droplet

(a) The wave is refracted at the entrance of the droplet, then reflected at the opposite
side and finally refracted back at the exit. The trajectory is shown in Fig. A.149.

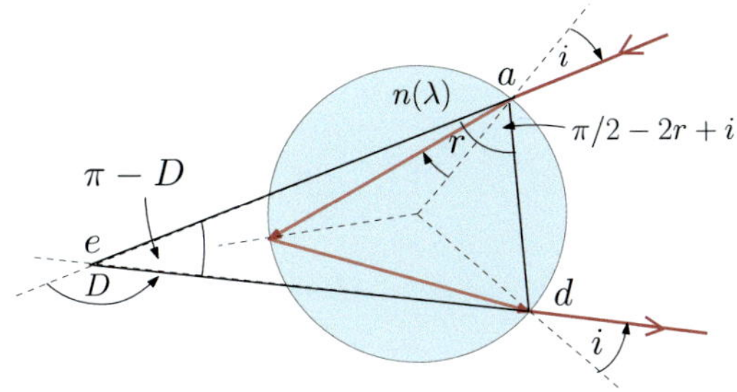
Fig. A.150 Triangle *aed* used to determine D

By Snell–Descartes law, $\sin i = n \sin r$, and since the triangle *abc* is isosceles, we have $\alpha = r$. By the law of reflection, $\beta = \alpha = r$, and the triangle *cbd* being also isosceles, we conclude that $r' = r$. Finally, since $\sin i' = n \sin r'$, we obtain $i' = i$. The deviation angle D can be obtained by considering the triangle *aed* shown in Fig. A.150, the sum of its angles reads

$$(\pi - D) + 2(\pi/2 - 2r + i) = \pi.$$

Hence,

$$D = \pi - 4r + 2i$$

and replacing r

$$D = \pi + 2i - 4 \sin^{-1} \left(\frac{\sin i}{n} \right). \tag{A.89}$$

(b) We look for i such that $(dD/di)_{i=i_m} = 0$, that is

$$\left(\frac{dD}{di} \right)_{i=i_m} = 2 - \frac{4 \cos i_m}{n \sqrt{1 - \left(\frac{\sin i_m}{n} \right)^2}} = 0,$$

which is resolved as

$$n \sqrt{1 - \left(\frac{\sin i_m}{n} \right)^2} = 2 \cos i_m = 2 \sqrt{1 - \sin^2 i_m}.$$

By squaring the above equation, we obtain

$$n^2 - \sin^2 i_m = 4 - 4 \sin^2 i_m$$

and replacing $\sin^2 i_m = 1 - \cos^2 i_m$, we find

$$i_m = \cos^{-1} \left(\frac{n^2 - 1}{3} \right).$$

Fig. A.151 In the case $n > 2$, D decreases with increasing i

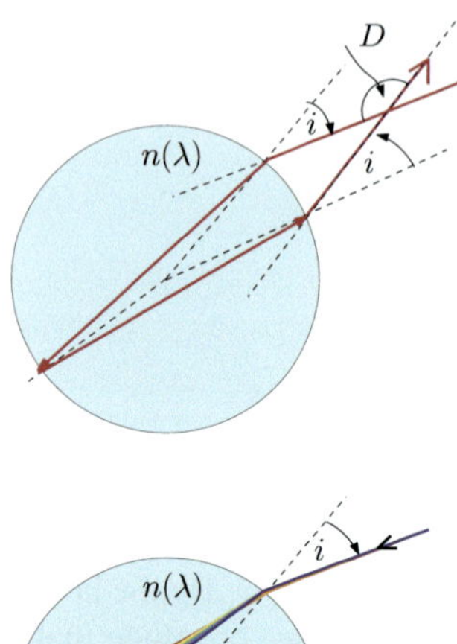

Fig. A.152 The deviation angle increases with n

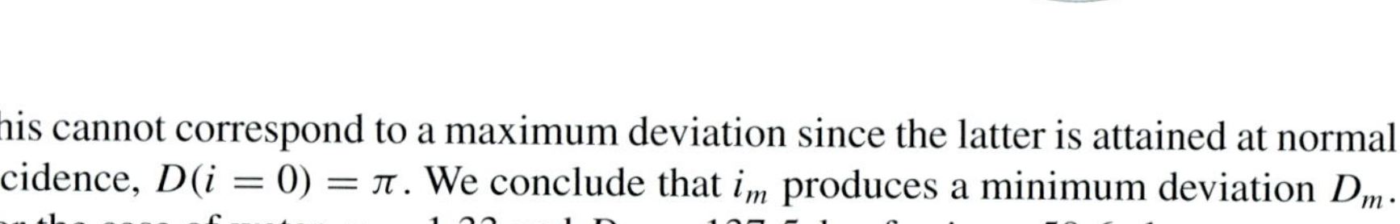

This cannot correspond to a maximum deviation since the latter is attained at normal incidence, $D(i = 0) = \pi$. We conclude that i_m produces a minimum deviation D_m. For the case of water, $n = 1.33$ and $D_m = 137.5$ deg for $i_m = 59.6$ deg.

Note finally that no solution is obtained if $n^2 - 1 > 3$, that is for $n > 2$. In this case there is a strong refraction toward the normal to the interface, and the deviation angle decreases monotonically with increasing i, but never reaching a local minimum. This is illustrated in Fig. A.151.

(c) From (A.89) we see that increasing n will yield a smaller $r = \sin^{-1}(\sin i / n)$ and therefore a larger deviation D. Blue light gets more deviated than red as shown in Fig. A.152.

15.4 The optical fiber

(a) In region i (with $i =$ core, cladding), $\underline{\mathbf{E}}_i$ must be a solution of d'Alembert's wave equation for a monochromatic wave,

$$\nabla^2 \underline{\mathbf{E}}_i + \frac{n_i^2 \omega^2}{c^2} \underline{\mathbf{E}}_i = 0.$$

Note that this is true provided that the refractive index in each medium is constant in space. Indeed when the refractive index is inhomogeneous, there is a non-zero polarization charge density. The right-hand side term in the wave equation is non-zero and the existence of longitudinal waves is possible, for which the electric field may have a non-zero component along $\mathbf{u}_z$.

Replacing the Laplacian in cylindrical coordinates and injecting a wave of the form $\underline{\mathbf{E}}_i(r, \theta, z, t) = E_i(r, \theta)e^{i(kz - \omega t)}\mathbf{u}$, we obtain

$$\frac{1}{r}\frac{\partial}{\partial r}\left(r\frac{\partial E_i(r,\theta)}{\partial r}\right) + \frac{1}{r^2}\frac{\partial^2 E_i(r,\theta)}{\partial \theta^2} - k^2 E_i(r,\theta) = -\frac{n_i^2(r)\omega^2}{c^2}E_i(r,\theta).$$

(b) Inserting a solution of the form $\mathcal{E}_i(r)e^{\pm il\theta}$ into the latter equation yields

$$\frac{1}{r}\left(\frac{d\mathcal{E}_i}{dr} + r\frac{d^2\mathcal{E}_i}{dr^2}\right)e^{\pm il\theta} - \left(\frac{l^2}{r^2} + k^2\right)\mathcal{E}_i(r)e^{\pm il\theta} = -\frac{n_i^2(r)\omega^2}{c^2}\mathcal{E}_i(r)e^{\pm il\theta}$$

and we find

$$r\frac{d\mathcal{E}_i}{dr} + r^2\frac{d^2\mathcal{E}_i}{dr^2} - \left(l^2 + r^2 k^2\right)\mathcal{E}_i(r) = -\frac{r^2 n_i^2(r)\omega^2}{c^2}\mathcal{E}_i(r),$$

which can be rewritten as

$$r\frac{d\mathcal{E}_i}{dr} + r^2\frac{d^2\mathcal{E}_i}{dr^2} + \left(\frac{r^2}{a^2}q_i^2(r) - l^2\right)\mathcal{E}_i(r) = 0,$$

with

$$q_i^2(r) = a^2\left(\frac{n_i^2\omega^2}{c^2} - k^2\right).$$

(c) A wave confined in the core of the fiber is obtained if $q_1^2 > 0$ (the field inside the core behaves like a Bessel function of the first kind) and if $q_2^2 < 0$ (the field in the cladding behaves like a Bessel function of the second kind, and therefore decays exponentially as r increases. This means

$$\frac{n_1^2\omega^2}{c^2} - k^2 > 0 \quad \text{and} \quad \frac{n_2^2\omega^2}{c^2} - k^2 < 0.$$

Hence, the modulus of the wavevector along the z-axis must satisfy

$$\frac{n_2^2\omega^2}{c^2} < k^2 < \frac{n_1^2\omega^2}{c^2}.$$

Recalling that $k_0 = \omega/c$ is the value of k associated with the wave in vacuum, finally the condition for a guided mode writes

$$n_2^2 k_0^2 < k^2 < n_1^2 k_0^2.$$

This guided electric field is the superposition of plane waves propagating in the core and reflected at the core-cladding interface. As such, $k = n_1 k_0 \cos\alpha = n_1 k_0 \sqrt{1 - \sin^2\alpha}$ where α is the angle formed by the plane waves with respect to the fiber axis. Snell–Descartes's law gives $\sin\alpha = \dfrac{\sin\theta}{n_1}$ and so we find

$$n_2^2 k_0^2 < k_0^2 (n_1^2 - \sin^2 \theta) < n_1^2 k_0^2,$$

leading to

$$0 < \sin^2 \theta < n_1^2 - n_2^2.$$

15.5 Reflection of a circularly polarized wave on a metallic conductor

(a) The incident wave is a plane wave of wavevector $\mathbf{k} = -k\mathbf{u}_x$ with $k = \omega/c$ and therefore the magnetic field can be easily obtained by writing Maxwell–Faraday law for the case of a plane wave,

$$\mathbf{B}_i = \frac{\mathbf{k}}{\omega} \times \mathbf{E}_i = -\frac{E_0}{c} \mathbf{u}_x \times (\cos(\omega t + kx)\mathbf{u}_y + \sin(\omega t + kx)\mathbf{u}_z)$$

$$= \frac{E_0}{c}(-\cos(\omega t + kx)\mathbf{u}_z + \sin(\omega t + kx)\mathbf{u}_y)$$

and the incident Poynting vector reads

$$\mathbf{\Pi}_i = \frac{\mathbf{E}_i \times \mathbf{B}_i}{\mu_0} = -\frac{E_0^2}{\mu_0 c}\mathbf{u}_x.$$

(b) The electric field is zero everywhere inside the conductor, assumed here to be perfect.

(c) Since the wave is totally reflected at the interface, the continuity of the tangential component of the electric field gives

$$\mathbf{E}_i(0) + \mathbf{E}_r(0) = \mathbf{0} \quad \rightarrow \quad \mathbf{E}_r(0) = -E_0(\cos \omega t \, \mathbf{u}_y + \sin \omega t \, \mathbf{u}_z).$$

and since the reflected wave propagates along the x-axis, we find

$$\mathbf{E}_r = -E_0(\cos(\omega t - kx)\mathbf{u}_y + \sin(\omega t - kx)\mathbf{u}_z).$$

This plane wave is characterized by a wavevector $\mathbf{k}_r = k\mathbf{u}_x = \dfrac{\omega}{c}\mathbf{u}_x$ and the reflected magnetic field is obtained by

$$\mathbf{B}_r = \frac{\mathbf{k}_r}{\omega} \times \mathbf{E}_r = \frac{1}{c}\mathbf{u}_x \times \mathbf{E}_r = \frac{E_0}{c}(-\cos(\omega t - kx)\,\mathbf{u}_z + \sin(\omega t - kx)\,\mathbf{u}_y).$$

(d) The total electric field $\mathbf{E} = \mathbf{E}_i + \mathbf{E}_r$ for $x > 0$ is given by

$$\mathbf{E} = E_0(\cos(\omega t + kx) - \cos(\omega t - kx))\mathbf{u}_y + E_0(\sin(\omega t + kx) - \sin(\omega t - kx))\mathbf{u}_z$$

$$= 2E_0 \sin(kx)\left(-\sin \omega t \mathbf{u}_y + \cos \omega t \mathbf{u}_z\right)$$

and similarly for the total magnetic field $\mathbf{B} = \mathbf{B}_i + \mathbf{B}_r$,

$$\mathbf{B} = \frac{E_0}{c}(-\cos(\omega t + kx) - \cos(\omega t - kx))\mathbf{u}_z + \frac{E_0}{c}(\sin(\omega t + kx) + \sin(\omega t - kx))\mathbf{u}_y$$

$$= \frac{2E_0}{c}\cos(kx)\left(-\cos\omega t\,\mathbf{u}_z + \sin\omega t\mathbf{u}_y\right).$$

From this, we obtain the total Poynting vector

$$\mathbf{\Pi} = \frac{\mathbf{E}\times\mathbf{B}}{\mu_0} = \mathbf{0},$$

so that there is no net energy flux through any plane at fixed z, the flux of the incident wave is canceled by that of the reflected wave.

(e) At $x < 0$ the electric field is zero and equally $\mathbf{E}(0^+) = 0$ so that the normal component of the electric field is continuous across $x = 0$, consistent with the fact no charge density is present at the interface. The magnetic field at $x = 0^+$ is given by

$$\mathbf{B}(0) = \frac{2E_0}{c}(\sin\omega t\,\mathbf{u}_y - \cos\omega t\,\mathbf{u}_z),$$

whereas inside the conductor $\mathbf{B} = 0$. The tangential component of the magnetic field is therefore discontinuous at $x = 0$. The jump condition

$$\mathbf{u}_x \times \left(\mathbf{B}(0^+) - \underbrace{\mathbf{B}(0^-)}_{=0}\right) = \mu_0\mathbf{j}_S$$

allows us to determine the surface current density $\mathbf{j}_S$, and so, the conductor carries a surface current

$$\mathbf{j}_S = \frac{2E_0}{\mu_0 c}(\sin\omega t\,\mathbf{u}_z + \cos\omega t\,\mathbf{u}_y)$$

which oscillates in phase with the incident electric field at the surface. The electromagnetic wave caused by this current cancels the incident one everywhere inside the conductor and produces the reflected wave at $x > 0$.

15.6 Electromagnetic cavity

(a) Inside the conductors, the electromagnetic wave vanishes and in particular $\mathbf{E} = \mathbf{0}$. By continuity of the tangential component of the electric field at an interface, we have $x = 0$

$$\underline{\mathbf{E}}(0, t) = \underline{E}_1 e^{-i\omega t}\mathbf{u}_y + \underline{E}_2 e^{-i\omega t}\mathbf{u}_y = 0 \quad \forall t$$

and we conclude $\underline{E}_2 = -\underline{E}_1$. Thus, the electric field can be rewritten as

$$\underline{\mathbf{E}}(x, t) = \underline{E}_1\left(e^{i(kx-\omega t)} - e^{i(-kx-\omega t)}\right)\mathbf{u}_y = 2i\underline{E}_1\sin(kx)e^{-i\omega t}\mathbf{u}_y.$$

The continuity of the tangential component of $\mathbf{E}$ at the interface $x = L$ yields

$$\sin kL = 0 \quad \rightarrow \quad k_n = \frac{n\pi}{L}, \quad n \in \mathbb{N}$$

and so

$$\underline{\mathbf{E}}_n(x, t) = 2i\,\underline{E}_1 \sin(k_n x)e^{-i\omega t}\mathbf{u}_y \; .$$

(b) The electric field $\underline{\mathbf{E}}_n$ must be a solution to d'Alembert's wave equation. Since it only depends on x,

$$\frac{\partial^2 \underline{\mathbf{E}}_n}{\partial x^2} - \frac{1}{c^2}\frac{\partial^2 \underline{\mathbf{E}}_n}{\partial t^2} = 0.$$

We have

$$\frac{\partial^2 \underline{\mathbf{E}}_n}{\partial x^2} = -2ik_n^2\underline{E}_1 \cos(k_n x)e^{-i\omega t}\mathbf{u}_y = -k_n\underline{\mathbf{E}}$$

$$\frac{\partial^2 \underline{\mathbf{E}}_n}{\partial t^2} = -2i\omega^2\underline{E}_1 \cos(k_n x)e^{-i\omega t}\mathbf{u}_y = -\omega^2\underline{\mathbf{E}}$$

and this gives the dispersion relation

$$-k_n^2 + \frac{\omega_n^2}{c^2} = 0 \quad \rightarrow \quad \omega_n = ck_n = \frac{n\pi c}{L}, \quad n \in \mathbb{N}^+.$$

(c) For the mode n, the real electric field reads

$$\mathbf{E}_n(x, t) = \mathrm{Re}\{2i\,\underline{E}_1 \sin(k_n x)e^{-i\omega_n t}\mathbf{u}_y\}$$

and by writing $\underline{E}_1 = E_1 e^{i\phi}$, we obtain

$$\mathbf{E}_n(x, t) = 2E_1 \sin(k_n x)\sin(\omega_n t - \phi)\mathbf{u}_y.$$

The wave clearly does not propagate, since it cannot be written as a function of $x \pm ct$. The spatial dependence is always the same, only the overall amplitude oscillates in time. This is called a standing wave.

(d) The node positions x_p are such that

$$\sin k_n x_p = 0 \quad \rightarrow \quad x_p = \frac{p\pi}{k_n} = \frac{pL}{n} = \frac{p\lambda_n}{2},$$

where $\lambda_n = 2\pi/k_n = 2L/n$. Two successive nodes are thus separated by $\lambda_n/2$.

(e) From Faraday–Maxwell's equation, we have

$$\underbrace{\nabla \times \mathbf{E}_n}_{\frac{\partial E_y}{\partial x}\mathbf{u}_z} = -\frac{\partial \mathbf{B}_n}{\partial t}$$

and so, $\mathbf{B}_n = B_z\mathbf{u}_z$, with

$$\frac{\partial B_z}{\partial t} = -2E_1 k_n \cos(k_n x)\sin(\omega_n t - \phi) \quad \rightarrow \quad B_z = \frac{2E_1}{c}\cos(k_n x)\cos(\omega_n t - \phi).$$

The nodes of the magnetic field are given by $\cos k_n x'_p = 0$, i.e.,

$$x'_p = \left(\frac{\pi}{2} + p\pi\right)\frac{1}{k_n} = \frac{\lambda_n}{4} + \frac{p\lambda_n}{2}$$

so that the electric and magnetic field nodes are separated by

$$x'_p - x_p = \frac{\lambda_n}{4}.$$

The most general solution for an electric field depending on x inside the cavity will be a superposition of modes for each of the two possible orthogonal polarizations along $\mathbf{e}_y$ and $\mathbf{e}_z$,

$$\mathbf{E}(x, t) = \sum_n A_n \sin(k_n x)\sin(\omega_n t - \phi_y)\mathbf{u}_y + \sum_n B_n \sin(k_n x)\sin(\omega_n t - \phi_z)\mathbf{u}_z.$$

15.7 Guided wave

(a) The wave is described by a function of $ky - \omega t = k(y - (\omega/k)t)$, so it propagates along the direction of increasing values of y at a speed ω/k. Using the identity $\sin A \cos B = 1/2\,(\sin(A - B) + \sin(A + B))$, we can rewrite the electric field as

$$\mathbf{E} = \frac{E_0}{2}\underbrace{\sin\left(ky + \frac{\pi x}{L} - \omega t\right)}_{\mathbf{k}_+ \cdot \mathbf{x} - \omega t}\mathbf{u}_z - \frac{E_0}{2}\underbrace{\sin\left(ky - \frac{\pi x}{L} - \omega t\right)}_{\mathbf{k}_- \cdot \mathbf{x} - \omega t}\mathbf{u}_z.$$

We recognize the superposition of two plane waves having the same frequency ω and wavevector

$$\mathbf{k}_+ = k\mathbf{u}_y + \frac{\pi}{L}\mathbf{u}_x \qquad \mathbf{k}_- = k\mathbf{u}_y - \frac{\pi}{L}\mathbf{u}_x,$$

so that the components of the wavevector along the x-axis represent counterpropagating waves. The $\mathbf{k}_+$ (resp. $\mathbf{k}_-$) wave plays the role of an incident (resp. reflected) wave for the plane wall at $x = L$, and vice versa for the plane wall at $x = 0$. Each wave propagates between the walls and bounces on them, ensuring under certain condition to be detailed below, that it is guided along the y direction. so that in the x-axis they are counterpropagating waves, shown in Fig. A.153.

Fig. A.153 The field is a superposition of plane waves propagating along $\mathbf{k}_+$ and $\mathbf{k}_-$

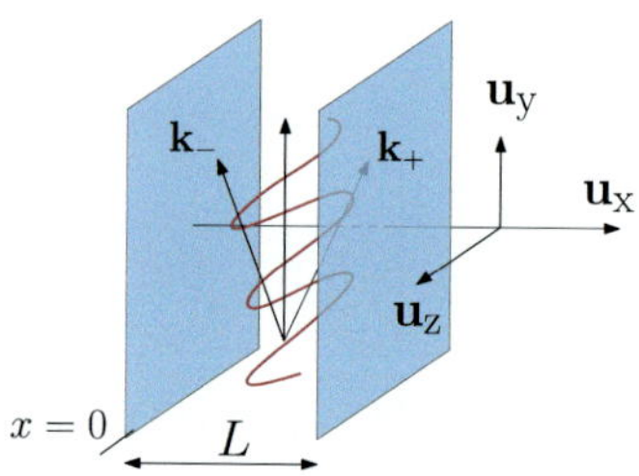

(b) Assuming perfect conductive plates, the wave vanishes everywhere inside them. If no charge is present at the interfaces, the electric field must be continuous (both its normal and tangential components), and so the field is zero at $x = 0$ and $x = L$.

(c) The electric field must be a solution to d'Alembert's wave equation

$$\nabla^2 \mathbf{E} - \frac{1}{c^2}\frac{\partial^2 \mathbf{E}}{\partial t^2} = 0.$$

We have

$$\nabla^2 \mathbf{E} = \frac{\partial^2 \mathbf{E}}{\partial x^2} + \frac{\partial^2 \mathbf{E}}{\partial x^y} = -\left(\frac{\pi^2}{L^2} + k^2\right)\mathbf{E}$$

and $\dfrac{\partial^2 \mathbf{E}}{\partial t^2} = -\omega^2 \mathbf{E}$ so that the dispersion relation reads

$$k^2 + \frac{\pi^2}{L^2} = \frac{\omega^2}{c^2}.$$

For the wave to propagate along the y-axis, the wavenumber k must be real. This gives a minimum frequency ω for the wave to be able to propagate along the axis of the waveguide

$$k^2 = \frac{\omega^2}{c^2} - \frac{\pi^2}{L^2} > 0$$

and so

$$\omega > \frac{c\pi}{L}.$$

(d) The magnetic field can be obtained by using Maxwell–Faraday law

$$\nabla \times \mathbf{E} = \frac{\partial \mathbf{E}_z}{\partial y}\mathbf{u}_x - \frac{\partial \mathbf{E}_z}{\partial x}\mathbf{u}_y = -\frac{\partial \mathbf{B}}{\partial t}.$$

Calculating and replacing $\nabla \times \mathbf{E}$, this yields

$$\frac{\partial \mathbf{B}}{\partial t} = k E_0 \sin\left(\frac{\pi}{L}x\right)\sin(ky - \omega t)\mathbf{u}_x + \frac{\pi}{L}E_0 \cos\left(\frac{\pi}{L}x\right)\cos(ky - \omega t)\mathbf{u}_y.$$

Integrating with respect to time, we find

$$\mathbf{B} = \frac{k}{\omega}E_0 \sin\left(\frac{\pi}{L}x\right)\cos(ky - \omega t)\mathbf{u}_x - \frac{\pi}{\omega L}E_0 \cos\left(\frac{\pi}{L}x\right)\sin(ky - \omega t)\mathbf{u}_y.$$

(e) The electromagnetic energy density u_{EM} writes

$$u_{EM} = \frac{\epsilon_0}{2}|\mathbf{E}|^2 + \frac{1}{2\mu_0}|\mathbf{B}|^2 = \frac{\epsilon_0}{2}(|\mathbf{E}|^2 + c^2|\mathbf{B}|^2)$$

$$= \frac{\epsilon_0 E_0^2}{2}\left[\left(1 + \frac{c^2 k^2}{\omega^2}\right)\sin^2\left(\frac{\pi x}{L}\right)\cos^2(ky - \omega t) + \frac{\pi^2 c^2}{\omega^2 L^2}\cos^2\left(\frac{\pi x}{L}\right)\sin^2(ky - \omega t)\right].$$

Using the fact that the average in time of a $\sin^2(t)$ or $\cos^2(t)$ function is equal to $1/2$, the time-averaged density is therefore

$$\langle u_{EM}\rangle_T = \frac{\epsilon_0 E_0^2}{4}\left(1 + \frac{c^2 k^2}{\omega^2}\right)\sin^2\left(\frac{\pi x}{L}\right) + \frac{\epsilon_0 \pi^2 c^2 E_0^2}{4\omega^2 L^2}\cos^2\left(\frac{\pi x}{L}\right).$$

Now we can average in the x direction over the length L of the cavity, noting that $\langle \sin^2(\pi x/L)\rangle = \langle \cos^2(\pi x/L)\rangle = 1/2$ and finally,

$$\langle\langle u_{EM}\rangle_T\rangle = \frac{\epsilon_0 E_0^2}{8}\left(1 + \underbrace{\frac{c^2 k^2}{\omega^2} + \frac{\pi^2 c^2}{\omega^2 L^2}}_{1}\right) = \frac{\epsilon_0 E_0^2}{4}.$$

The Poynting vector is given by

$$\mathbf{\Pi} = \frac{\mathbf{E}\times\mathbf{B}}{\mu_0} = \epsilon_0 c^2 \mathbf{E}\times\mathbf{B}$$

$$= \epsilon_0 c^2 \frac{E_0^2 k}{\omega}\sin^2\left(\frac{\pi x}{L}\right)\cos^2(ky - \omega t)\mathbf{u}_y + \epsilon_0 c^2 \frac{E_0^2 \pi}{2\omega L}\sin\left(\frac{2\pi x}{L}\right)\cos(2(ky - \omega t))\mathbf{u}_x$$

whose average over time and space gives

$$\langle\langle\mathbf{\Pi}\rangle_T\rangle = \epsilon_0 c^2 \frac{E_0^2 k}{4\omega}\mathbf{u}_y = \frac{c^2 k}{\omega}\langle u_{EM}\rangle_T\mathbf{u}_y.$$

(f) Let S_y be a surface perpendicular to the propagation direction. The flux of the Poynting vector through S_y represents a power, that is, a quantity of energy crossing the surface during a time interval Δt, expressed as

$$\Delta E = |\langle\langle\mathbf{\Pi}\rangle_T\rangle| S_y \Delta t = v_e \langle\langle u_{EM}\rangle_T\rangle S_y \Delta t,$$

where v_e is the speed at which the energy propagates. The right-hand side corresponds to the quantity of energy in the cylinder of bases S_y and height $v_e \Delta t$ with density $\langle\langle u_{EM}\rangle_T\rangle$, i.e., the energy that will cross the surface S_y during the time interval Δt. Since $|\langle\langle\mathbf{\Pi}\rangle_T\rangle| = \dfrac{c^2 k}{\omega} \langle\langle u_{EM}\rangle_T\rangle$, we obtain

$$\mathbf{v}_e = \frac{c^2 k}{\omega} \mathbf{u}_y = c\sqrt{1 - \left(\frac{\pi c}{\omega L}\right)^2}\,\mathbf{u}_y.$$

15.8 Reflexion and transmission coefficients of a thin dielectric plate: the Fabry–Pérot resonator

(a) The incident wave will be partially reflected by and partially transmitted through the entrance surface at $x = 0$, and the same will happen at the exit surface $x = e$.

(b) In the complex representation, the incident field writes

$$\underline{\mathbf{E}}_i = \underline{E}_0 e^{i(n_1 kx - \omega t)} \mathbf{u}_z, \quad x < 0,$$

the reflected wave,

$$\underline{\mathbf{E}}_r = \underline{r}\, \underline{E}_0 e^{i(-n_1 kx - \omega t)} \mathbf{u}_z, \quad x < 0$$

and the transmitted wave,

$$\underline{\mathbf{E}}_t = \underline{t}\, \underline{E}_0 e^{i(n_1 kx - \omega t)} \mathbf{u}_z \quad x > e.$$

Finally, inside the dielectric plate, there will be an electric field resulting from the superposition of the transmitted wave through the entrance surface and the reflected wave by the exit surface. It is a superposition of a progressive and a regressive plane wave with wavenumber $n_2 k$, which reads

$$\underline{\mathbf{E}}_{\text{plate}} = \underline{E}_1 e^{i(n_2 kx - \omega t)} \mathbf{u}_z + \underline{E}_2 e^{i(-n_2 kx - \omega t)} \mathbf{u}_z.$$

The incident, reflected and transmitted magnetic fields write

$$\underline{\mathbf{B}}_i = \frac{n_1 k}{\omega} \mathbf{u}_x \times \underline{\mathbf{E}}_i = -\frac{n_1 \underline{E}_0}{c} e^{i(n_1 kx - \omega t)} \mathbf{u}_y,$$

$$\underline{\mathbf{B}}_r = -\frac{n_1 k}{\omega} \mathbf{u}_x \times \underline{\mathbf{E}}_r = \underline{r}\,\frac{n_1 \underline{E}_0}{c} e^{i(-n_1 kx - \omega t)} \mathbf{u}_y,$$

$$\underline{\mathbf{B}}_t = \frac{n_1 k}{\omega} \mathbf{u}_x \times \underline{\mathbf{E}}_t = -\underline{t}\,\frac{n_1 \underline{E}_0}{c} e^{i(n_1 kx - \omega t)} \mathbf{u}_y.$$

and for the magnetic field inside the plate, we can write

$$\mathbf{B}_{\text{plate}} = \frac{n_2 k \mathbf{u}_x}{\omega} \times \underline{E}_1 e^{i(n_2 kx - \omega t)} \mathbf{u}_z + \frac{-n_2 k \mathbf{u}_x}{\omega} \underline{E}_2 e^{i(-n_2 kx - \omega t)} \mathbf{u}_z$$

$$= -\frac{n_2 \underline{E}_1}{c} e^{i(n_2 kx - \omega t)} \mathbf{u}_y + \frac{n_2 \underline{E}_2}{c} e^{i(-n_2 kx - \omega t)} \mathbf{u}_y.$$

(c) At $x = 0$, the tangential components of the electric and magnetic field are continuous (we assume the absence of a current surface at the interface), which yields $\left(\underline{\mathbf{E}}_i + \underline{\mathbf{E}}_r\right)|_{x=0} = \underline{\mathbf{E}}_{\text{plate}}|_{x=0}$ and so

$$\underline{E}_0 \left(1 + \underline{r}\right) = \underline{E}_1 + \underline{E}_2, \tag{A.90}$$

and $\left(\underline{\mathbf{B}}_i + \underline{\mathbf{B}}_r\right)|_{x=0} = \underline{\mathbf{B}}_{\text{plate}}|_{x=0}$ gives

$$n_1 \underline{E}_0 \left(\underline{r} - 1\right) = n_2 \left(\underline{E}_2 - \underline{E}_1\right). \tag{A.91}$$

The continuity conditions are now applied to the exit surface $x = e$, $\underline{\mathbf{E}}_t|_{x=e} = \underline{\mathbf{E}}_{\text{plate}}|_{x=e}$, so

$$\underline{t}\, \underline{E}_0 e^{i n_1 k e} = \underline{E}_1 e^{i n_2 k e} + \underline{E}_2 e^{-i n_2 k e} \tag{A.92}$$

and $\underline{\mathbf{B}}_t|_{x=e} = \underline{\mathbf{B}}_{\text{plate}}|_{x=e}$, i.e.,

$$-\underline{t}\, n_1 \underline{E}_0 e^{i n_1 k e} = n_2 \left(\underline{E}_2 e^{-i n_2 k e} - \underline{E}_1 e^{i n_2 k e}\right). \tag{A.93}$$

Multiplying both sides of (A.90) by n_2 and combining with (A.91), we get

$$\underline{E}_1 = \frac{\underline{E}_0}{2} \left(\frac{n_1 + n_2}{n_2} + \underline{r}\frac{n_2 - n_1}{n_2}\right),$$

$$\underline{E}_2 = \frac{\underline{E}_0}{2} \left(\underline{r}\frac{n_2 + n_1}{n_2} + \frac{n_2 - n_1}{n_2}\right),$$

and recalling Fresnel's coefficients at normal incidence for the two interfaces $t_{12} = \dfrac{2n_1}{n_1 + n_2}$, $t_{21} = \dfrac{2n_2}{n_1 + n_2}$, $r_{12} = \dfrac{n_1 - n_2}{n_1 + n_2}$ and $r_{21} = \dfrac{n_2 - n_1}{n_1 + n_2} = -r_{12}$, the result can be rewritten as

$$\underline{E}_1 = \underline{E}_0 \left(\frac{1}{t_{21}} + \underline{r}\frac{r_{21}}{t_{21}}\right),$$

$$\underline{E}_2 = \underline{E}_0 \left(\underline{r}\frac{1}{t_{21}} + \frac{r_{21}}{t_{21}}\right).$$

Equations (A.92) and (A.93) then become

$$\underline{t}e^{in_1ke} = \frac{1}{t_{21}}e^{in_2ke} + \frac{r_{21}}{t_{21}}e^{-in_2ke} + \underline{r}\left(\frac{r_{21}}{t_{21}}e^{in_2ke} + \frac{1}{t_{21}}e^{-in_2ke}\right),$$

$$\underline{t}e^{in_1ke} = \left(\frac{1}{t_{12}}e^{in_2ke} - \frac{r_{21}}{t_{12}}e^{-in_2ke}\right) + \underline{r}\left(\frac{r_{21}}{t_{12}}e^{in_2ke} - \frac{1}{t_{12}}e^{-in_2ke}\right).$$

where we have used the relation $\dfrac{n_2}{n_1 t_{21}} = \dfrac{1}{t_{12}}$. The latter system can be solved by using $t_{12} + t_{21} = 2$, $t_{12} - t_{21} = 2r_{12}$ and $1 - r_{21}^2 = t_{12}t_{21}$. We find

$$\boxed{\begin{aligned} \underline{r} &= \frac{r_{21}(e^{2in_2ke} - 1)}{1 - r_{21}^2 e^{2in_2ke}}. \\[2ex] \underline{t} &= \frac{t_{12}t_{21}e^{i(n_2-n_1)ke}}{1 - r_{21}^2 e^{2in_2ke}}. \end{aligned}}$$

(d) The transmission through the dielectric plate is given by the transmittance

$$T = |\underline{t}|^2 = \frac{|t_{12}|^2 |t_{21}|^2}{|1 - Re^{\Delta\phi}|^2}$$

where $R = r_{12}^2 = r_{21}^2$ is the reflectance of each interface. Since $|t_{12}|^2|t_{21}|^2 = (1-R)^2$ and $|1 - Re^{i\Delta\phi}|^2 = (1 - Re^{i\Delta\phi})(1 - Re^{-i\Delta\phi})$,

$$T = \frac{(1-R)^2}{1 + R^2 - 2R\cos\Delta\phi}.$$

Finally, since $\cos\Delta\phi = 1 - 2\sin^2(\Delta\phi/2)$, we find

$$T = \frac{(1-R)^2}{1 + R^2 - 2R + 4R\sin^2(\Delta\phi/2)} = \frac{1}{1 + \underbrace{\frac{4R}{(1-R)^2}}_{\mathcal{F}^2}\sin^2(\Delta\phi/2)},$$

where $\mathcal{F} = \dfrac{2\sqrt{R}}{1-R}$ is called the cavity finesse coefficient. We see that the transmission through the dielectric plate varies periodically with $\Delta\phi$ and attains a maximum value of $T_{\max} = 1$ for $\Delta\phi = 2p\pi$ with $p \in \mathbb{N}$, as shown in Fig. A.154. This is a remarkable result since there are always particular values of $\Delta\phi$ for which the wave is completely transmitted through the plate, even though each interface has a non-zero reflectivity! This is due to constructive interferences of all the waves being reflected multiple times inside the cavity that eventually add up at the exit of the plate. The minimum value of T is simply $T_{\min} = 1/(1 + \mathcal{F}^2)$. If the finesse tends to zero, that is for a very poor contrast index between the plate and its surroundings, then the transmittance

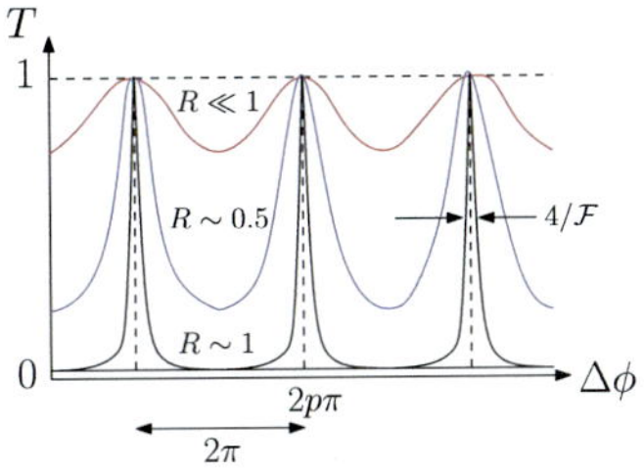

Fig. A.154 Transmission
coefficient v/s $\Delta\phi$ for
different values of R

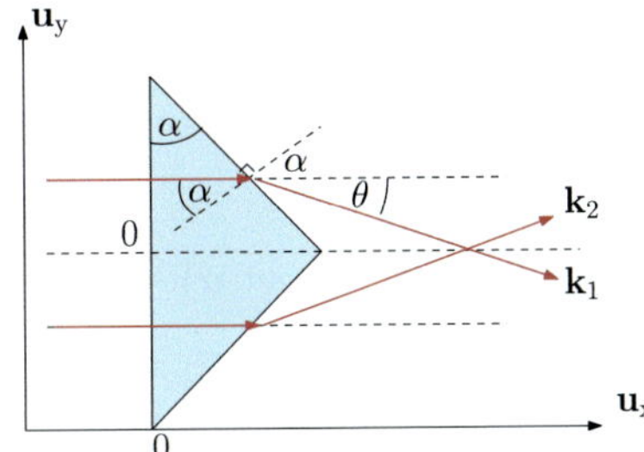

Fig. A.155 The beams form
an angle of α at the prism-air
interface

varies weakly and slowly around its maximum of 1. This corresponds to the case of an almost transparent dielectric, such as a glass plate.

However, for values of R approaching unity, the finesse tends to infinity and the transmittance oscillates between ~ 0 and 1, with very narrow resonances whose widths vary as $1/\mathcal{F}$. Having highly reflective surfaces (and therefore high finesse $\mathcal{F}$) helps to build up the interferences between the multiple beams at the exit of the waveplate, leading to a perfect transmittance if they all arrive in phase at the exit of the slit.

Problems of Chap. 16

16.1 Fresnel's biprism

(a) After the passage through the first interface ($x = 0$) the wavevector is not deviated since the waves arrive at normal incidence. Only its modulus change because of the refractive index of the prism, that is $\mathbf{k}' = nk\mathbf{u}_x$.

By symmetry, if $\mathbf{k}_1 = k(\cos\theta\mathbf{u}_x - \sin\theta\mathbf{u}_y)$ is the wavevector at the output of the upper half of the biprism, then the wavevector $\mathbf{k}_2$ at the output of the lower part will be the symmetric of $\mathbf{k}_1$ with respect to the plane $y = 0$, i.e., $\mathbf{k}_2 = k(\cos\theta\mathbf{u}_x + \sin\theta\mathbf{u}_y)$. The angle θ can be obtained by applying Snell–Descartes law after noticing that the waves are incident at the second prism-air interface at an angle α with respect to the normal, as shown in Fig. A.155.

$$n \sin\alpha = \sin(\alpha + \theta).$$

Fig. A.156 Interference pattern produced by the biprism

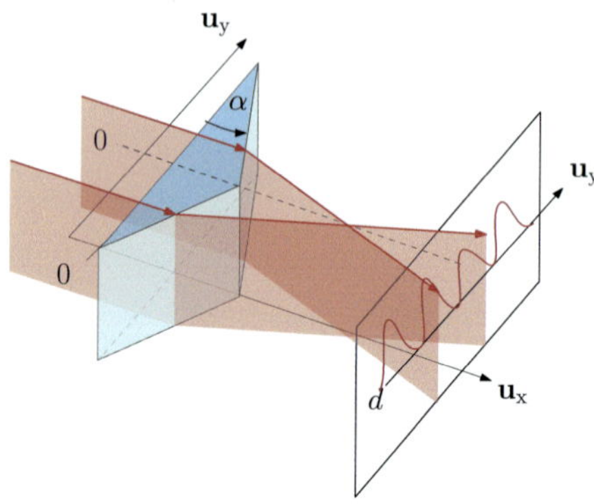

so that $\theta = \sin^{-1}(n \sin \alpha) - \alpha$ and

$$\sin \theta = \sin(\sin^{-1}(n \sin \alpha)) \cos \alpha - \sin \alpha \cos(\sin^{-1}(n \sin \alpha)),$$
$$\cos \theta = \cos(\sin^{-1}(n \sin \alpha)) \cos \alpha + \sin \alpha \sin(\sin^{-1}(n \sin \alpha)).$$

Using $\cos(\sin^{-1}(x)) = \sqrt{1 - x^2}$, we obtain

$$\sin \theta = n \sin \alpha \cos \alpha - \sin \alpha \sqrt{1 - n^2 \sin^2 \alpha},$$
$$\cos \theta = n \sin \alpha \sin \alpha + \cos \alpha \sqrt{1 - n^2 \sin^2 \alpha}.$$

(b) The total electric field at a point $P = (x = d, y, z)$ on the screen will be the superposition of $\underline{\mathbf{E}}_1(d, y, z)$ and $\underline{\mathbf{E}}_2(d, y, z)$, that is

$$\underline{\mathbf{E}}(y, z) = \underline{\mathbf{E}}_0 e^{i(k \cos \theta d - \omega t)} \left(e^{-ik \sin \theta y} + e^{ik \sin \theta y} \right) = 2\underline{\mathbf{E}}_0 e^{i(k \cos \theta d - \omega t)} \cos(k \cos \theta y).$$

Finally, by writing $\underline{\mathbf{E}}_0 = \mathbf{E}_0 e^{i\phi}$, we find

$$\mathbf{E}(y, z) = 2\mathbf{E}_0 \cos(k \cos \theta d - \omega t + \phi) \cos(k \cos \theta y).$$

(c) The intensity being proportional to the squared modulus of the total electric field, we see that it only depends on the y coordinate and therefore,

$$\mathcal{I}(y) = 4\mathcal{I}_0 \cos^2(k \cos \theta y),$$

where $\mathcal{I}_0$ corresponds to the maximum intensity that would be obtained by keeping only half of the incoming beams. The intensity then varies on the screen periodically between $4\mathcal{I}_0$ and 0 along the y-axis, due to the interference between the two output beams coming from the biprism. This is shown in Fig. A.156. The observed pattern is a sequence of bright and dark vertical fringes, spaced by a distance $\Delta y = \lambda/(2 \sin \theta) = \lambda/(2n \sin \alpha \cos \alpha - \sin \alpha \sqrt{1 - n^2 \sin^2 \alpha})$.

16.2 Degree of coherence of a doublet

(a) The field's amplitude at a given position reads

$$\underline{E}(t) = \underline{E}_1 e^{-i\omega_1 t} + \underline{E}_2 e^{-i\omega_2 t}$$

and so

$$\underline{E}^*(t)\underline{E}(t-\tau) = |\underline{E}_1|^2 e^{i\omega_1\tau} + |\underline{E}_1|^2 e^{i\omega_2\tau} + \underline{E}_1^*\underline{E}_2 e^{-i(\omega_2-\omega_1)t} e^{i\omega_2\tau} + \underline{E}_1\underline{E}_2^* e^{i(\omega_2-\omega_1)t} e^{i\omega_1\tau} \, .$$

Since $T_{\text{det}} \gg \frac{2\pi}{|\omega_1-\omega_2|}$, the average of the terms oscillating at frequency $\omega_2 - \omega_1$ is zero. If both monochromatic waves have equal amplitude, we obtain

$$g(\tau) = \frac{\langle \underline{E}^*(t)\underline{E}(t-\tau)\rangle_{T_{\text{det}}}}{\langle |\underline{E}(t)|^2\rangle_{T_{\text{det}}}} = \frac{1}{2}(e^{-i\omega_1\tau} + e^{-i\omega_2\tau}) \, .$$

The intensity measured by the interferometer is then given by

$$\mathcal{I}(dl) = 2\mathcal{I}_0 \left(1 + \frac{1}{2}\cos\omega_1\frac{dl}{c} + \frac{1}{2}\cos\omega_2\frac{dl}{c} \right) \tag{A.94}$$

which is simply the sum of the interference patterns produced by each of the monochromatic waves separately, consistent with the fact that two non-synchronized waves are mutually incoherent. By using the identity $\cos a + \cos b = 2\cos((a+b)/2)\cos((a-b)/2)$ we find

$$\mathcal{I}(dl) = 2\mathcal{I}_0 \left(1 + \cos\left(\frac{\Delta\omega}{2}\frac{dl}{c}\right)\cos\left(\omega_0\frac{dl}{c}\right) \right) \, .$$

From Eq. (16.6) we see that the contrast depends on the path length dl and it is given by

$$C(dl) = \cos\left(\frac{\Delta\omega}{2}\frac{dl}{c}\right) \, .$$

(b) The path length is in this case $dl = ax/D$, and so the interference pattern at position x on the screen reads

$$\mathcal{I}(x) = 2\mathcal{I}_0 \left(1 + \cos\left(\frac{\Delta\omega}{2}\frac{ax}{cD}\right)\cos\left(\omega_0\frac{ax}{cD}\right) \right) \, .$$

In terms of wavelength, we have $\lambda_1 = 2\pi c/\omega_1$, $\lambda_2 = 2\pi c/\omega_2$, $\lambda_0 \sim 2\pi c/\omega_0$ and

$$\Delta\lambda = \lambda_1 - \lambda_2 = \frac{\Delta\omega\lambda_1\lambda_2}{2\pi c} \sim \frac{\Delta\omega\lambda_0^2}{2\pi c}$$

Fig. A.157 Interference
pattern produced by a
doublet placed in front of a
double slit

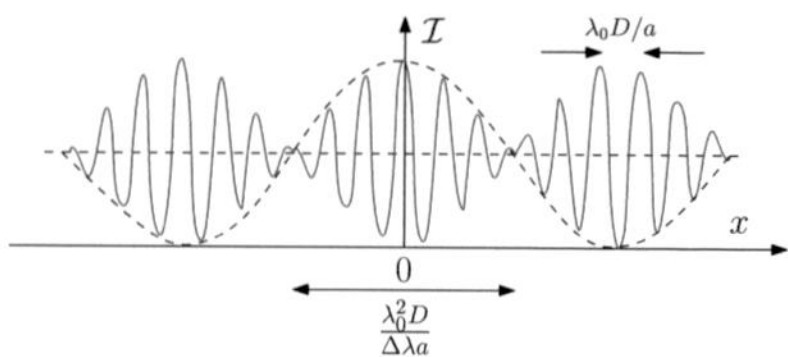

so that

$$\mathcal{I}(x) = 2\mathcal{I}_0 \left(1 + \cos\left(\pi \frac{\Delta\lambda a x}{\lambda_0^2 D} \right) \cos\left(2\pi \frac{a x}{\lambda_0 D} \right) \right).$$

This is illustrated in Fig. A.157. It corresponds to the interference pattern of a
monochromatic source of wavelength λ_0 modulated by an envelope function that
changes the contrast locally on the screen. The width of the envelope is inversely
proportional to the difference in wavelength $\Delta\lambda$ of the doublet.

16.3 A stellar interferometer

The angle subtended by the double slit from Betelgeuse is a/D', where $D' = 600$
lightyears. The fringes are expected to disappear when α reaches a critical value
given by the coherence angle

$$\theta_c = \frac{\lambda}{L}.$$

Since one lightyear corresponds to 1 ly $= 3 \times 10^8 \times 60 \times 60 \times 24 \times 365 = 9.46 \times 10^{15}$ m, the critical angle is then

$$\theta_c = \frac{3}{600 \times 9.46 \times 10^{15}} = 5.28 \times 10^{-19}$$

and for a red wavelength, $\lambda = 0.6\,\mu$m, we can estimate the diameter of Betelgeuse

$$L = \frac{\lambda}{\theta_c} = 1.13 \times 10^{12} \text{ m}$$

that is, about 1000 times larger than the Sun.

16.4 Interferences with sunlight

(a) When viewed from Earth, the sun looks like a disc of radius R, whose coherence
angle $\theta_c = \lambda/2R$ for the central part of the visible spectrum ($\lambda = 500$ nm) reads

$$\theta_c = \frac{500 \times 10^{-9}}{2 \times 6.96 \times 10^8} = 3.6 \times 10^{-16}$$

whereas the angle α subtended by the slit when looked from the sun is $\alpha = a/D'$.
For the fringes to appear on the screen, this angle must be smaller than the coherence
angle, so that

$$\alpha = \frac{a}{D'} < \theta_c$$

and we obtain

$$a < \theta_c D' = 50 \ \mu\text{m} .$$

We see that the observation of fringes requires $a < 0.05$ mm, and so a 2 mm separation between the apertures is too large to observe interferences with sunlight.

(b) A $\phi = 1$ mm aperture at distance l from the double aperture will act as an effective source of light with a coherence angle

$$\theta_c = \frac{\lambda}{\phi} = \frac{500 \times 10^{-9}}{10^{-3}} = 5 \times 10^{-4}$$

that should be larger than $\alpha = a/l$, and so the distance l at which the pinhole should be placed is such that

$$l > \frac{a}{\theta_c} = \frac{2 \times 10^{-3}}{5 \times 10^{-4}} = 4 \ \text{m} .$$

(c) The fringes on the screen have a periodicity of

$$\Delta x = \frac{\lambda D}{a} = \frac{500 \times 10^{-9} \times 2}{2 \times 10^{-3}} = 0.5 \ \text{mm} .$$

If the pinhole is centered with respect to the two apertures, at $x = 0$ one has a bright fringe, and the first dark fringe is observed at $x = \pm \Delta x/2 = 0.25$ mm, corresponding to a path difference between the beams of

$$dl = \frac{ax}{D} = \frac{2 \times 10^{-3} \times 2.5 \times 10^{-4}}{2} = 0.25 \ \mu\text{m}$$

which is a bit more than $1/4$ of the sunlight's coherence length. This means that one can see at most 3 dark fringes on the two sides of the central bright fringe. In total, a number of 6 dark fringes is observed inside a spot of size $6 \times \Delta x = 3$ mm.

Problems of Chap. 17

17.1 Fermat's principle and Snell–Descartes law

Let $P = (x, 0, z)$ be the point at which the ray intersects the interface separating the two media, as shown in Fig. A.158. The time t_{AB} that it takes for light to go from A to B is then

$$t_{AB} = n_1 \frac{AP}{c} + n_2 \frac{PB}{c} .$$

Fig. A.158 Trajectory of the
optical ray

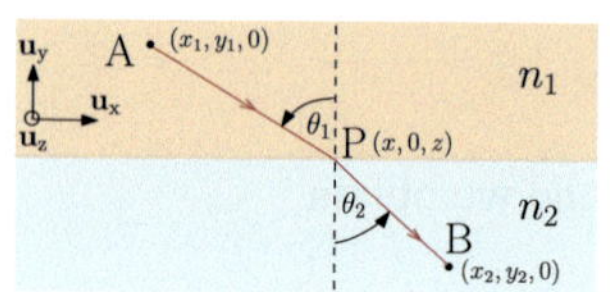

Replacing in terms of x and z

$$t_{AB}(x, z) = \frac{n_1}{c}\sqrt{(x - x_1)^2 + y_1^2 + z^2} + \frac{n_2}{c}\sqrt{(x - x_2)^2 + y_2^2 + z^2}.$$

When this time is stationary, the partial derivatives with respect to x and z vanish.
Then

$$\frac{\partial t_{AB}(x, z)}{\partial z} = \frac{n_1}{c}\frac{z}{AP} + \frac{n_2}{c}\frac{z}{PB} = 0$$

and we conclude, as in the previous question, that $z = 0$ so that A, P, and B belong
to the same (incidence) plane $z = 0$. The derivative with respect to x gives

$$\frac{\partial t_{AB}(x, z)}{\partial x} = \frac{n_1}{c}\frac{(x - x_1)}{AP} + \frac{n_2}{c}\frac{(x - x_2)}{PB} = 0$$

and we conclude that $x_1 < x < x_2$ and that

$$n_1\frac{(x - x_1)}{AP} = n_2\frac{x_2 - x}{PB}$$

which can be written in terms of the incident and refracted angles as

$$n_1 \sin\theta_1 = n_2 \sin\theta_2$$

which corresponds to Snell–Descartes law.

17.2 Deviation of light by the atmosphere

(a) Taking the cross product with $\mathbf{r}$ on both sides, we have

$$\mathbf{r} \times \left(\frac{d}{ds}n\mathbf{t}\right) = \mathbf{r} \times \left(\frac{dn(r)}{dr}\mathbf{u}_r\right) = 0$$

and since, by definition, $\dfrac{d\mathbf{r}}{ds} = \mathbf{t}$,

$$\mathbf{r} \times \left(\frac{d}{ds}n\mathbf{t}\right) = \frac{d}{ds}(\mathbf{r} \times n\mathbf{t}) = \mathbf{0}.$$

Fig. A.159 A distant object is seen at an apparent angle θ_0 with respect to the horizon

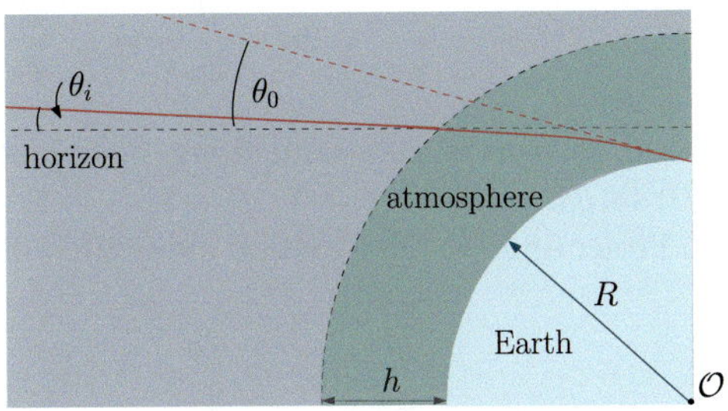

We conclude that $\mathbf{r} \times n\mathbf{t} = \mathbf{A}$, where $\mathbf{A}$ is a constant vector. The trajectory lies in a plane perpendicular to $\mathbf{A}$.

(b) From the previous result

$$|\mathbf{A}| = rn(r) \sin \phi = \text{Cst.}$$

(c) The situation is shown in Fig. A.159. Since $rn(r) \sin \phi$ is a constant, and $h \ll R$, we have

$$Rn(R) \sin \left(\frac{\pi}{2} - \theta_0\right) = (R + h)n(R + h) \sin \left(\frac{\pi}{2} - \theta_i\right)$$

and we conclude by replacing $n(R + h) = 1$ that the light rays coming from the star seem to form an angle of θ_0 with respect to the horizon, where θ_0 is given by

$$Rn_0 \cos \theta_0 = (R + h) \cos \theta_i.$$

(d) We have $\cos \theta_0 = \cos(\theta_i + \Delta\theta) = \cos \theta_i \cos \Delta\theta - \sin \Delta\theta \sin \theta_i \approx \cos \theta_i - \Delta\theta \sin \theta_i$ and so,

$$(R + h) \cos \theta_i \approx Rn_0 \left(\cos \theta_i - \Delta\theta \sin \theta_i\right).$$

Assuming $\theta_i \neq \pi/2$ (otherwise $\Delta\theta = 0$), we have $\cos \theta_i \neq 0$ and we obtain

$$\Delta\theta = \left(1 - \frac{R + h}{Rn_0}\right) \frac{1}{\tan \theta_i}.$$

17.3 **The mirage effect**

(a) We write the ray equation

$$\frac{d}{ds}(n\mathbf{t}) = \nabla n = \frac{dn}{dz}\mathbf{u}_z,$$

from which we conclude that

$$\mathbf{u}_z \times \frac{d}{ds}(n\mathbf{t}) = \frac{d}{ds}(\mathbf{u}_z \times n\mathbf{t}) = 0$$

and so, $\mathbf{u}_z \times n\mathbf{t}$ is a constant along the trajectory, hence, $|\mathbf{u}_z \times n\mathbf{t}| = n(z)\sin\theta(z) = n(0)\sin\theta_0$.

(b) Projecting the ray equation along the z-axis,

$$\frac{d}{ds}(nt_z) = \frac{dn}{dz}, \qquad \rightarrow \qquad \frac{d}{ds}\left(n\frac{dz}{ds}\right) = \frac{dn}{dz}$$

and using the chain rule

$$\frac{dn}{ds}\frac{dz}{ds} + n\frac{d^2z}{ds^2} = \frac{dn}{dz},$$

and so,

$$\frac{dn}{dz}\left(\frac{dz}{ds}\right)^2 + n\frac{d^2z}{ds^2} = \frac{dn}{dz}.$$

Since $\dfrac{dz}{ds} = t_z = -\cos\theta$, we find

$$\frac{dn}{dz}\cos^2\theta + n\frac{d^2z}{ds^2} = \frac{dn}{dz}, \quad \text{thus,} \quad \frac{d^2z}{ds^2} = \frac{\sin^2\theta}{n}\frac{dn}{dz}.$$

Similarly, the projection of the ray equation along the x-axis yields

$$\frac{d}{ds}(nt_x) = \frac{d}{ds}\left(n\frac{dx}{ds}\right) = 0$$

and so,

$$\underbrace{\frac{dn}{ds}\frac{dx}{ds}}_{\frac{dn}{dz}\frac{dz}{ds}} + n\frac{d^2x}{ds^2} = 0.$$

Since $\dfrac{dx}{ds} = \sin\theta$, we find

$$\frac{d^2x}{ds^2} = \frac{\sin\theta\cos\theta}{n}\frac{dn}{dz}.$$

(c) We have $\dfrac{dz}{ds} = \dfrac{dz}{dx}\dfrac{dx}{ds}$ and so, $\frac{dz}{dx} = -\dfrac{\cos\theta}{\sin\theta}$. In addition,

$$\frac{d^2z}{ds^2} = \frac{d}{ds}\left(\frac{dz}{dx}\frac{dx}{ds}\right) = \frac{d^2z}{dx^2}\underbrace{\left(\frac{dx}{ds}\right)^2}_{\sin^2\theta} + \frac{dz}{dx}\underbrace{\frac{d^2x}{ds^2}}_{-\frac{\cos\theta}{\sin\theta}},$$

Fig. A.160 The mirage
effect

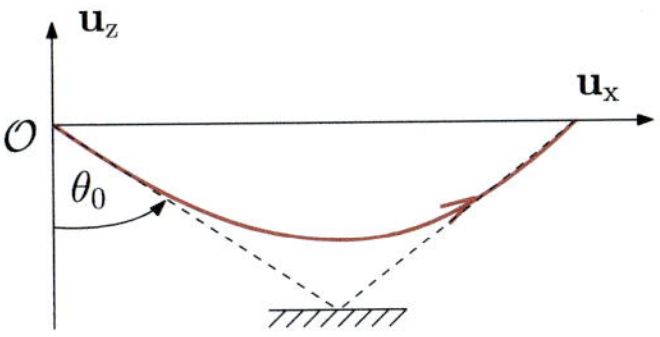

hence,

$$\frac{dz^2}{dx^2} = \frac{1}{\sin^2 \theta} \left(\frac{d^2 z}{ds^2} + \frac{\cos \theta}{\sin \theta} \frac{d^2 x}{ds^2} \right).$$

Replacing $\dfrac{d^2 z}{ds^2}$ and $\dfrac{d^2 x}{ds^2}$ from the previous question, we obtain

$$\frac{dz^2}{dx^2} = \frac{1}{\sin^2 \theta} \left(\frac{\sin^2 \theta}{n} + \frac{\cos \theta}{\sin \theta} \frac{\sin \theta \cos \theta}{n} \right) \frac{dn}{dz} = \frac{1}{n \sin^2 \theta} \frac{dn}{dz} \left(\sin^2 \theta + \cos^2 \theta \right).$$

Recalling that $z = an^2 + b$, we have $1 = 2an \dfrac{dn}{dz}$ and since $A = n \sin \theta$, we find

$$\frac{dz^2}{dx^2} = \frac{1}{2an^2 \sin^2 \theta} = \frac{1}{2a A^2}.$$

(d) We have $n(0) \sin \theta_0 = A$. Integrating the equation found in the previous question,
we obtain

$$z(x) = \frac{1}{4a A^2} x^2 + Cx + D.$$

Imposing the boundary conditions

$$\begin{cases} z(x = 0) = 0, \\ \left(\dfrac{dz}{dx} \right)_{x=0} = -\dfrac{1}{\tan \theta_0}, \end{cases} \quad \rightarrow \quad \begin{cases} D = 0, \\ C = -\dfrac{1}{\tan \theta_0}. \end{cases}$$

The trajectory is given by the parabola of equation

$$z(x) = \frac{x^2}{4a A^2} - \frac{x}{\tan \theta_0}.$$

The ray eventually bends upwards and gives the impression of being reflected by a
mirror-like surface, as shown in Fig. A.160. This mirage effect is typically observed
in a road overheated by the Sun, or in a desertic zone, and gives the illusion to an
observer of the presence of water on the road due to the virtual image of the sky
produced by the bending of optical rays.

Fig. A.161 Snell–Descartes law applied to the ray passing through B, O, and B'

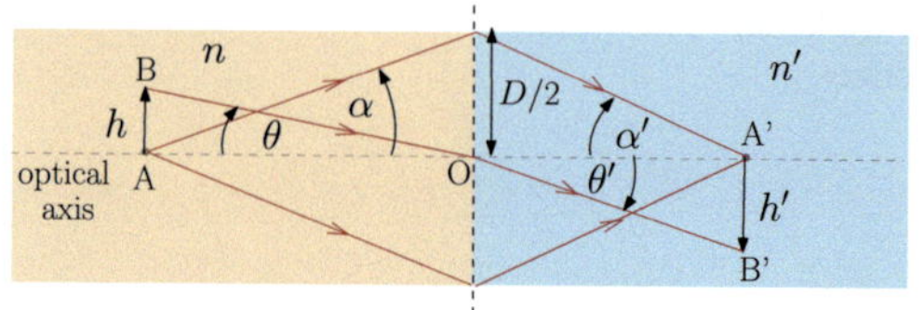

17.4 The Lagrange–Helmholtz invariant

If D is the aperture diameter of the system, we have

$$\frac{D}{2} = \alpha\,\overline{\text{AO}} = \alpha'\,\overline{\text{OA'}}.$$

On the other hand, Snell–Descartes law for the ray shown in Fig. A.161 passing through B, O, and B' gives, in the paraxial approximation

$$n\theta \approx n'\theta'.$$

Replacing $\theta \approx h/\overline{\text{AO}}$ and $\theta' \approx h'/\overline{\text{OA'}}$, this rewrites as

$$n\frac{h}{\overline{\text{AO}}} \approx n'\frac{h'}{\overline{\text{OA'}}}$$

and so

$$\alpha\frac{nh}{n'h'}\overline{\text{OA'}} = \alpha'\overline{\text{OA'}}$$

and we conclude that

$$nh\alpha = n'h'\alpha'.$$

17.5 Ray tracing with spherical mirrors

The system is supposed to be aplanatic, so that in order to find the image of the segment AB, it is enough to find the position of the image B' of B. For this, we look for the intersection of any pair of rays stemming from B and reflected by the mirror. Firstly, a horizontal ray coming from B will pass through (or seem to come from) the image focal point F'. This is shown in Fig. A.162.

In addition, the ray coming from B such that the ray (or its prolongation) passes through F' will exit the system horizontally. We find the position of B' as the intersection of the latter two rays (or their prolongations), as shown in Fig. A.163.

For the concave mirror, we have therefore

- For $\overline{OA} < \overline{OC} = 2f'$, the image is real, inverted, and diminished.
- For $2f' < \overline{OA} < \overline{OF'} = f'$, the image is real, inverted, and magnified.
- For $f' < \overline{OA} < 0$, the image is virtual, upright, and magnified.

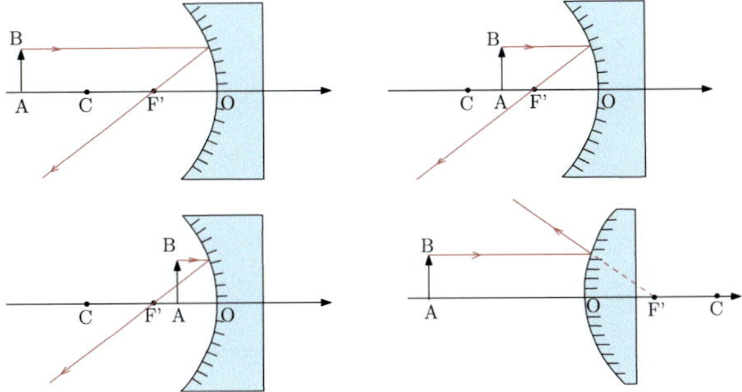

Fig. A.162 A horizontal ray will pass through (or seem to come) from the image focal point

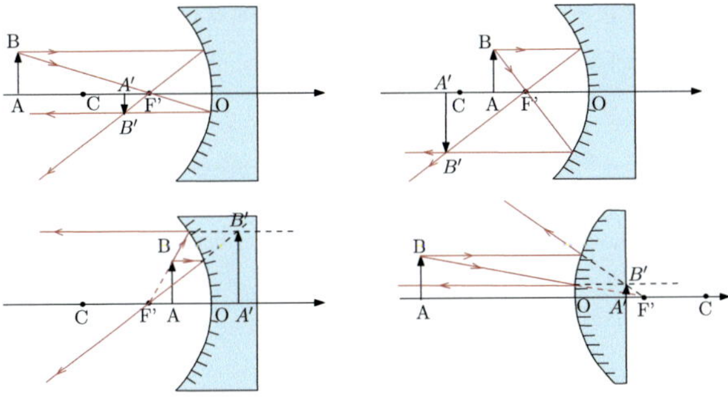

Fig. A.163 A ray passing through F' will exit the system horizontally

For the convex mirror, the image is virtual, upright, and diminished.

17.6 Image formation with a convergent thin lens

We need to find for each case the position of the image B' of point B. For this, we look for the intersection of any two rays stemming from B after refraction by the lens. Here, we will show three of them for completeness. Firstly, a horizontal ray coming from or aimed at B will pass through the image focal point F' as shown in Fig. A.164.

The ray coming from or aimed toward B and such that the ray (or its prolongation) passes through the object focal point F will exit the system horizontally, as shown in Fig. A.165

A ray coming or directed toward B and passing through the center O of the lens will not be deviated, as shown in Fig. A.166. All these rays (or their prolongations) meet at a single point B', the image of B. We find A' as the projection of B' onto the optical axis.

For the convergent thin lens, we have therefore

- For a real object with $\overline{OA} < \overline{OF} = f$, the image is real, inverted, and diminished (although it can be magnified if $\overline{OA} > 2\,\overline{OF}$).
- For a real object with $f < \overline{OA} < 0$, the image is virtual, upright, and magnified.

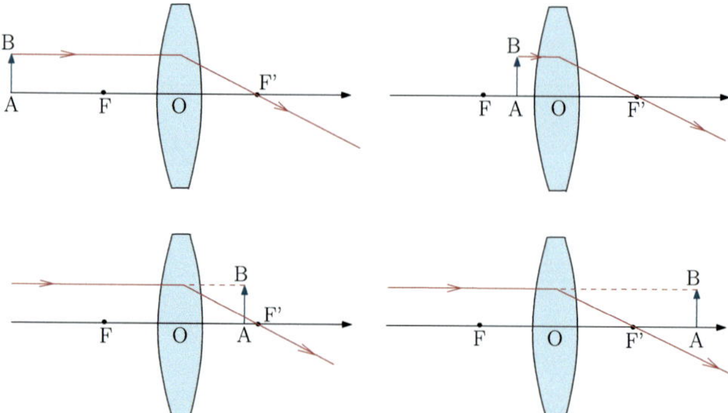

Fig. A.164 A horizontal ray will pass through the image focal point

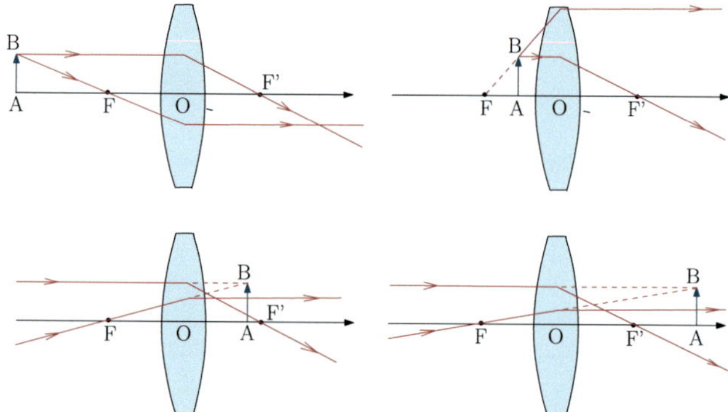

Fig. A.165 Rays passing through F will leave the system horizontally

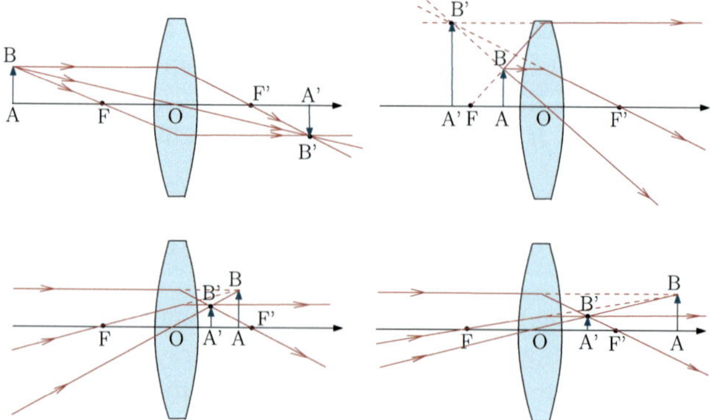

Fig. A.166 Rays passing through the center will not be deviated

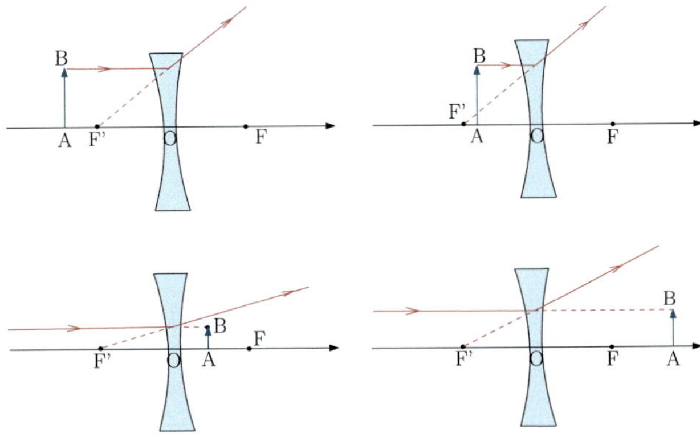

Fig. A.167 A horizontal ray will pass through (or seem to come from) the image focal point

- For a virtual object with $0 < \overline{OA} < \overline{OF'} = f'$, the image is real, upright, and diminished.
- For a virtual object with $f' < \overline{OA}$, the image is real, upright, and diminished.

17.7 Image formation with a divergent thin lens

We need to find the image B' of point B since the system is aplanatic, implying that A' is the projection of B along the optical axis. To find the position of B', we look for the intersection of any two rays stemming from B. Here, we will show three of them for completeness. Firstly, a horizontal ray coming from B (or directed toward B when the latter is a virtual object) will seem to come from the image focal point F', as shown in Fig. A.167.

The ray coming or seeming to converge to B and whose prolongation passes through the object focal point F will exit the system horizontally, as shown in Fig. A.168. A ray coming from or aimed at B and passing through the center O will not be deviated by the lens. All the rays mentioned above should meet at a single point B', the image of B. We find A' as the projection of B' onto the optical axis, as shown in Fig. A.169. For the divergent thin lens, we have therefore

- For a real object with $\overline{OA} < \overline{OF'} = f$, the image is virtual, upright, and diminished.
- For a real object with $f < \overline{OA} < 0$, the image is virtual, upright, and diminished.
- For a virtual object with $0 < \overline{OA} < \overline{OF} = f'$, the image is real, upright, and magnified.
- For a virtual object with $f' < \overline{OA}$, the image is virtual, inverted, and magnified (it can be diminished if $\overline{OA} > 2\overline{OF}$.

17.8 The human eye and its resolution

(a) Let A be the position of an object located on the axis of the eye at a distance $d_{\min} = 20\,\text{cm}$ from it, and let A' be the position of its image by the eye in the retina. The thin lens equation gives

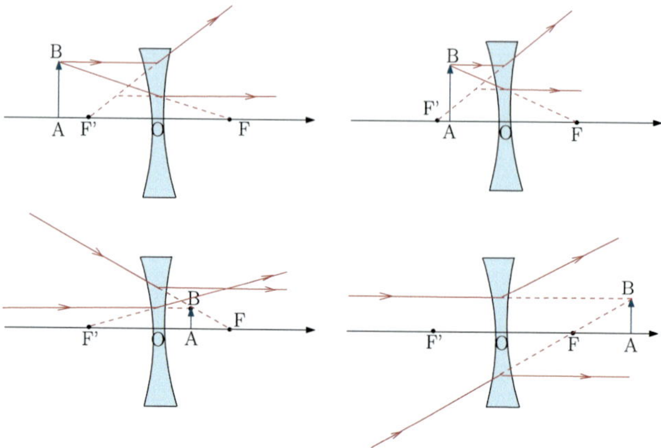

Fig. A.168 A ray passing through F will exit horizontally

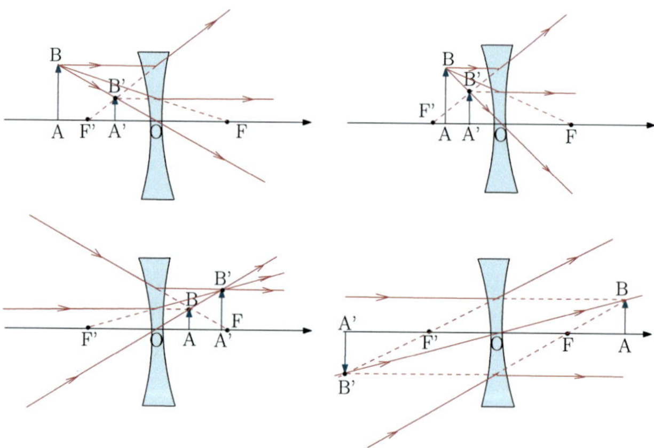

Fig. A.169 A ray passing through the center is not deviated

Fig. A.170 An object at A imaged by the eye into the retina

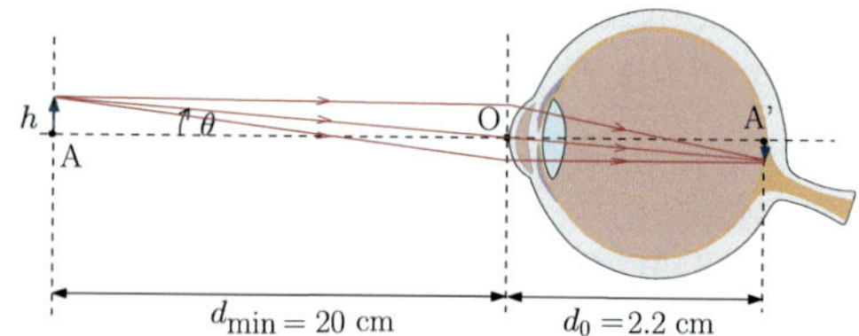

$$\frac{1}{\overline{OA'}} - \frac{1}{\overline{OA}} = \frac{1}{f'}.$$

Replacing $\overline{OA'} = d_0 = 2.2\,\text{cm}$ and $\overline{OA} = -d_{\min} = -20\,\text{cm}$, we obtain the image focal length of the eye

Fig. A.171 The maximum angle at which the eye can see an object is $\theta_{\max}$

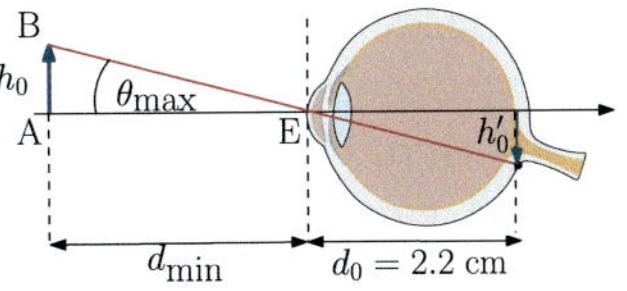

$$f' = 1.982 \,\text{cm}.$$

A small object of size h is seen by the eye under an angle

$$\theta = \tan^{-1} \frac{h}{d_{\min}} \approx \frac{h}{d_{\min}}.$$

Ray tracing can be used to show the image formation by the eye, as shown in Fig. A.170.

(b) An object seen under an angle $\theta_{\min} \ll 1$ will have a physical size h' on the retina, given by

$$h' \approx d_0 \theta_{\min} = 6.6 \,\mu\text{m}.$$

The eye can resolve the smallest features on an object when the latter is placed at $d_{min} = 20\,\text{cm}$ from the eye. Given the minimum angular separation $\theta_{\min}$ allowing to separate two points on the retina, we conclude that the smallest feature that the eye can resolve has a size h of

$$h \approx \theta_{\min} d_{min} = 60 \,\mu\text{m},$$

which is close to the typical diameter of a human hair.

17.9 The magnifying glass

(a) The maximum angle $\theta_{\max}$ at which the naked eye can see an object of size h_0 is the angle subtended by the object when it is located at the *punctum proximum* of the eye, that is, in the paraxial approximation,

$$\theta_{max} \approx h_0 / d_{\min}.$$

as shown in Fig. A.171.

The maximum physical size of the image in the retina is then

$$h_0' = \theta_{max} d_0 = h_0 \frac{d_0}{d_{\min}} = 0.088 \, h_0.$$

(b) To find the image of AB by the magnifying lens, we need to find the position of the image B' of B. For this, we find the point at which the rays coming from B intersect after refraction by the lens. A horizontal ray coming from B will pass through the lens image focal point F', and the ray coming from B and passing through the center

O of the lens will not be deviated. The prolongations of these two rays intersect at B', and we conclude that the image of AB by the lens is virtual, upright, and magnified, as shown in Fig. A.172. To find the position at which the virtual image of the lens is formed, we can use the Descartes relation

$$\frac{1}{\overline{OA'}} - \frac{1}{\overline{OA}} = \frac{1}{f'}$$

and so,

$$\overline{OA'} = -\frac{f'd}{f'-d}.$$

The image is formed at a distance d_i from the eye given by

$$d_i = \overline{A'E} = \overline{A'O} + \overline{OE} = \frac{f'd}{f'-d} + l.$$

Finally, the image $A'B'$ subtends an angle θ from the eye given by

$$\theta = \frac{\overline{A'B'}}{d_i} = \frac{\overline{A'B'}(f'-d)}{f'(d+l)-ld}$$

and since the transverse magnification of the lens is $M = \overline{A'B'}/\overline{AB} = \overline{OA'}/\overline{OA}$

$$\theta = \frac{h_0(f'-d)}{f'(d+l)-ld} \frac{f'}{(f'-d)} = \frac{h_0 f'}{f'd+l(f'-d)}.$$

(c) The angular magnification reads

$$\gamma = \frac{\theta}{\theta_{\max}} = \frac{d_{\min} f'}{f'd+l(f'-d)}.$$

(d) When $l = 0$, the angular magnification reads

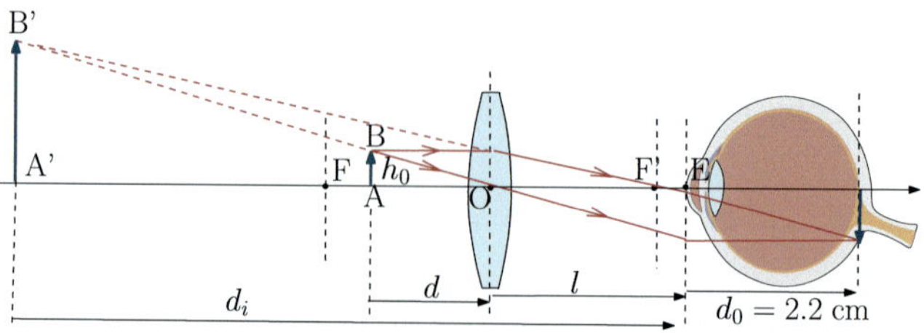

Fig. A.172 Ray tracing showing image formation by the magnifying lens

$$\gamma = \frac{d_{\min}}{d},$$

so that the smaller d is, the larger the magnification. However, d cannot be arbitrarily small, since the virtual image of the object by the lens should be at least at a distance $d_i = d_{\min}$ from the eye. Imposing $\overline{OA'} = \overline{EA'} = -d_i < -d_{\min}$ yields

$$d_{\min} < \frac{f'd}{f'-d} \qquad \rightarrow \qquad d > \frac{f'd_{\min}}{f'+d_{\min}}$$

and so the maximum magnification is

$$\gamma_{\max} = \frac{d_{\min}}{d} = \frac{f'+d_{\min}}{f'} = 1 + \frac{d_{\min}}{f'}.$$

On the other hand, the maximum possible value of d is $d = f'$, so that the virtual image $A'B'$ formed by the lens is at infinity. This corresponds to the most comfortable situation for the observer, since the eye does not need to accommodate its lens. The angular magnification in this case becomes

$$\gamma_{\min} = \frac{d_{\min}}{f'}$$

and we conclude that

$$\frac{d_{\min}}{f'} \leq \gamma \leq 1 + \frac{d_{\min}}{f'}.$$

Evaluating numerically, $5 \leq \gamma \leq 6$. Note finally that to obtain high magnifications, a lens with a small focal length should be employed. It is thus required to hold very close to our eyes a lens with small radii of curvature, which is not very convenient. To overcome this drawback, a system formed by two lenses, also called microscope, can be used.

17.10 The microscope

(a) A horizontal ray coming from B will pass through the objective image focal point F_1', and the ray coming from B passing through the center of the objective will not be deviated. The intersection of these two rays allows us to find the position of the image h_1 of h_0 by the objective, as shown in Fig. A.173. Since $d_0 > f_1'$, the latter is real, inverted, and magnified (although it can also be reduced if the object is farther than $2f_1'$ from the objective). The image h_2 of h_1 by the eyepiece can be found in a similar way by considering two rays coming from the extremal point of h_1: a horizontal ray that will pass through F_2' and the ray passing through the center of the eyepiece without deviation. If $d_i > L - f_2'$, the image h_1 is between the eyepiece object focal point F_2 and the eyepiece, and so the resulting image h_2 is virtual, magnified, and upright (but still inverted with respect to h_0), as shown in the figure below.

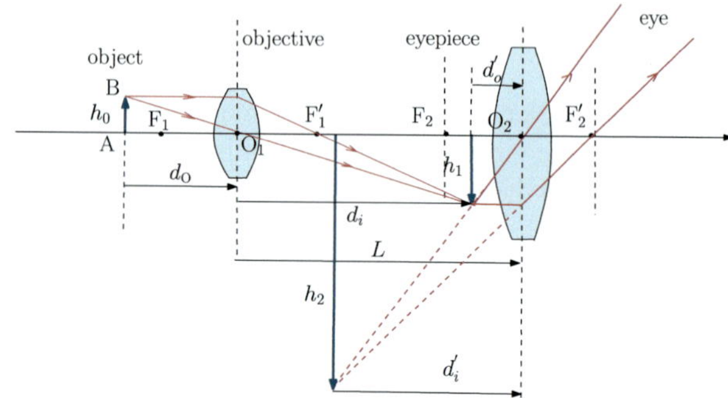

Fig. A.173 Ray tracing for the objective-eyepiece ensemble

(b) Note that the eyepiece acts as a magnifying glass for the object h_1 generated by the objective, which is a magnified version of h_0 by a quantity $M_1 = -\frac{d_i}{d_o}$. We conclude that the relevant magnification is therefore $\gamma = M_1 \theta / \theta_{\text{max}}$, where θ is the angle at which the image h_2 is looked at by the eye, and where $\theta_{\text{max}} = h_0/d_{\text{min}}$ is the maximum angle at which the object is observed by the naked eye, that is, when the latter is positioned at the *punctum proxium* at a distance d_{min} from the eye. This angular magnification is finally transformed into a transverse magnification by the eye lens, which forms an image of size $f'\theta$ on the retina, where f' is the eye focal length and θ the angle at which the imaged object is looked at.

Based on the result of Exercise 17.9, when the image h_2 is located at infinity, that is, when the eye is relaxed, the magnification is

$$\gamma = -\frac{d_i}{d_o}\frac{d_{\text{min}}}{f_2'}$$

and corresponds to the case in which h_1 is at the object focal point F_2 of the eyepiece, so that $d_i = L - f_2'$ and d_i can be found by using the thin lens equation

$$\frac{1}{d_i} + \frac{1}{d_o} = \frac{1}{f_1'}.$$

This yields $d_o = f_1'(L - f_2')/(L - f_1' - f_2')$ and we obtain

$$\gamma = -\frac{L - (f_1' + f_2')}{f_1'}\frac{d_{\text{min}}}{f_2'}.$$

17.11 Projecting an image on a screen

(a) For a real object, only the convergent lens is capable of producing a real image, provided that the object is at a distance larger than f' from the lens (see Exercise 17.6). For a divergent lens, the image of a real object is always virtual (see Exercise 17.7).

(b) Let O be the center of the lens. Given that the image of the object is formed at the position A' of the screen, the thin lens equation writes

Fig. A.174 Ray tracing shows that the image is reversed, with no net magnification

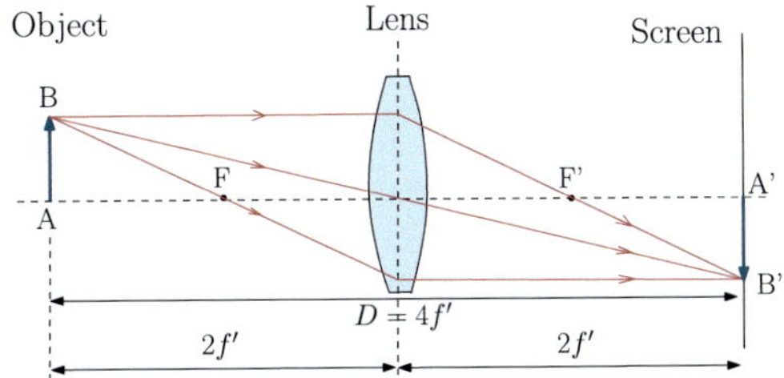

$$\underbrace{\frac{1}{\overline{OA'}}}_{D-x} - \underbrace{\frac{1}{\overline{OA}}}_{-x} = \frac{1}{f'},$$

which results in the quadratic equation for x,

$$x^2 - Dx + f'D = 0,$$

whose general solution writes

$$x = \frac{D \pm \sqrt{D^2 - 4Df'}}{2} = \frac{D}{2}\left(1 \pm \sqrt{1 - \frac{4f'}{D}}\right).$$

From this, we conclude that real solutions exist provided that $1 - 4f'/D \geq 0$, or in other words if the screen is placed sufficiently far from the object, that is, $D \geq 4f'$. Under these conditions, two solutions arise

$$x_1 = \frac{D}{2}\left(1 + \sqrt{1 - \frac{4f'}{D}}\right) < D \quad \text{and} \quad x_2 = \frac{D}{2}\left(1 - \sqrt{1 - \frac{4f'}{D}}\right) < x_1.$$

(c) If $D = 4f'$, only one solution for x is obtained and given by $x = D/2$. The lens is therefore placed at the middle point between the object and the screen, at a distance $2f'$ from each. The magnification on the screen is therefore

$$M = \frac{\overline{A'B'}}{\overline{AB}} = \frac{\overline{OA'}}{\overline{OA}} = -\frac{2f'}{2f'} = -1.$$

Ray tracing and simple geometrical arguments allow us to confirm that the image is simply reversed, without a net magnification, as shown in Fig. A.174.

(d) For $D > 4f'$, we obtain the two solutions for x mentioned in question b. Since $M = \frac{\overline{OA'}}{\overline{OA}} = -\frac{D-x}{x}$, we conclude that for $x_1 = \frac{D}{2}\left(1 + \sqrt{1 - \frac{4f'}{D}}\right)$ and $x_2 =$

$$\frac{D}{2}\left(1 - \sqrt{1 - \frac{4f'}{D}}\right)$$ the magnification is given by

$$M_1 = -\frac{1 - \sqrt{1 - \dfrac{4f'}{D}}}{1 + \sqrt{1 - \dfrac{4f'}{D}}} \qquad M_2 = -\frac{1 + \sqrt{1 - \dfrac{4f'}{D}}}{1 - \sqrt{1 - \dfrac{4f'}{D}}}.$$

Since $|M_2| > 1$ and $M_1 M_2 = 1$, we find that the solution which gives a magnified version of the object on the screen is the one obtained when the lens is placed closer to the object, at a distance x_2. The two cases are shown in Fig. A.175 for $D = 5f'$.

17.12 Longitudinal magnification of a thin lens

Let $x = -\overline{OA}$ and $y = \overline{OA'}$ as shown in Fig. A.176. The thin lens equation then writes

$$\frac{1}{y} + \frac{1}{x} = \frac{1}{f'},$$

where f' is the image focal length of the lens. Differentiating we find

$$-\frac{1}{y^2}dy - \frac{1}{x^2}dx = 0,$$

so that

$$\frac{dy}{dx} = -\left(\frac{y}{x}\right)^2.$$

The minus sign tells us that if point A is displaced by an infinitesimal quantity dx along the optical axis, its image moves by an infinitesimal quantity dy in the same direction. For $\overline{AB} = -\Delta x \ll x$, we can then write $\overline{A'B'} = \Delta y \ll y$ and we have

$$\frac{\overline{A'B'}}{\overline{AB}} = -\frac{\Delta y}{\Delta x} \approx \left(\frac{y}{x}\right)^2 = \left(\frac{\overline{OA'}}{\overline{OA}}\right)^2.$$

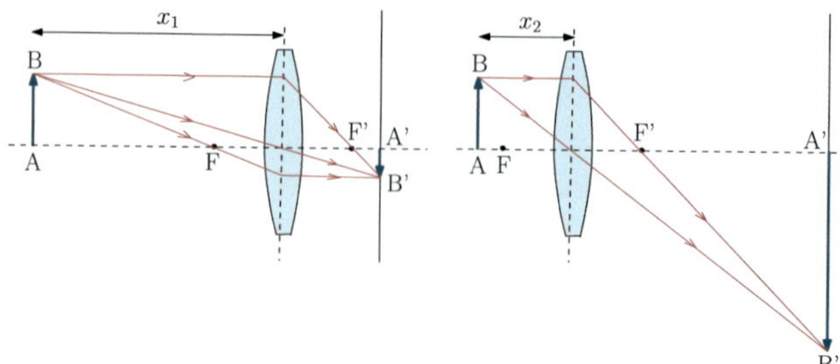

Fig. A.175 Ray tracing for the two solutions previously found

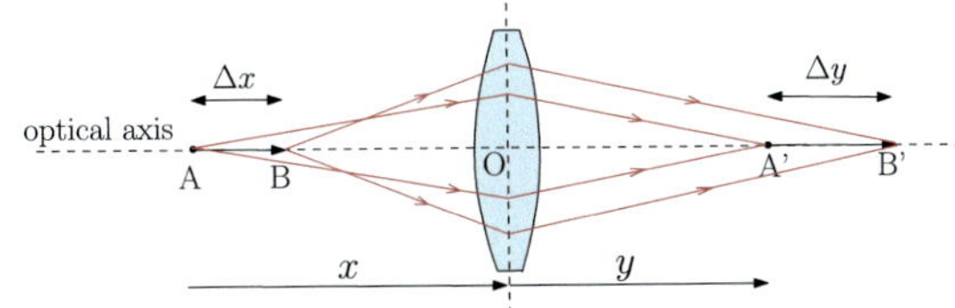

Fig. A.176 Definition of x and y

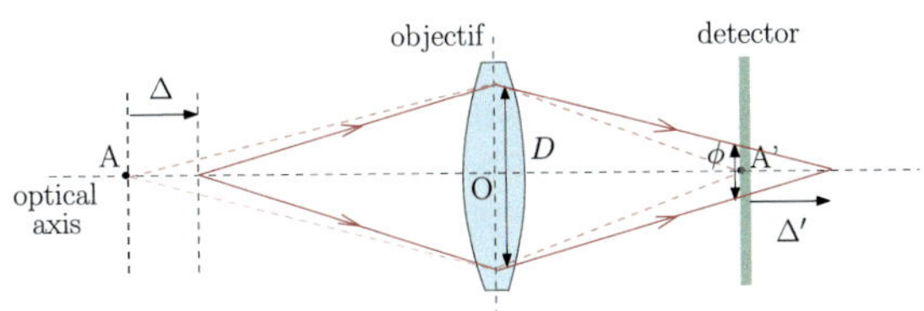

Fig. A.177 A displacement of an object by Δ produces a displacement of Δ' of its image by the lens

On the other hand, the transverse magnification of a thin lens is $M = -\overline{OA'}/\overline{OA}$ from which we conclude that $M_l = M^2$.

17.13 The field of depth of an objective lens

Supposing that a point object is displaced by a small quantity Δ with respect to A, its image will be displaced by Δ' with respect to A', as shown in Fig. A.177. Both are related by the longitudinal magnification M_l of a thin lens (see Exercise 17.12),

$$M_l = \frac{\Delta'}{\Delta} = M^2 = \left(\frac{\overline{OA'}}{\overline{OA}}\right)^2 .$$

On the detector, the image of the point object will be a circle of diameter ϕ, determined by the maximal rays passing through the lens. We have

$$\phi \approx 2\Delta' \frac{D/2}{\overline{OA'}},$$

imposing $\phi < \phi_0$ in order to have a sharp image on the detector. This yields

$$\Delta' \frac{D}{\overline{OA'}} = D\Delta \frac{\overline{OA'}}{\overline{OA}^2} < \phi_0$$

and so, the field of depth is given by the maximum value of Δ allowing for a sharp image, that is,

$$\Delta_{\max} = \frac{\phi_0 \overline{OA}^2}{D\overline{OA'}} .$$

If the object is far from the objective (at a distance much larger than f'), the lens equation gives $\overline{OA'} \approx f'$ so that

$$\Delta_{\max} \approx \frac{\phi_0 \overline{OA}^2}{D f'}.$$

17.14 The Galilean telescope

(a) All the rays coming from the extremity of the object, which is at infinity, form an angle α with respect to the optical axis. The system being afocal, all the rays leaving the eyepiece are parallel to each other and form an angle α' with respect to the optical axis, as shown in Fig. A.178. To find this angle, we can trace the trajectory of an incident ray passing through the object focal point F_1 of the objective, as shown in the figure below. This ray will exit the objective lens horizontally and so, after passing through the eyepiece, it will seem to come from its image focal point F_2'. We have

$$\tan \alpha = \frac{h}{f_1'} \quad \text{and} \quad \tan \alpha' = -\frac{h}{f_2'}.$$

In the paraxial approximation, the angles are small so that the angular magnification reads

$$\gamma = \frac{\alpha'}{\alpha} = -\frac{f_1'}{f_2'}.$$

(b) The angle under which Io is observed with the naked eye is simply

$$\alpha_1 = \frac{\phi}{D} = \frac{3600}{628.3 \times 10^6} = 5.7 \times 10^{-6}$$

Galileo's telescope had an angular magnification of $\gamma = -f_1'/f_2' = 98/5 = 19.6$, and so, when looked under the telescope, Io appears to subtend an angle

$$\alpha_2 = \gamma \alpha_1 = 1.11 \times 10^{-4}.$$

(c) An object placed at infinity, looked under an angle $\alpha \ll 1$ by the eye will form an image of transverse size

$$x = f' \alpha,$$

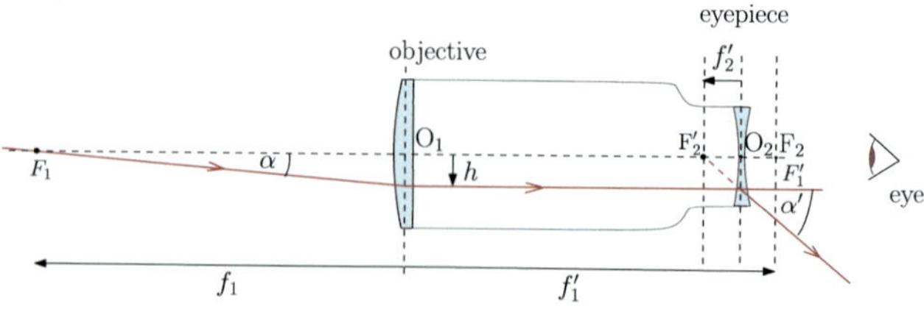

Fig. A.178 Trajectory of a ray entering the system at an angle α

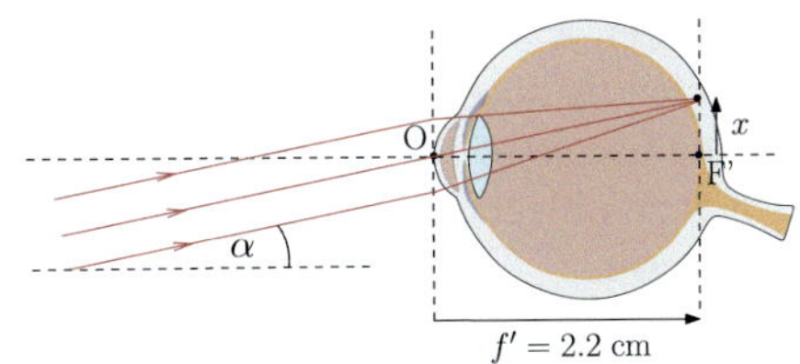

Fig. A.179 An object at infinity is focused by the eye in the retina

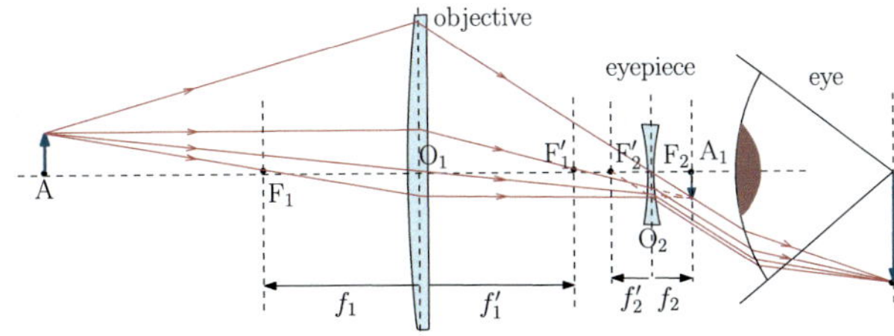

Fig. A.180 Ray tracing showing the image of an object at A at the eye's retina

where $f' = 2.2\,\text{cm}$ is the focal length of the eye without accommodation, see Fig. A.179. Io's size on the retina when looked with the naked eye is therefore

$$x_1 = f'\alpha_1 = 2.2 \times 10^{-2} \times 5.7 \times 10^{-6} = 1.25 \times 10^{-7}\,\text{m},$$

which is much smaller than the distance between cones in the fovea and so, it cannot be resolved. With the telescope, the image is magnified by $\gamma = 19.6$ so that

$$x_2 = f'\alpha_2 = 2.46 \times 10^{-6}\,\text{m},$$

which now falls in the resolution limit of the eye.

(d) Let A be the position of an object on the optical axis, and A_1 the image of A by the objective. The image A_2 of A_1 formed by the eyepiece should be located at infinity, in which case the eyes of the observer will form a neat image of A on the retina without accommodation, as shown in Fig. A.180. The position of A_1 should thus coincide with the object focal point F_2 of the eyepiece. Applying the lens equation, the position of the image A_1 of A by the first lens is given by

$$\frac{1}{\overline{O_1 A_1}} = \frac{1}{\overline{O_1 A}} + \frac{1}{f_1'}$$

and replacing $\overline{O_1 A} = -50\,\text{m}$ and $f_1 = 0.98\,\text{m}$, we find $\overline{O_1 A_1} = 0.9996\,\text{m}$. Since for an object at infinity we had $\overline{O_1 A_1} = f_1' = 98\,\text{cm}$, we conclude that the eyepiece should be displaced by

$$\Delta = (99.96 - 98)\,\text{cm} = 1.96\,\text{cm}$$

in the direction of light propagation.

(e) Let A be the position on the optical axis of the closest object that can be observed by accommodating the eye lens, A_1 the image of A by the objective, and A_2 the image of A_1 by the eyepiece. If $d_{\min} = 20\,$cm is the minimum distance at which an object can be imaged by the eye, modeled by a thin lens located at E, we must have

$$\overline{A_2 E} = \overline{A_2 O_2} + \underbrace{\overline{O_2 E}}_{15\,\text{mm}} = d_{\min} = 20\,\text{mm}.$$

Hence, A_2 must be a virtual image formed by the eyepiece, such that $\overline{O_2 A_2} = -18.5\,$cm. This is shown in Fig. A.181. From the lens equation for the eyepiece

$$\frac{1}{\overline{O_2 A_2}} = \frac{1}{\overline{O_2 A_1}} + \frac{1}{f_2'},$$

we obtain

$$\overline{O_2 A_1} = \frac{f_2'\,\overline{O_2 A_2}}{f_2' - \overline{O_2 A_2}} = 6.85\,\text{cm}.$$

Since

$$\overline{O_1 A_1} = \overline{O_2 A_1} + \underbrace{\overline{O_1 O_2}}_{L = f_1' + f_2' + \Delta = 94.96\,\text{cm}},$$

we obtain

$$\overline{O_1 A_1} < L + 6.85\,\text{cm} = 101.81\,\text{cm}$$

and from the lens equation for the objective

$$\frac{1}{\overline{O_1 A_1}} = \frac{1}{\overline{O_1 A}} + \frac{1}{f_1'},$$

we find

$$\overline{O_1 A} = \frac{f_1'\,\overline{O_1 A_1}}{f_1' - \overline{O_1 A_1}} = -26.2\,\text{m}.$$

17.15 The achromatic doublet

(a) Equation 17.17 with air ($n = 1$) on both sides of the lens gives

$$\frac{1}{f_1'} = (n_1 - 1)\left(\frac{1}{R_1} - \frac{1}{R_2} + \frac{e_1(n_1 - 1)}{R_1^2 n_1}\right).$$

Evaluating numerically, we obtain $f_1' = 99.779$ mm. Since $e_1 \ll R_1, R_2$, we may use the thin lens approximation which gives instead

$$\frac{1}{f_1'} = (n_1 - 1)\left(\frac{1}{R_1} - \frac{1}{R_2}\right) \tag{A.95}$$

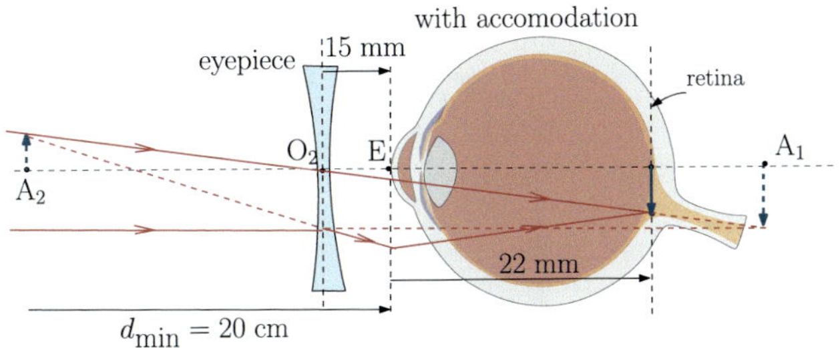

Fig. A.181 A_2 is the virtual image of A_1 by the eyepiece

and we obtain $f_1' = 98.61$ mm, thus an error of only 1%.

(b) Differentiating both sides of (A.95) with respect to n_1 gives

$$-\frac{1}{f_1'^2}\frac{df_1'}{dn_1} = \left(\frac{1}{R_1} - \frac{1}{R_2}\right)$$

and so

$$df_1' = -dn_1 f_1'^2 \left(\frac{1}{R_1} - \frac{1}{R_2}\right) = -dn_1 \frac{f_1'}{n_1 - 1}$$

for $f_1' = 98.61$ mm, $n_1 = 1.651$ and $dn_1 = 0.01165$, we find

$$df_1' = -1.76 \text{ mm}$$

so that the focal length for blue light is shorter by 1.76 mm with respect to that of red light.

(c) The plano-concave lens has an input surface of radius of curvature $R_2 = -142$ mm and an output surface with infinite radius of curvature. Both radii are thus orders of magnitude larger than e_2, so that the focal length f_2' of this lens (alone in air) can also be obtained by using the thin lens approximation

$$\frac{1}{f_2'} = \frac{(n_2 - 1)}{R_2} \tag{A.96}$$

so that $f_2' = -195.19$ mm. When the two lenses are glued together, the effective focal length f' of the hole system is given by Gullstrand's formula (17.25)

$$\frac{1}{f'} = \frac{1}{f_1'} + \frac{1}{f_2'} \tag{A.97}$$

and so $f' = 199.3$ mm.

(d) The change in f_2' due to a change in n_2 can be obtained by differentiating (A.96)

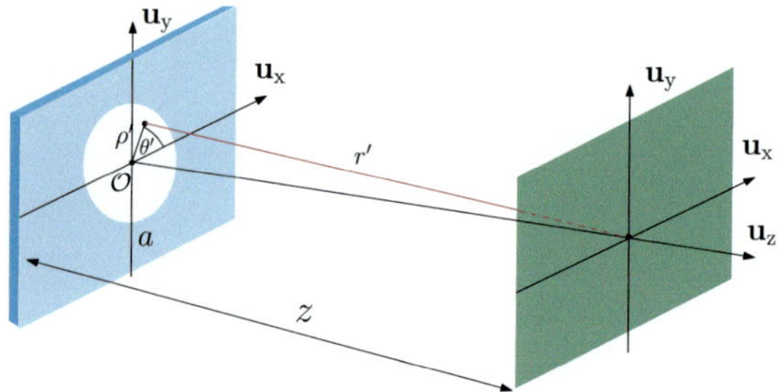

Fig. A.182 Cylindrical coordinates used for the calculation of the Huygens–Fresnel integral

$$-\frac{1}{f_2'^2}\frac{df_2'}{dn_2} = -\frac{1}{R_1} = \frac{1}{f_2'(n_2-1)}$$

and so

$$df_2' = -dn_2\frac{f_2'}{(n_2-1)} = 6.83 \text{ mm.}$$

Finally, from (A.97) we conclude that the change in f' due to dispersion is given by

$$df' = f'^2\left(\frac{df_1'}{f_1'^2} + \frac{df_2'}{f_2'^2}\right) = -80 \ \mu\text{m.}$$

We see that between blue and red light, the relative change in focal length is $df'/f' = -4 \times 10^{-4}$ for the doublet, whereas for the single lenses we have relative changes which are two orders of magnitude higher, of $df_1'/f_1' = -0.0179$ and $df_2'/f_2' = -0.035$.

Problems of Chap. 18

18.1 Diffraction along the axis of a circular aperture

(a) Using cylindrical coordinates (ρ', θ') in the plane $z = 0$, shown in Fig. A.182, the Huygens–Fresnel integral (18.8) at a point $P = (0, 0, z)$ in the axis can be written as

$$\underline{E}(z, t) = -\frac{ie^{-i\omega t}}{\lambda} \int_0^{2\pi} \int_0^\infty \underline{E}_0\, \underline{t}(\rho', \theta') \times \frac{e^{ikr'}}{r'} \frac{z}{r'}\rho'd\rho'd\theta'$$

with $r' = \sqrt{z^2 + \rho'^2}$ and where the transmission factor of the aperture is given by $\underline{t}(\rho', \theta') = 1$ for $\rho' \leq a$ and zero otherwise.

Due to the rotational symmetry of the aperture along the Oz axis, the integral over θ' is simply 2π and we find

$$\underline{E}(z, t) = -\frac{2\pi i \underline{E}_0 z e^{-i\omega t}}{\lambda} \int_0^a \frac{e^{ik\sqrt{\rho'^2 + z^2}}}{z^2 + \rho'^2} \rho' d\rho' .$$

A change of variables $v^2 = \rho'^2 + z^2$ leads finally to

$$\underline{E}(z, t) = -\frac{2\pi i \underline{E}_0 z e^{-i\omega t}}{\lambda} \int_z^{\sqrt{a^2 + z^2}} \frac{e^{ikv}}{v} dv .$$

(b) Using Sommerfeld's lemma and recalling that $k = 2\pi/\lambda$, we obtain

$$\underline{E}(z, t) \approx -\underline{E}_0 z e^{-i\omega t} \left(\frac{e^{ik\sqrt{a^2 + z^2}}}{\sqrt{a^2 + z^2}} - \frac{e^{ikz}}{z} \right) = -\underline{E}_0 e^{-i\omega t} \frac{z e^{ik\sqrt{a^2 + z^2}} - \sqrt{a^2 + z^2} e^{ikz}}{\sqrt{a^2 + z^2}}$$

that is

$$\underline{E}(z, t) \approx \underline{E}_0 e^{i(kz - \omega t)} - \frac{z \underline{E}_0}{\sqrt{z^2 + a^2}} e^{ik(\sqrt{a^2 + z^2} - \omega t)} . \tag{A.98}$$

This can be interpreted as the superposition of the incident plane wave $\underline{E}_0 e^{i(kz - \omega t)}$ with a diffracted wave of amplitude $\underline{E}_0 z / \sqrt{z^2 + a^2}$ having traveled a path $\sqrt{a^2 + z^2}$, which corresponds to the distance between the point z in the axis and the border of the aperture.

(c) The intensity being proportional to $|\underline{E}|^2$, we can write

$$\mathcal{I}(z) = \frac{K|\underline{E}_0|^2}{a^2 + z^2} \left(z^2 + a^2 + z^2 - \frac{2z}{\sqrt{a^2 + z^2}} \cos(k(\sqrt{a^2 + z^2} - z)) \right) \tag{A.99}$$

where K is a proportionality constant. Finally, writing the incident intensity as $\mathcal{I}_0 = K|\underline{E}_0|^2$, we find that along the axis of the aperture, the diffracted intensity varies according to

$$\mathcal{I}(z) = \mathcal{I}_0 \left(1 + \frac{z^2}{a^2 + z^2} \right) - \frac{2\mathcal{I}_0 z}{\sqrt{a^2 + z^2}} \cos\left(k(\sqrt{a^2 + z^2} - z) \right) . \tag{A.100}$$

In the limit $z \gg a^2/\lambda \gg a$, we can neglect a with respect to z outside the cos function and keep the first-order expansion $\sqrt{a^2 + z^2} \approx z + \frac{a^2}{2z}$ inside its argument, so that

$$\mathcal{I}(z) \approx 2\mathcal{I}_0 \left(1 - \cos\left(\frac{ka^2}{2z} \right) \right)$$

and since $ka^2/2z = \pi a^2/\lambda z \ll 1$, we find

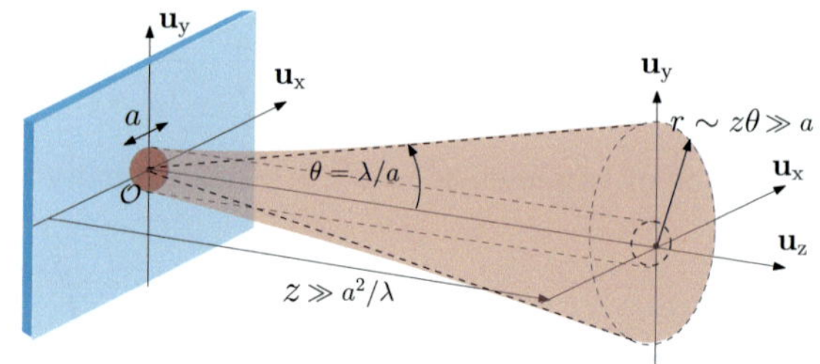

Fig. A.183 The diffracted beam has a minimum divergence of λ/a

$$I(z) \approx 2I_0 \left(1 - 1 + \frac{k^2 a^2}{8z^2}\right) = I_0 \frac{k^2 a^2}{4z^2} .$$

We see that the intensity decreases as $1/z^2$ far from the aperture, in contrast with the prediction of geometrical optics according to which the beam should conserve its size and its uniform intensity after passing through the aperture, since light rays travel in straight paths. In reality, the transverse size of any light beam will eventually increase due to diffraction.

Here, a uniform field distributed over a circle of radius $a \gg \lambda$ at $z = 0$ will diverge with an angle $\theta = \lambda/a$, as shown in Fig. A.183. The intensity is thus distributed over a larger area as z increases, and so its value in the axis decreases to zero as $1/z^2$. This effect is visible when the radius r of the beam becomes much larger than the aperture's size a or, in other words, when the size of the beam becomes much larger than the size a of the beam predicted by geometrical optics. We then expect this decay in intensity to become significant when

$$r = \theta z = \frac{z\lambda}{a} \gg a$$

which indeed corresponds to $z \gg a^2/\lambda$.

(d) The intensity in the axis is modulated by a cosine term, and it reaches a minimum whenever $\cos k(\sqrt{a^2 + z^2} - z) = 1$, that is

$$k(\sqrt{a^2 + z^2} - z) = p2\pi \qquad p \in \mathbb{N}^*$$

and since $k = 2\pi/\lambda$, this rewrites

$$\sqrt{a^2 + z^2} - z = p\lambda \quad p \in \mathbb{N}^* .$$

This tells us that, in order to obtain a destructive interference between the incident wave and the wave diffracted by the aperture's border, the optical path difference should be a multiple of the wavelength. Figure A.184 shows the intensity variation along the axis.

Fig. A.184 The diffracted
beam has a minimum
divergence of λ/a

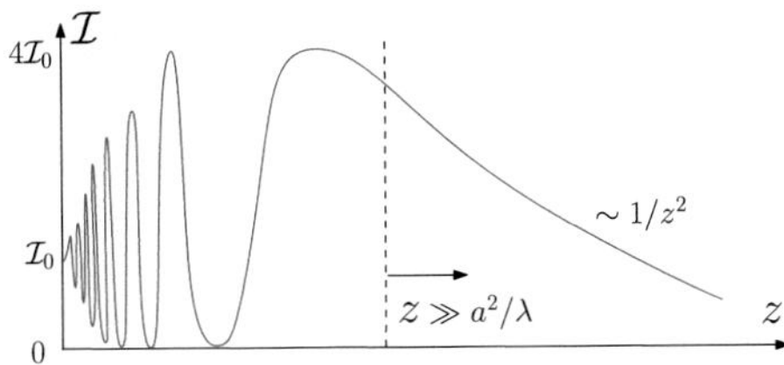

18.2 The Fresnel–Arago spot

(a) The Huygens–Fresnel integral (18.8) allows us to write the electric field at point P as

$$\underline{E}(\rho, \theta, z, t) = -\frac{ie^{-i\omega t}}{\lambda} \times \int_0^{2\pi} \int_0^{\infty} \underline{E}_0(\rho', \theta')\underline{t}(\rho', \theta')\frac{e^{ikr'}}{r'}\frac{z}{r'}\rho'd\rho'd\theta'$$

where

$$r' = \sqrt{\rho^2 + \rho'^2 - 2\rho\rho' \cos(\theta - \theta') + z^2}.$$

For a uniform incident wave $\underline{E}_0(\rho', \theta') = \underline{E}_0$, and since the transmission factor of the disc is given by $\underline{t}(\rho', \theta') = 1$ if $\rho' > a$ and 0 otherwise, the above integral becomes

$$\underline{E}(\rho, \theta, z, t) = -\frac{i\underline{E}_0 z e^{-i\omega t}}{\lambda} \times \int_0^{2\pi} \int_a^{\infty} \frac{e^{ik\sqrt{\rho^2+\rho'^2-2\rho\rho'\cos(\theta-\theta')+z^2}}\rho'd\rho'}{\rho^2 + \rho'^2 - 2\rho\rho' \cos(\theta - \theta') + z^2}d\theta' \; .$$

Now we make the substitution $r^2 = \rho^2 + \rho'^2 - 2\rho\rho' \cos(\theta - \theta') + z^2$, so that

$$2r\,dr = 2\rho'd\rho' - 2\rho \cos(\theta - \theta')$$

and for $\rho \ll a \leq \rho'$, we can write $r\,dr \approx \rho'd\rho'$ and the field is given by

$$\underline{E}(\rho, \theta, z, t) = -\frac{i\underline{E}_0 z e^{-i\omega t}}{\lambda} \times \int_0^{2\pi} \int_{R(\rho',\theta',z)}^{\infty} \frac{e^{ikr}dr}{r}d\theta'$$

with

$$R(\rho', \theta', z) = \sqrt{z^2 + \rho^2 + a^2 - 2a\rho \cos(\theta - \theta')} \; .$$

Since $\rho^2 \ll a^2 + z^2$, we can expand R up to first order in $\rho^2/(a^2 + z^2)$ and so

$$R \approx \sqrt{a^2 + z^2} + \frac{\rho^2}{2\sqrt{z^2 + a^2}} - \frac{\rho a \cos(\theta - \theta')}{\sqrt{z^2 + a^2}} \; .$$

(b) Using Sommerfeld's lemma, we can approximate the field by

Fig. A.185 Diffraction pattern in a plane perpendicular to the axis of the disc, at large distance

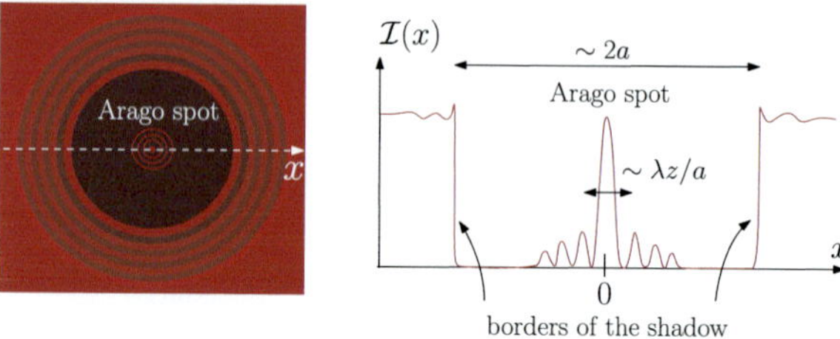

$$\underline{E}(\rho, \theta, z, t) = \frac{i\,\underline{E}_0 z e^{-i\omega t}}{ik\lambda} \int_0^{2\pi} \frac{e^{ikR}}{R} d\theta' .$$

In the denominator, we approximate R by $R \approx \sqrt{a^2 + z^2}$ and we only consider the θ' and ρ' dependence of R in the phase of the exponential. This yields

$$\underline{E}(\rho, \theta, z, t) = \frac{E_0 z e^{i(k\sqrt{a^2+z^2}-\omega t)}}{2\pi\sqrt{a^2+z^2}} e^{i\frac{k\rho^2}{2\sqrt{z^2+a^2}}} \times \underbrace{\int_0^{2\pi} e^{-i\frac{k\rho a \cos(\theta-\theta')}{\sqrt{z^2+a^2}}} d\theta'}_{2\pi J_0\left(\frac{k\rho a}{\sqrt{z^2+a^2}}\right)} .$$

Finally, the intensity $\mathcal{I} \propto |\underline{E}|^2$ reads

$$\mathcal{I}(\rho, z) = \frac{\mathcal{I}_0 z^2}{z^2 + a^2} J_0^2\left(\frac{2\pi\rho a}{\lambda\sqrt{z^2 + a^2}}\right) .$$

The bessel function J_0 is maximum at $\rho = 0$ and presents several oscillations as ρ increases, with a first zero located at $\rho = \rho_1$ such that

$$\frac{2\pi\rho_1 a}{\lambda\sqrt{z^2 + a^2}} = 2.4 .$$

For $z \gg a$, this corresponds to $\rho_1 \approx \frac{2.4\lambda z}{2\pi a}$ and so the Arago spot has a typical diameter of $\sim \lambda z/a$. The diffracted intensity in a plane perpendicular to the z-axis is thus characterized by a central spot surrounded by concentric rings which correspond to the local maxima of J_0^2, whose relative intensities rapidly decrease as a function of the distance to the axis. Eventually, the intensity increases again abruptly at the edges of the shadow. This is illustrated in Fig. A.185.

(c) The Arago spot has zero intensity at $z = 0$ (immediately behind the disc) and the latter increases up to $\sim \mathcal{I}_0$ for $z \sim a$. For $z \gg a$, the Arago spot size increases linearly with distance as $\lambda z/a = \theta z$ (here, we recognize the minimum divergence angle $\theta = \lambda/a$ of a propagating beam in free space). The Arago spot merges with the transmitted wave when its size becomes comparable to the shadow's radius, that is $\lambda z/a \geq a$, and so for distances such that

$$z \geq \frac{a^2}{\lambda} \, .$$

For a disc of radius $a = 50$ μm and a red laser of wavelength $\lambda = 0.65$ μm, the Arago spot will cease to be visible at distances such that $z \geq 3.8$ mm.

18.3 The fundamental Gaussian mode

(a) We have

$$\frac{\partial \mathcal{E}(x, y, z)}{\partial x} = \frac{ikx}{q(z)} A(z) e^{ik \frac{x^2+y^2}{2q(z)}}$$

$$\frac{\partial^2 \mathcal{E}(x, y, z)}{\partial x^2} = \left(\frac{ik}{q(z)} - \frac{k^2 x^2}{q^2(z)} \right) A(z) e^{ik \frac{x^2+y^2}{2q(z)}}$$

and the same result is obtained for $\frac{\partial^2 \mathcal{E}(x,y,z)}{\partial y^2}$ by interchanging x and y. For the derivative with respect to z

$$\frac{\partial \mathcal{E}(x, y, z)}{\partial z} = \frac{dA(z)}{dz} e^{ik \frac{x^2+y^2}{2q(z)}} + A(z) \left(-\frac{ik(x^2 + y^2)}{2q(z)^2} \frac{dq(z)}{dz} \right) e^{ik \frac{x^2+y^2}{2q(z)}}$$

and so

$$\left(-1 + \frac{dq(z)}{dz} \right) \frac{A(z)k^2}{q(z)^2} (x^2 + y^2) + 2ik \left(\frac{A(z)}{q(z)} + \frac{dA(z)}{dz} \right) = 0 \, .$$

(b) Since the latter equality must hold for every x and y, we conclude that

$$\frac{dq(z)}{dz} = 1 \quad \rightarrow \quad q(z) = q_0 + (z - z_0)$$

and replacing in the paraxial wave equation

$$\frac{A(z)}{q(z)} + \frac{dA(z)}{dz} = 0 \quad \rightarrow \quad \frac{1}{A(z)} \frac{dA(z)}{dz} = -\frac{1}{q(z)} \, .$$

Integrating the latter equality

$$\frac{A(z)}{A(z_0)} = \frac{q(z_0)}{q} = \frac{q_0}{q(z)} \, .$$

(c) Replacing q and A into the proposed solution, we obtain

$$\mathcal{E}(x, y, z) = A_0 \frac{q_0}{q(z)} e^{ik \frac{x^2+y^2}{2q(z)}}$$

and writing $1/q(z) = 1/R(z) + i \frac{\lambda}{\pi \omega(z)^2}$

$$\mathcal{E}(x, y, z) = A_0 \frac{q_0}{q(z)} e^{ik \frac{x^2+y^2}{2R(z)}} e^{-k \frac{\lambda(x^2+y^2)}{2\pi \omega(z)^2}} .$$

Finally, since $k\lambda = 2\pi$, we have

$$\mathcal{E}(x, y, z) = A_0 \frac{q_0}{q(z)} e^{ik \frac{x^2+y^2}{2R(z)}} e^{-\frac{(x^2+y^2)}{\omega(z)^2}} .$$

We see that $R(z)$ represents the radius of curvature of the wavefront at position z, whereas $\omega(z)$ defines the characteristic distance over which the field intensity decays as a function of the distance $x^2 + y^2$ with respect to the propagation axis. Since $q(z) = (z - z_0) - iz_R$,

$$\begin{aligned}
\frac{1}{q(z)} &= \frac{1}{(z - z_0) - iz_R} \\
&= \frac{z - z_0}{(z - z_0)^2 + z_R^2} + i \frac{z_R}{(z - z_0)^2 + z_R^2} \\
&= \frac{1}{R(z)} + i \frac{\omega_0^2}{\omega(z)^2 z_R} .
\end{aligned}$$

From this, we obtain the radius of curvature and the beam radius as a function of the position z on the axis:

$$R(z) = z - z_0 + \frac{z_R^2}{(z - z_0)} , \tag{A.101}$$

$$\omega(z)^2 z_R = \omega_0^2 \frac{(z - z_0)^2 + z_R^2}{z_R} ,$$

that is

$$\omega(z) = \omega_0 \sqrt{1 + \left(\frac{z - z_0}{z_R} \right)^2} . \tag{A.102}$$

(d) The field intensity $\mathcal{I}(x, y, z) \propto |\mathcal{E}(x, y, z)|^2$ is a Gaussian function:

$$\mathcal{I}(x, y, z) \propto \frac{1}{|q(z)|^2} e^{-2 \frac{x^2+y^2}{\omega(z)^2}}$$

$$\mathcal{I}(x, y, z) = \frac{2\mathcal{I}_0}{\pi \omega(z)^2} e^{-2 \frac{x^2+y^2}{\omega^2(z)}} .$$

The intensity decays by $1/e^2$ at a distance from the axis given by the radius $\omega(z)$. This is shown in Fig. A.186.

$$\omega(z) \approx \omega_0 \left(1 + \frac{(z - z_0)^2}{2z_R^2} \right) .$$

Fig. A.186 The intensity is a Gaussian function of x

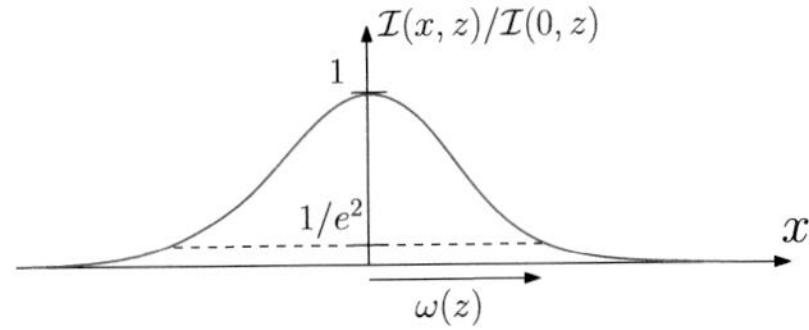

18.4 The Descartes relation

(a) At a point (x', y') in the plane $z = 0$ and just before the lens, the electric field writes

$$\underline{E}_i(x', y', t) = \underline{E}_0 \frac{e^{i(k\sqrt{d_1^2+x'^2+y'^2}-\omega t)}}{\sqrt{d_1^2 + x'^2 + y'^2}} \ .$$

In the paraxial approximation, $x'^2 + y'^2 \ll d_1^2$ so that we can approximate the denominator by d_1 and keep the (x', y') dependence of the phase in the exponential up to first order in $(x'^2 + y'^2)/d_1^2$, which gives

$$\underline{E}_i(x', y', t) \approx \underline{E}_0 \frac{e^{i(kd_1 - \omega t)}}{d_1} e^{ik\frac{x'^2+y'^2}{2d_1}} \ .$$

The field immediately after the lens is simply $\underline{E}_0(x', y', 0, t) = \underline{t}(x', y', t)\underline{E}_i(x', y', t)$, where $\underline{t}(x', y') = \underline{t}_0 e^{-ik\frac{x'^2+y'^2}{2f'}}$ is the transmission factor of the lens. We have

$$\underline{E}_0(x', y', 0, t) = \underline{E}_0\underline{t}_0 \frac{e^{i(kd_1 - \omega t)}}{d_1} \times e^{ik\frac{(x'^2+y'^2)}{2}\left(\frac{1}{d_1}-\frac{1}{f'}\right)} \ .$$

(b) The field can be rewritten as

$$\underline{E}_0(x', y', 0, t) = \underline{A}e^{-ik\frac{(x'^2+y'^2)}{2d_2}}$$

where $\underline{A}$ is a constant and d_2 is given by $1/d_2 = 1/f' - 1/d_1$. It corresponds, in the paraxial approximation, to the field created by a point source located at $\mathbf{x}_i = d_2\mathbf{u}_z$. If $d_1 > f'$, $d_2 > 0$ so that the wave converges toward the point $\mathbf{x}_i$ after the lens, corresponding therefore to a real image. If $d_1 < f'$, $d_2 < 0$, and the field after the lens is identical to the one generated by a virtual image before the lens. The position of the object and its image by the lens are therefore related by the Descartes relation

$$\frac{1}{d_1} + \frac{1}{d_2} = \frac{1}{f'} \ .$$

18.5 The human eye

For a very distance point-like source, the radius of the Airy disc in the retina is given by

$$r^* = 1.22 \frac{\lambda f'}{n' \phi}$$

where $\phi = 2 - 8$ mm is the pupil's diameter, $f' = 2.2$ cm the focal length of the eye at rest, and $n' = 1.33$ the refractive index of the medium between the eye's lens and the retina (vitreous humor). For $\lambda = 500$ nm, we obtain an Airy spot size of

$$r^* = 1.26 - 5.04 \ \mu\text{m} \ .$$

Since the distance between cone cells in the region of the retina with the highest resolution (fovea) is around 2.5 μm, we conclude that in general the eye's resolution is close to the diffraction limit and that it is actually diffraction-limited for small diameters of the pupil.

18.6 Divergence of a Gaussian beam

(a) The Fraunhofer diffraction is valid in the far field limit $z \gg a^2/\lambda$ where a is the typical size of the field distribution at $z = 0$. In this case, this corresponds to distances much larger than the Rayleigh length of the beam $z_R = w_0^2/\pi\lambda$. We must then have

$$z \gg \frac{w_0^2}{\lambda} = 1.6 \ \text{m} \ .$$

A few meters from the laser tube, it is valid to use the formula of Fraunhofer diffraction.

(b) The field at distance z is given by Eq. (18.29)

$$\underline{E}(x, y, z, t) = \frac{2\pi k}{iz} e^{i(kz - \omega t)} e^{ik \frac{(x^2 + y^2)}{2z}} \times \mathcal{F}\{\underline{E}_0 e^{-\frac{(x'^2 + y'^2)}{w_0^2}}\} \left(\frac{kx}{z}, \frac{ky}{z} \right) .$$

The Fourier transform of a Gaussian is also a Gaussian

$$\mathcal{F}\{e^{-\alpha x^2}\}(u) = \frac{1}{\sqrt{4\pi\alpha}} e^{-\frac{k^2}{4\alpha}}$$

so that

$$\mathcal{F}\{e^{-\frac{x'^2 + y'^2}{w_0^2}}\}(u, v) = \frac{w_0^2}{4\pi} e^{-\frac{w_0^2(u^2 + v^2)}{4}} \ .$$

The Gaussian beam remains Gaussian while propagating. The same is true for the intensity profile, which reads

$$I(x, y, z) = I_0 \left| \mathcal{F}\{e^{-\frac{(x'^2 + y'^2)}{w_0^2}}\} \left(\frac{kx}{z}, \frac{ky}{z} \right) \right|^2 = I_0 e^{-\frac{k^2 w_0^2 (x^2 + y^2)}{2z^2}} = I_0 e^{-\frac{2(x^2 + y^2)}{w(z)^2}}$$

whose spatial extent is given by the radius

$$w(z) = 2\frac{z}{kw_0} = \frac{\lambda z}{\pi w_0}$$

which increases linearly as a function of distance. This linear increase describes well the evolution of the laser profile for distances in the far field limit $z \gg w_0^2/\lambda$.

(c) The angular divergence of the beam in the far field is simply given by

$$\theta = \tan^{-1}\left(\frac{w(z)}{z}\right) = \tan^{-1}\left(\frac{\lambda}{\pi w_0}\right)$$

and since $w_0 \gg \lambda$, $\sin\theta \approx \tan\theta \approx \theta$ and therefore

$$\theta \approx \frac{\lambda}{\pi w_0} \, .$$

The divergence of a Gaussian beam, illustrated in Fig. A.187, is therefore the minimal possible divergence limited by diffraction. For the He-Ne laser with $w_0 = 1$ mm and $\lambda = 632.8$ nm, we obtain $\theta \sim 0.2$ mrad.

Problems of Chap. 19

19.1 The Lorenz gauge
Let $\mathbf{A}' = \mathbf{A} + \nabla\phi$ and $V' = V - \frac{\partial\phi}{\partial t}$. The electric and magnetic fields generated by $\mathbf{A}'$ and V' are the same as the ones generated by $\mathbf{A}$ and V, regardless of the choice of the scalar field ϕ. Indeed:

$$\mathbf{B}' = \nabla \times \mathbf{A}' = \nabla \times \mathbf{A} + \underbrace{\nabla \times \nabla\phi}_{=0} = \mathbf{B}$$

and

$$\mathbf{E}' = -\frac{\partial\mathbf{A}'}{\partial t} - \nabla V' = -\frac{\partial\mathbf{A}}{\partial t} - \nabla V \underbrace{- \frac{\partial}{\partial t}\nabla\phi + \nabla\frac{\partial\phi}{\partial t}}_{=0} = \mathbf{E}\,.$$

Let us now impose the Lorenz gauge condition

$$\frac{1}{c^2}\frac{\partial V'}{\partial t} + \nabla \cdot \mathbf{A}' = 0\,.$$

This is

$$\frac{1}{c^2}\frac{\partial V}{\partial t} - \frac{1}{c^2}\frac{\partial^2\phi}{\partial t^2} + \nabla \cdot \mathbf{A} + \nabla \cdot \nabla\phi = 0\,.$$

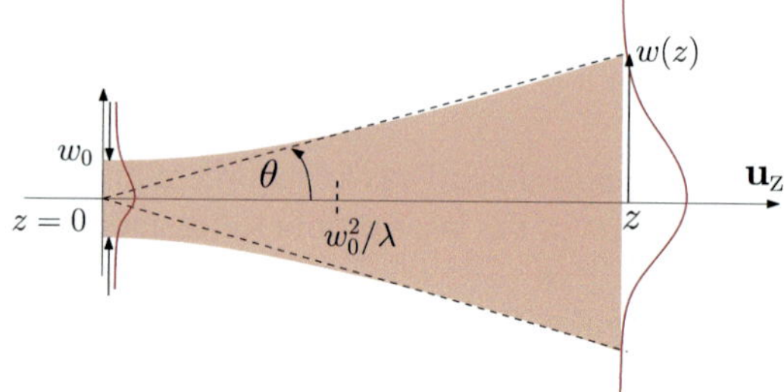

Fig. A.187 The intensity is a Gaussian function of x

We find that ϕ must be a solution of the propagation equation with $\frac{1}{c^2}\frac{\partial V}{\partial t} + \boldsymbol{\nabla}\cdot\mathbf{A}$ as source term

$$\boldsymbol{\nabla}^2\phi - \frac{1}{c^2}\frac{\partial^2\phi}{\partial t^2} = -\frac{1}{c^2}\frac{\partial V}{\partial t} - \boldsymbol{\nabla}\cdot\mathbf{A}.$$

19.2 Derivation of the radiated magnetic field from the vector potential

For the magnetic field, we start from the radiation vector potential

$$\mathbf{A}_{\mathrm{rad}}(\mathbf{x}, t) = \frac{\mu_0}{4\pi r} \iiint_\Omega d^3x' \mathbf{J}(\mathbf{x}', t^*)$$

where

$$t^* = t - \frac{r}{c} + \frac{\mathbf{u}_r \cdot \mathbf{x}'}{c}.$$

First use the following vector calculus identity[2]

$$\boldsymbol{\nabla} \times \left(\frac{\iiint_\Omega d^3x' \mathbf{J}(\mathbf{x}', t^*)}{r} \right) = \boldsymbol{\nabla}\left(\frac{1}{r}\right) \times \iiint_\Omega d^3x' \mathbf{J}(\mathbf{x}', t^*) + \frac{1}{r}\boldsymbol{\nabla} \times \iiint_\Omega d^3x' \mathbf{J}(\mathbf{x}', t^*).$$

The first term decays as $(1/r^2)$, while for the second,

$$\boldsymbol{\nabla} \times \iiint_\Omega d^3x' \mathbf{J}(\mathbf{x}', t^*) = \iiint_\Omega d^3x' \boldsymbol{\nabla} \times \mathbf{J}(\mathbf{x}', t^*) = \iiint_\Omega d^3x' \boldsymbol{\nabla} t^* \times \frac{\partial}{\partial t}\mathbf{J}(\mathbf{x}', t^*).$$

We therefore need to calculate

$$\boldsymbol{\nabla} t^* = -\frac{1}{c}\left[\boldsymbol{\nabla} r - \boldsymbol{\nabla}(\mathbf{u}_r \cdot \mathbf{x}')\right] = -\frac{\mathbf{u}_r}{c} + \frac{\mathbf{x}'}{cr} - \frac{\mathbf{x}\cdot\mathbf{x}'}{cr^2}\mathbf{u}_r$$

and we retain only the dominant (first) term on the right-hand side:

$$\boxed{\boldsymbol{\nabla} t^* \approx -\frac{\mathbf{u}_r}{c}}$$

---

[2] For any scalar a and vector $\mathbf{b}$, $\boldsymbol{\nabla} \times a\mathbf{b} = (\boldsymbol{\nabla} a) \times \mathbf{b} + a\boldsymbol{\nabla} \times \mathbf{b}$.

$$\nabla \times \left(\frac{\dot{\mathbf{p}}(\tau)}{r} \right) = -\frac{1}{rc} \mathbf{u}_r \times \ddot{\mathbf{p}}(\tau) + O\left(\frac{1}{r^2} \right).$$

We finally compute the integral and differentiation with respect to time:

$$\nabla \times \iiint_\Omega d^3x' \mathbf{J}(\mathbf{x}', t^*) = -\frac{\mathbf{u}_r}{c} \times \iiint_\Omega d^3x' \frac{\partial}{\partial t} \mathbf{J}(\mathbf{x}', t^*) = -\frac{\mathbf{u}_r}{c} \times \frac{\partial}{\partial t} \iiint_\Omega d^3x' \mathbf{J}(\mathbf{x}', t^*),$$

which yields the radiated magnetic field Eq. (19.31)

$$\mathbf{B}_{\mathrm{rad}}(\mathbf{x}, t) = -\frac{\mu_0}{4\pi cr} \mathbf{u}_r \times \frac{\partial}{\partial t} \iiint_\Omega d^3x' \mathbf{J}(\mathbf{x}', t^*).$$

19.3 Derivation of the radiated magnetic field from the potentials

We start with the gradient of the scalar potential:

$$V_{\mathrm{rad}}(\mathbf{x}, t) = \frac{1}{4\pi \epsilon_0 r} \iiint_\Omega d^3x' \varrho(\mathbf{x}', t^*),$$

where

$$t^* = t - \frac{t}{c} + \frac{\mathbf{u}_r \cdot \mathbf{x}'}{c}.$$

Applying the vector calculus identity for the gradient of a product,[3] we find

$$\nabla V_{\mathrm{rad}}(\mathbf{x}, t) = \frac{1}{4\pi \epsilon_0} \left[\frac{1}{r} \nabla \left(\iiint_\Omega d^3x' \varrho(\mathbf{x}', t^*) \right) - \left(\frac{\mathbf{u}_r}{r^2} \right) \iiint_\Omega d^3x' \varrho(\mathbf{x}', t^*) \right].$$

The second term on the right-hand side generates a field that decays as $1/r^2$, and therefore does not contribute to radiation. For the first term, using the chain rule, we have

$$\nabla \left(\iiint_\Omega d^3x' \varrho(\mathbf{x}', t^*) \right) = \iiint_\Omega d^3x' \nabla \varrho(\mathbf{x}', t^*) = \iiint_\Omega d^3x' \nabla t^* \frac{\partial}{\partial t} \varrho(\mathbf{x}', t^*)$$

$$\approx \frac{\mathbf{u}_r}{c} \iiint_\Omega d^3x' \frac{\partial}{\partial t} \varrho(\mathbf{x}', t^*).$$

Then we use charge conservation to eliminate the charge density in favor of the current density:

$$\frac{\partial}{\partial t} \varrho(\mathbf{x}', t^*) = \left[\frac{\partial}{\partial t'} \varrho(\mathbf{x}', t') \right]_{t'=t^*} = -\left[\nabla' \cdot \mathbf{J}(\mathbf{x}', t') \right]_{t'=t^*}.$$

On the other hand, we can use the chain rule to calculate

---

[3] For two scalar fields a and b, $\nabla(ab) = a\nabla b + b\nabla a$.

$$\nabla' \cdot \mathbf{J}(\mathbf{x}', t^*) = \left[\nabla' \cdot \mathbf{J}(\mathbf{x}', t')\right]_{t'=t^*} + \nabla' t^* \cdot \frac{\partial}{\partial t^*} \mathbf{J}(\mathbf{x}', t^*)$$

and $\nabla' t^* = \mathbf{u}_r/c$. Thus,

$$\frac{\partial}{\partial t}\varrho(\mathbf{x}', t^*) = -\nabla' \cdot \mathbf{J}(\mathbf{x}', t^*) + \frac{\mathbf{u}_r}{c} \cdot \frac{\partial}{\partial t}\mathbf{J}(\mathbf{x}', t^*).$$

When we perform the integral over the volume Ω, the term $-\nabla' \cdot \mathbf{J}(\mathbf{x}', t^*)$ involving the divergence is transformed into a surface integral which vanishes because $\mathbf{J}(\mathbf{x}', t^*) = 0$ for $\mathbf{x}' \in \partial\Omega$. Finally,

$$\nabla \left(\iiint_\Omega d^3x' \varrho(\mathbf{x}', t^*) \right) = -\frac{1}{c^2}\mathbf{u}_r \left(\mathbf{u}_r \cdot \boldsymbol{\alpha}(\mathbf{x}, t) \right),$$

where

$$\boldsymbol{\alpha}(\mathbf{x}, t) \equiv \frac{\partial}{\partial t} \iiint_\Omega d^3x' \mathbf{J}(\mathbf{x}', t^*), \quad \text{with} \quad t^* = t - r/c + \mathbf{u}_r \cdot \mathbf{x}'/c.$$

Another contribution to the radiated magnetic field comes from

$$\frac{\partial \mathbf{A}_{\text{rad}}}{\partial t} = \frac{\mu_0}{4\pi r}\frac{\partial}{\partial t} \iiint_\Omega d^3x' \mathbf{J}(\mathbf{x}', t^*) = \frac{\mu_0}{4\pi r}\boldsymbol{\alpha}(\mathbf{x}, t).$$

Collecting the two contributions, we find

$$\nabla V_{\text{rad}} + \frac{\partial \mathbf{A}_{\text{rad}}}{\partial t} = \frac{\mu_0}{4\pi r} \left\{ -\mathbf{u}_r \left(\mathbf{u}_r \cdot \boldsymbol{\alpha}(\mathbf{x}, t) \right) + \boldsymbol{\alpha}(\mathbf{x}, t) \right\}.$$

We recognize the triple product expansion: $\mathbf{u}_r \times (\mathbf{u}_r \times \boldsymbol{\alpha}) = \mathbf{u}_r(\mathbf{u}_r \cdot \boldsymbol{\alpha}) - \boldsymbol{\alpha}$, hence

$$\mathbf{E}_{\text{rad}}(\mathbf{x}, t) = \frac{\mu_0}{4\pi r}\mathbf{u}_r \times (\mathbf{u}_r \times \boldsymbol{\alpha}(\mathbf{x}, t)) = \frac{\mu_0}{4\pi r}\mathbf{u}_r \times \left(\mathbf{u}_r \times \frac{\partial}{\partial t} \iiint_\Omega d^3x' \mathbf{J}(\mathbf{x}', t^*) \right).$$

19.4 Energy radiated by a charged particle

The dipole moment of the charge q with respect to the origin is given by

$$\mathbf{p}(t) = qr(t)\mathbf{u}_r.$$

Then

$$\ddot{\mathbf{p}}(t) = q\frac{d^2r(t)}{dt^2}\mathbf{u}_r.$$

The total power radiated by the charge at an instant t will be

$$P(t) = \left(\frac{\mu_0}{6\pi c}\right) q^2 \{\ddot{r}(t)\}^2 \ .$$

Solving this problem exactly is really complex. However, we can assume that the total radiated energy is much smaller than the initial energy of the particle. In this way, we can assume that the trajectory is that of a particle that does not lose energy under the action of the potential. We can then determine $r_{\min}$ satisfying

$$\frac{1}{2}mv_0^2 = \frac{q\alpha}{r_{\min}}.$$

Thus

$$r_{\min} = \frac{2q\alpha}{mv_0^2}.$$

Newton's equation also yields

$$m\ddot{r}(t) = -\frac{q\alpha}{r^2}$$

so that

$$P(t) = \left(\frac{\mu_0}{6\pi c}\right) \frac{q^4\alpha^2}{m^2 r^4(t)}.$$

In addition, given the simplification of the problem, we can obtain the total radiated energy as twice the radiated energy on the path $r_{\min} \to \infty$. Thus

$$E = 2 \int_{t_0}^{t_f} dt \, P(t),$$

with

$$dt = \frac{dr}{v}$$

where

$$\frac{1}{2}mv^2 + \frac{q\alpha}{r} = \frac{1}{2}mv_0^2.$$

Then

$$v(r) = \sqrt{v_0^2 - \frac{2q\alpha}{mr}}.$$

Finally

$$E = 2 \int_{r_{\min}}^{\infty} dr \left(\frac{\mu_0}{6\pi c}\right) \frac{q^4\alpha^2}{m^2 r^4 \sqrt{v_0^2 - \frac{2q\alpha}{mr}}}$$

$$= 2\left(\frac{\mu_0 q^4 \alpha^2}{6\pi m^2 c}\right) \int_{r_{\min}}^{\infty} dr \, \frac{1}{r^4 \sqrt{v_0^2 - \dfrac{2q\alpha}{mr}}}$$

$$= 2\left(\frac{\mu_0 q^4 \alpha^2}{6\pi m^2 c}\right) \int_{r_{\min}}^{\infty} dr \, \frac{1}{r^4 \sqrt{\dfrac{2q\alpha}{m}} \sqrt{\dfrac{1}{r_{\min}} - \dfrac{1}{r}}}$$

$$= 2\left(\frac{\mu_0 q^4 \alpha^2}{6\pi m^2 c}\right) \sqrt{\frac{m}{2q\alpha}} \int_{r_{\min}}^{\infty} dr \, \frac{1}{r^4 \sqrt{\dfrac{1}{r_{\min}} - \dfrac{1}{r}}} \, .$$

This last integral can be evaluated numerically, leading to

$$E = 2\left(\frac{\mu_0 q^4 \alpha^2}{6\pi m^2 c}\right) \sqrt{\frac{m}{2q\alpha}} \frac{16}{15 r_{\min}^{5/2}} \, .$$

Replacing the value of $r_{\min}$,

$$E = \left(\frac{\mu_0 q m}{3\pi \alpha c}\right) \frac{2 v_0^5}{15} \, .$$

Note that this result is valid while

$$E \ll \frac{1}{2} m v_0^2 \, .$$

19.5 Electron in a gravitational field

The acceleration of the electron in the terrestrial gravitational field is equal to $a = g = 9.81 \text{ m/s}^2$, so that the radiated power is

$$P = \left(\frac{\mu_0 e^2}{6\pi c}\right) g^2 \, .$$

The time T that the electron takes to fall a distance d is given by

$$d = \frac{g T^2}{2}, \quad \text{i.e.,} \quad T = \sqrt{\frac{2d}{g}}$$

so that the energy radiated over a distance d is

$$E_{\text{rad}} = PT = \left(\frac{\mu_0 e^2}{6\pi c}\right) g^2 \sqrt{\frac{2d}{g}} \, .$$

If $d = 1$ cm, we find $T = 4.51 \times 10^{-2}$ s, and

$$E_{\text{rad}} = 5.49 \times 10^{-52} \text{ J} = 3.42 \times 10^{-33} \text{ eV}.$$

On the other hand, the loss of potential energy after falling a distance d is

$$\Delta U = -mgd = -8.94 \times 10^{-32} \text{ J} = -5.6 \times 10^{-13} \text{ eV}.$$

We see that the fraction of potential energy that is transferred in the form of radiation is negligible compared to ΔU:

$$\frac{E_{\text{rad}}}{\Delta U} \sim 10^{-22}.$$

Then, virtually all the potential energy is converted into kinetic energy.

19.6 Charge suspended by a spring

Assuming that the charge oscillates in the z direction, the equation of motion, without taking into account any friction due to radiation, is given by

$$m\frac{d^2z}{dt^2} = -K(z - z_0),$$

where z_0 is the equilibrium position. The solution is a harmonic motion of amplitude A_0 around z_0:

$$z(t) - z_0 = A_0 \cos(\omega t)$$

with $\omega = \sqrt{K/m}$. Assuming that the energy radiated during a cycle is much smaller than the mechanical energy of the system, we can consider that the acceleration corresponds at all times to $\mathbf{A}(t) = -A_0\omega^2 \cos(\omega t)\mathbf{u}_z$ and then the average power radiated during a cycle is

$$\langle P \rangle = \left(\frac{\mu_0 e^2}{6\pi c}\right)\langle a^2 \rangle = \left(\frac{\mu_0 e^2}{6\pi c}\right)\frac{\omega^4 A_0^2}{2} = -\frac{dE}{dt}.$$

The mechanical energy of the system is given by $E(t) = \dfrac{1}{2}KA^2(t) = \dfrac{1}{2}m\omega^2 A^2(t)$, then:

$$\langle P \rangle = \left(\frac{\mu_0\omega^2 e^2}{6\pi cm}\right)E(t) = -\frac{dE(t)}{dt}$$

so that the mechanical energy decays exponentially as $E(t) = E_0 e^{-t/\tau}$, with a time constant given by

$$\tau = \frac{6\pi mc}{\mu_0\omega^2 e^2}.$$

Finally, the mechanical energy will be equal to half of its initial value for $t = t^*$ such that

$$e^{-t^*/\tau} = \frac{1}{2} \quad \rightarrow \quad t^* = \tau \ln(2) = \frac{6\pi mc}{\mu_0\omega^2 e^2}\ln(2).$$

19.7 Cyclotron radiation

The radius of the orbit is given by the balance between the acceleration and the Lorentz force:

$$m\frac{v^2}{R} = qvB \quad \rightarrow \quad R = \frac{mv}{qB}.$$

Assuming that the radius of the orbit is approximately constant during a revolution, we have $a = v^2/R$ and the power radiated during a revolution is

$$\langle P \rangle = \left(\frac{\mu_0 q^2}{6\pi c}\right)\frac{v^4}{R^2} = -\frac{dE}{dt}.$$

The kinetic energy at time t is given by

$$E(t) = \frac{1}{2}mv(t)^2 = \frac{1}{2m}R(t)^2 q^2 B^2,$$

so that

$$\frac{dE}{dt} = R\frac{dR}{dt}\frac{q^2 B^2}{m}.$$

Then

$$-\frac{dE}{dt} = -\frac{q^2 B^2}{m}R\frac{dR}{dt} = \left(\frac{\mu_0 q^2}{6\pi c}\right)\frac{v^4}{R^2}$$
$$= \left(\frac{\mu_0 e^2}{6\pi c}\right)\frac{R^4 q^4 B^4}{m^4}.$$

Finally

$$\frac{dR}{dt} = -\left(\frac{q^4 B^2 \mu_0}{6\pi c m^3}\right)R,$$

the solution of which is

$$R(t) = R_0 e^{-t/\tau}$$

with

$$\tau = \frac{6\pi c m^3}{q^4 B^2 \mu_0}.$$

As the radiated power can be written as a function of the radius as

$$P = \left(\frac{q^6 B^4 \mu_0}{6\pi c m^4}\right)R^2 = \left(\frac{q^6 B^4 \mu_0}{6\pi c m^4}\right)R_0^2 e^{-2t/\tau},$$

the total radiated energy is given by

$$\Delta E = \left(\frac{q^6 B^4 \mu_0}{6\pi c m^4}\right) R_0^2 \int_0^\infty e^{-2t/\tau}\, dt = \left(\frac{q^6 B^4 \mu_0}{6\pi c m^4}\right) R_0^2 \frac{\tau}{2}\,.$$

Introducing the expression for τ in the latter equation leads to

$$\Delta E = \left(\frac{q^2 B^2}{2m}\right) R_0^2 = \frac{1}{2} m v_0^2$$

which corresponds to the initial kinetic energy of the charge q.

19.8 Radiation power and polarization of a rotating dipole

(a) The dipole moment has modulus p_0 and its direction makes an angle ωt with respect to the y-axis (at $t = 0$, it is assumed to be aligned with the y-axis). The dipole moment thus reads, component-wise,

$$\mathbf{p}(t) = \begin{cases} p_x = 0 \\ p_y = p_0 \cos \omega t \\ p_z = p_0 \sin \omega t \end{cases} \Rightarrow \quad \ddot{\mathbf{p}}(t) = \begin{cases} 0 \\ -\omega^2 p_0 \cos \omega t \\ -\omega^2 p_0 \sin \omega t \end{cases}.$$

In the system of spherical coordinates with basis $(\mathbf{u}_r, \mathbf{u}_\theta, \mathbf{u}_z)$, we have $\mathbf{u}_y = \sin\theta\,\mathbf{u}_r + \cos\theta\,\mathbf{u}_\theta$, hence

$$\ddot{\mathbf{p}}(t) = -\omega^2 p_0 (\sin\theta \cos\omega t\, \mathbf{u}_r + \cos\theta \cos\omega t\, \mathbf{u}_\theta + \sin\omega t\, \mathbf{u}_z)\,.$$

The magnetic field radiated at large distances from the rotating dipole thus reads

$$\mathbf{B}(M, t) = \frac{\mu_0}{4\pi cr}\ddot{\mathbf{p}}(\tau) \times \mathbf{u}_r = -\frac{\mu_0 \omega^2 p_0}{4\pi cr} \begin{bmatrix} \sin\theta \cos\omega\tau \\ \cos\theta \cos\omega\tau \\ \sin\omega\tau \end{bmatrix} \times \begin{bmatrix} 1 \\ 0 \\ 0 \end{bmatrix}$$

$$= \frac{\mu_0 \omega^2 p_0}{4\pi cr} \begin{bmatrix} 0 \\ -\sin\omega\tau \\ \cos\theta \cos\omega\tau \end{bmatrix},$$

where $\tau = t - r/c$. The radiated electric field reads

$$\mathbf{E} = c\mathbf{B} \times \mathbf{u}_r = \frac{\mu_0 \omega^2 p_0}{4\pi r} \begin{bmatrix} 0 \\ -\sin\omega\tau \\ \cos\theta \cos\omega\tau \end{bmatrix} \times \begin{bmatrix} 1 \\ 0 \\ 0 \end{bmatrix} = \frac{\mu_0 \omega^2 p_0}{4\pi r} \begin{bmatrix} 0 \\ \cos\theta \cos\omega\tau \\ \sin\omega\tau \end{bmatrix}.$$

Thus,

$$\mathbf{E}(\mathbf{r}, t) = \frac{\mu_0 \omega^2 p_0}{4\pi r} \left[\cos\theta \cos\omega\left(t - \frac{r}{c}\right) \mathbf{u}_\theta + \sin\omega\left(t - \frac{r}{c}\right) \mathbf{u}_z\right].$$

(b) The modulus of **B** reads

$$|\mathbf{B}| \frac{\mu_0 p_0 \omega^2}{4\pi c} \frac{1}{r} \sqrt{\sin^2 \omega\tau + \cos^2\theta \cos^2 \omega\tau}\,.$$

Thus, using $\langle \cos^2 \omega\tau \rangle = \langle \sin^2 \omega\tau \rangle = 1/2$, we find

$$\langle \mathbf{B}^2 \rangle = \frac{\mu_0^2 p_0^2 \omega^4}{32\pi^2 c^2} \frac{1}{r^2} (1 + \cos^2\theta)\,.$$

(c) The propagation direction of the electromagnetic radiation is that of the Poynting vector, hence it is in the radial direction:

$$\mathbf{\Pi} = \frac{\mathbf{E} \times \mathbf{B}}{\mu_0} = \frac{c B^2}{\mu_0}\mathbf{u}_r\,.$$

In the solid angle $d\Omega = dS/r^2$ around the observation direction making an angle θ with the x-axis, the average radiated power reads

$$\langle d\mathcal{P} \rangle = \langle \mathbf{\Pi} \rangle \cdot d\mathbf{S} = \langle |\mathbf{\Pi}| \rangle r^2 d\Omega$$

where the modulus of the average Poynting vector reads

$$\langle |\mathbf{\Pi}| \rangle = \frac{c}{\mu_0}\langle B^2 \rangle = \frac{\mu_0 p_0^2 \omega^4}{32\pi^2 c}\frac{1}{r^2}(1 + \cos^2\theta)\,.$$

Hence

$$\frac{\langle d\mathcal{P} \rangle}{d\Omega} = \langle |\mathbf{\Pi}| \rangle r^2 = \frac{\mu_0 p_0^2 \omega^4}{32\pi^2 c}(1 + \cos^2\theta)\,.$$

The radiated power per unit solid angle is maximum for $\theta = 0$ and $\theta = \pi$, i.e., along the direction of the rotation axis of the dipole. It is minimum for $\theta = \pi/2$, in the orthogonal plane to the x-axis.

(d) The average total power radiated by the rotating dipole is obtained by integration over the solid angle: $d\Omega = 2\pi \sin\theta d\theta$, thus

$$\langle \mathcal{P} \rangle = \int \langle d\mathcal{P} \rangle = \frac{\mu_0 p_0^2 \omega^4}{32\pi^2 c} 2\pi \int_0^\pi (1 + \cos^2\theta) \sin\theta d\theta$$

$$= \frac{\mu_0 p_0^2 \omega^4}{16\pi c}\left[-\cos\theta - \frac{\cos^3\theta}{3} \right]_0^\pi$$

$$= \langle \mathcal{P} \rangle = \frac{\mu_0 p_0^2 \omega^4}{6\pi c}\,.$$

The radiated power is proportional to the fourth power of the rotation frequency of the dipole.

(e) The radiation electromagnetic wave is linearly polarized if the tip of the **E**-vector describes a line when time evolves. This is realized if $\cos\theta = 0$, i.e., $\theta = \pi/2$. Then the polarization is along the z-axis:

$$\mathbf{E} = \frac{\mu_0 p_0 \omega^2}{4\pi r} \sin\omega\left(t - \frac{r}{c}\right)\mathbf{u}_z .$$

The polarization is circular if the components E_θ and E_z of **E** have the same amplitude and a quadrature $(\pi/2)$ phase difference. This is realized if $\cos\theta = \pm 1$, hence, if $\theta = 0$ or $\theta = \pi$.

For all other values of θ, the tip of the **E** vector describes an ellipse in the plane $\mathbf{u}_\theta$ and $\mathbf{u}_z$. The polarization is elliptic.

In conclusion the polarization state depends on the observation direction θ with respect to the rotation axis of the dipole.

Problems of Chap. 20

20.1 Global Positioning System and relativity

Since the satellite travels at a speed of $v = 14000$ km/h, the Lorentz factor is

$$\gamma = \frac{1}{\sqrt{1 - \left(\frac{14000/3.6}{3\times 10^8}\right)^2}} = 1 + 8.4 \times 10^{-11} .$$

As a result, the time error between the satellite clock and the receiver clock after $T = 12\,\text{h}$ would be

$$\delta t = T' - T = (\gamma - 1)\,T = 8.4 \times 10^{-11} \times 12 \times 3600 = 3.6\,\mu\text{s}$$

and this corresponds to a positioning error of

$$\delta x = c\delta t = 1.1\,\text{km} .$$

20.2 Acceleration to relativistic speeds

Since the electron is initially at rest and the force has only an x component, the motion is one dimensional. We have

$$\frac{dp}{dt} = F_0 \quad \rightarrow \quad p(t) = F_0 t .$$

The energy is given by

Fig. A.188 An observer at
O will receive the radiation
emitted between points A
and B

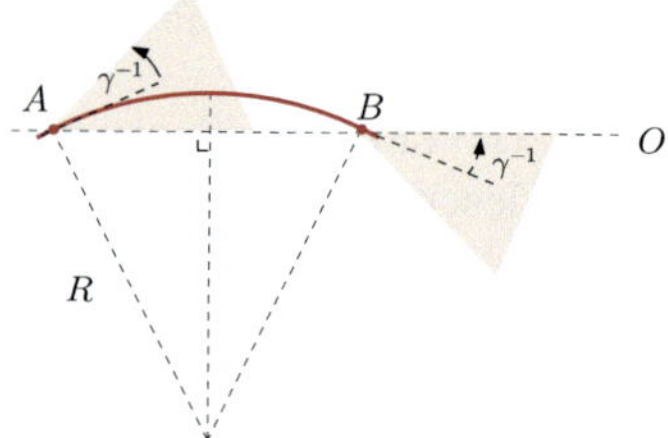

$$E(t) = \sqrt{m^2 c^4 + c^2 p(t)^2} = mc^2 \sqrt{1 + \frac{F_0^2 t^2}{m^2 c^2}}$$

and the electron velocity can be obtained from

$$v(t) = \frac{p(t)}{\gamma(t) m} = \frac{p(t)}{E(t)/c^2} = \frac{F_0 t}{m\sqrt{1 + \frac{F_0^2 t^2}{m^2 c^2}}} \, .$$

If the electron is accelerated to an energy of $E = 10\,\text{MeV}$, then the Lorentz factor
will be

$$\gamma = \frac{E}{mc^2} = \frac{10^7 \times 1.6 \; 10^{-19}}{9 \; 10^{-31} \times (3 \times 10^8)^2} = 19.75$$

which translate for the speed

$$v = c \left(1 - \frac{1}{\gamma^2} \right)^{1/2} = 0.9988c \, .$$

20.3 Duration of a synchrotron pulse

An observer at position O will only receive radiation if it is located within the cone
of emission which, according to the relativistic beaming effect, is defined by an angle
$1/\gamma$ with respect to the particle's propagation direction. As shown in Fig. A.188, this
will happen once per turn, every time the particle is within points A and B.

The period of the detected signal is simply the period of the charge's circular
motion. If the speed is $v \approx c$, then

$$T \approx \frac{2\pi R}{c} \, .$$

On the other hand, the arc A and B subtends an angle $2/\gamma \ll 1$ so that $(AB) \approx \frac{2R}{\gamma}$.
The duration ΔT of the pulse is given by the difference in arrival times at O of
the light emitted at A with respect to the light emitted at B. Since the charge takes
a time $(AB)/v$ to go from A to B, the light that was emitted at A is at a distance
$c(AB)/v - (AB)$ from B when the particle is at B. We conclude that

$$\Delta T = \frac{\left(\frac{c}{v} - 1\right)(AB)}{c} = \frac{2R\frac{v}{c}\left(1 - \frac{v}{c}\right)}{c\gamma}$$

In the highly relativistic regime, $\gamma^{-2} = \left(1 - \frac{v^2}{c^2}\right) = \left(1 + \frac{v}{c}\right)\left(1 - \frac{v}{c}\right) \approx 2\left(1 - \frac{v}{c}\right)$

$$\Delta T \approx \frac{R}{c\gamma^3}$$

We see that

$$\frac{\Delta T}{T} = \frac{1}{2\pi\gamma^3} \ll 1$$

20.4 Relativistic composition of velocities

Using the general transformation (20.3) for the four-velocity $\mathcal{U} = (\gamma(v)\mathbf{v}, \gamma(v)ic)$, we get for the fourth-component

$$i\gamma(v)c = \gamma(u)\left(i\gamma(v')c + i\frac{\mathbf{u}}{c} \cdot \gamma(v')\mathbf{v}'_{\parallel}\right)$$

which gives the transformation of the Lorentz factors associated with the velocity of the particle in each frame. Solving for $\gamma(v')$

$$\gamma(v') = \frac{\gamma(v)}{\gamma(u)\left(1 + \frac{\mathbf{u}\cdot\mathbf{v}'_{\parallel}}{c^2}\right)}. \qquad (A.103)$$

The Lorentz transformation for the parallel component reads

$$\gamma(v)\mathbf{v}_{\parallel} = \gamma(u)\left(\gamma(v')\mathbf{v}'_{\parallel} - i\frac{\mathbf{u}}{c}ic\gamma(v')\right)$$

and using (A.103)

$$\mathbf{v}_{\parallel} = \frac{\gamma(u)\gamma(v')}{\gamma(v)}\left(\mathbf{v}'_{\parallel} + \mathbf{u}\right) = \frac{1}{1 + \frac{\mathbf{u}\cdot\mathbf{v}'_{\parallel}}{c^2}}\left(\mathbf{v}'_{\parallel} + \mathbf{u}\right)$$

and since the perpendicular component of a four-vector is unchanged

$$\gamma(v)\mathbf{v}_{\perp} = \gamma(v')\mathbf{v}'_{\perp},$$

we get

$$\mathbf{v}_{\perp} = \frac{1}{\gamma(u)\left(1 + \frac{\mathbf{u}\cdot\mathbf{v}'_{\parallel}}{c^2}\right)}\mathbf{v}'_{\perp}.$$

Finally

$$v = \frac{1}{1 + \frac{\mathbf{u}\cdot\mathbf{v}'_\parallel}{c^2}}\left(\mathbf{v}'_\parallel + \mathbf{u} + \frac{1}{\gamma(u)}\mathbf{v}'_\perp\right)$$

but $\mathbf{v}'_\parallel + \mathbf{u} + \frac{\mathbf{v}'_\perp}{\gamma(u)} = \frac{\mathbf{v}'}{\gamma(u)} + \mathbf{u} + \left(1 - \frac{1}{\gamma(u)}\right)\mathbf{v}'_\parallel$, and writing $\mathbf{v}'_\parallel = (\mathbf{v}'\cdot\mathbf{u})\mathbf{u}/u^2$

$$v = \frac{1}{1 + \frac{\mathbf{u}\cdot\mathbf{v}'_\parallel}{c^2}}\left(\frac{\mathbf{v}'}{\gamma(u)} + \mathbf{u} + \left(1 - \frac{1}{\gamma(u)}\right)\frac{\mathbf{v}'\cdot\mathbf{u}}{u^2}\mathbf{u}\right).$$

20.5 Magnetic interaction and relativity

Using the Lorentz transformation (20.3) for the quadriforce $\mathcal{F} = \gamma(\mathbf{F}, i\frac{\mathbf{v}\cdot\mathbf{F}}{c})$, we get

$$\gamma(v)\mathbf{F}_\perp = \gamma(v')\mathbf{F}'_\perp, \tag{A.104}$$

$$\gamma(v)\mathbf{F}_\parallel = \gamma(u)\left(\gamma(v')\mathbf{F}'_\parallel + \gamma(v')\frac{\mathbf{u}}{c^2}\mathbf{v}'\cdot\mathbf{F}'\right), \tag{A.105}$$

and

$$\gamma(v)\mathbf{v}\cdot\mathbf{F} = \gamma(u)\left(\gamma(v')\mathbf{v}'\cdot\mathbf{F}' + \gamma(v')\mathbf{u}\cdot\mathbf{F}'_\parallel\right). \tag{A.106}$$

The relationship between the Lorentz factors $\gamma(v)$ and $\gamma(v')$ can be retrieved by writing the Lorentz transformation of the fourth-component of the four-velocity $i\gamma(v')c = \gamma(u)\left(i\gamma(v)c - i\frac{\mathbf{u}}{c}\cdot\gamma(v)\mathbf{v}_\parallel\right)$ which gives

$$\gamma(v') = \gamma(u)\gamma(v)\left(1 - \frac{\mathbf{u}}{c^2}\cdot\mathbf{v}_\parallel\right).$$

From (A.104) we can solve for the perpendicular component of the force

$$\mathbf{F}_\perp = \frac{\gamma(v')}{\gamma(v)}\mathbf{F}'_\perp = \gamma(u)\left(1 - \frac{\mathbf{u}}{c^2}\cdot\mathbf{v}_\parallel\right)\mathbf{F}'_\perp.$$

From (A.105)

$$\mathbf{F}_\parallel = \frac{\gamma(u)\gamma(v')}{\gamma(v)}\left(\mathbf{F}'_\parallel + \frac{\mathbf{u}}{c^2}\mathbf{v}'\cdot\mathbf{F}'\right)$$

$$= \gamma^2(u)\left(1 - \frac{\mathbf{u}}{c^2}\cdot\mathbf{v}_\parallel\right)\left(\mathbf{F}'_\parallel + \frac{\mathbf{u}}{c^2}\mathbf{v}'\cdot\mathbf{F}'\right).$$

From (A.106) we can solve for $\mathbf{v}'\cdot\mathbf{F}' = \frac{\mathbf{v}\cdot\mathbf{F}}{\gamma^2(u)\left(1 - \frac{\mathbf{u}\cdot\mathbf{v}_\parallel}{c^2}\right)} - \mathbf{u}\cdot\mathbf{F}'_\parallel$ and so

$$\mathbf{F}_\parallel = \gamma^2(u)\left(1 - \frac{\mathbf{u}}{c^2}\cdot\mathbf{v}_\parallel\right)\left(\mathbf{F}'_\parallel - \frac{\mathbf{u}}{c^2}\mathbf{u}\cdot\mathbf{F}'_\parallel\right) + \frac{\mathbf{u}}{c^2}\mathbf{v}\cdot\mathbf{F}.$$

The total force acting on the charge q_2 in the reference $\mathcal{R}$ is

$$\mathbf{F} = \mathbf{F}_\perp + \mathbf{F}_\parallel = \gamma(u)\left(1 - \frac{\mathbf{u}}{c^2}\cdot\mathbf{v}\right)\left(\mathbf{F}'_\perp + \gamma(u)\mathbf{F}'_\parallel - \gamma(u)\frac{\mathbf{u}}{c^2}\mathbf{u}\cdot\mathbf{F}'_\parallel\right) + \frac{\mathbf{u}}{c^2}\mathbf{v}\cdot\mathbf{F}.$$

$$\mathbf{F} = \mathbf{F}_\perp + \mathbf{F}_\parallel = \gamma(u)\left(1 - \frac{\mathbf{u}}{c^2}\cdot\mathbf{v}\right)\left(\mathbf{F}'_\perp + \frac{1}{\gamma(u)}\mathbf{F}'_\parallel\right) + \frac{\mathbf{u}}{c^2}\mathbf{v}\cdot\mathbf{F}.$$

Taking the dot product with $\mathbf{v}$ at both sides, we get

$$\mathbf{F}\cdot\mathbf{v} = \gamma(u)\left(1 - \frac{\mathbf{u}}{c^2}\cdot\mathbf{v}\right)\left(\mathbf{F}'_\perp + \frac{1}{\gamma(u)}\mathbf{F}'_\parallel\right)\cdot\mathbf{v} + \frac{\mathbf{u}\cdot\mathbf{v}}{c^2}\mathbf{v}\cdot\mathbf{F},$$

which simplifies to

$$\mathbf{F}\cdot\mathbf{v} = \gamma(u)\left(\mathbf{F}'_\perp + \frac{1}{\gamma(u)}\mathbf{F}'_\parallel\right)\cdot\mathbf{v}.$$

Now, since $(\mathbf{r}'_2 - \mathbf{r}'_1)_\perp = (\mathbf{r}_2 - \mathbf{r}_1)_\perp$ and $(\mathbf{r}'_2 - \mathbf{r}'_1)_\parallel = \gamma(u)(\mathbf{r}_2 - \mathbf{r}_1)_\parallel$,

$$\frac{1}{|\mathbf{r}'_2 - \mathbf{r}'_1|^3} = \frac{1}{|\mathbf{r}_2 - \mathbf{r}_1|^3 \gamma^3(u)\left(1 - \frac{u^2}{c^2}\sin^2\alpha\right)^{3/2}}$$

where α is the angle between $(\mathbf{r}_2 - \mathbf{r}_1)$ and $\mathbf{u}$. We find

$$\mathbf{F}\cdot\mathbf{v} = \frac{q_1 q_2}{4\pi\epsilon_0|\mathbf{r}_2 - \mathbf{r}_1|^3}\frac{(\mathbf{r}_2 - \mathbf{r}_1)\cdot\mathbf{v}}{\gamma^2(u)\left(1 - \frac{u^2}{c^2}\sin^2\alpha\right)^{3/2}}$$

and finally

$$\mathbf{F} = \frac{q_1 q_2}{4\pi\epsilon_0|\mathbf{r}_2 - \mathbf{r}_1|^3}\frac{1}{\gamma^2(u)\left(1 - \frac{u^2}{c^2}\sin^2\alpha\right)^{3/2}}\underbrace{\left(\left(1 - \frac{\mathbf{u}}{c^2}\cdot\mathbf{v}\right)(\mathbf{r}_2 - \mathbf{r}_1) + \frac{\mathbf{u}}{c^2}(\mathbf{r}_2 - \mathbf{r}_1)\cdot\mathbf{v}\right)}_{(\mathbf{r}_2 - \mathbf{r}_1) + \frac{\mathbf{v}\times(\mathbf{u}\times(\mathbf{r}_2 - \mathbf{r}_1))}{c^2}}.$$

The force takes the form of a Lorentz force $\mathbf{F} = q_2\mathbf{E}(\mathbf{r}_2) + q_2\mathbf{v}\times\mathbf{B}(\mathbf{r}_2)$, where $\mathbf{E}$ and $\mathbf{B}$ are the Liénard–Wiechert fields of a point charge $\mathbf{q}_1$ at position $\mathbf{r}_1$ moving at constant speed $\mathbf{u}$ in $\mathcal{R}$:

$$\mathbf{E}(\mathbf{r}) = \frac{q_1}{4\pi\epsilon_0|\mathbf{r} - \mathbf{r}_1|^3}\frac{\mathbf{r} - \mathbf{r}_1}{\gamma^2(u)\left(1 - \frac{u^2}{c^2}\sin^2\alpha\right)^{3/2}},$$

$$\mathbf{B}(\mathbf{r}) = \frac{q_1\mu_0}{4\pi|\mathbf{r} - \mathbf{r}_1|^3}\frac{\mathbf{u}\times(\mathbf{r} - \mathbf{r}_1)}{\gamma^2(u)\left(1 - \frac{u^2}{c^2}\sin^2\alpha\right)^{3/2}}.$$

Index